MATHEMATICS AND STATISTICS FOR ECONOMICS

Y THE SAME AUTHOR AND PUBLISHER:

- *Mathematics for Mangement and Econonmics*

MATHEMATICS AND STATISTICS FOR ECONOMICS

(Second Revised Edition)

G.S. MONGA
M.A., M.Sc., D.STAT.
Dean
Invertis Institute of Management Studies,
Civil Lines, Barelley

VIKAS® PUBLISHING HOUSE PVT LTD

VIKAS® PUBLISHING HOUSE PVT LTD
E-28, Sector-8, Noida-201301 (UP) India
Phone: +91-120-4078900 • Fax: +91-120-4078999
VIKAS® Registered Office: 576, Masjid Road, Jangpura, New Delhi-110014. India
E-mail: ***helpline@vikaspublishing.com*** • Website: ***www.vikaspublishing.com***

- Ahmedabad : 305, Grand Monarch, 100 ft Shyamal Road, Near Seema Hall, Ahmedabad-380051 • Ph. +91-79-65254204
- Bengaluru : First Floor, N S Bhawan, 4th Cross, 4th Main, Gandhi Nagar, Bengaluru-560009 • Ph. +91-80-22281254, 22204639
- Chennai : E-12, Nelson Chambers, 115, Nelson Manickam Road, Aminjikarai, Chennai-600029 • Ph. +91-44-23744547, 23746090
- Hyderabad : Aashray Mansion, Flat-G (G.F.), 3-6-361/8, Street No. 20, Himayath Nagar, Hyderabad-500029 • Ph. +91-40-23269992 • Fax. +91-40-23269993
- Kolkata : 82, Park Street, Kolkata-700017 • Ph. +91-33-22837880
- Mumbai : 67/68, 3rd Floor, Aditya Industrial Estate, Chincholi Bunder, Behind Balaji International School & Evershine Mall, Malad (West), Mumbai-400064 • Ph. +91-22-28772545, 28768301
- Patna : Flat No. 101, Sri Ram Tower, Beside Chiraiyatand Over Bridge, Kankarbagh Main Rd., Kankarbagh, Patna-800020 • Ph. +91-612-2351147

Mathematics and Statistics for Economics
ISBN: 978-81259-0982-8

Printed in India.

PREFACE TO THE SECOND EDITION

The purpose of this book is to provide important and vital aspects of mathematical and statistical methods widely used by economists and managers in various contexts. The number of topics required to be studied in mathematics and statistics has increased. It is also playing an increasingly vital role in the solution of problems encountered in social sciences. This book includes considerably more material than that required in most courses and therefore, should be able to generate greater interest in the subject.

The contents and methodology of this book have been tested over the past three decades and more. Each chapter gives a concise presentation of concepts and theories which are illustrated and amplified with the help of examples. It is hoped that students will receive a complete and rigorous introduction to the basic principles of the subject and discover for themselves, sources of creativity in during this exercise.

I take this opportunity to thank friends and colleagues for invaluable suggestions and inspiring support.

G.S. MONGA

PREFACE TO THE FIRST EDITION

This book is an introduction to the basic ideas and techniques of mathematics and statistics and is intended to prepare the students of economics, for a better understanding of current economic literature.

It is now generally accepted that a study of modern economics is incomplete without a knowledge of mathematics and statistics. In view of the increasing complexity and variety of problems in economic theory, a student without a fair grounding in these tools, may not be able to cope and hence, remain unfamiliar with many aspects of economic theory. Due to unfounded fear of the so called 'difficult' mathematics, students tend to shy away from the subject and often, it may be difficult to persuade them to take special courses in mathematics. But the fact is that, learning to use mathematical and statistical techniques does not require elaborate mathematical preparation and the student need not be dismayed by some mathematical manipulations.

One learns through practice. Mathematics is best learnt if the student handles it systematically. He is bound to find it rewarding and exciting, while following the logical arguments and steps involved.

This book gives a simple and clear presentation of the logic and meaning of mathematics and statistics which will be useful to beginners. There is a fairly self-contained development of the topics and the student will find here, a starting point which will help gain self-confidence and familiarity with the subject. A knowledge of elementary economics is assumed and no attempt has been made to explain simple economic terms. The student will attain proficiency in both economics and mathematics simultaneously, while he proceeds further with different aspects of his studies.

A wide range of topics is covered in the book although, the treatment is limited to the elementary level. There are solved, unsolved and party solved exercises with the every section, to make the student familiar with methods introduced.

The book is divided into four parts which are independent of one another. Part One, beginning with a review chapter, contains a treatment of some mathematical topics with reference to problems in economics. Part Two considers the elementary aspects of linear programming and related topics. Part Three provides a basic course in probability and here, we enter the realm of statistics. Part Four deals with index numbers, time series and survey methods. These are familiar topics in economics.

It is a pleasure to acknowledge my gratitude to the many colleagues and students who have been kind enough, to offer their valuable suggestions and constructive criticisms which have contributed significantly to the improvement of the book.

G.S. MONGA

CONTENTS

CHAPTER 1

Review of Some Elementary Mathematics

This chapter is a brief survey of some selected elementary mathematical concepts which will be found to be useful in subsequent chapters.

1.1 ALGEBRAIC EXPRESSIONS

Arithmetic is concerned with numbers. Algebra employs symbols which are treated as unknown numbers. The laws of combination for algebraic expressions are the same as those of arithmetic.

An algebraic *term* is a product of numbers and symbols. $5xy$ is the product of 5, x and y each of which is a *factor.*

An algebraic expression is a combination of algebraic terms separated by + or – signs.

An expression containing one term is called a *monomial;* with two terms we have a *binomial.* With more than one term, we have, in general, *a polynomial.*

To *factorise* a polynomial means to find such polynomials which when multiplied together produce the given polynomial.

Ex. 1. $5 \times x \times y = 5xy$; $(x + y)(x - y) = x^2 - y^2$.

An expression is said to be *prime* when it has no factor other than itself and unity.

Ex. 2. 7 and 1 are the only factors of 7, a prime number.

$a + b$ and 1 are the only factors of $a + b$.

A polynomial is *factorised* when it has been resolved into its prime factors.

Ex. 3. $x + y$ and $x - y$ are the prime factors of $x^2 - y^2$.

If the same factor is contained in two or more polynomials, it is a *common* factor of the polynomials.

Ex. 4. $x - y$ is a common factor of $x^2 - y^2$ and $x^3 - y^3$.

The *degree* of a polynomial in certain letters is that of its terms of the highest degree in those letters.

Ex. 5. $5x^3y^6 + 7x^5y - 20xy^7$ is a polynomial of fifth degree in x, seventh degree in y and ninth degree in x, y.

The product of all the common factors of two or more polynomials is called the highest common factor (H.C.F.).

The lowest common factor (L.C.M.) of two or more polynomials is the product of all the different prime factors, each factor taken the greatest number of times it occurs in any one of the polynomials.

The following results in elementary algebra are well known :

(i) $x(y + z) = xy + xz$ (ii) $(x + y)^2 = x^2 + 2xy + y^2$

(iii) $(x - y)^2 = x^2 - 2xy + y^2$ (iv) $(x + a)(x + b) = x^2 + ax + bx + ab$

(v) $(x + y)(x - y) = x^2 - y^2$ (vi) $(x + y)^3 = x^3 + 3x^2y + 3xy^2 + y^3$

(vii) $(x - y)^3 = x^3 - 3x^2y + 3xy^2 - y^3$

(viii) $(x + y)(x^2 - xy + y^2) = x^3 + y^3$

(ix) $(x - y)(x^2 + xy + y^2) = x^3 - y^3$

(x) $(x + y + z)^2 = x^2 + y^2 + z^2 + 2xy + 2yz + 2zx$.

Ex. 6. Factorise : (i) $1 - x^2$; (ii) $1 - x^3$; (iii) $1 + x - 2x^2$

Ans. (i) $1 - x^2 = (1 + x)(1 - x)$

(ii) $1 - x^3 = (1 - x)(1 + x + x^2)$

(iii) $1 + x - 2x^2 = (1 - x)(1 + 2x)$.

Ex. 7. Find H.C.F. and L.C.M. for the three expressions in Ex. 6.

H.C.F. $= 1 - x$ L.C.M. $= (1 - x)(1 + x)(1 + 2x)(1 + x + x^2)$.

Ex. 8. Factorise and find H.C.F. and L.C.M. for the following expressions:

(i) $x^2 + 5x + 6,\ 5x + 10,\ x^2 - 4$; (ii) $x^4 + x^2 + 1,\ x^3 - x^2 + x$;

(iii) $x^3 - y^3,\ x^4 - y^4,\ x^2 - y + x - xy$; (iv) $a^3 - 27,\ 3ab - 6a,\ ab + b - 2a - 2$

Ans. (i) $x + 2$; $5(x + 2)(x + 3)(x - 2)$

(ii) $x^2 - x + 1$; $y(x^2 - x + 1)(x^2 + x + 1)$

(iii) $x - y$; $(x^2 - y^2)(x^2 + y^2)(x^2 + xy + y^2)(x + 1)$

(iv) No HCF LCM $= 3a(a - 3)(a + 1)(b - 2)(a^2 + 3a + 9)$

Algebraic fractions are subject to the same rules as those of arithmetic. A fraction is defined when its denominator is not zero.

Ex. 9. $\dfrac{x^2 + 5x + 7}{x - 1}$ is not defined when $x = 1$.

Addition, subtraction, multiplication, division and simplification of fractions can be illustrated by means of examples.

Ex. 10. $\dfrac{x}{2} + \dfrac{y}{6} = \dfrac{3x + y}{6}$.

Ex. 11. $$\frac{x+1}{x} + \frac{x}{x-3} - \frac{1}{x+3} = \frac{(x+1)(x^2-9) + x^2(x+3) - x(x-3)}{x(x-3)(x+3)}$$

$$= \frac{2x^3 + 3x^2 - 6x - 9}{x(x-3)(x+3)} = \frac{(2x+3)(x^2-3)}{x(x-3)(x+3)}$$

Ex. 12. $$\frac{x^4-1}{x^4} \div \frac{x^2-1}{x^2} = \frac{x^2+1}{x^2}$$

Ex. 13. $$\frac{5x}{1-\dfrac{x-1}{x+1}} = \frac{5x(x+1)}{2}.$$

Ex. 14. $$\frac{x^2-xy-2y^2}{x^4-9xy^2} \div \frac{x-2y}{x+3y} = \frac{x+y}{x(x-3y)}.$$

1.2 PROPERTIES OF EXPONENTS : INDICES

If n is a positive integer then a^n, *the* n^{th} *power of a* represents the product of n factors :

$$a^n = a \times a \times \ldots \times a \ (n \text{ factors})$$

The integer n is called the exponent of a^n to the base a.

Let a be any real number, then :

if $n = 1$ $\quad a^n = a$

if $n > 1$ $\quad a^n = a \cdot a^{n-1} = a^2 \cdot a^{n-2}$ etc.

if $a \neq 0$ $\quad a^{-n} = \dfrac{1}{a^n}$

if $a = 0$ $\quad a^0 = 1$

The basic laws of exponents are given below. If m and n are real numbers, then :

(*i*) $a^m \cdot a^n = a^{m+n}$ (*ii*) $\dfrac{a^m}{a^n} = a^{m-n} = \dfrac{1}{a^{n-m}}$

(*iii*) $(a^m)^n = a^{mn}$ (*iv*) $(ab)^m = a^m b^m$

(*v*) $\left(\dfrac{a}{b}\right)^m = \dfrac{a^m}{b^m}$

Ex. 15. (*i*) $a^5 \times a^2 = a^7;\ a^5 \times a^{-2} = a^3$

(*ii*) $\dfrac{a^5}{a^2} = a^3;\ \dfrac{a^5}{a^{-2}} = a^7$ (*iii*) $(a^5)^2 = a^{10},\ (a^5)^{-2} = a^{-10} = \dfrac{1}{a^{10}}$

(*iv*) $(3a)^2 = 3^2 \cdot a^2 = 9a^2$ (*v*) $\left(\dfrac{3}{a}\right)^2 = \dfrac{9}{a^2},\ \left(\dfrac{9}{a^2}\right)^{\frac{1}{2}} = \dfrac{9^{\frac{1}{2}}}{a}$

Ex. 16. $1 = \dfrac{a^m}{a^m} = a^{m-m} = a^0 \quad \therefore a^0 = 1$. Any quantity (other than zero) raised to the power zero equals unity.

$$a^{\frac{1}{2}} \cdot a^{\frac{1}{2}} = a = \sqrt{a}\sqrt{a} \qquad \therefore\ a^{\frac{1}{2}} = \sqrt{a}$$

Similarly $\quad a^{\frac{1}{3}} = \sqrt[3]{a},\ a^{\frac{1}{n}} = \sqrt[n]{a}$

In general $\quad a^{\frac{m}{n}} = \sqrt[n]{a^m}$, *i.e.* nth root of a^m

Ex. 17. $\dfrac{1}{a^m} = \dfrac{a^0}{a^m} = a^{0-m}$ $\quad\therefore \dfrac{1}{a^m} = a^{-m}$

Similarly $\dfrac{1}{a^{-m}} = a^m$.

a^{-m} is the reciprocal of a^m

a^m is the reciprocal of a^{-m}

Meaning of $a^{m/n}$

If $a > 0$; m, n are positive integers, then

$$a^{m/n} = (a^m)^{1/n} = n\text{th root of } a^m$$

$$= (a^{\frac{1}{n}})^m = m\text{th power of } a^{\frac{1}{n}}$$

Ex. 18. $\left(\dfrac{1}{25}\right)^{-\frac{3}{2}} = \left(\dfrac{1}{5^2}\right)^{-\frac{3}{2}} = (5^2)^{+\frac{3}{2}} = 5^{2.\frac{3}{2}} = 5^3 = 125.$

Ex. 19. $\left(a^{\frac{1}{3}}b^{\frac{2}{3}}\right)^{\frac{3}{4}} = a^{\frac{1}{4}}b^{\frac{1}{2}}.$

Ex. 20. $\dfrac{(ab^2)^{-2}}{(a/b)^3} = \dfrac{a^{-2}b^{-4}}{a^3/b^3} = a^{-2-3} \cdot b^{-4+3} = a^{-5}b^{-1} = \dfrac{1}{a^5b}.$

Ex. 21. $\sqrt{\left(\dfrac{a^{-2}}{-2}\right)^4} = \sqrt{\dfrac{a^{-8}}{2^4}} = \left(\dfrac{a^{-8}}{2^4}\right)^{\frac{1}{2}} = \dfrac{a^{-4}}{2^2} = \dfrac{1}{4a^4}.$

Ex. 22. If $a^m \cdot a^n = (a^m)^n$ show that $m(n-2) + n(m-2) = 0$.

Ex. 23. Show that (*i*) $2x^{-\frac{3}{2}} \cdot px^{-\frac{5}{2}} = \dfrac{2p}{x^4}$

(*ii*) $\left(\dfrac{2x^{-2}}{9y}\right)^{-2} = \dfrac{81x^2y^2}{4}$

(*iii*) $(xy)^{\frac{2}{3}} (xy)^{\frac{3}{2}} = x^{\frac{13}{6}} y^{\frac{13}{6}}$

(*iv*) $(px^{\frac{1}{2}} y^{-\frac{1}{2}})(qx^{-\frac{3}{4}} y^{\frac{1}{4}}) = pqx^{-\frac{1}{4}} y^{-\frac{1}{4}}$

(*v*) $\dfrac{(a^{p-4n})^p}{a^{p(p-5n)} \cdot a^{n(p-6n)}} = a^{6n^2}$

Ex. 24. Show that

(*i*) $\dfrac{(x^{\frac{1}{2}})^{-\frac{1}{2}} (y^{\frac{2}{3}})^{-\frac{3}{4}}}{(x^{-\frac{1}{2}})^{-\frac{1}{3}} (y^{-\frac{3}{4}})^{-\frac{2}{3}}} = x^{-1}y^{-2}$ $\qquad$ (*ii*) $\left(\dfrac{a^{n^2+1}}{a^{1-n}}\right)^{\frac{1}{n+1}} = a^n$

(*iii*) $\left(\dfrac{q+r}{x^r - p}\right)^{\frac{1}{p-q}} \left(\dfrac{r+p}{x^{p-q}}\right)^{\frac{1}{q-r}} \left(\dfrac{p+q}{x^{q-r}}\right)^{\frac{1}{r-p}} = 1$

(*iv*) $\dfrac{7 \cdot 8^3 \cdot 2^{4n-1}}{2^6 \cdot 2^{4n+7} - 2^{10} \cdot 2^{4n}} = 1$

Ex. 25. Solve for x : (*i*) $x^3 = 8^2$ (*ii*) $x^{-4} = 81$ (*iii*) $x^{\frac{3}{2}} = 64$

(*iv*) $3^x + 3^{x+1} + 3^{x+2} + 3^{x+3} = 120$ [L.H.S. $= 3^x(1 + 3 + 3^2 + 3^3)$]

(*v*) $x^{\sqrt{x}} = (x\sqrt{x})^x$. $\left[\text{Hint} : x^{x^{1/2}} = x^{\frac{3}{2}x} \quad \therefore \quad x^{1/2} = \frac{3}{2}x\right]$

Ans. (*i*) 4 (*ii*) $\frac{1}{3}$ (*iii*) 16 (*iv*) 1 (*v*) $\frac{4}{9}$

Ex. 26. If $\sqrt[m]{3^{m2} \div (3^m)^2} = 81$ find m. **Ans.** 6

Ex. 27. If $a = 10^x$, $b = 10^y$, $(a^y/b^x)^x = 100$. Show that $xyz = 1$.

Ex. 28. If $2^x = 3^y = 12^z$, show that $\dfrac{1}{z} = \dfrac{1}{y} + \dfrac{2}{x}$.

Ex. 29. Solve for x, y : (*i*) $9^{x-y} = 81$, $9^{x+y} = 74$

$9^{x-y} = 9^2$ $\quad$ $9^{x+y} = 9^3$ $\quad$ Solving

$\therefore x - y = 2$ $\quad$ $x + y = 3$ $\quad$ $x = 5/2 \;\; y = 1/2$

(*ii*) $7^{x+7} = 49^{4x-7}$

i.e., $7^{x+7} = 7^{2(4x-7)}$ or $x + 7 = 2(4x - 7)$

$\therefore x = 3$

(*iii*) $4^x\, 5^y = 20$, $4^{2x} \cdot 5^y = 80$. **Ans.** $x = 1 = y$

(*iv*) $2^{x+y} = 2^{2x-y} = \sqrt{8}$. **Ans.** $x = 1$, $y = 1/2$

Ex. 30. If $3^{1/3} + 3^{-1/3} = 0$ then $3a^3 - 9a - 10 = 0$.

Ex. 31. (*i*) If $p^q = q^p$ then $\left(\dfrac{p}{q}\right)^{p/q} = p^{\frac{p}{q}-1}$

(*ii*) If $a = xy^{p-1}$, $b = xy^{q-1}$, $c = xy^{r-1}$ then $a^{q-r}b^{r-p}c^{p-q} = 1$.

Ex. 32. Solve the equation

$$3^x + 3^{x+1} - 3^{x+2} + 3^{x+3} = 120 \times 3^{\sqrt{2}}$$

LHS $= 3^x(1 + 3 + 9 + 27)$ or $3^x + 40 = 120 \times 3^{\sqrt{2}}$

$$3^x = \frac{120}{40} \cdot 3^{\sqrt{2}} = 3 \cdot 3^{\sqrt{2}} = 3^{\sqrt{2}+1}$$

$\therefore \quad x = \sqrt{2} + 1$

Ex. 33. Solve $3^{x+5} = 3^{x+3} + 8/3$

$$3^x(3^5 - 3^3) = \frac{8}{3}$$

$$\therefore \qquad 3^x = \frac{8}{3}\cdot\frac{1}{216} = \frac{1}{81} = 3^{-4} \text{ or } x = -4$$

Ex. 34. If $(a^n)^m = a^{n^m}$ then $n^{m-1}\cdot m^{n-1} = m^n$

Since $\qquad (a^n)^m = a^{n^m}$ implies $nm = n^m$

$$\text{L.H.S.} = n^{m-1}\, m^{n-1} = \frac{n^m}{n}\cdot\frac{m^n}{m} = \frac{nm}{n}\cdot\frac{m^n}{m} = m^n$$

1.3 SURDS

A rational number can be expressed as a ratio of two integers in the form m/n where $n \neq 0$. Such numbers as are not rational may be irrational. Thus $\sqrt{2}$, $\sqrt[3]{7}$, $\sqrt{2+\sqrt{3}}$, π etc. are irrational numbers because they cannot be expressed in the form m/n.

Definition. A Surd is an irrational root of a rational number.

If x is a prime number then $\sqrt[n]{x}$ is always a surd.

e.g., $\sqrt{2}$, $\sqrt{3}$, $\sqrt[3]{7}$, are surds since 2, 3, 7 are prime numbers. If x is not prime and also not the perfect nth power of any rational number then $\sqrt[n]{x}$ is a surd.

e.g., $\sqrt{4}$ is not a surd but $\sqrt[3]{4}$ is a surd.

$\sqrt[3]{16} = 2\sqrt[3]{2}$ is a surd.

In the surd $\sqrt[n]{x}$, n is the *order* or *index* and x the *radicand.*

Ex. 35. $\sqrt{9} = 3$ is not a surd

$\sqrt[4]{9}$ is a surd

$\sqrt{4/9}$ is not a surd

$\sqrt{3+\sqrt{5}}$ is a surd.

Ex. 36. $\sqrt{2}$ is a surd of order 2. It is called a quadratic surd.

$\sqrt[3]{2}$ is a cubic surd of order 3.

If in a given case, $\sqrt[n]{a}$ can be found exactly then it is not a surd.

Types of Surds

Surds of the type $\sqrt{2}$, $\sqrt[3]{3}$ are called pure surds.

$5\sqrt{2}$, $7\sqrt[4]{2}$ are cases of mixed surds.

If we write $5\sqrt{2} = \sqrt{50}$ it becomes a pure surd.

If we write $\sqrt{50} = 5\sqrt{2}$ it becomes a mixed surd.

Similar Surds

Surds having the same irrational factors are called Similar Surds.

$2\sqrt{3}$, $5\sqrt{3}$, $7\sqrt{3}$ are similar surds.

Ex. 37. $\sqrt{27}$, $\sqrt{12}$ $\sqrt{75}$ are similar because they can be written $3\sqrt{3}$, $2\sqrt{3}$, $5\sqrt{3}$.

Ex. 38. $\sqrt[5]{2}$, $\sqrt[4]{3}$, $\sqrt[3]{5}$ are not of the same order but can be made of the same order as follows: LCM of 5, 4, 3 is 60.

Write $\sqrt[5]{2} = 2^{\frac{1}{5}} = 2^{12/60} = (2^{12})^{1/60} = \sqrt[60]{2^{12}}$

$\sqrt[4]{3} = 3^{\frac{1}{4}} = 3^{15/60} = (3^{15})^{1/60} = \sqrt[60]{3^{15}}$

$\sqrt[3]{5} = 5^{\frac{1}{3}} = 5^{20/60} = (5^{20})^{1/60} = \sqrt[60]{5^{20}}$

We can arrange the three surds in decreasing order of magnitude by reducing them to surds of the same order.

Ex. 39. $\sqrt[3]{2} = 2^{1/3} = (2^2)^{1/6} = 4^{1/6}$ $\qquad \sqrt{3} = 3^{\frac{1}{2}} = (3^3)^{1/6} = 27^{1/6}$

$\therefore \qquad \sqrt{3} > \sqrt[3]{2}$

Compound surds are sums of two or more simple surds

e.g., $\sqrt{5} + \sqrt[3]{7} + \sqrt[4]{11}$

Compound surds are called binomial if they have only two terms as follows.

Surds of the type $a + b\sqrt{c}$ are called binomial surds. Here one part is real, the other a surd.

If such a surd is equal to another the rational part of one equals the rational part of the other and the surd part of one equals the surd part of the other.

If $\qquad a + b\sqrt{c} = d + e\sqrt{f}$

then $\qquad a = d,\ b\sqrt{c} = e\sqrt{f}$

Binomial Quadratic Surds

A surd of the type $a + b\sqrt{c}$ is binomial quadratic surd.

$\sqrt{2}$, $\sqrt{3}$ are examples of monomial quadratic surds.

Conjugate Surds

$a + b\sqrt{c}$ and $a - b\sqrt{c}$ are conjugate surds.

$\sqrt{2}$ and $-\sqrt{2}$ are conjugate surds.

The sum of two conjugate surds $a + \sqrt{b}$ and $a - \sqrt{b}$ is the rational number $2a$ and their product $a^2 - b^2$.

Some Rules

(i) $\sqrt[n]{x}\,\sqrt[n]{y} = \sqrt[n]{xy}$

(ii) $\sqrt[n]{x}\,/\sqrt[n]{y} = \sqrt[n]{x/y}$

(iii) $\sqrt[m]{\sqrt[n]{x}} = \sqrt[mn]{x} = \sqrt[n]{\sqrt[m]{x}}$

(iv) $\sqrt[m]{x^n} = \sqrt[mp]{x^{np}}$

(v) $(\sqrt[m]{x})^n = \sqrt[m]{x^n}$

Essentially the rules are similar to those for indices.

This is obvious since a surd can be expressed in the exponent form.

Thus $\sqrt[n]{x} = x^{\frac{1}{n}},\ \sqrt[n]{x^m} = x^{\frac{m}{n}}$ etc.

Ex. 40. (i) $\sqrt[3]{5}\cdot\sqrt[3]{7} = \sqrt[3]{35}$

or $\sqrt[3]{5}\cdot\sqrt[3]{7} = 5^{\frac{1}{3}}\cdot 7^{\frac{1}{3}} = 35^{\frac{1}{3}} = \sqrt[3]{35}$

(ii) $\sqrt[4]{128} = \sqrt[4]{2^5} = \sqrt[4]{2^4\cdot 2} = 2\sqrt[4]{2}$

or $\sqrt[4]{128} = 2^{5/4} = 2\cdot 2^{\frac{1}{4}} = 2\sqrt[4]{2}$

(iii) $7\sqrt[4]{25} \div 21\sqrt[6]{125}$

The expression $= \dfrac{7\sqrt{5}}{21\sqrt{5}} = \dfrac{1}{3}$

$$\frac{7\times 25^{\frac{1}{4}}}{21\times 125^{\frac{1}{6}}} = \frac{1}{3}\,\frac{5^{2/4}}{5^{3/6}} = \frac{1}{3}\,\frac{5^{1/2}}{5^{1/2}} = \frac{1}{3}$$

Some Properties of Surds

Addition and Subtraction

$$A\sqrt[m]{x} + B\sqrt[m]{x} = (A+B)\sqrt[m]{x}$$

Ex. 41. $\sqrt{18} + \sqrt{50} - 4\sqrt{2} = 3\sqrt{2} + 5\sqrt{2} - 4\sqrt{2}$

$= (3+5-4)\sqrt{2} = 4\sqrt{2}$

Ex. 42. $\sqrt{2},\ \sqrt[3]{9},\ \sqrt[4]{12}$

have a common order 12 and so may be written

$\sqrt[12]{2^6},\ \sqrt[12]{9^4},\ \sqrt[12]{12^3}$

or $\sqrt[12]{64},\ \sqrt[12]{6561},\ \sqrt[12]{1728}$

In the decreasing order of magnitude these surds may be arranged as follows :

$$\sqrt[3]{9},\ \sqrt[4]{12},\ \sqrt{2}$$

Simplifying Surd Expressions

This is done by simplifying and combining different terms suitably.

Ex. 43. Simplify : $(\sqrt{a}+\sqrt{b})\ (\sqrt[4]{a}+\sqrt[4]{b})(\sqrt[4]{a}-\sqrt[4]{b})$

The expression $= (\sqrt{a}+\sqrt{b})(\sqrt{a}-\sqrt{b}) = (a-b)$

Alt. the expression $= \left(a^{\frac{1}{2}}+b^{\frac{1}{2}}\right)\left(a^{\frac{1}{4}}+b^{\frac{1}{4}}\right)\left(a^{\frac{1}{4}}-b^{\frac{1}{4}}\right)$

$$= \left(a^{\frac{1}{2}}+b^{\frac{1}{2}}\right)\left(a^{\frac{1}{2}}-b^{\frac{1}{2}}\right)$$

$$= (a-b)$$

Ex. 44. Simplify by multiplication

$$(\sqrt{11}-\sqrt{5}-\sqrt{2})\ (\sqrt{11}-\sqrt{5}+\sqrt{2})$$

Let $x = 11 - \sqrt{5}$, then the expression becomes

$$(x-\sqrt{2})\ (x+\sqrt{2}) = x^2-2$$

$$= (11-\sqrt{2})^2 - 2 = 11 + 5 - 2\sqrt{55} - 2$$

$$= 14 - 2\sqrt{55}$$

Additional Properties

Suppose a, c are rational and b, d are surds,

If $a+\sqrt{b} = c+\sqrt{d}$

then $a = c,\ b = d$

If $\sqrt{b} = c+\sqrt{d}$

then $c = 0,\ b = d$

If $a+\sqrt{b} = c+\sqrt{d}$

then $a-\sqrt{b} = c-\sqrt{d}$

Rationalisation

If we multiply a surd by a suitable expression so that the product is rational the multiplying factor is called a rationalising factor. The method is useful in simplifying surds.

Ex. 45. To simplify $\dfrac{3\sqrt{2}}{\sqrt{3}+\sqrt{6}}$, a suitable rationalising factor is $\sqrt{3}-\sqrt{6}$. We may write the expression

$$= \frac{3\sqrt{2}}{\sqrt{3}+\sqrt{6}} \cdot \frac{\sqrt{3}-\sqrt{6}}{\sqrt{3}-\sqrt{6}}$$

$$= \frac{3\sqrt{2}(\sqrt{3}-\sqrt{6})}{3-6}$$

$$= \frac{3\sqrt{2}}{-3}(\sqrt{3}-\sqrt{6}) = -\sqrt{2}(\sqrt{3}-\sqrt{6})$$

$$= -\sqrt{2}\sqrt{3}(1-\sqrt{2})$$

$$= \sqrt{6}(\sqrt{2}-1)$$

If Y is a rationalising factor of X then X is a rationalising factor of Y.

Two conjugate binomial quadratic surds are rationalising factors of each other. Thus $a + b\sqrt{c}$ and $a - b\sqrt{c}$ rationalise each other.

Ex. 46. $a\sqrt{b} + c\sqrt{d}$ and $a\sqrt{b} - c\sqrt{d}$ are rationalising factors of each other. Their product is $a^2b - c^2d$.

Ex. 47. $\sqrt{7}+\sqrt{3}$ and $\sqrt{7}-\sqrt{3}$ rationalise each other. The product is 4.

Ex. 48. Show that $\left(\frac{\sqrt{26-15\sqrt{3}}}{5\sqrt{2}-\sqrt{38+5\sqrt{3}}}\right)^2$ is a rational number.

Ex. 49. If $a = \frac{\sqrt{3}-\sqrt{2}}{\sqrt{3}+\sqrt{2}}$, $b = \frac{\sqrt{3}+\sqrt{2}}{\sqrt{3}-\sqrt{2}}$, prove $a^2 + ab + b^2 = 99$.

Ex. 50. Find the square root of

(*i*) $49 + 20\sqrt{6}$ (*ii*) $\sqrt{32}-\sqrt{24}$ (*iii*) $16 + 6\sqrt{7}$ (*iv*) $26 - 15\sqrt{3}$

(*i*) Write the expression as $25 + 24 + 2.\sqrt{600} = (\sqrt{25}+\sqrt{24})^2$;

Sq. rt. $= \sqrt{25}+\sqrt{24}$

(*ii*) $\sqrt{32}-\sqrt{24} = \sqrt{2}(\sqrt{16}-\sqrt{12})$

$= \sqrt{2}(4-2\sqrt{3}) = \sqrt{2}(3+1-2\sqrt{3})$

$= \sqrt{2}(\sqrt{3}-1)^2$

Sq. rt. $= 2^{\frac{1}{4}}(\sqrt{3-1})$

Ans. (*iii*) $3+\sqrt{7}$ (*iv*) $\frac{1}{\sqrt{2}}(3\sqrt{3}-5)$

Ex. 51. Find the square root of $a + x + \sqrt{2ax + x^2}$.

Ans. $\frac{1}{\sqrt{2}}(\sqrt{2a+x}+\sqrt{x})$

1.4 COMPLEX NUMBERS *See 4.6*

1.5 CONSTANTS AND VARIABLES

Constants : A symbol which retains the same value throughout a particular problem is called a *constant.* Constants are usually denoted by letters at the beginning of the alphabet: *a, b, c* etc. Constants may be absolute or arbitrary.

Absolute Constants : An absolute or numerical constant retains the same value in all problems and operations. 2, – 5, π, *e* are absolute constants.

Parameters : An arbitrary or parametric constant is a symbol which is constant for the purpose of any particular problem but may assume different values in different problems. The value of a parameter depends on the particular situation represented in the problem.

Variables : A symbol which assumes different values in a given problem is called a variable. A variable is usually denoted by the last letters of the alphabet: *x, y, z,* etc. The set of values which a variable assumes is its *range*. A variable may be continuous or discrete.

Continuous Variables : A continuous variables *x* assumes any value of an aggregate of all real values between a given number *a* to another given number *b*. There is an infinity of points in the range of a continuous variable. This implies a noncountable or continuous number of points.

The concept of a continuous variable is useful in mathematical analysis although in practice all variables can be measured only to a finite accuracy.

Discrete Variables : A discrete variable assumes values only within a countable or discrete range.

Real Variables : If a variable *x* does not take all the real values between *a* and *b* but takes only values of an aggregate selected from an infinity of values between *a* and *b*, then we say that *x* is a real variable.

Variate : This term is used for a typical measurement of a variable. The distinction between a variable and a variate is a fine one often ignored in practice.

1.6 RATIO, PROPORTION, VARIATION

Ratio : A ratio expresses a comparison between two quantities— it is a relation between two like numbers or values.

The ratio of *a* to *b* is written

(*i*) with the colon sign $a : b$

or (*ii*) as a fraction $\dfrac{a}{b}$

or (*iii*) as a division $a \div b$

It can be seen that a ratio is an indicated division. It gives an index of comparison.

The first term, *a*, in the ratio $\dfrac{a}{b}$ is called the *antecedent* and the second term, *b,* is called the *consequent.*

The quantities that are being compared should be expressed in the same units—they should be dimensionally balanced.

Some rules may be noted.

(1) A ratio can be reduced to its lowest terms:

$$\frac{4}{8}=\frac{1}{2} \qquad \frac{a^2}{ab}=\frac{a}{b}$$

(2) A ratio does not change if both its terms are multiplied by the same number:

$$\frac{2}{3}=\frac{4}{6} \qquad \frac{x}{y}=\frac{ax}{ay}$$

(3) The total of any quantity can be divided into parts corresponding to a given ratio by taking the fractional part of each term in the ratio of the whole.

To divide 10 coins between two children in the ratio $\frac{2}{3}$ we take the fractional parts $\frac{2}{2+3}=\frac{2}{5}$ and $\frac{3}{2+3}=\frac{3}{5}$ and give $10\times\frac{2}{5}=4$ coins to the first child and $10\times\frac{3}{5}=6$ coins to the second child.

Proportion: An equation expressing the equality between two ratios is called a *proportion.*

The equality of the ratios $\frac{a}{b}$ and $\frac{c}{d}$ may be written

(*i*) $a : b : : c : d$

or (*ii*) $\frac{a}{b}=\frac{c}{d}$

or (*iii*) We may say that a, b, c, d are in proportion.

The end terms a, d are called the *extremes* of the proportion. The middle terms b, c are called the *means* of the proportion.

If $\frac{a}{b}=\frac{c}{d}$ then a, b, c and d are called the first, second, third and fourth proportional respectively.

It may be noted that

(*i*) The product of the means is equal to the product of the extremes.

If $a : b : : c : d$ then $ad = bc$

$1 : 2 : : 4 : 8$ then $1 \times 8 = 2 \times 4$

(*ii*) Either the extremes or means of both can be interchanged.

If $\frac{a}{b}=\frac{c}{d}$ then $\frac{b}{a}=\frac{d}{c}$

$$\frac{a}{c}=\frac{b}{d} \qquad \frac{d}{b}=\frac{c}{a}$$

Direct Proportion and Inverse Proportion : A *direct* proportion indicates a relationship between two quantities such that an increase in one causes a corresponding increase in the other and a decrease in one causes a corresponding decrease in the other, *e.g.*, in Fisher's quantity theory of money, other things remaining equal, in the equation $MV = PT$, the larger the quantity of money, the higher the price and the smaller the quantity of money, the lower the price. We say

P is directly proportional to M

An *inverse* proportion indicates a relationship between two quantities such that an increase in one causes a corresponding decrease in the other and a decrease in one causes a corresponding increase in the other, *e.g.*, the larger the number of competitors, the lower the price (of course in general, and up to a certain limit).

Again, the larger the number of workers, the lower the wages (according to wages fund theory only).

Ex. 52. Solve the proportion $\dfrac{3x}{4} = \dfrac{7}{12}$

(This problem may also be stated: find the first proportional in $3x : 3 :: 7 : 12$)

$3x = 4 \times \dfrac{7}{12}$ or $x = \dfrac{4}{3} \times \dfrac{7}{12} = \dfrac{7}{9}$.

Ex. 53. Solve the proportions

(*i*) $\dfrac{2}{5} = \dfrac{x}{10}$ (*ii*) $\dfrac{7}{y} = \dfrac{y}{7}$ (*iii*) $\dfrac{5}{x} = \dfrac{15}{x+6}$

(*iv*) $\dfrac{a+2}{4} = \dfrac{2a-5}{6}$ (*v*) $\dfrac{x}{3} = \dfrac{12}{x}$

Ans. (*i*) $x = 4$ (*ii*) $y = 7$ (*iii*) $x = 3$ (*iv*) $a = 16$ (*v*) $x = \pm 6$

Ex. 54. Two numbers are in the ratio $\frac{2}{3}$. Find them if their sum is 40.

Ans. 16, 24

Ex. 55. Simplify:

(*i*) $\dfrac{a}{x-y} : \dfrac{b}{x-y}$ (*ii*) $\dfrac{-x}{ab} : \dfrac{-xy}{abc}$

(*iii*) $\dfrac{-1}{a} : \dfrac{a}{-1}$ (*iv*) $\dfrac{a}{a+b} : \dfrac{b}{a-b}$

Ans. (*i*) $\dfrac{a}{b}$ (*ii*) $\dfrac{c}{y}$ (*iii*) $\dfrac{1}{a^2}$ (*iv*) $\dfrac{a(a-b)}{b(a+b)}$

Ex. 56. 40 men finish a job in 10 days. How many men are required to finish the job in 5 days?

Ans. $x : 40 :: 10 : 5 \quad \therefore x = 80$

Ex. 57. In a camp there are provisions for 500 men for 20 days. How long will the provisions last if 750 men are camped there? (Inverse proportion)

Ans. $\dfrac{750}{500} = \dfrac{20}{x} \quad \therefore x = 13\frac{1}{3}$ days

Ex. 58. Equal sums of money are invested at 3% and 6% respectively. If the income at 3% is 250, what is the income at 6%? **Ans.** 500

Ex. 59. Find x, y, z so that the sets $\{8, x, 20, y\}$ and $\{3, 4, z, 12\}$ are proportional. (Here $8/3 = x/4 = 20/z = y/12$) **Ans.** 32/3, 32, 15/2

Ex. 60. If $\frac{a}{b} = \frac{c}{d}$, add 1 on both the sides, *i.e.*, $\frac{a}{b} + 1 = \frac{c}{d} + 1$ and prove that,

(*i*) $\frac{a+b}{b} = \frac{c+d}{d}$ similarly (*ii*) $\frac{a}{a+b} = \frac{c}{c+d}$.

Ex. 61. If $\frac{a}{b} = \frac{c}{d} = \frac{e}{f}$ prove that

(*i*) $\frac{a+b}{a} = \frac{c+d}{c}$ (**Hint :** Write $\frac{b}{a} + 1 = \frac{d}{c} + 1$)

(*ii*) $\frac{a-b}{a} = \frac{c-d}{d}$ (*iii*) $\frac{a}{a-b} = \frac{c}{c-d}$

(*iv*) $\frac{a-b}{a+b} = \frac{c-d}{c+d}$ (*v*) $\frac{a+b}{a-b} = \frac{c+d}{c-d}$

(*vi*) $\frac{a+c}{b+d} = \frac{a}{b} = \frac{c}{d}$ (*vii*) $\frac{a+c}{b+d} = \frac{c+e}{d+f} = \frac{a+e}{b+f} = \frac{a+c+e}{b+d+f}$

(To get the last ratio add the numerators and denominators, respectively, of the first three ratios).

Ex. 62. If $\frac{a}{b} = \frac{2}{3}$ find the value of $\frac{3a+4b}{6a-5b}$.

Ans. $$\frac{3a+4b}{6a-5b} = \frac{\frac{3a}{b}+4}{\frac{6a}{b}-5} = \frac{3\times\frac{2}{3}\times 4}{6\times\frac{2}{3}-5} = \frac{2+4}{4-5} = \frac{6}{-1} = -6$$

Ex. 63. If $\frac{a}{b} = \frac{2}{3}$ find the value of $\frac{2a^2+3b^2}{a^2+b^2}$. **Ans.** $\frac{35}{13}$

Ex. 64. Two numbers are in the ratio $\frac{3}{4}$. If 4 is subtracted from each they are in the ratio $\frac{5}{7}$. Find the numbers.

Let the numbers be $3a$ and $4a$. Then $\frac{3a-4}{4a-4} = \frac{5}{7}$

$\therefore$ $21a - 23 = 20a - 20$ $\therefore$ $a = 8$

$\therefore$ The numbers are $3a$, $4a$ *i.e.*, 24 and 32.

Ex. 65. Find the number which when subtracted from each of the terms of the ratio $\frac{19}{31}$ makes it $\frac{1}{4}$. **Ans.** 15

1.7 RATIO OF DIVISION OF A LINE SEGMENT

Let R be a point on the line segment PQ. Then we can say that PQ is divided at R in the ratio $\frac{PR}{RQ}$.

It is a convention to regard distances measured along the right as positive and those along the left as negative.

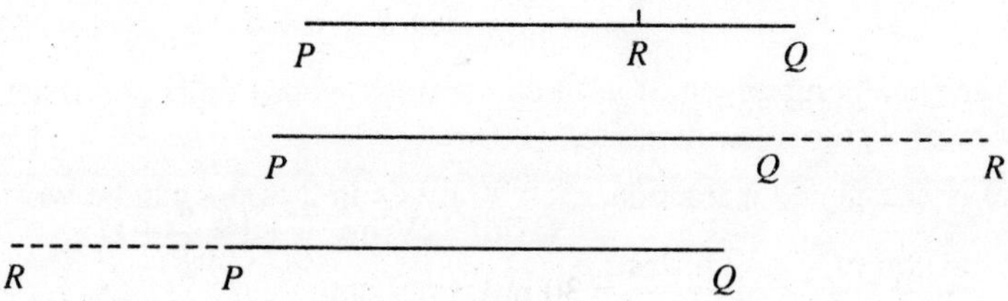

If R lies between P, Q so that PR, PQ are both positive, then PR/PQ is positive and this is a case of internal division.

If R lies to the right of Q so that PR is positive and RQ is negative then $\frac{PR}{RQ}$ is negative and this is a case of external division.

If R lies to the left of P so that PR is negative and RQ is positive then $\frac{PR}{RQ}$ is negative and this is also a case of external division.

In each case if $PR = k_1$, $PQ = k_2$ then $\frac{PR}{RQ} = \frac{k_1}{k_2}$ for internal division and

$\frac{PR}{RQ} = -\frac{k_1}{k_2}$ for external division.

Percentage : A percentage is the numerator of a fraction whose denominator is 100. In other words it is a fractional or decimal part of 100.

Two per cent may be expressed in any one of the following ways:

$$2\%,\ 0.02,\ \frac{2}{100} \text{ or } \frac{1}{50}$$

Addition, subtraction, multiplication and division of percentages are done appropriately.

A per cent figure can be written as a decimal or fraction, 50 per cent can be written 0.50 or $\frac{50}{100}$ or $\frac{1}{2}$.

A decimal or fraction can be written as a per cent.

$$0.25 \text{ means } 25\%,\ \frac{10}{100} \text{ means } 10\%$$

Ex. 66. Find 15% of 500. **Ans.** $500 \times \frac{15}{100} = 75$

Ex. 67. What per cent is 200 of 500? **Ans.** $\frac{200}{500} \times 100 = 40$ per cent

Ex. 68. What is the quantity whose 20% is 100.

Ans. $100 \div 20\%$ or $100 \div \frac{1}{5}$ *i.e.*, 500

Rate : A *rate* is a comparison of unlike quantities whose units are given, *e.g.*, miles per hour, number of lectures per day, etc.

With a rate is associated a fraction, *e.g.*, 60 miles in 2 hours can be written

$$\frac{60 \text{ miles}}{2 \text{ hours}} = \frac{30 \text{ miles}}{1 \text{ hour}} = 30 \text{ miles per hour.}$$

$$= \frac{44 \text{ ft}}{1 \text{ sec}} = 44 \text{ ft. per sec.}$$

$$\frac{\$\,100}{5 \text{ kg}} = \frac{\$\,20}{1 \text{ kg}} = \frac{2 \text{ cents}}{1 \text{ gram}}$$

Ex. 69. 20 articles were sold for Rs 100. What was the rate at which they were sold?

Ans. 5

Ex. 70. A train covers 200 miles in 5 hours. What is the rate at which it is travelling?

Ans. 40 mph

Direct Variation : A quantity y is said to vary directly as another quantity x if changes in x bring about changes in y in the same ratio.

This relation between the two variables x and y can be written

$$\frac{y}{x} = k \text{ where } k \text{ is a constant quantity}$$

The relation is usually described by saying

y varies directly as x

or symbolically, $y \propto x$

When y varies as x it follows that x varies as y.

Thus if $\frac{y}{x} = k$

then $x = \frac{1}{k} \cdot y$

Hence, x varies directly as y since $\frac{1}{k}$ is a constant quantity.

Ex. 71. Price level (P) varies directly as quantity of money (M). What will happen if quantity of money is

(*i*) doubled? (*ii*) trebled? (*iii*) halved?

Since $P \propto M$

$\therefore \quad P = kM \quad \text{or} \quad \frac{P}{M} = k$, a constant quantity

If M is doubled, P is doubled and so on.

Ex. 72. Supply (S) varies directly as price (P)

$$S \propto P$$

This statement implies that if price rises, supply rises in the same proportion.

Inverse Variation: If y varies as the reciprocal of x we say that y varies inversely as x.

$$y \propto \frac{1}{x} \text{ or } y = \frac{k}{x} \text{ where } k \text{ is a constant.}$$

Ex. 73. Demand (D) varies inversely as price (p) :

$$D \propto \frac{1}{p}$$

This statement implies that if price rises, demand falls in the same proportion.

Joint Variation: If y varies jointly as u and v, we write

y varies as the product of u and v

i.e., $y \propto uv$ or $y = kuv$

This means that if y varies as u when v is constant and y varies as v when u is constant then y varies jointly as u and v when both u and v vary.

If $y = k\frac{u}{v}$ it means that y varies directly as u and inversely as v, *i.e.,* $y \propto \frac{u}{v}$.

Ex. 74. If k is the constant of proportionality we can write

(*i*) y varies as the cube of x

$$y = kx^3$$

(*ii*) x varies as the square of p

$$x = kp^2$$

(*iii*) y varies directly as x and inversely as the sum of x and u

$$y = \frac{kx}{x + u}.$$

Ex. 75. If (x_1, y_1), (x_2, y_2) are points on the graph of $xy = k$ then

$$x_1y_1 = k, x_2y_2 = k$$

$$\therefore \quad x_1y_1 = x_2y_2$$

$$\therefore \quad \frac{x_1}{x_2} = \frac{y_2}{y_1} \text{ i.e., } y \text{ varies inversely as } x.$$

Ex. 76. If u varies inversely as v find v_2 when

$$u_1 = 5, v_1 = 10, u_2 = 8; u_1v_1 = u_2v_2$$

$$\frac{u_1}{u_2} = \frac{v_2}{v_1} \quad \therefore \quad \frac{5}{8} = \frac{v_2}{10}$$

$$\therefore \quad v_2 = \frac{50}{8} = 6\frac{1}{4}$$

Ex. 77. If income tax (T) varies as the square of the excess (E) of a man's income over Rs. 3,000 and on an income of Rs. 6,000 it amounts to Rs. 100 find how much income tax will a man with an income of Rs. 9,000 pay?

$T = kE^2$. Now when $E = 6000 - 3000 = 3000$ then $T = 100$

$$\therefore \quad 100 = k\,(3000)^2 \text{ or } k = \frac{1}{90000}$$

When $E = 9000 - 3000 = 6000$, the income tax to be paid is

$$T = \frac{1}{90000}E^2 = \frac{1}{90000}(6000)^2 = 400.$$

Ex. 78. If (x_1, y_1), (x_2, y_2) are points on a line $y = mx$ ...(1)

then
$$\frac{y_1}{x_1} = m, \quad \frac{y_2}{x_2} = m$$

therefore we get the proportional relation

$$\frac{y_1}{x_1} = \frac{y_2}{x_2} \qquad \text{...(2)}$$

$$\frac{y_1}{y_2} = \frac{x_1}{x_2} \qquad \text{...(3)}$$

(1), (2) and (3) are different ways of saying that y is directly proportional to x.

m is the constant of proportionality.

Ex. 79. If $y \propto x$ and $y = 8$ when $x = 15$, find y when $x = 10$.

$$y = kx \quad \therefore \quad 8 = k\,15 \quad \therefore \quad k = \frac{15}{8}$$

$$\therefore \quad y = \frac{15}{8}x,$$

When $x = 10,\ y = \dfrac{15}{8} \cdot 10 = \dfrac{75}{4}$.

Ex. 80. y varies directly as x_1 and inversely as x_2. It is known that when $y = 20$, $x_1 = 30$, $x_2 = 12$. Find y when $x_1 = 12$, $x_2 = 4$.

We have $y = k\dfrac{x_1}{x_2}$ or $20 = k\dfrac{30}{12}$ $\therefore$ $k = 8$

$\therefore \qquad y = 8 \cdot \frac{x_1}{x_2}$ Substituting the new values

$$y = 8 \cdot \frac{12}{4} = 24.$$

Ex. 81. If $x \propto y$ prove that $x^2 + y^2 \propto x^2 - y^2$

$$\frac{x}{y} = k \quad \therefore \quad \frac{x^2}{y^2} = k^2$$

$$\therefore \qquad \frac{x^2 + y^2}{x^2 - y^2} = \frac{k^2 + 1}{k^2 - 1}, \text{ a constant quantity}$$

$$\therefore \qquad x^2 + y^2 \propto x^2 - y^2.$$

Ex. 82. If $x \propto u$ and $y \propto u$ prove that

$$x^2 - y^2 \propto u^2$$

Ex. 83. If $x + y \propto x - y$ show that $x^2 + y^2 \propto x^2 - y^2$

(Hint: $\frac{x + y}{x - y} = k_1$ where k_1 is a constant

$\therefore \qquad \frac{y}{x} = k \quad$ where $k = \frac{k_1 + 1}{k_1 - 1}$, another constant)

Ex. 84. If a, b, c, d, are constants prove that $ax + by \propto cx + dy$.

Ex. 85. Two persons A and B are awarded pensions in proportion to the number of years they served. A served 8 years longer than B and received Rs. 400 p.m. more in pension. If A had served 12 years longer than B his pension would have been double that of B. How long did each of them serve and what were their pensions?

Ans. A 20 years, Rs. 12000
B 12 years, Rs. 7200

1.8 SEQUENCES

A sequence is a set of numbers arranged in a definite order corresponding to the order of natural numbers 1, 2, 3,....

The numbers in the range of the sequence are called the *elements* of the sequence.

a_n is the *general* term of the sequence $a_1, a_2, a_3, \ldots, a_n, \ldots$

Often, this sequence is written $\{a_n\}$. A sequence is known if its nth term is known.

If $f(n)$ is a function, it gives the sequence $f(1), f(2), f(3) \ldots,$

A sequence is said to be *finite* if it has a first and a last number.

4, 6, 8, 10, 12 is a finite sequence. A sequence is said to be *infinite* if every term in the sequence is followed by another. The statement "$n \to \infty$" or "n tends to infinity" expresses unending numbers coming.

$\frac{1}{2}, \frac{2}{3}, \frac{3}{4}, \frac{4}{5}$ ...... is an infinite sequence.

A sequence $\{a_n\}$ is said to be

(*i*) increasing if for each positive integer n,

$$a_n < a_{n+1}$$

(*ii*) decreasing if for each positive integer n,

$$a_n > a_{n+1}.$$

$1^2, 2^2, 3^2, 4^2, \ldots\ldots$ is an increasing sequence.

$1, \frac{1}{2}, \frac{1}{4}, \frac{1}{8}, \ldots\ldots$ is a decreasing sequence.

$0, -1, -2, -3, \ldots\ldots$ is a decreasing sequence.

All these three sequences are infinite sequences. The first tends to $+\infty$ and the last tends to $-\infty$. The second tends to zero.

A sequence may *oscillate* without tending to any particular value.

The sequences $+1, -1, +1, -1, +1, -1, \ldots\ldots$ and $1, \frac{1}{2}, 1, \frac{1}{3}, 1, \frac{1}{4}, \ldots$ are oscillatory.

Limit of a sequence

A sequence $\{a_n\}$ is said to have a limit l if there is a number l such that a_n can be made as close to l as we wish by taking n sufficiently large. We write this

$\lim_{n\to\infty} a_n = l$ which can be read :

Limit of a_n as n tends to infinity is l.

A sequence which has a limit is said to be convergent. A sequence which has no limit is said to be divergent.

Ex. 86. Given the sequence $a_n = \frac{1}{n} + 2$

As $n \to \infty, \frac{1}{n} \to 0$ $\quad\therefore\quad \lim_{n\to\infty} a_n = 2.$

Ex. 87. Prove that $\lim_{n\to\infty} \frac{1}{1+1/n} = 1$ where $a_n = \frac{1}{1+1/n}$.

Ex. 88. Prove that $\lim_{n\to\infty} \frac{3}{n^2} = 0$ where $a_n = \frac{3}{n^2}$.

Ex. 89. Prove that 1 is not the limit of the sequence $a_n = (-1)^n$.

Ex. 90. $$\lim_{n\to\infty} \frac{n^2 - 5n + 2}{3n^2 + n + 1} = \lim_{n\to\infty} \frac{\frac{n^2}{n^2} - \frac{5n}{n^2} + \frac{2}{n^2}}{\frac{3n^2}{n^2} + \frac{n}{n^2} + \frac{1}{n^2}} = \lim_{n\to\infty} \frac{1 - \frac{5}{n} + \frac{2}{n^2}}{3 + \frac{1}{n} + \frac{1}{n^2}}$$

(dividing the numerator and denominator by n^2)

$$= \frac{1}{3}$$

because the terms with n in the denominator vanish.

Ex. 91. $\lim_{n\to\infty} \frac{n^2 + 4n}{5n^3} = \frac{1}{5}$.

The nth Term

We have seen that a sequence is a set of numbers arranged according to some definite rule *i.e.,* arranged in a definite order according to some law; often there is some regular pattern of growth or decline or something else.

(*i*) The numbers 2, 4, 6, 8, 10, form a sequence in which each term is obtained by adding 2 to the preceding term.

(*ii*) The numbers 2, 4, 8, 16, 32, ... (*i.e.,* 2, 2^2, 2^3, 2^4, 2^5, ...) form a sequence in which each term is obtained by multiplying the preceding term by 2.

The terms of a sequence are denoted by T_1, T_2, T_3, T_4, ...

In case (*i*), the second term is $T_2 = 4 = 2 \times 2$. We can write $T_3 = 6 = 2 \times 3$, $T_4 = 8 =$ $\times$ 4 etc. The nth term is $T_n = 2n$.

In case (*ii*), $T_1 = 2$, $T_2 = 4 = 2^2$ $T_3 = 8 = 2^3$, ..., $T_n = 2^n$.

The nth term T_n is the general term of a sequence. Any term can be written by putting the appropriate value of n in the expression for T_n.

Ex. 92. If $T_n = n^2$, $T_1 = 1^2$, $T_2 = 2^2$, $T_3 = 3^2$ etc.

Ex. 93. If $T_n = (2n + 1)^3$, $T_1 = (2 \times 1 + 1)^3 = 27$

$$T_2 = (2 \times 2 + 1)^3 = 125 \text{ etc.}$$

Ex. 94. If the sequence is 1, 3, 5, 7, ...

Can you find the nth term?

$$T_1 = 1 = 2 \times 1 - 1$$

$$T_2 = 3 = 2 \times 2 - 1$$

$$T_3 = 5 = 2 \times 3 - 1$$

....

$$T_n = 2 \times n - 1 = 2n - 1$$

Ex. 95. Find the nth term for the sequence

2^2, 5^2, 8^2, ...**Ans.** $T_n = (3n - 1)^2$

Finite and infinite sequences. A sequence is finite if it has a finite number of terms.

A sequence is an infinite sequence if it has an indefinite number of terms.

The sequence 2, 5, 8, ..., 32 is finite.

The sequence 2, 5, 8,...... is infinite.

Ex. 96. Write the terms of the sequence which has

$$T_n = \frac{6n^2}{(2n-1)\,(n+1)}.$$ **Ans.** $3, \frac{8}{3}, \frac{27}{10}, \frac{96}{35}, \ldots$

Ex. 97. Write the terms of the sequence

$T_n = 2n + 1$ if n is odd

$= 2n + 2$ if n is even

$T_1 = 2 \times 1 + 1 = 3,$ $T_2 = 2 \times 2 + 2 = 6$

$T_3 = 2 \times 3 + 1 = 7,$ $T_4 = 2 \times 4 + 2 = 10$

$T_5 = 2 \times 5 + 1 = 11,$ $T_6 = 2 \times 6 + 2 = 14$ etc.

Ex. 98. Write the terms of $T_n = \frac{n^3}{(2n-1)^2}$. **Ans.** $1, \frac{8}{9}, \frac{27}{25}, \ldots\ldots$

1.9 ARITHMETIC, GEOMETRIC AND HARMONIC PROGRESSIONS

An *arithmetic* progression is a sequence in which each term is obtained from the preceding term by adding to it a fixed number called the *common difference.*

A *geometric* progression is one in which each term is obtained from the preceding term by multiplying it by a fixed number called the *common ratio.*

The terms of a *harmonic progression* are reciprocals of the terms of an arithmetic progression.

Ex. 99. 2, 4, 6, 8, 10,... is an arithmetic progression with *common difference* (C.D.) 2.

Ex. 100. 2, 4, 8, 16, 32, ... is a geometric progression with *common ratio* (C.R.) 2.

Ex. 101. 10, 9, 8, 7, ... is an arithmetic progression with C.D. – 1.

Ex. 102. $1, \frac{1}{2}, \frac{1}{4}, \frac{1}{8}, \ldots\ldots$ is a geometric progression with C.R. $\frac{1}{2}$.

Ex. 103. 2, 5, 6, 8, 18, 11, 54 ...... is a mixed series with alternative terms in arithmetic and geometric progression.

Ex. 104. $1, \frac{1}{2}, \frac{1}{3}, \ldots\ldots, \frac{1}{n}$ ...... is a harmonic progression.

1.10 ARITHMETIC PROGRESSION (AP)

In an AP, the difference between any term and the preceding term remains constant. This constant is called *common difference*, denoted by d.

The rth term of the AP is denoted by T_r.

The 1st term T_1 is denoted by a

The 2nd term, $T_2 = a + d$

The 3rd term, $T_3 = a + 2d$

The nth term, or the last term $T_n = a + (n-1)d = l$ say

The sum of the first n terms of the AP is written

$$S_n = a + (a + d) + (a + 2d) + \ldots\ldots + (a + (n - 1)d)$$

$$= \frac{n}{2}\{2a + (n - 1)d\} = \frac{n}{2}[a + l]$$

Some Characteristics of AP

(*i*) If a, b, c are in AP then $b - a = c - b$

$$\therefore \quad 2b = a + c \quad \text{or} \quad b = \frac{a + c}{2}$$

If a, b, c, d are in AP then $a + d = b + c$

(*ii*) If each term of an AP is multiplied (or divided) by a constant, the series continues to remains an AP.

If a constant is added to (or subtracted from) an AP, the series continues to remain an AP.

(*iii*) If 3 numbers are in AP, take them as $a - d, a, a + d$

If 5 numbers are is AP, take $a - 2d, a - d, a, a + d, a + 2d$

These are odd number of terms. When there is an even number of terms *e.g.*, when 4 terms are in AP, take $a - 3d, a - d, a + d, a + 3d$ and so on.

Ex. 105. The sum of the first n natural number is $\frac{n(n + 1)}{2}$.

Put $a = 1,\ d = 1,\ n = n$ in the formula above for S_n

$$S_n = \frac{n}{2}\{2a + (n - 1)d\} = \frac{n}{2}\{2 \times 1 + (n - 1)1\} = \frac{n(n + 1)}{2}$$

Ex. 106. Find the sum to n terms for the AP 2 + 4 + 6 +......

Here $a = 2, d = 2$

$$S_n = \frac{n}{2}\{2 \times 2 + (n - 1)2\} = n(n + 1)$$

Ex. 107. Sum to 10 terms : $S_{10} = 3 + 6 + \ldots\ldots$ upto 10 terms

$$= \frac{10}{2}\{2 \times 3 + (10 - 1)3\} = 165$$

Ex. 108. The sum of the first 10 natural numbers is

$$S = \frac{10 \times (10 + 1)}{2} = 55.$$

Ex. 109. The sum of the arithmetic progression

5 + 10 + 15 + 20 + 25 + 30 is given by

$$S_a = \frac{n}{2}(2a + (n - 1)d\} \text{ where } n = 6,\ a = 5,\ d = 5$$

$$= \frac{6}{2}\{10 + (6 - 1)5\} = 105.$$

Ex. 110. A loan of Rs. 3250 is to be repaid by paying Rs. 20 in the first month, increasing per month by Rs. 15 in subsequent months. How many months it will take to clear the loan?

$$a = 20,\ d = 15,\ S_n = \frac{n}{2}\{2a + (n-1)\,d\}$$

where n is the number of months required to clear the loan.

$$3250 = \frac{n}{2}\{2 \times 20 + (n-1)\,15\}$$

Simplifying $3n^2 + 5n - 1300 = 0$

or $(n - 20)(3n + 65) = 0$

Since n should be positive, the negative answer is discarded. Hence $n = 20$ months.

Ex. 111. A firm produces 1000 units of its product in the first year and produced 14500 units in 10 years. Find the increase per annum and also find the output in the fifteenth year.

Here $a = 1000, \quad n = 10$

$\therefore \quad 14500 = \frac{10}{2}\{2000 + 9d\}$ using $S_n = \frac{n}{2}\{2a + (n-1)d\}$

$\therefore \quad 2900 - 2000 = 9d$ or $d = 100$

The increase per annum is 100 units.

In 15 years, output is given by $\frac{15}{2}\{2000 + 1400\} = 25500$

In 14 years, output is $\frac{14}{2}\{2000 + 1300\} = 23100$

Output in the 15th year $25500 - 23100 = 2400$

Alternatively use $T_n = a + (n-1)d$, the nth term

$$T_{15} = a + 14d$$
$$= 1000 + 14 \times 100 = 2400$$

Ex. 112. A firm produces 600 units in the third year and 700 units in the seventh year. Find the production that was in the (*a*) first year (*b*) fifth year (*c*) in the first five years.

Ans. 550, 660, 3000

Ex. 113. Find three numbers is AP whose sum is 12 and the sum of whose squares is 56.

Let the numbers be $a - d,\ a,\ a + d$.

$$a - d + a + a + d = 12 \quad i.e.,\ 3a = 12 \text{ or } a = 4$$
$$(a - d)^2 + a^2 + (a + d)^2 = 56 \quad i.e.,\ 3a^2 + 2d^2 = 56$$

$\therefore \quad 3 \times 16 + 2d^2 = 56$

or $\quad 2d^2 = 8$ or $d \pm 2$

If $d = +2$ the numbers are $4 - 2, 4, 4 + 2$ or 2, 4, 6

If $d = -2$ the numbers are $4 + 2, 4, 4 - 2$ or 6, 4, 2.

Ex. 114. Find four numbers in AP such that their sum is 16 and their product is 105.

Let the numbers be $a - 3d, a - d, a + d, a + 3d$

Note. Here the common difference is $2d$. The number of terms is even.

$$a - 3d + a - d + a + d + a + 3d = 16 \text{ i.e., } 4a = 16$$

$\therefore$ $a = 4$

$$(a - 3d)(a - d)(a + d)(a + 3d) = 105$$

$$(a^2 - 9d^2)(a^2 - d^2) = 105$$

$\therefore$ $(16 - 9d^2)(16 - d^2) = 105$

$$9d^4 - 160d^2 + 151 = 0$$

or $(d^2 - 1)(9d^2 - 151) = 0$

The possible answer is $d^2 = 1$ or $d = \pm 1$ (we reject $d^2 = 151/9$)

If $d = +1$ the numbers are $a - 3d, a + d, a + d, a + 3d$

i.e., $4 - 3, 4 - 1, 4 + 1, 4 + 3$ or $1, 3, 5, 7$

If $d = -1$, the numbers are 7, 5, 3, 1.

Ex. 115. Find (*i*) the sum of the first n odd natural numbers. **Ans.** n^2

(*ii*) the sum of odd integers between 50 and 150. **Ans.** 4950

Hint: Find $\{1 + 3 + \ldots\ldots + 149\} - \{1 + 3 + \ldots\ldots + 49\}$

Ex. 116. The sum of n terms of an AP is $n^2 + 2n$. Find T_n and d.

Ans. $T_n = 2n + 1, d = 2$

Ex. 117. If a, b, c are in AP then show that the following are in AP:

(*i*) $b + c, c + a, a + b$

(*ii*) $\dfrac{a(b+c)}{bc}, \dfrac{b(c+a)}{ca}, \dfrac{c(a+b)}{ab}$ (*iii*) $\dfrac{1}{bc}, \dfrac{1}{ca}, \dfrac{1}{ab}$

Ex. 118. If a^2, b^2, c^2 are in AP show $\dfrac{1}{b+c}, \dfrac{1}{c+a}, \dfrac{1}{a+b}$ are in AP

Hint: Use $2b^2 = a^2 + c^2$

Ex. 119. How many terms of 54, 51, 48, ... should be taken to have a total of 513?

$$a = 54, \quad d = -3$$

$$513 = \frac{n}{2}\{2 \times 54 + (n - 1)(-3)\}$$ **Ans.** $n = 18$ or 19

Ex. 120. The sum of m terms of an AP is equal to the sum of its n terms. Show that the sum of $m + n$ terms is zero.

$$S_m = S_n \text{ i.e., } \frac{m}{2}\{2a + (m - 1)d\} = \frac{n}{2}\{2a + (n - 1)d\}$$

$\therefore$ $-2a = (m + n - 1)d$

$$S_{m+n} = \frac{m+n}{2}\{2a + (m + n - 1)d\} = \frac{m+n}{2}\{2a - 2a\} = 0$$

Ex. 121. The sums of n terms of two AP's are in the ratio $7n + 1 : 4n + 27$. Find the ratio of their nth terms.

If a_1, a_2 are their first terms and d_1, d_2 their common differences, the ratio of sum to n terms is

$$\frac{S_1}{S_2} = \frac{7n+1}{4n+27} \quad i.e., \quad \frac{\frac{n}{2}\{2a_1+(n-1)d_1\}}{\frac{n}{2}(2a_2+(n-1)d_2\}} = \frac{7n+1}{4n+27}$$

We have $$\frac{2a_1+(n-1)d_1}{2a_2+(n-1)d_2} = \frac{7n+1}{4n+27} \qquad ...(1)$$

To find $$\frac{T_{n1}}{T_{n2}} = \frac{a_1+(n-1)d_1}{a_2+(n-1)d_2}$$

put in (1) $\quad n - 1 = 2(p - 1)$ so that $$\frac{2a_1+2(p-1)d_1}{2a_2+2(p-1)d_2} = \frac{7(2p-1)+1}{4(2p-1)+1}$$

$$\therefore \quad \frac{a_1+(p-1)d_1}{a_2+(p-1)d_2} = \frac{14p-6}{8p+23}$$

Replacing p by n we have $$\frac{T_{n1}}{T_{n2}} = \frac{14n-6}{8n+23}$$

Ex. 122. If S_1, S_2, S_2 are sums to n terms of 3 APs, if the first term of each is 1 and their common differences are 1, 2, 3, show that $2S_2 = S_1 + S_3$.

Ex. 123. If a, b, c are is AP, show

$$(a - c)^2 = 4\,(b^2 - ac)$$

Ex. 124. In an AP, if $S_p = a$, $S_q = b$, $S_r = c$, show

$$\frac{a}{p}(q-r)+\frac{b}{q}(r-p)+\frac{c}{r}(p-q)=0$$

Ex. 125. The sums of n terms of two AP's are in the ratio $\frac{4n-10}{3n+13}$, find the ratio of their 12th terms. **Ans.** 1

Ex. 126. If S_n, S_{2n}, S_{3n} are the sums of the first n, $2n$, $3n$ terms of an AP show

$$S_{3n} = 3(S_{2n} - S_n)$$

Ex. 127. In any AP, show that the sum of the second half of $2n$ terms is $\frac{1}{3}$ of the sum of $3n$ terms.

Ex. 128. $S_1, S_2, \ldots\ldots, S_m$ are sums to n terms of m APs whose first term and common difference are $1, 2, 3, \ldots\ldots, m$. Show $S_1 + S_2 + \ldots + S_m = \frac{mn}{4}(m + 1)(n + 1)$.

1.11 GEOMETRIC PROGRESSION (GP)

In a G.P., the ratio of any term to the preceding term remains constant. This constant is called the Common Ratio (CR), denoted by r.

The nth term of a GP is $T_n = ar^{n-1}$

The 1st term T_1 is denoted by a

The 2nd term $T_2 = ar$

The 3rd term $T_3 = ar^2$

The nth or last term $T_n = ar^{n-1}$

The sum of n terms of the series is written

$$S_n = a + ar + ar^2 + \ldots\ldots + ar^{n-1}$$

$$= a\frac{r^n - 1}{r - 1} \text{ if } r > 1$$

$$= a\frac{1 - r^n}{1 - r} \text{ if } r < 1$$

If $r < 1$ and the number of terms is infinite *i.e.*, $n \to \infty$ then

$$S_n = \frac{a}{1 - r}$$

Some characteristics of GP

(*a*) If a, b, c are in GP, $\frac{b}{a} = \frac{c}{b}$ *i.e.*, $b^2 = ac$

If a, b, c, d are in GP, $ad = bc$ or $b/a = d/c$

(*b*) If every term of a GP is multiplied or divided by a constant or is raised to a given power, the series continues to remain a GP.

The reciprocals of terms of a GP continue to be in GP.

(*c*) If there are 3 numbers in GP take them as $\frac{a}{r}$, a, ar

If 5 numbers are in GP take $\frac{a}{r^2}$, $\frac{a}{r}$, a, ar, ar^2

These are odd number of terms. With even number of terms, say 4 terms, take $\frac{a}{r^3}$, $\frac{a}{r}$, ar, ar^3.

Ex. 129. $S_n = 1 + \frac{1}{2} + \frac{1}{2^2} + \frac{1}{2^3} + \ldots + \frac{1}{2^{n-1}} = 1\frac{1 - \left(\frac{1}{2}\right)^n}{1 - \frac{1}{2}} = 2\left[1 - \frac{1}{2^n}\right]$

Ex. 130. Sum to 4 terms, $S_4 = 2 + 4 + 8 + 16$

$$a = 4,\ r = 2,\ n = 4,\ r > 1 \ \therefore\ S_4 = a\,\frac{r^4 - 1}{r - 1} = 2 \times \frac{2^4 - 1}{2 - 1} = 30$$

Ex. 131. $S_5 = 1 + \frac{1}{3} + \frac{1}{3^2} + \frac{1}{3^3} + \frac{1}{3^4} = 1 \times \frac{1-(1/3)^5}{1-1/3} = \frac{3}{2}\left[1 - 1/3^5\right] = \frac{121}{81}$

Ex. 132. The sum of the geometric progression

$$1 + \frac{1}{2} + \frac{1}{4} + \frac{1}{8} + \frac{1}{16} \text{ is given by}$$

$$S_5 = \frac{a(1-r^n)}{1-r} \text{ where, } a = 1, \ r = \tfrac{1}{2}, \ n = 5$$

$$= \frac{1[1-(\frac{1}{2})^5]}{1-\frac{1}{2}} = \frac{1-\frac{1}{32}}{\frac{1}{2}} = \frac{31}{16}.$$

Ex. 133. The sum of the infinite series

$$1 + \tfrac{1}{2} + \tfrac{1}{4} + \ldots\ldots \text{ is given by } \frac{1}{1-\frac{1}{2}} = 2. \left(\text{Use } S_n = \frac{a}{1-r}\right)$$

Ex. 134. The *r*th term of 72, – 18, 9/2,... is $(-1)^{n-1} \frac{9}{2^{2n-5}}$.

Ex. 135. If *a, b, c,* are the *p*th, *q*th, *r*th terms of a GP, show that $a^{q-r} b^{r-p} c^{p-q} = 1$.

If f is the first term $T_p = fg^{p-1} = a$

where g is the common ratio.

Similarly $T_q = fg^{q-1} = b$, $T_r = fg^{r-1} = c$

Substituting for *a, b, c*

$$(fg^{p-1})^{q-r} (fg^{q-1}) (fg^{r-1})^{p-q}$$

$$= f^{q-r+r-p+p-q} g^{(q-r)(p-1)+(r-p)(q-1)+(p-q)(r-1)} = f^0 g^0 = 1$$

Ex. 136. If *a, b, c* are in GP and $a^{1/x} = b^{1/y} = c^{1/z}$ then *x, y, z* are in AP.

(**Hint:** Use $b^2 = ac$)

Ex. 137. Find three numbers *a, b, c* between 2 and 18 such that

(*i*) their sum is 25.

(*ii*) 2, *a, b* are the consecutive term of an AP.

(*iii*) *b, c,* 18 are the consecutive terms of a GP.

We have $a + b + c = 25$, $2a = 2 + b$, $c^2 = 18b$

Solve by substituting, $c = -24, 12$

$\therefore$ $c = 12$ ($\because -24$ does not lie between 2, 18)

$b = 8$

$a = 5$

Ex. 138. Find the sum to *n* terms : 6 + 66 + 666 +...

Write $S_n = 6(1 + 11 + 111 + \ldots\ldots)$

$$= \frac{6}{9}(9 + 99 + 999 + \ldots\ldots)$$

$$= \frac{6}{9}\{(10-1) + (10^2-1) + (10^3-1) + \ldots\ldots n \text{ term}\}$$

$$= \frac{6}{9}\,(10 + 10^2 + 10^3 + \ldots\ldots n \text{ terms } - 1 - 1\ldots\ldots n \text{ terms})$$

$$= \frac{6}{9}\left[10\frac{10^n - 1}{10 - 1} - n\right] = \frac{60}{81}(10^n - 1) - \frac{6n}{9}$$

Ex. 139. Similarly $0.4 + 0.44 + 0.444 + \ldots\ldots = \frac{4}{9}\,[0.9 + 0.99 + 0.999 + \ldots\ldots]$

Ex. 140. 0.125 125 125 ......

$$= \frac{125}{10^3} + \frac{125}{10^6} + \frac{125}{10^9} + \ldots = \frac{125}{10^3}\left[1 + \frac{1}{10^3} + \frac{1}{10^6} + \ldots\right]$$

$$= \frac{125}{10^3}\;\frac{1}{1 - \frac{1}{10^3}} = \frac{125}{999}$$

Ex. 141. If the pth, qth, rth terms of an AP are in GP show that the common ratio of the GP is $\frac{q-r}{p-q}$.

1.12 HARMONIC PROGRESSION (H.P.)

If the reciprocals of a sequence form an arithmetic progression then the sequence is in Harmonic Progression.

$1, \frac{1}{2}, \frac{1}{3}, \frac{1}{4}, \ldots\ldots$ is in H.P. since the series of reciprocals 1, 2, 3, 4, ... is in A.P.

Similarly $\frac{1}{8}, \frac{1}{13}, \frac{1}{18}, \ldots\ldots$ is in H.P.

The general form of an H.P. may be written

$$\frac{1}{a}, \frac{1}{a+d}, \frac{1}{a+2d}, \ldots\ldots\ldots$$

There is no direct formula for the sum to n terms of an H.P.

Ex. 142. If $a, b, c,$ are three consecutive terms of an H.P.

then $\frac{1}{a}, \frac{1}{b}, \frac{1}{c}$ are in A.P.

$$\therefore \quad \frac{1}{b} - \frac{1}{a} = \frac{1}{c} - \frac{1}{b}$$

$$\therefore \quad \frac{a-b}{ab} = \frac{b-c}{bc} \qquad \therefore \quad \frac{a-b}{b-c} = \frac{a}{c}$$

Some useful definitions

(*i*) If x, A, y are in A.P. then A is the arithmetic mean of x, y.

(*ii*) If x, G, y are in G.P. then G is the geometric mean of x, y.

(*iii*) If x, H, y are in H.P. then H is the harmonic mean of x, y.

In case (*i*) $A - x = y - A \quad \therefore \; 2A = x + y$ or $A = \dfrac{x+y}{2}$

In case (*ii*) $\dfrac{x}{G} = \dfrac{G}{y} \quad \therefore \; G^2 = xy$ or $G = \sqrt{xy}$

In case (*iii*) $\dfrac{1}{H} - \dfrac{1}{x} = \dfrac{1}{y} - \dfrac{1}{H} \quad \therefore \; \dfrac{2}{H} = \dfrac{x+y}{xy}$ or $H = \dfrac{2xy}{x+y}$

From the above it can be seen that $G^2 = AH$

$$AH = \frac{x+y}{2} \cdot \frac{2xy}{x+y} = xy = G^2$$

It can be seen that $A > G > H$

$$\text{Proof} \quad A - G = \frac{x+y}{2} - \sqrt{xy}$$

$$= \frac{1}{2}\,(x + y - 2\sqrt{xy})$$

$$= \frac{1}{2}\,(\sqrt{x} - \sqrt{y})^2 > 0 \quad \therefore \; A > G$$

since the squared term is always positive.

Also $\quad AH = G^2$ or $\dfrac{A}{G} = \dfrac{G}{H}$

Since $\quad A > G,\; \dfrac{A}{G} > 1 \quad \therefore \; \dfrac{G}{H} > 1 \; \therefore \; G > H$

$\therefore \quad A > G > H$

Thus for any given series, the A.M. exceeds the G.M. and the G.M. exceeds H.M.

Ex. 143. If a, b, c, d, e are five numbers such that a, b, c, are in A.P., b, c, d are in G.P. and c, d, e are in H.P., show that a, c, e are in G.P.

We have $b = \dfrac{a+c}{2},\; c^2 = bd,\; d = \dfrac{2ce}{c+e}$

$\therefore \quad c^2 = bd = \dfrac{a+c}{2} \cdot \dfrac{2ce}{c+e}$ or $c = \dfrac{(a+c)e}{c+e}$

$\therefore \quad c\,(c + e) = (a + c)e$

$\therefore \quad c^2 = ae$

So that a, c, e are in G.P.

Ex. 144. If a, b, c, d are in HP and a, b, c, d are all positive show that $ad > bc$

Hint: If a, b, c are in GP, the geometric mean of a, c is $\sqrt{ac} = b$

Since $GM > HM$

$$\sqrt{ac} > b \quad (\because b \text{ is HM, given})$$

Similarly $\sqrt{bd} > c$

$\therefore \quad ac > b^2, bd > c^2$

$\therefore \quad acbd > b^2c^2$ or $ad > bc.$

Ex. 145. If a, b, c are in HP show (*i*) bc, ca, ab are in AP.

(*ii*) $\dfrac{a}{b+c-a}, \dfrac{b}{c+a-b}, \dfrac{c}{a+b-c}$ are in AP

(*iii*) $\dfrac{bc}{ca+ab}, \dfrac{ca}{bc+ab}, \dfrac{ab}{bc+ca}$ are in AP

Ex. 146. If a, b, c are in HP. Show $a^3 + c^3 > 2b^3$.

Ex. 147. If a, x, y, z, b are in AP, $x + y + z = 15$

If a, x, y, z, b are in HP, $\dfrac{1}{x}+\dfrac{1}{y}+\dfrac{1}{z}=\dfrac{5}{2}$

Find a, b. **Ans.** 9, 1 or 1, 9

1.13. ARITHMETICO GEOMETRIC PROGRESSION

Here, each term is a product of the corresponding terms of an AP and a GP, as in $a, (a + d)r, (a + 2d)r^2, \ldots\ldots, \{a + (n - 1)d\}r^{n-1}$

It can be shown that for this series

$$S_n = \frac{a}{1-r} + \frac{rd}{(1-r)^2} - \frac{r^n d}{(1-r)^2} - \{a + (n-1)d\}\frac{r^n}{1-r}$$

$$= \frac{a}{1-r} + \frac{rd}{(1-r)^2} \text{ if } n \to \infty$$

To find the nth Term of any Series. This can be done by

(*i*) Inspection. The nth term of 1.3.5 + 3.5.7 + 5.7.9 +... is $(2n - 1)(2n + 1)(2n + 3)$

The nth term of

$$1 + (1 + 2) + (1 + 2 + 3) + \ldots$$

is $\quad 1 + 2 + 3 + \ldots + n$

(*ii*) Method of difference. To find the nth term of

$$2 + 4 + 7 + 11 + \ldots$$

Write $\quad S = 2 + 4 + 7 + 11 + \ldots + T_n$

Again

$$S = 2 + 4 + 7 + \ldots + T_{n-1} + T_n$$

Subtracting $0 = 2 + 2 + 3 + 4 + \ldots + n + T_n$

$$\therefore \qquad 1 + \frac{n(n+1)}{2} = T_n \text{ or } T_n = \frac{n^2 + n + 2}{2}.$$

Ex. 148. Show that $T_n = \frac{(n+1)^2}{4}$ for the series

$$\frac{1^2}{1} + \frac{1^2 + 2^2}{1+3} + \frac{1^2 + 2^2 + 5^2}{1+3+5} + \ldots$$

Ex. 149. Show that $T_n = 4n^3 + 4n^2 + n$ for the series

$$1.3^2 + 2.5^2 + 3.7^2 + \ldots$$

1.14 GRAPHS

A graph is a picture or diagram showing the mathematical relation between the quantities represented by it. The positions of points on the graph can be described with reference to two intersecting lines called *Coordinate Axes.*

The horizontal axis $X'X$ is called the x axis and the vertical axis $Y'Y$, the y-axis. They intersect at O, the point called the origin.

Distances measured to the right of $X'X$ are positive and to the left of $X'X$ are negative.

Distances measured upward from $Y'Y$ are positive and those measured downward are negative.

The two axes divide the plane into four quadrants: I, II, III, IV.

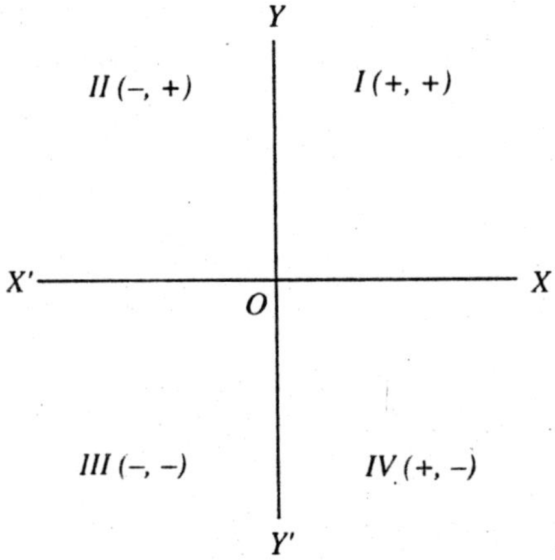

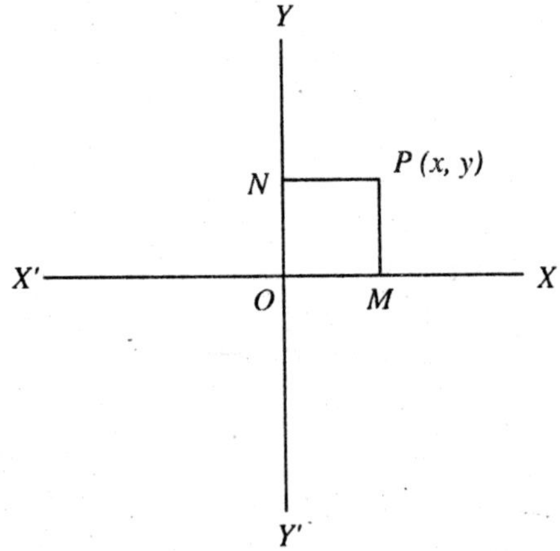

The position of a point P in the plane is shown by (x, y), called the rectangular coordinates of the point.

x is called the abscissa and is equal to OM or NP.

y is called the ordinate and is equal to MP or ON.

The coordinates of the origin O are (O, O).

To draw a graph, units of suitable size are taken along the axes on a graph paper.

Ex. 150. The graph shows the following points:

$A : (3, 0), B : (0, 2)$
$C : (-2, 1)$
$D : (-3, -2)$
$E : (1, -2)$
$F : (0, -1)$

We can draw the graph of any equation by writing pairs of coordinates and joining the points thus obtained.

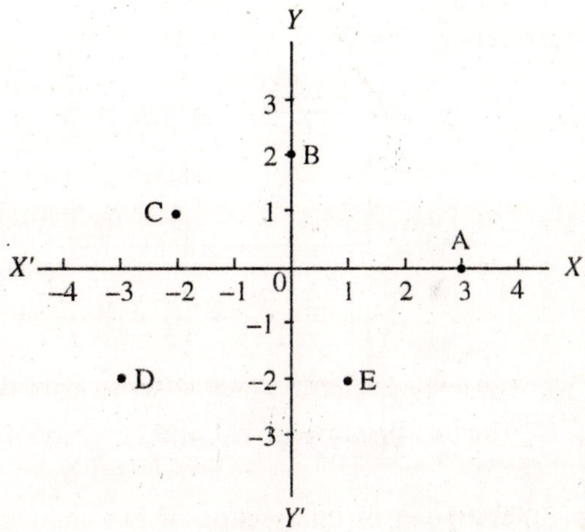

Ex. 151. Draw the graph of $y = x$.

If $x = 1, y = 1$

If $x = 2, y = 2$

If $x = -1, y = -1$

If $x = 0, y = 0$

If we plot the four points (1, 1), (2, 2), (–1, –1), (0, 0) and join them we get a straight line passing through the origin.

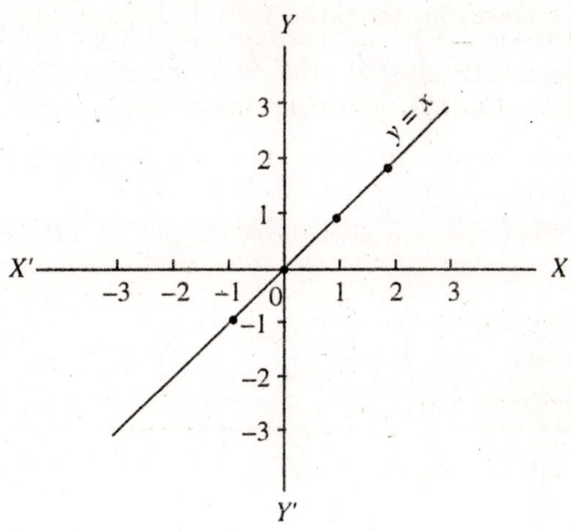

Ex. 152. Draw the graph of

$y = 2x + 1.$

We can form a table:

x	1	2	–1	–2
$y = 2x + 1$	2	5	–1	–3

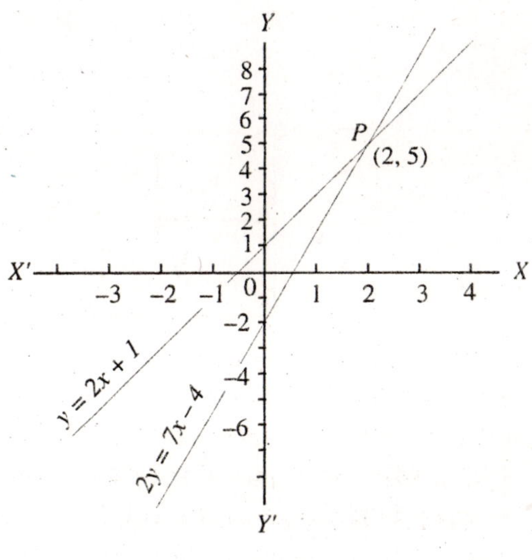

Ex. 153. Draw the graph of

$$2y = 7x - 4 \quad i.e., \quad y = \frac{7x}{2} - 2$$

x	0	2	–1	1
$y = \frac{7x}{2} - 2$	–2	5	$-\frac{11}{2}$	$\frac{3}{2}$

The two lines $y = 2x + 1$ and $2y = 7x - 4$ when drawn are found to intersect at a point P (2, 5).

The point of intersection of two lines gives the solution of simultaneous equations representing the lines. Thus if we solve these equations we find,

$$x = 2, y = 5.$$

Ex. 154. $y = x^2$ is the equation of a parabola.

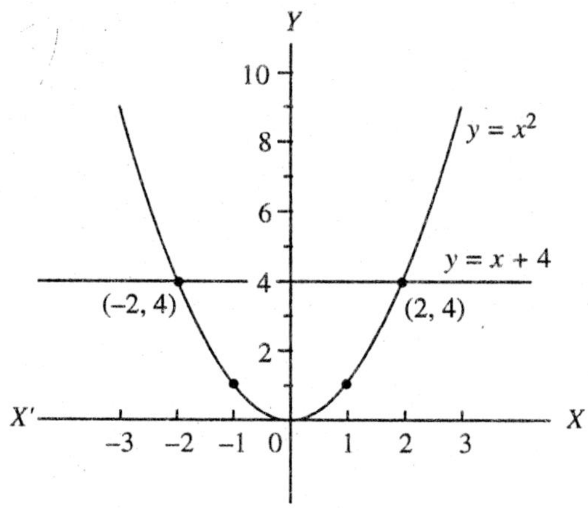

x	0	1	2	3	–1	–2	–3
$y = x^2$	0	1	4	9	1	4	9

If we draw the line $y = x + 4$ on the same graph, there are two points of intersection, (–2, 4) and (2, 4).

These are the solutions of the equations: $y = x^2$, $y = x + 4$.

Ex. 155. Draw the graphs of $y = x^2 - 2x + 1$, and $2y = x + 4$.

Find the points of intersection. **Ans.** (2.8, 3.4), (–3.5, 1.8)

Ex. 156. Find the solutions of the following simultaneous equations graphically:

(*i*) $3x + 4y = 7$ (*ii*) $5x - 9y = 1$ (*iii*) $y = \frac{x}{2} + 1$

$2x - y = 1$ $9x - 5y = 13$ $x = \frac{y}{3} + 8$

Ans. (*i*) (1, 1) (*ii*) (2, 1) (*iii*) (10, 6)

Ex. 157. Find the equilibrium price and quantity, given the

demand curve : $x = 25 - 3p$.

supply curve : $x = 2p + 10$.

In this case the demand curve and the supply curve are straight lines. In the equilibrium position the quantity demanded equals quantity supplied (we have used x for quantity demanded as well as quantity supplied) at the price p acceptable to buyers as well as sellers. The answer will therefore be obtained by solving the simultaneous equations given or by drawing their graphs and noting the point of intersection which in this case will be $x = 16$, $p = 3$.

Ex. 158. Show in the first quadrant the following demand, supply and cost curves:

$p = x^2 - 6x + 10$, $p = \sqrt{6 + x}$, $c = \sqrt{3x + 4}$

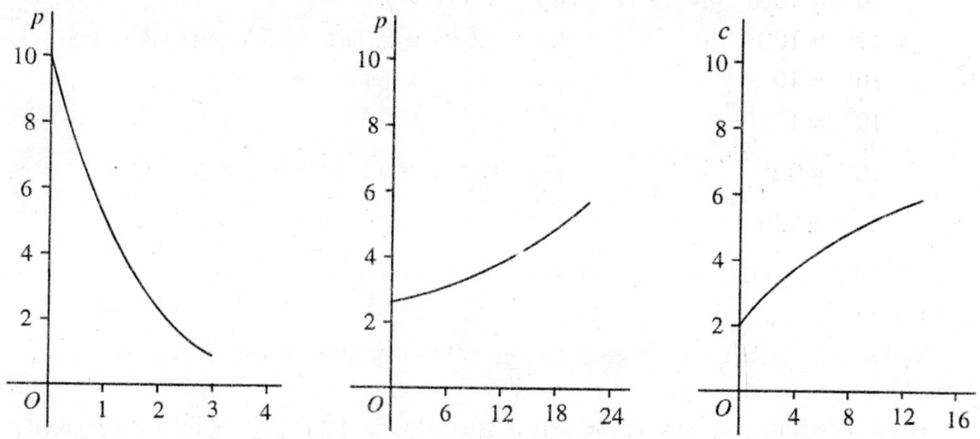

All curves are in the first quadrants because all quantities are positive.

The graphs show the required curves by arms of parabolas in the first quadrant.

The revenue (R) curve and average cost (AC) curve for the above cases can be found by using the fact that

$$R = \text{price} \times \text{quantity sold} = px$$
$$= (x^2 - 6x + 10)\,x$$

$$AC = \frac{\text{Total cost}}{\text{Quantity}} = \frac{C}{x} = \frac{\sqrt{3x+4}}{x} = \sqrt{\frac{3}{x} + \frac{4}{x^2}}$$

Ex. 159. Sketch the following demand, supply and cost curves in the first quadrant.

	Demand	Supply	Cost
(*i*)	$x = 36 - p^2$	$x = \frac{p^2}{4} - p$	$C = \sqrt{6x + 11}$
(*ii*)	$x = \sqrt{\frac{48 - p}{3}}$	$p = 2\sqrt{x + 4}$	$C = x + \frac{x^2}{12}$
(*iii*)	$p = x^2 - 10x + 25$	$x = \frac{p^2}{2} + p - 4$	$C = \sqrt{9x - 25}$

Also plot the revenue and average cost curves.

1.15 COMMON LOGARITHMS

The logarithm of a number is the exponent of the power to which a constant number must be raised to produce the given number.

The constant number is called the base. The logarithms to the base 10 are called common logarithms and can be used for computational purposes. Note the following results.

$10^3 = 1000$ gives $\log_{10} 1000 = 3$

$10^2 = 100$ $\log 100 = 2$ (base 10 need not be written)

$10^1 = 10$ $\log 10 = 1$

$10^0 = 1$ $\log 1 = 0$

$10^{-1} = 0.1$ $\log 0.1 = -1$, usually written $\bar{1}.0$

$10^{-2} = 0.01$ $\log 0.01 = -2$, written $\bar{2}.0$

$10^{-3} = 0.001$ $\log 0.001 = -3$, written $\bar{3}.0$

and so on.

The number after = sign in each case gives the logarithm of the number preceding it. (above, right)

If the logarithm, *i.e.*, the second part is given, the first part is called the antilogarithm and can be found out. Thus antilog 3 = 1000, antilog $\bar{2}$ = 0.01.

There are tables available for logarithms and antilogarithms of numbers.

Ex. 160. From the tables we find

$$\log 1.6 = 0.2041$$

Similarly log 1.63 = 0.2122. In this case, 0, the number before the decimal, is called the *characteristic* and the positive decimal part .2122 is called the *mantissa*.

In general if $a^y = x$, then

$$y \log a = x = \text{an integer part + a decimal part}$$

The integer part is the *characteristic*,

The decimal part is the *mantissa*.

How to use the tables?

To find the logarithm of a number first look into the tables for the decimal part or mantissa of the logarithm, irrespective of the decimal place in the given number.

The mantissa for 5.71 or 57.1 or 571 is the same *viz.* 0.7559.

The characteristic is equal to one less than the number of digits *before the decimal point*. Thus

$$\log 571 = 2.7559 \quad \text{(2 because 571 has 3 digits)}$$
$$\log 57.1 = 1.7559 \quad \text{(1 because 57.1 has 2 digits)}$$
$$\log 5.71 = 0.7559 \quad \text{(0 because 5.71 has 1 digit)}$$
$$\log 0.571 = \bar{1}.7559,$$

In the last case there is no digit before the decimal point in the number so that the characteristic is –1, written $\bar{1}$ since the mantissa is always positive. It may be noted that

$$\log 0.0571 = \bar{2}.7559$$
$$\log 0.00571 = \bar{3}.7559$$

and so on, the rule being that the negative characteristic is equal to one more than the number of zeros immediately following the decimal point in the given number.

Ex. 161. Find the product 24×1.2

Let $y = 24 \times 1.2$

$\log y = \log 24 + \log 1.2 = 1.3802 + 0.0792 = 1.4594$

To find y take antilogarithm of 1.4594

From the tables of antilogarithms, we have

Antilog 1.4594 = 28.80

The rule of finding antilogarithms from the tables is exactly the reverse of that for logarithms. In the table of antilogarithms we look for the value against the mantissa only, 0.4594 in this case.

The characteristic of the given logarithm determines the position of the decimal point in the antilogarithm.

The characteristic being 1 in this example, the decimal point will come after the second digit.

It can be seen that

$$\text{Antilog } 2.7559 = 571$$
$$\text{Antilog } 1.7559 = 57.1$$
$$\text{Antilog } 0.7559 = 5.71$$
$$\text{Antilog } \bar{1}.7559 = 0.571 \text{ and so on.}$$

Ex. 162. $y = \dfrac{3.21^2 \times 0.9573^{1/2}}{98.32}$

$$\log y = 2 \log 3.21 + \tfrac{1}{2} \log 0.9573 - \log 98.32$$
$$= 2 \times .5065 + \tfrac{1}{2} \times (\bar{1}.9810) - 1.9927$$

$$= 1.0130 + \tfrac{1}{2}(2.9810 + 1) - 1.9927$$

$$= 1.0130 + \bar{1}.4905 + 0.5 - 1.9927$$

$$= \bar{1}.0108$$

$$y = \text{Antilog } \bar{1}.0108 = 0.1025.$$

Ex. 163. Evaluate : (*i*) $\dfrac{17.2 \times 24.5}{43.15}$; (*ii*) $\sqrt{\dfrac{628}{523} \div \dfrac{16}{27}}$; (*iii*) $\dfrac{85.76 - \sqrt[3]{0.0007}}{\sqrt{1067}}$.

(*i*) $x = \dfrac{17.2 \times 24.5}{43.15}$

$\log x = \log 17.2 + \log 24.5 - \log 43.15$

$= 1.2355 + 1.38.92 - 1.6350 = 0.9897$

$\therefore$ $x = 9.766$ using antilogarithm tables

(*ii*) $x = \sqrt{\dfrac{628}{523} \div \dfrac{16}{27}} = \sqrt{\dfrac{628 \times 27}{523 \times 16}}$

$\log x = 1/2\,[\log 628 + \log 27 - \log 523 - \log 16]$

$= 1/2\,[2.7980 + 1.4314 - 2.7185 - 1.2041]$

$= 1.2 \times 0.3068 = 0.1534$

$x = 1.423$

(*iii*) We cannot take logarithms till +ve or –ve sign in the expression is removed. We first simplify the number

Let $y = \sqrt[3]{.0007}$ so that $\log y = 1/3 \log 0.0007$

$\therefore$ $\log y = 1/3 \times \bar{4}.8451 = 1/3\,(-3.1594) = -1.0516$

$= \bar{2}.9484$ [since $-1.0516 = -2 + (1 - .0516)$]

Taking antilogarithms, $y = 0.0888$

The given expression $= x = \dfrac{85.76 - 0.0888}{\sqrt{1067}} = \dfrac{85.6712}{\sqrt{1067}}$

$\log x = \log 85.6712 - 1/2 \log 1067 = 1.9329 - 1/2\,(3.0281)$

$= 1.9329 - 1.5140 = 0.4189$

$\therefore$ $x = 2.623$

Ex. 164. Prove by using logarithms :

(*i*) $\dfrac{273 \times 943 \times 975}{254 \times 760} = 1300$

(*ii*) $6.31^6 = 63100$

(*iii*) $(0.0123)^5 = 2.82 \times 10^{-10}$

(*iv*) $\dfrac{\sqrt{3850}}{286} = 11.6$

(*v*) $10^{-2.0804} = 0.00831$

Ex. 165. Solve for x :

		Ans.
(*i*)	$4^{x+3} = 7^{x-1}$	(*i*) 10.92.
(*ii*)	$3^{x+1} = 27$	(*ii*) 2.
(*iii*)	$5\,(6^x) = 21^{x-2}$	(*iii*) 6.14.
(*iv*)	$\log (x + 2) - \log 2x = \log 12$	(*iv*) $\frac{2}{23}$.
(*v*)	$9^{x-3}\ 27^{x+4}$	(*v*) – 18.
(*vi*)	$\log (x^2 - 4) - \log (x + 2) = \log 16.$	(*vi*) 18.

Ex. 166. $3^{2x} = \frac{323}{5}$

$\therefore \quad 2x \log 3 = \log 323 - \log 5$

$\therefore \quad x = \frac{\log 323 - \log 5}{2 \log 3} = 1.9$ approx.

Ex. 167. Given $\log_e 2 \log_x 625 = \log_{10} 16 \log_e 10$. Find x.

Taking common base $\frac{\log 2}{\log e} \cdot \frac{\log 625}{\log x} = \frac{\log 16}{\log 10}\ \frac{\log 10}{\log e}$ (See Rules below)

$$\therefore \quad \log x = \frac{\log 2 \cdot \log 625}{\log 16}$$

$$= \frac{\log 2 \cdot (4 \log 5)}{4 \log 2}$$

$$= \log 5$$

$$\therefore \quad x = 5$$

Ex. 168. Show $\log_a x \log_b x + \log_b x \log_c x + \log_c x \log_a x$

$$= \frac{\log_a x \log_b x \log_c x}{\log_{abc} x}$$ (Using Base Change Rule given below)

Ex. 169. If $\log_{\sqrt{8}} x = \frac{10}{3}$, find x. **Ans.** $x = 32$

Ex. 170. If a, b, c are is GP show $\log_a x$, $\log_b x$, $\log_c x$ are in HP.

Ex. 171. If $\log (m + n) = \log m + \log n$ show $m = \frac{n}{n-1}$.

Ex. 172. If $\log \frac{m+n}{2} = \frac{1}{2} (\log m + \log n)$ show $m = n$.

Rules for Logarithms

Multiplication Rule :

$\log (MN) = \log M + \log N$

$\log (MNL) = \log M + \log N + \log L$

$\log M(N + L) = \log M + \log (N + L)$

$\log (245 \times 0.3967) = \log 245 + \log 0.3967$

$\log [45 (397 - 222)] = \log 45 + \log (397 - 222) = \log 45 + \log 175$

Division Rule :

$$\log \frac{M}{N} = \log M - \log N$$

$$\log \frac{M}{NL} = \log M - \log (NL) = \log M - (\log N + \log L) = \log M - \log N - \log L.$$

$$\log \frac{MN}{L} = \log M + \log N - \log L$$

$$\log \frac{M}{N} \cdot \frac{L}{K} = \log \frac{M}{N} + \log \frac{L}{K} = \log M - \log N + \log L - \log K$$

or $$= \log ML - \log NK = \log M + \log L - \log N - \log K$$

$$\log \frac{M - N}{L} = \log (M - N) - \log L.$$ Note that $\log (M - N) \neq \log M - \log N$.

$$\log \frac{45}{2793} = \log 45 - \log 2793$$

$$\log \frac{855}{3782 - 5.79} = \log 855 - \log (3782 - 5.79)$$

$$= \log 855 - \log 3776.21$$

Power Rule :

$\log M^N = N \log M$

$\log 16 = \log 4^2 = 2 \log 4$; $\log 16 = \log 2^4 = 4 \log 2$

$\log 1000 = \log 10^3 = 3 \log 10 = 3 \times 1 = 3.$

Base Change :

$$\log_N M = \frac{\log_x M}{\log_x N}, \quad \log_{100} 10 = \frac{\log_x 10}{\log_x 100}$$

where x is any quantity. Usually we take $x = 10$.

i.e., we reduce the problem to base 10. (See Ex. 167)

1.16 NATURAL LOGARITHMS

Common logarithms have the base 10. If the base is e (see next section) we have what are known as natural logarithms, briefly written *ln* instead of *log*.

There can be other bases besides 10 and e, but these two are in common use.

Natural logarithms can be converted into common logarithm and *vice versa* by using the fact that $e = 2.7183$ approx. and $\log e = 0.4343$ approx.

$$\text{Thus } \log_e u = \frac{\log_{10} u}{\log_{10} e} = \frac{1}{\log_{10} e}\log_{10} u = \frac{1}{.4343}\log_{10} u$$

$$= 2.303 \log_{10} u.$$

Ex. 173. ln 3.24 = $\log_e 3.24 = 2.303 \log 3.24 - 2.303 \times 0.5105 = 1.175$.

1.17 THE NUMBER *e*

The number e is one of the fundamental constants of mathematical analysis.

The number e is defined as a limit

$$e = \lim_{h \to 0} (1+h)^{1/h}$$

which is the same thing as

$$e = \lim_{a \to \infty}\left(1+\frac{1}{a}\right)^a$$

This is a convergent series having a positive limit. It can be shown that

$$e = 1 + \frac{1}{1!} + \frac{1}{2!} + \frac{1}{3!} + \ldots\ldots$$

$$= 2.71828\ldots, \text{ an irrational number.}$$

In practical work we use $e = 2.7183$ up to the fourth place of decimal.

The number e is a *transcendental* number.

Another example is π.

Such a number cannot be expressed as the root of any polynomial with integer coefficients.

The Exponential e^x

The exponential e^x is the inverse of the natural logarithm defined by

$x = ln\ y$ where $ln\ y = \log_e y$

It follows that $e^{\ln x} = x$ and $ln\ e^x = x$

Some properties

1. $e^a = 1$
2. $e^{a+b} = e^a \cdot e^b$
3. $e^{a-b} = \dfrac{e^a}{e^b}$
4. $e^{ar} = (e^a)^r$
5. $a^x = e^{x \ln a}$
6. $e^x = 1 + x + \dfrac{x^2}{2!} + \dfrac{x^3}{3!} + \ldots$
7. If $x < 0$ then $0 < e^x < 1$

   If $x = 0$ then $e^x = 1$

   If $x > 0$ then $e^x > 1$.

Ex. 174. Verify the following results:

$$ln\ e^2 = \log_e e^2 = 2$$

$$ln\ e = \log_e e = 1$$

$$ln\ 1 = \log_e 1 = 0$$

$$ln\ \frac{1}{e} = \log_e e^{-1} = -1$$

Ex. 175. If $x = -0.4$ find e^x.

If $y = e^{-0.4}$ then $\log y = -0.4 \log e = -0.4 \log 2.7183$

$$= -0.4\ (0.4343) = -0.17372 = \bar{1}.9263$$

$$y = \text{Antilog } \bar{1}.9263 = 0.1684$$

The answer can be found directly from natural logarithm tables.

Ex. 176. Find e^x and e^{-x} when (*i*) $x = 1$ (*ii*) $x = 2$ (*iii*) $x = 0.6$

Ans. (*i*) 2.71828 (*ii*) 7.38905 (*iii*) 1.82211
0.36787 0.13533 0.54881.

Ex. 177. Verify the following results:

$$\lim_{h \to 0} (1 + xh)^{\frac{1}{h}} = e^x$$

$$\lim_{hx \to 0} (1 + xh)^{\frac{1}{hx}} = e.$$

1.18 MATHEMATICS OF MONEY: COMPOUND INTEREST

Interest is the sum paid as compensation to those who lend money and who thus give up the chance of using it themselves and, therefore, expect to be reimbursed.

If P = principal *i.e.*, initial balance or investment
r = rate of interest per period
t = number of periods

We have simple interest $I = Prt$

The total amount A is the principal plus the interest.

Ex. 178. The simple interest on Rs. 500 for 2 years at 3% is

$$I = 500 \cdot \frac{3}{100} \cdot 2 = 30$$

When interest is added to the principal at the end of each period it is converted into principal. The total amount due at the end of the period is called the *compound amount.*

Compound interest is the difference between this compound amount and the original principal.

If the initial investments is P, after one period it becomes $P(1 + r)$. After two periods it is $P(1 + r)(1 + r) = P(1 + r)^2$.

The amount of money accumulated at the end of t compounding periods is
$A = P(1 + r)^t$.

Ex. 179. The compounded amount of Rs. 500 after 5 years at 5% is

$$A = 500\ (1 + 0.05)^5 = 500 \times 1.05^5$$

$$\log A = \log 500 + 5 \log 1.05 = 2.6990 + 5 \times 0.0212$$

$$= 2.8050$$

Whence A = Rs. 638.30.

Ex. 180. The compounded amount of Rs. 3000 after 4 years at 3% interest each half year, compounded semi-annually is given by

$$A = P(1 + r)^t = 3000\ (1.03)^8$$

since there are 8 period of 6 months each.

Using logarithms, we have

$$\log A = \log 3000 + 8 \log 1.03$$

$$= 3.4771 + 8 \times 0.0128 = 3.5795$$

$\therefore$ A = Rs. 3800.

A modification

A principal P compounded m times a year for n years at the annual interest rate r grows to an amount

$$A = P\left(1+\frac{r}{m}\right)^{mn}$$

1.19 CONTINUOUS COMPOUNDING

Suppose the interest from an investment is obtained as well as reinvested immediately after it is earned. This means that the compounding takes place m times a year where m tends to infinity. Now we know that

$$e = \lim_{m\to\infty}\left(1+\frac{1}{m}\right)^m \text{ and } e^h = \lim_{m\to\infty}\left(1+\frac{h}{m}\right)^m$$

so that we have the formula for continuous compounding

$$e^r = \lim_{m\to\infty}\left(1+\frac{r}{m}\right)^m \text{ in one year.}$$

In t years the compounded amount of the principal P becomes: $A = Pe^{rt}$

This is an example of the process of exponential growth and is equally applicable to population, wealth, etc. r may be called *instantaneous* rate of growth.

Equivalent rate of interest

Interest rates resulting in equal amounts in the same length of time on the same principal are called equivalent rates.

Let i be the effective rate equivalent to a given nominal rate r compounded m times a year. Since equivalent rates result in equal amounts in the same length of time, we have

$$P(1 + i) = P\left(1+\frac{r}{m}\right)^m$$

or $$1 + i = \left(1+\frac{r}{m}\right)^m$$

$$i = \left(1+\frac{r}{m}\right)^m - 1$$

In the case of continuous compounding $i = e^r - 1$.

Ex. 181. The equivalent of 8% compounded quarterly is obtained by writing

$$m = 4, \ \frac{r}{m} = \frac{0.08}{4} = 0.02. \text{ We have}$$

$$i = (1.02)^4 - 1 = 0.0824 \text{ or } 8.24\%$$

Ex. 182. Find: (*i*) simple and (*ii*) compound interest at 8% converted quarterly on Rs. 2000 invested for $1\frac{3}{4}$ years. **Ans.** 290, 297.37

Ex. 183. How many years it will take for Rs. 100 to double itself if invested at 4% compounded semi-annually? **Ans.** 35

Ex. 184. For $n > 1$ show that $(1 + r)^n > 1 + nr$

Hence compare amounts at compound and simple interests.

Ex. 185. The equivalent of continuous compounding at 6% is

$$i = e^r - 1 = e^{0.06} - 1$$

We may write $i + 1 = e^{0.06}$

$$\log (i + 1) = 0.06 \log e = 0.06 \times 0.43429 = 0.0260574$$

$$i + 1 = 1.06184$$

$$\therefore \quad i = 6.184\%.$$

Ex. 186. Find the effective rate equivalent to 6% nominal converted (*i*) semi-annually, (*ii*) quarterly, (*iii*) monthly, (*iv*) continuously.

Ans. 6.09, 6.1364, 6.1678, 6.1837

Ex. 187. What is the effective rate equivalent to a nominal rate of 2% compounded continuously? **Ans.** 2.02

Ex. 188. If $(1 + r)^{85} = 10.0330$ show that $(1 + r)^{170} = 100.661$.

Ex. 189. If money is invested for 8 years what should be the rate of interest for this money (*i*) to treble (*ii*) to quintuple? One may do this by writing $3P = P(1 + r)^8$ and solving for r where $(1 + r)^8 = 3$.

Alternatively write $3P = Pe^{8r}$ or $e^{8r} = 3$

$$\therefore \quad 8r = ln\ 3 = 1.0986$$

$$\therefore \quad r = 0.1373 \text{ i.e., } 13.73\%$$

(*ii*) $5 = e^{8r}$ *i.e.*, $8r = ln\ 5 = 1.60944$

$$\therefore \quad r = 0.20118 \text{ i.e., } 20.12\%$$

Ex. 190. Find the *time* for the money to treble if compounded semiannually at 8%.

(Semiannually $\frac{8}{2} = 4\% = .04$).

$$3P = P(1 + .04)^{2t} \text{ (semiannually, hence } 2t)$$

$\therefore$ $$1.04^{2t} = 3 \text{ or } 2t \ ln\ 1.04 = ln\ 3$$

i.e., $$2t(0.03922) = 1.09861$$

$$2t = \frac{1.09861}{0.03922} \text{ or } t = 13.95 \text{ years.}$$

1.20 PRESENT VALUE AT COMPOUND INTEREST

The present value of an amount A due in n periods is the principal P which invested now at r per cent amounts to A in n periods. Since $A = P(1 + r)^n$, we have

$$P = \frac{A}{(1+r)^n} = A(1+r)^{-n}$$

Ex. 191. The present worth of Rs. 1000 due in 5 years at 3% compounded annually is

$$P = 1000\ (1.03)^{-5}$$

$$\log P = \log 1000 - 5 \log 1.03 = 3 - 5 \times 0.0128 = 2.9360$$

$$P = 863.00$$

Ex. 192. Show that the present value of Rs. 500 due in 4 years at 3% compounded semi-annually is Rs. 444 approximately.

Ex. 193. At what nominal rate, compounded semi-annually, will Rs. 840 amount to Rs. 2100 in 20 years? **Ans.** 4.634% compounded semi-annually.

1.21 PARTIAL FRACTIONS

$R(x) = \dfrac{f(x)}{g(x)}$ is a rational function if $f(x)$, $g(x)$ are polynomials

It is a *proper* function if the degree of the denominator is greater than that of the numerator.

Two or more rational functions can be combined to make one composite rational function.

A composite rational function can be decomposed into simple rational functions.

Ex. 194. $\dfrac{1+x+x^2}{x(1-x^2)}$ has numerator degree = 2, denominator degree = 3.

It can be made into partial fractions.

Procedure : Write $\dfrac{1+x+x^2}{x(1+x)(1-x)} = \dfrac{A}{x} + \dfrac{B}{1+x} + \dfrac{C}{1-x}$

$\therefore$ $$\frac{1+x+x^2}{x(1+x)(1-x)} = \frac{A(1-x^2)+B(x-x^2)+C(x+x^2)}{x(1+x)(1-x)}$$

The Denominators being identical, the numerators can be equated. Put $x = 0$ in

$$1 + x + x^2 = A(1 - x^2) + B(x - x^2) + C(x + x^2)$$

$\therefore$ $A = 1$

Put $1 + x = 0$ *i.e.*, $x = -1$ so that $1 - 1 + 1 = B(-1 - 1)$

$\therefore$ $B = -\dfrac{1}{2}$

Put $1 - x = 0$ *i.e.*, $x = 1$ so that $1 + 1 + 1 = 2C$

$\therefore$ $C = \dfrac{3}{2}$

$$\therefore \quad \frac{1 + x + x^2}{x(1 - x^2)} = \frac{1}{x} + \frac{-1/2}{1 + x} + \frac{3/2}{1 - x}$$

Ex. 195. $\dfrac{5x - 18}{2x^2 + x - 6} = \dfrac{4}{x + 2} - \dfrac{3}{2x - 3}$ (Similar to Ex. 194)

Note. 1. $\dfrac{1}{(x - a)\,(x - b)} = \dfrac{1}{a - b}\left(\dfrac{1}{x - a} - \dfrac{1}{x - b}\right)$

2. If the denominator has some linear factors repeated as in

$$\frac{ax^3 + bx + c}{(x + 1)^3\,(x - 2)}$$

the procedure is to write it as

$$\frac{A}{x + 1} + \frac{B}{(x + 1)^2} + \frac{C}{(x + 1)^3} + \frac{D}{x - 2}$$

3. If the denominator is of the form given in

$$\frac{1}{x^3 + 8} = \frac{1}{(x + 2)\,(x^2 - 4x + 4)}$$

the procedure is to write it as

$$\frac{A}{x + 2} + \frac{Bx + C}{x^2 - 4x + 4}$$

4. If the denominator has repeated quadraic expressions as in

$$\frac{ax + b}{(x + 1)(x^2 + 1)^2}$$

the procedure is to write it as

$$\frac{A}{x + 1} + \frac{Bx + C}{x^2 + 1} + \frac{Dx + E}{(x^2 + 1)^2}$$

See the chapter on integration.

1.2 THE SIGMA (Σ) NOTATION AND USE OF SUBSCRIPTS

The Greek letter Σ is used to facilitate the writing of sums. Thus the sum of four numbers a_1, a_2, a_3, a_4 is written

$$\sum_{i=1}^{4} a_i = a_1 + a_2 + a_3 + a_4$$

If $A = \{a_1, a_2, a_3, \ldots, a_n\}$ is a set of n elements

$$\sum_{i=1}^{n} a_i = a_1 + a_2 + a_3 + \ldots\ldots + a_n$$

Here a_i is any element of the set and as i takes values from 1 to n, every element is considered.

i is the subscript, often used to identify the position of an element.

Ex. 196. $\sum_{i=1}^{5} i = 1 + 2 + 3 + 4 + 5$ 5 terms

$\sum_{j=1}^{4} j^2 = 1^2 + 2^2 + 3^2 + 4^2$ 4 terms

$\sum_{k=1}^{6} \frac{1}{k} = \frac{1}{1} + \frac{1}{2} + \frac{1}{3} + \frac{1}{4} + \frac{1}{5} + \frac{1}{6}$ 6 terms

$\sum_{i=1}^{3} x_i^3 = x_1^3 + x_2^3 + x_3^3$

In general $\sum_{i=m}^{n} f(i) = f(m) + f(m+1) + f(m+2) + \ldots + f(n)$

where m, n are integers and $m < n$.

There are $n - m + 1$ terms in this sum. m is called the *lower limit*, n the *upper limit* and i the *index* of summation.

Ex. 197. $\sum_{i=1}^{4} i = \sum_{j=1}^{4} j$. It shows the index is a mere dummy.

Ex. 198. $\sum_{k=1}^{3} k(a_k + b_k) = 1(a_1 + b_1) + 2(a_2 + b_2) + 3(a_3 + b_3)$.

Ex. 199. $\sum_{r=1}^{4} (r+1) = (1+1) + (2+1) + (3+1) + (4+1) = 14$

Ex. 200. $\sum_{k=1}^{3} x_k + \sum_{j=2}^{4} x_i = (x_1 + x_2 + x_3) + (x_2 + x_3 + x_4)$

$$= x_1 + 2x_2 + 2x_3 + 2x_4.$$

Ex. 201. Find the value of

(*i*) $\sum_{r=2}^{5} r^2$, (*ii*) $\sum_{r=0}^{3} x_{i+1}$, (*iii*) $\sum_{r=i}^{4} (-1)^r x^{-r}$,

Ans. (*i*) $2^2 + 3^2 + 4^2 + 5^2$, (*ii*) $x_1 + x_2 + x_3 + x_4$,
(*iii*) $-x^{-1} + x^{-2} - x^{-3} + x^{-4}$

Ex. 202. Write in sigma notation

(*i*) $x_1 + x_2 + x_3 + x_4$. (*ii*) $\frac{1}{n}(x_1 + x_2 + \ldots + x_n)$

(*iii*) $\frac{1}{2} + \frac{2}{3} + \frac{3}{4} + \frac{4}{5} + \frac{5}{6} + \frac{6}{7}$ (*iv*) $10^2 + 11^2 + 12^2 + 13^2 + 14^2$

(*v*) $1 + 3 + 5 + 7 + 9 + 11$. (*vi*) $2 + 4 + 6 + \ldots + 2n$.

(*vii*) $8 + 9 + 10 + 11 - 1 - 2 - 3 - 4$.

Ans. (*i*) $\sum_{i=1}^{4} x_i$ (*ii*) $\frac{1}{n}\sum_{i=1}^{n} x_i$ (*iii*) $\sum_{i=1}^{6} \frac{i}{i+1}$

(*iv*) $\sum_{i=10}^{14} i^2$ (*v*) $\sum_{k=0}^{5} (2k+1)$ (*vi*) $\sum_{h=1}^{n} 2h$

(*vii*) $\sum_{i=8}^{11} x_i - \sum_{j=1}^{4} x_j$

Some Rules of Σ *notation*

I. If c is a constant $\sum_{i=1}^{n} c = cn$.

Ex. 203. $\sum_{i=1}^{3} k = 3k$; $\sum_{j=1}^{3} 5 = 5 \times 3$; $\sum_{r=m}^{n} 1 = n - m + 1$; $\sum_{i=1}^{n} 1 = n$.

II. If c is a constant $\sum_{i=1}^{n} cx_i = c\sum_{i=1}^{n} x_i$

Ex. 204. $\sum_{i=1}^{3} 5a_i = 5\sum_{i=1}^{3} a_i$

III. $\sum_{i=1}^{n} (x_i + y_i) = \sum_{i=1}^{n} x_i + \sum_{i=1}^{n} y_i$.

Ex. 205. $\sum_{i=1}^{7} (cx_i + lu_i) = c\sum_{i=1}^{7} x_i + l\sum_{i=1}^{7} u_i$.

IV. $\sum_{i=1}^{n} (x_i - x_{i-1}) = \sum_{i=1}^{n} x_i - \sum_{i=1}^{n} x_{i-1}$

$$= (x_1 + x_2 + \ldots + x_n) - (x_0 + x_1 + \ldots + x_{n-1}) = x_n - x_0.$$

V. $\sum_{i=m}^{n} x_i = \sum_{i=0}^{n-m} x_{i+m}$.

Ex. 206. $\sum_{i=3}^{8} a_i = \sum_{i=0}^{8-3} a_{i+3} = \sum_{i=0}^{5} a_{i+3}$

Each equals $a_3 + a_4 + a_5 + a_6 + a_7 + a_8$.

VI. $\sum_{i=0}^{n} x_i = x_0 + \sum_{i=1}^{n} x_i = x_0 + x_1 + \sum_{i=2}^{n} x_i$ etc.

Ex. 207. The following results are useful and may be proved.

$$\sum_{i=1}^{n} i = \frac{n(n+1)}{2}$$

$$\sum_{i=1}^{n} i^2 = \frac{n(n+1)(2n+1)}{6}.$$

$$\sum_{i=1}^{n} i^3 = \frac{n^2(n+1)^2}{4}.$$

Ex. 208. Use sigma notation for the following:

(*i*) 1.2 + 2.3 + 3.4 +...*n* terms (*ii*) 1.2 + 3.4 + 5.6 +...*n* terms

(*iii*) $f_1x_1 + f_2x_2 + f_3x_3 + \ldots + f_nx_n$ (*iv*) 1 – 2 + 3 – 4 +...*n* terms

(*v*) $f_1x_1^2 + f_2x_2^2 + f_3x_3^2 + \ldots n$ terms

Ans. (*i*) $\sum_{i=1}^{n} i.(i+1)$ (*ii*) $\sum_{i=1}^{n} (2i-1)2i$ (*iii*) $\sum_{i=1}^{n} f_ix_i$

(*iv*) $\sum_{r=1}^{n} (-1)^{r+1} \cdot r$ (*v*) $\sum_{r=1}^{n} f_ix_i^2$.

Ex. 209. The quantity $\dfrac{f_ix_1 + f_2x_2 + f_3x_3 + \ldots + f_nx_n}{f_1 + f_2 + f_3 + \ldots + f_n}$

may be denoted by $\overline{x}$ and written

$$\overline{x} = \frac{\Sigma f_ix_i}{\Sigma f_i}; \text{ more concisely } \overline{x} = \frac{\Sigma fx}{\Sigma f}.$$

This is the *arithmetic mean* of a set of observations having values

$$x_1, x_2, x_3, \ldots, x_n$$

occurring, respectively, $f_1, f_2, f_3, \ldots, f_n$ times.

Ex. 210. Write the following in full

(*i*) $\sum_{r=0}^{n} \frac{x^r}{r!}$, (*ii*) $\sum_{i=1}^{n} (x_i - \overline{x})$, (*iii*) $\sum_{i=1}^{n} f_i(x_i - \overline{x})^2$.

Ans. (*i*) $\dfrac{x^0}{0!} + \dfrac{x^1}{1!} + \dfrac{x^2}{2!} + \ldots + \dfrac{x^n}{n!}$

$$(ii) \quad (x_1 - \bar{x}) + (x_2 - \bar{x}) + \ldots + (x_n - \bar{x}) = \sum_{i=1}^{n} x_i - n\bar{x}$$

$$(iii) \quad f_1(x_1 - \bar{x})^2 + f_2(x_2 - \bar{x}_2)^2 + \ldots + f_n(x_n - \bar{x})^2.$$

The quantity $\frac{1}{n} \sum f_i(x_i - \bar{x})^2$ is called the *variance* of $x_1, x_2, \ldots, x_n$.

Ex. 211. Expand

$$(i) \quad \sum_{i=1}^{n} (x_i - \bar{x}))(y_i - \bar{y}) \qquad (ii) \quad \sum_{l=1}^{9} (\sqrt{l+1} - \sqrt{l})$$

Ans. (*i*) $(x_1 - \bar{x})(y_1 - \bar{y}) + (x_2 - \bar{x})(y_2 - \bar{y}) + \ldots + (x_n - \bar{x})(y_n - \bar{y})$.

(*ii*) $(\sqrt{2} - \sqrt{1}) + (\sqrt{3} - \sqrt{2}) + (\sqrt{4} - \sqrt{3}) + \ldots + (\sqrt{9} - \sqrt{8})$

Also prove that (*i*) equals $\sum_{i=1}^{n} x_i y_i - n\bar{x}\bar{y}$ and (*ii*) equals 2.

Ex. 212. Is it correct to say that

$$(i) \quad \sum_{n=0}^{100} n^2 = \sum_{n=1}^{100} n^2 \quad \text{(Yes)} \qquad (ii) \quad \sum_{n=0}^{100} (n+2) = \left(\sum_{n=0}^{100} n\right) + 2 \quad \text{(No)}$$

$$(iii) \quad \sum_{n=0}^{99} (n+1) = \sum_{n=1}^{100} n^2 \quad \text{(No)} \qquad (iv) \quad \sum_{n=0}^{100} n^2 = \left(\sum_{n=0}^{100} n\right)^2 \quad \text{(No)}$$

$$(v) \quad \sum_{n=2}^{5} \frac{n-1}{n+1} = 2.1 \quad \text{(Yes)} \qquad (vi) \quad \sum_{n=3}^{6} n(n-2) = 50? \quad \text{(Yes)}$$

1.23 PERMUTATIONS AND COMBINATIONS; FACTORIAL NOTATION

A study of permutations and combinations is concerned with the number of ways in which it is possible to arrange things subject to certain given conditions.

Suppose there are four cars to be taken by four persons. In how many different ways can these cars be taken?

Any one of the four cars can be taken by the first person in four ways. There are now three choices left for the second person so that there are 4 × 3 = 12 ways of taking two cars. Similarly there are 4 × 3 × 2 = 24 ways of taking three cars. There remains now one person and one car which can be taken by this last person in one way only.

The total number of different arrangements is 4. 3. 2. 1 = 24

We write this in the *factorial* notation :

4 ! = 4. 3. 2. 1 = 24 (read 4 factorial)

We have similarly

5 ! = 5. 4. 2. 1 = 120

6 ! = 6. 5. 4. 3. 2. 1 = 720

$n! = n(n-1)(n-2) \ldots\ldots\ldots 3.\ 2.\ 1$

The symbol n! represents the product of all whole numbers from 1 to n.

We define 0! = 1.

Ex. 213. $\frac{7!}{4!} = \frac{7.6.5.4.3.2.1}{4.3.2.1} = 7.6.5 = 210$

Ex. 214. $\frac{n!}{(n-1)!} = \frac{n(n-1)(n-2)\ldots3.2.1}{(n-1)(n-2)\ldots3.2.1} = n.$

Ex. 215. $6.5! = 6!.$

Ex. 216. $n(n-1)! = n!.$

Ex. 217. $n(n-1)! = n(n-1)\,(n-2)!.$

Ex. 218. $\frac{6!}{2!\,4!} = \frac{6.\ 5.\ 4.\ 3.\ 2.\ 1}{2.\ 1.\ 4.\ 3.\ 2.\ 1} = 15.$

Ex. 219. $\frac{n!}{r!(n-r)!} = \frac{n(n-1)\ldots\ldots\ldots\ldots 3.2.1}{r(r-1)\ldots3.2.1(n-r)(n-r-1)\ldots3.2.1}$

$$= \frac{n(n-1)(n-2)\ldots\ldots(n-r+1)}{r!}$$

Ex. 220. $\frac{(n+2)!}{n!} = (n+1)(n+2)$

Ex. 221. $\frac{n!}{r!(n-r)!} + \frac{n!}{(r-1)!(n-r+1)!} = \frac{(n+1)!}{r!(n-r+1)!}$

Permutations

A permutation of a set of distinguishable objects is any arrangement of some of these objects in a definite order. In such an arrangement, order is essential. A different order yields a different permutation.

Ex. 222. If there are three letters *a, b, c* there are six possible permutations:

abc *bca* *cab* *acb* *bac* *cba*

This can be done by means of a *tree* diagram where the total number of paths as shown gives the number of distinct permutations:

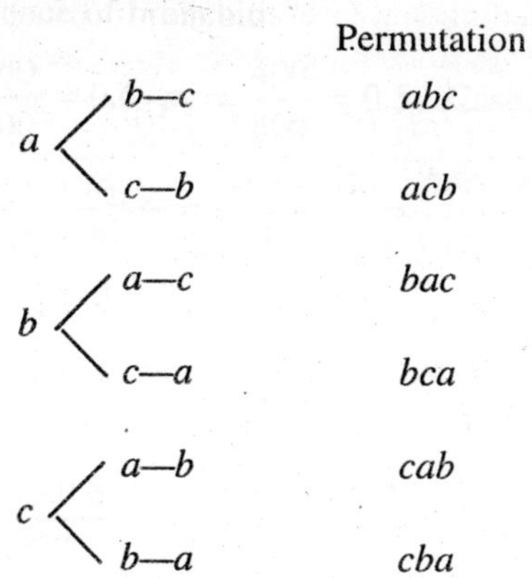

We can use the following *multiplication principle* to find the number of permutations:

If an event can occur in n different ways and for each of these n ways, a second event can occur in m different ways, then the two events can occur in nm ways. The argument can be extended to more events.

When the events are *mutually exclusive i.e.,* not occurring simultaneously, the total number of permutations in such a situation can be found by the following *addition principle* :

If two events are mutually exclusive and the first can occur in n different ways and the second in m different ways then one event *or* the other can occur in $n + m$ ways.

Ex. 223. Of the three letters a, b, c, any one letter can be taken in 3 ways. The second letter can now be taken in 2 ways and the third letter in 1 way.

The total number of permutations is $3 \times 2 \times 1 = 6$.

This is usually written

$$^3P_3 = 3 \times 2 \times 1 = 6$$

$^3P_3$ represent the number of ways of selecting 3 objects taken 3 at a time.

The symbol ^nP_r

^nP_r denotes the total number of permutations of a set of n different objects taken r at a time.

$$^nP_r = n(n-1)(n-2)......(n-r+1)$$

$$= \frac{n(n-1)(n-2)......3.2.1}{(n-r)(n-r-1)...3.2.1} = \frac{n!}{(n-r)!}$$

This is the formula to find the number of ways in which r objects can be placed in a linear order if there are n objects from which to select.

We may argue as follows:

The number of permutations of n objects taken r at a time is the same as the number of ways in which r seats can be occupied by n persons.

There are n ways of occupying the first seat. There being $n - 1$ seats left now, there are $n - 1$ ways of occupying the second seat so that the first two seats can be occupied in $n(n - 1)$ ways. Proceeding similarly, we note that the number of ways in which all the seats can be occupied is

$$n(n-1)(n-2).........(n-r+1)$$

which is the definition of ^nP_r.

Ex. 224. The number of permutations of 7 objects taken 2 at a time is

$$^7P_2 = 7.6 = 42.$$

Ex. 225. (*i*) The number of permutations of n objects taken n at a time is

$$^nP_n = n\,!.$$

(*ii*) Show $^nP_r = {}^{n-1}P_r + r\,{}^{n-1}P_{r-1}$

(*iii*) If $\frac{^{96}P_{r+6}}{^{54}P_{r+3}} = \frac{30800}{1}$ show $r = 41$

Ex. 226. There are 5 entrances to a hall and 2 extra doors to leave (*i.e.*, these doors open only from the inside). In how many different ways can a person enter and leave the hall?

Ans. $5 \times 7 = 35$ ways

Ex. 227. Show that there are 15600 ways of putting 3 letters without repetition on the licence plates of road vehicles.

If repetition is allowed the number of ways is 26^3.

Ex. 228. If repetition is allowed, the number of arrangements of n objects taken r at a time is n^r.

Ex. 229. Three persons have 4 coats, 5 shirts and 6 ties. Find the number of ways in which they can put on their clothes.

Ans. 1956

Ex. 230. (*i*) If no digit is repeated in any number, how many four digit numbers can be formed out of the digits 1, 2 3, 4, 5, 6?

(*ii*) How many of these numbers will exceed 4000?

Ans. (*i*) ${}^6P_4 = 360$; (*ii*) In this case the digit in the thousand's place has to be 4, 5 or 6. The remaining 3 digits can be chosen from the remaining 5 digits in ${}^5P_3 = 60$ ways. The answer, therefore, is $3 \times {}^5P_3 = 180$ ways.

Ex. 231. The offices of chairman, secretary and treasurer are to be filled out of eight eligible candidates for the college union council. In how many ways can this be done?

Ans. 336

Ex. 232. In how many ways can 6 men and 4 women be seated in a straight line so that no two women are seated together?

Ans. 604800

Ex. 233. In how many ways can the letters of the word M O U N T be arranged?

Ans. The total number of words that can be formed is ${}^5P_5 = 5! = 120$

Suppose, here, that all the words are required to begin with the letter O. Then there are only 4 letters to be arranged and the total number of words, all beginning with O, is $4! = 24$.

If now, it is required that all the words begin with O and end with U, the total number of words that can be formed is $3! = 6$.

If the beginning or end is with O *or* U this can be done in ${}^2P_2 = 2 \times 1 =$ ways. The remaining 3 letters give us 6 words and the total number of possibilities is $6 \times 2 = 12$.

Ex. 234. (*i*) In how many ways m men and w women can sit in a row?

Ans. $(m + w)!$ ways

(*ii*) In how many ways can 4 men and 2 women sit in a row?

Ans. $6! = 720$ ways

(*iii*) In how many ways can they be seated so that men must sit together and women must sit together?

Ans. Men can sit in 4! ways

Women can sit in 2! ways

There can be first men MMMMWW

or there can be first women WWMMMM

The total number of ways is $4!\,2! + 2!\,4! = 2 \times 4!\,2! = 96$ ways (*iv*) If only the women are to sit together there are five possible cases: MMMMWW, MMMWWM, MMWWMM, MWWMMM, WWMMMM

Men can sit in 4! ways.

Women can sit in 2! ways.

The total number of ways is $5 \times 4! \times 2! = 240$

Ex. 235. An urn contain 7 balls. Of these 3 balls can be drawn in (*i*) $7^3 = 343$ ways with replacement and (*ii*) in 7.6.5 = 210 ways without replacement.

Permutations when some objects are alike

There are six possible permutations with the letters *a, b, c*.

If the letters are *a, b, a* there are only three permutations, *viz. aba, aab, baa*. This can be found by dividing 3! by 2! which is the number of ways in which the *a*'s can be permuted. Thus, with a letter repeated twice the number of permutations of 3 letters is $\frac{3!}{2!} = 3$.

The number of permutations of n objects taken n at a time in which n_1 objects of one kind are alike, n_2 of another kind are alike, ..., n_k of the kth kind are alike, is given by

$$\frac{n!}{n_1!\, n_2! \ldots n_k!} \text{ (where } n_1 + n_2 + \ldots + n_k = n).$$

Ex. 236. In how many ways can the letters of the following words be written?

(*i*) E C O N O M I C S (*ii*) S T A T I S T I C I A N

Ans. (*i*) C and O occur twice each. Hence the required number of ways is $\frac{9!}{2!\,2!}$

(*ii*) $\frac{12!}{2!\; 2!\; 3!\; 3!}$ (Since S, A occur twice and T, I thrice each)

Ex. 237. In how many different ways can the letters of (*i*) INDIA (*ii*) BHARAT (*iii*) HINDUSTAN be arranged? **Ans.** (*i*) 60 (*ii*) 360 (*iii*) 1,81,440

Circular Permutations

Circular permutations depend on the relative positions of objects. If there are *n* objects we fix the position of one object and arrange the remaining $n - 1$ objects in all possible ways. We therefore conclude that the number of distinct permutations of *n* objects taken *n* at a time on a circle is $(n - 1)!$.

Ex. 238. In how many ways can four men and four women be seated at a round table so that boys and girls occupy alternate seats? If one couple is allowed to sit together what is the number of arrangements? **Ans.** 144, 72

Combinations

Arrangements in which order is not a consideration are called combinations.

A combination of a set of distinct objects is any selection of some of these objects without regard to their order. The order of arrangements is immaterial.

ab and *ba* represent the same combination.

abc, bca, cab, acb, cba, bac represent different permutations but the same combination of three letters *a, b, c* taken three at a time.

Ex. 239. There are 4 combinations and 24 permutations of the letters *a, b, c, d* taken 3 at a time.

The symbol ^nC_r is used to denote the number of combinations of *n* objects taken *r* at a time. It can be shown that

$$^nC_r = \frac{n!}{r!(n-r)!} = \frac{^nP_r}{r!}$$

The following results may be verified.

(*i*) $^nP_r = \frac{n!}{(n-r)!} = r!\ ^nC_r.$

(*ii*) $^nC_r = {}^nC_{n-r}$. This shows that the number of combinations of *n* distinct objects taken *r* at a time equals the number of combinations of these *n* objects taken $n - r$ at a time.

(*iii*) $^nC_n = {}^nC_0$. This follows from (*ii*) $^nC_0 = 1,\ ^nC_n = 1$

(*iv*) $^nC_{r-1} = \frac{r \cdot {}^nC_r}{n-r+1}$. Hence $^nC_r + {}^nC_{r-1} = {}^{n+1}C_r$

(*v*) If $^nC_r = {}^nC_s$ then $r = s$ or $n - r = s$

(*vi*) The number of ways in which $m + n$ items can be divided into two groups of *m* and *n* items is $^{m+n}C_m = \frac{(m+n)!}{m!\ n!}$

Ex. 240. Show :

(*i*) If $^{15}C_{3r} = {}^{15}C_{r+3}$ then $r = 3$

(*ii*) If $^nC_{r-1} = 36,\ ^nC_r = 84,\ ^nC_{r+1} = 126$

Find *n, r*. **Ans.** $n = 9, r = 3$

Ex. 241. From 10 different letters of the alphabet, words with 5 letters are formed. Find the number of words which have at least one letter repeated.

When a letter is allowed repetition any number of times then the number of 5 letter words = 10^5 = 100000.

Number of words when all 5 letters are different

$= {}^{10}C_5 = 252$

Required number = 100000 – 252 = 99748

Ex. 242. (*i*) In how many ways can 3 boys be selected out of 8 boys? (*ii*) If a particular boy is to be always selected in how many ways can this be done?

Ans. (*i*) $^8C_3 = \frac{8.7.6}{1.2.3} = 56$ (*ii*) $^7C_2 = \frac{7.6}{1.2} = 21$

Ex. 243. In how many ways can 2 books be selected out of 10 different books?

Ans. 45 ways

Ex. 244. Out of 6 ruling and 5 opposition party members 4 are to be selected. In how many ways can this be done so as to include, (*i*) exactly one ruling party member, (*ii*) at least one opposition member?

Ans. (*i*) $^6C_1 \,.\, ^5C_3 = 60.$

(*ii*) This means selecting 1, 2, 3 or 4 opposition members. The number of ways required, therefore, is $^5C_1 \cdot {}^6C_3 + {}^5C_2 \cdot {}^6C_2 + {}^5C_3 \cdot {}^6C_1 + {}^5C_4 = 315$

Ex. 245. From a group of 10 democrats and 8 socialists how many committees of seven can be formed so as to include, (*i*) exactly 4 democrats, (*ii*) at most 4 democrats?

Ans. 11760, 19968

Ex. 246. The number of ways of selecting *r* objects from *n* + 1 objects is

$$^{n+1}C_r = {}^nC_r + {}^nC_{r-1}.$$

Ex. 247. Find the value of

(*i*) $^6P_2$ **Ans.** (*i*) $\dfrac{6!}{(6-2)!} = \dfrac{6!}{4!} = \dfrac{6.5.4.3.2.1}{4.3.2.1} = 30.$

(*ii*) $^6P_6$ (*ii*) $^6P_6 = 6! = 6.5.4.3.2.1 = 720.$

(*iii*) $^6C_2$ (*iii*) $^6C_2 = \dfrac{6!}{2!\ 4!} = \dfrac{^6P_2}{2!} = \dfrac{30}{2} = 15$

(*iv*) $^6C_4$ (*iv*) $^6C_4 = {}^6C_2 = 15.$

(*v*) $^6C_6$ (*v*) $^6C_6 = {}^6C_0 = 1.$

Ex. 248. In how many ways can two major and two minor tests be given to a class of 10 students? **Ans.** 10.9.8.7 = 5040 ways

Ex. 249. There are 8 different colours. How many selections of 3 different colours are possible? **Ans.** 336

Ex. 250. Show that from the digits 1, 2, 3, 4, 5, 6 we can form 60 even numbers, each having three digits, none of the digits being repeated in a number. (There are three digits 2, 4, 6 for the unit's place. The digit in ten's place can be filled in 5 ways after the digit in the unit's place. The digit in the hundred's place can now be filled in 4 ways. We have 3.5.4 = 60.)

Ex. 251. In how many ways may 2 dice show up? **Ans.** $6 \times 6 = 36$ ways

Ex. 252. In how many ways may 3 dice show up? **Ans.** $6 \times 6 \times 6 = 216$ ways

Ex. 253. In how many ways can three or more people be selected out of 5?

Ans. $^5C_3 + {}^5C_4 + {}^5C_5 = 16$

Ex. 254. A straight line is determined by two points. How many straight lines can be determined by 12 points no two of which lie in a straight line? **Ans.** 66

Ex. 255. In how many ways can a committee of 3 people be formed from a group of 7 people? **Ans.** 35

Ex. 256. If $^{18}C_r = {}^{18}C_{r+2}$ find *r*. **Ans.** $r = 8$

Ex. 257. If $^nC_{n-4} = 15$ find *n*. **Ans.** $n = 6$

Ex. 258. Find the number of four letter arrangements of the letters of the word BRASS. How many of them begin with S? **Ans.** 60, 24

Ex. 259. In a question paper, 16 questions are to be attempted out of 20. In how many ways can this be done? How often may any particular question be selected?

Ans. 4845, 3876

Note. Each factor in ${}^nC_r = \dfrac{n(n-1)......(n-r+1)}{r!}$ is positive for each value of r.

$${}^nC_r > {}^nC_{r-1} \text{ provided } \frac{n-r+1}{r} > 1$$

i.e., $$n - r + 1 > r \text{ or } r < \frac{n+1}{2}$$

$\therefore$ The numbers ${}^nC_1$, ${}^nC_2$, ${}^nC_3$,...... increase steadily till a maximum is reached. Thereafter the numbers start diminishing.

Case 1 If n is an odd integer, $n + 1$ even, then $\dfrac{n+1}{2}$ is an integer.

If $$r = \frac{n+1}{2},\ {}^nC_r = {}^nC_{r+1}$$

i.e., $${}^nC_{\frac{n+1}{2}} = {}^nC_{\frac{n-1}{2}}$$

Thus there are two maximum terms in this case

Case 2 If n is an even integer, $\dfrac{n+1}{2}$ is not an integer

$\therefore$ ${}^nC_r$ steadily increases till $r = n/2$ when

${}^nC_r$ is maximum.

Thereafter ${}^nC_r$ steadily decreases up to the value ${}^nC_n = 1$.

1.24 THE BINOMIAL THEOREM

The binomial theorem is used to carry out operations of the form $(a + 2b)^7$, $(2x - 3y)^5$ etc.

To find the expansion of $(x + y)^n$, let us consider the following expansions:

$$(x + y)^1 = x + y$$

$$(x + y)^2 = x^2 + 2xy + y^2 = {}^2C_0x^2 + {}^2C_1xy + {}^2C_2y^2$$

$$(x + y)^3 = x^3 + 3x^2y + 3xy^2 + y^3 = {}^3C_0x^3 + {}^3C_1x^2y + {}^3C_2xy^2 + {}^3C_3y^3$$

$$(x + y)^4 = x^4 + 4x^3y + 6x^2y^2 + 4xy^3 + y^4$$

$$= {}^4C_0x^4 + {}^4C_1x^3y + {}^4C_2x^2y^2 + {}^4C_3xy^3 + {}^4C_4y^4.$$

and so on. We have, by induction,

$$(x + y)^n = {}^nC_0x^n + {}^nC_1x^{n-1}y + {}^nC_2x^{n-2}y^2 + \ldots + {}^nC_rx^{n-r}y^r + \ldots + {}^nC_ny^n$$

$$= \sum_{r=0}^{n} {}^nC_rx^{n-r}y^r \cdot [\text{Here } {}^nC_rx^{n-r}y^r \text{ is the general term})$$

This is the *binomial theorem.* The numbers ${}^nC_r$ are called *binomial coefficients.*

Ex. 260. $(a + 3b)^4 = {}^4C_0a^4 + {}^4C_1a^3(3b) + {}^4C_2a^2(3b)^2 + {}^4C_3a(3b)^3 + {}^4C_4(3b)^4$

$= a^4 + 12a^3b + 54a^2b^2 + 108ab^3 + 81b^4.$

Ex. 261. $(x^2 - y^3)^5 = {}^5C_0(x^2)^5 - {}^5C_1(x^2)^4(y^3) + {}^5C_2(x^2)^3(y^3)^2$

$- {}^5C_3(x^2)^2(y^3)^3 + {}^5C_4(x^2)(y^3)^4 - {}^5C_5(y^3)^5$

$= x^{10} - 5x^8y^3 + 10x^6y^6 - 10x^4y^9 + 5x^2y^{12} - y^{15}.$

Ex. 262. Find $\left(a + \frac{1}{a}\right)^6$. **Ans.** $a^6 + 6a^4 + 15a^2 + 20 + \frac{15}{a^2} + \frac{16}{a^4} + \frac{1}{a^6}$.

Ex. 263. Find $(10.1)^5$

Ans. $(10.1)^5 = (10 + 0.1)^5 = 10^5 + {}^5C_1 \cdot 10^4 \cdot (0.1) + \ldots$etc.

$= 105101.00501.$

Ex. 264. Show that $(1.02)^8$ is approximately 1.171.

Notes. (*a*) *General Term in the Expansion of* $(x + y)^n$

The $(r + 1)$th term is the general term denoted by

$$T_{r+1} = {}^nC_rx^{n-r}y^r$$

(*b*) *Middle Term* (*i*) If n is even, then the middle term is

$$T_{\frac{n}{2}+1} = {}^nC_{n/2}x^{n/2}y^{n/2}$$

Ex. 265. If $n = 10$, the middle is $T_{\frac{10}{2}+1}$ *i.e.,* T_6

The middle term of $(x + y)^{10}$ is $T_6 = {}^{10}C_5\,x^5y^5$

(*ii*) If n is odd there are two middle terms.

If $n = 5$ as in $(x + y)^5$, the two middle terms are

$$T_3 = {}^5C_2x^3y^2\ T_4 = {}^5C_3x^2y^3$$

Thus when n is odd we have two middle terms

(1) $T_{\frac{n}{2}+\frac{1}{2}} = {}^nC_{\frac{n}{2}-\frac{1}{2}}\,x^{\frac{n}{2}+\frac{1}{2}}\,y^{\frac{n}{2}-\frac{1}{2}}$

(2) $T_{\frac{n}{2}+\frac{3}{2}} = {}^nC_{\frac{n}{2}+\frac{1}{2}}\,x^{\frac{n}{2}-\frac{1}{2}}\,y^{\frac{n}{2}+\frac{1}{2}}$

Ex. 266. The general term in the expansion of $(1 + x)^n$ is

$$T_{r+1} = {}^nC_rx^r$$

The general term in $(1 - x)^n$

$$T_{r+1} = (-1)^r\,{}^nC_rx^r$$

Ex. 267. Find the coefficient of (*i*) x^9 (*ii*) x^0 in $(x^2 - 1/3x)^9$

The general term is $T_{r+1} = {}^9C_r(x^2)^{9-r}\left(-\frac{1}{3x}\right)^r = {}^9C_r(-1)^r3^{-r}x^{18-3r}$

(*i*) If $18 - 3r = 9$, $r = 3$

The required term is $T_4 = (-1)^3 (3^{-3}) \cdot {}^9C_3 = \frac{-28}{9}$

(*ii*) If $18 - 3r = 0$, $r = 6$

The required term is $T_7 = (-1)^6\, 3^{-7} \cdot {}^9C_6 = \frac{28}{243}$

Ex. 268. (*i*) In the expansion of $\left(ax + \frac{1}{x}\right)^n$ if the fourth term is 5/2 find a, n.

(*ii*) Prove $101^{50} > 100^{50} + 99^{50}$ **Ans.** (*i*) $\frac{1}{2}$, 6

Ex. 269. Find the term independent of x in (*i*) $\left(x + \frac{1}{x}\right)^{2n}$ **Ans.** $\frac{(2n)!}{(n!)^2}$

(*ii*) $\left(\sqrt{\frac{x}{3}} + \frac{3}{2x^2}\right)^{10}$ **Ans.** 5/4

Notes. (*a*) In any binomial expansion $(x + y)^n$, the total number of terms is $n + 1$ which is one more than n, the binomial index.

The sum of the indices of x and y in any term is always n.

(*b*) In the expansion.

$$(x + y)^n$$
$$= {}^nC_0x^n + {}^nC_1x^{n-1}y^1 + {}^nC_2x^{n-2}y^2 + \ldots + {}^nC_rx^{n-r}y^n + \ldots + {}^nC_ny^n$$

The coefficients of the terms equidistant from the two sides are equal.

This is because ${}^nC_r = {}^nC_{n-r}$

Thus ${}^nC_0 = {}^nC_n$, ${}^nC_1 = {}^nC_{n-1}$ etc.

If we substitute $-y$ for y, it can be seen that

$$(x - y)^n = {}^nC_0x^n - {}^nC_1x^{n-1}y + {}^nC_2x^{n-2}y^2 \_\ldots + (-1)^ny^n$$

The terms are +ve and –ve alternately. The last term is +ve if n is even and –ve if n is odd.

Ex. 270. $(1 - x)^n = 1 - {}^nC_1x + {}^nC_2x^2 - {}^nC_3x^3 + \ldots + (-1)^nx^n$

$$\left(x - \frac{1}{x}\right)^4 = x^4 - 4x^3\left(\frac{1}{x}\right) + 6x^2\left(\frac{1}{x}\right)^2 - 4x\left(\frac{1}{x}\right)^3 + \left(\frac{1}{x}\right)^4$$

$$= x^4 - 4x^2 + 6 - 4/x^2 + 1/x^4$$

Ex. 271. In the binomial expansion of $(x + y)^n$, the $n + 1$ coefficients

$${}^nC_0, {}^nC_1, {}^nC_2, \ldots. {}^nC_r, \ldots {}^nC_n$$

are called the binomial coefficients.

Since $(1 + x)^n = {}^nC_0 + {}^nC_1x + {}^nC_2x^2 + \ldots. + {}^nC_nx^n$,

we have by putting $x = 1$,

$$(1 + 1)^n = {}^nC_0 + {}^nC_1 + {}^nC_2 + {}^nC_2 + \ldots. + {}^nC_n$$

Thus the sum of binomial coefficients

$${}^nC_0 + {}^nC_1 + {}^nC_2 + \ldots + {}^nC_n = 2^n$$

By giving suitable values to x, y and n in the expansion of $(x + y)^n$ it can be seen that

$$(1+x)^{-n} = 1 - nx + \frac{n(n-1)}{1\times 2}x^2 - \frac{n(n-1)(n-2)}{1\times 2\times 3}x^3 + \ldots\ldots$$

$$(1-x)^{-n} = 1 + nx + \frac{n(n-1)}{1\times 2}x^2 + \frac{n(n-1)(n-2)}{1\times 2\times 3}x^3 + \ldots\ldots$$

$$(1+x)^{-1} = \frac{1}{1+x} = 1 - x + x^2 - x^3 + \ldots + (-1)^r x^r + \ldots\ldots \quad |x| < |$$

$$(1-x)^{-1} = \frac{1}{1+x} = 1 + x + x^2 + x^3 + \ldots + x^r + \ldots\ldots \quad |x| < |$$

$$(1+x)^{-2} = \frac{1}{(1+x)^2} = 1 - 2x + 3x^2 - 4x^3 \ldots\ldots \quad |x| < |$$

Mathematical Induction

To prove that a proposition or statement $P(n)$ is true we can use the method of induction which consists of the following steps.

1. Verify if the statement is true for $n = 1$, $n = 2$
2. Assume it is true for $n = k$
3. Prove it is true for $n = k + 1$

Ex. 272. $P(n) = 1^2 + 2^2 + \ldots + n^2 = \dfrac{1}{6}n\,(n + 1)\,(2n + 1)$

Step (1) For $n = 1$, L.H.S. $= 1^2 = 1 =$ RHS, $P(1)$ is true

For $n = 2$, $1^2 + 2^2 = \dfrac{1}{6} \times 2(2 + 1)\,(2 \times 2 + 1) = 5$, $P(2)$ is true

Step (2) For some natural number $n = k$

$P(k) = 1^2 + 2^2 + \ldots + k^2 = \dfrac{1}{6}k\,(k + 1)\,(k + 2)$ is true

Step (3) $P(k) + (k + 1)^2 = 1^2 + 2^2 + \ldots + k^2 + (k + 1)^2$

$$= \left\{\frac{1}{6}k(k+1)(2k+1)\right\} + (k+1)^2 = \frac{1}{6}(k+1)(2k^2 + k + 6)$$

$$= \frac{1}{6}(k+1)\,(k+2)\,(2k+3)$$

$= P(k + 1)$ $\because$ k is replaced by $k + 1$.

$\therefore$ Since $P(1)$ is true and $P(k)$ leads to $P(k + 1)$

$\therefore$ $P(k)$ is true by induction.

Ex. 273. $\frac{1}{3.7}+\frac{1}{7.11}+\ldots+\frac{1}{(4n-1)(4n+3)}=\frac{n}{3(4n+3)}$

$P(1) = \frac{1}{21}$ (Putting $n = 1$ on LHS and RHS)

$$P(k)=\frac{1}{3.7}+\frac{1}{7.11}+\ldots+\frac{1}{(4k-1)(4k+3)}=\frac{k}{3(4k+3)}$$

Add $(k+1)$th term *i.e.*, $\frac{1}{(4k+3)(4k+7)}$ to $P(k)$

$$\therefore \left\{\frac{1}{3.7}+\frac{1}{7.11}+\ldots+\frac{1}{(4k-1)(4k+3)}+\ldots+\frac{1}{(4k-1)(4k+3)}\right\}+\frac{1}{(4k+3)(4k+7)}$$

$$=\frac{k}{3(4k+3)}+\frac{1}{(4k+3)(4k+7)}=\frac{k+1}{3(4k+7)}=P(k+1)$$

where k is replaced by $k + 1$ in $P(k)$

$\therefore$ $P(k + 1)$ is true when $P(k)$ is true

$\therefore$ $P(n)$ is true by induction.

Ex. 274. Prove that $7^{2n} + 2^{3n-3}\, 3^{n-1}$ is divisible by 25

When $n = 1$, $7^{2n} + 2^{3n-3} \times 3^{n-1} = 7^2 + 2^\circ \cdot 3^\circ = 49 + 1 = 50$

When $n = k$, let $7^{2k} + 2^{3n-3} \cdot 3^{n-1} = 25k$

When $n = k + 1$

$$7^{2(k+1)} + 2^{3(k+1)-3}\, 3^{k+1-1}$$

$$= 7^{2k}\, 49 + 2^{3k-3} \cdot 8 \cdot 3^{k-1} \cdot 3$$

(Add $7^{2k} + 2^{3k-3} \cdot 3^{k-1}$ and also subtract)

$$= 7^{2k}(49 + 1) + 2^{3k-3}\, 3^{k-1}\, (24 + 1) - 7^{2k} + 2^{3k-3} \cdot 3^{k-1}$$

$$= 25\, [2.7^{2k} + 2^{3k-3} \cdot 3^{k-1}] - 25k$$

This is divisible by 25.

Ex. 275. Prove the sum to n term S_n is true for $n \geq 1$ where

$$S_n = \sum_{r=0}^{n} r^2\, {}^nC_r = n(n+1)2^{n-2}$$

$S_1 = 0 + 1^2 \cdot {}^1C_1 = 1$ (sum of 2 terms)

$S_2 = 0 + 1^2 \cdot {}^2C_1 + 2^2 \cdot {}^2C_2 = 6$ (sum of 3 terms)

S_k = Sum of $k + 1$ term

Let $S_k = k(k + 1)\, 2^{k-2}$

$$S_{k+1} = \sum_{r=0}^{n}(r^2)\,{}^{k+1}C_r + (k+1)^2 \cdot {}^{k+1}C_{k+1}$$

$$= \sum_{r=0}^{k} r^2[{}^kC_r + {}^kC_{r-1}] + (k+1)^2 \quad [\because {}^kC_r + {}^kC_{r-1} = {}^{k+1}C_r]$$

$$= \sum_{r=0}^{k} r^2\, {}^kC_r + \sum_{r=0}^{k} r^2\, {}^kC_{r-1} + (k+1)^2$$

$$= k(k+1)2^{k-2} + \sum_{r=0}^{k} r^2\, {}^kC_{r-1} + (k+1)^2$$

$= k(k + 1)\ (k + 2)\ 2^{k-1}$ which is the same as S_k with k replaced by $k + 1$.

CHAPTER 2

Mathematical Reasoning

2.1 INDUCTIVE AND DEDUCTIVE REASONING

Mathematics has a well defined language and rules of operation. It provides one of the best methods of dealing with complex analytical problems. Mathematics deals with abstract general models : it is concerned with their internal logic and not with their applications. Mathematics is a tool which can be *used* by each expert in his own field. Every idea developed in mathematics need not necessarily have a direct application to the science under consideration but it can be relevant to the development of some principles.

Mathematics enables us to see the structure of an economic or a social science model in a clear, precise and compact form. In any such study, mathematics begins with the empirical content of the problem, makes use of deductive reasoning and arrives at empirical conclusions whose validity depends on the original empirical statement. Mathematics is uncommitted in the choice of the model but it is very much there in discovering the fitness of the model to the problem under consideration.

Reasoning, inductive or deductive, is an essential element in all mathematical work.

Inductive reasoning consists of conclusions drawn on the basis of observations. Deductive reasoning proceeds from assumptions and not from observations.

Induction is often the road to some discovery. In the course of inductive reasoning, one guesses, makes observations and may be led to certain general truths or principles which cannot be accepted outright. They must be proved.

Proofs are most often arrived at through deductive reasoning. The deductive arguments used in proofs must be consistent and convincing.

Mathematical work requires the use of both deductive and inductive reasoning. Pure deduction is impossible in a social science which deals with a certain aspect of reality. Judgements about reality can be based on induction and hence on what may be called intuition in social sciences.

2.2 AXIOM SYSTEMS

Mathematics is built upon a system of propositions or axioms involving undefined and defined concepts. Theorems are proved on the basis of these axioms.

A mathematical system or structure is the result of the application of the postulational method of logic to a set of elements, relations and operations. The choice of postulates determines the mathematical system completely. The set of postulates has to be noncontradictory or consistent. Different valid mathematical systems arise depending on the choice of elements, relations and operations, depending upon needs and interests. Algebra, Geometry, Calculus are examples of different mathematical systems.

In the process of reasoning in deductive logic, any proof involves some hypothesis and conclusion. The reasoning behind any deductive logical system can be understood by the properties of the system and the operations performed therein.

A deductive logical system or *an axiomatic system* is any logical system having primitive terms and postulates. It consists of four essential parts.

1. Undefined Terms
2. Defined Terms
3. Axioms
4. Theorems

A mathematical system is an example of a deductive logical system.

1. *Undefined Terms* : An axiomatic system is a formal system based ultimately on primitive terms which have no definite meaning. The primitive terms are *undefined technical terms* including elements, relations and operations which can be performed on elements.

All basic terms cannot be defined. We have to use some undefined terms to define other terms.

2. *Defined Terms* : All non-primitive technical terms used in the system are *defined terms*, described by the use of undefined terms.

A *model* of the axiomatic system is obtained by assigning terms of definite meaning in such a way as to make the postulates true statements.

A good *definition* should

(*a*) be consistent *i.e.*, convey exactly the same meaning in every possible case.

(*b*) be clear and precise without redundant words.

(*c*) use only undefined or previously defined terms.

(*d*) be broad enough to include all objects in the given set and yet be detailed enough to exclude any that do not belong to the set.

3. *Axioms* : An axiom (assumption, postulate, or primitive statement) is any statement that is assumed to be true for the system under consideration. It is an unproved statement containing only primitive terms. It is an assumption of some basic property about the elements of the system. Thus axioms are neither self evident nor necessarily true. They are assumptions, sometimes proper, sometimes not.

The axiom set of an axiomatic system should be consistent *i.e.*, non-contradictory.

A concrete model based on real world objects, relations and operations and relying on noncontradictory statements is absolutely consistent because contradictions in the real world are impossible. A derived model is relatively consistent.

Axioms become more comprehensible with mental pictures which, in turn, can be easily formed if the formulation of the axioms is precise.

4. *Theorems* : A theorem is a statement which must be logically deduced or proved by using undefined terms, defined terms, axioms and statements whose truth value has been previously established. The truth of a statement is a relative idea depending on the particular set of axioms.

2.3 LOGIC

Mathematical logic or symbolic logic is the art of reasoning which helps us to systematise the principles of valid reasoning and which may be used to judge the correctness of a chain of reasoning. Mathematical reasoning is formal: it is based on the form of the sequence of statements. It gives us a mathematical theory as a logical system.

Philosophers and logicians believe that mathematics is a part of logic. Mathematicians feel the opposite. In either case the importance of logic cannot be denied. Logical reasoning is a must in science. Intuition and accidental results may be prominent at times but even in such cases logic can play an important role. The ability to reason logically reflects intelligence. This ability can be improved and cultivated, though sometimes a genius may have it as a natural gift.

Real world observations and undefined and defined motives of the behaving units give rise to propositions. The complex relationships in the economic behaviour of men and of nations along with these propositions lead to predictions, policy recommendations and various types of suggestions. A consideration of economic phenomena in terms of variables like prices, production, taxes etc., although subject to severe limitations and practical difficulties, is helpful in giving a systematic formulation of the multifarious activities of the economic system. Logic brings economic analysis and mathematical reasoning together to help draw conclusions from a set of postulates. Logical reasoning, used along with the characteristics of the economic system can be useful in understanding complex economic phenomena more clearly.

We can use the language of sets in the study of logic. The notion of sets is essential in any mathematical thinking. Almost all the sciences, physical or social, make use of the set theory, implicitly or explicitly.

2.4 STATEMENTS, TRUTH TABLES AND QUANTIFIERS

Logic, commonly described as the science of the formal principles of reasoning, studies the forms of valid arguments and different types of propositions. It considers the methods and principles used in distinguishing between good and bad arguments.

Statements : A statement (also called an assertion, a proposition or a declarative sentence), either written or oral, is a sentence which is either true or false. It cannot be true and false at the same time.

The statement *Economics is a science* has a *truth value. The* truth value of a statement is its truthfulness (T) or falsehood (F). The truth value is a relation between the subject and the predicate of the sentence.

A statement is denoted by letters like p, q, etc.

Compound Statements : A statement which is the combination of two or more simple statements is a *compound* (or composite) statement. Simple statements are combined by *connectives.*

The truth value of a statement is completely determined by the truth value of each of its substatements and the way they are connected.

If we write all possible truth values of a given statement for all possible combinations of truth values of its components we get a device known as the *truth table* which is useful in considering the validity of the statement.

Open Sentences

An open sentence is not a statement but is a sentence which contains a variable and which can be converted into a statement by replacing the variable by the name of a specific number or object or person and then quantifying the variable.

A sentence in mathematics does not include exclamatory, interrogative and imperative sentences.

Quantifiers

A quantifier is a specific collection of numbers, objects or persons.

Conclusions in logic are drawn mostly by the use of the following quantifiers.

(*i*) *All*, the *universal* quantifier, written $\forall x$ and read, *for all x* or *for every x*

(*ii*) *Some*, the *existential* quantifier, written $\exists x$ and read, *for some x* or *there exists an x*.

(*iii*) The quantifier *no* (or none) can be expressed in terms of *all* or *some*.

A single sentence may contain several quantifiers of each kind. Often, the order in which quantifiers occur, may make a difference in meaning.

2.5 TYPES OF STATEMENTS

Conjunction $p \wedge q$

Conjunction is the joining of two statements p and q by the connective *and*.

The statement "p and q" written $p \wedge q$ is called the conjunction of p and q.

If p is true and q is true then $p \wedge q$ is true. Otherwise $p \wedge q$ is false.

There are four cases, which can be shown in the *truth table* which exhibits all the information about $p \wedge q$.

	p	q	$p \wedge q$
(*i*) p true, q true	T	T	T
(*ii*) p true, q false	T	F	F
(*iii*) p false q true	F	T	F
(*iv*) p false, q false	F	F	F

Disjunction $p \vee q$

Disjunction is the joining of two statements by the connective *or* used in the inclusive sense.

$p \vee q$ (i.e., p or q) is called the inclusive disjunction of p and q.

If both p, q are false, $p \vee q$ is false; otherwise $p \vee q$ is true. There are four cases as shown in the truth table.

If we wish to assert that either p or q is true but not both p and q are true we have the exclusive disjunction $p \vee q$ for which the truth table is given below.

Inclusive disjunction

p	q	$p \vee q$
T	T	T
T	F	T
F	T	T
F	F	F

Exclusive disjunction

p	q	$p \vee q$
T	T	F
T	F	T
F	T	T
F	F	F

Ex. 1. If p = Rent is high, q = Corn is low

then $p \wedge q$ = Rent is high and corn is low

$p \vee q$ = Rent is high or corn is low.

Negation $\sim p$

Conjunction and disjunction are *binary* compositions, combining two statements to give a new one. Negation or denial just turns a statement into a new one; it is *unary*. Negation consists in adding to or removing from a sentence the word 'not'.

If p is a statement $\sim p$ (*i.e.*, not p) is a statement such that

if p is true then $\sim p$ is false,

if p is false then $\sim p$ is true.

p	$\sim p$
T	F
F	T

The truth value of the negation of a statement is just the opposite of the statement.

Ex. 2.

p : Rent is high $\sim p$: Rent is not high.

p : Sometimes rent is high $\sim p$: Rent is never high.

p : Some rents are not high $\sim p$: All rents are high.

p : Rent and corn are high $\sim p$: Rent or corn is not high.

Conditional $p \to q$

A conditional has two components : the *antecedent* or hypothesis and the *consequent*.

If p, q are statements then $p \to q$ (i.e., if p then q) is called an *implication* or a *conditional*.

$p \to q$ is false when p is true and q is false.

Otherwise it is true.

p	q	$p \to q$
T	T	T
T	F	F
F	T	T
F	F	T

Converse $q \to p$

The implication $q \to p$ (*i.e.*, if q then p) is called the converse of the implication $p \to q$.

Inverse $\sim p \to \sim q$

The implication $\sim p \to \sim q$ (*i.e.*, if not p then not q) is called the inverse of the implication $p \to q$.

Contrapositive $\sim q \to \sim p$

The implication $\sim q \to \sim p$ (*i.e.*, if not q then not p) is called the contrapositive of the implication $p \to q$.

The contrapositive method of proof begins by assuming the false result to be true and arrives at a contradiction. This is the famous *reductio ad absurdum* method of proving results in an indirect manner.

Equivalent Statement $\Leftrightarrow$

Equivalent statements are true together. They have identical truth tables.

Ex. 3. (*i*) $p \Leftrightarrow \sim(\sim p)$ (*ii*) $\sim(p \vee q) \Leftrightarrow (\sim p \wedge \sim q)$

(*iii*) $(p \vee p) \Leftrightarrow p$ (*iv*) $(p \vee q) \Leftrightarrow (q \vee p)$

(*v*) $(p \wedge p) \Leftrightarrow p$ (*vi*) $(p \wedge q) \Leftrightarrow (q \wedge p)$

The statement $p \to q$ and its contrapositive $\sim q \to \sim p$ are equivalent. This is verified from the truth table.

The converse $q \to p$ and the inverse $\sim p \to \sim q$ are equivalent. The truth table shows this.

				Implication	Converse	Inverse	Contrapositive
p	q	$\sim p$	$\sim q$	$p \to q$	$q \to p$	$\sim p \to \sim q$	$\sim q \to \sim p$
T	T	F	F	T	T	T	T
T	F	F	T	F	T	T	F
F	T	T	F	T	F	F	T
F	F	T	T	T	T	T	T

Ex. 4.

Implication	: If price is high then demand is low.
Converse	: If demand is low then price is high.
Inverse	: If price is not high then demand is not low.
Contrapositive	: If demand is not low then price is not high.

Tautology

A statement logically true in all possible cases is a tautology. Instances of tautology: $p \to p$, $p \to [q \to (p \text{ and } q)]$

$[p \vee (q \text{ and } r)] \Leftrightarrow [(p \vee q) \text{ and } (p \vee r)]$

Contradiction : A statement logically false in all possible cases is a contradiction. Instances of contradiction : $p \wedge \sim p$, $p \to \sim p$.

Ex. 5. Prove that the inverse and the contrapositive of an implication are not equivalent. (Use Truth Tables)

Ex. 6. Prove that the converse and the contrapositive of an implication are not equivalent.

Ex. 7. State the inverse, converse and contrapositive of the following implications:

(*i*) If x is an integer then $2x$ is an even integer.

(*ii*) If $x = 3$ then $x^2 + 1 = 10$.

(*iii*) If $x^2 = 2x$ then $x = 2$ or $x = 0$.

Ans. (*i*) Inverse : If x is not an integer then $2x$ is not an even integer.
Converse : If $2x$ is an even integer then x is an integer.
Contrapositive : If $2x$ is not an even integer then x is not an integer.

(*ii*) Inverse : If $x \neq 3$ then $x^2 + 1 \neq 10$
Converse : If $x^2 + 1 = 10$ then $x = 3$.
Contrapositive : If $x^2 + 1 \neq 10$ then $x \neq 3$

(*iii*) Inverse : If $x^2 \neq 2x$ then $x \neq 2$ and $x \neq 0$
Converse : If $x = 2$ or $x = 0$ then $x^2 = 2x$
Contrapositive : If $x \neq 2$ and $x \neq 0$ then $x^2 \neq 2x$.

Ex. 8. Write the above answer in symbolic form.

Ex. 9. Given the statements :

p : The P.C.I. is low $\quad$ q : The country is poor,

give verbal representations of the following symbolic forms:

(*i*) $p \wedge q$ $\quad$ (*ii*) $p \vee q$ $\quad$ (*iii*) $p \wedge \sim q$

(*iv*) $\sim p \wedge q$ $\quad$ (*v*) $(\sim p \wedge q) \vee (p \wedge \sim q)$

Ans. (*i*) P.C.I. is low and the country is poor.

(*ii*) P.C.I. is low or the country is poor.

(*iii*) P.C.I. is low and the country is not poor.

(*iv*) P.C.I. is not low and the country is poor.

(*v*) P.C.I. is not low and the country is poor.

or P.C.I. is low and the country is not poor.

Ex. 10. Construct truth tables for the following:

(*i*) $\sim(p \vee q)$ (*ii*) $\sim p \wedge \sim q$

Are they equivalent?

Ans.

p	q	$p \vee q$	$\sim(p \vee q)$	$\sim p$	$\sim q$	$\sim p \wedge \sim q$
T	T	T	F	F	F	F
T	F	T	F	F	T	F
F	T	T	F	T	F	T
F	F	F	T	T	T	T

They are equivalent.

Ex. 11. Prove that the following are equivalent:

$p, \sim(\sim p)$.

Ex. 12. Construct truth tables for the following:

(*i*) $(p \vee q) \vee (\sim p)$

(*ii*) $(p \vee \sim q) \wedge \sim p$

(*iii*) $\sim(p \vee q) \vee \sim(q \vee p)$.

Ans. These may be constructed in stages.

p	q	$p \vee q$	$(p \vee q \vee (\sim p)$	$p \vee \sim q$	$(p \vee \sim q) \wedge \sim p$
T	T	T	T	T	F
T	F	T	T	T	F
F	T	T	T	F	F
F	F	F	T	T	T

p	q	$\sim(p \vee q)$	$\sim(q \vee p)$	$\sim(p \vee q) \vee \sim(q \vee p)$
T	T	F	F	F
T	F	F	F	F
F	T	F	F	F
F	T	T	T	T

Note that (*ii*) and (*iii*) are equivalent.

Ex. 13. Are the following statements equivalent? Construct the truth tables:

(*i*) $p \wedge (q \vee r)$ (*ii*) $(p \wedge q) \vee (p . \wedge r)$

These are examples of iterated compositions. An iterated composition consists of application of truth functional compositions a number of times as done in Ex. 12.

A good example of an iterated composition is the biconditional statement $(p \to q) \wedge (q \to p)$.

Biconditionals $p \leftrightarrow q$

The statements $p \leftrightarrow q$ (*i.e.*, p if and only if q) is a *biconditional statement*. The connective 'if and only if' is called the biconditional connective.

$p \leftrightarrow q$ is logically equivalent to two one way implications $p \to q$ and $q \to p$.

If p, q are both true, $p \to q$, $q \to p$ are both true.

If p, q are both false, $p \to p$, $q \to p$ are both true.

We may say, therefore, that the biconditional statement is true when both p, q are true and when both p, q are false. The biconditional is false when p is true and q is false or *vice versa*.

p	q	$p \leftrightarrow q$
T	T	T
T	F	F
F	T	F
F	F	T

Ex. 14. Construct truth tables for $q \leftrightarrow p$ and show that $p \leftrightarrow q$ and $q \leftrightarrow p$ are equivalent.

Ex. 15. Construct truth tables for the following:

(*i*) $p \sim \leftrightarrow q$ (*ii*) $q \leftrightarrow \sim p$

(*iii*) $(\sim p \leftrightarrow q) \wedge (q \to \sim p)$.

Ans.

p	q	$\sim p \leftrightarrow q$	$q \leftrightarrow \sim p$	$(\sim p \to q) \wedge (q \to \sim p)$
T	T	F	F	F
T	F	T	T	T
F	T	T	T	T
F	F	F	F	F

Note that (*i*), (*ii*), (*iii*) are equivalent statements.

Ex. 16. Given the statement: "If I pass the test then I shall invite you", judge under the following conditions if this statement is true (T) or false (F).

(*a*) I pass the test and invite you.

(*b*) I pass the test and do not invite you.

(*c*) I do not pass the test and invite you.

(*d*) I do not pass the test and do not invite you. **Ans.** T, F, T, T

Ex. 17. Prove that the contrapositive of an implication is the inverse of the converse of the implication or it is the converse of the inverse of the implication.

This has been proved above by the method of truth tables.

The result means that the hypothesis and conclusion may be interchanged and then negated or the hypothesis and conclusion may first be negated and then interchanged.

Implication	...	Converse
$p \to q$		$q \to p$
⋮		⋮
Inverse	...	Contrapositive
$\sim p \to \sim q$		$\sim q \to \sim p$

A formal proof may be given.

Since $p \to q$, it means, either p is true or $\sim p$ is true.

If p is true then $\sim q$ cannot be true.

If $\sim q$ is true then p cannot be true.

Hence the result.

Ex. 18. Prove similarly : If $r \to \sim p, \sim r \to s$ then $p \to s$.

Ex. 19. Prove that the contrapositive of a theorem is the theorem itself.

Ex. 20. Prove that $[(p \to q)$ and $(q \to r)] \to (p \to r)$

(This is the transitive property of implications. It can be generalised. A series of implications thus gives us a deductive sequence.)

A summary of the Laws of Operations governing statements

We have considered several basic laws concerning statements in logic. Given below is a summary of the same.

Let p, q, r be the given statements.

(*i*) **Idempotent laws**

$(p \vee p) \Leftrightarrow p$ $\qquad$ $(p \wedge p) \Leftrightarrow p$

(*ii*) **Conjunction and Disjunction**

$(p \vee q) \Leftrightarrow (q \vee p)$ $\qquad$ $(p \wedge q) \Leftrightarrow (q \wedge p)$

$[(p \vee q) \vee r] \Leftrightarrow [p \vee (q \vee r)]$ $\qquad$ $[(p \wedge q) \wedge r \Leftrightarrow p \wedge (q \wedge r)]$

$p \vee \sim p$ is true $\qquad$ $p \wedge \sim p$ is false.

(*iii*) **De Morgan's Laws**

$\sim (p \vee q) \Leftrightarrow (\sim p \wedge \sim q)$ $\qquad$ $\sim (p \wedge q) \Leftrightarrow (\sim p \vee \sim q)$

2.6 VALIDITY, TRUTH AND PROOF

An argument in which a conclusion is reached correctly is a *valid* argument. The validity of an argument is not concerned with the truth of premises or conclusions.

A valid argument results when the conjunction of the hypotheses implies the conclusion.

To prove the validity of an argument we may use the truth table analysis.

Consider the truth table for $[p \text{ and } (p \to q)] \to q$. .

p	q	$p \to q$	p and $(p \to q)$	$[p \text{ and } (p \to q)] \to q$
T	T	T	T	T
T	F	F	F	T
F	T	T	F	T
F	F	T	F	T

From the truth table we see that the statement $[p \text{ and } (p \to q)] \to q$ is always true when both p and q are false or true. It means that if we accept p and $p \to q$ then the conclusion is that we must accept q.

The deductive statement $[p \text{ and } (p \to q) \to q]$ is a syllogism.

p is a minor premise

$p \to q$ is a major premise

q is the conclusion of the syllogism.

The major premise is a statement which is always of a form such that given the minor premise, the conclusion necessarily follows.

Indirect Proof

Sometimes to prove that a statement is true we may first assume that it is untrue and then by arguing logically and proceeding we can arrive at an explicit contradiction. A correct argument will lead to the conclusion that the original statement was not untrue.

Thus to prove that a statement q is true, suppose q is a consequence of p which is known to be true. To show $p \to q$ by the *reductio ad absurdum* we assume q to be false and arrive at some contradictory results and conclude $\sim q \to \sim p$, which being the same as $p \to q$, the proof is complete.

This reasoning is commonly used. Thus someone may say: If there were deficit financing prices would have gone up. But there has been no deficit financing it follows that prices have not gone up.

Reductio ad absurdum is an indirect method of proof. It is a proof by contradiction and hence is related to the contraposition.

2.7 MEANING OF NECESSARY AND SUFFICIENT CONDITIONS

In the implication $p \to q$

(*i*) if p is satisfied then q is a true statement. Statement p is said to be a sufficient condition for q because the knowledge that p is true is sufficient for knowing q is true.

(*ii*) p cannot be true unless q is satisfied. Statement q is said to be a necessary condition for p.

q may be necessary for p but may not be sufficient for p. Brains may be necessary to win an election but may not be sufficient for it.

If q is both a necessary and sufficient condition for p, then the converse of the original implication is true. Thus for a necessary and sufficient condition both the implication and its converse must be true i.e., $p \rightarrow q$. For this purpose we attach the phrase 'if and only if' to a statement.

'if' conveys the direct implication and 'only if' conveys the converse.

Ex. 21. Consider the statement:

Only if a student is active he should be the leader of the class.

This means: if a student is the leader of the class he should be active. Being active is a necessary condition to being a leader but it is not a sufficient condition.

Ex. 22. Comment on the statement.

A professor is good if and only if he imparts knowledge to his students.

Ex. 23. Consider the argument.

If prices are high then wages are high. Prices are high or there are price controls. If there are price controls then there is not an inflation. There is an inflation. Therefore, wages are high. (Robert R. Stoll).

We are in agreement with each of the first four statements (the premises). Must we accept the conclusion in the fifth statement?

The statements can be written: (Notation self explanatory)

$$\begin{array}{l} P \rightarrow W \\ P \vee C \\ C \rightarrow \sim I \\ I \\ \hline W \end{array}$$

If the premises have truth value T then the conclusion must have value T.

Since I and $C \rightarrow \sim I$ have value T

$\therefore$ The value of C is F according to the truth table for the conditional.

Since $P \vee C$ has value T

$\therefore$ P has value T

Since $P \rightarrow W$ has value T

$\therefore$ W has value T

The conclusion is : Wages are high.

Ex. 24. In the above argument consider the conjunction

$$(P \vee C) \wedge (C \rightarrow \sim I)$$

Its truth value depends on the assignments made to the prime statements, depending further on periods of changing economic conditions.

We may, therefore, consider the truth table for the composite statement for all possible (*i.e.* $2 \times 2 \times 2 = 8$) assignments of truth values to P, C, I.

P	C	I	$(P \vee C) \wedge (C \rightarrow \sim I)$
T	T	T	F
T	T	F	T
T	F	T	T
T	F	F	T
F	T	T	F
F	T	F	T
F	F	T	F
F	F	F	F

Ex. 25. Comment on the statement:

If either labour or management is stubborn, then the strike will be settled if and only if the government obtains an injunction, but troops are not sent into the mills.

Show that if we use the symbols

L : Labour is stubborn

M : Management is stubborn

S : The strike will be settled

G : The government obtains an injunction

R : Troops are sent into the mills.

then the given statement may be written

$$L \vee M \rightarrow (S \rightarrow G \wedge \sim R)$$

Ex. 26. By a truth value analysis, determine whether the above statement is true or false under each of the following assumptions:

(*i*) Labour is stubborn, management is not, the strike will be settled, the government obtains an injunction and troops are sent into mills.

(*ii*) Both labour and management are stubborn, the strike will not be settled, the government obtains an injunction and troops are sent into the mills.

(*iii*) If the government obtains an injunction then troops will be sent into the mills and the strike will not be settled.

Ex. 27. Every firm wants to maximise its profits and every firm wants to minimise its losses. But every firm will sell some items at a loss and will not seek to maximise its profits.

Symbolise the statement. Construct the truth table and comment.

Ans. Contradiction

Ex. 28. If the commodity market is perfect, the prices of units of a particular commodity are equal. But it always happens that prices for such units are not equal. Therefore the commodity market is not perfect.

Comment on the validity of the argument and construct the truth table.

Ans. Validity may be shown by showing the resultant conditional to be a tautology.

Ex. 29. Comment on the validity of the following statements:

(*i*) If our exports do not exceed faster than our imports, then foreign exchange will continue to flow out. Foreign exchange stops flowing. Thus exports have increased faster than imports.

(*ii*) Either businessmen are honest or there is a parallel economy. There is no parallel economy. Therefore businessmen are honest.

(*iii*) Either the country should import more goods or it should start manufacturing goods. The country is not at all thinking of imports. Therefore, the country is going to start manufacturing goods.

CHAPTER 3

Sets and Relations

3.1 SETS

In mathematics, the idea of a set is fundamental and all mathematical objects and constructions go back to the set theory. In social sciences and business analysis we have sets of data, sets of items produced, sets of outcomes of decisions and so on.

Words like groups, family, flock, crowd, association, etc. are often used to convey the idea of a set in everyday life. Enumeration and measurement lead to numbers and sets whose study gives us a good insight into the depths of a subject.

We can use symbols to represent sets and operations on these sets. We can solve many problems by means of set diagrams.

Notation and Elements

A set is a collection of any type of objects, things or numbers.

The constituent members of a set are called *elements* of that set.

The elements may be either listed by enclosing them in brackets or they may be described by giving a statement of the properties of the elements.

A set is known by its elements. A set is well defined if and only if it can be decided that a given object is an element of the set.

Capital letters A, B, C, etc. are usually employed to denote sets. Elements of sets are usually denoted by lower case letters a, b, c, etc. or shown by description.

Ex. 1. $A = \{2, 4, 6, 8\}$ is a set of the even numbers 2, 4, 6, 8. We may say A is a set of all even numbers between 1 and 9; or 2, 4, 6, 8 are elements of A.

Ex. 1 gives the *Roster* or *tabulation method* to describe a set. It consists of listing of each element of the set within the brace.

Another method is the *descriptive phrase method* which consists of placing a phrase describing the elements of the set within the brace. Thus

$A = \{\text{even numbers between 1 and 9}\}$.

This method may be used when there is a large number of elements or when all the elements cannot be named.

A third method known as the *rule method* or *set builder* method consists of enclosing within the brace a general element and describing it. Thus

$$A = \{x/x \text{ is an even integer, } 1 < x < 9\}$$

or $$A = \{x \text{ even integer})/1 < x < 9\}$$

A set is determined by its elements and not by the method of description. The set A described in any of these three different ways, remains the same.

The idea of a set is both obvious and ancient. But what looks obvious must be proved and the proof may be either insufficient to justify the obvious or it may lead to other possibilities including the opposite of the obvious.

The algebra of sets provides a student with laws which can be used to prove important results in any science.

Diagrammatic Representation of Sets

Sets can be represented by circular or similarly enclosed curves commonly known as Venn diagrams. Every element in a set is represented by a point.

Ex. 2. $V = \{a, e, i, o, u\}$

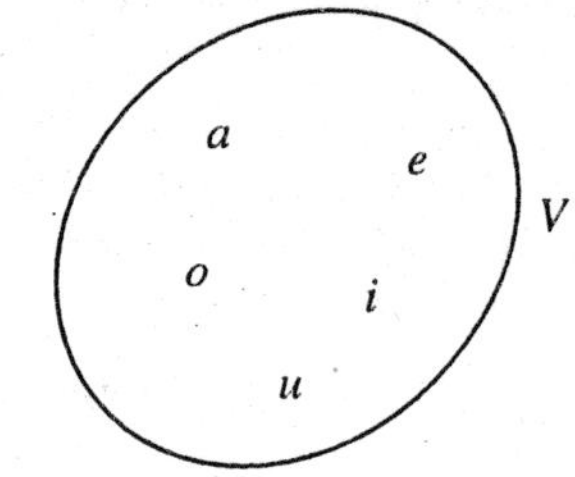

The Venn diagram shows how the vowels which are the elements of the set V belong to the set V.

We shall use symbols and diagrams to represents and understand sets.

It should be remembered that a set remains unchanged if the order of its elements is changed. Thus, we can write

$$V = \{e, o, i, a, u\}.$$

Ex. 3. $B = \{1, 2, 3, 4, 5\}$ is the set of counting numbers less than 6. Alternatively, $B = \{x \text{ (integer)} \mid 0 < x < 6\}$.

Ex. 4. $C = \{$Physics, Chemistry, Sociology, Economics$\}$ is a set of four sciences.

Ex. 5. How will you describe the set of the two smallest integers larger than 15?

Ans. $\{16, 17\}$

Ex. 6. How will you express the following in the set notation?

(*i*) the letters in the word 'internal'.
(*ii*) integers greater than 100 but less than 102.
(*iii*) integers less than —5 and greater than or equal to —10.
(*iv*) letters of the English alphabet.
(*v*) odd numbers upto 99.

Ans. (*i*) $\{i, n, t, e, r, v, a, l\}$ (*ii*) $\{101\}$
(*iii*) $\{-6, -7, -8, -9, -10\}$.
(*iv*) $\{a, b, c, \ldots, z\}$
(*v*) $\{1, 3, 5, \ldots, 99\}$

These sets may be expressed in different forms.

Ex. 7. What do the following sets mean?

(*i*) $A = \{x/x$ is a person weighing more than 60 kg$\}$
(*ii*) $B = \{b/b > 2\}$
(*iii*) $C = \{c \text{ (integer)}/c < 0\}$.

Ans. (*i*) A consists of the elements x such that x is a person weighing more than 60 kg.

(*ii*) B is the set of all numbers greater than 2.

(*iii*) C is the set of all integers less than 0 *i.e.*, C is the set of all negative integers.

Finite and Infinite Sets

A set is finite, if it has a finite number of elements. The elements of such a set can be counted by a finite number. The number of elements in a finite set A is denoted by $n(A)$. Here n is a finite positive integer.

If a set has an infinite number of elements it is an infinite set. The elements of such a set cannot be counted by a finite number.

A set of points along a line or in a plane is called a point set. A finite set has a finite subset. An infinite set may have an infinite subset.

Ex. 8. $F = \{a, b, x, z, 0, 18, p\}$

F is a finite set. $n(A) = 7$.

Ex. 9. $S = \{x/x$ is a grain of sand on the sea shore$\}$

S is a finite set.

Ex. 10. $R = \{x/x$ is a natural number$\}$

$= \{1, 2, 3, \ldots\}$. R is an infinite set.

Ex. 11. $A = \{x/x$ is a number between 2 and 3$\}$

A is an infinite set.

Ex. 12. Is $D = \{x/x$ is a multiple of 2$\}$ an infinite set? **Ans.** Yes.

Ex. 13. Do the letters of the English alphabet give an infinite set? **Ans.** No.

Ex. 14. Is $m = \{x/x$ is a currency note in India$\}$ an infinite set? **Ans.** No.

Ex. 15. The set of points between 10 and 20 is a point set. There is an infinite number of points between 10 and 20. This is an example of an infinite set.

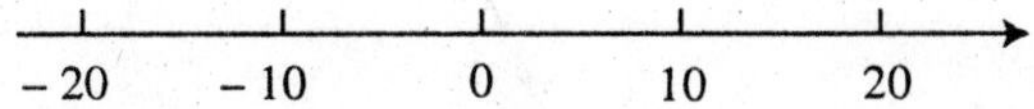

Ex 16. The set of all points along a line is an infinite set. There is an infinity of points on a line. A member of the set of points along a line can be represented on the line by a variable x.

Ex. 17. A point set in a plane is shown by the set of points in a plane figure. Is it an infinite set? **Ans.** Yes.

Null Set or Empty Set

The null set is the set which contains no elements. A null set is denoted by ϕ.

Ex. 18. $A = \{x/x$ is a married bachelor$\} = \phi$.

Ex. 19. $B = \{x/x$ is a planet visiting the earth$\} = \phi$.

Ex. 20. $C = \{x/x$ is a potato growing on a mango tree$\} = \phi$.

Singleton or Unit Set

A unit set contains only one element.

Ex. 21. $A = \{a\}$ is a unit set.

$X = \{0\}$ is a unit set containing a single element 0. X is not a null set.

Ex. 22. Although a is an element of $\{a\}$ it is not true that $\{a\}$ is an element of a. Explain.

Universal Set

The universal set is the totality of elements under consideration as elements of any set. It is the set of all objects relevant to a particular application.

A universal set is denoted by E. E is the universe or universal set including all possible elements and is often drawn as a rectangle.

Ex. 23. $E = \{a, b, c, ... z\}$, the set of letters of the English alphabet is a universal set.

Ex. 24. E = {head, tail} is a universal set obtained by the throw of a coin.

Ex. 25. E = {rise in demand, constant demand, fall in demand} is a universal set resulting from a change in the price of a commodity. It covers all the possibilities.

Disjoint Sets

If two sets have no elements in common then they are disjoint sets.

Ex. 26. If A = {all odd numbers}

B = {all even numbers}

then A, B are disjoint sets.

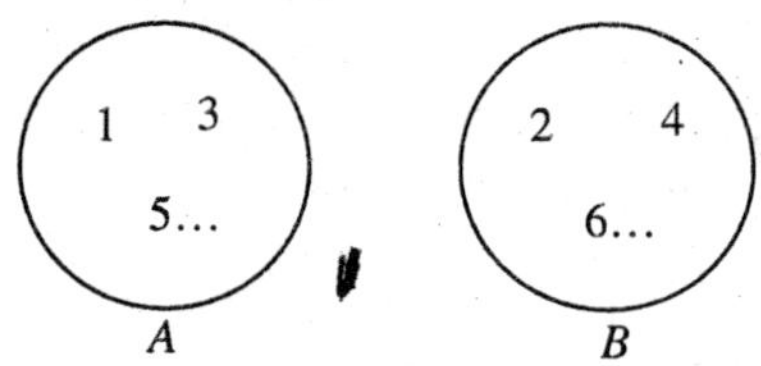

Ex. 27. If $C = \{0, 1, 2, 3, 4\}$

and $D = \{5, 6\}$

then C and D are disjoint sets. The Venn diagram illustrates this fact.

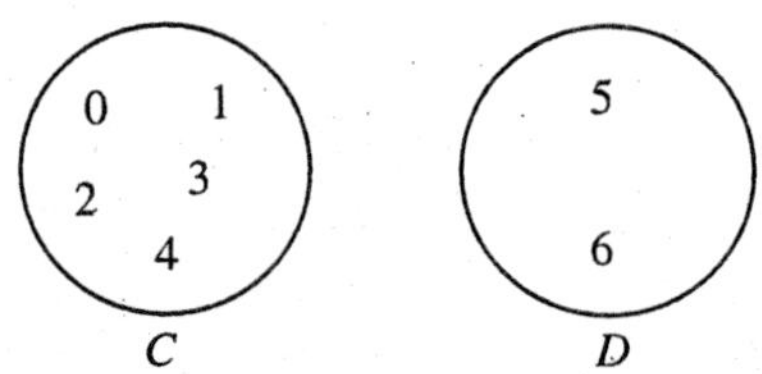

Overlapping Sets

When two sets overlap, the overlapping portion will include the common points between the two sets.

Ex. 28. $C = \{0, 2, 3, 4\}$

and $A = \{2, 4, 6\}$

The Venn diagram shows the common points 2,4 in the overlapping position.

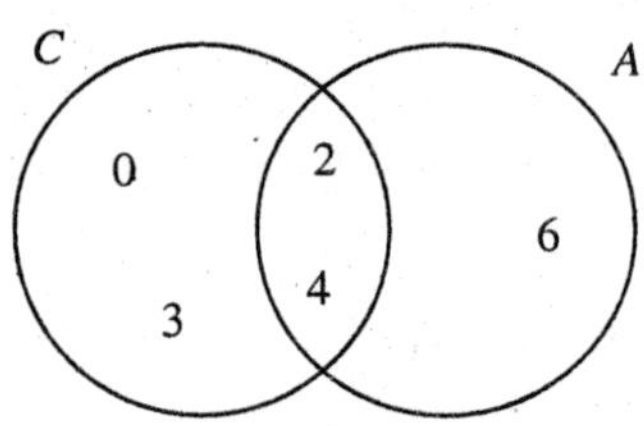

Membership of a Set

If an element x belongs to a set A, we write $x \in A$.

Read x is an element of A.

If an element x does not belong to a set A, we write $x \notin A$.

We say that x is not an element of A.

If an element x belongs to A it does not mean that the set containing x belongs to A.

$x \in X$ does not mean $\{x\} \in X$

$x \in C, x \notin D$ means that x belongs to C but not to D

$x \in P, x \in Q$ means that x belongs to P and also to Q.

$x \notin A, x \notin B$ means that x does not belong to A or to B.

Ex. 29. In the set $X = \{x/x^2 - 5x + 6 = 0\}$, $2 \in X$, $3 \in X$ $\because$ 2, 3 are solutions of $x^2 - 5x + 6 = 0$ $\therefore x = \{2, 3\}$

Subsets

If from a set we want to select some elements having a special property it means we are interested in a subset of the set.

A subset of a set X is a set which consists of some or all of the elements X. The set Y is a subset of X if and only if every element of Y is an element of X and we write $Y \subseteq X$. $a \in Y$ implies $a \in X$.

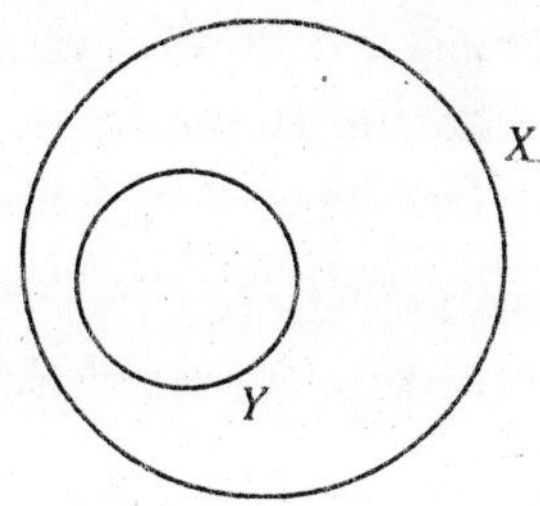

This is called the *inclusion relation.* Every element of Y is also an element of X *i.e.,* Y is included in X or X is a superset of Y.

A set Y is a *proper subset* of a set X if and only if $Y \subset X$ and $Y \neq X$ and $Y \neq \phi$.

Every set is a subset of itself. Thus $X \subseteq X$.

Ex. 30. $Y = \{2, 4, 6, 8, 10\}$ is a subset of $X = \{1, 2, 3,\ldots, 10\}$.

$V = \{a, e, i, o, u\}$ is a subset of the set of English alphabets, $A = \{a, b,\ldots,z\}$.

The set of positive integers is a subset of the set of all real numbers.

If A, B are equal sets then $A \subset B$ and $B \subset A$.

A finite set has a finite subset. An infinite set may have an infinite subset.

If a set Y is not a subset of a set X we write

$$Y \not\subset X$$

Here the set Y may have some points which are also in X but not all the points of Y are in X.

If a set Y is a subset of set X, all the elements of Y must be in X. A subset can be formed by taking a certain number of elements from the given set.

If $Y = \{y/0 \leq y \leq 5\}$

and $X = \{y/0 \leq y \leq 10\}$

then $Y \subset X$

If $Y = \{a, b, c\}, X = \{a, b, c\}$

then $Y \subset X$ anc $X \subset Y$

Ex. 31. State whether the following are true or false. Give reasons.

(*i*) $E \subseteq E$	(*ii*) $\phi \subseteq \phi$	(*iii*) $\phi \in \phi$
(*iv*) $\phi \subseteq E$	(*v*) $\phi \subset E$	(*vi*) $\phi = \{\phi\}$
(*vii*) $\phi \subseteq \{\phi\}$	(*viii*) $\phi \subseteq A$	(*ix*) $\phi = \{0\}$

Ans. *T, T, F, T, T, F, T, T, F*

Ex. 32. There are six members in a committee. Any decision can be made by a majority vote. What are the decision making votes?

(**Hint:** All subsets of at least 4 votes)

Ex. 33. If $A = \{a, b, c\}$, is A a subset of A? **Ans.** Yes

Ex. 34. {Economics, Sociology} is a subset of {Physics, Chemistry, Economics, Sociology}.

Ex. 35. If $A = \{\text{Money, Banking, Public Finance}\}$

and $B = \{x/x = \text{all subjects in Economics}\}$

then $A \subseteq B$.

Ex. 36. The empty set is a subset of every set.

Thus $\phi \subseteq A$ for *any* set A.

It means all elements of ϕ are elements of A.

(And ϕ has no elements at all).

Equal or Identical Sets

If every element of one set is also an element of another set then the two sets are said to be equal.

Equal sets have exactly the same elements.

If sets A, B are equal we write $A = B$.

It follows that if $A \subset B$ and $B \subset A$ are both true, then $A = B$.

Ex. 37. If $A = \{2, 4, 6, 8\}$, $B = \{6, 4, 8, 2\}$

then $A = B$.

Also $\{2, 4, 6, 8\} = \{2, 2, 4, 4, 6, 8, 8\}$

since the set on either side contains only four distinct elements, *viz.* 2, 4, 6, 8.

Ex. 38. If $A = \{x/x \text{ is a letter of the word 'equal'}\}$

and $B = \{q, a, e, u, l\}$

then $A = B$.

Ex. 39. $A = \{x/x^2 - 5x + 6 = 0\}$ and $B = \{2, 3\}$ are equal sets.

Ex. 40. If $A = \{10, 12, 14, 16, 18, 20\}$

$B = \{\text{all even numbers between 9 and 21}\}$

then $A = B$.

If we use the symbol $\Rightarrow$ to stand for "implies" we may describe the equality of two sets symbolically.

If $x \in A \Rightarrow x \in B$

and $x \in B \Rightarrow x \in A$

then $A = B$

It means that x as a member of A is also a member of B and *vice versa*.

Equivalent Sets

Two sets are said to be equivalent if there is a *one-to-one correspondence* between the elements of the two sets. The order in which the elements appear in each set is immaterial.

It may be noted that a one-to-one correspondence exists between two sets A, B if it is possible to associate the elements of A with the elements of B in such a way that each element of each set is associated with exactly one element of the other.

Ex. 41. If $A = \{a, b, c\}$

$B = \{\alpha, \beta, \gamma\}$

then A and B are equivalent sets.

We may express equivalence symbolically by writing

$$A \equiv B$$

or

$$A \leftrightarrow B.$$

Equivalent sets have the same number of elements. In other words, their cardinal number is the same.

Thus in Ex. 41, $n(A) = n(B) = 3$.

Ex. 42. Is the set of letters in the word 'economics' equivalent to the set {1, 2, 3, 4, 5, 6, 7}?

Note that two sets are equivalent if they have the same number of elements and no element is repeated when the members of a set are listed.

$$\{1, 2, 3, 4, 5, 6, 7\} = \{e, c, o, n, m, i, s\}$$

Power Set

The class of *all* subsets of a set A is called the power set of A. It may be written $P(A)$.

Ex. 43. If $A = \{0, 1\}$ then

ϕ, {0}, {1}, {0, 1} are the subsets of A:

We can write in this case, the power set

$$P(A) = [\phi, \{0\}, \{1\}, \{0, 1\}]$$

Ex. 44. If $S = \{a, b, c\}$

$$P(S) = [\phi, \{a\}, \{b\}, \{c\}, \{a, b\}, \{b, c\}, \{c, a\}, \{a, b, c\}]$$

Ex. 45. Designate the following sets in words:

(*i*) {2, 4, 6, 8, 10} (*ii*) $\{a, b, c, d, e, f\}$

(*iii*) $\left\{\frac{1}{2}, \frac{1}{2^2}, \frac{1}{2^3}, \ldots\right\}$ (*iv*) {2, 4, 8, 16,...}

Which of these are finite sets? **Ans.** (*i*) and (*ii*)

Which of these are infinite sets? **Ans.** (*iii*) and (*iv*)

Ex. 46. Express the following symbolically:

(*i*) 5 is an element of A

(*ii*) Set A is a subset of itself

(*iii*) The set of all odd integral multiples of 3

(*iv*) The set of even natural numbers less than 16.

Ans. (*i*) $5 \subset A$ (*ii*) $A \subseteq A$ (*iii*) $\{3, 9, 15, 21, ...\}$

(*iv*) $\{x/x$ is an even natural number less than $16\}$

Ex. 47. In a class 75% students pass an examination.

Is this a proper subset of the class?

Ans. $A = \{75\%$ of students$\}$, $B = \{$students in class$\}$. Then $A \subset B$ and $A \neq B$. Therefore it is a case of a proper subset.

Ex. 48. Can a one-to-one correspondence be established between the set of real numbers and the set of people in India?

Ans. No. The set of real numbers is an infinite set.

Ex. 49. Give some examples of the null set.

3.2. OPERATIONS ON SETS

Sets may be combined and operated on in various ways to form new sets.

The basic operations on sets are complementation, union, difference and intersection.

Complement of a Set

If A is a subset of the universal set E then the collection of elements belonging to E but not belonging to A is called the complement of A.

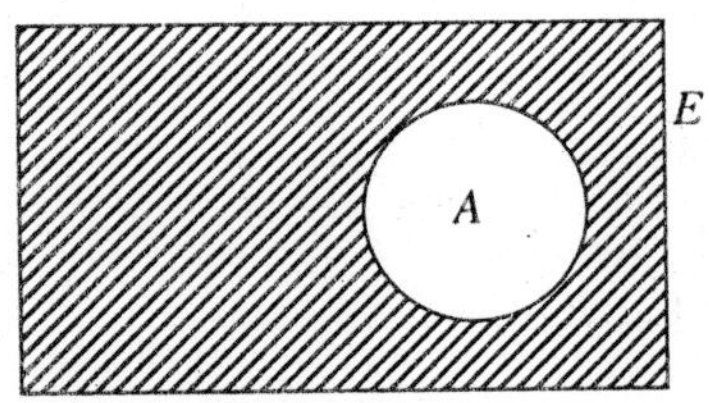

The shaded portion in the Venn diagram stands for the complement of A.

We thus note that the complement of a set A to a universal set E is the set of all elements of E which are not elements of A.

The complement of A is written A'.

$$A' = \{x/x \notin A\}$$

Ex. 50. $V' = \{a, e, i, o, u\}$ is the complement of the set

$$V = \{x/x = \text{any consonant in the alphabet}\}$$

If A has a complement A' then A' has a complement A.

$$(A')' = A$$

The complement of the complement of A is A.

Two sets are complements of each other if all the elements of the two together constitute the universal set.

The complement of the universal set E contains no elements, *i.e.*

$$E' = \phi$$

Similarly $$\phi' = E.$$

Ex. 51. If $E = \{1, 2, 3, 4, 5\}$

$$A = \{1, 3, 5\}$$

then the complement of A is

$$A' = \{2, 4\}$$

Ex. 52. If E is the set of all positive integers show that the complement of the set

$$A = \{2, 4, 6, 8,...\} \text{ is}$$

$$A' = \{1, 3, 5,...\}$$

Draw the Venn diagram.

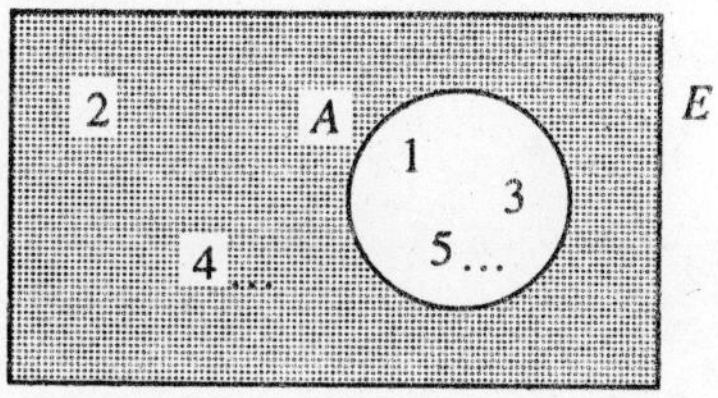

Union $A \cup B$

The union of two sets A and B is the set of all elements belonging to A or B or to both A and B.

Symbolically

$$A \cup B = \{x/x \in A, x \in B\}$$

or $\{x \in \text{both } A, B\}$.

The shaded area in the Venn diagram represents $A \cup B$ which shows the set of objects that are elements of at least one of the sets A and B.

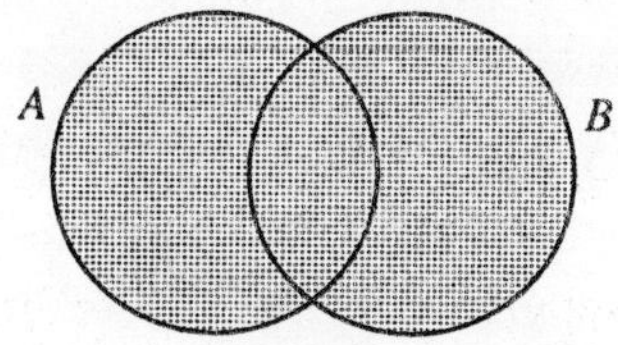

$A \cup B$

Ex. 53. The union of $\{a, b, c, d\}$ and $\{c, d, e\}$ is $\{a, b, c, d, e\}$ shown by the boundary of the Venn diagram.

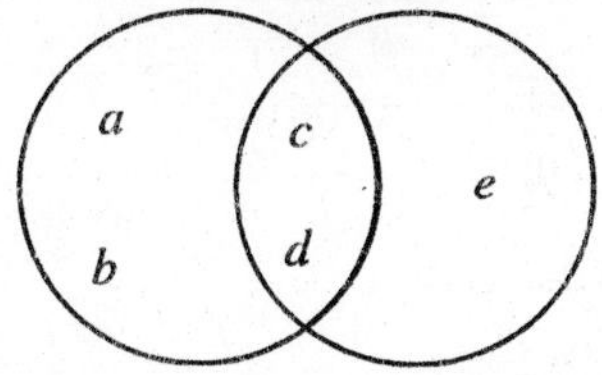

Ex. 54. How are the following relations implied in the definition of a union?

(*i*) $A \cup A = A$ (*ii*) $A \cup A' = E$

(*iii*) $A \cup \phi = A$ (*iv*) $A \cup E = E$

(*v*) $A \cup B = B \cup A$

Venn diagrams should be used here.

Ex. 55. If A and B are nonempty sets, then $A \cup B$ is nonempty. Three possibilities are shown in the Venn diagrams.

(*i*) A, B are overlapping (*ii*) A, B are disjoint (*iii*) $B \subset A$.

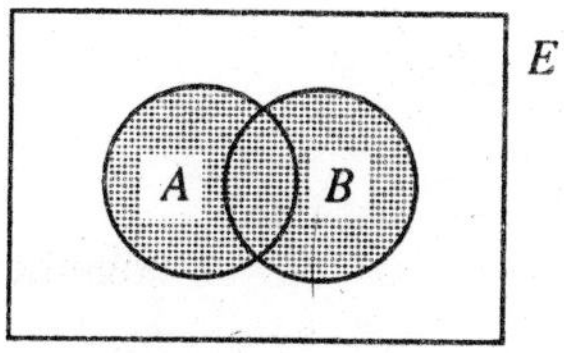

(*i*)

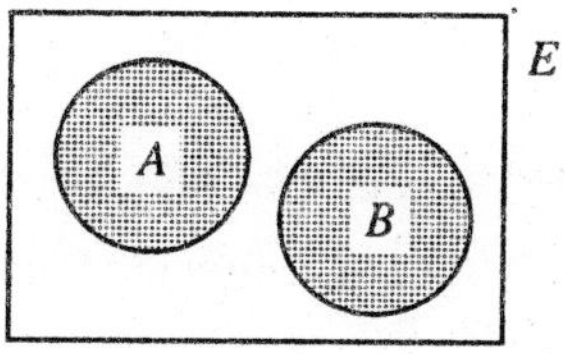

(*ii*)

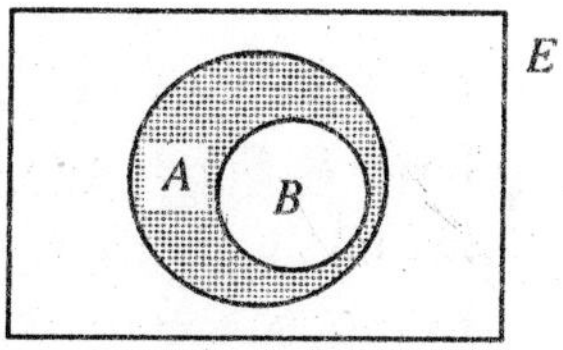

(*iii*)

Difference $A - B$

The difference of two sets A and B is the set of all elements belonging to A but not to B.

$$A - B = \{x/x \in A, x \notin B\}$$
$$B - A = \{x/x \in B, x \notin A\}$$

Ex. 56. If $A = \{a, b, x, y\}$

and $B = \{c, d, x, y\}$

then $A - B = \{a, b\}$.

What is $B - A$? **Ans.** $\{c, d\}$

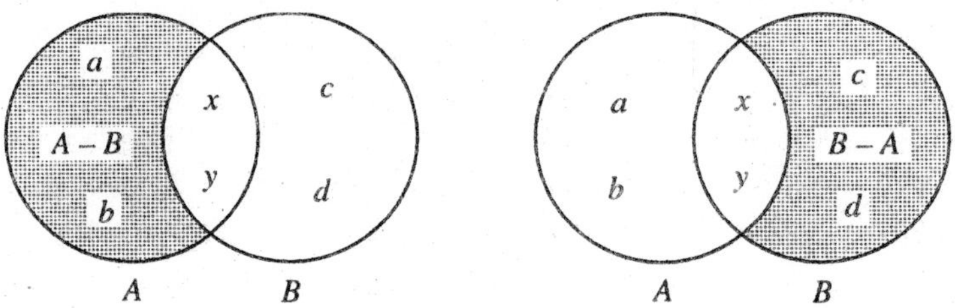

Ex. 57. How are the following relations implied in the definition of difference?

(*i*) $A - A = \phi$ (*ii*) $E - A = A'$ (*iii*) $(A - B) \cap B = \phi$.

Ex. 58. If $A = \{a/0 \le a \le 10\}$ and $B = \{a/5 \le a \le 15\}$

then $A - B = \{a/0 \le a \le 5\}$, $B - A = \{a/10 \le a \le 15\}$

Is it correct to say $A - B = B - A$? **Ans.** Only if $A = B$

Ex. 59. If A and B are non-empty sets then $A - B$ is nonempty when

(*i*) $B \subset A$

(*ii*) A, B are disjoint

(*iii*) A, B are overlapping.

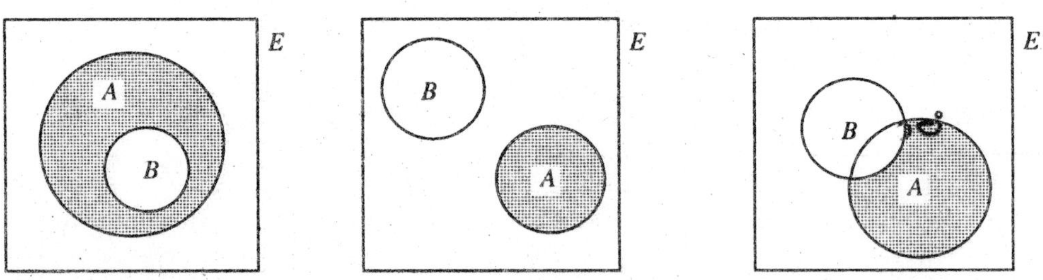

(*i*) (*ii*) (*iii*)

If $A - B$ is non-empty then

$$A - B \subset A$$

$\therefore$ any element $x \in A, x \notin B$

If $A - B$, $B - A$, $A \cap B$ are non-empty then they are disjoint

Ex. 60. If $A = \{8, 9, 10\}$, $B = \{1, 2, 3\}$, find $A - B$. **Ans.** ϕ

Ex. 61. If $A \subset B$ what is $A - B$? **Ans.** ϕ

Intersection $A \cap B$

The intersection of two sets A and B is the set of all elements which belong to both A and B. The shaded area in the Venn diagram shows $A \cap B$.

Symbolically

$$A \cap B = \{x/x \in A \text{ and } x \in B\}.$$

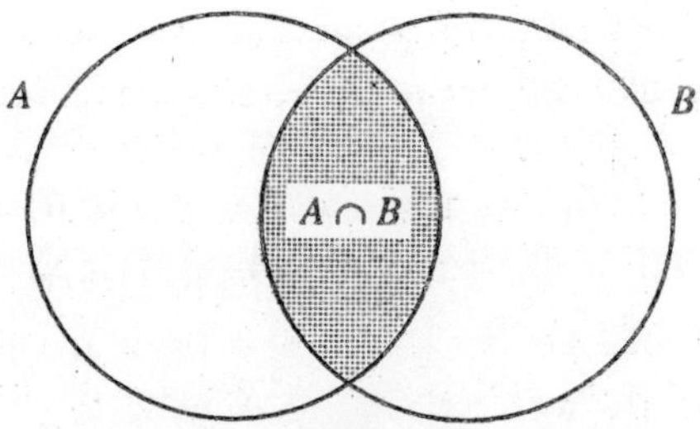

$A \cap B$

Ex. 62. If $A = \{a, b, c, d\}$

and $B = \{c, d, e\}$

then $A \cap B = \{c, d\}$.

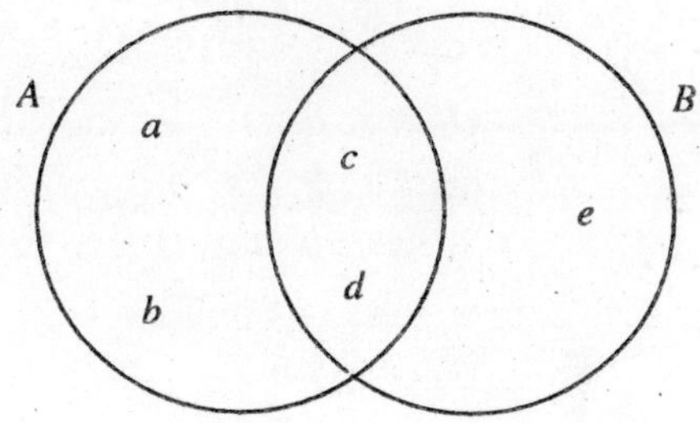

Ex. 63. How are the following relations implied in the definition of intersection?

(*i*) $A \cap A = A$ (*ii*) $A \cap A' = \phi$

(*iii*) $A \cap \phi = \phi$ (*iv*) $A \cap E = A$

(*v*) $A \cap B = B \cap A$.

Ex. 64. Find $A \cup B$ and $A \cap B$, given

$A = \{0, 1, 2\}$ $B = \{2, 3, 4\}$

Ans. $\{0, 1, 2, 3, 4\}$, $\{2\}$

Ex. 65. Find $A \cup B$, $A - B$ and $A \cap B$, given

(*i*) $A = \{c, f\}$, $B = \{c, d, g\}$

(*ii*) $A = \{c, d\}$, $B = \{e, f, g\}$.

Ans.	$A \cup B$	$A - B$	$A \cap B$
(*i*)	$\{c, d, f, g\}$	$\{f\}$	$\{c\}$
(*ii*)	$\{c, d, e, f, g\}$	$\{c, d\}$	ϕ

Ex. 66. If A and B are non-empty sets, $A \cap B$ is non-empty then

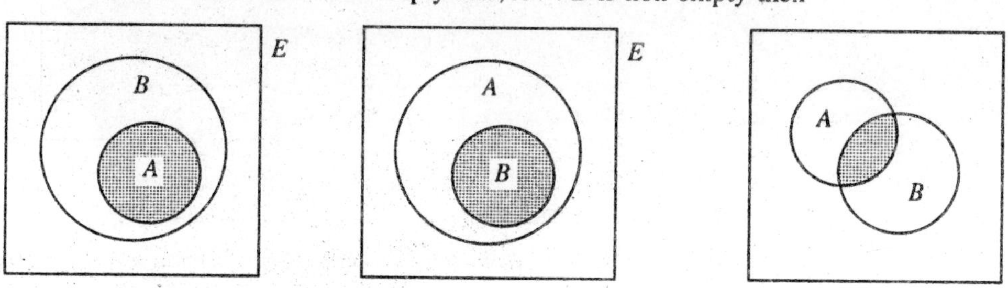

(*i*) (*ii*) (*iii*)

(*i*) $A \subset B$ (*ii*) $B \subset A$ (*iii*) A, B are overlapping.

If A, B are disjoint

$$A \cap B = \phi.$$

Ex. 67. Given the universal set $E = \{1, 2, 3, 4, 5, 6, 7, 8, 9, 10\}$ what is the complement of

(*i*) $\{1, 2, 3, 4, 5\}$ (*ii*) $\{1\}$

(*iii*) $\{\ \}$ (*iv*) $\{1, 10\}$?

Ans. (*i*) $\{6, 7, 8, 9, 10\}$ (*ii*) $\{2, 3,..., 10\}$.

(*iii*) $\{1, 2, 3,..., 10\}$ (*iv*) $\{2, 3, 4, 5, 6, 7, 8, 9\}$

Ex. 68. $A = \{1, 2, 3, 4, 5, 6\}$

$B = \{1, 5, 9, 15, 21\}$

$C = \{13, 14, 15, 16, 17, 18, 19, 20, 21\}$

Find (*i*) $A \cup B$ (*ii*) $B \cup C$ (*iii*) $C \cup A$

(*iv*) $A \cap B$ (*v*) $B \cap C$ (*vi*) $C \cap A$

(*vii*) $A \cup B \cup C$ (*viii*) $A \cap B \cap C$

(*ix*) $(A \cup B) \cap C$ (*x*) $A \cup (B \cap C)$

(*xi*) $A \cup B'$ (*xii*) $B \cup A'$

(*xiii*) $A - B$ (*xiv*) $(A \cup B) - C$.

Ex. 69. Show (*i*) $A \cup B = B \cup A$ (*ii*) $B \cap A = A \cap B$.

Three Sets

The intersection and union of three sets gives rise to several interesting results.

We consider here some Venn diagrams for three sets.

$A \cup B \cup C$ is the same as $(A \cup B) \cup C$ or $A \cup (B \cup C)$ since we get the same result by joining the union of any two of A, B, C to the third. An element in $A \cup B \cup C$ belongs to at least one of A, B, C.

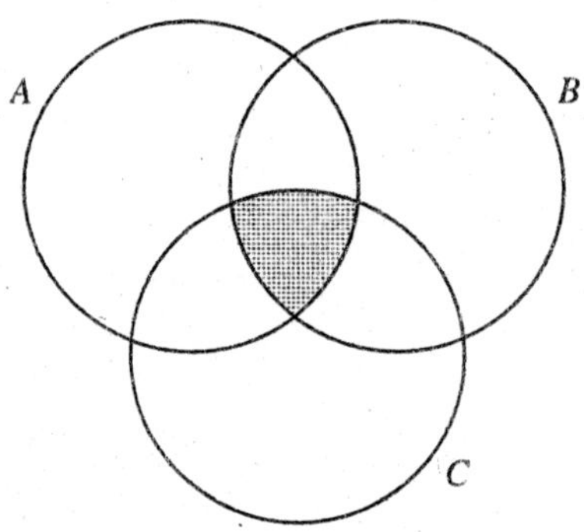

$A \cap B \cap C$

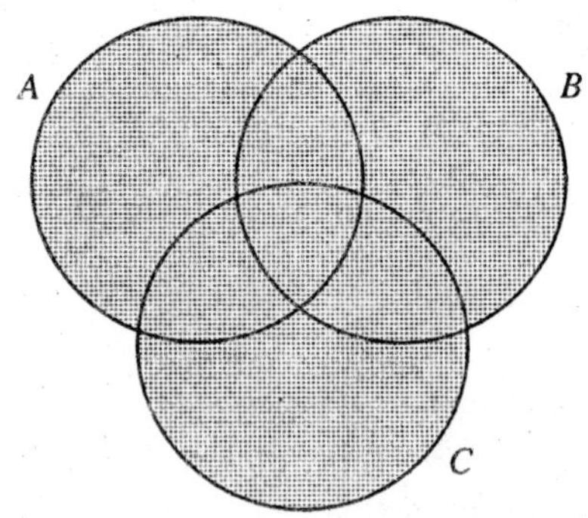

$A \cup B \cup C$

Similarly $A \cap B \cap C$ is the same as $A \cap (B \cap C)$ or $(A \cap B) \cap C$. Thus an element in $A \cap B \cap C$ is common to all the three sets A, B, C.

Ex. 70. Let $A = \{1, 2, 3, 4\}$

$B = \{1, 3, 5, 7\}$

$C = \{1, 5, 6, 8\}$

Then $A \cup B \cup C = \{1, 2, 3, 4, 5, 6, 7, 8\}$

$A \cap B \cap C = \{1\}$.

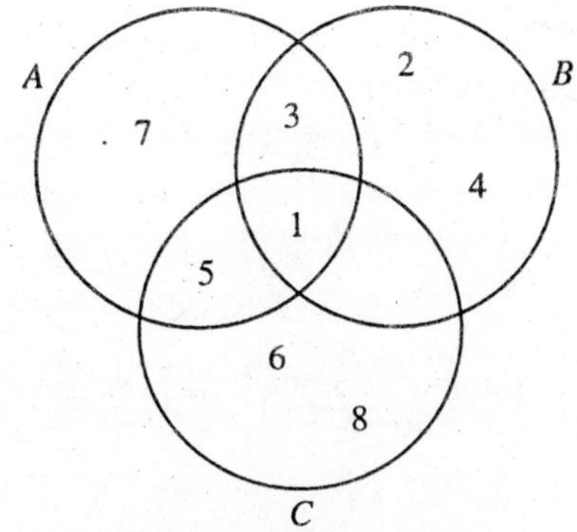

Ex. 71. $A \cap (B \cup C)$ is the intersection of A with the union of B and C.

$(A \cap B) \cup (A \cap C)$ is the union of $A \cap B$ and $A \cap C$. We get the same diagram as for $A \cap (B \cup C)$.

$\therefore A \cap (B \cup C) = (A \cap B) \cup (A \cap C)$.

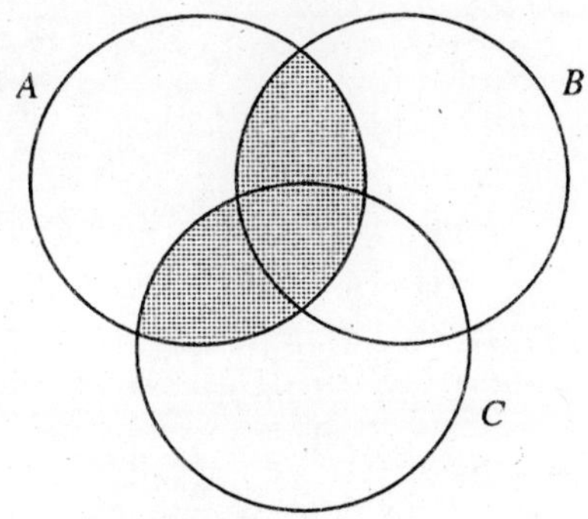

$A \cap (B \cup C)$ or $(A \cap B) \cup (A \cap C)$

$A \cup (B \cap C)$ is the union of A with the intersection of B, C. It may be noted that

$A \cup (B \cap C) = (A \cup B) \cap (A \cup C)$.

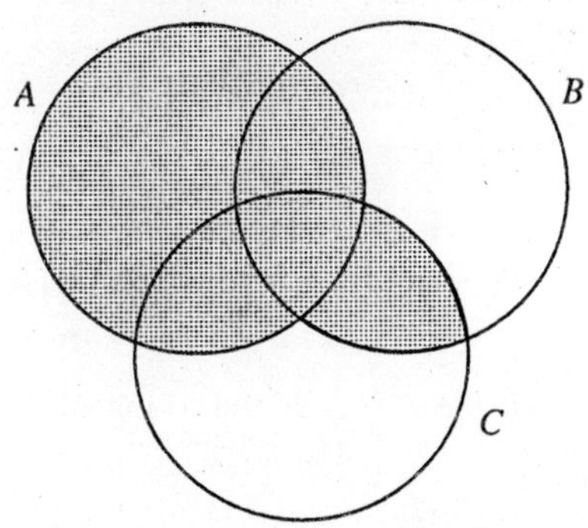

$A \cup (B \cap C)$ or $(A \cup B) \cap (A \cup C)$

Ex. 72. $(A \cup B \cup C)$

The shaded portion is the complement of $A \cup B \cup C$. It is the same as

$A' \cap B' \cap C'$.

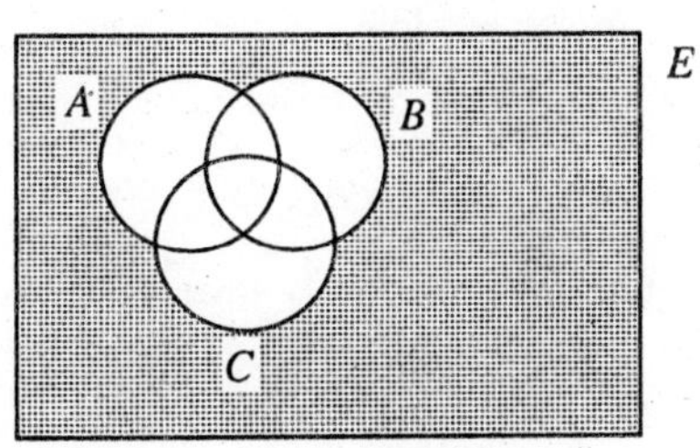

Ex. 73. $(A \cup B)' \cap C$

The shaded portion indicates the intersection of the complement of $(A \cup B)$ and C.

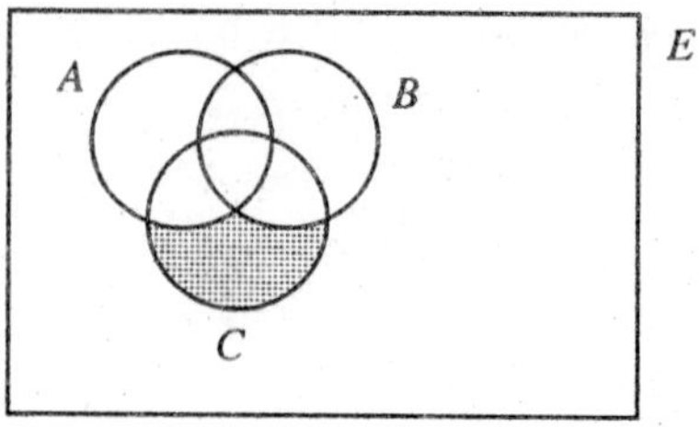

Ex. 74. Prove the following identities:

(*i*) $(A \cap B) \cup (B' \cap A) = A$

(*ii*) $A \cup (A' \cap B) = A \cup B$

(*iii*) $(A \cup E) \cap (A \cap \phi) = \phi$

(*iv*) $A \cap (A \cup B) = A$

(*v*) $A \cup (A' \cap B) = A \cup (A \cup B')$

(*vi*) $[A \cup B \cap B]' = A' \cap B'$.

(*vii*) $(A - B) - C = [A - C] \cap (B - C)$

(*viii*) $(A - B) \cap C = (A \cap C) - (B \cap C)$

(*ix*) $(A - B)' = A' \cup B$

Hint: (*i*) The proof for these may be given by means of diagrams. Or as follows. For (*i*) $(A \cap B) \cup (A \cap B') = A \cap (B \cup B')$. (Distributive law. See below)

$B \cup B' = E$ the universal set. (Complement law.)

$\therefore \quad (A \cap B) \cup (A \cap B') = A \cap E = A$

Ex. 75. Simplify

(*i*) $[A' \cap (B \cup C)]'$

(*ii*) $(A \cup B \cup C)'$

(*iii*) $(A \cup B') \cap (A' \cup C) \cap (B \cup C')$

(*iv*) $(A \cap B) \cup (A' \cap C) \cup (B \cap C)$

(*v*) $A \cap (A' \cup B) \cup B \cap (B \cup C) \cup B$

Ex. 76. Prove the result $(A \cup B) \cup C = A \cup (B \cup C)$

Proof. *L.H.S.* is the union of $A \cup B$ and C

i.e., it is the union of A, B and C.

Similarly *R.H.S.* is the union of A, B and C.

In the same way $(A \cap B) \cap C = A \cap (B \cap C)$.

Laws of Operations

There are seven basic laws of set algebra. They provide the basis for manipulating sets.

Every law is associated with a dual. The principle of duality is that if, in any true statement involving only union, intersection or complementation, $\cup$, $\cap$ are interchanged and E, ϕ are interchanged, a true statement results.

These laws can be tested by means of Venn diagrams.

Let A, B, C be any subsets in the universal set E. Then we have the following laws of operations on sets:

(*i*) *Idempotent Laws or Laws of Inclusion or Tautology*

$A \cup A = A$ $\quad$ $A \cap A = A$

(*ii*) *Commutative Laws*

$A \cup B = B \cup A$ $\quad$ $A \cap B = B \cap A$

(*iii*) *Associative Laws*

$(A \cup B) \cup C = A \cup (B \cup C)$ $\quad$ $(A \cap B) \cap C = A \cap (B \cap C)$

(*iv*) *Distributive Laws*

$A \cup (B \cap C) = (A \cup B) \cap (A \cup C)$ $\quad$ $A \cap (B \cup C) = (A \cap B) \cup (A \cap C)$

(*v*) *Identity Laws*

$A \cup \phi = A, A \cup E = E$ $\quad$ $A \cap E = A, A \cap \phi = \phi$

(*iv*) *Complement Laws*

$A \cup A' = E, (A')' = A$ $\quad$ $A \cap A' = \phi, E' = \phi, \phi' = E$

(*vii*) *De Morgan's Laws*

$(A \cup B)' = A' \cap B'$ $\quad$ $(A \cap B)' = A' \cup B'$.

Some Additional Properties

(*i*) If A, B, C are three sets such that

$A \subset B$ and $B \subset C$ then $A \subset C$.

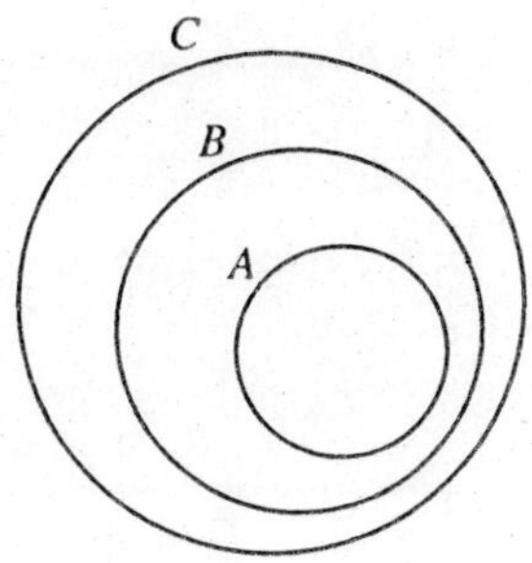

Since each element of A is an element of $B(\because A \subset B)$ and since each element of B is an element of C ($\because B \subset C$), it follows that each element of A is an element of C.

(*ii*) If $A \not\subset B, B \not\subset C$ then $A \not\subset C$

(*iii*) If $A \subset B, B \subset C$ then $A \subset (B \cap C)$

(*iv*) If $A \subset B$ then $A \subset (B \cup C)$ for all C.

Ex. 77. Draw a Venn diagram to represent $(P \cup Q) \subset Q$.

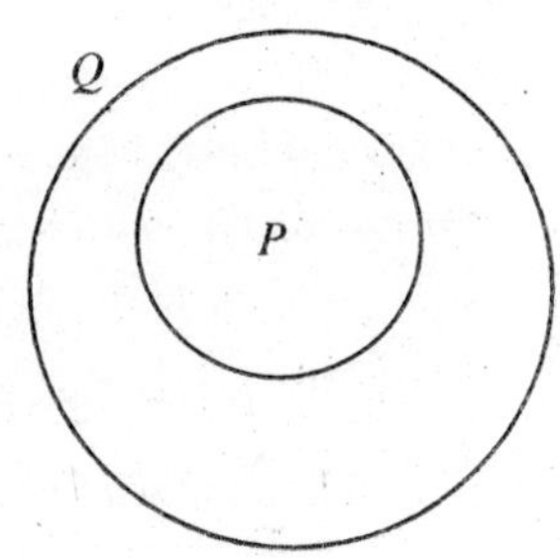

3.3 THE NUMBER OF ELEMENTS IN A SET

Let $n(A)$ be the number of elements in set A and and $n(B)$ in B. Then

$$n(A \cup B) = n(A) + n(B) - n(A \cap B)$$

$$n(A \cup B) = n(A) + n(B) \text{ if } A, B \text{ are disjoint.}$$

We can extend the formula to 3 sets.

$$\begin{aligned} n(A \cup B \cup C) &= n(A) + n(B) + n(C) \\ &\quad - n(A \cap B) - n(B \cap C) - n(C \cap A) \\ &\quad + n(A \cap B \cap C). \end{aligned}$$

Ex. 78. Show that

$$n(A) = n(A \cap B') + n(A \cap B)$$

$$n(B) = n(A' \cap B) + n(A \cap B)$$

$$n(A) = n(B') = n(A \cap B') + n(A' \cap B) + 2n(A \cap B).$$

Ex. 79. If $A \cap B = \phi$ then $n(A \cap B) = 0$.

Ex. 80. A firm has 40 workers working in its factory premises, 30 working in its office and 20 working in both the places. How many workers are there in the firm? How many are working exclusively in the (*i*) factory, (*ii*) office?

We have $n(F) = 40, \quad n(O) = 30, \quad n(F \cap O) = 20$

$$n(F \cup O) = n(F) + n(O) - n(F \cap O)$$

$$= 40 + 30 - 20 = 50, \text{ total number of workers}$$

The number working in the factory only

$$= n(F) - n(F \cap O)$$

$$= 40 - 20 = 20$$

The number working in the office only

$$= n(O) - n(F \cap O) = 30 - 20 = 10.$$

Ex. 81. In a class, 64% students have taken politics (P) and 56% students have taken History (H). How many students have taken both the subjects?

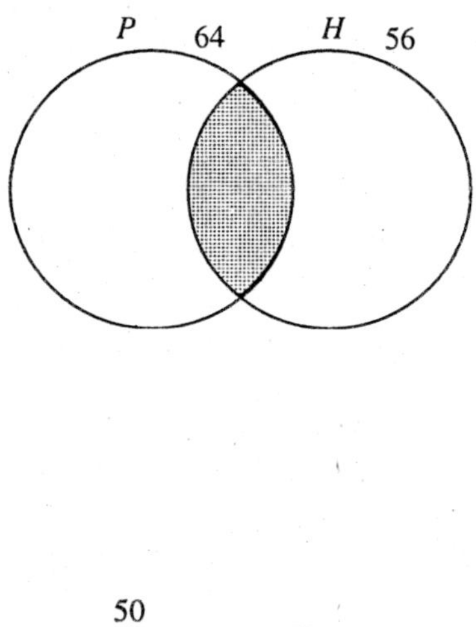

If the number of students in the universal set E is 100,

$$n(P \cup H) = 100.$$

The shaded area shows the students who have taken both the subjects. The number is 64 + 56 – 100 = 20. Thus 20% students have taken both the subjects.

Alternatively, we may use the formula

$$n(P \cup H) = n(P) + n(H) - n(P \cap H)$$

$$\therefore \quad 100 = 64 + 56 - n(P \cap H)$$

$$\therefore \quad n(P \cap H) = 20.$$

Ex. 82. In a survey of 100 students it was found that 50 used the college library, 40 had their own library and 30 borrowed books. Of these, 20 used both the college library and their own. 15 used their own and borrowed books and 10 used the college library and borrowed books. How many students used all the three sources of books?

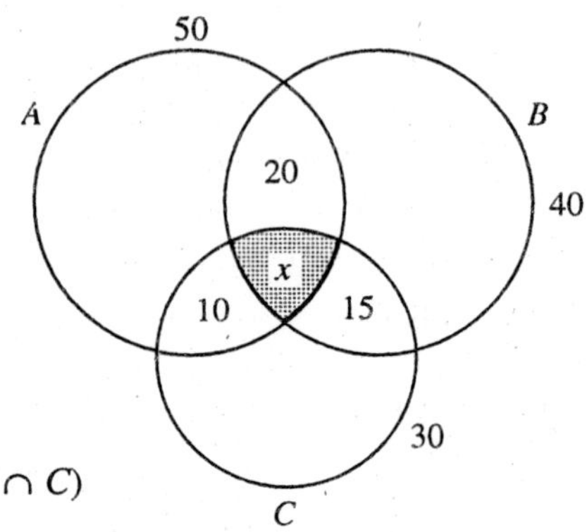

Suppose x students use all the three sources

$$n(A \cup B \cup C) = n(A) + n(B) + n(C)$$

$$- n(A \cap B) - n(B \cap C) - n(C \cap A) + n(A \cap B \cap C)$$

Here $\quad n(A \cap B \cap C) = x$

By using this formula,

$$n(A \cup B \cup C) = 50 + 40 + 30 - 20 - 15 - 10 + x = 100 \quad \therefore\ x = 25.$$

Ex. 83. Given the set

$A = \{0, 1, 2, 3, \ldots\}$. Assume x a positive integer.

State whether the following statements are true or false: **Ans.**

				Ans.
(i)	$10 \in A$	(ii)	$101 \in A$	T, T
(iii)	$0.5 \in A$	(iv)	$1.5 \notin A$	F, T
(v)	$0 \notin A$	(vi)	$-10 \notin A$	F, T
(vii)	$\{1, 2) \subset A$	(viii)	$\{0, 1, 4) \not\subset A$	T, T
(ix)	$\{-1, 0, 1\} \not\subset A$	(x)	$\{x \subset A/x < 5\}$	T, F
(xi)	$\{x \in A/0 \le x \le 10\}$	(xii)	$\{x \in A/x > 101\}$	T, T
(xiii)	$\{x \in A/x < 0\} = \phi$	(xiv)	$\{0\} = 0$	T, F
(xv)	$\{1\} = 1$	(xvi)	$\{\ \} \neq 0$	F, T
(xvii)	$\phi = 0$	(xviii)	$\{0, 1, 2\} \cup \{1, 2, 3\} = \{1, 2, 3\}$	F, F
(xix)	$\{10, 11, 12, 13\} \cup \{8, 9, 10, 11\} = \{8, 9, 10, 11, 12, 13\}$			T
(xx)	$\{0, 1, 2\} \cap \{1, 2, 3\} = \{1, 2\}$.			T

Ex. 84. If $B = \{\ldots, -3, -2, -1, 0, 1, 2, \ldots\}$

and $A = \{0, 1, 2, \ldots\}$,

are the following statements true or false? **Ans.**

				Ans.
(i)	$B \subset A$	(ii)	$\{x \in B/x > -1\}$	F, T
(iii)	$A \cup \phi = A$	(iv)	$B \cup \phi = B$	T, T
(v)	$B \cap \phi = B$	(vi)	$A \cup B = B$	F, T
(vii)	$A \cap B = A$	(viii)	$A \subset B$.	T, T

Ex. 85. In a certain examination 53 per cent students passed in Economics, 61 per cent in Politics, 60 per cent in History, 24 per cent in Economics and Politics, 35 per cent in Politics and History, 27 per cent in Economics and History and 5 per cent students passed in none of these subjects. How many students passed in all the three subjects?

Ans. 7 per cent

Ex. 86. In an examination, 32 per cent students failed in Economics, 30 per cent in Politics, 46 per cent in History, 12 per cent in Economics and Politics, 9 per cent in Politics and History, 10 per cent in Economics and History and 3 per cent in all the three, how many students passed in all the three subjects? How many students failed in exactly one subject? **Ans.** 20, 55 per cent

Ex. 87. Shade the following in a Venn diagram.

(i) B' (ii) $A' \cap B'$ (iii) $(A \cup B')$ (iv) $(A - B')$

(v) $A \cup B$ (vi) $A \cap B$ (vii) $A' \cup B$ (viii) $A' - B'$

Ex. 88. Draw a Venn diagram showing the proper relationship between the following sets:

(i) E : the set of all students

A : the set of all students of Bombay University

B : the set of all postgraduate students of Bombay University

(*ii*) E : the set of all students

A : the set of all school students

B : the set of all college students

C : the set of all male students.

(*iii*) E : the set of students appearing for B.A. examination

A : the set of students with Economics and Statistics as their subjects for *B.A.*

B : the set of students with Philosophy and Psychology as their subjects for *B.A.*

C : the set of students with Economics, Philosophy, Sociology, and History as their subjects for B.A.

Ex. 89. Define *A, B* as follows:

$A = \{$firms producing identical products$\}$

$B = \{$industries with few firms$\}$

Show in a Venn diagram the position of the following: (shown alongside)

(*i*) A firm in perfect competition

(*ii*) A firm in monopolistic competition

(*iii*) A pure oligopoly firm

(*iv*) A differentiated oligopoly firm.

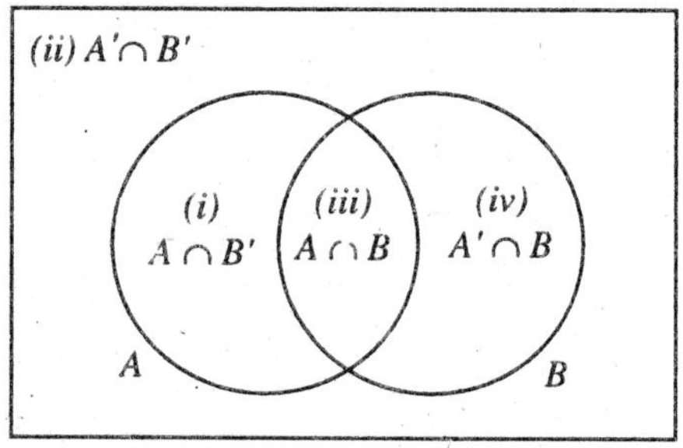

Ex. 90. Show that

(*i*) $A - B = A \cap B' = \{x/x \in A \text{ and } x \notin B\}$

(*ii*) $B - A = B \cap A' = \{x/x \in B \text{ and } x \notin A\}$

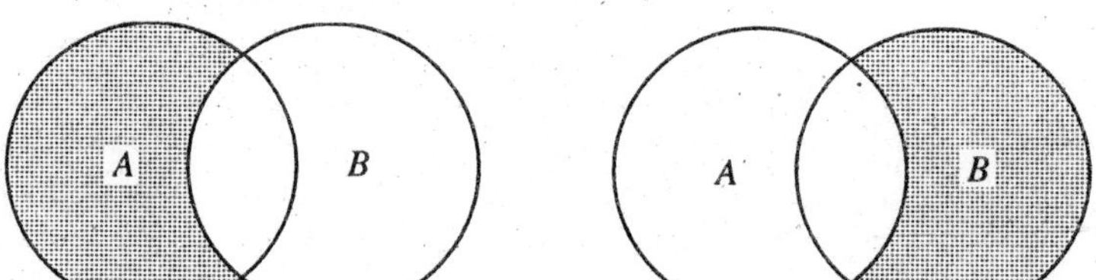

(*i*) (*ii*)

The set $A - B$ is the set of elements in A but not in B.

The set $B - A$ is the set of elements in B but not in A.

Ex. 91. We define *symmetric difference* of two sets *A, B* as the set of elements either in *A* or in *B* but not in both *A, B.*

If we write the symmetric difference of *A, B* as $A \nabla B$ then show that

(*i*) $A \nabla B = \{x/x \in A \cup B, x \notin A \cap B\}$

$= (A \cup B) - (A \cap B)$

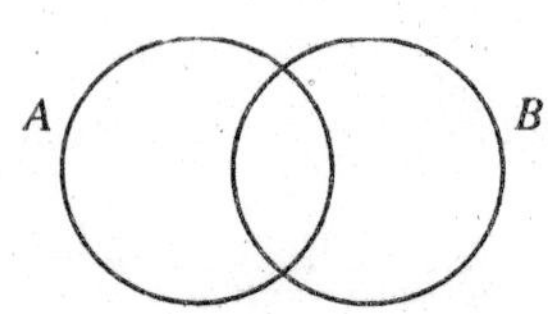

(*ii*) $A \nabla B = (A - B) \cup (B - A)$

(*iii*) $A \nabla B = B \nabla A$

[**Hint:** $(A - B) \cup (B - A)$ is the same $(B - A) \cup (A - B)$]

Ex. 92. Represent the following by Venn diagrams.

(*i*) $Q' \cap (P \cup R)$

(*ii*) $R \cap (P \cup Q')$

(*iii*) $R \cup (P \cap Q')$

Which of these represents the shaded portion in the Venn diagram shown here? **Ans.** (*i*)

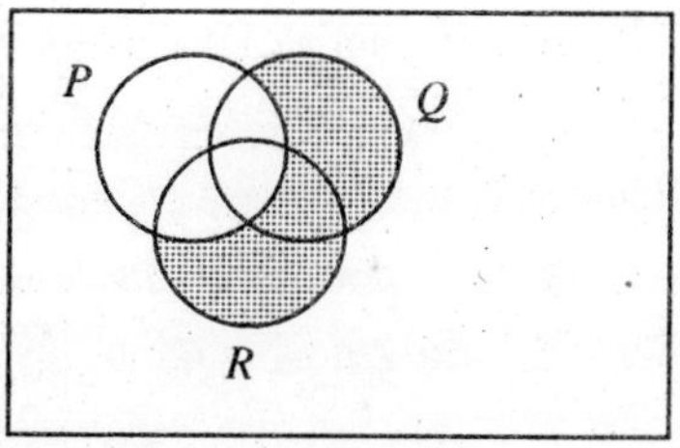

Ex. 93. Find which of the diagrams given here corresponds to the following properties:

(*i*) $P \subset (Q \cap R)$, $Q \subset R, Q \neq R, P \neq R$

(*ii*) $P \subset R, P \neq R$, $Q \cap R = \phi$

(*iii*) $P \subset Q, R \not\subset Q$, $P \cap R \neq \phi$

(*iv*) $P \subset Q, R \subset Q$, $P \cap R = \phi$

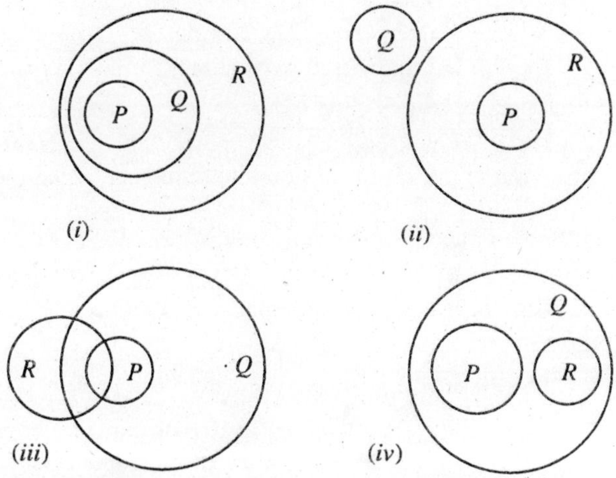

3.4 ORDERED PAIRS

The elements of a set $\{a, b\}$ need not be listed in any particular order because $\{a, b\} = \{b, a\}$. In the case of *ordered* pairs, the order in which the elements are given has a special meaning. This is because, often, the way of arrangement or order of elements is very important.

Let A, B be two sets and let $a \in A, b \in B$.

Then (a, b) is an *ordered pair* consisting of the *first component a* and the *second component b.*

In general $(a, b) \neq (b, a)$

If $A = B = \{b\}$ then the only ordered pair is (b, b).

3.5 CARTESIAN PRODUCT

The set of all possible pairs (a, b) where $a \in A$ and $b \in B$, is called the *Cartesian product* of A and B and is written:

$$A \times B = \{(a, b)/a \in A, b \in B\}$$

Thus $A \times B$ contains all the ordered pairs of elements in A and B. It gives the most general or universal relation between A and B. For three sets we write similarly, $A \times B \times C = \{(a, b, c)/a \in A, b \in B, c \in C\}$.

Ex. 94. $A = \{1, 2\}, B = \{a, b\}$

$$A \times B = \{(1, a), (1, b), (2, a), (2, b)\}$$

$$B \times A = \{(a, 1), (a, 2), (b, 1) (b, 2)\}$$

$$A \times A = \{(1, 1), (1, 2), (2, 1), (2, 2)\}$$

What is $B \times B$?

Ex. 95. $A = \{H, T\}, B = \{1, 2, 3\}$

$$A \times B = \{(H, 1), (H, 2), (H, 3), (T, 1), (T, 2), (T, 3)\}$$

$$B \times A = \{1, H), (2, H), (3, H), (1, T), (2, T), (3, T)\}$$

$$B \times B = \{(1, 1), (1, 2), (1, 3), (2, 1), (2, 2), (2, 3), (3, 1), (3, 2), (3, 3)\}.$$

What is $A \times A$?

Ex. 96. In the ordered pairs (h, w) for heights and weights of students in a class, will any (h_i, w_i) be equal to (w_i, h_i)?

Ex. 97. Let A be the set of the possible outcomes 1, 2, 3, 4, 5, 6 for the throw of a die. What is the set of all possible outcomes when the die is tossed (*i*) twice (*ii*) thrice?

Ans. (*i*) $A \times A = \{(1, 1), (1, 2), \ldots (6, 6)\}$. There are 36 cases.

(*ii*) $A \times A \times A = \{(1, 1, 1), \ldots, (6, 6, 6)\}$. There are 216 cases.

Ex. 98. If there are 4 elements in A and 5 in B how many ordered pairs are there in $A \times B$? **Ans.** 20

Ex. 99. If $n(A) = r$, $n(B) = s$, find $n(A \times B)$.

Ans. *r.s.* If there are r elements in A and s elements in B, the number of ordered pairs is $r \times s$

Ex. 100. Is it possible to have $(a, b) = (c, d)$?

Ans. Yes, if $a = b$, $c = d$

Ex. 101. The Cartesian product of

(*i*) $A = \{(x/x = 3, 5\}$, $B = \{(y/y = x + 1\}$ is obtained by first writing $A = \{3, 5\}$, $B = \{3 + 1, 5 + 1\} = \{4, 6\}$.

Thus $A \times B = (3, 4), (3, 6), (5, 4), (5, 6)\}$.

Find similarly (*i*) $A \times B$ where $A = \{x/x = 2, 4, 6\}$, $B = \{y/y = x^2\}$

(*ii*) $A \times B$ where $A = \{x/x = 0, 1, -1\}$, $B = \{y/y = 2x - 1\}$.

3.6 RELATIONS

A relation is an association between two or more things. It may or may not be true.

Relations are a necessary part of a mathematical system and we use them intuitively in mathematical work.

Relations are classified according to the number of elements associated. When a relation suggests a correspondence or an association between the elements of two sets it is a *binary* or *dyadic* relation. For three elements it is *ternary* or *triadic* and so on.

Most relations in mathematics are binary. Equality, identity, equivalence, <, > are examples of binary relations.

Ex. 102. (*i*)

(*i*) $x \to y$	(*ii*) $x \to y$
$1 \to 3$	$1 \to 2$
$2 \to 6$	$2 \to 5$
$3 \to 9$	$3 \to 8$
. . .	. . .
$x \to 3x$	$x \to 3x - 1$

In (*i*) the correspondence between the x and y columns is given by the relation $y = 3x$. In (*ii*) it is $y = 3x - 1$.

Ex. 103. Consider the = relation for the set {1, 2, 3}. The relation = is completely defined by the set of pairs (1 = 1), (2 = 2), (3 = 3).

Ex. 104. Let A = {Energee, Gold Spot, Pepsi Cola}

B = {Aarey, Parle, Pepsico}.

Here the relationship is "is bottled by". A reversal of the order will be meaningless for the relation.

Domain and Range

The set of elements from which the first component of the ordered pair is drawn is the *domain* of the relation.

The *range* of the relation is the set of elements to which the second component of the ordered pair belongs.

Definition. A relation is a set of ordered pairs.

Any subset of a set of ordered pairs is also a relation. This follows from the definition of the Cartesian product set.

The number of relations in any Cartesian product depends on the number of ordered pairs in that particular Cartesian product. If the number of ordered pairs is n the number of relations is 2^n.

A relation R has a *complement* R' with respect to a given set. R' consists of all the elements of the set that are not elements of R.

A *universal* relation is such that all ordered pairs under consideration belong to it.

An *empty* relation is such that no ordered pair belongs to it.

The *inverse* of relation R is written R^{-1}. The relations > and < are inverses of each other. If $A = \{(a, b)/a, b \in A\}$ then $A^{-1} = \{(b, a)/a, b \in A\}$

Functions

Let there be two sets A and B. If to each element of A, a unique element of B is assigned in some manner we have a *function* or a *functional relation* from the set A to the set B. Such an assignment, which is a special kind of relation, is usually written

$f : A \to B$ or f is a function of A into B.

It implies that if $(x, y) \in f$ and $(x, z) \in f$ then $y = z$. In other words, for each $x \in A$ there is at most one $y \in B$ with $(x, y) \in f$.

To represent the member of the range which corresponds to a particular x in the given domain we use the notation $y = f(x)$.

A function may be described by means of a graph. The function is the truth set of the equation $y = f(x)$ and the graph of the function is the graph of the truth set.

Properties of Relations

Let R be any relation in a set A. Let a, b, c be any elements of the set. The *sentence* xRy means two objects x and y are in the relation R with each other.

Reflexivity : R is *reflexive* if every element in A is related to itself *i.e.,* for every $x \in A$, xRx is true.

R is *irreflexive* if and only if aRa is false for all $a \in A$,

A *nonreflexive* relation is neither reflexive nor irreflexive.

Symmetry : If $(a, b) \in R$ implies $(b, a) \in R$ then R is symmetric. This means $xRy \equiv yRx$.

R is *asymmetric* if and only if for every $x \in A$, $y \in A$, xRy excludes yRx.

A *nonsymmetric* relation is neither symmetric nor asymmetric.

Transitivity : If a is related to b and b is related to c then a is related to c. This gives a *transitive* relation. It can be written

(xRy) and $(yRz) \to (xRz)$ for all x, y, z

R *is intransitive* if and only if xRy, yRz together exclude xRz.

A *nontransitive* relation is neither transitive nor intransitive.

Connectedness : R is *connected* if and only if for every $x \in A$, $y \in A$

If $x \neq y$ then either xRy or yRx or both.

Ex. 105. Which properties of relations are satisfied by the following?

$A_1 = \{(1, 1), (2, 2), (3, 3)\}$. **Ans.** Ref; Tra; Sym

$A_2 = \{(1, 1), (2, 2), (3, 3), (2, 1)\}$. **Ans.** Ref; Tra; Not Sym as $(1, 2) \notin A_2$.

$A_3 = \{(1, 1), (2, 2), (3, 3), (1, 3), (3, 2)$.

Ans. Ref; Not Tra as $(1, 3), (3, 2) \in A_3$ but not $(3, 1)$; Not Sym as $(1, 3) \in A_3$.

$A_4 = \{(1, 1), (2, 2), (3, 3), (1, 2), (2, 1), (2, 3), (3, 2)\}$

Ans. Ref; Not Tra as $(1, 3) \notin A_4$; Sym.

$A_5 = \{(1, 2), (2, 3), (1, 3)\}$ **Ans.** Not Ref as $(1, 1) \notin A_5$, Tra; Non Sym as $(2, 1) \notin A_5$ and $(1, 2) \in A_5$

$A_6 = \{(1, 1), (1, 3), (3, 1)$. **Ans.** Not Ref as $(2, 2) \notin A_6$; Not Tra

$A_7 = \{(1, 3), (3, 2)\}$. **Ans.** Not Ref as $(1, 1) \notin A_7$; Not Tra as $(1, 2) \notin A_7$; Not Sym as $(3, 1) \notin A_7$.

$A_8 = \{(1, 4)\}$. **Ans.** Not Ref; Tra; Not Sym

$A_9 = \phi$ **Ans.** Not Ref; Tra; Sym.

Ex. 106. What are the characteristics of the equivalence relation = ?

Ans. Let R be the = relation in a set A. Then

(i) R is reflexive if for each $a \in A$ we have $a = a$ for all a

(ii) R is transitive if $a = b$, $b = c$ results in $a = c$

(iii) R is symmetric if $a = b$ implies $b = a$.

This is a case of *equivalence relations* which are reflexive, symmetric and transitive. Other cases are congruence, identity etc.

Linear order relations, like $<$, $>$ are never reflexive, never symmetric and are transitive.

Ordering relations are those binary relations which are transitive and antisymmetric.

If an ordering is irreflexive it is referred to as *strong* ordering.

$>$, $<$ are cases of strong ordering.

$\geq$, $\leq$, "is a subset of" are cases of weak ordering because they are transitive and reflexive.

If R is transitive and reflexive R is said to be weak ordering or *quasi-ordering* or *preordering.*

The ordering R is said to be *complete* if it is transitive, antisymmetric and also connected. An ordering that is not complete is called *partial* ordering.

3.7 CHOICE PROBLEMS

To attain certain goals or *results*, an *individual* (or a group) makes, more or less *consciously*, a *choice* for which he performs some *activities*. He might as well make other choices with results valued differently.

There may be any number of feasible activities. The choice set consists of all results following any activity of the feasible activities.

Economic theory is concerned with a number of choice problems.

Maximum satisfaction : The best choice of a consumer, with a given income, between different combinations of commodities at given prices.

Minimum cost : The choice of a producer, desirous of producing a certain quantity of some commodity for the process with the minimum cost.

Maximum returns : The choice, out of various investment projects, of the project with the highest rate of return.

Maximum social advantage : The choice of the most favoured government budget out of all possible combinations of taxes and expenditures.

Let us consider the first of the above cases.

Consumer Preference

A complete quasi-ordering relation is one which is transitive, reflexive and connected. Such relations are available in economics in the theory of consumer choice.

A preference relation or indifference relation is a complete quasi-ordering of the sets of states in which a consumer can find himself.

If a consumer is ordering a finite number of alternatives he may have a *strong ordering* which means that each alternate will have a place of its own in the order and to each number in the order, one and only one item will correspond.

A consumer, however, may face alternatives between which he is indifferent. He will then have a *weak ordering* which means that there is the possibility of some items which are incapable of being arranged in front of one another. There may be division into groups which are themselves strongly ordered but there may be no ordering within the groups.

Let P, Q, R be distinct items for consumer preference.

Strong ordering depends on :

(*i*) *Preliminary conditions* : Items other than P are either before P or after P giving two classes which must overlap.

(*ii*) *Consistency conditions* : It Q is to the left of P, P must be to the right of Q. If Q is to the right of P, P must be to the left of Q.

(*iii*) *Transitivity conditions* : If Q is to the left of P and R is to the left of Q then R is to the left of P.

If there are more items the process can be continued until the whole set of items has been ordered.

Weak ordering may take two forms. With basis P the remaining items are classified into two possibly overlapping classes:

Negative form : Items not to the right of P

Items not to the left of P

Positive form : Items to the left of P

Items to the right of P

Items neither to the left nor right of P.

Order Preserving Functions

The utility function is an order preserving function in choice theory. Production function, index numbers etc. are examples of other order preserving functions. A utility function *indicates order*; it does not measure intensities.

Ex. 107. Consider an ideal consumer who is purchasing two commodities and who has a fixed scale of preference.

Initially the consumer is confronted with given prices (p_1 and p_2) for the two commodities (quantities x_1 and x_2) and has a fixed income y. He can buy as much of any commodity as he likes at the given price (*i.e.*, the market is perfectly competitive from the consumer's point of view).

(*i*) What will be his behaviour if prices change?

(*ii*) Consider the case of (*a*) three goods (*b*) more goods.

(The binary relation here is the preference relation. A consumer's ranking of the available alternatives is a very complex phenomenon. In deciding what to buy an individual consumer ignores any effect that decision may have on market prices. The initial constraint is ($p_1x_1 + p_2x_2 \leq y$).

Ex. 108. Comment on the following three Axioms describing possible patterns of consumer reaction.

Strong : A consumer always prefers more to less.

Weaker : A consumer never prefers less to more and always prefers more to less with respect to at least one commodity.

Weakest : A consumer never prefers less to more.

Ex. 109. Which of the following are equivalence relations?

(*i*) {(1, 1), (2, 2), (1, 2)} (*ii*) {(1, 1), (2, 1), (1, 2)}

(*iii*) {(1, 1), (2, 2), (1, 2), (2, 1)}

Ans. (*i*) Not an equivalence relation because the symmetric property does not hold.

(*ii*) Also not, because the reflexive property is not satisfied.

(*iii*) It is an equivalence relation.

Ex. 110. What are the domain and range of a relation?

Find the domain and range of the relation whose members are $\{(-a_1, a_2), (-b_1, b_2), (-c_1, c_2)\}$

The domain of a relation is the set of first members of all the pairs in a relation.

The range of a relation is the set of second members.

In the above case, we have,

Domain : $(-a_1, -b_1, -c_1)$

Range : (a_2, b_2, c_2)

Ex. 111. Do the following give binary relations?

(*i*) the perpendicular relation (*ii*) the converse relation

(*iii*) the relation of being a factor. Give reasons. **Ans.** Yes.

Ex. 112. Comment on (*i*) {(3, 3), (4, 4), (5, 5)}

(*ii*) {(3, 3) (4, 4), (3, 4)}.

Ans. (*i*) It describes an equivalence relation. In each pair every element is related to itself showing the reflexive property.

(*ii*) It is not an equivalence relation. The symmetry property does not hold.

Ex. 113. Comment on the following relations:

(*i*) Income	Consumption	(*ii*) Price	Demand
20	15	5	20
25	21	4	25
26.	20	3	27
27	22	4	28

Ans. These are lists of ordered pairs showing the relations (income, consumption) and (price, demand). In (*i*) incomes form the *domain* and the consumptions form the *range* of the relation. In (*ii*) price and demand form the domain and the range respectively. For the ordered pair (*a, b*) belonging to the relation *R*, any numbers that *a* can possibly represent are said to form the *domain* and any numbers that *b* can possibly represent give the *range* of the relation.

Ex. 114. What are the properties of the relation "less than" and "greater than"?

The properties for "less than" are :

(*i*) $a < a$ never holds (irreflexive)

(*ii*) If $a < b$ then $b < a$ never holds (asymmetric)

(*iii*) If $a < b$, $b < c$ then $a < c$ always holds (transitive)

Similarly for "greater than".

These are cases of an order relation which is transitive but not reflexive or symmetric.

Ex. 115. What are the properties of the composite relation $a \leq b$? What is weak ordering?

$a \leq b$ means either $a < b$ or $a = b$

(*i*) One of the two component relations always holds. Therefore the composite relation is reflexive.

(*ii*) The composite relation is neither symmetric nor asymmetric because (*a*) there is no symmetry if a, b are unequal, (*b*) there is no asymmetry if a, b are equal.

But if $a \leq b$ and $b \leq a$ then $a = b$ gives us the property of *antisymmetry.*

(*iii*) The composite relation is transitive because both component relations $a < b$ and $a = b$ are transitive.

The relation "$\leq$" gives us a weak ordering as compared with the relation "$<$" which defines a strong ordering. In the strong ordering no ties are permitted but in the weak ordering they are permitted.

Ex. 116. Examine the following statements for symmetry (*s*), antisymmetry (*a*), transitivity (*t*) and reflexitivity (*r*).

(*i*) $\{(a, a), (a, b)\}$

(*ii*) $\{(1, 1), (2, 3), (4, 1)\}$

(*iii*) $\{(1, 3), (2, 4)\}$

(*iv*) $\{(a, a), (b, b), (c, c)\}$

(*v*) $A \times A$

(*vi*) a is greater than b

(*vii*) a is a multiple of b

(*viii*) a times b is the square of a number.

Ans. (*i*) *t* (*ii*) *t* (*iv*) *s*, *t* (*v*) *s* and not *a*; *t*, *r* (*vi*) *a*, *t* (*vii*) *a*, *t*, *r* (*viii*) *s*, *t*, *r*

Ex. 117. Examine the following relations for strong and weak ordering.

(*i*) The NI (national income) of country b is higher than that of a.

(*ii*) a and b have the same goods.

(*iii*) a is below b in the matter of economic development.

(*iv*) a is in a position to "take off" straight to the level of b.

(*v*) Agricultural production in a is not less than that in b.

(*vi*) a generates at least as much power as b.

Ex. 118. Let aRb mean that the commodity b is not preferred to a by a consistent consumer. Define bRc similarly. Show that R is reflexive, transitive, connected and is completely weakly ordered.

(aRb implies indifference so that R is reflexive. Since aRb and bRc imply aRc, R is transitive. R is connected in the set of commodities a, b, c).

Ex. 119. Let aRb mean that neither a nor b is preferred to the other. Then show that R is an equivalence relation implying equal utilities in a and b.

Ex. 120. If a is strongly preferred to b, without any element of indifference, then show that R becomes a complete ordering of the above equivalence class.

Chapter 4

The Number System

4.1 NUMBERS

A study of the order and regularity among numbers is an important feature of mathematics. Discovery of patterns in numbers can help predict future behaviour through chains of reasoning. For this a keen perception of values and number relations is essential.

Natural Numbers

The most familiar numbers are the counting numbers or the natural numbers

1, 2, 3......

They are also called cardinal numbers or positive integers. They are associated with any process of counting objects through one-to-one correspondence though often this one-to-one correspodence may be used without counting and *vice versa.*

Integers

The set of integers contains all natural numbers, negatives of natural numbers and zero

{......–20, –19,......, –1, 0, 1, 2, 3,......, 105,......}

An integer is a whole number and does not have a fractional part.

Measurement

We can associate the concepts of distance and measurement with numbers.

Take a point 0 on a line. On the right of 0 mark all positive integers with a certain unit of measurement. With the same unit mark all negative integers on the left of 0.

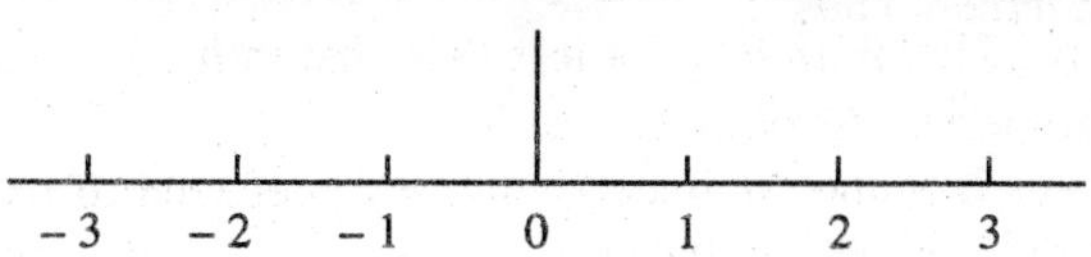

The distance between 0 and 1 is a unit which is the distance between any two consecutive points on either side of 0. There is one-to-one correspondence between certain points on the number line and the integers. To each natural number r there corresponds one and only one point with coordinate r on this line, *i.e.,* r units to the right of 0. Similarly a point r units to the left of 0 will have the coordinate $-r$.

A manipulation of integers by adding, subtracting multiplying, dividing etc., gives rise to integers as well as fractions, decimals and irrational numbers.

Rational Numbers

A rational number is a ratio of two integers, positive or negative, the denominator in the quotient being nonzero.

If a, b are integers and $b \neq 0$ then $\frac{a}{b}$ is a rational number.

All integers are rational numbers but not all rational numbers are integers. The fraction $\frac{6}{3}$ is a rational number. Being equal to 2 it is an integer. But the fraction $\frac{1}{3}$ is not an integer.

$$\frac{a}{b} \text{ and } \frac{ka}{kb},\; k \neq 0 \text{ represent the same rational number.}$$

If $k = 0$, $\frac{a}{b} \cdot \frac{0}{0}$ has no meaning.

A rational number $\frac{a}{b}$ can be represented by a point on the number line shown as a distance from the origin 0. There is an infinite number of rational points on any line segment.

The set of rational numbers contains all decimal numbers of the form $\frac{a}{b}$ where a, b are integers and $b \neq 0$.

$$\frac{1}{4} = 0.25,\ \frac{1}{3} = 0.333,\ \ldots,\ \frac{1}{6} = 0.1616\ldots$$

are examples of rational numbers. A rational number can thus be expressed as a terminating decimal or as a repeating decimal.

Irrational Numbers

Rational numbers do not constitute the set of all the numbers. There are numbers whose distances from the origin, on the number line, cannot be described by the ratios of two integers.

A number which cannot be expressed exactly by using integers, fractions or decimals is called an irrational number. Thus

$$\sqrt{2} = 1.4142\ldots\ldots$$

is an irrational number, as it cannot be expressed as a fraction and the process of finding the square root in decimals never terminates.

Other examples are $\sqrt[3]{7}$, $\sqrt[5]{2}$, π etc.

Numbers like 0.1616... although having infinite decimal expansions are not irrational since they can be expressed as fractions of two integers.

An irrational number does not have a rational coordinate on the number line. The coordinate for an irrational number exists but its distance from the origin cannot be given as the ratio of two integers.

Real Numbers

Rationals and irrationals together constitute the real number system.

Any real number is either rational or irrational. Any point on the number line is related to some corresponding real number uniquely and each real number is related to some corresponding point on the number line uniquely.

The classification of the set of all real numbers is as follows:

- Real Numbers
  - Rational Numbers
    - Positive Integers
    - Negative Integers
    - Zero
    - Positive Fractions
    - Negative Fractions
  - Irrational Numbers

The real number system can be presented as a deductive logical system. By the use of the properties of this system and the operations performed therein we can understand the reasoning behind any steps in mathematical methods connected with real numbers. The operations on real numbers are applicable to rational and irrational numbers.

4.2 AXIOMS OF REAL NUMBERS

The real number system is fundamental to mathematical analysis. It may be defined as a complete, simply ordered field thus suggesting that it is a system which is (*i*) a field (*ii*) simply ordered, and (*iii*) complete.

We divide the axioms for real numbers into three groups.

A. The axioms of the field.

B. The axioms of order.

C. The axioms of completeness.

The meanings of fields, simple order and completeness will be clear from the axioms which are discussed below and which are used in all the familiar algebraic manipulations. In practice these laws are used with great rapidity so that often we are not even aware of the fact that we are using them.

The Axioms of the Field

A field is a set of numbers which contains the sum, the difference, the product and the quotient (except division by zero) of any two numbers in the set. When subjected to these operations the resulting numbers continue to remain in the field.

There are six axioms of a field for addition and multiplication.

Let a, b, c be real numbers. Then we have the following axioms.

(*i*) *The Axiom of Closure* : There is a *unique* real number $a + b$ which is the result of addition of b to a. Similarly there is a *unique* real number $a \cdot b$ which is obtained by multiplying a by b.

The sum $a + b$ is a unique real number

The product $a \cdot b$ is a unique real number.

(*ii*) *The Axiom of Commutation* : $a + b$ and $b + a$ are the same real number. Similarly $a \cdot b$ and $b \cdot a$ are the same real number.

$$a + b = b + a, \; a + (b + c) = (b + c) + a \quad -6 + 11 = +11 - 6$$

$$a \cdot b = b \cdot a, \qquad a(bc) = (bc)a, \qquad 3.5 = 5.3$$

(*iii*) *The Axiom of Association* : Any two of a, b, c, may be added to the third and the result will always be the same, *viz.*, $a + b + c$ which is uniquely defined.

The product of any two of a, b, c, by the third is the same *viz.*, abc which is uniquely defined.

$$a + (b + c) = (a + b) + c = (a + c) + b = a + b + c$$

$$a(bc) = (ab)c = (ac)b = abc$$

$$5 + (3 + 4) = 3 + 4 + 5, \; 5(3.4) = 3.4.5$$

(*iv*) *The Axiom of Distribution* : The distribution law connects the two operations of addition and multiplication.

The product of a and $b + c$ is the same real number as the sum of ab and ac.

$$a(b + c) = ab + ac$$

$$(a + b)(c + d) = ac + ad + bc + bd$$

(*v*) *The Axiom of Identity* : The addition of zero to any number does not change the number. Zero is called the identity element for addition as it leaves the number unchanged.

$$a + 0 = 0 + a = a$$

The multiplication by 1 leaves a number unchanged. 1 is the identity element for multiplication. For each and every a, $a \cdot 1$ and a are the same real number.

$$a \cdot 1 = 1 \cdot a = a$$

(*vi*) *The Inverse Axiom* : For all a,

$$a + (-a) = 0$$

$-a$ is called the additive inverse of $+a$

$+a$ is called the additive inverse of $-a$

For all a, $a \neq 0$

$$a \cdot \frac{1}{a} = 1.$$

$\frac{1}{a}$ is called the multiplicative inverse of a

a is called the multiplicative inverse of $\frac{1}{a}$.

The following statements may be regarded as theorems. Proofs are given in some cases.

1. If $a + a = a$ then a is equal to 0

Proof. $a + a = a$ — hypothesis

$\therefore \quad a + a + (-a) = a + (-a)$

$\therefore \quad a + [a + (-a)] = a + (-a)$ — association

$\therefore \quad a + 0 = 0$

$\therefore \quad a = 0$

2. If a is a real number then $a \cdot 0 = 0$

Proof. $0 + 0 = 0$

$\therefore \quad a(0 + 0) = a \cdot 0$

$\therefore \quad a \cdot 0 + a \cdot 0 = a \cdot 0$ distribution

$\therefore \quad a \cdot 0 = 0$

3. If $a = b$ then $-a = -b$

4. There is no real number a which is a reciprocal of 0.

Proof. Let there be a real number a which is a reciprocal of 0, then $a \cdot 0 = 1$. But $a \cdot 0 = 0$ so that $0 = 1$ which is absurd (Every real number a, other than zero has a unique reciprocal $1/a$).

5. If $a + c = b + c$ then $a = b$

Proof. $a + c + (-c) = b + c + (-c)$

$\therefore \quad a + [c + (-c)] = b + [c + (-c)]$ association

$\therefore \quad a + 0 = b + 0$

$\therefore \quad a = b.$

6. If $ac = bc$ and $c \neq b$ then $a = b$

Proof. $(ac)\dfrac{1}{c} = (bc)\dfrac{1}{c}$

$\therefore \quad a\left(c \cdot \dfrac{1}{c}\right) = b\left(c \cdot \dfrac{1}{c}\right)$ association

$\therefore \quad a \cdot 1 = b \cdot 1$

$\therefore \quad a = b$

7. If $a + b = 0$ then $b = -a$

Proof. $-a + (a + b) = -a + 0$

$\therefore \quad (-a + a) + b = -a + 0$

$\therefore \quad 0 + b = -a + 0$

$\therefore \quad b = -a.$

8. If $ab = 1$, $a \neq 0$ then $b = \dfrac{1}{a}$

Proof. $(ab)\dfrac{1}{a} = 1 \cdot \dfrac{1}{a}$

$\therefore \quad b\left(a \cdot \dfrac{1}{a}\right) = 1 \cdot \dfrac{1}{a}$

$\therefore \quad b \cdot 1 = 1 \cdot \dfrac{1}{a}$

$\therefore \quad b = \dfrac{1}{a}$

9. $b(-a) = -(ab).$

10. $(-a)(-b) = ab.$

11. $a(b-c) = ab - ac$

Proof $a(b-c) = a[b + (-c)]$

$= a \cdot b + a(-c)]$

$= ab - ac.$

12. $a - 0 = a.$

13. $a + a = 2a.$

14. $\frac{a}{b} \cdot \frac{c}{d} = \frac{ac}{bd}$ if $b \neq 0,\ d \neq 0.$

15. $-(a+b) = -a - b.$

16. $b + (a - b) = a.$

17. $-\frac{a}{b} = -\frac{-a}{b} = \frac{a}{-b}.$

18. $\frac{a+b}{c} = \frac{a}{c} + \frac{b}{c}.$

19. $-a + (-a) = (-2) \cdot a$

20. If $b = ax$ then $x = \frac{b}{a}$ $a \neq 0$

21. $(ab)^{-1} = a^{-1} b^{-1}$ *i.e.*, the reciprocal of the product is the product of the reciprocals.

Proof. $(ab)(a^{-1} b^{-1}) = a[b(a^{-1} b^{-1})] = a[(bb^{-1})a^{-1}]$

$= a[1 \cdot a^{-1}] = a \cdot a^{-1} = 1.$ Hence the result.

22. If $\frac{a}{b} = \frac{c}{b}$ then $a = c$ $b \neq 0$

23. $-(a+b) = (-a) + (-b)$

24. $(x-a)(x+a) = x^2 - a^2$

25. $(a+b)^2 = a^2 + 2ab + b^2$

The Axioms of Order

A field is said to be *ordered* if it satisfies the properties of trichotomy, transitivity, addition and multiplication. This means that real numbers can be thought of as being arranged on the number line.

A real number a is positive if $a > 0$ and negative if $a < 0$.

Zero is neither positive nor negative.

If a, b are real then $a < b$ if and only if $b > a$.

There are six axioms of an ordered field.

Let a, b, c be real numbers.

(*i*) *The Axiom of Trichotomy.* If a, b are real then one and only one of the following statements is true:

$a < b,\ a = b,\ a > b$

(*ii*) *The Axiom of Transitivity* :

If $a > b$ and $b > c$ then $a > c$

If $a < b$ and $b < c$ then $a < c$

(*iii*) *The Axiom of Addition* :

If $a > b$ then $a + c > b + c$

(*iv*) *The Axiom of Multiplication* :

If $a > b$ and $c > 0$ then $ac > bc$

(*v*) *The Axiom of Extent* : If $a < b$ there exist real numbers c, d such that

$$c < a,\ b < d$$

(*vi*) *The Axiom of Density* : If $a < b$ there exists a real number c such that

$$a < c < b$$

i.e., between any two real numbers, however close to each other, there is always a real number.

26. $a > b$ if and only if $-a < -b$

Proof. $a > b$ $\therefore$ $a + [(-a) + -(b)] > b + [(-a) + (-b)]$

$\therefore$ $[a + (-a)] + (-b) > [b + (-b)] + (-a)$

$\therefore$ $0 + (-b) > 0 + (-a)$ or $-b > -a$.

27. If $a > 0$ then $-a < 0$.

28. If $a > 0$ and $b > 0$ then $ab > 0$

29. If $a > b$ and $c > d$ then $a + c > b + d$.

Proof. $a + c > b + c$, $\quad b + c > b + d$.

$\therefore$ $a + c > b + d$.

30. If $a > b$ and $c < 0$ then $ac < bc$

Proof. $a(-c) > b(-c)$

$\therefore$ $-ac > -bc$

$\therefore$ $ac < bc$.

An inequality is reversed if we multiply both sides by the same negative number.

Thus $\quad -5 > -7$ gives $7 > 5$ or $5 < 7$

Absolute Value or Modulus

Some quantities like price, demand, supply never take negative values while other quantities like profit, utility etc. may be either positive or negative. Often we need to use a quantity regardless of its sign. It is called the absolute value of the quantity.

The absolute value of a real number a is denoted by $|a|$. Thus $|-2| = 2, |2| = 2$. The absolute value of a real number a is defined by the following conditions.

$$|a| = a \quad \text{if} \quad a > 0$$

$$|a| = -a \quad \text{if} \quad a < 0$$

$$|a| = 0 \quad \text{if} \quad a = 0$$

The square of the absolute value of a number is the square of the number.

$|a|^2 = |a^2|, |-a|^2 = a^2$

31. For every a, $|-a| \leq a \leq |a|$
32. $|ab| = |a|\,|b|$ Thus $|5.3| = |5|\,|3| = 15$.
33. $|a+b| \leq |a| + |b|$. Thus $|5+3| \leq |5| + |3| = 8$.

and $|5+(-3)| < |5| + |-3| =$ *i.e.* < 8.

34. $|a| \geq a$ for every a.
35. $|-a| = |a|$ for every a.
36. Let $a > 0$ then $|x| < a$ if and only if $-a < x < a$.

Proof. If $x \geq 0$ then $x < a$ *i.e.,* $|x| < a$ when $-a < x < a$

If $x < 0$ then $-a < x$ *i.e.,* $|x| < a$ then $-a < x < 0$

Combining we get $-a < x < a$

$$|x| < a \text{ implies } -a < x < a$$

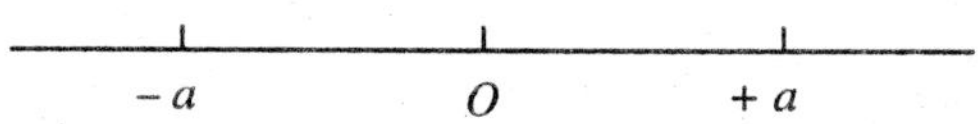

$|x| < a$

The Axiom of completeness

This axiom is a subtle characterising property of the real number system. It is sometimes called the least upper bound axiom. It states that every set which has an upper bound has a least upper bound. Every set which has a lower bound has a greatest upper bound.

Consider the approximate values of $\sqrt{3}$. There are several approximations : 1.7, 1.73, 1.732, ...

Since $1.7 < 1.73 < 1.732 < \ldots < \sqrt{3}$

i.e., each value is less than $\sqrt{3}$ and none equals $\sqrt{3}$,

We say $\sqrt{3}$ is an upper bound of the set.

$$\{1.7, 1.73, 1.732, \ldots\}$$

There are other upper bounds but $\sqrt{3}$ is the best. It is the least upper bound.

An upper bound t of a set is said to be the least upper bound if no upper bound is less than t.

A lower bound v is the greatest lower bound if no lower bound is greater than v.

The set of positive integers has no upper bound.

The set of negative integers has no lower bound.

Ex. 1. Perform the indicated operations:

(*i*) $5(-3) + 4(6)$ (*ii*) $(-3)(-7) + (-6+3)$

(*iii*) $2^2 \times 2^3 \times 2^4$ (*iv*) $2^5 \div 2^3$

(v) $|-9|-|2|+|-6|$

(vi) $|-8|+|-2|\cdot|-3|$

Ex. 2. Prove that

(i) $ab > 0$ if and only if a, b are both positive or both negative.

(ii) $-3 < -1$ (iii) $a > \frac{a+b}{2} > 1$ if $a > b$

(iv) $0 < \frac{1}{2} < 1$ (v) $a > \sqrt{ab} > b$ if $a > b > 0$

(vi) $(-4)(-2) > (-2)(-2)$ (vii) $3 + (-3) < 4 + (-3)$

Ex. 3. Solve the following equations:

(i) $|x| = 2$ (ii) $\sqrt{x^2} = 7$

(iii) $x = |2|$ (iv) $x = |\frac{1}{2}| + |-\frac{5}{2}|$

(v) $\sqrt{(2-x)^2} = 4$ (vi) $|x-2| = -5$.

Ans (i) 2 (ii) ± 7 (iii) 2
(iv) 3 (v) 6, 2 (vi) No solutions

Ex. 4. Express the following rational numbers as repeating decimals:

(i) $\frac{1}{7}$ (ii) $\frac{1}{8}$ (iii) $\frac{2}{3}$ (iv) $\frac{1}{5}$

Ans. (i) $0.\dot{1}\dot{4}\dot{2}\ \dot{8}\dot{5}\dot{7}$ (ii) $\frac{1}{3} = 0.\dot{3}\dot{3}$ (iii) $0.\dot{6}\dot{6}\dot{6}$ (iv) 0.20

Ex. 5. Give the multiplicative inverse of (i) 3 (ii) -5 and additive inverse of (iii) -8 (iv) 10 (v) 0.

Ans. (i) $\frac{1}{2}$ (ii) $-\frac{1}{5}$ (iii) $+8$ (iv) -10 (v) 0

Ex. 6. Solve the following inequalities:

(i) $|x-3| > 2$ (ii) $\left|\frac{x-3}{5}\right| > \frac{1}{2}$ (iii) $|\pi - x| \le 2$.

Ans. (i) $x > 5, x < 1$ (ii) $x > \frac{11}{2}, x < \frac{1}{2}$ (iii) $\pi + 2 \ge x \ge \pi - 2$

Ex. 7. Show that (i) $|\bar{x} - \mu| < d$ means $\mu - d < \bar{x} < \mu + d$

(ii) $|\mu| \ge d$ means $\mu \ge d$ and $\mu \le -d$ if $d > 0$.

Ex. 8. Solve the following inequalities

(i) $|3x-2| < 5$ (ii) $|15-2x| < \frac{1}{2}$ (iii) $|x-2| \le 3$

Ans. (i) $x < \frac{7}{3}, x > -1$ i.e., $-1 < x < \frac{7}{3}$

(ii) $-\frac{1}{2} < |5-2x| < \frac{1}{2}$ (iii) $-3 \le x-2 \le 3$

Ex. 9. If $x + 2 > 5$ then $x + 7 > ?$

If $x > 5$ then $5x > ?$

If $x + 2 < 5$ then $x + 3 < ?$

If $x < 5$ then $5x < ?$

If $x + 5 > 10$ and $y + 10 > 15$ then $x + y + 15 > ?$

Ex. 10. Which of the alternative answers is correct in the following cases?

		α β	Ans.
(*i*)	$a - b > a - 5$	then $b > 5/b < 5$	β
(*ii*)	$5 - a > b - a$	then $b > 5/b < 5$	β
(*iii*)	$2a + 5 < 2a + b$	then $b > 5/b < 5$	α
(*iv*)	$2a + b < 2b + a$	then $a < b/b < a$	α
(*v*)	$a + b + c > a + 2b$	then $b < c/b > c$	α
(*vi*)	$2a - b < 2b - a$	then $a < b/b > a$	α
(*vii*)	$a - b > a - 2b + 10$	then $b > 10/b < 10$	α

4.3 INTERVAL NOTATION

An interval is a special type of point set. Since there is an infinite number of real numbers between any two real numbers, intervals of real numbers are also infinite sets.

A *closed interval* is written $[a, b]$ or $a \leq x \leq b$ where $a < b$. The end points a, b of a closed interval are elements of the interval *i.e.*, a, b belong to the interval. Thus x can take the values a, b and all the values between a, b.

An *open interval* is written (a, b) or $a < x < b$ where $a < b$. The end points a, b do not belong to the interval. Here x can take any values between the numbers a, b.

A partially open *i.e.*, half open, half closed interval is written $a < x \leq b$ or $(a, b]$.

A partially open *i.e.*, half closed, half open interval is written $a \leq x < b$ or $[a, b)$.

An interval $a \leq x \leq b$ where $a = b$ is called a *degenerate* interval. It reduces to a single point on a line.

An interval $a \leq x \leq b$ where $a > b$ is called a null interval. The null interval contains no points. For example, $4 \leq x \leq 0$ is null since both the inequalities $4 \leq x$ and $x \leq 0$ cannot be satisfied for any number x.

An interval may be bounded or unbounded. If an interval contains all points less than a certain number b, it is said to be unbounded from below. It is written $-\infty < x \leq b$ in the closed form and written $-\infty < x < b$ in the open form.

If an interval contains all points more than a certain number a, it is said to be unbounded from above. It is written $a \leq x < \infty$ in the closed form and written $a < x < \infty$ in the open form.

If a, b are given numbers, an interval $a \leq x \leq b$ is said to be bounded.

If an interval extends to infinity on either side it is said to be unbounded and written $-\infty < x < \infty$.

4.4 MATHEMATICAL INDUCTION: A PROPERTY OF NATURAL NUMBERS

Mathematical induction is an important method of proving certain kinds of theorems. These theorems involve the argument that some fact is true for every positive integer n or for every such n greater than some fixed positive integer m.

The basic axioms used in mathematical induction are that, in the set S of natural numbers.

(*i*) 1 is an element of the set.

(*ii*) Every natural number n has a successor which is a natural number.

(*iii*) No two natural numbers have the same successor.

(*iv*) 1 is not the successor of any number.

In deductive reasoning we begin with a basic assumption : the hypothesis, from which a valid statement is deduced by using established axioms, definitions and theorems.

A sequence of valid statements, each established with the help of the previous statement, is organised and concludes with the statement to be proved by induction.

Mathematical induction is a form of deductive reasoning in which conclusions are established beyond any doubt. It is not the same as inductive logic which considers the formulation of generalisation by observing many specific cases in experimental sciences.

Mathematical induction proves a general proposition p by considering the constituent sequence of propositions $p_1, p_2, p_3, \ldots$

First the proposition p_1 is verified for $n = 1$.

If the proposition p_k is assumed to be true for $n = k$ then p_{k+1} is checked to be true for $n = k + 1$.

Then the general proposition p is true for all natural numbers n.

4.5 SUMMARY OF PROPERTIES OF ALGEBRAIC OPERATIONS ON REAL NUMBERS

Let a, b, c be real numbers.

Properties of +

(*i*)	The sum $a + b$ is unique	*Closure and uniqueness*
(*ii*)	$a + b = b + a$	*Commutation*
(*iii*)	$(a + b) + c = a + (b + c)$	*Association*
(*iv*)	$a + 0 = a$ for all a	*Identity*
(*v*)	$a + (-a) = 0$ for all a	*Inverse*

Properties of ×

(*i*)	The product $a \cdot b$ is unique	*Closure and uniqueness*
(*ii*)	$a \cdot b = b \cdot a$	*Commutation*
(*iii*)	$(a \cdot b)c = a\,(b \cdot c)$	*Association*
(*iv*)	$a \cdot 1 = a$ for all a	*Identity*
(*v*)	$a \cdot \frac{1}{a} = 1$ for all a $(\neq 0)$	*Inverse*

Properties of =

(*i*)	$a = a$	*Reflexivity*
(*ii*)	$a = b$ implies $b = a$	*Symmetry*
(*iii*)	$a = b$ and $b = c$ implies $a = c$	*Transitivity*

Properties of <

(*i*) One and only one of the following is true : $a = b$, $a < b$, $b < a$ — *Trichotomy*

(*ii*) If $a < b$ and $b < c$ then $a < c$ — *Transitivity*

(*iii*) If $a < b$ then $a + c < b + c$ — *Addition*

(*iv*) If $a < b$ and $c > 0$ then $ac < bc$ — *Multiplication*

(*v*) For real a, b however close there exists a real c such that $a < c < b$ — *Density*

Similarly for *properties* of >

Other properties and derived results :

(*i*) $a \cdot (b + c) = a \cdot b + a \cdot c$ — *Distribution*

(*ii*) Replacement of any part of an algebraic expression by an equal leaves the expression unaltered
Substitution

(*iii*) Subtraction $a - b = a + (-b)$

(*iv*) Division $a \div b = a \cdot 1/b \;\; (b \neq 0)$

(*v*) $a > b$ *implies* $b < a$

(*vi*) $a \leq b$ *implies* $a < b$ or $a = b$

(*vii*) $a \geq b$ *implies* $b < a$ or $b = a$

(*viii*) Absolute value $|a| = a$ if $a \geq 0$
$= -a$ if $a < 0$

(*ix*) a is positive *implies* $a > 0$
a is negative *implies* $a < 0$

(*x*) Exponent : $a^m = a \cdot a \cdot a \cdot a \ldots\ldots\ldots\ldots a$, m times.

4.6 COMPLEX NUMBERS

We have noticed man's imaginative skill in the case of three invented systems of numbers:

(*i*) The set of all integers, constructed from the counting numbers.

(*ii*) The set of all rational numbers, constructed from the integers.

(*iii*) The set of all real numbers, constructed from the rationals.

Each system is richer than and contains the previous systems. The corresponding equations are of the form

(*i*) $x + a = 0$ where a is an integer

(*ii*) $ax + b = 0$ where a, b are rationals and $a \neq 0$

(*iii*) $ax^2 + bx + c = 0$ where $a \neq 0$ and $b^2 > 4\,ac$

We consider now the fourth system resulting from an equation like $x^2 + 1 = 0$.

If x is any real number, $x^2 \geq 0$ so that the equation gives the impossible situation, $x^2 = -1$.

We, therefore, define a new set of numbers given by

$x = \sqrt{-1} = i$, an imaginary number.

If a, b are real, a number of the form $a + bi$ is called a complex number, a is the real part and bi is the imaginary part of $a + bi$ which is sometimes written (a, b).

The imaginary number can be used by remembering its value.

$$i = \sqrt{-1},\ i^2 = -1,\ i^3 = -i$$

$$i^4 = 1,\ i^5 = i,\ i^6 = -1 \text{ and so on.}$$

The laws of operations of complex numbers may be briefly stated as follows:

Equality : $a + ib = c + id$ if $a = c$, $b = d$

Addition : $(a + ib) + (c + id) = (a + c) + i(b + d)$

Multiplication : $(a + ib)(c + id) = (ac - bd) + i(ad + bc)$

$$a(c + id) = (ac) + i(cd)$$

Division
$$\frac{c+id}{a+ib} = \frac{(c+id)(a-ib)}{(a+ib)(a-ib)}$$

$$= \frac{(ac+bd)+i(ad-bc)}{a^2+b^2} = x + iy$$

$$\text{where } x = \frac{ac+bd}{a^2+b^2},\ y = \frac{ad-bc}{a^2+b^2}$$

Absolute value : $|a + ib| = \sqrt{a^2 + b^2}$

A polynomial of nth degree with real coefficients has n roots. The roots may be real or complex. Some roots may coincide.

Complex roots, if any, always occur in *conjugate* pairs, in the form $a + ib$ and $a - ib$. The *sum* $a + ib + a - ib = 2a$ and the *product* $(a + ib)(a - ib) = a^2 + b^2$ of these complex conjugates are real.

Note 1. A complex number is the sum of a real number x and imaginary number iy

$$z = x + iy$$

If
$$z_1 = x_1 + iy_1,\ z_2 = x_2 + iy_2$$

$$z_1 + z_2 = (x_1 + x_2) + i(y_1 + y_2)$$

$$z_1 z_2 = x_1x_2 + x_1y_1i + x_2y_1i + i^2y_1y_2$$

$$= x_1x_2 - y_1y_2 + i(x_1y_2 + x_2y_1) \quad \because i^2 = -1$$

$$\frac{z_1}{z_2} = \frac{x_1+iy_1}{x_2+iy_2} = \frac{x_1+iy_1}{x_2+iy_2} \cdot \frac{x_2-iy_2}{x_2-iy_2}$$

$$= \frac{x_1x_2+y_1y_2}{x_2^2+y_2^2} + i\frac{y_1y_2-x_1x_2}{x_2^2+y_2^2}$$

Note 2. The complex conjugate

The conjugate of $z = x + iy$ is $\bar{z} = x - iy$

$\bar{z}$ is obtained by replacing i by $-i$. If z is a conjugate of $\bar{z}$, then $\bar{z}$ is a conjugate of z.

Note 3. A complex number can be written in a convenient form by multiplying by a suitable conjugate.

Note 4. If $a + bi = c + di$ where a, b, c, d are real numbers then $a = c$, $b = d$

Ex. 11. $\dfrac{1+2i}{2+i}$ can be written

$$\frac{(1+2i)(2-i)}{(2+i)(2-i)} = \frac{2+3i-2i^2}{4-i^2} \quad \text{(This is called rationalisation)}$$

$$= \frac{2+3i-2(-1)}{4+1} = \frac{4+3i}{5} = \frac{4}{5} + i\frac{3}{5}$$

Ex. 12. $$\left(\frac{1+2i}{2+i}\right)^2 = \left(\frac{4}{5} + i\frac{3}{5}\right)^2 = \left(\frac{4}{5}\right)^2 - \left(\frac{3}{5}\right)^2 + 2i \cdot \frac{12}{25} \quad (\because i^2 = -1)$$

$$= \frac{7}{25} + i\frac{24}{25}$$

or $$\left(\frac{1+2i}{2+i}\right)^2 = \frac{(1+2i)^2}{(2+i)^2} = \frac{1-4+4i}{4-1+4i} = \frac{-3+4i}{3+4i}$$

$$= \frac{(-3+4i)(3-4i)}{(3+4i)(3-4i)} = \frac{-9+16+24i}{9+16}$$

$$= \frac{-7+24i}{25}$$

Ex. 13. Find x, y if $\dfrac{x-4}{4+i} + \dfrac{y}{4-i} = i$

$(x-4)(4-i) + y(4+i) = (4+i)(4-i)i$

$4x - 16 - xi + 4i + 4y + yi = (16+1)i$

$4x + 4y - 16 = 17i + xi - 4i - yi$

$4(x+y) - i(x-y) = 16 - 13i$

Equate real and imaginary parts.

$4(x+y) = 16$

$x - y = 13$ Solving simultaneously $x = \dfrac{-9}{2}$, $y = \dfrac{17}{2}$

Ex. 14. If $x = 2 + 3i$ find the value of $x^4 - 4x^2 + 8x + 35$

$x - 2 = i3$

$\therefore \quad (x-2)^2 = -9$ or $x^2 - 4x + 7 = 0$

$$x^4 - 4x^2 + 8x + 35 = x^4 - 4x^3 + 7x^2 + 4x^3 - 16x^2 + 28x + 5x^2 - 20x + 35$$
$$= x^2(x^2 - 4x + 7) + 4x(x^2 - 4x + 7) + 5(x^2 - 4x + 7)$$
$$= x^2 \cdot 0 + 4x \cdot 0 + 5.0 = 0$$

Ex. 15. If $x = 4 + i7$ show that $x^3 - 4x^2 - 9x + 97 = -342 + 721i$

Ex. 16. $\sqrt{5+12i}$

$$= \sqrt{9-4+12i}$$
$$= \sqrt{3^2 + (2i)^2 + 2.3.2i} = \sqrt{(3+2i)^2} = 3 + 2i$$

Similarly $\sqrt{5-12i} = 3 - 2i$.

Ex. 17. A quadratic equation has two roots.

A cubic equation has three roots.

Find the roots of $x^3 - 1 = 0$

This is the same as finding the *cube roots of unity.*

$$x^3 - 1 = (x-1)(x^2 + x + 1) = 0$$

$\therefore$ $x = 1$ and the other two roots are those from

$$x^2 + x + 1 = 0$$
$$x^2 + x + \frac{1}{4} + \frac{3}{4} = 0$$
$$\therefore \quad \left(x + \frac{1}{2}\right)^2 = -\frac{3}{4} = \left(\frac{\sqrt{3}}{2}i\right)^2$$
$$\therefore \quad x = -\frac{1}{2} \pm \frac{\sqrt{3}}{2}i$$

$\therefore$ $1, -\frac{1}{2} + \frac{\sqrt{3}}{2}i, -\frac{1}{2} - \frac{\sqrt{3}}{2}i$ are the cube roots of 1. These are denoted by $1, w, w^2$

Check $w^3 = \left(-\frac{1}{2} + \frac{\sqrt{3}}{2}\right)^3 = \frac{1}{8}((-1)^3 + (\sqrt{3})^3 - 3\sqrt{3} + 3.3)$

$$= \frac{1}{8}(-1 + 3\sqrt{3} - 3\sqrt{3} + 9) = \frac{8}{8} = 1$$

Similarly $(w^2)^3 = 1$

Note $1 + w + w^2 = 0$

Ex. 18. $x + iy = \sqrt{\frac{a+ib}{c+id}}$ prove $(x^2 + y^2)^2 = \frac{a^2 + b^2}{c^2 + d^2}$ using the conjugate property.

$$x - iy = \sqrt{\frac{a-ib}{c-id}}$$

$$\therefore \quad (x+iy)(x-iy) = \sqrt{\frac{(a+ib)(a-ib)}{(c+id)(c-id)}} = \sqrt{\frac{a^2+b^2}{c^2+d^2}}$$

$$\therefore \quad x^2+y^2 = \sqrt{\frac{a^2+b^2}{c^2+d^2}}$$

Ex. 19. If 1, w, w^2 are the cube roots of unity, find the roots of $(x-1)^3+8=0$

Ans. $-1,\ 1-2w,\ 1-2w^2$

Ex. 20. If $\left(\frac{1+i}{1-i}\right)^{98} = p+iq$, find p, q

$$\frac{1+i}{1-i} = \frac{(1+i)(1+i)}{(1-i)(1+i)} = \frac{1+2i+i^2}{1+1} = \frac{2i}{2} = i$$

$$\left(\frac{1+i}{1-i}\right)^{98} = i^{98} = (i^2)^{49} = -1$$

Ex. 21. Show $(1+w)(1+w^2)(1+w^4)(1+w^8) = 1$

LHS $= (1+w)(1+w^2)(1+w)(1+w^2)$ $(\because w^3 = 1)$

$= (1+w^2)(1+w^2)$

Note $1+w+w^2 = 0$

$= w^6 = 1$

Ex. 22. Show $\dfrac{\sqrt{5+12i}+\sqrt{5-12i}}{\sqrt{5+12i}-\sqrt{5-12i}} = -\dfrac{3}{2}i.$

Ex. 23. If $(a+ib) = (x+iy)^{1/3}$ prove $4(a^2-b^2) = \dfrac{x}{a}+\dfrac{y}{b}$

Ex. 24. Show $\left(\dfrac{1}{1-2i}+\dfrac{3}{1+i}\right)\left(\dfrac{3+4i}{2-4i}\right) = -\dfrac{14}{13}$

Ex. 25. $\sqrt{-8-6i} = \sqrt{-9+1-6i} = \sqrt{(3i)^2+1-2.3i.1.} = 3i-1$

Ex. 26. Show $(1-w+w^2)^5+(1+w-w^2)^5 = 32.$

Other Examples

Ex. 27. Change the following into interval notation

(*i*) $\{x/3 < x < 7\}$ (*ii*) $\{y/10 \geq y > 0\}$ (*iii*) $\{a/a \leq 0\}$.

Ans. (*i*) (3, 7) (*ii*) (0, 10] (*iii*) $(-\infty, 0]$.

Ex. 28. Change the following into inequalities

(*i*) (2, 3] (*ii*) (10, 15) (*iii*) $(0, \infty)$ (*iv*) $(-\infty, \infty)$

(*v*) $[k, \infty)$ (*vi*) $[c, d]$.

Ex. 29. Show that the product of two odd integers is odd.

Ex. 30. Show that the square of an even integer is itself even. (An even number can be written $2k$ whose square is $4k^2$ which is divisible by 2.)

Ex. 31. If the square root of an even integer is an integer then it is an even integer.

Ex. 32. Prove that $\sqrt{2}$ is irrational.

Ans. If $\sqrt{2}$ is not irrational then it should be equal to a rational number p/q where p, q have no factors in common

$\therefore$ $(p/q)^2 = p^2/q^2 = 2$ or $p^2 = 2q^2$

$\therefore$ p^2 is even which means p is also even.

If $p = 2k$ where k is a rational

then $p^2 = 4k^2 = 2q^2$ or $q^2 = 2\,k^2$

$\therefore$ q^2 is even which means q is also even

$\therefore$ p, q both have a common factor 2.

This contradicts the assumption that p, q are prime. The contradictory result fallows from the assumption that the theorem is false. Therefore the theorem must be true.

Ex. 33. Show $\sqrt{2}$, e, π on the real number line

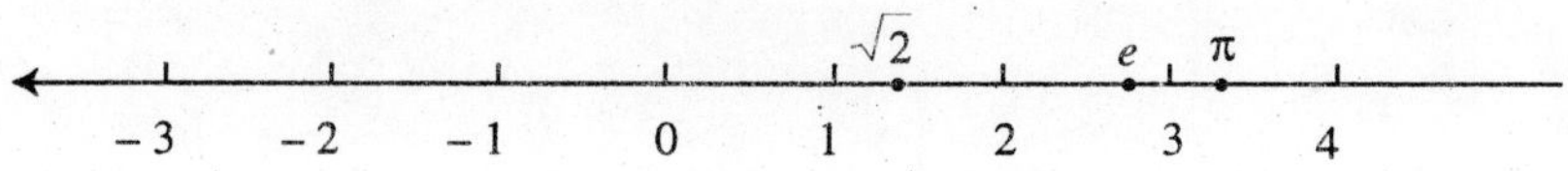

Ex. 34. State which of the three systems of integers, rational numbers and real numbers is infinite, simply ordered or complete.

Ans. The system of integers is infinite, simply ordered and complete but it is not a field

The rational number system is a field which is infinite and simply ordered but not complete.

The real number system is a field which is infinite, simply ordered and complete.

Ex. 35. State whether the following statements are true or false. Give reasons for your answer:

(*i*) Every integer is a natural number.

(*ii*) Every natural number is rational.

(*iii*) Every real number is rational.

(*iv*) The sum of two even integers is an even integer.

(*v*) The product of two positive integers is a positive integer.

(*vi*) There is a greatest rational number less than 2.

(*vii*) There is a least integer greater than 2.

(*viii*) The product of two rational numbers is rational.

(*ix*) The sum of two irrational numbers is irrational.

(*x*) If $x < y$ then $x^2 < y^2$.

(*xi*) Given any real number x we can find a real number y such that $xy = 1$.

(*xii*) A real number is either rational or irrational.

(*xiii*) There can be a real number which is both rational and irrational.

(*xiv*) The product of two equal integers is positive.

Ans. (*i*) F (*ii*) T (*iii*) F (*iv*) T (*v*) T (*vi*) F (*vii*) T (*viii*) T (*ix*) F (*x*) F (*xi*) F (*xii*) T (*xiii*) F (*xiv*) F

CHAPTER 5

Systems of Equations and Inequalities Market Equilibrium

5.1 EQUATIONS AND IDENTITIES

Relations between economic quantities can be expressed in the form of equations. A better understanding of the economic system can be obtained by a study of these relations expressed as systems of equations. A solution of two or more simultaneous equations will be the main subject of our study in this chapter.

An *equation* is a statement showing the equality of two expressions.

$x + 2 = 7$ is an equation with an unknown variable x.

When a value of the variable is substituted in the equation making the equality true we have a solution or a root of the equation.

$x = 5$ is a root of the above equation.

The roots of an equation can be found by applying suitable algebraic operations.

If an equality is true for all values of the variable it is called an *identity.* The identity $2x = 5x - 3x$ is true for all values of x.

If an equality is not true for any value whatsoever of the variable it is a null equation.

If two equations have the same solution set they are equivalent equations.

5.2 DEGREE OF AN EQUATION : SOLUTIONS

The highest power of the unknown variable gives the degree of the equation.

(*i*) A first degree or *linear equation* is a polynomial in one variable occurring only to the first power. In the linear equation

$$ax + b = 0$$

a, b are constants and the solution is obtained by writing $ax = -b$ and getting

$$x = -\frac{b}{a} \qquad a \neq 0.$$

(*ii*) A second degree or quadratic equation is of the general form

$$ax^2 + bx + c = 0.$$

Where a, b, c are constants. We can get the solution by factorising the left side. It can be shown that the solution of the general quadratic equation is given by

$$x = \frac{-b \pm \sqrt{b^2 - 4ac}}{2a} \qquad a \neq 0$$

This gives us two roots or solutions.

(*iii*) A cubic equation of the general form

$$ax^3 + bx^2 + cx + d = 0$$

has three solutions.

(*iv*) In general an equation of the *n*th degree has *n* solutions.

Ex. 1. Solve for x

(*i*) $5 - 3x = -1$ (*ii*) $2x^2 - 5x + 3 = 0$

Ans. (*i*) $5 + 1 = 3x$ or $3x = 6$ $\therefore x = 2$

(*ii*) By factorising $(x - 1)(2x - 3) = 0$

$\therefore x - 1 = 0$ or $2x - 3 = 0$ $\therefore x = 1$ or $x = \frac{3}{2}$

Alternatively, by using the formula, $a = 2$, $b = -5$, $c = 3$,

$$x = \frac{-b \pm \sqrt{b^2 - 4ac}}{2a} = \frac{-(-5) \pm \sqrt{25 - 4 \times 2 \times 3}}{2 \times 2}$$

$$= \frac{5 \pm \sqrt{1}}{4}$$

$$\therefore \qquad x = \frac{5+1}{4} = \frac{3}{2} \text{ or } x = \frac{5-1}{4} = 1.$$

Ex. 2. Solve for x : $\frac{k}{x} - 7 = m.$ **Ans.** $x = \frac{k}{m+7}$

Ex. 3. A market demand curve is given by

$$D = 36 - 3p$$

(*i*) Find p if $D = 24$. (*ii*) Find D if $p = 10$

Ans. (*i*) $24 = 36 - 3p$.

$\therefore 3p = 36 - 24$ or $3p = 12$ $\therefore p = 4$

(*ii*) $D = 36 - 3 \times 0 = 36 - 30 = 6.$

Ex. 4. (*i*) Find S, given $p = 2$ and (*ii*) find p, given $S = 18$ for the supply curve

$$S = 5p + 2p^2$$

Ans. (*i*) $S = 5p + 2p^2 = 5 \times 2 + 2 \times 4 = 18$

(*ii*) $18 = 5p + 2p^2$ or $2p^2 + 5p - 18 = 0.$

$$p = \frac{-5 \pm \sqrt{25 - 4 \times 2 \times (-18)}}{2 \times 2} = \frac{-5 \pm \sqrt{169}}{4}$$

$$\therefore \qquad p = \frac{-5+13}{4} = 2 \text{ or } p = \frac{-5-13}{4} = -\frac{9}{2}.$$

We may discard the negative value of p, the price.

Ex. 5. Solve for x :

(*i*) $x^2 = k$ (*ii*) $3x^2 + 7 = 10$

(*iii*) $x^2 - 9x + 14 = 0$ (*iv*) $a^2x^2 = 9$.

Ans. (*i*) $\pm k$ (*ii*) ± 1 (*iii*) 7, 2 (*iv*) $\pm 3/a$.

5.3 SYSTEMS OF SIMULTANEOUS EQUATIONS

A linear equation in two unknown x, y is of the general form

$$ax + by = c$$

A solution of such an equation is an ordered pair (x_1, y_1) such that

$$ax_1 + by_1 = c$$

which means (x_1, y_1) satisfies the given equation.

An equation with two unknowns may have no solution or any number of solutions.

If there are two unknowns, two equations are necessary to arrive at the solution.

Three equations are necessary for three unknowns and n equations for n unknowns.

A simultaneous solution of two or more equations gives a solution of each of the given equations at the same time.

There may be different types of combinations of simultaneous equations including equations of higher degree polynomials.

Solutions of Simultaneous Linear Equations

Solutions can be found for simultaneous equations only if the equations are consistent and independent.

Simultaneous equations should not be

(*i*) inconsistent as in $x + y = 3$, $x + y = 7$

(*ii*) dependent as in $x + y = 3$, $2x + 2y = 6$. The second equation is derived from the first and does not provide any additional information about x, y.

When there are three or more equations we first try any two equations. If the solution is unique and the remaining equations are satisfied with it this may lead to the solution of the system. If the first two equations are found to be coincident, one may be discarded. If the two equations have no common solution the system has no solution.

Ex. 6. In the system $2x + 3y = 5$, $5x - 4y = 1$, $7x - 3y = 4$ the first two equations have the solution (1, 1) which satisfies the third equation and hence is unique.

Ex. 7. In the system $3x + 7y = 20$, $6x + 14y = 40$, $7x - y = 12$, $8x + y = 18$, the first two equations are coincident. We may discard one and proceed. (2, 2) is the solution.

Ex. 8. In the system $2x + 3y = 8$, $4x + 6y = 15$, $7x - y = 5$, $5x + y = 7$, the first two equations are inconsistent so that the system as a whole is inconsistent.

Ex. 9. Consider the nature of solutions of the following systems:

(*i*) $2x + y = 7$, $3x - y = 8$, $4x + 3y = 15$

(*ii*) $x + 3y = 10$, $2x + 6y = 20$, $7x - y = 4$, $5x + y = 8$

(*iii*) $5x - y = 9$, $10x - 2y$, $7x - 2y = 12$.

Ans. (*i*) (3, 1), (*ii*) (1, 3) (*iii*) inconsistent

Ex. 10. To solve the system of equations

$$x + 6y - z = 10$$
$$2x + 3y + 3z = 17$$
$$3x - 3y - 2z = -9$$
$$4x - y + 3z = 11$$

We may take any three equations and see if the solution satisfies the fourth. Thus the first three equations give $x = 1$, $y = 2$, $z = 3$. These values satisfy the last equation, it follows that (1, 2, 3) is the solution of the system.

If, however, the fourth equation were, instead, $2x + 3y - z = 8$ and we substitute the solution here, we get $2.1 + 3.2 - 3 = 8$ which is impossible. In this case the system has no solution.

Ex. 11. Solve the system of equations

(*i*) $x - y + 2z = 6$, $x + y + 3z = 8$, $2z + y + 2z = 7$, $x + 2y + 3z = 9$.

(*ii*) $x - y - 2z = 3$, $2x + y + 3z = 1$, $x + 2y + 3z = -4$, $3x + y - 2z = -3$

(*iii*) $2x + y - 2z = 6$, $-x - 3y + 4z = 2$, $x - 2y + 2z = 5$.

Ans. (*i*) and (*ii*) (1, – 4, 1), (*iii*) No solution

Note the Following:

1. A solution for consistent equations is unique if and only if
   No. of independent equations = No. of variables.
2. There are multiple solutions if and only if
   No. of independent equations < No. of variables.
3. The equations are inconsistent if
   No. of independent equations > No. of variables.

In 1 the statement will not hold if the equations are inconsistent.

In 2 there may not be multiple solutions if the equations are inconsistent.

There may be no solution at all in the situation 3.

The General Case

In a system of m equations in n unknowns

$$a_{11}x_1 + a_{12}x_2 + \ldots + a_{1n}x_n = k_1$$
$$a_{21}x_1 + a_{22}x_2 + \ldots + a_{2n}x_n = k_2$$
$$\ldots \quad \ldots \quad \ldots \quad \ldots \quad \ldots$$
$$a_{m1}x_1 + a_{m2}x_2 + \ldots + a_{mn}x_n = k_m$$

if $m \le n$, the system usually has solutions

if $m > n$, it usually does not have solutions

if $m = n$, the system usually has a unique solution.

If in the above system each $k_i = 0$ then the system is said to be homogeneous. If any k_i is not zero, it is a nonhomogeneous system of equations.

There are several methods of solving simultaneous equations.

(A) Elimination by addition and subtraction.

Ex. 12. $3x + 2y = 7$ ...(1)

$5x - y = 3$ ...(2)

Multiply (2) by 2 and add to (1)

$$\begin{array}{r} 3x + 2y = 7 \\ 10x - 2y = 6 \\ \hline 13x \quad = 13 \end{array} \qquad \therefore\ x = 1$$

Substituting in (1), $2y = 7 - 3 = 4$ $\therefore\ y = 2$

(B) Elimination by substitution

Consider the pair of equations given above.

From (2) substitute $y = 5x - 3$ in (1)

$$3x + 2(5x - 3) = 7$$

$$\therefore\quad 3x + 10x - 6 = 7$$

$$\therefore\quad 13x = 13 \qquad \therefore\ x = 1$$

By substitution in (1) $y = 2$

(C) By using determinants and matrices

These methods are discussed later.

(D) Graphing methods.

If there are two simultaneous equations in two unknowns, we can draw their graphs on the same rectangular coordinate system. The point of intersection of the two curves gives the solution now.

Three Equations

Ex. 13. $x + 6y - z = 10$ ...(1)

$2x + 3y + 3z = 17$ ...(2)

$3x - 3y - 2z = -9$ ...(3)

Adding (2) and (3), $5x + z = 8$ ...(4)

Adding (3) × 2 and (1).

we get $7x - 5z = -8$ ...(5)

Solve (4) and (5) now. We get $x = 1$, $z = 3$

Substituting in (1), $1 + 6y - 3 = 10$ and we get $y = 2$

The solution is $x = 1$, $y = 2$, $z = 3$.

Ex. 14. Solve the following systems of equations:

(*i*)
$$4x + 3y = 9$$
$$2x + 4y + z = 0$$
$$3x - 2y + 4z = 3$$

(*ii*)
$$x + z = 6$$
$$y + z = 7$$
$$y - x = 1.$$

Ans. (*i*) 3, –1, –2 (*ii*) Inconsistent

Square and Triangular Systems of Equations

When the number of equations and the number of unknowns are equal we have a *square system* of equations.

Ex. 13 gives a square system of three equations with three unknowns.

If we rewrite these equations by subtracting $2 \times (1)$ from (2) and $3 \times (1)$ from (3) we get

$$x + 6y - z = 10; \; -9y + 5z = -3; \; -21y + z = -39$$

The second and third equations here form a square system and if we eliminate y from these two equations we have

$$\begin{aligned} x + 6y - z &= 10 \\ y - \tfrac{5}{9}z &= \tfrac{1}{3} \\ z &= 3 \end{aligned}$$

This is the triangular form of equations. By continuing elimination the solution can be written in the *diagonal* form

$$\begin{aligned} x &= 1 \\ y &= 2 \\ z &= 3 \end{aligned}$$

A square system of equations is redundant if it does not satisfy the conditions of consistency and independence. In such a case the solution is not unique.

Rectangular Systems of Equations

When the number of unknowns is greater than the number of equations we have a rectangular system of equations

$$\begin{aligned} x + 6y - z + 3u &= 10 \\ 2x + 3y + 3z - 2u &= 17 \\ 3x - 3y - 2z + u &= -9 \end{aligned}$$

These equations are obtained by adding a fourth variable u to the system in Ex. 13. These equations can be solved by treating u as an independent variable and getting the values of x, y, z in terms of u and the constant quantities. In other words we are considering the system as a square system :

$$\begin{aligned} x + 6y - z &= 10 - 3u \\ 2x + 3y + 3z &= 17 + 2u \\ 3x - 3y - 2z &= -9 - u \end{aligned}$$

Different values assigned to u give different solutions for x, y, z.

We come across such problems in linear programming.

It may be noted here that if we have a system with more equations than unknowns then the system may be redundant or unsolvable unless the extra equations reduce to any one of the others, keeping the number of equations equal to the number of unknowns.

Systems of equations, in practical work, may not always be simple. Often the amount of arithmetic involved is prohibitive and possibilities of errors cannot be ruled out. When the task is laborious it is preferable to use the methods of determinants and matrices given later.

5.4 GRAPHS

If the two lines given by

$$l_1 : a_1x + b_1y + c_1 = 0$$
$$l_2 : a_2x + b_2y + c_2 = 0$$

(*i*) are parallel they have no points in common

(*ii*) intersect, they have exactly one point in common

(*iii*) are coincident, they represent the same line.

The common point, if any, is the solution of the equation system. There are three possibilities for the given system of linear equations.

In case (*i*) the system has no solution and there is no value (x, y) which satisfies both equations and the system is inconsistent.

In case (*ii*), a *unique* (x, y) satisfies both equations and (x, y) is the *solution* of the system.

In case (*iii*), there is an infinity of solutions.

Ex. 16. Consider the following systems of equations for the nature of their solutions and draw their graphs.

(*i*) $2x + 3y = 5$, $5x - 4y = 1$ (*ii*) $3x + 7y = 20$, $6x + 14y = 40$ (*iii*) $2x + 3y = 8$, $4x + 6y = 15$

Ans. (*i*) unique (1, 1);
(*ii*) coincident, infinite, solutions;
(*iii*) parallel, no solution

Three or More Lines

Generally, three lines (*a*) meet at three different points forming a triangle. Sometimes they may (*b*) intersect at a single point; (*c*) be parallel or (*d*) be coincident.

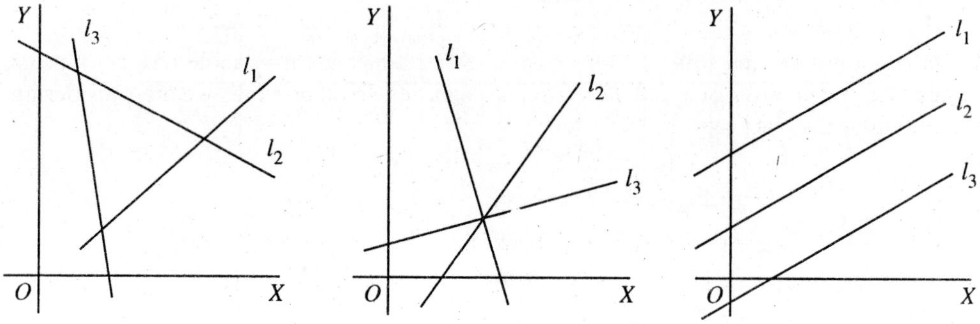

(*a*) (*b*) (*c*)

5.5 LINEAR EQUATIONS

The graph of a linear equation like

$$y = a + bx$$

requires only two points to be plotted.

The dependence on output x of cost C can be expressed as a linear relation

$$C = 400 + 0.5\,x$$

Any two points satisfying this equation will suffice to draw the line. Preferably, we may take the points

$x = 0$, $C = 400$ and $x = 200$, $C = 500$

The graph shows a fixed cost of 400 and with every additional quantity x produced, the increased cost is $0.5\,x$.

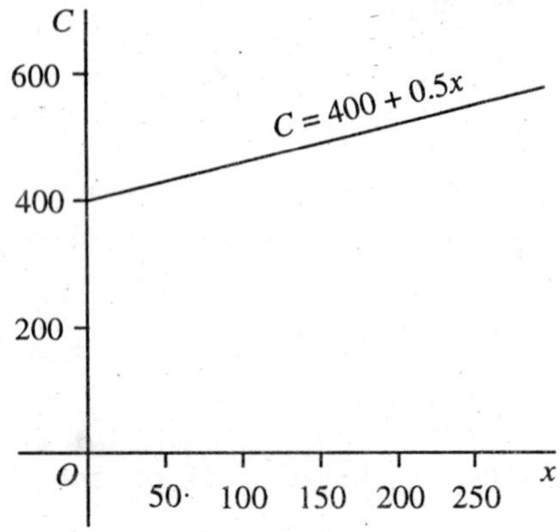

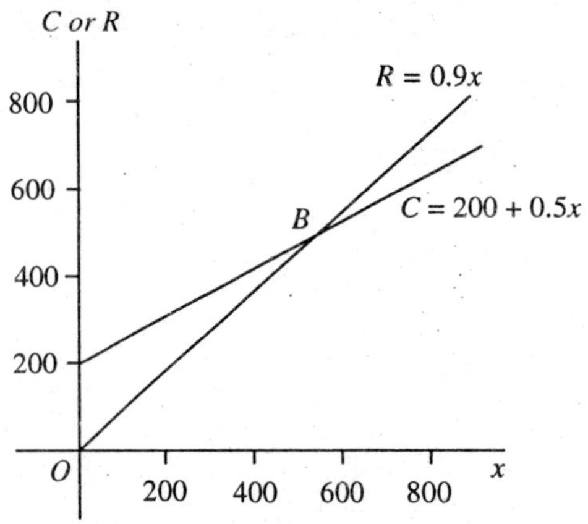

Ex. 17. The revenue equation $R = 0.9\,x$ can be plotted along with the cost equation $C = 200 + 0.5\,x$ on the same graph.

The intersection point B is the break even point. The shaded portion below B is loss ($R < C$) and the unshaded portion above B shows profit ($R > C$).

5.6 THE EXPONENTIAL

We come across exponential equations like $y = ab^x$ in studies on growth or decay. Several values of x should be plotted in such cases to get a fair idea of the trend.

If $y = 3 \times 2^x$ we first prepare the table

x	0	1	2	3	–3	–2	–1
$y = 3 \times 2^x$	3	6	12	24	3/8	3/4	3/2

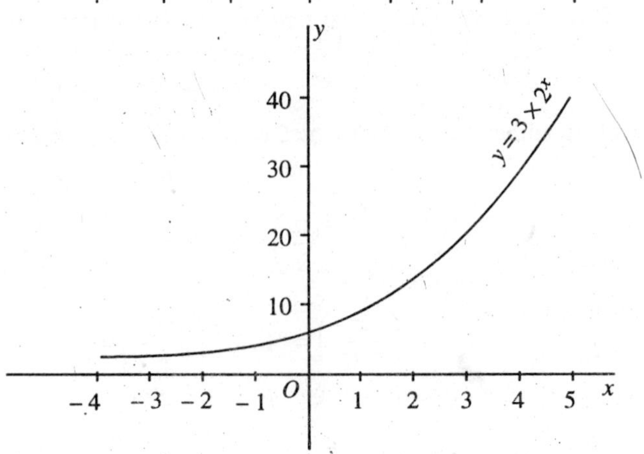

We can similarly plot $y = 3 \times 2^{-x}$

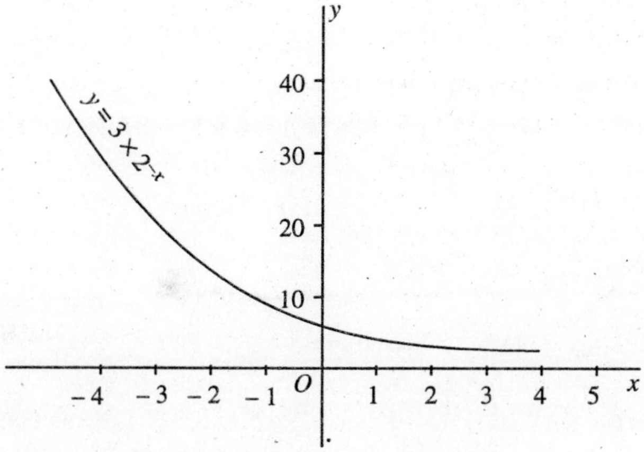

The exponential equation $y = ab^x$ can be linearised by a logarithmic transformation by writing $\ln y = \ln a + x \ln b$.

$$y = 3 \times 2^x \text{ can be written}$$

$$\ln y = \ln 3 + x \ln 2$$

and gives rise to a linear curve if $\ln y$ is plotted against x.

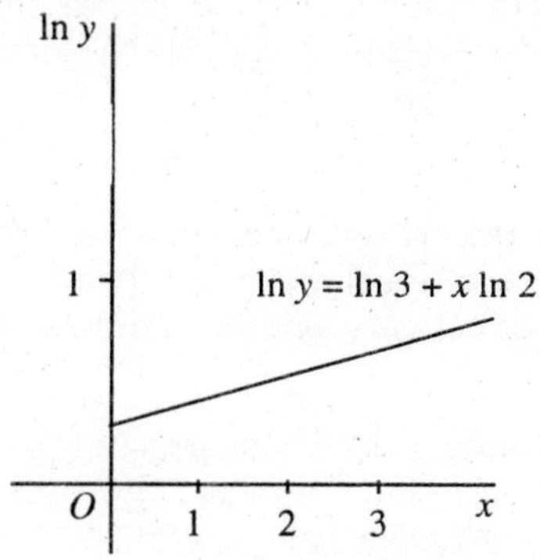

x	0	1	2	3
$\ln y = \ln 3 + x \ln 2$	0.497	0.778	1.079	1.380

This is a case of *semi-logarithmic* transformation since it involves the logarithm of one variable only. We may have logarithms of both the variables appearing in the transformation in some cases. Thus, on taking logarithms,

$$y = ax^b \text{ becomes}$$

$$\ln y = \ln a + b \ln x$$

or

$$Y = A + bX$$

where

$$Y = \ln y,\ X = \ln x,\ A = \ln a$$

In all these transformations, generally, natural logarithms are employed.

The equation $y = e^{ax + b}$ can be linearised by writing.

$$\ln y = ax + b$$

The equation $y = \dfrac{a}{x + b}$ can be linearised by first writing

$$\frac{1}{y} = \frac{x + b}{a} = \frac{x}{a} + \frac{b}{a}$$

and then substituting $Y = \dfrac{1}{y},\ A = \dfrac{1}{a},\ B = \dfrac{b}{a}$ so that $Y = Ax + B$

5.7 GRAPHS OF QUADRATIC EQUATIONS

The graph of a quadratic function is a parabola with a vertex and a turning point. In the quadratic

$$y = ax^2 + bx + c$$

If $a > 0$ the graph is concave upwards. If $a < 0$ it is concave downwards.

$a > 0$ $a < 0$

If the graph cuts the x-axis the roots are real, unequal. For the concave-up case the roots are OA_1, OB_1; for the concave-down case, the roots are OA_2, OB_2. The Minimum at L_1 or the Maximum at L_2 is the same for both. It is $\frac{-b}{2a}$.

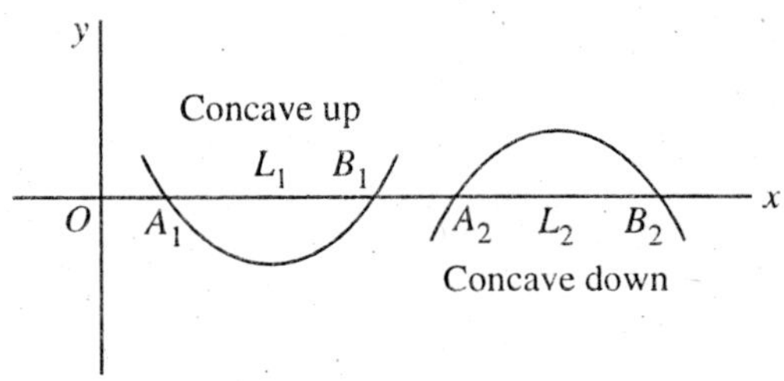

If the curve is tangent to the x-axis the roots are real and equal. If the curve does not cut the x-axis anywhere the roots are imaginary.

Some relationships, linear or otherwise, often represented by means of graphs are given below:

Relationship $y = f(x)$	x	y
Demand curve for a commodity	Demand	Price
Cost curve of a firm	Output	Cost
Supply curve of a commodity	Output	Price
Revenue from sales	Sales	Revenue
Average cost curve of a firm	Output	Average Cost
Consumption function	Income	Consumption
Profit function	Output	Profit
Marginal efficiency of investment	Interest rate	Investment

Ex. 18. A person with Rs. 50,000 is desirous of getting an average rate of return of 15 per cent from three alternative investments, with rates of return 15, 10 and 18 per cent. He wants to put in 40 per cent of his capital in the high return investment. Determine the strategy to meet this requirement.

Ans. 18,000 at 15%, 12,000 at 10%, 20,000 at 18%

Ex. 19. To produce a certain desired milk food we require 6 units of nutrient A, 10 units of B and 8 units of C. A jar of milk food type I contains 1 unit of A, 3 units of B and 4 units of C.

A jar of type II contains 3 units each of A, B, C.

A jar of type III contains 2 units of A and 5 units of B.

Find out how much of each type of milk food should be combined to produce the desired variety?

Suppose we require for the purpose x units of type I, y units of type II and z units of type III.

Then
$$x + y + z = 24$$

Also
$$\frac{x}{8} + \frac{y}{3} + \frac{2z}{7} = 6 \qquad \text{Nutrient A}$$
$$\frac{3x}{8} + \frac{y}{3} + \frac{5z}{7} = 10 \qquad \text{Nutrient B}$$
$$\frac{x}{2} + \frac{y}{3} = 8 \qquad \text{Nutrient C}$$

Solve the equations as usual. Alternatively, the augmented matrix may be written

$$\begin{bmatrix} 1 & 1 & 1 & 24 \\ 1/8 & 1/3 & 2/7 & 6 \\ 3/8 & 1/3 & 5/7 & 10 \\ 1/2 & 1/3 & 0 & 8 \end{bmatrix} \qquad \text{(See chapter 7)}$$

Using the elimination process we find
$$x = 112/13,\ y = 114/13,\ z = 56/13.$$

Ex. 20. Find the linear relation between x units of a product and its total cost C, given that for 500 units, $C = 600$ and for 1000 units, $C = 9000$. Find the fixed cost (F) and variable costs (V) and the number of units that should be produced when profit is (a) 0 (b) 1000 (c) – 600 at the selling price of 10 per units.

Let
$$C = F + V = F + bx$$

Since
$$\left.\begin{aligned} 6000 &= F + 500\,b \\ 9000 &= F + 1000\,b \end{aligned}\right\} \quad \begin{aligned} F &= 3000 \\ b &= 6 \end{aligned}$$

The cost line is $C = 3000 + 6\,x$

where the slope $b = 6$ is the marginal cost.

(a) When $p = 10$, revenue $R = 10\,x$

For zero profit $C = R$.

$$3000 + 6\,x = 10\,x \quad \therefore\ 4x = 3000 \text{ or } x = 750$$

(b) For profit 1000

$$3000 + 6x - 10x = 1000 \text{ or } 4x = 2000 \text{ or } x = 500$$

(c) For profit – 600

$$3000 + 6x - 10x = -600 \text{ or } 4x = 3600 \text{ or } x = 900$$

Ex. 21. 200 units of a product cost Rs. 800 and 400 units cost Rs. 1200. The cost curve is a straight line. Find the fixed cost and the cost of producing 800 units?

Ans. $\left.\begin{aligned}800 &= F + 200x\\ 1200 &= F + 400x\end{aligned}\right\}$ $F = 400,\ C = 400 + 2x$; Cost of 800 units $= 2000$

Ex. 22. For 100 units of a product the cost is 1200 and for 400 units, the cost is 2700. Find the number of units that could result in zero profit, 300 profit and – 300 profit if the selling price is 7 per unit. **Ans.** 350, 500, 200

Ex. 23. National Income Analysis

Let Y, C stand for the endogenous variables national income and consumption. Then the Keynesian national income model is

$$Y = C + I_0 + G_0 \qquad \text{(National Income identity)}$$

$$C = a + bY,\ a > 0,\ 0 < b < 1 \qquad \text{(Consumption function)}$$

Here I_0, G_0 are exogenously determined investment and government expenditures and the parameters a, b stand respectively for autonomous consumption and marginal propensity to consume.

These are independent, consistent relations which can be solved simultaneously to get the equilibrium values of the endogeneous variables :

Income $\quad \overline{Y} = \dfrac{a + I_0 + G_0}{1 - b}$

Consumption $\quad \overline{C} = \dfrac{a + b(I_0 + G_0)}{1 - b}$

The relations can also be written

$$Y - C = I_0 + G_0$$

$$-bY + C = a$$

Ex. 24. In a closed economy, the consumption and investment functions are given by

$$C = 15 + 0.9Y,\ I = 20 + 0.005Y$$

Find the equilibrium values of C, I, Y where Y is total income and the government expenditure is given by (*a*) $G = 25$ (*b*) $G = 0$

Assume that the total national expenditure is given by $E = C + I + G$.

(*a*) For equilibrium $E \equiv Y = C + I + G$

$\therefore \quad Y = (15 + 0.9Y) + (20 + 0.05Y) + 25$

$\therefore \quad Y(1 - 0.95) = 25 + 20 + 15 = 60$

$\therefore \quad$ Equilibrium $Y = \dfrac{60}{0.005} = 1200$

$\therefore \quad C = 15 + 0.9Y = 1095$

$\quad I = 20 + 0.05Y = 80$

(*b*) When $G = 0$, $Y = (15 + 0.9Y) + (20 + 0.05Y)$

$\therefore \quad Y(1 - 0.95) = 35$

$\therefore \quad Y = \dfrac{35}{0.05} = 700$

$\left.\begin{aligned}C &= 645\\ I &= 55\end{aligned}\right\}$ are the new values if $G = 0$

5.8 BREAKEVEN POINT

Breakeven analysis is concerned with the profitability of a firm. It helps one in identifying the level of operation or output at which there is zero profit. This level of operation is the breakeven (*BE*) point. Before this point there are losses, after this point profits start coming in.

Breakeven analysis is useful as a planning tool to commence or expand a business or introduce a new product.

The analysis makes use of revenue and cost functions which can have any shape but here they are assumed to be *linear.*

A linear cost function implies that average variable cost (*AVC*) is constant and total cost (*C*) is a function of output *x i.e.*, $C = C(x)$ which is a linear function.

A linear revenue function $R(x)$ assumes that selling price *p* is constant. It is also assumed that *p* is greater than *AVC*.

The breakeven (*BE*) point can be determined in terms of volume of output, sales or percentage of production capacity. The breakeven value is obtained by equating $R(x)$ and $C(x)$ implying that

$$\text{Profit } \pi(x) = R(x) - C(x) = 0$$

at the breakeven point.

Ex. 25. To produce good quality plastic buckets, a firm estimates that its fixed cost comes to Rs. 250,000 and average variable cost Rs. 20. What should be the number of buckets produced to break even if each bucket is sold at Rs. 30? If 50,000 buckets are likely to be sold with the present equipment what is the expected final profit?

$$R(x) = 30x \text{ for } x \text{ buckets sold}$$

$$C(x) = 250{,}000 + 20x$$

$$R(x) = C(x) \text{ or } 30x = 250{,}000 + 20x$$

The solution of this equation gives the BE value $x_B = 25{,}000$

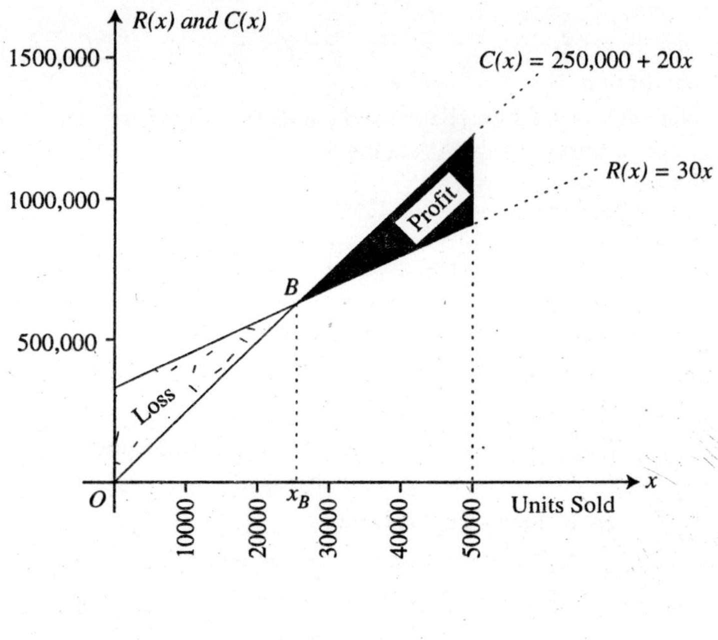

It can be seen that at the BE point x_B,

$$R(x) = 30x_B = 30 \times 25{,}000 = 750{,}000$$

$$C(x) = 250{,}000 + 20x_B = 250{,}000 \div 500{,}000 = 750{,}000$$

At x_B, Profit $\pi\ (x) = R(x) - C(x) = 0$

If expected sales $x = 50{,}000$ buckets,

$$\begin{aligned}\pi &= R(50{,}000) - C(50{,}000)\\ &= [30 \times 50{,}000] - [250{,}000 + 20 \times 50{,}000]\\ &= 1500{,}000 - 1250{,}000\\ &= 250{,}000.\end{aligned}$$

Ex. 26. A toothpaste company A can manufacture toothpaste (a) in its own factory at a cost of Rs. 1.25 per unit or (b) can have it manufactured by B who charges Rs. 3000 for a year initially and Re. 0.95 per unit ordered or (c) can get them from C who is prepared to handle the job for an initial annual payment of Rs. 15000 and Re. 0.65 per unit produced. Compare the alternatives.

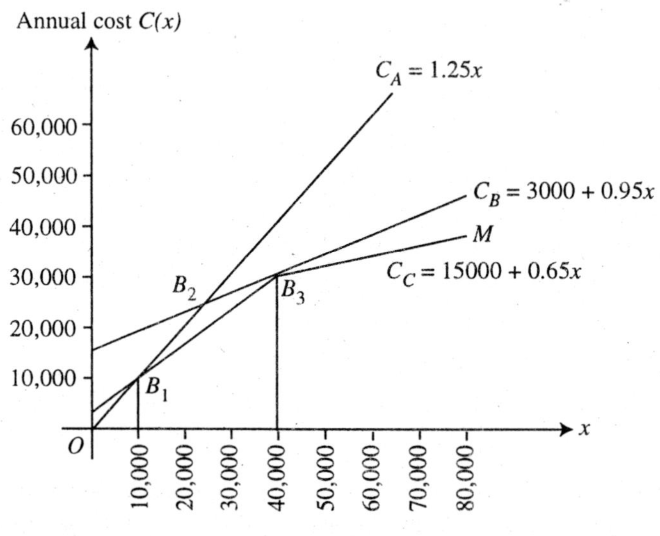

We can obtain any of the three breakeven points by solving for x, the corresponding cost equations: For instance, compare B and C we find

$$3000 + 0.95\,x = 15000 + 0.65\,x$$

or $$x = 40{,}000$$

If production exceeds 40,000 units, option C is less costly. For less than 40,000, B is better. Similarly, for less than 10,000 units, A is least costly. For units between 10,000 and 40,000 B is to be preferred.

The line OB_1B_2M is the least cost option at any level of production.

Note 1. The BE point changes when

(a) The price changes. The BE point shifts to the left when price increases. The BE point shifts to the right when price decreases.

(b) The fixed cost increases or decreases.

(c) The average variable cost increases or decreases.

Note 2. As long as the price p exceeds v, the average variable cost (AVC), the sale of each unit contributes to the profit margin $= p - v$.

At lower levels of output total profit margin for all units sold is less than total fixed cost. This is a loss situation which vanishes when the BE sales take place. After the BE point, when fixed costs (FC) are recovered, profits start coming. We can also calculate the BE point by using the formula

$$\text{BE output level} = \frac{FC}{p - v}$$

Note 3. Two or more alternatives may be compared to discover potential advantages or disadvantages. Similarly, two or more products may be analysed using the BE technique.

Ex. 27. The AVC of producing an instrument comes to Rs. 20 per unit and fixed cost Rs. 90,000. Find the breakeven point if the instrument sells for Rs. 20 per unit.

Ans. 7500 units

Ex. 28. An automobile factory can buy switches at Rs. 10 each. (Alt. A). It can also manufacture them at a cost of Rs. 8 each after investing in a machine costing Rs. 80,000. (Alt. B). Alternatively it can produce the same at Rs. 5 each if it buys a superior machine for Rs. 120,000. (Alt. C) Graph the cost functions and compare the alternatives.

Ans. If $x < 24000$ use Alt. A

If $x > 24000$ use Alt. C

Ex. 29. Product Mix Case. A firm produces three items A, B, C in the ratio 3 : 2 : 1 *i.e.*, 1 unit of the product mix consists of 3 units of A, 2 of B, 1 of C. The prices per unit of A, B, C are Rs. 40, Rs. 30 and Rs. 55 and the AVC of A, B, C are Rs. 30, Rs. 21 and Rs. 43 respectively. Find the BE point if FC = Rs. 240,000.

Profit margin for A Rs. 10, B Rs. 9, C Rs. 12

Profit margin per unit of product mix is

$$3 \times 10 + 2 \times 9 + 1 \times 12 = 60$$

$$\pi(x) = 60x - 240{,}000 = 0 \text{ for BE}$$

$x = 4000$ product mix units for BE *i.e.*, A 12,000 units, B 8000 units, C 4000 units.

Ex. 30. Sensitivity Analysis. This is used to assess how much the solution of a problem might change if there are changes in some of the parameters.

(*a*) A firm requires 20,000 ball pens which it can buy from outside at Rs. 2 a piece or assemble itself at a cost of Rs. 1.25 a piece after investing in a small machine costing Rs. 12,500. How many ball pens would make the costs of the two alternative equal? (This is the BE point and it can be found to be $x = 16{,}666.67$)

Ans. For producing 20,000 ball pens, self assembling will cost Rs. 37,500; buying outside will cost Rs. 40,000

(*b*) In (*a*) what should the self assembling cost per unit be to make the two alternatives equally costly? **Ans.** Rs. 1.375 per unit

(*c*) If outside cost (at Rs. 2) fluctuates by 20 per cent up or down, recalculate the BE point when the cost is 20 per cent higher. Also under the same condition and the additional information that self assembling variable cost may go up by 30 per cent find the combined effect of these possibilities.

5.9 REVENUE AND PROFIT FUNCTIONS

The revenue function is usually taken to be linear

$$R = px \qquad (p \text{ price, } x \text{ quantity})$$

If there are n products and $p_1, p_2,..., p_n$ are the prices and $x_1, x_2,..., x_n$ units of these products are sold then

$$R = p_1x_1 + p_2x_2 +...+ p_n x_n$$

Where both R, C are linear functions of the same variable x, the profit function is also linear

$$\pi (x) = R (x) - C (x)$$

Ex. 31. A firm has three plants with fixed costs of Rs. 2500, 3750, 1250 respectively. It produces three items 1, 2, 3 in these plants, their quantities being x_1, x_2, x_3. The average costs a_1, a_2, a_3 are Rs. 9, 11, 7.50 and the average revenues r_1, r_2, r_3 are Rs. 13, 16.50, 12 respectively. Find the profit function if the firm has an annual fixed cost of Rs. 12,500 of its own.

$$\begin{aligned}\pi(x) &= R(x_1, x_2, x_3) - C(x_1, x_2, x_3)\\ &= (r_1x_1 + r_2x_2 + r_3x_3) - \{2500 + c_1x_1 + 3750 + c_2x_2 + 1250 + c_3x_3\} - 12500\\ &= 13x_1 + 16.5x_2 + 12x_3 - \{2500 + 9x_1 + 3750 + 11x_2 + 1250 + 7.5x_3) - 12500\\ &= 4x_1 + 5.5x_2 + 4.5x_3 - 20000\end{aligned}$$

Ans.

5.10 LINEAR DEPRECIATION

There are many methods of estimating depreciation. A simple method is that of straight line depreciation. In the linear depreciation function, the rate of depreciation is assumed constant over time. It can be written

$$v = C - rt$$

where v is the value of the asset at the end of t years, r being the rate of depreciation and C is the purchase price.

If the purchase price is Rs. 20000 and the depreciation rate ten per cent *i.e.*, Rs. 2000 per year then at the end of t years the book value is

$$v = 20000 - 2000\ t$$

This will be zero at the end of 10 years.

If the asset, after having been used for 10 years has a salvage or resale value of say, Rs. 4000. The annual depreciation rate is

$$\frac{C - 4000}{\text{asset life in years}} = \frac{16000}{10} = \text{Rs. } 1600.$$

Ex. 32. A machine costing Rs. 4000 will be fully depreciated after 20 years. Represent this by a linear depreciation function. What happens if there is salvage value of Rs. 1000?

Ans. $v = 4000 - 20\ t$; With salvage value,

$$\text{the depreciation rate} = \frac{4000 - 1000}{20} = \text{Rs. } 150.$$

5.11 MARKET EQUILIBRIUM FOR ONE COMMODITY

In a market there is a tendency for the price to reach an *equilibrium* level where the demand for the commodity equals its supply. A high price is forced down by low demand. A low price is pulled up by low supply.

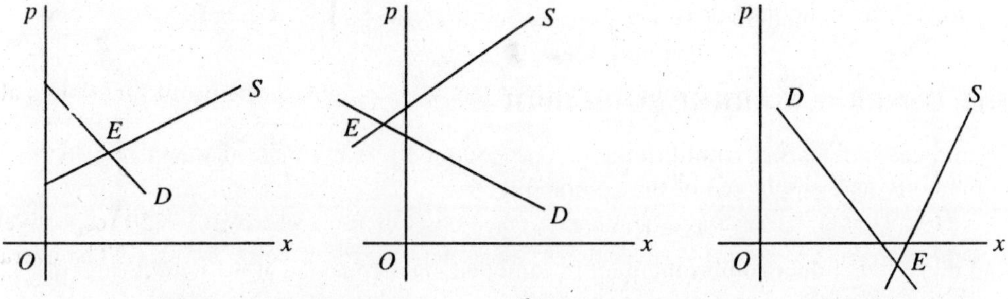

Market equilibrium corresponds to the point of intersection of the demand curve and the supply curve on a graph paper. It may be obtained by solving the demand and supply equations. Equilibrium price and equilibrium quantities are obtained simultaneously and are meaningful only if the point of intersection E lies in the first quadrant as in (a) and not as in (b) or (c) *i.e.*, price p and quantity x both are greater than zero.

Ex. 33. Find the equilibrium price (p) and quantity (x) if the demand law is $2p = 14 - x$ and the supply law is $12p = 14 + x$.

Ans. Either solve the two equations to get the equilibrium $p = 2$ and equilibrium $x = 10$ or note the equilibrium point (of intersection) by drawing the lines.

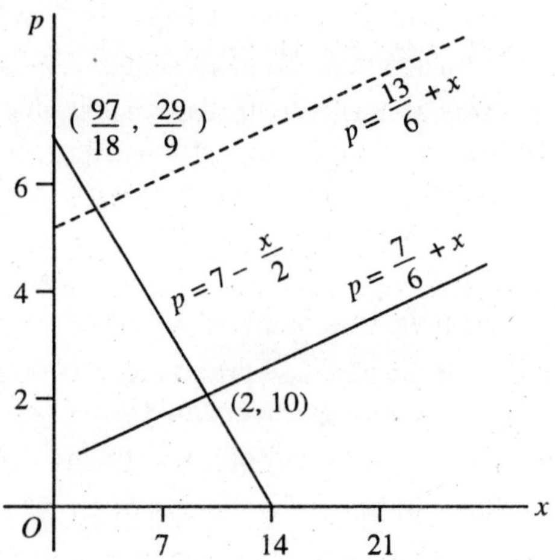

Ex. 34. The two equations of Ex. 33 can be written as

$$\text{demand} : p = 7 - \frac{x}{2}, \quad \text{supply} : p = \frac{7}{6} + x$$

If there is a tax of 1 per unit sold, the new equations become

$$\text{demand} : p = 7 - \frac{x}{2} \text{ (same)},$$

$$\text{supply} : p = \frac{7}{6} + x + 1 \text{ } i.e., p = \frac{13}{6} + x.$$

From the graph or by solving the new equations

$$p = \frac{97}{18}, \quad x = \frac{29}{9}$$

Ex. 35. If instead of a tax of 1, a subsidy of $\frac{1}{4}$ per unit is given in the above case what will be the result?

Ex. 36. What tax do you suggest in the following case so that price rises by 1.5 units.

demand law : $p = 5 - \frac{x}{2}$, supply law : $6\,p = 6 + x$ **Ans.** 2

Ex. 37. Show that if a tax k per unit is imposed when the demand and supply laws are $p = \frac{27}{8} - \frac{2}{3}x$ and $3p = 2x + 2$ the new equilibrium price is $\frac{2}{3} + \frac{9}{13}k$.

5.12 GENERAL MARKET EQUILIBRIUM

IN the case of market equilibrium for one commodity we equate demand (x_d) for the commodity and supply (x_s) of the commodity,

$$x_d = x_s$$

and thus predict the equilibrium quantity supplied and demanded at the equilibrium price.

Usually the price and quantity of a commodity influence and are influenced by the prices and quantities of other commodities. In cases like these, the equilibrium prices and quantities can be obtained by equating the quantity supplied and the quantity demanded of each commodity separately.

There will be two equations for two commodities.

$$\text{Commodity 1} \ : \ x_{d1} = x_{s1}$$

$$\text{Commodity 2} \ : \ x_{d2} = x_{s2}$$

The two simultaneous equations can be solved.

If there are n commodities the general market equilibrium or *Walrasian equilibrium* will involve n equations, there being n demand functions and n supply functions

$$x_{di} = x_{si} \qquad i = 1, 2, \ldots\ldots\ldots, n.$$

If there are n simultaneous equations in n unknowns we can solve them.

We shall consider only linear equations.

Ex. 38. Let p_1, p_2 be the prices of two commodities whose demand and supply functions are as follows:

$$x_{d_1} = 20 - p_1 + p_2 \qquad x_{s_1} = 16 - p_1 + 3p_2$$

$$x_{d_2} = 36 - 5p_1 + 4p_2 \qquad x_{s_2} = 10 - p_1 + 9p_2$$

By equating the demand and supply for each commodity,

$$x_{d_1} = x_{s_1} \text{ gives} \qquad 20 - p_1 + p_2 = 16 - p_1 + 3p_2$$

$$x_{d_2} = x_{s_2} \text{ gives} \qquad 36 - 5p_1 + 4p_2 = 10 - p_1 + 9p_2$$

We solve the two simultaneous equations to get equilibrium prices

$$p_1 = 4, \qquad p_2 = 4$$

Equilibrium quantities are obtained by substituting these equilibrium prices in the original demand and supply equations. We have

$$x_{d_1} = x_{s_1} = 18, \qquad x_{d_2} = x_{s_2} = 24.$$

Ex. 39. Find equilibrium prices and quantities for the two commodity market model

(*i*) $x_{d_1} = 6 - p_1 - p_2$ $\qquad x_{s_1} = -2 + p_2$

$x_{d_2} = 10 - p_1 - 2p_2$ $\qquad x_{s_2} = -3 + p_1 + p_2$

Ans. $p_1 = 2, p_2 = 3, x_1 = 1, x_2 = 2$

(*ii*) $x_{d_1} = a_0 + a_1p_1 + a_2p_2$ $\qquad x_{s_1} = b_0 + b_1p_1 + b_2p_2$

$x_{d_2} = g_0 + g_1p_1 + g_2p_2$ $\qquad x_{s_2} = h_0 + h_1p_1 + h_2p_2$

Ans. $$p_1 = \frac{(a_2 - b_2)(g_0 - h_0) - (a_0 - b_0)(g_2 - h_2)}{(a_1 - b_1)(g_2 - h_2) - (a_2 - b_2)(g_1 - h_1)}$$

$$p_2 = \frac{(a_0 - b_0)(g_1 - h_1) - (a_1 - b_1)(g_0 - h_0)}{(a_1 - b_1)(g_2 - h_2) - (a_2 - b_2)}$$

Ex. 40. A commodity is produced by using 3 units of labour and 2 units of capital. The total cost comes to 62. If the commodity is produced by using 4 units of labour and 1 unit of capital, the cost comes to 56. What is the cost per unit of labour and capital?

Ans. $3L + 2K = 62$ $\qquad$ where L, K = cost per unit

$4L + K = 56$ $\qquad$ of labour, capital respectively

The solution is $L = 10, K = 16$

Ex. 41. A man's income from interest and wages is 500. He doubles his investments and also gets an increase of 50% in wages and his income increases to Rs. 800. What was his original income separately in terms of interest (i) and wages (w)?

Ans. $i + w = 500$

$2i + \frac{3}{2}w = 800$

The solution is $i = 100, w = 400$

Ex. 42. Get the values of equilibrium price and quantity by solving the equations representing the supply and demand schedules:

Supply schedule : $\qquad x = 500 + 20\,p$

Demand schedule : $\qquad x = -\,100 + 40\,p$

Show that the equilibrium price $p = 30$

and the equilibrium output $\qquad x = 1100$.

Ex. 43. If there are two commodities with prices p_1, p_2, demands D_1, D_2 and supplies S_1, S_2 and we have the demand and supply schedules

$D_1 = 10 - p_1 + p_2$ $\qquad S_1 = 6 + p_1 + 2p_2$

$D_2 = 12 + 2p_1 - p_2$ $\qquad S_2 = 19 + 3p_1 - 5p_2$

then to solve these equations we write $D_1 = S_1, D_2 = S_2$

$D_1 = S_1$ gives $10 - p_1 + p_2 = 6 + p_1 + p_2$

i.e., $\qquad 2p_1 + p_2 = 4 \qquad \ldots(1)$

Similarly $D_2 = S_2$ gives $p_1 - 4p_2 = 7 \qquad \ldots(2)$

and we get $p_1 = 1, p_2 = 2$ from (1), (2). We have the equilibrium values by substitution

$D_1 = S_1 = 10 - 1 + 2 = 11$

$D_2 = S_2 = 12 + 2 - 2 = 12.$

5.13 GRAPHS OF LINEAR INEQUALITIES

In mathematical economics, game theory and linear programming, inequalities play an important role. They are useful when the solution of a problem is required to satisfy certain minimum or maximum requirements.

Inequalities are *relations*, not functions. They can be solved and graphed like equations.

Ex. 44. The graph of $x \leq 3$ is the set of all points

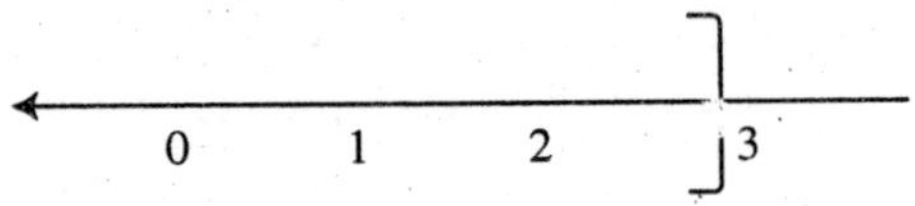

to the left of and including 3. This is the graph of $(-\infty, 3]$.

Ex. 45. The graph of $x \leq 3$ and $x > -2$, *i.e.*, the graph of $-2 < x \leq 3$ is the set of all points to the left of and including 3 and to the right of but not including -2.

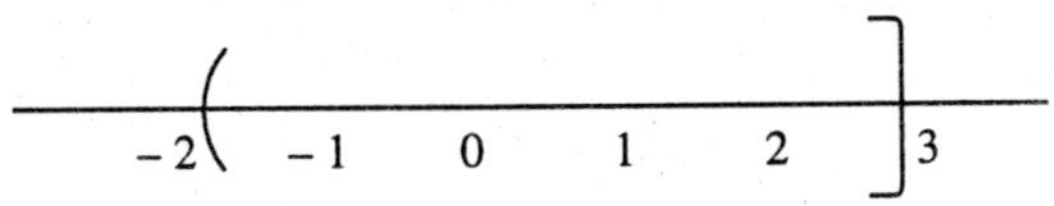

This is the graph of the interval $(-2 < x \leq 3)$ or $(-2, 3]$

Similarly we have the following graphs.

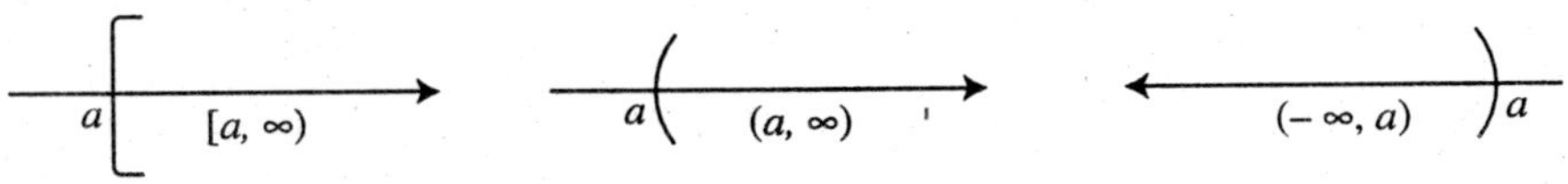

Ex. 46. Draw the graphs of

$$(2 \leq x \leq 4), (-3 < x \leq 2), (-\infty, 0), (0, \infty)$$

Ex. 47. The graph of $|x| < 3$ is the same as the graph of $-3 < x < 3$. Draw it.

Ex. 48. Draw the graph of $|3x - 2| < 8$.

Ex. 49. Consider a pair of inequalities

$$x + 3 > 0, x - 2 < 0$$

Their simultaneous solution is given by

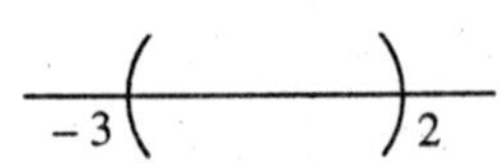

the open interval $(-3 < x < 2)$.

Ex. 50. Find the simultaneous solutions for the following inequalities.

(*i*) $2x - 10 \leq 0,$ $\quad -2x + 4 \leq 0$

(*ii*) $x > 0,$ $\quad x - 3 \leq 0$

Linear Inequalities in two Variables

To graph a linear inequality in two variables we change the inequality sign to an equality sign and graph the equation. This line will give the boundary below or above which the set of points corresponding to the inequality will lie.

Ex. 51. (*i*) The graph of the inequality $x + 2y > 6$ is the shaded portion above the line $x + 2y = 6$. This does not include the points on the line. The graph of $x + 2y \geq 6$ includes the points on the line.

(*ii*) The graph of $x + 2y < 6$ is below the line $x + 2y = 6$ and does not include the points on the line. The graph of $x + 2y \leq 6$ would include the points on the line.

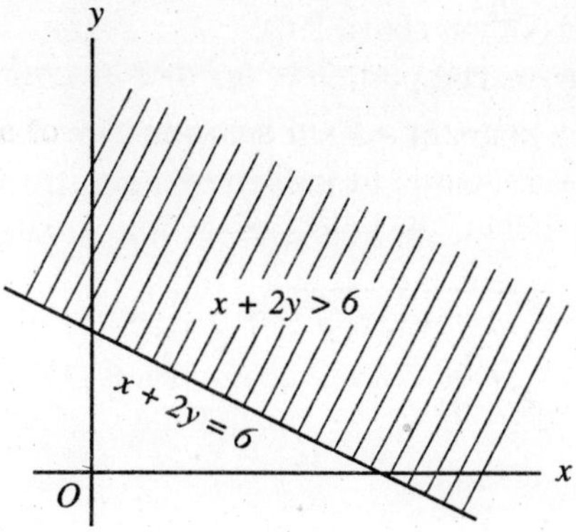

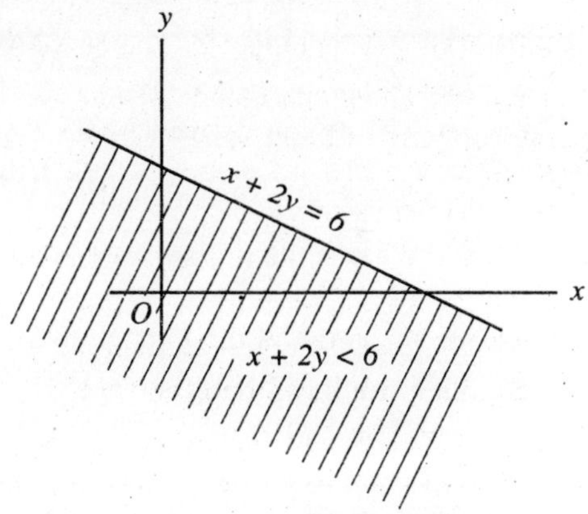

Ex. 52. Graph : (*i*) $x + y > 1$ (*ii*) $x - y > 0$, (*iii*) $y \leq x$
(*iv*) $x < y$, (*v*) $2x - y > 1$, (*vi*) $2x - y \leq 2$.

5.14 SYSTEMS OF LINEAR INEQUALITIES

Consistent simultaneous linear equations have unique solutions but a pair of linear inequalities does not have a unique solution.

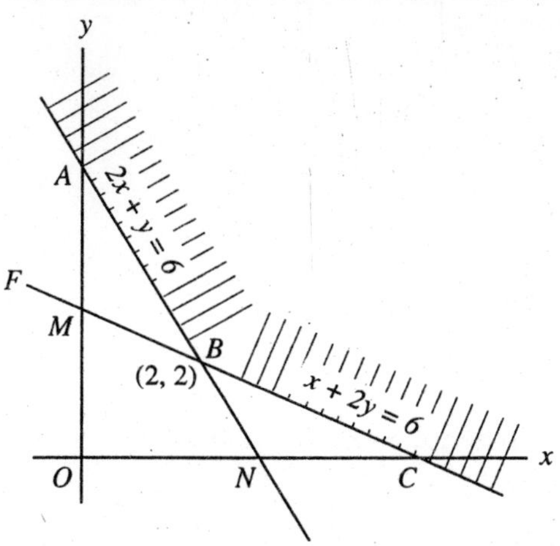

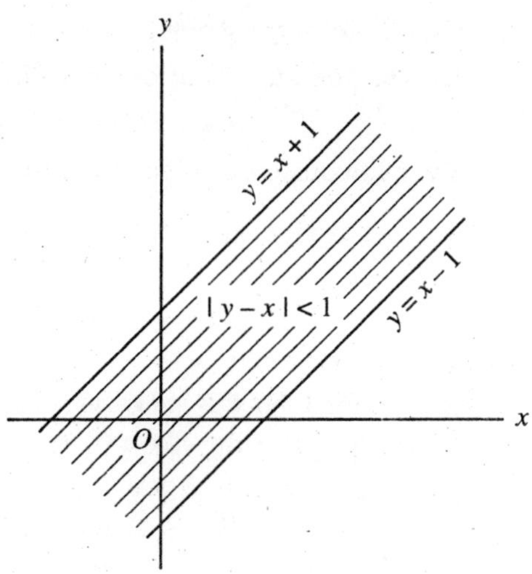

The intersection of the graphs of two or more inequalities is a subset of the plane. It may be empty, as in the case of parallel lines.

If there are more than two variables, graphical representation becomes difficult.

Ex. 53. The region defined by the inequalities $2x + y > 6$, $x + 2y > 6$ is the shaded region to the right of *ABC*. If $2x + y \geq 6$, $x + 2y \geq 6$, the points on the lines are also included in the region. (See the graph above)

Ex. 54. The region defined by $|y - x| < 1$ can be found by writing $y < x + 1$ and $y > x - 1$. It is the set of points between the lines $y = x + 1$, $y = x - 1$.

The region defined by $y > x + 1$ and $y < x - 1$ is an empty set.

Ex. 55. What is the region defined by the inequalities

$2x + y > 6$, $x + 2y > 6$, $x > 0$, $y > 0$? (Graph above left)

Ans. It is the shaded portion in the *first quadrant* ($\because x > 0, y > 0$) to the right of *yABCx*

Ex. 56. The region defined by $2x + y < 6$, $x + 2y < 6$ is to the left of *FBN*. If the inequalities are $2x + y \leq 6$ and $x + 2y < 6$ the region include the points on the line *AB* also but not the points on *BC*.

Ex. 57. What is the region defined by $2x + y < 6$, $x + 2y < 6$, $x > 0$, $y > 0$?

Ans. It is the convex quadrilateral *OMBN*

Ex. 58. Find the region defined by $x + y > 1$, $x - y > 0$.

Ex. 59. Find the solution set of

(*i*) $|x - y| > 2$,

(*ii*) $x + y > 1$, $x - y > 0$

(*iii*) $(x + y) \geq 1$, $x - y < 0$, $x + 2y < 1$

(*iv*) $2x - y > 1$, $x + y \leq 4$, $y > 1$

(*v*) $y \geq 0$, $x < 3$, $x + y \leq 6$

(*vi*) $|x + 2| \geq 3$, $x + y \leq 4$

(*vii*) $x - 3y \geq 6$, $2x - 6y \leq 8$ (This is an empty set)

(*viii*) $x < 2$, $x > -3$.

Ex. 60. Find the solution set determined by

$$y - 2x - 4 < 0$$
$$2y - x - 4 < 0$$
$$6x + 5y - 30 < 0$$
$$y - 4 < 0.$$

Ans. The shaded quadrilateral

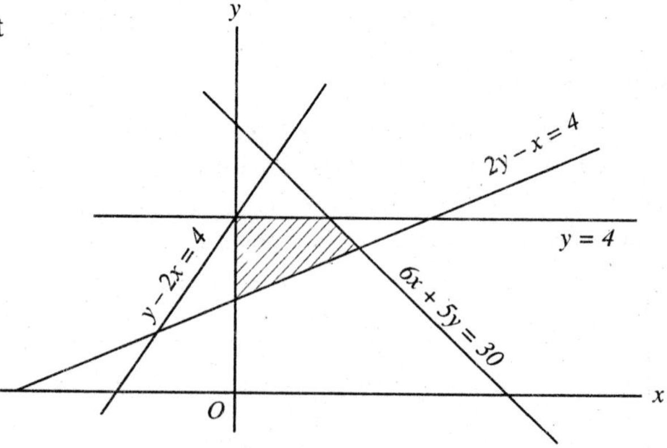

Ex. 61. Graph the following and find the solution set

(*i*) $2x + 3y - 24 < 0$, $3x + y < 0$, $x - 2y + 2 < 0$,
$3x + y - 8 > 0$,

(*ii*) $x \geq -3$, $y \leq 2$, $x < 3$, $y > -4$

Chapter 6

Trigonometry

6.1 PERIODIC FUNCTIONS—TRIGONOMETRIC FUNCTIONS

Trigonometry deals with the problem of measurement and solution of triangles and periodic functions. The applications of trigonometry to business cycles and other situations do not involve the triangle specifically. They are concerned with the properties and applications of circular or periodic functions.

Periodic Functions

A function is periodic with period p ($p \neq 0$) if $f(x + p) = f(x)$.

The cyclical curve shows a periodic curve with period p.

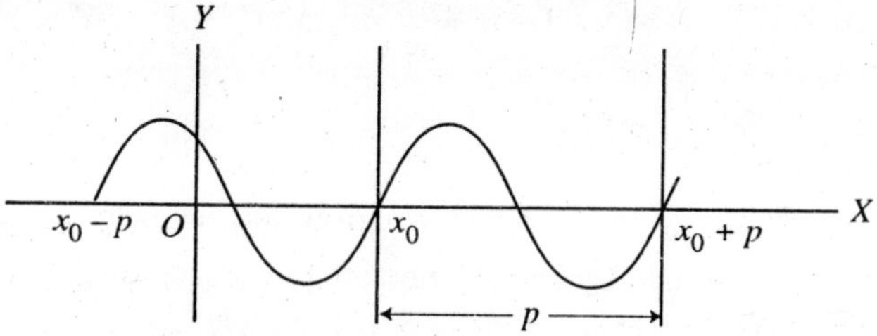

Radian Measure of an Angle

$\angle$ *LON* is called a directed angle which is measured by the rotation about its vertex *O*. The angle is positive if the rotation is anticlockwise and negative if the rotation is clockwise.

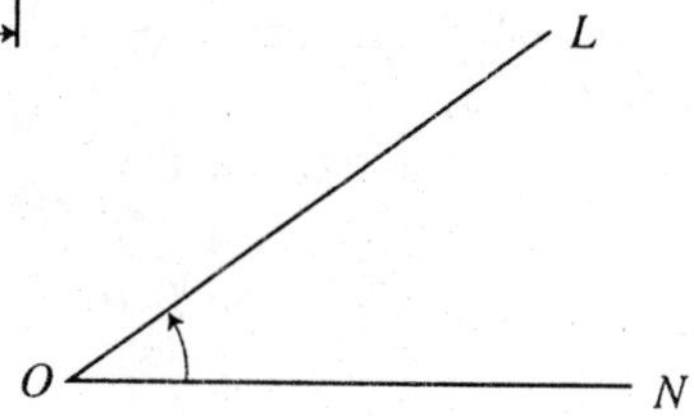

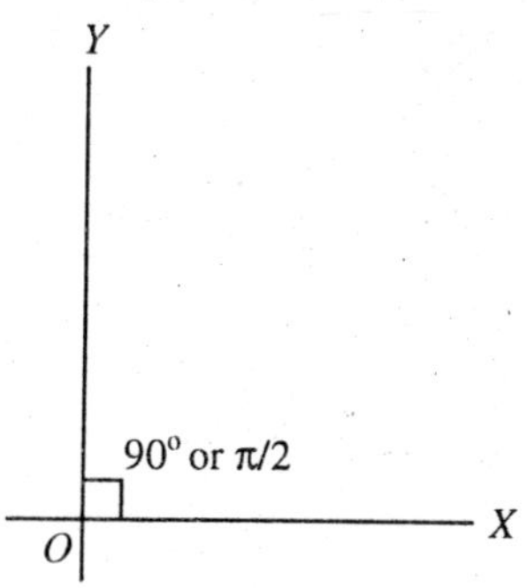

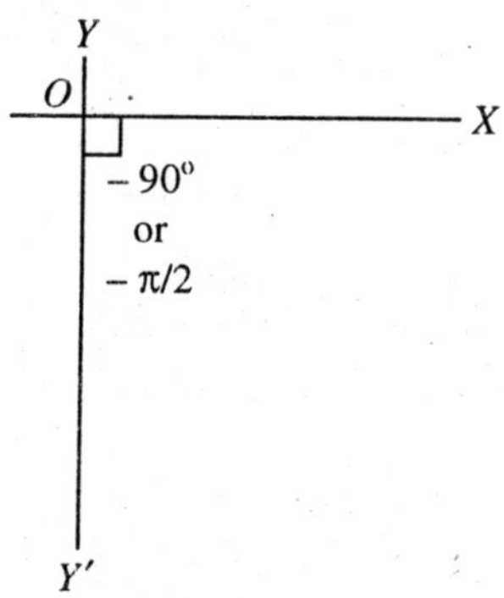

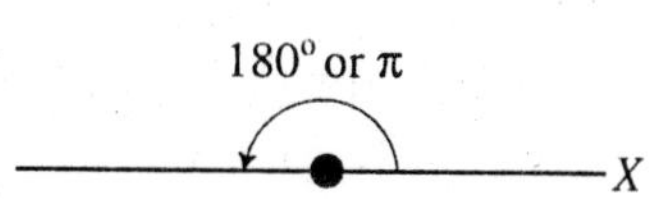

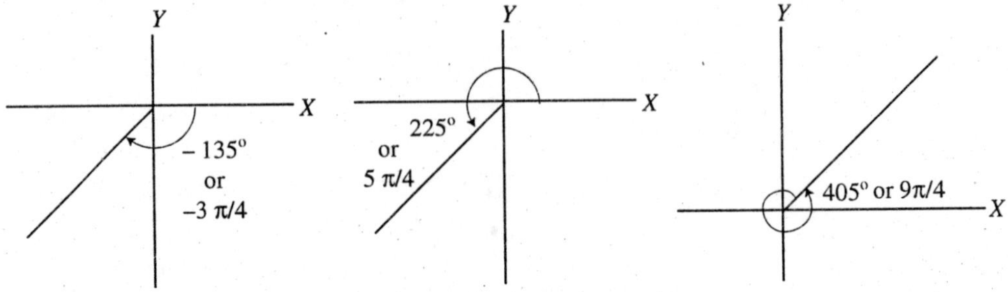

The radian measure of an angle expresses degrees in terms of radians. π radians, written π^c, corresponds to 180°. Similarly, $2\pi^c$ corresponds to 360°. Usually, c is not written to express radian measures.

An angle $\alpha°$ can be converted into radians by multiplying it by π/180. Thus

$$\alpha° = \alpha\frac{\pi}{180} \text{ radians}$$

Conversely θ radians can be written

$$\theta^c = \theta\frac{180}{\pi} \text{ degrees}$$

Ex. 1. (*i*) $45° = \frac{\pi^c}{4}$ (*ii*) $\frac{\pi^c}{2} = 90°$ (*iii*) $\frac{\pi^c}{6} = 30°$ (*iv*) $60° = \frac{\pi^c}{3}$ (*v*) $\frac{5\pi^c}{12} = 75°$

(*vi*) $\frac{4\pi^c}{3} = 240°$ (*vii*) $\frac{4\pi^c}{3} = 240°$ (*viii*) $\frac{-5\pi^c}{6} = -150°$ (*ix*) $-30° = -\frac{\pi^c}{6}$

Note

Measurement of Angles

Two systems of measuring and comparing angles may be considered.

In the *Sexagesimal system* a right angle (obtained by a quarter revolution) is divided into 90 equal parts and each part equals one degree, written 1°.

$$1° = 60 \text{ minutes written } 60'$$

$$1' = 60 \text{ seconds written } 60''$$

In the *circular measure* system, the unit is a *radian.* A radian is the measure of the angle made at the centre of a circle by an arc whose length equals the radius of the circle.

The notation used is: 1 radian = 1^c

The circumference of a circle equals $2\pi r$ where r is the radius of the circle. It can be shown that:

$$\frac{1}{2}\pi \text{ radians} = \left(\frac{\pi}{2}\right)^c = 90°$$

$$\therefore \quad 1^c = \left(\frac{180}{\pi}\right) \text{ or } 1° = \left(\frac{\pi}{180}\right)^c$$

Since $\pi = \frac{22}{7}$ or 3.142, $1^c = \left(57\frac{3}{11}\right) = 57° \; 17' \; 45''$

$180° = \pi$ radians, $360° = 2\pi$ radians

$\frac{\pi}{10}$ radians $= 18°$, $\frac{\pi}{6}$ radians $= 30°$

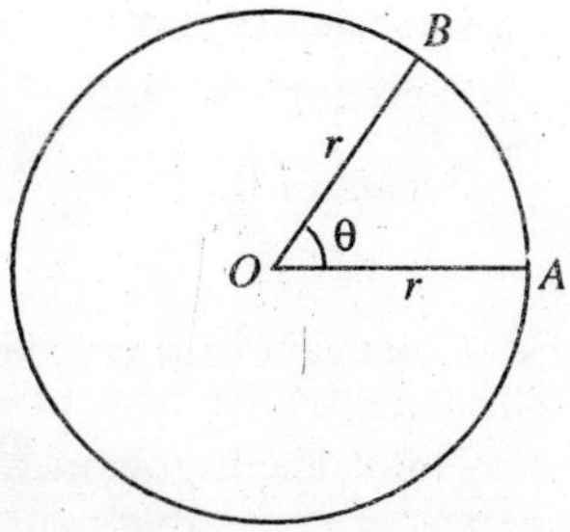

Area of a Circle

The area of a circle with radius r is πr^2.

The area of a sector AOB subtending an angle θ at the centre is $\frac{1}{2} r^2 \theta$.

Length of the arc $AB = r\,\theta$.

Ex. 2. An arc $AB = 55$ cms. subtends an angle of 150° at the centre of a circle. Find the area of the sector AOB.

$$\theta = 150° = 150 \times \frac{\pi}{180} = \frac{5}{6} \times \frac{22}{7} = \left(\frac{55}{21}\right)^c$$

Arc $\quad AB = r\theta = r \times \frac{55}{21} = 55 \qquad \therefore \quad r = 21$

$\therefore$ Area of $AOB = \frac{1}{2} r^2 \theta = \frac{1}{2} \times 21 \times 21 \times \frac{55}{21} = \frac{1155}{2}$ sq. cms.

Area of circle $= \pi r^2 = \frac{22}{7} \times 21 \times 21 = 1386$ sq. cms.

Ex. 3. Find the measure of the angle between the hour hand and the minute hand of a clock at a quarter past one.

Ans. $\left(\frac{7\pi}{24}\right)^c$

Trigonometric Ratios

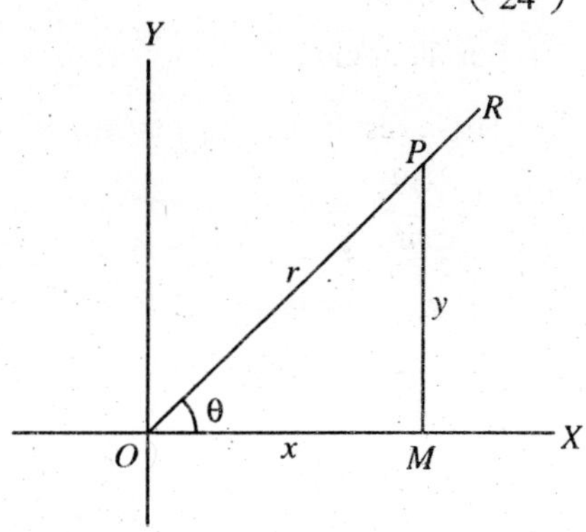

Take a point $P(x, y)$ on the line OR which makes an angle θ with the x axis.

In $\Delta\, OPM$, $OM = x$, and $MP = y$.

The hypotenuse $OP = r$

OM is the side adjacent to angle θ

MP is the side opposite angle θ

There are six trigonometric ratios

Sine of θ written $\quad \sin\theta = \frac{\text{Opp. Side}}{\text{Hypotenuse}} = \frac{MP}{OP} = \frac{y}{r}$

Cosine of θ written $\quad \cos\theta = \frac{\text{Adjacent side}}{\text{Hypotenuse}} = \frac{OM}{OP} = \frac{x}{r}$

Tangent of θ written $\tan\theta = \dfrac{\text{Opp. side}}{\text{Adj. side}} = \dfrac{MP}{OM} = \dfrac{y}{x}$

Cosecant of θ or $\operatorname{cosec}\theta = \dfrac{r}{y} = \dfrac{1}{\sin\theta}$

Secant of θ or $\sec\theta = \dfrac{r}{x} = \dfrac{1}{\cos\theta}$

Cotangent of θ or $\cot\theta = \dfrac{x}{y} = \dfrac{1}{\tan\theta}$

Note: The trigonometric ratios remain the same wherever the point P is taken on OR.

Values of Trigonometric Functions of Standard Angles

We can find the numerical values of sin 45°, cos 30° etc. and use them whenever required. Tables are available to find the values of trigonometric functions of all angles but it is useful to remember some standard values.

To find sin 45° or $\sin\dfrac{\pi}{4}$ note that in the right angled triangle POM

$$OP^2 = OM^2 + PM^2$$

If $\angle POM = 45° = \angle MOP,\ OM = MP = x$

If $OP = a$ then $a^2 = 2x^2$ or $a = \sqrt{2}\,x$ or $x = a/\sqrt{2}$

$$\sin 45° = \frac{PM}{OP} = \frac{x}{a} = \frac{a/\sqrt{2}}{a} = \frac{1}{\sqrt{2}}$$

$\therefore \operatorname{cosec} 45° = \sqrt{2}$

Similarly $\cos 45° = \dfrac{1}{\sqrt{2}}$, $\sec 45° = \sqrt{2}$

$\tan 45° = 1$, $\cot 45° = 1$

For an angle of 30° we can prove that $\sin 30° = \dfrac{1}{2}$ etc.

These results can be put in a tabular form and remembered.

$\sin 0° = 0$	$\cos 0° = 1$	$\tan 0° = 0$
$\sin 30° = 1/2$	$\cos 30° = \sqrt{3}/2$	$\tan 30° = 1/\sqrt{3}$
$\sin 45° = 1/\sqrt{2}$	$\cos 45° = 1/\sqrt{2}$	$\tan 45° = 1$
$\sin 60° = \sqrt{3}/2$	$\cos 60° = 1/2$	$\tan 60° = \sqrt{3}$
$\sin 90° = 1$	$\cos 90° = 0$	$\tan 90° = \infty$

The trigonometric ratios for other angles can be found by suitable methods. Tables for trigonometric ratios are available. It may be remembered that

$$\sin(90° - \theta) = \cos\theta,\ \cos(90° - \theta) = \sin\theta \text{ and } \tan(90° - \theta) = \cot\theta.$$

Note the following identities which can be easily proved.

$$\frac{\sin\theta}{\cos\theta} = \tan\theta,\quad \frac{\cos\theta}{\sin\theta} = \cot\theta \qquad \sin^2\theta + \cos^2\theta = 1$$

$$\sec^2\theta - \tan^2\theta = 1 \qquad \operatorname{cosec}^2\theta - \cot^2\theta = 1$$

Ex. 4. $2 \tan^2 45° + 3 \tan^2 60°$

$$= 2 \times (1)^2 + 3\,(\sqrt{3})^2 = 2 + 9 = 11$$

Ex. 5. $4 \sin^2 30° + 16 \cos^2 60°$

$$= 4\left(\frac{1}{2}\right)^2 + 16\left(\frac{1}{2}\right)^2 = 1 + 4 = 5$$

Ex. 6. $4 \cos^3 30° - 3 \cos 30° = 0$

$$\tan 45° - 2 \sin^2 30° - \cos 60° = 0$$

The trigonometric ratios are useful in finding the angles and lengths of the sides of a triangle and for a variety of other purposes.

Signs of Trigonometric Ratios

As θ increases from 0 to 2π,
$\sin \theta$ rises from 0 to 1 in Quadrant I
falls from 1 to 0 in Quadrant II
falls from 0 to – 1 in Quadrant III
rises from – 1 to 0 in Quadrant IV

Similarly $\cos \theta$	falls from 1 to 0	in Q. I
	falls from 0 to – 1	in Q. II
	rises from – 1 to 0	in Q. III
	rises from 0 to 1	in Q. IV
$\tan \theta$	rises from 0 to ∞	in Q. I
	rises from $-\infty$ to 0	in Q. II
	rises from 0 to ∞	in Q. III
	rises from $-\infty$ to 0	in Q. IV

All trigonometric ratios are positive in Q.I
$\sin \theta$ and $\operatorname{cosec} \theta$ are positive in Q.II
$\tan \theta$ and $\cot \theta$ are positive in Q.III
$\cos \theta$ and $\sec \theta$ are positive in Q.IV

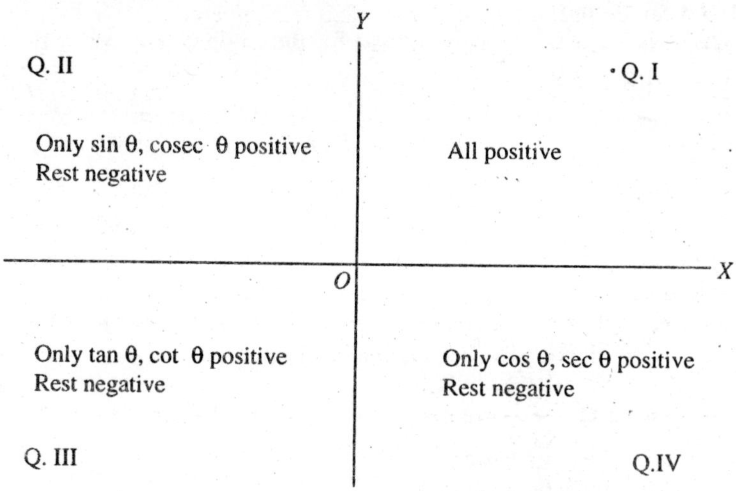

The following results can be verified and used when required.

$$\sin(-\theta) = -\sin\theta$$

$$\cos(-\theta) = \cos\theta$$

$$\tan(-\theta) = \tan\theta$$

The angle $-\theta$ is in the fourth quadrant in which only $\cos(-\theta)$ is positive.

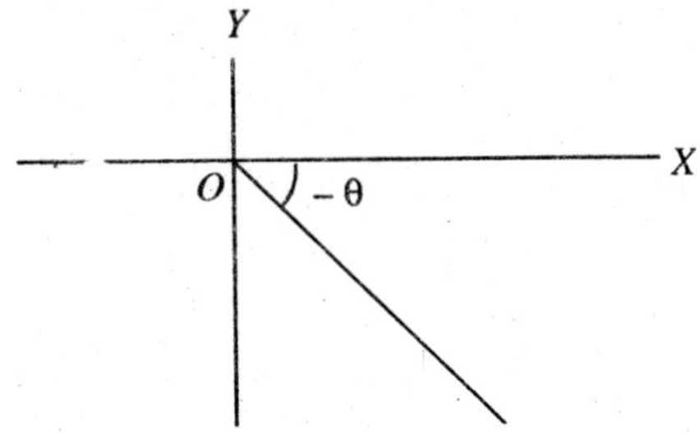

$\sin(90-\theta) = \cos\theta$
$\cos(90-\theta) = \sin\theta$
$\tan(90-\theta) = \cot\theta$

The angle $90-\theta$ is in the first quadrant. All ratios are positive.

$\sin(90+\theta) = \cos\theta$
$\cos(90+\theta) = -\sin\theta$
$\tan(90+\theta) = -\cot\theta$

The angle $90+\theta$ is in the second quadrant. Only $\sin(90+\theta)$ is positive.

$\sin(180-\theta) = \sin\theta$
$\cos(180-\theta) = -\cos\theta$
$\tan(180-\theta) = -\tan\theta$

The angle $180-\theta$ is in the second quadrant. Only $\sin(180-\theta)$ is positive.

$\sin(180+\theta) = -\sin\theta$
$\cos(180+\theta) = -\cos\theta$
$\tan(180+\theta) = +\tan\theta$

The angle $180+\theta$ is in the third quadrant. Only $\tan(180+\theta)$ is positive.

$\sin(360-\theta) = -\sin\theta$
$\cos(360-\theta) = \cos\theta$
$\tan(360-\theta) = -\tan\theta$

Same rule as for the angle $-\theta$.

$\sin(360+\theta) = \sin\theta$
$\cos(360+\theta) = \cos\theta$
$\tan(360+\theta) = \tan\theta$

Same rule as for the angle θ

Ratios of other angles can be expressed in terms of the above ratios.

Ex. 7. $\sin\left(3\frac{\pi}{2}-\theta\right) = \sin(270-\theta) = \sin(90+180-\theta)$

$$= \cos(180-\theta) = -\cos\theta$$

or $\sin(270-\theta) = \sin(180+90-\theta)$

$$= -\sin(90-\theta) = -\cos\theta$$

or $\sin(270-\theta) = \sin(270-\theta-360)$

$$= \sin(-90-\theta) = \sin\{-(90+\theta)\}$$

$$= -\sin(90+\theta) = -\cos\theta$$

or $\sin(270 - \theta) = -\sin(\theta - 270)$

$= -\sin(\theta - 270 + 360) = -\sin(90 + \theta)$

$= -\cos\theta$

or $\sin(270 - \theta) = -\cos[90 + (270 - \theta)]$

$= -\cos(360 - \theta) = -\cos\theta$

Ex. 8. $\sin\frac{3}{2}\pi = \sin(\pi + \pi/2) = -\sin\frac{\pi}{2} = -1$

$\cos\frac{5}{4}\pi = -\frac{1}{\sqrt{2}}$, $\quad \sin\frac{5}{4}\pi = -\frac{1}{\sqrt{2}}$

$\cos\frac{7}{4}\pi = \frac{1}{\sqrt{2}}$, $\quad \sin\frac{7}{4}\pi = -\frac{1}{\sqrt{2}}$

$\cos\frac{7}{6}\pi = -\frac{\sqrt{3}}{2}$, $\quad \sin\frac{7}{6}\pi = -\frac{1}{2}$

$\cos\frac{11}{6}\pi = \frac{\sqrt{3}}{2}$, $\quad \sin\frac{11}{6}\pi = -\frac{1}{2}$

$\cos\frac{2}{3}\pi = -\frac{1}{2}$ $\quad \sin\frac{5}{3}\pi = -\frac{\sqrt{3}}{2}$

$\cos\left(-\frac{3}{4}\pi\right) = -\frac{1}{\sqrt{2}}$, $\quad \sin\left(-\frac{4}{3}\pi\right) = \frac{\sqrt{3}}{2}$

Ex. 9. If $\sin\theta = -\frac{3}{5}$, then $\cos\theta = \frac{4}{5}$, $\quad \tan\theta = -\frac{3}{4}$

$\operatorname{cosec}\theta = -\frac{5}{3}$, $\quad \sec\theta = \frac{5}{4}$, $\quad \cot\theta = -\frac{4}{3}$.

Inclination and Slope of a Line

The inclination of a line corresponds to the angle made by the line with the horizontal axis.

If a line is horizontal or parallel to the x axis its inclination is zero *i.e.*, $\theta = 0$. If a line is perpendicular to the x axis, $\theta = 90°$.

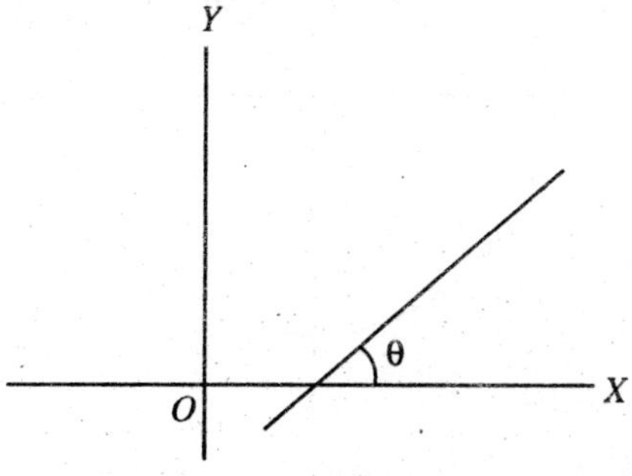

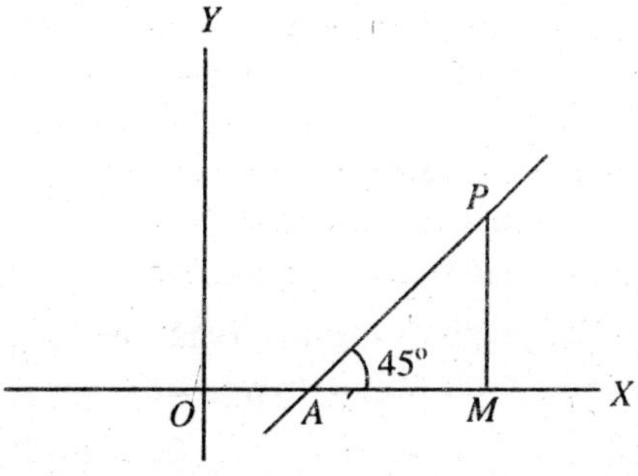

If the inclination of a line is θ, the slope of the line is $\tan\theta$.

The slope of a line with 45° inclination is $\tan 45° = 1$. Alternatively. since $\angle APM = 45°$, $AM = MP$ and $\tan 45° = MP/AM = 1$

Ex. 10. The slope of a line with inclination (*i*) 30° (*ii*) 60° (*iii*) 0° (*iv*) 90° is

(*i*) $\tan 30° = \dfrac{1}{\sqrt{3}}$ (*ii*) $\tan 60° = \sqrt{3}$ (*iii*) $\tan 0 = 0$ (*iv*) $\tan 90° = \infty$

Note that parallel lines have the same inclination and hence equal slopes.

The slope of a curve at a point *P* is the slope of the tangent drawn at the point *P*.

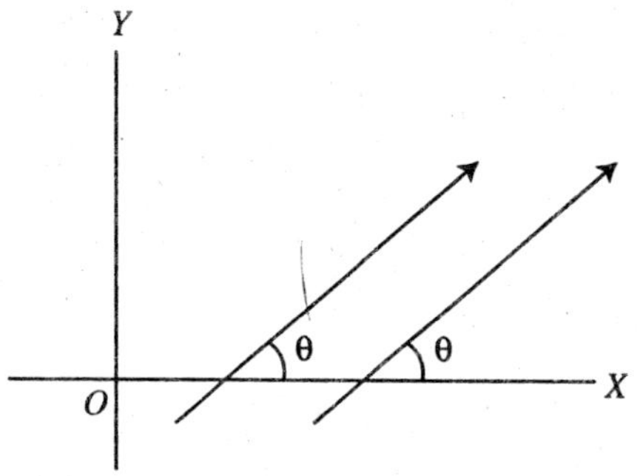

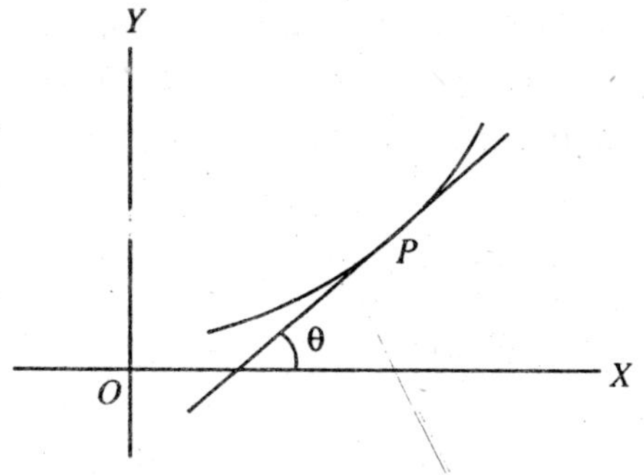

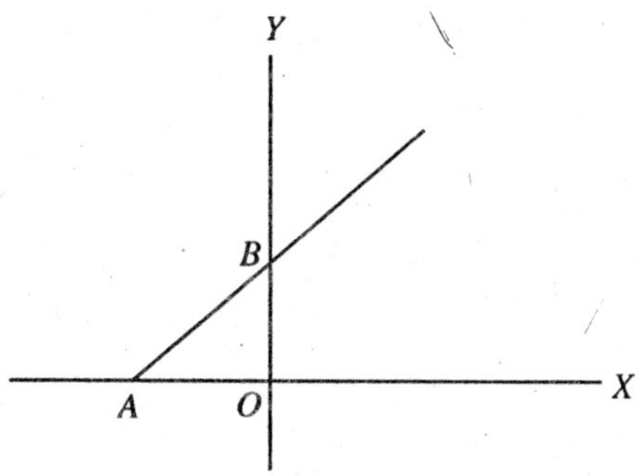

The intercept of a line on the *y* axis is the distance from the origin to the point of intersection on the *y* axis.

OB is the *y* intercept.

Similarly *OA* is the *x* intercept.

The intercept may be positive or negative. Zero intercept implies that the line passes through the origin.

Note. We know the trigonometric ratios

$$\sin\theta = \frac{MP}{OP}, \ \cos\theta = \frac{OM}{OP}, \tan\theta = \frac{MP}{OM}.$$

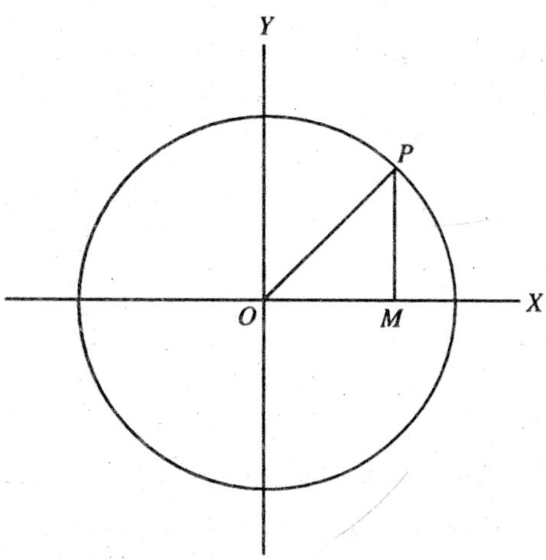

As *OP* rotates in a counterclockwise direction, the angle θ changes. While *OP* remains constant, *MP* and *OM* change. Thus the three ratios are functions of θ. The functions $\sin\theta$, $\cos\theta$, $\tan\theta$ being related to a circle are called circular functions. Being also related to a triangle, they are often referred to as trigonometric functions. They are periodic with the fundamental period 2π.

The graph of the sine function $f(\theta) = \sin\theta$ repeats itself at intervals of 2π

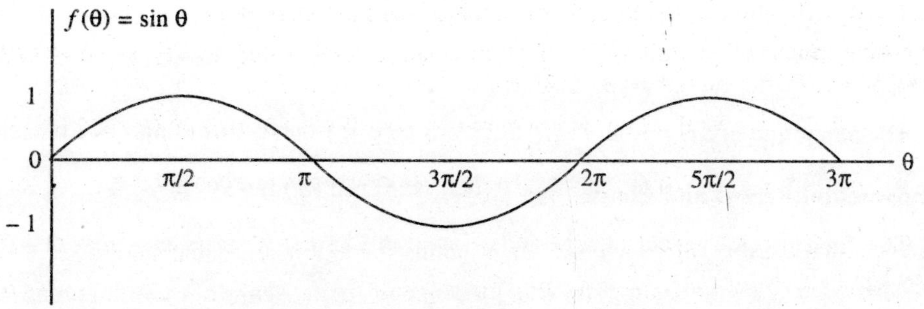

Since $\sin(\theta + 2\pi) = \sin(\theta + 360°) = \sin\theta$

$$\cos(\theta + 360) = \cos\theta$$

it means sin θ, cos θ are periodic functions each with period 2π.

In general, $\sin(\theta + 2n\pi) = \sin\theta$

Similarly, $\cos(\theta + 2n\pi) = \cos\theta$

where n is any integer.

Note: As θ increases from (*i*) 0 to $\frac{\pi}{2}$, sin θ rises from 0 to 1

(*ii*) $\frac{\pi}{2}$ to π, sin θ falls from 1 to 0

(*iii*) π to 3 $\frac{\pi}{2}$, sin θ falls from 0 to − 1

(*iv*) $\frac{3\pi}{2}$ to 2π, sin θ rises from − 1 to 0

θ	0	π/4	π/2	3π/4	π	5π/4	3π/2	7π/4	2π
sin θ	0	$1/\sqrt{2}$	1	$1/\sqrt{2}$	0	$-1/\sqrt{2}$	−1	$-1/\sqrt{2}$	0

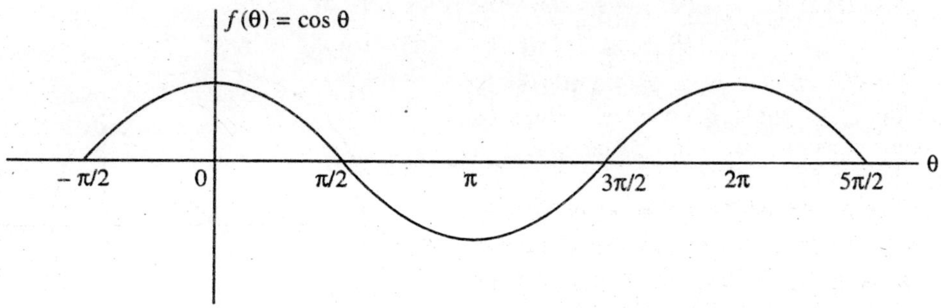

The graph of $f(\theta) = \cos\theta$ is also periodic.

However, cos 0 = 1 and cos π/2 = 0

The *amplitude* of a sine wave is the absolute value of one half of the difference between the greatest and the least ordinate of the wave. The amplitude in each of the sine and cosine graphs is

$$1/2\ \{1 - (-1)\} = 1$$

The amplitude of $f(t) = 5 \sin t$ is 5. Each ordinate is 5 times the ordinate of $f(t) = \sin t$.

Fundamental Period and Phase

The fundamental period of a periodic function like $f(t) = \sin bt$ or $g(t) = \cos bt$ is given by

$$T = \frac{2\pi}{|b|}$$

$$\begin{aligned}\sin bt &= \sin (bt + 2\pi) \text{ because the period is } 2\pi\\ &= \sin b\left(t + \frac{2\pi}{b}\right)\\ &= f\left(t + \frac{2\pi}{b}\right)\end{aligned}$$

The phase of $f(t) = \sin (t - p)$ or $f(t) = \cos (t - p)$ is defined by $|p|$

The phase of $f(t) = \sin\left(t + \frac{1}{2}\pi\right)$ is given by $|p| = \left|-\frac{1}{2}\pi\right| = \frac{\pi}{2}$

Ex. 11. For the function $\quad f(t) = 4 \cos (2t - \pi/2)$

$$= 4 \cos 2(t - \pi/4)$$

the amplitude is 4 and phase $\frac{\pi}{4}$.

The fundamental period is $\frac{2\pi}{2} = \pi$.

Ex. 12. The amplitude of $2 \sin \frac{1}{2}\pi t$ is 2 and the fundamental period is $\frac{2\pi}{(1/2)\pi} = 4$.

Ex. 13. If $f(x) = \sin kx$ is periodic what is the pericd?

$$\begin{aligned}f(x) &= \sin (kx) = \sin (kx + 2\pi) = \sin k(x + 2\pi/k)\\ &= f(x + 2\pi/k).\end{aligned}$$

The period is $2\pi/k$

This is also true for $f(x) = \cos kx$. The period is $2\pi/k$.

The period for $f(x) = \tan kx$ is π/k.

Ex. 14. Find the period, if any, for $f(x) = \sin x^2$.

If the period is t, $f(x + t) = \sin (x + t)^2$.

Also $f(x) = \sin x^2 = \sin (x^2 + 2\pi)$

$\therefore \qquad (x + t)^2 = x^2 + 2\pi$

$\therefore \qquad 2tx + t^2 = 2\pi \qquad$ which does not give a fixed value of t.

$f(x) = \sin x^2 \qquad$ This function is not periodic.

Periodic Functions and Graphs

Ex. 15. Graph of $f(\theta) = \sin\theta + 2\cos\theta$

$$f(0) = 0 + 2 = 2 \qquad f\left(\frac{\pi}{4}\right) = \frac{1}{\sqrt{2}} + 2\cdot\frac{1}{\sqrt{2}} = 2.1$$

$$f(\pi) = 0 - 2 = -2 \qquad f\left(\frac{3}{4}\pi\right) = \frac{1}{\sqrt{2}} - \frac{2}{\sqrt{2}} = -0.7$$

$$f(2\pi) = 0 + 2 = 2 \qquad f\left(\frac{5}{4}\pi\right) = -\frac{1}{\sqrt{2}} - \frac{2}{\sqrt{2}} = -2.1$$

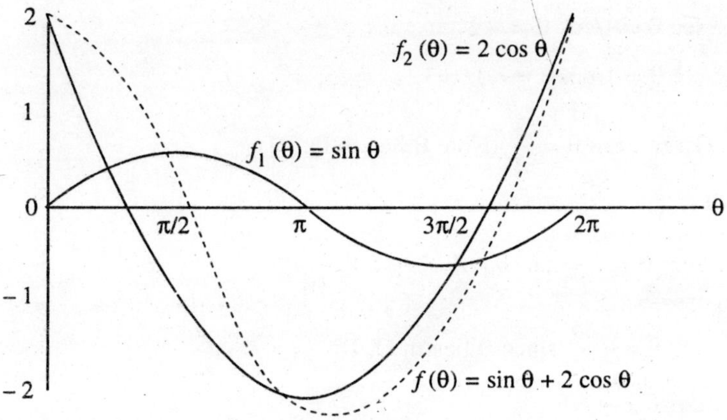

Ex. 16. Graph of $f(t) = e^{-t}\sin 2t \quad t \geq 0$

$|f(t)| \leq e^{-t}$ because $|\sin 2t| \leq 1$ and $e^{-t} > 0$ for all t

i.e., $-e^{-t} \leq f(t) \leq e^{-t}$ *i.e.*, the graph of $f(t)$ lies between the graphs of $f_1(t) = -e^{-t}$ and $f_2(t) = e^{-t}$.

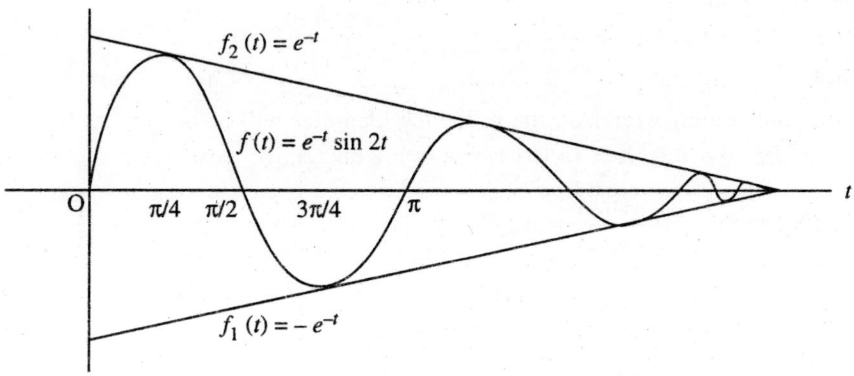

There are damped oscillations. In the limit, the amplitude is zero.

Trigonometric Ratios

Given any one trigonometric ratio, the other ratios can be found by using the identities given earlier.

Ex. 17. $\cos\theta = -\dfrac{1}{2}$ (Note θ lies in Q. II)

To find other ratios, use

$$\sin\theta = \sqrt{1-\cos^2\theta} = \sqrt{1-\frac{1}{4}} = \sqrt{\frac{3}{4}} = \frac{\sqrt{3}}{2}$$

We have taken the value + $\sqrt{3}/2$ since θ lies in Q. II

$$\tan\theta = \sin\theta/\cos\theta = (\sqrt{3}/2)/(-1/2) = -\sqrt{3}$$

$$\text{cosec}\,\theta = 1/\sin\theta = 2/\sqrt{3}$$

$$\sec\theta = 1/\cos\theta = -2$$

$$\cot\theta = 1/\tan\theta = -1/\sqrt{3}$$

Ex. 18. Given : $\tan\theta = \dfrac{3}{4}$ (Note θ lies in Q. III)

We have

$$\sec^2\theta = 1 + \tan^2\theta = 1 + 9/16 = \frac{25}{16}$$

$$\sec\theta = -\frac{5}{4} \text{ since } \theta \text{ lies in Q. III}$$

$$\cos\theta = -4/5$$

$$\sin\theta = \sqrt{1-\cos^2\theta} = \sqrt{1-\frac{16}{25}} = \sqrt{\frac{9}{2}} = -\frac{3}{5} \text{ (Q. III)}$$

$$\cot\theta = 4/3,\ \text{cosec}\,\theta = -5/3$$

Ex. 19. Show $4\cot^2 45° - \sec^2 60° + \sin^2 30° = \dfrac{1}{8}$

Identities

In the forthcoming exercises, the following identities will be useful.

(*a*) $\sin^2 A + \cos^2 A = 1$, (*b*) $\sec^2 A = 1 + \tan^2 A$, (*c*) $\text{cosec}^2 A = 1 + \cot^2 A$

Ex. 20. Prove : $\sqrt{\dfrac{1-\sin A}{1+\sin A}} = \sec A - \tan A$

$$\text{LHS} = \sqrt{\frac{(1-\sin A)(1-\sin A)}{(1+\sin A)(1-\sin A)}} = \sqrt{\frac{(1-\sin A)^2}{1-\sin^2 A}}$$

$$= \sqrt{\frac{(1-\sin A)^2}{\cos^2 A}} = \frac{1-\sin A}{\cos A} = \frac{1}{\cos A} - \frac{\sin A}{\cos A} = \sec A - \tan A$$

Ex. 21. Prove : $\dfrac{\cos A}{1-\tan A} + \dfrac{\sin A}{1-\cot A} = \sin A + \cos A$

Put $\tan A = \sin A/\cos A$ and $\cot A = \cos A/\sin A$

$$\text{LHS} = \frac{\cos A \cdot \cos A}{\cos A - \sin A} + \frac{\sin A \sin A}{\sin A - \cos A}$$

$$= \frac{\cos^2 A}{\cos A - \sin A} - \frac{\sin^2 A}{\cos A - \sin A}$$

$$= \frac{\cos^2 A - \sin^2 A}{\cos A - \sin A} = \frac{(\cos A + \sin A)(\cos A - \sin A)}{\cos A - \sin A}$$

$$= \cos A + \sin A$$

Ex. 22. Show (*i*) $(\sin^6 A + \cos^6 A) = 3(\sin^4 A + \cos^4 A) - 1$

(*ii*) $\dfrac{\tan A + \sec A - 1}{\tan A - \sec A + 1} = \dfrac{1 + \sin A}{\cos A}$

(*iii*) $\dfrac{\cot A + \tan B}{\cot B + \tan A} = \cot A \tan B$

(*iv*) $(\operatorname{cosec} A - \sin A)(\sec A - \cos A)(\tan A + \cot A) = 1$

(*v*) $\left(\dfrac{1 + \sin\theta - \cos\theta}{1 + \sin\theta + \cos\theta}\right)^2 = \dfrac{1 - \cos\theta}{1 + \cos\theta}$

Ex. 23. (*i*) If $\tan\theta = \dfrac{p}{q}$ show $\dfrac{p\sin\theta - q\cos\theta}{p\sin\theta + q\cos\theta} = \dfrac{p^2 - q^2}{p^2 + q^2}$

(*ii*) Show $\dfrac{1 + \sin A}{1 + \cos A}\,\dfrac{1 + \sec A}{1 + \operatorname{cosec} A} = \tan A$

Compound Angles

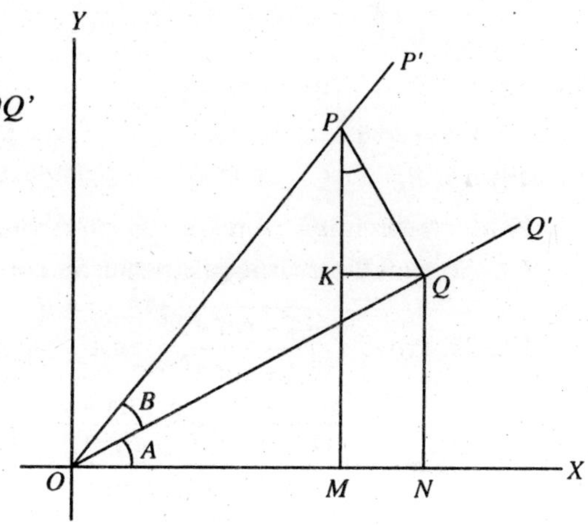

To find $\sin(A + B)$.

From a point P on OP' draw $PQ \perp OQ'$

From Q draw $QK \perp PM$.

$$PM = QN + PK$$

$$\frac{PM}{OP} = \frac{QN}{OP} + \frac{PK}{OP}$$

$$= \frac{QN}{OQ} \cdot \frac{OQ}{OP} + \frac{PK}{PQ} \cdot \frac{PQ}{OP}$$

$$= \sin A \cos B + \cos A \sin B$$

Since $\angle KPQ = \angle A$

$$\sin(A + B) = \frac{PM}{OP} = \sin A \cos B + \cos A \sin B \qquad \ldots(1)$$

Similarly $\cos(A + B) = \dfrac{OM}{OP} = \dfrac{ON}{OP} - \dfrac{MN}{OP}$

$$= \frac{ON}{OQ} \cdot \frac{OQ}{OP} - \frac{KQ}{PQ} \cdot \frac{PQ}{OP}$$

$\therefore$ $\cos(A + B) = \cos A \cos B - \sin A \sin B.$ ...(2)

Replacing B by $-B$ in the above results

$$\sin(A - B) = \sin A \cos B - \cos A \sin B \quad ...(3)$$

$$\cos(A - B) = \cos A \cos B + \sin A \sin B \quad ...(4)$$

Also $\tan(A + B) = \dfrac{\sin(A + B)}{\cos(A + B)} = \dfrac{\sin A \cos B + \cos A \sin B}{\cos A \cos B - \sin A \sin B}$

Dividing the numerator and denominator by $\cos A \cos B$, we have

$$\tan(A + B) = \frac{\frac{\sin A}{\cos A} + \frac{\sin B}{\cos B}}{1 - \frac{\sin A}{\cos A}\frac{\sin B}{\cos B}} = \frac{\tan A + \tan B}{1 - \tan A \tan B}. \quad ...(5)$$

Similarly $\tan(A - B) = \dfrac{\tan A - \tan B}{1 + \tan A \tan B}.$ ...(6)

It can be seen by putting $A + B = C$, $A - B = D$ that

$$\sin C + \sin D = 2 \sin \frac{C + D}{2} \cos \frac{C - D}{2};$$

$$\sin C - \sin D = 2 \cos \frac{C + D}{2} \sin \frac{C - D}{2};$$

$$\cos C + \cos D = 2 \cos \frac{C + D}{2} \cos \frac{C - D}{2};$$

$$\cos C - \cos D = -2 \sin \frac{C + D}{2} \sin \frac{C - D}{2}.$$

...(7)

We know that $\sin(A + B) = \sin A \cos B + \cos A \sin B.$

Put $A = B$, then $\sin(2A) = \sin A \cos A + \cos A \sin A.$

We have the result $\sin 2A = 2 \sin A \cos A.$ ...(8)

On the same lines using the result for $\cos(A + B)$ we can prove

$$\cos 2A = \cos^2 - \sin^2 A$$

$$= 2\cos^2 A - 1 = 1 - 2\sin^2 A \quad ...(9)$$

$$\sin 3A = \sin(2A + A)$$

$$= \sin 2A \cos A + \cos 2A \sin A$$

$$= 2 \sin A \cos A + (1 - 2\sin^2 A) \sin A$$

$$= 2 \sin A(1 - \sin^2 A) + (1 - 2\sin^2 A) \sin A$$

$$= 2 \sin A - 2\sin^3 A + \sin A - 2\sin^3 A$$

$\therefore$ $\sin 3A = 3 \sin A - 4 \sin^3 A.$ ...(10)

Similarly $\cos 3A = \cos(2A + A) = 4\cos^3 A - 3 \cos A$ ...(11)

$$\tan 3A = \tan(2A + A)$$

$$= \frac{\tan 2A + \tan A}{1 - \tan 2A \tan A} = \frac{\dfrac{2\tan A}{1-\tan^2 A} + \tan A}{1 - \dfrac{2\tan A}{1-\tan^2 A}\tan A}$$

$$= \frac{2\tan A + \tan A - \tan^3 A}{1 - \tan^2 A - 2\tan^2 A} = \frac{3\tan A - \tan^3 A}{1 - 3\tan^2 A} \qquad \ldots(12)$$

In the result $\cos 2A = \cos^2 A - \sin^2 A$ if we replace A by $\frac{A}{2}$ we have

$$\cos A = \cos^2\frac{A}{2} - \sin^2\frac{A}{2} = 2\cos^2\frac{A}{2} - 1 = 1 - 2\sin^2\frac{A}{2} \qquad \ldots(13)$$

Similarly $\quad \sin A = 2\sin\frac{A}{2}\cos\frac{A}{2} \qquad \ldots(14)$

$$\tan A = \frac{2\tan A/2}{1-\tan^2 A/2} = \frac{2t}{1-t^2} \text{ where } t = \tan\frac{A}{2}. \qquad \ldots(15)$$

The following additional results are useful and may be derived:

Since $\quad \cos 2A = 2\cos^2 A - 1 = 1 - 2\sin^2 A$

$$\therefore \quad \cos^2 A = \frac{1+\cos 2A}{2} \text{ or } \cos A = \pm\frac{\sqrt{1+\cos 2A}}{2}$$

$$\sin^2 A = \frac{1-\cos 2A}{2} \text{ or } \sin A = \pm\frac{\sqrt{1-\cos 2A}}{2}$$

A Summary of Some Important Results

I. $\sin(A \pm B) = \sin A \cos B \pm \cos A \sin B$

$\cos(A \pm B) = \cos A \cos B \mp \sin A \sin B$

$\sin(A + B + C) = \sin A \cos B \cos C + \cos A \sin B \cos C + \cos A \cos B \sin C - \sin A \sin B \sin C$

$\cos(A + B + C) = \cos A \cos B \cos C - \cos A \sin B \sin C - \sin A \cos B \sin C - \sin A \sin B \cos C$

$$\tan(A+B) = \frac{\tan A + \tan B}{1 - \tan A \tan B}, \quad \tan(A-B) = \frac{\tan A - \tan B}{1 + \tan A \tan B}$$

$$\tan(A+B+C) = \frac{\tan A + \tan B + \tan C - \tan A \tan B \tan C}{1 - \tan B \tan C - \tan C \tan A - \tan A \tan B}$$

$$\cot(A+B) = \frac{\cot A \cot B - 1}{\cot B + \cot A} \text{ and } \cot(A-B) = \frac{\cot A \cot B + 1}{\cot B - \cot A}$$

II. $\sin 2A = 2\sin A \cos A = \dfrac{2\tan A}{1+\tan^2 A} \qquad \sin A = \dfrac{2\tan A/2}{1+\tan^2 A/2}$

$$\cos 2A = \left\{\begin{array}{l} = \cos^2 A - \sin^2 A \\ = 1 - 2\sin^2 A \\ = 2\cos^2 A - 1 \end{array}\right\} = \frac{1 - \tan^2 A}{1 + \tan^2 A}; \qquad \cos A = \frac{1 - \tan^2 A/2}{1 + \tan^2 A/2}$$

$$\tan 2A = \frac{2\tan A}{1 - \tan^2 A} = \frac{1 - \cos^2 A}{1 + \cos^2 A}; \qquad \tan A = \frac{2\tan A/2}{1 - \tan^2 A/2}$$

$$1 - \cos 2A = 2\sin^2 A$$

$$1 + \cos 2A = 2\cos^2 A$$

$$\sin 3A = 3\sin A - 4\sin^3 A \qquad \tan 3A = \frac{3\tan A - \tan^3 A}{1 - 3\tan^2 A}$$

$$\cos 3A = 4\cos^3 A - 3\cos A$$

III. Conversion of Product into Sum

$$\sin A \cos B = \frac{1}{2}\{\sin(A + B) + \sin(A - B)\}$$

$$\cos A \sin B = \frac{1}{2}\{\sin(A + B) - \sin(A - B)\}$$

$$\cos A \cos B = \frac{1}{2}\{\cos(A + B) + \cos(A - B)\}$$

$$\sin A \sin B = -\frac{1}{2}\{\cos(A + B) - \cos(A - B)\}$$

IV. Conversion of Sum into Product

$$\sin C + \sin D = 2\sin\frac{C + D}{2}\cos\frac{C - D}{2}$$

$$\sin C - \sin D = 2\cos\frac{C + D}{2}\sin\frac{C - D}{2}$$

$$\cos C + \cos D = 2\cos\frac{C + D}{2}\cos\frac{C - D}{2}$$

$$\cos C - \cos D = -2\sin\frac{C + D}{2}\sin\frac{C - D}{2}$$

V.
$$\cos A/2 = \pm\sqrt{\frac{1 + \cos A}{2}} \qquad \text{...(16)}$$

$$\sin A/2 = \pm\sqrt{\frac{1 - \cos A}{2}} \qquad \text{...(17)}$$

$$\tan A/2 = \pm\sqrt{\frac{1 - \cos A}{1 + \cos A}}. \qquad \text{...(18)}$$

Also if we write $\tan A/2 = t$ we have

$$\sin A = \frac{2t}{1 + t^2}, \quad \cos A = \frac{1 - t^2}{1 + t^2}, \tan A = \frac{2t}{1 - t^2} \qquad \text{...(19)}$$

$$\tan\frac{A}{2} = \frac{\sin A/2}{\cos A/2} = \frac{\sin A/2 \cdot 2\cos A/2}{\cos A/2 \cdot 2\cos A/2}$$

$$= \frac{\sin A}{2\cos^2 A/2} = \frac{\sin A}{1+\cos A}.$$

Ex. 24. Prove : (*i*) $\tan^2 A \operatorname{cosec} A = \dfrac{\sin A}{1-\sin^2 A}$

(*ii*) $\dfrac{\sec^2 A - 1}{\sin^2 A} = \dfrac{1}{\cos^2 A}$

Ans. (*i*) $\left(\dfrac{\sin A}{\cos A}\right)^2 \dfrac{1}{\sin A} = \dfrac{\sin^2 A}{\cos^2 A}\cdot\dfrac{1}{\sin A} = \dfrac{\sin A}{\cos^2 A} = \dfrac{\sin A}{1-\sin^2 A}.$

(*ii*) $\dfrac{\sec^2 A - 1}{\sin^2 A} = \dfrac{\tan^2 A}{\sin^2 A} = \dfrac{\sin^2 A}{\cos^2 A}\cdot\dfrac{1}{\sin^2 A} = \dfrac{1}{\cos^2 A}$

Ex. 25. Prove:

(*i*) $\dfrac{\sec x + \operatorname{cosec} x}{1+\tan^2 x} = \operatorname{cosec} x.$

(*ii*) $(1 + \sec\theta)(1 - \cos\theta) = \tan\theta \sin\theta$

$\left(\textbf{Hint : } \text{LHS} = \left(1+\dfrac{1}{\cos\theta}\right)(1-\cos\theta) = \dfrac{1-\cos^2\theta}{\cos\theta} = \dfrac{\sin^2\theta}{\cos\theta} = \text{RHS}\right).$

(*iii*) $\dfrac{1-\cot^2 x}{1+\cot^2 x} = \sin^2 x - \cos^2 x.$

(*iv*) $(\tan x + \cos x)^2 = \sec^2 x + \operatorname{cosec}^2 x.$

(**Hint :** LHS $= \tan^2 x + \cot^2 x + 2 = \tan^2 x + 1 + \cot^2 x + 1 =$ RHS).

(*v*) $\dfrac{\sin^3\theta + \cos^3\theta}{\sin\theta + \cos\theta} = 1 - \sin\theta\cos\theta.$

$\left(\textbf{Hint : } \text{LHS} = \dfrac{(\sin\theta+\cos\theta)(\sin^2\theta+\cos^2\theta-\sin\theta\cos\theta)}{\sin\theta+\cos\theta}\right)$

(*vi*) $\sec x + \tan x = \dfrac{1}{\sec x - \tan x}.$

(*vii*) $\dfrac{\cos x}{1+\sin x} = \dfrac{1-\sin x}{\cos x}\left[\textbf{Hint : } \text{LHS} = \dfrac{\cos x\,(1-\sin x)}{1-\sin^2 x}\right]$

(*viii*) $\sin^3 t + \cos^3 t + \sin t \cos^2 t + \sin^2 \cos t = \sin t + \cos t.$

Ex. 26. Prove the following identities

(*i*) $\dfrac{\cos 2A + \cos 4A}{\sin 4A - \sin 2A} = \cot A.$

$$\left(\text{LHS} = \frac{2\cos\frac{2A+4A}{2}\cos\frac{2A-4A}{2}}{2\cos\frac{4A+2A}{2}\sin\frac{4A-2A}{2}} = \frac{2\cos 3A\cos(-A)}{2\cos 3A\sin A} = \text{RHS}\right). \because \cos(-A) = \cos A$$

(*ii*) $\dfrac{\sin 2A - \sin 2B}{\sin 2A + \sin 2B} = \dfrac{\tan (A - B)}{\tan (A + B)}$.

(*iii*) $\sin^3 x \cos^3 x = \dfrac{3\sin 2x - \sin 6x}{32}$.

(*iv*) $\sin 2\alpha + \sin 2\beta + \sin 2\gamma = 4 \cos \alpha \cos \beta \cos \gamma$ where $\alpha + \beta + \gamma = \pi$.

(See Ex. 37)

(*v*) $\dfrac{1 - \cos 8x}{8} = \sin^2 2x \cos^2 2x$.

(*vi*) $\dfrac{1 - \tan^2 \theta}{1 + \tan^2 \theta} = \cos 2\theta$.

(*vii*) $\tan \dfrac{t}{2} = \dfrac{1 - \cos t}{\sin t} = \dfrac{\sin t}{1 + \cos t}$.

$\left(\textbf{Hint :}\ \tan \dfrac{t}{2} = \dfrac{\sin t/2}{\cos t/2} = \dfrac{2\sin^2 t/2}{2\sin t/2 \cos t/2} = \dfrac{1 - \cos t}{\sin t}\ \text{See Ex. 25 } (vii)\right)$.

(*viii*) If $u = \tan \dfrac{x}{2}$ then $\sin x = \dfrac{2u}{1 + u^2}$, $\cos x = \dfrac{1 - u^2}{1 + u^2}$.

$\left(\textbf{Hint :}\ \sin x = 2\sin \dfrac{x}{2} \cos \dfrac{x}{2} = \dfrac{2\sin x/2}{\cos x/2} \cos^2 x/2 = 2\tan \dfrac{x}{2} \cdot \dfrac{1}{\sec^2 x/2}\right)$.

(*ix*) $\dfrac{2}{1 + \cos 2t} = \sec^2 t$.

(*x*) $\cos^4\theta - \sin^4\theta = \cos 2\theta$.

(*ix*) $\dfrac{1 + \sin 2A + \cos 2A}{1 + \sin 2A - \cos 2A} = \cot A$.

Ex. 27. Given $q \cos \theta = p \sin \theta$ find the value of $\dfrac{p\cos\theta + q\sin\theta}{p\cos\theta - q\sin\theta}$.

Since $p \cos \theta = \dfrac{p^2}{q} \sin \theta$, the expression becomes

$$\frac{\dfrac{p^2}{q}\sin\theta + q\sin\theta}{\dfrac{p^2}{q}\sin\theta - q\sin\theta} = \frac{\left(\dfrac{p^2}{q} + q\right)\sin\theta}{\left(\dfrac{p^2}{q} - q\right)\sin\theta} = \frac{p^2 + q^2}{p^2 - q^2}$$

Ex. 28. $\sin(A+B)\sin(A-B) = \sin^2 A - \sin^2 B$

$$\begin{aligned} LHS &= (\sin A \cos B + \cos A \sin B)(\sin A \cos B - \cos A \sin B) \\ &= \sin^2 A \cos^2 B - \cos^2 A \sin^2 B \\ &= \sin^2 A (1 - \sin^2 B) - (1 - \sin^2 A)\sin^2 B \\ &= \sin^2 A - \sin^2 A \sin^2 B - \sin^2 B + \sin^2 A \sin^2 B \\ &= \sin^2 A - \sin^2 B \end{aligned}$$

Ex. 29. $\cos(A+B)\cos(A-B) = \cos^2 A - \sin^2 B$

Ex. 30. $\dfrac{\tan(A+B)}{\cot(A-B)} = \dfrac{\sin^2 A - \sin^2 B}{\cos^2 A - \sin^2 B}$ (Use Ex. 28, 29)

Ex. 31. $\dfrac{1+\sin 2A - \cos 2A}{1+\sin 2A + \cos 2A} = \tan A$

Hint: $\text{LHS} = \dfrac{1 + 2\sin A\cos A - (1 - 2\sin^2 A)}{1 + 2\sin A \cos A + (2\cos^2 A - 1)}$

Ex. 32. If $\tan\theta = \dfrac{b}{a}$ then $a\cos^2\theta + b/2 \sin 2\theta = a$

Ex. 33. $a\cos\theta + b\sin\theta = \sqrt{a^2+b^2}\left(\dfrac{a\cos\theta}{\sqrt{a^2+b^2}} + \dfrac{b\sin\theta}{\sqrt{a^2+b^2}}\right)$

$$= \sqrt{a^2+b^2}\,(\sin\alpha\cos\theta + \cos\alpha\sin\theta)$$

$$= \sqrt{a^2+b^2}\,\sin(\theta+\alpha) \text{ where } \tan\alpha = \frac{b}{a}$$

Ex. 34. $\dfrac{\sec 8A - 1}{\sec 4A - 1} = \dfrac{\tan 8A}{\tan 2A}$ (**Note:** $\sin 8A = 2\sin 4A \cos 4A$)

Hint: $\text{LHS} = \dfrac{1-\cos 8A}{\cos 8A} \; \dfrac{\cos 4A}{1 - \cos 4A} = \dfrac{2\sin^2 4A \cos 4A}{\cos 8A \cdot 2\sin^2 2A}$

$$= \frac{\sin 4A \cdot 2\sin 4A \cdot \cos 4A}{\cos 8A \cdot 2\sin^2 2A} \text{ etc.}$$

Ex. 35. $\dfrac{\cos 9^\circ + \sin 9^\circ}{\cos 9^\circ - \sin 9^\circ} = \tan 54^\circ$

Divide Num. & Den. by $\cos 9^\circ$

$$\text{LHS} = \frac{1+\tan 9^\circ}{1-\tan 9^\circ} = \frac{\tan 45^\circ + \tan 9^\circ}{1 - \tan 45^\circ \tan 9^\circ} \quad (\because \tan 45^\circ = 1)$$

$$= \tan(45^\circ + 9^\circ) = \tan 54^\circ$$

Ex. 36. $8\sin\dfrac{\pi}{14}\sin\dfrac{3\pi}{14}\sin\dfrac{5\pi}{14} = 1$

$$\text{LHS} = 8\cos\left(\frac{\pi}{2} - \frac{\pi}{14}\right)\cos\left(\frac{\pi}{2} - \frac{3\pi}{14}\right)\cos\left(\frac{\pi}{2} - \frac{5\pi}{14}\right)$$

$$= 8\cos\frac{3\pi}{7}\cos\frac{2\pi}{7}\cos\frac{\pi}{7} = \frac{8}{2\sin\frac{\pi}{7}}\cdot 2\sin\frac{\pi}{7}\cos\frac{\pi}{7}\cos\frac{2\pi}{7}\cos\frac{3\pi}{7}$$

Note: $2\sin A\cos A = \sin 2A$

$$= \frac{8}{2\sin\pi/7}\sin\frac{2\pi}{7}\cos\frac{2\pi}{7}\cos\frac{3\pi}{7} = \frac{8}{4\sin\frac{\pi}{7}}\sin\frac{4\pi}{7}\cos\frac{3\pi}{7}$$

$$= \frac{8}{8\sin\pi/7}\left(\sin\pi + \sin\frac{\pi}{7}\right) = \frac{1}{\sin\pi/7}(0 + \sin\pi/7) = 1$$

Ex. 37. If $A + B + C = \pi$ show

(*i*) $\cos 2A + \cos 2B + \cos 2C = -1 - 4\cos A\cos B\cos C$

LHS $= (2\cos^2 A - 1) + 2\cos(B + C)\cos(B - C)$

Since $B + C = \pi - A$, $\cos(B + C) = \cos(\pi - A) = -\cos A$

LHS $= -1 + 2\cos^2 A - 2\cos A\cos(B - C)$

$= -1 + 2\cos A(\cos A - \cos(B - C))$

$= -1 - 2\cos A(-\cos(B + C) - \cos(B - C))$

$= -1 - 2\cos A\, 2\cos B\cos C$

$= -1 - 4\cos A\cos B\cos C$

(*ii*) $\cos^2 A + \cos^2 B + \cos^2 C = 1 - 2\cos A\cos B\cos C$

(*iii*) $\sin^2\frac{A}{2} + \sin^2\frac{B}{2} + \sin^2\frac{C}{2} = 1 - 2\sin\frac{A}{2}\sin\frac{B}{2}\sin\frac{C}{2}$.

(*iv*) $\sin(B + C - A) + \sin(C + A - B) + \sin(A + B - C) = 4\sin A\sin B\sin C$.

Ex. 38. If $A + B + C = \pi/2$ show

$\sin^2 A + \sin^2 B + \sin^2 C = 1 - 2\sin A\sin B\sin C$

Ex. 39. If $A + B = C$ then show

$\cos^2 A + \cos^2 B + \cos^2 C = 1 + 2\cos A\cos B\cos C$

Ex. 40. If $\cos\theta + \sin\theta = \sqrt{2}\cos\theta$ show $\cos\theta - \sin\theta = \sqrt{2}\sin\theta$

Dividing by $\cos\theta$ we have

$1 + \tan\theta = \sqrt{2}$ or $\tan\theta = \sqrt{2} - 1 < 1$, *i.e.*, $\sin\theta < \cos\theta$

Squaring $\cos\theta + \sin\theta = \sqrt{2}\cos\theta$

$\cos^2\theta + \sin^2\theta + 2\cos\theta\sin\theta = 2\cos^2\theta$

$\therefore$ $\cos^2\theta - \sin^2\theta - 2\cos\theta\sin\theta = 0$

$\therefore$ $(\cos\theta - \sin\theta)^2 = 2\sin^2\theta$

$\therefore$ $\cos\theta - \sin\theta = \pm\sqrt{2}\sin\theta$

$\therefore$ $\cos\theta - \sin\theta = \sqrt{2}\sin\theta$

($-\sqrt{2}\sin\theta$ not possible because $\tan\theta < 1$)

Ex. 41. Show $\sin 12° \sin 48° \sin 45° = \frac{1}{8}$

Ex. 42. $\dfrac{\cos 6A + 6\cos 4A + 15\cos 2A + 10}{\cos 5A + 5\cos 3A + 10\cos A} = 2\cos A$

Hint: $2 \cos A (\cos 5A + 5 \cos 3A + 10 \cos A)$

$= (\cos 6A + \cos 4A) + 5(\cos 4A + \cos 2A) + 10 (\cos 2A + 1)$

$= \cos 6A + 6 \cos 4A + 15 \cos 2A + 10$

Ex. 43. Given $a = \sin A + \sin B$, $b = \cos A + \cos B$,

show that $\tan \dfrac{A-B}{2} = \sqrt{\dfrac{4-a^2-b^2}{a^2+b^2}}$

We have

$$\begin{aligned} a^2 + b^2 &= (\sin A + \sin B)^2 + (\cos A + \cos B)^2 \\ &= \sin^2 A + 2 \sin A \sin B + \sin^2 B \\ &\quad + \cos^2 A + 2 \cos A \cos B + \cos^2 B \\ &= 1 + 2 (\sin A \sin B + \cos A \cos B) + 1 \\ &= 2 + 2 \cos (A - B) \end{aligned}$$

$$\frac{4-a^2-b^2}{a^2+b^2} = \frac{4}{a^2+b^2} - 1 = \frac{4}{2\,(1+\cos(A-B))} - 1$$

$$= \frac{1-\cos(A-B)}{1+\cos(A-B)} = \frac{2\sin^2\frac{A-B}{2}}{2\cos^2\frac{A-B}{2}} = \tan^2\frac{A-B}{2}.$$

$$\therefore \quad \sqrt{\frac{4-a^2-b^2}{a^2+b^2}} = \tan\frac{A-B}{2}.$$

Ex. 44. Show $\left(\dfrac{\cos A + \cos B}{\sin A - \sin B}\right)^n + \left(\dfrac{\sin A + \sin B}{\cos A - \cos B}\right)^n = 2\cot^n \dfrac{A-B}{2}$ or 0.

$$\text{LHS} = \left(\frac{2\cos\frac{A+B}{2}\cos\frac{A-B}{2}}{2\cos\frac{A+B}{2}\sin\frac{A-B}{2}}\right)^n + \left(\frac{2\sin\frac{A+B}{2}\cos\frac{A-B}{2}}{-2\sin\frac{A+B}{2}\sin\frac{A-B}{2}}\right)^n \quad \text{use reasult (7)}$$

$= 2\cot^n \dfrac{A-B}{2}$ if n is even

$= 0$ if n is old.

Ex. 45. If $\sin\theta + \operatorname{cosec}\theta = 2$ show that $\sin^5\theta \sec^5\theta = 2$

(*Hint:* $\sin\theta + \dfrac{1}{\sin\theta} = 2$ $\quad\therefore \sin^2\theta - 2\sin\theta + 1 = 0$.)

Ex. 46. If $\sin\alpha \sin\beta - \cos\alpha\cos\beta + 1 = 0$ show that $1 + \cot\alpha \tan\beta = 0$.

6.2 INVERSE TRIGONOMETRIC FUNCTIONS

The inverse sine function is written

$y = \sin^{-1}x$ which implies $x = \sin y$ $\quad -\frac{1}{2}\pi \le y \le \frac{1}{2}\pi.$

Note: $\sin^{-1}x \neq (\sin x)^{-1}$

$$(\sin x)^{-1} = \frac{1}{\sin x}. \text{ and } \sin(\sin^{-1}x) = x$$

$$\cos(\cos^{-1}x) = x,\ \cos^{-1}(\cos x) = x.$$

Ex. 47. $\sin^{-1}\left(\frac{1}{\sqrt{2}}\right) = \frac{\pi}{4}$ since $\sin\frac{\pi}{4} = \frac{1}{\sqrt{2}}$

$$\sin^{-1}\left(-\frac{1}{\sqrt{2}}\right) = -\frac{\pi}{4} \text{ since } \sin\left(-\frac{\pi}{4}\right) = -\sin\frac{\pi}{4} = -\frac{1}{\sqrt{2}}.$$

Other trigonometric inverse functions are $\cos^{-1}x$, $\tan^{-1}x$, $\sec^{-1}x$, $\text{cosec}^{-1}x$, $\cot^{-1}x$.

Ex. 48. $\cos^{-1}\left(\frac{1}{\sqrt{2}}\right) = \frac{\pi}{4}$ since $\cos\frac{\pi}{4} = \frac{1}{\sqrt{2}}$

Ex. 49. Prove $\cos^{-1}x = \frac{1}{2}\pi - \sin^{-1}x$ for $|x| \le 1$

(Let $\cos^{-1}x = y$ $\quad\therefore\quad$ $\cos y = x$ or $\sin\left(\frac{\pi}{2} - y\right) = x$

$\therefore\quad \sin^{-1}x = \frac{\pi}{2} - y = \frac{\pi}{2} - \cos^{-1}x$. Hence the result).

Ex. 50. Prove $\cot^{-1}x = \frac{1}{2}\pi - \tan^{-1}x$.

Ex. 51. Prove (*i*) $\sec^{-1} 2 = \frac{\pi}{2}$ (*ii*) $\text{cosec}^{-1}(-1) = \frac{3}{2}\pi$ (*iii*) $\sec^{-1}(-2) = \frac{4}{3}\pi$

(*iv*) $\cos^{-1}\left(\frac{3}{\sqrt{5}}\right) + \cos^{-1}\left(\frac{2}{\sqrt{5}}\right) = \frac{\pi}{4}$ (*v*) $2\tan^{-1}\frac{1}{3}\tan^{-1}\left(-\frac{1}{7}\right) = \frac{\pi}{4}$.

Ex. 52. Plot the Graphs of $y = \sin^{-1}x$, $\cos^{-1}x$, $\tan^{-1}x$.

If	$x = 0$,	$\sin^{-1}x = \sin^{-1} 0 = 0$,	$\cos^{-1} 0 = \pi/2$.
If	$x = \frac{1}{2}$,	$\sin^{-1}1/2 = \pi/6$,	$\cos^{-1} 1/2 = \pi/3$
If	$x = -1/2$	$\sin^{-1}(-1/2) = -\pi/6$,	$\cos^{-1}(-1/2) = \pi/3$.
If	$x = 1$.	$\sin^{-1} 1 = \pi/2$,	$\cos^{-1} 1 = 0$
If	$x = -1$,	$\sin^{-1}(-1) = -\pi/2$,	$\cos^{-1}(-1) = \pi$.

Graphs of Inverse Functions

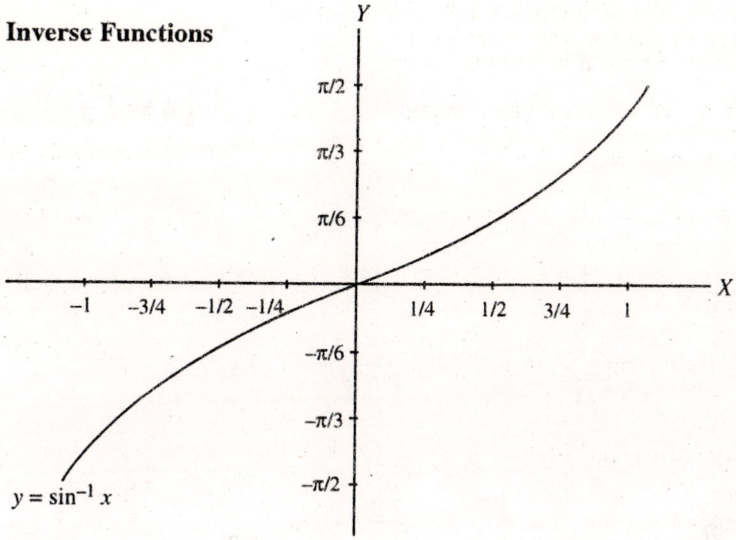

$y = \sin^{-1} x$

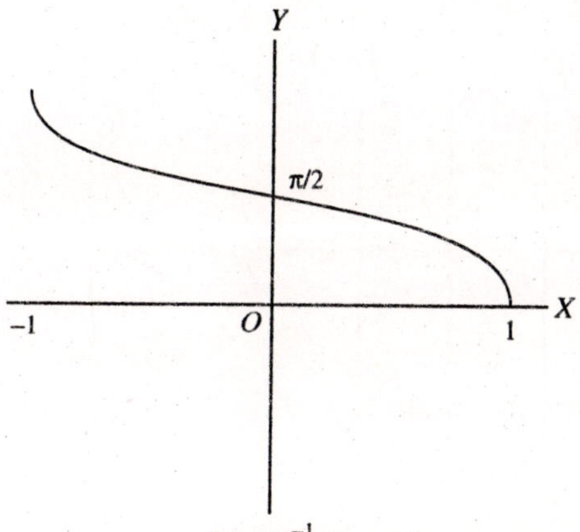

$y = \cos^{-1} x$

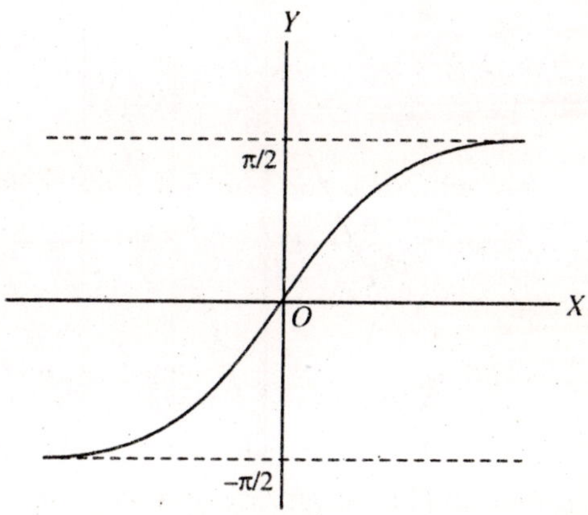

$y = \tan^{-1} x$

Range of Trigonometric Functions

$y = \sin x$

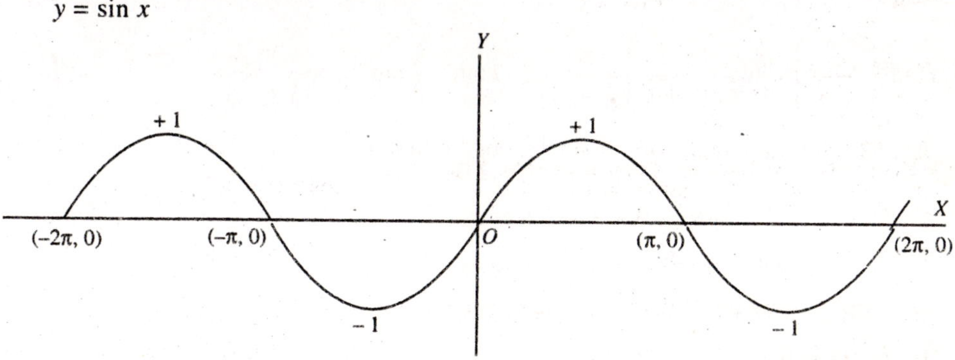

The range of $y = \sin x$ is $-1 \le y \le 1$. The function increases strictly from -1 to $+1$ as x increases from $-\pi/2$ to $\pi/2$ and decreases strictly from $+1$ to -1 as x increases from $\pi/2$ to $3\pi/2$ and so on.

The range of $y = \cos x$ is $-1 \leq y \leq 1$

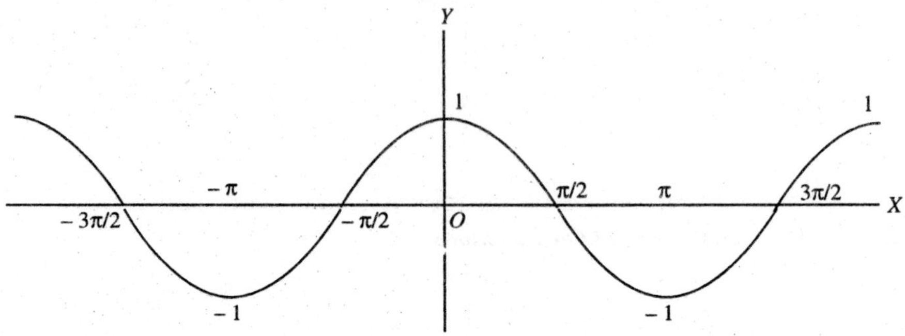

The range of $y = \tan x$ is $-\infty \leq y \leq \infty$

The range of $y = \tan^{-1}x$ is $-\dfrac{\pi}{2} \leq y \leq \dfrac{\pi}{2}$

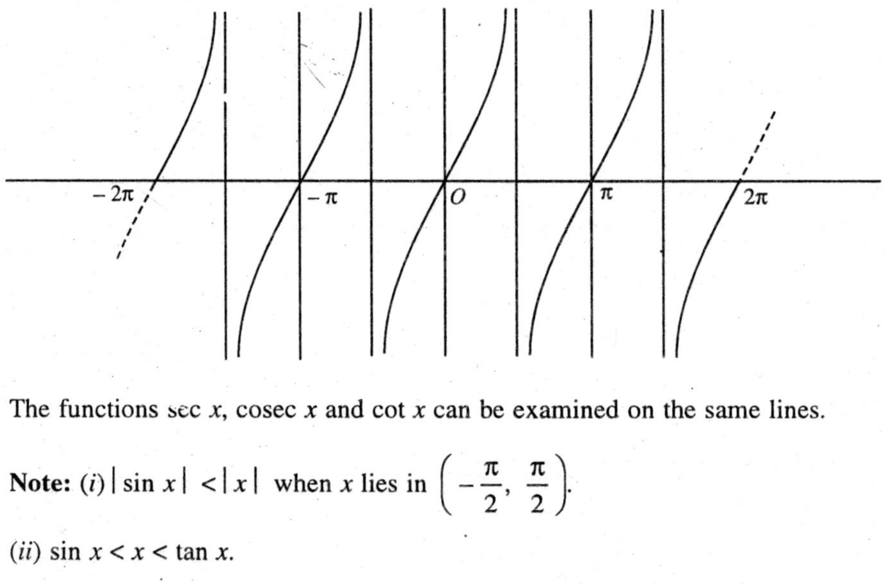

The functions sec x, cosec x and cot x can be examined on the same lines.

Note: (*i*) $|\sin x| < |x|$ when x lies in $\left(-\dfrac{\pi}{2}, \dfrac{\pi}{2}\right)$.

(*ii*) $\sin x < x < \tan x$.

Some Properties of Inverse Trigonometric Functions

I. If $\quad \sin^{-1}x = y, \sin y = x$

$\therefore \quad \sin^{-1}(\sin y) = y, \cos^{-1}(\cos y) = y$ etc.

Also $\quad \sin(\sin^{-1}y) = y, \cos(\cos^{-1}y) = y$ etc.

II. *Reciprocity*

$$\sin^{-1}\frac{1}{x} = \text{cosec}^{-1}x$$

(Since $\sin^{-1}\frac{1}{x} = y$ implies $\sin y = \frac{1}{x}$ or $x = \frac{1}{\sin y}$

$\therefore \quad x = \text{cosec } y$ or $y = \text{cosec}^{-1}x$)

Similarly $\quad \cos^{-1}\frac{1}{x} = \sec^{-1}x, \qquad \sec^{-1}\frac{1}{x} = \cos^{-1}x$

$$\tan^{-1}\frac{1}{x} = \cot^{-1}x, \qquad \cot^{-1}\frac{1}{x} = \tan^{-1}x \qquad \ldots(20)$$

III. *Inverse Functions are Odd Functions*

$$\sin^{-1}(-x) = -\sin^{-1}x$$

$$[\sin^{-1}(-x) = y \text{ implies } -x = \sin y \text{ or } x = -\sin y]$$

$$\tan^{-1}(-x) = -\tan^{-1}x$$

$$\text{cosec}^{-1}(-x) = -\text{cosec}^{-1}x$$

IV. *Other Properties*

$$\sin^{-1}(-x) = -\sin^{-1}x \qquad -1 \le x \le 1$$

$$\text{cosec}^{-1}(-x) = -\text{cosec}^{-1}x \qquad -1 \ge x, x \ge 1$$

$$\cos^{-1}(-x) = \pi - \cos^{-1}x \qquad -1 \le x \le 1$$

$$\sec^{-1}(-x) = \pi - \sec^{-1}x \qquad -1 \ge x, x \ge 1$$

$$\tan^{-1}(-x) = -\tan^{-1}x \qquad -\infty < x < \infty$$

$$\cot^{-1}(-x) = \pi - \cot^{-1}x \qquad -\infty < x < \infty \qquad \ldots(21)$$

$$\sin^{-1}x + \cos^{-1}x = \frac{\pi}{2}$$

$$\tan^{-1}x + \cot^{-1}x = \frac{\pi}{2}$$

$$\text{cosec}^{-1}x + \sec^{-1}x = \frac{\pi}{2} \qquad \ldots(22)$$

$$\tan^{-1}x + \tan^{-1}y = \tan^{-1}\frac{x+y}{1-xy} \text{ where } xy < 1, x > 0, y > 0$$

$$\tan^{-1}x - \tan^{-1}y = \tan^{-1}\frac{x-y}{1+xy}. \qquad \ldots(23)$$

$$x, y > 0$$

e.g., If $\quad \sin^{-1}x = \theta$

$$x = \sin\theta$$

$$= \cos\left(\frac{\pi}{2} - \theta\right)$$

$\therefore \quad \cos^{-1}x = \frac{\pi}{2} - \theta$

$\therefore \quad \sin^{-1}x + \cos^{-1}x = \frac{\pi}{2}$

Ex. 54. Find

(*i*) $\cos(\tan^{-1} 3/4)$, (*ii*) $\sin \cot^{-1}x$

(*i*) If $\tan^{-1}\frac{3}{4} = x$, $\tan x = \frac{3}{4}$, $\cos x = \frac{4}{5}$ $\therefore x = \cos^{-1}\frac{4}{5}$

$$\cos(\tan^{-1} 3/4) = \cos\left(\cos^{-1}\frac{4}{5}\right) = \frac{4}{5}.$$

(*ii*) If $\cot^{-1}x = y$ then $\cot y = x$

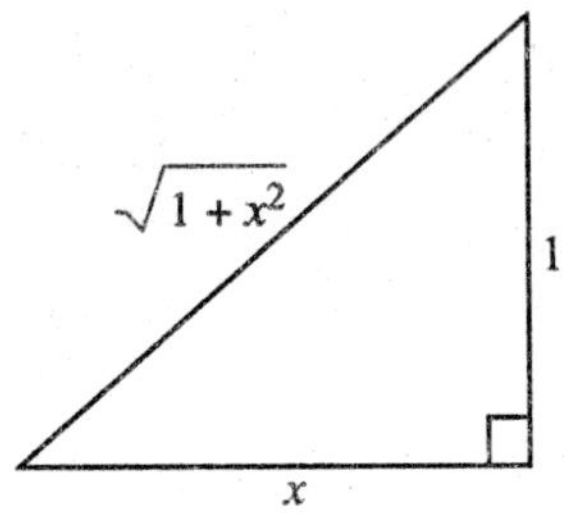

$$\therefore \quad \sin y = \frac{1}{\sqrt{1+x^2}}$$

$$y = \sin^{-1}\frac{1}{\sqrt{1+x^2}}$$

$$\therefore \quad \sin\cot^{-1}x = \sin y$$

$$= \sin\sin^{-1}\frac{1}{\sqrt{1+x^2}} = \frac{1}{\sqrt{1+x^2}}$$

Ex. 55. Simplify: $y = \tan^{-1}\left(\frac{3a^2x - x^3}{a(a^2 - 3x^2)}\right)$

Let $x = a\tan\theta$ so that $\theta = \tan^{-1}\frac{x}{a}$

$$\therefore \quad y = \tan^{-1}\left(\frac{3a^2 a\tan\theta - a^3\tan^3\theta}{a(a^2 - 3a^2\tan^2\theta)}\right)$$

$$= \tan^{-1}\frac{3\tan\theta - \tan^3\theta}{1 - 3\tan^2\theta} = \tan^{-1}(\tan 3\theta) = 3\theta$$

$$= 3\tan^{-1}\frac{x}{a}.$$

Ex. 56. $\cot^{-1} 3 + \cot^{-1} 4$

$$= \tan^{-1}\frac{1}{3} + \tan^{-1}\frac{1}{4}$$

$$= \tan^{-1}\frac{\frac{1}{3}+\frac{1}{4}}{1-\frac{1}{3}\cdot\frac{1}{4}} \quad \text{(Rule 23)}$$

$$= \tan^{-1}\frac{7}{11}$$

Ex. 57. $\cos\left(\frac{\pi}{6} + \cos^{-1}\frac{\sqrt{3}}{2}\right)$

$$= \cos\left(\frac{\pi}{6} + \pi - \cos^{-1}\frac{\sqrt{3}}{2}\right) \quad \text{(Rule 21)}$$

$$= \cos\left(\frac{\pi}{6} + \pi - \frac{\pi}{6}\right)$$

$$= \cos \pi = -1$$

Ex. 58. Prove: $2 \tan^{-1}\frac{1}{2} + \tan^{-1}\frac{1}{7} = \tan^{-1}\frac{31}{17}$

Ex. 59. Prove: $\tan^{-1}\frac{\sqrt{1+x^2}-\sqrt{1-x^2}}{\sqrt{1+x^2}+\sqrt{1-x^2}} = \frac{\pi}{4} + \frac{1}{2}\cos^{-1}x^2.$

Ex. 60. Show: $\sec^{-1}\frac{1}{2x^2-1} = 2\cos^{-1}x.$

Ex. 61. Given: $x = 0.2$, $0 < \cos^{-1}x < \pi$, $-\frac{\pi}{2} < \sin^{-1}x < \frac{\pi}{2}$

show that $\cos(2\cos^{-1}x + \sin^{-1}x) = -\frac{2\sqrt{6}}{5}$

Ex. 62. Show $\sin\left(\sin^{-1}\frac{1}{2} + \cos^{-1}\frac{1}{2}\right) = 1.$

Ex. 63. If $\cos^{-1}\frac{x}{a} + \cos^{-1}\frac{y}{b} = \alpha.$

Show that $\frac{x^2}{a^2} + \frac{y^2}{b^2} - \frac{2xy\cos\alpha}{ab} = \sin^2\alpha.$

Ex. 64. Prove: (*i*) $\sin^{-1}\frac{3}{5} + \sin^{-1}\frac{18}{17} = \cos^{-1}\frac{36}{85}$

(*ii*) $\cos[2\tan^{-1}1/7] = \sin[4\tan^{-1}1/3]$

(*iii*) $\tan^{-1}\frac{1}{2} + \tan^{-1}\frac{1}{5} + \tan^{-1}\frac{1}{8} = \pi/4$

Ex. 65. Solve: $\tan^{-1}2x + \tan^{-1}3x = \pi/4$

Since $\frac{\pi}{4} = \tan^{-1}1$

$$\tan^{-1}2x + \tan^{-1}3x = \tan^{-1}\frac{2x+3x}{1-6x^2} = \tan^{-1}1$$

$\therefore \quad 5x = 1 - 6x^2$ or $6x^2 - 5x + 1 = 0$

$\therefore \quad (x+1)(6x-1) = 0$

$\therefore \quad x = -1, x = 1/6$

Substitute $x = -1$ in the original sum : $\tan^{-1}(-2) + \tan^{-1}(-3) \neq \frac{\pi}{4}$

$\therefore \quad x = -1$ is not a solution **Ans.** $x = 1/6$

Ex. 66. Show $\tan^{-1}x + \cot^{-1}(x + 1) = \tan^{-1}(x^2 + x + 1)$

Solve for z : $\sin^{-1}\dfrac{2x}{1+x^2} + \sin^{-1}\dfrac{24}{1+y^2} = 2\tan^{-1}z$ **Ans.** $z = \dfrac{x+y}{1-xy}$

Ex. 67. If $\cos^{-1}x + \cos^{-1}y + \cos^{-1}z = \pi$ show $x^2 + y^2 + z^2 + 2xyz = 1$.

6.3 PROPERTIES OF THE TRIANGLE

We shall denote by $2s$ the perimeter of the triangle ABC. If $AC = b$, $BC = a$, $AC = b$

$$25 = a + b + c$$

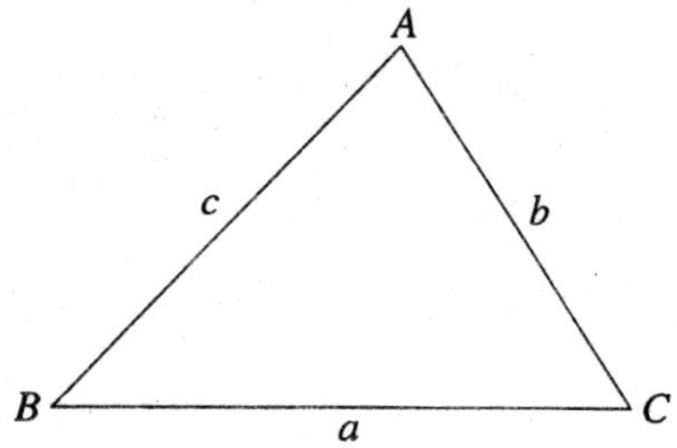

The area of the triangle is written Δ

The incentre is the point where the three internal bisectors of the triangle meet.

r = radius of the incircle

The perpendicular bisectors of the triangle meet at the circumcentre.

R = radius of the circumcircle.

The following results can be put to good use. (some proofs given later)

I. **Sine Rule :** $\dfrac{a}{\sin A} = \dfrac{b}{\sin B} = \dfrac{c}{\sin C} = 2R$

$a = 2R \sin A$, $b = 2R \sin B$, $c = 2R \sin C$

If two sides and one angle are known, the remaining side and angles can be found *e.g.*, if $a = 3$, $b = 4$, $\angle C = 90°$

$$\frac{a}{\sin A} = \frac{c}{\sin C} \quad \therefore \quad \frac{3}{\sin A} = \frac{5}{\sin 90} \quad (c^2 = a^2 + b^2 = 25,\ c = 5)$$

$$\sin A = \frac{3}{5} \text{ and } \sin C = \frac{4}{5}$$

II. **Cosine Rule :** $\cos A = \dfrac{b^2 + c^2 - a^2}{2bc}$ *i.e.*, $a^2 = b^2 + c^2 - 2bc \cos A$

$\cos B = \dfrac{c^2 + a^2 - b^2}{2ca}$ *i.e.*, $b^2 = c^2 + a^2 - 2ca \cos B$

$\cos C = \dfrac{a^2 + b^2 - c^2}{2ab}$ *i.e.*, $c^2 = a^2 + b^2 - 2ab \cos C$

III. **Projection Rule :** $a = b \cos C + c \cos B$

$b = c \cos A + a \cos C$

$c = a \cos B + b \cos A$

IV. **Half Angles**

$$\sin\frac{A}{2} = \sqrt{\frac{(s-b)(s-c)}{bc}},\ \sin\frac{B}{2} = \sqrt{\frac{(s-c)(s-a)}{ca}},\ \sin\frac{C}{2} = \sqrt{\frac{(s-a)(s-b)}{ab}}$$

$$\cos\frac{A}{2} = \sqrt{\frac{s(s-a)}{bc}},\ \cos\frac{B}{2} = \sqrt{\frac{s(s-b)}{ca}},\ \cos\frac{C}{2} = \sqrt{\frac{s(s-c)}{ab}}$$

$$\tan\frac{A}{2} = \sqrt{\frac{(s-b)(s-c)}{s(s-a)}},\ \tan\frac{B}{2}\sqrt{\frac{(s-c)(s-a)}{s(s-b)}},\ \tan\frac{C}{2} = \sqrt{\frac{(s-a)(s-b)}{s(s-c)}}$$

V. **Area of Triangle** *ABC*

$$\Delta = \frac{1}{2}bc\sin A = \frac{1}{2}ca\sin B = \frac{1}{2}ab\sin C = \frac{abc}{4R} = 2R^2\sin A\sin B\sin C = rs$$

Also $\Delta = \sqrt{s(s-a)(s-b)(s-c)}$

VI. $\tan\dfrac{B-C}{2} = \dfrac{b-c}{b+c}\cot\dfrac{A}{2}$ $\qquad \sin\dfrac{B-C}{2} = \dfrac{b-c}{a}\cos\dfrac{A}{2}$

$\tan\dfrac{C-A}{2} = \dfrac{c-a}{c+a}\cot\dfrac{B}{2}$ $\qquad \sin\dfrac{C-A}{2} = \dfrac{c-a}{b}\cos\dfrac{B}{2}$

$\tan\dfrac{A-B}{2} = \dfrac{a-b}{a+b}\cot\dfrac{C}{2}$ $\qquad \sin\dfrac{A-B}{2} = \dfrac{a-b}{c}\cos\dfrac{C}{2}$

$$\frac{a\sin(B-C)}{b^2-c^2} = \frac{b\sin(C-A)}{c^2-a^2} = \frac{c\sin(A-B)}{a^2-b^2}$$

$(a + b + c) = (b + c)\cos A + (c + a)\cos B + (a + b)\cos C$

Ex. 68. Show that if $\angle C = 60°$, $\dfrac{1}{a+c} + \dfrac{1}{b+c} = \dfrac{3}{a+b+c}$

Since $c^2 = a^2 + b^2 - 2ab\cos 60° = a^2 + b^2 - ab$ *i.e.*, $a^2 + b^2 - c^2 - ab = 0$

$$\text{LHS} = \frac{a+b+2c}{(a+c)(b+c)} = \frac{a+b+2c}{(a+c)(b+c)}\,\frac{a+b+c}{a+b+c} = \frac{3(a+c)(b+c)}{(a+c)(b+c)(a+b+c)} = \frac{3}{a+b+c}$$

Note. Using the result $a^2 + b^2 - c^2 = ab$

$$(a + b + c)(a + b + 2c) = a^2 + 2ab + b^2 + 2c^2 + ac + bc + 2ac + 2bc$$
$$= 3(a + c)(b + c)$$

Ex. 69. Prove the sine law OR

Prove that: In ΔABC, the sides *a, b, c* are proportional to the sines of the angles opposite to the sides.

$$\frac{a}{\sin A} = \frac{b}{\sin B} = \frac{C}{\sin C}$$

All the three angles may be acute as in (*i*)

One may be obtuse and the others acute as in (*ii*)

One a right angle and the others acute as in (*iii*)

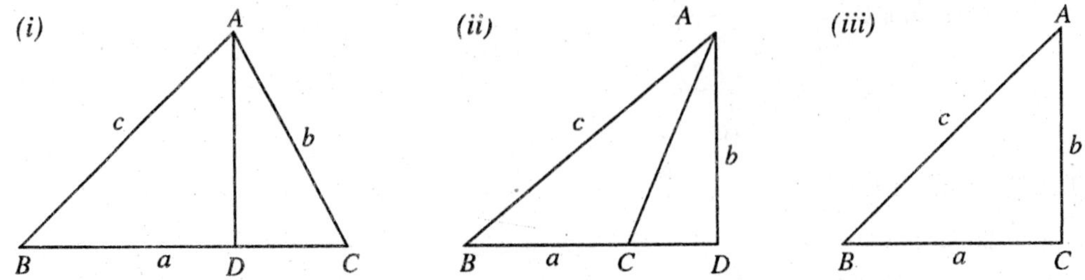

$AB = c \quad BC = a \quad CA = b$

AD is perpendicular to BC.

$$\frac{AD}{AB} = \sin B \text{ and } \frac{AD}{AC} = \sin C$$

$$\therefore \quad AD = AB \sin B = AC \sin C$$

$$c \sin B = b \sin C \text{ or } \frac{c}{\sin C} = \frac{b}{\sin B}$$

Similarly it can be shown that $\dfrac{a}{\sin A} = \dfrac{b}{\sin B}$

$$\frac{a}{\sin A} = \frac{b}{\sin B} = \frac{c}{\sin B}$$

We can also write $\dfrac{\sin A}{a} = \dfrac{\sin B}{b} = \dfrac{\sin C}{c}$

Ex. 70. If $a = b$ then $\sin A = \sin B$ or $A = B$

If $a = 3$, $b = 4$, and $\sin A = \dfrac{3}{4}$, find B.

$$\frac{3}{\sin A} = \frac{4}{\sin B} \quad \therefore \quad \sin B = \frac{4}{3}\sin A = \frac{4}{3} \cdot \frac{3}{4} = 1$$

$$\therefore \quad B = 90.$$

Ex. 71. Prove the cosine law. (Any diagram can be used)

In diagram (*i*) since $AD = b \sin C$, $CD = a - BD = a - b \cos C$

We have $AB^2 = AD^2 + BD^2 = (b \sin C)^2 + (a - b \cos C)^2$

$$\therefore \quad c^2 = b^2 \sin^2 C + a^2 - 2ab \cos C + b^2 \cos^2 C$$

$$= b^2 (\sin^2 C + \cos^2 C) + a^2 - 2ab \cos C$$

$$\therefore \quad c^2 = a^2 + b^2 - 2ab \cos C$$

Similarly $\quad b^2 = c^2 + a^2 - 2ca \cos B$

$$a^2 = b^2 + c^2 - 2bc \cos A$$

These results give the cosine law and can be proved for diagrams (*ii*) and (*iii*) also.

The cosine law can be written in the alternative form

$$\cos A = \frac{b^2 + c^2 - a^2}{2bc}$$

$$\cos B = \frac{c^2 + a^2 - b^2}{2ca}$$

$$\cos C = \frac{a^2 + b^2 - c^2}{2ab}$$

The Projection Law

In the above diagram (*i*)

$$BC = BD + DC = c \cos B + b \cos C$$

Note that $\frac{BD}{c} = \cos B$ and $\frac{DC}{b} = \cos C$

$\therefore \quad BC = a = b \cos C + c \cos B$

Similarly $CA = b = c \cos A + a \cos C$

$AB = c = a \cos B + b \cos C$

This can be proved from diagrams (*ii*) and (*iii*) also.

e.g., in diagram (*iii*) we have *a* right angled triangle

$a = BC = BD$

$= c \cos B + 0 \quad (\because \cos C = \cos 90° = 0)$

$= c \cos B + b \cos C \quad \because b \cos C = 0$

In diagram (*ii*)

$BC = BD - DC$

$\therefore \quad a = c \cos B - b \cos (\pi - C) \quad \because C$ is obtuse

$= \cos B + b \cos C$

Ex. 72. Find the formula for the area of a triangle.

$\Delta\, ABC$ has sides

$AB = c,$

$BC = a,$

$CA = b$

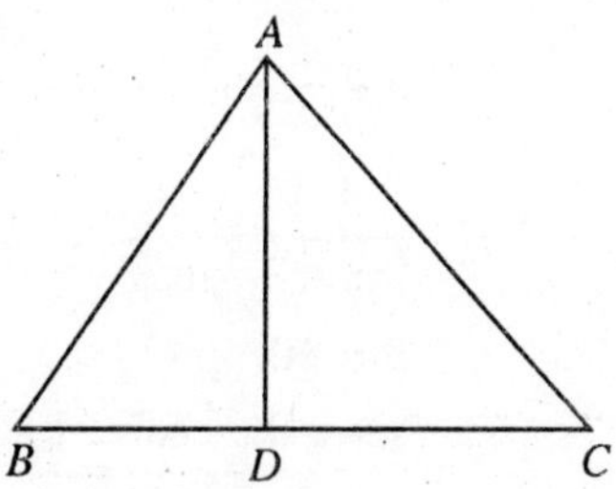

If $AD = h$ is the altitude from A to BC then the area of the triangle

$$\Delta = \frac{1}{2} BC \cdot AD$$

$$= \frac{1}{2} ab \sin C$$

If can be shown that $\Delta = \frac{1}{2} bc \sin A = \frac{1}{2} ca \sin B$

If we write $2s = a + b + c$ or $s = \frac{a+b+c}{2}$

Then $s - a = \frac{1}{2}(b + c - a),\ s - b = \frac{1}{2}(c + a - b),\ s - c = \frac{1}{2}(a + b - c)$

$$\Delta^2 = \left(\frac{1}{2}bc\sin A\right)^2 = \frac{1}{4}b^2c^2(1-\cos^2 A)$$

$$= \frac{1}{2}bc(1+\cos A)\cdot\frac{1}{2}bc(1-\cos A)$$

$$= \frac{2bc+b^2+c^2-a^2}{4}\cdot\frac{2bc-b^2-c^2+a^2}{4} \quad \left[\because \cos A = \frac{b^2+c^2-a^2}{2bc}\right]$$

$$= \frac{(b+c)^2-a^2}{4}\cdot\frac{a^2-(b-c)^2}{4}$$

$$= \frac{(b+c+a)}{2}\,\frac{(b+c-a)}{2}\,\frac{(a-b+c)}{2}\,\frac{(a+b-c)}{2}$$

$$= s(s-a)(s-b)(s-c)$$

$$\therefore \quad \Delta = \sqrt{s(s-a)(s-b)(s-c)}$$

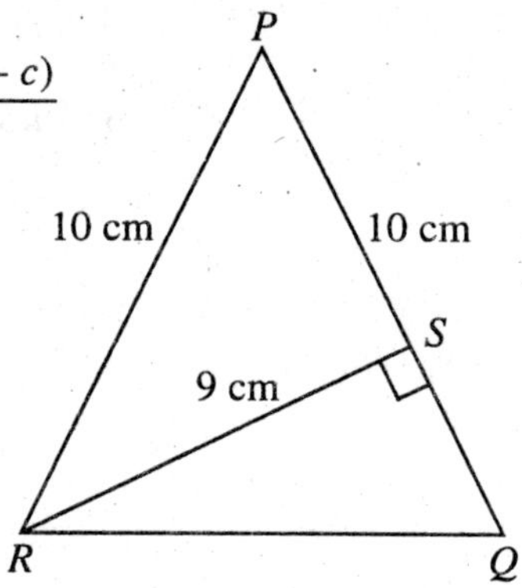

Ex. 73(*i*). RS is perpendicular to PQ. Find the area of the isosceles $\Delta\, PQR$

$\sin P = 9/10$,

$$\Delta = \frac{1}{2}PR\cdot PQ\sin P = \frac{1}{2}\cdot 10\cdot 10\cdot\frac{9}{10} = 45 \text{ cm}^2$$

Ex. 73(*ii*). A triangle has sides $5x + 4y$, $4x + 5y$ and $7x + 7y$ units (x, y positive). Show the triangle is obtuse angled.

Ex. 74. $\tan\dfrac{B-C}{2} = \dfrac{b-c}{b+c}\cot\dfrac{A}{2}$

or $\quad \tan\dfrac{A}{2} = \dfrac{b-c}{b+c}\cot\dfrac{B-C}{2}$

or $\quad$ If $B = 90°$, $\tan\dfrac{A}{2} = \sqrt{\dfrac{b-c}{b+c}}$

$$\frac{b-c}{b+c} = \frac{\frac{b}{c}-1}{\frac{c}{b}+1} = \frac{\frac{\sin B}{\sin C}-1}{\frac{\sin B}{\sin C}-1} = \frac{\sin B-\sin C}{\sin B+\sin C}$$

$$\therefore \quad \frac{b-c}{b+c} = \frac{2\cos\frac{B+C}{2}\sin\frac{B-C}{2}}{2\sin\frac{B+C}{2}\cos\frac{B-C}{2}} = \cot\frac{B+C}{2}\tan\frac{B-C}{2}$$

$$\therefore \quad \tan\frac{B-C}{2} = \frac{b-c}{b+c}\tan\frac{B+C}{2} = \frac{b-c}{b+c}\cot\frac{A}{2}$$

$$\left(\because \frac{A}{2}+\frac{B}{2}+\frac{C}{2} = 90° \text{ and } \tan\frac{B+C}{2} = \cot\frac{A}{2}\right)$$

Similarly $\tan\dfrac{C-A}{2} = \dfrac{c-a}{c+a}\cot\dfrac{B}{2}$

$$\tan\frac{A-B}{2} = \frac{a-b}{a+b}\cot\frac{C}{2}$$

If $B = 90°$ then $\tan\dfrac{B-C}{2} = \tan\dfrac{90-(90-A)}{2} = \tan\dfrac{A}{2}$

$(C = 180 - A - B = 180 - A - 90 = 90 - A)$

Thus $\tan\dfrac{B-C}{2} = \dfrac{b-c}{b+c}\cot\dfrac{A}{2}$

$$\therefore \quad \tan^2\frac{A}{2} = \frac{b-c}{b+c}$$

or
$$\tan\frac{A}{2} = \sqrt{\frac{b-c}{b+c}}$$

Ex. 75. Show: (*i*) $\tan\dfrac{A-B}{2} = \dfrac{a-b}{a+b} + \dfrac{c}{2}\cot\dfrac{c}{2}$

$$= \frac{a-b}{a+b} \text{ if } C = 90°$$

(*ii*) $\tan\dfrac{C-A}{2} = \dfrac{c-a}{c+a}\cot\dfrac{B}{2}$

Ex. 76. Prove: $\dfrac{a-b\cos C}{c-b\cos A} = \dfrac{\sin C}{\sin A}$

$$\text{LHS} = \frac{b\cos C + c\cos B - b\cos C}{a\cos B + b\cos A - b\cos A} = \frac{c\cos B}{a\cos B}$$

$$= \frac{c}{a} = \frac{\sin C}{\sin A}$$

Ex. 77. If A, B, C are in AP and $\dfrac{b}{c} = \sqrt{\dfrac{3}{2}}$, show $A = 75°$.

Since $A + C = 2B = 180 - B$ ($\because A + B + C = 180°$)

$$\therefore \quad 3B = 180° \text{ or } B = 60°$$

$$\therefore \quad \frac{b}{c} = \frac{\sqrt{3}}{\sqrt{2}} = \frac{\sin B}{\sin C} = \frac{\sin 60°}{\sin C}$$

$$\sin C = \sin 60°\frac{\sqrt{2}}{\sqrt{3}} = \frac{\sqrt{3}}{2}\,\frac{\sqrt{2}}{\sqrt{3}} = \frac{1}{\sqrt{2}}$$

$$\therefore \quad C = 45°$$

$$A = 180° - B - C = 180° - 60° - 45° = 75°$$

Ex. 78. Angles A, B, C are in the ratio 1 : 2 : 7. Show that the ratio of the largest side to the smallest side is $\dfrac{\sqrt{5}+1}{\sqrt{5}-1}$.

$$x + 2x + 7x = 180 \quad \therefore\ A = 18°,\ B = 36°,\ C = 126°$$

$$\frac{\text{Largest side}}{\text{Smallest side}} = \frac{\sin 126°}{\sin 18°} = \frac{\sin 54°}{\sin 18°} = \frac{(\sqrt{5}+1)/4}{(\sqrt{5}-1)/4}$$

Ex. 79. If $\dfrac{\sin A}{\sin C} = \dfrac{\sin(A-B)}{\sin(B-C)}$ show a^2, b^2, c^2 are in AP.

$$\frac{\sin A}{\sin C} = \frac{a}{c} = \frac{\sin A \cos B - \cos A \sin B}{\sin B \cos C - \cos B \sin C}$$

We can replace $\dfrac{\sin A}{\sin B}$ by $\dfrac{a}{b}$ after dividing the numerator and denominator by $\sin B$

$$\therefore \quad \frac{a}{c} = \frac{a\cos B - b\cos A}{b\cos C - c\cos C}$$

$$\therefore \quad ab \cos C - ac \cos C = ac \cos B - b \cos A$$

$$\therefore \quad b(a \cos C + c \cos A) = 2ca \cos B = 2ca \cdot \frac{c^2 + a^2 - b^2}{2ca}$$

$$\therefore \quad b \cdot b = c^2 + a^2 - b^2$$

or $$c^2 + a^2 = 2b^2$$

Ex. 80. Prove $a \sin(B - C) + b \sin(C - A) + c \sin(A - B) = 0$

LHS $= 2R \sin A \sin(B - C) + 2R \sin B \sin(C - A) + 2R \sin C \sin(A - B)$

Since $a = 2R \sin A$

LHS $= 2R [\sin(B + C) \sin(B - C) + \sin(C - A) \sin(C + A) + \sin(A - B) \sin(A + B)]$

$= 2R[\cos 2C - \cos 2B + \cos 2B - \cos 2A + \cos 2A - \cos 2C] = 0$

Ex. 81. Prove $\quad c^2 = (a-b)^2 \cos^2 \dfrac{C}{2} + (a+b)^2 \sin^2 \dfrac{C}{2}$

Since $\quad c^2 = a^2 + b^2 - 2ab \cos C$

$$= (a^2 + b^2)\left(\cos^2 \frac{C}{2} + \sin^2 \frac{C}{2}\right) - 2ab\left(\cos^2 \frac{C}{2} - \sin^2 \frac{C}{2}\right)$$

$$= \cos^2 \frac{C}{2}(a^2 + b^2 - 2ab) + \sin^2 \frac{C}{2}(a^2 + b^2 + 2ab)$$

$$= (a-b)^2 \cos^2 \frac{C}{2} + (a+b)^2 \sin^2 \frac{C}{2}$$

Ex. 82. (*i*) Show $\dfrac{(a+b+c)^2}{a^2+b^2+c^2} = \dfrac{\cot A/2 + \cot B/2 + \cot C/2}{\cot A + \cot B + \cot C}$

(*ii*) If the angles A, B, C are in A.P. show

$$2\cos\frac{A-C}{2} = \frac{a+c}{\sqrt{a^2+c^2-ac}}$$

(*iii*) If $A = 45°$, $B = 75°$ show $a + c\sqrt{2} - 2b = 0$

Ex. 83. The sides of a triangle are in AP and the greatest angle exceeds the smallest by 90°. Prove that the sides are in the proportion $\sqrt{7}+1 : \sqrt{7} : \sqrt{7}-1$

Ex. 84. If $\cos A = \dfrac{\sin B}{2\sin C}$ prove the triangle is isosceles

Ex. 85. If $\cot A + \cot B + \cot C = \sqrt{3}$, the triangle is equilateral.

Trigonometric Equations

A trigonometric equation involves trigonometric expressions like $\sin\theta$, $\cos\theta$, etc. where θ is an unknown. The solution of the equation is the value (or sometimes values) of θ that satisfies the equation.

Since the number of solutions of a trigonometric equation is infinite, the numerically smallest angle α is of importance. e.g. the equation $\sin\theta = k$ has solutions

$$\theta = n\pi + (-1)^n\alpha \begin{cases} n \in I \text{ i.e.,} \\ n = \ldots\ldots -2, -1, 0, 1, 2, 3, \ldots \\ \text{(all integers } - \text{ or } +) \end{cases}$$

where α is numerically the smallest angle for which $\sin\alpha = k$

$\cos\theta = k$ has solutions $\theta = 2n\pi \pm \alpha$ $(n \in I)$

where α is numerically the smallest angle for which $\cos\alpha = k$.

Similarly for $\tan\theta = k$, the solutions are $\theta = n\pi + \alpha$ $(n \in I)$

where α is numerically the smallest angle for which $\tan\alpha = k$

Ex. 86. $2\sin^2\theta + \sqrt{3}\cos\theta + 1 = 0$ $\qquad 0 \le \theta \le \pi$

$2(1 - \cos^2\theta) + \sqrt{3}\cos\theta + 1 = 0$ $\qquad$ *i.e.,* $2\cos^2\theta - \sqrt{3}\cos\theta + 3 = 0$

$\therefore \qquad (\cos\theta - \sqrt{3})(2\cos\theta + \sqrt{3}) = 0$

$\cos\theta = \sqrt{3}$ not possible; $\cos\theta = -\dfrac{\sqrt{3}}{2}$ or $\theta = \alpha = \dfrac{5\pi}{6}$ or $\dfrac{7\pi}{6}$

Ex. 87. $\sin^2\theta = \sin^2\alpha$

$\therefore \qquad \frac{1}{2}(1 - \cos 2\theta) = \frac{1}{2}(1 - \cos 2\alpha)$ or $\cos 2\theta = \cos 2\alpha$

$\therefore \qquad 2\theta = 2n\pi \pm 2\alpha \qquad$ *i.e.,* $\theta = n\pi \pm \alpha$.

Ex. 88. Given the usual trigonometric ratios find the trigonometric ratios of 18°, 36°, 54°, 72°.

Let $A = 18°$ then $2A = 36° = 90° - 3A$

$$\sin 2A = \sin (90 - 3A) = \cos 3A$$

i.e., $$2 \sin A \cos A = 4 \cos^3 A - 3 \cos A$$

$\therefore$ $$\cos A (2 \sin A - 4 \cos^2 A - 3) = 0$$

Since $\cos A \neq 0$, $2 \sin A - 4 \cos^2 A - 3 = 0$

$\therefore$ $$2 \sin A - 4 (1 - \sin^2 A) - 3 = 0$$

$\therefore$ $4 \sin^2 A + 2 \sin A - 1 = 0$ (compare $ax^2 + bx + c = 0$)

$\therefore$ $$\sin A = \sin 18° = \frac{-2 \pm 2\sqrt{5}}{8}$$

$\therefore$ $\sin 18° = \dfrac{-1 + \sqrt{5}}{4}$, a positive value

or $= \dfrac{-1 - \sqrt{5}}{4}$, a negative value which is not possible in the first quadrant

$\therefore$ $$\cos 18° = \sqrt{1 - \sin^2 18°} = \frac{\sqrt{10 + 2\sqrt{5}}}{4}$$

$$\tan 18° = \frac{\sqrt{5} - 1}{\sqrt{10 + 2\sqrt{5}}}$$

$$\sin 36° = 2 \sin 18° \cos 18° = \frac{2}{16}(\sqrt{5} - 1)(\sqrt{10 + 2\sqrt{5}})$$

$= \dfrac{\sqrt{10 - 2\sqrt{5}}}{4}$ on simplifying

$$\cos 36° = \frac{\sqrt{5} + 1}{4}$$

$$\tan 36° = \frac{\sqrt{5} + 1}{\sqrt{10 - 2\sqrt{5}}}$$

Since $\cos (90 - 2A) = \sin 2A$

i.e., $\cos 54° = \sin 36° = \dfrac{\sqrt{10 - 2\sqrt{5}}}{4}$

$$\sin 54° = \cos 36° = \frac{\sqrt{5} + 1}{4}$$

$$\sin 72° = \cos 18° = \frac{\sqrt{10 + \sqrt{5}}}{4}$$

$$\cos 72° = \sin 18° = \frac{\sqrt{5} - 1}{4}$$

Ex. 89. Solve (*i*) $2 \sin^2 \theta = 3 \cos \theta$ **Ans.** $\theta = 2n\pi \pm \frac{\pi}{3}$, $\alpha = \frac{\pi}{3}$

(*ii*) $4 \sin^4 \theta + 12 \cos^2 \theta = 7$, **Ans.** $\theta = n\pi + (-1)^n \left(-\frac{\pi}{4}\right)$, $\alpha = -\frac{\pi}{4}$

(*iii*) $(1 - \tan \theta)(1 + \sin 2\theta) = 1 + \tan \theta$ **Ans.** $\theta = n\pi - \frac{\pi}{4}$, $\alpha = -\frac{\pi}{4}$

Ex. 90. (*i*) Solve $\cos 6\theta + \cos 4\theta + \cos 2\theta + 1 = 0$ $0 \le \theta \le \pi$

Ans. $\begin{cases} \theta = \pi/2 \\ \theta = \pi/4, 3\pi/4 \\ \theta = \pi/6,\ 3\pi/6,\ 5\pi/6 \end{cases}$

(*ii*) $\sec 4\theta - \sec \theta = 2$ **Ans.** $\theta = n\pi + \pi/10$

(*iii*) $\cos(\pi/2 + 5\theta) + \sin \theta - 2 \cos 3\theta = 0$ **Ans.** $\theta = n\pi - \pi/4$

(*iv*) $\sin 3\alpha = 4 \sin \theta \sin(\theta + \alpha) \sin(\theta - \alpha)$ **Ans.** $\theta = n\pi \pm \frac{\pi}{3}$

CHAPTER 7

Determinants and Matrices

7.1 DETERMINANTS

A determinant is a symbol expressing the difference between two or more products. It is represented by a group of numbers enclosed by two vertical lines. a, b, c, d are the *elements* of the determinant $\begin{vmatrix} a & b \\ c & d \end{vmatrix}$ and its value is $ad - bc$.

This determinant is of the second order as it has two rows and two columns. Its value is obtained by taking the difference between the products of the elements of the two diagonals.

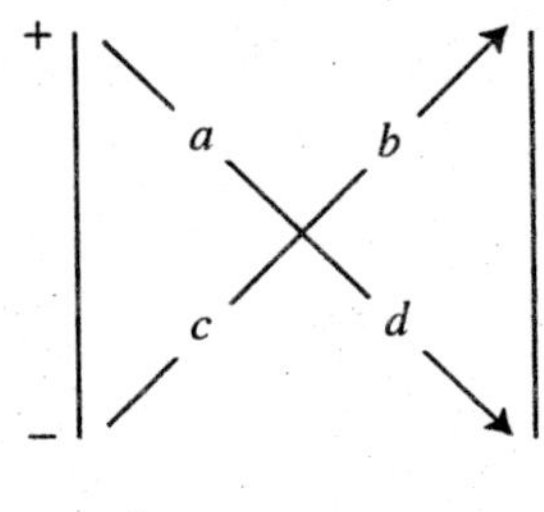

Ex. 1. $\begin{vmatrix} 5 & 7 \\ 2 & 4 \end{vmatrix} = 5 \times 4 - 7 \times 2 = 6;$ $\quad \begin{vmatrix} 2 & -4 \\ 1 & 0 \end{vmatrix} = 2 \times 0 - (-4) = 4$

$$\begin{vmatrix} m_1 & b_1 \\ m_2 & b_2 \end{vmatrix} = m_1 b_2 - m_2 b_1; \quad \begin{vmatrix} a_{11} & a_{12} \\ a_{21} & a_{22} \end{vmatrix} = a_{11}a_{22} - a_{12}\, a_{21}$$

7.2 SOLUTION OF TWO SIMULTANEOUS EQUATIONS BY DETERMINANTS

To solve the equations $a_1 x + b_1 y = m_1$ and $a_2 x + b_2 y = m_2$ multiply the first by b_2 and the second by b_1.

(*i*) $a_1 b_2 x + b_1 b_2\, y = m_1 b_2$ (*ii*) $a_2 b_1 x + b_1 b_2 y = m_2 b_1$

Subtracting (*ii*) from (*i*), $x(a_1 b_2 - a_2 b_1) = m_1 b_2 - m_2 b_1$

Subtracting (*i*) from (*ii*), $y(a_1 b_2 - a_2 b_1) = m_1 a_2 - m_2 a_1$

We have the solution

$$x = \frac{m_1 b_2 - m_2 b_1}{a_1 b_2 - a_2 b_1} = \frac{\begin{vmatrix} m_1 & b_1 \\ m_2 & b_2 \end{vmatrix}}{\begin{vmatrix} a_1 & b_1 \\ a_2 & b_2 \end{vmatrix}},$$

$(a_1 b_2 \neq a_2 b_1)$

$$y = \frac{a_1 m_2 - a_2 m_1}{a_1 b_2 - a_2 b_1} = \frac{\begin{vmatrix} a_1 & m_1 \\ a_2 & m_2 \end{vmatrix}}{\begin{vmatrix} a_1 & b_1 \\ a_2 & b_2 \end{vmatrix}}$$

This is known as Cramer's Rule. The formula can be easily used to solve simultaneous equations.

Ex. 2. In the equations $3x + 2y = 7$, $5x - y = 3$ we have

$a_1 = 3,\ b_1 = 2,\ m_1 = 7$ $\qquad a_2 = 5,\ b_2 = -1,\ m_2 = 3$

$$x = \frac{\begin{vmatrix} 7 & 2 \\ 3 & -1 \end{vmatrix}}{\begin{vmatrix} 3 & 2 \\ 5 & -1 \end{vmatrix}} = \frac{-7-6}{-3-10} = \frac{-13}{-13} = 1;$$

$$y = \frac{\begin{vmatrix} 3 & 7 \\ 5 & 3 \end{vmatrix}}{\begin{vmatrix} 3 & 2 \\ 5 & -1 \end{vmatrix}} = \frac{9-35}{-3-10} = 2.$$

Ex. 3. Solve by using determinants

(*i*) $4x + 2y = 2$, $\quad 3x - 5y = 21$ **Ans.** (*i*) $2, -3$

(*ii*) $5x + 3y = 1$, $\quad 2x - 3y = -8$. (*ii*) $-1, 2$.

7.3 HIGHER ORDER DETERMINANTS

A third order determinant has 3 rows and 3 columns.

$$\begin{vmatrix} a_1 & b_1 & c_1 \\ a_2 & b_2 & c_2 \\ a_3 & b_3 & c_3 \end{vmatrix} \qquad \begin{vmatrix} a_{11} & b_{12} & c_{13} \\ a_{21} & b_{22} & c_{23} \\ a_{31} & b_{32} & c_{33} \end{vmatrix}$$

are determinants of order 3.

A determinant of order n has n rows and n columns.

$$\begin{vmatrix} a_{11} & a_{12} & \cdots & a_{1n} \\ a_{21} & a_{22} & \cdots & a_{2n} \\ \cdots & \cdots & \cdots & \cdots \\ a_{n1} & a_{n2} & \cdots & a_{nn} \end{vmatrix}$$

is a determinant of order n. The element in the ith row and j^{th} column is a_{ij}. The first subscript in a_{ij} is the row subscript and the second, the column subscript.

Minors and Cofactors

With each element of a determinant are associated two quantities: the minor and the cofactor of that element.

The *minor* of an element of a determinant is the determinant of one lower order, obtained by deleting the row and the column containing that element. In the nth order determinant given above the minor of a_{ij} is the $(n-1)^{th}$ order determinant obtained by deleting the ith row and j^{th} column.

The *cofactor* of the element a_{ij} is the minor of a_{ij} multiplied by $(-1)^{i+j}$. Thus to get the cofactor multiply the minor by + 1 or – 1 according as $i + j$ (row number + column number) is even or odd.

In the third order determinant

$$\begin{vmatrix} a_1 & b_1 & c_1 \\ a_2 & b_2 & c_2 \\ a_3 & b_3 & c_3 \end{vmatrix}$$

the minors of a_1, b_1, c_2, c_3 are, respectively

$$\begin{vmatrix} b_2 & c_2 \\ b_3 & c_3 \end{vmatrix} = b_2c_3 - b_3c_2 \text{ (minor of } a_1), \qquad \begin{vmatrix} a_2 & c_2 \\ a_3 & c_3 \end{vmatrix} = a_2c_3 - a_3c_2 \text{ (minor of } b_1)$$

$$\begin{vmatrix} a_1 & b_1 \\ a_3 & b_3 \end{vmatrix} = a_1b_3 - a_3b_1 \text{ (minor of } c_2), \qquad \begin{vmatrix} a_1 & b_1 \\ a_2 & b_2 \end{vmatrix} = a_1b_2 - a_2b \text{ (minor of } c_3)_1$$

To find the corresponding cofactors note the row and column number of the element. Thus a_1 is in row 1, column 1. Here $i + j = 1 + 1 = 2$, an even number so that the cofactor of a_1 has + sign.

c_2 is in row 2, column 3. Since $2 + 3 = 5$ is odd the cofactor of c_2 has – sign.

The cofactor of an element is denoted by the corresponding capital letter of the element. Given below are the cofactors of some of the elements of the third order determinant given above.

Element	*Cofactor*
a_1	$A_1 = +(b_2c_3 - b_3c_2)$
b_1	$B_1 = -(a_2c_3 - b_3c_2)$
c_2	$C_2 = -(a_1b_3 - a_3b_1)$
c_3	$C_3 = +(a_1b_2 - a_2b_1)$ etc.

Expansion by Cofactors

The value of a determinant can be obtained by expansion by cofactors. This is done by adding together the product of the elements of any row or column of the determinant with their respective cofactors. Thus in terms of its first row the third order determinant may be evaluated as follows:

Steps

$$\begin{vmatrix} a_1 & b_1 & c_1 \\ a_2 & b_2 & c_2 \\ a_3 & b_3 & c_3 \end{vmatrix} = a_1A_1 + b_1B_1 + c_1C_1 \qquad (1)$$

$$= a_1\begin{vmatrix} b_2 & c_2 \\ b_3 & c_3 \end{vmatrix} - b_1\begin{vmatrix} a_2 & c_2 \\ a_3 & c_3 \end{vmatrix} + c_1\begin{vmatrix} a_2 & b_2 \\ a_3 & b_3 \end{vmatrix} \qquad (2)$$

$$= a_1(b_2c_3 - b_3c_2) - b_1(a_2c_3 - a_3c_2) + c_1(a_2b_3 - a_3b_2) \qquad (3)$$

$$= a_1b_2c_3 - a_1b_3c_2 - a_2b_1c_3 + a_3b_1c_2 + a_2b_3c_1 - a_3b_2c_1. \qquad (4)$$

Expanding the determinant in terms of the first *column* we get $a_1A_1 + a_2A_2 + a_3A_3$ which can be seen to be equal to $a_1A_1 + b_1B_1 + c_1C_1$.

If the elements of any row (or column) are multiplied by the cofactors of any other row (or column) the result is zero.

Thus $\quad a_1B_1 + a_2B_2 + a_3B_3 = 0,\ b_1C_1 + b_2C_2 + b_3C_3 = 0.$

Ex. 4. $\quad a_1A_1 + b_1B_1 + c_1C_1 = b_1B_1 + b_2B_2 + b_3B_3$

$$= c_1C_1 + c_2C_2 + c_3C_3$$

Note: To expand a determinant, one may straightaway proceed to steps (2), (3), (4) without using cofactors.

Ex. 5. $\begin{vmatrix} 1 & 3 & 2 \\ 0 & 2 & -1 \\ 4 & 1 & 5 \end{vmatrix} = 1\begin{vmatrix} 2 & -1 \\ 1 & 5 \end{vmatrix} - 3\begin{vmatrix} 0 & -1 \\ 4 & 5 \end{vmatrix} + 2\begin{vmatrix} 0 & 2 \\ 4 & 1 \end{vmatrix}$

$$= 1\,(10 + 1) - 3\,(0 + 4) + 2\,(0 - 8) = -17$$

Here we have expanded the determinant using the first row. If we expand using the second column, we have the same result since

$$-3\begin{vmatrix} 0 & -1 \\ 4 & 5 \end{vmatrix} + 2\begin{vmatrix} 1 & 2 \\ 4 & 5 \end{vmatrix} - 1\begin{vmatrix} 1 & 2 \\ 0 & -1 \end{vmatrix}$$

$$= -3\,(0 + 4) + 2\,(5 - 8) - 1\,(-1 - 0) = -17$$

Ex. 6. Find the value of $\begin{vmatrix} 2 & 1 & 0 \\ 3 & 2 & 4 \\ 1 & 2 & 3 \end{vmatrix}$ expanding by the first row, second column and third column. Show that in each case the result is -9.

7.4 SOME PROPERTIES OF DETERMINANTS

Property 1. If the rows of a determinant are changed into columns and columns into rows, the value of the determinant remains unchanged.

$$\begin{vmatrix} a_1 & b_1 \\ a_2 & b_2 \end{vmatrix} = \begin{vmatrix} a_1 & a_2 \\ b_1 & b_2 \end{vmatrix},\quad \begin{vmatrix} a_1 & b_1 & c_1 \\ a_2 & b_2 & c_2 \\ a_3 & b_3 & c_3 \end{vmatrix} = \begin{vmatrix} a_1 & a_2 & a_3 \\ b_1 & b_2 & b_3 \\ c_1 & c_2 & c_3 \end{vmatrix}$$

Property 2. If any two rows (columns) of a determinant are *interchanged* the sign of the determinant changes.

$$\begin{vmatrix} a_1 & b_1 & c_1 \\ a_2 & b_2 & c_2 \\ a_3 & b_3 & c_3 \end{vmatrix} = -\begin{vmatrix} b_1 & a_1 & c_1 \\ b_2 & a_2 & c_2 \\ b_3 & a_3 & c_3 \end{vmatrix}$$

Property 3. If the elements of any row (or column) of a determinant are multiplied by k, the value of the determinant is multiplied by k.

$$\begin{vmatrix} ka_1 & b_1 & c_1 \\ ka_2 & b_2 & c_2 \\ ka_3 & b_3 & c_3 \end{vmatrix} = k\begin{vmatrix} a_1 & b_1 & c_1 \\ a_2 & a_2 & c_2 \\ a_3 & a_3 & c_3 \end{vmatrix}$$

Property 4. If any row (or column) or a multiple of any row (or column) is added to or subtracted from another row (or column) the value of the determinant remains unchanged.

$$\begin{vmatrix} a_1 & b_1 & c_1 \\ a_2 & b_2 & c_2 \\ a_3 & b_3 & c_3 \end{vmatrix} = \begin{vmatrix} a_1 - kb_1 & b_1 & c_1 \\ a_2 - kb_2 & b_2 & c_2 \\ a_3 - kb_3 & b_3 & c_3 \end{vmatrix}.$$

Property 5. If any two rows (or columns) of a determinant are equal, the value of the determinant is zero.

$$\begin{vmatrix} a_1 & a_1 & c_1 \\ a_2 & a_2 & c_2 \\ a_3 & a_3 & c_3 \end{vmatrix} = 0$$

This may be checked either by expansion or by using property 4 in which case,

$$\begin{vmatrix} a_1 - a_1 & a_1 & c_1 \\ a_2 - a_2 & a_2 & c_2 \\ a_3 - a_3 & a_3 & c_3 \end{vmatrix} = \begin{vmatrix} 0 & a_1 & c_1 \\ 0 & a_2 & c_2 \\ 0 & a_3 & a_3 \end{vmatrix} = 0$$

If all the elements of any row (or column) are zero, the value of the determinant is zero.

These properties can be used to convert some elements of determinants to zero and simplify the calculation work.

Ex. 7. $$\begin{vmatrix} 1 & 6 & -1 \\ 2 & 3 & 3 \\ 3 & -3 & -2 \end{vmatrix} = \begin{vmatrix} 0 & 7 & -1 \\ 5 & 0 & 3 \\ 1 & -1 & -2 \end{vmatrix}$$ (Add col. 3 to col. 1 and substract col. 3 from col. 2) *i.e.* $C_1 + C_3,\ C_2 - C_3$

$$= 0\,(0 + 3) - 7\,(-10 - 3) - 1\,(-5 - 0)$$

$$= 0 + 91 + 5 = 96.$$

Ex. 8. $$\begin{vmatrix} 10 & 6 & -1 \\ 17 & 3 & 3 \\ -9 & -3 & -2 \end{vmatrix} = \begin{vmatrix} -8 & 0 & -5 \\ 8 & 0 & 1 \\ -9 & -3 & -2 \end{vmatrix}$$ (Here we have added row 3 to row 2 and twice row 3 to row 1) *i.e.* $R_2 + R_3,\ R_1 + 2R_3$

(Now add row 2 to row 1) i.e. $R_1 + R_2$

$$= \begin{vmatrix} 0 & 0 & -4 \\ 8 & 0 & 1 \\ -9 & -3 & -2 \end{vmatrix}$$

$$= -4(-24 - 0) = 96.$$

7.5 CRAMER'S RULE FOR THREE LINEAR EQUATIONS IN THREE UNKNOWNS (Extension of Sec. 7.2)

The system of equations

$$a_1x + b_1y + c_1z = m_1$$
$$a_2x + b_2y + c_2z = m_2$$
$$a_3x + b_3y + c_3z = m_3$$

has the unique solution

$$x = \frac{\begin{vmatrix} m_1 & b_1 & c_1 \\ m_2 & b_2 & c_2 \\ m_3 & b_3 & c_3 \end{vmatrix}}{\begin{vmatrix} a_1 & b_1 & c_1 \\ a_2 & b_2 & c_2 \\ a_3 & b_3 & c_3 \end{vmatrix}}, \quad y = \frac{\begin{vmatrix} a_1 & m_1 & c_1 \\ a_2 & m_2 & c_2 \\ a_3 & m_3 & c_3 \end{vmatrix}}{\begin{vmatrix} a_1 & b_1 & c_1 \\ a_2 & b_2 & c_2 \\ a_3 & b_3 & c_3 \end{vmatrix}}, \quad z = \frac{\begin{vmatrix} a_1 & b_1 & m_1 \\ a_2 & b_2 & m_2 \\ a_3 & b_3 & m_3 \end{vmatrix}}{\begin{vmatrix} a_1 & b_1 & c_1 \\ a_2 & b_2 & c_2 \\ a_3 & b_3 & c_3 \end{vmatrix}}$$

No solution exists if the denominator is zero.

Ex. 9. Solve by Cramer's Rule:

$$x + 6y - z = 10$$
$$2x + 3y + 3z = 17$$
$$3x - 3y - 2z = -9$$

$$x = \frac{\begin{vmatrix} 10 & 6 & -1 \\ 17 & 3 & 3 \\ -9 & -3 & -2 \end{vmatrix}}{\begin{vmatrix} 1 & 6 & -1 \\ 2 & 3 & 3 \\ 3 & -3 & -2 \end{vmatrix}} = \frac{96}{96} = 1 \text{ (Using the results of Ex. 7, 8)}$$

$$y = \frac{\begin{vmatrix} 1 & 10 & -1 \\ 2 & 17 & 3 \\ 3 & -9 & -2 \end{vmatrix}}{\begin{vmatrix} 1 & 6 & -1 \\ 2 & 3 & 3 \\ 3 & -3 & -2 \end{vmatrix}} = \frac{192}{96} = 2, \quad z = \frac{\begin{vmatrix} 1 & 6 & 10 \\ 2 & 3 & 17 \\ 3 & -3 & -9 \end{vmatrix}}{\begin{vmatrix} 1 & 6 & -1 \\ 2 & 3 & 3 \\ 3 & -3 & -2 \end{vmatrix}} = \frac{288}{96} = 3.$$

Ex. 10. Solve by determinants, the following equations:

$$3x + 3y - z = 11$$
$$2x - y + 2z = 9$$
$$4x + 3y + 2z = 25.$$

Ans. $x = 2, y = 3, z = 4$.

Ex. 11. It is known that consumption (C) and savings (S) are functions of income (Y). Also $C + S = Y$. If an economy may be described by $C = 100 + 0.5Y$, $S = 50 + 0.35Y$, find the equilibrium values of C, S and Y.

Ans. $C = 600$, $S = 400$, $Y = 1000$

Ex. 12. A man with an income of 1200 is indifferent about the following combinations of his purchases of goods X and Y as he gets the same satisfaction from either.

(*i*) 10 units of X and 20 units of Y (**Hint.** $10p_x + 20p_y = 120$

(*ii*) 15 units of X and 10 units of Y $15p_x + 10p_y = 1200$)

What are the prices of X and Y ? **Ans.** 60, 30

Ex. 13. Evaluate the following determinants

(*i*) $\begin{vmatrix} 1 & 1 & 1 \\ a & b & c \\ a^2 & b^2 & c^2 \end{vmatrix}$ (*ii*) $\begin{vmatrix} 1-a & a & 0 \\ -2 & 1+a & 1-a \\ 2+a & a & 2-a \end{vmatrix}$

(*iii*) $\begin{vmatrix} 2 & -1 & 5 \\ 3 & 0 & -2 \\ -5 & -2 & 1 \end{vmatrix}$ (*iv*) $\begin{vmatrix} 0 & b & c \\ -b & 0 & a \\ -c & -a & 0 \end{vmatrix}$ (**Hint.** Change rows into cols. and take $(-1)^3$ out)

Ans. (*i*) $ab^2 + bc^2 + ca^2 - a^2b - b^2c - c^2a$
(*ii*) $-5a^2 + 5a + 2$ (*iii*) 45 (*iv*) 0

Ex. 14. Show that the equation $\begin{vmatrix} x & y & 1 \\ a_1 & b_1 & 1 \\ a_2 & b_2 & 1 \end{vmatrix} = 0$ (**Hint.** Expand and substitute $x = a_1$, $y = b_1$ etc.)

is that of a straight line through (a_1, b_1), (a_2, b_2).

Ex. 15. Solve the following equations by the method of determinants:

(*i*) $x - 3y + z = 4$ (*ii*) $5a + 7b + 9c = 0$

$3x - y + 2z = 3$ $3a + 4b + 5c = 0$

$2x + 2y + z = 2$ $2a + 3b + 4x = 0.$

Ans. (*i*) Inconsistent equations, check by adding
(*ii*) If $a = p$, $b = -2p$, $c = p$

Ex. 16. Show that $\begin{vmatrix} 1 & a & a^3 \\ 1 & b & b^3 \\ 1 & c & c^3 \end{vmatrix} = (a - b)(b - c)(c - a)(a + b + c)$

Hence or otherwise show that $\begin{vmatrix} 1 & 1 & 1 \\ 1 & 2 & 8 \\ 1 & 3 & 27 \end{vmatrix} = 12$ (**Hint.** Subtract row 2 from row 1 and row 3 from row 2 and expand after taking out common factors)

Ex. 17. $\begin{vmatrix} 2 & -1 & 1 & 2 \\ 1 & 2 & 0 & -1 \\ -2 & -3 & -4 & 1 \\ 1 & -1 & 1 & 2 \end{vmatrix} = 2\begin{vmatrix} 2 & 0 & -1 \\ -3 & -4 & 1 \\ -1 & 1 & 2 \end{vmatrix}$

$$-(-1)\begin{vmatrix} 1 & 0 & -1 \\ -2 & -4 & 1 \\ 1 & 1 & 2 \end{vmatrix} + 1\begin{vmatrix} 1 & 2 & -1 \\ -2 & -3 & 1 \\ 1 & -1 & 2 \end{vmatrix}$$

$$-2\begin{vmatrix} 1 & 2 & 0 \\ -2 & -3 & 4 \\ 1 & -1 & 1 \end{vmatrix}$$

$$= 2(-11) - (-1)(-11) + 1(0) - 2(-11) = -11$$

Instead of expanding, it is better to use the properties of determinants. In this case if we add col. 2 to col. 3, and add twice col. 2 to col. 1 and col. 4, we have

$$\begin{vmatrix} 0 & -1 & 0 & 0 \\ 5 & 2 & 2 & 3 \\ -8 & -3 & -7 & -5 \\ -1 & -1 & 0 & 0 \end{vmatrix} = -(-1)\begin{vmatrix} 5 & 2 & 3 \\ -8 & -7 & -5 \\ -1 & 0 & 0 \end{vmatrix}$$

$$= -1\begin{vmatrix} 2 & 3 \\ -7 & -5 \end{vmatrix} = -11.$$

Ex. 18. Show that

(*i*) $\begin{vmatrix} a^2 + 2ab & a & 1 \\ 3b^2 & b & 1 \\ c^2 + 2bc & c & 1 \end{vmatrix} = \begin{vmatrix} a^2 & a & 1 \\ b^2 & b & 1 \\ c^2 & c & 1 \end{vmatrix}$, (**Hint.** Subtract $2b$ times col. 2 from col. 1)

(*ii*) $\begin{vmatrix} 4 & 11 & 10 \\ 7 & 2 & 6 \\ 1 & 5 & 4 \end{vmatrix} = 3$ (*iii*) $\begin{vmatrix} a-b-c & 2a & 2a \\ 2b & b-c-a & 2b \\ 2c & 2c & c-a-b \end{vmatrix} = (a+b+c)^2$,

(*iv*) $\begin{vmatrix} b+c & c & b \\ c & c+a & a \\ b & a & a+b \end{vmatrix} = 4abc.$ (**Hint.** (*iii*) & (*iv*) Add row 2 and row 3 to row 1)

Ex. 19. Find the value of

(*i*) $\begin{vmatrix} 1 & -2 & 4 & 2 \\ 3 & 0 & 5 & -2 \\ 2 & 7 & 3 & -1 \\ -4 & -3 & 1 & 2 \end{vmatrix}$ (*ii*) $\begin{vmatrix} 0 & a & b & c \\ a & 0 & c & b \\ b & c & 0 & a \\ c & b & a & 0 \end{vmatrix}$

(*iii*) $\begin{vmatrix} 1 & 1 & 1 & 1 \\ a & b & c & d \\ a^2 & b^2 & c^2 & d^2 \\ a^3 & b^3 & c^3 & d^3 \end{vmatrix}$ (*iv*) $\begin{vmatrix} 1 & ab & c^2 \\ 1 & bc & a^2 \\ 1 & ca & b^2 \end{vmatrix}$

Ans. (*i*) -461

(*ii*) $(a+b+c)(a+b-c)(a-b+c)(-a+b+c)$

(*iii*) $(a-b)(b-c)(c-d)(a-d)(a-c)(b-d)$

(*iv*) $(b-c)(c-a)(b-a)(a+b+c)$.

Ex. 20. Show that $\begin{vmatrix} 7 & 13 & 10 & 6 \\ -5 & -9 & -7 & -4 \\ 8 & 12 & 11 & 7 \\ -4 & -10 & -6 & -3 \end{vmatrix} = 0$

Ex. 21. If $\begin{vmatrix} x^3+1 & x^2 & x \\ y^3+1 & y^2 & y \\ z^3+1 & z^2 & z \end{vmatrix} = 0$ show $xyz = -1$ given $x \neq y \neq z$

In this the first column can be split as follows:

$$\text{LHS} = \begin{vmatrix} x^3 & x^2 & x \\ y^3 & y^2 & y \\ z^3 & z^2 & z \end{vmatrix} + \begin{vmatrix} 1 & x^2 & x \\ 1 & y^2 & y \\ 1 & z^2 & z \end{vmatrix}$$

$$= xyz \begin{vmatrix} x^2 & x & 1 \\ y^2 & y & 1 \\ z^2 & z & 1 \end{vmatrix} + \begin{vmatrix} 1 & x^2 & x \\ 1 & y^2 & y \\ 1 & z^2 & z \end{vmatrix} = (xyz+1) \begin{vmatrix} x^2 & x & 1 \\ y^2 & y & 1 \\ z^2 & z & 1 \end{vmatrix} = 0$$

$\therefore (xyz + 1)(x - y)(y - z)(z - x) = 0$

Since $x \neq y \neq z$ $\therefore xyz = -1$

Ex. 22. Prove $\begin{vmatrix} bc & bc_1 + b_1c & b_1c_1 \\ ca & ca_1 + c_1a & c_1a_1 \\ ab & ab_1 + a_1b & a_1b_1 \end{vmatrix} = (bc_1 - b_1c)(ca_1 - c_1a)(ab_1 - a_1b)$

Ex. 23. Given $a + b + c = 0$, solve for x :

$$\begin{vmatrix} a-x & c & b \\ c & b-x & a \\ b & a & c-x \end{vmatrix} = 0$$

Ans. $x = 0$ or $\pm\sqrt{\frac{3}{2}(a^2+b^2+c^2)}$

7.6 MATRICES

A matrix is a rectangular array of numbers, represented by enclosing the array by brackets. The numbers of rows and columns are the dimensions or order of the matrix.

$\begin{bmatrix} 2 & 4 & 3 \\ 6 & -1 & 0 \end{bmatrix}$ is a matrix of dimensions or order 2×3. It has 2 rows and 3 columns.

$\begin{bmatrix} a_{11} & a_{12} \\ a_{21} & a_{22} \end{bmatrix}$ is a 2×2 matrix. It has 2 rows and 2 columns.

$$A = \begin{bmatrix} a_{11} & a_{12} & \cdots & a_{1n} \\ a_{21} & a_{22} & \cdots & a_{2n} \\ \cdots & \cdots & \cdots & \cdots \\ a_{m1} & a_{m2} & \cdots & a_{mn} \end{bmatrix}$$

is a matrix of order $m \times n$. It has m rows and n columns. The element in the ith row and j^{th} column is denoted by a_{ij}. Briefly we write $A = [a_{ij}]$

If $m = n$, *i.e.*, if the number of rows equals the number of columns, we have a *square* matrix. An $n \times n$ matrix is also called an nth order matrix.

Vectors

A vector is a special type of matrix in which there is only one row or one column. A $1 \times n$ matrix $[a_1 a_2 \ldots a_n]$ is called a *row vector.*

An $m \times 1$ matrix $\begin{bmatrix} a_1 \\ a_2 \\ \vdots \\ a_m \end{bmatrix}$ is called a *column vector.*

We note that a row vector is an ordered collection of numbers written in a row. A column vector is an ordered collection of numbers written in a column. Two row (or column) vectors are equal if and only if the corresponding elements of the vectors are equal. If the vectors given are

$$u = [a, b],\ v = \begin{bmatrix} a \\ b \end{bmatrix},\ u_1 = [a\ b],\ v_1 = \begin{bmatrix} b \\ a \end{bmatrix} \text{ then } u = u_1,\ u \neq v \neq v_1.$$

The *sum* of two vectors of the same type is a vector whose components are the sums of the corresponding elements of the given vectors:

$$[u_1 u_2 \ldots u_m] + [v_1 v_2 \ldots v_m] = [u_1 + v_1\ u_2 + v_2 \ldots u_m + v_m].$$

The *product* of a vector by a scalar k i.e. the numerical multiplication of a vector by a number k, gives a vector whose components are k times the components of the given vector.

$$\begin{aligned} k[u_1\ u_2 \ldots u_m] &= [k\,u_1\ k\,u_2 \ldots k\,u_m] \\ &= [u_1\ u_2 \ldots u_m]k. \end{aligned}$$

Ex. 24. If $u = [1, 2], v = [3, 5],\ w = [4, 7]$

$u - v = [-2, -3],\ 2u + 3w = [14\ \ 25]$

The *associative law of addition* holds for vectors

$$[u + v] + w = u + [v + w]$$

The *product* of a row vector $u = [u_1\ u_2 \ldots u_m]$

and a column vector $v = \begin{bmatrix} v_1 \\ v_2 \\ \vdots \\ v_m \end{bmatrix}$, each having the *same* number m of components, is defined by

$$uv = [u_1v_1 + u_2v_2 + \ldots + u_m\ v_m]$$

The row vector u is written first and the column vector v written after u and the number of components is the same in each.

Ex. 25. $[1\ 2\ 3]\begin{bmatrix} 2 \\ 0 \\ -1 \end{bmatrix} = [1.2 + 2.0 + 3\,(-1)] = -1$

$$[1\ 0]\begin{bmatrix} 0 \\ 1 \end{bmatrix} = [1.0 + 0.1] = 0.$$

Ex. 26. Show that if $u = \begin{bmatrix} 2 \\ 4 \\ 3 \end{bmatrix}$, $x = \begin{bmatrix} 1 \\ 2 \\ 0 \end{bmatrix}$

$$v = [-2\ \ 0\ \ 1],\ y = [0\ \ 0\ \ 1]$$

then $$uv - yx = -1$$

Also find $2\,vu + 3\,yx$, $yu - vx$. **Ans.** -2, 1

7.7 ELEMENTARY ALGEBRA OF MATRICES

Equality : Two matrices are equal if and only if they have the same number of rows and the same number of columns and the corresponding elements in the two are equal.

$$\begin{bmatrix} a & b \\ c & d \end{bmatrix} = \begin{bmatrix} p & q \\ r & s \end{bmatrix} \text{ if } \begin{matrix} a = p, & b = q \\ c = r, & d = s \end{matrix}$$

In symbols, if $A = [a_{ij}]$, $B = [b_{ij}]$ then $A = B$ if A, B have the same dimensions and $a_{ij} = b_{ij}$ for each (i, j).

Addition. The sum of two matrices with the same dimensions is a matrix with elements which are the sums of the corresponding elements of the given matrices.

$$\begin{bmatrix} a & b \\ c & d \end{bmatrix} = \begin{bmatrix} p & q \\ r & s \end{bmatrix} = \begin{bmatrix} a+p & b+q \\ c+r & d+s \end{bmatrix}$$

If $A = [a_{ij}]$, $B = [b_{ij}]$ have the same dimensions, then $A + B = [c_{ij}]$ where $c_{ij} = a_{ij} + b_{ij}$.

If two matrices have different dimensions their addition is not defined.

Ex. 27. $\begin{bmatrix} 1 & 2 \\ 4 & 7 \end{bmatrix} + \begin{bmatrix} 2 & 0 & 5 \\ 3 & 4 & 7 \end{bmatrix}$ is not defined.

The following laws of addition for matrices may be noted :

Associative law $A + B = B + A$

Commutative $A + (B + C) = (A + B) + C$

Scalar Multiplication

The product of a matrix by a scalar k is a matrix with new elements k times the elements of the given matrix. (Compare with scalar multiplication of a determinant where only one row or column of elements is affected).

$$k\begin{bmatrix} a & b \\ c & d \end{bmatrix} = \begin{bmatrix} ka & kb \\ kc & kd \end{bmatrix} = \begin{bmatrix} a & b \\ c & d \end{bmatrix}k$$

$$kA = k[a_{ij}] = Ak$$

Ex. 28. $2\begin{bmatrix} 3 & -2 \\ 0 & 1 \end{bmatrix} = \begin{bmatrix} 6 & -4 \\ 0 & 2 \end{bmatrix}$

Note the following:

$$-A = (-1)A,\ A - B = A + (-1)B$$

$$k_1 k_2 A = k_1(k_2 A),\ k(A + B) = kA + kB$$

$$(k_1 + k_2)A = k_1 A + k_2 A,\ A + A = 2A.$$

Multiplication of Matrices

Consider the product

$$\begin{bmatrix} a_1 & b_1 & c_1 \\ a_3 & b_2 & c_2 \end{bmatrix}\begin{bmatrix} p \\ q \\ r \end{bmatrix} = \begin{bmatrix} a_1p + b_1q + c_1r \\ a_2p + b_2q + c_2r \end{bmatrix}$$

$$2 \times 3 \qquad 3 \times 1 \qquad 2 \times 1$$

Each row of the 2×3 matrix, taken as a row vector is multiplied by the 3×1 column vector. This gives us two numbers which are the elements of a 2×1 matrix. The procedure would be similar for higher dimension matrices.

Ex. 29. $\begin{bmatrix} a_1 & b_1 \\ a_2 & b_2 \end{bmatrix}\begin{bmatrix} p_1 & p_2 \\ q_1 & q_2 \end{bmatrix} = \begin{bmatrix} a_1p_1 + b_1q_1 & a_1p_2 + b_1q_2 \\ a_2p_1 + b_2q_1 & a_2p_2 + b_2q_2 \end{bmatrix}$

Ex. 30. $\begin{bmatrix} 1 & 0 & 2 & 3 \\ 2 & -1 & 0 & 3 \\ 3 & 2 & -2 & 0 \end{bmatrix}\begin{bmatrix} 2 & 1 \\ 0 & 2 \\ 1 & 4 \\ 2 & 0 \end{bmatrix}$

$$= \begin{bmatrix} 1.2 + 0.0 + 2.1 + 3.2 & 1.1 + 0.2 + 2.4 + 3.0 \\ 2.2 - 1.0 + 0.1 + 3.2 & 2.1 - 1.2 + 0.4 + 3.0 \\ 3.2 + 2.0 - 2.1 + 0.2 & 3.1 + 2.2 - 2.4 + 0.0 \end{bmatrix}$$

$$= \begin{bmatrix} 10 & 9 \\ 10 & 0 \\ 4 & -1 \end{bmatrix}$$

Ex. 31. Show that (*i*) $\begin{bmatrix} 2 & 4 \\ 6 & 8 \end{bmatrix}\begin{bmatrix} 1 & -1 \\ 2 & 0 \end{bmatrix} = \begin{bmatrix} 10 & -2 \\ 22 & -6 \end{bmatrix}$

(*ii*) $[3 - 2]\begin{bmatrix} 2 & -1 & -2 \\ 6 & 3 & 4 \end{bmatrix} = [-6 - 9 - 14]$

(*iii*) $\begin{bmatrix} 1 & 2 & -3 \\ 0 & -2 & 4 \\ 1 & -3 & 1 \end{bmatrix}\begin{bmatrix} 5 & 2 & 1 \\ -3 & 1 & 7 \\ 0 & 1 & 2 \end{bmatrix} = \begin{bmatrix} -1 & 1 & 9 \\ 6 & 2 & -6 \\ 4 & 0 & -18 \end{bmatrix}$

(*iv*) $\begin{bmatrix} 2 & 2 & -1 & 3 \\ 1 & 0 & 1 & 0 \\ 0 & -1 & 4 & 0 \end{bmatrix}\begin{bmatrix} 3 & 4 \\ 0 & -1 \\ 1 & 0 \\ 2 & -2 \end{bmatrix} = \begin{bmatrix} 11 & 0 \\ 4 & 4 \\ 4 & 1 \end{bmatrix}$

Note. 1. If two matrices A, B are such that the number of columns in A is the same as the number of rows in B they are *conformable* for multiplication.

2. If A is of order $p \times q$ and B is of order $q \times r$ then A, B are conformable for multiplication and $A \times B$ is a matrix of order $p \times r$.

This does not mean we can also find $B \times A$ necessarily. B, A are not conformable for multiplication unless the number of columns in B is the same as the number of rows in A.

Rule. Let A be a $p \times q$ matrix and B, a $q \times r$ matrix then their product $AB = C$ is a $p \times r$ matrix whose $(ij)^{th}$ element c_{ij} is the scalar product of the ith row vector of A and the jth column vector of B

$$AB = [c_{ij}] \text{ where } c_{ij} = \sum_{k=l}^{q} a_{ik}\, b_{kj} \qquad \begin{matrix} i = 1, 2 \ldots p \\ j = 1, 2 \ldots r \end{matrix}$$

3. *The following may be noted:*

Associative law $(AB)C = A(BC)$

Distributive law $A(B + C) = AB + AC$, $(B + C)A = BA + CA$

Scalar product $k(AB) = (kA)B = A(kB)$

In general $AB \neq BA$

Sometimes AB exists and BA may not be defined.

4. Algebraic product and factorising are not, in general, valid for matrices. Thus for square matrices A and B of the same order, we have to write

$$[A + B]^2 = A^2 + AB + BA + B^2 \ (not\ A^2 + 2AB + B^2)$$

$$[A + B]\,[A - B] = A^2 - AB + BA - B^2 \ (not\ A^2 - B^2)$$

Ex. 32. If $A = \begin{bmatrix} -3 & 2 \\ 5 & 0 \end{bmatrix}$ $\quad B = \begin{bmatrix} -k & k \\ 2k & 1 \end{bmatrix}$

$C = \begin{bmatrix} 0 & 1 \\ 0 & 1 \end{bmatrix}$, $\quad D = \begin{bmatrix} 1 & 0 \\ 0 & 0 \end{bmatrix}$, $\quad E = \begin{bmatrix} 6 & 1 \\ 7 & -4 \end{bmatrix}$

Show that $BD = \begin{bmatrix} -k & 0 \\ 2k & 0 \end{bmatrix}$ $CD = \begin{bmatrix} 0 & 0 \\ 0 & 0 \end{bmatrix}$

$$2A + C = \begin{bmatrix} -6 & 5 \\ 10 & 1 \end{bmatrix}, \qquad DE + B = \begin{bmatrix} 6-k & k \\ 7+2k & 1 \end{bmatrix}$$

Ex. 33. Let $A = [a_1\ a_2\ a_3], x = \begin{bmatrix} x_1 \\ x_2 \\ x_3 \end{bmatrix}$

then $Ax = [a_1x_1 + a_2x_2 + a_3x_3]$

Ex. 34. If $A = [4\ 3\ 0 - 2], x = \begin{bmatrix} 2 \\ 3 \\ 1 \\ -2 \end{bmatrix}$

$$Ax = [4 \times 2 + 3 \times 3 + 0 \times 1 + (-2)(-2)] = [21]$$

The element of the product matrix Ax is obtained by adding the products of row elements of A by the corresponding column elements of x.

Ex. 35. If $A = [a_1\ a_2],\ B = \begin{bmatrix} b_1 & c_1 \\ b_2 & c_2 \end{bmatrix}$

A, B are conformable for multiplication

$$A = [a_1b_1 + a_2b_2 \qquad a_1c_1 + a_2c_2]$$

AB is of order 1×2.

Ex. 36. If $A = [a_1\ a_2],\ B = \begin{bmatrix} b_1 & c_1 & d_1 \\ b_2 & c_2 & d_2 \end{bmatrix}$

$AB = [a_1b_1 + a_2b_2\ \ a_1c_1 + a_2a_2\ \ a_1d_1 + a_2d_2]$ is of order 1×3.

Ex. 37. $\begin{bmatrix} 2 & 1 & 0 \\ 0 & 2 & 3 \end{bmatrix} \begin{bmatrix} 3 \\ 4 \\ 2 \end{bmatrix} = \begin{bmatrix} 2\times3+1\times4+0\times2 \\ 0\times3+2\times4+3\times2 \end{bmatrix} = \begin{bmatrix} 10 \\ 14 \end{bmatrix}$

The product of a 2×3 and 3×1 matrix is a 2×1 matrix.

Ex. 38. $\begin{bmatrix} 3 & 2 & 4 \\ -1 & 0 & 1 \end{bmatrix} \times \begin{bmatrix} 2 & -1 \\ 4 & -2 \\ 3 & 0 \end{bmatrix}$ $(2 \times 3$ and $3 \times 2)$

$$= \begin{bmatrix} 3\times2+2\times4+4\times3 & 3(-1)+2(-2)+4\times0 \\ -1\times2+0\times4+1\times3 & -1(-1)+0(-2)+1\times0 \end{bmatrix}$$

$= \begin{bmatrix} 26 & -7 \\ 1 & 1 \end{bmatrix}$. This is a 2×2 matrix.

Ex. 39. If $A = \begin{bmatrix} 2 & 4 \\ 0 & 3 \end{bmatrix}$, a 2×2 matrix.

and $B = \begin{bmatrix} -1 & 1 \\ 0 & -1 \end{bmatrix}$, a 2×2 matrix.

They are conformable for multiplying

$$AB = \begin{bmatrix} 2(-1)+4(0) & 2\times1+4(-1) \\ 0(-1)+3\times0 & 0\times1+3(-1) \end{bmatrix} = \begin{bmatrix} -2 & -2 \\ 0 & -3 \end{bmatrix}$$

$$BA = \begin{bmatrix} -1\times2+1\times0 & -1\times4+1\times3 \\ 0\times2+(-1)\times0 & 0\times4+(-1)3 \end{bmatrix} = \begin{bmatrix} -2 & -1 \\ 0 & -3 \end{bmatrix}$$

AB, BA are both of order 2×2. In general $AB \neq BA$.

Ex. 40. Let $A = \begin{bmatrix} 3 & 2 \\ 10 & -2 \end{bmatrix}$, $x = \begin{bmatrix} x_1 \\ x_2 \end{bmatrix}$, $b = \begin{bmatrix} 7 \\ 3 \end{bmatrix}$

$$Ax = \begin{bmatrix} 3x_1 + 2x_2 \\ 10x_1 - 2x_2 \end{bmatrix}, \text{ a } 2 \times 1 \text{ matrix.}$$

If $Ax = b$, we can write $\begin{bmatrix} 3x_1 + 2x_2 \\ 10x_1 - 2x_2 \end{bmatrix} = \begin{bmatrix} 7 \\ 3 \end{bmatrix}$

which means $3x_1 + 2x_2 = 7$ This is a pair of
$10x_1 - 2x_2 = 3$ simultaneous equations.

7.8 SOME TYPES OF MATRICES

Zero or Null Matrix

A matrix, with every element zero, is called a null matrix. It is denoted by *O*. It need not be square. In matrix theory it plays the role of zero.

$$O_{2\times3} = \begin{bmatrix} 0 & 0 & 0 \\ 0 & 0 & 0 \end{bmatrix} \text{ is a } 2 \times 3 \text{ null matrix.}$$

Diagonal Matrix

A square matrix all of whose elements are zero except those in the leading diagonal is called a diagonal matrix.

$$D = \begin{bmatrix} d_1 & 0 & 0 \\ 0 & d_2 & 0 \\ 0 & 0 & d_3 \end{bmatrix} \text{ is a } 3 \times 3 \text{ diagonal matrix.}$$

Ex. 41. If $G = \begin{bmatrix} g_1 & 0 & 0 \\ 0 & g_2 & 0 \\ 0 & 0 & g_3 \end{bmatrix}$ then $DG = GD = \begin{bmatrix} d_1g_1 & 0 & 0 \\ 0 & d_2g_2 & 0 \\ 0 & 0 & d_3g_3 \end{bmatrix}$

Note that DG is also a diagonal matrix, its diagonal elements being the products of the corresponding elements of the given matrices. The multiplication of diagonal matrices is commutative.

Upper and Lower Triangular Matrices

A square matrix all of whose elements below the main diagonal are zero is called *upper triangular.* If all elements above the main diagonal are zero it is *lower triangular.*

$$\begin{bmatrix} a_{11} & a_{12} & a_{13} \\ 0 & a_{22} & a_{23} \\ 0 & 0 & a_{33} \end{bmatrix} \text{ and } \begin{bmatrix} a_{11} & 0 & 0 \\ a_{21} & a_{22} & 0 \\ a_{31} & a_{32} & a_{33} \end{bmatrix}$$

are, respectively, upper and lower triangular matrices.

The Identity Matrix

If in the diagonal matrix D, each diagonal element $d_i = k$, it behaves like a scalar matrix. An *identity* matrix has 1 for every diagonal element and zero elsewhere.

$I_3 = \begin{bmatrix} 1 & 0 & 0 \\ 0 & 1 & 0 \\ 0 & 0 & 1 \end{bmatrix}$ is an identity matrix of order 3. The identity matrix plays the role of the number 1.

The Transpose of a Matrix

If we interchange the rows and columns of an $m \times n$ matrix A we get an $n \times m$ matrix A' which is called the transpose of A.

If $A_{mn} = [a_{ij}]$ then $A'_{nm} = [a_{ji}]$

Ex. 42. If $A = \begin{bmatrix} a_{11} & a_{12} & a_{13} \\ a_{21} & a_{22} & a_{23} \end{bmatrix}$ then $A' = \begin{bmatrix} a_{11} & a_{12} \\ a_{12} & a_{22} \\ a_{13} & a_{23} \end{bmatrix}$

Ex. 43. The transpose of a column vector $x = \begin{bmatrix} x_1 \\ x_2 \\ x_3 \end{bmatrix}$

is the row vector $x' = [x_1\ x_2\ x_3]$

Note the following results

The transpose of the transpose of a matrix is the original matrix. Thus $(A')' = A$.

If k is a scalar $(kA)' = kA'$

The transpose of a sum : $(A + B)' = A' + B'$

The transpose of a product : $(AB)' = B'A'$

Symmetric Matrices

A square matrix A such that $A' = A$ is a symmetric matrix.

A square matrix A such that $A' = -A$ is skew symmetric. Its leading diagonal has all zeros.

$$\begin{bmatrix} 5 & a \\ a & 5 \end{bmatrix}, \begin{bmatrix} r_{11} & r_{12} & r_{13} \\ r_{12} & r_{22} & r_{23} \\ r_{13} & r_{23} & r_{33} \end{bmatrix} \text{ are symmetric matrices}$$

$$\begin{bmatrix} 0 & -1 & 2 \\ 1 & 0 & 3 \\ -2 & -3 & 0 \end{bmatrix} \text{ is skew symmetric.}$$

Note that if A is a square matrix then, (*i*) $A + A'$ is symmetric (*ii*) $A - A'$ is skew symmetric, (*iii*) The square matrix A is the sum of the symmetric matrix $\frac{A+A'}{2}$ and the skew symmetric matrix $\frac{A-A'}{2}$. These results can be easily proved.

If A, B are square and $AB = BA$ then A, B are *commutative*. If $AB = -BA$ then A, B are *anti-commutative*.

If $A^2 = A$ then A is called *idempotent.*

Determinant of a Matrix

A square matrix A has a uniquely defined determinant $|A|$ associated with the matrix. The determinant of

$$A = \begin{bmatrix} a_{11} & a_{12} \\ a_{21} & a_{22} \end{bmatrix} \text{ is } |A| = \begin{vmatrix} a_{11} & a_{12} \\ a_{21} & a_{22} \end{vmatrix} = a_{11}\,a_{22} - a_{12}\,a_{21}$$

The determinant of the product of two matrices is the product of their determinants.

$$|AB| = |A|\,|B|$$

Ex. 44. Verify the above results for

$$(i) \quad A = \begin{bmatrix} 1 & 1 & 1 \\ 4 & -1 & 1 \\ 0 & 1 & -1 \end{bmatrix}, \qquad B = \begin{bmatrix} 2 & 2 & 1 \\ 1 & 0 & 2 \\ 2 & 1 & 2 \end{bmatrix}$$

$$(ii) \quad A = \begin{bmatrix} 1 & 0 & 0 \\ 0 & 1 & 0 \\ 0 & 0 & 1 \end{bmatrix}, \qquad B = \begin{bmatrix} a & b & c \\ d & e & f \\ g & h & i \end{bmatrix}$$

Singular and Non-singular Matrices

A square matrix A is (*a*) singular if $|A| = 0$

(*b*) non-singular if $|A| \neq 0$

Ex. 45. $A = \begin{bmatrix} 12 & 3 \\ 20 & 5 \end{bmatrix}$ is singular because $|A| = 60 - 60 = 0$

Adjoint Matrix

The adjoint matrix of A is obtained by replacing the elements of A by their respective cofactors and then transposing.

If $A = [a_{ij}]$ and $B = [A_{ij}]$ where A_{ij} is the cofactor of a_{ij} in A then we have the adjoint matrix of A, written

$$\text{adj } A = [A_{ij}]' = [A_{ji}]$$

Ex. 46. If $A = \begin{bmatrix} a_{11} & a_{12} & a_{13} \\ a_{21} & a_{22} & a_{23} \\ a_{31} & a_{32} & a_{33} \end{bmatrix}$, $\text{adj } A = \begin{bmatrix} A_{11} & A_{21} & A_{31} \\ A_{12} & A_{22} & A_{32} \\ A_{13} & A_{23} & A_{33} \end{bmatrix}$

where $A_{11} = +\begin{bmatrix} a_{22} & a_{23} \\ a_{32} & a_{33} \end{bmatrix}$, $A_{12} = -\begin{bmatrix} a_{21} & a_{23} \\ a_{31} & a_{33} \end{bmatrix}$ etc.

7.9 LINEAR DEPENDENCE AND THE RANK IDEA

Let $u_1, u_2, \ldots\ldots, u_m$ be m column vectors of the same order. Let $c_1, c_2, \ldots\ldots, c_m$ be scalars, not all zero, such that the linear combination $c_1u_1 + c_2u_2 + \ldots + c_mu_m = 0$ then the vectors $u_1, u_2, \ldots\ldots, u_m$ are said to be *linearly dependent.*

$u_1, u_2, \ldots\ldots, u_m$ are said to be *linearly independent* if no such scalars are available such that $c_1u_1 + c_2u_2 + \ldots + c_mu_m = 0$.

Ex. 47. $u_1 = \begin{bmatrix} 1 \\ 2 \\ 3 \end{bmatrix}$, $u_2 = \begin{bmatrix} -2 \\ 4 \\ 0 \end{bmatrix}$, $u_3 = \begin{bmatrix} 0 \\ 2 \\ 3 \end{bmatrix}$

are linearly dependent since $2u_1 + u_2 - 4u_3 = 0$

Ex. 48. $\begin{bmatrix} 1 \\ 0 \\ 0 \end{bmatrix}, \begin{bmatrix} 0 \\ 1 \\ 0 \end{bmatrix}, \begin{bmatrix} 0 \\ 0 \\ 1 \end{bmatrix}$ are linearly independent.

Ex. 49. Show that the vectors $\begin{bmatrix} -6 \\ 0 \\ 4 \end{bmatrix}, \begin{bmatrix} 3 \\ -1 \\ -6 \end{bmatrix}, \begin{bmatrix} 6 \\ 2 \\ 4 \end{bmatrix}$ are linearly dependent.

Note that if $u_1, u_2, \ldots\ldots, u_m$ are linearly dependent then either any vector, say, u_1 is linearly dependent on $u_2, \ldots\ldots, u_m$ or $u_2, u_3, \ldots\ldots, u_m$ are themselves linearly dependent.

In a matrix, the number of linearly independent rows is the same as the number of linearly independent columns.

The Rank of a Matrix

The *rank* of a matrix is the maximum number of linearly independent rows (or columns) in the matrix. In other words the rank of a matrix A is the order of the largest nonzero minor of $|A|$.

The rank of a matrix is a unique number.

Ex. 50. The rank of $A = \begin{bmatrix} -4 & 0 & 5 \\ 1 & 2 & 3 \end{bmatrix}$ is 2 since

$\begin{vmatrix} -4 & 0 \\ 1 & 2 \end{vmatrix} \neq 0$ and this is a minor of the highest order

Ex. 51. Show that the rank of $\begin{bmatrix} 2 & 3 & 5 & 1 \\ 1 & 2 & 3 & 2 \\ 1 & 3 & 4 & 5 \end{bmatrix}$ is 2.

Ex. 52. The rank of $A = \begin{bmatrix} 1 & 4 & 0 \\ 2 & 5 & 0 \\ 3 & 6 & 0 \end{bmatrix}$ is 1 since $|A| = 0$. Each of the second order minors is zero but every element is not zero.

Ex. 53. Show that any matrix A and its transpose A' have the same rank.

7.10 INVERSE OF A MATRIX

If A is a non-singular square matrix, then it has an *inverse* denoted by A^{-1}.

The product of a matrix and its inverse gives the identity matrix.

$$AA^{-1} = \text{I} = A^{-1}A$$

A singular matrix does not have an inverse. A rectangular matrix cannot have an inverse.

If A, B are square matrices of order n such that $AB = I_n = BA$, then A and B are the multiplicative inverses of each other.

If A has a finite inverse then $|A| \neq 0$

$$(kA)^{-1} = \frac{1}{k}A^{-1} \quad \text{where } k \neq 0; \text{ Also } (A')^{-1} = (A^{-1})'$$

It can be shown that $(AB)^{-1} = B^{-1}A^{-1}$

since $\qquad (AB)(B^{-1}A^{-1}) = A(BB^{-1})A^{-1} = AIA^{-1} = AA^{-1} = \text{I}$

Similarly $(ABC)^{-1} = C^{-1}B^{-1}A^{-1}$

There are several procedures for computing the inverse. We shall find the inverse by using the adjoint matrix remembering that $A^{-1} = \dfrac{1}{|A|}[A_{ji}]$.

For the 2×2 matrix $A = \begin{bmatrix} a & b \\ c & d \end{bmatrix}$, the cofactors of a, b, c, d are, respectively, $d, -c, -b, a$ so that the inverse is given by $\dfrac{\text{adj } A}{|A|} = \dfrac{[A_{ji}]}{|A|}$. Thus

$$A^{-1} = \frac{1}{|A|}\begin{bmatrix} d & -b \\ -c & a \end{bmatrix} \text{ where } |A| = ad - bc$$

For the 3×3 matrix

$$A = [a_{ij}] = \begin{bmatrix} a_{11} & a_{12} & a_{23} \\ a_{21} & a_{22} & a_{23} \\ a_{31} & a_{32} & a_{33} \end{bmatrix},$$ we have the matrix of cofactors

$$[A_{ij}] = \begin{bmatrix} A_{11} & A_{12} & A_{13} \\ A_{21} & A_{22} & A_{23} \\ A_{31} & A_{32} & A_{33} \end{bmatrix}$$ whose transpose is

$$[A_{ji}] = \begin{bmatrix} A_{11} & A_{21} & A_{31} \\ A_{12} & A_{22} & A_{32} \\ A_{13} & A_{23} & A_{33} \end{bmatrix}$$

The inverse of A is, therefore, $A^{-1} = \frac{1}{|A|}[A_{ji}]$

Ex. 54. Check this result by showing that $AA^{-1} = I$.

Ex. 55. Show that $\begin{bmatrix} 7 & -11 & 16 \\ -3 & 5 & -7 \\ 1 & -2 & 3 \end{bmatrix}$ and $\begin{bmatrix} 1 & 1 & -3 \\ 2 & 5 & 1 \\ 1 & 3 & 2 \end{bmatrix}$ are inverses of each other.

This is so because the product is $\begin{bmatrix} 1 & 0 & 0 \\ 0 & 1 & 0 \\ 0 & 0 & 1 \end{bmatrix}$

Ex. 56. Show that the matrix $A = \begin{bmatrix} 2 & -1 & 1 \\ -1 & 2 & -1 \\ 1 & -1 & 2 \end{bmatrix}$ satisfies the equation $A^3 - 6A^2 +$ $9A - 4I = 0$

Hence find A^{-1}.

Hint. Find $A^2 = AA = \begin{bmatrix} 6 & -5 & 5 \\ -5 & 6 & -5 \\ 5 & -5 & 6 \end{bmatrix}$

$$A^3 = A^2A = \begin{bmatrix} 22 & -21 & 21 \\ -21 & 22 & -21 \\ 21 & -21 & 22 \end{bmatrix}$$

Substitute and show $A^3 - 6A^2 + 9A = 4I$

Alternatively, premultiply by A^{-1} then $A^2 - 6A + 9I = 4A^{-1}$

Show $A^{-1} = \frac{1}{4}\begin{bmatrix} 3 & 1 & -1 \\ 1 & 3 & 1 \\ -1 & 1 & 3 \end{bmatrix}$

Steps

$A^{-1}(A^3 - 6A^2 + 9A - 4I) = 0$

$\therefore A^2 - 6A + 9I - 4A^{-1} = 0$

Substitute for A^2, A.

Ex. 57. If $A = \begin{bmatrix} 2 & 2 & 0 \\ 2 & 1 & 1 \\ -7 & 2 & -3 \end{bmatrix}$ verify that $A^3 - 13A + 12I = 0$. Hence find A^{-1}.

Verify $AA^{-1} = I$

Ex. 58. If $A = \begin{bmatrix} 7 & 3 & -3 \\ 2 & 4 & 1 \end{bmatrix}$

Verify $(A'A)\,(A'A)^{-1} = I$.

Ex. 59. Corresponding to the matrix

$$A = \begin{bmatrix} 2 & 3 & 4 \\ 4 & 3 & 1 \\ 1 & 2 & 4 \end{bmatrix} \text{ we have the matrix of cofactors:}$$

$$[A_{ij}] = \begin{bmatrix} 10 & -15 & +5 \\ -4 & +4 & -1 \\ -9 & +14 & -6 \end{bmatrix}$$

Since $|A| = -5$ we have

$$A^{-1} = \frac{1}{-5}[A_{ji}] = -\frac{1}{5}\begin{bmatrix} 10 & -4 & -9 \\ -15 & 4 & 14 \\ 5 & -1 & -6 \end{bmatrix} = \begin{bmatrix} -2 & 0.8 & 1.8 \\ 3 & -0.8 & -2.8 \\ 2 & 0 & 1 \end{bmatrix}$$

Ex. 60. Show that

if $$A = \begin{bmatrix} 1 & 2 & 3 \\ 1 & 3 & 3 \\ 2 & 4 & 3 \end{bmatrix} \text{ then } A^{-1} = \frac{1}{3}\begin{bmatrix} 3 & -6 & 3 \\ -3 & 3 & 0 \\ 2 & 0 & 1 \end{bmatrix}$$

Ex. 61. For any matrix A

$$A(\text{adj } A) = |A| \cdot I$$
$$(\text{adj } A)A = |A| \cdot I$$

so that if $$|A| \neq 0,\ A^{-1} = \frac{\text{adj } A}{|A|}$$

7.11 APPLICATION OF MATRICES TO LINEAR SYSTEMS

A system of equations

$$a_{11}x_1 + a_{12}x_2 = b_1$$
$$a_{21}x_1 + a_{22}x_2 = b_2$$

can be expressed in the matrix form

$$\begin{bmatrix} a_{11} & a_{12} \\ a_{21} & a_{22} \end{bmatrix}\begin{bmatrix} x_1 \\ x_2 \end{bmatrix} = \begin{bmatrix} b_1 \\ b_2 \end{bmatrix}$$

or in the condensed form $AX = B$

where A is the matrix of coefficients.

X is the column vector of variables $\begin{bmatrix} x_1 \\ x_2 \end{bmatrix}$

B is the column vector $\begin{bmatrix} b_1 \\ b_2 \end{bmatrix}$

We may also write the equation as $X' A' = B'$

For three equations we may use a similar notation

$$a_{11}x_1 + a_{12}x_2 + a_{13}x_3 = b_1$$
$$a_{21}x_1 + a_{22}x_2 + a_{23}x_3 = b_2$$
$$a_{31}x_1 + a_{32}x_2 + a_{33}x_3 = b_3$$

i.e., $$\begin{bmatrix} a_{11} & a_{12} & a_{31} \\ a_{21} & a_{22} & a_{23} \\ a_{31} & a_{32} & a_{33} \end{bmatrix} \begin{bmatrix} x_1 \\ x_2 \\ x_3 \end{bmatrix} = \begin{bmatrix} b_1 \\ b_2 \\ b_3 \end{bmatrix} \quad \text{or } AX = B$$

and so on.

Now if $$AX = B$$

then $$A^{-1}AX = A^{-1}B \quad \text{or} \quad X = A^{-1}B \text{ since } A^{-1}A = 1$$

i.e., to solve the system of equations we must find A^{-1} and then calculate $A^{-1}B$.

Ex. 62. The coefficient matrix for the system

$2x_1 + 3x_2 = 9,\ 3x_1 + 5x_2 = 12$ is given by

$$A = \begin{bmatrix} 2 & 3 \\ 3 & 5 \end{bmatrix}. \text{ Since } A^{-1} = \frac{1}{1}\begin{bmatrix} 5 & -3 \\ -3 & 2 \end{bmatrix}$$

$$\therefore \quad \begin{bmatrix} x_1 \\ x_2 \end{bmatrix} = X = A^{-1}B = \begin{bmatrix} 5 & -3 \\ -3 & 2 \end{bmatrix}\begin{bmatrix} 9 \\ 12 \end{bmatrix} = \begin{bmatrix} 9 \\ -3 \end{bmatrix}$$

The solution is $x_1 = 9,\ x_2 = -3$

Ex. 63. In the system
$$2x_1 - 4x_2 + 3x_3 = 3$$
$$4x_1 - 6x_2 + 5x_3 = 2$$
$$-2x_1 + x_2 - x_3 = 1$$

$$A = \begin{bmatrix} 2 & -4 & 3 \\ 4 & -6 & 5 \\ -2 & 1 & -1 \end{bmatrix},\ A^{-1} = \begin{bmatrix} 0.5 & -0.5 & -1 \\ -3 & 2 & 1 \\ -4 & 3 & 2 \end{bmatrix}$$

$$\begin{bmatrix} x_1 \\ x_2 \\ x_3 \end{bmatrix} = X = A^{-1}B \begin{bmatrix} 0.5 & -0.5 & -1 \\ -3 & 2 & 1 \\ -4 & 3 & 2 \end{bmatrix}\begin{bmatrix} 3 \\ 2 \\ 1 \end{bmatrix} = \begin{bmatrix} -0.5 \\ -4 \\ -4 \end{bmatrix}$$

The solution is $x_1 = -0.5,\ x_2 = -4,\ x_3 = -4$.

Ex. 64. Solve the systems:

(*i*)
$$\begin{aligned} x_1 + x_2 + x_3 &= 10 \\ 2x_1 - 2x_2 + x_3 &= 1 \\ 3x_1 + 2x_2 - x_3 &= -16 \end{aligned}$$

(*ii*)
$$\begin{aligned} x_1 + x_2 + x_3 &= 6 \\ 4x_1 - x_2 + x_3 &= 5 \\ x_2 - x_3 &= -1 \end{aligned}$$

(*iii*)
$$\begin{aligned} x_1 - x_3 &= 4 \\ 2x_2 - x_3 &= 6 \\ x_1 + x_2 &= 10 \end{aligned}$$

Ans. (*i*) – 3, 2, 11, (*ii*) 1, 2, 3 (*iii*) 6, 4, 2

The above problems can also be solved by using the concept of an *augmented* matrix. It amounts to reducing the coefficient matrix to the diagonal form as shown in the following exercise, by performing suitable operations.

Ex. 65.
$$\begin{aligned} x_1 + 6x_2 - x_3 &= 10 \\ 2x_1 + 3x_2 + 3x_3 &= 17 \\ 3x_1 - 3x_2 - 2x_3 &= -9 \end{aligned}$$

For this system, the augmented matrix is

$$\begin{bmatrix} 1 & 6 & -1 & 10 \\ 2 & 3 & 3 & 17 \\ 3 & -3 & -2 & -9 \end{bmatrix}$$

$$= \begin{bmatrix} 1 & 6 & -1 & 10 \\ 0 & -9 & 5 & -3 \\ 0 & -21 & 1 & -39 \end{bmatrix} \quad \begin{matrix} \text{Row } 2 - 2\text{ (Row 1)} \\ \text{Row } 3 - 3\text{ (Row 1)} \end{matrix}$$

$$= \begin{bmatrix} 1 & 0 & \frac{7}{3} & 8 \\ 0 & -9 & 5 & -3 \\ 0 & 0 & -\frac{32}{3} & -32 \end{bmatrix} \quad \begin{matrix} \text{Row } 1 + 3/2\text{ (Row 2)} \\ \text{Row } 3 - 7/3\text{ (Row 2)} \end{matrix}$$

$$= \begin{bmatrix} 1 & 0 & 0 & 1 \\ 0 & -9 & 0 & -18 \\ 0 & 0 & -\frac{32}{3} & -32 \end{bmatrix} \quad \begin{matrix} \text{Row } 1 + 7/32\text{ (Row 2)} \\ \text{Row } 1 + 15/32\text{ (Row 3)} \end{matrix}$$

The solution is $x_1 = 1$ using first row

$-9x_2 = -18$ or $x_2 = 2$ " second row

$-32x_3 = -32$ or $x_3 = 3.$ " third row

Apply the technique to Ex. 64.

Ex. 66. If $A = \begin{bmatrix} 3 & 1 \\ 5 & 2 \end{bmatrix}$, $|A| = 3 \times 2 - 1 \times 5 = 1$

$$A^{-1} = \frac{1}{1}\begin{bmatrix} 2 & -1 \\ -5 & 3 \end{bmatrix} = \begin{bmatrix} 2 & -1 \\ -5 & 3 \end{bmatrix}$$

Check: $AA^{-1} = \begin{bmatrix} 3 & 1 \\ 5 & 2 \end{bmatrix}\begin{bmatrix} 2 & -1 \\ -5 & 3 \end{bmatrix} = \begin{bmatrix} 1 & 0 \\ 0 & 1 \end{bmatrix}$, a unit matrix.

Ex. 67. If $A = \begin{bmatrix} 2 & -1 \\ 7 & 4 \end{bmatrix}$; $|A| = 8 - (-7) = 15$

$$A^{-1} = \frac{1}{15}\begin{bmatrix} 4 & 1 \\ -7 & 2 \end{bmatrix} = \begin{bmatrix} \frac{4}{15} & \frac{1}{15} \\ -\frac{7}{15} & \frac{8}{15} \end{bmatrix}$$

Ex. 68. If $A = \begin{bmatrix} 2 & -3 \\ -6 & 9 \end{bmatrix}$, the inverse does not exist since

$|A| = 0.$

Ex. 69. Show that for any matrix

$$A = \begin{bmatrix} a_1 & a_2 \\ b_1 & b_2 \end{bmatrix} \text{ and the identity matrix } I = \begin{bmatrix} 1 & 0 \\ 0 & 1 \end{bmatrix}$$

(*i*) $IA = AI = A$ (*ii*) $I^2 = I$.

Ex. 70. If $P = \begin{bmatrix} -1 & 0 \\ 0 & 1 \end{bmatrix}$, $Q = \begin{bmatrix} 1 & 1 \\ -1 & 1 \end{bmatrix}$

show that (*i*) $P^2 = I$, (*ii*) $Q^2 = \begin{bmatrix} 0 & 2 \\ -2 & 0 \end{bmatrix}$,

(*iii*) $Q^4 = \begin{bmatrix} -4 & 0 \\ 0 & -4 \end{bmatrix}$, (*iv*) $PQ \neq QP$

Ex. 71. Show that

$$\begin{bmatrix} 2 & 1 & -3 \\ 0 & 2 & -2 \\ -1 & -1 & 3 \\ 3 & 0 & 1 \end{bmatrix} \times \begin{bmatrix} 3 & 0 \\ 2 & 4 \\ 2 & -1 \end{bmatrix} = \begin{bmatrix} 2 & 7 \\ 0 & 10 \\ 1 & -7 \\ 8 & -1 \end{bmatrix}$$

Ex. 72. Given a matrix A, the product

(*i*) $A \cdot O = O$ (*ii*) $O \cdot A = O$

where O is a zero matrix (or null matrix).

The products of two nonzero matrices can sometimes equal a zero matrix

Thus $$\begin{bmatrix} a & b \\ 0 & 0 \end{bmatrix}\begin{bmatrix} b & 0 \\ -a & 0 \end{bmatrix} = \begin{bmatrix} 0 & 0 \\ 0 & 0 \end{bmatrix}$$

Ex. 73. Given $A = \begin{bmatrix} 1 & 1 \\ 0 & 1 \end{bmatrix}$ show that

$$A^2 = \begin{bmatrix} 1 & 2 \\ 0 & 1 \end{bmatrix}, A^3 = \begin{bmatrix} 1 & 3 \\ 0 & 1 \end{bmatrix}, \ldots, A^n = \begin{bmatrix} 1 & n \\ 0 & 1 \end{bmatrix}$$

Ex. 74. Show that $\begin{bmatrix} 1 & 2 \\ 2 & 4 \end{bmatrix} \times \begin{bmatrix} -4 & 6 \\ 2 & -3 \end{bmatrix}$ gives a zero matrix.

Ex. 75. Show that $\begin{bmatrix} 3 & 4 \\ 2 & 3 \end{bmatrix} \times \begin{bmatrix} 3 & -4 \\ -2 & 3 \end{bmatrix}$ gives a unit matrix.

Ex. 76. Solve the following matrix equation by finding the inverse.

$$A = \begin{bmatrix} 3 & -4 \\ 9 & 2 \end{bmatrix} \begin{bmatrix} x_1 \\ x_2 \end{bmatrix} = \begin{bmatrix} -4 \\ 9 \end{bmatrix}$$

If $\quad A = \begin{bmatrix} 3 & -4 \\ 9 & 2 \end{bmatrix}$ then $|A| = 42$

$$\therefore \quad A^{-1} = \frac{1}{42}\begin{bmatrix} 2 & 4 \\ -9 & 3 \end{bmatrix} = \begin{bmatrix} \frac{4}{42} & \frac{2}{42} \\ \frac{9}{42} & \frac{3}{42} \end{bmatrix}$$

$$\therefore \quad \begin{bmatrix} x_1 \\ x_2 \end{bmatrix} = \begin{bmatrix} \frac{2}{42} & \frac{4}{42} \\ -\frac{9}{42} & \frac{3}{42} \end{bmatrix} \begin{bmatrix} -4 \\ 9 \end{bmatrix} = \begin{bmatrix} \frac{2}{3} \\ \frac{3}{2} \end{bmatrix}$$

$$\therefore \quad x_1 = \tfrac{2}{3}, \; x_2 = \tfrac{3}{2}.$$

Ex. 77. If

$$A = \begin{bmatrix} 2 & 0 \\ 1 & 3 \end{bmatrix}, \; B = \begin{bmatrix} 4 & 0 \\ 0 & 1 \end{bmatrix}, \; x = \begin{bmatrix} x_1 \\ x_2 \end{bmatrix}, \; y = \begin{bmatrix} y_1 \\ y_2 \end{bmatrix}$$

Solve the equations

(*i*) $Y = AX$ (*ii*) $Y = BX$ (*iii*) $Y = ABX$ (*iv*) $Y = BAX$ (*v*) $Y = (A + B)X$

Ex. 78. If $\quad A = \begin{bmatrix} a_1 & b_1 & c_1 \\ a_2 & b_2 & c_2 \end{bmatrix}$ show that

$$AA' = \begin{bmatrix} a_1^2 + b_1^2 + c_1^2 & a_1a_2 + b_1b_2 + c_1c_2 \\ a_1a_2 + b_1b_2 + c_1c_2 & a_2^2 + b_2^2 + c_2^2 \end{bmatrix}$$

$$A'A = \begin{bmatrix} a_1^2 + a_2^2 & a_1b_1 + a_2b_2 & a_1c_1 + a_2c_2 \\ a_1b_1 + a_2b_2 & b_1^2 + b_2^2 & b_1c_1 + b_2c_2 \\ a_1c_1 + a_2c_2 & b_1c_1 + b_2c_2 & c_1^2 + c_2^2 \end{bmatrix}$$

Also find $AA'A$

Ex. 79. Show that the inverse of the matrix

$$(i) \quad A = \begin{bmatrix} 0 & 4 & 2 \\ -1 & 6 & 1 \\ 3 & -2 & -5 \end{bmatrix} \text{ is } A^{-1} = \begin{bmatrix} 0.70 & -0.40 & 0.20 \\ 0.05 & 0.15 & 0.05 \\ 0.40 & -0.30 & -0.10 \end{bmatrix}$$

(ii) $D = \begin{bmatrix} d_1 & 0 & 0 \\ 0 & d_2 & 0 \\ 0 & 0 & d_3 \end{bmatrix}$ is $D^{-1} = \begin{bmatrix} \frac{1}{d_1} & 0 & 0 \\ 0 & \frac{1}{d_2} & 0 \\ 0 & 0 & \frac{1}{d_3} \end{bmatrix}$

Ex. 80. If $A = \begin{bmatrix} 1 & 0 & 1 \\ 2 & 1 & 3 \\ 1 & 4 & 0 \end{bmatrix}$ show that

$\text{adj } A = \begin{bmatrix} -12 & 4 & -1 \\ 3 & -1 & -1 \\ 7 & -4 & 1 \end{bmatrix}$. Hence find A^{-1}

Ex. 81. Find the unknown quantities in the following equations.

(i) $\begin{bmatrix} 3 & x & w+3 \\ 4y & 2 & 5z \end{bmatrix} = \begin{bmatrix} 3 & 2w-1 & 2x-19 \\ 20 & 2 & z-12 \end{bmatrix}$

(ii) $\begin{bmatrix} 3x & 4 \\ 2y+3 & z \\ 3w-2 & y \end{bmatrix} = \begin{bmatrix} 12+x & 4 \\ 11 & 3z-6 \\ 5w+7 & 4 \end{bmatrix}$.

Ans. (i) $x = 15$, $w = 8$, $y = 5$, $z = -3$

(ii) $x = 6$, $y = 4$, $z = 3$, $w = -4\frac{1}{2}$

Ex. 82. A firm produces five qualities of its product which needs the following materials.

Quality	*Materials needed*			
	M_1	M_2	M_3	M_4
A_1	6	6	10	8
A_2	3	4	12	6
A_3	4	5	15	8
A_4	2	2	12	5
A_5	3	2	10	

If the firm has to produce, respectively, 3, 22, 20, 12 and 7 units of the five qualities find the amounts of different materials required by writing their requirements as a row vector. **Ans.** (169, 194, 658, 324)

Ex. 83. In an opinion poll 30% are found to favour party A and 70% opposed to A. A month later 20% of those originally in favour are now opposed to A and 30% of those originally opposed are now in favour of A. How have the odds changed?

We can write the $\begin{pmatrix} \text{For} \\ \text{Against} \end{pmatrix}$ vector as $X = \begin{pmatrix} 30 \\ 70 \end{pmatrix}$

This was originally under the operation of

$$I = \begin{bmatrix} 1 & 0 \\ 0 & 1 \end{bmatrix} \text{ as may be verified : } \begin{bmatrix} 1 & 0 \\ 0 & 1 \end{bmatrix} \begin{bmatrix} 30 \\ 70 \end{bmatrix} = \begin{bmatrix} 30 \\ 70 \end{bmatrix}$$

The new poll gives the new operator $Y = \begin{bmatrix} 1-0.2 & 0+0.3 \\ 0+0.2 & 1-0.3 \end{bmatrix}$

$$YX = \begin{bmatrix} 0.8 & 0.3 \\ 0.2 & 0.7 \end{bmatrix} \begin{bmatrix} 30 \\ 70 \end{bmatrix} = \begin{bmatrix} 45 \\ 55 \end{bmatrix}$$

The new poll has resulted in 45% for, 55% against.

Ex. 84. In an economy consisting of 3 industries *A, B, C,* the demands by each industry for the products *A, B, C* are given by the vectors $D_A = (0\ 1\ 4)$, $D_B = (20\ 0\ 8)$, $D_C = (30\ 5\ 0)$ *i.e.,* each industry demands the products of the other industries only.

If there are only three consumers *a, b, c* having demands $D_a = (3, 2\ 5)$, $D_b = (0\ 17\ 1)$, $D_c = (4\ 6\ 12)$, show that the overall demand is given by the vector (57 31 30).

If the prices of *A, B, C* are 1, 2, 4 per unit respectively, find the profits made by the three industries. **Ans.** (39 10 80)

Ex. 85. Show that for the 3 × 3 matrix $A = [A_{ij}]$ if *d* and *D* stand for the determinants of *A* and adj *A* then $dD = \begin{vmatrix} d & 0 & 0 \\ 0 & d & 0 \\ 0 & 0 & d \end{vmatrix} = d^3$ so that $D = d^{3-1}$

In general for the $n \times n$ matrix *A*, $D = d^{n-1}$

Also show that the cofactor of A_{11} in *D* is $a_{11}\, d$.

Ex. 86. Leontieff inverse. We know that

$$(1-x)^{-1} = \frac{1}{1-x} = 1 + x + x^2 + \ldots\ldots = \sum_{r=0}^{\infty} x^r \text{ or } (1-x)\sum_{r=0}^{\infty} x^r = 1$$

Consider the matrix $I - A$ which occurs in input output models. We can write

$$(I-A)^{-1} = I + A + A^2 + \ldots\ldots = \sum_{k=0}^{\infty} A^k$$

$$(I-A)\sum_{k=0}^{\infty} A^k = I$$

If the series converges, $(I - A)^{-1}$ may be evaluated by selecting a finite number of terms in the series expansion.

Ex. 87. Show, by taking a suitable example that $(A + B)^{-1} = A^{-1} + B^{-1}$.

Ex. 88. Under what conditions do the following hold?

$$(A + B)^2 = A^2 + 2AB + B^2$$

$$(A + B)(A - B) = A^2 - B^2$$

$$(A + B)\,C = AC + BC$$

Ans. If $AB = BA$

Applications and Advantages of Matrix Algebra

We have considered some manipulations which are possible with the help of matrix notation which has a number of applications in many fields.

The matrix notation has the merit of being compact and precise. It is time saving and suitable for systematic mechanical calculations.

Any accounting system can be conveniently represented in the form of matrices. Thus national income accounting can be done in terms of matrix notation. Linear programming makes a wide use of matrix methods.

CHAPTER 8

Analytical Geometry

8.1 POINTS AND LINES

Analytical Geometry (also called algebraic or coordinate geometry) is that branch of geometry which considers the properties of plane figures by means of coordinates of points. To do this we make use of the notations and operations of algebra and analyse the problems discussed in pure geometry systematically to arrive at their solutions.

Relationships in general and economic relationships in particular are usually described by more than one variable. Analytical geometry is one of several methods of studying these relationships.

We shall consider here the concepts of the point, lines and conics in analytical geometry.

The Distance Formula

To find the distance between two points $P_1(x_1, y_1)$ and $P_2(x_2, y_2)$.

Draw $P_1M_1 \perp Ox$, $P_2M_2 \perp Ox$,

$P_1N \perp P_2M_2$.

$P_1N = OM_2 - OM_1 = x_2 - x_1$

$P_2N = P_2M_2 - P_1M_1 = y_2 - y_1$

In the right angled triangle P_1P_2N we have, by Pythagoras' theorem

$$d^2 = P_1P_2^2 = P_1N^2 + P_2N^2$$

$$= (x_2 - x_1)^2 + (y_2 - y_1)^2$$

The distance P_1P_2 is given by

$$d = \sqrt{(x_2 - x_1)^2 + (y_2 - y_1)^2}$$

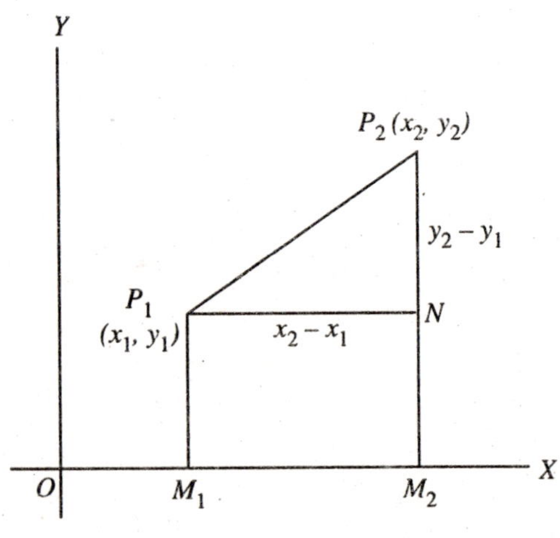

Ex. 1. The length of the segment between the points (2, – 1), (– 3, 4) is given by

$$d = \sqrt{(-3-2)^2 + (4-(-1))^2}$$

$$= \sqrt{25+25} = \sqrt{50} = 5\sqrt{2}.$$

Ex. 2. The distance between (a, o), (o, b) is given by

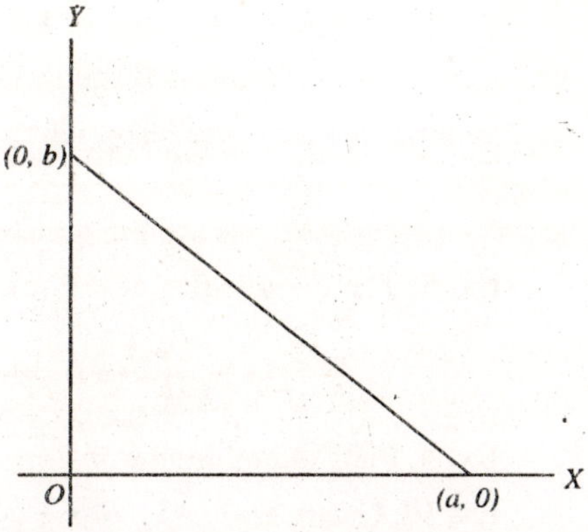

$$d = \sqrt{(o-a)^2 + (b-o)^2}$$

$$= \sqrt{a^2 + b^2}$$

which can be proved by Pythagoras' theorem.

Ex. 3. If the distance of a point $P(x, y)$ from the origin is twice that from (a, b) what is the relation between x, y, a, b?

Distance of P from (0, 0) is

$$\sqrt{(x-0)^2 + (y-0)^2} = \sqrt{x^2 + y^2}$$

Distance of P from (a, b) is $\sqrt{(x-a)^2 + (y-b)^2}$

We have $\sqrt{x^2 + y^2} = 2\sqrt{(x-a)^2 + (y-b)^2}$

$\therefore \quad x^2 + y^2 = 4(x-a)^2 + 4(y-b)^2$

This is the required relation which may be simplified further.

Ex. 4. Find the lengths of the sides and diagonals of the rectangle formed by the four points (0, 0) (a, b), $(0, b)$, $(a, 0)$. **Ans.** a, b, a, b; $\sqrt{a^2 + b^2}$, $\sqrt{a^2 + b^2}$

Ex. 5. If a circle has the centre (– 5, 1) and passes through the point (– 4, – 3) what is its radius? **Ans.** $\sqrt{17}$

Ex. 6. Find the third vertex if two vertices of an equilateral triangle are given to be (0, 0), (– 4, 3).

Midpoint of a Line Segment

Let $P_1(x_1, y_1)$, $P_2(x_2, y_2)$ be the end points on the line segment P_1P_2 whose midpoint is $P(x, y)$.

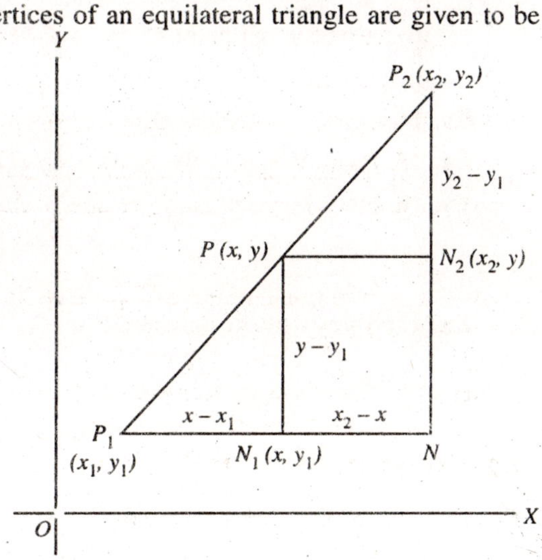

We can easily see that

$$x - x_1 = x_2 - x$$

i.e., $$2x = x_2 + x_1$$

$$\therefore \quad x = \frac{x_1 + x_2}{2}$$

Similarly $$y - y_1 = y_2 - y$$

$$\therefore \quad y = \frac{y_1 + y_2}{2}$$

The coordinates (x, y) of the midpoint P can be written

$$\left(\frac{x_1 + x_2}{2}, \frac{y_1 + y_2}{2}\right)$$

Ex. 7. Find the coordinates of the centroid G of the triangle whose vertices are (3, 2), (– 1, – 4), (– 5, 6). [The centroid is the meeting point of the three medians. The centroid divides each median in the ratio 2 : 1]. Prove $G = \left(\frac{x_1 + x_2 + x_3}{3}, \frac{y_1 + y_2 + y_3}{3}\right)$ where $(x_1, y_1), (x_2, y_2), (x_3, y_3)$ are the vertices of the triangle. **Ans.** (– 1, 4/3)

Ex. 8. The coordinates of the midpoint of the line segment joining (3, 4), (5, 6) are

$$x = \frac{3+5}{2} = 4, \quad y = \frac{4+6}{2} = 5$$

Ex. 9. Find the midpoint of $(-a, 0), (a, 0)$ [**Ans.** (0, 0)]

Ex. 10. Given $A(-1, -1)$, $B(3, -3)$, $C(2, 3)$, find the lengths of the sides and the medians of $\Delta\ ABC$. Also find the coordinates of the midpoints of the medians.

Ex. 11. The segments joining the midpoints of the sides of a quadri-lateral form a parallelogram.

Hint. Find the lengths of sides joining the midpoints. (See diagram).

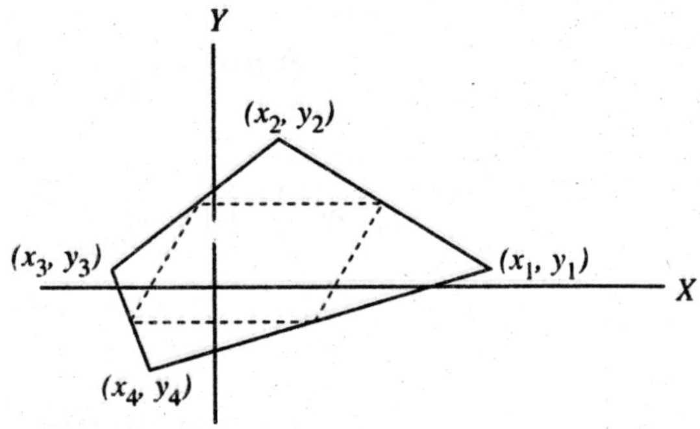

Ex. 12. The diagonals of a square are equal.

Ex. 13. Given $P(3, 2)$, $Q(5, 8)$, find the coordinates of R on PQ so that

(*i*) P is the midpoint of QR

(*ii*) Q is the midpoint of PR

(*iii*) R is the midpoint of PQ

[**Ans.** (*i*) (1, – 4); (*ii*) (7, 14); (*iii*) (4, 5)]

Ex. 14. Prove that the diagonals of a parallelogram bisect each other.

Ex. 15. Prove that $A(0, 0)$, $B(10, -4)$ $C(2, 5)$ form a right triangle.

Ex. 16. Find the point (x, y) on the y-axis equidistant from (3, 2), (– 5 – 2).

[**Ans.** (0, – 2)]

Ex. 17. Show that the distance from (1, 1) to $\left(\frac{2t^2}{1+t^2}, \frac{(1-t)^2}{1+t^2}\right)$ is the same for all values of t. (Show the answer does not contain t)

8.2 DIVISION OF A LINE

Coordinates of the Point Dividing a Line

There are two possibilities

(*a*) P lies on the segment P_1P_2.

(*b*) P lies externally on P_1P_2 extended

Case (a) Internal Division

Let the line segment P_1P_2 formed by joining $P_1(x_1, y_1)$ and $P_2(x_2, y_2)$ be divided at $P(x, y)$. Given that

$$\frac{P_1P}{PP_2} = \frac{k_1}{k_2}$$

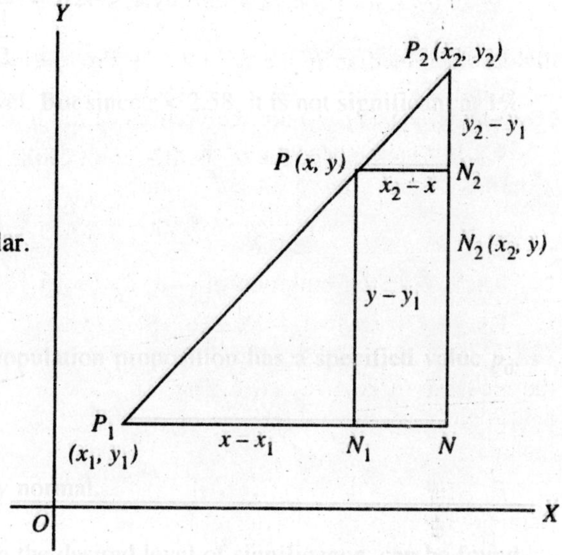

Triangles PP_1N_1, P_2PN_2 are similar.

$$\therefore \quad \frac{P_1P}{PP_2} = \frac{x - x_1}{x_2 - x}$$

$$\therefore \quad \frac{k_1}{k_2} = \frac{x - x_1}{x_2 - x}$$

$$\therefore \quad k_1x_2 - k_1x = k_2x - k_2x_1$$

or $$k_1x + k_2x = k_1x_2 + k_2x_1$$

$$\therefore \quad x = \frac{k_1x_2 + k_2x_1}{k_1 + k_2}$$

Similarly since $\dfrac{P_1P}{PP_2} = \dfrac{y - y_1}{y_2 - y}$ we have $y = \dfrac{k_1y_2 + k_2y_2}{k_1 + k_2}$

Ex. 18. The coordinates of $P(x, y)$ dividing the line segment joining (1, 2), (3, 4) in the ratio 1 : 3 are given by

$$x = \frac{k_1x_2 + k_2x_1}{k_1 + k_2} = \frac{1\times 3 + 3\times 1}{1+3} = \frac{3}{2}$$

$$y = \frac{k_1y_2 + k_2y_1}{k_1 + k_2} = \frac{1\times 4 + 0\times 2}{1+3} = \frac{5}{2}$$

Case (b) External Division

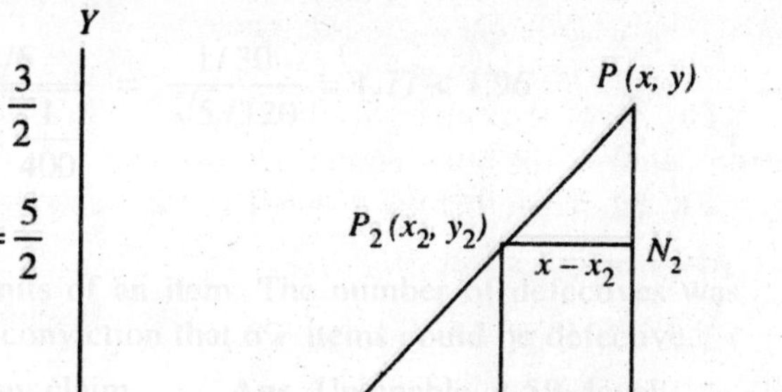

Here $\dfrac{P_1P}{PP_2} = -\dfrac{k_1}{k_2}$

Triangles PP_1N, PP_2N_2 are similar

$$\therefore \quad \frac{P_1P}{PP_2} = \frac{P_1N}{N_2P_2}$$

$$\therefore \quad \frac{k_1}{k_2} = \frac{x - x_1}{x_2 - x}$$

[Note N_2P_2 is negative and hence equals $-(x - x_2) = x_2 - x$]

$$\therefore \quad -k_1(x_2 - x) = k_2(x - x_1)$$

$$\therefore \quad x(k_1 - k_2) = k_1x_2 - k_2x_1$$

Thus we have $$x = \frac{k_1x_2 - k_2x_1}{k_1 - k_2}$$

Similarly $$y = \frac{k_1y_2 - k_2y_1}{k_1 - k_2}$$

Ex. 19. The coordinates of $P(x, y)$ dividing the line segment joining (1, 2), (3, 4) externally in the ratio 5 : 4 are given by

$$x = \frac{5 \times 3 - 4 \times 1}{5 - 1} = 11$$

$$y = \frac{5 \times 4 - 4 \times 2}{5 - 4} = 12$$

To sum up:

The coordinates of internal division are

$$\left(\frac{k_1x_2 + k_1x_1}{k_1 + k_2}, \frac{k_1y_1 + k_2y_1}{k_1 + k_2}\right)$$

The coordinates of external division are

$$\left(\frac{k_1x_2 - k_2x_1}{k_1 - k_2}, \frac{k_1y_2 - k_2y_1}{k_1 - k_2}\right), \text{ or } \left(\frac{k_2x_1 - k_1x_2}{k_2 - k_1}, \frac{k_2y_1 - k_1y_2}{k_2 - k_1}\right)$$

Ex. 20. In what ratio does the origin divide the join of (6, 0), (– 3, 0)?

Let the ratio be k : 1. Then the coordinates of the dividing point, *i.e.*, (0, 0) are

$$x = \frac{k(-3) + 1.6}{k - 1} = 0, \quad y = 0$$

$\therefore$ $-3k + 6 = 0$ or $k = 2$ the ratio is 2 : 1

Ex. 21. If one end of the diameter of a circle with centre $C(-4, 1)$ is $P(2, 6)$, find the coordinates of the other end of the diameter. **Ans.** (– 10 – 4)

Ex. 22. Write the coordinates of the point of trisection of the line segment joining (1, 0), (8, 10). **Ans.** $\left(\frac{10}{3}, \frac{14}{3}\right), \left(\frac{17}{3}, \frac{22}{3}\right)$

Ex. 23. Find the ratio in which the y-axis divides the join of (3, 5) (6, 7).

Ans. External division – 1 : 2

8.3 AREA OF A TRIANGLE

Let the coordinates of the vertices be given by

$A(x_1, y_1)$, $B(x_2, y_2)$, $C(x_3, y_3)$.

It may be remembered that the area of a trapezium = $\frac{1}{2}$ (sum of parallel sides) × perpendicular distance between them.

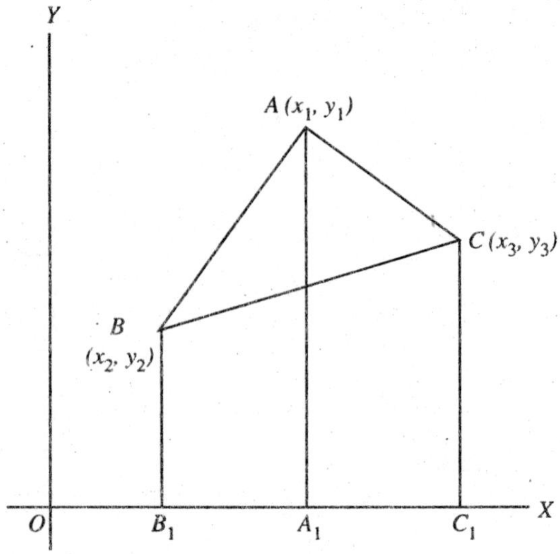

Thus $\quad \text{Trap. } AA_1B_1B = \frac{1}{2}(AA_1 + BB_1)A_1B_1$

$$= \frac{1}{2}(y_1 + y_2)(x_1 - x_2)$$

$$\Delta ABC = \text{Trap. } AA_1B_1B + \text{Trap. } ACC_1A_1 - \text{Trap. } BCC_1B_1$$

$$= \frac{1}{2}(y_1 + y_2)(x_1 + x_2) + \frac{1}{2}(y_1 + y_3)(x_3 - x_1) - \frac{1}{2}(y_2 + y_3)(x_3 - x_2)$$

$$= \frac{1}{2}(x_1y_2 - x_2y_1 + x_2y_3 - x_3y_2 + x_3y_1 - x_1y_3)$$

If three points lie on a straight line the area of the triangle formed by them is zero *i.e.*, $\frac{1}{2}(x_1 y_2 - x_2y_1 + x_2y_3 - x_3y_2 + x_3y_1 - x_1y_3) = 0$. In fact, this is the condition for three points to be on a line. (Collinearity)

Ex. 24. Show that if the vertex C in a triangle is the origin (0, 0), the area of the triangle is $\frac{1}{2}(x_1y_2 - x_2y_1)$.

Ex. 25. Show that the area of a quadrilateral $A(x_1, y_1)$, $B(x_2, y_2)$ $C(x_3, y_3)$, $D(x_4, y_4)$ is $\frac{1}{2}(x_1y_2 - x_2y_1 + x_2y_3 - x_3y_2 + x_3y_4 - x_4y_3 + x_4y_1 - x_1y_4)$. (Sum of two triangles)

Ex. 26. Find the area of a triangle with vertices

(*i*) (10, 7), (– 2, 3), (0, 0); (*ii*) (1, 3), (5, 6), (2, 4) **Ans.** (*i*) 22 (*ii*) $\frac{1}{2}$

Ex. 27. Show that points (0, 2), (1, 5), (– 1, – 1) are collinear.

Ex. 28. If (0, 4), (– 1, 3), (k, 5) are collinear, find k. (Use Ex. 23) **Ans.** 1

8.4 LINEAR FUNCTIONS

Equations of a Line

The graph of a linear function $ax + by + c = 0$ where a, b, c are real numbers is a straight line.

Every equation of the first degree in x, y represents a straight line. Two points are enough to determine a line.

The equation of a straight line is often written in the form

$$y = mx + c$$

Slope of a line

Let $P_1(x_1, y_1)$, $P_2(x_2, y_2)$ be any two points on a line.

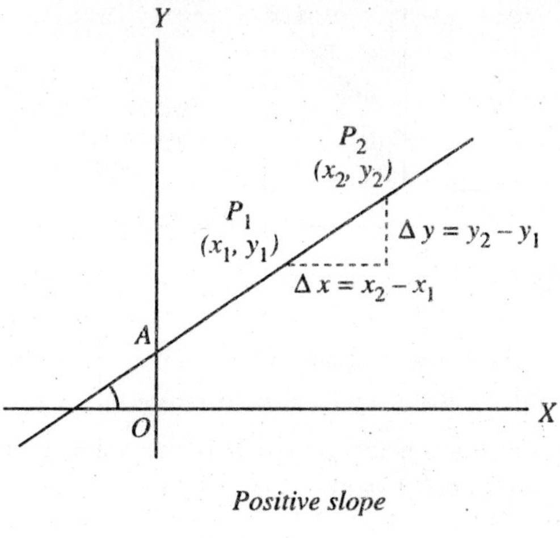

Positive slope

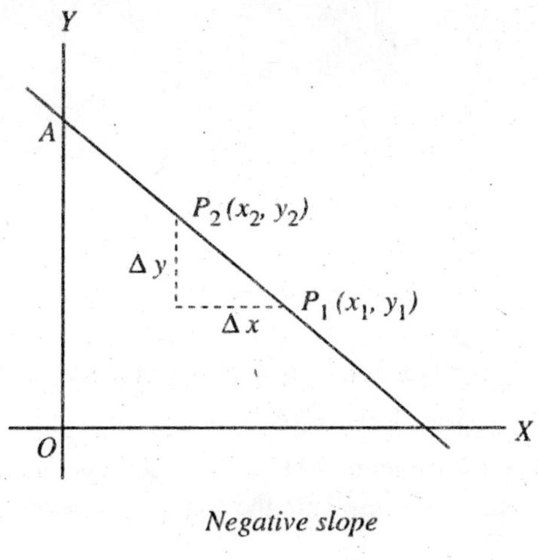

Negative slope

Then the *slope m of the line* is defined by

$$m = \frac{\Delta y}{\Delta x} = \frac{y_2 - y_1}{x_2 - x_1}$$

$$= \frac{\text{Vertical change}}{\text{Horizontal change}}$$

The slope is positive if the line rises to the right. The slope is negative if the line rises to the left.

Ex. 29. The slope of a line through (4, 7), (2, 1) is

$$m = \frac{y_2 - y_1}{x_2 - x_1} = \frac{1 - 7}{2 - 4} = 3.$$

Ex. 30. If (4, 2) is a point on a line passing through the origin (0, 0) its slope is $m = \frac{-2 - 0}{4 - 0} = -\frac{1}{2}$.

Ex. 31. Find the slopes of the lines passing through

(*i*) $(-1, 0), (1, 0)$ (*ii*) $(a, b), (2a, 2b)$

(*iii*) $(0, 0), (2, 5)$ (*iv*) $(a + k, b + k), (a + m, b + m)$

(*v*) $(-a, -b), (-b, -a)$.

Ans. (*i*) 0, (*ii*) b/a, (*iii*) 5/2, (*iv*) 1, (*v*) -1.

The Line $y = mx + c$

This is the *slope intercept form* of a line.

Slope. Wherever we take two points on the line the slope will be the same quantity m.

Let (x_1, y_1) be on the line

$$y = mx + c.$$

Then

$$y_1 = mx_1 + c_1.$$

If (x_2, y_2) is on the same line then

$$y_2 = mx_2 + c_2$$

Subtracting

$$y_2 - y_1 = mx_2 - mx_1$$
$$= m(x_2 - x_1)$$

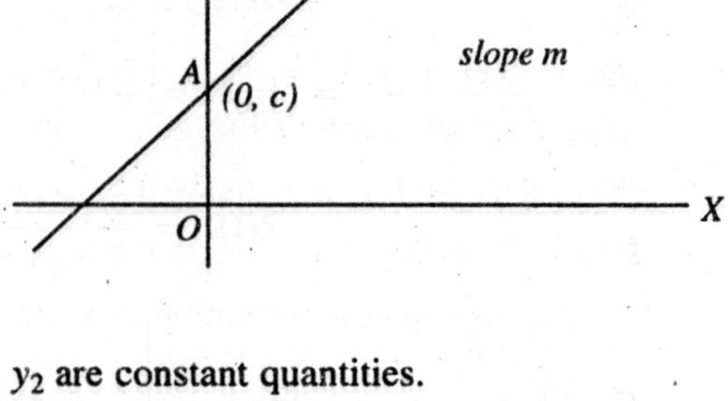

We have the slope $m = \frac{y_2 - y_1}{x_2 - x_1}$

which is a constant quantity since x_1, y_1, x_2, y_2 are constant quantities.

The slope of the line $y = mx + c$ is the value of the *coefficient* of x. (Here m)

Intercept. OA is the intercept made by the line on the y-axis. It is the value of y where the line cuts the y-axis *i.e.*, where $x = 0$. (Here $OA = c$)

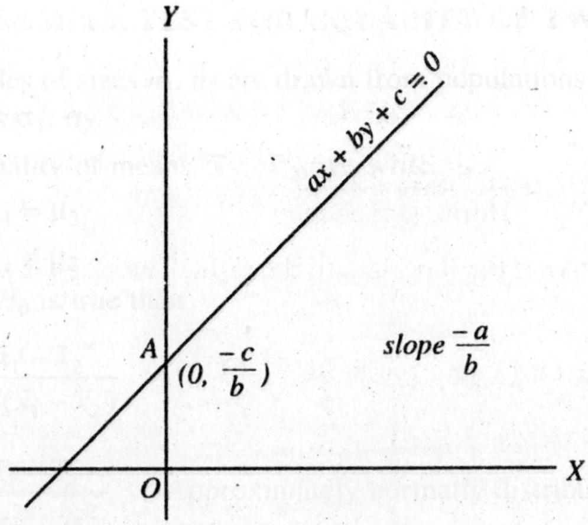

For any equation $ax + by + c = 0$, we can find the slope by writing it as

$y = -(a/b)x - c/b$

This is of the form $y = mx + c$.

$\therefore$ slope $= -\dfrac{a}{b}$ and the intercept $= -\dfrac{c}{b}$.

Ex. 32. If $P_1(4, 7)$, $P_2(2, 1)$ are points on a line, find the slope of the line.

$$m = \frac{y_2 - y_1}{x_2 - x_1} = \frac{1 - 7}{2 - 4} = \frac{-6}{-2} = 3.$$

Ex. 33. If $P(4, 2)$ is a point on a line passing through the origin $(0, 0)$ the slope is given by

$$m = \frac{2 - 0}{4 - 0} = \frac{1}{2}.$$

Ex. 34. The slope of the line $3y = 9x - 2$ is obtained by writing the line as

$$y = \frac{9x}{3} - \frac{2}{3} \quad \therefore \quad m = 3.$$

Ex. 35. The slope of the line $x = -2y - 7$ is obtained by writing the line as

$$y = -\tfrac{1}{2}x - \tfrac{7}{2} \quad \therefore \quad m = -\tfrac{1}{2}.$$

Ex. 36. The slope of $x = -y$ is the same as the slope of $y = -x$ which is -1.

Ex. 37. The slope of each of the following lines is zero

(*i*) $y = 0$, (*ii*) $y = k$, (*iii*) $y = -k$.

(Note that $y = 0$ is the equation of x-axis)

Ex. 38. The slope of lines (*i*) $x = 0$, (*ii*) $x = k$, (*iii*) $x = -k$ is infinite.

($x = 0$ is the equation of y-axis)

Ex. 39. The slope of the line $\dfrac{x}{2} + \dfrac{y}{3} = 1$ is obtained by writing it as

$$y = -\frac{3}{2}x + 3 \quad \therefore \quad m = -\frac{3}{2}.$$

Note 1. (*i*) $y = mx + c$

This is the equation of a line with slope = m and intercept c on the y-axis.

Let the line cut the y-axis at A. At A, $x = 0$ so that by substitution,

$$y = m \times 0 + c, \quad \therefore \quad y = c \quad \therefore \quad \text{The coordinates of } A \text{ are } (0, c)$$

At B, $y = 0$ so that $0 = mx + c$

$$\therefore \quad x = -\frac{c}{m} \quad \therefore \quad \text{The coordinates of } B \text{ are } \left(-\frac{c}{m}, 0\right)$$

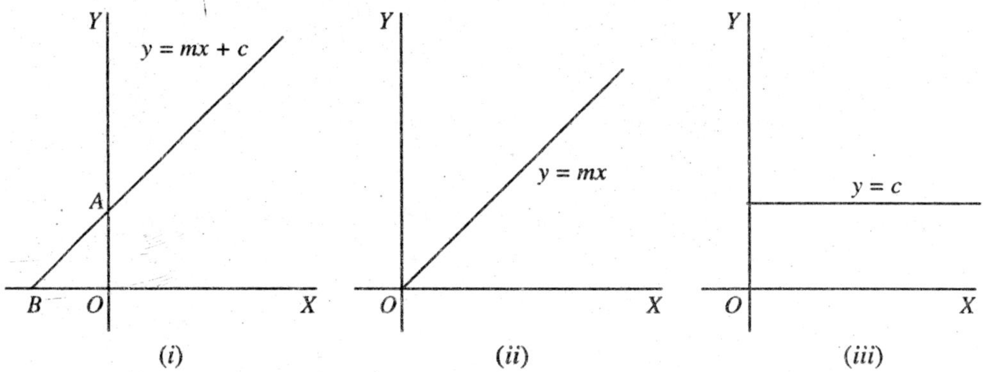

(*ii*) If the line passes through the origin there is no intercept *i.e.*, $c = 0$

$y = mx$ is the equation of a line with slope m, passing through the origin. The slope is

$$m = \frac{y}{x}.$$

If m is positive, the line makes an acute angle with the x-axis.

If m is negative, the angle is obtuse.

Note : 2. Any straight line can be written in the general abstract form of an equation:

$$y = mx + c$$

A line is a simple, particular form of association or *relation* between two quantities.

x and y are *variables* whose variations are under consideration.

m is the slope of the line. The slope of the line shows the increase in y for a unit increase in x. A line has a constant slope.

Also, $\quad c = OA$ intercept on y-axis

$$-\frac{c}{m} = OB \text{ intercept on } x\text{-axis}$$

If $c = 0$ there is no intercept on the y-axis (nor on the x-axis) and the equation reduces to $y = mx$. This is a line passing through the origin.

(*iii*) If $m = 0$, the resulting line $y = c$ is parallel to the x-axis, at a distance c from the x-axis. Its slope is zero.

In the same manner, the line $x = b$ is parallel to the y-axis, at a distance b from the y-axis.

m and c are called *parameters* of the equation. They are constant for a given straight line. Different values of m and c will give lines with different slopes and intercepts.

Note: 3.

The relation $y = mx + c$

is often used in economics as an *approximate* linear model because, in practice, generally, we do not have exactly linear relations.

Equation of a line with slope m and passing through (x_1, y_1)

Let (x, y) be any point on the line

The slope of the line is given by

$$m = \frac{y - y_1}{x - x_1}$$

$$\therefore \quad y - y_1 = m(x - x_1)$$

This is the equation of the line passing through (x_1, y_1).

Equation of a line having intercept a on x-axis and intercept b on y-axis

The slope of this line is

$$\frac{b-0}{0-a} = -\frac{b}{a}$$

If (x, y) is any point on the line we can also write the slope as

$$\frac{y-0}{x-a} = \frac{y}{x-a}$$

$$\therefore \quad \frac{y}{x-a} = -\frac{b}{a}$$

$$\frac{y}{b} = -\frac{x}{a} + 1$$

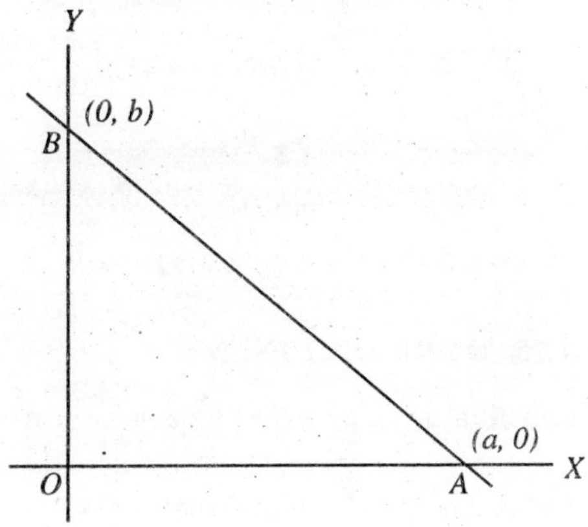

$\therefore \quad \dfrac{x}{a} + \dfrac{y}{b} = 1$ is the equation of the line having intercepts a, b on the axes

Two lines $y = m_1x + c$, $\quad y = m_2x + c_2$

(*i*) *Intersecting lines.* If the two lines intersect there is a value of x which satisfies the two simultaneously.

$$m_1x + c_1 = m_2x + c_2$$

$$\therefore \quad x = \frac{c_2 - c_1}{m_1 - m_2} (m_1 \neq m_2)$$

If θ is the angle between two intersecting lines, $\tan\theta = \dfrac{m_2 - m_1}{1 + m_1m_2}$

(*ii*) *Parallel lines.* If $m_1 = m_2$ and $c_1 \neq c_2$ the lines are parallel. The slopes of parallel lines are equal.

(*iii*) *Identical lines.* If $m_1 = m_2$ and $c_1 = c_2$ the two lines coincide. The slopes as well as intercepts of identical lines are equal.

(*iv*) *Perpendicular lines.* If $m_1 m_2 = -1$, the lines are perpendicular.

The product of the slopes of perpendicular lines is -1.

Alternative forms of two lines:

$$a_1 x + b_1 y + c_1 = 0 \qquad a_2 x + b_2 y + c_2 = 0$$

Their slopes are $m_1 = -\dfrac{a_1}{b_1}$, $\quad m_2 = -\dfrac{a_2}{b_2}$

The lines *intersect* if $-\dfrac{a_1}{b_1} \neq -\dfrac{a_2}{b_2}$

The lines are *parallel* if $-\dfrac{a_1}{b_1} = -\dfrac{a_2}{b_2}$ *i.e.*, if $\dfrac{a_1}{b_1} = \dfrac{a_2}{b_2}$

The lines are *identical* if $\dfrac{a_1}{b_1} = \dfrac{a_2}{b_2} = \dfrac{c_1}{c_2}$

The lines are *perpendicular* if $\left(-\dfrac{a_1}{b_1}\right)\left(-\dfrac{a_2}{b_2}\right) = -1$

i.e., $\dfrac{a_1 a_2}{b_1 b_2} = -1$ or $a_1 a_2 = -b_1 b_2$

We can write this as $a_1 a_2 + b_1 b_2 = 0$.

8.15 MORE ON LINES

Equation of a line *AB* at a distance *p* from the origin.

Let $P(x, y)$ be a point on the line. *PM* makes $\angle \theta$ with the line.

(*PM* is perpendicular to the *x*-axis $OQ = p$ is perpendicular to the line)

$$OQ = OA \cos\theta$$
$$= (OM + MA)\cos\theta$$
$$\therefore \quad p = (x + y\tan\theta)\cos\theta$$
$$\therefore \quad x\cos\theta + y\sin\theta = p$$

is the equation of a line making $\angle \theta$ with the *x*-axis.

Put $x = 0$, the *y* intercept $= p/\sin\theta$

Put $y = 0$, the *x* intercept $= p/\cos\theta$

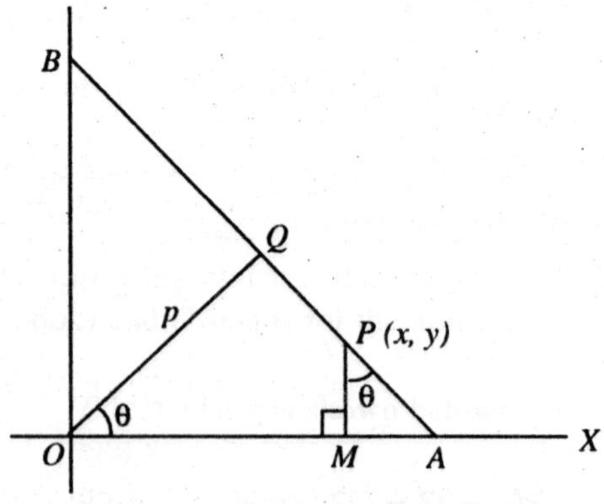

Ex. 40. Find the equation of the line through $A(x_1, y_1)$ and making $\angle \theta$ with the *x*-axis.

Let $P(x, y)$ be on the line and let $AP = r$ (See diagram below)

$$\cos\theta = \frac{AB}{AP} = \frac{x - x_1}{r} \quad \text{or} \quad \frac{x - x_1}{\cos\theta} = r$$

$$\sin\theta = \frac{BP}{AP} = \frac{y - y_1}{r}$$

or $$\frac{y - y_1}{\cos\theta} = r$$

$\therefore$ The required equation is

$$\frac{x - x_1}{\cos\theta} = \frac{y - y_1}{\sin\theta} = r$$

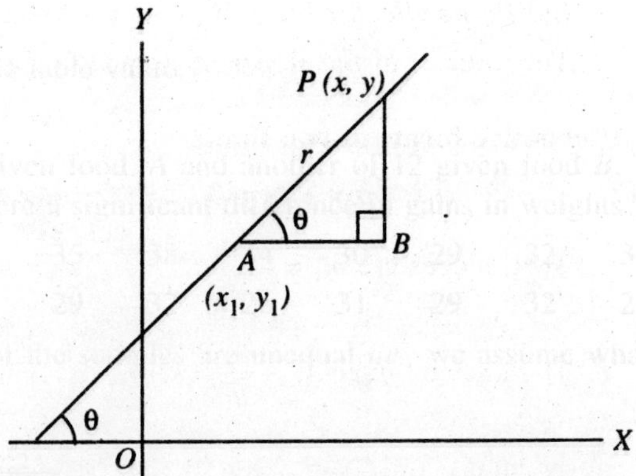

Ex. 41. $x = x_1 + r\cos\theta$ and $y = y_1 + r\sin\theta$ are the coordinates of any point on the above line at a distance r from A.

8.6 ANGLE θ BETWEEN LINES

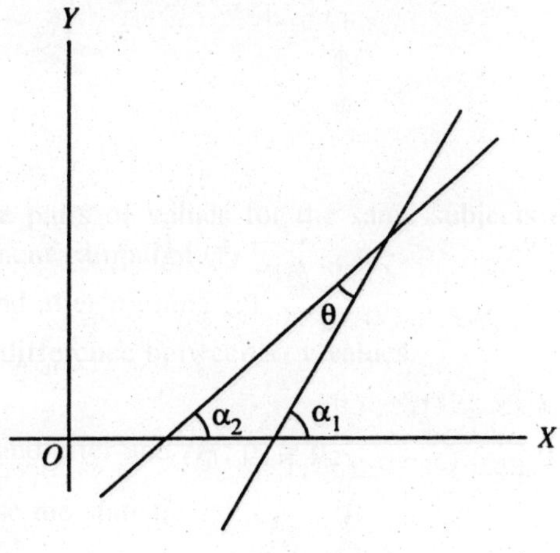

Given $y = m_1x + c_1$, and $y = m_2x + c_2$ making angles α_1, α_2 with the x-axis

Slopes $\quad \tan\alpha_1 = m_1, \quad \tan\alpha_2 = m_2$

$$\alpha_1 - \alpha_2 = \theta \text{ or } \pi - \theta$$

$$\therefore \quad \tan(\alpha_1 - \alpha_2) = \tan\theta$$

$$\text{or } \tan(\pi - \theta) \text{ i.e., } -\tan\theta$$

$$\therefore \quad \tan\theta = \pm\frac{\tan\alpha_1 - \tan\alpha_2}{1 + \tan\alpha_1\tan\alpha_2}$$

$$= \pm\frac{m_1 - m_2}{1 + m_1m_2}$$

Ex. 42. If two lines are $x\cos\alpha + y\sin\alpha = p$, $x\cos\beta + y\sin\beta = q$

$$m_1 = -\cos\alpha/\sin\alpha, \; m_2 = -\cos\beta/\sin\beta$$

$$\tan\theta = \pm\frac{-\cos\alpha/\sin\alpha + \cos\beta/\sin\beta}{1 + (-\cos\alpha/\sin\alpha)(-\cos\beta/\sin\beta)}$$

$$= \pm\frac{\sin\alpha\cos\beta - \cos\alpha\sin\beta}{\cos\alpha\cos\beta + \sin\alpha\sin\beta} = \pm\frac{\sin(\alpha - \beta)}{\cos(\alpha - \beta)}$$

$$= \pm\tan(\alpha - \beta) \text{ i.e., } \theta = \alpha - \beta \text{ or } \beta - \alpha$$

Note : (1) Two lines are parallel if $\tan\theta = 0$ *i.e.,* $m_1 = m_2$

(2) They are perpendicular if $\tan\theta = \infty$ *i.e.,* $m_1m_2 = -1$

(3) The lines $a_1x + b_1y + c_1 = 0$ and $a_2x + b_2y + c_2 = 0$ are parallel if $m_1 = -a_1/b_1 = -a_2/b_2 = m_2$ and perpendicular if $\left(-\frac{a_1}{b_1}\right)\left(-\frac{a_2}{b_2}\right) = -1$ *i.e.,* $a_1a_2 + b_1b_2 = 0$

Ex. 43. What is the equation of a line passing through $(a\cos^3\theta, a\sin^3\theta)$ and perpendicular to the line $x\sec\theta + y\,\text{cosec}\,\theta = a$.

Slope of the line $= -\dfrac{\sec\theta}{\operatorname{cosec}\theta}$

The required line perpendicular to it has slope $= \dfrac{\operatorname{cosec}\theta}{\sec\theta}$

$\therefore$ Its equation is $y - a\sin^3\theta = \dfrac{\operatorname{cosec}\theta}{\sec\theta}(x - a\cos^3\theta)$

$$\therefore \quad x\cos\theta - y\sin\theta = a(\cos^4\theta - \sin^4\theta)$$
$$= a(\cos^2\theta - \sin^2\theta)(\cos^2\theta + \sin^2\theta)$$
$$= a\cos 2\theta$$

Ex. 44. Find the equation of the line through (a, b) which makes $\angle\alpha$ with $y = mx + c$.

If this line PQ has slope m_1

$$\tan\alpha = \pm\frac{m - m_1}{1 + mm_1}$$

$$\therefore \quad m_1 = \frac{m \pm \tan\alpha}{1 + m\tan\alpha}$$

There are two lines

$$y - b = \frac{m + \tan\alpha}{1 + m\tan\alpha}(x - a)$$

$$y - b = \frac{m - \tan\alpha}{1 + m\tan\alpha}(x - a)$$

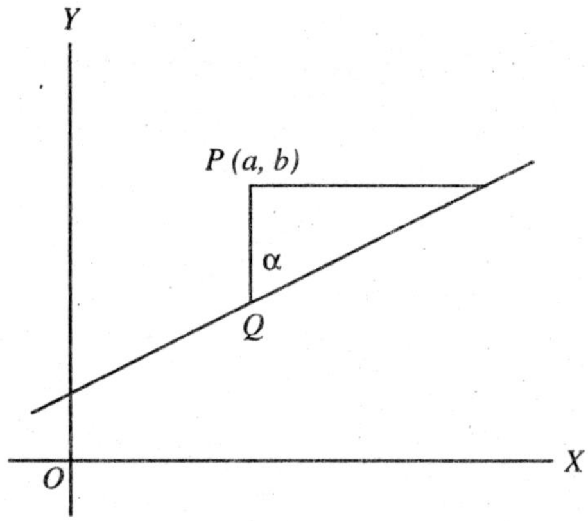

Families of Parallel Lines

A family of parallel lines can be expressed generally by the equation $y = mx + c$ where c can take varying values $c_1, c_2, \ldots$etc.

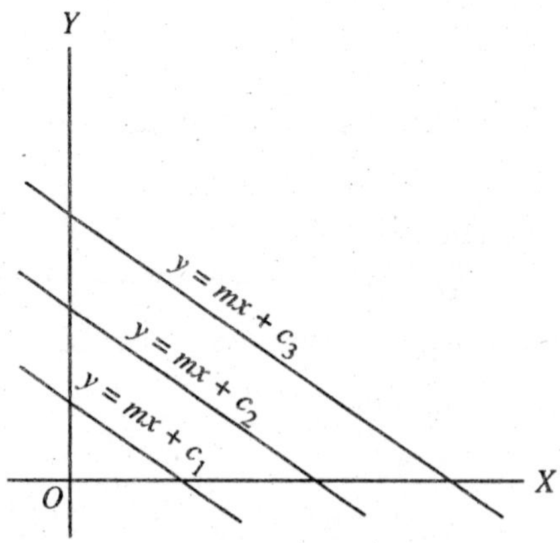

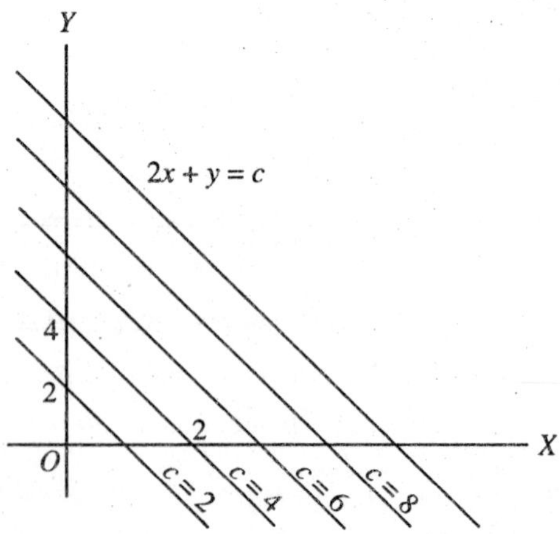

If some constraints are imposed on the equation it is possible to find the highest value of the parameter c in a given situation.

Ex. 45. If $2x + y = c$, the line giving the highest value of c passes through the upper right vertex of the quadrilateral which is given by $0 \le x \le 2, 0 \le y \le 4$.

This corresponds to $c = 8$ $2x + y = 8$.

Ex. 46. What is the equation of a line which meets the x-axis at (3, 0) and y-axis at (0, – 5). **Ans.** $\frac{x}{3} + \frac{y}{-5} = 1$ or $\frac{x}{3} - \frac{y}{5} = 1$ or $5x - 3y = 15$

Ex. 47. What is the equation of a line with intercepts – 2 and – 5 on the axes?

$$\frac{x}{-2} + \frac{y}{-5} = 10$$

$$\therefore \quad -5x - 2y = 100$$

or $$5x + 2y + 100 = 0$$

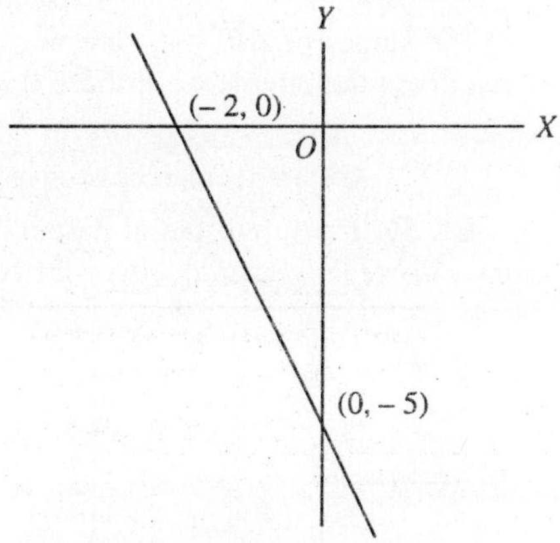

Ex. 48. What is the equation of a line passing through (2, 3) and having an intercept on the x-axis twice the intercept on the y-axis?

Ans. $x + 2y - 8 = 0$

Ex. 49. Show that $\left(-\frac{3}{2}, -1\right)$ lies on the join of (7, – 5) and (– 4, 3).

Ex. 50. What is the equation of a line having (*i*) equal intercepts on the axes?

Ans. If a is the intercept, the equation of the line is given by $\frac{x}{a} + \frac{y}{a} = 1$ or $x + y = a$

Ex. 51. Find the slopes of the following lines

(*i*) $\frac{x+2}{3} - 4y = \frac{1}{2}$ (*ii*) $x - \sqrt{3y} = 6$ (*iii*) $\sqrt{3x} + y = 12$

(*iv*) $y - 12x = 0$ (*v*) $x = y$ (*vi*) $y\sqrt{2} = x\sqrt{6}$

Are there any parallel or perpendicular lines among these?

Draw the graphs and check.

Ans. $\frac{1}{12}, \frac{1}{\sqrt{3}}, \frac{1}{\sqrt{3}}$, 12, 1, 3. No lines || or perpendicular

Ex. 52. Find the equation of a line passing through (2, 6) and (5, 3). Find its slope and intercepts.

Ans. $x + y = 8$; – 1; 8, 8

Ex. 53. For what values of c will the point (7, c) lie on the line passing through (3, 6) and (– 5, 2)? **Ans.** 8

Ex. 54. Two vertices of an equilateral triangle are (– 4, 0), (4, 0). Find the third vertex. **Ans.** $(0, \pm 4\sqrt{3})$

Ex. 55. Consider the slopes of the lines

AB, AB_1, AB_2, AB_3

Ay, AO, AB_4, AB_5

The slope of AB is zero. The slope goes on increasing and the slope of Ay is ∞.

The line AO is the same as Ay

The slopes of AB_4, AB_5 are negative. What about the intercepts of these lines?

Ans. Intercept OA on y-axis; different intercepts on x-axis.

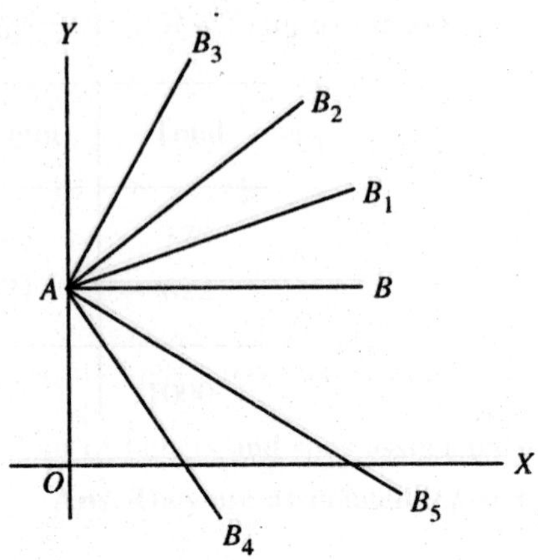

Ex. 56. The intercepts of parallel lines are different but the slopes are the same. Can lines with negative and positive intercepts on the same axis be parallel? **Ans.** Yes

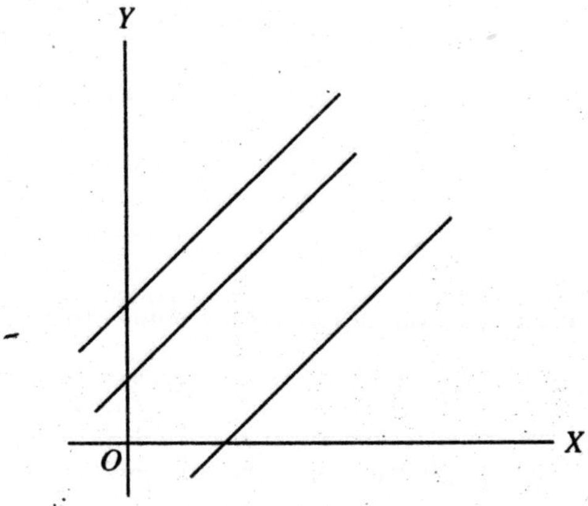

Ex. 57. Prove that the following pairs of lines are parallel:

(*i*) $y = 3x + 7$, $y = 3x - 5$

(*ii*) $y = 7x$, $y = 7x + \frac{1}{2}$

(*iii*) $y = x$, $y = x + a$

(*iv*) $ax + by + c = 0$, $bx + k = -\dfrac{b^2}{a}y$

(*v*) $x = k$, $x = 0$

(*vi*) $y = 5$, $y = -5.$

Ex. 58. Show that (0, 0), $(a, 0)$, $a + (\sqrt{a^2 - b^2}, b)$, $(\sqrt{a^2 - b^2}, b)$ are vertices of a rhombus $(a > b > 0)$. Show that the diagonals are perpendicular.

Ex. 59. Show that $y = 7x + 2$, $y = -\frac{1}{7}x + 4$ are perpendicular lines.

Ans. $m_1 m_2 = 7 \times \frac{1}{7} = -1$

Ex. 60. Prove that the following pairs of lines are perpendicular:

(*i*) $y = 3x + 7$, $x = 5 - 3y$

(*ii*) $y = 2x$, $y = \frac{3}{4} - \frac{x}{2}$

(*iii*) $\frac{x}{a} + \frac{y}{b} = 1$, $ax - by = 10$

(*iv*) $ax + by + c = 0$, $ay = bx$

(*v*) $y = x$, $x = -y.$

Ex. 61. Prove that the figure formed by the points $A(3, 1)$, $B(6, 0)$, $C(4, 4)$ is a right triangle.

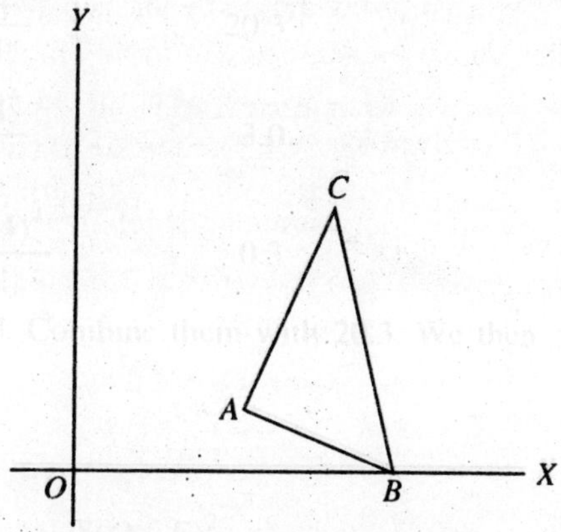

$$\text{Slope of } AB = \frac{0-1}{6-3} = -\frac{1}{3}$$

$$\text{Slope of } AC = \frac{4-1}{4-3} = 3$$

$$-\tfrac{1}{3} \times 3 = -1$$

$\therefore$ $AB \perp AC$

By finding the lengths of the lines we can prove that

$$BC^2 = AB^2 + AC^2$$

Ex. 62. Show that $(-3, -2)$, $(7, 4)$, $(1, 14)$ are the vertices of an isosceles right triangle, find the length of the perpendicular from the vertex of the right angle to the hypotenuse. **Ans.** $2\sqrt{17}$

Ex. 63. Show that $(3, -2)$, $(4, 3)$, $(-1, 1)$, $(-2, -4)$ are the vertices of a parallelogram. Find the lengths of the diagonals. **Ans.** 5, $\sqrt{85}$

(Find the slopes of opposite pairs of sides and show they are parallel and equal).

Ex. 64. (*a*) Prove that the lines joining the midpoints of opposite sides of a quadrilateral bisect each other.

(*b*) Prove that the diagonals of a rhombus are perpendicular.

Ex. 65. (*a*) $(2, 4)$, $(6, 2)$, $(8, 6)$ are three vertices of a square. Find the fourth vertex. **Ans.** $(4, 8)$

(*b*) If $(3, 6)$, $(-5, 2)$, $(7, y)$ are points on the same line find y.

Ans. The slopes of the lines joining any two of these three points will have the same value so that $y = 8$

Ex. 66. (*a*) Find the equation of the line parallel to the line $5x - 7y - 10 = 0$ and passing through $(3, -4)$.

(*b*) Find the equations of the medians of a triangle with vertices $(1, 6)$, $(3, -4)$, $(-5, -1)$. **Ans.** $17x - 14y + 7 = 10$, $13x + 10y + 1 = 0$, $2x - 7y + 3 = 0$

Ex. 67. Find the equation of the line passing through $(8, 3)$ and through the point of intersection of the lines $5x - 2y + 15 = 0$, $3x + y = 13$. **Ans.** $x + y = 11$

Ex. 68. Show that $\left(x_1, -\frac{c + ax_1}{b}\right), \left(x_2, -\frac{c + ax_2}{b}\right)$ are points on the line $ax + by + c = 0$. (Substitute each point in the equation and see if it is satisfied).

Ex. 69. The lines $\left.\begin{array}{l} 5x - ky + 3 = 0 \\ ax - 2y + 6 = 0 \end{array}\right\}$ are identical if

$$\frac{5}{a} = \frac{-k}{-2} = \frac{3}{6} \qquad \therefore \quad a = 10, \ k = 1.$$

Ex. 70. Show that the lines $3x + by + 5 = 0$, $cx - 3y - 2 = 0$ are perpendicular if $3c - 3b = 0$ or $b = c$.

Ex. 71. The equation of a line through (4, 5) and perpendicular to the line $4x - 3y + 7 = 0$ is given by $3x + 4y - 32 = 0$.

A summary and some additional results

(*i*) $ax + by + c = 0$ is a linear equation. It is of degree one in x and y. It can be geometrically represented by a straight line.

(*ii*) If (x_1, y_1), (x_2, y_2) are points on a line, its slope is defined by

$$\tan\theta = m = \frac{y_2 - y_1}{x_2 - x_1}$$

where θ is the smallest positive angle between the lines and the positive x-axis. The rate of change of a linear function $ax + by + c = 0$ is constant and equals the slope of the line.

(*iii*) There are many forms of the equation of a straight line:

The slope intercept from $\quad y = mx + c$

The point slope form $\quad y - y_1 = m(x - x_1)$

The two point form $\quad \frac{y - y_1}{y_2 - y_1} = \frac{x - x_1}{x_2 - x_1}$

The intercept form $\quad \frac{x}{a} + \frac{y}{b} = 1$

The line parallel to y-axis $\quad x = k$

The line parallel to x-axis $\quad y = k$

(*iv*) In the general form $\quad ax + by + c = 0$

The slope $m = -\frac{a}{b}$ and the intercepts are $-\frac{c}{a}$ on the x-axis and $-\frac{c}{b}$ on the y-axis

if $a = 0$, the line is parallel to x-axis.

if $b = 0$, the line is parallel to y-axis.

(*v*) If two lines with slopes m_1, m_2 are parallel, $m_1 = m_2$

If they are perpendicular $m_1 m_2 = -1$

If θ is the angle between them then $\tan\theta = \left|\frac{m_2 - m_1}{1 + m_1 m_2}\right|$

(*vi*) *Intersection of two lines.* The coordinates of the points of intersection, obtained by solving the equations, satisfy the equations of both the lines.

(*vii*) *Concurrent lines.* Three given lines have a common point of intersection if the coordinates of the intersection point on any two lines satisfy the equation of the third line.

(*viii*) *Normal form.* The normal form of the equation of the line $ax + by + c = 0$ is given by

$$\frac{a}{\sqrt{a^2+b^2}}x + \frac{b}{\sqrt{a^2+b^2}}y + \frac{c}{\sqrt{a^2+b^2}} = 0.$$

(*ix*) The perpendicular distance of a given point (x_1, y_1) from a given line $ax + by + c = 0$ is given by

$$p = \frac{ax_1 + by_1 + c}{\sqrt{a^2+b^2}}$$

The perpendicular from the origin to this line is $p = \dfrac{c}{\sqrt{a^2+b^2}}$

(*x*) *Distance between parallel lines.* It is the perpendicular distance of any point on one of the lines to the other line.

(*xi*) *Any line through the intersection of two given lines,*

$a_1x + b_1y + c_1 =$ and $a_2x + b_2y + c_2 = 0$ is given by

$a_1x + b_1y + c_1 + k(a_2x + b_2y + c_2) = 0$

which represents the family of all lines passing through the point of intersection of the given lines.

8.7 ISOPROFIT AND ISOCOST LINES FOR TWO PRODUCTS

An isoprofit line shows different combinations of two products X_1, X_2 which will yield the same total profit.

If x_1, x_2 are the quantities of the two products, the profit function describing the isoprofit line is given by

$$\pi = a_1x_1 + a_2x_2$$

where π is profit and a_1, a_2 are known values.

The slope of the profit line is found by fixing the value of π say at π_1 thus

$$a_2x_2 = \pi_1 - a_1x_1 \quad \text{or} \quad x_2 = -\frac{a_1}{a_2}x_1 + \frac{\pi_1}{a_2}$$

The slope is $-\dfrac{a_1}{a_2}$. The intercept on the x_1-axis is $x_1 = \dfrac{\pi_1}{a_1}$; that on the x_2-axis is $x_2 = \dfrac{\pi_1}{a_2}$.

A family of isoprofit lines can be drawn by assigning different values to the profit constant. The slopes of all isoprofit lines for a given problem are equal.

An *isocost* line shows different combinations of two products X_1, X_2 which will involve the same total cost. The total cost function is given by

$$C = b_1x_1 + b_2x_2$$

where b_1, b_2 are constants. If $C = C_1$, the slope is $-\frac{b_1}{b_2}$ since

$$x_2 = -\frac{b_1}{b_2}x_1 + \frac{C}{b_2}$$

A family of isocost lines can be drawn by assigning different values to the cost.

8.8 CHANGE OF ORIGIN: TRANSLATION OF AXES

If the coordinates axes are changed, the coordinates of a point would change. The point remains in the same place.

Suppose the coordinates of a point P in the *old* coordinate system (OX, OY) are (x, y)

Let the *new* coordinates system be $(O'X', O'Y')$ with the new origin $O'(h, k)$

There is a *shift* or change of origin from O to O'. In other words, there is a translation of axes to a new point $O'(h, k)$.

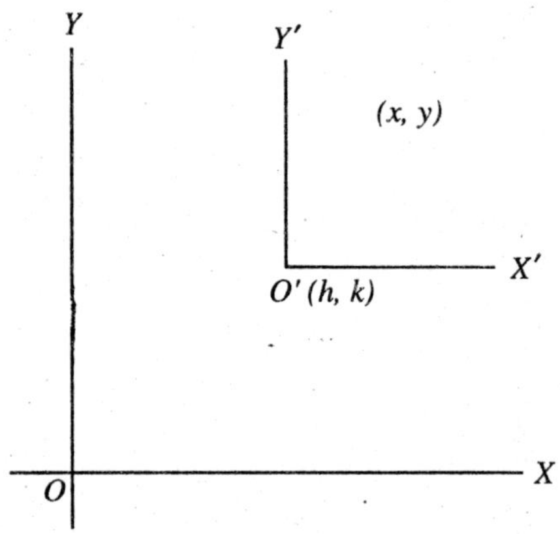

This means that

$$x = h + x'$$
$$y = k + y'$$

Thus if the new origin is $O'(h, k)$, the new coordinates of P are given by

$$x' = x - h$$
$$y' = y - k$$

If in a problem, new coordinates are known we can return to the old coordinate system by using

$$x = x' + h$$
$$y = y' + k$$

Ex. 72. (*a*) If the origin is shifted to $(-5, 1)$, the coordinates of a point $P(-5, 10)$ with reference to new axes can be found as follows. Here $(h, k) = (-5, 1)$; $(x, y) = (-5, 10)$

$$x' = x - h = -5 - (-5) = 0$$
$$y' = y - k = 1 - 10 = 9$$

Thus $\quad (x', y') = (0, -9)$

(*b*) if by a change of origin, $P(3, -5)$ becomes $(4, 2)$, find the new origin.

$$x' = x - h \quad \therefore \quad h = x - x' = 3 - 4 = -1$$
$$y' = y - k \quad \therefore \quad k = y - y' = -5 - 2 = -7$$

Hence $\quad (h, k) = (-1, -7)$

(*c*) If there is a change of the coordinates system from O to O' (α, β) and $P(x, y)$ becomes $P(x', y')$ then

$$x = x' + \alpha$$
$$y = y' + \beta$$

(old in terms of new)

or
$$x' = x - \alpha$$
$$y' = y - \beta$$

(new in terms of old)

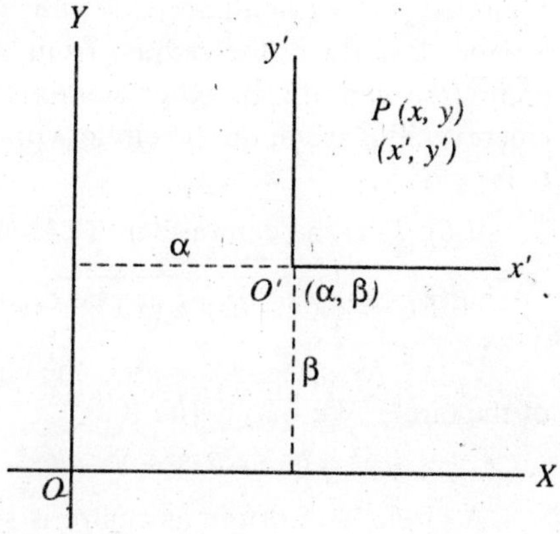

The line $ax + by + c = 0$, by shifting the origin to O' becomes

$$a(x' + \alpha) + b(y' + \beta) + c = 0$$

i.e., $ax' + by' + (a\alpha + b\beta) + c = 0)$

(*d*) If the new axes are perpendicular and are through the same origin but at an angle θ

Then
$$x = x' \cos\theta - y' \sin\theta$$
$$y = x' \sin\theta + y' \cos\theta$$

or
$$x' = x\cos\theta + y\sin\theta$$
$$y' = -x\sin\theta + \cos\theta$$

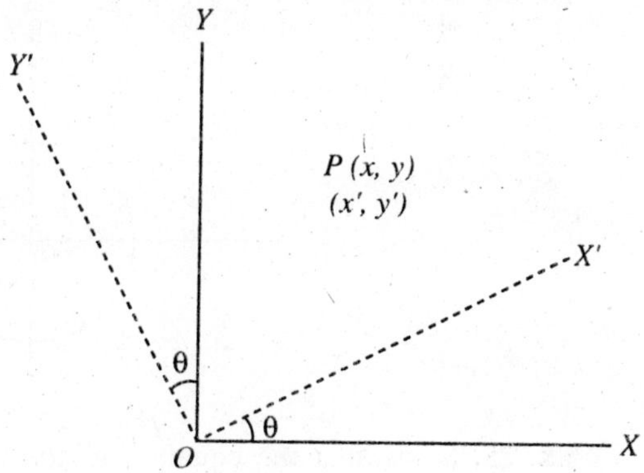

Ex. 73. In some cases it is possible to find an appropriate origin O' (h, k) such that the new equation assumes a simple form. If in the equation

$$(x - 3)^2 + (y + 4)^2 = 36$$

the origin is shifted to $(3, -4)$, the new equation becomes

$$x'^2 + y'^2 = 36$$

Ex. 74. (*a*) Show that by shifting the origin suitably the equation $y^2 - 10x - 6y + 149 = 0$ takes the new form $y'^2 = 20x'$

If we factorise the equation we get $(y - 3)^2 = 20(x - 7)$. Now shift the origin to $(7, 3)$.

(*b*) Show that by shifting the origin suitably the equation $x^2 - 4x - 16y = 14$ takes the form $x'^2 = 16y'$

(*c*) If we shift the origin to $(-2, 3)$ what form does the equation $x^2 + y^2 + 4x - 6y = 12$ take?

Ans. $x'^2 + y'^2 = 25$

8.9 CONIC SECTIONS

The first degree equation in x and y represents a straight line. The graphs of second degree equations are called conic sections. We can easily use the rectangular coordinate system to study the geometry of four conic sections: the circle, the parabola, the ellipse and the hyperbola.

8.10 THE CIRCLE

A circle is the set of all points P which are at a fixed distance r (the radius) from a fixed point (the centre). In other words $PC = r$ wherever P is taken on the circle with centre C.

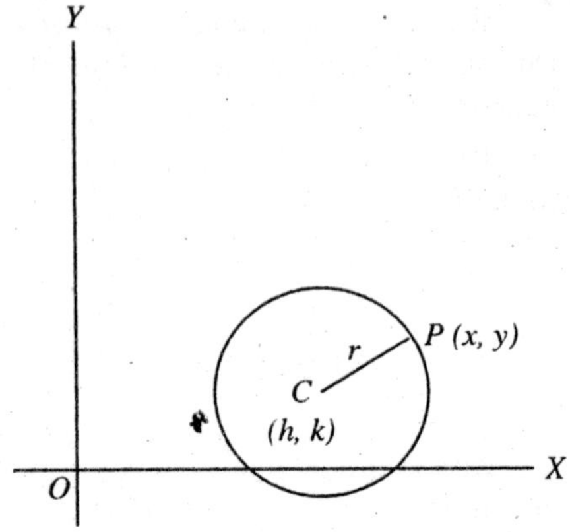

If (h, k) is the centre then we have

$$PC = \sqrt{(x-h)^2 + (y-k)^2} = r$$

$\therefore (x-h)^2 + (y-k)^2 = r^2$ is the equation of the circle. We may write it as

$$x^2 + y^2 - 2hx - 2ky + h^2 + k^2 - r^2 = 0$$

A circle with origin as centre is symmetric with respect to the origin and the axes. The equation is $x^2 + y^2 = r^2$.

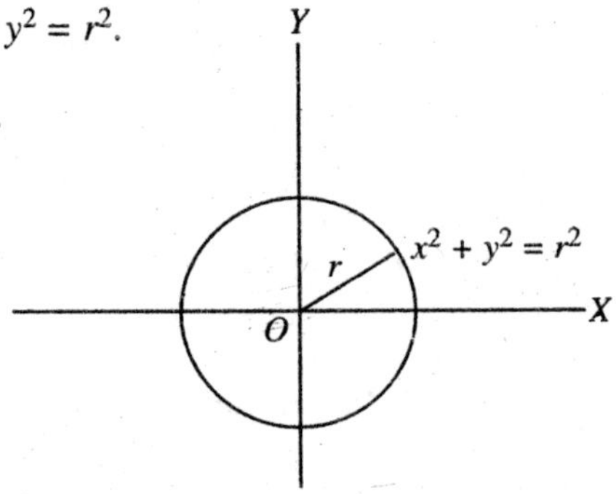

Ex. 75. Show that the equation of the circle with centre (2, – 3) and radius 5 is $(x-2)^2 + (y+3)^2 = 25$ or $x^2 + y^2 - 4x + 6y - 12 = 0$.

Ex. 76. Find the equation of the circle with (*i*) centre (2, 7), $r = 8$ (*ii*) centre (– 5, 3), $r = 3$.

Sketch the circles. **Ans.** $x^2 + y^2 - 4x - 14y - 11 = 0$; $x^2 + y^2 + 10x - 6 + 25 = 0$

Ex. 77. Find the centre and radius of the following circles.

(*i*) $(x-8)^2 + (y-5)^2 = 25$ (*ii*) $(x+2)^2 + y^2 = 9$

(*iii*) $x^2 + y^2 + 4x - 6y - 5 = 0$ (*iv*) $x^2 + y^2 - 3x + 4y + 4 = 0$

Ans. (*i*) (8, 5), 5. (*ii*) (– 2, 0), 3 (*iii*) (– 2, 3), $3\sqrt{2}$ (*iv*) (3/2, – 2), 3/2

Ex. 78. Find the circle

(*i*) of radius 3 and concentric with $(x-1)^2 + y^2 = 1$

(*ii*) of radius 5 passing through (3, – 2) with centre on $2x - y + 2 = 0$

(*iii*) touching the x-axis, centre (3, 7)

(*iv*) with (4, – 3), (6, 2) as ends of a diameter

Ans. (*i*) $(x-1)^2 + y^2 = 9$ (*ii*) $(x+2)^2 + (y+2)^2 = 25$ or $x^2 + (y-2)^2 = 25$

(*iii*) $x^2 + y^2 - 6x - 14y + 9 = 0$ (*iv*) $x^3 + y^2 - 10x + y + 18 = 0$

Ex. 79. Find the circle passing through the three points (5, 4), (3, 2), (– 3, 0). (Substitute these values in the general equation $(x - h)^2 + (y - k)^2 = r^2$ and solve the three equations for h, k, r). **Ans.** $(x + 3)^2 + (y - 10)^2 = 100$

Ex. 80. Analyse the equation: $x^2 + y^2 + 4x - 6y - 12 = 0$. We may write this equation in the form

$$(x^2 - 4x + 4) + (y^2 - 6y + 9) = 25$$

$\therefore$ $(x + 2)^2 + (y - 3)^3 = 25$ is the equation of a circle with centre (–2, 3) and radius 5.

Ex. 81. Analyse the equation: $x^2 + y^2 + 4x + 4y + 9 = 0$

Ans. Circle with centre (– 2, – 2); radius 1

Ex. 82. Show that the length of the tangent drawn from an exterior point (x_1, y_1) to the circle $(x - h)^2 + (y - k)^2 = r^2$ is given by

$$l = \sqrt{(x_1 - h^2) + (y_1 - k)^2 - r^2}$$

8.11 TANGENT AND CHORD

The equation of second degree

$$x^2 + y^2 + 2gx + 2fy + c = 0$$

represents a circle with centre $(-g, -f)$

and radius $r = \sqrt{g^2 + f^2 - c}$

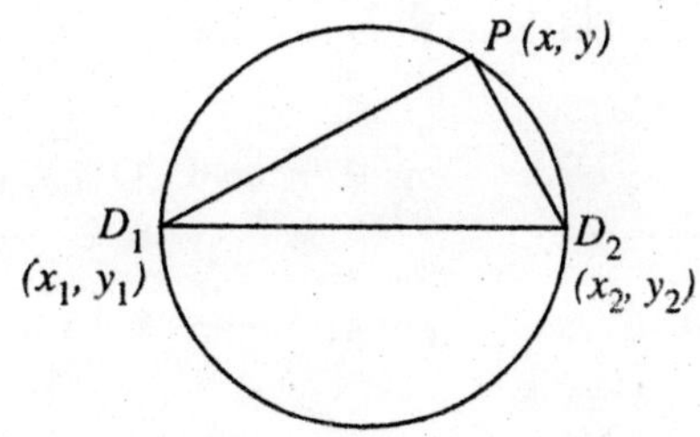

The equation can be written

$$x^2 + 2gx + g^2 + y^2 + 2fy + f^2 = g^2 + f^2 - c$$

$$\therefore \quad (x + g)^2 + (y + f)^2 = (\sqrt{g^2 + f^2 - c})^2$$

$\therefore$ Centre $\equiv (-g, -f)$, $r = \sqrt{g^2 + f^2 - c}$

Note: The equation of a circle is always such that the coefficients of x^2, y^2 are equal and the coefficient of xy is zero.

If $ax^2 + 2hxy + by^2 + 2gx + 2fy + c = 0$ represents a circle, then $a = b$, $h = 0$.

Circle on a Given Diameter

Let $P(x, y)$ be any point on the circle with diameter extremities $D_1(x_1, y_1)$, $D_2(x_2, x_2)$. The slopes of PD_1, PD_2 are m_1, m_2. The angle D_1PD_2 is a right angle.

Product of slopes of D_1P, $D_2P = -1$

$$\therefore \quad m_1 m_2 = \frac{(y - y_1)(y - y_2)}{(x - x_1)(x - x_2)} = -1$$

$$\therefore \quad (x - x_1)(x - x_2) + (y - y_1)(y - y_2) = 0$$

is the equation of the circle with the given diameter.

Tangent

The slope of the tangent to the circle $x^2 + y^2 + 2gx + 2fy + c = 0$ is found by differentiation thus:

$$2x + 2y\frac{dy}{dx} + 2g + 2f\frac{dy}{dx} = 0$$

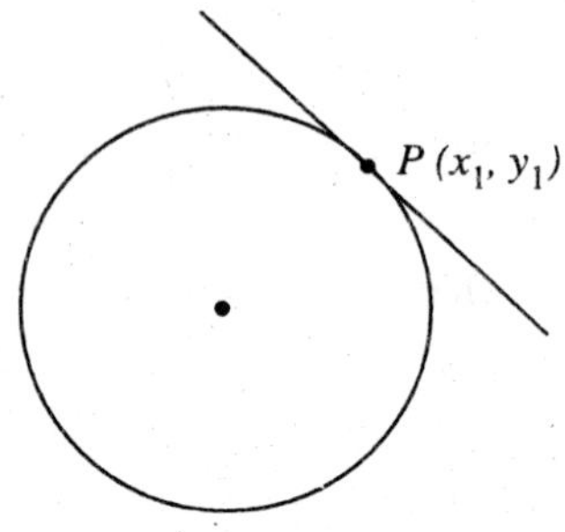

$$\therefore \quad \frac{dy}{dx} = -\frac{x+g}{y+f} = -\frac{x_1+g}{y_1+f} \text{ at the point } P(x_1, y_1)$$

$\therefore$ The equation of the tangent is

$$y - y_1 = -\frac{x_1+g}{y_1+f}(x - x_1)$$

Simplify and note that since $x_1^2 + y_1^2 + 2gx_1 + 2fy_1 + c = 0$, the required equation of the tangent is $xx_1 + yy_1 + g(x + x_1) + f(y + y_1) + c = 0$

Tangent to $x^2 + y^2 = r^2$

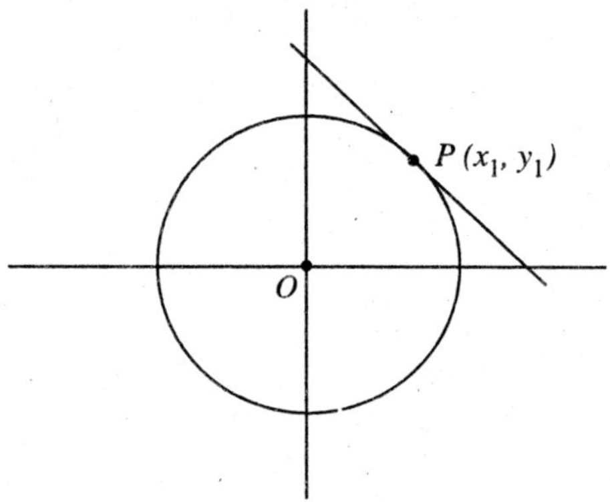

Differentiating, $\quad 2x + 2y\frac{dy}{dx} = 0$

$$\frac{dy}{dx} = -\frac{x_1}{y_1} \text{ at } P(x_1, y_1)$$

$$y - y_1 = -\frac{x_1}{y_1}(x - x_1)$$

Noting that $x_1^2 + y_1^2 = r^2$, the required equation of the tangent is $xx_1 + yy_1 = r^2$

Normal

The normal at a point on the circle is perpendicular to the tangent at that point. For $x^2 + y^2 = r^2$, the slope of the normal is $+\frac{y_1}{x_1}$ so that the equation of the normal is

$$y - y_1 = \frac{y_1}{x_1}(x - x_1)$$

Note: 1. *Condition for the Line* $y = mx + c$ *to be a Tangent to the Circle.*

If the line touches $x^2 + y^2 = a^2$, substitute $y = mx + c$ in the equation

$$x^2 + (mx + c)^2 = a^2$$

$$(1 + m^2)x^2 + 2mcx + c^2 - a^2 = 0$$

If the two points coincide, the two roots of this equation are equal for tangency.

$$\therefore \quad 4m^2c^2 - 4(1 + m^2)(c^2 - a^2) = 0$$

$$\therefore \quad c^2 = a^2(1 + m^2) \text{ or } c = a\sqrt{1+m^2}$$

is the *condition* for $y = mx + c$ to touch the circle. This is the same as saying that the perpendicular from the centre to the tangent equals the radius.

Note: 2. From any point (x_1, y_1) two tangents can be drawn to a circle $x^2 + y^2 = a^2$.

The tangent is $y = mx + a\sqrt{1+m^2}$. It passes through (x_1, y_1).

$$\therefore \quad y_1 - mx_1 = a\sqrt{1+m^2}$$

Square and simplify to get

$$m^2(x_1^2 - a^2) - 2mx_1y_1 + y_1^2 - a^2$$

There are two roots. Correspondingly, the two tangents are

(*i*) Real if $4x_1^2y_1^2 - 4(x_1^2 - a^2)(y_1^2 - a^2) > 0$

i.e., $x_1^2 + y_1^2 - a^2 >$ *i.e.,* (x_1, y_1) is outside the circle

(*ii*) Coincident if $x_1^2 + y_1^2 = a^2$, *i.e.,* (x_1, y_1) is on the circle

(*iii*) Imaginary if $x_1^2 + y_1^2 < a^2$, *i.e.,* (x_1, y_1) is inside the circle

Note: 3. The line joining the two points of contact is called the chord of contact $(C_2\, C_3)$

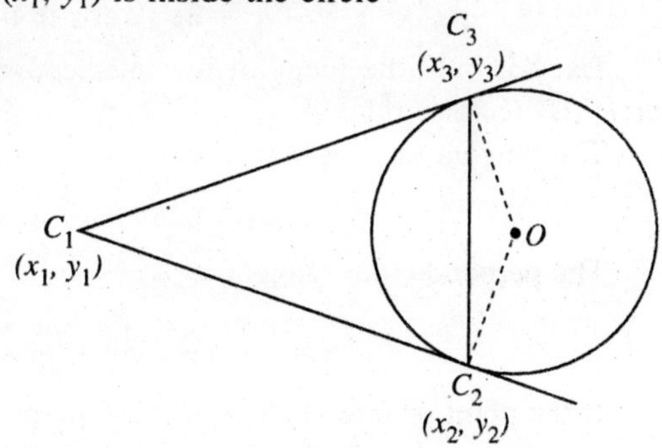

The equations of the tangents are

$$xx_2 + yy_2 = a^2,\ xx_3 + yy_3 = a^2$$

Since they pass through (x_1, y_1)

$$\therefore \quad x_1x_2 + y_1y_2 = a^2,\ x_1x_3 + y_1y_3 = a^2$$

$\therefore$ The equation of the chord of contact of $x^2 + y^2 = a^2$ is

$$x_1x + y_1y = a^2$$

Ex. 83. For the circle $x^2 + y^2 + 2gx + 2fy + c = 0$ the chord of contact subtends a right angle at the centre. The combined equation of C_1O and C_2O passing through the points of intersection of $x_1x + y_1y = a^2$ and $x^2f + y^2 = a^2$ is

$$x^2f + y^2 - a^2[(xx_1 + yy_1)/a^2]^2 = 0$$

For the lines to be perpendicular

$$x_1^2 + y_1^2 - 2a^2 = 0$$

8.12 PARAMETRIC FORM OF THE CIRCLE

The parametric equations of $x^2 + y^2 = a^2$ are

$$x = a \cos \theta,\ \ y = a \sin \theta$$

If there are two points $\theta_1 \equiv (a \cos \theta_1, a \sin \theta_1)$, $\theta_2 \equiv (a \cos \theta_2, a \sin \theta_2)$ on the circle, the equation of the line joining θ_1, θ_2 is

$$y - a \sin \theta_1 = \frac{a\sin\theta_2 - a\sin\theta_1}{a\cos\theta_2 - a\cos\theta_1}(x - a\cos\theta_1)$$

$$= -\frac{\cos(\theta_1+\theta_2)/2}{\sin(\theta_1+\theta_2)/2}(x - a\cos\theta_1)$$

i.e.,

$$x\cos\frac{\theta_1+\theta_2}{2} + y\sin\frac{\theta_1+\theta_2}{2} = a\cos\frac{\theta_1-\theta_2}{2}$$

Ex. 84. The circle drawn on the chord $x \cos \alpha + y \sin \alpha = p$ of the circle $S_1 \equiv x^2 + y^2 = a^2$ as diameter is $x^2 + y^2 - a^2 - 2p(x \cos \alpha + y \sin \alpha - p) = 0$

The equation of the circle through the intersection of S_1 and the line is

$$x^2 + y^2 - a^2 + \lambda(x \cos \alpha + y \sin \alpha - p) = 0$$

whose centre $\left(-\frac{\lambda \cos \alpha}{2}, -\frac{\lambda \sin \alpha}{2}\right)$ lies on the line $x \cos \alpha + y \sin \alpha = p$

i.e., $$\left(-\frac{\lambda \cos \alpha}{2}\right)\cos \alpha + \left(-\frac{\lambda \sin \alpha}{2}\right)\sin \alpha = p$$

$$\therefore \quad -\frac{\lambda}{2}(\cos^2 \alpha + \sin^2 \alpha) = p \text{ or } \lambda = -2p$$

The required equation is

$$x^2 + y^2 - a^2 - 2p(x \cos \alpha + y \sin \alpha - p) = 0$$

Ex. 85. Find the locus of the intersection point of two perpendicular tangents to a circle. (Director circle)

The tangent to $x^2 + y^2 = a^2$ is

$$y = mx + a\sqrt{1 + m^2} \qquad \ldots(1)$$

The perpendicular tangent is

$$y = -\tfrac{1}{m}x + a\sqrt{1 + \left(-\tfrac{1}{m}\right)^2} \qquad \ldots(2)$$

If the point of intersection of these perpendicular tangents is (x_1, y_1) then

$$y_1 - mx_1 = a\sqrt{1 + m^2}$$

$$my_1 + x_1 = a\sqrt{1 + m^2}$$

Square and add to get

$$x_1^2 + y_1^2 = 2a^2$$

This locus $x^2 + y^2 = 2a^2$ is called the *director circle.*

8.13 THE PARABOLA

A parabola is a set of points each of which is equidistant from a given point called the focus (S) and from a given line called the directrix (ZM)

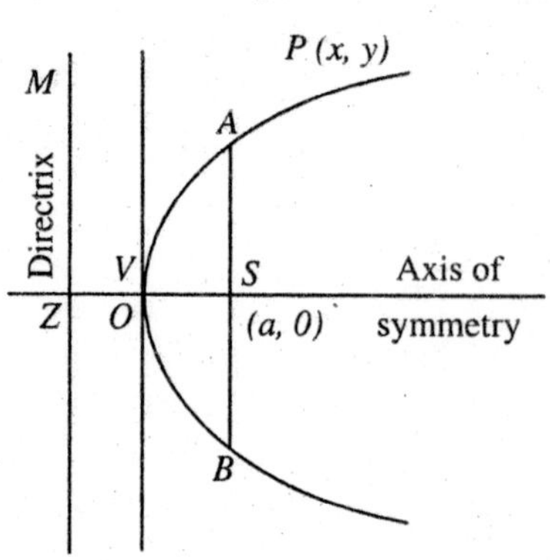

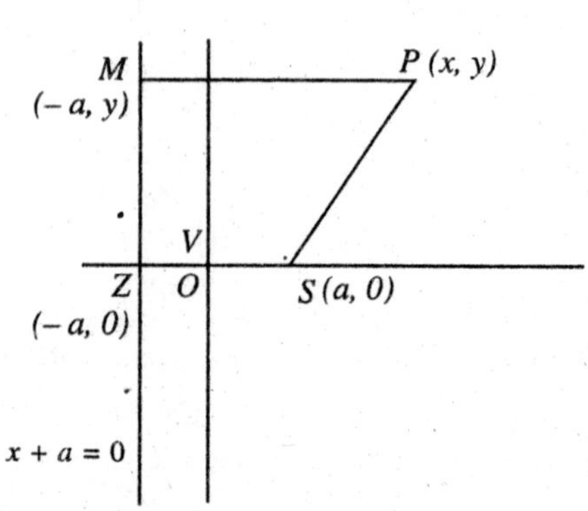

The directrix is parallel to and at a distance $-a$ from the y-axis. Its equation is $x + a = 0$

Since $PS = PM$ $\qquad (x - a)^2 + y^2 = (x + a)^2$

Simplifying this we get the equation of the parabola with the vertex at the origin:

$$y^2 = 4ax$$

The line through the focus perpendicular to the directrix is called the *axis of symmetry*. The line segment AB with end points on the parabola perpendicular to the x-axis at the focus is the *latus rectum*. V is the *vertex* of the parabola.

Note the following cases of parabolas in different positions along with their directrices. Each of these has its vertex at the origin.

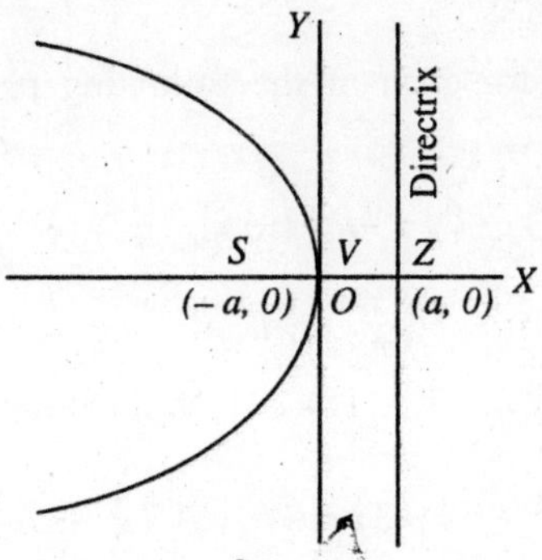

$y^2 = -4ax$

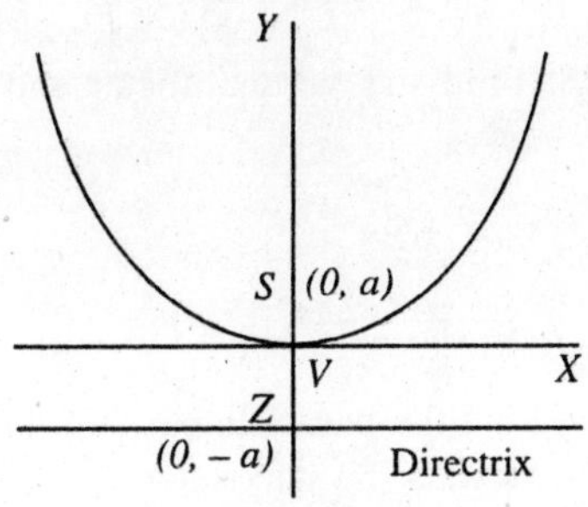

$x^2 = 4ay$

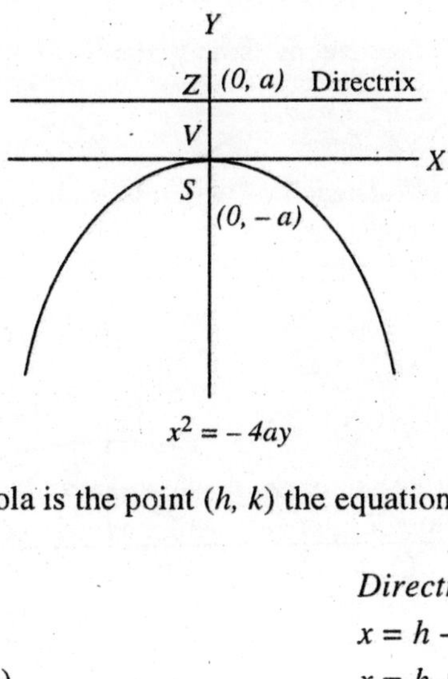

$x^2 = -4ay$

If the vertex V of the parabola is the point (h, k) the equations in the above four cases become

	Equation	*Directrix*
(*i*)	$(y - k)^2 = 4a(x - h)$	$x = h - c$
(*ii*)	$(y - k)^2 = -4a(x - h)$	$x = h + c$
(*iii*)	$(x - h)^2 = 4a(y - k)$	$y = k - c$
(*iv*)	$(x - h)^2 = -4a(y - k)$	$y = k + c$

Ex. 86. Find the focus and directrix of the parabola $y^2 = -8x$. Sketch the curve. Comparing this with $y^2 = -4ax$ we have $4a = -8$ or $a = -2$.

The focus is $S(-4, 0)$ and the directrix $x - 2 = 0$.

Note that the latus rectum $AB = 4a = 8$.

Ex. 87. Find the standard form for the parabola

$$y^2 + 2y - 12x - 3 = 0$$

Rewriting the equation, $y^2 + 2y + 1 = 12x + 4$

or $\quad (y + 1)^2 = 12\left(x + \frac{1}{3}\right)$. This is the case (*i*) above

The vertex is $\left(-\frac{1}{3}, -1\right)$

Since $4a = 12, \quad \therefore\ a = 3;$

the directrix: $x = -\frac{1}{3} - 3 = -\frac{10}{3}$ or $x + \frac{10}{3} = 0$

the focus: $\left(-\frac{1}{3} + 3, -1\right) = \left(\frac{8}{3}, -1\right)$.

Ex. 88. Find the vertex, focus and directrix for each of the following parabolas. Sketch the curves.

(*i*) $y = x^2 + 4x$ **Ans.** $V(-2, -4)$, $S\left(-2, -\frac{15}{4}\right)$, $D: y = -\frac{71}{4}$

(*ii*) $2x^2 + 5y - 3x + 4 = 0$ **Ans.** $V\left(\frac{3}{4}, \frac{23}{40}\right)$, $S\left(\frac{3}{4}, \frac{6}{5}\right)$, $D: x = \frac{3}{4}$

Ex. 89. Find the parabola passing through $(-1, 2)$, $(1, -1)$, $(2, 1)$ and having its axis parallel to (*i*) x-axis (*ii*) y-axis.

Ans. (*i*) $7y^2 + 6x - 3y = 16$ (*ii*) $7x^2 - 9x - 6y = 4$

Ex. 90. Find the points of intersection of $y^2 = 4x$ and $x^2 = y$.

Ans. $(0, 0)$, $(4^{1/3}, 4^{2/3})$

8.14 THE ELLIPSE

An ellipse is the set of all points each of which is such that the sum of its distances from two points, called foci, is a constant.

S_1, S_2 are the foci. C, the midpoint of S_1S_2, is called the centre of the ellipse. The ellipse is symmetric with respect to each of two perpendicular lines that intersect at its centre. In the diagram the centre is at the origin. V_1, V_2 are the vertices on the line of symmetry which is the x-axis here. V_3, V_4 are the vertices on the other line of symmetry, here the y-axis.

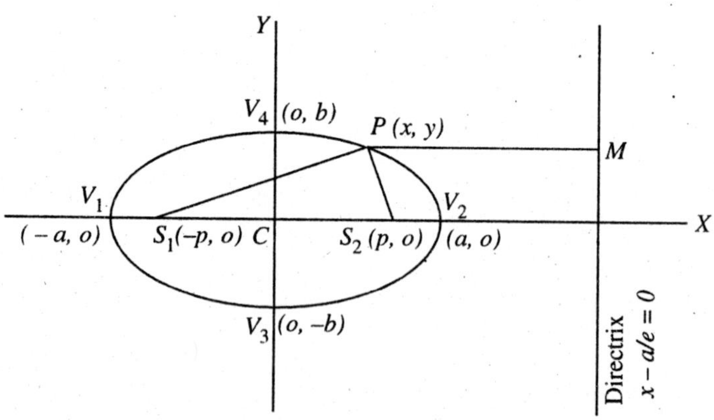

The longer segment V_1V_2, determined by the vertices is called the *major axis* and the shorter segment, V_3V_4, the *minor axis*.

Let $P(x, y)$ be a point on the ellipse.

Let $PS_1 + PS_2 = 2a$, a constant. $S_1 \equiv (-p, o)$, $S_2 \equiv (p, o)$

Then $\sqrt{(x+p)^2 + y^2} + \sqrt{(x-p)^2 + y^2} = 2a$ for an ellipse

Simplifying and substituting $a^2 - p^2 = b^2$; we get from this the equation of the ellipse:

$$\frac{x^2}{a^2} + \frac{y^2}{b^2} = 1$$

Here the major axis $V_1V_2 = 2a$ and the minor axis $V_3V_4 = 2b$

An ellipse is not a circle. If b is equal to a, the resulting circle may be considered as a special case of the ellipse whose foci coincide with the centre of the circle with radius a.

Ex. 91. Find the vertices and the foci of the ellipse $x^2 + 4y^2 = 4$. Writing the equation in the form $\frac{x^2}{4} + \frac{y^2}{1} = 1$, we have the vertices V_1 $(-2, 0)$, V_2 $(2, 0)$, V_3 $(0, -1)$, V_4 $(0, 1)$.

Since $p^2 = a^2 - b^2 = 4 - 1 = 3$ and $p = \sqrt{3}$, the coordinates of the foci are S_1 $(-\sqrt{3}, 0)$, S_2 $(\sqrt{3}, 0)$.

Ex. 92. Find the coordinates of the vertices and foci for

(*i*) $\frac{x^2}{100} + \frac{y^2}{81} = 1$ (*ii*) $2x^2 + 3y^2 = 6$ (*iii*) $\frac{x^2}{25} + \frac{y^2}{49} = 1$

Ans. (*i*) $V(\pm 10, 0)$, $S(\pm \sqrt{19}, 0)$ (*ii*) $V(\pm \sqrt{3}, 0)$, $S(\pm 1, 0)$, (*iii*) $V(0, \pm 7)$, $S(0 \pm 2\sqrt{6})$

Ex. 93. Show that the equation of the ellipse with foci $(\pm 1, 0)$ and major axis $2a = 5$ is given by $\frac{x^2}{25} + \frac{y^2}{21} = \frac{1}{4}$. $\left(\text{Since } p = 1, b^2 = a^2 - p^2 = \frac{21}{4}\right)$

Ex. 94. Find the equation of the ellipse, given

(*i*) $S(\pm 2, 0)$, major axis $2a = 6$

(*ii*) $S(\pm 2, 0)$, minor axis $2b = 6$

Ans. (*i*) $5x^2 + 9y^2 = 45$ (*ii*) $9x^2 + 13y^2 = 117$

Ex. 95. Show that an ellipse is the locus of all points P each of which is such that the ratio of the distances of P from a fixed point S and a fixed line (called the directrix) is a constant e (where $e < 1$); e is called the eccentricity of the ellipse.

Given $\frac{PS}{PM} = e$, we have $PM = \frac{p}{e^2} - x$. (see diagram above)

The equation of the directrix is $x = -\frac{a}{e}$ and $PS = e.\,PM$ or $PS^2 = e^2\,PM^2$

$$\therefore \quad (x - (-ae))^2 + y^2 = e^2(x + a/e)^2$$

Simplifying we get the equation $\frac{x^2}{a^2} + \frac{y^2}{a^2(1-e^2)} = 1$ or $\frac{x^2}{a^2} + \frac{y^2}{b^2} = 1$.

where $\quad b^2 = a^2(1 - e^2)$

Ex. 96. If the major axis is horizontal the equation of the ellipse with centre (h, k) where $b^2 = a^2 (1 - e^2)$ is given by

$$\frac{(x-h)^2}{a^2} + \frac{(y-k)^2}{b^2} = 1$$

where $a > b$

If the major axis is vertical

$$\frac{(x-h)^2}{b^2} + \frac{(y-k)^2}{a^2} = 1$$

where $b > a$.

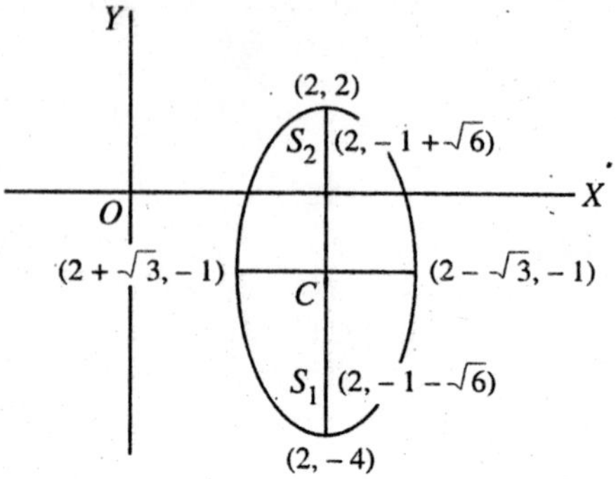

Ex. 97. Find the centre, foci and eccentricity of the ellipse

$$y^2 - 12x + 2y + 4 = 0$$

[write the equations as $3(x-2)^2 + (y+1)^2 = 9$ (see diagram)

or $$\frac{(x-2)^2}{3} + \frac{(y+1)^2}{9} = 1]$$

Ans. $C(2, -1), e = \sqrt{\frac{2}{3}}$

8.15 THE HYPERBOLA

A hyperbola is the set of points each of which is such that the absolute value of the difference of the distance from two points, called foci, is a constant.

The line S_1S_2 determined by the foci is the line of symmetry. C, the midpoint of S_1S_2 is the centre of the hyperbola. $V_1(-a, 0)$, $V_2(a, 0)$ are the vertices and V_1V_2 $(= 2a)$ is called the transverse axis. Let $P(x, y)$ be a point on the hyperbola.

The difference $|PS_1 - PS_2|$ $= 2a$, a constant.

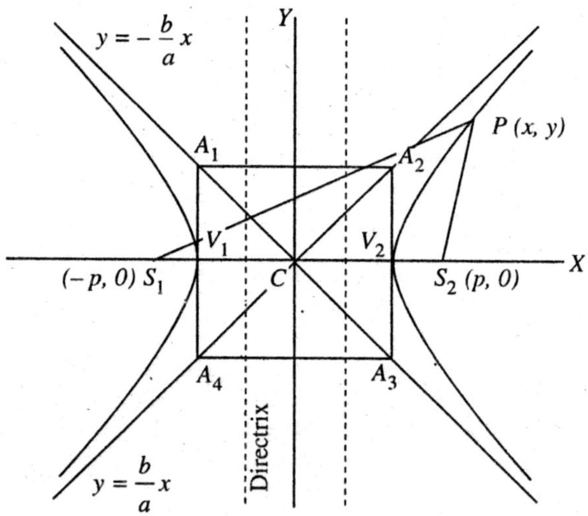

$$\left|\sqrt{(x+p)^2 + y^2} - \sqrt{(x-p)^2 + y^2}\right| = 2a$$

Simplifying and substituting

$$a^2 - p^2 = -b^2$$

we get the equation of the hyperbola,

$$\frac{x^2}{a^2} - \frac{y^2}{b^2} = 1$$

The graph of the hyperbola is symmetric with respect to the axes. The hyperbola does not intersect the y-axis. If we write the equation

$$\frac{x^2}{a^2}-\frac{y^2}{b^2}=1 \text{ as } y^2=\frac{b^2}{a^2}(x^2-a^2)$$

then
$$y=\pm\frac{bx}{a}\sqrt{1-\frac{a^2}{x^2}}$$

For large x, $y=\pm\dfrac{bx}{a}$ *i.e.*, the hyperbola approaches the lines whose equations are

$$y=\frac{bx}{a} \text{ and } y=-\frac{bx}{a}$$

There lines are called the *asymptotes* of the hyperbola.

An asymptote is a line which a curve approaches as a point on the curve moves towards infinity. An asymptote is not a part of the hyperbola but is useful for other purposes and for drawing the hyperbola. We first draw a rectangle $A_1A_2A_3A_4$ with the sides $2a$ and $2b$ whereafter it becomes easy to draw the asymptotes and the hyperbola.

Sometimes the asymptotes are parallel to or coincident with the coordinate axes.

Ex. 98. For the hyperbola $\dfrac{x^2}{16}-\dfrac{y^2}{25}=1$ the vertices are $(\pm 4, 0)$

$$p^2=a^2+b^2=41,\ p=\sqrt{41} \text{ so that the foci are } (\pm\sqrt{41},\ 0)$$

The asymptotes are $y=\pm\dfrac{bx}{a}$ or $4y=\pm 5x$

Ex. 99. Find the vertices, foci and the asymptotes of

(*i*) $\dfrac{x^2}{4}-\dfrac{y^2}{5}=1$ (*ii*) $\dfrac{x^2}{16}-\dfrac{y^2}{9}=1$

Ans. (*i*) $V(\pm 2, 0)$, $S(\pm 3, 0)$, Asymptotes: $y=\pm\frac{5}{2}x$

(*ii*) $V(\pm 4, 0)$, $S(\pm 5, 0)$, Asymptotes: $y=\pm\frac{3}{4}x$

Ex. 100. Show that a hyperbola is the set of all points P each of which is such that the ratio of the distances of P from a fixed point S and from a fixed line (the directrix) is a constant e (where $e > 1$). [As in the case of the ellipse, the equation turns out to be

$$\frac{x^2}{a^2}-\frac{y^2}{a^2(e^2-1)}=1 \quad \text{Here } b^2=a^2(e^2-1)]$$

Ex. 101. If the transverse axis is on the y-axis the equation of the hyperbola with centre (h, k) is given by

$$\frac{(x-h)^2}{a^2}-\frac{(y-k)^2}{b^2}=1$$

Ex. 102. Find the centre, foci and the asymptotes of the hyperbola

(*i*) $25x^2 - 9y^2 + 150x + 18y - 9 = 0$ (*ii*) $-2x^2 + y^2 + 12x + 4y + 4 = 0$

Ans. (*i*) $(-3, 1)$; $(-3 \pm \sqrt{34}, 1)$; $y = 1 \pm \frac{5}{3}(x + 3)$

(*ii*) $(3, -2)$; $(3 \pm 3\sqrt{3}\ -2)$; $y = -2 \pm \sqrt{2}(x - 2)$

EQUILATERAL OR RECTANGULAR HYPERBOLA

If $a = b$ we get the rectangular hyperbola $\frac{x^2}{a^2} - \frac{y^2}{a^2} = 1$

or $x^2 - y^2 = a^2$ whose asymptotes are perpendicular and have the equations $x = a \pm y$

If $\frac{x^2}{a^2} - \frac{y^2}{b^2} = 0$ we have two intersecting lines $-\ \frac{x}{a} = \pm\frac{y}{b}$

The general equation of the equilateral hyperbola is $(x - h)(y - k) = c$ where (h, k) is the centre of the hyperbola.

If the coordinate axes themselves are asymptotes the standard form of the equation of the hyperbola is $xy = c$. It shows an inversely proportional relation between x and y.

8.16 EQUATION AFTER TRANSLATION OF AXES

By selecting a new origin (h, k) appropriately we can often simplify an equation in x, y.

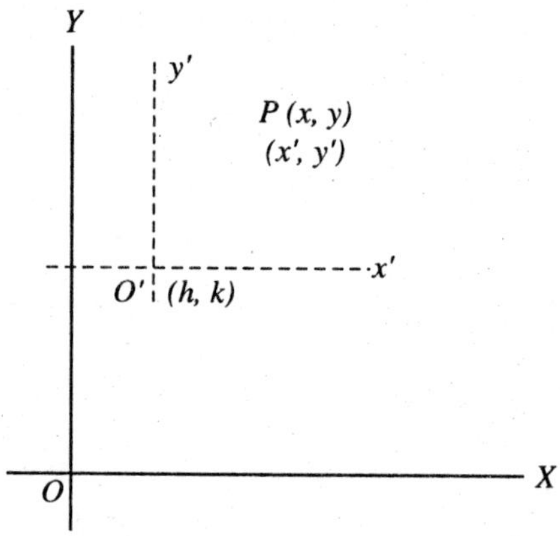

Ex. 103. With the new set of coordinate axes above, the equation of the circle $(x - h)^2 + (y - k)^2 = r^2$ becomes $x^2 + y^2 = r^2$

Ex. 104. With the origin translated to $(1, -2)$ the ellipse

$$\frac{(x-1)^2}{9} + \frac{(y+2)^2}{25}\ 1 \text{ can be written}$$

$$\frac{x'^2}{9} + \frac{y'^2}{25} = 1.$$

Ex. 105. Show that $2x'^2 + 3y'^2 = 12$ is the equation obtained from $8x^2 + 12y^2 - 16x + 36y - 13 = 0$ by translating the origin to $(1, -3/2)$.

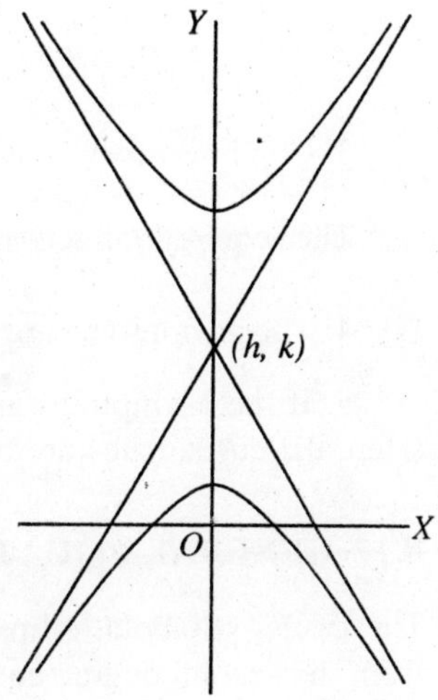

Ex. 106. A translation to the new origin (h, k) changes a given hyperbola

$$\frac{(y-k)^2}{a^2} - \frac{(x-h)^2}{b} = 1$$

to the new form $\dfrac{y'^2}{a^2} - \dfrac{x'^2}{b^2} = 1.$

The equations of tangents, chord to the parabola, the ellipse and the hyperbola can be found in the same way as done earlier for the circle. A summary is given below.

Parabola

1. The tangent at (x_1, y_1) is $yy_1 = 2a(x + x_1)$
2. At $(at^2, 2at)$ it is $yt = x + at^2$
3. Condition for $y = mx + c$ to be tangent is $c = a/m$

The required tangent is $y = mx + a/m$

4. The chord joining the points $(at_1^2, 2at_1)$ and $(at_2^2, 2at_2)$ is $(t_1 + t_2) = 2(x - at_1t_2)$
5. The normal at (x_1, y_1) is $y - y_1 = -\dfrac{y_1}{2a}(x - x_1)$

Ellipse

1. The tangent at (x_1, y_1) is $\dfrac{xx_1}{a^2} - \dfrac{yy_1}{b^2} = 1$
2. At $(a\cos\theta, b\sin\theta)$ it is $\dfrac{x}{a}\cos\theta + \dfrac{y}{b}\sec\theta = 1$
3. Condition for $y = mx + c$ to be tangent is $c = \sqrt{a^2m^2 + b^2}$

The required tangent is $y = mx + \sqrt{a^2m^2 + b^2}$

4. The chord joining $(a\cos\theta_1, b\sin\theta_1)$, $(a\cos\theta_2, b\sin\theta_2)$ is

$$\frac{x}{a}\cos\frac{\theta_1+\theta_2}{2} + \frac{y}{b}\sin\frac{\theta_1+\theta_2}{2} = \cos\frac{\theta_1-\theta_2}{2}$$

5. The auxiliary circle is the circle with radius a on the major axis of the ellipse as diameter

Hyperbola

1. The tangent at (x_1, y_1) is $\dfrac{xx_1}{a^2} - \dfrac{yy_1}{b^2} = 1$

2. At ($a \sec\theta$, $b \tan\theta$) it is $\frac{x}{a}\sec\theta - \frac{y}{b}\tan\theta = 1$

3. Condition for $y = m + c$ to be tangent is $c = \sqrt{a^2m^2 - b^2}$

The required tangent is $y = mx + \sqrt{a^2m^2 - b^2}$

4. The asymptotes are $y = \pm \frac{b}{a}x$

5. If the asymptotes are at right angles, we have the rectangular hyperbola $xy = c^2$ (Here the coordinates are the axymplotes).

8.17 GENERAL EQUATION OF THE SECOND DEGREE

The circle, parabola, ellipse and hyperbola are all special cases of and can be derived from the second degree equation

$$ax^2 + by^2 + 2gx + 2fy + c = 0$$

(*i*) If $a = b \neq 0$ we have a circle.

(*ii*) If $ab = 0$ *i.e.*, if the equation is a quadratic in one variable, we have a parabola.

(*iii*) If $ab > 0$ *i.e.*, a, b are both positive or both negative, we have an ellipse

(*iv*) If $ab < 0$ *i.e.*, if a, b have opposite signs, we have a hyperbola

(*v*) If $a = b = 0$ and not both of f and g vanish we have a straight line.

8.18 INVARIANTS

In the more general quadratic equation

$$ax^2 + 2hxy + by^2 + 2gx + 2fy + c = 0$$

it can be shown that the quantity $h^2 - ab$ remains *invariant* under any rotation of coordinate axes. The quantity $h^2 - ab$ known as the *discriminant* is useful in discriminating between different curves. In the above equation if

$h^2 - ab < 0$ we have an ellipse

$h^2 - ab = 0$ we have a parabola

$h^2 - ab > 0$ we have a hyperbola.

Another invariant is the sum $a + b$.

8.19 HIGHER DEGREE CURVES: SKETCHING CURVES

The graphs of first and second degree curves can be sketched fairly easily but not those of higher degree curves. There are, however, some cases in which we can solve for y and get some points on the curve which can then be sketched.

Ex. 107. Sketch the curves

(*i*) $y^2 = ax^3$ (*ii*) $y^2 = x(x - 1)(x - 2)$ (*iii*) $y^2 = x(x - 1)^3$

(*iv*) $y = x^2y - x^2$ (*v*) $y = \frac{x - 1}{(x - 2)(x - 3)}$

Some General Methods of Sketching Curves

One or more of the following methods may be found useful in sketching or getting some hints about the same.

(*i*) *Factorisation:* If the equation of the curve can be written as the product of two or more factors then the graph consists of two ore more curves corresponding to those factors.

Ex. 108. $3x^2y + xy^2 + 9x + 3y = 0$ can be written

$(xy + 3)(3x + y) = 0$. Here the graph consists of two curves $xy + 3 = 0$ and $3x + y = 0$.

Ex. 109. $3x^2 - 8xy - 3y^2 = 0$ gives two lines $x - 3y = 0$, $3x + y = 0$.

(*ii*) *Finding Intercepts* : The intercepts are points at which a curve cuts the axes.

To obtain x-intercepts, put $y = 0$ and solve for x.

To obtain y-intercepts, put $x = 0$ and solve for y.

Ex. 110. c is the y-intercept and $-\dfrac{c}{m}$ the x-intercept made by the line $y = mx + c$.

Ex. 111. The intercepts made by the ellipse $\dfrac{x^2}{a^2} + \dfrac{y^2}{b^2} = 1$ on the axes are $\pm a$ and $\pm b$ respectively.

(*iii*) *Symmetry:* Generally a curve is symmetric to the x-axis if it has only even powers in y and symmetric to the y-axis if it has only even powers in x. The curves $y^2 = 4px$, $x^2 = 4py$ provide the respective examples.

(x, y), $(-x, y)$ both lie on a curve symmetric to the x-axis and (x, y), $(x, -y)$ both lie on a curve symmetric to the y-axis.

If both (x, y), $(-x, -y)$ lie on a curve it is symmetric to the origin as in the case of the circle $x^2 + y^2 = r^2$.

Ex. 112. $4x^2 - 3xy + x^3 = 0$ is not symmetric to the origin or the axes. The curve $x^3y + 2xy = 0$ is symmetric to the origin and the axes.

$4x^2y + y + x^3 = 0$ is symmetric to the origin but not to the axes.

(*iv*) *Asymptotes* : The movement of the curve with respect to the asymptotes may be examined by letting x, y approach $\pm\infty$. This may suggest possible limitations on the extent of the curve.

The procedure consists in solving the equation first for one variable and then for another and noticing the behaviour of the curve.

(*v*) Use of derivatives, maxima and minima translation of axes and various other techniques may be found useful in some cases.

Ex. 113. Sketch the curves (*i*) $xy = 1$, (*ii*) $(x - 2)\left(y - \frac{3}{2}\right) = 1$

For the curve $xy = 1$ the asymptotes are the axes $x = 0$, $y = 0$ and the curve lies in the first and third quadrants. The centre is at the origin.

The curve $(x - 2)(y - 3/2) = 1$ can be written $x'y' = 1$ by translating the origin to $(2, 3/2)$. The asymptotes are $x = 2$ and $y = 3/2$. The centre is $(2, 3/2)$. The shape of (*ii*) is similar to that of (*i*) as can be seen from the diagrams.

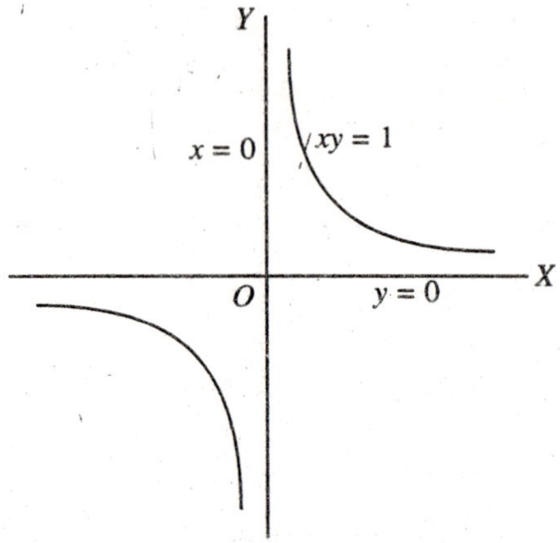

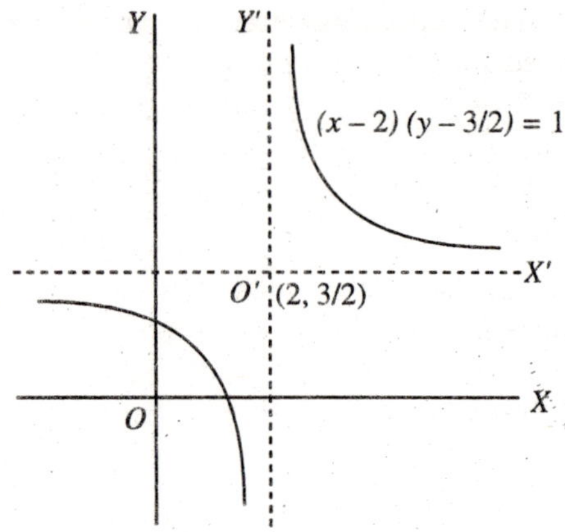

Ex. 114. Show by sketching that the curves $(x - 3)\, y = -6$ and $x(y - 4) = -3$ are rectangular hyperbolas lying in the second and fourth quadrants.

Ex. 115. Sketch (*i*) the parabola $y = -x^2 + 2x + 3$ (*ii*) the ellipse $x^2 + 9y^2 - 8x + 7 = 0$.

[Find the standard forms: (*i*) $(y - 4) = -(x - 1)^2$, vertex (1, 4) and (*ii*) $\dfrac{(x-4)^2}{9} + \dfrac{y^2}{1}$ $= 1$, centre (4, 0) and the semiaxes 3 and 1]

Ex. 116. Sketch the parabola $5x^2 + 4y = 12$.

Ex. 117. Sketch the curve (*i*) $y = \dfrac{x^2y - x^2}{4}$ (*ii*) $y = \dfrac{x^2 - x^2y}{4}$

(Note zero intercepts and symmetry to *y*-axis).

The equations can be written

$$(i)\ x = \pm 2\sqrt{\frac{y}{y-1}} \text{ and } (ii)\ x = \pm 2\sqrt{\frac{y}{1-y}}$$

There are three asymptotes for (*i*) $x = \pm 2$ and $y = 1$ and only one asymptote for (*ii*) $y = 1$ since x is not defined for negative values of $\dfrac{y}{1-y}$

Ex. 118. Sketch the curves (*i*) $y = x - 4x^3$ (*ii*) $y = x^3 (x - 2)^2$ (*iii*) $y = 1/9x^3$ (*iv*) $y = x^2 - x^4$.

8.20 MARKET EQUILIBRIUM

If demand and supply curves are given the market equilibrium values of price and quantity are obtained by solving the given equations or noting the points of intersection geometrically.

Ex. 119. To find the equilibrium price and quantity for the demand and supply curves $p^2 + x + 4p = 32$, $20p = x + 20$ we should consider only the *relevant portions* of the curves in the *first quadrant* since price and output are non-negative quantities.

The equations may be either solved simultaneously to give $x = 20$, $p = 2$ or may be sketched to get the point of intersection.

The line $20p = x + 20$ is easy to draw.

For the parabola note that the x-intercept 32 and the p-intercepts -8, 4 are obtained by putting $p = 0$ and $x = 0$ respectively.

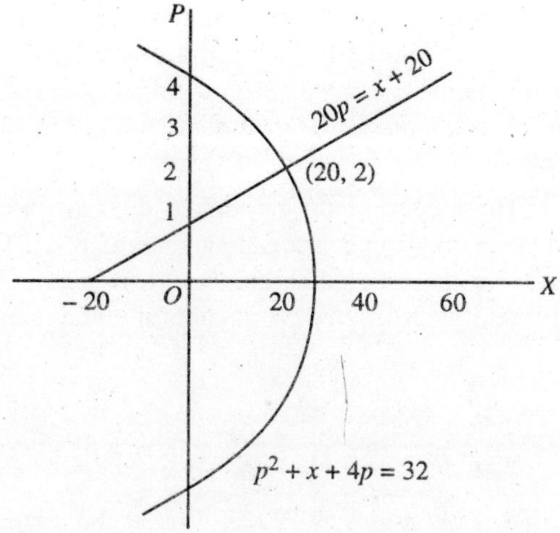

Ex. 120. Find the equilibrium price (y) and quantity (x) for the demand and supply curves given below.

(*i*) $2y = 16 - x$, $y^2 = 4(x - y)$

(*ii*) $(x + 16)(y + 12) = 480$, $y = 2x + 4$

(*iii*) $x = 64 - 8y - 2y^2$, $x = 10y + 5y^2$

(*iv*) $x = \sqrt{36 - y}$, $4y = 24 + x^2$

Ans. (*i*) 8, 4 (*ii*) 4, 12 (*iii*) 40, 2 (*iv*) $(2\sqrt{6},\ 12)$

8.21 INDIFFERENCE CURVES AND PRODUCT TRANSFORMATION CURVES

If there are two commodities x, y a particular level of satisfaction can be attained from a large number of combinations of x, y and from an infinite number of such combinations in the case of continuous utility functions. The curve joining all commodity combinations giving the consumer the same level of satisfaction is called an *indifference curve.* An *indifference map* is a collection of indifference curves at different levels of satisfaction. No two indifference curves can intersect. An indifference curve is convex to the origin.

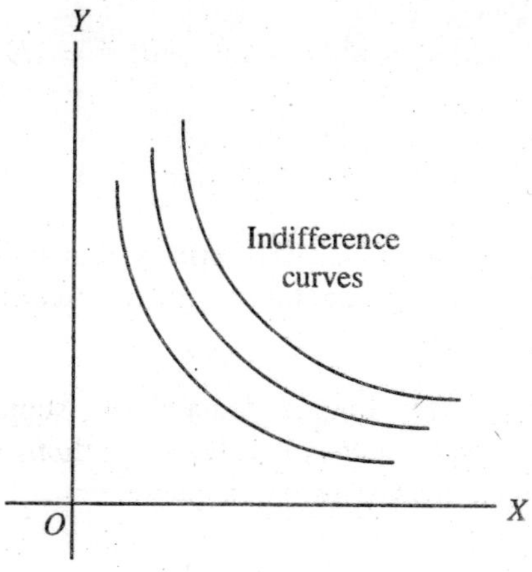

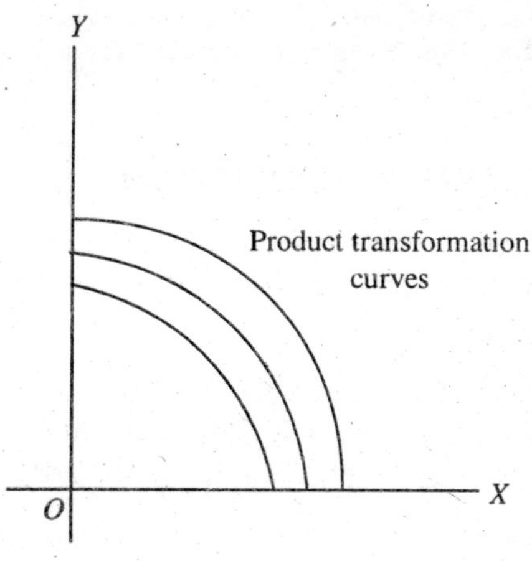

A *product transformation curve* gives the relationship between the quantities of two different commodities x, y produced by the same firm. It is a curve joining all output combinations available from a given quantity of resources. A family of transformation curves is obtained from different inputs. Transformation curves are a form of indifference curves.

These curves may be looked upon as parts of a parabola, an ellipse, a hyperbola, a circle or any other convenient curve. In each case only the appropriate branch of the curve in the first quadrant should be taken. A family of curves is obtained by giving different values to the parameters in the equation of the curve.

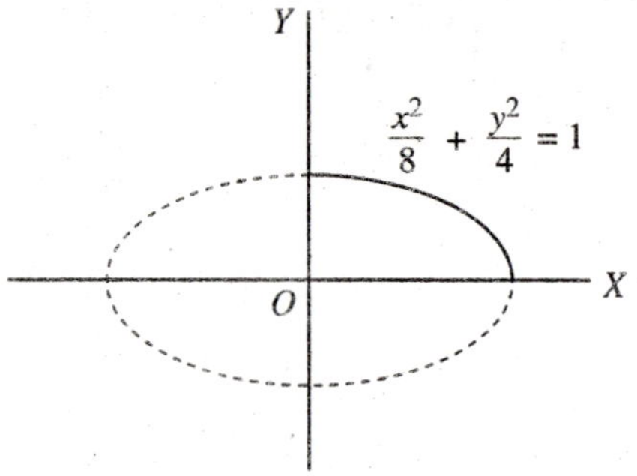

unbroken portions of curves shown here can be used

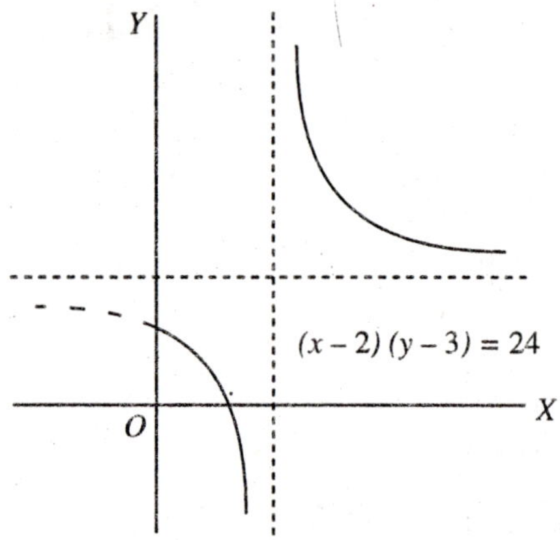

Ex. 121. Sketch the indifference curves

(*i*) $(x + 20)(y + 30) = k$ $\qquad 0 \le x \le \dfrac{k}{30} - 20$

(*ii*) $(x + y - k)^2 = 2xy$ $\qquad 0 \le x \le k.$

Draw the indifference map by giving different values to the parameter k.

Ex. 122. Sketch the transformation curves (*i*) $y^2 = 20 - x - 4y$

(*ii*) $x^2 = 100 - 5y$

CHAPTER 9

Functions and Limits

9.1 FUNCTIONS

In economics and other sciences we have to deal with more than one variable. The changes in any variable are not of great importance unless they are associated with changes in related variables. The way in which one variable depends on other variables is described by means of *functions* which pervade all mathematics and its applications. A function is a relation which associates any given number with another number. Functions can be defined in several ways.

We define a function from the set X into the set Y as a set of ordered pairs (x, y) where x is an element of X and y is an element of Y such that for each x in X there is only one ordered pair (x, y) in the function f. The notation used is

$$f : X \rightarrow Y \text{ or } x \rightarrow f(x) \text{ or } y = f(x)$$

A function is a mapping or transformation of x into y or $f(x)$. The variable x presents elements of the *domain* and is called the *independent* variable. The variable y representing elements of the *range* is called the *dependent* variable. The function $y = f(x)$ is often called a *single valued function* since there is a unique y in the range for each specified x. The converse may not necessarily be true.

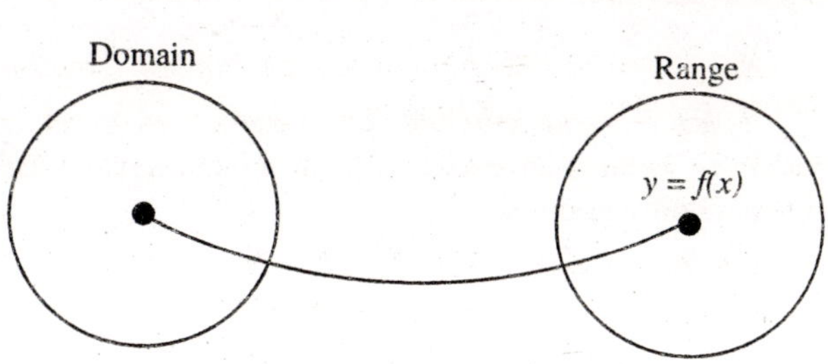

$y = f(x)$ is the *image* of x

A function whose domain and range are sets of real numbers is called a *real valued* function of a real variable.

A function is a *constant* function if the range consists of a single element. It may be written $y = k$ or $f(x) = k$ where k is a constant.

If $f(x) = x$ for all x it is an *identity* function.

$f(x) = |x|$ is the *absolute value* function.

Often a function depends on several independent variables. If there are n independent variables $x_1, x_2, \ldots\ldots, x_n$ and the range is the set of all possible values of y corresponding to the domain of $(x_1, x_2, \ldots, x_n)$ we say that y is a function of x_i's:

$$y = f(x_1, x_2, \ldots, x_n)$$

Letters other than f may be used to represent a function.

Equality of functions. Two functions f and g are equal if and only if they have the same domain and $f(x) = g(x)$ for all x in the domain.

Ex. 1. If p is the price and x the quantity of a commodity demanded by the consumers we write the demand function:

$$x = f(p)$$

Ex. 2. If S stands for saving, Y for income

$$S = f(Y)$$

conveys that saving depends on income.

Ex. 3. If saving depends on income and also on rate of interest r, saving is a function of two variables r, Y.

$$S = f(r, Y)$$

Ex. 4. If u is a function of three variables x, y, z we write

$$u = f(x, y, z)$$

Ex. 5. If production (x) is a function of land (D), labour (L), capital (K), organisation (O) we can write

$$x = f(D, L, K, O)$$

as the production function.

Ex. 6. If utility is a function of n commodities whose quantities are represented by $x_1, x_2, \ldots, x_n$, then the utility function can be written

$$U = U(x_1, x_2, \ldots x_n)$$

Explicit and Implicit Functions

If one variable is expressed directly in terms of other variables, it is a case of an explicit function.

Ex. 7. $y = 3x + 2$, $y = 4x^2 + 9$ are explicit functions.

If the relation between the variables is given by an equation containing all the variables we have an implicit function which may not distinguish between dependent and independent variables.

Ex. 8.

$$\begin{aligned} x^2 + 7xy + 9y^2 + 8 &= 0 \\ y - x \log y &= x^3 \\ f(x, y) &= 0 \end{aligned}$$

are implicit functions.

Graphs of Functions

The graph of a function $f(x)$ is a set of points in the plane.

We can picture a function $y = f(x)$ as a graph showing the relation between x and y.

Every point (x, y) which satisfies the equation $y = f(x)$ is on the curve.

The coordinates of every point (x, y) on the curve satisfy the equation $y = f(x)$. Thus the graph of a function consists of all points in the form $[x, f(x)]$ where x is a number in the range of $f(x)$.

9.2 VALUE OF A FUNCTION AT A POINT

$f(a)$ is the value of the function $f(x)$ when x is replaced by a.

Ex. 9. If $y = f(x) = \sqrt{x+1}$

$f(0) = \sqrt{1}$ $\qquad f(a) = \sqrt{a+1}$

$f(1) = \sqrt{2}$ $\qquad f(a^2 + 2a) = \sqrt{a^2+2a+1} = \pm(a+1)$

$f(-1) = 0$ $\qquad f(a-1) = \sqrt{a-1+1} = \sqrt{a}$

Ex. 10. If $f(x) = |x - 3|$ then

$f(0) = 3$ $\qquad f(a+3) = |a+3-3| = |a|$

$f(3) = 0$ $\qquad f(3-x) = |3-x-3| = |x|$

$f(-3) = 6$ $\qquad f(1/x) = |1/x - 3|$

Ex. 11. If $f(x) = 2$ then

$f(a) = 2, f(0) = 2, f(5) = 2$

Ex. 12. If $f(x) = x^2 + x + 1$ find $f(a+b)$. **Ans.** $a^2 + 2ab + b^2 + a + b + 1$

Ex. 13. If $g(x) = x^{\frac{1}{2}}$ find $g(0), g(2), g(4)$. **Ans.** $0, 2^{\frac{1}{2}}, 2$

Ex. 14. If $h(x) = x^3$ show that $h(-x) = -h(x)$.

Ex. 15. If $h(x) = x^2$ show that $h(-x) = h(x)$.

Ex. 16. Find $\frac{f(x)-f(a)}{x-a}$ $(x \neq a)$ for the following:

(i) $f(x) = x$ $\qquad$ **Ans.** (i) $\frac{f(x)-f(a)}{x-a} = \frac{x-a}{x-a} = 1$

(ii) $f(x) = x^2$ $\qquad$ (ii) $\frac{x^2-a^2}{x-a} = \frac{(x+a)(x-a)}{x-a} = x+a$

(iii) $f(x) = 1/x$ $\qquad$ (iii) $-\frac{(a-x)^2}{ax}$

(iv) $f(x) = \sqrt{x}$ $\qquad$ (iv) $\frac{1}{\sqrt{x}+\sqrt{a}}$

Ex. 17. If $f(x) = |x-2|$ find $f(0), f(2), f(-2)$. **Ans.** 2, 0, 4

9.3 NATURE OF SOME FUNCTIONS

Even and Odd Functions

A function $f(x)$ is said to be an *even* function if for every x in a certain range,

$$f(-x) = f(x)$$

A function $f(x)$ is an odd function if for every x in a certain range,

$$f(-x) = -f(x)$$ *See* Ex. 14, 15

Ex. 18. $f(x) = x^2$ is an even function.

Ex. 19. $f(x) = x^3$ is an odd function.

Ex. 20. Prove that if $f(x)$, $g(x)$ are both odd functions then

$f(x) + g(x)$ and $f(x) - g(x)$ are also odd functions.

Single and multivalued functions

If to each value of x there corresponds one and only one value of y, the function is said to be single valued or uniform, *e.g.*,

$$y = x(x - 1)$$

If more than one value of y corresponds to one value of x the function is said to be multivalued or multiform, *e.g.*,

$$y = \sqrt{x}$$

This is a two valued function because any value of x, say 4, gives two values of y (*i.e.*, + 2 and − 2).

Ex. 21. Given the implicit function between the variables x, y.

$$3x - y = 0$$

this can be written as

(*i*) $3x = y$ or $x = y/3$ making x the dependent variable and y the independent variable.

(*ii*) $y = 3x$ making y, x the dependent and independent variables respectively.

In this example we have described an implicit function in an explicit form thus getting two relationships, or two explicit functions.

Ex. 22. If $x^2 + y^2 = 9$ we have the explicit functions

(*i*) $x^2 = 9 - y^2$ or $x = \pm\sqrt{9 - y^2}$

(*ii*) $y^2 = 9 - x^2$ or $y = \pm\sqrt{9 - x^2}$

Ex. 23. $x^5y^2 + 3x^2y + 2 = 0$

This implicit function cannot be conveniently expressed in the form of explicit functions.

Ex. 24. If $f(x) = x^2 - 2x + 1/x$

Show that $f(a) + f(-a) = 0$

Ex. 25. If $f(x) = x^4 - 3x^2 + 10$ show that $f(x) = f(-x)$

Also find the value of $f(2x) - 2f(x)$.

$$f(2x) = (2x^4) - 3(2x)^2 + 10 = 16x^4 - 12x^2 + 10$$

$$2f(x) = 2x^4 - 6x^2 + 20$$

$$\therefore \quad f(2x) - 2f(x) = 14x^4 - 18x^2 - 10$$

Ex. 26. Given $f(x) = 3 - 2x + x^2$.

Find $f(0), f(-2), f(5), f(-1)$

In $f(x)$, putting $x = 0$, $f(0) = 3 - 2 \times 0 + 0 = 3$

Similarly $f(-2) = 3 - 2(-2) + (-2)^2 = 3 + 4 + 4 = 11$

$$f(5) = 3 - 2 \times 5 + 25 = 18$$

$$f(-1) = 3 - 3 - 2(-1) + (-1)^2 = 6$$

Ex. 27. Let $y = f(x) = ax + b$ where a, b are constants.

Find $f(a), f(-a), f\left(\frac{a}{b}\right), f(k), f(o), f\left(-\frac{b}{a}\right)$

$$f(x) = ax + b$$

$$\therefore \quad f(a) = a \cdot a + b = a^2 + b$$

$$f(-a) = -a \cdot a + b = -a^2 + b$$

$$f\left(\frac{a}{b}\right) = a \cdot \frac{a}{b} + b = \frac{a^2}{b} + b$$

$$f(k) = a \cdot k + b$$

$$f(o) = a \cdot o + b = b$$

$$f\left(-\frac{b}{a}\right) = -a \cdot \frac{b}{a} + b = -b + b = 0$$

Ex. 28. For $f(x) = 2^x + 3$ show that (*i*) $f(2) = 2f(-1)$.

(*ii*) $f(3) = f(0) + f(1) + f(2) - 5$

Monotone Functions

A function is *monotone* if it is either increasing or decreasing.

A function is increasing if $f(x_1) < f(x_2)$ for $x_1 < x_2$

A function is non-decreasing if $f(x_1) \le f(x_2)$ for $x_1 \le x_2$

A function is decreasing if $f(x_1) > f(x_2)$ for $x_1 < x_2$

A function is non-increasing if $f(x_1) \ge f(x_2)$ for $x_1 \le x_2$

Inverse Functions

If there is an implicit function between x and y we can derive two explicit functions from it.

(*i*) y as an explicit function of x, *i.e.*, $y = f(x)$

(*ii*) x as an explicit function of y, *i.e.*, $x = g(y)$

The functions $y = f(x)$ and $x = g(y)$

are inverse functions. Each is inverse to the other

If $f(x)$, $g(x)$ are two functions so related that

$f[g(x)] = x$ for every x in the domain of $g(x)$.

and $g[f(x)] = x$ for every x in the domain of $f(x)$

then $f(x)$ and $g(x)$ are called inverses of each other.

The inverse of a function $f(x)$ is often denoted by $f^{-1}(x)$ (read, f inverse or the inverse of f) so that

$$f\{f^{-1}(x)\} = x$$
$$f^{-1}\{f(x)\} = x$$

whenever the inverse exists.

Every function does not necessarily have an inverse.

Ex. 29. $$f(x) = 5x - 6,\ g(x) = \frac{x+6}{5}$$

are inverses of each other because

$$f[g(x)] = 5\left(\frac{y+6}{5}\right) - 6 = x$$

or $$g[f(x)] = \frac{5x-6+6}{5} = x$$

Ex. 30. Find the inverse of

(*i*) $\dfrac{x^2-1}{x^2}$ (*ii*) $x^2 - 4x$ (*iii*) $\dfrac{x}{x^2-1}$ (*iv*) $\dfrac{x^2-1}{x}$

Ans. (*i*) $\pm\dfrac{1}{\sqrt{1-x}}$ (*ii*) $2 \pm \sqrt{x+4}$

(*iii*) $\dfrac{1 \pm \sqrt{1+4x_2}}{2x}$ (*iv*) No inverse.

Note that $f(x)$ and $f_{-1}(x)$ are inversely related functions of x. The subscript -1 is a symbol for an inverse function, not an exponent. It is true, however, that

$$[f(x)]^{-1} = \frac{1}{f(x)}$$

but this has nothing to do with an inverse function.

9.4 ALGEBRA OF FUNCTIONS

We can combine two (or more) functions under the operations of addition, subtraction, multiplication and division to form new functions.

Let $f(x)$ and $g(x)$ be two functions

Their *sum* is the function $y = f(x) + g(x)$

Their *difference* is the function $y = f(x) - g(x)$

Their *product* is the function $y = f(x) \cdot g(x)$

Their *quotient* is the function $y = f(x)/g(x)$ where $g(x) \neq 0$

The *composite* function of f by g is $f\{g(x)\}$

These functions are defined for all those values of x for which f and g are defined. The commutative, associative and distributive laws hold for functions.

9.5 ALGEBRAIC AND TRANSCENDENTAL FUNCTIONS

Algebraic functions are obtained through a finite number of algebraic operation like addition, subtraction, multiplication, division and through solving a finite number of algebraic equations. Polynomials, rational and radical functions are algebraic.

A function that is not algebraic is a *transcendental* function. Trigonometric, exponential and logarithmic functions are transcendental.

A *polynomial* function is of the form

$f(x) = a_0x^n + a_1x^{n-1} + \ldots\ldots + a_{n-1}x + a_n$ where n is a positive integer and $a_0, a_1, \ldots, a_n$ are real numbers.

This is a polynomial function of degree n in x.

Constant Functions

A zero degree polynomial function is a constant function.

Ex. 31. $f(x) = 3$, $g(x) = \pi$, $h(x) = -k$ are constant functions

Linear Functions

A polynomial function of degree 1 is a linear function.

Ex. 32. $f(x) = mx + c$ is a linear function.

The linear function $f(x) = x$ is called the *identity* function

The graph of a linear function is a straight line.

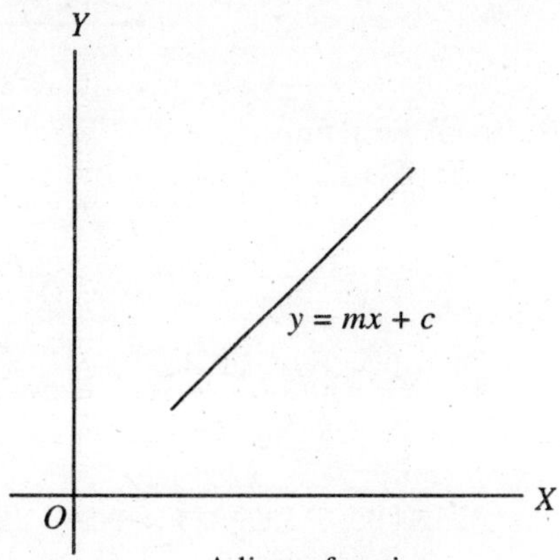

A linear function

Quadratic Functions

A polynomial function of degree 2 is a quadratic function. We can similarly have functions of higher degrees.

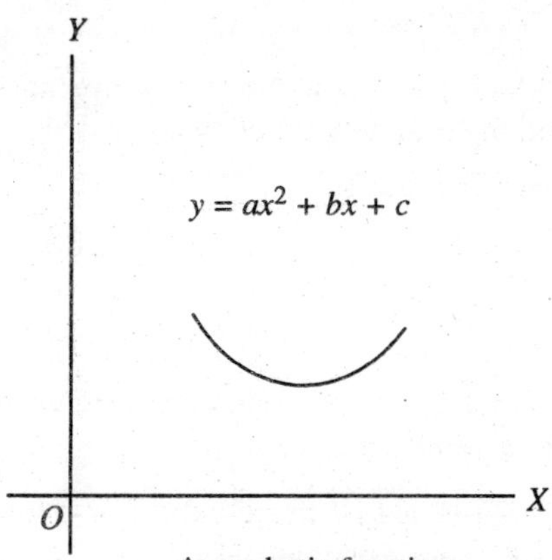

A quadratic function

Ex. 33. $f(x) = ax^2 + bx + c$ is a quadratic function. Its graph is a parabola.

Ex. 34. $f(x) = 3x^3 - 7x^2 + 2x - 1$ is a cubic function.

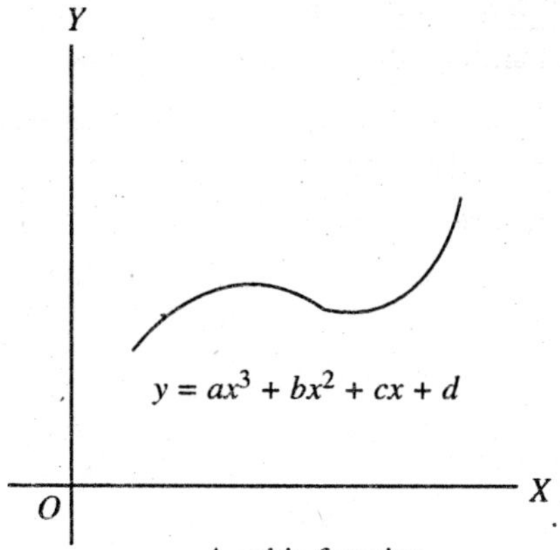

A cubic function

Rational Functions

A rational function is of the form

$$y = \frac{f(x)}{g(x)} \text{ where } f(x),\ g(x) \text{ are polynomials and } g(x) \neq 0$$

Ex. 35. $y = \dfrac{2x - 1}{3x^2 + 2x - 5}$ is a rational function.

9.6 PERIODIC FUNCTIONS—TRIGONOMETRIC FUNCTIONS

Trigonometry deals with the problem of measurement and solution of triangles. Many of the important applications of trigonometry to business cycles and elsewhere do not involve triangles specifically. They deal with the properties and applications of circular functions.

Periodic Functions

A function is periodic with period $p (p \neq 0)$ if $f(x + p) = f(x)$ (see diagram)

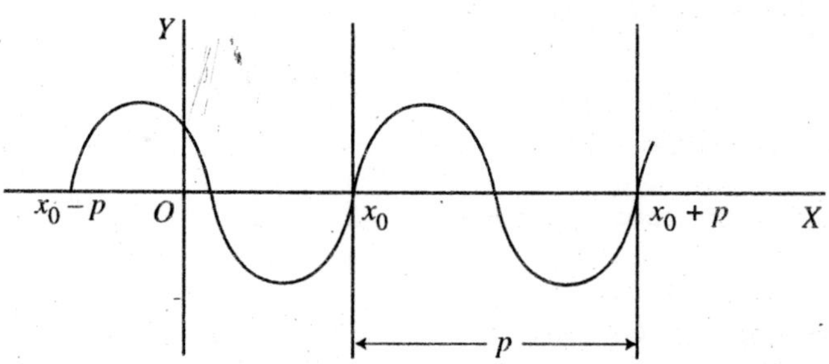

Trigonometric or Circular Functions

In the angle *PAX*, *CB* is drawn perpendicular to *AX*, forming the right angled triangle *ABC*. The point *C* may be taken anywhere on *AP*.

We define the following trigonometric functions.

Sine of angle θ written as

$$\sin\theta = \frac{\text{opp. side}}{\text{hypotenuse}} = \frac{BC}{AC}$$

Cosine of angle θ

$$\cos\theta = \frac{\text{adjacent side}}{\text{hypotenuse}} = \frac{AB}{AC}$$

Tangent of angle θ

$$\tan\theta = \frac{\text{opp. side}}{\text{adjacent side}} = \frac{BC}{AB}$$

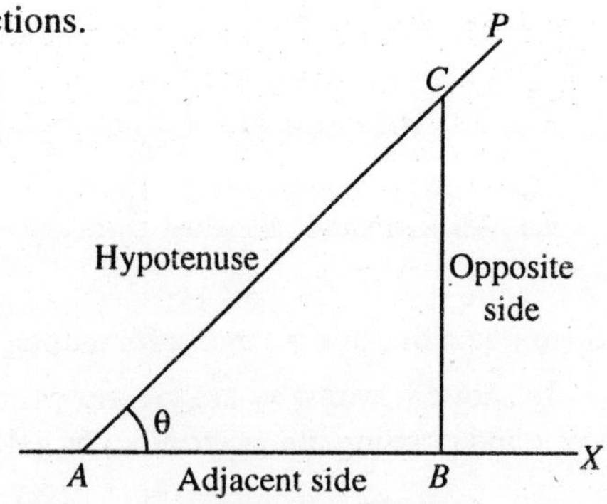

These trigonometric functions are the most important periodic functions.

There are three more trigonometric functions—secant, cosecant and cotangent defined by

$\sec\theta = 1/\cos\theta$, $\operatorname{cosec}\theta = 1/\sin\theta$, $\cot\theta = 1/\tan\theta$

The following identities follow easily.

$$\sin^2\theta + \cos^2\theta = 1$$
$$\sec^2\theta - \tan^2\theta = 1$$
$$\operatorname{cosec}^2\theta - \cot^2\theta = 1$$
$$\frac{\sin\theta}{\cos\theta} = \tan\theta,\quad \frac{\cos\theta}{\sin\theta} = \cot\theta.$$

9.7 INCLINATION AND SLOPE

If a line AP is not horizontal, its angle of inclination is the least positive angle through which the positive x-axis must be rotated to coincide with AP. θ is the angle of inclination of AP. If AP is horizontal, $\theta = 0$. If AP is vertical $\theta = 90°$.

The slope of a line PQ is defined by $m = \tan\theta$.

Ex. 36. What is the slope of a line with inclination 45° which passes through the origin?

Take a point P on the line. Draw $PM \perp OX$. Then the slope is given by

$\tan 45° = \dfrac{MP}{OM} = 1$ by geometry since $MP = OM$.

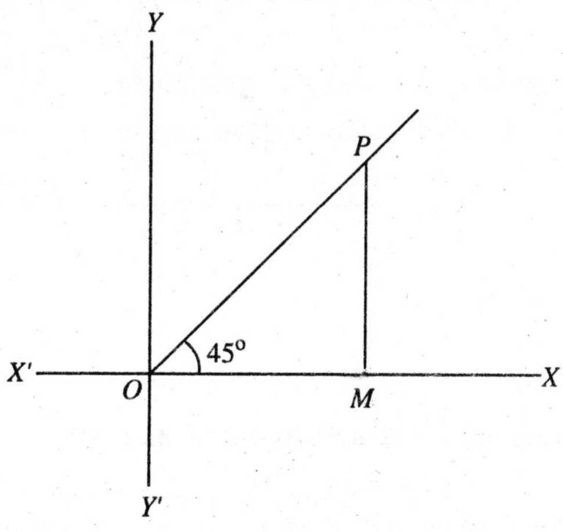

Ex. 37. The slope of the lines with inclination 30°, 60° respectively can be proved to be

$$\tan 30° = \frac{1}{\sqrt{3}},\ \text{and}\ \tan 60° = \sqrt{3}$$

Ex. 38. The slope of a horizontal line is zero.

Ex. 39. The slope of a vertical line is infinite.

Ex. 40. Consider the cost function

$$c = 2x$$

where c stands for cost and x for output.

Its slope is found by taking any point P on it and drawing the perpendicular PM.

$$PM = 5, \; OM = \frac{5}{2}$$

$$\therefore \text{ Slope} = \tan\theta = \frac{PM}{OM} = \frac{5}{\frac{5}{2}} = 2$$

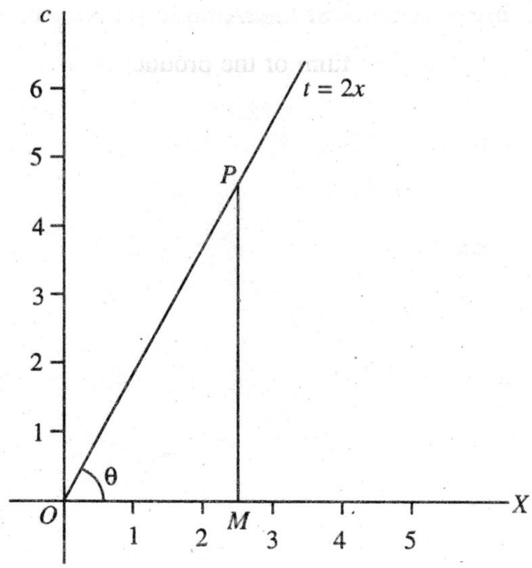

The following values of tan θ corresponding to different values of θ may be noted:

θ	0°	30°	45°	60°	90°	180°	270°	360°
tan θ	0	$\frac{1}{\sqrt{3}}$	1	$\sqrt{3}$	∞	0	∞	0

9.8 LOGARITHMIC AND EXPONENTIAL FUNCTIONS

In the equation

$$a^y = x$$

the exponent y is called the *logarithm* of x to the base a. We write the equation in the alternative form

$$y = \log_a x \quad x > 0, \quad a > 0, \quad a \neq 1$$

This is the *logarithmic function.* The expressions

$$x = a^y \text{ and } y = \log_a x$$

are equivalent and give the same relation.

x may be called the antilogarithm of y to the base a.

Ex. 41.

$\log_2 16 = 4$ since $2^4 = 16$

$\log_{10} 100 = 2$ since $10^2 = 100$

$\log_a 1 = 0$ since $a^0 = 1$

$\log_5\left(\frac{1}{125}\right) = -3$ since $5^{-3} = \frac{1}{5^3} = \frac{1}{125}$

$\log_x x = 1$ since $x^1 = x$

Some properties of logarithmic functions

I. The logarithm of the product of two quantities equals the sum of their logarithms

$$\log_a uv = \log_a u + \log_a v$$

It is true because if $u = a^x$, $v = a^y$ then $uv = a^{x+y}$

$\therefore \qquad \log_a uv = x + y = \log_a u + \log_a v$

Ex. 42. $\log_{10} (16 \times 49) = \log_{10} 16 + \log_{10} 49$

Ex. 43. $\log_{10} (15 \times 29 \times 47) = \log_{10} 15 + \log_{10} 29 + \log_{10} 47$

II. The logarithm of the quotient of two quantities equals the logarithm of the numerator minus the logarithm of the denominator.

$$\log_a \frac{u}{v} = \log_a u - \log_a v$$

Since $\qquad \dfrac{u}{v} = a^{x-y}$

$\therefore \qquad \log_a \dfrac{u}{v} = x - y = \log_a u - \log_a v$

Ex. 44. $\log_{10} \frac{80}{16} = \log_{10} 80 - \log_{10} 16$

Ex. 45. $\log_{10} \left(\dfrac{47 \times 32}{10 \times 16}\right) = \log_{10} (47 \times 32) - \log_{10} (10 \times 16)$

$$= \log_{10} 47 + \log_{10} 32 - \log_{10} 10 - \log_{10} 16$$

III. The logarithm of a power of a quantity is the power times the logarithm of the quantity

$$\log_a u^m = m \log_a u$$

If $\qquad x = \log_a u,\ a^x = u,\ (a^x)^m = a^{xm} = u^m$

Now $\qquad u^m = a^{xm}$

$\therefore \qquad \log_a u^m = mx = m \log_a u$

Ex. 46. $\qquad \log_{10} 5^2 = 2 \log_{10} 5$

$\qquad \log_a a^2 = 2 \log_a a = 2$

Ex. 47. (*a*) Simplify

(*i*) $\log_m a + \log_m b - \log_m c$ (*ii*) $\log_{\sqrt{5}} 25 - \log_{\sqrt{5}} 5$

(*iii*) $\log_b (a^2 + 2) - \log_a (a^2 - 2)$ (*iv*) $\log_3 \frac{2}{3} - \log_3 \frac{1}{3}$

(*b*) Find the value of x

(*v*) $\log_x 32 = 5$ (*vi*) $x = \log_9 \frac{1}{9}$

(*vii*) $\log_{\sqrt{5}} 25 = x$ (*viii*) $x = \log_2 4^7$

(*ix*) $\log_{10} 1000 = x$ (*x*) $\log_{10} 0.10 = x$

Ans. (*i*) $\log_m \dfrac{ab}{c}$ (*ii*) $\log_{\sqrt{5}} 5$ (*iii*) $\log_b \dfrac{a^2+2}{a^2-2}$

(*iv*) $\log_3 2$ (*v*) 2 (*vi*) -1

(*vii*) 4 (*viii*) 14 (*ix*) 3 (*x*) -1

Ex. 48. Show (*i*) $\log_{10} [28^6 \times \sqrt{240} \times \frac{1}{3}] = 6 \log_{10} 28 + \frac{1}{2} \log_{10} 240 - \log_{10} 3$

$$(ii) \quad \log_{10} \frac{\sqrt{37} \times 47^5 \times \sqrt[5]{32}}{0.04 \times \frac{1}{2} \times 100} = \tfrac{1}{2}\log_{10} 37 + 5\log_{10} 47$$

Ex. 49. Show (*i*) $\log_a x^{1/m} = 1/m \log_a x$

(*ii*) $\log_a x^{-1} = - \log_a x$

IV. The base of a logarithm can be changed to any other base as follows:

$$\log_v u = \frac{\log_a u}{\log_a v}$$

It is true because if $x = \log_v u$, then $v^x = u$

$$\therefore \quad \frac{\log_a u}{\log_a v} = \frac{\log_a v^x}{\log_a v} = \frac{x \log_a v}{\log_a v} = x = \log_v u$$

Ex. 50. (*i*) $$\log_{20} 40 = \frac{\log_{10} 40}{\log_{10} 20}$$

(*ii*) $$\log_{100} 1000 = \frac{\log_{10} 1000}{\log_{10} 100} = \frac{3}{2}$$

$$\log_v a = \frac{\log_a a}{\log_a v} = \frac{1}{\log_a v}$$

From this it follows that since

$$\log_v a = \frac{1}{\log_a v}$$

$$\log_v a \times \log_a v = 1$$

Ex. 51. $$\frac{\log_5 15}{1 + \log_5 50} = \frac{\log_5 15}{\log_5 5 + \log_5 50} = \frac{\log_5 15}{\log_5 250}$$

$$= \log_{250} 15 = \frac{1}{\log_{15} 250}$$

Exponential Functions

A simple exponential function with base a is defined by the equation $x = a^y$, $a > 0$

The logarithmic function $y = \log_a x$ is the inverse of the exponential function $x = a^y$.

It should be noted that exponential functions are different from

(*i*) Linear functions *e.g.* $y = ax$, $x = ky$

(*ii*) Power functions *e.g.* $y = x^a$, $x = y^k$

An exponential function is of the form

$$y = k^x, \quad x = z^y$$

The laws of growth and decay in economics and other sciences provide an application of the exponential function. This occurs when the rate of change of the variable under consideration with respect to time is proportional to the amount of the variable present at a given instant of time.

Ex. 52. Let $y = f(x)$ be an exponential function with base a. Show that for any two real numbers x_1, x_2

$$f(x_1 + x_2) = f(x_1)\, f(x_2)$$

Ans. If $f(x) = a^x$, $a^{x_1+x_2} = a^{x_1} \cdot a^{x_2}$ etc.

Ex. 53. Compare the properties of $f(x) = \log_3 x$ with those of $f(x) = \log_{1/3} (x)$. What is the difference?

$$\log_3 x = \frac{\log x}{\log 3}$$

$$\log_{1/3} x = \frac{\log x}{\log 1/3} = \frac{\log x}{-\log 3}$$

$$\log_3 x = -\log_{1/3} x$$

Ex. 54. Show that $x = 0$ is the solution of

$$\log_3 (x + 1) + \log_3 (x + 3) = 1.$$

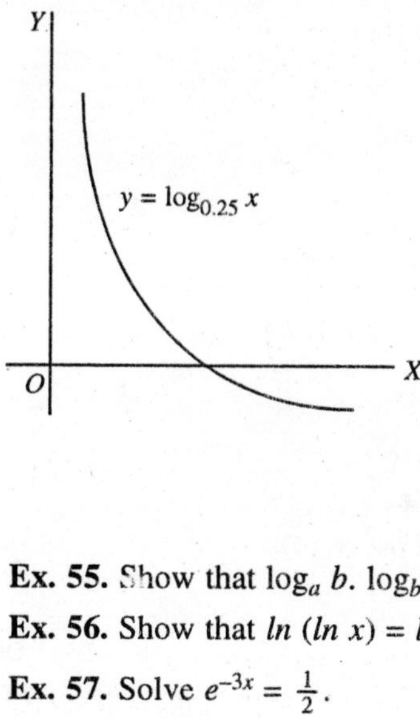

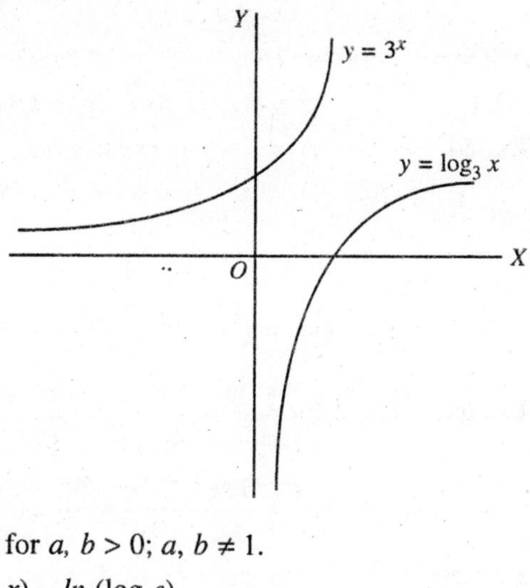

Ex. 55. Show that $\log_a b. \log_b a = 1$ for $a, b > 0$; $a, b \neq 1$.

Ex. 56. Show that $ln\ (ln\ x) = ln\ (\log x) - ln\ (\log e)$.

Ex. 57. Solve $e^{-3x} = \frac{1}{2}$. **Ans.** 0.231

Ex. 58. Draw the graph of $y = \log_{0.25} x$.

(It is a continuous and decreasing function. For $x = 0.5$, $y = 0.5$; $x = 1$; $y = 0$; $x = 2$, $y = -0.5$ etc. See the diagram).

Ex. 59. Draw the graph of $f(x) = y = \log_3 x$ (See the diagram).

Also draw the graph of its inverse $g(x) = 3^x$ (Both the functions are continuous and increasing. For $x = 3$, $y = 1$; $x = 1$, $y = 0$ etc.).

Ex. 60. Draw the graphs of $f(x) = \log_2 x$ and $g(x) = 2^x$.

9.9 HOMOGENEOUS FUNCTIONS

A homogeneous functions is a special type of function frequently used in economics.

A function is homogeneous of degree n if when each of its variables is replaced by k times the variable, the new function is k^n times the original function.

If $z = f(x, y)$ is homogeneous of degree n,

then $$f(kx, ky) = k^n z = k^n f(x, y)$$

Ex. 61. $f(x, y) = x^2 - 3xy + 5y^2$ is a homogeneous function of degree 2 because

$$f(kx, ky) = k^2x^2 - 3kx \cdot ky + 5k^2y^2$$

$$= k^2(x^2 - 3xy + 5y^2) = k^2 f(x, y)$$

Ex. 62. $f(L, K)$ is homogeneous of degree α if

$$f(aL, aK) = a^{\alpha} f(L, K)$$

Ex. 63. $f(x, y, z) = \dfrac{y^2}{z} - \dfrac{3x^2}{y}$ is homogeneous of degree one.

$$f(kx, ky, kz) = \frac{k^2y^2}{kx} - \frac{3k^2x^2}{ky} = k\left(\frac{y^2}{z} - \frac{3x^2}{y}\right)$$

$$= kf(x, y, z)$$

A linearly homogeneous function of degree one is such that if

$$z = f(x, y), f(kx, ky) = kf(x, y)$$

Ex. 64. $f(x, y) = x + y$ is linearly homogeneous.

(*i*) $f(lx, ly) = lx + ly = l(x + y) = lf(x, y)$

(*ii*) $f(x, y) = \sqrt{ax^2 + by^2}$ is linearly homogeneous

$$f(kx, ky) = \sqrt{ak^2x^2 + bk^2y^2} = k\sqrt{ax^2 \quad by^2} = kf(x, y)$$

Ex. 65. $f(x, y, z) = \dfrac{3z}{x} - \dfrac{5x}{y}$

$$f(kx, ky, kz) = \frac{3kz}{kx} - \frac{3kx}{ky} = \frac{3z}{y} - \frac{5x}{y} = k^{\circ} f(x, y, z)$$

Here $f(x, y, z)$ is homogeneous of degree zero.

Ex. 66. Show that the following functions are linearly homogeneous

(*i*) $z = ax + by$ (*ii*) $z = \sqrt{ax^2 + bxy + cy^2}$

(*iii*) $z = (ax^2 + bxy + cy^2)/(x + y)$

9.10 COBB DOUGLAS PRODUCTION FUNCTION

If L stands for labour, K for capital and α, β are constants such that $\alpha + \beta = 1$ then in the Cobb Douglas production function, replacing L, K by pL, pK

$$Y = aL^{\alpha} K^{\beta}$$

We have $a \cdot (pL)^\alpha \cdot (pK)^\beta = a \cdot p^\alpha L^\alpha \cdot p^\beta K^\beta$

$= a \cdot p^{\alpha+\beta} L^\alpha K^\beta = p \cdot aL^\alpha K^\beta = pY$ since $\alpha + \beta = 1$

Hence the Cobb Douglas production function is *linearly homogeneous.*

Ex. 67. Show that (*i*) $Y = \left(\frac{a}{L^r} + \frac{\beta}{K^r}\right)^{-1/r}$

(*ii*) $Y = a[\delta K^{-r} + (1 - \delta)L^{-r}]^{-1/r}$

are homogeneous of degree 1.

Returns to scale

Let $Y = f(K, L)$ be a linearly homogeneous production function in labour L and capital K.

Raising all inputs of L, K c-fold will raise the output c-fold.

since $f(cK, cL) = cY$

In other words, a linearly homogeneous production function implies constant returns to scale in production. If no input is used there is no output. If inputs are doubled, output is doubled.

9.11 DEMAND FUNCTIONS AND SUPPLY FUNCTIONS

Let x denote the quantity of a commodity demanded and p its price. x and p being variables we may write the demand function

$x = f(p)$ showing dependence of x on p or

$p = F(x)$ showing dependence of p on x

These are the explicit forms of the implicit demand function $g(x, p) = 0$.

The variables, in the case of a demand function, as in the case of other functions in economics, are hypothetical quantities and not actual observable quantities. Changes in the values of parameters cause shifts in the demand curve.

The arguments given above apply to a supply function also if x stands for the variable supply. The slope of a supply curve is positive and that of a demand curve is negative, in general.

Ex. 68. Construct the market demand schedule and plot the demand curves given by

(*i*) $x = 10 - 2p$ (*ii*) $x = 10/p$

(*iii*) $x = -3p^2 + p + 65$ (*iv*) $x = 5\sqrt{p}$

Ex. 69. Draw the supply curves (*i*) $x = 3^p - 3$ (*ii*) $x = 3p + 3$

9.12 COST FUNCTIONS AND PRODUCTION FUNCTIONS

If x is the quantity produced by a firm at total cost C, we have the total cost function

$C = f(x)$ explicitly. We may write this in the implicit form:

$$g(C, x) = 0$$

Average cost of production or cost per unit is obtained by dividing total cost by the quantity produced.

$$AC = C/x \text{ or } C = AC \cdot x$$

Cost curves can be obtained from a knowledge of *production functions.* A production relationship is described by inputs associated with specified amounts of outputs. In the production function

$$Y = f(L, K)$$

L, K are quantities of labour and capital respectively required to produce Y.

If p_L is the wage rate or price of labour, p_K the price per unit of capital we can write the total cost as the sum of all inputs (*i.e.*, L, K in this case) times their respective prices:

$$C = LP_L + KP_K$$

Ex. 70. Plot the total cost and average cost curves for the cost function.

$$C = x^3 - 3x^2 + 4x + 60$$

Ex. 71. The liquidity preference function can be written

$$M = f(r)$$

where M is the total quantity of money and r the interest rate.

Plot the curve $M = \dfrac{500}{r-2}$ and show that the interest rate of 2% is the *liquidity trap* in this case.

9.13 LIMITS

The concept of limits is useful in developing some mathematical techniques and also in analysing various problems in economics.

We know, for instance, that the rate of interest r can never fall to zero, however large be the quantity of capital available. There is a minimum below which r cannot fall. Let this minimum be 2 per cent. Then if K stands for capital and a is a positive constant, we may write, roughly,

$$r = 2 + \frac{a}{K}$$

If K is small r is large.

If K is large r is small.

However large K may become r will never be less than 2. The limit of r as K increases is 2. We write this

$$\lim_{K \to \infty} r = 2$$

In a similar manner, we can see that there is a limit below which average cost will not go however large the quantity of output.

A variable x may increase without bound in such a way that x is larger than any large number N. Then we say that x tends to infinity. We write this

$$x \to \infty$$

A function $f(x)$ may become infinite as x tends to some value c. We write this

$$\lim_{x \to c} f(x) = \infty$$

A function may attain a finite value l as x tends to a value k. Then we write

$$\lim_{x \to k} f(x) = l$$

This statement means that provided $f(x)$ is defined for all values of x in some range and x approaches a given number k in that range then $f(x)$ approaches a limit l.

In other words (*i*) if the difference between x and k is made as small as we please (*i.e.*, if $x \to k$) then (*ii*) the difference between $f(x)$ and l can be made as small as we please (*i.e.*, $f(x) \to l$)

Note: $x = 1 \quad \therefore \quad 2x^2 = 2$

$x \to 1 \quad \therefore \quad 2x^2 \to 2$

These two are not identical statements. Although the answers are the same the procedures of arriving at the answers are different. Moreover the answers need not necessarily be the same always.

Consider the function $f(x) = 1/x$

As x tends to infinity, $f(x)$ tends to zero.

We write

$$\lim_{x \to \infty} \frac{1}{x} = 0$$

If $f(x) = x^2 + 3$, $f(x)$ approaches 7 as x approaches 2.

We write

$$\lim_{x \to 2} (x^2 + 3) = 7$$

Definition of a limit. Let $f(x)$ be a function.

If there exists a number l such that $f(x)$ may be made as close to l as we please by choosing for x any number close to a given number k, we say that the limit of $f(x)$ as x approaches k is l.

We write this symbolically

$$\lim_{x \to k} f(x) = l$$

This statement means that $f(x)$ is close to l if x tends to (but is not equal to) k. If we take x sufficiently close to k we can have $f(x)$ as close to l as we like.

The definitions of the limit of a sequence and the limit of a function are identical but while n is restricted to positive integers in a sequence, the functions $f(x)$ is defined for all real numbers in a given range.

9.14 THEOREMS ON LIMITS OF FUNCTIONS

1. If a is a constant

$$\lim_{x \to k} a = a$$

2. If a, b are constants

$$\lim_{x\to k}(ax+b)=ak+b$$

3. If $\lim_{x\to k} f(x)=l,\qquad \lim_{x\to k} f(x)=m$

(i) $\lim_{x\to k}[f(x)+g(x)]=l+m$

(ii) $\lim_{x\to k}[f(x)\cdot g(x)]=l\cdot m$

(iii) $\lim_{x\to k}\dfrac{f(x)}{g(x)}=\dfrac{l}{m},\quad m\neq 0$

(iv) $\lim_{x\to k}\sqrt[n]{f(x)}=\sqrt[n]{l}.$

(v) If $\lim_{x\to k} f(x)=l,\ \lim_{x\to k} g(x)=0$

$\lim_{x\to k} f(x)\cdot g(x)=0$

Now consider the following:

$$\lim_{x\to 0}\frac{k}{x}=\infty;\qquad \lim_{x\to\infty} kx=\infty;\qquad \lim_{x\to\infty}\frac{x}{k}=\infty$$

These are some convenient though pedogogically bad methods of writing limits when division and multiplication by infinity or division by zero is involved. It is preferable to talk of no limits in such cases.

In evaluating a limit we let the variable approach the limiting value. Often this may not be feasible in which case factorisation may remove the difficulties. Sometimes a convenient division of the terms in the numerator and the denominator by certain quantities is found useful.

Ex. 72. $\lim_{x\to a} x=a$

We have not set $x=a$, this is not the idea of a limit. We have expressed that, in the limit, x approaches a.

Ex. 73. $\lim_{x\to 2}(5x-7)=\lim_{x\to 2}5x-\lim_{x\to 2}7=10-7=3$

Ex. 74. $\lim_{x\to 2}\dfrac{x^2-4}{x-2}=\lim_{x\to 2}\dfrac{(x+2)\,(x-2)}{x-2}=\lim(x+2)=4$

Here factorisation is useful in finding the limit.

Ex. 75. $\lim_{x\to 2}\dfrac{x^2+1}{x^2-1}=\lim_{x\to 2}\dfrac{x^2/x^2-1/x^2}{x^2/x^2-1/x^2}=\lim_{x\to 2}\dfrac{1+1/x^2}{1-1/x^2}=1$

(Dividing numerator and denominator by x^2)

Since $\dfrac{1}{x^2}\to 0$ as $x\to\infty$

Here the division of numerator and denominator by x^2 is useful.

Ex. 76. Prove the following results:

(*i*) $\lim_{x \to -2} x^2 = 4$ (*ii*) $\lim_{x \to 0} (7 - x - x^2) = 7$

(*iii*) $\lim_{x \to 1} \frac{x^2 - 4x + 3}{x^2 + 2x - 3} = -\frac{1}{2}$ (*iv*) $\lim_{x \to 0} \frac{x^4 - a^4}{x - a} = 4a^3$

(*v*) $\lim_{x \to \infty} \frac{2x + 3x^{-1}}{3x + 2x^{-1}} = \frac{2}{3}$ (*vi*) $\lim_{x \to 0} \frac{\sqrt{1+x} - \sqrt{1-x}}{x} = 1$

(*vii*) $\lim_{x \to \infty} \frac{a + bx}{x^2} = 0$ (*viii*) $\lim_{p \to 2} \frac{p^3 - 8}{p - 2} = 12.$

(*ix*) $\lim_{r \to 2} \frac{r^3 - 5}{2r^3 + 6} = -\frac{1}{22}$ (*x*) $\lim_{x \to 1} \frac{x^2 + x - 2}{x^2 + 6x - 7} = \frac{2}{3}$

(*xii*) $\lim_{x \to -1} \frac{2x + 1}{x^2 - 3x + 4} = -\frac{1}{8}$ (*xii*) $\lim_{h \to 0} \frac{(x+h)^2 - x^2}{h} = 2x$

(*xiii*) $\lim_{h \to 0} \frac{(x+h)^3 - x^3}{h} = 3x^2$ (*xiv*) $\lim_{h \to 0} \frac{1}{h}\left\{\frac{1}{x+h} + \frac{1}{x}\right\} = \frac{-1}{x^2}$

Ex. 77. Evaluate the limits if they exist:

(*i*) $\lim_{x \to 1} \frac{x^2}{x - 1}$ (*ii*) $\lim_{x \to 2} \frac{3x - 7}{x^2 - 4}$

(*iii*) $\lim_{x \to 0} \frac{5}{x}$ (*iv*) $\lim_{x \to 0} \left(x - \frac{3}{x}\right)$

(*v*) $\lim_{p \to 1} \frac{p}{p^2 - 1}$ (*vi*) $\lim_{a \to 2} \frac{a - 2}{a^2 - 4}$

(*vii*) $\lim_{x \to -4} \left(\frac{2x}{x+4} + \frac{8}{x+4}\right)$ (*viii*) $\lim_{y \to 1} \frac{y^2 - y}{y - 1}$

(*ix*) $\lim_{x \to -1} \frac{6x^3 - 4x^2 + 8}{x^2 - x}$ (*x*) $\lim_{x \to 1} \frac{|2x^5 - 1|}{4}$

Ans. (*i*) – (*v*) no limit (*vi*) $\frac{1}{4}$ (*vii*) 2 (*viii*) 1 (*ix*) – 1 (*x*) $\frac{1}{4}$

Ex. 78. $\lim_{x \to 4} \frac{(x^3 - 64)}{\sqrt{x^2 + 9} - 5} = \lim_{x \to 4} \frac{x(x-4)(x^2 + 4x + 16)(\sqrt{x^2 + 9} + 5)}{(\sqrt{x^2 + 9} - 5)(\sqrt{x^2 + 9} + 5)}$

$$= \lim_{x \to 4} \frac{x(x^2 + 4x + 16)(\sqrt{x^2 + 9} + 5)}{x + 4}$$

$$= 240$$

Ex. 79. $\lim\limits_{x\to4} \dfrac{3-\sqrt{5+x}}{1+\sqrt{5-x}} = \lim\limits_{x-4} \dfrac{(3-\sqrt{5+x})(3+\sqrt{5+x})(1+\sqrt{5-x})}{(1-\sqrt{5-x})(1+\sqrt{5+x})(3+\sqrt{5+x})}$)

$$= \lim_{x\to4} \frac{\{9-(5+x)\}(1+\sqrt{5-x})}{\{1-(5-x)\}(3+\sqrt{6+x})}$$

$$= \lim_{x\to4} \frac{4-x}{-(4-x)} \frac{1+\sqrt{5-x}}{3+\sqrt{5+x}} = -\frac{1}{3}$$

Ex. 80. $\lim\limits_{x\to3} \dfrac{x(x-2)-3}{x(x-3)(x-2)} = \dfrac{4}{3}$

Ex. 81. $\lim\limits_{x\to1} \dfrac{\cos \pi x/2}{1-x}$ Put $y = x-1$
As $x \to 1, y \to 0$

$$= \lim_{y\to0} \frac{\cos\left(\frac{\pi}{2}+\frac{\pi y}{2}\right)}{-y} = \lim_{y\to0} \frac{-\sin \pi y/2}{-\frac{2}{\pi}\cdot\frac{\pi y}{2}} = \frac{\pi}{2}$$

Ex. 82. $\lim\limits_{x\to0} \dfrac{\sqrt{2}-\sqrt{1+\cos x}}{\sin^2 x} = \dfrac{1}{4\sqrt{2}}$

Ex. 83. $\lim\limits_{x\to1} \dfrac{\tan (x^2-1)}{x-1} = 2$

9.15 LIMITS AND CONTINUITY OF FUNCTIONS

$\lim\limits_{x\to a} f(x)$ may or may not be the same as $f(a)$.

If $\lim\limits_{x\to a} f(x) = f(a)$, a finite quantity, then $f(x)$ is said to be a continuous function of x at $x = a$.

If any of the two conditions, viz.

(*i*) $\lim\limits_{x\to a} f(x) = f(a)$, a finite quantity

(*ii*) $f(a)$ exists

is not satisfied for any value of x, $f(x)$ is said to be discontinuous for that value of x.

The discussion of a limit is necessary to get a precise concept of continuity. The concepts of limit and continuity are essential in the study of calculus.

Ex. 84. $f(x) = \dfrac{1}{x-2}$ and $\lim\limits_{x\to a} f(x) = \dfrac{1}{a-2}$

If $x \to 2$, $f(x)$ does not have a finite value

$\therefore$ $f(x)$ is continuous for any value of a except 2

Ex. 85. $f(x) = \sqrt{(x-1)(2-x)}$

This function is continuous only in the interval (1, 2)

Ex. 86. $f(x) = \dfrac{1}{x^2 - a^2}$

This function is discontinuous at $x = a$, $x = -a$.

9.16 A NOTE ON FINITE DISCONTINUITY AND INFINITE DISCONTINUITY

When a limit exists it means it exists finitely.

If $\lim_{x \to a} f(x) = \infty$, it means that as $x \to a$, $f(x)$ becomes larger and larger and not that $f(x)$ equals ∞ which is not a number at all.

$f(x)$ is continuous on an interval if it is continuous at every point of the interval.

If $y = f(x)$ is continuous in the interval $[a, b]$ then the range of y is the interval $[a, b]$ and $f(x)$ cannot pass from one value to another value in the interval without assuming at least once each intermediate value.

The graph of a function which is continuous in an interval $[a, b]$, is a continuous unbroken curve.

The concept of a limit should not be applied to a discrete function which is defined for a finite number of values in an interval and is discontinuous at all other points.

Price, cost, demand, supply and other functions in economics are often discrete but are treated as continuous for convenience in the applications of mathematical analysis.

A function $f(x)$ has a finite discontinuity at $x = a$ if $f(x)$ exists but changes abruptly at $x = a$. Thus $f(x) = (e^{1/x} + 5)^{-1}$ has a finite discontinuity at $x = 0$ though at all other points it is continuous.

Infinite discontinuity at $x = a$ implies that as $x \to a$, $f(x)$ becomes infinite. Thus at $x = 5$, $f(x) = \dfrac{1}{x-5}$ has an infinite discontinuity since $f(5)$ does not exist though $f(x)$ exists at all points other than $x - 5$.

CHAPTER 10

Differential Calculus: Functions of One Variable

10.1 DIFFERENTIATION

Calculus, which may be called the mathematics of change, motion and growth, provides the means to great advances in exact thinking and has applications in practically all scientific work. The precision of various concepts in economics depends on the use of calculus.

The basic operations of calculus are : *differentiation,* which involves finding the rate at which a variable quantity is changing and *integration,* which involves finding a function whose rate of change is given.

Differential calculus, concerned with finding the *instantaneous* rate of change of a given function, is very useful in business and economics which involve changes in the form of growth, decay, increasing and decreasing returns and variations in costs, profits, etc. Prominent applications of the concept are in the fields of marginal analysis and problems of optimisation.

The *average rate of change* is the average value of a variable under consideration over an interval. The *marginal rate of change* is the instantaneous rate of change *i.e.,* a change over a particular small change in a related variable. Thus, marginal cost may be looked upon as the rate of change of total cost following a change in the quantity produced. Marginal revenue is, similarly, the rate of change of total revenue with quantity. To express the rate of change of any functional relationship, we use the concept of the derivative which involves infinitesimally small changes in the independent and dependent variables. These two forms of variables are always present in any problem of differential calculus. The dependent variable is a continuous function of the independent variable.

Differential calculus attempts to find a function derived from a given relationship between two variables to express the idea of change. This derived function is called the *derivative* of the given function. The process of obtaining a derivative is called *differentiation* of the function. We say that a function has a derivative if it is differentiable.

Average Rate of Change

If a point $P_1(x_1, y_1)$ moves to a position $P_2(x_2, y_2)$ we define the increments in x, y as

$$\Delta x = x_2 - x_1, \qquad \Delta y = y_2 - y_1$$

which may be any real quantities, positive, negative or zero.

If there is a functional relation between x and y, the average rate of change of y with respect to x is given by

$$\frac{\Delta y}{\Delta x} = \frac{y_2 - y_1}{x_2 - x_1}.$$

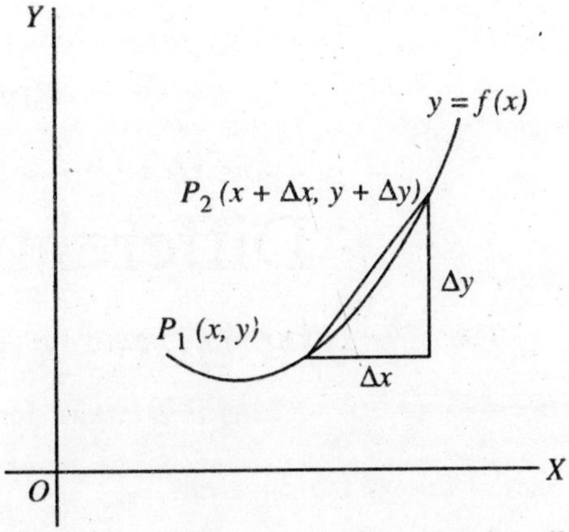

Let $y = x^2$. If x changes from 1 to 2 *i.e.*, $\Delta x = 1$ then y changes from 1 to 4 *i.e.*, $\Delta y = 3$ so that $\frac{\Delta y}{\Delta x} = \frac{3}{1}$.

If x changes from 2 to 3 *i.e.*, $\Delta x = 1$ then y changes from 4 to 9 *i.e.*, $\Delta y = 5$ so that $\frac{\Delta y}{\Delta x} = 5$.

This means that $\frac{\Delta y}{\Delta x}$, the average rate of change of y with x, is not the same at all points.

Definition of the Derivative: First Principles

The derivative, or the rate of change of $y = f(x)$ with respect to x is written $\frac{dy}{dx}$ and is defined as a limit :

$$\frac{dy}{dx} = \lim_{\Delta x \to 0} \frac{\Delta y}{\Delta x} = \lim_{\Delta x \to 0} \frac{f(x + \Delta x) - f(x)}{\Delta x}$$

provided such a limit exists.

The derivatives of different functions can be found with the help of this limit and use may be made of *differential by first principles*, as follows.

Ex. 1. If $y = f(x) = x^2$ then $f(x + \Delta x) = (x + \Delta x)^2$

$$\frac{dy}{dx} = \lim_{\Delta x \to 0} \frac{(x + \Delta x)^2 - x^2}{\Delta x}$$

$$= \lim_{\Delta x \to 0} \frac{x^2 + 2x\ \Delta x + \Delta^2 x - x^2}{\Delta x} = \lim_{\Delta x \to 0} \frac{2x\,\Delta x + \Delta^2 x}{\Delta x}$$

$$= \lim_{\Delta x \to 0} \frac{\Delta x(2x + \Delta x)}{\Delta x} = \lim_{\Delta x \to 0} (2x + \Delta x) = 2x.$$

Ex. 2. If $y = f(x) = \frac{1}{x}$ then $f(x + \Delta x) = \frac{1}{x + \Delta x}$

$$\frac{dy}{dx} = \lim_{\Delta x \to 0} \frac{\frac{1}{x + \Delta x} - \frac{1}{x}}{\Delta x} = \lim_{\Delta x \to 0} \frac{\frac{x - (x + \Delta x)}{x(x + \Delta x)}}{\Delta x} = \lim_{\Delta x \to 0} \frac{-\Delta x}{x(x + \Delta x)\Delta x}$$

$$= \lim_{\Delta x \to 0} \frac{-1}{x(x + \Delta x)} = \frac{-1}{x^2}.$$

Ex. 3. Find $\frac{dy}{dx}$ for (*i*) $y = x^3$ (*ii*) $y = x^{-2}$ (*iii*) $y = x^{1/2}$

(*iv*) $y = \frac{3}{2x-5}$ (*v*) $y = 7x^2 - \frac{1}{x^2}$

Ans. (*i*) $3x^2$ (*ii*) $-2x^{-3}$ (*iii*) $\frac{1}{2}x^{-1/2}$ (*iv*) $\frac{-6}{(2x-5)^2}$ (*v*) $14x + 2x^{-3}$

Slope

The slope of a line joining the points $P_1(x_1, y_1)$ and $P_2(x_2, y_2)$ is a constant $m = \frac{y_2 - y_1}{x_2 - x_1} = \frac{\Delta y}{\Delta x}$. This is constant for a line because the line, throughout its length, rises or falls at the same rate.

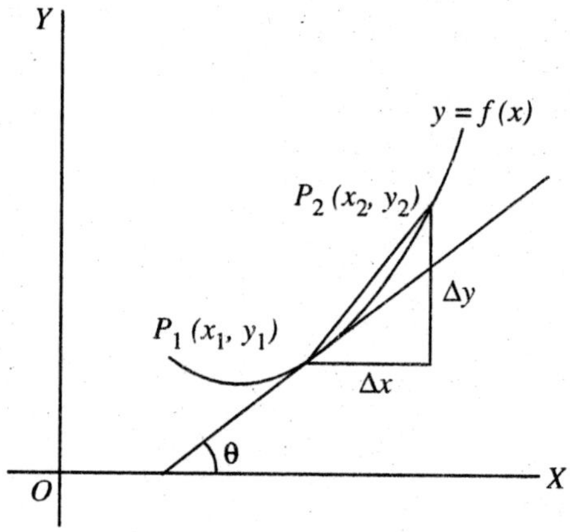

A curve does not rise or fall at the same rate throughout its length. The slope of a curve at any point $P_1(x_1, y_1)$ is the slope of the tangent to the curve at P_1. This is found as follows.

Let P_2 (x_2, y_2) be any other point on the curve. Keeping P_1 fixed and moving P_2 along P_1 and closer and closer to P_1, the slope of the secant P_1P_2 approaches a constant limiting value which is the slope of the tangent at P_1.

$$\lim_{P_2 \to P_1} \frac{\Delta y}{\Delta x} = \lim_{\Delta x \to 0} \frac{\Delta y}{\Delta x} = \lim_{\Delta x \to 0} \frac{f(x + \Delta x) - f(x)}{\Delta x}$$

which, we know, is written $\frac{dy}{dx}$. This means that the slope of a curve at a point P_1 is the *rate of change* of the ordinate (*i.e.*, the y coordinate) with respect to the abscissa (*i.e.*, the x coordinate) at P_1.

The slope of the tangent is also written $\frac{dy}{dx} = \tan\theta$ where θ is the angle of inclination of the tangent to the x-axis.

Ex. 4. We have seen that for the curve $y = x^2$, $\frac{dy}{dx} = 2x$ which is the slope of the given curve at any point (x, y).

The slope at any particular point (2, 4) on the curve can be found by substitution.

At (2, 4), $\frac{dy}{dx} = 2x = 2 \times 2 = 4$

At (3, 4), $\frac{dy}{dx} = 2x = 2 \times 3 = 6$

Ex. 5. Find the slopes of the following curves:

(*i*) $2x + 3y = 5$ at any point

(*ii*) $y = x^3 - 12x + 13$ at (1, 2)

(*iii*) $y = \dfrac{x^3+1}{x}$ at (1, 2) and (2, 4.5)

(*iv*) $y = 1/x$ at (1, 1)

Ans. (*i*) $-\dfrac{2}{3}$ (*ii*) – 9 (*iii*) 1, 3.75 (*iv*) – 1

Ex. 6. Differentiate the following functions:

(*i*) $2x^3$ (*ii*) $2x^2 + x$ (*iii*) $\dfrac{1}{x+1}$ (*iv*) $5 - x^2$ (*v*) $\dfrac{x^4}{3}$ (*vi*) $\sqrt{x}$ (*vii*) $\dfrac{1}{\sqrt{x}}$ (*viii*) $\left(\dfrac{x+1}{x}\right)^2$

Ans. (*i*) $6x^2$ (*ii*) $4x + 1$ (*iii*) $\dfrac{-1}{(x+1)^2}$ (*iv*) $-2x$ (*v*) $\frac{4}{3}x^3$

(*vi*) $\dfrac{1}{2\sqrt{x}}$ (*vii*) $\dfrac{-1}{x^{3/2}}$ (*viii*) $-2\left(\dfrac{x+1}{x^3}\right)$

It is not necessary in later work to find derivatives from the first principles. Some rules can be framed.

10.2 RULES OF DIFFERENTION

There are some useful rules for differentiating particular types of functions.

Polynomial Functions

(*a*) If $y = x^n$, $\dfrac{dy}{dx} = nx^{n-1}$ *e.g.*, if $y = x^5$, $\dfrac{dy}{dx} = 5x^{5-1} = 5x^4$

This holds at all points except at $x = 0$. At $x = 0$ it holds if $n \geq 1$

(*b*) If $y = c$, *a* constant, then $\dfrac{dy}{dx} = 0$ *e.g.*, if $y = 7$, $\dfrac{dy}{dx} = 0$

(*c*) If $y = cu$, where $u = f(x)$

$$\frac{dy}{dx} = c\frac{du}{dx}$$

(*d*) If $y = u + v$ *e.g.*, if $y = 7x^5 + 3$

where $u = f(x)$, $v = g(x)$ $\dfrac{dy}{dx} = 35x^4 + 0 = 35x^4$

$$\frac{dy}{dx} = \frac{du}{dx} + \frac{dv}{dx}$$

(*e*) If $y = u_1 + u_2 + \ldots + u_m$ where $u_1 = f_1(x)$, $u_2 = f_2(x)$, etc.

$$\frac{dy}{dx} = \frac{du_1}{dx} + \frac{du_2}{dx} + \ldots + \frac{du_m}{dx}.$$

Ex. 7. (*i*) $y = x^2$

$$\frac{dy}{dx} = 2x$$

(*ii*) $y = x^3$

$$\frac{dy}{dx} = 3x^2$$

(*iii*) $y = x$

$$\frac{dy}{dx} = 1x^0 = 1$$

(*iv*) $y = \frac{1}{x} = x^{-1}$,

$$\frac{dy}{dx} = -1x^{-2} = -\frac{1}{x^2}$$

(*v*) $y = x^{-n}$

$$\frac{dy}{dx} = -nx^{-n-1}$$

(*vi*) $y = x^{1/2}$

$$\frac{dy}{dx} = \frac{1}{2}x^{-1/2} = \frac{1}{2x^{1/2}}$$

(*vii*) $y = 5$

$$\frac{dy}{dx} = 0$$

(*viii*) $y = 5x$

$$\frac{dy}{dx} = 5$$

(*ix*) $y = 7x^4$

$$\frac{dy}{dx} = 28x^3$$

(*x*) $y = \frac{7}{x^{3/5}}$

$$\frac{dy}{dx} = -\frac{21}{5}x^{-8/5}$$

(*ix*) $y = 4x^3 + 3x$

$$\frac{dy}{dx} = 12x^2 + 3$$

(*xii*) $y = 7x^4 - 3x^3 + 2x^2 + 2$

$$\frac{dy}{dx} = 28x^3 - 9x^2 + 4x$$

(*xiii*) $y = C_0x^n + C_1x^{n-1} + C_2x^{n-2} + \ldots + C_{n-1}x + C_n$

$$\frac{dy}{dx} = C_0nx^{n-1} + C_1(n-1)x^{n-2} + C_2(n-2)x^{n-3} + \ldots + 1C_{n-1}$$

Ex. 8. Differentiate with respect to x :

(*i*) $x^2 - 4x + 3$ (*ii*) $3x^7 - 2$

(*iii*) $\frac{x^5}{5} + \frac{x^4}{4} - \frac{x^3}{3} - \frac{x^2}{2} + x - 1$ (*iv*) $(x - 2)(x - 3)$

Ans. (*i*) $2x - 4$ (*ii*) $21x^6$ (*iii*) $x^4 + x^3 - x^2 - x + 1$ (*iv*) $2x - 5$

Ex. 9. Find the slope of the curve $y = x^3 - 6x + 2$ where it cuts the y-axis.

Ans. $\frac{dy}{dx} = 3x^2 - 6 = -6 \quad \because$ on the y-axis, $x = 0$

Rational Functions: Products and Quotients

(*a*) If y is a product of two functions *i.e.*, if $y = uv$ where u, v are function of x

$$\frac{dy}{dx} = u\frac{dv}{dx} + v\frac{du}{dx}$$

This can be proved by the usual method of limits. (see Ex. 14)

Ex. 10. (*i*) $y = 2x^2 (5x + 3)$

Let $u = 2x^2 \quad \therefore \quad \frac{du}{dx} = 4x$

Let $v = 5x + 3 \quad \therefore \quad \frac{dv}{dx} = 5$

$$\frac{dy}{dx} = u\frac{dv}{dx} + v\frac{du}{dx}$$
$$= 2x^2 \cdot 5 + (5x + 3)\, 4x$$
$$= 10x^2 + 20x^2 + 12x$$
$$= 30x^2 + 12x.$$

Ex. 11. $y = (x + 5)(x^2 + 3)$ Let $u = x + 5,\ v = x^2 + 3$

$$\frac{dy}{dx} = u\frac{dv}{dx} + v\frac{du}{dx} \qquad \frac{du}{dx} = 1,\ \frac{dv}{dx} = 2x$$
$$= (x + 5)\, 2x + (x^2 + 3)\, 1$$
$$= 3x^2 + 10x + 3$$

Alternatively $y = x^3 + 5x^2 + 3x + 15$

$$\frac{dy}{dx} = 3x^2 + 10x + 3.$$

Ex. 12. $y = \frac{1}{x\sqrt{x}} = \frac{1}{x} \cdot \frac{1}{x^{1/2}}$

Let $u = \frac{1}{x} = x^{-1},\ v = \frac{1}{x^{1/2}} = x^{-1/2}$

$$\frac{du}{dx} = -1x^{-2} = -\frac{1}{x^2},\ \frac{dv}{dx} = -\frac{1}{2}x^{-3/2} = -\frac{1}{2x^{3/2}}.$$
$$\frac{dy}{dx} = u\frac{dv}{dx} + v\frac{du}{dx}$$
$$= \frac{1}{x}\left(-\frac{1}{2x^{3/2}}\right) + \frac{1}{x^{1/2}}\left(\frac{1}{x^2}\right)$$
$$= -\frac{1}{2x^{5/2}} - \frac{1}{x^{5/2}} = -\frac{3}{2x^{5/2}}.$$

Alternatively $y = \frac{1}{x^{3/2}} = x^{-3/2}$

$$\frac{dy}{dx} = -\frac{3}{2}x^{-\frac{3}{2}-1} = -\frac{3}{2}x^{-5/2} = -\frac{3}{2x^{5/2}}$$

Ex. 13. $y = (x + 1)(2x + 3)(3x + 4)$

Let $u = (x + 1)(2x + 3)$
$$= 2x^2 + 5x + 3$$

$$\frac{du}{dx} = 4x + 5$$

$$v = (3x + 4), \frac{dv}{dx} = 3$$

$$\frac{dy}{dx} = (x + 1)(2x + 3).\ 3 + (3x + 4)(4x + 5)$$

$$= 6x^2 + 15x + 9 + 12x^2 + 31x + 20$$

$$= 18x^2 + 46x + 29$$

Alternatively if $y = u\ v\ w$

$$\frac{dy}{dx} = uv\frac{dw}{dx} + vw\frac{du}{dx} + wu\frac{dv}{dx}.$$

This will give the same result.

Ex. 14. If u, v are differentiable functions of x and $y = uv$ then prove that

$$\frac{dy}{dx} = \frac{d(uv)}{dx} = u\frac{dv}{dx} + v\frac{du}{dx}$$

Since $\qquad y = uv$

$$y + \Delta y = (u + \Delta u)(v + \Delta v)$$

$$= uv + u\ \Delta v + v\ \Delta u + \Delta u\ \Delta v$$

$\therefore \qquad \Delta y = (y + \Delta y) - y = u\ \Delta v + v\ \Delta u + \Delta u\ \Delta v$

$$\frac{dy}{dx} = \lim_{\Delta x \to 0} \frac{\Delta y}{\Delta x} = \lim_{\Delta x \to 0} \frac{u\Delta v + v\Delta u + \Delta u\,\Delta v}{\Delta x}$$

$$= u \lim_{\Delta x \to 0} \frac{\Delta v}{\Delta x} + v \lim_{\Delta x \to 0} \frac{\Delta u}{\Delta x} + \left(\lim_{\Delta x \to 0} \frac{\Delta u}{\Delta x}\right)\left(\lim_{\Delta x \to 0} \Delta v\right)$$

$$= u\frac{dv}{dx} + v\frac{du}{dx} + \frac{du}{dx} \times 0$$

$\therefore \qquad \dfrac{dy}{dx} = u\dfrac{dv}{dx} + v\dfrac{du}{dx}.$

Ex. 15. If $\quad y = u \cdot \dfrac{1}{v}$ then (see Ex. 18)

$$\frac{dy}{dx} = u \cdot \frac{d}{dx}\left(\frac{1}{v}\right) + \left(\frac{1}{v}\right)\frac{du}{dx}$$

$$= u\left(-\frac{1}{v^2}\right)\frac{dv}{dx} + \frac{1}{v}\frac{du}{dx}$$

$$= \frac{-u\dfrac{dv}{dx} + v\dfrac{du}{dx}}{v^2}$$

$$= \left\{v\frac{du}{dx} - u\frac{dv}{dx}\right\} \Big/ v^2$$

Ex. 16. If $y = \frac{2x^2+3}{x}$

Let $u = 2x^2 + 3,\ v = x$

$$\frac{du}{dx} = 4x, \frac{dv}{dx} = 1.$$

$$\frac{dy}{dx} = \frac{v\frac{du}{dx} - u\frac{dv}{dx}}{v^2}$$

$$= \frac{x(4x) - (2x^2+3).1}{x^2}$$

$$= \frac{2x^2-3}{x^2}$$

Ex. 17. Are the following correct?

(*i*) $\frac{d(uv)}{dx} = \frac{du}{dx}\frac{dv}{dx}$ **Ans.** No

(*ii*) $\frac{d(u/v)}{dx} = \left(\frac{du}{dx}\right) \Big/ \left(\frac{dv}{dx}\right)$ **Ans.** No

(*iii*) $\frac{d}{dx}(u+v) = \frac{du}{dx} + \frac{dv}{dx}$ **Ans.** Yes

The derivative of a product of terms is not the product of the derivatives of the terms. The derivative of a quotient is not the quotient of derivatives.

The derivative of the sum of expressions is the sum of the derivatives of these expressions.

Ex. 18. Prove that if $y = \frac{u}{v}$, $\frac{dy}{dx} = \frac{v\frac{du}{dx} - u\frac{dv}{dx}}{x^2}$. We may use the product rule by writing $y = \frac{u}{v} = (u)\left(\frac{1}{v}\right)$. as done earlier.

Alternatively we may make use of the first principles.

$$y = \frac{u}{v}, y + \Delta y = \frac{u+\Delta u}{v+\Delta v}$$

$$\Delta y = \frac{u+\Delta u}{v+\Delta v} - \frac{u}{v} = \frac{uv + v\Delta u - uv - u\Delta v}{v(v+\Delta v)}$$

$$\therefore \quad \Delta y = \frac{v\Delta u - u\Delta v}{(v^2 + v\Delta v)}$$

$$\frac{dy}{dx} = \lim_{\Delta x \to 0} \frac{v\Delta u - u\Delta v}{(v^2 + v\Delta v)\Delta x}$$

$$= \lim_{\Delta x \to 0} \frac{v\frac{\Delta u}{\Delta x} - u\frac{\Delta v}{\Delta x}}{v^2 + v\Delta v}$$

$$= \frac{v \lim\limits_{\Delta x \to 0} \frac{\Delta u}{\Delta x} - u \lim\limits_{\Delta x \to 0} \frac{\Delta v}{\Delta x}}{v^2 + v \lim\limits_{\Delta x \to 0} \Delta v}$$

$$= \frac{v\frac{du}{dx} - u\frac{dv}{dx}}{v^2 + v \times 0}$$

$$\therefore \quad \frac{dy}{dx} = \frac{v\frac{du}{dx} - u\frac{dv}{dx}}{v^2}.$$

Ex. 19. If $y = x^n$ where $n = p/q$, q any positive integer and p any integer, positive negative or zero, then

we may write $\frac{dy}{dx} = \frac{p}{q} x^{p/q-1}$

Alternatively, write $y = x^{p/q}$

$$\therefore \quad y^q = x^p \text{ and } \frac{d}{dx} y^q = \frac{d}{dx} x^p$$

$$\therefore \quad qy^{q-1} \frac{dy}{dx} = p\, x^{p-1} \text{ or}$$

$$\frac{dy}{dx} = \frac{px^{p-1}}{qy^{q-1}} = \frac{p}{q} \frac{x^{p-1}}{(x^{p/q})^{q-1}} = \frac{p}{q} x^{p/q-1}.$$

Ex. 20. If $y = u^n$ where $u = f(x)$

$$\frac{dy}{dx} = nu^{n-1} \frac{du}{dx}.$$

Ex. 21. Show that if (*i*) $y = (x^2 + 1)(2x^3 + 3x^2)$

(*ii*) $y = \frac{x^2+1}{x^2-1}$, then

(*i*) $\frac{dy}{dx} = 6(x^2+1)^3 + (2x+3)$ (*ii*) $\frac{dy}{dx} = \frac{-4x}{(x-1)^2}$.

Ex. 22. Differentiate with respect to x :

(*i*) $(x^2 + 1)^3 (x^3 - 1)^2$ (*ii*) $(x^2 + 1)^2 (x + 1)^{-3}$

(*iii*) $\frac{x}{x^2+1}$ (*iv*) $\left(x + \frac{1}{x}\right)^2$

(*v*) $\frac{x^2}{1+x}$

Ans. (*i*) $6x(x^2+1)^2(x^3-1)(2x^3+x-1)$

(*ii*) $-2(2x^2+3x-1)(x+1)(x^2+1)^{-4}$ (*iii*) $(1-x^2)(x^2+1)^{-2}$

(*iv*) $2\left(x+\frac{1}{x}\right)\left(1-\frac{1}{x^2}\right)=2\left(x-\frac{1}{x^3}\right)$ (*v*) $\frac{x^2+2x}{(1+x)^2}$

Ex. 23. If $y=\frac{1+x}{(1-x)^{1/3}}$, $\frac{dy}{dx}=(1-x)^{-1/3}+\frac{1+x}{3(1-x)^{-4/3}}$

Ex. 24. Differentiate with respect to x :

$$y=\frac{x+1}{\sqrt{x}}$$

We can differentiate this function in two ways :

(1) By Quotient Rule

$$y=\frac{x+1}{\sqrt{x}}=\frac{u}{v} \text{ where } u=x+1,\ v=\sqrt{x}.$$

$$\frac{dy}{dx}=\frac{1}{v^2}\left(v\frac{du}{dx}-u\frac{dv}{dx}\right)$$

$$=\frac{1}{(\sqrt{x})^2}\left(\sqrt{x}\frac{d}{dx}(x+1)-(x+1)\frac{d\sqrt{x}}{dx}\right)$$

$$=\frac{1}{x}\left\{\sqrt{x}.1-(x+1)\times\tfrac{1}{2}x^{-1/2}\right\}$$

$$=\frac{1}{x}\left(\sqrt{x}-\frac{x+1}{2\sqrt{x}}\right)=\frac{1}{x}\frac{2x-x-1}{2\sqrt{x}}=\frac{x-1}{2x\sqrt{x}}.$$

(2) Alternatively $y=\frac{x+1}{\sqrt{x}}=\sqrt{x}+\frac{1}{\sqrt{x}}$

$$\frac{dy}{dx}=\frac{1}{2}x^{-1/2}-\frac{1}{2}x^{-3/2}=\frac{x-1}{2x\sqrt{x}}.$$

Ex. 25. $y=\frac{x-1}{\sqrt{x+1}}$

Let $u=x-1;\ \frac{du}{dx}=1;\ v=(x+1)^{1/2},$

$$\frac{dv}{dx}=\frac{1}{2}(x+1)^{-1/2}=\frac{1}{2(x+1)^{1/2}}$$

$$\frac{dy}{dx} = \frac{v\frac{du}{dx} - u\frac{dv}{dx}}{u^2}$$

$$= \frac{(x+1)^{\frac{1}{2}} \cdot 1 - (x-1)\frac{1}{2}(x+1)^{-1/2}}{x+1}$$

$$= \frac{x+3}{2(x+1)^{3/2}}$$

Ex. 26. $y = \sqrt{x^3 - 5x + 3}$

Let $u = x^3 - 5x + 3, \ \frac{du}{dx} = 3x^2 - 5, \ \frac{dy}{du} = \frac{1}{2}u^{-1/2}$

We know that

If $y = u^n, \ \frac{dy}{dx} = n u^{n-1} \frac{du}{dx}$

$\therefore$ If $y = u^{1/2}, \ \frac{dy}{dx} = \left(\frac{1}{2}u^{-1/2}\right)\left(\frac{du}{dx}\right)$

$\therefore$ $$\frac{dy}{dx} = \left(\frac{1}{2\sqrt{x^3 - 5x + 3}}\right)(3x^2 - 5)$$

$$= \frac{3x^2 - 5}{2\sqrt{(x^3 - 5x + 3)}}$$

Ex. 27. If $y = (x^2 - 5x + 6)^{1/3}(x^2 + 1)^{2/3}$

$$\frac{dy}{dx} = \frac{6x^3 - 25x^2 + 26x - 5}{3(x^2+1)^{1/3}(x^2 - 5x + 6)^{2/3}}.$$

Ex. 28. If $y = \frac{3 - x^2}{\sqrt{x^2 - 6x + 2}}, \ \frac{dy}{dx} = \frac{9 - 7x + 9x^2 - x^3}{(x^2 - 6x + 2)^{3/2}}$

Ex. 29. If $y = \sqrt[3]{(1 + x^2)^2}, \ \frac{dy}{dx} = \frac{4x}{3}(1 + x^2)^{-1/3}.$

Ex. 30. If $y = \frac{\sqrt{x}}{2x + 3}, \ \frac{dy}{dx} = \frac{3 - 2x}{2\sqrt{x}} \frac{1}{(2x + 3)^2}.$

Ex. 31. If $y = \frac{x^2 + x + 1}{x^2 - x + 1}, \ \frac{dy}{dx} = \frac{2 - 2x^2}{(x^2 - x + 1)^2}$

Ex. 32. Find $\frac{dx}{dy}$ for some of the above cases where you can solve for x and verify $\frac{dy}{dx} \cdot \frac{dx}{dy} = 1.$

Composite Functions

Differentiation of expressions involving functions of different variables is carried out by the *chain rule.*

Let $y = f(x)$ where y is a function of x

$z = g(y)$ z is a function of y

then $$\frac{dz}{dx} = \frac{dz}{dy} \cdot \frac{dy}{dx}$$

Here $z = g(y) = g[f(x)]$ is a composite function. This is also written $(g \circ f)(x)$.

We can extend this result to the case of three or more variables.

If $y = f(x),\ z = g(y),\ w = h(z)$

then $$\frac{dw}{dx} = \frac{dw}{dz} \cdot \frac{dz}{dy} \cdot \frac{dy}{dx}.$$

Ex. 33. $y = 5x^2$, $z = 7y + 3$

$$\frac{dy}{dx} = 10x, \quad \frac{dz}{dy} = 7$$

$$\therefore \quad \frac{dz}{dx} = \frac{dz}{dy} \cdot \frac{dy}{dx} = 7 \times 10x = 70x.$$

Ex. 34. $x = 10p - p^2$, $p = 0.7y - 2$

$$\frac{dx}{dp} = 10 - 2p \quad \frac{dp}{dy} = 0.7$$

$$\therefore \quad \frac{dx}{dp} = 0.7\ (10 - 2p).$$

Ex. 35. If $x = at^2$, $y = 2at$ find $\frac{dy}{dx}$ **Ans.** $1/t$

Ex. 36. If $y = x^3 - 3x^2 + 5x - 4$ where $x = z + z^2$

Show that $\frac{dy}{dz} = [3(z + z^2)^2 - 6(z + z^2) + 5](2z + 1)$.

Ex. 37. If $x = 2t + 3$, $y = t^2 - 1$ show that $\frac{dy}{dx} = \frac{x-3}{2}$.

Ex. 38. $y = (1 + x^2)^2$, we may put $u = 1 + x^2$ so that

$$y = u^2, \ \frac{du}{dx} = 2u, \ \frac{dy}{dx} = 3x^2$$

$$\frac{dy}{dx} = \frac{dy}{du} \cdot \frac{du}{dx} = (2u)\ (3x^2) = 2(1+x^2).\ 3x^2 = 6x^2\ (1+x^2)$$

Ex. 39. If $y = f(u)$, $u = g(x)$, $y = f(g(x)$ then

$$\frac{dy}{dx} = \frac{dy}{du}\frac{du}{dx} = f'(u)g'(x)$$

Note. $(f \circ g') = (f' \circ g)g'$ and $f'(g(x')) = f'(g(x))\ g'(x))$.

Ex. 40. If $y = (3x^2 + 6x - 5)^{3/2}$, put $u = 3x^2 + 6x - 5$

$$\frac{dy}{du} = \frac{du^{3/2}}{du} = \frac{3}{2}u^{1/2} \text{ and } \frac{du}{dx} = 6x + 6$$

$$\frac{dy}{dx} = \frac{3}{2}(3x^2 + 6x - 5)^{1/2}(6x + 6) = 9(3x^2 + 6x - 5)^{1/2}.$$

Ex. 41. If $y = (3x^2 + 6x - 5)^{2/3}$, $\dfrac{dy}{dx} = \dfrac{4(x+1)}{(3x^2 + 6x - 5)^{1/3}}$.

Ex. 42. Differentiate (*i*) $x^2 (1 - 2x)^{1/3}$ (*ii*) $\dfrac{x}{\sqrt{4 - 9x^2}}$. w.r.t. x.

Ans. (*i*) $\dfrac{2x(3 - 7x)}{3(1 - 2x^2)^{2/3}}$ (*ii*) $\dfrac{4}{(4 - 9x^2)^{3/2}}$

The Inverse Relation *dx*/*dy*

If $x = F(y)$ we may find $y = f(x)$ and obtain $\dfrac{dy}{dx}$

Alternatively, we may find

$$\frac{dx}{dy} \text{ and obtain } \frac{dy}{dx} = \frac{1}{dx/dy}$$

If $x = y^2$ then $y^2 = x$ or $2y\dfrac{dy}{dx} = 1$,

$$\therefore \quad \frac{dy}{dx} = \frac{1}{2y} = \frac{1}{2\sqrt{x}}$$

Alternatively, $\dfrac{dx}{dy} = 2y = 2\sqrt{x}$

$$\therefore \quad \frac{dy}{dx} = \frac{1}{2\sqrt{x}}.$$

Ex. 43. If $x^2 + y^2 = 9$

We have $y = \pm\sqrt{9 - x^2}$ and $\dfrac{dy}{dx} = \mp\dfrac{x}{\sqrt{9 - x^2}}$

Alternatively $x^2 = 9 - y^2$ so that $2x\dfrac{dx}{dy} = 2y$ or $\dfrac{dx}{dy} = -\dfrac{y}{x} = \mp\dfrac{\sqrt{9 - x^2}}{x}$

and

$$\frac{dy}{dx} = \mp\frac{x}{\sqrt{9 - x^2}}$$

Ex. 44. If $xy = 1$ find $\dfrac{dy}{dx}$ **Ans.** $-1/x^2$

10.3 IMPLICIT DIFFERENTIATION

It may sometimes be difficult to solve an equation explicitly for y in terms of x.

Given the implicit function $f(x, y) = 0$, we may differentiate it term by term and then solve the resulting equation for $\frac{dy}{dx}$.

Ex. 45. $x^2 + y^2 = 1$. We differentiate this term by term

$$2x + 2y\frac{dy}{dx} = 0 \quad \therefore \quad \frac{dy}{dx} = -\frac{x}{y}.$$

Ex. 46. $x^2y^3 + 2xy + y + 2 = 0$

$$x^2 \cdot 3y^2\frac{dy}{dx} + y^3 \cdot 2x + \left(2x\frac{dy}{dx} + 2y\right) + \frac{dy}{dx} = 0$$

$$\therefore \quad \frac{dy}{dx}(3x^2y^2 + 2x + 1) + 2xy^3 + 2y = 0$$

$$\therefore \quad \frac{dy}{dx} = -\frac{2xy^3 + 2y}{3x^2y^2 + 2x + 1}.$$

Ex. 47. Find $\frac{dy}{dx}$ by implicit differentiation

(*i*) $x^2 + y^2 = a^2$ (*ii*) $4x^3 + x^2y - 3xy^2 - 5y^3 = 7$

(*iii*) $6x^4 - 3xy^3 + 2xy - x + 7y = 10$ (*iv*) $6x^2 = y^3$

(*v*) $x^{1/2} + y^{1/2} = a^{1/2}$

Ans. (*i*) $-\frac{y}{x}$ (*ii*) $-\frac{12x^2 + 2xy - 3y^2}{x^2 - 6xy - 15y^2}$ (*iii*) $\frac{24x^3 - 3y^3 + 2y - 1}{9xy^2 - 2x - 7}$

(*iv*) $4x/y^2$ (*v*) $-(y/x)^{1/2}$

Find $\frac{dx}{dy}$ in the above cases and verify that $\frac{dy}{dx} \cdot \frac{dx}{dy} = 1$.

Ex. 48. For the relations (*i*) $(x - y)^2 = (2x + 3)^2$ (*ii*) $6x\sqrt{x-2} - 4(x - 2)^{3/3} = y$

Show $\frac{dy}{dx}\frac{dx}{dy} = 1$. **Ans.** (*i*) $2(x - y)\left(1 - \frac{dy}{dx}\right) = 2\,(2x + 3).2$

$$\frac{x - y}{4x + 6}\left(1 - \frac{dy}{dx}\right) = 1.$$

$$\therefore \quad \frac{dy}{dx} = \frac{4x + 6}{x - y} - 1 \text{ or } \frac{dy}{dx} = \frac{3x + y + 6}{y - x}$$

We can similarly find dx/dy

$$2(x - y)\left(\frac{dx}{dy} - 1\right) = 2(2x + 3)2\frac{dx}{dy}.$$

$$(x-y)\frac{dx}{dy} - 2(2x+3)\frac{dx}{dy} = x-y$$

$$(-3x-y-6)\frac{dx}{dy} = x-y \quad \therefore \quad \frac{dx}{dy} = \frac{y-x}{3x+y+6}$$

$$\therefore \quad \frac{dy}{dx}\frac{dx}{dy} = 1.$$

Ex. 49. At the point (2, – 2) on the curve $x^2 + xy + y^2 = y$ find the slopes of the tangent and the normal. (*i.e.*, dy/dx and $-dx/dy$) **Ans.** 1, – 1

Ex. 50. For the curve $\sqrt{x} + \sqrt{y} = \sqrt{a}$ show that the sum of intercepts of the tangent at (p, q) on the axes is constant

The slope at (p, q) is $\frac{dy}{dx} = -\frac{\sqrt{q}}{\sqrt{p}}$ and the tangent is $\frac{x}{\sqrt{p}} + \frac{y}{\sqrt{q}} = \sqrt{a}$

Sum of intercepts = $\sqrt{ap} + \sqrt{aq}$ = constant.

Ex. 51. Let the sum to n terms of a GP be $y_n = a\,\frac{r^n - 1}{r-1}$ where r is the common ratio.

Show that $(r-1)\,\frac{dy_n}{dr} = (n-1)\,y_n - ny_{n-1}$

$$\frac{dy_n}{dr} = a\frac{(r-1)nr^{n-1} - (r^n - 1)}{(r-1)^2} = \frac{a}{(r-1)^2}[nr^n - nr^{n-1} - r^n + 1]$$

$$= \frac{a}{(r-1)^2}[(n-1)\,(r^n - 1) - n(r^n - 1)] = \frac{1}{r-1}[(n-1)y_n - ny_{n-1}].$$

Parametric Form

The derivatives of functions in parametric form can be found either by eliminating the parameters or using the chain rule.

If $\quad x = f(t),\ y = g(t)$ where t is a parameter,

then the elimination of t gives $h(x, y) = 0$ whence dy/dx can be obtained. Alternatively

$$\frac{dy}{dx} = \frac{dy}{dt}\Big/\frac{dx}{dt}.$$

Ex. 52. Find $\frac{dy}{dx}$ when $x = \frac{1}{1-2t}$ and $y = \sqrt{2t+1}$ at $t = 2$

$$\frac{dx}{dt} = \frac{+2}{(1-2t)^2},\ \frac{dy}{dt} = \frac{1}{\sqrt{2t+1}}$$

$$\frac{dy}{dx} = \frac{dy}{dt} \div \frac{dx}{dt} = \frac{1}{2}\frac{(1-2t)^2}{\sqrt{1+2t}} = \frac{1}{2}\frac{(1-2\times 2)^2}{\sqrt{1+4}} = \frac{9}{2\sqrt{5}} \text{ at } t = 2.$$

Ex. 53. $x = t + \frac{1}{t}$, $y = t - \frac{1}{t}$

$$\frac{dx}{dt} = 1 - \frac{1}{t^2}, \quad \frac{dy}{dt} = 1 + \frac{1}{t^2} \quad \frac{dy}{dx} = \frac{t^2+1}{t^2-1}.$$

Ex. 54. $x = \frac{1}{t+1}$, $y = \sqrt{5-t^2}$, $\frac{dy}{dx} = \sqrt{\frac{t^2}{5-t^2}}\,(1+t^2)$.

Other Forms

We are now in a position to find the derivatives of :

(*a*) Polynomial Functions in the form

$$y = a_0 x^n + a_1 x^{n-1} + \ldots + a_n$$

or

$$y = (a_0 x^n + a_1 x^{n-1} + \ldots + a_n)(b_0 x^m + b_1 x^{m-1} + \ldots + b_m)$$

which possess derivatives for all values of x.

(*b*) Rational Functions of the form

$$y = \frac{a_0 x^n + a_1 x^{n-1} + \ldots + a_n}{b_0 x^m + b_1 x^{m-1} + \ldots + b_m}$$

which possesses derivatives except when the denominator becomes zero.

(*c*) Implicit or algebraic functions like

$$f_0(x)\, y^n + f_1(x) y^{n-1} + f_2(x) y^{n-2} + \ldots + f_n(x) = 0$$

where $f_i(x)$ are polynomial functions.

Other types of functions that will be considered are

(*i*) Exponential and Logarithmic Functions

(*ii*) Trigonometric Functions

(*iii*) Hyperbolic Functions

10.4 LOGARITHMIC FUNCTIONS

The natural logarithm is written $ln\, x$ for all positive x. Its derivative is $1/x$.

(*a*) If $y = ln\, x$, $\frac{dy}{dx} = \frac{1}{x}$

(*b*) If $y = ln\, u$ where u is a positive differentiable function of x then

$$\frac{d}{dx}(ln\, u) = \frac{d}{du}(ln\, u) \cdot \frac{du}{dx} = \frac{1}{u}\frac{du}{dx}.$$

(*c*) If $y = ln\, f(x)\, g(x) = ln\, f(x) + ln\, g(x)$

$$\frac{dy}{dx} = \frac{d}{dx} ln\, f(x) + \frac{d}{dx} ln\, g(x)$$

$$= \frac{1}{f(x)} f'(x) + \frac{1}{g(x)} g'(x)$$

$$= \frac{f'(x)}{f(x)} + \frac{g'(x)}{g(x)}$$

(*d*) If $y = ln\dfrac{f(x)}{g(x)} = ln\ f(x) - ln\ g(x)$

$$\frac{dy}{dx} = \frac{f'(x)}{f(x)} - \frac{g'(x)}{g(x)}.$$

(*e*) If $y = \{ln\ f(x)\}^n = n\ ln\ f(x)$

$$\frac{dy}{dx} = \frac{n}{f(x)} f'(x) = n\frac{f'(x)}{f(x)}.$$

(*f*) If $y = \log_a u$, we can write $y = \dfrac{ln\ u}{ln\ a}$

$$\frac{dy}{dx} = \frac{1}{ln\ a}\frac{1}{u}\frac{du}{dx}$$

The derivative of a logarithmic function can be written only if the base of the logarithm is *e*.

Ex. 55. $y = ln\ x^2,\ \dfrac{dy}{dx} = \left(\dfrac{1}{x^2}\right)\cdot 2x = \dfrac{2}{x}$

or since $y = 2\,ln\ x,\ \dfrac{dy}{dx} = 2.\dfrac{1}{x} = \dfrac{2}{x}.$

Ex. 56. $y = \log_a x^2 = \dfrac{1}{ln\ a} ln\ x^2$

$$\frac{dy}{dx} = \left(\frac{1}{ln\ a}\right)\frac{2}{x}$$

Ex. 57. $y = \log_{10} x = \dfrac{\log_e x}{\log_e 10} = \dfrac{ln\ x}{ln\ 10}$

$$\frac{dy}{dx} = \frac{1}{ln\ 10}\cdot\frac{1}{x}$$

Ex. 58. $y = ln\dfrac{\sqrt{1+x} - \sqrt{1-x}}{\sqrt{1+x} + \sqrt{1-x}}$. The devivative may be found by using the $\dfrac{u}{v}$ formula. Alternatively,

Rationalising $y = ln\dfrac{(\sqrt{1+x} - \sqrt{1-x})^2}{(\sqrt{1+x} + \sqrt{1-x})\,(\sqrt{1+x} - \sqrt{1-x})}$

$$= ln\frac{2 - 2\sqrt{1-x^2}}{2x} = ln\ (1 - \sqrt{1-x^2}) - ln\ x$$

$$\therefore \quad \frac{dy}{dx} = \frac{-\dfrac{1}{2\sqrt{1-x^2}}\dfrac{d}{dx}(1-x^2)}{1-\sqrt{1-x^2}} - \frac{1}{x} = \frac{\dfrac{-2x}{-2\sqrt{1-x^2}}}{1-\sqrt{1-x^2}} - \frac{1}{x}$$

$$= \frac{x}{\sqrt{1-x^2}\,(1-\sqrt{1-x^2})} - \frac{1}{x} = \frac{1}{x\sqrt{1-x^2}}.$$

Ex. 59. $y = \ln \ln x, \ \dfrac{dy}{dx} = \dfrac{1}{\ln x} \cdot \dfrac{1}{x} \cdot = \dfrac{1}{x \ln x}$

Ex. 60. $y = 2 \ln (1 + \sqrt{x}), \ \dfrac{dy}{dx} = \dfrac{1}{x + \sqrt{x}}.$

Ex. 61. $y = \ln \ln \ln x, \ \dfrac{dy}{dx} = \ln x \ln (\ln x).$

Ex. 62. From the first principles if $y = \log_a x$

$$\Delta y = \log_a (x + \Delta x) - \log_a (x)$$

$$= \log_a\left(\frac{x+\Delta x}{\Delta x}\right) = \log_a\left(1 + \frac{\Delta x}{x}\right)$$

$$\frac{\Delta y}{\Delta x} = \frac{1}{\Delta x}\log_a\left(1+\frac{\Delta x}{x}\right) = \frac{1}{x}\frac{x}{\Delta x}\log_a\left(1+\frac{\Delta x}{x}\right)$$

$$= \frac{1}{x}\log_a\left(1+\frac{\Delta x}{x}\right)^{x/\Delta x}$$

$$\lim_{\Delta x \to 0} \frac{\Delta y}{\Delta x} = \frac{1}{x}\log_a e.$$

Ex. 63. If $y = x^3 \log_a 2x = u \cdot v$

$$\frac{dy}{dx} = x^3 \frac{2}{2x}\log_a e + (\log_a 2x)3x^2$$

$$= x^2 (3 \log_a 2x + \log_a e).$$

Ex. 64. (*i*) $y = e^{(2x+1)/x}$ (*ii*) $y = x^2 e^x$

Ans. (*i*) $\dfrac{dy}{dx} = \dfrac{-e^{(2x+1)/x}}{x^2}$ (*ii*) $\dfrac{dy}{dx} = (x^2 + 2x)e^x$

Ex. 65. If $y = e^{-x}, \ \dfrac{dy}{dx} + y = 0.$

Ex. 66. Let $x^y = e^{x-y}$ then $x^y e^y = e^x$.

$\therefore \quad (ex)^y = e^x$ and $y \ln e + y \ln x = x$ or $y + y \ln x = x$

$$\therefore \quad \frac{dy}{dx} + \left(y\frac{1}{x} + \ln x\frac{dy}{dx}\right) = 1$$

Substituting for y, $\dfrac{dy}{dx} = \ln x\,(1 + \ln x)^{-2}$

Exponential Functions

The inverse of $y = \ln x$ is the exponential, written

$x = e^y$ or exp y.

The graph of $y = e^x$ is the inverse image of the graph of $y = \ln x$, the mirror being the line $y = x$. Every point (α, β) on one graph has the image (β, α) in the other.

Since $y = e^x$

The domain of e^x is the range of $\ln x$

The range of e^x is the domain of $\ln x$.

The Exponential

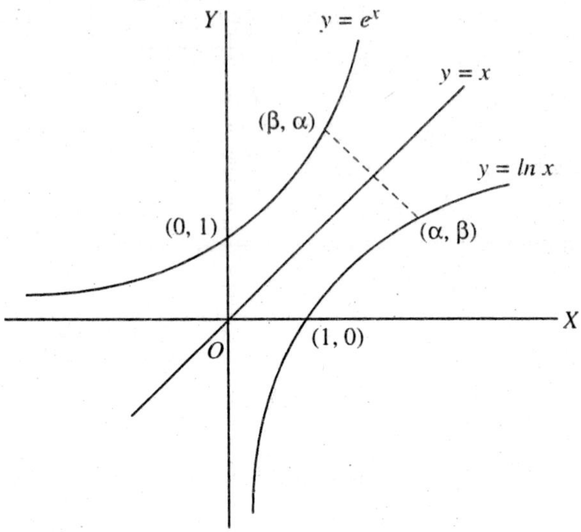

(g) If $y = \ln x$ for all $x > 0$

$$\frac{dy}{dx} = \frac{1}{x} > 0 \text{ for all } x > 0$$

The derivative of $\ln x$ is always positive

$\therefore$ $\ln x$ is an increasing function of x.

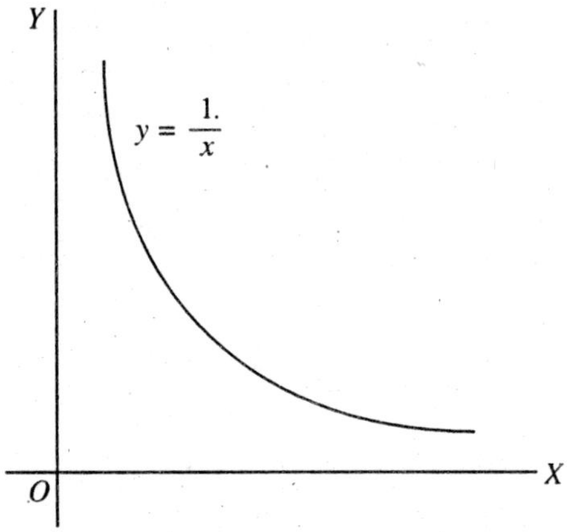

The graph of

$$y = \ln x$$

meets the x-axis at (1, 0)

where $$\frac{dy}{dx} = \frac{1}{x} = \frac{1}{1} = 1$$

$\therefore$ The slope of $y = \ln x$ at (1, 0) is 1 *i.e.*, The tangent at (1, 0) is at 45° to the x-axis.

As $x \to \infty$, $\ln x \to \infty$.

As $x \to 0$ through positive values

$$\ln x = -\ln \frac{1}{x} \to -\infty$$

The domain is $(0, \infty)$

The range is $(-\infty, \infty)$.

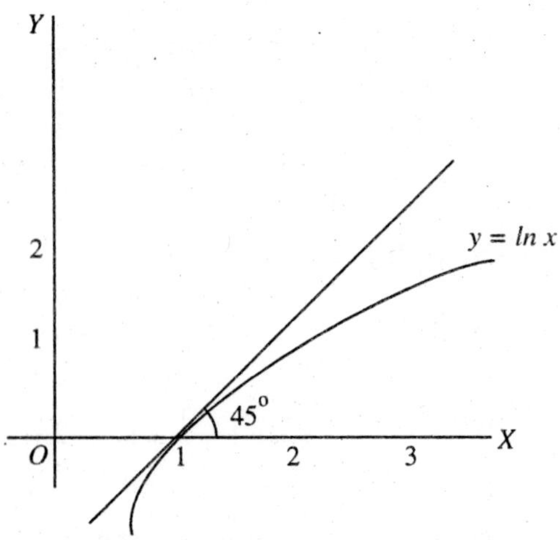

Other Exponentials

(*h*) Let $y = a^x$ where a is any real positive number and x is any real number

We can see that $y = a^x = e^{x \ln a}$

and $\ln y = x \ln a$

If y is any positive number then $\ln y = x \ln a$ takes all real values between $-\infty$ and ∞ (a being a positive number $\neq 1$).

The mapping of $y = a^x$ $(-\infty < x < \infty)$ is one to one from the domain of all real numbers x onto the range of positive a^x.

For each positive real $y = a^x$ there exists one real number x when $a > 0$ $(\neq 1)$.

Again $y = \ln_a x = \dfrac{\ln x}{\ln a}$ if a is positive real, $a \neq 1$, x real.

Also
$$\lim_{x\to 0} \frac{a^x - 1}{x} = \ln a.$$

Since
$$\lim_{x\to 0} \frac{a^x - 1}{x} = \lim_{x\to 0} \frac{z}{\log_a (1+z)} \quad \text{where } z = a^x - 1$$

$$= \lim_{z\to 0} \frac{1}{\frac{1}{z}\log_a (1+z)} = \lim_{z\to 0} \frac{1}{\log_a (1+z)^{1/z}} = \frac{1}{\log_a e} = \ln a$$

(*i*) $y = e^x$ implies $x = \ln y$

Since
$$\frac{dx}{dy} = \frac{1}{y} = \frac{1}{e^x}$$

$$\therefore \quad \frac{dy}{dx} = e^x.$$

(*j*) If $y = e^u, \dfrac{dy}{dx} = e^u \dfrac{du}{dx}.$

(*k*) If $y = a^u = e^{u \ln a}$, $\dfrac{dy}{dx} = e^{u \ln a} \dfrac{d}{dx}(u \ln a)$

$$= e^{u \ln a}(\ln a)\frac{du}{dx}.$$

(*l*) If $y = a^x = e^{x \ln a}$, $\dfrac{dy}{dx} = e^{x \ln a} \dfrac{d}{dx}(x \ln a)$

$$= e^{x \ln a} (\ln a) = a^x \ln a.$$

(*m*) If $y = e^{a x}$, $\dfrac{dy}{dx} = ae^{a x}.$

Ex. 67. Find the derivative of $y = a^x$ using the first principles.

Since $\quad y + \Delta y = a^{x + \Delta x}$, $\Delta y = a^{x + \Delta x} - a^x = a^x (a^{\Delta x} - 1)$

$$\frac{\Delta y}{\Delta x} = a^x \frac{a^{\Delta x} - 1}{\Delta x}$$

$$\lim_{\Delta x \to 0} \frac{\Delta y}{\Delta x} = a^x \lim_{\Delta x \to 0} \frac{a^{\Delta x} - 1}{\Delta x} = a^x \cdot ln\ a.$$

Similarly if $y = e^x$, $\Delta y = e^{x + \Delta x} - e^x = e^x(e^{\Delta x} - 1)$

$$\frac{\Delta y}{\Delta x} = e^x \left(\frac{e^{\Delta x} - 1}{\Delta x} \right)$$

$$\frac{dy}{dx} = e^x \lim_{\Delta x \to 0} \frac{e^{\Delta x} - 1}{\Delta x} = e^x\ ln\ e = e^x.$$

Ex. 68. If $y = u^v$ where u, v are function of x

$$ln\ y = v\ ln\ u$$

$$\therefore \quad \frac{1}{y}\frac{dy}{dx} = v \cdot \frac{d}{dx} ln\ u + ln\ u \cdot \frac{dv}{dx}$$

$$= v \cdot \frac{1}{u} \cdot \frac{du}{dx} + ln\ u \cdot \frac{dv}{dx}$$

$$\therefore \quad \frac{dy}{dx} = u^v \left(\frac{v}{u}\frac{du}{dx} + ln\ u \frac{dv}{dx} \right)$$

$$= v u^{v-1} \frac{du}{dx} + u^v ln\ u \frac{dv}{dx}.$$

Ex. 69. If $y = u^x$ where u is a function of x

$$ln\ y = x\ ln\ u$$

$$\therefore \quad \frac{1}{y}\frac{dy}{dx} = x \cdot \frac{1}{u}\frac{du}{dx} + ln\ u$$

$$\therefore \quad \frac{dy}{dx} = u^x \left(\frac{x}{u}\frac{du}{dx} + ln\, u \right)$$

$$= xu^{x-1} + u^x\ ln\ u.$$

Ex. 70. If $y = x^u$, $ln\ y = u\ ln\ x$

$$\frac{1}{y}\frac{dy}{dx} = u \cdot \frac{1}{x} + ln\, x \cdot \frac{du}{dx}$$

$$\therefore \quad \frac{dy}{dx} = x^u \left(\frac{u}{x} + ln\, x \frac{du}{dx} \right)$$

$$= ux^{u-1} + x^u\ ln\, x \frac{du}{dx}.$$

Ex. 71. Find $\frac{dy}{dx}$ when (*i*) $y = (1 + x^2)\, e^{ax}$ (*ii*) $y = 2x\ ln\ x^2$ Take $y = uv$

(*i*) $$\frac{dy}{dx} = (1 + x^2)\frac{d}{dx}e^{ax} + e^{ax}\frac{d}{dx}(1 + x^2)$$

$$= (1 + x^2)\, ae^{ax} + e^{ax} \cdot 2x.$$

(*ii*) We can write $y = 4x \ln x$

$$\frac{dy}{dx} = 4\left[x \cdot \frac{1}{x} + \ln x \cdot 1\right]$$

$$= 4\,[1 + \ln x]$$

Ex. 72. Differentiate with respect to x

(*i*) $\ln\sqrt{x^2 - 1}$ (*ii*) $\ln\frac{x}{x+1}$ (*iii*) $\sqrt{\ln x}$

(*iv*) $a^x a^a$ (*v*) $x^2 e^x$ (*vi*) $(1 + x)e^{-2x}$

(*vii*) $\frac{e^x - e^{-x}}{e^x + e^{-x}}$ (*viii*) $\frac{ax-1}{a^2}e^{ax}$ (*ix*) $e^x \ln x$

Ans. (*i*) $\frac{x}{x^2-1}$ (*ii*) $\frac{1}{x(x+1)}$ (*iii*) $\frac{1}{2x\sqrt{\ln x}}$ (*iv*) $a^x a^{a-1}(a + x \ln a)$

(*v*) $(x^2 + 2x)e^x$ (*vi*) $-4xe^{-2x}$ (*vii*) $\frac{4}{(e^x + e^{-x})^2}$ (*viii*) xe^{ax} (*ix*) $e^x\left(\ln x + \frac{1}{x}\right)$

Ex. 73. If $y = \ln(u^3 - 1)$ and $u = x^2 - 1$

$$\frac{dy}{dx} = \frac{6x(x^2-1)^2}{(x^2-1)^3 - 1}.$$

Ex. 74. If $v = \ln(y + 4)$ and $y = x^2$

then
$$\frac{dv}{dx} = \frac{2x}{x^2+4}.$$

Ex. 75. If
$$y = x^{x(2x^3+1)}$$

then
$$\frac{dy}{dx} = x^{x(2x^3+1)}(2x^3 + 1 + \ln x + 8x^3 \ln x).$$

10.5 DERIVATIVES OF TRIGONOMETRIC FUNCTIONS

We shall begin with the use of the First Principles.

(*i*) $y = \sin x,\ y + \Delta y = \sin(x + \Delta x)$

$$\Delta y = \sin(x + \Delta x) - \sin x = 2\cos\left(x + \frac{\Delta x}{2}\right)\sin\frac{\Delta x}{2}$$

$$\lim_{\Delta x \to 0}\frac{\Delta y}{\Delta x} = \lim_{\Delta x \to 0} 2\cos\left(x + \frac{\Delta x}{2}\right) \cdot \frac{\sin \Delta x/2}{\Delta x}$$

$$= \lim_{\Delta x \to 0}\cos\left(x + \frac{\Delta x}{2}\right) \cdot \lim_{\Delta x \to 0}\frac{\sin \Delta x/2}{\Delta x/2}$$

$$= \cos x.\ 1 = \cos x \quad \left(\text{Using the rule } \lim_{z \to 0}\frac{\sin z}{z} = 1\right)$$

$$\therefore \quad \frac{dy}{dx} = \cos x.$$

(*ii*) $y = \cos x$ so that $\Delta y = \cos(x + \Delta x) - \cos x = -2\sin\left(x + \frac{\Delta x}{2}\right)\sin\frac{\Delta x}{2}$

$$\lim_{\Delta\to 0}\frac{\Delta y}{\Delta x} = \lim_{\Delta x\to 0}\left\{-2\sin\left(x+\frac{\Delta x}{2}\right)\frac{\sin\Delta\, x/2}{\Delta x}\right\}$$

$$= \lim_{\Delta x\to 0}\sin\left(x+\frac{\Delta x}{2}\right)\cdot\lim_{\Delta x\to 0}\frac{\sin\Delta\, x/2}{\Delta\, x/2}$$

$$= -\sin x\cdot 1$$

$$\frac{dy}{dx} = -\sin x.$$

(*iii*) $y = \tan x$, $\Delta y = \tan(x + \Delta x) - \tan x$

We know that $\tan(A - B) = \dfrac{\tan A - \tan B}{1 + \tan A\tan B}$

$\therefore\ \tan A - \tan B = \tan(A - B)(1 + \tan A\tan B)$

Put $A = x + \Delta x$, $B = x$ so that

$$\tan(x + \Delta x) - \tan x = \tan\Delta x\,[1 + \tan(x + \Delta x)\tan x]$$

$$\lim_{\Delta x\to 0}\frac{\Delta y}{\Delta x} = \lim_{\Delta x\to 0}\left[\frac{\tan\Delta x}{\Delta x}\{1+\tan(x+\Delta x)\tan x\}\right]\quad\left(\textbf{Note.}\ \lim_{z\to 0}\frac{\tan z}{z} = 1\right)$$

$$= \left\{\lim_{\Delta x\to 0}\frac{\tan\Delta x}{\Delta x}\right\}\lim_{\Delta x\to 0}\{1+\tan(x+\Delta x)\tan x\} = 1.\ \{1+\tan^2 x\} = \sec^2 x$$

(*iv*) $y = \operatorname{cosec} x$ (*v*) $y = \sec x$ (*vi*) $y = \cot x$

Ans. (*iv*) $\dfrac{dy}{dx} = -\operatorname{cosec} x\cot x$; (*v*) $\dfrac{dy}{dx} = \sec x\tan x$; (*vi*) $\dfrac{dy}{dx} = -\operatorname{cosec}^2 x$

Ex. 76. $y = \cos\sqrt{x}$ and $\Delta y = \cos\sqrt{x+\Delta x} - \cos\sqrt{x}$

$$\frac{dy}{dx} = \lim_{\Delta x\to 0}\frac{\cos\sqrt{x+\Delta x} - \cos\sqrt{x}}{\Delta x}$$

$$= \lim_{\Delta x\to 0}\frac{1}{\Delta x}\left[-2\sin\frac{\sqrt{x+\Delta x}+\sqrt{x}}{2}\sin\frac{\sqrt{x+\Delta x}-\sqrt{x}}{2}\right]$$

$$= \lim_{\Delta x\to 0}\left[-2\sin\frac{\sqrt{x+\Delta x}+\sqrt{x}}{2}\cdot\frac{\sin\dfrac{\sqrt{x+\Delta x}-\sqrt{x}}{2}}{\left(\dfrac{\sqrt{x+\Delta x}-\sqrt{x}}{2}\right)}\cdot\left(\frac{\sqrt{x+\Delta x}-\sqrt{x}}{2}\right)\right]\frac{1}{\Delta x}$$

$$= \lim_{\Delta x\to 0}\left\{-2\sin\frac{\sqrt{x+\Delta x}+\sqrt{x}}{2}\right\}\lim_{\Delta x\to 0}\frac{\sin\dfrac{\sqrt{x+\Delta x}-\sqrt{x}}{2}}{\dfrac{\sqrt{x+\Delta x}-\sqrt{x}}{2}}\lim_{\Delta x\to 0}\frac{\Delta x}{\sqrt{x+\Delta x}+\sqrt{x}}\cdot\frac{1}{\Delta x}$$

$$\because \quad \frac{\sqrt{x+\Delta x}-\sqrt{x}}{2} = \frac{\sqrt{x+\Delta x}-\sqrt{x}\,(\sqrt{x+\Delta x}-\sqrt{x})}{2(2(\sqrt{x+\Delta x}+\sqrt{x}))}$$

$$= \frac{\Delta x}{2(\sqrt{x+\Delta x}+\sqrt{x})}$$

$$\therefore \quad \frac{dy}{dx} = -2\sin\sqrt{x}\cdot 1\cdot\frac{1}{2}\lim_{\Delta x\to 0}\frac{1}{\sqrt{x+\Delta x}+\sqrt{x}}$$

$$= -2\sin\sqrt{x}\,\frac{1}{2}\,\frac{1}{2\sqrt{x}} = -\frac{\sin\sqrt{x}}{2\sqrt{x}}.$$

Ex. 77. $\quad y = x\sin x$

$$\frac{\Delta y}{\Delta x} = \frac{(x+\Delta x)\sin(x+\Delta x) - x\sin x}{\Delta x}$$

$$= \frac{1}{\Delta x}\{x\sin(x+\Delta x) - x\sin x + \Delta x\sin(x+\Delta x)\}$$

$$= \frac{1}{\Delta x}\left\{x\cdot 2\cos\left(x+\frac{\Delta x}{2}\right)\sin\frac{\Delta x}{2} + \Delta x\sin(x+\Delta x)\right\}$$

$$\frac{dy}{dx} = \lim_{\Delta x\to 0}\left[x\cos\left(x+\frac{\Delta x}{2}\right)\frac{\sin\Delta x/2}{\Delta x/2} + \sin(x+\Delta x)\right] = x\cos x.1 + \sin x.$$

Ex. 78. (*i*) $y = \dfrac{\sin x}{x}$ (*ii*) $y = \tan(ax+b)$ (*iii*) $y = x\sin x\cos x$.

Ans. $\dfrac{dy}{dx} =$ (*i*) $\dfrac{x\cos x - \sin x}{x^2}$ (*ii*) $a\sec^2(ax+b)$ (*iii*) $\dfrac{1}{2}\sin 2x + x\cos 2x$

Trigonometric Functions: Domain and Range

Domain	*Functions*	*Range*
All	sin x, cos x	$-1 \le x \le 1$
All except $(2n+1)(\pi/2)$	tan x	All
All except $n\pi$	cot x	All
All except $(2n+1)(\pi/2)$	sec x	All ≥ 1 or ≤ -1
All except $n\pi$	cosec x	All ≥ 1 or ≤ -1

Note: All implies "All real numbers." n is an integer.

Ex. 79. Differentiate w.r.t x

(*i*) $\cos\sqrt{1-x}$ **Ans.** $\dfrac{1}{2}\dfrac{\sin\sqrt{1-x}}{\sqrt{1-x}}$

(*ii*) $\dfrac{1+\cos 2x}{1-\cos 2x}$ **Ans.** $-\operatorname{cosec}^2 x$

(*iii*) $\operatorname{cosec}\dfrac{1-x}{1+x}$ **Ans.** $\dfrac{2}{(1+x)^2}\operatorname{cosec}\dfrac{1-x}{1+x}\cot\dfrac{1-x}{1+x}$

(*iv*) $\sin x°$ (convert degrees into radians first, $x° = \dfrac{\pi}{180}x$) **Ans.** $\dfrac{\pi}{180}\cos x°$

(*v*) $x = a\cos^3 t$

$y = a\sin^3 t$ **Ans.** $-\tan t$

$$\left[\frac{dx}{dt} = 3a\cos^2 t(-\sin t)\right.$$

$$\left.\frac{dy}{dt} = 3a\sin^2 t(\cos t)\right]$$

(*vi*) $x = 4(1-\sin t)$ **Ans.** $1/2 \sec t - \cos t$

$y = \sin 2t$

Ex. 80. If $x\sin(a+y) = \sin y$

then $$\frac{dy}{dx} = \frac{\sin^2(a+y)}{\sin a}$$

10.6 DERIVATIVE OF A DERIVATIVE

Derivatives of Higher Orders

$\dfrac{dy}{dx}$ is the first derivative of y with respect to x. The second derivative is obtained by differentiating with respect to x twice and is written

$$\frac{d^2y}{dx^2} = \frac{d}{dx}\left[\frac{dy}{dx}\right] \text{ or } y''.$$

Derivatives of Higher order can be defined similarly.

Ex. 81. $y = 5x^4 + 2x^3, \quad \dfrac{dy}{dx} = 20x^3 + 6x^2$

$$\frac{d^2y}{dx^2} = \frac{d}{dx}(20x^3+6x^2) = 60x^2 + 12x$$

$$\frac{d^3y}{dx^3} = \frac{d}{dx}\left(\frac{d^2y}{dx^2}\right) = \frac{d}{dx}(60x^2+12x) = 12x + 12.$$

Ex. 82. $y = \dfrac{2}{x}, \dfrac{dy}{dx} = -\dfrac{2}{x^2}, \dfrac{d^2y}{dx^2} = \dfrac{d}{dx}\left(-\dfrac{2}{x^2}\right) = \dfrac{4}{x^3}.$

Ex. 83. If $x = \dfrac{t^2}{1+t},\ y = \dfrac{t^3}{1+t}$.

$$\frac{dy}{dx} = \frac{dy}{dt}\cdot\frac{dt}{dx} = \frac{t(3+2t)}{2+t}$$

$$\frac{d^2y}{dx^2} = \frac{d(dy/dx)/dt}{dx/dt} = \frac{(t+3)\,(t+1)^3}{t(2+t)^3}$$

Ex. 84. If $x = t - t^2,\ y = t - t^3$

$$\frac{dy}{dx} = \frac{1-3t^2}{1-2t},\ \frac{d^2y}{dx^2} = \frac{2-6t+6t^2}{(1-2t)^2}.$$

Ex. 85. $y = \sqrt{1+x^2},\ \dfrac{d^2y}{dx^2} = (x^2+1)^{-3/2},\ \dfrac{d^3y}{dx^3} = -3x(x^2+1)^{-5/2}$

$\dfrac{d^4y}{dx^4} = 3\,(4x^2-1)\,(x^2+1)^{-7/2}$ and so on.

Ex. 86. Find the slope of the tangent to the following functions at the indicated points.

(*i*) $y = x^4 - 3x^3 + 2x^2 + 1$ at (0, 1), (2, 1)

(*ii*) $y = (x-1)\,(x+1)^9$ at (0, 1), (1, 0)

(*iii*) $y = x^{2/3}$ at $(\sqrt{8}, 2)$, (0, 0) **Ans.** (*i*) 0, 4 (*ii*) – 8, 2^9 (*iii*) $\dfrac{\sqrt{2}}{3}$, undefined

Ex. 87. Find $\dfrac{dy}{dx}$ for the following functions :

(*i*) $y = (x^2 + x - 1)^{-1/2}$ (*ii*) $\sqrt{xy} = 1$

(*iii*) $x = y/(1-x^2)$ (*iv*) $(x+2y)^2 + 2xy^2 - 6 = 0$

(*v*) $x^2y + xy^2 = 6(x^2+y^2)$ (*vi*) $y = x^2 + t^2,\ t = \dfrac{x}{1+x^2}$

(*vii*) $y = \sqrt{x^2+5} + \sqrt{x^2-5}$

Ans. (*i*) $-\dfrac{1}{2}\,(2x+1)\,(x^2+1)^{3/2}$ (*ii*) $-\dfrac{y}{x}$ (*iii*) $(1-x^2)^{-3/2}$

(*iv*) $-\dfrac{y^2+2y+x}{2(xy+2y+x)}$ (*v*) $\dfrac{12x-12xy-y^2}{x^2+2xy-12y}$ (*vi*) $2x + \dfrac{2x(1-x)^2}{(1+x^2)^3}$

(*vii*) $x[(x^2+5)^{-1/2} + (x^2-5)^{-1/2}]$

Also find d^2y/dx^2 for the above functions.

Ex. 88. If $y = \sin x$ show $\dfrac{d^2y}{dx^2} + y = 0$.

Ex. 89. If $y = (\sin^{-1} x)^2$ then $(1 - x^2)\dfrac{d^2y}{dx^2} - x\dfrac{dy}{dx} = 2.$

Ex. 90. If $ax^2 + 2hxy + by^2 = 0$ then $\dfrac{d^2y}{dx^2} = 0$

Hint: $$2ax + 2h\left(y + x\frac{dy}{dx}\right) + 2by\frac{dy}{dx} = 0$$

$$\therefore \quad \frac{dy}{dx} = -\frac{ax + hy}{hx + by}$$

Since $ax^2 + 2hxy + by^2 = (ax^2 + hxy) + (hxy + by^2)$

$$= x(ax + hy) + y(hx + by) = 0$$

$$\therefore \quad \frac{ax + hy}{hx + by} = -\frac{y}{x}$$

$$\therefore \quad \frac{dy}{dx} = \frac{y}{x}$$

$$\frac{d^2y}{dx^2} = \frac{x\frac{dy}{dx} - y}{x^2} = \frac{x\left(\frac{y}{x}\right) - y}{x^2} = \frac{y - y}{x^2} = 0.$$

Ex. 91. If $x = a(\cos\theta + \theta\sin\theta)$, $y = a(\sin\theta - \theta\cos\theta)$

$$\frac{dx}{d\theta} = a\theta\cos\theta, \frac{dy}{d\theta} = a\theta\sin\theta, \frac{dy}{dx} = \tan\theta$$

$$\frac{d^2y}{dx^2} = \frac{d}{dx}\left(\frac{dy}{dx}\right) = \frac{d}{dx}(\tan\theta) = \frac{d\tan\theta}{d\theta}\cdot\frac{d\theta}{dx}$$

$$= \frac{\sec^2\theta}{dx/d\theta} = \frac{\sec^2\theta}{a\theta\cos\theta} = \frac{1}{a\theta\cos^3\theta}.$$

10.7 THE DIFFERENTIAL

If for the function $y = f(x)$ we write $\dfrac{dy}{dx} = f'(x)$, we define dx as the *differential* of x, which represents any real number. Also dy, the differential of y, which is a function of x and dx, is given by $dy = f'(x)\,dx$.

$$\frac{dy}{dx} = \frac{\text{the differential of } y}{\text{the differential of } x}$$

The rules for finding differentials are obtained from those of derivatives.

For a constant c, $dc = 0$, $d(cu) = c\,du$.

Similarly $du^n = nu^{n-1}$, $d(u + v) = du + dv$

$$d(uv) = u\,dv + v\,du$$

$$d(u/v) = \frac{v\,du - u\,dv}{v^2}.$$

Ex. 92. If $y = x^4 + 5x^3 - 3x^2 + 2$

$$dy = (4x^3 + 15x^2 - 6x)dx.$$

Ex. 93. If $xy^2 + x^2y = 4$, $\quad dy = -\dfrac{y^2 + 2xy}{x^2 + 2xy}dx.$

Ex. 94. If $y = \dfrac{1 + x - x^2}{1 - x}$, $\quad dy = \dfrac{2 - 2x + x^2}{(1 - x)^2}dx.$

Chain Rule. If $x = g(t)$ is a function of the variable t we can write the function $y = f(x)$ as

$$y = f[g(t)]$$

Since

$$dy = f'(x)\, dx$$

$$= f'[g(t)]\, dg(t)$$

$\therefore$

$$dy = f'[g(t)]\, g'(t)\, dt$$

We thus have the chain rule in terms of differentials. Compare

$$\frac{dy}{dt} = f'(g(t))g'(t)$$

where $y = f(x) = f[g(t)]$

or

$$\frac{dy}{dt} = \frac{dy}{dx} \cdot \frac{dx}{dt}.$$

Differential Approximation

For the function $y = f(x)$, we have

$$\frac{dy}{dx} = \frac{df}{dx} = f'(x) \text{ or } df = f'(x)\, dx$$

df is defined as the differential of f.

In the notation $\dfrac{dy}{dx} = f'(x)$, set

$$dx = h \text{ and } dy = df.$$

The differential of y can be written

$$dy = f'(x)dx = hf'(x)$$

Here dx, the differential of x is an independent variable. The ratio $\dfrac{dy}{dx}$ is the derivative which can be written in terms of differentials.

Also $df(x, h) = f'(x) \cdot h$ where h has replaced dx

This represents the value of the function when x and h are substituted into the right side.

Ex. 95. For the function $f(x) = x^4 + 3x^2 - 2x$, we have

$$f'(x) = 4x^3 + 6x - 2$$

$$df(x, h) = (4x^3 + 6x - 2)h$$

Let $\quad x = 1,\ h = 0.1$ then

$$df(1, 0.1) = (4 + 6 - 2)\,0.1 = 0.8$$

Ex. 96. Since $\lim\limits_{h\to 0} \dfrac{f(x+h) - f(x)}{h} = f'(x)$

we can write approximately, after removing the limit part, $f(x + h) - f(x) = hf'(x)$ or using Δx for h,

$$f(x + \Delta x) \simeq f(x) + f'(x)\,\Delta x$$

Using this, find the value of $\sqrt{3.99}$ approximately.

Let $\quad y = f(x) = \sqrt{x}$ and $\Delta x = -0.01$ where $x = 4$

$$\sqrt{3.99} = f(4 + (-0.01)) = f(4) - 0.01\,f'(4)$$

$$f'(x) = \frac{1}{2\sqrt{x}},\ f(4) = \sqrt{4} = 2,\ f'(4) = \left.\frac{1}{2\sqrt{x}}\right]_{x=4} = \frac{1}{2\sqrt{4}} = \frac{1}{4}$$

$\therefore$ Using (1) $\quad \sqrt{3.99} = 2 - 0.01 \times \dfrac{1}{4} = 2 - 0.0025 = 1.9975.$

Ex. 97. $82^{1/4} = 81^{1/4} + \left.\dfrac{dy}{dx}\right]_{x=81} \cdot 1 \left(\text{where } y = x^{1/4} = f(x),\ f'(x) = \dfrac{1}{4}x^{-3/4};\ \Delta x = 1\right)$

$$= 3 + \left.\frac{1}{4}x^{-3/4}\right]_{x=81} = 3 + \frac{1}{4 \times 27} = 3.00906. \quad \left(\because 81^{-3/4} = 3^{4\left(-\frac{3}{4}\right)} = 3^{-3}\right)$$

Use of Derivatives to Find Percentage Change and Roots

If α_0 is an approximate root of $f(x) = 0$, a closer approximation is $\alpha = \alpha_0 - f(\alpha_0)/f'(\alpha_0)$. This can be improved by repetition *i.e.* by using α as the new α_0 and repeating for improved values of the root.

Ex. 98. To find the root of $f(x) = e^{-x} - x/2 = 0$, we have

$$f(0.8) = 0.0493,\ f(0.9) = 0 - 0.0434$$

$f(0.85) = -0.0025$ which is nearly zero, *i.e.* 0.85 is an approximate root

For a better approximation write

$$f'(x) = -e^{-x} - \frac{1}{2} \quad \text{or } f'(0.85) = 0.9274$$

$$\alpha = \alpha_0 - \frac{f(\alpha_0)}{f'(\alpha_0)} = 0.85 - \frac{-0.0025}{0.9274} = 0.853.$$

Ex. 99. Let x increase by 1 per cent in $y^3 = 2x^2$ Find the percentage increase in y.

$\dfrac{dy}{dx} = \dfrac{4x}{3y^2},\ \Delta x = x/100$; change in y is Δy which equals $(dy/dx)\Delta x$

The percentage change in y is

$$100\frac{\Delta y}{y} = \left(\frac{100}{y}\right)\left(\frac{dy}{dx}\right)\Delta x = \left(\frac{100}{y}\right)\left(\frac{4x}{3y^2}\right)\frac{x}{100}$$

$$= \frac{400\,x^2}{3.2x^2 100} = \frac{2}{3}\% \quad (\text{since } y^3 = 2x^2)$$

Ex. 100. Find $\tan^{-1} 0.999$.

Let $f(x) = \tan^{-1} x$, $f'(x) = \frac{1}{1+x^2}$

$$h = -0.001, f(a) = f(1) = \tan^{-1} 1 = \frac{\pi}{4}, f'(a) = \frac{1}{2}$$

$$f(a + h) = f(a) + hf'(a)$$

$$\therefore \quad \tan^{-1} 0.999 = f(0.999) = f(1) - 0.001\, f'(1)\cdot$$

$$= \frac{\pi}{4} - 0.001 \times \frac{1}{2} = \frac{\pi}{4} - 0.0005$$

Ex. 101. Show (*i*) $(80.7)^{\frac{1}{4}} = 2.9972$ (*ii*) $\sqrt[3]{26.96} = 2.9986$.

Ex. 102. $\sin 31° = 0.51516$ given $1° = 0.0175$ radian and $\cos 30° = 0.8660$.

Ex. 103. (*i*) $ln\ 101 = 4.6152$ given that $ln\ 10 = 2.3026$

(*ii*) $ln\ 9.01 = 2.1983$ given that $ln\ 3 = 1.0986$.

Ex. 104. If there is an error of 0.3% in the measurement of the radius of a sphere, find the percentage error in the calculation of its volume.

$$V = \frac{4}{3}\pi r^3, \quad \Delta V = \frac{dV}{dr}\Delta r = 4\pi r^2 \cdot \Delta r$$

$$\text{Percentage error} = \frac{\Delta V}{V} \times 100 = \frac{4\pi r^2 \Delta r}{\frac{4}{3}\pi r^3} \times 100$$

$$= 3 \times 100 \times \left(\frac{\Delta r}{r}\right) = 3 \times 100 \times 0.003$$

$$= 0.9\%$$

Ex. 105. The side of a square of size 5 metres is incorrectly measured as 5.11 metres. Find the error in the calculation of the area. **Ans.** 1.1 sq. metres

Miscellaneous

Ex. 106. (*i*) Find dy/dx given $y = (ln\ x)^x + (\sin^{-1} x)^{\sin x}$

Ans. $(ln\ x)^x\ (ln\ ln\ x + 1/ln\ x) + (\sin^{-1} x)^{\sin x}\left(\cos x\ ln \sin^{-1} x + \frac{\sin x}{\sqrt{1-x^2}\sin^{-1} x}\right)$

(*ii*) $y = \cos^{-1}\dfrac{1-x^2}{1+x^2}$ **Ans.** $\dfrac{2}{1+x^2}$

(*iii*) $y = (x^x)^x$ **Ans.** $x^{x^2+1}\ ln\ ex^2$

(*iv*) $y = x^{(x^x)}$ **Ans.** $[1 + x\ lnx\ ln\ ex]\ x^{(x^x)+x-1}$

(*v*) $y = e^{ax}\cos(b\tan^{-1}x)$

Ans. $e^{ax}\left[a\cos(b\tan^{-1}x) - b\sin(b\tan^{-1}x)\cdot\dfrac{1}{1+x^2}\right]$

Ex. 107. (*i*) Find $\dfrac{dy}{dx}$ when

$$x = a\sqrt{\frac{t^2-1}{t^2+1}},\quad y = \text{at}\sqrt{\frac{t^2-1}{t^2+1}}$$ **Ans.** $\dfrac{2\text{ at }(t^2-1)^{-1/2}}{(t^2+1)^3/2}$

(*ii*) $x = 3\cos t - 2\cos^3 t$

$y = 3\sin t - 2\sin^3 t$ **Ans.** $\cot t$

Ex. 108. If $y = f(x)$ is differentiable such that the inverse function $x = f^{-1}(y)$ is defined then show that

$$\frac{dx}{dy} = \frac{1}{dy/dx}\left(\frac{dy}{dx} \neq 0\right)$$

We have $x + \Delta x = f^{-1}(y + \Delta y)$

$\therefore$ $y + \Delta y = f(x + \Delta x)$

$$\frac{dx}{dy} = \lim_{\Delta y\to 0}\frac{\Delta x}{\Delta y} = \frac{1}{\lim\limits_{\Delta y\to 0}\dfrac{\Delta y}{\Delta x}} = \frac{1}{\lim\limits_{\Delta x\to 0}\dfrac{\Delta y}{\Delta x}}\quad (\text{Since as } \Delta y \to 0,\ \Delta x \to 0)$$

$$= \frac{1}{dy/dx}$$

Ex. 109. If $x = f(t)$, $y = g(t)$ are differentiable then

$$\frac{dy}{dx} = \frac{dy/dt}{dx/dt}$$

$$\frac{dy}{dx} = \lim_{\Delta x\to 0}\frac{\Delta y}{\Delta x} = \lim_{\Delta x\to 0}\frac{\Delta y/\Delta t}{\Delta x/\Delta t} = \lim_{\Delta x\to 0}\left(\frac{\Delta y/\Delta t}{\Delta x/\Delta t}\right)\left(\begin{array}{r}\text{As } \Delta x \to 0\\ \Delta t \to 0\end{array}\right)$$

$$= \frac{\lim\limits_{\Delta t\to 0}(\Delta y/\Delta t)}{\lim\limits_{\Delta t\to 0}(\Delta x/\Delta t)} = \frac{dy/dt}{dx/dt}.$$

Ex. 110. $y = \tan^{-1}\dfrac{\sqrt{1+x^2}-1}{x}$ then $\tan y = \dfrac{\sqrt{1+x^2}-1}{x}$

Hence find $dy/dx = \dfrac{1}{2\,(1+x^2)}$

Alternatively put $x = \tan t$

$$\therefore \quad y = \tan^{-1} \frac{\sqrt{1+\tan^2 t} - 1}{\tan t} = \tan^{-1} \frac{\sec t - 1}{\tan t}$$

$$= \tan^{-1} \frac{1 - \cos t}{\sin t} = \tan^{-1} \frac{2 \sin^2 t/2}{2 \sin \frac{t}{2} \cos \frac{t}{2}} = \tan^{-1} \tan \frac{t}{2}$$

$$= \frac{t}{2} = \frac{1}{2} \tan^{-1} x$$

$$\therefore \quad \frac{dy}{dx} = \frac{1}{2} \frac{1}{1+x^2}.$$

Ex. 111. $y = \sin^{-1} \dfrac{1-x^2}{1+x}$, $\dfrac{dy}{dx} = \dfrac{-2}{1+x^2}$ (**Hint.** Put $x = \tan t$)

Ex. 112. $y = \cos^{-1} (2x\sqrt{1-x^2})$, $\dfrac{dy}{dx} = \dfrac{-2}{\sqrt{1-x^2}}$ (**Hint.** Put $x = \sin t$)

Ex. 113. $y = \cos^{-1} (\sin x)$, $\dfrac{dy}{dx} = -1.$

Ex. 114. $y = \tan^{-1} \dfrac{\cos x}{1 + \sin x}$, $\dfrac{dy}{dx} = -\dfrac{1}{2}.$

Ex. 115. Find $\dfrac{dy}{dx}$ for the following:

(*i*) $y = \tan^{-1} 5x$, $\dfrac{dy}{dx} = \dfrac{1}{1+25x^2} \cdot \dfrac{d}{dx} (5x) = \dfrac{5}{1+25x^2}$

(*ii*) $y = \dfrac{x}{\sqrt{1-x^2}} \cos^{-1} x$, $\dfrac{dy}{dx} = \dfrac{\cos^{-1} x - x\sqrt{1-x^2}}{(1-x^2)^{3/2}}$

(*iii*) $y = x \sin^{-1} x + \sqrt{1-x^2}$, $\dfrac{dy}{dx} = \dfrac{x}{\sqrt{1-x^2}} + \sin^{-1} x \dfrac{x}{\sqrt{1-x^2}}$

(*iv*) $y = \tan^{-1} \dfrac{x}{2} + \cot^{-2} \dfrac{2}{x} = \tan^{-1} \dfrac{x}{2} + \tan^{-1} \dfrac{x}{2} = 2 \tan^{-1} \dfrac{x}{2}$

Since $\cot^{-1} z = \tan^{-1} 1/z$

$$\frac{dy}{dx} = \frac{2}{1+\left(\frac{x}{2}\right)^2} \cdot \frac{1}{2} = \frac{4}{4+x^2}$$

(*v*) $x = \cos^{-1}\left(x\dfrac{1-y}{1+y}\right), \dfrac{dy}{dx} = \dfrac{2(x\sin x + \cos x)}{(x+\cos x)^2}, (x + \cos x \neq 0)$

(*vi*) $y = x\cos^{-1} 2x - \dfrac{1}{2}\sqrt{1-x^2}, \dfrac{dy}{dx} = \cos^{-1} 2x$

(*vii*) $y = \tan^{-1}\dfrac{1+x}{1-x}, \dfrac{dy}{dx} = \dfrac{1}{1+x^2}.$

Ex. 116. $y = ln\left[e^x\left(\dfrac{x-2}{x+2}\right)^{3/4}\right]$

$$\frac{dy}{dx} = 1 + \frac{3}{4(x-2)} - \frac{3}{4(x+3)}$$

Ex. 117. $y = ln\,(\sec x + \tan x), \dfrac{dy}{dx} = \sec x$

Ex. 118. $y = \dfrac{\cot x}{1+e^x}, \dfrac{dy}{dx} = \dfrac{-(1+e^x)\text{cosec}^2 x - e^x \cot x}{(1+e^x)^2}.$

Ex. 119. If $y\sqrt{x} = \sin^{-1}\sqrt{x}$

prove that $(4x - 4x^2)\dfrac{d^2y}{dx^2} + (6-8x)\dfrac{dy}{dx} = y$

Differentiating,

$$\sqrt{x}\frac{dy}{dx} + y\frac{1}{2\sqrt{x}} = \frac{1}{\sqrt{1-x}}\cdot\frac{1}{2\sqrt{x}}$$

$$\therefore \quad 2x\frac{dy}{dx} + y = \frac{1}{\sqrt{1-x}}$$

Differentiating again,

$$2x\cdot\frac{d^2y}{dx^2} + 2\frac{dy}{dx} + \frac{dy}{dx} = \frac{1}{2(1-x)^{3/2}} = \frac{1}{2(1-x)}\frac{1}{\sqrt{1-x}}$$

$$\therefore \quad 2x\frac{d^2y}{dx} + 3\frac{dy}{dx} = \frac{1}{2(1-x)}\left(2x\frac{dy}{dx} + y\right)$$

$$\therefore \quad 4x(1-x)\frac{d^2y}{dx^2} + (6-8x)\frac{dy}{dx} = y.$$

Ex. 120. If $x^p y^q = (x+y)^{p+q}$ show that $\dfrac{dy}{dx} = \dfrac{y}{x}.$

Ex. 121. If $y = x^{x^{x^{\cdot^{\cdot^{\cdot\infty}}}}}$ show that $x\dfrac{dy}{dx} = \dfrac{y^2}{1 - y\,ln\,x}$

We can write $y = x^{x^{x^{\cdot^{\cdot^{\cdot^{\infty}}}}}} = x^y$

$\therefore \quad \ln y = y \ln x$

$\therefore \quad \dfrac{1}{y}\dfrac{dy}{dx} = y \cdot \dfrac{1}{x} + \ln x \dfrac{dy}{dx}$

$\therefore \quad \dfrac{dy}{dx}\left(\dfrac{1}{y} - \ln x\right) = \dfrac{y}{x}$ or $\dfrac{dy}{dx} = \dfrac{y^2}{x(1 - y \ln x)}$

$\therefore \quad x\dfrac{dy}{dx} = \dfrac{y^2}{1 - y \ln x}.$

Ex. 122. If $\sqrt{y/x} + \sqrt{x/y} = 6$ show that $\dfrac{dy}{dx} = \dfrac{x - 17y}{17x - y}.$

Ex. 123. If $e^x + e^y = e^{x+y}$ show that $\dfrac{dy}{dx} = -e^{-(y+x)}$

Ex. 124. If $y = (\sqrt{x})^{(\sqrt{x})^{(\sqrt{x})\ldots\infty}}$ show that $\dfrac{dy}{dx} = \dfrac{y^2}{2 - y \ln x}.$

Ex. 125. If $\sin y = x \sin (a + y)$ show $\dfrac{dy}{dx} = \dfrac{\sin^2 (a + y)}{\sin a}.$

Ex. 126. $y = (\sin x)^x + x^{\sin x} = u + v$ say

$\ln u = x \ln \sin x, \ \ln v = \sin x \ln x$

$\dfrac{1}{u}\dfrac{du}{dx} = x.\dfrac{1}{\sin x}\cos x + \ln \sin x = x \cot x + \ln \sin x$

$\therefore \quad \dfrac{du}{dx} = (\sin x)^x (x \cot x + \ln \sin x)$

Similarly $dv/dx = x^{\sin x}\left(\dfrac{\sin x}{x} + \ln x \cdot \cos x\right)$

$\therefore \quad \dfrac{dy}{dx} = (\sin x)^x (x \cot x + \ln \sin x) + x^{\sin x}\left(\dfrac{\sin x}{x} + \cos x \ln x\right)$

Ex. 127. $y = (\cos x)^{\sec^{-1} x} + (\ln x)^{\tan x} = u + v$ say

$\ln u = \sec^{-1} x \ln \cos x, \ \ln v = \tan x \ln \ln x$

$\dfrac{1}{u}\dfrac{du}{dx} = \sec^{-1} x \cdot \dfrac{1}{\cos x}(-\sin x) + \ln \cos x \cdot \dfrac{1}{x\sqrt{x^2 - 1}}$

$\therefore \quad \dfrac{du}{dx} = (\cos x)^{\sec^{-1} x}\left[\dfrac{\ln \cos x}{x\sqrt{x^2 - 1}} - \tan x \cdot \sec^{-1} x\right]$

$$\frac{1}{v}\frac{dv}{dx} = \frac{\tan x}{ln\, x}\cdot\frac{1}{x} + (ln\ ln\ x)\sec^2 x$$

$$\therefore \quad \frac{dv}{dx} = (ln\, x)^{\tan x}\left[\frac{\tan x}{x\, ln\, x} + \sec^2 x\, ln\, ln\ x\right]$$

$$\frac{dy}{dx} = \frac{du}{dx} + \frac{dv}{dx}.$$

Ex. 128. $y = (4x^2)^x + x^{4x^2} = u + v$ say

$$ln\ u = x\ ln\ 4x^2,\ ln\ v = 4x^2\ ln\ x$$

$$\frac{1}{u}\frac{du}{dx} = x\cdot\frac{1}{4x^2}\ 8x + ln\ 4x^2$$

$$\therefore \quad \frac{du}{dx} = (4x^2)^x\ [2 + ln\ 4x^2]$$

Similarly $\quad \dfrac{dv}{dx} = 4x^{4x^2+1}\ (1 + 2\ ln\ x)$

$$\frac{dy}{dx} = \frac{du}{dx} + \frac{dv}{dx}$$

Ex. 129. (*i*) $y = \dfrac{x^2 + \sin x}{1+x}$, $\dfrac{dy}{dx} = (x^2 + 2x + x\cos x - \sin x + \cos x)/(1 + x)2$

(*ii*) $y = x^{-3}\ 3^x$, $\dfrac{dy}{dx} = 3^x x^{-4}\ (x\ ln\ \ 3 - 3)$

(*iii*) $y = \dfrac{\cot x}{1+x^2}$, $\dfrac{dy}{dx} = \dfrac{(1+x^2)\ \text{cosec}^2 + 2x\cot x}{-(1+x^2)^2}$

(*iv*) $y = \sqrt{\dfrac{1+\sin x}{1-\sin x}}$, $\dfrac{dy}{dx} = \sec x\,(\sec x + \tan x)$

(*v*) $y = ln\ (e^{mx} + e^{-mx})$, $\dfrac{dy}{dx} = m\dfrac{e^{mx} - e^{-mx}}{e^{mx} + + e^{-mx}}$

(*vi*) $y = (1 + x)^x$, $\dfrac{dy}{dx} = (1+x)^x\left[\dfrac{x}{1+x} + ln\,(1+x)\right]$

(*vii*) $y = x^{\sqrt{x}}$, $\dfrac{dy}{dx} = x^{\sqrt{x}}\left(\dfrac{2 + ln\, x}{\sqrt{x}}\right)$

(*viii*) $y = (ln\ x)^x + x^{\cos x}$,

$$\frac{dy}{dx} = (ln\, x)^x\left[\frac{1}{ln\, x} + ln\, ln\, x\right] + x^{\cos x}\left[\frac{\cos x}{x} - \sin x\, ln\, x\right]$$

(*ix*) $y = x^x + a^x + x^a + a^a + e^a + a^e$

$$\frac{dy}{dx} = x^x\ (1 + ln\ x) + a^x\ ln\ a + ax^{a-1}$$

(x) $y = \dfrac{(\cos x)^x}{1+x^2}, \; \dfrac{dy}{dx} = \dfrac{(\cos x)^x}{1+x^2}\left(ln\cos x - x\tan x - \dfrac{2x}{1+x^2}\right)$

(xi) $y = \sqrt{\dfrac{a-x}{a+x}}(x^2+b)$

$$\frac{dy}{dx} = \sqrt{\frac{a-x}{a+x}}(x^2+b)\left[\frac{2x}{x^2+b} - \frac{1}{2(a+x)} - \frac{1}{2(a-x)}\right]$$

(xii) $y = (x+2)^{x^3+1}$

$$\frac{dy}{dx} = (x+2)^{x^3+1}\left[\frac{x^3+1}{x+2} + 3x^2 ln\,(x+2)\right].$$

Ex. 130. If $x = y\sin(2y+5) + \sqrt{1+y^2}$

$$\frac{dx}{dy} = 2y\cos(2y+5) + \sin(2y+5) + \frac{y}{\sqrt{1+y^2}}.$$

Ex. 131. $f(t) = t^2 + 3t + 7$ where $t = 2x + 3$

$$\frac{d\,f(t)}{dx} = 8x + 18$$

Ex. 132. (i) $y = \tan^{-1}\dfrac{a+b\cos x}{b-a\cos x}, \; \dfrac{dy}{dx} = \dfrac{-\sin x}{1+\cos^2 x}$

(ii) $y = x^3\tan^{-1}x, \; \dfrac{dy}{dx} = x^2\left(3\tan^{-1}x + \dfrac{x}{1+x^2}\right)$

(iii) $y = \tan^{-1}\dfrac{1}{x}, \; \dfrac{dy}{dx} = -\dfrac{1}{1+x^2}$

(iv) $y = \sin^{-1}\sqrt{x} + \cos^{-1}\sqrt{1-x}, \; \dfrac{dy}{dx} = \dfrac{1}{\sqrt{x}\sqrt{1-x}}$

Hint. Put $x = \sin^2 t$

(v) $y = \sin^{-1}\dfrac{2-3x}{1+6x}, \; \dfrac{dy}{dx} = \dfrac{-3}{1+9x^2}.$

Ex. 133. (i) $\sqrt{x} + \sqrt{y} = \sqrt{a}, \dfrac{dy}{dx} = -\sqrt{y}/\sqrt{x}$

(ii) $xy + y\sec^{-1}x = 1$

$$\frac{dy}{dx} = -\left(1 + \frac{1}{x\sqrt{x^2-1}}\right)y^2$$

(iii) $x = \cos xy, \; \dfrac{dy}{dx} = \dfrac{-(1+y\sqrt{1-x^2})}{x\sqrt{1-x^2}}$

(*iv*) $\sin y = ln\,(x + y),\ \frac{dy}{dx} = \frac{1}{(x+y)\cos y - 1}$

(*v*) $x^p y^q = (x + y)^{p+q},\ \frac{dy}{dx} = \frac{y}{x}$

(*vi*) $y = x e^{xy},\ \frac{dy}{dx} = \frac{y(1+xy)}{x(1-xy)}$

(*vii*) $ax^2 + 2\,hxy + by^2 + 2gx + 2fy + c = 0$

$$\frac{dy}{dx} = -\frac{ax + hy + g}{hx + by + f}$$

(*ix*) $y = x \sin y,\ \frac{dy}{dx} = \frac{y}{x(1 - x\cos y)}$

(*x*) $x^{2/3} + y^{2/3} = a^{2/3},\ \frac{dy}{dx} = -\sqrt{y/x}.$

Ex. 134. Show that if $y = ln\sqrt{\frac{(1+x)e^{-2x}}{1-x}}$

$$\frac{dy}{dx} = \frac{x^2}{1-x^2}$$

Ex. 135. If $\tan y = 2\ ln\ x$ show (*i*) $x\frac{dy}{dx} = 2\cos^2 y$

(*ii*) $x^2\frac{d^2y}{dx^2} + 2(1 + 2\sin y)\cos^2 y = 0$

Ex. 136. Find dy/dx when (*i*) $y = ln \sin 2x + e^{1/x}$

(*ii*) $x = \sin^2\theta,\ y = \sin 2\theta$

(*iii*) $y = e^{\tan x}$ **Ans.** (*i*) $2\cot 2x - e^{1/x}/x^2$ (*ii*) $2\cot 2\theta$ (*iii*) $\sec^2 x e^{\tan x}$.

Ex. 137. If $y = b + ax + 54x^2 + x^3 + x^4/144$ find a if $(d^2y/dx^2)^2 = y$ for all values of x.

Ans. 1296

Ex. 138. If $x = \cot\theta,\ y = \sin^2\theta$ show $\frac{d^2y}{dx^2} = 2\sin 3\theta \sin^3\theta.$

Ex. 139. If $y = \frac{\sin x - \cos x}{\sin x + \cos x}$ show $\frac{dy}{dx} = 1 + y^2$ and $\frac{d^2y}{dx^2} = 0$ when $y = 0$.

Ex. 140. If $y = ae^{-kt}\cos(ct + f)$

Show that $\frac{d^2y}{dt^2} + 2k\frac{dy}{dt} + (c^2 + k^2)y = 0$

Hint. First find $\frac{dy}{dt} = -ky - c\,ae^{-kt}\sin(ct + f)$

Then $\frac{d^2y}{dt^2} = -2k\frac{dy}{dt} - (c^2 + k^2)y$

Ex. 141. If $a^2 \cos^2 \theta + b^2 \sin^2 \theta = p^2$ find $\dfrac{d^2p}{d\theta^2}$.

Hint. $$p^2 = (a^2 - b^2) \cos^2 \theta + b^2$$

$$p\frac{dp}{d\theta} = (b^2 - a^2)\sin\theta\cos\theta$$

$$p\frac{d^2p}{d\theta^2} + \left(\frac{dp}{d\theta}\right)^2 = (b^2 - a^2)(\cos^2\theta - \sin^2\theta)$$

Substitute for $\dfrac{dp}{d\theta}$ and get $\dfrac{d^2p}{d\theta^2}$

Ex. 142. If $y = \sin^m x$ show that

$$\sin^2 x\frac{d^2y}{dx^2} = (m^2\cos^2 x - m)y.$$

Ex. 143. If $x^{\frac{1}{m}} + y^{\frac{1}{m}} = 2x$ show that

$$(x^2 - 1)\frac{d^2y}{dx^2} + x\frac{dy}{dx} - m^2y = 0.$$

Ex. 144. If $y = \dfrac{1}{2}(\sin^{-1} x)^2$ show that

$$(1 - x^2)\frac{d^2y}{dx^2} - x\cdot\frac{dy}{dx} = 1$$

Since $2y = (\sin^{-1}x)^2$

$$\therefore \quad 2\frac{dy}{dx} = \frac{2(\sin^{-1}x)}{\sqrt{1-x^2}} \text{ or } (1-x^2)\left(\frac{dy}{dx}\right)^2 = (\sin^{-1}x)^2 = 2y$$

Differentiating, $(1 - x^2)\, 2\left(\dfrac{dy}{dx}\right)\dfrac{d^2y}{dx^2} - 2x\left(\dfrac{dy}{dx}\right)^2 = 2\dfrac{dy}{dx}$

Rearranging and cancelling out we get the result.

Ex. 145. If $x = (a + bx)\, e^{y/x}$ show that

$$x^3\left(\frac{d^2y}{dx^2}\right) = \left(x\frac{dy}{dx} - y\right)^2$$

Ex. 146. If $y = a \cos \log x + b \sin \log x$ show that

$$x^2 = \frac{d^2y}{dx^2} + x\frac{dy}{dx} + y = 0.$$

10.8 ECONOMIC VARIABLES

Variables like demand, price, employment etc. cannot change by infinitesimally small quantities as desired in calculus which is concerned with the behaviour of continuous variables. But under certain conditions it is feasible to treat certain economic variables as continuous. For instance, a change of 10 numbers in the employment level of one million is relatively extremely small and the calculus techniques can be employed. Strictly speaking, economic variables are discontinuous but this shortcoming can usually be ignored.

It is possible to get some useful results by translating various economic laws into the language of calculus.

The relations between economic variables can be expressed by means of functions and curves. The derivatives obtained from these functions give us the marginal values in economics. All decision processes in economics make use of the concept of the marginal. Calculus along with other mathematical tools can be usefully employed here.

We make use of symbols to represent economic variables. Some symbols may be used for different purposes. But the meaning will be clear from the context.

u : Utility	S : Saving
x : Quantity, output;	I : Investment
demand, supply, scales	Y : Income
C : Cost/Consumption	π : Profit
R : Revenue	r : Rate of Interest
p : Price or average revenue	K : Capital
	L : Labour etc.

The common marginal concepts in economics are :

Marginal utility $\frac{du}{dx}$; Marginal cost $\frac{dC}{dx}$; Marginal revenue $\frac{dR}{dx}$; Marginal product of the factor labour $\frac{dx}{dL}$; Marginal return on investment $\frac{d\pi}{dI}$ etc.

In macroeconomics, the symbol C is used for consumption and $\frac{dC}{dY}$ denotes marginal propensity to consume (or *mpc*). Similarly $\frac{dS}{dY}$ denotes marginal propensity to save (mps).

The concept of the average is the familiar one. Thus average cost per unit of output is defined as the ratio of total cost C to output x.

Average cost $AC = C/x$, Average Revenue $AR = R/x$ and so on.

10.9 POINT ELASTICITY

Elasticity of demand is the relative change in demand in response to a relative change in price. If these changes are small and the demand function $x = f(p)$ is a continuous function of price p, we define the elasticity of demand

$$\eta_d = \frac{dx/x}{dp/p} = \frac{dx}{dp}\cdot\frac{p}{x}, \text{ often denoted by } \eta.$$

Elasticity of supply is the relative change in supply in response to a relative change in price. If now x stands for supply and the supply function is written $x = g(p)$, the formula for elasticity of supply retains the same form as that for η_d.

$$\eta_s = \frac{p}{x}\frac{dx}{dp} \text{ where } x \text{ is the supply function.}$$

In general, the slope of the demand curve is negative and hence η_d is *negative.*

Similarly, the slope of the supply curve is generally positive and hence η_s is *positive.*

It is conventional to use the absolute value of elasticity to consider if a function is elastic at a point.

Ex. 147. Find η_d for the demand function $x = y - 2p$ when

(*i*) $p = 1$ (*ii*) $p = 2$

Here we have $\frac{dx}{dp} = -2$

$$\eta_d = \frac{p}{x}\cdot\frac{dx}{dp} = \frac{p}{7-2p}(-2) = \frac{-2p}{7-2p}$$

(*i*) For $p = 1$, $\eta_d = \frac{-2\times 1}{7-2\times 1} = -\frac{2}{5}$

(*ii*) For $p = 2$, $\eta_d = \frac{-2\times 2}{7-2\times 2} = -\frac{4}{3}$

Ex. 148. Find η_s for the supply function $x = 5 + 2p^2$

since $\frac{dx}{dp} = 4p$

$$\eta_s = \frac{p}{x}\frac{dx}{dp} = \frac{p}{5+2p^2}\cdot 4p = \frac{4p^2}{5+2p^2}$$

(*i*) For $p = 2$, $\eta_s = \frac{4\times 4}{5+2\times 4} = \frac{16}{13}$

(*ii*) For $p = 3$ $\eta_s = \frac{4\times 9}{5+2\times 9} = \frac{36}{23}$

Ex. 149. Prove that for the demand curve $xp^{-a} = b$ where a, b are constants, $\eta_d = -a$, a constant. (Note that $xp^{-a} = b$ is the equation of the rectangular hyperbola).

Ex. 150. Prove that elasticity is the ratio of the marginal function and the average function.

$$\eta = \frac{p}{x}\cdot\frac{dx}{dp} = \frac{dx}{dp} \div \frac{x}{p} = \frac{\frac{dx}{dp}}{\frac{x}{p}} = \frac{\text{marginal value}}{\text{average value}}$$

Ex. 151. Show that
(*i*) $\eta_d = 0$ when demand is perfectly inelastic
(*ii*) $\eta_d = \infty$ when demand is perfectly elastic
(*iii*) $\eta_d > 1$ when demand is elastic
(*iv*) $\eta_d < 1$ when demand is inelastic

10.10 ELASTICITY OF SUBSTITUTION

Consider the production function $Y = f(K, L)$

where output Y is a function of capital K and labour L. For an optimal input combination of L and K we must have

$$\frac{p_L}{p_K} = \frac{MP_L}{MP_K}$$

where p_L, p_K are prices per unit of L, K, and MP_L, MP_K, the marginal products of L, K respectively.

$$\frac{MP_L}{p_L} = \frac{MP_K}{p_K}$$

$$\therefore \quad \frac{1}{p_L}\frac{dY}{dL} = \frac{1}{p_K}\frac{dY}{dK}$$

$$\therefore \quad \frac{dL}{dK} = \frac{p_K}{p_L}$$

If s is the marginal rate of substitution between L, K,

$$s = \frac{dL}{dK} = \frac{p_K}{p_L}$$

We may define elasticity of sulutibution when two factor inputs, L K are given, as follows:

The elasticity of substitution σ measures the extent to which factor K can be substituted for factor L when the price of K falls relative to the price of L. If we write $x = \frac{L}{K}$ we have

$$\sigma = \frac{dx}{x} \div \frac{ds}{s} = \frac{dx}{ds}\frac{s}{x}$$

A similar definition of elasticity of substitution can be given in the case of a consumer's preference scale at two points of time.

Ex. 152. Show that for the Cobb Douglas production function $Y = aL^{\alpha}K^{1-\alpha}$ we have $\sigma = 1$ at all points on every production isoquant. Show that if $\sigma > 1$ the share of receipts from a relative price fall increases and *vice versa*.

10.11 REVENUE

Total Revenue $R = p \cdot x$ = price × quantity sold

Marginal Revenue $MR = \dfrac{dR}{dx} = p + x\dfrac{dp}{dx} = p\left(1 + \dfrac{x}{p}\dfrac{dp}{dx}\right)$

$= p\left(1 - \dfrac{1}{\eta_d}\right)$ since η_d is a negative quantity

i.e., $MR = AR\left(1 - \dfrac{1}{\eta_d}\right)$ or $AR = MR\dfrac{\eta_d}{\eta_d - 1}$

It follows from this that when

(*i*) $\eta_d = 1$, TR remains constant with a fall in price

(*ii*) $\eta_d > 1$, TR rises with a fall in price

(*iii*) $\eta_d < 1$, TR falls with a fall in price

10.12 COST

Total cost C is a function of output x : $C = f(x)$

Average cost $AC = \dfrac{\text{Total cost}}{\text{Total output}} = \dfrac{C}{x}$

Total Cost $C = AC \times x$

Marginal cost is the rate of change of cost with output.

Marginal cost $MC = \dfrac{dC}{dx}$

Ex. 153. Let $C = 10 + 2x + 3x^2$

then $AC = \dfrac{C}{x} = \dfrac{10}{x} + 2 + 3x$

$$MC = \frac{dC}{dx} = 2 + 6x$$

The slope of a curve is obtained by finding its derivative. The slope of the average cost curve is obtained by finding the derivative of C/x with respect to x.

Ex. 154. The slope of AC curve,

$\dfrac{C}{x} = \dfrac{10}{x} + 2 + 3x$ is given by

$$\frac{d}{dx}\left(\frac{C}{x}\right) = \frac{d}{dx}\left(\frac{10}{x} + 2 + 3x\right) = -\frac{10}{x^2} + 3.$$

In general if $AC = \dfrac{C}{x}$, the slope is given by

$$\frac{d}{dx}\left(\frac{C}{x}\right) = \frac{x\dfrac{dC}{dx} - C}{x^2} = \frac{1}{x}\left(\frac{dC}{dx} - \frac{C}{x}\right) = \frac{1}{x}(MC - AC)$$

Slope of AC curve $= \frac{1}{x}(MC - AC)$. This

(*i*) Slope < 0 if $MC < AC$ *i.e.*, the curve slopes downwards

(*ii*) Slope > 0 if $MC > AC$ *i.e.*, the curve slopes upwards

(*iii*) Slope $= 0$ if $MC = AC$ which gives the lowest point of the curve where MC curve cuts AC curve.

Cost Elasticity

The elasticity of total cost function $C = C(x)$ is given by

$$\eta_c = \frac{dC}{dx}\frac{x}{C} = \frac{C'}{A} \text{ where } A = \frac{C}{x}$$

$$= \frac{\text{Marginal cost}}{\text{Average cost}}$$

The elasticity of average cost $A = C/x$

$$\eta_a = \frac{dA}{dx}\frac{x}{A} = \frac{x^2}{C}\frac{x\frac{dC}{dx} - C}{x^2} = \frac{x}{C}\frac{dC}{dx} - 1 = \eta_c - 1$$

This can be used to infer the behaviour of average cost function and hence the nature of returns to scale and the expansion path.

The elasticity η_m of marginal cost C' is obtained from the relation $C' = A\eta_c$.

$$\eta_m = \frac{dC'}{dx}\frac{x}{C'} = \left(A\frac{d\eta_c}{dx} + \eta_c\frac{dA}{dx}\right)\frac{x}{C'}$$

$$= \left(A\frac{d\eta_c}{dx} + \eta_c\frac{dA}{dx}\right)\frac{x}{A\eta_c} = \frac{x}{\eta_c}\frac{d\eta_c}{dx} + \frac{x}{A}\frac{dA}{dx}$$

$$= \eta_\eta + \eta_a = \eta_\eta + \eta_c - 1$$

where η_η is the elasticity of η_c.

Ex. 155. For $C = a_1x^2 + a_2x + a_3$, $\eta_c = \frac{x(2a_1x + a_2)}{a_1x^2 + a_2x + a_3}$,

$$\eta_a = \eta_c - 1 = \frac{a_1x^2 - a_3}{a_1x^2 + a_2x + a_3}$$

When an average falls, the marginal falls faster. When an average rises, the marginal rises faster. If we take the example of the cost function $C = f(x)$ we note that the MC curve rises or falls faster than the AC curve. When AC is the minimum, $AC = MC$. We can derive these results.

Let $\quad AC = A = \frac{C}{x}$ then $\frac{dA}{dx} = \frac{1}{x}\left(\frac{dC}{dx} - \frac{C}{x}\right)$.

(*i*) For minimum AC, $\frac{dA}{dx} = 0 \ \therefore\ \frac{dC}{dx} = \frac{C}{x}$

i.e., MC curve cuts the AC curve and $MC = AC$

(*ii*) When $\frac{dC}{dx} < \frac{C}{x}$ *i.e.*, $MC < \dot{A}C$, $\frac{dA}{dx} < 0$ *i.e.*, AC curve is falling.

(*iii*) When $\frac{dC}{dx} > \frac{C}{x}$, $\frac{dA}{dx} > 0$ *i.e.*, AC curve is rising.

Ex. 156. For the cost function $C = (ax + b)^{1/2}$, the elasticity of cost, η_c rises but remains less than unity as x rises.

$$\eta_c = \frac{dC}{dx}\frac{x}{C} = \frac{d\ln C}{d\ln x} = \frac{d\left\{\frac{1}{2}ln(ax+b)\right\}}{d\{ln\, x\}} = \frac{a}{2(a+b)}\Big/\frac{1}{x}$$

$$= \frac{ax}{2(ax+b)} < 1 \text{ as } x \text{ increases}$$

$$\frac{d\eta_c}{dx} = \frac{a(ax+b) - ax \cdot x}{2(ax+b)^2} = \frac{ab}{2(ax+b)^2} > 0 \text{ if } a, b > 0$$

$\therefore$ η_c rises as x rises.

Ex. 157. The demands x_1, x_2 for two goods at prices p_1, p_2 have the relation $p_2(a_1 + b_1x_1) = p_1(a_2 + b_2x_2)$. Given the consumer's income $M = p_1x_1 + p_2x_2$, find the demand function for each good.

Substitute $M = p_1x_1 + p_2x_2$ or $x_1 = \frac{M - p_2x_2}{p_1}$

into the relation so that

$$p_2\left[a_1 + b_1\left(\frac{M - p_2x_2}{p_1}\right)\right] = p_1(a_2 + b_2x_2) - b_1\frac{p_2}{p_1}x_2 - p_1p_2x_2$$

$$= p_1a_1 - p_2a_1 - p_2\frac{b_1}{p_1}M$$

Simplifying $x_2 = \frac{M b_1 p_2 + a_1 p_1 p_2 - a_2 p_1^2}{b_2 p_1^2 + a_1 p_2^2}$ demand function for good 2.

Similarly for x_1.

10.13 PRODUCTION FUNCTION AND MARGINAL REVENUE PRODUCT

For simplicity, let us consider production (Y) as a function of a single factor, say quantity of capital (K) used to produce Y :

$$Y = f(K)$$

AP = Average Product $= \frac{Y}{K}$

The Marginal Product $= \frac{dY}{dK}$ is the slope of the product curve. It is written MPK.

Thus if $Y = 100 + 10K - K^2$

Marginal product of capital is given by $\frac{dY}{dK} = 10 - 2K$

Now consider output as a function of labour, L.

$$Y = f(L)$$

Since revenue R is a function of output

$$R = g(Y) = g[f(L)]$$

$$\frac{dR}{dL} = \frac{dR}{dY}\frac{dY}{dL}$$

which means $MRP = MR \times MPP$

where $MRP = \frac{dR}{dL}$ = marginal revenue product of labour

$MR = \frac{dR}{dY}$ = marginal revenue

MPP = marginal physical product of labour

Let $R = 5x + 12$, $Y = 3L^2 - 2L$ then

$$\frac{dR}{dY} = 5, \frac{dY}{dL} = 6L - 2$$

$$\frac{dR}{dL} = \frac{dR}{dY}\cdot\frac{dY}{dL} = 5(6L - 2).$$

If output is a function of labour only, $Y = f(L)$, then we have price $p = h(Y)$ and revenue $R = Yg(Y)$.

$$\frac{dR}{dL} = \frac{dR}{dY}\frac{dY}{dL} = \{Yh'(Y) + h(Y)\}\frac{dY}{dL}$$

$$= \left(Y\frac{dp}{dY} + p\right)\frac{dY}{dL}$$

$$= MR \cdot MPP$$

Since labour is the only input with wage rate w, cost $C = wL$ and marginal cost of labour $dC/dL = w$.

The marginal cost with respect to output Y is

$$C_y = \frac{dC}{dY} = \frac{dC}{dL}\frac{dL}{dY} = \frac{w}{dY/dL} = \frac{MC \text{ of variable input}}{MPP}$$

The slope of MC curve is given by

$$\frac{d^2C}{dY^2} = \frac{d}{dY}\left(\frac{w}{dY/dL}\right) = -\frac{w}{(dY/dL)^2}\cdot\frac{d^2Y}{dL^2}\cdot\frac{dL}{dY} = \frac{-w(d^2Y/dL^2)}{(dY/dL)^3}$$

> 0 if $d^2Y/dL^2 < 0$ *i.e.*, if MPP is diminishing.

Ex. 158. For the demand curve $p = a + bY$ find MRP if the production function is $Y = c_1L + c_2L^2$

$$\frac{dR}{dL} = \left(Y\frac{dp}{dY} + p \right)\frac{dY}{dL} = \{(c_1L + c_2L^2)b + a + b\,(c_1L + c_2L^2)\}\,(c_1 + 2c_2L)$$

$$= (a + 2bc_1L + 2bc_2L^2)\,(c_1 + 2c_2L)$$

Ex. 159. Find the elasticity of demand and MR and plot the functions.

(*i*) $p = 100 - x - x^2$ at $p = 70$ (*ii*) $x = 50 - 5p - p^2$ at $p = 2$

(*iii*) $x = 2p - p^2$ at $p = 1$ (*iv*) $px = 1$ (*v*) $x = 20 - 2p^2$ at $p = 2$

(*vi*) $p = \frac{7}{x}$ (*vii*) $p = \frac{10}{3+x}$ at $x = 0$, $x = 9$

Ans. (*i*) At $p = 70$, $x = 5$ $\therefore \eta_d = \frac{100 - x - x^2}{x(1+2x)} = \frac{14}{11}$ (*ii*) $\frac{1}{2}$ (*iii*) 0 (*iv*) 1

(*v*) 1.33 (*vi*) 1 (*vii*) undefined; $\frac{4}{3}$

MR : *e.g.* (*i*) $R = px = 100x - x^2 - x^3$ $\therefore$ $MR = dR/dx = 100 - 2x - 3x^2$

(*ii*) $R = px = 50p - 5p^2 - p^3$ $\therefore$ $\frac{dR}{dx} = \frac{dR}{dp}\cdot\frac{dp}{dx} = \frac{3p^2 + 10p - 50}{5 + 2p}$

Ex. 160. Find MC if

(*i*) $C = 1000 + 2x - 5x^2$ (*ii*) $C = 500 - 3x + 2x^2$

(*iii*) $C = 11 + x$ (*iv*) $C = \frac{5}{x} + 10 + 0.5x$

Ans. (*i*) $2 - 10x$, (*ii*) $-3 + 4x$, (*iii*) 1, (*iv*) $-\frac{5}{x^2} + 0.5$

Ex. 161. Find MC and AC if

$$C = 1000 + 100x - 10x^2 + x^3$$

Ans. MC : $10C - 20x + 3x^2$; AC : $\frac{1000}{x} + 100 - 10x + x^2$

Ex. 162. If total production is given as a function of labour,

$$TP : Y = 10L + 15L^2 - 3L^3$$

We have the functions for the average and marginal product,

$$AP : \frac{Y}{L} = 10 + 15L - 3L^2;\ MP : \frac{dY}{dL} = 10 + 30L - 9L^2$$

These functions can be presented in tables or in diagrams.

Units of labour L	*Total product* Y	AP $\frac{Y}{L}$	MP $\frac{dY}{dL}$
0	0	0	0
1	22	22	31
2	56	28	34
3	111	28	19
4	88	22	– 14
5	50	10	– 65

Ex. 163. If the curves shown here correspond to the production function $Y = f(L)$, what do the points P, M_T, M_A, F stand for?

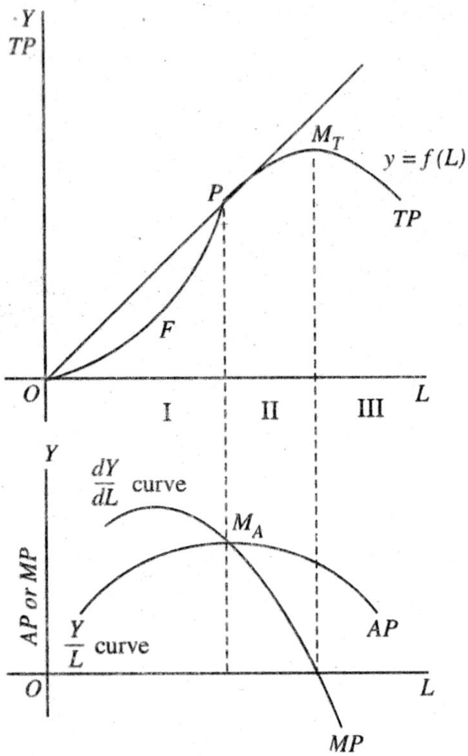

$$\frac{Y}{L} = AP$$

= Average Productivity

$$\frac{dY}{dL} = MP$$

= Marginal Productivity

M_T corresponds to the maximum production when MP is zero.

M_A corresponds to the maximum AP, the point where MP curve cuts AP curve.

F is the point of inflection of the production curve. In the diagram we can notice : Stage

I. increasing returns when $\frac{dY}{dL} > \frac{Y}{L}$

II. decreasing returns when $\frac{dY}{dL} < \frac{Y}{L}$

III. constant returns when $\frac{dY}{dL} = \frac{Y}{L}$

10.14 OUTPUT ELASTICITY AND AGGREGATE DEMAND

If p is the general price level, Y the aggregate demand, Q the level of output, then the equilibrium level of Y is given by

$$Y = pQ$$

If we assume the aggregate supply function

$$p = AQ^{\alpha-1}, \alpha > 1$$

then $$Y = AQ^{\alpha} \text{ or } Q = (Y/A)^{1/\alpha}$$

Here α is the *inflation coefficient*. It α is large, inflation is high requiring a larger output. We can write $p = Y^{\frac{\alpha-1}{\alpha}} A^{\frac{1}{\alpha}}$

The output elasticity with respect to aggregate demand is given by

$$\eta_{QY} = \frac{dQ}{dY}\frac{Y}{Q} = \frac{1}{\alpha A}\left(\frac{Y}{A}\right)^{\frac{1}{\alpha}-1} \cdot \frac{Y}{Q} = \frac{1}{\alpha}$$

Similarly $$\eta_{pY} = \frac{dp}{dY}\frac{Y}{p} = \frac{\alpha-1}{\alpha}\left(\frac{A}{Y}\right)^{1/\alpha} = 1 - \frac{1}{\alpha} = 1 - \eta_{QY}$$

$$\eta_{Qp} = \frac{1}{\alpha - 1}, \quad \eta_{pQ} = \alpha - 1$$

Ex. 164. Given the supply function $x = 2\sqrt{p-7}$, examine how the elasticity of supply η_s changes with price.

Note. $\eta_s = \dfrac{dx}{dp}\dfrac{p}{x} = \dfrac{1}{\sqrt{p-7}}\dfrac{p}{2\sqrt{p-7}} = \dfrac{2p}{x^2}$ which decreases as x increases.

10.15 THE LOGISTIC FUNCTION

This function has many uses in economics and other fields. Its general form is

$$y = \frac{k}{1+e^{P(x)}}$$

where k is a constant and $P(x)$ is a polynomial in x. Usually x stands for time periods. The curve can be used to represent adoption of new products by the market, growth of animal populations, industrial products etc.

If $P(x) = -ax$ or $P(x) = -(-b + ax)$ the curve has a positive slope and the value of k determines the asymptotic limit of the relation.

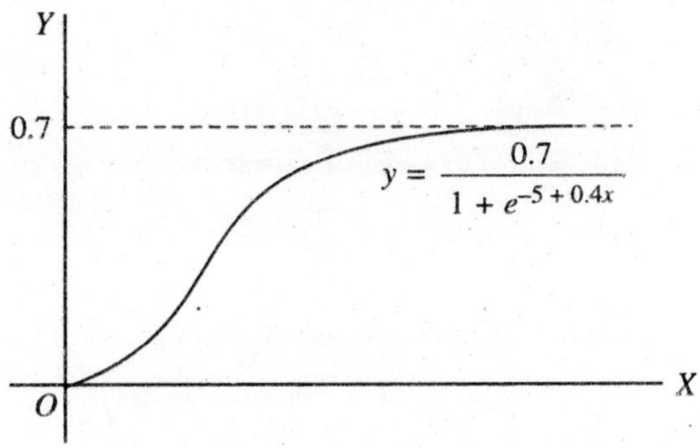

The function $y = \dfrac{100}{1+e^{-0.5x}}$ approaches the value 100 as x becomes large.

Similarly $y = \dfrac{0.7}{1+e^{-(5+0.4x)}}$ approaches 0.7 as x increases.

Ex. 165. If y is the percentage of people who have acquired a washing machine in time period x (measured in years, $x \geq 1$) and if $a = -0.2$, $k = 0.8$ when $P(x) = -0.2x$ we have, for

$$x = 1,\ y = 0.8\ (1 + e^{-0.2 \times 1})^{-1} = 0.44$$
$$x = 2,\ y = 0.8\ (1 + e^{-0.2 \times 2})^{-1} = 0.48$$

and so in.

$$x = 20,\ y = 0.8\ (1 + e^{-2 \times 20})^{-1} = 0.78$$

Since $x \geq 1$, this means one year following the introduction of the washing machine in the market; after 1 year, 44% people bought a set; after 2 years, 48% people bought a set; after 20 years, 78% people bought a set. If $x \to \infty$, $y \to 0.8$ *i.e.*, the maximum number of sets bought will be 80%.

Ex. 166. Show that the logistic curve

$$y = \frac{k}{1+e^{-a+bx}}$$

has a positive slope. **Ans.** $\dfrac{dy}{dx} = \dfrac{kbe^{-(a+bx)}}{[1+e^{-(a+bx)}]^2}$

The slope is positive and $\to 0$ as $x \to \infty$. Then the curve moves parallel to the x-axis.

Ex. 167. Show $y = \dfrac{100}{1+e^{0.3x}}$ approaches 0 as $x \to \infty$.

10.16 RATE OF GROWTH

The rate of growth of any function $y = f(t)$ with reference to time t is defined by

$$g = \frac{dy/dt}{y} = \frac{\text{marginal function}}{\text{total function}}$$

provided the function $f(t)$ is continuous and differentiable.

The rate of change of population N under normal conditions may be written

$$g = \frac{dN/dt}{N}$$

For exponential change if N_o is the initial population,

$$N = N_o e^{gt}$$

If g is positive it is a case of exponential growth.

If g is negative it is exponential decay.

The rate of growth of a function $y = f(t)$ may be written.

$$\frac{d\,ln\,y}{dt} = \frac{dy/dt}{y}$$

The rate of growth of a product $y = u(t)\ v\ (t)$ is

$$\frac{d\,ln\,y}{dt} = \frac{d\,ln\,u}{dt} + \frac{d\,ln\,v}{dt}$$

$= \text{Rate of growth of } u + \text{Rate of growth of } v.$

The rate of growth of $y = \dfrac{u}{v}$ is the difference between the rate of growth of u and v.

For the function $y = a^t$, the rate of growth is

$$\frac{d\,ln\,y}{dt} = \frac{d}{dt}(t\,ln\,a) = ln\,a.$$

If a, b, c are the rates of growth of population (N), national income (Y) and per capita income (y) respectively, then since $y = Y/N$,

$$\frac{d\,ln\,y}{dt} = \frac{d\,ln\,Y}{dt} - \frac{d\,ln\,N}{dt} \quad \text{or } c = b - a.$$

Growth and Decay

If r is the instantaneous rate of growth, P, the initial value of a quantity and A the final value in a continuous compounding process then

$$A = Pe^{rt}$$

$$\textit{Rate of change of } A = \frac{dA}{dt} = Pre^{rt} = rA$$

If P stands for money and r for rate of interest it means that at any time t, money is increasing at interest rate times the principle *at that time*.

Rate of growth of A is the relative quantity,

$$g = \frac{\frac{dA}{dt}}{A} = \frac{rA}{A} = r.$$

This rate of growth being with reference to a specific point of time t, is an instantaneous rate of growth. In this case the rate of growth, r, happens to be a constant though it may not be so always.

Although the discussion here is with reference to continuous growth it is found to be useful in practical cases where growth is discrete or discontinuous *i.e.*, where changes occur once or a few times during the period. For instance, in the usual (discontinuous) compound interest pattern of growth $P(1 + r)^t$, where t stands for the number of compounding periods, we may suitably replace $1 + r$ by e^r so that

$$P(1 + r)^t = Pe^{rt}$$

In the opposite case of discounting, to find the present value P of a given sum A due in t years from now we have

$P = A(1 + r)^{-t}$ in the discontinuous case

$P = Ae^{-rt}$ in the continuous case.

Here $-r$ is the instaneous rate of negative growth or decay.

Ex. 168. How long will it take for money invested at 5.5 per cent compounded daily to triple? **Ans.** 20 years

(This can be taken as a case of continuous compounding)

10.17 THE INVESTMENT MULTIPLIER

Macroeconomics makes use of identities and technical relations for models. It we write Y for national income or national expenditure, C for consumption, G for government expenditure, I for investment, X and M for exports and imports, we have the relation.

$$Y = C + I + G + X - M.$$

If C includes all expenditure and I all autonomous investment we may write the simplified version.

$$Y = C + I$$

Suppose that consumption is a linear function of income so that $C = a + bY$. We get by rearranging.

$$Y = \frac{a + I}{1 - b}.$$

If an autonomous increase in investment induces an increase in income, we define the investment multiplier.

$$\frac{dY}{dI} = \frac{1}{1 - b}$$

where b, the marginal propensity to consume (mpc), is less than one since a community tends to spend a certain proportion of any additional income.

If S stands for savings, $C + S = Y$ so that

$$\frac{dC}{dY} + \frac{dS}{dY} = \frac{dY}{dY} = 1 \text{ i.e., mpc + mps = 1.}$$

We have written mps for marginal propensity to save.

Ex. 169. For the continuous linear consumption function

$$C = 20 + \frac{2}{3}Y, \quad \text{mpc} = \frac{dC}{dY} = \frac{2}{3}, \quad \text{mps} = 1 - \text{mpc} = \frac{1}{3}.$$

Average propensity to consume $apc = \dfrac{C}{Y}$.

If $Y = 300$, $C = 20 + 200 = 220$ then apc $= \dfrac{220}{300}$.

Ex. 170. Find apc, mpc, mps for the consumption function $C = 10 + 0.5\ Y$ when Y is (*i*) 8 (*ii*) 10 (*iii*) 20. **Ans.** (*i*) 1.75, 0.5, 0.5 (*ii*) 1.5, 0.5, 0.5 (*iii*) 1, 0.5, 0.5

10.18 BALANCED BUDGET MULTIPLIER

Let C, I, G stand for consumption, investment, and government expenditure. Let the disposable income be $Y_d = Y - t$ where t is the level of net tax yield and let consumption be a linear function of disposable income:

$$C = a + bY_d = a + b(Y - t)$$

If we write the equation for the market equilibrium.

$$Y = C + I + G$$

$$= \frac{a - bt + I + G}{1 - b} \text{ by substitution}$$

we have

the investment multiplier $\dfrac{dY}{dI} = \dfrac{1}{1-b}$

the government purchase multiplier $\dfrac{dY}{dG} = \dfrac{1}{1-b}$

the tax multiplier $\dfrac{dY}{dt} = \dfrac{-b}{1-b}$

It can be seen that $\dfrac{dY}{dG} + \dfrac{dY}{dt} = 1$, irrespective of the value of b, the marginal propensity to consume. This result is known as the balanced budget multiplier when a change in government expenditure is balanced by an equal increase in tax revenue.

Ex. 171. Show that $\dfrac{dC}{dG} = \dfrac{b}{1-b} = b\left(\dfrac{dY}{dG} - \dfrac{dt}{dG}\right)$

which may be written $dC = b\ (dY - dt)$.

Ex. 172. If $t = a_1 + b_1 Y$ so that $C = a + b\,(Y - t) = a - ba_1 + b(1 - b_1)\,Y$ show that

$$dY = \frac{dI}{1 - b(1 - b_1)},\quad dt = \frac{b_1 dI}{1 - b(1 - b_1)}$$

(If investment declines by dI, the level of income falls by dY and an induced budgetary deficit dt results).

Ex. 173. Show that if $C = ax^b + k,\ a > 0,\ b > 0,\ k \geq 0$

then $MC = abx^{b-1}$

and marginal average cost $= a\,(b - 1)x^{b-2} - k/x^2$

Also show that $MC = AC$ if $x = \dfrac{k^{1/b}}{a^{1/b}\,(b - 1)^{1/b}}$

Ex. 174. Show that MC and AC are equal for the function

$$C = ae^{bx} \text{ if } x = 1/b$$

Ex. 175. Show that if $x = f(p)$ we may write the elasticity of x with respect to p as

$$\eta = \frac{dlnx/dp}{dlnp/dp}$$

$$\left(\text{Note that } \frac{dlnx}{dp} = \frac{1}{x}\frac{dx}{dp} \text{ and } \frac{dlnp}{dp} = \frac{1}{p}\right).$$

Ex. 176. If the disposable income $Y_d = Y - t$ where Y is income, t the amount of tax and utility u is a function of income, write

$$u_1 = u(Y),\ u_2 = u(Y - t)$$

then $u_1 - u_2 = \text{constant}.$

Show that the change in tax with respect to income that is necessary for equal absolute sacrifice is given by

$$\frac{dt}{dY} = 1 - \frac{du_1/dY}{du_2/dY} = 1 - \frac{du_1}{du_2}$$

Ex. 177. For the tax yield function $z = xt$ find the change in tax yield when the demand function $x = \dfrac{5}{p^{3/2}}$ changes to $x = \dfrac{5}{p^{11/10}}$. What happens when $z = xtp$?

$$z_1 = xt = \frac{5t}{p^{3/2}} \text{ become } z_2 = \frac{5t}{p^{11/10}} = p^{0.4}\,z_1.$$

This is a case of per unit tax.

When $z = xtp,\ z_1 = \dfrac{5tp}{p^{3/2}}$ becomes $z_2 = \dfrac{5tp}{p^{11/10}} = p^{0.4}\,z_1$

This is a case of tax on total revenue $R = xp$.

Ex. 178. Given total cost $C = 100 + 1.2\,x - 0.0001\,x^2$, is it a case of a perfectly competitive market? Are there decreasing returns?

$dC/dx = MC = 1.2 - 0.0002\,x$ is a decreasing function of x. In a perfectly competitive market with constant price, the firm will go on producing more and more.

No diminishing returns because diminishing marginal product is necessary and sufficient for increasing marginal cost.

10.19 IMPLICIT FUNCTIONS AND POLICY MODELS

From the implicit relation $f(x, y) = 0$ one may want to know x in terms of y or y in terms of x. For example, if x, p stand for demand and price in the function $f(x, p) = 0$, one may write demand as a function of price or *vice versa* depending on the situation.

In policy models, target variables like national income may be expressed in terms of policy variables like development expenditure. Conversely, for policy formulation, we may want to express policy variables explicitly for consistency with target variables.

Income Determination: Implicit Differentiation with Income Identity

Ex. 179. When two variables are simultaneously determined by equilibrium mechanism, the causality of a variable depends on the situational convenience. In the simple model below, income Y is an aggregate of consumption C (which is a function of Y), investment I, government expenditure G, net exports or exports X minus imports M.

$$Y = C(Y) + I + G + (X - M)$$

If I, G, X, M are assumed constant we have

$$\frac{dY}{dY} = \frac{dC}{dY} + \frac{dI}{dY} \text{ or } \frac{dI}{dY} = 1 - \frac{dC}{dY}$$

i.e.,
$$\frac{dY}{dI} = \frac{1}{1 - dC/dY}, \text{ the investment multiplier}$$

Here if $C = a + bY$ then the average propensity to consume, $apc = C/Y = b + a/Y$ and marginal propensity to consume $mpc = dC/dY = b$.

If $a > 0$ then $apc > mpc$

For large Y, $apc \simeq mpc$

If $C = bY$, $Y = C + I$, $I = I_o$, a constant then

$$Y = bY + I = \frac{1}{1-b} I \text{ so that } \frac{dY}{dI} = \frac{1}{1-b} > 1$$

For $C = bY$, $\dfrac{dY}{dI} = \dfrac{1}{1-b}(I + a)$ *i.e.*, all autonomous expenditure as consumption or investment can be lumped together.

Ex. 180. Let Go, Xo, M stand for autonomous government expenditure, constant export demand and import demand respectively. Show that the model

$$Y = C + I + Go + Xo - M$$
$$C = a + bY$$
$$I = m + vY$$
$$M = f + gY$$

can be written $\begin{bmatrix} 1 & -1 & -1 & 1 \\ -b & 1 & 0 & 0 \\ -v & 0 & 0 & 0 \\ -g & 0 & 0 & 1 \end{bmatrix} \begin{bmatrix} Y \\ C \\ I \\ M \end{bmatrix} = \begin{bmatrix} Go + Xo \\ a \\ m \\ f \end{bmatrix}$

Ex. 181. In the model $Y = C + I_o$, $C = a + bY$, it is given that the exogenous investment I_o is 50, $a = 100$ and $b = 0.8$. Show that

$$\begin{bmatrix} Y \\ C \end{bmatrix} = \begin{bmatrix} 1 & -1 \\ -0.8 & 1 \end{bmatrix}^{-1} \begin{bmatrix} 50 \\ 100 \end{bmatrix} = \begin{bmatrix} 750 \\ 700 \end{bmatrix}$$

What will be the values of Y and C if $I_o = 200$

and $a = 440$, $b = 0.6$? **Ans.** $Y = 1600$, $C = 1400$

Ex. 182. By substituting the equation $C = a + bY$ into $Y = C + I_o$ we have,

$$Y = \left(\frac{1}{1-b}\right)a + \left(\frac{1}{1-b}\right)I_o$$

We have the investment multiplier given by

$$\frac{dY}{dI_o} = \frac{1}{1-b} = k$$

If we write, $C = \left(\frac{1}{1-b}\right)a + \left(\frac{b}{1-b}\right)I_o$

we have the consumption investment multiplier, $\frac{dC}{dI_o} = \frac{b}{1-b}$.

Changes in I_o, C

$k = \frac{1}{1-b}$ is the change in Y corresponding to a change in autonomous investment I_o.

$\frac{dC}{dI_o} = \frac{b}{1-b}$ represents a change in C corresponding to a change in autonomous investment.

Similarly $\frac{dY}{da} = \frac{1}{1-b}$ represents a change in Y corresponding to a change in autonomous consumption expenditure a.

This is the same as the change resulting from I_o.

Finally $\frac{dC}{da} = \frac{1}{1-b}$ shows the impact on C of a change in a.

Ex. 183. Show that for the model $Y = C + G_o$, $C = a + bY$ the equilibrium values are $Y = \frac{G_o + a}{1-b}$, $C = \frac{G_o b + a}{1-b}$ and the multipliers are

$$\begin{bmatrix} \frac{1}{1-b} & \frac{1}{1-b} \\ \frac{b}{1-b} & \frac{1}{1-b} \end{bmatrix}$$

Ex. 184. Find the trade off rate between change in income, government expenditure and deficit.

If Y_d is disposable income and G the autonomous government expenditure, $Z = tY$ the total tax yield, then we have,

$$C = bY_d$$
$$Y = C + I + G$$
$$Y_d = Y - tY$$

so that

$$Y = bY(1-t) + I + G$$
$$= \frac{I+G}{1-b(1-t)}$$

The multiplier is

$$k = \frac{1}{1-b(1-t)}$$
$$< \frac{1}{1-b} \text{ if } t > 0$$

For $k > 1$, $1 - b(1-t) < 1$ or $bt < b$

i.e., $t < 1$

i.e., for k to exceed unity the tax rate should be less than one and the smaller is t the greater will be the multiplier.

Again, since $\dfrac{dY}{dG} = \dfrac{1}{1-b(1-t)}$, $dY = k\, dG$

The deficit $dH = dG - dZ$
$$= dG - tk\, dG$$
$$= (1 - tk)\, dG$$

A rise dG in government expenditure results in income increase greater than itself but an increase in budget deficit results in an increase smaller than itself. The trade off rate between dY, dH, is

$$\varphi = \frac{dY}{dH} = \frac{k}{1-tk}$$

Ex. 185. Show that an increase in deficit is accompanied by a larger increase in income. *i.e.*, show that $d\varphi/dt = \dfrac{k^2(1-b)}{(1-tk)^2} > 0$

Hint: $dk/dt = -bk^2$ and $\dfrac{d}{dt}(1-tk) = k(btk-1)$

Ex. 186. If E is the autonomous expenditure and the monotonically increasing function of tax yield is given by $z = tY = tk\,E$ where $k = \dfrac{1}{1-b(1-t)}$ (as given above) find the rate of change of $\xi = dH/dG$ with tax rate t. Show how you will maximise tax yield.

Hint: Since $\dfrac{dk}{dt} = -bk^2, \dfrac{dz}{dt} = \dfrac{d}{dt}(t\,k\,E)$

For maximum Z, $\dfrac{dZ}{dt} = \left(t + \dfrac{dk}{dt} + k\right)E = (-tbk^2 + k)E = 0$

$\therefore \quad t = 1/bk$

$$\frac{dH}{dG} = 1 - tk \quad \therefore \quad \frac{d\xi}{dt} = -t\frac{dk}{dt} - k = k(btk - 1)$$

10.20 IS-LM EQUATION

Given the money supply $M_s = M$, the linear demand function for money as a function of income Y and interest rate r is given by

$$M_d = a_1 + a_2Y + a_3r \qquad a_1, a_2 > 0, a_3 < 0$$

The goods and money markets are:

Goods (IS)		*Money (LM)*	
$C = bY$	consumption	$M_d = a_1 + a_2Y + a_3r$	demand
$I = i_1 + i_2r$	investment $i_2 < 0 < i_1$	$M_s = M$	supply
$Y = C + I$	equilibrium	$M_d = M_s$	equilibrium
$r = -\dfrac{i_i}{i_2} + \dfrac{1-b_1}{b_2}Y$	(1)	$r = \dfrac{M - a_1}{a_3} - \dfrac{a_2}{a_3}Y$	(2)

IS curve slopes downwards because $\dfrac{dr}{dY} = \dfrac{1-b_1}{b_2} < 0$

An increase in Y induces a smaller increase in C

LM curve slopes upwards because $\dfrac{dr}{dY} = \dfrac{-a_2}{b_3} > 0$

An increase in Y increases M_d and r must rise sufficiently to leave M_d unchanged

Solving (1), (2) for Y and r

$$Y = \frac{i_1}{(1-b) + a_2i_2/a_3} + \frac{i_2/a_3}{(1-b) + a_2i_2/a_3}(M - a_1)$$

$$r = \frac{as\,i_1/a_3}{(1-b) + a_2i_2/a_3} + \frac{(1-b)/a_3}{(1-b) + a_2i_2/a_3}(M - a_1)$$

Ex. 187. $\dfrac{dY}{dM} = [a_2 + (1-b)a_3/i_2]^{-1} > 0$

10.21 DUAL NATURE OF INVESTMENT

Harrod Domar Model is concerned with the dual nature of investment. It finds an equilibrium growth pattern consistent with increased aggregate demand.

Output-capital ratio Y/K is the maximum output attainable per unit of capital. The model assumes $\frac{Y}{K}$ to be constant for all values of K so that

$$0 = d\left(\frac{Y}{K}\right) = \frac{dY}{K} - \frac{Y}{K^2}dK$$

$$\therefore \quad \frac{dY}{dK} = \frac{Y}{K}$$

This gives the equality of marginal and average output-capital ratios. If we equate the change in K *i.e.*, dK to net investment I then

$$dY = \frac{Y}{K}dK = \frac{Y}{K}I$$

Assuming the general price level $p = 1$, aggregate demand $AD = pY = Y$

$$AD = Y = C + I = bY + I$$

$$\therefore \quad Y = \frac{1}{1-b}I,\ dY = \frac{dI}{1-b}$$

where b is the marginal propensity to consume.

Since $\quad dY = \frac{Y}{K}I,\ \frac{dI}{I} = (1-b)\frac{Y}{K} = s\frac{Y}{K}$

and $1 - b = s$, marginal propensity to save

$$\therefore \quad \frac{dI}{I} = s\frac{Y}{K}$$

The proportionate rate of investment growth must equal mps times marginal output capital ratio.

Ex. 188. If capital stock is to be fully utilised without deficit or excess demand, net investment must grow by a constant percentage equal to mps times marginal output capital ratio. (Proved above)

Ex. 189. Show that in the Solow Growth Model, the proportionate change in K/L equals the proportion of total production saved divided by the existing capital stock.

The model, for a single commodity and with no technical progress, assumes full employment and shows that growth paths converge to a unique path with full employment and a natural rate of growth in which income growth rate equals labour growth rate. The annual output rate in the economy is $Y(t) = F(K, L)$ with the proportion of $Y(t)$ saved being s and the accumulated stock of capital K. Labour $L(t)$ grows at a constant rate n *i.e.*, $L(t) = Loe^{nt}$.

We try to find if capital labour ratio $r = K/L$ can always ensure full employment and if this ratio approaches some equilibrium path.

Since $\quad K = rL,\ dK/dt = \left(nr + \frac{dr}{dt}\right)L_oe^{nt}$ (since $L = L_oe^{nt}$)

Also $\quad \frac{dK}{dY} = sY = sF(K, L) = sF(K/L, 1)L = sF(r, 1)L$

$$\therefore \qquad nr + \frac{dr}{dt} = sF(r, 1) \text{ or } \frac{dr}{dt} = sF(r, 1) - nr$$

(*i*) If r is constant, $sF(r, 1) = nr$

(*ii*) If r changes, there is no capital accumulation, $\frac{dr}{dt} = -nr$

i.e., $\frac{1}{r}\frac{dr}{dt} = -n$ *i.e.*, the proportionate change in capital labour ratio is the negative of rate of change of L.

(*iii*) If L is constant or $n = 0$, $\frac{1}{r}\frac{dr}{dt} = \frac{1}{r}sF(r, 1) = \frac{s}{K}F(K, L)$

i.e., the proportionate change in capital labour ratio is the proportion of total production saved divided by the existing capital stock.

Ex. 190. The cost of a machine is Rs. 272 (amount in '000). Its value after t years is $100\ e^{0.5\sqrt{t}}$. (*a*) Show that the present value of the machine is the largest if it is disposed of after 25 years.

(Use 5 per cent compound interest rate)

(*b*) Also show that after 16 years, the maximum continuously compounded interest rate is 6.25 per cent.

(*a*) Present value $y = A(t) \cdot e^{-rt}$ where $r = 0.05$

$$= (100\ e^{0.5\sqrt{t}})\ (e^{-0.05t})$$

$$lny = ln\ 100 + 0.5\sqrt{t} - 0.05t$$

For max. $$\frac{1}{y}\frac{dy}{dt} = \frac{0.5}{2\sqrt{t}} - 0.05 = 0 \qquad \text{(See 10.22)}$$

or $$\sqrt{t} = \frac{0.5}{2 \times 0.5} = 5$$

or $$t = 25 \text{ years.}$$

(*b*) If i is the required interest rate after t years.

$$272 = A(t)\ e^{-it}$$

$$= 100\ e^{0.5\sqrt{t}}\ e^{-it}$$

$$\therefore \qquad 2.72 = e^{0.5\sqrt{t}-it} \text{ where } t = 16$$

$$ln\ 2.72 = 0.5\sqrt{16} - i(16)$$

$$1 = 4 \times 0.5 - 16i \text{ since } ln\ 2.72 \simeq 1$$

$$\therefore \qquad i = \frac{1}{16} = \frac{1}{16} \times 100\% = 6.25\%$$

10.22 MAXIMA AND MINIMA

Under certain given conditions, we may want to find how small or large a given expression may become. We may wish to discover the minimum or maximum value of a function or to find the minimum or maximum points on the curve corresponding to the function. In economics and management we are required to maximise the utility obtained from goods and services, maximise profits, minimise costs and so on.

The term optimisation may be used, in general, both for maximisation and minimisation since, in either case it implies the best result under the available circumstances. In mathematics, the maximum and minimum values are not used in the *best* sense and the term commonly applied to either of them is the *extremum* which conveys the sense of an extreme value, the highest or the lowest.

Equilibrium

The term *optimal position* is often applied to the equilibrium position resulting from the balancing of the forces of, say, demand and supply and having no further tendency to change. This kind of optimal position is the best under the given conditions; it may not be, on the face of it, a result of any conscious effort by anybody. It may be impersonal and automatic in spite of the presence of and the working of numerous variables in the background. Such an equilibrium position and a stable maximum or minimum are not necessarily identical and would give rise to different interpretations. However, the two may be related under some conditions when certain qualitative properties of equilibrium equations are known.

Some problems may not seem to involve maximum and minimum positions but may be converted into equivalent maximum and minimum problems for technical convenience.

We can study the maximising or minimising behaviour of an individual, a firm or an economy. We shall consider here the mathematics and economics of the subject.

Maximum and Minimum Points on a Curve

A *maximum* value of a function $y = f(x)$ is a value which is greater than any value of the function immediately preceding or following the given *value*. A maximum point on a curve is higher than all the *neighbouring* points on the curve.

A *minimum* value of a function $y = f(x)$ is a value less than any value of the function immediately preceding or following the given value. A minimum point on a curve is lower than all the *neighbouring* points on the curve.

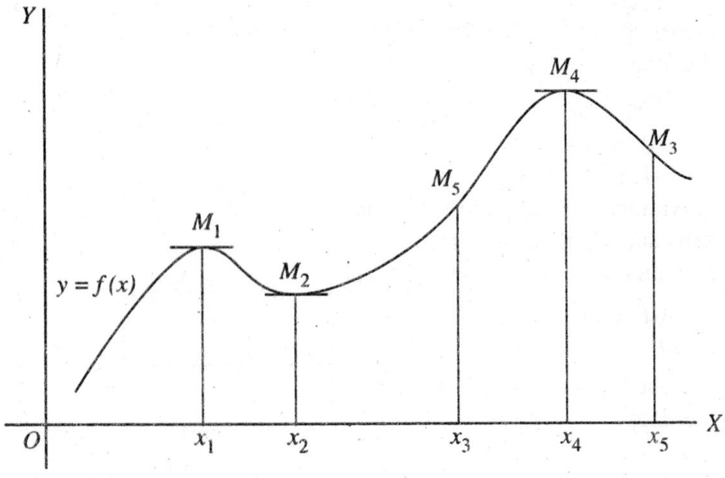

A function may have several maxima and minima.

If there are several points of maxima (or minima) in *various neighbourhoods,* the highest of them is called the *global* maximum (or minimum). Each of these points is a relative or local maximum (or minimum) in the sense that each represents a maximum (or minimum) in the immediate neighbourhood of that point only.

Rising and Falling Curves

A curve is said to be *rising* if its ordinates increase algebraically when x increases. A rising curve has a positive slope.

A curve is said to be falling if its ordinates decrease algebraically when x increases. A falling curve has a negative slope.

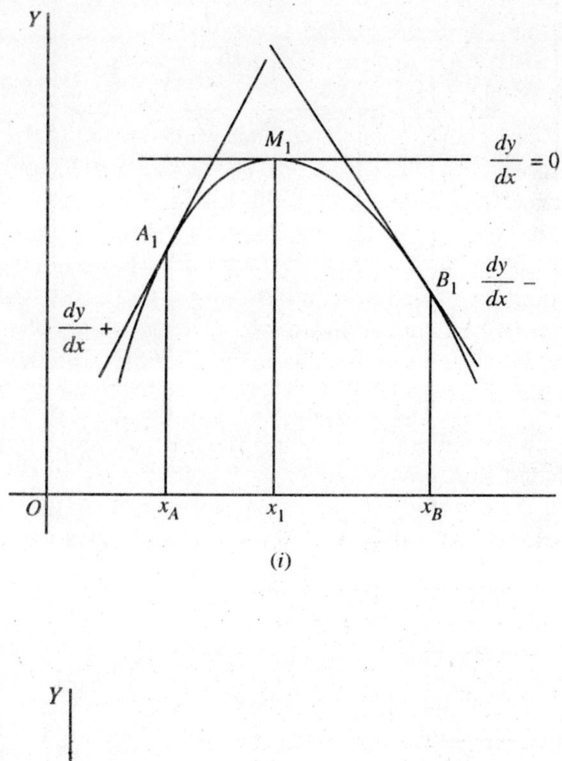

(i)

Turning Points

A turning point of the graph of a function $y = f(x)$ is a point where the tangent is parallel to the x-axis. This means that the slope of the curve at a turning point is zero.

There are various ways as at M_1, M_2, M_3, M_4, M_5 in which a turning point can occur. At each of these $dy/dx = 0$.

Points like M_1 and M_4 are the maximum turning points, M_2 is a minimum turning point and M_3, M_5 are points of inflexion.

(i) M_1 : *Maximum Turning Point*

A_1, B_1 are in the neighbourhood of M_1. The curve is rising at A_1 and the slope is positive.

The slope is zero at M_1.

At B_1 the curve is falling and the slope is negative. At the maximum turning point the slope is zero and changes sign from + to – as x increases from x_A to x_1 to x_B.

(ii) M_2 : *Minimum Turning Point*

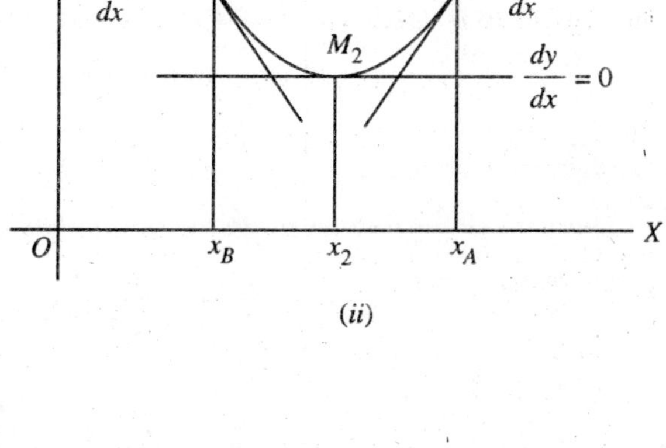

(ii)

At B_2, the curve is falling and the slope is negative. At A_2, the curve is rising and the slope is positive. At M_2, the slope is zero and changes sign as x increases from x_B to x_2 to x_A.

(iii) M_3, M_5 : Points of inflextion

The tangent crosses the curve at the point of inflexion M_3 (or M_5).

At A_3 as well as B_3 the curve is falling (or rising) and there is no change of sign of the slope as x increases from x_A to x_3 to x_B or x_A to x_5 to x_B.

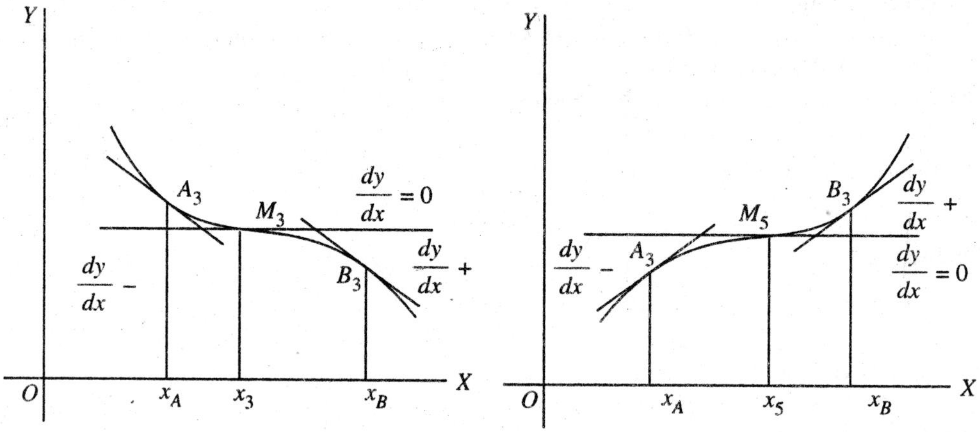

(iii)

An inflexion makes a change in curvature, the curve changing from convex to concave or vice versa. The diagram shows at the point of inflexion a tangent parallel to x-axis though it need not be so in all cases. It may slope upwards or downwards giving rise to different situations.

The First Derivative Test for Extreme Values : Necessary Condition

At any turning point $dy/dx = 0$. This is not a decisive test. It only gives an idea as to how the curve is changing.

At a maximum dy/dx changes sign from + to −

At a minimum dy/dx changes sign from − to +

At an an inflexion dy/dx does not change sign.

$\frac{dy}{dx} = 0$ is a necessary but not sufficient condition for an extreme value. We call it the first order condition.

The Second Derivative Test : Sufficiency Condition

In the case of a maximum the curve is rising on the left at A_1, at a decreasing rate so that $\frac{d^2y}{dx^2} < 0$.

Similarly on the right at B_1 the curve is falling at an increasing rate so that

$$\frac{d^2y}{dx^2} < 0.$$

In either case $\qquad \dfrac{d^2y}{dx^2} < 0$

(1) ∴ For a maximum $\qquad \dfrac{dy}{dx} = 0, \dfrac{d^2y}{dx^2} < 0$

(2) Similarly, for a minimum $\qquad \dfrac{dy}{dx} = 0, \dfrac{d^2y}{dx^2} > 0$

The second derivative test is less general than the first derivative test. The necessary and sufficient conditions are both needed simultaneously.

(3) At a point of inflexion, a necessary condition is that $\dfrac{d^2y}{dx^2} = 0$. A sufficient condition is that $\dfrac{d^2y}{dx^2}$ changes sign as x increases through the point. The direction of sign change indicates the type of inflexion.

(4) If $\dfrac{d^3y}{dx^3} < 0$, the function changes from concave from above to concave from below.

(5) If $\dfrac{d^3y}{dx^3} > 0$, it changes from concave from below to concave from above.

We shall use the following rule:

To find the maximum or minimum value of $y = f(x)$ write $\dfrac{dy}{dx} = 0$, and solve the equation. If at a given point,

$$\frac{d^2y}{dx^2} > 0 \text{ then } f(x) \text{ is a minimum}$$

$$\frac{d^2y}{dx^2} < 0 \text{ then } f(x) \text{ is a maximum.}$$

The second derivative test may fail sometimes when $\dfrac{d^2y}{dx^2} = 0$ and the result may be inconclusive.

In such a case, take values of x close to the extreme value.

If, for x just less than the extreme value, $\dfrac{dy}{dx}$ is negative and for x just more than it, $\dfrac{dy}{dx}$ is positive then this is a case of a minimum. In the opposite case, we have a maximum.

Ex. 191. $y = x^2 - 3x + 2$

$$\frac{dy}{dx} = 2x - 3 = 0 \qquad \therefore \quad x = \frac{3}{2}$$

The function has a minimum or maximum at $x = \frac{3}{2}$.

$$\frac{d^2y}{dx^2} = \frac{d}{dx}(2x - 3) = 2, \text{ a positive value at } x = 3/2.$$

$\therefore$ The function has a minimum at $x = \frac{3}{2}$

Ex. 192. $y = 3x - 12x^2$

$$\frac{dy}{dx} = 3 - 24x = 0 \text{ or } x = \frac{1}{8}$$

$$\frac{d^2y}{dx^2} = -24, \text{ a negative value}$$

$\therefore$ The function has a maximum at $x = \frac{1}{8}$.

Ex. 193. $y = x^3 - 3x$

$$\therefore \quad \frac{dy}{dx} = 3x^2 - 3 = 0$$

$$\therefore \quad 3(x - 1)(x + 1) = 0$$

$$\therefore \quad x = 1 \text{ or } x = -1$$

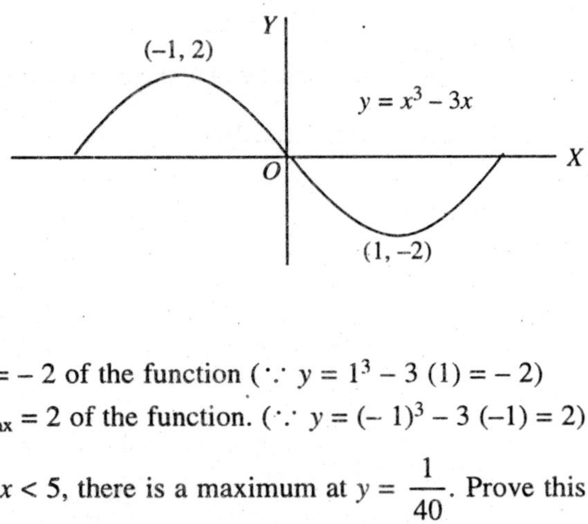

At $x = 1,\ \frac{d^2y}{dx^2} = 6x = 6 > 0.$

At $x = -1,\ \frac{d^2y}{dx^2} = -6 < 0.$

$x = 1$ gives a minimum value $y_{min} = -2$ of the function ($\because\ y = 1^3 - 3(1) = -2$)
$x = -1$ gives a maximum value $y_{max} = 2$ of the function. ($\because\ y = (-1)^3 - 3(-1) = 2$)

Ex. 194. (*i*) If $y = \sqrt{x} - 10x,\ 0 < x < 5$, there is a maximum at $y = \frac{1}{40}$. Prove this result.

(*ii*) If $y = 0.2x^2 - 4x + 40$, there is a minimum at $y = 20$.

$$\frac{dy}{dx} = 0.4x - 4 = 0 \text{ or } x = 10$$

$$\frac{d^2y}{dx^2} = 0.4, \text{ a positive value}$$

$$\therefore \quad y = 0.2 \times 10^2 - 4 \times 10 + 40 = 20$$

Hence, the point (10, 20) gives a minimum on the curve.

(*iii*) $y = \dfrac{ax}{x^2 + a^2}$ has a max. $= \dfrac{1}{2}$ at $x = a$ and min. $= -\dfrac{1}{2}$ at $x = -a$

(*iv*) $y = x^4 - 4x^2 + 4$ has a max. = 4 at $x = 0$ and min. = 0 at $x = \pm\sqrt{2}$

(*v*) $y = ax^4$, $a > 0$

$$\frac{dy}{dx} = 4ax^3 = 0 \quad \therefore\ x = 0$$

$$\frac{d^2y}{dx^2} = 12ax^2 = 0 \text{ at } x = 0$$

The second derivative test fails.

If $x < 0$, $\dfrac{dy}{dx} = 4ax^3$ is negative

If $x > 0$, it is positive

which means the function is decreasing. It attain a minimum and then increases. This shows that $x = 0$ gives a minimum for

$$y = ax^4.$$

Ex. 195. Examine the following functions for their maxima and minima:

(*i*) $y = 2x^2 - x^3$ **Ans.** (*i*) Min at $x = 0$, max at $x = 4/3$

(*ii*) $y = 3x^2 - 12x + 12$ (*ii*) min at (2, 0)

(*iii*) $y = ax^2 + bx + c$

(*iii*) min at $\left(-\dfrac{b}{2a}, \dfrac{b^2}{4a} + c\right)$ if $a > 0$; max at $\left(-\dfrac{b}{2a}, -\dfrac{b^2}{4a} + c\right)$ if $a < 0$

(*iv*) $y = \dfrac{6x}{x^2 + 4}$ (*iv*) min at $\left(-2, -\dfrac{3}{2}\right)$ max at $\left(2, \dfrac{3}{2}\right)$

(*v*) $y = 3x^2 - x + 1$ (*v*) min at $x = 1/6$

(*vi*) $y = -2x^2 + 3x - 1$ (*vi*) min at $x = 3/4$, no maximum.

(*vii*) $y = x^3$ (*vii*) $x = 0$, point of inflexion

Draw rough diagrams.

Ex. 196. Find the point of inflexion for the curves

(*a*) $y = \dfrac{a^2x}{a^2 + x^2}$ (*b*) $y = (ln\ x)^3$ (*c*) $y = x(x+1)^2$ (*d*) $y = be^{-\left(\frac{x}{2}\right)^2}$

Ans. (*a*) $\dfrac{dy}{dx} = \dfrac{a^2(a^2 - x^2)}{(a^2 + x^2)^2}$ and $\dfrac{d^2y}{dx^2} = \dfrac{-2a^2x(3a^2 - x^2)}{(a^2 + x^2)^2} = 0$

$\therefore$ $x(3a^2 - x^2) = 0$ gives the points of inflexion.

$$x(\sqrt{3}a - x)(\sqrt{3}a + x) = 0$$

The points of inflexion are $x = 0, \sqrt{3}\,a, -\sqrt{3}\,a$

(*b*) $\frac{dy}{dx} = 3\,(ln\ x)^2 \frac{1}{x}$

$$\frac{d^2y}{dx^2} = \frac{3lnx}{x^2}\,(2 - ln\ x) = 0$$

$\therefore$ $ln\ x = 0$ or $ln\ x = 2$ *i.e.*, $x = 1$, $x = e^2$ are points of inflexion.

(*c*) $x = \frac{1}{3}, +\frac{4}{3\sqrt{3}}, \frac{-4}{3\sqrt{3}}$

(*d*) $x = \frac{a}{\sqrt{2}}, -\frac{a}{\sqrt{2}}$

Ex. 197. Find the maximum and minimum points for (*i*) the hyperbola $y = (x - 3)^{-1}$, (*ii*) $y = 3x + (x + 2)^{3/5}$

(*i*) This curve has the x axis *i.e.*, $y = 0$ and $x = 3$ as asymptotes.

$$\frac{dy}{dx} = y' = -\frac{1}{(x-3)^2}$$

is not defined for $x = 3$ so that there is no maximum or minimum value.

$y' < 0$ for $x \neq 3$. Hence y is decreasing over the interval $x < 3$ and $x > 3$.

(*ii*) $y = 3x + (x + 2)^{3/5}$, $y' = 3\left(1 + \frac{1}{5}(x+2)^{-2/5}\right)$

$$y'' = -\frac{6}{25}(x+2)^{-7/5}$$

We find the point of inflection is $x = -2$ by noting that

y'' is not defined here at $x = -2$

y'' is changing sign through $x = -2$

When $x < -2$, y'' is positive, the curve is concave upward

When $x > -2$, y'' is negative, the curve is concave downward.

Ex. 198. Examine for critical values

(*i*) $y = -x^2 + 2x + 3$

Ans. max = 4 at (1, 4); no point of inflection, concave downward.

(*ii*) $y = (2 - x)^3$

Ans. No real max or min; point of inflection at $x = 2$; concave downward $x > 2$; concave upward $x < 2$

Ex. 199. Show that $y = \sum_{i=1}^{n} (a_i - x)^2$ has a relative minimum

when $x = \sum_{i=1}^{n} a_i / n$

Ex. 200. Critical points on the Normal Curve

$$y = \frac{1}{\sigma\sqrt{2\pi}} e^{-\frac{1}{2}\left(\frac{x-\mu}{\sigma}\right)^2}$$

We shall consider the properties of the normal curve written in the form

$$y = k\, e^{-b^2x^2}, \; k > 0$$

This curve has the x-axis as its horizontal asymptote since $y \to 0$ as $x \to \infty$.

The highest point on the curve is at (o, k)

Since $\frac{dy}{dx} = 2kb^2x\, e^{-b^2x^2} = 0$

gives $x = 0$

which corresponds to a maximum since

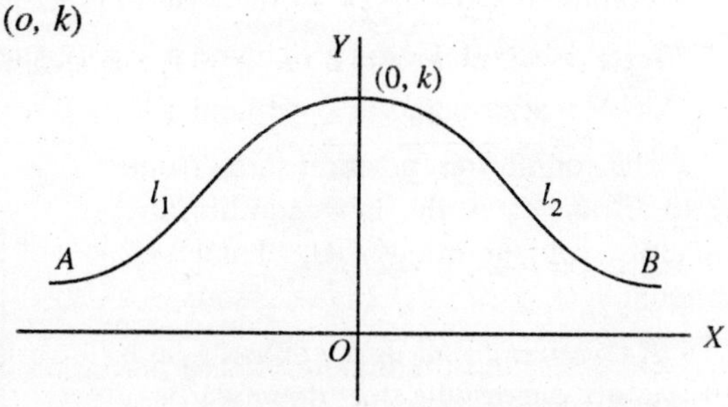

$$\frac{d^2y}{dx^2} = 2kb^2(2b^2x^2 - 1)\, e^{-b^2x^2} < 0$$

$\frac{d^2y}{dx^2} = 0$ gives $x = \pm\frac{1}{\sqrt{2}b}$, two points (I_1, I_2) of inflection on the curve.

When $\frac{d^2y}{dx^2} > 0$ on the left of O or on the right of O it is concave upward at A (before the inflection point I_1) and at B (after the inflection point I_2).

After I_1 and before I_2, $\frac{d^2y}{dx^2} < 0$ and the curve is concave downward.

10.23 MINIMISATION OF COST

If the cost function is given by $C = f(x)$ we may minimise the average cost $AC = \frac{C}{x}$ so that

$$\frac{dAC}{dx} = \frac{x\frac{dC}{dx} - C}{x^2} = 0 \text{ or } x\frac{dC}{dx} = C$$

Thus for minimum average cost $\frac{C}{x} = \frac{dC}{dx}$

i.e., $AC = MC$ which means that AC and MC curves intersect at the point of minimum average cost.

The equation $AC = MC$ gives us the quantity of output corresponding to minimum AC.

10.24 MAXIMISATION OF REVENUE

If the demand function $p = f(x)$ for a commodity is known, giving $R = px$, we can obtain the maximum revenue by writing

$$MR = \frac{dR}{dx} = 0$$

This gives the output corresponding to maximum revenue if it is found that $\frac{d^2R}{dx^2} < 0.$

Revenue from Taxation : Under perfect competition conditions, supposing that a tax t on a commodity results in an increase in price and decrease in demand. We have

(*i*) before taxation : demand $x = f(p)$, supply $x = g(p)$;

(*ii*) after taxation : demand $x = f(p)$, supply $x_t = g(p) + t$.

The equilibrium position shifts from E to E_t and x_t is the new equilibrium quantity sold at price E_tM_t. Total tax amount = tx_t.

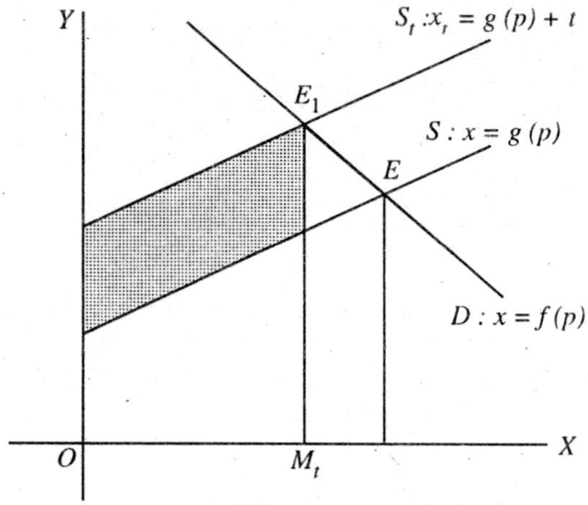

A negative t will imply subsidy and the supply curve will shift downwards.

To maximise revenue δ from taxation, the government should maximise

$$\delta = tx_t \text{ so that } \frac{d\delta}{dt} = 0, \quad \frac{d^2\delta}{dt^2} < 0$$

Sometimes, we may, instead, employ $\frac{d\delta}{dx}$ and $\frac{d^2\delta}{dx^2}$ for ease in calculation work.

Ex. 201. If the demand function for a commodity is given by $p = 12\,e^{-x/4}$ then we can write the revenue function

$$R = px = 12xe^{-\frac{1}{4}x}$$

$$MR = \frac{dR}{dx} = -3xe^{-x/4} + 12e^{-x/4} = 3e^{-x/4}(4 - x) = 0.$$

$$\therefore \quad x = 4 \text{ in which case } R = \frac{48}{e}$$

$$\frac{d^2R}{dx^2} = -3e^{-x/4} + 3(4 - x)\left(-\frac{1}{4}e^{-x/4}\right)$$

$$= e^{-x/4}\left(\frac{3}{4}x - 6\right)$$

$$= 0 \text{ if } x = 8$$

$$< 0 \text{ if } x = 4$$

If follows that $x = 4$ is the level of production for maximum revenue.

It may be proved that $x = 8$ gives a point of inflexion.

Ex. 202. If the AC function is given by $AC = 25 - 8x + x^2$, we have

$$C = 25x - 8x^2 + x^3$$

The marginal cost is given by

$$MC : \frac{dC}{dx} = 25 - 16x + 3x^2 \qquad \text{Solve } \frac{dC}{dx} = 0$$

$\therefore$ $x = 4$ is the output corresponding to the minimum AC which equals $25 - (8 \times 4) + 16$.

Ex. 203. Determine the tax that the government should impose for maximum tax revenue if the demand and supply functions are:

before taxation, demand : $p = 25 - 2x^2$; supply : $p = 5 + x$

after taxation, demand : $p = 25 - 2x^2$; supply : $p = 5 + x + t$

The new equilibrium position gives

$$25 - 2x^2 = 5 + x + t \text{ or } t = 20 - x - 2x^2$$

We have to maximise the total tax revenue.

$$\delta = tx = (20 - x - 2x^2)\, x = 20x - x^2 - 2x^3$$

$$\frac{d\delta}{dx} = 20 - 2x - 6x^2 = 0$$

i.e.,

$$(3x - 5)(x + 2) = 0$$

$\therefore$ $x = 5/3$ since $x = -2$ is not admissible

$$\frac{d^2\delta}{dx^2} = -2 - 12x = -22 < 0 \quad \text{at } x = 5/3$$

$\therefore$ $x = 5/3$ gives a maximum which corresponds to

$$t = 20 - x - 2x^2$$

$$= 20 - 5/3 - 2 \cdot \frac{25}{9} = \frac{115}{9}.$$ This is the tax rate which leads to

$$\text{Maximum tax revenue } \delta_{max} = tx = \frac{115}{9} \cdot \frac{5}{3} = \frac{575}{27}$$

Ex. 204. Find the minimum average cost for

(*i*) $AC = 10 - 4x^3 + 3x^4$

(*ii*) $AC = 6x + 7 + \dfrac{36}{x}$

(*iii*) $AC = 2e^x + e^{-x}$ **Ans.** $9,\ 7 + 12\sqrt{6},\ 2\sqrt{2}$

Ex. 205. Find the output for maximum revenue, given the demand functions

(*i*) $p = \sqrt{12 - x}$ (*ii*) $p = (x - 8)^2$ **Ans.** 8, 8/3

Ex. 206. Find the maximum revenue from a tax t given the functions

(*i*) demand $x = \dfrac{\sqrt{25 - p}}{2}$, supply $x = p - 5$ **Ans.** $t = 12.8$, $tx = 21.3$

(*ii*) demand $p = 30 - 2x^2$, supply $p = 3 + x^2$ **Ans.** $t = 18$, $tx = 18\sqrt{3}$

10.25 PROFIT MAXIMISATION UNDER MONOPOLY

Given the demand function, the monopolist fixes the output leaving price to be determined by demand conditions or, he fixes the price leaving the output to be determined by demand conditions.

Let revenue (R) and cost (C) be functions of output (x). Then the profit of a firm whose revenue function is $R(x)$ and cost function $C(x)$ is given by

$$\pi(x) = R(x) - C(x)$$

To get the condition for maximum profit write

$$\frac{d\pi}{dx} = 0$$

so that $$\frac{dR}{dx} - \frac{dC}{dx} = 0$$

$$\therefore \quad \frac{dR}{dx} = \frac{dC}{dx}$$

i.e., $MR = MC$ which is the condition for the equilibrium of a firm. It also means that the slopes of tangents to the revenue and cost curves at the point of maximum profit are equal.

$\frac{dR}{dx} = \frac{dC}{dx}$ is only a necessary condition. A sufficient condition for profit maximisation is

$$\frac{d^2\pi}{dx^2} = \frac{d^2R}{dx^2} - \frac{d^2C}{dx^2} < 0$$

$$\therefore \quad \frac{d^2R}{dx^2} < \frac{d^2C}{dx^2} \quad \text{or} \quad \frac{dMR}{dx} < \frac{dMC}{dx}$$

$\therefore$ for the profit to be maximum,

$\frac{dMR}{dx}$ (*i.e.*, the rate of change of MR) should be less than

$\frac{dMC}{dx}$ (*i.e.*, the rate of change of MC) at the equilibrium output where $MR = MC$.

Ex. 207. Let the demand function of a monopolist be $p = 15 - 2x$ then the revenue function is $R(x) = xp = x(15 - 2x) = 15x - 2x^2$

The MR function $\frac{dR}{dx} = 15 - 4x$

Let the total cost function be given by $C = x^2 + 2x$

Then the average cost function is $\frac{C}{x} = x + 2$

The MC function is given by

$$MC: \quad \frac{dC}{dx} = 2x + 2$$

All the four curves AR, MR, AC, MC can be sketched.

For equilibrium: $MR = MC$

$$15 - 4x = 2x + 2$$

$\therefore \quad x = 2\frac{1}{6}$ is the equilibrium output,

Alternatively, we may write

$$\pi = R - C = 15x - 2x^2 - (x^2 + 2x)$$

i.e.,

$$\pi = -3x^2 + 13x$$

$$\frac{d\pi}{dx} = -6x + 13 = 0$$

$\therefore \; x = \frac{13}{6}$ is the necessary condition for profit maximisation.

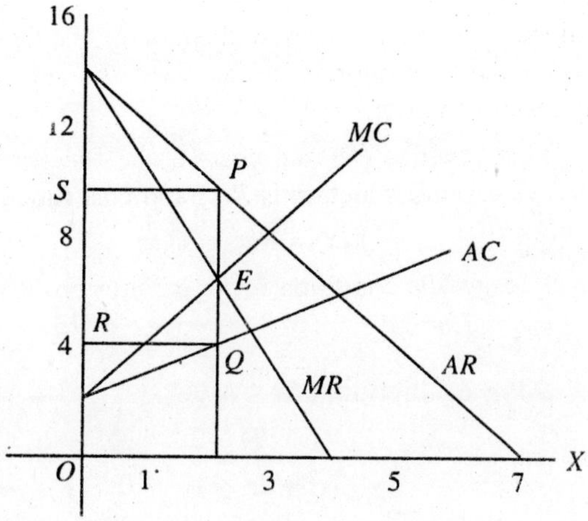

To get the sufficient condition, we have

$\frac{d^2\pi}{dx^2} = -6$, a negative quantity, which means, the profit is maximum at $x = 2\frac{1}{6}$.

This corresponds to the point E. The rectangle PQRS shows the maximum profit.

At the equilibrium point which corresponds to maximum profit,

$$AR = p = 15 - 2x = 15 - 2 \times \tfrac{13}{6}$$

$$= \tfrac{32}{3} = 10\tfrac{2}{3}$$

$$AC = x + 2 = \tfrac{13}{6} + 2 = 4\tfrac{1}{6}$$

$$MR = 15 - 4x = 15 - 4 \times \tfrac{13}{9} = \tfrac{19}{3}$$

$$= 6\tfrac{1}{3}$$

$$MR = MC = 2x + 2 = 2 \times \tfrac{13}{6} + 2$$

$$= 6\tfrac{1}{3}$$

Profit $= -3x^2 + 13x = 14\frac{1}{12}$

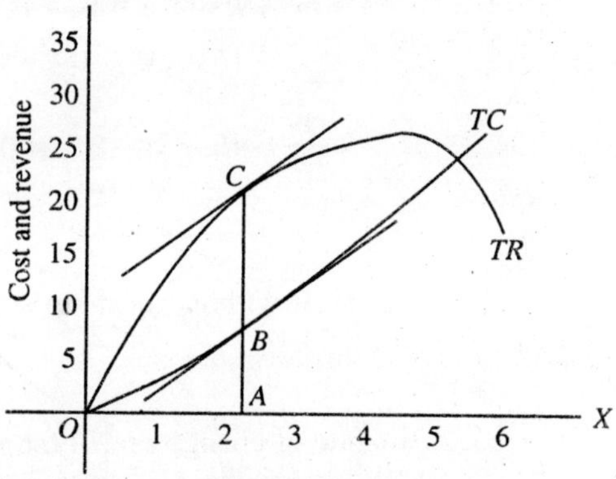

We can find the maximum profit from total revenue and total cost. (TR, TC)

We know $x = 2\frac{1}{6}$ is the necessary condition for profit maximisation. This gives us the point A from where a perpendicular is drawn to cut TC and TR curves at B, C. The tangents drawn at B, C are parallel because at these points $\frac{dR}{dx} = \frac{dC}{dx}$ and the profit BC is maximum.

Ex. 208. A firm has the demand function $p = 12 - 3x$ and the cost function $C = x^2 + 2x$. Find AC, AR, MC, MR and profit for the firm at the point of equilibrium. What is the condition for profit maximisation? Draw the graphs.

Ans. $x = \frac{5}{4}, \; \pi_{max} = \frac{25}{4}$

Ex. 209. Let the revenue function be given by $R = 14x - x^2$ and the cost function $C = x(x^2 - 2)$.

Then
$$AC = x^2 - 2$$
$$MC = \frac{dC}{dx} = \frac{d}{dx}(x^3 - 2x) = 3x^2 - 2$$
$$AR = \frac{R}{x} = 14 - x$$
$$MR = \frac{dR}{dx} = \frac{d}{dx}(14x - x^2) = 14 - 2x$$

For equilibrium $MR = MC$
$$14 - 2x = 3x^2 - 2$$
$$\therefore \quad 3x^2 + 2x - 16 = 0$$
$$\therefore \quad (3x + 8)(x - 2) = 0$$

$\therefore\ x = -\frac{8}{3}$ or $x = 2$. The first value being negative is rejected and the second of these two values gives the equilibrium position. Alternatively, we can get the result by putting
$$\frac{d\pi}{dx} = 0$$

We have,
$$\pi = R - C = 14x - x^2 - (x^3 - 2x)$$
i.e.,
$$\pi = 16x - x^2 - x^3$$
$$\frac{d\pi}{dx} = 16 - 2x - 3x^2 = 0$$
$$\therefore \quad x = -\frac{8}{3} \text{ or } x = 2$$

$x = -\frac{8}{3}$ is a result without meaning so that to get the point of maximum profit, we find, at $x = 2$
$$\frac{d^2\pi}{dx^2} = \frac{d}{dx}\left(\frac{d\pi}{dx}\right) = -2 - 6x = -2 - 6 \times 2 = -14$$

$\therefore$ The output $x = 2$ gives the maximum profit

Maximum profit $= 16x - x^2 - x^3 = 16 \times 2 - 4 - 8 = 20$.

Ex. 210. If A is the cost per unit and p is the selling price, a seller finds that he can sell the quantity
$$x = \frac{k_1}{p - A} + \frac{k_2}{100 - p}$$
where k_1, k_2, are constants. Show that for maximum profit, $p = \dfrac{100 + A}{2}$

Hint: $R = x(p - A)$ and $dR/dx = 0$

Ex. 211. Given a monopolist's cost and demand functions
$$C = ax^2 + bx + c,\ p = \beta - \alpha x^2,$$

Show that for maximum profit, output is given by

$$x = \frac{1}{3\alpha}[a^2 + 3\alpha(\beta - b)]^{\frac{1}{2}} - \frac{a}{3\alpha}$$

Ex. 212. If the cost of processing x applications is given by $C = 60x - 5x \ln x + 0.001x^3$, find the minimum AC.

Hint: $$\frac{dAC}{dx} = \frac{d}{dx}(60 - 5\ln x + 0.001x^2) = -\frac{5}{x} + 0.002x = 0 \Rightarrow x = 50$$

$$\frac{d^2AC}{dx^2} = \frac{5}{x^2} + .002 > 0$$

AC_{min} (at $x = 50$) is 42.9

Ex. 213. A monopolist sells two commodities with demand functions given by $p_x = g(x)$ and $p_y = h(y)$. If $y = f(x)$ show that $f'(x)$ is minus one times the ratio of marginal revenues of x and y $\left[\text{Note } \frac{dy}{dx} = f'(x)\right]$

Total revenue R from both $= xp_x + yp_y$

$$R = xg(x) + yh(y) = xg(x) + f(x)\,h(y)$$

$$\frac{dR}{dx} = \{xg'(x) + g(x)\} + \left\{fh'(y)\frac{dy}{dx} + h(y)f'(x)\right\} = 0$$

$$\therefore \quad xg'(x) + g(x) + f'(x)\{fh'(y) + h(y)\} = 0$$

$$\therefore \quad f'(x) = \frac{xg'(x) + g(x)}{yh'(y) + h(y)}$$

Since $R_x = xp_x = xg(x) \quad \therefore \quad R_y = yp_y = yh.(y)$

$$\therefore \quad MR_x = \frac{dR_x}{dx} = xg'(x) + g(x)$$

$$MR_y = \frac{dR_y}{dy} = yh'(y) + h(y)$$

$$\therefore \quad f'(x) = -\frac{MR_x}{MR_y}$$

Ex. 214. Given a monopolist's cost function $C = ax^2 + bx + c$ and the demand law $p = \beta - \alpha x$ show that the imposition of a tax t per unit maximises the tax revenue if $2t = \beta - b$ and the increase in p is less than t.

$$\pi = px - C = (\beta - \alpha x)x - (ax^2 + bx + c + tx) \text{ and } \frac{d\pi}{dx} = 0 \Rightarrow \beta - 2\alpha x - 2ax - b - t = 0$$

i.e., $x(2a + 2\alpha) = \beta - b - t$ or $x_e = \dfrac{\beta - b - t}{2a + 2\alpha}$ (equilibrium quantity)

$$\frac{d^2\pi}{dx^2} = -2a - 2\alpha < 0.$$

Equilibrium price $p_e = \beta - \alpha x_e = \beta - \alpha \dfrac{(\beta - b - t)}{2a + 2\alpha}$

Tax revenue $T = tx = t\dfrac{\beta - b - t}{2a + 2\alpha} = \dfrac{\beta t - t^2 - bt}{2a + 2\alpha}$

$\dfrac{dT}{dt} = \dfrac{\beta - 2t - b}{2a + 2\alpha} = 0$ *i.e.,* $2t = \beta - b$ for T_{max}

Increase in price $= \left\{\beta - \dfrac{\alpha(\beta - b - t)}{2a + 2\alpha}\right\} - \left\{\beta - \dfrac{\alpha(\beta - b)}{2a + 2\alpha}\right\}$

$= \dfrac{\alpha t}{2(a + \alpha)} < t$ since numerator < denominator

Ex. 215. Given the following revenue and cost functions for a firm:

(*i*) $R = 200x - .01x^2$ (*ii*) $R = x(100 - x)$

$C = 50 + 20000$ $C = x^3 - \dfrac{57}{2}x^2$

Find the maximum profit and the level of output for maximum profit.

Ans. (*i*) 5000, 20 (*ii*) 5425, 7500

Ex. 216. Find the minimum average cost if the cost function is given by $C = 36x - 10x^2 + 2x^3$. Find the marginal cost at the point.

The average cost function is given by $\dfrac{C}{x} = AC = 36 - 10x + 2x^2$

For an extremum $\dfrac{dAC}{dx} = -20 + 4x = 0$ or $x = \frac{5}{2}$

Now $\dfrac{d^2AC}{dx^2} = 4 > 0$

$\therefore$ $x = \frac{5}{2}$ gives the minimum value of the average cost.

At $x = \frac{5}{2}$, the minimum AC is given by

$$AC_{min} = 36 - 10x + 2x^2$$

$$= 36 - 10 \times \tfrac{5}{2} + 2 \times \left(\tfrac{5}{2}\right)^2 = 23.5$$

$$MC = \frac{dC}{dx} = 36 - 20x + 6x^2$$

$$= 36 - 20 \times \tfrac{5}{2} + 6\left(\tfrac{5}{2}\right)^2 \text{ at } x = \frac{5}{2}$$

$$= 23.5$$

10.26 DUOPOLY

A duopoly is characterised by the interdependence of the sellers' actions. The price-quantity combination and profit of a duopolist depend on the actions of the other duopolist.

Depending on the behaviour assumptions, different solutions to the problems are obtained. We consider here the *Cournot solution.* Other solutions may be considered similarly.

Suppose two duopolists supply quantities x_1, x_2 of a good required by the market whose demand is $x = x_1 + x_2$. Let $C_1(x_1)$, $C_2(x_2)$ be their cost functions.

$\frac{dx_1}{dx_2}$ is called the *conjectural variation* of the first *i.e.*, the adjustment which the first duopolist assumes to have been made by the second in view of a change in x_1.

$\frac{dx_2}{dx_1}$ is the conjectural variation of the second.

$x_1 = f_1(x_2)$, $x_2 = f_2(x_1)$ are the *reaction curves* of the two.

The shapes of the reaction curves depend on the cost functions. In the case of Cournot solution $dx_2/dx_1 = 0$ and $dx_1/dx_2 = 0$.

To maximise the profit function of the first duopolist

$$\pi_1(x_1) = px_1 - C_1(x_1)$$

we have $\frac{d}{dx_1}[px_1 - C_1(x_1)] = 0$

$$\therefore \qquad p + x_1 \frac{dp}{dx}\frac{dx}{dx_1} = \frac{dC_1(x)}{dx_1}$$

$$i.e., \qquad p + x_1 \frac{dp}{dx}\left(\frac{dx_1}{dx_1} + \frac{dx_2}{dx_1}\right) = \frac{dC_1}{dx} = MC_1$$

$$i.e., \qquad p + x_1 \frac{dp}{dx} = MC_1 \text{ since } \frac{dx_2}{dx_1} = 0$$

Similarly for the second duopolist $p + x_2 \frac{dp}{dx} = MC_2$

If $\frac{dx_2}{dx_1} \neq 0$ we have for the

first
$$p + x_1 \frac{dp}{dx}\left(1 + \frac{dx_2}{dx_1}\right) = MC_1,$$

second
$$p + x_2 \frac{dp}{dx}\left(1 + \frac{dx_1}{dx_2}\right) = MC_2.$$

10.27 MONOPSONY

Let the output $x(L)$ and wages $w(L)$ of a monopsonist be functions of the input labour L only. A monopsonist has monopoly in buying

We have to find the labour input which maximises his profit function

$$\pi = px(L) - Lw(L) - F$$

where F is the monopsonist's fixed costs. p is the buying price.

We have $$\frac{d\pi}{dL} = p\frac{dx}{dL} - L\frac{dw}{dL} - w = 0$$

i.e., *MRP* equals marginal factor cost.

We have $$\frac{d^2\pi}{dL^2} = p\frac{d^2x}{dL^2} - \frac{d}{dL}\left(L\frac{dw}{dL} - w\right) < 0 \quad \text{(Second order condition)}$$

If we write $f(L) = L\frac{dw}{dL} - w$ for the labour cost function, the second order condition can be written

$$p\frac{d^2x}{dL^2} < \frac{df}{dL}$$

i.e., for profit to be maximum, the increase in the marginal factor cost should not exceed *MRP*.

Under competitive conditions $w(L) = w_1$, a constant, so that to maximise

$$\pi = px(L) - w_1L - F$$

we have $$\frac{d\pi}{dL} = p\frac{dx}{dL} - w_1 = 0 \text{ or } p\frac{dx}{dL} = w_1$$ *i.e.*, *MRP* equals factor cost w_1

and $$\frac{d^2\pi}{dL^2} = p\frac{d^2x}{dL^2} < 0$$ *i.e.*, *MRP* should be a decreasing function of *L*.

10.28 BILATERAL MONOPOLY

This is a market situation in which a single seller *S* faces a single buyer *B*, each being a monopolist in his own realm.

The monopsonist uses the monopolist's output of good x_2 to produce the good x_1 so that $x_1 = f_1(x_2)$ which he sells at price p_1. He maximises his profit

$$\pi_B = p_1f_1(x_2) - p_2x_2$$

$$\therefore \quad \frac{d\pi_B}{dx_2} = p_1f'_1(x_2) - p_2 = 0$$

This gives the monopsonist's demand function $p_2 = p_1f'_1(x_2)$ which is used by the monopolist to maximise his profit:

$$\pi_s = [p_1f'_1(x_2)]x_2 - cf_2(x_2)$$

where $f_2(x_2)$ is the production function of the monopolist whose cost per unit is *c*.

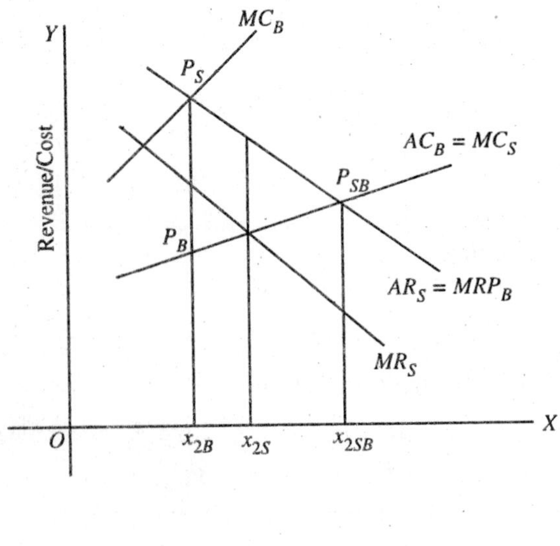

$$\frac{d\pi_s}{dx_2} = 0 \text{ gives } p_1\,[f_1'(x_2) + x_2 f_2''(x_2)] = cf_2'\,(x_2)$$

which is the monopolist's supply function. This is used by the monopsonist to maximise his profit.

$$\pi_B = p_1 f_1\,(x_2) - p_2 x_2 = p_1 f_1\,(x_2) - cx_2 f'2\,(x_2)$$

$$\frac{d\pi_B}{dx_2} = 0 \text{ gives } p_1 f'_1\,(x_2) = c\,[f_2\,(x_2) + x_2 f''_2\,(x_2)]$$

The solution x_{2s} gives the monopoly side solution with dominant seller and x_{2B} gives the monopsony side solution with dominant buyer. The competitive solution x_{sB} is obtained by writing

$$p_1 f'_1\,(x_2) = cf'_2\,(x_2)$$

when the competitive price is p_{sB} which always lies between p_s and p_B.

Ex. 217. Two duopolists sell differentiated products. The follower is confined to a one fourth share of the market. The leader's cost and demand functions are $C_1 = 2x_1^2$, $p_1 = 50 - 50x_1 - 2x_2$. Those of the follower are $C_2 = 3x_2^2$, $p_2 = 60 - 3x_1 - 4x_2$ find the output and prices of each.

$$x_2 = \frac{kx_1}{1-k} = \frac{\frac{1}{4}x_1}{1-\frac{1}{4}} = \frac{x_1}{3}$$

$$\pi_1 = R_1 - C_1 = p_1 x_1 - C_1 = (50 - 5x_1 - 2x_2)x_1 - 2x_1^2$$

$$= 50x_1 - \frac{23}{3}x_1^2$$

$$\frac{d\pi_1}{dx_1} = 0 \text{ gives } x_2 = \frac{75}{23}, \frac{x_2}{3} = \frac{75}{69}$$

$$p_1 = 50 - 5x_1 - 2x_2,\ p_2 = 60 - 3x_1 - 4x_2$$

Ex. 218. Given the Demand and Supply functions

$$D = 40 - 2p,\ S = 2p - 20$$

if the producers adjust the supply function to include a tax t per unit in the supply calculate the tax rate to maximise tax revenue T and the maximum T.

Ans. $p = 20 - 0.5D$, $p = 10 + 0.5S + t$

In equilibrium $t = 10 - x$ where $x = D = S$

To maximise $T = (10 - x)x$

$$\frac{dT}{dx} = 0 \text{ or } x = 5,\ \frac{d^2T}{dx^2} < 0$$

$$t = 10 - x = 10 - 5 = 5, \text{ the tax rate for } T_{max}$$

$$T_{max} = tx = 5 \times 5 = 25$$

Ex. 219. A monopolist has a demand function $p = 15 - 6x$ and cost function $C = 2 + 3x - 3x^2 + 2x^3$. What is the effect of a 25% profit tax on (*i*) output (*ii*) price (*iii*) profit? What will be the effect with a 100% profits tax? Without tax $\pi_{max} = 5$.

$$\pi = R - C - 0.25\,(R - C) = 0.75\,(12x - 3x^2 - 3x^3 - 2)$$

$$\frac{d\pi}{dx} = 0 \text{ gives } x = 1, p = 9, \pi_{max} = 0.75 \times 5 = 3.75$$

The price and output remain the same but the profit goes down by 25%

With a hundred per cent profit tax, we must first have $R \geq C$ or simplifying $12x - 3x^2 - 2x^3 - 2 \geq 0$

Since for many values like $x = \frac{1}{4}, \frac{1}{2}, 1, \frac{3}{2}$, etc. the inequality holds, the solution is indeterminate.

Ex 220. Given the Cost Function $C = 5x$ and the demand function $p = 29 - 4x$.

Find the optimum price and quantity before and after a tax of one rupee per unit is imposed. Find the tax rate that will maximise tax revenue. What will be the tax revenue if an additional tax of 20 per cent is imposed?

$$R = x\,(29 - 4x) \qquad \therefore\ MR = 29 - 8x$$

$$C = 5x \qquad \therefore\ MC = 5$$

$$MR = MC \text{ gives } 29 - 8x \text{ or } x = 3, p = 17$$

With a one rupee tax, $C = 5x + 1x = 6x$

New $MC = MR$ means $6 = 29 - 8x$ or $x = \frac{23}{8}$, $p = \frac{35}{2}$

If the tax revenue maximising tax rate is t,

New $MC = MR$ or $5 + t = 29 - 8x$

$$\therefore \qquad x = \frac{24 - t}{8}$$

$$p = 29 - 4x = 29 - 4\left(\frac{24 - t}{8}\right) = 17 + \frac{t}{2}$$

= pretax price + half the tax rate imposed.

$$\pi_{max} = (29x - 4x^2) - (5x + tx) = (24 - t)\,x - 4x^2$$

$$= (24 - t)\frac{24 - t}{8} - 4\left(\frac{24 - t}{8}\right)^2 = \frac{(24 - t)^2}{16}$$

$$T = tx = t\left(\frac{24 - t}{8}\right) = \frac{24t - t^2}{8}$$

T_{max} requires $\frac{dT}{dt} = 0$ or $24 - 2t = 0$ or $t = 12$.

$$T_{max} = tx = 12 \cdot \frac{24 - 8}{8} = 24$$

If there is a tax of 20 per cent, the new R will be

$$R = \frac{80}{100}(29 - 4x)x \text{ or } MR = 0.8(29 - 8x)$$

$$MR = MC \text{ gives } 0.8\ (29 - 8x) = 5x \text{ or } x = \frac{116}{57}$$

When a tax t is imposed

$$\begin{aligned} \pi &= R - C \\ &= (29x - 4x^2) - (5x + tx) \\ &= (24 - t - 4x)x \\ &= \left(24 - t - 4\cdot\frac{24-t}{8}\right)\frac{24-t}{8} \\ &= \frac{(24-t)^2}{16} \end{aligned}$$

$$T = \text{Tax revenue } tx = t\left(\frac{24-t}{8}\right) = \frac{24-t^2}{8}$$

For $\quad T_{max}, \dfrac{dT}{dt} = 0 \text{ or } \dfrac{24-2t}{8} = 0 \text{ or } t = t_0 = 12$

$\therefore \quad T_{max} = t_0 \cdot x = 12\left(\dfrac{24-12}{12}\right) = 12$

Ex. 221. The tax rate t for a good is related to its consumption quantity q by the relation $t = 18 - 3q^2$. Find the optimal tax rate.

Revenue $R = tq = 18q - 3q^3$

For R_{max}, $\dfrac{dR}{dq} = 18 - 9q^2 = 0 \quad i.e., q = \sqrt{2}$ (Note $d^2R/dq^2 < 0$)

Substituting we can also get maximum value of R as $R_{max} = 18\sqrt{2} - 3(\sqrt{2})^3 =$ 16.9

Ex. 222. The average cost of production is given by $A = \dfrac{3 + 48L - 192L}{2L - 8}$

where L is the number of units employed. Find L for A_{min}.

$$\frac{dA}{dL} = \frac{-3}{2(L-4)^2} + 24 = 0$$

$\therefore \quad (L - 4)^2 = \dfrac{1}{16}$

$\therefore \quad L = 4\dfrac{1}{4}$

Question. $L = 4$ or $L = 5$?

If $L = 4$, $A = \infty$ (by substituting in the A expression)

If $L = 5$, $A = 121\frac{1}{2}$

Answer, $L = 5$

Ex. 223. A discriminating monopolist has two distinct markets described by

I. $x_1 = 16 - 0.2p_1$

II. $p_2 = 180 - 20x_2$

If the total cost function is $C = 20x + 20$ where $x = x_1 + x_2$, find the prices the monopolist will charge in each market while maximising his profits. Find the profits with and without price discrimination. **Note:** In each market $MC = MR_1 = MR_2$

Since $R_1 = p_1x_1 = 80x_1 - 5x_1^2$, $R_2 = 180x_2 - 20x_2^2$

$MR_1 = MC$ gives $80 - 10x_1 = 20$ or $x_1 = 6$, $p_1 = 50$

$MR_2 = MC$ gives $180 - 40x_1 = 20$ or $x_2 = 4$, $p_2 = 100$

Profit with price discrimination $\pi_d = R_1 + R_2 - C$

$\therefore$ max $\pi_d = (300 + 400) - [20 + 20(x_1 + x_2)] = 480$

where $x_1 = 6$, $x_2 = 4$, $p_1 = 50$, $p_2 = 100$, $R_1 = 300$, $R_2 = 400$

For profit without price discrimination π_w the overall demand function is given by $x = x_1 + x_2$ (with $p_1 = p_2$) (**Note:** $x_2 = 9 - 0.05p$)

$$x = 16 - 0.2p + 9 - 0.05p \text{ or } p = 100 - 4x$$

$$\therefore \quad R = (100 - 4x)x \text{ and } \pi_w = 100x - 4x^2 - 20x - 20$$

For max π_w, $\frac{d\pi_w}{dx} = 0$ or $-8x + 80 = 0$, $\frac{d^2\pi_w}{dx^2} < 0$

$$\therefore \quad x = 10,\ \pi_{w\ max} = -4(100) + 80(10) - 20 = 380$$

This is less than max π_d,

Ex. 224. Let A be the advertising rate and $R = R(t)$ the change in sales rate per unit spent on advertising. If S is a saturation constant *i.e.*, the upper limit of sales rate whatever the expenditure on advertising then we can write

$$\frac{dS}{dt} = RA - \lambda S$$

where λ is the decay constant. Find the predicted sales rate.

Let the zero sales response be r *i.e.*, $S = 0$ when $R = r$. We have the relation between sales and change in sales rate due to advertising given by

$$\frac{R}{r} + \frac{S}{M} = 1$$

Then we have $\frac{dS}{dt} = rA\left(1 - \frac{S}{M}\right) - \lambda S$

If $\frac{dS}{dt} = 0$ *i.e.*, if $A = \frac{\lambda SM}{r(M - S)} = \frac{\lambda M}{r} \cdot \frac{S}{M - S}$

we have constant sales.

If $\quad A(t) = A$ (constant) $0 < t < T$

$= 0, \ t > T$

If a is the total advertising expenditure $a = AT$ where $0 < t < T$ so that

$$\frac{dS}{dt} + \left(\frac{ar}{TM} + \lambda\right)S = \frac{ar}{T}$$

whose general solution is $S = ke^{-bt} + \dfrac{ar}{bT}$

where $\quad b = \dfrac{ar}{TM} + \lambda$

For $t > T$, $\dfrac{dS}{dt} = -\lambda S$ or $S = ke^{-\lambda T} = ke^{-\lambda T}$ when $t = T$

∴ The predicted sales rate is

$$S(t) = \frac{ar}{bT}(1 - e^{-bt}) + S_0 e^{-bt} \text{ for } 0 < t < T$$

$$= S(T)\, e^{\lambda(T-t)} \text{ for } t > T$$

Ex. 225. Two doupolists have identical demand and average cost functions:

$$p = 50 - \frac{x}{3}, \quad AC = \frac{100}{x} + 3 + \frac{x}{25}$$

Find the total equilibrium output.

Total output x is equally divided between the two.

$$R = px = \left(50 - \frac{x}{3}\right)\frac{x}{2} = 25x - \frac{x^2}{6}$$

$$C = 100 + 3x + x^2/25$$

$$\pi = R - C = -100 + 22x - \frac{28}{75} - x^2$$

$$\frac{d\pi}{dx} = 0$$

$$\therefore \quad 22 - \frac{56}{75}x = 0 \text{ or } x = \frac{22 \times 75}{56} \simeq 23$$

Ex. 226. Given the average cost

$$\frac{C}{x} = \frac{10}{x} + 10 - \frac{9}{10}x + \frac{1}{25}x^2$$

Find MC and the slopes of AC, MC. When is the AC a minimum?

$$C = 10 + 10x - \frac{9}{10}x^2 + \frac{1}{25}x^3$$

$$MC = \frac{dC}{dx} = 10 - \frac{18}{10}x + \frac{3}{25}x^2$$

Slope of $AC = \frac{dAC}{dx} = -\frac{10}{x^2} - \frac{9}{10} + \frac{2x}{25}$

Slope of $MC = \frac{dMC}{dx} = -1.8 + \frac{6}{25}x$

For minimum AC, slope of AC is zero.

$$\therefore \quad -\frac{10}{x^2} - \frac{9}{10} + \frac{2x}{25} = 0$$

Ex. 227. A firm's fixed cost is Rs. 28000 and the demand function for its product is $x = 30000 - 1500p$. The labour and other cost per unit is Rs. 8.

What should be the price for maximum profit?

Total Revenue $R = px = p\,(30000 - 1500p)$

Total Cost $C = 28000 + 8x = 28000 + 8\,(30000 - 1500p)$

Profit $\pi = R - C = -\,1500p^2 + 42000p - 268000$

$$\frac{d\pi}{dp} = -3000p + 42000$$

For maximum profit $\frac{d\pi}{dp} = 0$ giving $p = 14$ and $x = 9000$

$$\frac{d^2\pi}{dp^2} = -\,3000 < 0$$

$$\pi_{max} = 26000 \text{ when } x = 9000$$

Ex. 228. For the Cost Function

$$C = 5 + 3x(x + 7)/(x + 8)$$

Show that marginal cost falls continuously with output.

Ex. 229. For the demand curve $x = p^a e^{-b(p + c)}$ show that as price rises, demand rises and becomes large as the price approaches the value $\frac{a}{b}$. Also find the effect on demand elasticity of any price greater than $\frac{a}{b}$.

Note. $\eta = -\frac{p}{x} \cdot \frac{dx}{dp} = -b\left(\frac{a}{b} - p\right)$. If $\frac{a}{b} < p$, then η changes sign.

Ex. 230. If L is the labour input find the number of units required for maximum input for the production function given by $Y = L\,(72 + 3L - L^2/3)$.

Find the maximum average (APL) and marginal (MPL) product of labour and show that when APL is maximum it equals MPL.

Note. $\frac{dY}{dL} = 72 + 6L - L^2 = 0 \Rightarrow L = 12, -6$

$\frac{d^2Y}{dL^2} < 0$ at $L = 12$, the maximum input

$$MPL = \frac{dY}{dL} = 72 + 6L - L^2;\ APL = 72 + 3L - L^2/3$$

To find the maximum MPL we have $\frac{d(MPL)}{dL} = 6 - 2L = 0$ or $L = 3$

$$\frac{d^2 MPL}{dL^2} = -2 < 0$$

MPL is maximum at $L = 3$.

To find the maximum APL, we have $\frac{dAPL}{dL} = 3 - 2L/3 = 0$ or $L = 9/2$

$$\frac{d^2 APL}{dL^2} = -\frac{2}{3} < 0$$

APL is maximum at $L = 9/2$

At $L = 9/2$, $MPL = 72 + 6 \cdot \frac{9}{2} - \left(\frac{9}{2}\right)^2 = 78.75$

At $L = 9/2$, max $APL = 72 + 3 \cdot \frac{9}{2} - \frac{1}{3}\left(\frac{9}{2}\right)^2 = 78.75$

Ex. 231. The price p of a good is given by

$$x = ae^{-bp}$$

where x is the quantity sold. The cost function is $C = l + mx$ where a, b, l, m are positive constants. Find the optimal price which maximises the profit. **Ans.** $p = (1 + mb)/b$

Ex. 232. Given a firm's cost function

$$C = 100{,}000 + 50x^2/(1000)^2$$

find the profit function if x stands for the quantity of good sold and $p = 10$ is the price per unit. If π stands for profit then $d\pi/dx = 0$ gives $x = 100{,}000$.

Note. $$\pi = 100{,}000 + \frac{50x^2}{(1000)^2} - 10x$$

$$\frac{d\pi}{dx} = \frac{100x}{(1000)^2} - 10$$

$$= 0 \text{ at } x = 100{,}000$$

Ex. 233. The total cost and demand functions of a monopolist are $C = ax^2 + bx + c$, $p = \beta - \alpha x$ where α, β, a, b, c are positive constants. Find the profit maximising output level and show that the output level is the same whether output or price is fixed.

Note. $$\pi = px - C = \beta x - \alpha x^2 - (ax^2 + bx + c)$$

$$\frac{d\pi}{dx} = 0 \Rightarrow x_{max} = \frac{\beta - b}{2(\alpha + a)} \text{ where } \frac{d^2\pi}{dx^2} < 0$$

Alternatively, we should write the profit function in terms of p

i.e., $$x = \frac{1}{\alpha}(\beta - p),$$

$$\pi = \frac{(\beta - b)(\beta - p)}{a} - \frac{(\alpha + a)(\beta - p)^2}{\alpha^2} - c$$

$$\frac{d\pi}{dp} = 0 \Rightarrow p = \frac{\alpha\beta + 2a\beta + \alpha b}{2(\alpha + a)} \text{ where } \frac{d^2\pi}{dp^2} < 0$$

At this value of p, the maximum output is

$$x_{\max} = \frac{\beta - p}{\alpha} = \frac{1}{\alpha}\left\{\beta - \frac{\alpha\beta + 2\alpha\beta + \alpha b}{2(\alpha + a)}\right\} = \frac{\beta - b}{2(\alpha + a)}$$

which was the value obtained earlier.

Ex. 234. The cost and revenue functions of a commodity are given by

$$C = \frac{x^2}{25} + 5x + 1000, \; R = 126.6x - 3x^2$$

If a tax t per unit is imposed what will be the change in maximum profit if C remains unchanged? Find how much of t is passed on to the consumers.

Find the effect on profit when $t = 6$

For $\pi_{\max}$, $\frac{d\pi}{dx} = 0$ where $\pi = R - C$

$$\pi = -3.04x^2 + 121.6x - 1000$$

$$\frac{d\pi}{dx} = -6.08x + 121.6 = 0 \text{ or } x = 20$$

$$\frac{d^2\pi}{dx^2} = -6.08 < 0$$

The profit maximising price $= 126.6 - 3x$ [$\because p = R/x = 126.6 - 3x$]

$$= 126.6 - 3 \times 20 = 66.6$$

$$\pi_{\max} = 3.04 \times 20^2 + 121.6 \times 20 - 1000 = 216$$

After the tax t,

$$\pi = (126.6x - 3x^2) - (1000 + 5x + 0.04x^2) - tx$$

$$= -3.04x^2 + (121.6 - t)x - 1000$$

$$\frac{d\pi}{dx} = -6.08x + 121.6 - t = 0 \text{ or } x = \frac{121.6 - t}{6.08}$$

Since $\frac{d^2\pi}{dx^2} < 0$ we have for $\pi_{\max}$, $x = 20 - \frac{t}{6.08}$

and $$p = 126.6 - 3\left(20 - \frac{t}{6.08}\right) = 66.6 + \frac{3t}{6.08}$$

There is a change in $\pi_{\max}$ which depends on the tax rate t. The amount of tax passed onto the consumer is $t/6.08$

When $t = 6, \; x = 20 - \frac{6}{6.08} \simeq 19$

$$\pi_{\max} = -3.04x^2 + (121.6 - 6)x - 1000 = 96$$

Difference in profit $= 216 - 96 = 120$

Ex. 235. The cost and demand functions of a firm are given by

$$C = 1064 + 5x + x^2/25 \text{ and } p = 157 - 3x$$

Find the maximum profit (*i*) without any tax (*ii*) with a tax of t per unit. Find the tax portion passed on to the consumer. (*iii*) If however t is fixed at 4 what will be the effect on the profit of the company?

(*i*) Profit $\pi = R - C = (157 - 3x)x - [1064 + 5x + (x^2/25)]$

$$= -1064 + 152x - 76x^2/25$$

$$\frac{d\pi}{dx} = 152 - 152x/25 = 0 \text{ or } x = 25 \qquad \frac{d^2\pi}{dx^2} < 0$$

$$\pi_{max} = -1064 + 152 \times 25 - 76 \times 625/25 = 836$$

(*ii*) With a tax t per unit,

$$\pi = (157 - 3x)x - \left(1064 + 5x + \frac{x^2}{25} + tx\right)$$

$$\frac{d\pi}{dx} = 152 - \frac{152}{25}x - t = 0$$

or

$$x = 25\left(1 - \frac{t}{152}\right) \qquad \frac{d^2\pi}{dx^2} < 0$$

$\pi_{max} = 82 + \frac{75t}{152}$. This includes $\frac{75t}{152}$, the quantum of tax passed on to the consumer.

(*iii*) At $t = 4x = 25\ (1 - 4/152) = 3700/152$, substitute this in

$$\pi_{max} = -1064 + 152x - tx - \frac{76x^2}{26} = \pi_0$$

Change in profit $= \pi_0 - 836$.

Ex. 236. The cost and demand functions of each of two doupolists are

$$C = \frac{x^2}{25} + 3x + 100 \text{ and } p = 25 - x/3.$$

Find their total equilibrium output.

If the outputs are x_1 and x_2 then $x = x_1 + x_2$ and $p = 25 - \frac{x}{3} = 25 - \frac{x_1 + x_2}{3}$

For the first doupolist

$$\pi_1 = R_1 - C_1 = \left(25 - \frac{x_1 + x_2}{3}\right)x_1 - \left(\frac{x_1^2}{25} + 3x_1 + 100\right)$$

$$\frac{d\pi_1}{dx_1} = 22 - \frac{56x_1}{75} - \frac{x_1}{3}\frac{dx_2}{dx_1} - \frac{x_2}{3}$$

$$= 22 - \frac{56x_1}{75} - \frac{x_2}{3}$$

if the conjectural variation $\frac{dx_2}{dx_1} = 0$

$$\therefore \quad \frac{d\pi_1}{dx_1} = 0 \Rightarrow 22 - \frac{56x_1}{75} - \frac{x_2}{3}$$

$$\text{Similarly} \quad \frac{d\pi_2}{dx_2} = 0 \Rightarrow 22 - \frac{56x_2}{75} - \frac{x_1}{3}$$

$$\text{Solving} \quad x_1 = x_2 = 20.4$$

$$x = x_1 + x_2 = 40.8$$

Ex. 237. Determine the optimal point in time when to replace a capital asset whose depreciation decreases over time and operating cost increases over time.

If $C_1(t)$ is the total capital cost over the time period t and $C_2(t)$ the average operation cost the average cost per unit time period of owning the asset for time period t is

$$C(t) = \frac{C_1(t)}{t} + C_2(t)$$

$$C'(t) = -\frac{C_1}{t^2} + C_2'(t) \text{ if } C_1 \text{ is a constant}$$

If $\quad C_1(t) = p - s =$ purchase price – salvage price

and $\quad C_2(t) = a + bt$ increases linearly over time

$$\frac{dC(t)}{dt} = -\frac{1}{t^2}(p - s) + b = 0 \text{ for optimal } t$$

$\therefore$ The optimal point of time t is given by,

$$t = \sqrt{\frac{p - s}{b}}$$

The second order condition is satisfied, since

$$\frac{d^2C}{dt^2} = \frac{2}{t^3}(p - s) > 0$$

10.29 ECONOMIC ORDER QUANTITY

A firm must maintain some inventory for continuity of sales till the next lot of inventories arrives. A very low inventory may sometimes lead to stoppage of work and a large inventory would mean incurring costs. A reasonably sized inventory must balance these two aspects.

(*i*) The firm has to reorder, for say, x units every time. If the sales are uniformly spaced over time, the average number of units with the firm is $\frac{x}{2}$ whose total cost of maintenance or storage cost is $\frac{cx}{2}$ where c is the annual cost of maintaining a unit for one year.

(*ii*) In reordering there is a fixed cost F which does not vary with the order size. Another part of the cost varies with the order size. If b is the additional cost per unit the total reordering cost for x units in $F + bx$.

If a total of Q units is ordered and x units are ordered each time, the total number of orders is $\frac{Q}{x}$ and total annual reordering cost is

$$(F + bx)\frac{Q}{x} \text{ or } \frac{FQ}{x} + bQ$$

Total cost (including storage) is given by

$$Z = \frac{cx}{2} + \frac{FQ}{x} + bQ$$

For minimum total cost, we have

$$\frac{dZ}{dx} = \frac{c}{2} - \frac{FQ}{x^2} = 0 \quad i.e., \quad x^2 = \frac{2FQ}{c}$$

This gives us the optimal order size $x = \sqrt{\frac{2FQ}{c}}$

and
$$\frac{d^2Z}{dx^2} = \frac{2FQ}{x^3} > 0; \quad F, Q, x > 0$$

10.30 PRESENT VALUE

If $p\ (t)$ is the present value of a project and $R\ (t)$ is discounted continuously at discount rate r we have $p\ (t) = R(t)e^{-rt}$

If the initial investment is I, the NPV (net present value) is

$$\begin{aligned} N\ (t) &= p\ (t) - I \\ &= R\ (t)\ e^{-rt} - I \end{aligned}$$

(If $N\ (t) < 0$, the investment should not be undertaken)

To maximise N

$$\frac{dN(t)}{dt} = e^{-rt}\frac{dR}{dt} - rR(t)e^{-rt} = 0$$

or
$$e^{-rt}\left(\frac{dR}{dt} - rR\right) = 0$$

i.e.,
$$\frac{dR}{dt} = rR$$

$$\frac{d^2N}{dt^2} = \left(\frac{d^2R}{dt^2} - r\frac{dR}{dt}\right)e^{-rt} < 0$$

The optimum value of t thus obtained is the absolute maximum.

Ex. 238. Show how you will maximise the IRR (Internal Rate of Return r)? Find the condition for the maximum.

The *IRR* is found from

$$N = R(t)e^{-rt} - I = 0$$

$$\therefore \quad \frac{R}{I} = e^{rt} \text{ or } ln\left(\frac{R}{I}\right) = rt \text{ so that}$$

$$r = \frac{1}{t} ln\left(\frac{R}{I}\right)$$

To find the condition, write

$$\frac{dr}{dt} = \frac{1}{t}\frac{R'}{R} - \frac{1}{t^2} ln\left(\frac{R}{I}\right) = 0 \quad (R' \text{ is derivative of } R)$$

$$\frac{R'}{R} = \frac{1}{t} ln \frac{R}{I} \quad \text{is the required condition}$$

Ex. 239. If the present value per unit of a good sold after t years with interest at r per cent added continuously is given by $p(t) = R(t)e^{-rt}$, maximise $p(t)$, assuming production cost to be constant.

Ans. $ln\ p = ln\ R - rt \quad \therefore \quad \frac{1}{p}\frac{dp}{dt} = \frac{1}{R(t)} R'(t) - r = 0$

i.e., $R'(t)/R(t) = r$

Also $d^2p/dt^2 < 0$ which means $\frac{d}{dt}\frac{R'(dt)}{R(t)} < 0$

Ex. 240. An individual has various incomes x this year and various incomes y next year, the interest rate being i. The transformation relation is given by $F(x, y) = 0$. Show that the incomes x, y are such that the marginal rate of return over cost equals i.

Note. The present value of y is $\frac{y}{1+i}$

To find the incomes which maximise the total present value

$$u = x + \frac{y}{1+i} \text{ subject to } F(x, y) = 0$$

$$\frac{dx}{dt} = 1 + \frac{1}{1+i}\frac{dy}{dx} = 0 \quad i.e., \frac{dy}{dx} = -(1+i)$$

$$\therefore \quad 1 + i = -\frac{dy}{dx} \text{ or } i = -\frac{dy}{dx} - 1$$

$$\left(\text{The second order condition } \frac{d^2u}{dx^2} = \frac{1}{1+i}\frac{d^2y}{dx^2} < 0 \text{ because } \frac{d^2y}{dx^2} < 0\right)$$

Ex. 241. (Logistic distribution). A population grows so that at time t its size is

$$y = \frac{y_0}{1 + Ce^{-kt}}$$

Show that the rate of growth of the population

$$\frac{dy}{dt} = -y_0 \frac{-kCe^{-kt}}{(1+Ce^{-kt})^2} = \frac{ky_0Ce^{-kt}}{(1+Ce^{-kt})^2}$$

Ex. 242. After spending Rs. x on advertis.ng, the number of units of a good sold is

$$1000\,(1 - e^{-x/1000})$$

If the profit π per unit is Rs. 10 find $d\pi/dx$ where (*i*) $x = 1000$ (*ii*) $x = 3000$

$$\pi = \{1000\,(1 - e^{-x/1000})\} - x$$

$$\frac{d\pi}{dx} = 10e^{-x/1000} - 1$$

$= 2.679$ when $x = 1000$ (net profit)

$= -0.902$ when $x = 3000$ (net loss).

spending more on advertising leads to losses.

10.31 TRANSACTIONS DEMAND FOR MONEY

Suppose y is the income of an individual at some initial period of time. He spends this income uniformly to reach zero balance at the end of a certain period. As he is not using all his money at the same time he buys bonds to earn interest and may sell them when he runs out of cash. His optimal schedule will be to buy bonds at the beginning of the period and not to sell bonds until he needs cash. If he makes n transactions, his optimal purchase at the beginning of the period is $\frac{(n-1)\,y}{n}$. His revenue during the period is

$$R = \frac{r(n-1)y}{2n}$$

where r is the rate of interest and $\frac{(n-1)\,y}{2n}$ is the average bond holding. To maximise the revenue, we have

$$\frac{dR}{dn} = \frac{ry}{2n^2} = 0. \text{ This happens when } n \to \infty$$

$\therefore$ R is maximum when $n \to \infty$ *i.e.*, when all income in the beginning is converted into bonds and the bonds are cashed continuously as required.

If the individual incurs a fixed cost c per transaction, his profit

$$\pi = R - cn = \frac{(n-1)\,ry}{2n} - cn \text{ is maximum when}$$

$$\frac{d\pi}{dn} = \frac{ry}{2n^2} - c = 0 \text{ or } n = \sqrt{\frac{ry}{2c}}$$

$$\frac{d^2\pi}{dn^2} = -\frac{ry}{n^3} < 0, \text{ suggesting that } n = \sqrt{\frac{ry}{2c}}$$

gives the number of transactions for maximum profits.. A rise in transaction cost reduces and a rise in interest rate or the initial cash balance increases the optimum number of transactions.

A rise in interest rate also leads to increased bond holdings so that the transactions demand for money is a decreasing function of interest.

Ex. 243. A monopolist's cost function is given by $C = x^2 + 4x$ and his average revenue is given by $AR = 28 - 5x$. Show that $x = 2$ for maximum profit which equals 24.

Suppose a tax t per unit is imposed on the monopolist so that his total cost $C_1 = C + tx$. The profit $R - C_1$ is maximum if

$$\frac{dR}{dx} = \frac{dC}{dx} \text{ and } \frac{d^2R}{dx^2} < \frac{dC_1^2}{dx^2}$$

Find the maximum profit if in the problem given above a tax of (*i*) t per unit (*ii*) 25 per cent is imposed.

Ex. 244. Show that the maximum profit of a monopolist whose cost and revenue functions are $C = 14 - 3x$ and $R = 14x - 3x^2$ is given by $(9 - t)^2/16$ where t is the tax per unit imposed.

Sales x_t after taxation are less than x before taxation.

If $ln\ f(x) = t^{1/2}\ ln\ 2 - kt$ then $f(x)$ is maximum when $t = \left(\frac{\ln 2}{2k}\right)^2$. Show that for $k = 0.1$, $t \simeq 12$.

Ex. 245. Suppose a monopsonist's output is a function of labour: $x = 15L^2 - \frac{1}{5}L^3$ and he sells his output at a fixed price 3 per unit. If his cost is a function of labour

$$C = 23\tfrac{2}{5}L^2 + 144L$$

show that $L = 20$ gives him the maximum profit.

CHAPTER 11

Differential Calculus : Functions of Two or More Variables

11.1 PARTIAL DERIVATIVES

We have considered the derivatives of functions of one independent variable. Often, we have to deal with cases involving several factors. We can extend the idea of a derivative to cover functions of two or more independent variables.

Consider a function of two independent variables x, y

$$u = f(x, y)$$

The derivative of a function $y = f(x)$ is a way of measuring how y changes with x.

When u is a function of two variables we can see how u changes with x or y or both. We can consider

(*i*) how u changes with x ; y held constant

(*ii*) how u changes with y ; x held constant.

The derivative of u with respect to x when y is held constant is called the *partial derivative of u with respect to x* and written

$$f_x = \frac{\partial u}{\partial x} = \operatorname{Lim}_{\Delta x \to 0} \frac{f(x+\Delta x, y) - f(x, y)}{\Delta x}$$

Similarly if x is held constant, the *partial derivative of u with respect to y* is written

$$f_y = \frac{\partial u}{\partial y} = \operatorname{Lim}_{\Delta y \to 0} \frac{f(x, y+\Delta y) - f(x, y)}{\Delta y}$$

Thus finding a partial derivative amounts to finding derivatives with respect to one independent variable holding other independent variables constant.

Ex. 1. $u = x^2y + xy^4$

$$\frac{\partial u}{\partial x} = y(2x) + y^4\,(1) = 2xy + y^4$$

$$\frac{\partial u}{\partial y} = x^2 + 4xy^3$$

Ex. 2. $u = e^{xy}, \frac{\partial u}{\partial x} = ye^{xy}; \frac{\partial u}{\partial y} = xe^{xy}$

Ex. 3. $u = \frac{x}{y}, \frac{\partial u}{\partial x} = \frac{1}{y}; \frac{\partial u}{\partial y} = -\frac{x}{y^2}$

Ex. 4. Find $\frac{\partial u}{\partial x}, \frac{\partial u}{\partial y}$ for

(*i*) $u = x^2y^2 + x^5 + y^5$ (*ii*) $u = (x^2 - y^2)^{\frac{7}{5}}$ (*iii*) $u = (x + y)^2 e^{\frac{x^2}{y^2}}$

(*iv*) $u = x^m y^n$ (*v*) $u = \log_e (x^2 + y^3)^{2/3}$ (*vi*) $u = x^2 \log_e y + ye^x$

Ans. (*i*) $2xy^2 + 5x^4, 2x^2y + 5y^4$

(*ii*) $(7/5)(x^2 - y^2)^{2/5} (2x), (7/5) (x^2 - y^2)^{2/5} (-2y)$

(*iii*) $2(x+y)e^{x^2/y^2}\left(\frac{x(x+y)}{y^2}+1\right), 2(x+y)e^{x^2/y^2}\left(\frac{-x^2(x+y)}{y^3}+1\right)$

(*iv*) $mx^{m-1}y^n, nx^m y^{n-1}$ (*v*) $\frac{4x}{3(x^2+y^2)}, \frac{4y}{3(x^2+y^2)}$

(*vi*) $2x\ln y + ye^x, \frac{x^2}{y} + e^x$

Ex. 5. For the production function

$$x = 10 L^{0.7} K^{0.3}$$

$$\frac{\partial x}{\partial L} = 10K^{0.3}\, 0.7L^{-0.3} = 7\frac{K^{0.3}}{L^{0.3}}$$

This is the marginal productivity of labour—with capital fixed.

$$\frac{\partial x}{\partial K} = 3L^{0.7} K^{-0.7} = 3\frac{L^{0.7}}{K^{0.7}}$$

This is the marginal productivity of capital—with labour fixed.

Ex. 6. Given $y = \frac{5u - 3v}{5u^2 - v^2}$ find $\frac{\partial y}{\partial u}, \frac{\partial y}{\partial v}$.

Ex. 7. For the utility function of two commodities

$$u = u(x_1, x_2) = (x_1 - 2)^2 (x_2 + 1)^3$$

$\frac{\partial u}{\partial x_1} = 2(x_2 + 1)^3 (x_1 - 2)$ is the marginal utility function of the first commodity.

$\frac{\partial u}{\partial x_2} = 3(x_1 - 2)^2 (x_2 + 1)^2$ is the marginal utility function of the second .

If $x_1 = 3, x_2 = 4$

$$\frac{\partial u}{\partial x_1} = 2 \times 125 \times 1 = 250$$

$$\frac{\partial u}{\partial x_2} = 3 \times 1 \times 25 = 75$$

For the production function

$$x = f(L, K)$$

$\dfrac{\partial x}{\partial L}$ is the rate of change of output with respect to labour while capital input is held constant.

$\dfrac{\partial x}{\partial K}$ is the rate of change of output with respect to capital while labour input is held constant.

$\dfrac{\partial x}{\partial L}, \dfrac{\partial x}{\partial K}$, therefore, give marginal physical products of labour and capital respectively.

Ex. 8. For the production function

$$x = K^3 - 8K^2 + 10\,KL + 5KL^2 - 3L^3$$

find *MPP* of *L* and *K* at $L = 2$, $K = 3$. **Ans.** 14, 19

Partial derivatives of higher order

For the function $u = f(x, y)$ we have four *second* order derivatives.

The direct partial derivatives, written

(*i*) f_{xx} or $\dfrac{\partial^2 u}{\partial x^2} = \dfrac{\partial}{\partial x}\left(\dfrac{\partial u}{\partial x}\right)$

(*ii*) f_{yy} or $\dfrac{\partial^2 u}{\partial y^2} = \dfrac{\partial}{\partial y}\left(\dfrac{\partial u}{\partial y}\right)$

The cross partial derivatives written

(*iii*) f_{xy} or $\dfrac{\partial^2 u}{\partial x\, \partial y} = \dfrac{\partial}{\partial x}\left(\dfrac{\partial u}{\partial y}\right)$

(*iv*) f_{yx} or $\dfrac{\partial^2 u}{\partial y\, \partial x} = \dfrac{\partial}{\partial y}\left(\dfrac{\partial u}{\partial x}\right)$

If $f(x, y)$ is a continuous function with continuous partial derivatives we have

$$f_{xy} = f_{yx} \text{ i.e., } \frac{\partial^2 u}{\partial x \partial y} = \frac{\partial^2 u}{\partial y \partial x}$$

Higher order partial derivatives can be defined similarly.

Ex. 9. If $u = xy$, $f_x = y$, $f_y = x$

$$f_{xx} = 0, f_{yy} = 0, f_{xy} = 1 = f_{yx}$$

Ex. 10. For $u = xy^2 - 3x - 5y$, $f_x = y^2 - 3$, $f_y = 2xy - 5$

$$f_{xx} = 0, f_{yy} = 2x, f_{xy} = 2y = f_{yx}$$

Ex. 11. Find the partial derivatives of second order for the following:

(*i*) $u = \log(3x + y)$ (*ii*) $u = \dfrac{x+y}{x-y}$ (*iii*) $u = x^2 - xy^3$

Ans. (*i*) $\dfrac{-9}{(3x+y)^2}, \dfrac{-1}{(3x+y)^2}, \dfrac{-3}{(3x+y)^2}$

(*ii*) $\dfrac{4y}{(x-y)^3}, \dfrac{-4x}{(x-y)^3}, \dfrac{-2(x+y)}{(x-y)^3}$

(*iii*) $2, -2x, -2y$

11.2 THE TOTAL DIFFERENTIAL

The differential of a function $u = f(x, y)$ is defined by

$$du = \frac{\partial u}{\partial x}dx + \frac{\partial u}{\partial y}dy = f_x\, dx + f_y\, dy$$

The separate terms on the right are called partial differentials and their sum, du, is called the total differential. The result can be generalised to any number of variables.

If, for the case given above, $x = x(p, q)$, $y = y(p, q)$, $z = z(p, q)$ and we have $u = F(p, q)$, we may write

$$du = \frac{\partial u}{\partial p}dp + \frac{\partial u}{\partial q}dq = F_p\, dp + F_q dq$$

The value of du in both the cases remains the same.

Note. If $u = r + s$, $du = dr + ds$ sum

If $u = r \cdot s$, $du = rds + sdr$ product

If $u = \dfrac{r}{s}$, $du = \dfrac{sdr - rds}{s^2}$ quotient

If $u = u(x)$, $z = z(u)$, $dz = \dfrac{dz(u)}{du} \dfrac{du(x)}{dx} \cdot dx$

Ex. 12.

$$\begin{aligned} u &= (x^2 + y^2)(2x^2 - y) \\ du &= (x^2 + y^2)\, d(2x^2 - y) + (2x^2 - y)\, d(x^2 + y^2) \\ &= (x^2 + y^2)(4xdx - dy) + (2x^2 - y)(2xdx + 2ydy) \\ &= (8x^3 + 4xy^2 - 2xy)dx - (x^2 + y^2 - 4x^2y + 2y^2)dy \\ &= (8x^3 + 4xy^2 - 2xy)dx + (4x^2y - x^2 - 2y^2)dy \end{aligned}$$

Ex. 13. $u = \dfrac{x^3 + y^2}{x - y}$

$$\begin{aligned} du &= \frac{(x-y)(dx^3 + dy^2) - (x^3 + y^2)(dx - dy)}{(x-y)^2} \\ &= \frac{(x-y)(3x^2dx + 2ydy) - (x^3 + y^2)(dx - dy)}{(x-y)^2} \\ &= \frac{(2x^3 + 3x^2y - y^2)dx + (2xy - y^2 + x^3)dy}{(x-y)^2} \end{aligned}$$

Ex. 14. Find the total differential of u for

(*i*) $u = x^2y^3$ (*ii*) $u = x^2 + 3y^2$ (*iii*) $x^{\frac{1}{2}}y^{\frac{1}{3}}$

(*iv*) $u = \log (x^2 - y^2)$ (*v*) $u = \log (1 - x^2y)$

(*vi*) $u = z^2, \quad z = x^2 - y^2$

(*vii*) $u = \dfrac{x + y}{2x^2}$ (*viii*) $u = \dfrac{2x^2}{x + y}$ (*ix*) $u = \dfrac{xy}{x + y}$

11.3 PARTIAL ELASTICITIES

If $x = f(p_1, p_2)$ is the demand function for a commodity dependent on two variables p_1, p_2, we may define the partial elasticity of demand with respect to p_1 while p_2 is kept constant, as

$$\eta_1 = \frac{p_1}{x}\frac{\partial x}{\partial p_1}$$

Similarly
$$\eta_2 = \frac{p_2}{x}\frac{\partial x}{\partial p_2}$$

Ex. 15. For the function $x = 10 - 2p_1 - 3p_2$ we have

$$\eta_1 = \frac{p_1}{x}(-2) = \frac{-2p_1}{10 - 2p_1 - 3p_2}$$

and
$$\eta_2 = \frac{p_2}{x}(-3) = \frac{-3p_2}{10 - 2p_1 - 3p_2}$$

Ex. 16. If saving (S) is a function of income (Y) and interest (r) *i.e.*, if $S = f(Y, r)$, the partial elasticities of saving with respect to income and interest may be written

$$\eta_y = \frac{Y}{S}\frac{\partial S}{\partial Y}, \quad \eta_r = \frac{r}{S}\frac{\partial S}{\partial r}$$

Ex. 17. Find the partial elasticities for the utility function

$$U = 3x^2 + 5xy + 4y^2 \qquad \textbf{Ans. } \frac{x(6x + 5y)}{3x^2 + 5xy + 4y^2}, \ \frac{y(5x + 8y)}{3x^2 + 5xy + 4y^2}$$

Ex. 18. Find the partial elasticities for

(*i*) $x = 20 - 3p_1 + 4p_2$ (*ii*) $x = 30 - 2p_1 - p_2$

at $p_1 = 1, p_2 = 2$ **Ans.** (*i*) –3/9, -8/9 (*ii*) –2/26, –2/26

Ex. 19. Elasticity of Substitution. Suppose a firm's output is a function of inputs L_1, L_2. Thus $x = f(L_1, L_2)$ and we define RTS, the rate of technical substitution as dL_2/dL_1. The elasticity of substitution, σ, is a pure number which is defined as the proportionate change of input ratio divided by the proportionate change of the RTS. Then

$$\sigma = \frac{d\ln(L_2/L_1)}{d\ln(f_1/f_2)} = \frac{f_1L_1}{f_2L_2}\frac{d(L_2/L_1)}{d(f_1/f_2)}$$

$$= \frac{f_1(f_1L_1 + f_2L_2)}{f_2L_1L_2\left[f_1\frac{\partial}{\partial L_2}\left(\frac{f_1}{f_2}\right) - f_2\frac{\partial}{\partial L_1}\left(\frac{f_1}{f_2}\right)\right]}$$

$$= \frac{f_1f_2(f_1L_1 + f_2L_2)}{L_1L_2(2f_1f_2f_{12} - f_1^2f_{22} - f_2^2f_{11})}$$

since $\quad \dfrac{dL_2}{dL_1} = -\dfrac{f_1}{f_2}$

11.4 EULER'S THEOREM

The value of a linearly homogeneous function is the sum of the products of the independent variables and the corresponding partial derivatives. If $x = x(u, v)$, is homogeneous of degree n, Euler's theorem states that

$$u\frac{\partial x}{\partial u} + v\frac{\partial x}{\partial v} = nx(u, v)$$

If $n = 1$, we have $u\dfrac{\partial x}{\partial u} + v\dfrac{\partial x}{\partial v} = x(u, v) = x$ say

The result can be extended to any number of variables. In economics, Euler's theorem is applied to marginal productivity theory and various other situations.

The Adding up Theorem

If conditions of constant returns to scale are available and if each input factor is given the amount of its marginal product, the total product x will be exhausted by the shares distributed among all the input factors.

If $x = x(K, L)$ is a homogeneous production function (degree one) of capital and labour, we have, by Euler's theorem

$$x = \frac{\partial x}{\partial K}K + \frac{\partial x}{\partial L}L$$

We can see that the sum of the shares just add up to the total product.

Ex. 20. For the Cobb Douglas production function

$$x = AL^{\alpha}K^{\beta} \quad \text{with} \quad \alpha + \beta = 1$$

$$\frac{\partial x}{\partial L} = A\alpha L^{\alpha-1}K^{\beta} = \frac{\alpha}{L}AL^{\alpha}K^{\beta} = \frac{\alpha}{L}x$$

$$\frac{\partial x}{\partial K} = A\beta L^{\alpha}K^{\beta-1} = \frac{\beta}{K}AL^{\alpha}K^{\beta} = \frac{\beta}{K}x$$

We have $L\dfrac{\partial x}{\partial L} + K\dfrac{\partial x}{\partial K} = \left(L\cdot\dfrac{\alpha}{L} + K\cdot\dfrac{\beta}{K}\right)x = x$ since $\alpha + \beta = 1$

The relative shares of labour and capital are, respectively

$$\alpha = \frac{L}{x}\,\frac{\partial x}{\partial L} \text{ and } \beta = \frac{K}{x}\,\frac{\partial x}{\partial K}$$

The total product is exhausted since $\alpha + \beta = 1$

If we multiply by the price of x, in the relation

$$x = L\,\frac{\partial x}{\partial L} + K\frac{\partial x}{\partial K}$$

we have total revenue $R = px = pL\,\dfrac{\partial x}{\partial L} + pK\dfrac{\partial x}{\partial K}$

The total revenue R is exhausted by returns to the factors L, K.

Ex. 21. Apply the above result to the production functions

(*i*) $x = AL + BK$ (*ii*) $\dfrac{LK}{L+K}$ (*iii*) $x = \left[\dfrac{\alpha}{L^\rho} + \dfrac{\beta}{K^\rho}\right]^{1-\frac{1}{\rho}}$

(*iv*) $x = L^2K \log \dfrac{L}{K}$ (*v*) $x = AL^2 + BK^2$

Ex. 22. For the CES production function

$$x = A[\delta K^{-\rho} + (1-\delta)L^{-\rho}]^{-\frac{v}{\rho}},\ A > o,\ \rho > -1,\ O < \delta < 1$$

Show that $\dfrac{\partial x}{\partial L} = v(1-\delta)A^{-\frac{\rho}{v}}\,x^{1+\frac{\rho}{v}}L^{-1-\rho}$

$$\frac{\partial x}{\partial K} = v\delta A^{1-\frac{\rho}{v}}\,x^{1+\frac{\rho}{v}}\,L^{-1-\delta}$$

Also show that RTS $= \dfrac{\delta}{1-\delta}\left(\dfrac{K}{L}\right)^{\rho+1}$. Find $\dfrac{\partial(\text{RTS})}{\partial L}$.

Ex. 23. For the CES production function, the elasticity of substitution $\sigma = \dfrac{1}{1+\rho}$. What does the function reduce to when $\rho = 0$?

Find σ for the new function. **Ans.** Cobb Douglas function; $\sigma = 1$

Ex. 24. Show that the marginal products rise before eventually falling in the case of the production function given by

$$x = AL^{\alpha_1}e^{\alpha_2 L}K^{\beta_1}e^{\beta_2 K}$$

11.5 MAXIMA AND MINIMA : STATIONARY POINTS

For $u = f(x, y)$ to have a local maximum or minimum, a necessary condition is that at these points,

$$f_x = \frac{\partial u}{\partial x} = 0 \text{ and } f_y = \frac{\partial u}{\partial y} = 0$$

This is a necessary but not a sufficient condition. Sufficiency conditions are
for maximum $f_{xx} < 0, f_{yy} < 0, f_{xx} f_{yy} > f_{xy}^2$
for minimum $f_{xx} > 0, f_{yy} > 0, f_{xx} f_{yy} > f_{xy}^2$

If $f_{xx} f_{yy} < f_{xy}^2$ the stationary point under consideration is called a *saddle* point.

Ex. 25. Show that $u = a$ gives the maximum for the function

$$u^2 = a^2 - (x^2 + y^2)$$

Ex. 26. Show that $x = 2$ and $x = -3$ give the minima and $x = 0$ gives the maximum for $u = 3x^4 + 4x^3 - 36x^2 - y^3 - 12xy + 9$.

11.6 PURE COMPETITION

Suppose that, under pure competition conditions, a firm sells two goods at prices p_1, p_2 respectively so that, the revenue function is

$$R = R(x_1, x_2) = p_1x_1 + p_2x_2$$

and the cost function

$$C = C(x_1, x_2)$$

The profit function of the firm is

$$R - C = \pi(x_1, x_2, p_1, p_2) = p_1x_1 + p_2x_2 - C$$

To maximise π the necessary conditions to be satisfied are

$$\frac{\partial \pi}{\partial x_1} = 0 \text{ and } \frac{\partial \pi}{\partial x_2} = 0$$

The sufficient conditions are

$$\frac{\partial^2 \pi}{\partial x_1^2} < 0, \frac{\partial^2 \pi}{\partial x_2^2} < 0, \frac{\partial^2 \pi}{\partial x_1^2}\frac{\partial^2 \pi}{\partial x_2^2} > \left(\frac{\partial^2 \pi}{\partial x_1 \partial x_2}\right)^2$$

Ex. 27. If the joint cost function is $C = x_1^2 + x_1x_2 + 3x_2^2$ for two commodities whose prices are 7 and 20 find the maximum profit and total cost.

$$\pi = 7x_1 + 20x_2 - x_1^2 - x_1x_2 - 3x_2^2$$

$$\frac{\partial \pi}{\partial x_1} = 0 \text{ gives } 7 - 2x_1 - x_2 = 0$$

$$\frac{\partial \pi}{\partial x_2} = 0 \text{ gives } 20 - x_1 - 6x_2 = 0$$

Solving we get $x_1 = 2, x_2 = 3$ so that

$$C = 37, \pi_{max} = 37$$

Ex. 28. For two commodities the prices are given by $p_1 = 108 - 5x_1$ and $p_2 = 35 - 2.5x_2$ and the total cost function is $C = x_1^2 + x_2^2$, show that for maximum profit $x_1 = 9$, $x_2 = 5$. Find π_{max}.

11.7 MONOPOLIST PRODUCING TWO COMMODITIES

If x_1, x_2 are the outputs, and p_1, p_2 the prices of two commodities and $C(x_1, x_2)$ is the cost function of the monopolist, we have $x_1 = f(p_1, p_2)$ $x_2 = g(p_1, p_2)$ and we have to maximise

$$\pi(p_1, p_2) = p_1x_1 + p_2x_2 - C(x_1, x_2)$$

We set $$\frac{\partial \pi}{\partial p_1} = 0, \quad \frac{\partial \pi}{\partial p_2} = 0$$

and solve the resulting equations for profit maximising.

Sufficiency conditions can be written:

$$\frac{\partial^2 \pi}{\partial p_1^2} < 0, \quad \frac{\partial^2 \pi}{\partial p_2^2} < 0, \quad \frac{\partial^2 \pi}{\partial p_1^2}\frac{\partial^2 \pi}{\partial p_2^2} > \left(\frac{\partial^2 \pi}{\partial p_1 \partial p_2}\right)^2$$

11.8 THE MULTIPLE PLANT MONOPOLIST

If x_1, x_2 are the output of a two plant monopolist and $x = x_1 + x_2$ we have to maximise

$$\pi = R - C(x_1) - C(x_2)$$

$$\frac{\partial \pi}{\partial x_1} = \frac{dR}{dx}\frac{\partial x}{\partial x_1} - \frac{dC(x_1)}{dx_1} = 0$$

$$\frac{\partial \pi}{\partial x_2} = \frac{dR}{dx}\frac{\partial x}{\partial x_2} - \frac{dC(x_2)}{dx_2} = 0$$

Since $$\frac{\partial x}{\partial x_1} = \frac{\partial x}{\partial x_2} = 1, \text{ we have}$$

$$MR(x) = MC_1 \text{ and } MR(x) = MC_2$$

i.e., $$MC_1 = MC_2 = MR(x)$$

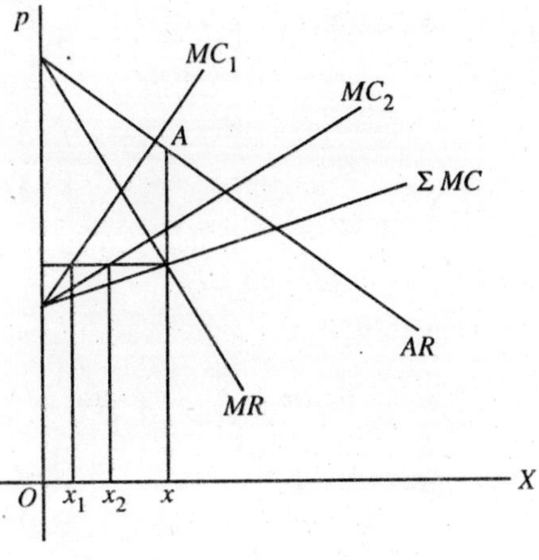

If x_I and x_{II} are outputs sold in markets *I* and *II* respectively, we can extend the argument to prove that

$$\Sigma MC = \Sigma MR$$

for maximum profits.

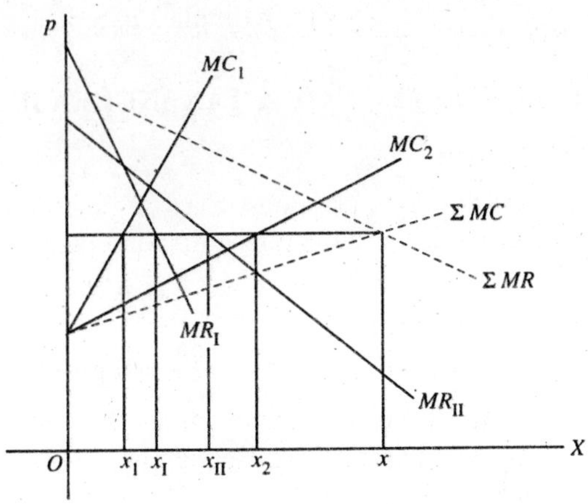

11.9 DISCRIMINATING MONOPOLIST

A discriminating monopolist, dealing with two isolated markets with different demand elasticities η_1, η_2 maximises his profit by selling quantities x_1, x_2 at prices p_1, p_2 in the two markets. Let $x = x_1 + x_2$.

To maximise $\pi = R_1(x_1) + R_2(x_2) - C(x)$ we write

$$\frac{\partial \pi}{\partial x_1} = 0, \qquad \frac{\partial \pi}{\partial x_2} = 0$$

We have $$\frac{\partial R_1}{\partial x_1} - \frac{\partial C}{\partial x} = 0 \qquad \text{i.e., } MR_1 = MC$$

$$\frac{\partial R_2}{\partial x_2} - \frac{\partial C}{\partial x} = 0 \qquad \text{i.e., } MR_2 = MC$$

$\therefore$ For maximum profit $MR_1 = MR_2$

i.e., $$p_1\left(1 - \frac{1}{\eta_1}\right) = p_2\left(1 - \frac{1}{\eta_2}\right)$$

$\therefore$ $$\frac{p_1}{p_2} = \frac{1 - 1/\eta_1}{1 - 1/\eta_2}$$

$\therefore$ If $\eta_1 > \eta_2$ *i.e.,* demand in market 1 is more elastic then $p_1 < p_2$ *i.e,* price in market 1 is lower and *vice versa.*

Generalising the above result to m markets we may write

$$p_1\left(1 - \frac{1}{\eta_1}\right) = p_2\left(1 - \frac{1}{\eta_2}\right) = \ldots\ldots\ldots\ldots\ldots\ldots = p_m\left(1 - \frac{1}{\eta_m}\right)$$

where $\eta_i > \eta_j$ implies $p_i < p_j$.

Ex. 29. Show that for monopsonistic discrimination

$$\eta_i > \eta_j \text{ implies } p_i > p_j.$$

11.10 EFFECT OF A TAX INCREASE ON OUTPUT OF A MONOPOLIST

Specific Tax

Let the price $p(x)$, cost $C(x)$ be functions of x. Let the specific tax be t per unit of output. We have to maximise the total profit

$$\pi = px - C - tx = R - C - tx$$

The first order condition $\dfrac{\partial \pi}{\partial x} = 0$ gives

$$\frac{\partial R}{\partial x} - \frac{\partial C}{\partial x} - t = 0$$

$\therefore$ $$MR = MC + t \text{ or } MC = MR - t$$

where t is a part of MC or MR.

The second order condition can help us to prove the obvious result that *as t increases demand for output falls.*

$$\frac{\partial^2 \pi}{\partial x^2} < 0 \text{ for maximum profits}$$

i.e., $$\frac{\partial^2}{\partial x^2}(R - C) < 0$$

Now since $\frac{\partial}{\partial x}(R - C) = t$, $\frac{\partial^2}{\partial x^2}(R - C) = \frac{\partial t}{\partial x}$

$$\therefore \quad \frac{\partial x}{\partial t} = \frac{1}{\frac{\partial^2}{\partial x^2}(R - C)} < 0 \text{ as the denominator is negative}$$

This gives the required result.

Effect of a Lump Sum Tax t_1

To maximise $\pi = px - C - t_1$, we have $\frac{\partial \pi}{\partial x} = 0$ which gives the usual condition for profit maximisation, *viz.* $\frac{\partial R}{\partial x} = \frac{\partial C}{\partial x}$

Effect of a Profits Tax t_2 per unit of profit

To maximise $\pi = R - C - t_2(R - C) \quad 0 < t_2 < 1$

we have $$\frac{\partial \pi}{\partial x} = (1 - t_2)\left(\frac{\partial R}{\partial x} - \frac{\partial C}{\partial x}\right) = 0$$

i.e., $\frac{\partial R}{\partial x} = \frac{\partial C}{\partial x}$ which is the usual expression for profit maximisation.

Sales Tax

The effect of a sales tax t_3 is similar to that of a specific tax.

To maximise $\pi = R - t_3R - C$,

we have $$\frac{\partial \pi}{\partial x} = (1 - t_3)\frac{\partial R}{\partial x} - \frac{\partial C}{\partial x} = 0$$

i.e., $(1 - t_3)MR = MC$

11.11 THE METHOD OF LEAST SQUARES

It is desired to fit a straight line $y = a + bx$ to a set of experimental observations

$$(x_1, y_1), (x_2, y_2), \ldots, (x_n, y_n)$$

For each value of x there are two values of y, the actual value y and the value y' predicted by the fitted straight line : $y = a + bx$.

Some of the deviations

$$d_i = y_i - y_i' = y_i - (a + bx_i)$$

are positive, others are negative. The squares are all positive. We consider that line as *the line of best fit* which minimises the sum of squared deviations

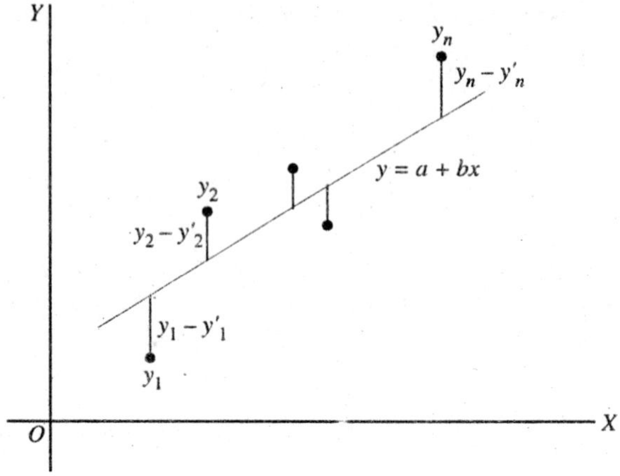

$$f(a,\ b) = \sum_{i=1}^{n} (y_i - a - bx_i)^2$$

To find suitable a, b for the purpose, we write

$$\frac{\partial f}{\partial a} = 0 \qquad i.e.,\ b \Sigma x_i + na = \Sigma y_i$$

$$\frac{\partial f}{\partial b} = 0 \qquad i.e.,\ b \Sigma x_i^2 + b \Sigma x_i = \Sigma x_i y_i$$

If we write $\bar{x} = \dfrac{\Sigma x_i}{n}$, $\bar{y} = \dfrac{\Sigma y_i}{n}$ we can, by solving these two equations see that, for the line of best fit

$$b = \frac{\Sigma x_i y_i - n\bar{x}\bar{y}}{\Sigma x_i^2 - n\bar{x}^2} \text{ and } a = \bar{y} - b\bar{x}$$

and that the point $(\bar{x}, \bar{y})$ lies on the line $y = a + bx$.

11.12 CONSTRAINED MAXIMA AND MINIMA

In mathematical and economic problems, the variables in a function are sometimes restricted by some relation or constraint. Thus the utility function $u = u(x_1, x_2 \ldots, x_n)$ may be subject to the budget constraint that income equals expenditure *i.e.*, $y = p_1x_1 + p_2x_2 + \ldots + p_nx_n$.

There are several methods of solving such problems. The method of direct elimination consists in substituting one of the variables of the constraint relation (expressed in terms of the remaining variables) in the given function and then dealing with the resulting variables as if they were independent of one another.

The method of *implicit functions* consists in similarly expressing one of the variables of the constraints in terms of the remaining variables and dealing with the implicit function obtained by substitution of this variable.

The method of Lagrange multiplier, given below, consists in forming an auxiliary function which results from the given function and its constraint.

To minimise (or maximise) $f(x, y)$ subject to $g(x, y) = 0$ construct the auxiliary function:

$$u(x, y) = f(x, y) - \lambda g(x, y)$$

where λ is an undetermined constant multiplier.

Now find values of x, y, λ for which the partial derivatives of x, y, λ are all zero.

$$u_x = 0 \text{ or } f_x - \lambda g_x = 0; \quad u_y = 0 \text{ or } f_y = \lambda g_y = 0; \quad u_\lambda = 0$$

i.e.,

$$f_x = \lambda g_x \qquad f_y = \lambda g_y$$

$$\therefore \qquad \lambda = \frac{f_x}{g_x} = \frac{f_y}{g_y}$$

Ex. 30. Consider two goods x_1, x_2; prices p_1, p_2. If the utility is proportional to the product of x_1 and x_2 *i.e.*, if $u = k\, x_1 x_2$, we have to maximise u subject to the budget constraint $y = p_1 x_1 + p_2 x_2$ *i.e.*, income equals expenditure.

By substitution
$$u = \frac{ky}{p_2} x_1 - \frac{kp_1}{p_2} x_1^2$$

The necessary condition for a maximum is $\dfrac{du}{dx_1} = 0$ *i.e.*, $x_1 = \dfrac{y}{2p_1}$.

This gives a maximum since $\dfrac{d^2u}{dx_1^2} < 0.$

The required values for maximum utility are

$$x_1 p_1 = \frac{y}{2}, \quad x_2 p_2 = \frac{y}{2}$$

which implies equal expenditure on the two commodities: $x_1 p_1 = x_2 p_2$

We can get the same result by using Lagrange multiplier method.

To maximise $kx_1x_2 - \lambda(p_1x_1 + p_2x_2 - y)$ we take partial derivatives

$$kx_2 - \lambda p_1 = 0, \; kx_1 - \lambda p_2 = 0$$

$$\therefore \qquad \lambda = \frac{kx_2}{p_1} = \frac{kx_1}{p_1} \text{ or } x_1 p_1 = x_2 p_2$$

Ex. 31. Maximise the production function $y = x_1 x_2$ subject to the budget constraint $x_2 = 6 - 2x_1$

Show that the maximum value of y is 8/9.

Ex. 32. Show that $x_1 = \frac{1}{6}$, $x_2 = \frac{1}{6}$, $x_3 = 2\frac{1}{2}$ maximises $u = x_1x_2x_3$ subject to $\dfrac{x_1x_2}{72} + \dfrac{x_2x_2}{96} + \dfrac{x_3x_1}{144} = 1.$

Ex. 33. Show that the condition to minimise (or maximise)

$u = f(x, y, z)$ subject to $g(x, y, z) = 0$ is $\lambda = \dfrac{f_x}{g_x} = \dfrac{f_y}{g_y} = \dfrac{f_z}{g_z}$

Ex. 34. Show that the minimum value of $x^2 + y^2 + z^2$ subject to $x + y + z = 1$ is given by $x = y = z = \frac{1}{3}$.

Ex. 35. Find the maximum value of $u = x^2 + 3xy - 5y^2$ subject to $2x + 3y = 6$.

Ex. 36. If F is fixed cost, w_i, the price of the variable factor v_i, minimise total cost

$C = F + \sum_{i=1}^{n} w_i v_i$ subject to the condition that total output $x = g(v_1, v_2, \ldots v_n)$. Deduce that

$\lambda = \dfrac{w_i}{\partial g/\partial v_i}$ *i.e.*, the equilibrium price of a factor is proportional to its marginal productivity.

Show that as the price of the ith factor changes with the prices of all other factors remaining unchanged, the demand for the ith factor will change in the opposite direction *i.e.*, show that $\dfrac{\partial v_i}{\partial w_i} < 0$. Also show that since λ equals marginal cost, under pure competition, the price of a factor equals its marginal product *i.e.*, $w_i = p\dfrac{\partial g_i}{\partial v_i}$.

11.13 LAW OF EQUIMARGINAL UTILITY

To maximise the utility function $u(x_1, x_2)$ subject to the budget constraint $y = p_1x_1 + p_2x_2$ we may maximise the function

$$z = u(x_1, x_2) - \lambda(p_1x_1 + p_2x_2 - y)$$

$$\frac{dz}{dx_1} = u_1 - \lambda p_1 = 0, \quad \frac{dz}{dx_2} = u_2 - \lambda p_2 = 0$$

where u_1, u_2 are marginal utilities of x_1, x_2 respectively.

It follows that $\lambda = \dfrac{u_1}{p_1} = \dfrac{u_2}{p_2}$ *i.e.*, marginal utilities should be proportional to prices for consumer equilibrium and hence for maximum utility.

The result can be extended to more goods.

11.14 LAW OF EQUIMARGINAL PRODUCTIVITY

We can find the conditions for maximum production given the production function $Q = f(x_1, x_2)$ subject to the budget restraint of cost

$$C = w_1x_1 + w_2x_2 + F$$

where x_1, x_2 are the factor inputs, w_1, w_2 their prices and F is the fixed cost.

We have $z = f(x_1, x_2) - \lambda(w_1x_1 + w_2x_2 + F - C)$

$$\frac{dz}{dx_1} = f_1 - \lambda w_1 = 0, \quad \frac{dz}{dx_2} = f_2 - \lambda w_2 = 0$$

We get $\dfrac{f_1}{w_1} = \dfrac{f_2}{w_2} = \lambda.$

i.e., marginal productivities should be proportional to the factor prices.

A corresponding problem is that of minimising cost $C = w_1x_1 + w_2x_2 + F$ subject to the production constraint $Q = f(x_1, x_2)$.

In this case, we get $\dfrac{w_1}{f_1} = \dfrac{w_2}{f_2} = \lambda_1$, say.

CHAPTER 12

Integral Calculus

12.1 THE INDEFINITE INTEGRAL

The inverse of the operation of differentiation is called integration. The integration process consists of finding a function whose derivative or differential is already known. The resulting function is an antiderivative or antidifferential, usually called an integral.

There are two fundamental concepts in integral calculus; indefinite integral and definite integral. We shall deal with indefinite integrals first.

Let $g(x)$ be a function of x such that

$$\frac{dg(x)}{dx} = f(x)$$

Then $g(x)$ is called an *integral of $f(x)$ with respect to x.* Symbolically

$$g(x) = \int f(x)\, dx$$

(*i*) $f(x)$ is the function to be integrated.

(*ii*) dx suggests the operation of integration is to be with respect to the variable x.

(*iii*) $\int$ is the sign of integration. It indicates the inverse of the differentiation process.

(*iv*) dx is the differential of x so that $f(x)\, dx$ may be considered as a differential of the primitive primary $g(x)$.

Thus $dg(x) = f(x)\, dx$ so that

$$g(x) = \int f(x)\, dx$$

We say, $g(x)$ is the integral of $f(x)$ with respect to x. If follows that

$$\frac{d}{dx}\left[\int f(x)dx\right] = f(x)$$

An integral of a given function is not unique.

If $\qquad y = x^2$ then $\dfrac{dy}{dx} = 2x$

$$y = x^2 + 3, \text{ then } \frac{dy}{dx} = 2x$$

$$y = x^2 - \pi, \text{ then } \frac{dy}{dx} = 2x$$

In general

If $y = x^2 + C$, then $\dfrac{dy}{dx} = 2x$ where C is any constant whatsoever.

If two differentiable functions $F(x)$, $G(x)$, have the same derivative then they differ only by a constant which means $G(x) = F(x) + C$. This is because the derivative of a constant is zero.

The functions $y = F(x)$ and $y = F(x) + C$ have the same derivative,

$$\frac{dy}{dx} = F'(x).$$

We write $\int f(x)\,dx = F(x) + C$ where C is an arbitrary constant.

This is the reason why $F(x) + C$ is called an indefinite integral. This value is not unique. Whenever we find an integral of a function we must add an arbitrary constant C to the result to get the most general value of the integral.

Ex. 1. $\dfrac{d}{dx}\left(\tfrac{1}{3}x^3\right) = \dfrac{3x^2}{3} = x^2 \qquad \therefore \int x^2 dx = \dfrac{x^3}{3} + C$

$$\frac{d}{dx}\left(\frac{x^4}{4}\right) = x^3 \qquad \therefore \int x^3 dx = \frac{x^4}{4} + C$$

$$\frac{d}{dx}\left(\frac{x^n}{n}\right) = x^{n-1} \qquad \therefore \int x^{n-1} dx = \frac{x^n}{n} + C$$

In general, $\int x^n dx = \dfrac{x^{n+1}}{n+1} + C$

This is the *power function rule* for all rational numbers $n \cdot (n \neq -1)$.

Ex. 2. $\int x\,dx = \dfrac{x^2}{2} + C$

$$\int dx = \int 1\,dx = \int x^0 dx = \frac{x^1}{1} + C = x + C$$

$$\int \sqrt{x^5}\,dx = \int x^{5/2} dx = \frac{x^{5/2+1}}{5/2+1} + C = \tfrac{2}{7}\,x^{7/2} + C$$

$$\int x^{-6} dx = \frac{x^{-6+1}}{-6+1} + C = \tfrac{1}{5}x^{-5} + C$$

$$\int \sqrt{x}\,dx = \int x^{1/2} dx = \frac{x^{1/2+1}}{\frac{1}{2}+1} + C = \tfrac{2}{3}\,x^{3/2} + C$$

$$\int \frac{1}{x^2} dx = \int x^{-2} dx = \frac{x^{-2+1}}{-2+1} + C = -\frac{1}{x} + C$$

$$\int \frac{1}{x} dx = \ln x + C \text{ (The power rule is } \textit{not} \text{ applicable in this case)}$$

Although $\frac{1}{x} = x^{-1}$ is a special form of the power function x^n with $n = -1$, the integral is different. Note the exception.

Standard result

If $y = \frac{1}{b(n+1)}(a+bx)^{n+1}$ where a, b, n are constants,

$$\frac{dy}{dx} = (a + bx)^n$$

$$\therefore \quad \int (a+bx)^n dx = \frac{1}{b(n+1)}(a+bx)^{n+1} + C$$

Ex. 3. $\int (1+3x)^2 dx = \frac{1}{3(2+1)}(1+3x)^3 + C_1$

(Here $a = 1$, $b = 3$ and C_1 is a constant)

$$= \tfrac{1}{9}(1 + 27x^2 + 9x + 27x^3) + C_1$$

$$= 3x^3 + 3x^2 + x + \tfrac{1}{9} + C_1$$

$$= 3x^3 + 3x^2 + x + C$$

where C is a new constant $C = \frac{1}{9} + C_1$

Ex. 4. $\int \frac{dx}{x+a} = \ln(x+a) + C.$

$$\int (3x^2 + 5/x) dx = x^3 + 5 \ln x + C$$

Ex. 5. Find the integrals of the following functions:

(*i*) $x^2 - 3x + 2$ (*ii*) $x^3 - \frac{1}{x^3}$

(*iii*) $x(x^2 - 1)^2$ (*iv*) $\frac{x-1}{3}$

(*v*) $1 - \frac{1}{u} + \frac{1}{u^2} - \frac{1}{u^3} + \frac{1}{u^4}$ (*vi*) $\sqrt[4]{t} - \sqrt[3]{t^2} + \sqrt[5]{t^3}$

(*vii*) $99x^{+99} + 100x^{-101}$

Ans. (*i*) $\frac{x^3}{3} - \frac{3x^2}{2} + 2x + C$ (*ii*) $x^4/4 - 4/x^4 + C$

(*iii*) $\frac{x^6}{6} - \frac{x^4}{2} + \frac{x^2}{2} + C$ (*iv*) $\frac{-1}{3}x + \frac{x^2}{6} + C$

(*v*) $u - \ln u - \frac{1}{u} + \frac{1}{2u^2} - \frac{1}{3u^3} + C$

(*vi*) $\frac{4}{5}t^{5/4} - \frac{3}{5}t^{5/3} + \frac{5}{8}t^{3/5} + C$

(*vii*) $0.99x^{100} - x^{-100} + C$

If these answers are differentiated, we will arrive at the original functions in each case.

Ex. 6. Determine the following integrals and check by differentiation:

(*i*) $\int (5 - 2x)\, dx$ (*ii*) $\frac{dx}{x\sqrt{x}}$

(*iii*) $\int \left(2\sqrt{x} - \frac{8}{\sqrt{x}}\right) dx$ (*iv*) $\int \left(\sqrt{x} - \frac{2}{x}\right)^2 dx$

(*v*) $\int (3 - 2x)(2x + 3)\, dx$ (*vi*) $\int (x - 1)(x^2 - 1/2)\, dx$

(*vii*) $\int \frac{x^{5/2} - x^{3/2} + 1}{x} dx$ (*viii*) $\int \frac{\sqrt[5]{4x}}{9} dx.$

Show that

Ans. (*i*) $5x - x^2 + C$ (*ii*) $-2x^{-12} + C$ (*iii*) $\frac{4}{3}x^{3/2} - 16x^{1/2} + C$

(*iv*) $\frac{x^2}{2} - \frac{4}{x} - 8x^{1/2} + C$ (*v*) $9x - \frac{4}{3}x^3 + C$ (*vi*) $\frac{x^4}{4} - \frac{x^3}{3} - \frac{x^2}{4} + \frac{x}{2} + C$

(*vii*) $\frac{2}{5}x^{5/2} - \frac{2}{3}x^{3/2} + \ln x + C$ (*viii*) $\frac{5 \times 4^{1/5}}{54}x^{6/5} + C$

Ex. 7. $\int 5x(1 - 2x^2)^{1/2} dx = -\frac{5}{6}(1 - 2x^2)^{3/2} + C.$

Ex. 8. $\int \frac{x + 2}{x - 1} dx = x + 3\ln(x - 1) + C.$

$\left(\textbf{Hint: } \frac{x + 2}{x - 1} = 1 + \frac{3}{x - 1}\right)$

Ex. 9. Check the following results

(*i*) $\int \frac{dx}{1 + x} = \ln (1 + x) + C$

(*ii*) $\int \frac{x}{\sqrt{1 + x}} dx = \frac{2}{3}(x + 1)^{3/2} - 2(x + 1)^{1/2} + C$

(*iii*) $\int \frac{dx}{x(3 + 4x)} = \frac{1}{3}\ln \frac{x}{3 + 4x} + C$

$$\left(\textbf{Hint}: \frac{1}{x(3+4x)} = \frac{4}{3}\left\{\frac{1}{4x} - \frac{1}{3+4x}\right\}\right)$$

Rules: The integral of the sum (or difference) of functions equals the sum (or difference) of their integrals.

$$\int [f_1(x) + f_2(x) + \ldots + f_n(x)]dx = \int f_1(x)dx + \int f_2(x)\,dx + \ldots + \int f_n(x)\,dx.$$

The integral of the product of a function by a constant equals that constant multiplied by the integral of the function

$$\int k\, f(x)dx = k \int f(x)\,dx.$$

12.2 SOME STANDARD FORMS

An indefinite integral has a symbolic meaning which suggests that the function whose derivative is $f(x)$ is of the form $F(x) + C$ where C is a constant. All the rules of integration, therefore, are derived from the corresponding results for differentiation. Often, the ability to guess the proper answer is an important requirement of integration.

Let u, v be differentiable functions of x. The following results may be used when necessary. Methods to prove these have been given in later sections.

1. $\int u^n du = \dfrac{u^{n+1}}{n+1} + C,\ n \neq -1$

2. $\int du = u + C$

3. $\int (du + dv) = \int du + \int dv$

4. $\int (du_1 + du_2 + \ldots + du_k) = \int du_1 + \int du_2 + \ldots + \int du_k$

   If $u = f(x)$, $v = g(x)$

5. $$\begin{aligned}\int \{f(x) + g(x)\}\, dx &= \int f(x)\, dx + \int g(x)\, dx \\ &= F(x) + C_1 + G(x) + C_2 \\ &= F(x) + G(x) + C \text{ where } C = C_1 + C_2\end{aligned}$$

6. $\int a\, du = a \int du,\ \int -\ du = -\int du$

   $$\begin{aligned}\int (x^4 - 2x^2 + 1)dx &= \int x^4 dx - \int 2x^2 dx + \int dx \\ &= \frac{x^5}{5} - \frac{2}{3}x^3 + x + C.\end{aligned}$$

   $$\begin{aligned}\int \frac{x^5 - 3x^3 + 2x}{x} dx &= \int x^4 dx - 3\int x^2 dx + 2\int dx \\ &= \frac{x^5}{5} - x^3 + 2x + C.\end{aligned}$$

7. $\int \frac{1}{x} dx = \ln x + C$ Check: $\dfrac{d}{dx}(\ln x) = \dfrac{1}{x}$

8. $\int e^x dx = e^x + C$ $\qquad \frac{d}{dx} e^x = e^x$

9. $\int e^{ax} dx = \frac{e^{ax}}{a} + C$ $\qquad \frac{d}{dx} e^{ax} = ae^{ax}$

10. $\int a^x dx = \frac{a^x}{\ln a} + C$ $\qquad \frac{d}{dx} a^x = a^x \ln a$

Similarly, check

11. $\int xe^x dx = (x-1)e^x + C$

12. $\int \frac{dx}{a^2 - x^2} = \frac{1}{2a} \ln \frac{a+x}{a-x} + C \text{ if } x^2 < a^2$

$\int \frac{dx}{x^2 - a^2} = \frac{1}{2a} \ln \frac{x+a}{x-a} + C \text{ if } x^2 > a^2$

13. $\int \frac{dx}{\sqrt{x^2 \pm a^2}} = \ln (x + \sqrt{x^2 \pm a^2}) + C$

14. $\int \sqrt{a^2 \pm x^2}\, dx = \frac{x}{2} \sqrt{a^2 \pm x^2} + \frac{a^2}{2} \ln (x + \sqrt{a^2 \pm x^2}) + C$

15. $\int \ln x dx = x \ln x - x + C$

16. $\int \frac{dx}{x \ln x} = \ln (\ln x) + C$

17. $\int \frac{dx}{x (\ln x)^2} = \frac{-1}{\ln x} + C$

18. $\int \frac{e^{-x} dx}{1 + e^{-x}} = \ln (e^x + 1) + C$, $\int xe^{-x} dx = -(x+1)e^{-x} + C$

19. If $\int f(x)dx = F(x) + C$ and $a (\neq 0)$ and b are constants

$\int f(x+b)\, dx = F(x+b) + C$

$\int f(ax) dx = \frac{1}{a} F(ax) + C$

$\int f(ax+b) = \frac{1}{a} F(ax+b) + C.$

20. $\int f(x)\, f'(x)\, dx = \frac{1}{2} \{f(x)\}^2 + C$

$\int \frac{f'(x)}{f(x)} dx = \ln f(x) + C$

$$\int \sqrt{f(x)}\, f'(x)\; dx = \frac{2}{3}[f(x)]^{3/2} + C$$

$$\int \frac{f'(x)}{\sqrt{f(x)}} dx = 2\sqrt{f(x)} + C$$

$$\int \{f(x)\}^n f'(x)\, dx = \frac{\{f(x)\}^{n+1}}{n+1} + C$$

$$\int f'(x) e^{f(x)}\, dx = e^{f(x)} + C$$

21. $\int x \ln x\, dx = \frac{x^2}{2} \ln x - \frac{x^2}{4} + C$

22. $\int x^n \ln ax\, dx = \frac{x^{n+1}}{n+1} \ln ax - \frac{x^{n+1}}{(n+1)^2} + C,\ n \neq -1$

23. $\int (\ln x)^n dx = x(\ln x)^n - n\int (\ln x)^{n-1} + C,\ n \neq -1$

24. $\int x^n \ln x\, dx = x^{n+1}\left[\frac{\ln x}{n+1} - \frac{1}{(n+1)^2}\right] + C,\ n \neq -1$

25. $\int \frac{(\ln x)^n}{x} dx = \frac{(\ln x)^{n+1}}{n+1} + C,\ n \neq -1$

26. $\int x e^{ax} dx = \frac{e^{ax}}{a^2}(ax - 1) + C$

27. $\int \frac{dx}{a + be^{gx}} = \frac{x}{a} - \frac{1}{ag}(a + be^{gx}) + C$

28. $\int x e^{-x^2}\, dx = -\frac{1}{2} e^{-x^2} + C$

Ex. 10. $\int \left(\sqrt{x} - \frac{x^2}{2} + \frac{2}{\sqrt{x}}\right) dx = \int \sqrt{x}\, dx - \int \frac{x^2}{2} dx + \int \frac{2}{\sqrt{x}} dx$

$$= \int x^{1/2} dx - \frac{1}{2}\int x^2\, dx + 2\int x^{-1/2}\, dx$$

$$= \frac{x^{3/2}}{3/2} - \frac{1}{2}\frac{x^3}{3} + 2\frac{x^{1/2}}{1/2} + C = \frac{2}{3}x^{3/2} - \frac{1}{6}x^3 + 4x^{1/2} + C$$

Ex. 11. $\int \left(x - \frac{1}{x}\right)^3 dx = \int \left(x^3 - 3x + \frac{3}{x} - \frac{1}{x^3}\right) dx$

$$= \int x^3 dx - \int 3x\, dx + \int \frac{3}{x} dx - \int \frac{1}{x^3} dx$$

$$= \int x^3 dx - 3\int x\, dx + 3 \int x^{-1} dx - \int x^{-3} dx$$

$$= \frac{x^4}{4} - 3\,\frac{x^2}{2} + 3\ln x - \frac{x^{-2}}{-2}$$

$$= \frac{1}{4}x^4 - \frac{3}{2}x^2 + 3\ln x + \frac{1}{2x^2} + C$$

Ex. 12. $\int \frac{a_1x^2 + a_2x^{-2} + a_3x^{-5} - a_4x^{-1/2}}{x^{-3}} dx$

$$= \int a_1 \frac{x^2}{x^{-3}} dx + a_2 \int \frac{x^{-2}}{x^{-3}} dx + a_3 \int \frac{x^{-5}}{x^{-3}} dx - a_4 \int \frac{x^{-1/2}}{x^{-3}} dx$$

$$= \int a_1 x^5 dx + a_2 \int x\, dx + a_3 \int x^{-2} dx - a_4 \int x^{5/2}\, dx$$

$$= a_1 \frac{x^6}{6} + a_2 \frac{x^2}{2} + a_3 \left(\frac{x^{-1}}{-1}\right) - a_4 \left(\frac{x^{7/2}}{7/2}\right)$$

$$= \frac{a_1}{6}x^6 + \frac{a_2}{2}x^2 - \frac{a_3}{x} - \frac{2a^4}{7}x^{7/2} + C$$

Ex. 13. $\int (1-3x)\,(1+x)\,dx = x - x^2 - x^3 + C$

Ex. 14. $\int \frac{x^5+1}{x} dx = \frac{x^5}{5} + \ln x + C$

Ex. 15. $\int (x^2-1)^2 dx = \frac{x^5}{5} - \frac{2x^3}{3} + x + C$

Ex. 16. $\int e^{2x}\, dx = \frac{1}{2} e^{2x} + C$

Ex. 17. $\int x\, e^x dx = xe^x - e^x + C$

Techniques of Integration

It may not always be very easy to integrate a given expression. Different types of complex situations require the use of special techniques.

12.3 INTEGRATION BY SUBSTITUTION

Often an integral $\int f(x)\,dx$ can be reduced to a simple integral $\int g(u)\,du$ by a convenient substitution of x by a new variable u such $x = g(u)$, so that $\frac{dx}{du} = g'(u)$ and $\int f(x)dx = \int f(g)g'(u)\,du$.

Ex. 18. To evaluate $\int x(x^2+1)^{3/2}dx$ let $u = x^2 + 1$. Then $du = 2xdx$ so that

$$\int x(x^2+1)^{3/2}dx = \int xu^{3/2}\frac{du}{2x} = \frac{1}{2}\int u^{3/2}du = \frac{1}{2}\cdot\frac{5}{2}u^{5/2} + C$$

$$= \frac{5}{4}(x^2+1)^{5/2} + C.$$

Ex. 19. $$\int \frac{xdx}{\sqrt{1+x^2}} = \int \frac{x}{\sqrt{1+x^2}}\frac{du}{2x} \text{ (Substituting } u = 1 + x^2,\ du = 2xdx)$$

$$= \frac{1}{2}\int \frac{du}{\sqrt{u}} = u^{1/2} + C = \sqrt{1+x^2} + C.$$

Ex. 20. $$\int 5a^{2x}dx = 5\int a^u \frac{du}{2} \text{ (Let } u = 2x,\ du = 2dx)$$

$$= \frac{5}{2}\frac{a^u}{\ln a} + C = \frac{5}{2}\frac{a^{2x}}{\ln 2a} + C$$

Ex. 21.

(*i*) $\int a^{kx}dx = \dfrac{a^{kx}}{k\ln a} + C$ (*ii*) $\int \dfrac{dx}{e^x} = -\dfrac{1}{e^x} + C$

(*iii*) $\int \dfrac{e^{\sqrt{x}}}{\sqrt{x}}dx = 2e^{\sqrt{x}} + C$ (*iv*) $\int a^x e^x dx = \dfrac{a^x e^x}{1+\ln a} + C$

(*v*) $\int \sqrt{e^x}dx = 2\sqrt{e^x} + C$ (*vi*) $\int \dfrac{dx}{x^2+4x+3} = \dfrac{1}{2}\ln\dfrac{x+1}{x+3} + C$

Ex. 22. $\int (x^3+4)^2\, 3x^2 dx$

Put $x^3 + 4 = u$

$du = 3x^2dx$

$\therefore$ $dx = \dfrac{du}{3x^2}$

The expression $= \int u^2 \cdot 3x^2 \cdot \dfrac{du}{3x^2}$

$$= \int u^2 du = \frac{u^2}{3} + C = \frac{(x^3+4)^3}{3} + C$$

Ex. 23. $\int \dfrac{x^5dx}{x^{12}+1}$

Put $x^6 = u$

$6x^5dx = du$

$\therefore$ $dx = \dfrac{du}{6x^5}$

The Expression $= \int \frac{x^5 du}{6x^5} \cdot \frac{1}{u^2+1}$

$= \frac{1}{6} \int \frac{du}{u^2+1} du$

This can be shown to be equal to $\tan^{-1}u + c$ or $\tan^{-1}x^6 + C$

Ex. 24. $\int (x+a)^n dx$

Put $x + a = u$

$\frac{dx}{du} = 1$

The Expression $= \int u^n du = \frac{u^{n+1}}{n+1} + C$

$= \frac{(x+a)^{n+1}}{n+1} + C$

12.4 INTEGRATION BY PARTS

Using the formula for the differential of a product

$$d(uv) = u\,dv + v\,du \text{ or } u\,dv = d(uv) - v\,du.$$

We may write the rule $\int u\,dv = uv - \int v\,du + C.$

This is known as integration by parts and expresses one integral $\int u\,dv$ in terms of another $\int v\,du$. Depending on individual cases u and v are suitably taken.

Ex. 25. $\int x e^x dx$. Let $u = x$, $dv = e^x dx$ so that $du = dx$, $v = e^x$

Using the rule $\int xe^x dx = xe^x - \int e^x dx = xe^x - e^x + C$

Ex. 26. Show that $\int x^2 e^x dx = x^2 e^x - 2xe^x + 2e^x + C$

(Let $u = x^2$, $dv = e^x dx$ so that $du = 2x$, $v = e^x$).

Ex. 27. $\int \ln x\,dx$. Let $u = \ln x$, $dv = dx$ so that $du = \frac{dx}{x}$, $v = x + C_1$

Using the rule $\int \ln x\,dx = (\ln x)(x + C_1) - \int (x + C_1)\frac{dx}{x} + C_2$

$= x \ln x + C_1 \ln x - \int dx - C_1 \int \frac{dx}{x} + C_2$

$= x \ln x + C_1 \ln x - x - C_1 \ln x + C_2$

$= x \ln x - x + C_2$.

The answer will be the same if instead of $v = x + C_1$, we write $v = x$ initially and proceed.

Ex. 28. Show that (*i*) $\int x \ln x\, dx = \frac{x^2}{2}\ln x - \frac{x^2}{4} + C$

(*ii*) $\int x^3 (\ln x)^2 dx = \frac{x^4}{32}[8\,(\ln x)^2 - 4\ln x + 1] + C$

(*iii*) $\int x^3 e^x dx = (x^3 - 3x^2 + 6x - 6)e^x + C.$

Ex. 29. $\int \sin^{-1} x dx = x\sin^{-1} x - \int x \frac{1}{\sqrt{1-x^2}} dx$ (Putt $u = \sin^{-1}x$, $dv = 1\, dx$)

LHS $= x\sin^{-1} x + \frac{1}{2}\int (-2x)\,(1-x^2)^{-1/2} dx$

$= x\sin^{-1} x + \frac{1}{2}\frac{(1-x^2)^{1/2}}{1/2} = x\sin^{-1} x + (1-x)^{1/2} + C$

Ex. 30. Similarly, $\int \sin^{-1}\frac{2x}{1+x^2} dx = x\sin^{-1}\frac{2x}{1+x^2} - \ln\,(1+x^2) + C$

Ex. 31(*i*). $I = \int e^{ax} \sin bx\, dx = \frac{e^{ax}}{a}\sin bx - \int \frac{e^{ax}}{a} b\cos bx\, dx$

$$= \frac{e^{ax}\sin bx}{a} - \frac{b}{a}\left\{\frac{e^{ax}}{a}\cos bx - \int \frac{e^{ax}}{a}\,(-b\sin bx)\,dx\right\}$$

$\therefore \quad I = \frac{e^{ax}\sin bx}{a} - \frac{b}{a^2}e^{ax}\cos bx - \frac{b^2}{a^2}$

$\therefore \quad (a^2 + b^2)I = e^{ax}\,(a \sin bx - b \cos bx)$

$\therefore \quad I = \frac{1}{a^2+b^2}(a\sin bx - b\cos bx) = \frac{e^{ax}}{\sqrt{a^2+b^2}}\sin\,(bx - \tan^{-1} b/a) + C$

(assuming $a = r\cos\theta$, $b = r\sin\theta$, $r = \sqrt{a^2+b^2}$)

(*ii*) $I = \int \sqrt{x^2-a^2}\, dx = x\sqrt{x^2-a^2} - \int x \cdot \frac{2x}{2\sqrt{x^2-a^2}} dx$

$= x\sqrt{x^2-a^2} - \int \frac{x^2-a^2+a^2}{\sqrt{x^2-a^2}}$

$= x\sqrt{x^2-a^2} - \left(\int \sqrt{x^2-a^2}\, dx + \int \frac{a^2}{\sqrt{x^2-a^2}} dx\right)$

$= x\sqrt{x^2-a^2} - I - a^2 \ln\,(x + \sqrt{x^2-a^2}) + C_1$

$$\therefore \quad 2I = x\sqrt{x^2 - a^2} - a^2 \ln (x + \sqrt{x^2 - a^2}) + C_1$$

$$\therefore \quad I = \frac{x}{2}\sqrt{x^2 - a^2} - a^2 \ln (x + \sqrt{x^2 - a^2}) + C$$

(*iii*) Replace $- a^2$ by $+ a^2$ in the above

$$\int \sqrt{x^2 + a^2}\,dx = \frac{x}{2}\sqrt{x^2 + a^2} + \frac{a^2}{2}\ln (x + \sqrt{x^2 + a^2}) + C$$

Ex. 32. $I = \int e^{ax} \cos bx\ dx$. Let $dv = \cos bx,\ u = e^{ax}$

$$I = e^{ax}\frac{\sin bx}{a} - \int \frac{\sin bx}{b} \cdot ae^{ax} dx$$

$$= \frac{1}{b} e^{ax} \sin bx - \frac{a}{b}\int e^{ax} \sin bx\ dx$$

$$= \frac{e^{ax}}{b}\sin bx - \frac{a}{b}\left[\frac{e^{ax}(-\cos bx)}{b} - \int \frac{-\cos bx}{b} ae^{ax} dx\right]$$

$$= \frac{e^{ax}}{b}\sin bx - \frac{a}{b}\left[\frac{-e^{ax}}{b}\cos bx + \frac{a}{b}\int e^{ax} \cos bx\ dx\right]$$

$$= \frac{e^{ax}}{b}\sin bx + \frac{a}{b^2}e^{ax}\cos bx - \frac{a^2}{b^2} I$$

$$\left(1 + \frac{a^2}{b^2}\right) I = \frac{e^{ax}}{b}\sin bx + \frac{a}{b^2} e^{ax} \cos bx + C_1$$

$$\therefore \quad I = \frac{e^{ax}}{a^2 + b^2}(a \cos bx + b \sin bx) + C$$

Ex. 33. $I = \int \frac{\ln \ln x}{x} dx$. Let $t = \ln x,\ dt = \frac{dx}{x}$

$$I = \int \frac{\ln t}{x} x\ dt = \int \ln t\, dt = t \ln t - t + C$$

$$= \ln x\ [\ln \ln x] - \ln x + C$$

Ex. 34. $I = \int \sin\sqrt{x}\, dx.$ Let $\sqrt{x} = t,\ \frac{dx}{2\sqrt{x}} = dt$

$$I = \int (\sin t) 2t\ dt$$

Let $dv = \sin t,\ u = t$

$$I = 2[t(-\cos t) - \int (-\cos t) \cdot 1\, dt]$$

$$= -2t \cos t + 2\int \cos t\, dt = -2t\cos t + 2\int \cos t\, dt + C$$

$$= -2\sqrt{x}\cos\sqrt{x} + 2\sin\sqrt{x} + C$$

Ex. 35. Prove that if u, v are differentiable functions of x,

(*i*) $\int \left(u\frac{dv}{dx} \right) dx = uv - \int v\frac{du}{dx} dx$

(*ii*) $\int uv\,dx = u\int v\,dx - \int \left\{ (v\,dx)\frac{du}{dx} \right\} dx$

(*i*) Since $\frac{duv}{dx} = u\frac{dv}{dx} + v\frac{du}{dx}$

Taking integrals on both sides

$$\int uv = \int \left(u\frac{dv}{dx} \right) dx + \int \left(v\frac{dv}{dx} \right) dx$$

or $$\int \left(u\frac{dv}{dx} \right) dx = uv - \int \left(v\frac{du}{dx} \right) dx$$

(*ii*) $\int uw = \int \left(u\frac{dw}{dx} \right) dx + \int \left(w\frac{du}{dx} \right) dx$ as done above

$$\therefore \quad \int uw - \int \left(w\frac{du}{dx} \right) dx = \int \left(u\frac{dw}{dx} \right) dx$$

Let $v = \frac{dw}{dx}$ so that $w = \int v\,dx$. On substituting

$$u\int v\,dx - \int \left\{ \left(\int v\,dx \right) \frac{du}{dx} \right\} dx = \int uv\,dx$$

The above two results can be put in an alternative form

(*i*) $\int f(x)g'(x)dx = f(x)\,g(x) - \int [f'(x)\,g(x)dx$

(*ii*) $\int f(x)\,g(x)dx = f(x)\int g(x)dx - \int [f'(x)\int g(x)dx]dx$

Ex. 36. Find $\int \sqrt{a^2 - x^2}\,dx$

(*i*) To integrate by parts let $dv = 1$, $u = \sqrt{a^2 - x^2}$

$$I = \int \sqrt{a^2 - x^2} \cdot 1\,dx = x\sqrt{a^2 - x^2} - \int x \cdot \frac{-2x}{2\sqrt{a^2 - x^2}} dx$$

$$= x\sqrt{a^2 - x^2} + \int \frac{x^2}{\sqrt{a^2 - x^2}} dx.$$

Now $-\int \frac{x^2}{\sqrt{a^2 - x^2}} dx = \int \frac{(a^2 - x^2) - a^2}{\sqrt{a^2 - x^2}} dx = \int (\sqrt{a^2 - x^2})\,dx - a^2 \int \frac{dx}{\sqrt{a^2 - x^2}}$

$$= I - a^2 \sin^{-1}\frac{x}{a} + C_1$$

$$\therefore \quad I = x\sqrt{a^2 - x^2} - \left(I - a^2 \sin^{-1}\frac{x}{a}\right) + C_1$$

$$\therefore \quad 2I = x\sqrt{a^2 - x^2} + a^2 \sin^{-1}\frac{x}{a} + C_1$$

$$\therefore \quad I = \frac{x}{2}\sqrt{a^2 - x^2} + \frac{a^2}{2}\sin^{-1}\frac{x}{a} + C \text{ where } C = 2C_1$$

Ex. 37. $\int 4^{1-x} dx$ (Put $y = 1 - x$, $dy = -dx$)

$$= -\int 4^y dy = -\frac{4^y}{\ln 4} + C = \frac{-4^{1-x}}{\ln 4} + C$$

Ex. 38. $\int 6x\, e^{x^2-1} dx$ (Put $y = x^2 - 1$, $dy = 2xdx$)

$$= \int 6x\, e^y \frac{dy}{2x} = 3\int e^y dy = 3e^y + C = 3e^{x^2-1} + C.$$

Ex. 39. To find $\int \frac{x^{1/2}}{x^{3/4}+1} dx$ it is convenient to substitute $y = x^{1/4}$ so that $x = y^4$, $dx = 4y^3 dy$

$$\int \frac{x^{1/2}}{x^{3/4}+1} dx = \int \frac{y^2 \cdot 4y^3 dy}{y^3+1} = \int \frac{4y^5}{y^3+1} dy$$

Now substitute again $y^3 + 1 = z$ so that $dz = 3y^2 dy$

$$4\int \frac{y^5}{y^3+1} dy = 4\int \frac{y^5}{z}\frac{dz}{3y^2} = \frac{4}{3}\int \frac{z-1}{z} dz = \frac{4}{3}\left(\int dz - \int \frac{dz}{z}\right)$$

$$= \frac{4}{3}(z - \ln z) + C_1 = \frac{4}{3}[(y^3+1) - \ln(y^3+1)] + C_1$$

$$= \frac{4}{3}y^3 - \frac{4}{3}\ln(y^3+1) + C$$

$$= \frac{4}{3}[x^{3/4} - \ln(x^{3/4}+1)] + C$$

Alternatively, write $x^{3/4} + 1 = z$ so that the integral becomes

$$\int \frac{x^{1/2}}{z}\frac{4}{3}x^{1/4} dz = \frac{4}{3}\int \frac{z-1}{z} dz$$

Ex. 40. To find $\int \frac{dx}{x\sqrt{x^2+x+1}}$ put $x = \frac{1}{y}$ and show that the value of the integral is

$$\ln\left\{\frac{x}{2 + x + \sqrt{1 + x + x^2}}\right\} + C$$

Ex. 41. Prove that (*i*) $\int \frac{dx}{x^2\sqrt{x^2+a^2}} = -\frac{\sqrt{x^2+a^2}}{a^2 x} + C$

(*ii*) $\int x(a+x)^{1/3} dx = \frac{3}{28}[(a+x)^{4/3}(4x-3a)] + C.$

12.5 INTEGRATION BY PARTIAL FRACTIONS

This method is applied to proper fractions in which the polynomial in the denominator is of higher degree than that in the numerator. The procedure consists in factorising the denominator and expressing the fraction as a sum of partial fractions as shown below.

Ex. 42. $\frac{x+5}{x^2+5x+6} = \frac{x+5}{(x+2)(x+3)} = \frac{A}{x+2} + \frac{B}{x+3}$

$\therefore \quad x + 5 = Ax + 3A + Bx + 2B = x(A + B) + 3A + 2B$

$\therefore \quad A + B = 1,\ 3A + 2B = 5$ by equating coefficients

$\therefore \quad A = 3,\ B = -2$

$$\int \frac{(x+5)\,dx}{x^2+5x+6} = \int \frac{3\,dx}{x+2} - \int \frac{2\,dx}{x+3}$$

$$= 3\ln(x+2) - 2\ln(x+3) + C = \ln\frac{(x+2)^3}{(x+3)^2} + C$$

While forming partial fractions the following rules are found useful if there are repeated factors. Use the rules and simplify to get the fractions.

Rules

If P stands for the numerator, we have

$$\frac{P}{(ax+b)^2} = \frac{A_1}{ax+b} + \frac{A_2}{(ax+b)^2}$$

$$\frac{P}{(ax+b)^3} = \frac{A_1}{ax+b} + \frac{A_2}{(ax+b)^2} + \frac{A_3}{(ax+b)^3}$$

$$\frac{P}{(ax^2+bx+c)^2} = \frac{A_1x+B_1}{ax^2+bx+c} + \frac{A_2x+B_2}{(ax^2+bx+c)^2}$$

Ex. 43. Show that (*i*) $\int \frac{x\,dx}{x^4+6x^2+5} = \frac{1}{8}\ln\left(\frac{x^2+1}{x^2+5}\right) + C$ (Factorise denominator)

(*ii*) $\int \frac{(4x^2+6)\,dx}{x^3+3x} = \ln x^2(x^2+3) + C$

Ex. 44. (*i*) $\int \frac{(1+x)^3}{(1-x)^3}\,dx$

$$\frac{(1+x)^3}{(1-x)^3} = \frac{1+3x^2+3x+x^3}{1+3x^2-3x-x^3} = \frac{6x^2+2}{1+3x^2-3x-x^3} - 1$$

$$= \frac{6x^2+2}{(1-x)^3} - 1$$

Let $\frac{6x^2+2}{(1-x)^3} = \frac{A}{1-x} + \frac{B}{(1-x)^2} + \frac{C}{(1-x)^3}$

$$= \frac{A(1-x)^2 + B(1-x) + C}{(1-x)^3}$$

$\therefore$ $6x^2 + 2 = A(1-x)^2 + B(1-x) + C$

Put $x = 1$, $C = 8$; put $x = 0$, $A + B + C = 2$, $x = -1$, $2A + B = 0$, $A = 6$, $B = -12$

$\therefore$ $\int \frac{(1+x)^3}{(1-x)^3}\,dx = \int\left(\frac{6}{1-x} - \frac{12}{(1-x)^2} + \frac{8}{(1-x)^3} - 1\right)dx$

$$= -6\ \ln(1-x) - \frac{12}{1-x} + \frac{4}{(1-x)^2} - x + C_1$$

(ii) $\int \frac{x}{1+x^3}dx$

$$\frac{x}{1+x^3} = \frac{x}{(1+x)(1-x+x^2)} = \frac{A}{1+x} + \frac{Bx+C}{1-x+x^2}$$

$\therefore$ $x = A(1 - x + x^2) + (Bx + C)(1 + x)$

Put $x = -1$, $A = -\frac{1}{3}$; Put $x = 0$, $C = \frac{1}{3} = A$; Put $x = 1$, $B = \frac{1}{3}$

$$\int \frac{x}{1+x^3}\,dx = \int\left[\frac{-1/3}{1+x} + \frac{x/3+1/3}{1-x+x^2}\right]dx$$

$$= \int\left[\frac{-1}{3(1+x)} + \frac{1}{6}\frac{2x-1}{1-x+x^2} + \frac{1}{2}\frac{1}{(x-1/2)^2+(\sqrt{3}/2)^2}\cdot\right]dx$$

$$= -\frac{1}{3}\ln(1+x) + \frac{1}{6}\ln(1-x+x^2) + \frac{1}{2}\frac{2}{\sqrt{3}}\tan^{-1}\frac{2x-1}{\sqrt{3}} + C_1$$

(iii) $\int \frac{x^2+1}{x^3+1}dx = \frac{2}{3}\ln(x+1) + \frac{1}{6}\ln(x^2-x+1) + \frac{1}{\sqrt{3}}\tan^{-1}\frac{2x-1}{\sqrt{3}}$

Note: If $f(x)$, $g(x)$ are polynomials, then $f(x)/g(x)$ is a rational fraction.

If the degree of $f(x)$ is greater than that of $g(x)$ then $f(x)/g(x)$ can be expressed as the sum of a polynomial and a proper rational fraction (in which the degree of $f(x) <$ the degree of $g(x)$. A proper rational fraction can be expressed as a sum of simpler fractions.

Ex. 45. $\int \frac{x^2+6x-25}{(x^2+4x)^2-25}dx = \int \frac{(x^2+6x-25)}{(x-1)(x+5)(x^2+4x+5)}dx$

Show this equals $\int \left[\frac{-3}{10(x-1)} + \frac{1}{2(x+5)} + \frac{-1/5x+5}{x^2+4x+5} \right] dx$

$$= -3\ln(x-1) + \frac{1}{2}\ln(x+8) - \frac{1}{10}\ln(x^2+4x+5) + \frac{17}{5}\tan^{-1}(x+2) + C$$

Ex. 46. $\int \frac{dx}{(x+1)^2(x^2+1)} = \int \left[\frac{A}{x+1} + \frac{B}{(x+1)^2} + \frac{Cx+D}{x^2+1} \right) dx$

Show $A = \frac{1}{2} = B,\ C = -\frac{1}{2}\quad D = 0$

Ans. $\frac{1}{2}\ln(x+1) - \frac{1}{4}\ln(x^2+1) - \frac{1}{2}\frac{1}{x+2} + C_1$

Ex. 47. $\int \frac{x}{1+x+x^2+x^3}dx = \int \frac{x}{(1+x)(1+x^2)}dx$

$$= \int \left[\frac{A}{x+1} + \frac{Bx+C}{x^2+1} \right] dx$$

where $A = -\frac{1}{2},\ B = C = \frac{1}{2}$ **Ans.** $\frac{1}{2}\ln\frac{(x^2+1)^{1/2}}{x+1} + \frac{1}{2}\tan^{-1}x + C_1$

Ex. 48. $\int \frac{x^4 dx}{(x^2-1)(x^2+4)} = x + \frac{1}{10}\ln\frac{x-1}{x+1} - \frac{8}{5}\tan^{-1}\frac{x}{2} + C.$

Ex. 49. $\int \frac{(x+a)dx}{x^2(x-a)(x^2+a^2)} = \frac{1}{a^3}\left(\frac{1}{x} + \ln\frac{(x-a)(x^2+a^2)^{1/2}}{x^2} \right) + C$

Ex. 50. $\int \frac{2x^2+1}{(x^2+1)^2}dx = \int \left[\frac{Ax+B}{x^2+1} + \frac{Cx+D}{(x^2+1)^2} \right] dx$

Show $C = A = 0,\ B = 2,\ D = 1$ **Ans.** $\frac{x}{2(x^2+1)} + \frac{5}{2}\tan^{-1}x + C_1$

Ex. 51. Evaluate the following integrals

(*i*) $\int (x^2 - \sqrt{x})dx$ (*ii*) $\int \frac{x\,dx}{x^2+1}$ (*iii*) $\int xe^{-x^2}dx$

(*iv*) $\int \frac{dx}{4+x}$ (*v*) $\int \frac{dx}{\sqrt{x^2+4}}$ (*vi*) $\int \frac{2x+3}{\sqrt{x^2+4}}dx$

(*vii*) $\int \frac{dx}{\sqrt{9+(2x+3)^2}}$ (*viii*) $\int \frac{x^3}{x+2}dx$

Ans. (*i*) (1/3) $(x^3 - 2x^{3/2}) + C$ (*ii*) (1/2) $\ln(x^2+1) + C$ (*iii*) $(-1/2)\ e^{-x^2} + C$

(*iv*) $\ln|4+x| + C$ (*v*) $\ln(x + \sqrt{4+x^2}) + C$ (*vi*) $\sqrt{x^2+4} + 3\ln(x + \sqrt{x^2+4}) + C$

(*vii*) $1/2\,[\ln 2x + 3 + \sqrt{(2x+3)^2+9}] + C$ (*viii*) $\frac{x^2}{2} - 2x + 4\ln(x+2) + C$

Ex. 52. Show that

(*i*) $\int (a+bx)^2 dx = \frac{(a+bx)^3}{3b} + C$

(*ii*) $\int x(a-bx^2)\,dx = \frac{2ax^2 - bx^4}{4} + C$

(*iii*) $\int x\sqrt{2x^2+3} = \frac{1}{6}(2x^2+3)^{3/2} + C$

(*iv*) $\int \frac{4x^2 dx}{\sqrt{x^3+8}} = \frac{8}{3}\sqrt{x^3+8} + C$

(*v*) $\int \frac{(\sqrt{k}-\sqrt{x})^2}{\sqrt{x}} dx = -\frac{2}{3}(\sqrt{k}-\sqrt{x})^3 + C$

(*vi*) $\int \frac{e^x}{a+be^x} dx = \frac{\ln(a+be^x)}{b} + C$

(*vii*) $\int \frac{e^{2x}}{e^{2x}+1} dx = \frac{1}{2}\ln(e^{2x}+1) + C$

(*viii*) $\int \frac{ae^x+b}{ae^x-b} dx = 2\ln(ae^x-b) - x + C$

(*ix*) $\int \sqrt{e^x}\,dx = 2\sqrt{e^x} + C$

(*x*) $\int (e^{5x}+a^{5x})\,dx = \frac{e^{5x}}{5} + \frac{a^{5x}}{5\ln a} + C$

Ex. 53. $\int e^{\tan x}\sec^2 x\,dx$

Put

$$\tan x = u$$
$$\sec^2 dx = du$$

$$\int e^u \sec^2 x \cdot \frac{du}{\sec^2 x} = \int e^u du$$
$$= e^u + C = e^{\tan x} + C$$

Ex. 54. $\int \frac{e^{m\tan^{-1}x}}{1+x^2} dx$

Put

$$m\tan^{-1}x = u$$

$$m\frac{1}{1+x^2}dx = du$$

$$dx = \frac{1+x^2}{m}du$$

$$\int \frac{e^u}{1+x^2}\cdot\frac{1+x^2}{m}du = \int \frac{e^u}{m}du$$

$$= \frac{e^u}{m} + C = \frac{1}{m}e^{m\tan^{-1}x} + C$$

Ex. 55. $\int [x\ln x \ln\ln x]^{-1}dx$

Put $\quad \ln\ln x = u$

$$\frac{1}{x\ln x}dx = du$$

$$= \int \frac{du}{u} = \ln u + C$$

$$= \ln\ln\ln(x) + C.$$

Ex. 56. $\int \sqrt{a^2 - x^2}\,dx = \int (\sqrt{a^2 - a^2\sin^2 u})\, a\cos u\, du$ (Putting $x = a\sin u$)

$$= a^2\int \cos^2 u\, du = a^2\int \frac{1+\cos 2u}{2}du = \frac{a^2}{2}\int du + \frac{a^2}{2}\int \cos^2 u\, du$$

$$= \frac{a^2}{2}u + \frac{a^2}{2}\frac{\sin 2u}{2} = \frac{a^2 u}{2} + \frac{a^2}{2}\sin u\cos u$$

$$= \frac{a^2}{2}\sin^{-1}\frac{x}{a} + \frac{a^2}{2}\frac{x}{a}\sqrt{1 - x^2/a^2}$$

$$\therefore \quad \int \sqrt{a^2 - x^2}\,dx = \frac{x\sqrt{a^2 - x^2}}{2} + \frac{a^2}{2}\sin^{-1}\frac{x}{a}$$

Ex. 57. $\int \sqrt{x^2 + a^2}\,dx = \int \sqrt{(a^2\sinh^2 u + a^2)}\, a\cosh u\, du$ (Putting $x = a\sinh u$)

$$= a^2\int \cosh^2 u\,du = \frac{1}{2}a^2\int (\cosh 2u + 1)\,du$$

(**Note.** This exercise uses hyperbolic functions and may be omitted. See Ex. 31(*iii*) above)

$$= \frac{a^2}{2}\int \cosh 2u\, du + \frac{a^2}{2}\int du$$

$$= \frac{a^2}{2}\sinh\frac{2u}{2} + \frac{a^2 u}{2} = \frac{a^2}{2}\sinh u\cosh u + \frac{a^2}{2}u$$

$$= \frac{a^2}{2}\cdot\frac{x}{2}\sqrt{1 + \frac{x^2}{a^2}} + \frac{a^2}{2}\sinh^{-1}\frac{x}{a}$$

$$= x\frac{\sqrt{x^2 + a^2}}{2} + \frac{a^2}{2}\ln\frac{x + \sqrt{x^2 + a^2}}{a}$$

Ex. 58. $\int \sqrt{x^2 - a^2}\,dx = \frac{x\sqrt{x^2 - a^2}}{2} - \frac{a^2}{2}\ln\frac{x + \sqrt{x^2 - a^2}}{a}$

Ex. 59. $\int \frac{dx}{a^2 - x^2} = \frac{1}{2a}\ln\frac{a + x}{a - x}$

Ex. 60. $\int \frac{dx}{x^{1/2} - x^{1/4}} = \int \frac{4u^3 dz}{u^2 - u}$ by putting $u = x^{1/4}$, $dx = 4u^3 du$

$$= 4\int \frac{u^2}{u - 1} du = 4\int \left(u + 1 + \frac{1}{u - 1} \right) du$$

$$= 4\left(\frac{u^2}{2} + u + \ln(u - 1) \right) = 2x^{1/2} + 4x^{1/4} + \ln(x^{1/4} - 1)^4 + C$$

Ex. 61. $I = \int \frac{\sin \ln x}{x^3} dx = \int \frac{\sin u\, du}{e^{2u}}$ where $u = \ln x$, $dx = x\, du$

$= \int \sin u\, e^{-2u} du$ (now integrate by parts)

$= e^{-2u}(-\cos u) - \int -2e^{-2u}(-\cos u)\, du$

$= -e^{-2u}\cos u - 2[e^{-2u}\sin u + 2\int e^{-2u}\sin u\, du]$

Show that $I = -\frac{1}{5x^4}\cos \ln x + 2\sin \ln x$

Ex. 62. $\int \sec(\tan x)\sec^2 x\, dx = \int \sec y\, dy$

$$= \ln \tan\left(\frac{\pi}{4} + \frac{y}{2} \right) = \ln\left[\tan\left(\frac{\pi}{4} + \frac{\tan x}{2} \right) \right] + C$$

(putting $y = \tan x$, $dy = \sec^2 x\, dx$)

Ex. 63. $\int x^{n-1} \operatorname{cosec} x^n\, dx = \frac{1}{n}\int \operatorname{cosec} y\, dy$

$$= \frac{1}{n}\ln \tan x^n / 2 + C$$

[putting $x^n = y$, $dy = nx^{n-1}\, dx$]

Ex. 64. $\int \frac{2\,dx}{\{2 - (2x + 3)^2\}^{1/2}} = \int \frac{\sqrt{2}\, dt}{\sqrt{2}(1 - t^2)^{1/2}}$

$$= \sin^{-1}(2x + 3) + C$$

(putting $2x + 3 = \sqrt{2}\,t$, $2xdx = \sqrt{2}\, dt$)

Ex. 65. $\int e^x \tan e^x \sec e^x = \int dt$

$$= t = \sec e^x + C$$

(putting $\sec e^x = t$, $\therefore$ $e^x \sec e^x \tan e^x\, dx = dt$

Ex. 66. $$I = \int \frac{dx}{\sqrt{x^2 + x + 1}}$$

$$I = \int \frac{dx}{\sqrt{\left(x + \frac{1}{2}\right)^2 + \left(\frac{\sqrt{3}}{2}\right)^2}}$$

$$= \int \frac{du}{\sqrt{u^2 + (\sqrt{3}/2)^2}} \qquad \text{where } u = x + \frac{1}{2}$$

$$= \sinh^{-1} \frac{u}{\sqrt{3/2}} = \sinh^{-1} \frac{2x+1}{\sqrt{3}}$$

12.6 IRRATIONAL FUNCTIONS

Types $\int \left(\frac{a_1 x + a_2}{b_1 x + b_2}\right)^{1/n} dx$ and $\int \frac{dx}{(a + bx)^n}$

These may be found by the substitution $u^n = \frac{a_1 x + a_2}{b_1 x + b_2}$ and $u = (a + bx)$.

Ex. 67. $\int \left(\frac{x+2}{2x+3}\right)^{1/2} \frac{dx}{x}$

$$\left[\textbf{Hint :} \text{ Let } u^2 = \frac{x+2}{2x+3} = \frac{1}{2}\left(1 + \frac{1}{2x+3}\right)\right.$$

$$\left.\therefore \quad 2udu = \frac{-dx}{(2x+3)^2}\right] \qquad \textbf{Ans. } \frac{1}{\sqrt{2}} \ln \frac{1+\sqrt{2}u}{1-\sqrt{2}u} + \frac{\sqrt{2}}{3} \ln \frac{\sqrt{3}y - \sqrt{2}}{\sqrt{3}y + \sqrt{x}} + C$$

Ex. 68. $\int (\sqrt{x} - 1)^{-1} x^{1/4} dx = \frac{4}{3} x^{3/4} + 4x^{1/4} + 2 \ln \frac{x^{1/4} - 1}{x^{1/4} + 1} + C$

Ex. 69. $\int \frac{\sqrt{1 + 1/x^2}}{\sqrt{1 - x^2}} dx = \frac{1}{2} \ln \frac{\sqrt{1+x^2} - \sqrt{1-x^2}}{\sqrt{1+x^2} + \sqrt{1-x^2}} + \tan^{-1} \frac{\sqrt{1+x^2}}{\sqrt{1-x^2}}$

Ex. 70. $\int \sqrt{ax^2 + bx + c}\, dx \quad (a < 0,\ b^2 - 4ac > 0)$

$$= \int (-\sqrt{a}) \sqrt{\frac{b^2 - 4ac}{4a^2} - \left(x + \frac{b}{2a}\right)^2}\, dx \quad \text{(Use Ex. 36)}$$

$$= \sqrt{-a} \left[\frac{x + b/2a}{2} \left(\frac{b^2 - 4ac}{4a^2} - \left(x + \frac{b}{2a}\right) \right)^2 \right.$$

$$\left. + \frac{b^2 - 4ac}{8a^2} \sin^{-1} \frac{x + b/2a}{\sqrt{(b^2 - 4ac)/4a^2}} + C \right]$$

Similarly $\int (ax^2 + bx + c)^{-1/2} dx = \frac{1}{\sqrt{-a}} \sin^{-1} \frac{x + b/2a}{(b^2 - 4ac)/4a^2} + C$

If, in these cases, the numerator is a linear function of x, it can be broken into two parts.

Ex. 71. $\int \frac{dx}{\sqrt{-x^2 + 5x - 6}} = \sin^{-1}(2x - 5) + C$

Ex. 72. $\int \frac{dx}{x + \sqrt{3x^2 + 2x - 1}} = -\sin^{-1} \frac{1 - x}{2x} + C$

Ex. 73. $\int \frac{dx}{x + \sqrt{x^2 + x + 2}}$

Put $u = \sqrt{x^2 + x + 2} + x$ or $x = \frac{u^2 - 2}{1 + 2u}$

$$dx = \frac{2(u^2 + u + 2)}{(1 + 2u)^2} \text{ and } \sqrt{x^2 + x + 2} = \frac{u^2 + u + 2}{1 + 2u}$$

$$\therefore \int \frac{dx}{x + \sqrt{x^2 + x + 2}} = 2\int \frac{du}{u^2 - 2} = \frac{1}{\sqrt{2}} \ln \frac{u - \sqrt{2}}{u + \sqrt{2}} + C \text{ where } u = \sqrt{x^2 + x + 2}$$

Ex. 74. $\int x^{1/2}(1 + x)^{-1} dx = 2x^{-1/2} - 2\tan^{-1}\sqrt{x} + C$

Ex. 75. $\int \sqrt{1 + \sqrt{x}}\, dx = \frac{4}{5}(1 + \sqrt{x})^{5/2} = \frac{4}{3}(1 + \sqrt{x})^{3/2} + C$

12.7 TRIGONOMETRIC FUNCTIONS

$$I = \int \frac{dx}{a + b\cos x} \text{ or } \int \frac{dx}{a + b\sin x}$$

For integrals of this type substitute $\sin x = \frac{2t}{1 + t^2}$, $\cos x = \frac{1 - t^2}{1 + t^2}$, $\tan x = \frac{2t}{1 - t^2}$ where $t = \tan x/2$. This makes the integral a rational function.

(*i*) $\int \frac{dx}{a + b\cos x} = \int \frac{dt \cdot 2/(1 + t^2)}{a + b\frac{1 - t^2}{1 + t^2}}$

$$\left[\text{Substitute } t = \tan\frac{x}{2}, \quad dt = \frac{1}{2}\sec^2\frac{x}{2}dx = \frac{1}{2}(1 + t^2)dx, \quad dx = \frac{2dt}{1 + t^2}\right]$$

$$\therefore \qquad \int \frac{dx}{a+b\cos x} = \frac{2}{a-b}\int \frac{dt}{t^2+\dfrac{a+b}{a-b}}$$

$$= \frac{2}{a-b}\int \frac{dt}{t^2+\left(\sqrt{\dfrac{a+b}{a-b}}\right)^2} \text{ if } a^2 > b^2$$

$$= \frac{2}{\sqrt{a^2-b^2}}\tan^{-1}\sqrt{\frac{a-b}{a+b}}\tan\frac{x}{2}+C$$

(*ii*) If $a^2 < b^2$ then

$$\int \frac{dx}{a+b\cos x} = \frac{2}{a-b}\int \frac{dt}{t^2-\dfrac{b+a}{b-a}} = \frac{2}{a-b}\int \frac{dt}{t^2-\left(\sqrt{\dfrac{b+a}{b-a}}\right)^2}$$

$$= \frac{2}{a-b}\cdot\frac{1}{2\sqrt{\dfrac{b+a}{b-a}}}\int\left[\frac{1}{t-\sqrt{\dfrac{b+a}{b-a}}}-\frac{1}{t+\sqrt{\dfrac{b-a}{b+a}}}\right]dx$$

$$= \frac{1}{a-b}\sqrt{\frac{b-a}{b+a}}\ln\frac{t-\sqrt{(a+b)/(b-a)}}{t+\sqrt{(b+a)/(b-a)}}+C$$

(*iii*) If $a^2 = b^2$, $a = b$ or $a = -b$. If $a = b$ then

$$\int \frac{dx}{a+b\cos x} = \frac{1}{2a}\int \sec^2\frac{x}{2}dx = \frac{1}{a}\tan\frac{x}{2}+C \qquad \left(\because \frac{1}{a(1+\cos x)} = \frac{1}{2a\cos^2 x/2}\right)$$

If $a = -b$, $\displaystyle\int \frac{dx}{a+b\cos x} = -\frac{1}{a}\cot\frac{x}{2}+C \qquad \left(\because \frac{1}{a(1-\cos x)} = \frac{1}{2a\sin^2 x/2}\right)$

Ex. 76. $\displaystyle\int \frac{\operatorname{cosec} x}{2+\operatorname{cosec} x}dx = \frac{1}{\sqrt{3}}\,\frac{\sin\frac{1}{2}x+(2-\sqrt{3})\cos\frac{1}{2}x}{\sin\frac{1}{2}x+(2+\sqrt{3})\cos\frac{1}{2}x}+C$

Ex. 77. $\displaystyle\int \frac{dx}{1-2\sin x} = \frac{1}{\sqrt{3}}\ln\left(\frac{\tan\frac{x}{2}-2-\sqrt{3}}{\tan\frac{x}{2}-2+\sqrt{3}}\right)+C$

Ex. 78. $\displaystyle\int \frac{\sin x\cos x\,dx}{1-\cos x} = \cos x + \ln(1-\cos x)+C$ [**Hint:** Let $u = \cos x$]

Ex. 79. $\int \sec x\, dx = \ln \tan\left(\frac{x}{2}+\frac{\pi}{4}\right) + C$ [**Hint:** Let $t = \tan x/2$]

Integration of $I = \dfrac{a_1 \cos x + a_2 \sin x + a_3}{b_1 \cos x + b_2 \sin x + b_3}$

This can be done by determining three constants k_1, k_2, k_3 such that

$$a_1 \cos x + a_2 \sin x + a_3 = k_1 (b_1 \cos x + b_2 \sin x + b_3) + k_2 (- b_1 \sin x + b_2 \cos x) + k_3$$

$$\therefore \quad a_1 = b_1k_1 + b_2k_2,\ a_2 = b_2k_1 - b_1k_2,\ a_3 = b_3k_1 + k_3$$

$$\therefore \quad I = \int k_1\, dx + k_2 \int \frac{-b_1 \sin x + b_2 \cos x}{b_1 \cos x + b_2 \sin x + b_3} dx + k_3 \int \frac{dx}{b_1 \cos x + b_2 \sin x + b_3}$$

$$= k_1x + k_2 \ln (b_1 \cos x + b_2 \sin x + b_3) + k_3I_1$$

where I_1 may be found by substituting $t = \tan \frac{x}{2}$.

If $a_3 = 0 = b_3$ in the original expression, we can write

$$I = k_4 \int \frac{-b_1 \sin x + b_2 \cos x}{b_1 \cos x + b_2 \sin x} dx + k_5 \int dx$$

$$= k_4 \ln (b_1 \cos x + b_2 \sin x) + k_5x$$

Here $k_4 = \dfrac{a_1b_2 - a_2b_1}{b_1^2 + b_2^2}$, $k_5 = \dfrac{a_1b_1 - a_2b_2}{b_1^2 + b_2^2}$

Ex. 80. $\int \frac{5\cos x + 6}{2\cos x + \sin x + 3} dx = 2x + \ln (2\cos x + \sin x + 3) + C$

Ex. 81. $\int (a + b\tan x)^{-1}\, dx = \frac{ax + b\ln (a\cos x + b\sin x)}{a^2 + b^2} + C$

Ex. 82. $\int \frac{dx}{\sin x + \sin 2x} = \frac{1}{3} \ln \frac{\sin x\ (1 + \cos x)}{1 + (2\cos x)^2} + C$

[**Hint:** This can be done by substituting $\cos x = u$, $- \sin x\, dx = du$]

Ex. 83. $I = \int \frac{e^{2\tan x} - 1}{e^{2\tan x} + 1} \sec^2 x\, dx = \int \frac{e^{\tan x} - e^{-\tan x}}{e^{\tan x} + e^{-\tan x}} \sec^2 x\, dx$

Put $\quad u = e^{\tan x} + e^{-\tan x}$

$\therefore \quad du = (e^{\tan x} + e^{-\tan x}) \sec^2 x\, dx$

$\therefore \quad I = \int \frac{du}{u} = \ln u + C = \ln (e^{\tan x} + e^{\tan x}) + C$

Ex. 84. $\int x \sin^{-1} x\, dx$ (use $\int u\, dv = uv - \int v\, du$)

$$= (\sin^{-1} x) \frac{x^2}{2} - \int \frac{1}{\sqrt{1 - x^2}} \frac{x^2}{2} dx$$

$$= \frac{1}{2}x^2 \sin^{-1} x + \frac{1}{2}\int \frac{1 - x^2 - 1}{\sqrt{1 - x^2}} dx$$

$$= \frac{1}{2}x^2 \sin^{-1} x + \frac{1}{2}\int \left(\sqrt{1 - x^2} - \frac{1}{\sqrt{1 - x^2}} \right) dx$$

$$= \frac{1}{2}x^2 \sin^{-1} x + \frac{1}{2}\left[\frac{x}{2}\sqrt{1 - x^2} + \frac{\sin^{-1} x}{2} - \sin^{-1} x \right]$$

$$= \left(\frac{1}{2}x^2 - \frac{1}{4} \right) \sin^{-1} x + \frac{x}{4}\sqrt{1 - x^2} + C$$

Ex. 85. $\int \sin^3 x \, dx$

$$= \int \frac{3\sin x - \sin 3x}{4} dx = -\frac{3}{4}\cos x + \frac{\cos 3x}{12} + C$$

or $\int \sin^3 x \, dx = \int (1 - \cos^2 x) \sin x = \int \sin x \, dx - \int \cos^2 x \sin x \, dx$

$$= -\cos x + \frac{1}{3}\cos^3 x + C$$

(*Note* : $\sin 3x - 3 \sin x - 4 \sin^3 x$)

Ex. 86. $\int \sin^3 3x \cos^5 3x \, dx = \int \sin^3 3x \, (1 - \sin^2 3x)^2 \cos 3x \, dx$

$$= \frac{\sin^4 3x}{12} - \frac{\sin^6 3x}{9} + \frac{\sin^8 3x}{24} + C.$$

Ex. 87. $\int (1 - \cos x)^{1/2} dx = \int \sqrt{2\sin^2 x/2} \, dx = \sqrt{2}\int \sin\frac{x}{2}$

$$= -2\sqrt{2}\cos\frac{x}{2} + C$$

Ex. 88. $\int \tan^4 x \, dx = \int \tan^2 x \tan^2 x \, dx$

$$= \tan^2 x \, (\sec^2 x - 1) dx = \int \tan^2 x \sec^2 x \, dx - \int \tan^2 x \, dx$$

$$= \frac{1}{3}\tan^3 x - (\tan x - x) + C$$

Ex. 89. $\int \cos 3x \cos 2x = \frac{1}{2}\sin x + \frac{1}{10}\sin 5x + C$

Ex. 90. $\int \tan^3 3x \sec 3x \, dx = \frac{1}{9}\sec^3 3x - \frac{1}{3}\sec^3 x + C$

Ex. 91. $\int \sec^n x\, dx = \dfrac{\sec^{n-2}}{n-1} \tan x + \dfrac{n-2}{n-1} \int \sec^{n-2} x\, dx + C$

$$\int \sec^3 x\, dx = \frac{1}{2} \sec x \tan x + \frac{1}{2} \ln\ (\sec x + \tan x) + C.$$

Ex. 92. $\int x(\cos^3 x^2 - \sin^3 x)^2\, dx = \dfrac{1}{12}\ (\sin x^2 + \cos x^2)\ (4 + \sin\ 2x^2) + C$

Ex. 93. $\int \sec^4 x (\tan x)^{-4}\, dx = -\dfrac{1}{3}\ (\tan x)^{-3} - \dfrac{1}{\tan x} + C$

Ex. 94. $\int \dfrac{dx}{\sqrt{1 - \sin 2x}} = -\dfrac{1}{\sqrt{2}} \ln\left[\ \text{cosec}\left(\dfrac{\pi}{4} - x\right) - \cot\left(\dfrac{\pi}{4} - x\right)\right] + C.$

12.8 REDUCTION FORMULAE

(1) $I = \int \sin^n x\, dx = \dfrac{-\cos x \sin^{n-1} x}{x} + \dfrac{n-1}{n} \int \sin^{n-2} x\, dx$ (n even)

Ex. 95. $\int \sin^4 x\, dx = -\dfrac{\cos x \sin^3 x}{4} - \dfrac{3}{8} \cos x \sin x + \dfrac{3}{8} x + C$

If the index n is odd as in $\int \sin^5 x\, dx$ we can integrate by substitution.

Ex. 96. $I = \int \sin^5 x\, dx$. Put $\cos x = t$, $-\sin x\, dx = dt$

$\therefore$ $I = -\int (1 - t^2)^2\, dt = -\int (1 - 2t^2 + t^4)\, dt$

$$= -\cos x + \frac{2}{3} \cos^3 x - \frac{1}{5} \cos^5 x + C$$

(2) $\int \cos^n x\, dx = \dfrac{\sin x \cos^{n-1} x}{n} + \dfrac{n-1}{n} \int \cos^{n-2} x\, dx$

(3) $\int \sin^p x \cos^q x\, dx = \dfrac{\cos^{q+1} x \sin^{p-1} x}{q+1} + \int (\cos^{q+1} x)\ \dfrac{p-1}{q+1} \sin^{p-2} x \cos x\, dx$

Ex. 97. $\int \sin^6 x (\cos x)^{-1} dx = \ln \tan\left(\dfrac{\pi}{4} + \dfrac{x}{2}\right) - \sin x - \dfrac{\sin^3 x}{3} - \dfrac{\sin^5 x}{5}$

(4) $\int \tan^n x\, dx = \dfrac{1}{n} \tan^{n-1} x - \int \tan^{n-2} x\, dx.$

(5) $\int x^m e^{ax} dx = \dfrac{x^m e^{ax}}{a} - \dfrac{m}{a} \int x^{m-1} e^{ax}\, dx$

(6) $\int (x^2 - a^2)^n dx = \dfrac{x(x^2 - a^2)^n}{2n+1} - \dfrac{2na^2}{2n+1} \int (x^2 - a^2)^{n-1} dx,\quad n \neq -\dfrac{1}{2}$

(7) $\int (a^2 \pm x^2)^{-n}\, dx = \dfrac{x/a^2}{(2n-2)\,(a^2 \pm x^2)^{n-1}} + \dfrac{2n-3}{2n-2} \int \dfrac{dx}{(a^2 + x^2)^{n-1}},\quad n \neq 1$

(8) $\int x^n \sin bx\, dx = -\dfrac{1}{b} x^n \cos bx + \dfrac{n}{b} \int x^{m-1} \cos bx\, dx$

(9) $\int x^n \cos bx\, dx = \dfrac{1}{b} \sin bx - \dfrac{n}{b} \int x^{n-1} \sin bx\, dx$

Ex. 98. $\int \dfrac{x\, dx}{\sqrt{a+bx}} = \dfrac{2(bx-2a)\sqrt{a+bx}}{3b^2}$

$$\int \sin \ln x\, dx = \frac{1}{2} x (\sin \ln x - \cos \ln x)$$

$$\int e^{ax} \sin bx\, dx = \frac{e^{ax}}{a^2+b^2} (a \sin bx - b \cos bx).$$

12.9 SOME IMPORTANT RESULTS AND EXERCISES

(1) $\int [f(x)]^n f'(x)\, dx = \dfrac{[f(x)]^{n+1}}{n+1}$

e.g., $\quad I = \int \tan^4 x \sec^2 x\, dx = \int \tan^4 x d \tan x$

Here $f(x) = \tan x$, $f'(x)\, dx = \sec^2 x\, dx = d \tan x$

$\therefore \quad I = \dfrac{1}{5} \tan^5 x + C.$

Ex. 99. $\int \dfrac{(\tan^{-1} x)^3}{1+x^2} dx = \dfrac{1}{4} (\tan^{-1} x)^4 + C$

$$\int \sqrt{2 + \sec^2 x}\, \sec^2 x \tan x\, dx = \frac{1}{3} (2 + \sec^2 x)^{3/2} + C$$

(2) $\int e^x (f(x) + f'(x))\, dx = e^x f(x) + C$

e.g., $\quad \int \dfrac{1 + x \ln x}{x} e^x dx = e^x \ln x + C$

Ex. 100. $\int e^x \dfrac{1 - \sin 2x}{1 + \cos 2x} dx = e^x \tan x + C$

(3) $\int e^{ax} \cos (bx + c)\, dx = \dfrac{e^{ax}}{\sqrt{a^2+b^2}} (\cos (bx+c) - \tan^{-1} b/a) + C$

$$\int e^{ax} \sin (bx + c)\, dx = \frac{e^{ax}}{\sqrt{a^2+b^2}} (\sin (bx+c) + \tan^{-1} b/a) + C$$

Ex. 101. (*i*) $\int \cos \ln x \, dx = \int \cos t \, e^t dt$ (Putting $x = e^t$)

$$= \frac{e^t}{\sqrt{2}} \cos\left(t - \frac{\pi}{4}\right) = \frac{x}{\sqrt{2}} \cos \ln\left(x - \frac{\pi}{4}\right) + C$$

(*ii*) $\int \frac{x e^{m \sin^{-1} x}}{\sqrt{1 - x^2}} dx = \frac{e^{mt}}{\sqrt{1 + m^2}} \sin (t - \cot^{-1} m) + C$

Hint: Put $t = \sin^{-1}x$

(4) $\int x^m \cos nx \, dx = x^m \frac{\sin nx}{n} - \int \frac{m}{n} x^{m-1} \sin nx \, dx$

Proceed further on the same lines.

Ex. 102. $\int x^3 \cos 3x \, dx = \frac{1}{3}(x^3 \sin 3x + x^2 \cos 3x) - \frac{2}{3} \int x \cos 3x \, dx$

(5) $\int x^m (1 - x)^{n-1} dx = (1 - x)^{n-1} \frac{x^{m+1}}{m+1} + \frac{n-1}{m+1} \int (1 - x)^{n-2} x^{m+1} dx$

(Integrate by parts)

(6) $\int \frac{dx}{(x^2 + a^2)^n} = \frac{x}{2a^2(x - 1)(x^2 + a^2)^{n-1}} + \frac{2n - 3}{2a^2(x - 1)} \int \frac{dx}{(x^2 + a^2)^{n-1}}$

(7) $\int \sin^n x \, dx = \frac{-1}{n} \cos x \sin^{n-1} x + \frac{n-1}{n} \int \sin^{n-2} x \, dx$

$$\int \cos^n x \, dx = \frac{1}{n} \sin x \cos^{n-1} x + \frac{n-1}{n} \int \cos^{n-2} x \, dx$$

$$\int \tan^n x \, dx = \frac{\tan^{n-1} x}{n-1} - \int \tan^{n-2} x \, dx$$

$$\int \sec^n x \, dx = \frac{\tan x \sec^{n-2} x}{n-1} + \frac{n-2}{n-1} \int \sec^{n-2} x \, dx$$

$$\int \operatorname{cosec}^n x \, dx = -\frac{\cot x \operatorname{cosec}^{n-2} x}{n-1} + \frac{n-2}{n-1} \int \operatorname{cosec}^{n-2} x \, dx$$

$$\int \cot^n x \, dx = -\frac{\cot^{n-1} x}{n-1} - \int \cot^{n-2} x \, dx$$

(8) $\int \frac{dx}{a + b \tan x} = \frac{1}{a^2 + b^2} [ax + b \ln (a \cos x + b \sin x)] + C$

Ex. 103. Find $\int e^x \frac{x+1}{(x+2)^2} dx$

Write $\frac{x+1}{(x+2)^2} = \frac{x+2-1}{(x+2)^2} = \frac{1}{x+2} - \frac{1}{(x+2)^2}$

If $$f(x)=\frac{1}{x+2},\ f'(x)=\frac{-1}{(x+2)^2}$$

$\therefore$ $$\int e^x\frac{x+1}{(x+2)^2}dx=\int e^x[f(x)+f'(x)]dx \text{ where } f(x)=\frac{1}{x+2}$$

$$=e^x f(x)+C=\frac{e^x}{x+2}+C. \text{ using result (2)}$$

Ex. 104. (*i*) $$I=\int\frac{dx}{1+\sin x-\cos x}=\int\frac{\frac{2\,dt}{1+t^2}}{1+\frac{2t}{1+t^2}-\frac{1-t^2}{1+t^2}}$$

By putting $t=\tan\dfrac{x}{2},\ dt=\dfrac{1}{2}\sec^2\dfrac{x}{2}$

$\therefore$ $$I=\int\frac{dt}{t(1+t)}=\ln t-\ln(1+t)+C$$

$$=\ln\tan\frac{x}{2}-\ln\left(1+\tan\frac{x}{2}\right)+C.$$

(*ii*) $$\int\sec x\,dx=\int\frac{1+t^2}{1-t^2}\frac{2\,dt}{1+t^2}=2\int\frac{dt}{1-t^2}=\ln\frac{1+\tan x/2}{1-\tan x/2}+C$$

$$=\ln\tan\left(\frac{x}{2}+\frac{\pi}{4}\right)+C$$

Alternative Answer $$\int\sec x\,dx=\int\frac{\sec x\,(\sec x+\tan x)}{\sec x+\tan x}dx$$

$$=\int\frac{\sec^2 x+\sec x\tan x}{\tan x+\sec x}dx=\ln(\tan x+\sec x)+C.$$

Ex. 105. $I=\int x^5\sqrt{1-x^3}\,dx=\int u(1-u^2)\,(-2/3u)\,du$

putting $u=\sqrt{1-x^3},\quad 3x^2dx=-2u\,du$

$\therefore$ $$I=-\frac{2}{3}\int(u^2-u^4)\,du=-\frac{2}{45}(1-x^3)^{3/2}\,(2+3x^3)+C$$

Ex. 106. $\int\sqrt{1+x^{1/2}}\,dx=\dfrac{4}{5}(1+x^{1/2})^{5/2}-\dfrac{4}{3}(1+x^{1/2})^{3/2}+C$

Ex. 107. $\int\dfrac{(e^x-2)e^x}{e^x+1}dx=e^x-3\ln(e^x+1)+C$

Ex. 108. (1) $\int\dfrac{x}{\sqrt{x-3}}dx=b\,(x-3)^{1/2}\times\dfrac{2}{3}(x-3)^{3/2}+C$

(2) $\int \frac{\ln x}{x^2} dx = -\frac{\ln x}{x} - \frac{1}{x} + C$

(3) $\int \frac{7-2x}{x^2-2x+1} dx = -2\ln(x-1) - \frac{5}{x-1} + C$

(4) $\int \frac{4x^2-2x-6}{x^3-x} dx = 6\ln x - 2\ln(x-1) + C$

(5) $\int (\ln x)^2 dx = (x \ln x)^2 - 2x \ln x + 2x + C$

(6) $\int e^{2.5x} dx = 0.4 e^{2.5x} + C$

(7) $\int \frac{dx}{5+3e^{2x}} = \frac{x}{5} - \frac{1}{10} \ln(5+3e^{2x}) + C$

Ex. 109. $I = \int \frac{\sqrt{1-x^{3/2}}}{x^{7/2}} dx$ $\left(\text{Put } x = \frac{1}{u},\ dx = -\frac{1}{u^2} du \right)$

$$I = \int \frac{\sqrt{x}\sqrt{1-x^{3/2}}}{\sqrt{x}\,x^{7/2}} dx = \int \frac{\sqrt{x-x^2}}{x^4} dx = -\int u\sqrt{u-1}\, du$$

Another substitution $u - 1 = v^2$ is necessary now

$\therefore$ $$-\int (v^2+1)v\,2v\,dv = -2\int (v^4+v^2)dv$$

$$= -2\left(\frac{v^5}{5} + \frac{v^3}{3}\right) = -\frac{2}{3}\left(\frac{1-x}{x}\right)^{5/2} - \frac{2}{3}\left(\frac{1-x}{x}\right)^{3/2} + C$$

Ex. 110. $\int \frac{dx}{(x+1)^{1/2}+(x+1)^{1/4}} = 2(x+1)^{1/2} - 4(x+1)^{1/4}$

$$+4\ln(1+(x+1)^{1/4}) + C.$$

Ex. 111. A socialist economy is being transformed into a liberal economy at a rate proportional to the number of remaining controlled firms whose original number is 50. At the period $t = 3$, the number of controlled firms is 25. When will 1/10 of the firms remain controlled?

It at time t, the number transformed is x then $\frac{dx}{dt} = \lambda(50 - x)$ where λ is a constant.

$\therefore$ $\frac{dx}{50-x} = \lambda\, dt$ or $\ln(50-x) = -\lambda t + \ln C$ (integrating)

$\therefore$ $50 - x = C e^{-\lambda t}$

when $t = 0$, $x = 0$, $50 - 0 = Ce^0$

$\therefore$ $C = 50$

$t = 3$, $50 - x = 25 = C e^{-3\lambda} = 50 e^{-3\lambda}$

$\therefore$ $e^{-3\lambda} = \frac{25}{50} = 0.5$ or $\lambda = 0.23$

$\therefore \qquad 50 - x = 50 e^{-0.23t}$

when the controlled firms remaining are $\frac{1}{10}$ *i.e.*, 5, $50 e^{-0.23t} = 5$ or $e^{-0.23t} = 0.1$

$\therefore \qquad t = 10.$

Ex. 112. $\displaystyle\int \frac{x^3 + 2x^2}{(x+1)(x^2+2x+2)} dx$

For partial fractions, note that the degree of the numerator here is the same as that of the denominator.

Let $$\frac{x^3 + 2x^2}{(x+1)(x^2+2x+2)} = 1 + \frac{A}{x+1} + \frac{Bx + C}{x^2 + 2x + 2}$$

Put $x = -1$, then $A = 1$. Take *LCM* and equate the coefficients of x^2 and the constant terms.

Then $B = -2$, $C = -4$

$$\therefore \qquad \int \left[1 + \frac{1}{x+1} - \frac{2x+4}{x^2+2x+2} \right] dx$$

$$= \int \left[1 + \frac{1}{x+1} - \left(\frac{2x+2}{x^2+2x+2} + \frac{2}{(x+1)^2 + 1} \right) \right] dx$$

$$= x + \ln(x+1) - \ln(x^2 + 2x + 2) - 2\tan^{-1}(x+1) + C_1$$

Ex. 113. $I = \displaystyle\int \frac{x}{(x-1)(x^2+1)^2} dx = \int \left[\frac{A}{x-1} + \frac{Bx + C}{x^2+1} + \frac{Dx + E}{(x^2+1)^2} \right] dx$

Note that $A = \frac{1}{4}, \; B = -\frac{1}{4}, \; C = -\frac{1}{4}, \; D = -\frac{1}{2}, \; E = \frac{1}{2}$

Show $I = \frac{1}{4}\ln(x-1) - \frac{1}{8}\ln(x^2+1) + \frac{1}{4(x^2+1)} + \frac{x}{4(1+x^2)} + C$

Ex. 114. $I = \displaystyle\int \frac{dx}{\sin x + \sin 2x}$ (Put $u = \cos x$)

$$= \int \frac{dx}{\sin x\,(1 + 2\cos x)} = \int \frac{-du}{(1-u)(1+u)(1+2u)}$$

$$= -\int \left[\frac{A}{1-u} + \frac{B}{1+u} + \frac{C}{1+2u} \right] dt$$

Note $A = \frac{1}{6}, \; B = -\frac{1}{2}, \; C = \frac{4}{3}$

Show $I = -\frac{1}{2}\ln(1+\cos x) - \frac{1}{6}\ln(1-\cos x) + \frac{2}{3}\ln(1+2\cos x) + C_1.$

Ex. 115. $\displaystyle\int \frac{dx}{1-\cos^4 x} = \int \frac{dx}{(1-\cos^2 x)(1+\cos^2 x)}$

$$= \frac{1}{2}\int \left[\frac{1}{1-\cos^2 x} + \frac{1}{1+\cos^2 x}\right] dx$$

$$= \frac{1}{2}\int \left(\text{cosec}^2 x + \frac{\sec^2 x}{\tan^2 x + 2}\right) dx$$

$$= \frac{1}{2\sqrt{2}} \tan^{-1} \frac{\tan x}{\sqrt{2}} - \frac{1}{2}\cot x + C$$

Ex. 116. $\displaystyle\int \frac{dx}{1+3e^x+2e^{2x}} dx = x + \ln(1+e^x) - 2\ln(1+2e^x) + C$

Ex. 117. $\displaystyle\int \frac{dx}{\sqrt{1+x}+\sqrt{x}} = \frac{2}{3}[(1+x)^{3/2} - x^{3/2}] + C$

12.10 THE DEFINITE INTEGRAL

Integration implies consolidation into a whole. Although it is a process which is the inverse of differentiation, it is also a process of summation and can be used to calculate areas by dividing them into infinitesimally small parts.

As indefinite integral possesses no definite numerical value. A definite integral has a numerical value independent of the constant C.

Let $f(x)$ be continuous on the interval (a, b). If we write $\int f(x)dx = F(x) + C$ then the value of the *definite* integral from a to b is written

$$\int_a^b f(x)dx = F(x)\Big]_a^b = F(b) - F(a)$$

Ex. 118. $\displaystyle\int_1^2 x^3 dx = \frac{x^4}{4}\Big]_1^2 = \frac{2^4}{4} - \frac{1^4}{4} = 4 - \frac{1}{4} = 3\frac{3}{4}.$

Ex. 119. $\displaystyle\int_0^1 e^x dx = e^x\Big]_0^1 = e^1 - e^0 = e - 1.$

Ex. 120. $\displaystyle\int_l^m (ax^2+bx+c)\,dx = \left[\frac{ax^3}{3} + \frac{bx^2}{2} + cx\right]_l^m$

$$= \frac{a}{3}(m^3 - l^3) + \frac{b}{2}(m^2 - l^2) + c(m - l).$$

Ex. 121. $\displaystyle\int_1^5 \left(x + \frac{4}{x^2}\right) dx = \left[\frac{x^2}{2} - \frac{4}{x}\right]_1^5 = \left(\frac{25}{2} - \frac{4}{5}\right) - \left(\frac{1}{4} - 4\right) = 15\tfrac{1}{5}.$

Ex. 122. $\int_0^2 \frac{5}{2+x} dx = 5\ln(2+x)\Big]_0^2 = 5\,(\ln 4 - \ln 2) = 5 \log 2.$

Ex. 123. Show that

(*i*) $\int_a^b 0\,dx = 0$

(*ii*) $\int_a^b x\,dx = \frac{b^2}{2} - \frac{a^2}{2}$

(*iii*) $\int_a^b x^2\,dx = \frac{b^3}{3} - \frac{a^3}{3}$

(*iv*) $\int_{-1}^1 u^3 du = 0$

(*v*) $\int_a^b (ax^2 + bx + c)\,dx = \frac{b-a}{6}\left[f(a) + 4f\left(\frac{a+b}{2}\right) + f(b) \right]$

where $f(x) = ax^2 + bx + c$

(*vi*) $\int_0^2 (x-1)\,(x^2 + x + 1)\,dx = 2$

(*vii*) $\int_0^{\pi/2} \cos x\; dx = [\sin x]_0^{\pi/2} = \sin\frac{\pi}{2} - \sin 0 = 1$

(*viii*) $\int_1^2 (x+7)^2\,dx + \int_2^1 (y+7)^2\,dy = 0$ (variable name immaterial)

(*ix*) $\int_0^{\pi/2} \sqrt{1+\sin x}\; dx = \int_0^{\pi/2} \left(\cos\frac{x}{2} + \sin\frac{x}{2} \right) dx = 2$

(*x*) $\int_1^0 (1 - r^2)\,dr = \frac{2}{3}$

(*xi*) $\int_1^2 u\,dv = uv\Big]_1^2 - \int_1^2 v\,du$

(*xii*) $\int_2^3 e^{2x} dx = \frac{e^4}{2}(e^2 - 1)$

12.11 THE AREA UNDER A CURVE AS A DEFINITE INTEGRAL: THE FUNDAMENTAL THEOREM OF INTEGRAL CALCULUS

Let $y = f(x)$ be a continuous function shown as a curve. To find the area under this curve in the interval (a, b) take a small strip of width $x_2 - x_1 = \Delta x_1$, its height being $f(x_1)$. The area of this strip is $f(x_1)\,\Delta x_1$. If we similarly take n strips of width Δx_i (i = 1, 2, ..., n) and n corresponding heights $f(x_i)$ (i = 1, 2, ..., n) we have n thin rectangles, each of area

$$f(x_i)\,\Delta x_i,\ i = 1, 2, \ldots, n$$

The total of all these areas is given by

$$\sum_{i=1}^{n} f(x)\,\Delta x_i$$

This is not exactly the area under the curve. But it can be if the widths of the rectangles are taken sufficiently small *i.e.*, if n is very large or as n tends to infinity. The rectangles will become thinner (almost lines) and we can write the area between (a, b) under the curve $y = f(x)$ as the limit

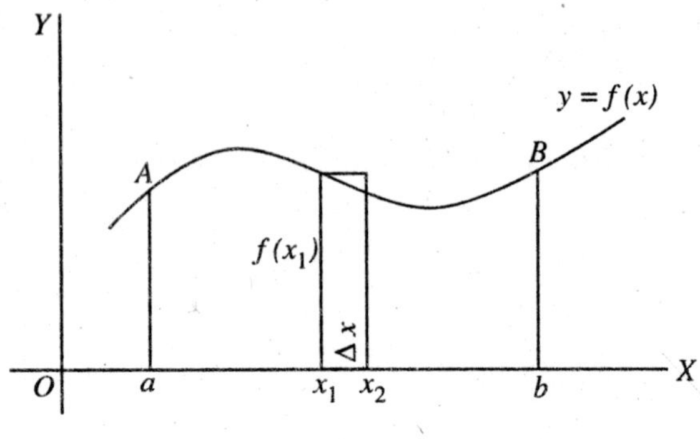

$$A = \lim_{n \to \infty} \sum_{i=1}^{n} f(x_i)\,\Delta x_i$$

The area under a curve is thus expressed as a *discrete* sum.

In the limit we can write the area in the *continuous* form

$$A = \int_a^b f(x)\,dx$$

The similarities between the expressions $\lim\limits_{n \to \infty} \Sigma f(x_i)\,\Delta x_i$ and $\int_a^b f(x)\,dx$ may be noted.

The discrete quantities $f(x_i)$ and Δx_i have their continuous counterparts $f(x)$ and dx and the discrete summation sign Σ is replaced by the continuous summation sign $\int$. The area under the curve $y = f(x)$ between the limits a, b can thus be written as a definite integral

$$\text{Area } abBA = \int_a^b f(x)\,dx = F(b) - F(a)$$

Ex. 124. Evaluate a definite integral as the limit of a sum by proving

$$\int_a^b e^x dx = e^x \Big]_a^b = e^b - e^a$$

The procedure from the first principles is as follows. Let the interval $[a, b]$ be divided into n subintervals each of size h at points a, $a + h$, $a + 2h$, ..., $a + nh$ where

$h = \dfrac{b-a}{n} \to 0$ as $n \to 0$.

Let $t_r = a + rh$, $f(t_r) = f(a + rh) = e^{a + rh} = e^a \cdot e^{rh}$

$$S_n = \sum_{r=1}^{n} f(t_t)\,\Delta_r = \sum_{r=1}^{n} (e^a \cdot e^{rh})h = e^a h \sum_{r=1}^{n} e^{rh}$$

$$= e^a h\,[e^h + e^{2h} + \ldots\ldots + e^{nh}]$$

$$= e^a \cdot h \cdot e^h \frac{(e^{nh}-1)}{(e^h-1)} = \frac{he^h}{e^h-1}(e^{b-a}-1) \quad \because b - a = nh$$

$$= (e^b - e^a)\,e^h \cdot \frac{h}{e^h - 1}$$

$$\text{As } \begin{matrix} h \to 0 \\ n \to \infty \end{matrix} \quad S_n = (e^b - e^a)\,(e^h)\left(\frac{h}{e^h - 1}\right) \to e^b - e^a$$

$$\left(\text{Since} \qquad \lim_{n \to 0} e^h = 1, \ \lim_{h \to 0} \frac{h}{e^h - 1} = 1.\right)$$

Ex. 125. $\int_a^b k\,dx = k(b-a)$. Let $h = \dfrac{b-a}{h}$

$$\therefore \qquad S_n = \sum_1^n k\,\Delta x = (k + k + \ldots\ldots k)h = nk\frac{b-a}{n} = k(b-a).$$

Ex. 126. $\int_0^2 (3x^2 + 7)\,dx = 22$. The intervals are at the points $0 + 0h$, $0 + 1h$, $0 + 2h$, ..., $0 + nh$ or $0, h, 2h, \ldots\ldots, nh\ (= 2)$ and $h = \dfrac{2-0}{n}$.

Ex. 127. Find the area bounded by the x-axis and the curve $y = x^2$ between $x = 1$ and $x = 3$

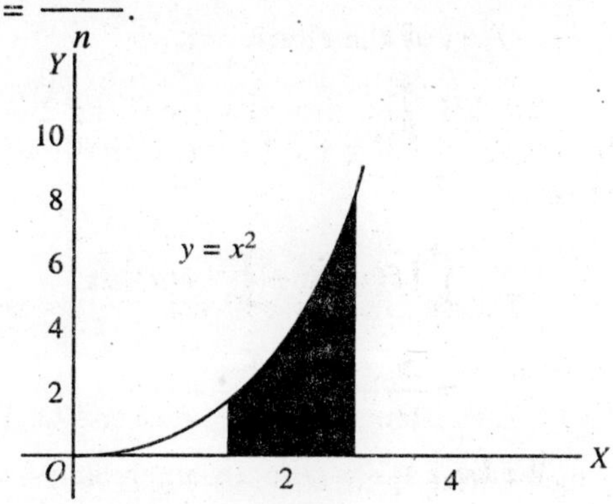

$$\int_1^3 x^2 dx = \frac{x^3}{3}\Bigg]_1^3 = \frac{3^3}{3} - \frac{1^3}{8} = 8\tfrac{2}{3}$$

Ex. 128. Find the area under the curve

$$y = \sqrt{x} \qquad 1 \le x \le 4$$

$$\int_1^4 \sqrt{x}\,dx = \frac{2}{3}x^{2/3}\Bigg]_1^4 = \frac{2}{3}[4^{3/2} - 1^{3/2}] = \frac{2}{3} \times 7 = \frac{14}{3}$$

Ex. 129. Find the area under the curve

(*i*) $y = x^2 + 4x + 5$ $\qquad -2 \le x \le 1$

(*ii*) $y = \dfrac{x^2}{2} + 1$ $\qquad 0 \le x \le 4$

(*iii*) $y = 9 - x^2$ $\qquad 1 \le x \le 3$

(*iv*) $y = a^2 - x^2$ $\qquad 0 \le x \le a$ $\qquad$ **Ans.** 12, 14/3, 28/3, $\pi \cdot \dfrac{2a^3}{3}$

Sign Convention

If the function $y = f(x)$ is positive in the interval (a, b) and the curve is above the x-axis then $\int_a^b f(x)\,dx$ is positive.

If $y = f(x)$ is negative in the interval (a, b) and the curve is below the x-axis then $\int_a^b f(x)\,dx$ is negative.

If $y = f(x)$ changes sign in the interval and the curve crosses the x axis, the area is the algebraic sum of a positive area and a negative area.

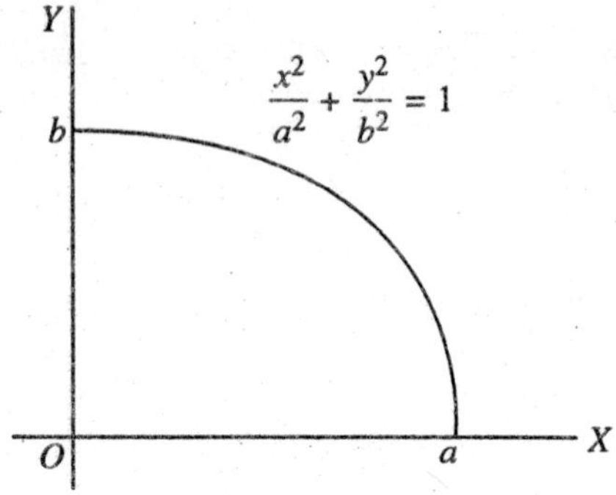

Ex. 130. To find the area of the ellipse $\frac{x^2}{a^2} + \frac{y^2}{b^2} = 1$ consider this ellipse divided into 4 equal parts. The area of part Oab is

$$\int_0^a y\,dx = \int_0^a \frac{b}{a}\sqrt{a^2 - x^2}\,dx$$

$$= \frac{b}{a}\,\frac{\pi a^2}{4} = \frac{\pi ab}{4}$$

$\therefore$ Area of the ellipse $= \pi ab$.

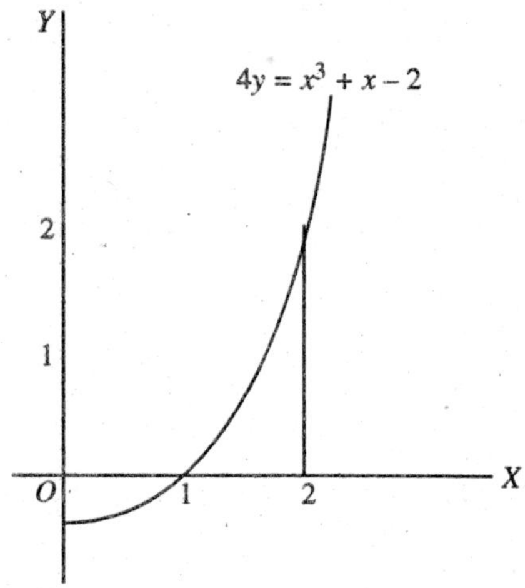

Ex. 131. The area between 0 and 2 of the curve $4y = x^3 + x - 2$ is the sum of two areas

$$\int_0^1 |f(x)|\,dx + \int_1^2 |f(x)|\,dx$$

$$= \frac{5}{16} + \frac{13}{16} = \frac{9}{8}$$

We have taken here the algebraic sum of a positive area and a negative area.

Ex. 132. Find the area bounded by the x-axis and the curve $y = 2x + x^2 - x^3$ between $x = 0$ and 1 **Ans.** 13/12

Ex. 133. Find the area bounded by (*i*) the x-axis and the curve $y = x^2 - 3x$ between $x = -1$ and $x = 4$ **Ans.** (*i*) 49/6

(*ii*) $y = \sqrt{x},\ y = \sqrt[3]{x}$ (*ii*) 5/12

(*iii*) $y = x(x^2 - 9),\ y = -5x$ (*iii*) 8

12.12 AREA BETWEEN TWO CURVES

The area between two curves $y = f_1(x)$ and $y = f_2(x)$ between a and b is given by

$$\int_a^b [f_1(x) - f_2(x)]\,dx$$

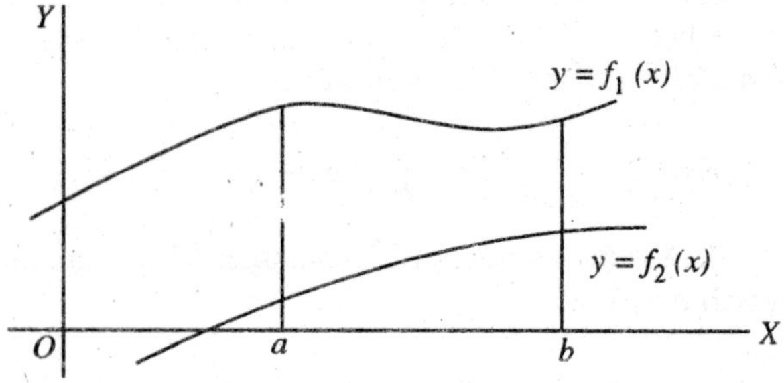

Ex. 134. The area between $y = x^3$ and $y = 9x^2$ can be found by first noting the points of intersection obtained by writing

$$y = x^3 = 9x^2$$

or $$x^2(x-9) = 0$$

$\therefore$ $$x = 0,\ x = 9$$

are the limits of integration.

The required area $= \int_0^9 (9x^2 - x^3)\,dx$

$$= \left[\frac{9x^3}{3} - \frac{x^4}{4}\right]_0^9 = 546.75$$

Ex. 135. Find the area between the y-axis and the curve

$$y^2 - 4y + 2x = 0$$

Hint: y axis means $x = 0$. **Ans.** 16/3

Ex. 136. Find the area bounded by (*i*) $4y = x^2$, $2y = x + 4$ and the axes, (*ii*) $y^2 = 2x$, $y = x - 4$, (*iii*) $y^3 = 4y + x$, $y^2 = 4 - x$ and x-axis. **Ans.** (*i*) 4 (*ii*) 1/2 (*iii*) 2

12.13 FUNDAMENTAL PROPERTIES OF DEFINITE INTEGRALS

It is assumed that the function $f(x)$ is integrable on the closed interval (a, b).

1. $\int_a^b kf(x)\,dx = k\int_a^b f(x)\,dx$

2. If $f_1(x)$, $f_2(x)$ are integrable on (a, b)

$$\int_a^b [f_1(x) + f_2(x)]\,dx = \int_a^b f_1(x)\,dx + \int_a^b f_2(x)\,dx$$

3. A definite integral can be expressed as a sum of definite integrals (additive property).

$$\int_a^b (f(x)\,dx = \int_a^c f(x)\,dx + \int_c^d f(x)\,dx, \qquad a < c < b$$

$$\int_a^b f(x)\,dx = \int_a^c f(x)\,dx + \int_c^d f(x)\,dx + \int_d^b f(x)\,dx, \qquad a < c < d < b$$

It means that the area under the curve between (a, b) is the sum of the areas under the curve between (a, c), (c, d) and (d, b). This property is useful in finding the areas under some discontinuous functions also.

Ex. 137. $\int_1^3 x^2\,dx = \int_1^2 x^2 dx + \int_2^3 x^2\,dx = \frac{19}{3} + \frac{7}{3} = 8\frac{2}{3}$

4. A definite integral is a *function* only *of its limits* a and b and not of the variable which may be changed.

$$\int_a^b f(x)\,dx = \int_a^b f(y)\,dy = \int_a^b f(z)\,dz$$

5. A definite integral equals zero when the limits of integration are identical:

$$\int_a^a F(x)\,dx = F(x)\Big]_a^a = F(a) - F(a) = 0$$

The area on a single point is zero because the width dx of the rectangle, is zero.

6. *The directed length* of the interval of integration is given by

$$\int_a^b dx = b - a$$

7. If the limits are interchanged the sign of the definite integral changes

$$\int_a^b f(x)\,dx = -\int_b^a f(x)\,dx$$

$$\int_2^4 x^2\,dx = -\int_4^2 x^2 dx$$

8. If one of the limits is the variable itself the definite integral becomes equal to the indefinite integral of the function

$$\int_a^x f(x)\,dx = F(x) - F(a) = F(x) + C$$

where $C = -F(a)$ is a constant.

Note. When one or both of the limits are infinite we have *improper* integrals. The concept of limits is to be used in such cases to find the value of the definite integral. (See 12.15)

Ex. 138. $\int_2^\infty \frac{dx}{x^2} = \lim_{n\to\infty} \int_2^n \frac{dx}{x^2} = \lim_{n\to\infty} \left[\frac{-1}{x}\right]_2^n = \lim_{n\to\infty}\left[-\frac{1}{n} + \frac{1}{2}\right] = \frac{1}{2}$

Ex. 139. Prove the following:

(*i*) $\int_0^{2a} f(x)\,dx = \int_0^a f(x)\,dx + \int_0^a f(2a - x)\,dx$

(*ii*) $\int_p^q f(x)\,dx = \int_p^q f(p + q - x)\,dx$

(*iii*) $\int_0^a f(a + x)\,dx = \int_0^a f(-x)\,dx$

Other useful properties of Definite Integrals

9. $\int_0^a f(x)\,dx = \int_0^a f(a - x)\,dx$

10. $\int_0^{2a} f(x)\,dx = \int_0^a f(x)\,dx + \int_0^a f(2a - x)\,dx$

$$= 2\int_0^a f(x)\,dx \text{ if } f(x) = f(2a - x)$$

$$= 0 \text{ if } f(x) = -f(2a - x)$$

11. $\int_a^b x f(x)\,dx = \frac{a+b}{2}\int_a^b f(x)\,dx$ (**Hint:** Put $a + b - u = x$, $dx = -du$)

12. $\int_0^a f(x)\,dx = 0$, if $f(a - x) = f(a + x)$ (**Hint:** Put $x = a + u$, $dx = du$)

Ex. 140. $\int_0^{\pi/2} \ln \sin x \, dx = -\frac{\pi}{2} \ln 2 = \int_0^{\pi/2} \ln \cos x \, dx$

Ex. 141. $I = \int_0^{\pi} \ln (1 + \cos \theta) d\theta = \int_0^{\pi} \ln (1 - \cos \theta) \, d\theta$

(Property 9. Note $\cos (\pi - \theta) = - \cos \theta$)

$$2I = \int_0^{\pi} \ln (1 - \cos^2 \theta) \, d\theta = 2 \int_0^{\pi} \ln \sin \theta \, d\theta$$

(Add: $\ln (1 + \cos \theta) + \ln (1 - \cos \theta) = (1 - \cos^2\theta)$)

$$= 2 \, [2 \int_0^{\pi/2} \ln \sin \theta \, d\theta] = 4 \left(-\frac{\pi}{2} \ln 2 \right)$$

$\therefore \quad I = -\pi \ln 2.$

Ex. 142. $\int_0^{\pi} \ln (1 + \tan x) dx = \frac{\pi}{8} \ln 2$

Ex. 143. $\int_0^{2a} x^{9/2} (2a - x)^{-1/2} \, dx$ (Put $x = 2a \sin^2 u$)

$$= 64 \, a^5 \int_0^{\pi/2} \sin^{10} u du = 64 \, a^5 \frac{9.7.5.3.1}{10.8.6.4.2} \frac{\pi}{2} \quad \text{(See Ex. 146 below)}$$

$$= \frac{63 \, a^5 \pi}{8}$$

Ex. 144. (*i*) $\int_0^{\pi} \frac{x \, dx}{b^2 \sin^2 x + a^2 \cos^2 x} = 2\pi \int_0^{\pi/2} \frac{dx}{a^2 \cos^2 x + b^2 \sin^2 x}$

(Substitute $t = \tan x$)

(*ii*) $I = \int_0^{\pi/3} (1 - \sin x)^{-1} \, dx$

Put $t = \tan \frac{x}{2}$, $dt = \frac{1}{2} \sec^2 \frac{x}{2} dx$

$$= \frac{1}{2} (1 + t^2) \, dx$$

Change the limits, $x = 0, t = 0$

$$x = \frac{\pi}{3}, \; t = \frac{1}{\sqrt{3}}$$

$$\therefore \quad I = \int_0^{1/\sqrt{3}} \frac{dt}{(1 - t)^2} = \left. \frac{2}{1 - t} \right]_0^{1/\sqrt{3}} = \frac{2}{1 - 1/\sqrt{3}} - 2 = \sqrt{3} + 1$$

Ex. 145. If $y = \frac{1}{2} x^2 - \frac{1}{4} \ln x$,

$$\int_1^4 (1 + (dy/dx)^2) \, dx = \frac{15 + \ln 2}{2}$$

Ex. 146. $\int_0^{\pi/2} \sin^n x \; dx = \frac{1.3.5...(n-1)}{2.4.6....n} \frac{\pi}{2}$ if n is even, $n > 0$

$= \frac{2.4.6....n-1}{1.3.5.... n}$ if n is odd, $n > 1$

Ex. 147. $\int_1^3 \ln(x + \sqrt{x^2 - 1}) \; dx = 3\ln(3 + 2\sqrt{2}) - 2\sqrt{2}$

Ex. 148. $\int_2^3 \left[\left(\frac{dx}{dt} \right)^2 + \left(\frac{dy}{dt} \right)^2 \right]^{1/2} dt = \sqrt{2}(e^3 - e^2)$

where $x = e^t \cos t, \; y = e^t \sin t$

Ex. 149. Find the area covered by $y = x\sqrt{1 - x^2}$. This curve when written $y^2 = x^2(1 - x^2)$ is symmetric and its area 4 times the area in quadrant one. The point of intersection on the x axis being at 1 we have

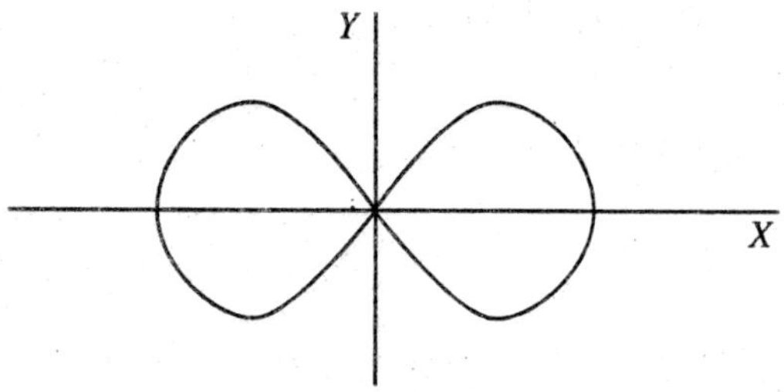

$$A = 4\int_0^1 x\sqrt{1 - x^2} dx = \frac{4}{3}$$

Ex. 150. Find the *average ordinate* of a parabola $f(x) = 4 - x^2$ from $x = a = -2$ to $x = b = 2$. The Average Ordinate (AO) over $a \le x \le b$ is given by

$$AO = \frac{\text{Area}}{\text{Base}} = \frac{\int_a^b f(x)\,dx}{b - a}$$

For $f(x) = 4 - x^2$, $AO = \frac{\int_{-2}^{2} (4 - x^2)\,dx}{4}$

$= \frac{1}{4} \times 2 \int_0^2 (4 - x^2)\,dx = 8/3.$

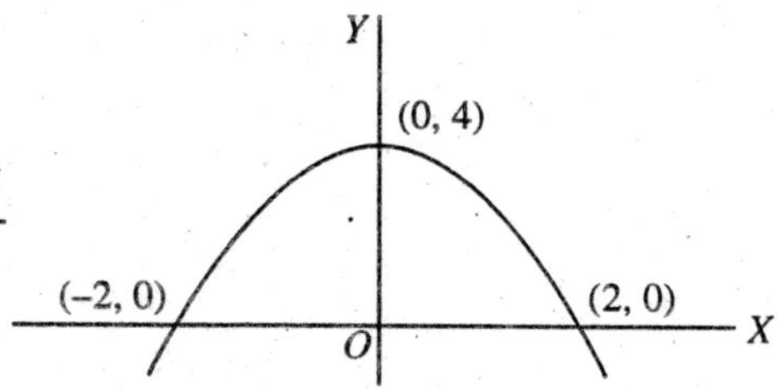

Ex. 151. For a semicircle, the base is $2r$ and the area $\frac{\pi r^2}{2}$.

Hence $AO = \frac{\pi r^2/2}{2r}$

$= \pi r/4.$

Ex. 152. Find (*i*) $I = \int_2^5 \frac{\sqrt{x}\,dx}{\sqrt{7-x} + \sqrt{x}}$

(*ii*) $I = \int_0^1 x^5 (1 - x^2)^{1/2} dx$

Use the rule $\int_a^b f(x)\ dx = \int_a^b f(a+b-x)dx$

$$I = \int_2^5 \frac{\sqrt{x}\, ax}{\sqrt{7-x}+\sqrt{x}} = \int_2^5 \frac{\sqrt{7-x}\, dx}{\sqrt{x}+\sqrt{7-x}}$$

$$\therefore \quad 2I = \int_2^5 \frac{\sqrt{x}+\sqrt{7-x}}{\sqrt{7-x}+\sqrt{x}} = \int_2^5 1\,dx = x\Big]_2^5 = 3$$

$$\therefore \quad I = 3/2.$$

(*ii*) Put $x = \sin\theta$, $dx = \cos\theta\, d\theta$

Limits: $x = 0$, $\sin\theta = 0$ or $\theta = 0$

$x = 1$, $\sin\theta = 1$ or $\theta = \pi/2$

$$\therefore \quad I = \int_0^{\pi/2} \sin^5\theta\,(1-\sin^2\theta)^{1/2}\cos\theta\, d\theta$$

$$= \int_0^{\pi/2} \sin^5 \cos^2\theta\, d\theta = \int_0^{\pi/2} (\sin^5\theta - \sin^7\theta)\, d\theta$$

$$= \int_0^{\pi/2} \sin^5\theta\, d\theta - \frac{6}{7}\int_0^{\pi/2} \sin^5\theta\, d\theta \left(\because \int_0^{\pi/2} \sin^7\theta\, d\theta = \frac{6}{7}\int_0^{\pi/2} \sin^5\theta\, d\theta\right)$$

(using reduction formula)

$$= \frac{1}{7}\int_0^{\pi/2} \sin^5\theta d\theta = \frac{1}{7}\cdot\frac{4}{5}\int_0^{\pi/2} \sin^3\theta d\theta$$

$$= \frac{1}{7}\cdot\frac{4}{5}\cdot\frac{2}{3}\int_0^{\pi/2} \sin\theta d\theta = \frac{8}{105}[-\cos\theta\,]_0^{\pi/2}$$

$$= \frac{8}{105}(0+1) = \frac{8}{105}.$$

Ex. 153. $\displaystyle\int_0^{\pi} \frac{x\,dx}{(1+\sin x)^2} = \int_0^{\pi} \frac{x\,dx}{\left(\cos\frac{x}{2}+\sin\frac{x}{2}\right)^2}$

$$= \int_0^{\pi} \frac{x\,dx}{2\left(\sin\frac{\pi}{4}\cos\frac{x}{2}+\cos\frac{\pi}{4}\sin\frac{x}{2}\right)^2} = \int_0^{\pi} \frac{x\ dx}{2\sin^2\left(\frac{\pi}{4}+\frac{x}{2}\right)}$$

$$= \frac{1}{2}\int_0^{\pi} x\,\text{cosec}^2\left(\frac{x}{2}+\frac{\pi}{4}\right)dx$$

$$= \left[-x\cos\left(\frac{x}{2}+\frac{\pi}{4}\right)+2\ln\sin\left(\frac{x}{2}+\frac{\pi}{4}\right)\right]_0^{\pi} = \pi + 2\left[\ln\sin\frac{3\pi}{4} - \ln\sin\frac{\pi}{4}\right] = \pi.$$

Ex. 154. Show $I = \int_0^{\pi/2} (\ln \tan x) \sin 2x\, dx = 0$

Hence show $\int_0^1 \ln\left(\frac{1-x}{x}\right) dx = 0$

Since $\int_0^a f(x)\,dx = \int_0^a f(a-x)\,dx$

$$I = \int_0^{\pi/2} \sin 2\left(\frac{\pi}{2} - x\right) \ln \tan\left(\frac{\pi}{2} - x\right) dx$$

$$= \int_0^{\pi/2} \sin 2x \ln \cot x\, dx$$

$$2I = \int_0^{\pi/2} [\sin 2x \ln \tan x + \sin 2x \ln \cot x]\, dx$$

$$= \int_0^{\pi/2} \sin 2x \ln \tan x \cot x\, dx$$

$$= \int_0^{\pi/2} \sin 2x \ln (1)\, dx = 0$$

If we put $x = \cos^2 u$

$$\int_0^1 \ln\left(\frac{1-x}{x}\right) dx = -\int_{\pi/2}^0 (\ln \tan^2 u) \sin 2u\, du$$

$$= 2\int_0^{\pi/2} \sin 2u \ln \tan u\, du = 0 \qquad \text{using the result above.}$$

Ex. 155. $\int_0^{2a} x^2 \sqrt{2ax - x^2}\, dx$ (Put $x = 2a \sin^2 u$)

$$= \int_0^{\pi/2} 32a^4 \sin^6 t\, dt = 32a^4 \times \frac{5.3.1.1}{8.6.4.2}\frac{\pi}{2} = \frac{5\pi a^4}{8}.$$

Ex. 156. $I = \int_0^{\pi} \frac{x \tan x}{\sec x + \tan x} dx = \int_0^{\pi} \frac{(\pi - x)\tan(\pi - x)}{\sec(\pi - x) + \tan(\pi - x)} dx$

$$= \int_0^{\pi} \frac{(\pi - x)\tan x}{\sec x + \tan x} dx = \int_0^{\pi} \frac{\pi \tan x}{\sec x + \tan x} dx - \int_0^{\pi} \frac{x \tan x}{\sec x + \tan x} dx$$

$$= \pi \int_0^{\pi} \frac{\tan x}{\sec x + \tan x} dx - I$$

$$\therefore \qquad 2I = \pi \int_0^{\pi} \frac{(\sec x - \tan x)\tan x}{(\sec x - \tan x)(\sec x + \tan x)} dx$$

$$= \pi \int_0^{\pi} (\sec x \tan x - \tan^2 x)\, dx$$

$$= \pi [\sec x - \tan x + x]_0^{\pi} = \pi(-1 + \pi - 1) = \pi(\pi - 2)$$

Ex. 157. $\int_{-\pi/2}^{\pi/2} \cos^3\theta(1+\sin\theta)^2\,d\theta$

$$=\int_{-\pi/2}^{\pi/2} (\cos^3\theta+2\cos^3\theta\sin\theta+\cos^3\theta\sin^2\theta)\,d\theta$$

$$=2\int_0^{\pi/2} (\cos^3\theta+2\cos^3\theta\sin^2\theta)\,d\theta$$

$$=2\left[\frac{2}{3\times1}+\frac{2\times1}{5\times3\times1}\right]=\frac{8}{5}.$$

Note: $\int_{-\pi/2}^{\pi/2} \cos^3\theta\sin\theta\,d\theta=0.$

Ex. 158. $\int_0^{\pi} x\sin^2 x\cos x\,dx$ (Integrate by parts)

$$=\left[\frac{x}{3}\sin^3 x\right]_0^{\pi}-\frac{1}{3}\int_0^{\pi}\sin^2 dx=-\frac{2}{3}\int_0^{\pi/2}\sin^3 x\,dx=-\frac{4}{9}$$

Ex. 159. $\int_0^1 \cot^{-1}(1-x+x^2)\,dx$ (Integrate by parts)

$$=\left[x\cot^{-1}(1-x+x^2)\right]_0^1+\int_0^1\frac{x(2x-1)}{1+(1-x+x^2)^2}dx$$

$$=\frac{\pi}{2}-\ln 2.$$

12.14 APPROXIMATE INTEGRATION

Sometimes it becomes necessary to find the approximate value of an integral. The approximation procedure is used in the following cases:

(*i*) when ordinary integration is difficult

(*ii*) when the function of the integrand is in the form of a table of numerical values

(*iii*) when the integral cannot be expanded as an elementary function

Some methods of approximate integration are given below.

If we divide the area under the curve $y = f(x)$ into n equal parts of width Δx we have the following rules.

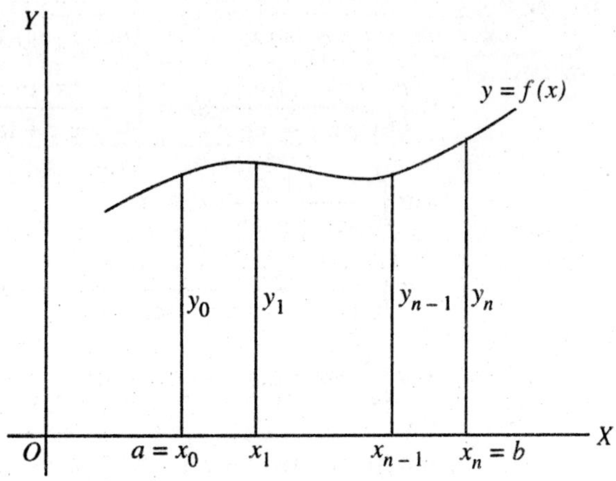

(*i*) The Trapezoidal rule:

$$\int_a^b f(x)\,dx \simeq \Delta x\left[\frac{y_0+y_n}{2}+\sum_{i=1}^{n-1} y_i\right]$$

(*ii*) Simpson rule

$$\int_a^b f(x)\,dx \simeq \frac{\Delta x}{3}(y_0+4y_1+2y_2+\ldots\ldots+2y_{n-2}+4y_{n-1}+y_n)$$

(*iii*) Taylor expansion

$$\int_a^b f(x)\,dx = \int_a^b\left(f(c)+\frac{x-a}{1!}f'(c)+\frac{(x-a)^2}{2!}f''(c)+\frac{(x-a)^3}{3!}f'''(c)+\ldots\right)dx$$

where $a \le c \le b$ and the series converges to $f(x)$ for all x in $[a, b]$

If we write $h=\dfrac{b-a}{n}$, the Trapezoidal rule may be put alternatively as

$$\int_a^b f(x)\,dx \sim \frac{h}{2}\,[f(a)+2f(a+h)+2f(a+2h)+\ldots\ldots+2f(a+\overline{n-1}h)+f(b)]$$

The Simpson rule alternative is

$$\int_a^b f(x)\,dx \sim \frac{h}{3}[f(a)+4f(a+h)+2f(a+2h)+4f(a+3h)+\ldots$$
$$+2f(a+\overline{n-2}h)+4f(a+\overline{n-1}h)+f(b)]$$

Ex. 160. $\displaystyle\int_0^{0.5}\frac{dx}{1+x^2}=\tan^{-1}x\Big]_0^{0.5}=\tan^{-1}0.5$

Here $f(x) = 1/(1 + x^2)$ whose integral can be found. But suppose it cannot be found. Then we can use e.g. the Trapezoidal rule. Let $n = 5$, $h = \dfrac{0.5-0}{5} = 0.1$,

$a = 0$, $a + h = 0.1$, $a + 2h = 0.2$, $a + 3h = 0.3$, $a + 4h = 0.4$, $b = 0.5$

$$\int_0^{0.5}\frac{dx}{1+x^2}=\frac{0.1}{2}\,[f(0)+2f(.1)+2f(.2)+2f(.3)+2f(.4)+f(.5)]$$

$$=\frac{0.1}{2}\left[1+\frac{2}{1.01}+\frac{2}{1.04}+\frac{2}{1.09}+\frac{2}{1.16}+\frac{1}{1.25}\right]=0.463.$$

Note: $\dfrac{1}{1+x^2}=\dfrac{1}{1+(.1)^2}=\dfrac{1}{1.01}$ for $x=.1$; $\dfrac{1}{1+.2^2}=\dfrac{1}{1.04}$ for $x=2$ etc.

Alt using Simpson's rule:

$$\int_0^{0.5}\frac{dx}{1+x^2}=\frac{0.1}{3}[f(0)+4f(.1)+2f(.2)+4f(.3)+2f(.4)+f(.5)]=0.437$$

Ex. 161. $\displaystyle\int_0^{0.5} e^{-x^2}dx=0.747$. Take $h=\dfrac{1}{4}$, $a=0$ by Simpson rule.

Ex. 162. $\displaystyle\int_0^1\sqrt{1+x^3}=1.115$. Take $n = 5$ with the Trapezoidal rule.

12.15 LIMITS OF INTEGRATION INFINITY

If $f(x)$ is continuous over $a \le x \le b$ we define $\int_a^\infty f(x)\,dx = \lim_{b\to\infty} \int_a^b f(x)\,dx$ provided the limit exists.

$$\text{Similarly} \quad \int_{-\infty}^{b} f(x)\,dx = \lim_{a\to-\infty} \int_a^b f(x)\,dx$$

$$\text{and} \quad \int_{-\infty}^{\infty} f(x)\,dx = \lim_{a\to-\infty} \int_a^{a_1} f(x)\,dx + \lim_{b\to\infty} \int_{a_1}^{b} f(x)\,dx.$$

If $f(x)$ has one or more points of discontinuity over $a \le x \le b$ OR at least one of the limits of integration is ∞ as in the above cases, we have an improper integral.

Ex. 163. $\int_0^\infty e^{-x} \sin x\,dx = \left.\frac{-e^{-x}}{2}(\sin x + \cos x)\right]_0^\infty$

This may be evaluated rigorously by writing

$$\lim_{b\to\infty} \frac{-e^{-x}}{2}[\sin x + \cos x]_0^b = -\frac{1}{2}(0-1) = \frac{1}{2}$$

or directly without going through this step, in practical cases.

Note: $e^{-\infty} = 0$, $\sin \infty$ or $\cos \infty$ lie between ± 1 and $e^0 = 1$

Ex. 164. (*i*) $\int_{-\infty}^{0} (4-x)^{-2}\,dx = \frac{1}{4}$; (*ii*) $\int_2^\infty \frac{dx}{x(\ln x)^2} = \frac{1}{\ln 2}$

12.16 CONSUMER SURPLUS (CS)

CS is the difference between the price that a consumer is willing to pay for a commodity and the actual price he pays for the commodity. The degree of satisfaction derived from a commodity is a subjective matter.

If DD_1 is the market demand curve then demand x_0 corresponds to the price p_0. The consumer surplus is

$$DD_1P_0 = \text{Area } DD_1x_0O - P_0D_1x_0O$$

$$= \int_0^{x_0} f(x)\,dx - p_0x_0$$

where $f(x)$ stands for the demand function.

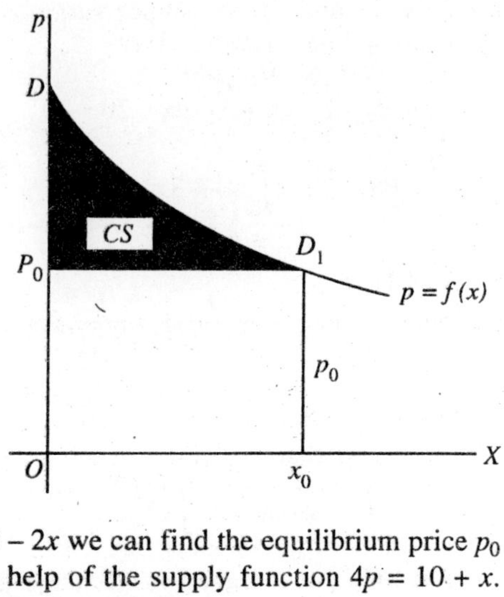

We have assumed that the area is defined when $x = 0$ and that satisfaction is measurable in terms of price for all the consumers. In other words, we assume that the utility function is the same for all the consumers and marginal utility of money is constant.

Ex. 165. Given the demand function $p = 25 - 2x$ we can find the equilibrium price p_0 and equilibrium quantity demanded x_0 with the help of the supply function $4p = 10 + x$. Thus $p_0 = 5$, $x_0 = 10$. (Solve the two equations simultaneously)

$$\text{Consumer surplus} = \int_0^{10} p\,dx - p_0x_0 = \int_0^{10}(25-2x)\,dx - 5\times 10$$

$$= [25x - x^2]_0^{10} - 50 = 100.$$

12.17 PRODUCER SURPLUS

If SS_1 is the market supply curve and if x_0 is the supply at market price p_0, the producer surplus PS is the area

$$SS_1P_0 = p_0x_0 - \int_0^{x_0} g(x)\,dx$$

where the supply function is given by

$$p = g(x)$$

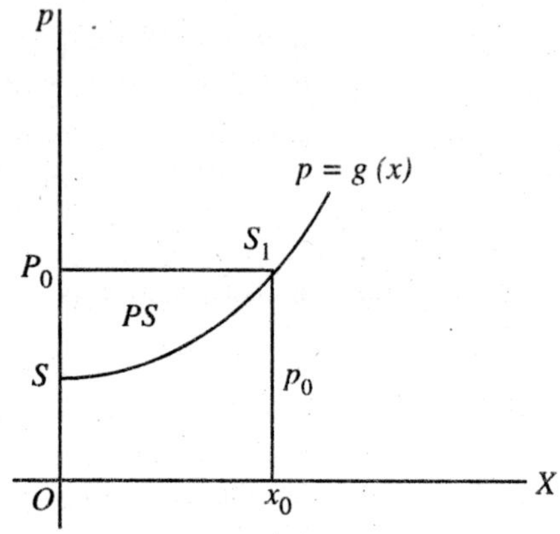

Ex. 166. Find the producer surplus for the supply function $p^2 - x = 9$ when $x_0 = 7$.

Ans. $3\frac{1}{3}$

Ex. 167. Show that if the demand and supply functions are given by $p = 36 - x^2$ and $4p - 6 = x^2$ the consumer and producer surpluses are $32\sqrt{6}$ and $8\sqrt{6}$.

Ex. 168. Given the demand function $p = 20 - 5x$ and the supply function $p = 4 + 3x$ find the consumer surplus (CS) and producer surplus (PS)

$$CS = \int_0^2 (20-5x)\,dx - 20 = 10$$

$$PS = 20 - \int_0^2 (4+3x)\,dx = 6.$$

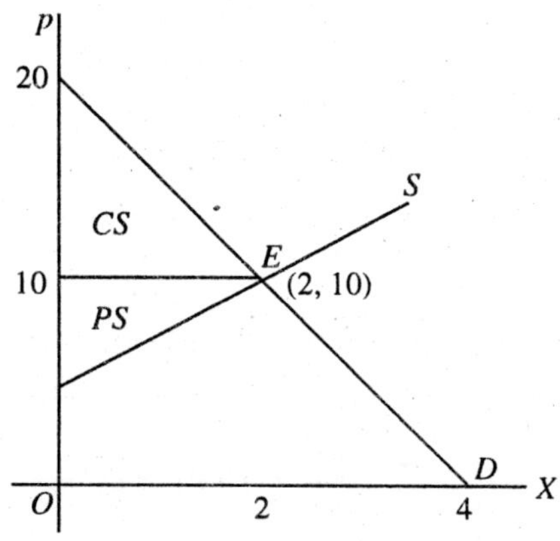

12.18 COST

The MC function is the derivative of the cost function:

$$\text{Marginal cost} = \frac{dC}{dx} = C', \text{ say}$$

If follows that the total cost function is the integral of the MC function.

$$\therefore \qquad \text{Cost } C = \int C'\,dx.$$

Ex. 169. The marginal cost function for a certain product is $5 + x^2$. Find the total cost and average cost functions if the fixed cost is 50.

$$\frac{dC}{dx} = 5 + x^2$$

$\therefore$ Cost $= \int (5 + x^2)\,dx \quad \therefore \quad C(x) = 5x + \frac{x^3}{3} + F$ where F is a constant.

For $x = 0$, $C(0) = 50$, given

$\therefore$ $50 = 5 \times 0 + 0 + F$ or $F = 50$.

Hence the cost function is given by

$$C = 5x + \frac{x^3}{3} + 50$$

The average cost function is given by

$$AC = \frac{C}{x} = 5 + \frac{x^2}{3} + \frac{50}{x}$$

Ex. 170. If the MC function is $3 + 2e^x$

$$C = \int (3 + 2e^x)\,dx = 3x + 2e^x + F \text{ where } F \text{ is a constant.}$$

If the fixed cost is 60 then $C = 3x + 2e^x + 60$, $AC = 3 + \frac{2}{x}e^x + \frac{60}{x}$.

Revenue

Marginal Revenue $= \frac{dR}{dx} = R'$ say

$\therefore$ Total Revenue $R = \int R'\,dx$

12.19 PROPENSITY TO CONSUME AND THE CONSUMPTION FUNCTION

If we know $b = mpc = \frac{dC}{dY}$ we can find the consumption function by writing (*Note* : In macroeconomics C stands for consumption, mpc = marginal propensity to consume)

$$\frac{dC}{dY} = f'(Y) \text{ or } dC = f'(Y)\,dY$$

$$\therefore \quad C = \int f'(Y)\,dY = f(Y) + C_1 \text{ where } C_1 \text{ is a constant.}$$

Since $s = mps = \frac{dS}{dY} = 1 - \frac{dC}{dY}$ (marginal propensity to save)

we can find the savings function similarly.

Ex. 171. If $mps = 2/5$ find the consumption function, given that consumption is 100 when income is zero.

$$mpc = 1 - \frac{2}{5} = \frac{3}{5}$$

$$C = \int \frac{3}{5}\,dY = \frac{3Y}{5} + C_1.$$

Using the given condition $100 = \frac{3}{5} \times 0 + C_1$

$\therefore \quad C_1 = 100$

The consumption function is $C = \frac{3Y}{5} + 100$.

Ex. 172. Find the consumption function if $mpc = \frac{1}{2} + \frac{1}{4\sqrt{Y}}$, given that consumption is 1000 when income is zero. **Ans.** $C = 1/2\ Y + 1/2\ \sqrt{Y} + 1000$

Ex. 173. Find the demand law given the marginal revenue function $MR = \frac{ab}{(x+b)^2} - \alpha$

Since $$\frac{dR}{dx} = \frac{ab}{(x+b)^2} - \alpha$$

$$R = \int dR = \int \left[\frac{ab}{(x+b)^2} - \alpha \right] dx$$

$$\therefore \quad R = -\frac{ab}{x+b} - \alpha x + C_1$$

When $x = 0$, $R = 0$

$$0 = \frac{-ab}{b} - \alpha \cdot 0 + C_1 \text{ or } C_1 = +a$$

$$R = -\frac{ab}{x+b} - \alpha x + a = \left(\frac{a}{x+b} - \alpha \right) x$$

$$\therefore \quad \frac{R}{x} = p = \frac{a}{x+b} - \alpha \text{ is the demand law}$$

Ex. 174. Given the marginal revenue function $MR = \frac{1}{2} x^{-1/2}$, find the demand function

$$R = \int \frac{1}{2} x^{-1/2} dx = x^{1/2} + C_1$$

when $x = 0$, $R = 0$ $\therefore$ $C_1 = 0$

$$\therefore \quad R = x^{1/2}$$

$$p = \frac{R}{x} = x^{-1/2} \text{ is the demand law}$$

Ex. 175. Given the MR function, $MR = \frac{6}{(x+2)^2} + 5$ find the demand function

Ans. $R = 5x - \frac{6}{x+2} + 3,\ p = \frac{3}{x+2} + 5$

Ex. 176. Find the area under $y = -\dfrac{x^3}{2}$ between 2.5 and 3.5

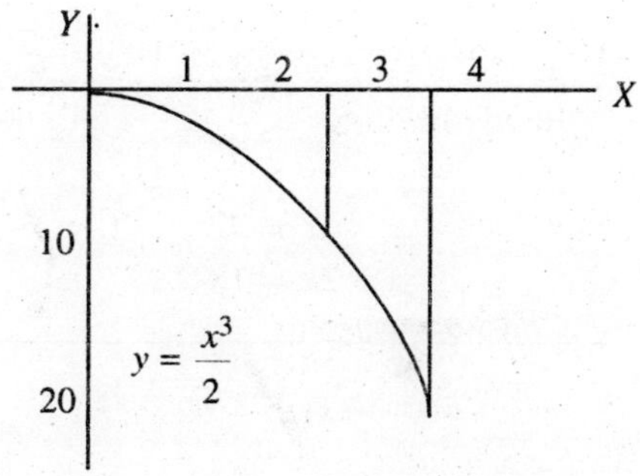

$$A = \int_{2.5}^{3.5} -\frac{x^3}{2}\,dx = \left.\frac{x^4}{8}\right]_{2.5}^{3.5}$$

$$= 13.875 \text{ (positive value).}$$

Ex. 177. A firm's daily expenditure E on a salesman is Rs. 1000 and the daily revenue function is given by $R(t) = 2000 - 10t^2$ where t stands for the number of days. Determine how long can the salesmen continue to maximise the net proceeds.

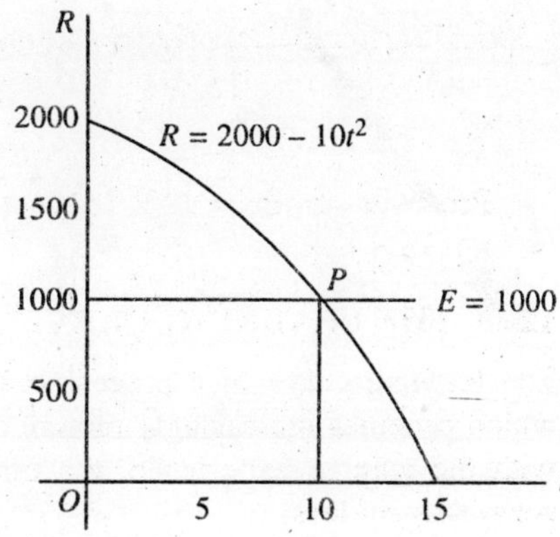

Net proceeds will be positive till $R = 1000$

$$2000 - 10t^2 = 1000$$

$$t = 10 \text{ days}$$

After 10 days expenditure will be more than revenue.

In 10 days, expenditure = $1000 \times 10 = 10000$

Total Revenue = Area between 0 and 10 under the revenue curve

$$= \int_0^{10} (2000 - 10t^2)\,dt$$

$$= \left[2000t - \frac{10t^3}{3}\right]_0^{10} = 16667$$

Net Revenue = 16667 – 10000 = 6667.

Ex. 178. Find the consumer and producer surplus in the sale of digital watches with the help of the following data.

Market price	*Actual Sales (units)*	*Available for sale*
400	20	200
100	120	50

Sales may be taken as demand

$$\text{Slope of demand curve} = \frac{400-100}{20-120} = -3$$

Demand curve $\quad p - 400 = -3(x - 20)$

or $\quad p = 460 - 3x$

Supply curve $\quad p - 400 = \dfrac{400-100}{200-50}(x - 200)$

$\therefore \quad p - 400 = 2(x - 200)$

or $\quad p = 2x$

In equilibrium

$$460 - 3x = 2x \quad (D = S)$$

$$\therefore \quad x = 92$$

$$p = 2x = 184.$$

Consumer surplus

$$= \int_0^{92} (460 - 3x)\,dx - 92 \times 184$$

$$= \left[460x - \frac{3x^2}{2}\right]_0^{92} - 16928 = 12696$$

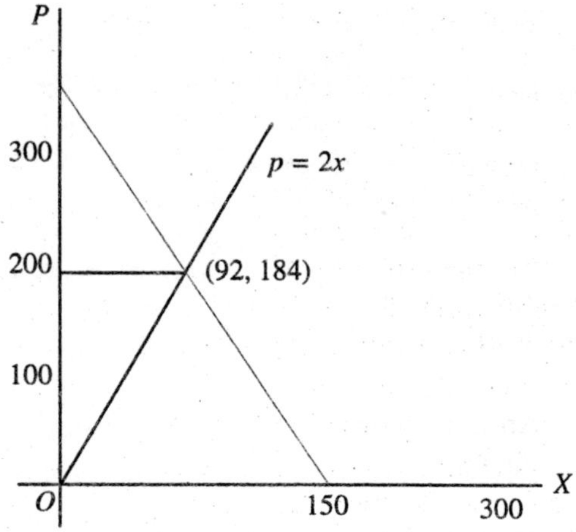

$$\text{Producer surplus} = 92 \times 184 - \int_0^{92} 2x\,dx = 16928 - x^2\Big]_0^{92} = 8464.$$

12.20 THE LEARNING CURVE

The learning curve of a process helps us in predicting the cost and time of a process which becomes increasingly mature over time and reaches a steady state. This is because with the gain in experience, the time needed per unit of production in later stages is progressively less.

Let y stand for average time over the life of the product and x the number of units produced. Then the learning curve can be written

$$y = Ax^{-\beta} \qquad A > 0$$

To procure the first unit *i.e.* for $x = 1$, the time required is $y = A$. We define

$$\beta = \frac{-\log\text{ (learning rate)}}{\log 2}$$

The average time per unit falls to a fixed percentage of the previous time.

Ex. 179. The first batch of 10 dolls is produced in 30 hours. Determine the time taken to produce the next 10 dolls and again 20 dolls using a 60 per cent learning curve. Estimate the time taken to produce the first doll.

Production of dolls	*Total hours*	*Increase*	*Average time*
10	*30*	*30*	*3*
20	$20\left(\frac{30 \times 60}{100}\right) = 36$	6	$3 \times \frac{60}{100} = 1.8$
40	$20\left(\frac{36 \times 60}{100}\right) = 43.2$	7.2	$1.8 \times \frac{60}{100} = 1.08$

To produce 20 dolls after the first 10 dolls takes 36 hours or an extra 6 hours.

To produce the next 20 dolls it takes an extra 7.2 hours. Learning rate $= \frac{1.08}{1.8} \times 100$ $= 60\% = 0.6$

From the learning curve $y = \alpha x^{-\beta}$

we have $\beta = \dfrac{-\log \text{(learning rate)}}{\log 2} = \dfrac{-\log 0.6}{\log 2} = 0.7369$

(*i*) when $x = 10$, $y = 3$ $\quad\therefore\quad 3 = \alpha \cdot 10^{-0.7369}$

$A = 16.38$ hours, the time required to produce the first doll.

Ex. 180. Because of learning experience, there is a reduction in labour requirement of a firm. Usually, we write the number of hours of labour required to produce the *x*th unit as (writing α for $-\beta$ in the learning curve)

$$f(x) = Ax^{\alpha} \quad A > 0, -1 \le \alpha < 0$$

After producing 36 units, a firm has a learning curve $f(x) = 1000x^{-1/2}$.

Find the labour hours H required to produce the next 28 units.

$$H = \int_{36}^{64} 1000x^{-1/2}\, dx$$

$$= 1000\,(2x^{1/2})\Big]_{36}^{64} = 2000x^{1/2}\Big]_{36}^{64}$$

$$= 2000\,(8 - 6) = 4000 \text{ hours.}$$

Ex. 181. A firm's monthly sales are Rs. 50000 which are expected to increase by 20 per cent per month. What will be the monthly sales in 10 months? (This is a simple growth problem)

$$\text{In 10 months, sales} = \int_0^{10} 50000\, e^{0.1t}\, dt$$

$$= 50000 \times \frac{1}{0.1} e^{0.1t}\Big]_0^{10}$$

$$= 500000\,[e - 1] = 500000 \times 1.35964$$

$$= \text{Rs. } 679{,}820.$$

Ex. 182. Rs. 1000 are deposited each year in a bank at 5 per cent per annum. Find the amount after 5 years.

$$\textbf{Ans. } A = \int_0^5 1000\, e^{0.05t}\, dt = 1000 \frac{e^{0.05t}}{0.05}\Big]_0^5 = 20000\,[e^{.25} - 1]$$

Ex. 183. A firm's learning curve after producing 100 units is $f(x) = 2400x^{-0.5}$ which is the rate of labour hours required to produce the *x*th unit. Find the hours needed to produce an additional 800 units.

$$\textbf{Ans. } \int_{100}^{900} f(x)\, dx = \int_{100}^{900} 2400x^{-0.5}\, dx = 2400 \frac{x^{0.5}}{0.5}\Big]_{100}^{900}$$

$$= 4800\,(30 - 10) = 96000$$

12.21 GINI COEFFICIENT OR LORENZ CURVE

The Lorenz curve is used to show the inequality of distribution of income, wealth etc. The cumulative percentages of a variable like incomes of individuals, are plotted against cumulative percentages of numbers of individuals. If these two percentages are equal everywhere (for instance, the lowest 40 per cent of individuals have 40 per cent of income and so on), the graph will be a 45° line. Otherwise we have the Lorenz curve OAO' which shows inequalities of income distribution. The degree of inequality is measured by the Gini Coefficient.

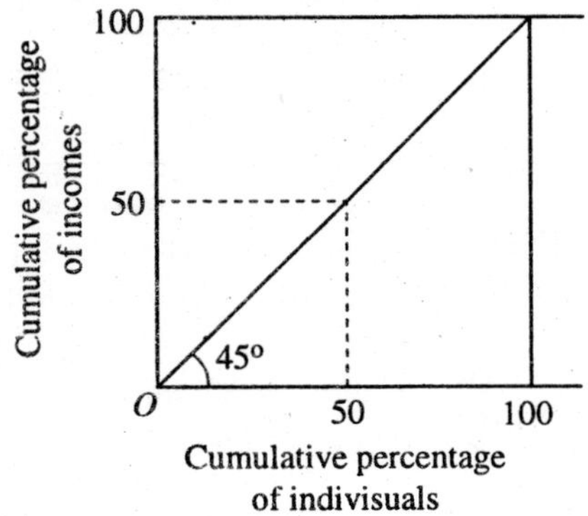

$$G = \frac{p}{p+q}$$

where p is the shaded area covered by the Lorenz curve and q is the remaining area on the right of OO'.

If the total area of the square is 1 then $p + q$ is half that area *i.e.*, $p + q = \frac{1}{2}$ so that

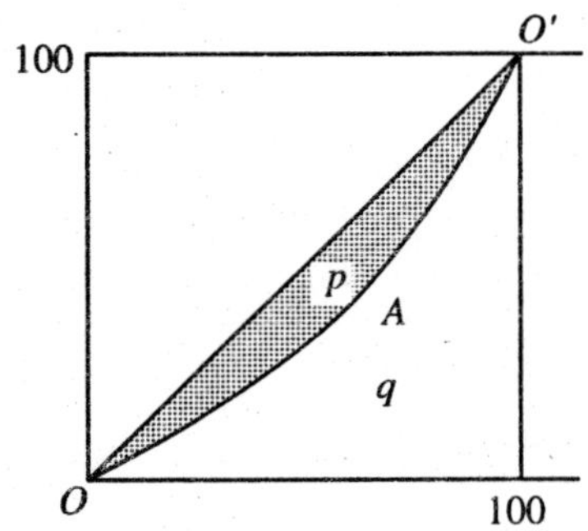

$$G = \frac{p}{p+q} = \frac{p+q-q}{p+q} = 1 - \frac{q}{p+q} = 1 - 2q.$$

If x stands for the cumulative percentage of individuals, the Lorenz function $f(x)$ can be used to find the Gini Coefficient G by using definite integrals.

The area of the square may also be written as $100 \times 100 = 10000$ so that we will have

$$G = \frac{\text{area of shaded region}}{(1/2)\text{ area of square}} = \frac{p}{5000}$$

$$p = 5000 - \int_0^{100} f(x)\,dx.$$

Ex. 184. For the Lorenz function $f(x) = \frac{x^2}{200} + \frac{x}{2}$ what is the percentage of income held by the highest cumulative 10% individuals? This is given by

$$f(100) - f(90) = 100 - 85.5 = 14.5\%$$

To find G first find the shaded region

$$p = 5000 - \int_0^{100} \left(\frac{x^2}{200} + \frac{x}{2}\right) dx$$

$$= 5000 - \frac{12500}{3} = \frac{2500}{3}$$

$$\therefore \quad G = \frac{2500/3}{5000} = \frac{1}{6}.$$

If the Lorenz curve is expressed as a function $f(x)$ of x, the cumulative percentage of individuals, the Gini Coefficient can be written as a definite integral

$$G = 1 - 2q = 1 - 2\int_0^{100} f(x)\,dx$$

Let $f(x) = \frac{x^2}{200} + \frac{x}{2}$. If $x = 100\%$ of individuals, $f(x) = \frac{100^2}{200} + \frac{100}{2} = 100.$

If $x = 90\%$, $f(x) = \frac{90^2}{200} + \frac{90}{2} = 85.5$

We can find the percentage of income held by the highest cumulative 10% individuals. This is found by subtracting. Thus

$$f(100) - f(90) = 100 - 85.5 = 14.5\%,$$

Area of the square = $100 \times 100 = 10000$.

Area of Lorenz curve is given by p where

$$2p = 10000 - 2q = 10000 - 2\int_0^{100}\left(\frac{x^2}{200} + \frac{x}{2}\right)dx$$

$$= 10000 - \frac{25000}{3}$$

$$= \frac{5000}{3} \text{ or } p = \frac{2500}{3}$$

$$G = \frac{p}{p+q} = \frac{2500/3}{5000} = \frac{1}{6}.$$

Ex. 185. Let the equation of the Lorenz curve drawn in a unit square be given by

$$y = \alpha x^2 + \beta x \qquad 0 \le x \le 1;\ \alpha, \beta < 1$$

∴ Area under the curve between (0, 1) is

$$\int_0^1 (\alpha x^2 + \beta x)\,dx = \left[\frac{\alpha x^3}{3} + \beta\frac{x^2}{2}\right]_0^1 = \frac{\alpha}{3} + \frac{\beta}{2}$$

Since the half area on one side of the diagonal is $\frac{1}{2}$, the area between the diagonal and the curve is $p = \frac{1}{2} - \left(\frac{\alpha}{3} + \frac{\beta}{2}\right)$

$$\text{Gini Coefficient} = \frac{\text{Shaded Area } p}{\text{Half Area}}$$

$$= \frac{1/2 - \alpha/3 - \beta/2}{1/2}$$

Ex. 186. Let $y = Nx^{-\mu}$ stand for the number of people earning an income more than or equal to x. It is believed that $\mu \simeq 1.5$ and N is a constant. Find the number of individuals with incomes between (*a*) 400 and 1600 (*b*) above 1600, given $N = 25000$.

This problem pertains to the Pareto distribution (see below)

$$y = Nx^{-\mu}$$

The number of people with income between x_0 and x_i is given by $y = \int_{x_0}^{x_i} Nx^{-\mu}\,dx$

$$\therefore \qquad y = \left.\frac{Nx^{1-\mu}}{1-\mu}\right]_{x_0}^{x_1} = \frac{N}{1-\mu}[x_1^{1-\mu} - x_0^{1-\mu}].$$

If $x_0 = 400$, $x_1 = 1600$, $\mu = 1.5$,

$$y = \frac{25000}{1-1.5}[1600^{-0.5} - 400^{-0.5}]$$

$$= -\frac{25000}{0.5}\left(\frac{1}{40} - \frac{1}{20}\right) = \frac{25000}{0.5}\left(\frac{1}{20} - \frac{1}{40}\right) = 2500.$$

For incomes above 1600, $y = \int_{1600}^{\infty} 25000x^{-1.5}dx$

$$\therefore \qquad y = 50000\,(1/40 - 0) = 1250.$$

12.22 PARETO INCOME DISTRIBUTION

In a population of size N, the number of individuals with incomes exceeding x is given by $y = Nx^{-\mu}$ where μ is a parameter which is nearly equal to 1.5.

The *number of individuals* having incomes between a and b is

$$N_1 = \int_a^b Nx^{-\mu}dx = \left.\frac{N}{-\mu+1}x^{-\mu+1}\right]_a^b = \frac{N}{1-\mu}[b^{1-\mu} - a^{1-\mu}] = 2N\left(\frac{1}{\sqrt{b}} - \frac{1}{\sqrt{a}}\right)$$

if $\mu = 1.5$.

The *total income exceeding* m is given by

$$\int_m^{\infty} x\mu\, Nx^{-\mu-1}dx = \int_m^{\infty} \mu\, Nx^{-\mu}dx = \frac{\mu N}{\mu - 1}m^{1-\mu}.$$

The average income cover m is $\dfrac{\mu}{\mu-1}Nm^{1-\mu}/Nm^{-\mu} = \dfrac{\mu m}{\mu - 1}$.

Ex. 187. What is the number of individuals falling in the income range (*i*) (C_1, C_2), (*ii*) (10, 200) (*iii*) above λ ?

Ans. (*i*) $\dfrac{N}{1-\mu}(C_2^{1-\mu} - C_1^{1-\eta})$ (*ii*) $\left[\dfrac{Nx^{1-\mu}}{1-\mu}\right]_{100}^{200} = \dfrac{N}{1-\mu}(200^{1-\mu} - 100^{1-\mu})$

(*iii*) The number with incomes above λ is given by

$$N_2 = \int_{\lambda}^{\infty} Nx^{-\mu}\,dx = \frac{2N}{\sqrt{\lambda}}$$

If $\mu = 1 + \beta$, this becomes $\frac{N}{\beta}\lambda^{-\beta}$.

Ex. 188. If $N = 5000\,x^{-1.5}$ the number with incomes (*i*) between 2000 and 5000 is 82 (*ii*) above 5000 is 141.

Ex. 189. Show that the Gini Coefficient is 0.53 if the Lorenz curve is given by $11y = 6x^2 + x$.

Ex. 190. Given $f(x) = \mu b^{\mu} x^{-\mu-1}$ $(b \le x \le \infty)$. Show that $\int_b^{\infty} f(x)\,dx = 1$.

Hint: $$\mu b^{\mu}\left.\frac{x^{-\mu}}{\mu}\right]_b^{\infty} = 0 - \frac{\mu b^{\mu} b^{-\mu}}{-\mu} = 1.$$

12.23 CAPITAL FORMATION

Let $K(t)$ stand for *stock* of capital K at a given time t. The net investment $I(t)$ is a flow concept per time period. We can write $\frac{dK}{dt} = I(t)$.

Given the investment function $I(t)$, capital accumulation over the time interval (t_1, t_2) is given by

$$\int_{t_1}^{t_2} I(t)\,dt = \int_{t_1}^{t_2} dK(t)$$
$$= K(t_2) - K(t_1).$$

Ex. 191. A firm has a current stock of machines worth Rs. 10 million and over the next 2 years, the investment function will be $I(t) = 5t + 0.3t^2 - 0.002t^3$. The total machine stock value after 2 years would be

$$K = 10 + \int_0^2 (5t + 0.3t^2 - 0.002t^3)\,dt$$

$$= 10 + \left[\frac{5t^2}{2} + \frac{0.3t^3}{3} - \frac{0.002t^4}{4}\right]_0^2$$

$$= 10 + [10 + .8 - .008]$$
$$= 20.792.$$

12.24 CAPITAL AND INVESTMENT

If the total capital invested at time t is K and that the requirement of capital at the equilibrium level is K_1 *i.e.*, capital is withdrawn if above K_1 and invested if below K_1 then the rate of investment is proportional to $K - K_1$ *i.e.,*

$$\frac{dK}{dt} = C(K - K_1) \text{ where } C \text{ is a constant.}$$

We may write $\dfrac{dK}{K - K_1} = C\,dt$ or $dK = C(K - K_1)\,dt$

Integrating, $\ln(K - K_1) = Ct + \ln C_1$ where $\ln C_1$ is a constant

$$K - K_1 = C_1 e^{Ct}$$

$$\therefore \qquad K = K_1 + C_1 e^{Ct}$$

Suppose the initial condition is that the investment at time $t = 0$ is $K_1/3$, then $K_1 + C_1 e^0 = K_1/3$ so that $C_1 = -\dfrac{2K_1}{3}$ and the solution is

$$K = K_1 - \frac{2}{3} K_1 e^{Ct}$$

Continuous Compounding

At the compound interest rate r an initial amount P_0 after t years becomes

$$P_t = P_0 (1 + r)^t$$

If compounding takes place m times per year,

$$P_t = P_0 \left(1 + \frac{r}{m}\right)^{mt}$$

If compounding is continuous *i.e.,* as $m \to \infty$, $P_t = P_0 e^{rt}$

The rate of increase of P_t over time is $dP_t/dt = P_0 r e^{rt} = r P_t$ *i.e.,* P_t increases at a rate proportional to P_t.

Conversely, given $\dfrac{dP_t}{dt} = rP_t$ *i.e.,* $\dfrac{dP_t}{P_t} = r\,dt$ or $\ln P_t = rt + \ln C$

$$\therefore \qquad P_t = Ce^{rt} = P_0 e^{rt} \text{ if the initial capital is } P_0$$

Also $P_0 = P_t e^{-\int_0^t r(t)\,dt}$

The present value at any time T can be found by using

$$P_T = P_t e^{-\int_T^t r(t)\,dt}$$

If $y_0, y_1, \ldots$ are expected yields of capital at time 0, 1,... the present value (PV) of the capital asset is

$$PV = \int_0^t y_t e^{-rt}\,dt$$

$$= \frac{y}{r}(1 - e^{-rt}) \text{ if } y_t = y = \text{constant}$$

$$= \frac{y}{r} \text{ as } t \to \infty.$$

This last case *i.e.,* $PV = \int_0^\infty y e^{-rt}\,dt = \dfrac{y}{r}$ is called capitalisation.

Present Value of Annuity

The present value of an annuity can be found by adding the present value of each of the individual annuity payment. If the amount paid at the end of each period is R the present value at the end of

period 1 is $\frac{R}{1+i}$ (i is discount rate)

period 2 is $\frac{R}{(1+i)^2}$

....

period n is $\frac{R}{(1+i)^n}$

The present value PV of the annuity is

$$PV = \frac{R}{1+i} + \frac{R}{(1+i)^2} + \ldots\ldots + \frac{R}{(1+i)^n}$$

This is a GP with common ratio $\frac{1}{1+i}$

$$\therefore \qquad PV = \frac{R}{1+i}\left[\frac{1-1/(1+i)^n}{1-1/(1+i)}\right] = R\,\frac{1-(1+i)^{-n}}{i}$$

If the annuity is for property, $n \to \infty$ and $PV = \frac{R}{i}$ (since $(1+i)^{-n} \to 0$).

If the amount paid at the end of each period is R_t

$$PV = \sum_{t=1}^{n} \frac{R_t}{(1+i)^t}$$

Net Present Value (NPV)

The NPV of an investment proposal with cash inflows R_t at the end of period t and initial investment y is

$$NPV = \sum_{t=1}^{n} \frac{R_t}{(1+i)^t} - y$$

where i is the discount rate.

IRR

The internal rate of return (IRR) of an investment is the discount rate which equates the PV of expected cash inflows to the initial investment y. It is the discount rate for which NPV is zero.

$$\sum_{t=1}^{n} \frac{R_t}{(1+i)^t} = y$$

By solving this equation we get i, the IRR which can be compared with the minimum acceptable return r. If $i > r$, the project is accepted. If $i < r$, the project is rejected.

It the equation is complicated, the trial error technique can help find i. Ready reckoner tables are also available.

If the income stream is assumed to be continuous and $n \to \infty$ we can write $R_t = f(t)$ and the capital value is

$$y_c = \int_0^T f(t)e^{-it}dt$$

If the income stream is constant at R

$$y_c = R\int_0^T e^{-it}dt = \frac{R}{i}(1 - e^{-iT})$$

If $R = 1000$, $i = 0.05$, $n = 2$, $y = \dfrac{1000}{0.05}(1 - e^{-0.1}) = 1904.$

The capital value y depends on R, i, T.

If amounts R_0, R_1, R_2,......R_n are invested is successive years, the total accumulation at the end of nth period is

$$y = \sum_{t=0}^{n} R_t(1+i)^{n-t}$$

or with continuous investment.

$$y_A = \int_0^T f(t)e^{i(T-t)}\,dt$$

If $f(t) = R$, a constant, $y = R\displaystyle\int_0^T e^{i(T-t)}\,dt = \frac{R}{i}(e^{iT} - 1).$

Since $y_A = e^{iT}\,y_c$ it follows that after T years, the accumulated value of the constant income stream is the present value itself accumulated at interest rate i.

12.25 INVERSE RETAIN BETWEEN CAPITAL AND INTEREST

If the production function is $x = f(L, K, t)$ there is a constant expenditure of $wL + rK$ per unit of time and the product x is realised after t years. The total accumulated investment is

$$K_A = \frac{wL + rK}{i}(e^{it} - 1) = \frac{(wL + rK)e^{it}}{i}(1 - e^{-it}) \text{ (using sec. 12.24)}$$

Under perfect competition $x = (wL + rK)\,e^{it}$

$$\therefore \qquad K_A = \frac{x}{i}(1 - e^{-it}).$$

Capital required varies inversely as the market interest rate.

Durable Goods. Assuming a constant stream of R per year from the durable good, the discounted value of all future product is

$$R\int_0^t e^{-it}\,dt = \frac{R}{i}(1 - e^{-it})$$

If we maximise the ratio of this to capital value of cost $f(t)$ we have

$$y = \frac{R}{i}(1 - e^{-it})/\,f(t)$$

$$\frac{dy}{dt} = \frac{R}{if^2}[ie^{-it}f - (1 - e^{it})f']$$

$$= \frac{Rf' e^{-it}}{if^2}\left[1 + e^{it} + i\frac{f}{f'}\right] = 0$$

i.e., $e^{it} = 1 + if/f'$ gives the optimum t.

Ex. 192. If $f(t) = C(1 - 1/(1 + t))$, the optimum value of t is given by

$$e^{it} = 1 + \frac{iCt/(1+t)}{C/(1+t)^2} = 1 + it + it^2$$

Ex. 193. If $f(t) = C\sqrt{t}$ show that $e^{it} = 1 + 2it$.

Ex. 194. If capital formation is regarded as a process continuous over time, the rate of capital formation or investment flow is

$$I(t) = \frac{dK}{dt}.$$

If it is known that $I(t) = 4t^{1/3}$ then the capital stock at time t is given by

$$\int dK = \int I(t)\,dt = \int 4t^{1/3}dt = 3t^{4/3} + K(0)$$

where K (0) is the initial amount of capital.

Show that if the capital flow is non-constant the capital stock at the end of the period t_0 is given by $\int_0^{t_0} 4t^{1/3}dt + K(0)$.

Ex. 195. If $I(t) = 1000e^{-0.5t}$ then capital formation in 2 years is

$$K(t) = \int_0^2 I(t)\,dt = \int_0^2 1000\,e^{0.5t}\,dt$$

$$= 1000\left.\frac{e^{0.5t}}{0.5}\right]_0^2 = \frac{1000}{0.5}\,[e^1 - e^0]$$

$$= 2000\,[2.7183 - 1] = 2000 \times 1.7183$$
$$= 3437.$$

Find in the above case K (t) from the remote past to the present:

$$K(t) = \int_{-\infty}^{0} I(t)\,dt = \frac{1000}{0.5}\,[e^{0.5t}]_{-\infty}^{0}$$
$$= 2000\,[e^0 - e^{-\infty}] = 2000.$$

Ex. 196. If $K(t)$ denotes an economy's (or a firm's) capital state at time t and $I(t)$ is investment net of depreciation then

$$\frac{dK}{dt} = I(t) - bK(t)$$

Show that this can be written as $\frac{d}{dt}[e^{bt}K\,b(t)] = e^{bt}I(t)$

If $I(t) = I_0$ for all t show that the solution is $K(t) = a\,e^{-bt} + I_0/b$

If $K(t)$ has exponential growth rate g show that $I(t) = (g + b)\,K(t)$.

12.26 POPULATION GROWTH

Suppose the human population is a function of the birth rate β and death rate δ. Suppose, further, that in a short period of time h, there occur βhN births and δhN deaths where $N = N(t)$ is assumed to be the population size as a continuous and differentiable function. We have

$$N(t+h) - N(t) = \beta hN - \delta hN$$

$$\therefore \quad \frac{N(t+h) - N(t)}{h} = \beta N - \delta N$$

hence
$$\frac{dN}{dt} = (\beta - \delta)N$$

or
$$\frac{dN}{N} = (\beta - \delta)dt$$

By integration $N = N_0 e^{(\alpha - \beta)t}$ where $N_0 = N(0)$.

If $\delta = \beta$, the population is constant.

If $\delta > \beta$, there is the possibility of extinction, *i.e.*, $N \to 0$.

If $\delta < \beta$, there is increase, slow at first, very fast later.

This is a simplified model. To get a better idea one must consider actual population size, age distribution and other factors.

Ex. 197. Suppose the population in a city was 5 million in 1961 and 10 million in 1971. What is the likely size of population in 1981?

$$N = N_0 e^{kt}$$

$$10 = 5e^{10k} \qquad \text{where } t = 10$$

$$\therefore \quad e^{10k} = 2$$

$$10k \log e = \log 2$$

$$10k \times 0.4343 = 0.3010$$

or
$$k = \frac{0.3010}{4.343} = 0.069$$

hence $N = 5e^{20k} = 5e^{1.38} = 20$ million app. (Alt. use natural logs)

Ex. 198. Let the population size be N at time t. If N_0, k are constants the exponential law of growth is

$$N = N_0 e^{kt}$$

If the population in 1950 ($t = 0$) was 100,000 and in 1976 ($t = 25$) was 271,828,

What is the expected population in 2000 ?

$$100{,}000 = N_0 e^0 \qquad \therefore \qquad N_0 = 100{,}000$$

$$271828 = N_0 e^{25k} \qquad \therefore \qquad \frac{271{,}828}{100{,}000} = e^{25k}$$

Since $e^{25k} = 2.71828$

$$\therefore \quad \ln e^{25k} = \ln 2.71828$$

$$\therefore \quad 25k = 1 \text{ or } k = \frac{1}{25}$$

In the year 2000, $t = 50$

$$N = 100{,}000\, e^{50/25} = 100{,}000\, e^2$$
$$\simeq 635{,}000 \text{ approx.}$$

12.27 LOGISTIC FUNCTION

Ex. 199. Because of environment factors like food supply, overcrowding etc., population size N is limited to a maximum number M and as $N \to M$, $dN/dt \to 0$. The corresponding population growth model is $\frac{dN}{dt} = kN\left(\frac{M-N}{M}\right)$. We will now get the logistic function.

(*Note* : If N is small, $\frac{M-N}{M} \simeq 1$ and as $N \to M$, $\frac{dN}{dt} \to 0$ since $M - N \to 0$)

If we write $k_1 = k/M$ then the equation becomes

$$\frac{dN}{dt} = k_1 N(M-N) \quad \therefore \quad \frac{dN}{N(M-N)} = k_1\, dt$$

Integrating $\int \frac{dN}{N(M-N)} = \int k_1 dt$ or $\ln \frac{N}{M-N} = Mk_1 t + Mk_2$

or $$\frac{N}{M-N} = e^{Mk_1 t + Mk_2} = (e^{Mk_2})e^{Mk_1 t} = Ae^{Mk_1 t}$$

$$\therefore \quad N = \frac{MAe^{Mk_1 t}}{Ae^{Mk_1 t} + 1} \quad \text{where } A = e^{Mk_2}.$$

The relation $N = \frac{MAe^{Mk_1 t}}{Ae^{Mk_1 t} + 1}$ can be written

$$N = \frac{M}{1 + \frac{1}{A}e^{-Mk_1 t}} = \frac{M}{1 + be^{-ct}} = \frac{1}{1 + bE^{-1}} \quad \text{with } e^c = E$$

where $b = \frac{1}{A}$, $c = Mk_1$ are constants.

This is the logistic function. The logistic curve shows that for $N = M$ there is a horizontal asymptote because

$$\lim_{t \to \infty} \frac{M}{1 + be^{-ct}} = \frac{M}{1 + b \times 0} = M$$

Since the growth rate is

$$G = \frac{dN}{dt} = kN(M-N)$$

the maximum growth rate occurs

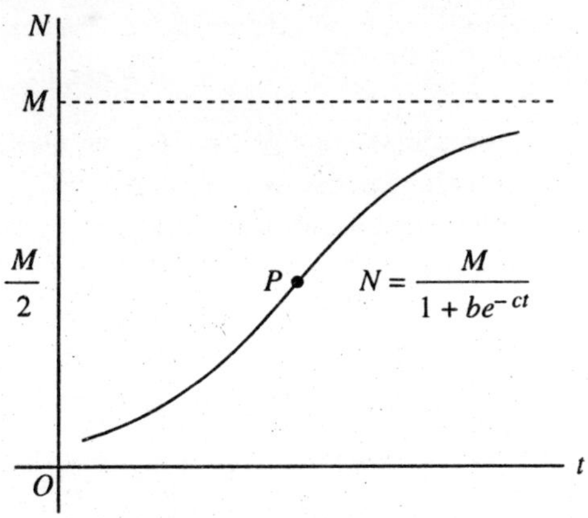

when $\frac{d}{dN}(kN(M-N)) = 0$ *i.e.*, $k(M - 2N) = 0$ or $N = \frac{M}{2}$.

Until the population size reaches $M/2$, the rate of growth increases. When the population size reaches $M/2$, the rate of growth decreases. $G_{max} = M/2$ corresponds to the point of uniform growth.

This occurs when $\frac{M}{2} = \frac{M}{1+be^{-ct}} \Rightarrow e^{ct} = b$ or $t = \frac{\ln b}{c}$.

MISCELLANEOUS

Ex. 200. The marginal revenue function of a product is given by $R' = 5 + 3x - x^2$. Find the total revenue and the average revenue, given that $R = 112$ when $x = 6$.

$$R = \int (5+3x-x^2)\,dx$$

$$= 5x + \frac{3}{2}x^2 - \frac{x^3}{3} + C$$

When $x = 6$ we have $112 = 5.6 + \frac{3}{2} \times 36 - \frac{6^3}{3} + C$

$\therefore \quad C = 100.$

Hence $\quad R = 5x + \frac{3}{2}x^2 - \frac{x^3}{3} + 100$

$$AR = 5 + \frac{3}{2}x - \frac{x^2}{3} + \frac{100}{x}.$$

Ex. 201. Find the cost function for a commodity whose MC function is given by $MC = 5 + 8x$. It is given that the fixed cost is 75. **Ans.** $75 + 5x + 4x^2$

Ex. 202 If for 2 units of production, total revenue is 200 find the total revenue function given that $MR = 10 + 20x - 3x^2$. **Ans.** $10x + 10x^2 - x^3 + 148$

Ex. 203. Let $MR = m$. Find the total revenue function when it is known that for zero production there is no revenue. Comment on the results.

$$R = \int m\,dx = mx + C \text{ where } C \text{ is a constant.}$$

Also $0 = 0 + C \therefore C = 0$ so that the revenue function is given by

$$R = mx \text{ and } AR = \frac{mx}{x} = m$$

We can see that $MR = AR$: the market is competitive.

Again, suppose $m = 0$; then $R = C_1$ where C_1 is a constant.

C_1 can be found if suitable conditions are given and then

$$AR = \frac{C_1}{x}.$$

Ex. 204. If the MR function for a product is given to be $MR = \frac{0.5}{x^{0.5}}$, what is the demand function of the commodity? **Ans.** $R = x^{1/2} + C_1$, $AR = x^{-\frac{1}{2}} + \frac{C_1}{x}$

Ex. 205. If $p_0 = 2$, $x_0 = 6$, find the consumer surplus, given the demand function $3p = 36 - 5x$. **Ans.** 30

Ex. 206. If the demand function for a commodity is $p = 50 - 2x - x^2$ find the consumer surplus if (*i*) $p_0 = 2$ (*ii*) $p_0 = 15$. **Ans.** 180, 108 1/3

Ex. 207. For a free good, $p_0 = 0$. Find the consumer surplus for this good if its demand function is given by $p = 64 - x^3$.

Ex. 208. The slope of a cost curve $C = f(x)$ passing through (2, 7) is $\frac{x}{2}$. Find the cost curve. **Ans.** $C = \frac{1}{4}x^2 + 6$

Ex. 209. Find the consumption function, given that when $Y = 125$, income equals consumption and that *mpc* is given by $0.5 + 0.1Y^{1/3}$.

Ans. $C = 0.5Y + 0.075Y^{4/3} + 15.6$

CHAPTER 13

Differential and Difference Equations

13.1 DIFFERENTIAL EQUATIONS

A differential equation is an equation which involves derivatives. It is an implicit functional relationship between variables and their differentials.

The *order* of a differential equation is the highest order of any of the derivatives contained in it. A differential equation is of order n if the nth derivative is the highest order derivative in it.

The highest power to which the derivative of the highest order occurs is the *degree* of a differential equation.

Ex. 1. $y = \frac{dy}{dx} + x^2 = x$ is a differential equation of order 1, degree 1

$\left(\frac{d^2y}{dx^2}\right)^3 - 6\left(\frac{dy}{dx}\right)^2 + xy$ is of order 2, degree 3.

A differential equation is said to be linear if the dependent variable is of the first degree.

$a_0 \frac{d^n y}{dx^n} + a_1 \frac{d^{n-1}y}{dx^{n-1}} + \ldots\ldots + a_n y = f(x)$ is a linear differential equation of order n.

If a_o, a_1, ..., a_n are constants and $f(x) = 0$ in the above equation we have a *linear homogeneous differential equation of order n* with constant coefficients.

Ordinary differential equations involve only one independent variable and its derivatives. When there are several independent variables we may have partial differential equations.

To solve a differential equation means to find all its solutions and in case it has no solution, to show it has no solution.

A solution of a differential equation is any function which satisfies the equation *i.e.*, reduces it to an identity when substituted into the equation.

A differential equation of order n has a *general solution* containing n arbitrary independent constants.

A *particular* solution may be obtained from the general solution by assigning to the arbitrary constants certain specific values which are obtained from any given information of the initial conditions given in the problem.

To solve differential equations we can recognise their different types and follow certain standard procedures some of which are considered below.

I. Direct Integration

To solve the differential equation of the type

$$\frac{dy}{dx} = f(x).$$

We can resort here to integration because, if $f(x)$ is the derivative of $F(x)$, we can write the above equation

$$dy = f(x)\, dx$$

$$\int dy = \int f(x)\, dx$$

$\therefore$ $y = F(x) + C.$ This is the required solution.

Ex. 2. $\frac{dy}{dx} = ay$ may be written $\frac{dy}{y} = dx$

Integrating $\quad ln\, y = x + c$

or $\quad y = e^{x+c}$ or $y = e^x \cdot e^c$

If we write $C_1 = e^c$ then $y = C_1 e^x$ is the solution

Ex. 3. $\frac{dy}{dx} = xe^x$ is equivalent to $dy = xe^x\, dx$ *i.e.*

$$y = \int xe^x dx = e^x (x - 1) + C, \text{ this is the general solution}$$

If the initial condition is that $y = 2$ when $x = 0$ then by substitution $2 = e^o (0 - 1) + C$ or $2 = -1 + C$

$\therefore$ $C = 3$

The particular solution is $y = e^x (x - 1) + 3$.

II. Separable Differential Equations

If M is a function of x and N is a function of y and we have a first order differential equation

$$M(x) + N(y)\frac{dy}{dx} = 0$$

This may be written

$$M(x)\, dx + N(y)\, dy = 0$$

The solution is obtained by writing

$$\int M(x)dx + \int N(y)dy = 0$$

Ex. 4. The equation $\frac{dy}{dx} = \frac{y}{x}$ may be written $\frac{dy}{y} = \frac{dx}{x}$

so that $\int \frac{dy}{y} = \int \frac{dx}{x}$

$\therefore$ ln y = ln x + ln C where ln C is an arbitrary constant.

This is the general solution.

If $y = 1$, when $x = 1$ then $C = 1$

Then the particular solution is $y = x$

Ex. 5. $dy/dx = x/y$ is a separable equation written $ydy = xdx$

$\therefore \int ydy - \int xdx = 0$ or $\frac{y^2}{2} - \frac{x^2}{2} = C$ where C is an arbitrary constant.

The solution is $\quad y^2 = 2C + x^2$

or $\quad y = \pm\sqrt{2C + x^2}$

Ex. 6. Solve the following separable differential equations.

(*i*) $dy/dx = 3xy$ (*ii*) $3x^2 + 2x - 3y\frac{dy}{dx} = 0$

(*iii*) $dy/dx = e^{x+y}$ (*iv*) $dy/dx = \frac{-y}{x}$

Ans. (*i*) $y = Ce^{3x^2/2}$ (*ii*) $y = \pm\sqrt{2/3(x^3 + x^2) + C}$

(*iii*) $e^x + e^{-y} = C$ (*iv*) $ln\ xy = C$

III. Exact Differential Equations

An equation of the form

$$M(x, y) + N(x, y)\frac{dy}{dx} = 0 \quad \text{where} \quad \frac{\partial M}{\partial y} = \frac{\partial N}{\partial x}$$

is called an exact differential equation

$\frac{\partial M}{\partial y} = \frac{\partial N}{\partial x}$ is a *necessary and sufficient* condition for a differential equation to be exact.

The solution of the exact differential equation can be shown to be

$$F(x, y) = \int Mdx + \int Ndy - \int\left(\frac{\partial}{\partial y}\int Mdx\right)dy$$

Ex. 7. $(xy^2 + y)dx = (x^2y + x)dx$. Here $M = xy^2 + y$, $N = x^2y + x$

$$F(x, y) = \int (xy^2 + y)\,dx + \int (x^2y + x)\,dy - \int\left[\frac{\partial}{\partial y}\int (xy^2 + y)\,dx\right]dy$$

$$= \frac{x^2y^2}{2} + xy + x^2y + xy - \int (x^2y + x)\,dy + C$$

$$= \frac{x^2y^2}{2} + xy + C$$

Ex. 8. Solve (*i*) $(x + 3y)dx + 3xdy = 0$ **Ans.** (*i*) $\frac{x^2}{2} + 3xy + C = 0$

(*ii*) $(2y + x^2)dy + (3x^2 + 2xy)dx$ (*ii*) $x^3 + x^2y + y^2 + C = 0$

(*iii*) $2y^3dx + 3xy^2\,dy = 0$ (*iii*) $x^2y^3 = C.$

IV. Homogeneous Differential Equations

If $M(x, y)$, $N(x, y)$ are homogeneous functions of degree n, then

$M(x, y)\,dx + N(x, y)\,dy = 0$ is a homogeneous differential equation of degree n.

A homogeneous equation is solved by reducing it to a separable equation after making an appropriate substitution.

Let $y = xv$, $dy = x\,dv + v\,dx$

The differential equation is transformed to

$M(x, xv)dx + N(x, xv)\,(xdv + vdx) = 0$

$$\frac{M(x, xv)}{N(x, xv)}dx + v\,dx + xdv = 0$$

M, N being homogeneous functions x can be taken out.

The equation now takes the separable form

$$\left(\frac{M(1, v)}{N(1, v)} + v\right)dx + x\,dv = 0$$

$$\frac{dx}{x} + \frac{N(1, v)\,dv}{M(1, v) + v\,N(1, v)} = 0$$

Ex. 9. $x\,dy = (x + y)\,dy$ is homogeneous of degree 1 in its coefficients. Let $y = xv$ and the equation becomes

$$x(xdv + vdx) = (x + xv)\,dx.$$

$\therefore\ dv = \frac{dx}{x}$ or $v = \ln x + \ln C$ where $\ln C$ is an arbitrary constant

Substituting we get, $y/x = \ln Cx$ or $y = x \ln Cx$.

Ex. 10. Solve (*i*) $(x + 2y)\frac{dy}{dx} + (x + y) = 0$ **Ans.** (*i*) $x^2 + 2xy + y^2 = C$

(*ii*) $(2x^2 - y^2) + x^2\frac{dy}{dx} = 0$ (*ii*) $y = \frac{2 + Cx^{1/3}}{2 - Cx^{1/3}}$

(*iii*) $(x^3 - y^2) + 2xy\frac{dy}{dx} = 0$ (*iii*) $y^2 = Cx - x^2$

V. Linear Differential Equations of First Order

$\frac{dy}{dx} + y\,P(x) = Q(x)$ is linear homogeneous of order 1.

If $Q(x) = 0$, it becomes homogeneous *i.e.*,

$\frac{dy}{dx} + y\,P(x) = 0$ whose solution is obtained by writing

$\frac{dy}{y} = -P(x)\,dx$ whence $\ln y = \int -P(x)\,dx$

$$\therefore \qquad y = Ce^{-\int P(x)\,dx}$$

If $Q(x) \neq 0$ we can prove that the solution of

$\frac{dy}{dx} + y\,P(x) = Q(x)$

is given by $y = e^{-\int P dx}\left(\int Qe^{\int P dx} + C\right)$

Ex. 11. The solution of $\frac{dy}{dx} + yx = x$ can be written ($P = x$, $Q = x$)

$$y = e^{-\int x dx}\left(\int x\, e^{\int x dx} + C\right)$$

$$= e^{-x^2/2}(xe^{x^2/2}\; dx + C)$$

$$= e^{-x^2/2}\;(e^{x^2/2} + C) = 1 + Ce^{-x^2/2}$$

Ex. 12. Solve (*i*) $x\frac{dy}{dx} + y = x^2$ **Ans.** (*i*) $\frac{x^3}{3} + \frac{C}{x}$

(*ii*) $x^2\frac{dy}{dx} = y^2 + xy - x^2$ (*ii*) $\frac{Cx^3 + x}{1 - Cx^2}$

(*iii*) $x\frac{dy}{dx} = y + xe^{y/x}$ (*iii*) $y = -x\ln\left(\ln\frac{1}{x} + C\right)$

(*iv*) $x(x-1)\frac{dy}{dx} + y = x^2$ (*iv*) $y = \frac{x^2 + Cx}{x - 1}$

(*v*) $(2xy + 1)\frac{dy}{dx} + (y^2 + 1) = 0$ (*v*) $x + y + xy^2 = C$

VI. Linear Differential Equations of Second Order

The solution of $\frac{d^2y}{dx^2}+b\frac{dy}{dx}+cy=f(x)$ is made of two parts. One part is called the *complementary function* y_c which is the solution of the homogeneous equation $\frac{d^2y}{dx^2}+b\frac{dy}{dx}+cy=0$

The other part is a *particular* solution y_p

The general solution is $y = y_c + y_p$

Let us assume $y = e^{mx}$ is a solution of $\frac{d^2y}{dx^2}+b\frac{dy}{dx}+cy=0.$

so that $m^2e^{mx} + bme^{mx} + ce^{mx} = 0$. Taking out e^{mx}

$m^2 + bm + c = 0$ which is called the *characteristic equation.*

This equation may have

(*i*) 2 real distinct roots in which case $y_c = C_1e^{m_1x} + C_2e^{m_2x}$

(*ii*) 2 real identical roots in which case $y_c = C_1e^{mx} + C_2xe^{mx}$

(*iii*) 2 imaginary roots $\alpha \pm i\beta$ in which case

$y_c = e^{\alpha x}(C_1 \cos \beta x + C_2 \sin \beta x)$ (See Ex. 13 below)

Ex. 13. (*i*) $\frac{d^2y}{dx^2}-3\frac{dy}{dx}+2y=0$ gives $m^2-3m+2=0$

$\therefore \quad m_1 = 1, m_2 = 2$ and $y_c = C_1e^x + C_2e^{2x}$

(*ii*) $\frac{d^2y}{dx^2}-4\frac{dy}{dx}+4y=0$ gives $m^2-4m+4=0$

$\therefore \quad m_1 = m_2 = 2$ and $y_c = C_1e^{2x} + C_2xe^{2x}$

(*iii*) $\frac{d^2y}{dx^2}-4\frac{dy}{dx}+5y=0$ gives $m^2-4m+5=0$

$\therefore \quad m_1, m_2 = 2 \pm i$

$\therefore \alpha = 2, \beta = 1$ and $y_c = C_1e^{2x}\cos x + C_2e^{-2x}\sin x.$

To find the particular solution y_p, we may assume that the general solution is given by $y = C_1P(x) + C_2Q(x)$. It can be shown that $y_p = P_1(x)\,P(x) + Q_1(x)\,Q(x)$ where

$$P_1(x)=\int\frac{-Qf(x)}{PQ'-QP'}dx,\quad Q_1(x)=\int\frac{Pf(x)}{PQ'-QP'}dx$$

Ex. 14. $\frac{d^2y}{dx^2}-\frac{5dy}{dx}+6y=e^{2x}.$

The characteristic equation $m^2 - 5m + 6 = 0$ has the roots $m_1 = 2, m_2 = 3$ so that the complementary solution is

$$y_c = C_1e^{2x} + C_2e^{3x}$$

Here $P(x) = e^{2x}$, $Q(x) = e^{3x}$, $f(x) = e^{2x}$ so that

$$P_1(x) = \int \frac{-e^{3x}\, e^{2x}\, dx}{3e^{2x}\, e^{3x} - 2e^{3x}\, e^{2x}} = -\int dx = -x + C_3$$

$$Q_1(x) = \int \frac{e^{2x}\, e^{2x}\, dx}{3e^{2x}\, e^{3x} - 2e^{3x}\, e^{2x}} = -\int e^{-x} dx = e^{-x} + C_4$$

$$\therefore \quad y_P = P_1P + Q_1Q = (-x + C_3)e^{2x} + (e^{-x} + C_4)e^{3x}$$

$$y = y_c + y_p = C_5e^{2x} + C_6e^{3x} - xe^{2x}.$$

Ex. 15. Solve (*i*) $\frac{d^2y}{dx^2} = 4y$

(*ii*) $\frac{d^2y}{dx^2} - 2\frac{dy}{dx} + \frac{dy}{dx} - x^{17}e^x = 0$

Ans. (*i*) $y = C_1e^{2x} + C_2e^{-2x}$ (*ii*) $\frac{1}{342}x^{19}e^x$

13.2 DOMAR ECONOMIC GROWTH MODEL

Investment has a dual character. It affects aggregate demand and promotes full employment. It also leads to expansion of capital stock and hence of supply of output. Domar tries to determine how rapidly must investment grow to maintain full capacity output.

Let b stand for marginal propensity to consume, s for marginal propensity to save and σ for the capital coefficient which represents the average as well as marginal productivity of capital.

If I, Y, S, and K stand for investment, income, savings and capital respectively, the multiplier $k = \frac{1}{1-b} = \frac{1}{s}$ and the production function is of the form $Y = \sigma K$, then we have

Increment to total demand = Increment to total supply.

Since increment to total demand

$$\frac{dY}{dt} = k\frac{dI}{dt} = \frac{1}{1-b}\frac{dI}{dt}$$

and increment to total supply

$$\frac{dY}{dt} = \sigma\frac{dK}{dt} = \sigma I$$

$$\therefore \quad \frac{1}{1-b}\frac{dI}{dt} = \sigma I \text{ or } \frac{dI}{I} = \sigma s dt.$$

Solving we get $I = I_o e^{\sigma st}$

This is the time path of the investment level required to maintain full capacity output and suggests that in this model, investment must grow exponentially at the rate σs. If σ and s are large, investment will require a correspondingly higher rate of growth.

It follows that since $Y = C + I$, $C = a + bY$, we have

$$Y = \frac{a}{s} + \frac{I_o}{s}\ e^{\sigma st}$$

If $a = 0$, *i.e.*, if consumption is proportional to income, we have

$Y = Y_o e^{\sigma st}$ showing exponential income growth.

Ex. 16. Instead of constant s and σ, let us have variable $s(t)$ and $\sigma(t)$. Then the differential equation can be written

$$\frac{dI}{I} = s(t)\sigma(t)\,dt.$$

Show that the solution is $I = I_o e^{\int s(t)\sigma(t)\,dt}$

Ex. 17. *Solow growth model.* In the usual growth model let us drop the assumption that capital (K) and labour (L) are combined in fixed proportions.

If (*i*) K, L are combined in varying proportions.

(*ii*) L grows exponentially, $L = L_o e^{\lambda t}$ where λ is the rate of growth of labour,

(*iii*) $\dfrac{K}{L} = V$ is the capital-labour ratio.

(*iv*) the production function is linearly homogeneous *i.e.*, $x = f(K, L) = Lf(K/L, 1) = L\ g(V)$, then, if a constant proportion of x is invested

$$\frac{dK}{dt} = sx = s.L_o e^{\lambda t}\ g(V)$$

since $K = VL$ we have the Solow model $\dfrac{dV}{dt} = sg(V) - \lambda V$

13.3 DOMAR DEBT MODEL : REDISTRIBUTIVE BURDEN OF NATIONAL DEBT

Suppose a constant fraction α of the income is borrowed each year *i.e.*, $\dfrac{dB}{dt} = \alpha Y$.

Case (*i*) Let $Y = Y_0$ = constant. Then the debt after t years is

$B = B_0 + \alpha Y_0 t$ where B_0 is debt at $t = 0$.

The ratio of debt B to income Y

$$\frac{B}{Y} = \frac{B_0}{Y_0} + \alpha t \to \infty \text{ as } t \to \infty$$

i.e., the redistributive burden becomes very large.

Case (*ii*) Let $Y = Y_0 + \beta t$ *i.e.*, national income increases at a constant rate β over time.

$$\frac{dB}{dt} = \alpha Y = \alpha Y_o + \alpha\beta t$$

or

$$B = B_0 + \alpha Y_o t + \tfrac{1}{2}\alpha\beta t^2$$

The ratio $\frac{B}{Y} = \frac{B}{Y_0 + \beta t} + \frac{\alpha Y_0 t}{Y_0 + \beta t} + \frac{\frac{1}{2}\alpha\beta t^2}{Y_0 + \beta t} \to \infty$ as $t \to \infty$

i.e., in this case the burden increases without limit.

Case (*iii*) Let $Y = Y_0 e^{gt}$ *i.e.*, income grows at a compound exponential rate and g is the exponential rate of growth.

$$\frac{dB}{dt} = \alpha Y_0 e^{gt}$$

so that
$$B = B_0 + \frac{\alpha}{g} Y_0 (e^{gt} - 1)$$

The ratio
$$\frac{B}{Y} = \frac{B_0}{Y_0 e^{gt}} + \frac{\alpha}{g}(1 - e^{-gt}) \to \frac{\alpha}{g}$$
as $t \to \infty$ *i.e.*, when income grows exponentially the debt-income ratio approaches a constant $\frac{\alpha}{g}$.

Ex. 18. Let the demand and supply functions be given by

$D(t) = a_0 + a_1 p(t)$, $S(t) = b_0 + b_1 p(t)$ and let the rate of change of price be proportional to excess demand so that

$$\frac{dp}{dt} = k\,(D - S)$$

Show that if at $t = 0$, $p = p_0$, the solution is $p = \bar{p} + (p_0 - \bar{p})\, e^{k(a_1 - b_1)}$

where
$$\bar{p} = \frac{a_0 - b_0}{b_1 - a_1}$$

Ex. 19. Show that if the labour-cost relation is given by

$$\frac{dC}{dL} + C(2 - Ce^{-L}) = 0$$

and if the fixed cost is 3 then $C = 3e^L$.

Ex. 20. The change in price with respect to demand for a commodity is given by

$$\frac{dp}{dx} = -\frac{2px + 24x}{x^2 + 16}$$

Show that if $x = 2$ when $p = 1$ then the demand function is

$$p = \frac{68 - 12x^2}{16 + x^2}$$

13.4 DIFFERENCE EQUATIONS

Let y_t be the value of a variable at time t. Suppose we know that the value of this variable at a previous time $t - 1$ is y_{t-1}.

The equation expressing the relation between the values of the variable at times t and $t - 1$ is called a difference equation.

If a, b, c are constants then

$y_t = ay_{t-1}$ is a linear *first* order homogeneous difference equation. There is no separate constant term independent of y. The time lag is 1.

$y_t = ay_{t-1} + b$ is a linear first order *non-homogeneous* difference equation. There is a separate constant term independent of y. The time lag is 1.

$y_t = ay_{t-1} + by_{t-2}$ is a linear *second* order homogeneous difference equation. The time lag is 2.

$y_t = ay_{t-1} + by_{t-k} + c$ is a linear kth order nonhomogeneous difference equation. The time lag is k.

In economics we often come across situations when variables occur with discrete time lags. We now consider the calculus of finite differences which deals with this subject. The use of *differences* as variables amounts to using lagged variables corresponding to previous time periods.

Differences

The difference $\Delta y_t = y_{t+1} - y_t$ is called the *first difference* of y_t and provides the rule for computing Δy_t. It may be noted that y is a function of t but that it takes only *integral* values. Thus y can be evaluated at intervals and not continuously.

The *second difference* is written

$$\begin{aligned}\Delta^2 y_t &= \Delta(\Delta y_t) = \Delta y_{t+1} - \Delta y_t \\ &= (y_{t+2} - y_{t+1}) - (y_{t+1} - y_t) \\ &= y_{t+2} - 2y_{t+1} + y_t\end{aligned}$$

The *third difference* is

$$\begin{aligned}\Delta^3 y_t &= \Delta^2 y_{t+1} - \Delta^2 y_t \\ &= (y_{t+3} - 2y_{t+2} + y_{t+1}) - (y_{t+2} - 2y_{t+1} + y_t) \\ &= y_{t+3} - 3y_{t+2} + 3y_{t+1} - y_t\end{aligned}$$

and so on.

An ordinary *difference equation* is an equation involving differences like Δy_t, $\Delta^2 y_t$, etc.

The *order* of a difference equation is the highest difference in the equation. The *degree* of a difference equation is the degree of the term of the highest order. A difference equation of degree one is said to be linear.

A difference equation may be written with or without the explicit presence of difference terms.

Ex. 21. (*i*) $y_{t+1} - y_t = 5$ or $\Delta y_t = 5$ are equivalent forms of a difference equation of degree one.

The subscript t may be taken arbitrarily. Thus t may be replaced by $t - 1$ or $t + 1$ throughout and this equation can be written in the alternative forms:

$$y_t - y_{t-1} = 5 \quad \text{or} \quad y_{t+2} - y_{t+1} = 5$$

(*ii*) $y_t - ky_{t-1} = a^{t-1}$ and $\Delta y_t - (a - 1)y_t = a^t$ are equivalent forms of a difference equation of order 1 degree 1.

Ex. 22. $y^3_{t+2} - 3y_t = 2$ is second order, third degree.

Ex. 23. $(\Delta y_t)^2 - ay_t + t^4 = 0$ is first order, second degree.

Ex. 24. Write the following equations with expressions of difference.

(*i*) $y_t - y_{t-1} + y_{t-2} = 0$ (*ii*) $y_t - 3y_{t-1} + 4y_{t-1} - 2y_{t-3} = 0$

Ans. (*i*) $\Delta^2 y_t + \Delta y_t + y_t = 0$ (*ii*) $\Delta^3 y_t + \Delta y_t = 0$

Ex. 25. Write the following equations without expressions of difference

(*i*) $8\Delta^2 y_t + 10\Delta y_t + y_t = 0$

(*ii*) $\Delta^2 y_t + 2\Delta y_t + (1 - k^2)y_t = at + 2a + b$

Ans. (*i*) $8y_t - 6y_{t-1} + y_{t-2} = 0$

(*ii*) $y_t - a^2 y_{t-2} = at + b$

The *general expression* for a difference equation of order n can be written

$$a_o y_t + a_1 y_{t-1} + \ldots\ldots + a_n y_{t-n} = f(t)$$

or

$$a_o \Delta^n y_t + a_1 \Delta^{n-1} y_t + \ldots\ldots + a_n y_t = f(t)$$

The coefficients are all constants, $a_n \neq 0$ and $f(t)$ is any function of t.

If $f(t) = 0$, it is a homogeneous linear equation with constant coefficients.

Solution of a Difference Equation

The solution of a difference (as of a differential) equation is a function. It is not a point. A solution of a difference equation is any function which satisfies the given equation. It means that such a solution, when inserted in the equation makes the relationship in the equation an identity.

A difference equation of the nth order has a solution containing n arbitrary constants. This is the general solution and is unique.

A particular solution is obtained from the general solution by assigning definite and specified values to the constants. Any particular solution is included in the general solution.

The arbitrary constants are specified by *initial* conditions. A particular solution requires the specification of n conditions in a difference equation of order n.

If $f(t)$ is a solution of a given difference equation then $Af(t)$, where A is *a* constant, is also a solution.

If $f_1(t), f_2(t)$ both satisfy a difference equation then $A_1 f_1(t) + A_2 f_2(t)$ also satisfies it.

Ex. 26. Show that for the equation $\Delta y_t = 1$, we have a solution $y_t = t + k$.

$\Delta y_t = 1$ can be written $y_{t+1} - y_t = 1$

Substituting $y_t = t + k$ we have $(t + 1 + k) - (t + k) = 1$

The equation is satisfied and $y_t = t + k$ is a solution.

Suppose we have an initial condition:

at $t = 0$ $\quad y_o = 5$, then $5 = y_o = 0 + k$

$\therefore$ $k = 5$

$\therefore$ When $y_o = 5$, a particular solution of the equation is $y_t = t + 5$

If we want to solve the above equation we may use the process of iteration *i.e.*, a repeated application of the difference equation:

$$y_{t+1} = y_t + 1$$

for $t = 1$, $\quad y_1 = y_o + 1$

for $t = 2$, $\quad y_2 = y_1 + 1 = y_o + 2$

for $t = 3$, $\quad y_3 = y_2 + 1 = y_o + 3.$

In general, $y_t = y_o + t = k + t$ if it is given that $y_o = k$. This is the required solution.

Ex. 27. Show that $y_t = k.\ 2^t - (t + 2)$ is a solution of $y_t - 2y_{t-1} = t$.

Substituting in the problem,

LHS $= k \cdot 2^t - (t + 2) - 2\,[k \cdot 2^{t-1} - (t + 1)]$

$= k2^t - t - 2 - k2^t + 2t + 2$

$= t$. Hence the result.

If $y_o = 10$, $10 = k2^0 - (0 + 2)$ or $k = 12$

Then the particular solution is $y_t = 12.2^t - (t + 2)$

Ex. 28. (*i*) Show that $y_t = \frac{1}{2}t(t-1)$ is a solution of $\Delta y_t = t$

(*ii*) Show that $y_t = \dfrac{k}{1+kt}$ is a solution of

$$y_{t+1} = y_t\,(1 - y_{t+1})$$

(*iii*) Show that $y_t = k_1 + k_2 2^t - t$ is a solution of

$$\Delta^2 y_t - y_{t+1} + y_t = 1.$$

What is the particular solution if $y_o = 0$, $y_1 = 2$?

$\Delta^2 y_t - y_{t+1} + y_t - 1 = 0$ can be written

$y_{t+2} - 3y_{t+1} + 2y_t - 1 = 0.$

Substituting, $k_1 + k_2\,2^{t+2} - t - 2 - 3\,(k_1 + k_2\,2^{t+1} - t - 1) + 2(k_1 + k_2 \cdot 2^t - t) - 1$

$= k_2\,(2^{t+2} - 3.2^{t+1} + 2.2^t) - (t + 2 - 3t - 3 + 2t) - 1$

$= k_2\,(2^{t+2} - 2^{t+2}) - (-1) - 1 = 0$

At $\quad t = 0,\ 0 = k_1 + k_2$

$t = 1,\ 1 = k_1 + 2k_2 - 1 \qquad \therefore\ k_2 = 2,\ k_1 = -2$

A particular solution is $y_t = -2 + 2\,2^t - t$.

Ex. 29. Show that $y_t = k_1 + k_2(-1)^t$ is a solution of $\Delta^2 y_t + 2\Delta y_t = 0$. Find a particular solution for $y_o = 1$, $y_1 = 3$.

Ex. 30. Consider the equation $y_t = at_{t-1}$

For $t = 1$, $y_1 = ay_o$

For $t = 2$, $y_2 = ay_1 = a^2 y_o$

For $t = 3$, $y_3 = ay_2 = a^2 y_1 = a^3 y_o$

Continuing thus we get, $y_t = a^t y_o$

$y_t = y_o a^t$ is the solution of the given linear first order homogeneous equation

Ex. 31. (*i*) If r is the interest rate compounded annually and P_t is the amount after t years then solve for P_t the difference equation

$$P_t = (1 + r) P_{t-1}$$ **Ans.** $P_o(1 + r)^t$

(*ii*) Show that the solution of the difference equation $ay_{t+1} - by_t = 0$

is given by $$y_t = (b/a)^t y_o$$

(*iii*) Solve: $$y_{t+1} - 2y_t = 0.$$ **Ans.** $y_t = y_o\, 2^t$

It is not necessary to use successive substitution or *iteration* every time. We can attempt a trial solution $y_t = ka^t$ where k and a are constants to be determined.

Let us try this substitution in the compound interest equation where a principal P_{t-1} carrying interest r becomes at the end of one year

$$P_t = (1 + r)P_{t-1}$$

To solve this difference equation let the trial solution be

$$P_t = k(1 + r)^t$$

Let the initial condition be that at time $t = 0$ the principle amount P_o was lent.

Then $P_o = k\,(1 + r)^o = k$ so that the complete solution is given by

$$P_t = P_o(1 + r)^t$$

Homogeneous Difference Equations

We have seen that the iterative procedure used above to solve a difference equation may be found useful in simple cases. But a more general approach is necessary in most cases. A parallel between the methods of solving differential and difference equations can be easily noted.

First Order Homogeneous Equations

The equation $y_t = by_{t-1}$ where b is a constant, has a solution of the form $y_t = b^t$ or $y_t = Ab^t$.

Since, by substitution we find $Ab^t = b\,(Ab^{t-1})$ which is true. This is called the *auxiliary* equation or *characteristic* equation.

If the initial condition is $y_o = C$, then by substitution

$$C = y_o = Ab^o = A$$

$\therefore\ y_t = Cb^t$ is the solution of the given equation

First Order Nonhomogeneous Equations

In the case of the nonhomogeneous equation $y_t = a + by_{t-1}$ where a, b are constants the general solution consists of two components: a particular solution y_p and a complementary function y_c which is the general solution of the reduced form $y_t = by_{t-1}$.

The general solution $y_t = y_c + y_p$

To find the particular solution we proceed as follows:

$$y_1 = a + by_o$$
$$y_2 = a + by_1 = a + b\,(a + by_o) = a + ab + b^2 y_o$$

$$y_3 = a + by_2 = a + ab + ab^2 + b^3 y_o$$

$$y_n = a + ab + ab^2 + \ldots + ab^{n-1} + b^n y_o$$

$$= b^n y_o + a \sum_{r=0}^{n-1} b^r = b^n y_o + a \frac{1-b^n}{1-b}$$

We have the solution $y_t = b^t y_o + a \dfrac{1-b^t}{1-b}$.

If the initial condition (at $t = 0$) is $y_o = C$, we have

$$y_t = Cb^t + a \frac{1-b^t}{1-b} \qquad \text{if } b \neq 1 \qquad \ldots(1)$$

$$= C + at \qquad \text{if } b = 1 \qquad \ldots(2)$$

Ex. 32. For the system $y_t = 4 + 3y_{t-1}$, the solution is

$$y_t = C \cdot 3^t + 4 \frac{1-3^t}{1-3} = C \cdot 3^t + 2(3^t - 1) \text{ if } y_o = C.$$

The solution of the system $y_t = 5 - 3y_{t-1}$ is

$$y_t = \tfrac{1}{4}[5 - (-3)^t].$$

Ex. 33. We may write the solution of the first order equation (1) as

$$y_t = Cb^t + \frac{a}{1-b} - \frac{ab^t}{1-b} = \frac{a}{1-b} + \left(C - \frac{a}{1-b} \right) b^t \qquad \ldots(1)$$

$$= a_0 + a_1 b^t \text{ where } a_o = \frac{a}{1-b},\ b_o = C - \frac{a}{1-b} \qquad b \neq 1$$

The time path of y_t, which depends on b shows the following behaviour pattern.

(*i*) If $b > 1$, $y_t \to \infty$ if $a_1 > 0$ and $y_t \to -\infty$ if $a_1 < 0$

(*ii*) If $b < -1$, y_t oscillates and diverges

(*iii*) If $0 < b < 1$, $y_t \to a_o$ since $b^t \to 0$

(*iv*) If $-1 < b < 0$, b^t oscillates to 0 and y_t oscillates to a_o.

Ex. 34. The complete solution of the difference equation

$$y_{t+1} - y_t = g_t$$

is given by $$y_t = y_o + \sum_{n=1}^{t} g_{n-1}$$

The complete solution of the general first order linear difference equation

$$y_{t+1} + ay_t = g_t$$

is given by $$y_t = Cy_t + y_t \sum_{n=1}^{t} \frac{g_{n-1}}{y_n}.$$

Second Order Homogeneous Equations

$y_t = ay_{t-1} + by_{t-2}$

The auxiliary equation is obtained by writing $y_t = r^t$ and substituting.

$$\therefore \qquad r^t - ar^{t-1} - br^{t-2} = 0$$

or
$$r^2 - ar - b = 0$$

There are three different cases for the solution. This characteristic equation may have

(*i*) 2 real distinct roots in which case $y_t = C_1 r_1{}^t + C_2 r_2{}^t$

(*ii*) 2 real identical roots in which case $y_t = C_1 r_1{}^t + C_2 t r_2{}^t \quad (r_1 = r_2)$

(*iii*) 2 imaginary roots in which case $y_t = p^t\,(C_1 \cos \theta t + C_2 \sin \theta t)$

where
$$r_1 = p(\cos \theta + i \sin \theta) = \alpha + i\beta$$
$$r_2 = p(\cos \theta - i \sin \theta) = \alpha - i\beta$$

and
$$p = \sqrt{\alpha^2 + \beta^2}$$

Compare this with the corresponding result for differential equations.

Ex. 35. (*i*) $y_t - 3y_{t-1} + 2y_{t-2} = 0$

$\therefore \qquad r^2 - 3r + 2 = 0 \qquad\qquad r_1 = 1, r_2 = 2$

The solution is $y_t = C_1 r_1{}^t + C_2 r_2{}^t = C_1 1^t + C_2\, 2^t$

(*ii*) $y_t - 4y_{t-1} + 4y_{t-2} = 0 \qquad \therefore \quad r^2 - 4t + 4 = 0,\ r_1 = 2,\ r_2 = 2$

The solution is $y_t = C_1 2^t + C_2 t\, 2^t$

(*iii*) $y_t - 4y_t + 5y_{t-2} = 0 \qquad \therefore \quad r^2 - 4r + 5 = 0,\ r_1, r_2 = 2 \pm i$

$$p = \sqrt{2^2 + 1^2} = \sqrt{5}$$

The solution is $\qquad (\sqrt{5})^t\,(C_1 \cos \theta t + C_2 \sin \theta t)$

Second Order Nonhomogeneous Equations

Depending on the form of the equation, different methods of finding the particular solution may be adopted as shown below:

(*i*) $y_{t+2} + a_1 y_t + a_2 y_t = g$ where g is a constant.

In such a case we may try a solution $y_t = k$. Then, substituting we get $k + a_1 k + a_2 k = g$ so that $y_p = k = \dfrac{g}{1 + a_1 + a_2}$ is the required particular solution.

(*ii*) $y_{t+2} + a_1 y_{t+1} + a_2 y_t = g^t$

In such a case we try a particular solution $y_t = Cg^t$

Ex. 36. $\quad y_{t+2} - 5y_{t+1} + 6y_t = 4^t$

We have $\qquad r^2 - 5r + 6 = 0 \qquad \therefore\ r_1 = 2,\ r_2 = 3$

Let $y_t = C \cdot 4^t$

$\therefore \quad C4^{t+2} - 5C4^{t+1} + 6C4^t = 4^t$

$\therefore \quad C(4^2 - 5.\,4 + 6) = 1$ or $C = \frac{1}{2}$

The general solution is $y_t = C_1 2^t + C_2 3^t + \frac{1}{2} 4^t$

(*iii*) If the complementary solution includes a term similar to g^t the particular solution is multiplied by t.

Ex. 37. $y_{t+2} - 5y_{t+1} + 6y_t = 3^t$. Since $r_1 = 2$, $r_2 = 3$, for the particular solution we take $y_t = Ct \cdot 3^t$

$$\therefore \quad C(t + 2)\, 3^{t+2} - 5C(t + 1)\, 3^{t+1} + 6Ct.\, 3^t = 3^t$$

$$\therefore \quad C\{3^2(t + 2) - 5 \times 3(t + 1) + 6t\} = 1$$

$$\therefore \quad C = 1/3$$

The general solution is $y_t = C_1 \cdot 2^t + C_2 \cdot 3^t + \frac{1}{3} t\, 3^t$.

(*iv*) If in the above case the roots are equal say 3 each, we take $y_t = Ct^2\, 3^t$ as the particular solution so that

$C = \dfrac{1}{18}$ and the general solution is

$$y_t = C_1 \cdot 3^t + C_2 t \cdot 3^t + \frac{1}{18} t^2 \cdot 3^t$$

13.5 THE COBWEB MODEL

Comparative Static Analysis. Consider the behavioural relations in the simple market model with *a, b, c, d* as parameters:

Demand as a function of price	$D = a + bp$
Supply as a function of price	$S = e + dp$
The equilibrium condition	$D = S$

Solving we get the equilibrium price and quantity,

$$\bar{p} = \frac{a - c}{d - b}, \quad \bar{x} = \frac{ad - bc}{d - b}$$

Such a comparative static analysis, which amounts to finding the solution of some simultaneous equations, does not give any indications of the time path of the movement of the equilibrium of the system.

Dynamic Analysis. Suppose, as in the case of agricultural commodities, demand in period t depends on the market price in period t:

$$D_t = a + bp_t \qquad b < 0$$

The adaptation of supply to price changes takes one period.

$$S_t = c + dp_{t-1} \qquad d > 0$$

The equilibrium condition $D_t = S_t$ gives

$$p_t = \frac{c - a}{b} + \frac{d}{b}\, p_{t-1}$$

If p_o is the initial price the solution may be written (Sec. 13.4)

$$p_t = p_o - \left(\frac{a - c}{d - b}\right)\left(\frac{d}{b}\right)^t + \frac{a - c}{d - b}$$

or $$p_t = (p_o - \overline{p}) \left(\frac{d}{b} \right)^t + \overline{p}$$

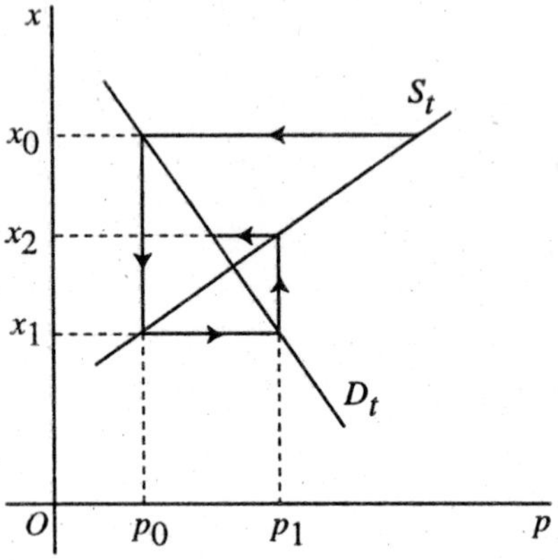

(i) *S* flatter than *D*
Stable equilibrium

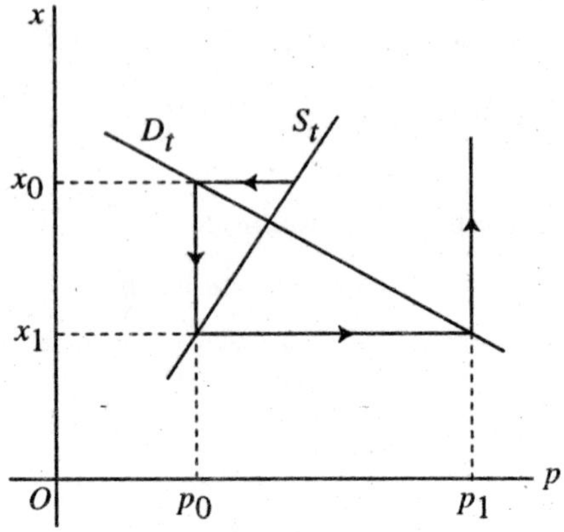

(ii) *D* flatter than *S*
Unstable equilibrium

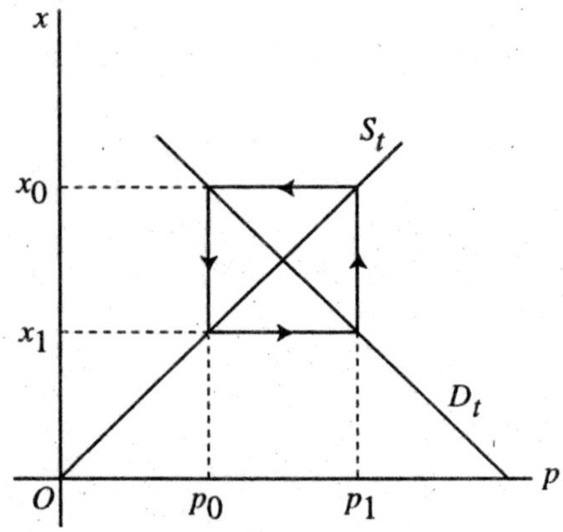

(iii) *D, S* having equal slopes
Uniform oscillations

$\frac{d}{b}$ is negative since $d > 0$, $b < 0$. This means that our model is oscillatory with three possible types of oscillations.

(*i*) If $\left| \frac{d}{b} \right| < 1$ *i.e.*, if $-1 < \frac{d}{b} < 1$ or $|d| < |b|$ (See diagrams)

$\left(\frac{d}{b} \right)^t \to 0$ as t increases and we have market stability with demand oscillations.

(*ii*) If $\left| \frac{d}{b} \right| > 1$, $\left(\frac{d}{b} \right)^t$ increases with t and we have unstable market with antidamped or explosive oscillations.

(*iii*) If $\left|\frac{d}{b}\right| = 1, \left(\frac{d}{b}\right)^t = 1$. This is the case of regular oscillations.

If m_d, m_s stand for the slopes of the demand and supply curves, we have in case
(*i*) $|m_d| > |m_s|$ (*ii*) $|m_d| < |m_s|$ (*iii*) $|m_d| = |m_s|$

From the time path of price p_t or otherwise we can also derive the time path of quantity x_t.

Ex. 38. Examine the stability of the following Cobweb models
(*i*) $D_t = 10 - 2p_t$, $S_t = -5 + 4p_{t-1}$ (*ii*) $D_t = 20 - 3p_t$, $S_t = 5 + p_{t-1}$.

13.6 EFFECT OF PUBLIC PREDICTION ON MARKET STABILITY

Devlotoglou examines this effect in the framework of the Cobweb model

$$D_t = a + bp_t,\quad S_t = c + dP_t,\quad D_t = S_t$$

where P_t is the expected price in period $t - 1$.

$$P_t = kF_t + (1 - k)\, p_{t-1}$$

where F_t is the forecast price and k, $1 - k$ are weights attached to the forecast and actual prices.

(*i*) $k = 0$, no confidence in forecast

(*ii*) $k = 1$, full confidence

If $F_t = p_t$ the forecast is correct

$$\therefore \quad P_t = kp_t + (1 - k)\, p_{t-1}$$

The solution is $p_t = (p_o - \bar{p})\left[\dfrac{d(1-k)}{b - dk}\right]^t + \bar{p}$

Stability conditions, as in the case of the Cobweb model, depend on $\dfrac{d(1-k)}{b-dk}$ which suggests stability when

$$-1 + \frac{2kd}{b} < \frac{d}{b} < 1$$

As confidence in forecast increases *i.e.*, as k increases, $-1 + 2kd/b$ decreases.

Ex. 39. If μ is the stock-induced-price-adjustment coefficient in the model

$$D_t = a + bp_t \qquad a > 0,\ b < 0$$

$$S_t = c + dp_t \qquad c < 0,\ d > 0$$

and $p_{t+1} - p_t = \mu(D_t - S_t)$ $\qquad \mu > 0$

Show that the solution is $p_t = (p_o - \bar{p})(1 - \mu b - \mu d)^t + \bar{p}$

Examine the stability conditions.

13.7 SAMUELSON MULTIPLIER ACCELERATOR MODEL

An explanation of business cycle fluctuations is provided by the multiplier accelerator principle. Let Y, C I, G stand respectively for national income, consumption, investment, and government expenditure. We have the acceleration principle that net investment in the economy during period t depends on the rate of change of aggregate income $Y_{t-1} - Y_{t-2}$ and consumption depends on the income of the preceding time period.

$$C_t = bY_{t-1},\ I_t = a(C_t - C_{t-1}) = ab(Y_{t-1} - Y_{t-2})$$

$$Y_t = C_t + I_t + G$$

$$\therefore \quad Y_t = b\,(1 + a)\,Y_{t-1} - ab\,Y_{t-2} + G$$

This second order nonhomogeneous equation may be written

$$Y_{t+2} - b(1 + a)\,Y_{t+1} + ab\,Y_t = G$$

The characteristics equation is $r^2 - b(1 + a)r + ab = 0$

The complete solution is

(*i*) $Y_t = C_1 r_1^t + C_2 r_2^t + \dfrac{G}{1-a}$ if the roots are real, unequal. No fluctuations but stability if $r_1, r_2 < 1$, $ac < 1$

(*ii*) $Y_t = (C_1 + C_2 t)\,r^t + \dfrac{G}{1-a}$ if the roots are real, equal. No fluctuations but stability if $r_1, r_2 < 1$, $ac < 1$ and instability if $r_1, r_2 > 1$, $ac > 1$

(*ii*) $Y_t = (C_1 + C_2 t) r^t + \dfrac{G}{1-a}$ if the roots are real, equal. No fluctuations but stability if $r < 1$, $a < 1$ and instability if $r > 1$, $a > 1$

(*iii*) $Y_t = p^t\,(C_1 \cos \theta t + C_2 \sin \theta t)$ if the roots are complex. Damped fluctuations if $p < 1$, $ac \leq 1$ and explosive or regular fluctuations if $p \geq 1$, $ac \geq 1$.

13.8 DUESENBERRY MULTIPLIER ACCELERATOR MODEL OF INCOME DETERMINATION

Assuming that (*i*) an excess of investment over saving generates income, (*ii*) the rate of investment is a function of the capital output ratio and (*iii*) investment expands capital stock and introducing a number of variables like business and house investments, stocks, profits, business debt etc., Duesenberry gives a simplified system of difference equations:

For income: $Y_t = (\delta + \rho)\,Y_{t-1} + (\eta + \mu)K_{t-1}$

For capital : $K_t = \delta y_{t-1} + [\eta + \chi(1 - \chi)]K_{t-1}$

where χ is a constant.

δ is the effect of changes in income on investment through changes in marginal efficiency of investments and in marginal cost,

η is the influence of capital stock through marginal efficiency and profit,

ρ is the mpc out of national income and also the profit increase resulting from an increase in income,

μ reflects the change in capital stock through the influence of profits on consumption.

Ex. 40. Show that for the Harrod model

$$S_t = aY_t,\ I_t = (Y_t - Y_{t-1}),\ S_t = I_t,\ a,\ b > 0$$

We have the difference equation $Y_t = \dfrac{b}{b-a}\ Y_{t-1}$

whose solution is $Y_t = \left(\dfrac{b}{b-a}\right)^t Y_0$

Show that Y_t, I_t, S_t are all diverging to infinity with t, *i.e.*, none of these variables attains equilibrium values.

Ex. 41. Solve the model : $Y_t = C_t + I_t,\ C_t = a + bY_{t-1}$.

Ex. 42. Show that for the difference equation $Y_t - Y_{t-1} = Y_t\, Y_{t+1}$ the solution (with $Y_0 = 1$) is $(1 + t)^{-1}$.

Ex. 43. Find a particular solution of

(*i*) $Y_t = 3Y_{t-1} - 2Y_{t-2} + 1 \qquad Y_0 = 2,\ Y_1 = 3$

Find the general solution of

(*ii*) $Y_t - 5Y_{t-1} + 4Y_{t-2} = 16$ (*iii*) $Y_t + 6Y_{t-1} + 9Y_{t-2} = 4$

(*iv*) $Y_{t+2} = 2Y_t$

Ans. (*i*) $Y_t = C_1 + C_2 \cdot 2^t - t = 2.2^t - t$ (*ii*) $C_1 4^t + C_2 + 5.5t$

(*iii*) $Y_t = C_1(-3)^t + C_2 t\,(-3)^t + \frac{1}{4}$ (*iv*) $C_1(\sqrt{2})^t + C_2(-\sqrt{2})^t$

Ex. 44. Show that the solution of $Y_{t+1} = t\ Y_{t+1} + t!$ is given by

$$Y_t = t! + C(t-1)!$$

Ex. 45. *Metzler Inventory Cycles.* If A stands for autonomous investment, C for sales of inventory goods and the stock held by producers at the end of period t is

$$K_t = hC_{t-1} - (C_t - C_{t-1})$$

the consumption function $C_t = bY_t$

and $Y_t = C_{t-1} + (hC_{t-1} - K_{t-1}) + A$,

show that the solution of the resulting equation

$$Y_t = bY_{t-1} + b(1+h)(Y_{t-1} - Y_{t-2}) + A_t$$

suggests oscillations or inventory cycles in output Y_t about some equilibrium path.

CHAPTER 14

Input Output Economics

14.1 INTRODUCTION

The Keynesian System based on aggregate quantities like consumption, investment, employment and income is concerned with the determinants of income and employment. The general equilibrium idea deals with relationships among the variables but not with economic interdependence or the structure of the economy.

Walras gave a model of interdependence among the producing sectors of the economy. The model has equations for incomes and expenditures and takes into account production costs in each sector as well as the competitive nature of demand and supply of commodities and factors. Early contributions to the idea of economic interdependence were made by Pareto, Cassel and others. Leontief's general theory of production, in the form of input output technique and tables, had the important ingredient of empirical content.

The input output technique or interindustry analysis has been used at various levels of aggregation in many countries irrespective of whether they rely on planning or on market forces for allocation of resources and income distribution.

The input output model has been used for single or multiple regions or sectors or companies and for whole economies and is useful in a variety of economic situations.

The input output model amounts to a general theory of production with all components of final demand treated as data. We are required to determine production levels of each sector to meet this final demand.

The model can be open or closed, static or dynamic, highly aggregated or disaggregated and can be applied to the analysis of development, planning and short term forecasting, to regions, sectors or whole economies following any kind of economic system. The input output table is a statistical description of certain features of an economy. Input output analysis can be used to predict the consequences of changes in the circumstances of an economy.

The static open model assumes fixed technical coefficients or proportionality and is based on current flows without allowing for substitution among inputs.

14.2 ECONOMIC INTERDEPENDENCE

The economic system may be thought of as a composite entity of mutually interrelated industries. In such a system, all industries may be reasonably assumed to be related to one another. In general, each industry buys its raw materials or inputs from and sells a part of its output to other industries in the system. There may be cases of some industries not buying from other industries and buying only from the primary producing sector. There may also be cases of industries selling their output not to other industries but only to the final consumption sector. But circular interdependence remains a feature of the economic system. Such a comprehensive system cannot be easily analysed.

A complete economic system represented by an input output model is called a Leontief model. Its analysis gives the formulation of general economic equilibrium in terms of industries. The method is suitable for empirical applications and is a simplified version of the production function equations of the general equilibrium system. The basic unit in Leontief analysis is an industry or a sector and not the firm.

Note. Input output analysis considers general economic equilibrium *empirically*, by studying the interdependence of the production plans of industries in the economy. An input output table presents at a point or over an interval of time, the structural characteristics of an economic system in the form of statistical information which is useful in analysing and understanding the system.

14.3 FINAL DEMAND

A typical industry in the system sells inputs to other industries, to consumers, governments and foreign countries. Those sales which are not used for production are called *final sales*. Correspondingly, the demand by the consumers, etc. is called final demand.

To simplify the model, sometimes the final goods sector is considered to be made of only consumers.

One can find what goes to the final goods sector and how much of each output is used to obtain these final goods. Alternatively if the final demand is known one wants to know how much output each industry should produce to meet this final demand.

14.4 APPLICATIONS OF INPUT OUTPUT METHODS

If we can tackle the problem of finding an appropriate relationship between the final goods sector and the outputs of various industries in an economy we may as well handle similar problems in other fields. We may use the method in planning, company transactions, social accounting, military mobilisation, transport, location, trade, regional economics etc.

Input output methods can be useful in the study of effects generated in the economy as a result of changes made in one or more sectors of the economy.

A comparison of input output tables for two or more countries or companies can help in determining the types of investment needed to stimulate growth and can be useful as a development tool.

In a situation of less than full employment resulting from aggregate demand deficiency, suitable changes in final demand in the input output table can suggest the level of economic activity necessary to achieve full employment. The policy makers can project full employment levels of demand with the help of input output tables.

Each row of the input output table presents the marketing profile of an industry or sector and each column present the input patterns. The market analysts can forecast from these the interindustry requirements for many products by projecting final sales.

14.5 ASSUMPTIONS OF INPUT OUTPUT ANALYSIS

A number of simplifying assumptions have to be made in the analysis.

1. Each industry or sector is assumed to produce a single, homogeneous product. To relax this assumption to some extent the single product may be considered as a composite product made of several items used in it in fixed proportions.

2. In a simple model there is only one primary input, *i.e.*, labour. The final demand is from consumers only. However, more primary inputs and final demand sectors can be introduced.

3. In any productive process all inputs are used in fixed proportions and any increase in inputs is in proportion to the level of output. Production takes place through processes with constant *technical coefficients.*

In other words, input output relations are assumed to be linear relations of direct proportionality. Production in each industry is subject to constant returns to scale.

4. The producing sector consists purely of competitive firms.

5. The output of any industry becomes either the input of another industry or the final demand or, it is added to the inventory of the industry. An industry may use some of its own product. Such use may be considered as a sale to the industry itself.

If, for any firm, total input equals total output, it is in equilibrium. If input exceeds output, inventories pile up, creating problems. Alternatively, if output exceeds input, inventories are exhausted.

6. The input output relations are technical in nature and not conditions of market equilibrium. The analysis is static, though easily extended to comparative statics *e.g.* by varying final demand.

7. All transactions may be considered in terms of money values since money is a suitable common unit for aggregating inputs and outputs of industries. Quantities and prices can also be considered for transactions.

8. In order that the total demand for the product of any industry be just sufficient we should know the level of output of each industry under consideration. Since the industries are interdependent the output levels must be consistent with the input requirements in the economy. This is to avoid difficulties in analysis.

9. Input output analysis assumes constant returns to scale which is a convenient assumption that simplifies conceptual and statistical matters. It is true, however, that in reality the production process may involve complex production relations.

10. The analysis assumes only one process without choice for each output. The input coefficients are assumed to be fixed, implying a fixed input per unit of output. This assumption is a great practical simplification though it is true that structural, technological changes as well as relative input price changes take place. The possibility of different production processes involving a variety of input combinations converts the input output problem into a linear programming problem involving optimisation. According to Samuelson, there is a unique, most efficient process of producing an output despite substitution possibilities. The fixed input proportion as an assumption is technologically quite convenient.

Note. Instead of treating final consumption as an exogenous factor it may be appropriate to devise a system of interrelationships to determine the consumption of various goods within the system.

14.6 THE TRANSACTIONS MATRIX

The basis of input output analysis is the *transactions table.* In a real transactions table there can be a large number of industries but for a quick understanding a small table with a few industries should be adequate. A transactions table presents the data pertaining to industries and sectors. An industry is an aggregate of firms producing similar products. The markets which industries serve are called sectors. These may be artificial definitions but in practice. for convenience, we can consider the industries on the basis of their principal products which are assumed to be homogeneous.

The transactions table pertaining to an economy describes the economy as well as the supply and demand relations of the economy in equilibrium. It neither sets an ideal for the economy nor is concerned with its efficiency. It shows the interindustry transactions required to satisfy a given final demand in addition to describing the structural interdependence of the economy.

The transactions matrix shows in monetary (or quantitative) terms, the transactions of the whole economy, in the form of the output of each industry as distributed among (*i*) other industries as intermediate products or raw materials and (*ii*) the ingradients of final demand (or autonomous) sector where demand for outputs is for consumption etc. and not for production.

If there are three industries X_1, X_2, X_3 we have the balancing equations of sales for industry X_1:

$$\begin{aligned}\text{Value of output of } X_1 &= \text{Value of sales to } X_1\\ &+ \text{Value of sales to } X_2\\ &+ \text{Value of sales to } X_3\\ &+ \text{Final sales to consumers}\end{aligned}$$

and so on for values of outputs of X_2 and X_3

An industry receives its inputs from other industries and labour services from the consuming sectors. We have the balancing equations of purchases, say, for X_1:

$$\begin{aligned}\text{Value of output of } X_1 &= \text{Value of purchases from } X_1\\ &+ \text{Value of purchases from } X_2\\ &+ \text{Value of purchases from } X_3\\ &+ \text{Value of services from the consuming sector}\end{aligned}$$

Here depreciation, interest, etc. are ignored to simplify the model. The process of capital formation through savings and investments has also not been considered.

Balancing equations can be recorded in an input output table or transactions matrix in which every industry appears once in a row as *seller to* other industries and once in the corresponding column as *purchaser from* other industries.

The *n* Industry Transaction Matrix

Output → / *Input* ↓		*Industry* 1	2	...	*n*	*Final Demand*	*Total Output*
Industry	1	x_{11}	x_{12}	...	x_{1n}	F_1	X_1
	2	x_{21}	x_{22}	...	x_{2n}	F_2	X_2
	...	...	...	...	...	...	...
	n	x_{n1}	x_{n2}	...	x_{nn}	F_n	X_n
Primary Input		y_1	y_2		y_n	—	y

The transactions matrix can be looked upon as a double table shown in quantitative terms though it could as well be in money value terms. There can be several columns for different types of final demands. There can be several rows for different primary inputs. Only one final demand and one primary input have been shown in this table.

Here x_{ij} is the interindustry flow from industry i to j and X_i the total output of the ith industry, F_i the final demand for the ith industry output and y_i the amount of primary inputs purchased by the ith industry. Thus the transactions matrix is an accounting technique or a record keeping device. Each entry is read twice, once in a column and once in a row. This is somewhat akin to the double entry system.

If the transactions matrix is in terms of quantities and has each row corresponding to an industry or sector with its own units, it is possible to add up the contents of each row but not those of each column because each amount would be in different units. Addition is possible if the transactions matrix is given in money values and the final demands or expenditures are in money terms also.

$$x_{11} + x_{12} + \ldots + x_{1n} + F_1 = X_1 \quad \text{or} \quad \sum_j x_{1j} + F_1 = X_1$$

$$x_{21} + x_{22} + \ldots + x_{2n} + F_2 = X_2 \quad \text{or} \quad \sum_j x_{2j} + F_2 = X_2 \text{ etc.}$$

$$y_1 + y_2 + \ldots + y_n = y \quad \text{or} \quad \sum_j y_j = y$$

Transactions Matrix in Compact Form

		n columns	*1 column*	*Total column*
n	rows	x_{ij}	F_i	X_i
1	row	y_j	0	y

(Total row not shown because column entries are in different units)

If
$$a_{ij} = \frac{x_{ij}}{X_j}, \qquad l_j = \frac{y_j}{X_j} \quad \text{(See Sec. 14.9)}$$

are the input coefficients, we have the equations, in a compact form,

$$\sum_j x_{ij} + F_i = X_i \quad i = 1, 2, \ldots n \quad \text{and} \quad \sum_j y_j = y$$

which can be written $\sum_j a_{ij} X_j + F_i = X_i$ and $\sum_j l_j X_j = y$

We have assumed pure competition and zero profit, *i.e.*, revenue equals cost.

14.7 STRUCTURE

If all entries in the input output table are nonzero *i.e.*, outputs of each industry become inputs of all the industries then we have a case of complete interdependence.

Zero entries in some columns imply non-drawing of inputs from the corresponding intersecting rows. If the reverse combination also has zero entries, then these will imply independent sectors.

A possible rearrangement of rows and columns may reveal a triangular array like a triangular matrix or block triangular matrix with all nonzero entries below and all zero entries above the diagonal. This implies that each industry in an upper row gets inputs from all industries below it (but not from those above it) and also fulfils the demands of those above and the final users. The bottommost row gets inputs from no other though it supplies its output to all others. The in-between industries get from those below and give to those above.

Block triangular matrices may be looked upon as having the characteristics of triangular matrices with blocks behaving like entries.

Basic industries are usually in the bottom rows and consumption goods industries in the upper rows.

If the data are given in value terms we can carry out the analysis in value terms. However, if the data are given in physical terms they can be converted into value terms by multiplying the quantities by their corresponding prices.

14.8 LEONTIEF OPEN AND CLOSED SYSTEMS

An input output table of interindustrial or intersectoral relations can be constructed either in the *open* or in the *closed* form.

In the closed system, we have a *closed* economy in which each industry purchases goods from other industries and services from the household "industry" which is assumed to be an additional industry. Similarly each industry sells outputs to other industries and to the household "industry". Because of the addition of this new "industry" all outputs become intermediate products. There is neither a final demand nor a primary input that could be considered separately.

An *open* system does not treat household as an industry. The transactions matrix above is a case of an open system. An open system is one in which besides other industries (or sectors) there is an additional sector like the final demand. There is thus an extra input which is not an output of any production process and a demand for outputs over and above their uses as inputs. The extra input is often identified as primary input labour.

When, in the closed system, the final demand becomes one of the sectors and is endogenous in the technology matrix, any change in the total system would be translated from disturbances in some of the coefficients. Output projections emanating from the envisaged changes do not require any additional exogenous information in such a closed system. But before using the technological matrix for projection purposes, it should be adjusted for changes in interrelated sectors. When some components like capital formation, government expenditure etc. do not seem to vary reasonably uniformly while the intermediate coefficients can be assumed to be stable, the open system may be recommended.

Ex. 1. We consider a simple *open* Leontief matrix of intersectoral relations. There are only two sectors: agriculture and industry. And, of course, the consumers.

Each row in the table gives the balancing equation of the output of an industry from the sales point of view and each column shows the input of an industry from the purchases point of view. All transactions are in money terms.

Output / *Input*		*Interindustry Sectors* (1) *Agri.*	(2) *Ind.*	*Final Demand Sector* (3) *Consumers*	*Total output (Sales)* (4)
Agriculture	(1)	300	600	100	1000
Industry	(2)	400	1200	400	2000
Consumers (Primary Input)	(2)	300	200	0	500
Total Input (Cost)	(4)	1000	2000	500	3500

Row 1, Column 1 : 300 is the amount consumed by the agricultural sector. (Out of its own total output of 1000).

Row 1, Col. 2 : 600 is the amount sold by agriculture to industrial sector.

Row 1,. Col. 3 : 100 is the amount sold by agriculture to household sector.

Row 2, Col. 1 : 400 is the amount sold by industry to agriculture and so on.

Row 2, Col. 2 : 1200 is the amount consumed by industry (out of its total output of 2000).

Row, 3 Column 3 : The consumer or household sector sells nothing to itself. However, it produces "labour" and services worth 300 (Row 3, Column 1) and 200 (Row 3, Column 2). The total of these two 300 + 200 = 500 (Row 3, Column 4) is, therefore, the sum of wages and salaries received by the consumers or household sector.

We can see that a transactions matrix is useful in the study of the structure of the producing sector. It shows the way in which the productive activity is distributed throughout the economy. When applied to the economic system it can help in predicting the response to any changes made in the system.

14.9 TECHNICAL COEFFICIENTS OR MARGINAL INPUT COEFFICIENTS OF PRODUCTION

The technical coefficient shows the number of units of any industry's output needed to produce one unit of another industry's output. In other words, it shows the amount of raw materials which are outputs of other industries needed by an industry from any other industry to produce a certain output.

A technical coefficient is obtained by dividing each sector's or industry's input (purchased from other sectors) by its total output.

Technical Coefficients (Ex. 1)

	Agriculture	*Industry*	*Explanation*
Agriculture	0.3	0.3	$\frac{300}{1000} = 0.3,\ \frac{600}{2000} = 0.3$
Industry	0.4	0.6	$\frac{400}{1000} = 0.4,\ \frac{1200}{2000} = 0.6$
Primary Input	0.3	0.1	$\frac{300}{1000} = 0.3,\ \frac{200}{2000} = 0.1$

Technical coefficients are structural parameters showing the linear relationship between the input of each industry (or sector) and its total output. These coefficients are technological relations so that the question of calculating them for the consumer sector does not arise. Thus we do not consider the technical coefficients for final demand.

We can see that a technical coefficient in an input output table implies the quantity of inputs required from an industry to produce one unit worth of output of a given industry. The units can be physical or monetary.

If, from one period to the next, the technical coefficients remain stable or can be suitably adjusted depending on the available information, they can be useful in obtaining an advance idea about purchases from supplying industries whenever there is a change in the output of an industry.

Stability Condition for Direct Technical Coefficients

An industry is not expected to pay for its inputs an amount exceeding the sale value of its output. This can be expressed by saying that in the table of technical coefficients, no column adds up to more than unity and at least one column adds up to less than unity. The fundamental conditions for the table of Direct and Indirect Requirements are contained in Hawkin Simon Conditions, given later.

In the table of direct and indirect requirements there can be no negative entries. This is because a negative entry for an industry would mean that if this industry expands its output it would be required to buy less from other industries.

14.10 TECHNOLOGY MATRIX

The table above gives the technology conditions in the interindustry relationship. The coefficients are usually put in the form of a matrix. If there are two industries (or sectors) the technology matrix may be written

$$A = \begin{bmatrix} a_{11} & a_{12} \\ a_{21} & a_{22} \\ 1 - a_{11} - a_{21} & 1 - a_{12} - a_{22} \end{bmatrix} = \begin{bmatrix} 0.3 & 0.3 \\ 0.4 & 0.6 \\ 0.3 & 0.1 \end{bmatrix}$$

where a_{ij} shows the input needed by the ith industry to produce each unit of the jth industry.

$a_{11} = 0.3$ means 30 paise worth of the agriculture output is required as an input to produce a rupee worth of the agriculture output.

$a_{21} = 0.4$ means 40 paise worth of the industry output is required as an input to produce a rupee worth of the agriculture output, and so on.

As we are dealing with an open model the sum of the coefficients of any column in the matrix must be less than or equal to one.

To sum up, column 1 of the matrix shows that for agriculture to produce output worth one rupee it needs inputs worth

30 paise from agriculture (*i.e.*, itself)

40 paise from industry

30 paise from households (*i.e.*, labour).

Column 2 shows that for industry to produce output worth one rupee it needs inputs worth 30 paise from agriculture, 60 paise from industry (*i.e.*, itself) and 10 paise from labour.

We may write these statements in the form of supply equations (X_1 = agriculture, X_2 = industry, L = labour).

$$X_1 = 0.3\, X_1 + 0.4\, X_2 + 0.3L$$

$$X_2 = 0.3\, X_1 + 0.6\, X_2 + 0.1L$$

Now consider the demand side. The demand for agriculture products is made up of 0.3 for internal use by agriculture itself, 0.3 by industry in addition to Rs. 100 worth of agriculture goods demanded by consumers. We may write

$$X_1 = 0.3\, X_1 + 0.3\, X_2 + 100 \quad \ldots(1)$$

Similarly for industry, $X_2 = 0.4\, X_1 + 0.6\, X_2 + 400 \quad \ldots(2)$

These are two simultaneous equations in two unknowns X_1, X_2. If by solving these equations we get values of X_1 and X_2 which do not exceed the available supply, that implies the solution and hence *feasible targets.*

The solution of (1), (2) in this case is obviously $X_1 = 1000$, $X_2 = 2000$.

Suppose now the final demand figures rise from 100 to 150 and 400 to 500 what is the new position? To get the answer, solve

$$X_1 = 0.3\, X_1 + 0.3\, X_2 + 150$$

$$X_2 = 0.4\, X_1 + 0.6\, X_2 + 500$$

The technical coefficients remain the same. And they remain the same even if we have different values of final demand. (Solve to get $X_1 = 1312.5$, $X_2 = 2562.5$)

Ex. 2. Find X_1, X_2 when final demands are 50, 300 in Ex. 1.

Ans. 687.5, 1437.5

Ex. 3. Find the technology matrix of the following economy and answer these questions:

What will be the level of output if the final demand becomes 65 for X_1 and 25 for X_2?

What will happen if the final demand is (*i*) doubled, (*ii*) halved? **Ans.** 97.1, 44.9

	X_1	X_2	*Final demand* *F*	*Total*
X_1	15	20	45	80
X_2	5	20	40	65
Primary Input	60	25		85
Total Input	80	65	85	230

14.11 INPUT OUTPUT TABLE WITH NET OUTPUT OF INDUSTRIES

In the input output table presented earlier x_{11}, x_{22}, x_{33} etc. (the diagonal elements in the input output matrix) stand for the consumption or inputs of industries X_1, X_2, X_3 etc. out of their own outputs. The *net output* of these industries, as available to other industries, can be obtained by subtracting these from total output. Accordingly the diagonal terms will all become zero as can be seen from the following three industry table. An input output table may be presented this way sometimes.

		Interindustry		*Sales*	*Final Demand*	*Total Output*
		1	2	3		
Interindustry Purchases	1	—	x_{12}	x_{13}	F_1	X_1
	2	x_{21}	—	x_{23}	F_2	X_2
	3	x_{31}	x_{32}	—	F_3	X_3
Primary Input		y_1	y_2	y_3		
Total Input		X_1	X_2	X_3		

Ex. 4. Interpret the following table: (Also see Ex. 6)

	Industry X_1	X_2	X_3	*Consumers* *F*	*Total*
X_1	0	5	3	1	9
X_2	3	0	4	3	10
X_3	4	2	0	5	11
P.I.	2	3	4	0	
Total	9	10	11		30

14.12 LEONTIEF OPEN SYSTEM FOR THREE INDUSTRIES

The case of three industries is adequate to illustrate the solution of a Leontief open system. It can be generalised easily.

Leontief Open System for 3 Industries

		Interindustry Sales			*Final Demand*	*Total Output*
		1	2	3		
Interindustry purchases	1	x_{11}	x_{12}	x_{13}	F_1	$X_1 = x_{11} + x_{12} + x_{13} + F_1$
	2	x_{21}	x_{22}	x_{23}	F_2	$X_2 = x_{21} + x_{22} + x_{23} + F_2$
	3	x_{31}	x_{32}	x_{33}	F_3	$X_3 = x_{31} + x_{32} + x_{33} + F_3$
Primary Input		y_1	y_2	y_3	$\Sigma F_i = \Sigma y_i$	
Total Input		$X_1 = x_{11} + x_{21} + x_{31} + y_1$	$X_2 = x_{12} + x_{22} + x_{32} + y_2$	$X_3 = x_{13} + x_{23} + x_{33} + y_3$		

Here x_{ij} is the output of industry i sold to industry j

F_i is the final demand for industry i

y_j is the primary input for industry j

X_j is the total output of industry j.

In a planning exercise the balance equations showing total output and total input in the table are accounting identities showing condition of equilibrium between demand and supply. They show the *flows* of outputs from an industry and the flows of inputs from other industries to an industry *i.e.*, they show the extent of economic activity.

Internal Consistency

The equations present the conditions of internal consistency of the plan. Without them the plan would not be feasible. If these equations were not satisfied there might be too much of some goods and too little of others. It may be noted that the internal consistency of an economic plan depends only on its structure and not on the scale of the plan.

Changes in consumption and investment either by the individuals or the government or due to increased exports or imports bring about changes in the interindustry system.

Production Function

Any industry, to produce the output X_j during a given period, needs the inputs from all the industries and the primary input, say labour. Thus each column gives the *production function* of an industry or the input output relation among the industries.

The technology matrix is a summary description of the technology conditions of the economy. The input coefficients $a_{ij} = \frac{x_{ij}}{X_j}$ are indicators of the amounts of inputs needed from all industries to produce a certain amount of output by any industry ($0 \le a_{ij} \le 1$). The relation $x_{ij} = a_{ij} X_j$ amounts to saying that sales of industry i to industry j are a constant proportion a_{ij} of the output of industry j.

The matrix of input coefficients may be written (in the 3 industry case)

$$A = \begin{bmatrix} a_{11} & a_{12} & a_{13} \\ a_{21} & a_{22} & a_{23} \\ a_{31} & a_{32} & a_{33} \end{bmatrix} \qquad \text{Note } a_{11} = \frac{x_{11}}{X_1},\ a_{12} = \frac{x_{12}}{X_2},$$

$$a_{21} = \frac{x_{21}}{X_1},\ a_{22} = \frac{x_{22}}{x_2}, \text{ etc.}$$

Replacing each x_{ij} by $a_{ij} X_j$ in the table we get the set of simultaneous linear equations

$$\begin{aligned} a_{11}X_1 + a_{12}X_2 + a_{13}X_3 + F_1 &= X_1 \\ a_{21}X_1 + a_{22}X_2 + a_{23}X_3 + F_2 &= X_2 \\ a_{31}X_1 + a_{32}X_2 + a_{33}X_3 + F_3 &= X_3 \end{aligned} \quad i.e. \quad \begin{bmatrix} a_{11} & a_{12} & a_{13} \\ a_{21} & a_{22} & a_{23} \\ a_{31} & a_{32} & a_{33} \end{bmatrix} \begin{bmatrix} X_1 \\ X_2 \\ X_3 \end{bmatrix} + \begin{bmatrix} F_1 \\ F_2 \\ F_3 \end{bmatrix} = \begin{bmatrix} X_1 \\ X_2 \\ X_3 \end{bmatrix}$$

which may be written in the matrix form .

$$AX + F = X$$

Rearranging, we may write

$$\begin{aligned} (1 - a_{11})X_1 - a_{12}X_2 - a_{13}X_3 &= F_1 \\ - a_{21}X_1 + (1 - a_{22})X_2 - a_{23}X_3 &= F_2 \\ - a_{31}X_1 - a_{32}X_2 + (1 - a_{33})X_3 &= F_3 \end{aligned}$$

This is the same as

$$\begin{bmatrix} 1 - a_{11} & -a_{12} & -a_{13} \\ -a_{21} & 1 - a_{22} & -a_{23} \\ -a_{31} & -a_{32} & 1 - a_{33} \end{bmatrix} \begin{bmatrix} X_1 \\ X_2 \\ X_3 \end{bmatrix} = \begin{bmatrix} F_1 \\ F_2 \\ F_3 \end{bmatrix}$$

or $(I - A)X = F$ where I is the 3×3 identity matrix.

The solution is $X = (I - A)^{-1}F$.

14.13 INPUT OUTPUT ANALYSIS : THE GENERAL CASE

The same notation will work if there are n industries

$$\begin{bmatrix} 1 - a_{11} & -a_{12} & \cdots & -a_{1n} \\ -a_{21} & 1 - a_{22} & \cdots & -a_{2n} \\ \cdots & \cdots & \cdots & \cdots \\ -a_{n1} & -a_{n2} & \cdots & 1 - a_{nn} \end{bmatrix} \begin{bmatrix} X_1 \\ X_2 \\ \cdots \\ X_n \end{bmatrix} = \begin{bmatrix} F_1 \\ F_2 \\ \cdots \\ F_n \end{bmatrix}$$

i.e. $(I - A)X = F$

To get the values of gross outputs we need to invert the Leontief matrix $I - A$ and get $X = (I - A)^{-1}F$.

Ex. 5.

		Interindustry Sales			*Final Demand*	*Total*
		1	*2*	*3*		
Interindustry	1	4	8	6	14	32
purchases	2	10	14	10	14	48
	3	6	4	8	22	40
Primary Input		12	22	16	—	50
Total		32	48	40	50	170

We have the technology matrix and the Leontief matrix

$$A = \begin{bmatrix} \frac{4}{32} & \frac{8}{48} & \frac{6}{40} \\ \frac{10}{32} & \frac{14}{48} & \frac{10}{40} \\ \frac{6}{32} & \frac{4}{48} & \frac{8}{40} \end{bmatrix}, \quad I - A = \begin{bmatrix} 1 - \frac{4}{32} & \frac{8}{48} & \frac{6}{40} \\ \frac{10}{32} & 1 - \frac{14}{48} & \frac{10}{40} \\ \frac{6}{32} & \frac{4}{48} & 1 - \frac{8}{49} \end{bmatrix}$$

Any suitable method can be used to find

$$(I - A)^{-1} = \begin{bmatrix} \frac{27}{20} & \frac{18}{50} & \frac{73}{200} \\ \frac{73}{100} & \frac{83}{50} & \frac{33}{50} \\ \frac{39}{100} & \frac{13}{50} & \frac{70}{50} \end{bmatrix}$$

We may verify the obvious result $(I - A)^{-1} F = X$ corresponding to the tabulated data.

$$\begin{bmatrix} \frac{27}{20} & \frac{18}{50} & \frac{73}{200} \\ \frac{73}{100} & \frac{83}{50} & \frac{33}{50} \\ \frac{39}{100} & \frac{13}{50} & \frac{70}{50} \end{bmatrix} \begin{bmatrix} 14 \\ 14 \\ 22 \end{bmatrix} = \begin{bmatrix} 32 \\ 48 \\ 40 \end{bmatrix}$$

If we want to find the effects of a change in one ore more final demand levels, we can use the above inverse matrix since the technology matrix remains the same.

Suppose the final demand targets are 10, 10, 20, then the new outputs will be given by

$$\begin{bmatrix} X_1 \\ X_2 \\ X_3 \end{bmatrix} = \begin{bmatrix} \frac{27}{20} & \frac{18}{50} & \frac{73}{200} \\ \frac{73}{100} & \frac{83}{50} & \frac{33}{50} \\ \frac{39}{100} & \frac{13}{50} & \frac{70}{50} \end{bmatrix} \begin{bmatrix} 10 \\ 10 \\ 20 \end{bmatrix} = \begin{bmatrix} 24.4 \\ 37.1 \\ 34.5 \end{bmatrix}$$

i.e., to satisfy the final demand target of 10, 10, 20 total outputs 24.4, 37.1, 34.5 must be produced by industries 1, 2, 3 respectively.

Ex. 6. Show that for Ex. 4 if the final demands are F_1, F_2, F_3 the solution may be written

$$\begin{bmatrix} X_1 \\ X_2 \\ X_3 \end{bmatrix} = \begin{bmatrix} 1 & -\frac{5}{10} & -\frac{3}{11} \\ -\frac{3}{9} & 1 & -\frac{4}{11} \\ -\frac{4}{9} & -\frac{2}{10} & 1 \end{bmatrix}^{-1} \begin{bmatrix} F_1 \\ F_2 \\ F_3 \end{bmatrix}$$

Determine the outputs if the final demands are 1, 2, 4 respectively *i.e.*, $F_1 = 1$, $F_2 = 2$, $F_3 = 4$.

Ex. 7. The following is the input output table for three industries X_1, X_2, X_3.

		Users				
		X_1	X_2	X_3	F	*Total output*
Producers	X_1	8	10	10	4	32
	X_2	8	20	6	6	40
	X_3	8	10	10	2	30

Show that the technology matrix is

$$A = \begin{bmatrix} 0.25 & 0.25 & 0.33 \\ 0.25 & 0.50 & 0.20 \\ 0.25 & 0.25 & 0.33 \end{bmatrix}$$

If the final demand for the outputs of industries X_1, X_2, X_3 becomes 6, 4 and 6 respectively find the outputs required of the industries.

Ex. 8. There are three industries X_1, X_2, X_3; X_1 requires 20% of the products of X_1, 40% of X_2 and 10% of X_3 (all in money terms). X_2 requires 30%, 10%, 30% of X_1, X_2, X_3.

X_3 requires 20%, 20%, 20% of X_1, X_2, X_3. Show that the technology matrix is

$$A = \begin{bmatrix} 0.2 & 0.3 & 0.2 \\ 0.4 & 0.1 & 0.2 \\ 0.1 & 0.3 & 0.2 \end{bmatrix}$$

Also find the outputs required of each industry if the find demand in money terms for the products of the three industries are Rs. 20, 5, 4 lakh respectively.

Ex. 9. Given the technology matrix

$$A = \begin{bmatrix} 0.1 & 0.4 & 0.1 & 0.1 \\ 0 & 0.1 & 0.4 & 0.2 \\ 0.2 & 0.1 & 0 & 0.2 \\ 0.3 & 0 & 0 & 0 \end{bmatrix}$$

if the final demands for the outputs of the four industries are 10, 100, 20, 5 respectively show that the output vector is given by

$$X = (I - A)^{-1} F = \begin{bmatrix} 1.31 & 0.62 & 0.38 & 0.33 \\ 0.25 & 1.28 & 0.54 & 0.39 \\ 0.36 & 0.29 & 1.15 & 0.39 \\ 0.39 & 0.19 & 0.11 & 1.10 \end{bmatrix} \begin{bmatrix} 10 \\ 100 \\ 20 \\ 5 \end{bmatrix}$$

Ex. 10. An economy has three activities producing phosphorus (X_1), tourism (X_2) and construction (X_3). Its technology matrix and final demand have been found to be

$$A = \begin{bmatrix} 0.05 & 0.02 & 0.40 \\ 0.10 & 0.70 & 0.25 \\ 0.05 & 0.10 & 0.10 \end{bmatrix}, \quad F = \begin{bmatrix} 15 \\ 3 \\ 2 \end{bmatrix}$$

Find the outputs in the three activities under these conditions.

Hint : Find

$$(I - A)^{-1} = \begin{bmatrix} 0.95 & -0.02 & -0.40 \\ -0.10 & 0.03 & -0.25 \\ -0.05 & -0.10 & 0.90 \end{bmatrix}^{-1} = \begin{bmatrix} 1.21 & 1.09 & 0.84 \\ 0.51 & 4.13 & 1.37 \\ 0.12 & 0.52 & 1.31 \end{bmatrix}$$

$$X = (I - A)^{-1} F = \begin{bmatrix} 1.21 & 1.09 & 0.84 \\ 0.51 & 4.13 & 1.37 \\ 0.12 & 0.52 & 1.31 \end{bmatrix} \begin{bmatrix} 15 \\ 3 \\ 2 \end{bmatrix} = \begin{bmatrix} 23.1 \\ 22.7 \\ 6.0 \end{bmatrix}$$

The required outputs are $X_1 = 23.1$, $X_2 = 22.7$, $X_3 = 6.0$

14.14 THE CLOSED MODEL

In the open model the open sector (say household) determines the final demand exogenously for each output. This open sector supplies the primary input (labour) not produced by industries so that the sum of any column element is less then one.

In the 3 industry technology matrix

$$A = \begin{bmatrix} a_{11} & a_{12} & a_{13} \\ a_{21} & a_{22} & a_{23} \\ a_{31} & a_{32} & a_{33} \end{bmatrix}$$

we have, the sum of elements of each column less than one, *e.g.*,

$$a_{11} + a_{21} + a_{31} < 1$$

Each column sum like this, in value terms, represents the partial input costs for one money unit worth of the output. The value of the primary input in that column (say the first) is $1 - a_{11} - a_{21} - a_{31}$

In general, if there are n industries, it is

$$1 - \sum_{i=1}^{n} a_{ij}$$

If the exogenous factor household is treated as an industry we have a closed model in which instead of the primary input we have another industry which is intermediate in nature.

When the final demand sector and primary input sector are considered as 'industries', we have in all $n + 1$ industries instead of n and we get a homogeneous system of equations. The result is that these new industries also have their technical coefficients so that the supply of the primary input is in a fixed proportion to the final demand 'industry' *i.e.*, the household sector now consumes the product of other industries in a fixed proportion to the quantity of labour supplied by it. In the closed model, none of the elements in the Leontief inverse matrix is zero. Moreover, these elements are larger than the corresponding elements of the open model.

The "supply" of primary input will now be in a fixed "technical proportion" to final demand *i.e.* household consumption will be in fixed proportion to the primary input supply. This will lead to a homogeneous equation system in which there is an equation corresponding to the primary input sector with subscript 0.

For the 3 industry case we then have the equation system:

$$\begin{bmatrix} 1 - a_{00} & -a_{01} & -a_{02} & -a_{03} \\ -a_{10} & 1 - a_{11} & -a_{12} & -a_{13} \\ -a_{20} & -a_{21} & 1 - a_{22} & -a_{23} \\ -a_{30} & -a_{31} & -a_{32} & 1 - a_{33} \end{bmatrix} \begin{bmatrix} F_0 \\ F_1 \\ F_2 \\ F_3 \end{bmatrix} = \begin{bmatrix} 0 \\ 0 \\ 0 \\ 0 \\ 0 \end{bmatrix}$$

This is a closed system and since the four rows are linearly not independent

$$|I - A| = 0$$

i.e., in a closed model, no unique solution for outputs can be obtained. However, output levels in a proportional form can be obtained.

14.15 HAWKIN SIMON CONDITIONS OF FEASIBILITY OF SOLUTIONS

For a unique set of output levels consistent with the final demands certain conditions have to be satisfied. First, that $(I-A)^{-1}$ exists, *i.e.*, it is nonsingular and its elements are all non-negative. Second, that each $a_{ii} \leq 1$ and no industry can use more of its output than it can produce. The determinant $|I-A|$, of the Leontief matrix, must be positive. Note that in the input output configuration the net output of a sector would be negative if its total output is less than the intermediate demand for it in the system.

The following factors may give rise to this situation:

(*a*) Output in the given sector is less than the total requirements because of technological shortcoming. There may be a nonviable technology in use.

(*b*) There may be an imbalance in the gross output levels. The situation can be improved by correcting the production structure.

If we are interested in a surplus producing system we should have an indecomposable technology matrix A which should satisfy certain conditions called Hawkin Simon Conditions.

What is Production Viability?

An industry is viable when its own input coefficient a_{ii} is less than one *i.e.*, it does not have a negative output.

For two industries

$$a_{11} < 1,\ a_{22} < 1$$

This can be written

$$1 - a_{11} > 0,\ 1 - a_{22} > 0$$

Hawkins Simon Conditions include these and the additional one

$$\frac{a_{21}}{1-a_{22}} < \frac{1-a_{11}}{a_{12}}$$

or

$$(1-a_{11})(1-a_{22}) - a_{12}\,a_{21} > 0$$

or

$$\begin{vmatrix} 1-a_{11} & -a_{12} \\ -a_{21} & 1-a_{22} \end{vmatrix} > 0$$

This is the main Hawkin Simon Condition for viability and is adequate for most purposes.

Generalisation : *n* sectors

This condition if fulfilled ascertains the avoidance of a nonviable situation like having negative outputs *e.g.*, when more than 1 unit of output is used to generate 1 unit of output. The condition states that all principal minors of $(I-A)$ are positive. It means that

(*a*) All diagonal elements of $(I-A)$ are strictly positive *i.e.*, in the n sector case, $1-a_{11}, 1-a_{22}, \ldots, 1-a_{nn}$ are positive.

Since $1 - a_{ii} > 0$, each $a_{ii} < 1$ $(i = 1, 2, \ldots n)$

(*b*) All principal minors of $(I-A)$ are positive.

The determinant of the matrix $(I-A)$ must be positive for the technology matrix A_{nn}. The conditions can be written.

$$|1-a_{11}|,\ \begin{vmatrix} 1-a_{11} & -a_{12} \\ -a_{21} & 1-a_{22} \end{vmatrix},\ \ldots\ldots,$$

$$\begin{vmatrix} 1-a_{11} & -a_{12} & \cdots & -a_{1n} \\ -a_{21} & 1-a_{22} & \cdots & -a_{2n} \\ \cdots & \cdots & \cdots & \cdots \\ -a_{n1} & -a_{n2} & \cdots & -a_{nn} \end{vmatrix} \text{ are all positive}$$

and the conditions become

$$1 - a_{11} > 0,\ 1 - a_{22} > 0,\ (1 - a_{11})(1 - a_{22}) - a_{12}a_{21} > 0 \text{ etc.}$$

Ex. 11. The matrix $A = \begin{bmatrix} .4 & .2 \\ .3 & .5 \end{bmatrix}$ with $[I - A] = \begin{bmatrix} .6 & -.2 \\ -.3 & .5 \end{bmatrix}$ is viable because

$$\begin{vmatrix} .6 & -.2 \\ -.3 & .5 \end{vmatrix} = 0.36$$

Ex. 12. $A = \begin{bmatrix} 2 & .3 \\ .2 & 3 \end{bmatrix}$

$$|I - A| = \begin{vmatrix} -1 & -.3 \\ -.2 & -2 \end{vmatrix} = 1.94$$

This system is not viable because $1 - a_{11}$ and $1 - a_{22}$ are negative.

Ex. 13. $A = \begin{bmatrix} 0.7 & 6 \\ 0.03 & 0.4 \end{bmatrix}$

$$|I - A| = \begin{bmatrix} 0.3 & -6 \\ -0.03 & 0.6 \end{bmatrix} = 0$$

This represents a self sustaining system.

Ex. 14. Give an economic interpretation of Hawkins Simon Conditions in the three industry case.

For the three sector (or industry) case,

$$I - A = \begin{bmatrix} 1-a_{11} & -a_{12} & -a_{13} \\ -a_{21} & 1-a_{22} & -a_{23} \\ -a_{31} & -a_{32} & 1-a_{33} \end{bmatrix},$$

the principal minor of $1 - a_{33}$ is

$$\begin{bmatrix} 1-a_{11} & -a_{12} \\ -a_{21} & 1-a_{22} \end{bmatrix} = (1 - a_{11})(1 - a_{22}) - a_{12}\, a_{21}.$$

This should be positive *i.e.*, $(1 - a_{11})(1 - a_{22}) > a_{12}\, a_{21}$

For an *economic interpretation* of this, note that $1 - a_{11}$ is the quantity from sector 1 for distribution among other sectors after satisfying its own needs. If this is not positive it would mean that each unit of production requires more than one unit to produce it which is not viable. Hence $1 - a_{11} > 0$. Similarly $1 - a_{22} > 0$ so that

$$(1 - a_{11})(1 - a_{22}) > 0$$

The net output of the two sectors must be available for use in other sectors and these two sectors must be capable of supplying more than their own needs. In the three sector case, given the third sector's requirement and the final demand, we can see that in practice the second condition will also be met *viz.*

$$(1 - a_{11})(1 - a_{22}) > a_{12}\, a_{21}$$

The interpretation is similar for other principal minors.

Ex. 15. Show that for the transactions matrix $\begin{bmatrix} 6 & 2 \\ 4 & 2 \end{bmatrix}$, with total output vector given by $X = \begin{bmatrix} 20 \\ 15 \end{bmatrix}$ and the final demand vector $\begin{bmatrix} 12 \\ 9 \end{bmatrix}$.

The Hawkin Simon Conditions are satisfied.

		Industry 1	2	F	X
Industry	1	6	2	12	20
	2	4	2	9	15
Primary Inputs		10	11		
Total		20	15		

$$A = \begin{bmatrix} 0.3 & 0.133 \\ 0.2 & 0.133 \end{bmatrix}$$

$$I - A = \begin{bmatrix} 0.7 & -0.133 \\ -0.2 & 0.867 \end{bmatrix}$$

$1 - a_{11} = 0.7$, $1 - a_{22} = 0.867$, Det $[I - A] = 0.58$. The conditions are satisfied since $1 - a_{11} > 0$, $1 - a_{22} > 0$, $|I - A| > 0$

Ex. 16. For the transactions matrix $\begin{bmatrix} 2.5 & 5 & 2.5 \\ 2.5 & 5 & 2.5 \\ 30 & 30 & 15 \end{bmatrix}$ with $X = \begin{bmatrix} 50 \\ 50 \\ 100 \end{bmatrix}$

show that Hawkin Simon Conditions are satisfied because all three principal and second order principal minors are positive and $|I - A|$ is also positive. Also find $(I - A)^{-1}$

Hint : $A = \begin{bmatrix} 0.05 & 0.20 & 0.025 \\ 0.05 & 0.10 & 0.025 \\ 0.60 & 0.60 & 0.150 \end{bmatrix}$, $(I - A)^{-1} = \begin{bmatrix} 1.09 & 0.27 & 0.04 \\ 0.08 & 1.15 & 0.04 \\ 0.83 & 1.01 & 1.23 \end{bmatrix}$

Ex. 17. Are the Hawkin Simon Conditions satisfied in the following cases?

(*a*) $A = \begin{bmatrix} 0.07 & 0.27 \\ 0.20 & 0.13 \end{bmatrix}$ (*b*) $A = \begin{bmatrix} 0.14 & 0.35 \\ 0.80 & 0.05 \end{bmatrix}$ **Ans.** Yes

Ex. 18. For the interindustry table

		Industry 1	2	3	Total
	1	35	0	0	100
Industry	2	5	25	15	50
	3	20	15	55	100

show that the Leontief inverse is given by

$$(I-A)^{-1} = \begin{bmatrix} 1.54 & 0 & 0 \\ 0.45 & 2.50 & 0.83 \\ 0.98 & 1.67 & 2.78 \end{bmatrix}. \text{ If } F = \begin{bmatrix} 130 \\ 10 \\ 20 \end{bmatrix} \text{ is the final demand vector show}$$

that the required outputs are $X_1 = 200, X_2 = 100, X_3 = 200$.

How does this show the linearity in the model?

[This final demand vector F is twice the original final demand vector $\begin{bmatrix} 65 \\ 5 \\ 10 \end{bmatrix}$ and the new outputs are all twice those of the original outputs. Hence the linearity.]

Ex. 19. Given the matrix of technical coefficients $A = \begin{bmatrix} 6/30 & 9/30 & 6/30 \\ 12/30 & 3/30 & 9/30 \\ 9/30 & 15/30 & 6/30 \end{bmatrix}$

and final demands 240, 480, 720

$$\text{Show that } (I-A)^{-1} = \begin{bmatrix} 57/24 & 34/24 & 27/24 \\ 41/24 & 58/24 & 32/24 \\ 47/24 & 49/24 & 60/24 \end{bmatrix}$$

$$\text{and } X = (I-A)^{-1} \begin{bmatrix} 240 \\ 480 \\ 720 \end{bmatrix} = \begin{bmatrix} 2060 \\ 2530 \\ 3250 \end{bmatrix}$$

Ex. 20. Given the closed matrix of production $A = \begin{bmatrix} P & Q & R & \\ 12/30 & 9/30 & 6/30 & P \\ 6/30 & 9/30 & 9/30 & Q \\ 12/30 & 12/30 & 15/30 & R \end{bmatrix}$

find the income ratios of P, Q, R who are consuming the production in the closed model.

If P, Q, R have incomes P_1, Q_1, R_1

P consumes (or spends on) $\frac{12}{30}$ of its own product *i.e.*, $\frac{12}{30} P_1$

P spends on $\frac{9}{30}$ of Q's product *i.e.* $\frac{9}{30}Q_1$

P spends on $\frac{6}{30}$ of R's product *i.e.* $\frac{6}{30}R_1$

P's income *i.e.*, $P_1 = \frac{12}{30}P_1 + \frac{9}{30}Q_1 + \frac{6}{30}R_1$

Similarly for Q *i.e.*, $Q_1 = \frac{6}{30}P_1 + \frac{9}{30}Q_1 + \frac{9}{30}R_1$

and for R *i.e.*, $R_1 = \frac{12}{30}P_1 + \frac{12}{30}Q_1 + \frac{15}{30}R_1$

$\therefore \frac{18}{30}P_1 - \frac{9}{30}Q_1 - \frac{6}{30}R_1 = 0$ These can be written as

$$-\frac{6}{30}P_1 + \frac{21}{30}P_1 - \frac{9}{30}R_1 = 0$$
$$-\frac{12}{30}P_1 - \frac{12}{30}Q_1 + \frac{15}{30}R_1 = 0$$

$$\frac{1}{30}\begin{bmatrix} 18 & -9 & -6 \\ -6 & 21 & -9 \\ -12 & -12 & 15 \end{bmatrix}\begin{bmatrix} P_1 \\ Q_1 \\ R_1 \end{bmatrix} = 0$$

If we add the second and third rows to the first row, the first row consists only of zeros. The first equations is redundant. The last two equations can be written

$$-6P_1 + 21Q_1 - 9R_1 = 0 \quad \text{or} \quad -6P_1 + 21Q_1 = 9R_1$$
$$-12P_1 - 12Q_1 + 15R_1 = 0 \qquad 12P_1 + 12Q_1 = 15R_1$$

Solving $\frac{Q_1}{R_1} = \frac{33}{54}, \ \frac{P_1}{R_1} = \frac{34.5}{54}$ *i.e.*, $P_1 : Q_1 : R_1 = 34.5 : 33 : 54$

Ex. 21. Find the income ratios for the matrix

$$\begin{bmatrix} 6/30 & 9/30 & 3/30 \\ 18/30 & 12/30 & 6/30 \\ 6/30 & 9/30 & 21/30 \end{bmatrix}$$

Ans. $3 : 5.5 : 7.5$

14.16 MORE ON PRODUCTION FUNCTIONS

The total output X_i of each industry in each row of the input output table equals the total of its parts used to meet other industry intermediate demands like x_{11}, x_{12}, x_{13}, etc.

This means that output X_1 is absorbed by industry 1, industry 2, industry 3 and final demand x_{01}. (in the three industry case)

Thus
$$X_1 = x_{11} + x_{12} + x_{13} + x_{01}$$
$$X_2 = x_{21} + x_{22} + x_{33} + x_{02}$$
$$X_3 = x_{31} + x^{32} + x_{33} + x_{03}$$
$$X_0 = x_{01} + x_{02} + x_{03}$$

The relations $X_i = \Phi(x_{i1}, x_{i2}, x_{i3}, x_{oi}) \qquad i = 1, 2, 3$

stand for production functions describing the input requirements for output X_i. Thus X_1 is a function of x_{11}, x_{12}, x_{13}, x_{o1} respectively from industries 1, 2, 3 and the primary input.

If there are n inputs we can write $X_1 = \varphi_1 (x_{11}, x_{12}, x_{13}, \ldots, x_{1n}, x_{01})$

$$X_2 = \varphi_2 (x_{21}, x_{22}, x_{23}, \ldots, x_{2n}, x_{02}) \text{ etc.}$$

Note: The Leontief production function is a fixed coefficient production function

$$Y = \min (aX_1, bX_2)$$

The Linear production function is given by

$$Y = a_0 + a_1X_1 + a_2X_2$$

The CES production function can be written

$$Y = (a_0 + a_1X_1^{\rho} + a_2X_2^{\rho})^{1/\rho}$$

The VES and other production functions can be similarly defined.

In the production function embodied in the input output model expressed as $X_j = \min\left(\frac{x_{1j}}{a_{1j}}, \frac{x_{2j}}{a_{2j}}, \ldots, \frac{x_{nj}}{a_{nj}}\right)$ the ratios $\frac{x_{ij}}{a_{ij}}$, $i = 1, 2, \ldots, n$ will all be the same and equal to X_j if their a_{ij} coefficients are not zero.

The definition of X_j implies a search for the smallest among the ratios.

The input output production function implies constant returns to scale.

14.17 ISOQUANTS

The isoquants in the two sector input output case are shown alongside those for the classical production function.

The classical isoquants with their convexity express the law of diminishing marginal productivity. In the input output production function if only one input is increased the output remains unchanged. In the two input case output will increase only when both inputs 1, 2 are increased together. If all inputs in the n input case are increased by 50 per cent and the nth input by 10 per cent then the minimum ratio is x_{nj}/a_{nj} and the output will increase by 10 per cent only.

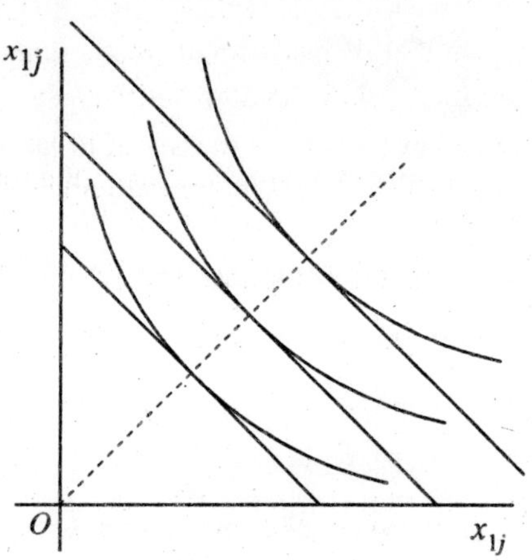

Classical isoquants

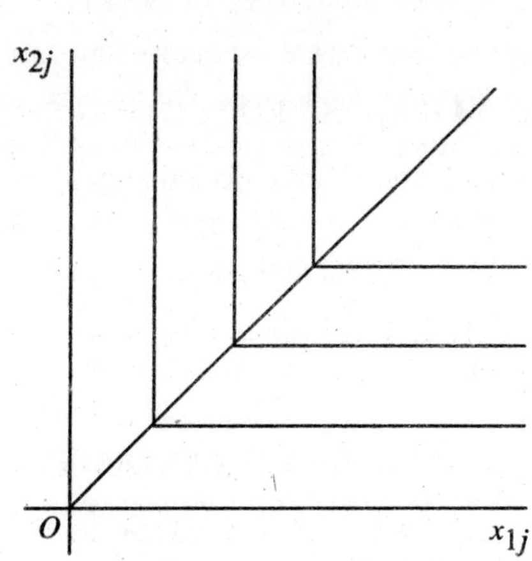

Input output isoquants

If x_{ij} is the flow of input from i to j and X_j the total output of sector j then the technical coefficient $a_{ij} = x_{ij}/X_j$ measures the fixed relationship between a sector's output and its inputs, ignoring economies of scale.

It can be seen that $$X_j = \frac{x_{1j}}{a_{1j}} = \frac{x_{2j}}{a_{2j}} = \ldots = \frac{x_{nj}}{a_{nj}}$$

To avoid the possibility of any a_{ij} being zero when input i is not used to produce j the input output production function is written

$$X_j = \min\left\{\frac{x_{1j}}{a_{1j}}, \frac{x_{2j}}{a_{2j}}, \ldots, \frac{x_{nj}}{a_{nj}}\right\}$$

and constant returns to scale are implied here.

14.18 GENERAL EQUILIBRIUM SYSTEM

If we think of the input output system in terms of the supply demand balance equations for the economy's productive system

$$\begin{aligned} X_1 &= x_{11} + x_{12} + \ldots + x_{1n} + F_1 \\ X_2 &= x_{21} + x_{22} + \ldots + x_{2n} + F_2 \\ \ldots & \qquad \ldots \qquad \ldots \\ X_n &= x_{n1} + x_{n2} + \ldots + x_{nn} + F_n \end{aligned}$$

then we have an empirically implementable general equilibrium system. Each x_{ij} is a function of X_j only so that

$$x_{ij} = f(X_j)$$

which when used in the above system makes it determinate in n unknowns, the final demand being given exogenously.

The elimination of prices may be implied to a certain extent.

(*a*) Prices can be ignored here because the relation links physical requirements of the ith input with the jth physical output. However physical measures are difficult to manage as for instance, in the case of services.

In the base year the quantity unit can be so chosen that its price equals the country's unit of account. Value data can be used as if it were made of physical quantities.

(*b*) If there is a single scarce factor say, labour and everything is measured in labour time then relative prices do not change. The argument will fail if there are more than one scarce factors.

(*c*) There is limited scope for substitution between inputs at the firm level.

(*d*) Price elimination may be justified if all cross-price elasticities are zero and the own price elasticity demand is unity.

14.19 DETERMINATE SYSTEMS

If the relation $x_{ij} = f(X_j)$ is linear, the form $x_{ij} = a_{ij}\, X_j$ provides constant returns to scale and a_{ij} becomes an average input output coefficient.

If we have the relation $x_{ij} = a_{ij} + c_{ij} X_j$, we get the marginal input output coefficient c_{ij} along with the average coefficient a_{ij}. Both the relations produce a system which is determinate in n unknowns.

$x_{ij} = a_{ij} X_j$ implies constant returns to scale. With nonconstant returns the relation

$$x_{ij} = a_{ij} + c_{ij} X_j$$

is preferable since the marginal coefficient would mean greater efficiency while a_{ij} will not be significantly different from zero.

In practice, however, the former relation is used.

Ex. 22. Show that the gross output level of an industry can be estimated as a function of final demand. Show how primary input demand can be derived and how the model may be closed to yield a solution for the income level.

Ex. 23. "The Leontief open model contains an implicit labour theory of value in that a set of prices proportional to total labour requirement coefficients forms an equilibrium set of prices for all final demands." Explain.

(The standard Leontief open model contains only one primary input labour. An equilibrium set of prices is such that the wage level just enables labour to purchase the net output of the economy so that for all industries, profits are exactly zero.)

Ex. 24. In a developing country, can be assumption of constancy of technical coefficients be tenable?

(The assumption of constant input coefficients is not very realistic, particularly in a developing country, where the constituents of the input output matrix are likely to change often and considerably.)

14.20 REGIONAL AND INTERREGIONAL INPUT OUTPUT ANALYSIS

The input output model is useful in regional and interregional applications. Regional models are similar to the national models but they cover a smaller geographical area. The interregional or multiregional models include more than one region. This can be done by aggregating a number of regional tables. A balanced regional model is obtained by suitably disaggregating a national model. Both the models are complementary and help in describing the economic system in terms of interdependent industries and regions.

In the absence of accurate data at the regional level, it is not unusual to utilise national coefficients to arrive at regional flow estimates as an approximate but not necessarily reliable alternative. There are variations in the industry mix of different regions.

Note. 1. Regional impact studies measure the direct, indirect and induced income effects of changes in the final demand in a region or the total impact of the location of a new industry in a region.

2. The treatment of purchases from and sales to other regions and other components of final demand as an exogenous factor does not permit the formulation of a comprehensive general equilibrium theory.

3. The difference between national and regional coefficients results from differences in technology, price structure, volume and pattern of interregional trade and industry mix.

To arrive at regional coefficients, adjustments in the national coefficients can be made by using one or more of the following methods:

(*a*) Location quotients as a proxy for regional coefficients (*b*) Cross industry quotients (*c*) Supply demand pool technique (*d*) RAS Technique. (*e*) Gravity model.

As an alternative, it would be proper to construct regional and national tables simultaneously for uniformity, using the survey method. It should be noted that fixed coefficients become redundant after a certain lapse of time. Moreover, often the use of national coefficients for measuring regional coefficients is not quite appropriate.

14.21 REGIONAL INPUT OUTPUT TABLES

Input Output models have been incorporated into Indian Planning from the Third Five Year Plan onwards. There are both national and regional input output tables but these cannot be well coordinated because of various dissimilarities. Since input output tables show the structural interdependence and structural basis of the economy they are useful in deriving regional intermediate demand, total production requirements and assessment of consistency of targets. They can help in assessing the worthwhileness of such objectives as poverty alleviation and reduction of inequalities, capital formation, competitive imports and exports and regional development.

14.22 LINKAGES

Degree of Interdependence Among Various Sectors

The degree of inter-indsutry dependence in advanced countries is higher than that in developing countries. Can we measure it? A knowledge of this degree can help us in assessing the extent of industrialisation of a country. Measures of this degree which is described as *linkages* are available.

Linkages

A new industrial set up is implied in the derived demand for inputs of other industries used in the new industry. This is Backward Linkage.

If the new industry does not cater exclusively to final demand its output will be used in other industries to a greater extent. This is Forward Linkage.

Both these linkages are likely to be higher in a closed economy than in an open one. Initially, the import content of a newly industrialising economy may be high but when inputs are later produced domestically, there are sizeable backward linkages.

The concepts are useful in studying the problems of development in growing economies.

Ex. 25. Is it possible to analyse the impact of changes in public and private expenditures in different sectors of the economy?

(**Hint.** The impact of an initial expenditure on an economic system depends not merely on the quantum of expenditure but also on interindustry linkages of sectors with the rest of the economy.)

Note: 1. *Factors of Production*. Input output analysis can be easily extended to consider demands for various factors of production. For instance, we may consider the quantities of labour and capital or additions to them to produce the final bill of goods. The procedure follows the same lines as these for final demand.

2. *Capital Flow Tables.* In a capital flow matrix, capital formation is the control total. This matrix shows the sale of capital producing industries across rows and purchases of capital consuming industries along columns.

3. *Projection of Input Output Tables.* We may project the table in such a way as to take into account changes in technology, material substitution and other factors. We may project the input coefficients of one year on to another year and so on.

14.23 EQUILIBRIUM PRICES: 3 INDUSTRY CASE

The following 3 industry case can be generalised to the n industry case

		Industry 1	2	3	*Final Demand*	*Row Total*
Industry	1	x_{11}	x_{12}	x_{13}	F_1	X_1
	2	x_{21}	x_{22}	x_{23}	F_2	X_2
	3	x_{31}	x_{32}	x_{33}	F_3	X_3
Primary input		y_1	y_2	y_3		

Suppose the prices of X_1, X_2, X_3 are p_1, p_2, p_3 and the wage rate for labour is w. Then assuming that outputs X_1, X_2, X_3 and final demands F_i are physical quantities to be valued in term of prices, the input output table becomes (using $x_{11} = a_1X_1$, $x_{21} = a_{21}X_2$, etc. and multiplying by p_j)

Industry	*1*	2	3	*F.D.*	*Total*
1	$a_{11}X_1p_1$	$a_{12}X_2p_1$	$a_{13}X_3p_1$	F_1p_1	X_1p_1
2	$a_{21}X_1p_2$	$a_{22}X_2p_2$	$a_{23}X_3p_2$	F_2p_2	X_2p_2
3	$a_{31}X_1p_3$	$a_{32}X_2p_3$	$a_{33}X_3p_3$	F_3p_3	X_3p_3
Primary Input	l_1X_1w	l_2X_2w	l_3X_3w		

Industry 1. Total Cost $= a_{11}X_1p_1 + a_{21}X_1p_2 + a_{21}X_1p_3 + l_1X_1w$

Total receipt $= X_1p_1$

Equating, $X_1p_1 = a_{11}X_1p_1 + a_{21}X_1p_2 + a_{31}X_1p_3 + l_1X_1w$

so that $p_1 = a_{11}p_1 + a_{21}p_2 + a_{31}p_3 + l_1w$

or $(1 - a_{11})p_1 - a_{21}p_2 - a_{31}p_3 = l_1w$ Industry 1

Similarly $-a_{21}p_1 + (1 - a_{22})p_2 - a_{32}p_3 = l_2w$ Industry 2

$-a_{31}p_1 - a_{32}p_2 + (1 - a_{33})p_3 = l_3w$ Industry 3

We have
$$\begin{bmatrix} 1-a_{11} & -a_{21} & -a_{31} \\ -a_{21} & 1-a_{22} & -a_{23} \\ -a_{31} & -a_{32} & 1-a_{33} \end{bmatrix} \begin{bmatrix} p_1 \\ p_2 \\ p_3 \end{bmatrix} = \begin{bmatrix} l_1w \\ l_2w \\ l_3w \end{bmatrix}$$

or $\quad (I - A)p = lw$ in matrix notation

$\qquad = V$ say (for value added lw)

then $\quad p = (I - A)^{-1}V$

Writing $(I - A)^{-1} = \begin{bmatrix} \gamma_{11} & \gamma_{12} & \gamma_{13} \\ \gamma_{21} & \gamma_{22} & \gamma_{23} \\ \gamma_{31} & \gamma_{32} & \gamma_{33} \end{bmatrix}$, we have ($\gamma_{ij}$ are elements of $(I - A)^{-1}$)

$$\therefore \quad p_1 = (\gamma_{11}l_1 + \gamma_{12}l_2 + \gamma_{13}l_3)w$$

$$p_2 = (\gamma_{21}l_1 + \gamma_{22}l_2 + \gamma_{23}l_3)w$$

$$p_3 = (\gamma_{31}l_1 + \gamma_{32}l_2 + \gamma_{33}l_3)w$$

Note. The demand y for a primary input like labour is related to find demand:

$$\frac{\partial y}{\partial F_1} = \gamma_{11}l_1 + \gamma_{12}l_2 + \gamma_{13}l_3$$

$$\therefore \quad p_1 = \frac{\partial y}{\partial F_1} w \text{ etc.}$$

i.e., MC of labour = (Marginal employment) × (wage rate).

Ex. 26. Derive the results for equilibrium prices in the two input case.

Industry	*1*	*2*	*Final Demand*	*Total receipts*
1	$a_{11}X_1p_1$	$a_{12}X_2p_1$	F_1p_1	X_1p_1
2	$a_{21}X_1p_2$	$a_{22}X_2p_2$	F_2p_2	X_2p_2
y	l_1X_1w	l_2X_2w		

Total cost $\quad a_{11}X_1p_1 + a_{21}X_1p_2 + l_1X_1w = X_1p_1$

$$a_{12}X_2p_1 + a_{22}X_2p_2 + l_2X_2w = X_2p_2$$

$$\therefore \quad a_{11}p_1 + a_{21}p_2 + l_1w = p_1$$

$$a_{12}p_2 + a_{22}p_2 + l_2w = p_2$$

$$(1 - a_{11})p_1 - a_{21}p_2 = l_1w$$

$$-a_{12}p_1 + (1 - a_{22})p_2 = l_2w$$

$$\therefore \quad \begin{bmatrix} p_1 \\ p_2 \end{bmatrix} = \begin{bmatrix} 1 - a_{11} & -a_{21} \\ -a_{12} & 1 - a_{22} \end{bmatrix}^{-1} \begin{bmatrix} l_1w \\ l_2w \end{bmatrix}$$

or $\quad p = (I - A)^{-1}V$ which yields

$$p_1 = (\gamma_{11}l_1 + \gamma_{12}l_2)w \qquad (\gamma_{ij} \text{ being elements of } (I - A)^{-1})$$

$$p_2 = (\gamma_{21}l_1 + \gamma_{22}l_2)w$$

Ex. 27. Given $A = \begin{bmatrix} 0.2 & 0.2 \\ 0.7 & 0.2 \end{bmatrix}$ and $V = \begin{bmatrix} 8 \\ 2 \end{bmatrix}$

$$\begin{bmatrix} p_1 \\ p_2 \end{bmatrix} = p = (I - A)^{-1} V = \begin{bmatrix} 1.6 & 0.4 \\ 1.4 & 1.6 \end{bmatrix} \begin{bmatrix} 8 \\ 2 \end{bmatrix} = \begin{bmatrix} 4.0 \\ 14.4 \end{bmatrix}$$

Ex. 28. Let W stand for the vector of wages per unit of output in different industries and Z stand for profits per unit of output in the industries. Write the price formation equation.

Since value added can be written as the sum of W and Z we have

$$p = (I - A)^{-1} V$$

$$\therefore \quad p = (I - A)^{-1} (W + Z)$$

$$= (I - A)^{-1} (wL + Z)$$

where L is the vector of labour coefficients and w is the uniform wage rate of all sectors.

Ex. 29. If the primary input is labour and l_1, l_2 are labour coefficients derive the labour requirement relation for the 2 input case.

Industry	*1*	2	*Final Demand*
1	a_{11}	a_{12}	F_1
2	a_{21}	a_{22}	F_2
Primary Input	l_1	l_2	

$$X = (I - A)^{-1} F = \begin{bmatrix} 1 - a_{11} & -a_{12} \\ -a_{21} & 1 - a_{22} \end{bmatrix}^{-1} \begin{bmatrix} F_1 \\ F_2 \end{bmatrix}$$

$$\begin{bmatrix} X_1 \\ X_2 \end{bmatrix} = \frac{1}{M} \begin{bmatrix} A_{11} & A_{21} \\ A_{12} & A_{22} \end{bmatrix} \begin{bmatrix} F_1 \\ F_2 \end{bmatrix}$$

where A_{ij} are cofactors of $[I - A]$ and M is the determinant of $[I - A]$.

$$X_1 = \frac{1}{M} [A_{11} F_1 + A_{21} F_2] = \gamma_{11} F_1 + \gamma_{21} F_2$$

$$X_2 = \frac{1}{M} [A_{12} F_1 + A_{22} F_2] = \gamma_{12} F_1 + \gamma_{22} F_2$$

The labour requirement is

$$L = l_1 X_1 + l_2 X_2$$

This can be written

$$L = [l_1 \; l_2] \begin{bmatrix} X_1 \\ X_2 \end{bmatrix} = lX$$

$\therefore$ $L = l(I - A)^{-1} F$. This can be generalised to the n input case.

Ex. 30. Derive the primary input requirement if there are two primary inputs L_1, L_2 in the above case.

Industry	*1*	2	*Final Demand*	*Total*
1	a_{11}	a_{12}	F_1	X_1
2	a_{21}	a_{22}	F_2	X_2
Primary	l_{11}	l_{12}		
Inputs	l_{21}	l_{22}		

$$\textbf{Ans. } L = \begin{bmatrix} L_1 \\ L_2 \end{bmatrix} = \begin{bmatrix} l_{11} & l_{12} \\ l_{21} & l_{22} \end{bmatrix} \begin{bmatrix} X_1 \\ X_2 \end{bmatrix} = lX = l(I - A)^{-1} F$$

Note. 1. We can translate final demand into demand for various primary inputs.

2. We can determine the quantity of any primary input required to lead to a certain given final demand.

3. We can find out what final demand will ensure full utilisation of a certain primary input.

4. We can determine the additional input requirement corresponding to a given rise in final demand.

5. In the case of excess capacity, we can try to absorb this excess by changes in final demand.

Three Industry, Two Input Case

For a 3-industry, 2 primary input case, we can write the total primary input use

$$L_1 = L_{11} + L_{12} + L_{13} \text{ for labour}$$

$$L_2 = L_{21} + L_{22} + L_{23} \text{ for capital}$$

L_{ij}'s are the quantities of inputs used in the three industries. The primary input coefficients are denoted by $l_{ij} = \dfrac{L_{ij}}{X_J}$ and can be shown in a matrix form

$$l = \begin{bmatrix} l_{11} & l_{12} & l_{13} \\ l_{21} & l_{22} & l_{23} \end{bmatrix}$$

Ex. 31. Show how to find the primary input vector in the three industry case.

$$L_1 = l_{11}X_1 + l_{12}X_2 + l_{13}X_3$$

$$L_2 = l_{21}X_1 + l_{22}X_2 + l_{23}X_3$$

$$\begin{bmatrix} L_1 \\ L_2 \end{bmatrix} = \begin{bmatrix} l_{11} & l_{12} & l_{13} \\ l_{21} & l_{22} & l_{23} \end{bmatrix} \begin{bmatrix} X_1 \\ X_2 \\ X_3 \end{bmatrix}$$

or $\quad L = lX$

Since $\quad X = (I - A)^{-1}F$

$\therefore \quad L = l(I - A)^{-1}F$

i.e., to find the primary input vector, the matrix of primary input coefficients should be postmultiplied by he Leontief inverse $(I - A)^{-1}$. The notation remains unchanged.

Ex. 32. Given the two industry technology matrix

Industry	X_1	X_2
X_1	0.3	0.3
X_2	0.4	0.6
Primary Input	0.3	0.1

Determine the output when final demands for the two industries are $F_1 = 90$, $F_2 = 520$. Also find the labour requirements.

$$X = \begin{bmatrix} X_1 \\ X_2 \end{bmatrix} = [I - A]^{-1} F \text{ where } I - A = \begin{bmatrix} 0.7 & -0.3 \\ -0.4 & 0.4 \end{bmatrix}$$

$$= \begin{bmatrix} 0.7 & -0.3 \\ -0.4 & 0.4 \end{bmatrix}^{-1} \begin{bmatrix} 90 \\ 520 \end{bmatrix}$$

$$\begin{bmatrix} X_1 \\ X_2 \end{bmatrix} = \frac{1}{0.16} \begin{bmatrix} 0.4 & 0.3 \\ 0.4 & 0.7 \end{bmatrix} \begin{bmatrix} 90 \\ 520 \end{bmatrix} = \begin{bmatrix} 1200 \\ 2500 \end{bmatrix}$$

Labour requirements $L = l_1X_1 + l_2X_2$ where l_1, l_2 are labour input coefficients.

$$L = 0.3 \times 1200 + 0.1 \times 5200 = 880$$

Ex. 33. Find outputs and labour requirements in the above example if final demands are 150, 600.

$$\textbf{Ans. } \begin{bmatrix} X_1 \\ X_2 \end{bmatrix} = \frac{1}{0.16} \begin{bmatrix} 0.4 & 0.3 \\ 0.4 & 0.7 \end{bmatrix} \begin{bmatrix} 150 \\ 600 \end{bmatrix} = \begin{bmatrix} 1500 \\ 3000 \end{bmatrix}$$

$$L = 0.3 \times 1500 + 0.1 \times 300 = 750$$

Ex. 34. Consider the input output table for the following system:

(Rs. '00 million)

Input / *Output*	*Agr.*	*Ind.*	*Final Demand*	*Total output*
Agr.	200	150	1050	1400
Ind.	—	750	1150	1900
Primary Input	1200	1000	—	2000
Total Input	1400	1900	2200	

Examine the extent to which the output level of agriculture will rise as a result of increase in its household consumption by Rs. 60 (in '00 million).

Hint: The technical coefficients are

$$a_{11} = 200/1400 = 1/7,\ a_{12} = 150/1900 = 15/190,$$
$$a_{21} = 0,\ a_{22} = 750/1900 = 75/190$$
$$a_{01} = 1200/1400 = 6/7,\ a_{02} = 100/1900 = 100/190$$

$$\left.\begin{array}{r} \frac{1}{7}X_1 + \frac{15}{190}X_2 + 1110 = X_1 \\ \frac{75}{190}X_2 + 1150 = X_2 \\ \frac{6}{7}X_1 + \frac{100}{190}X_2 = X_0 \end{array}\right\} \begin{array}{l} X_1 = 1470 \\ X_2 = 1900 \\ X_0 = 2260 \end{array}$$

Rise in agriculture output = 1470 – 1400 = 70

Ex. 35. Find the output requirements of three industries with the following technology matrix A when the final demands are 7000, 6000 and 9000. Also find the labour requirements and the change in output when final demand falls by 10 per cent

$$A = \begin{bmatrix} 0.2 & 0.4 & 0.5 \\ 0.5 & 0.1 & 0.1 \\ 0.1 & 0.2 & 0.1 \end{bmatrix}$$

Note. The labour coefficients are $l_1 = 0.2$, $l_2 = 0.3$, $l_3 = 0.3$

Ex. 36. Given the technology matrix

$$A = \begin{bmatrix} 0.3 & 0.3 & 0.2 \\ 0.1 & 0.2 & 0.3 \\ 0.2 & 0.1 & 0.4 \end{bmatrix}$$

find the equilibrium levels of outputs when final demands for the goods of the three industries are given to be $F_1 = 50{,}000$; $F_2 = 30{,}000$; $F_3 = 60{,}000$.

Ans. First find $(I - A)^{-1} = \begin{bmatrix} 1.837 & 0.816 & 1.020 \\ 0.490 & 1.551 & 0.939 \\ 0.694 & 0.531 & 2.163 \end{bmatrix}$

and show that

$$X = (I - A)^{-1}F = \begin{bmatrix} 177530 \\ 127370 \\ 180410 \end{bmatrix}$$

Ex. 37. Derive the interindustry matrix and labour requirements in the last exercise

Since $X_1 = 177530,\ X_2 = 127370,\ X_3 = 180410$

We can derive various ingredients of interindustry demand.

$$a_{11}X_1 = 0.3 \times 177530 = 53259$$
$$a_{12}X_2 = 0.3 \times 127370 = 38211 \text{ } etc.$$

The interindustry matrix is

Industry	*1*	*2*	*3*	*F*	*Total*
1	53259	38211	36082	50000	177530
2	17753	25474	54123	30000	127370
3	35506	12737	72164	60000	180410

The labour coefficients are $l_1 = 0.4, l_2 = 0.4, l_3 = 0.1$

The labour requirement is

$$L = l_1X_1 + l_2X_2 + l_3X_3$$
$$= 0.4 \times 177530 + 0.4 \times 127370 + 0.1 \times 180410$$
$$= 140000.$$

Ex. 38. In the 2 industry case, given $A = \begin{bmatrix} 0.2 & 0.2 \\ 0.7 & 0.2 \end{bmatrix}$

determine the outputs required to meet the final demand set 500, 5000. Also find the corresponding labour requirement. **Ans.** $[I - A]^{-1} = \frac{1}{0.5}\begin{bmatrix} 0.8 & 0.2 \\ 0.7 & 0.8 \end{bmatrix} = \begin{bmatrix} 1.6 & 0.4 \\ 1.4 & 1.6 \end{bmatrix}$

$$\therefore \quad X = \begin{bmatrix} 1.6 & 0.4 \\ 1.4 & 1.6 \end{bmatrix}\begin{bmatrix} 500 \\ 5000 \end{bmatrix} = \begin{bmatrix} 2800 \\ 8700 \end{bmatrix}$$

Labour Requirement $L = 0.3 \times 2800 + 0.6 \times 8700$
$= 840 + 5220 = 6040$

14.24 TRIANGULAR TECHNOLOGY MATRIX

If A is a triangular matrix, $(I - A)^{-1}$ is also triangular. If A is upper triangular in a 3 industry input output model then output

(*i*) X_3 depends on final demand F_3

(*ii*) X_2 depends on final demand F_2, F_3

(*iii*) X_1 depends on F_1, F_2, F_3

Ex. 39. For the triangular matrix $A = \begin{bmatrix} 0.3 & 0.2. & 0.3 \\ 0 & 0.2 & 0.3 \\ 0 & 0 & 0.2 \end{bmatrix}$

Find $(I - A)^{-1}$

Ex. 40. Use the value of $(I - A)^{-1}$ in the above case to determine the output levels consistent with the final demands $F_1 = 300$, $F_2 = 200$, $F_3 = 100$.

$$\textbf{Ans.} \begin{bmatrix} X_1 \\ X_2 \\ X_3 \end{bmatrix} = (I - A)^{-1}\begin{pmatrix} 300 \\ 200 \\ 100 \end{pmatrix} = \begin{bmatrix} 1.11 & 0.42 & 0.28 \\ 0 & 1.25 & 0.36 \\ 0 & 0 & 1.43 \end{bmatrix}\begin{bmatrix} 300 \\ 200 \\ 100 \end{bmatrix} = \begin{bmatrix} 445 \\ 286 \\ 143 \end{bmatrix}$$

Ex. 41. In the above case if there is 10 per cent change in final demands what will be the change in outputs of different industries?

Since $X = (I - A)^{-1}F$

$\therefore$ The change $dX = (I - A)^{-1}dF$

$$= \begin{bmatrix} 1.11 & 0.42 & 0.28 \\ 0 & 1.25 & 0.36 \\ 0 & 0 & 1.43 \end{bmatrix} \begin{bmatrix} 30 \\ 20 \\ 10 \end{bmatrix}$$

(Since 10% of 300 = 30, of 200 = 20, of 100 = 10)

$$\therefore \quad dX = \begin{bmatrix} 44.5 \\ 28.6 \\ 14.3 \end{bmatrix} \text{ i.e., } \begin{aligned} dX_1 &= 44.5 \\ dX_2 &= 28.6 \\ dX_3 &= 14.3 \end{aligned}$$

14.25 SERIES EXPANSION (See Sec. 14.26)

Ex. 42. Using the $(I - A)^{-1}$ expansion find the (*a*) unit costs of intermediate purchases, (*b*) prices in terms of expenditure on primary inputs.

Intermediate Purchases

In the two input cases let the price vector be

$$p = [p_1 p_2] \text{ and } A = \begin{bmatrix} a_{11} & a_{12} \\ a_{21} & a_{22} \end{bmatrix}$$

The costs of intermediate purchases are

$$PA = [p_1\ p_2] \begin{bmatrix} a_{11} & a_{12} \\ a_{21} & a_{22} \end{bmatrix} = [p_1 a_{11} + p_2 a_{21} \quad p_1 a_{12} + p_2 a_{22}]$$

If $$A = \begin{bmatrix} 1/2 & 1/3 \\ 1/4 & 1/2 \end{bmatrix}, \quad PA = \begin{bmatrix} \frac{p_1}{2} + \frac{p_2}{4} & \frac{p_1}{3} + \frac{p_2}{2} \end{bmatrix}$$

i.e., (1/2 unit of X_1) + (1/4 unit of X_2) produces 1 unit of X_1

(1/2 unit of X_1) + (1/2 unit of X_2) produces 1 unit of X_2

$\therefore$ Costs of intermediate purchases are

$$\frac{p_1}{2} + \frac{p_2}{4} \text{ per unit of } X_1$$

$$\frac{p_1}{3} + \frac{p_2}{2} \text{ per unit of } X_2$$

If V stands for the vector of expenditures on the primary inputs per unit of gross output then

$$pA + V = p \text{ or } V = p(I - A)$$

Also $p = V(I - A)^{-1} \simeq V[I + A + A^2 + ... + A^{m-1}]$
$= V$ (*i.e.* cost of primary inputs in producing finished goods)
$+ VA$ (*i.e.* cost of material consumed in producing)
$+ VA^2$ (*i.e.* cost of primary inputs to create raw materials) and so on.

Ex. 43. Show that for the two input homogeneous case

$$p_1 a_{11} + p_2 a_{21} = p_1$$
$$p_1 a_{12} + p_2 a_{22} = p_2$$

and hence the price ratio $\dfrac{p_1}{p_2} = \dfrac{1 - a_{22}}{a_{12}} = \dfrac{a_{21}}{1 - a_{11}}$

Ex. 44. (*a*) Can we get unique values of p_1, p_2 if there is a condition given that $\Sigma p_i = 1$.

Ans. Yes

(*b*) Can we get unique vlaues of p_1, p_2 if the wage rate is given?

Ans. The relations $\begin{matrix} p_1 a_{11} + p_2 a_{21} + l_1 w = p_1 \\ p_1 a_{12} + p_2 a_{22} + l_2 w = p_2 \end{matrix}$ give p's in terms of w

Ex. 45. If the profit rate is assumed to be equal to interest rate r then the above relations become

$$(1 + r)(p_1 a_{11} + p_2 a_{21}) + l_1 w = p_1$$
$$(1 + r)(p_1 a_{12} + p_2 a_{22}) + l_2 w = p_2$$

Can these be solved uniquely?

Ans. No, unless some additional stipulations are given

Ex. 46. Set out the equilibrium conditions when there are 2 industries. Find F_1, F_2. What is the inverse of the technology matrix in this case if $a_{11} = 0$, $a_{22} = 0$?

Ans. $\begin{bmatrix} \dfrac{1}{1 - a_{12} a_{21}} & \dfrac{a_{12}}{1 - a_{12} a_{21}} \\ \dfrac{a_{21}}{1 - a_{12} a_{21}} & \dfrac{1}{1 - a_{12} a_{21}} \end{bmatrix}$

Ex. 47. What is interindustry economics?

(As a form of applied economics, interindustry analysis deals with the quantitative analysis of the interdependence of producing units in an economy. It is used where methods of national income analysis and partial equilibrium analysis fail, though the former is closer to it. Input output analysis and linear programming can be looked upon as variations of interindustry economics.

Interindustry economics may be used to study structural analysis of an economy, formulation of programmes when various alternative policies are available, prediction of future events, projection of the economic structure like an overall development of the economy and so on.)

Ex. 48. "The input output system is derived from assumptions about economic behaviour and definitions of the variables used in the analysis." Discuss.

Ex. 49. What is structural disequilibrium in underdeveloped economies? How can input output analysis help in the matter?

[(*i*) Inefficient, improper use of available resources, shortage of entrepreneurs, underemployment, scarcity of capital and foreign exchange, etc. lead to structural disequilibrium.

(*ii*) Input output analysis may be utilised for the projection of growth of various sectors, balancing of resources, allocation of investment funds, projection of production, consumption and balance of payments etc.]

Ex. 50. Show that the Leontief inverse when

$$A = \begin{bmatrix} 0.1 & 0.3 & 0.1 \\ 0 & 0.2 & 0.2 \\ 0 & 0 & 0.3 \end{bmatrix} \text{ is } [I - A]^{-1} = \begin{bmatrix} 1.11 & 0.42 & 0.28 \\ 0 & 1.25 & 0.36 \\ 0 & 0 & 1.43 \end{bmatrix}$$

Also show $X_3 = 1.43F_3$, $X_2 = 1.25\, F_2 + 0.036\, F_3$ and $X_1 = 1.11\, F_1 + 0.42\, F_2 + 0.28\, F_3$.

Show that by using the power series

$$(I - A)^{-1} = I + A + A^2 + \ldots\ldots$$

up to three or four terms we get the same result as that found above.

Ex. 51. Let $A = \begin{bmatrix} 1/2 & 0 \\ 0 & 1/4 \end{bmatrix}$ then $A^2 = \begin{bmatrix} (1/2)^2 & 0 \\ 0 & (1/4)^2 \end{bmatrix}$

$$A^3 = \begin{bmatrix} (1/2)^3 & 0 \\ 0 & (1/4)^3 \end{bmatrix}, \ldots\ldots, A^m = \begin{bmatrix} (1/2)^m & 0 \\ 0 & (1/4)^m \end{bmatrix}$$

As $m \to \infty, \left(\frac{1}{2}\right)^m \to 0, \left(\frac{1}{4}\right)^m \to 0$

$$\therefore \quad \lim_{m\to\infty} A^m = \begin{bmatrix} 0 & 0 \\ 0 & 0 \end{bmatrix} = O \text{ a null matrix}$$

Note. Power series of $(I - A)^{-1}$

If a coefficient matrix A has the property that $\lim_{m\to\infty} A^m = 0$ we can use the approximation

$$R = (I - A)^{-1} \simeq I + A + A^2 + \ldots + A^{m-1}$$

Ex. 52. If $A = \begin{bmatrix} 1/2 & 0 & 1/3 \\ 0 & 2/3 & 0 \\ 0 & 0 & 1/2 \end{bmatrix}$, $A^2 = \begin{bmatrix} 1/4 & 0 & 1/3 \\ 0 & 4/9 & 0 \\ 0 & 0 & 1/4 \end{bmatrix}$,

$$A^3 = \begin{bmatrix} 0 & 0 & 1/4 \\ 1/8 & 8/27 & 0 \\ 0 & 0 & 1/8 \end{bmatrix}$$

$$(I - A)^{-1} \approx I + A + A^2 + A^3 = \begin{bmatrix} 15/8 & 0 & 11/12 \\ 0 & 65/27 & 0 \\ 0 & 0 & 15/8 \end{bmatrix}$$

14.26 MULTIPLIER CHAIN: ECONOMIC INTERPRETATION OF LEONTIEF INVERSE

Ex. 53. Can the expansion $(I - A)^{-1}$ be used as a multiplier chain? **Ans.** Yes.

Since $X = (I - A)^{-1} F \simeq (I + A + A^2 + \ldots + A^{m-1})F$

$\therefore$ $X = F + AF + A^2F + \ldots + A^{m-1} F$ assuming $A^m \to$ as $m \to \infty$

$\therefore$ Gross output X must be at least equal to

F + gross output AF to produce F

+ gross output A^2F to produce AF ... and so on

The multiplier chain $AF + A^2F + \ldots$ gives the inputs indirectly consumed in production.

14.27 SOME OTHER TECHNICAL COEFFICIENTS

Besides the technical coefficients which the producers of inputs require from each industry to produce a money unit worth of output in a given industry, there are other coefficients in use which are expressed in physical or value terms.

Capital Coefficient

In the static input output model, capital forms part of final demand. In contrast to the flows of the basic transactions, the capital coefficient is a stock concept and is the ratio of the quantity of capital required per unit of capacity in an industry *i.e.*, the ratio of a given type of capital to the maximum output of an industry.

The average capital coefficient, used in structural analysis, is the total amount of fixed capital per unit of capacity. It relates the existing capital stock to existing capacity and presents the capital structure of an economy at a given time.

The incremental capital coefficient gives the marginal capital structure of an industry. Used in dynamic analysis, it is the ratio of capital increment to capacity increment. It is usually based on relatively new plant data and differs from the average capital coefficients because of technological change. It can be evaluated even for a yet not started plant. It is a link between static and dynamic input output analysis.

Labour Input Coefficient

This coefficient stands for physical labour input in an industry per unit of output and can be used to estimate direct, indirect and induced effects of any given changes in final demand. If there are no major changes in product mix this coefficient is likely to remain constant in the short run though it may be variable in the long run, depending on any substitution of capital or other changes. The changes in the labour coefficient are likely to be gradual and in the same direction so that they can be used for employment projections while allowing for the effects of technological progress upon labour requirements.

Inventory Coefficient

This is an estimate of the stock of an input to be held in the economy per unit of output and is thus a measure of working capital. The coefficient is likely to be of the average rather than incremental variety because technological change is not likely to affect the normal level of inventories in the short period.

Dynamic Models

Although the static input output model has proved to be vary useful and handy in a number of situations, it is limited to the flow of current transactions, has fixed technical coefficients and is meant for short run analysis. The dynamic model is experimental in nature and is linked with the static model through incremental capital coefficients. It requires a full description of the capital structure of the economy to supplement the flow of current transactions.

14.28 DYNAMIC INPUT OUTPUT ANALYSIS

The dynamic case is an extension of static input output analysis. An output may be used as an intermediate input or may satisfy final demand. Alternatively, it may be used for capital formation and thus connect the past with the future through machinery, inventories etc. all of which ascertain productivity rise, growth, smoother economic activity and better sales.

Static input output analysis includes capital formation as one of the ingredients of final demand.

In the dynamic system with moving equilibrium, the acceleration principle is used to relate output to capital stock S.

(*i*) The flow x_{ij} of ith product is a fixed proportion a_{ij} of the corresponding flow of X_j.

(*ii*) The jth industry does not use its own product so that $a_{jj} = -1$.

(*iii*) Similar to a_{ij}, there are capital coefficients

$$b_{ij} = \frac{S_{ij}}{X_j}, \qquad i, \ j = 1, \ 2, \ \ldots, \ n$$

(*iv*) S_{ij} is the stock of the ith product held by the jth industry, including capital formation from inventories, plants, equipments etc.

(*v*) The structure of the economy is represented by the Flow Structure

$$A = \begin{bmatrix} 1 & -a_{12} & \ldots & -a_{1n} \\ -a_{21} & 1 & \ldots & -a_{2n} \\ -a_{n1} & -a_{n2} & \ldots & -a_{nn} \end{bmatrix}$$

and the capital structure

$$B = \begin{bmatrix} b_{11} & b_{12} & \ldots & b_{1n} \\ b_{21} & b_{22} & \ldots & b_{2n} \\ \ldots & \ldots & \ldots & \ldots \\ b_{n1} & b_{n2} & \ldots & b_{nn} \end{bmatrix}$$

(*vi*) a_{ij} is in ith product units per unit of the jth product

b_{ij} is in ith product units held per unit flow of the jth product.

(*vii*) Each capital coefficient b_{ij} in the relation $S_{ij} = b_{ij} X_j$ is the constant of an accelerator:

$$\frac{d S_{ij}}{dt} = b_{ij} \frac{dX_j}{dt}$$

i.e. Investment in ith product $= b_{ij} \times$ (change in output of jth industry).

Open System

In the open system, final demand F_i is a given function of time. We have

$$X_i = \sum_j x_{ij} + F_i + \sum_i \frac{dS_{ij}}{dt} \qquad i = 1, 2, \ldots, n$$

	input for			
=	other	+ final	+	other
	industries	demand		industries

The term $\dfrac{dS_{ij}}{dt}$ is positive or negative depending on whether inventories are being built or running down. The resulting equilibrium equation

$$\sum_j a_{ij} X_j + F_i + \sum_j b_{ij} \frac{dX_j}{dt} = 0 \qquad i = 1, 2, \ldots n$$

is an extended version of the static case.

Closed System

In the closed model, final demand is the nth "industry" so that the equilibrium condition is

$$\sum_j a_{ij} X_j + \sum_j b_{ij} \frac{dX_j}{dt} = 0 \qquad i = 1, 2, \ldots n.$$

The solutions of these differential equation sets provide the output path over time.

This dynamic system has possibilities similar to those of the trade cycle. It may or may not show oscillations or dampening.

In the two input open system case we have two equations:

$$a_{11} X_1 - a_{12} X_2 - b_{11} \frac{dX_1}{dt} - b_{12} \frac{dX_2}{dt} = F_1$$

$$-a_{21} X_1 - a_{22} X_2 - b_{21} \frac{dX_1}{dt} - b_{22} \frac{dX_2}{dt} = F_2$$

The closed system is a particular case in which $F_1 = 0$, $F_2 = 0$. The general solution is given by

$$X_1 = k_1 e^{-\theta_1 t} + k_2 e^{-\theta_2 t}$$

$$X_2 = k_1 m_1 e^{-\theta_1 t} + k_2 m_2 e^{-\theta_2 t}$$

Ex. 54. Can the *LP* model be used to arrive at an optimal trade pattern?

If input output matrices are used as technological constraints, a linear programming model can be used to determine the optimal trade pattern, without assuming a fixed trading relation between regions and without a given spatial distribution of industries. Other constraints relating to various aspects of socioeconomic reality can also be introduced. The procedure requires the optimisation of the objectives of (*a*) a desired output level and (*b*) a desirable trading pattern.

14.29 SOME LIMITATIONS

Input output analysis is useful in the study of technology of interdependence of various industries or sectors in the economy under given conditions of demand. But the optimum solution of problems involving explicit constraints is not possible with the tools of input output analysis. For that we make use of the method of linear programming.

1. Input output economics does not allow for substitution among inputs. It assumes fixed technical coefficients. This amounts to proportionality or constant returns to scale for all sectors. However, this criticism can be dealt with the help of empirical investigation and appeal to facts.

2. The technique has been questioned as a predictive tool though several projections for industries, sectors and national aggregates made by input output analysis have been found to be quite satisfactory.

As a description of the structure of an economy at a given time, the static analysis may not be very useful as a prediction device but the dynamic input output analysis can be employed in such cases.

3. The economy is not solely a function of coefficients. The classical postulate of one to one correspondence between measures of utility and values are not tenable because of alternative distributive shares and their relation with values in a coefficient matrix which, therefore, cannot result in a unique set of relative values. Also, a set of relative values does not imply a unique coefficient matrix even under the constant costs assumption. We can thus say that an economy is not a mere function of some coefficients. However, these arguments may not be fully applicable to certain sectors or processes. So far as the postulated goal of the society's utility maximisation goes, there are difficulties in synchronising this with the fixed input output technical coefficients.

4. There are difficulties in the induction of the effect of technical choices and technological changes in dynamic analysis.

5. Input output economics, despite its increasing popularity has not been able to displace the use of macro-econometric models whose popularity instead of waning has multiplied over the years.

At the same time it should be conceded that input output economics is being successfully utilised for a variety of purposes and the disillusionment with some aspects of its methodology has not come in the way of its increasing use.

14.30 FORECASTING

Forecasting is done by using time series and econometric models. It is also possible to make use of input output tables for consistent forecasting. The projections for individual industries or sectors can be added. The consistency of forecast here implies that when an input output table is projected, the output of each industry is consistent with the final demand and other industries' requirements for its products.

For short periods of two or three years the structural coefficients may be assumed to remain unchanged so that short term forecasting may be considered reasonable. But for longer periods it is better to make use of a dynamic input output model.

For short term forecasts, static input output models have been found reasonably effective.

Consistent forecasting from an input output table is useful in public policy as well as in the management of a private enterprise to estimate the changes in its share of the total market or to anticipate sales.

Consistent forecasting does not necessarily require the use of planning of any type in a country. Consistent forecasts are useful in sensitivity analysis to determine the more sensitive components of the economy in its pattern of growth, to develop detailed employment estimates and to evaluate effects of long range government programmes. They are based on the interrelationship of demand, production, income, prices and resources.

The relation $X = (I - A)^{-1} F$ can be easily used as a forecasting tool, preferably if the technology is assumed to have remained unchanged over time. In case

at time t $\qquad X_t = (I - A_t)^{-1}F$

at time $t + T$ $\qquad X_{t+T} = (I - A_t)^{-1}F_{t+T}$

where F_{t+T} is the desired demand level at time $t + T$. This result will be tenable if the intervening period is such as to allow the assumption of constant technology.

In using A after a certain period some errors are likely to creep in and these errors may have an inherent trend. Usually, the technology matrix A is constructed every 5 to 6 years. In between one may use what is known as Stone's RAS method to update technical coefficients.

Input Output Analysis in Capitalist and Socialist Economies. The objective of input output analysis is to measure the impact of autonomous changes in final demand on the economy and not to provide any form of economic organisation. The technique is equally applicable in a free market economy, a centrally or partially planned economy or any other system independently of any political, social or economic beliefs. It is a value free technique not concerned with any current situation in the matter of allocation of resources or income distribution and describes an economy as it is. The input output system is an analytical tool which shows, for a given final demand, the level of activity required within the endogenous sector of an economy, irrespective of the causes of demand change.

14.31 SOME OBSERVATIONS

The first application of input output analysis was in the U.S. (not a planned economy). Later it was applied in other countries. In the French indicative planning, use was made of the input output model for long term projections of final demand and to provide the corresponding production target. Demographic trends and other factors were also utilised in the process.

The input output technique has been very useful in partially planned economies. In the then centrally planned economy of the Soviet Union under a number of severe ideological constraints, the input output technique was used with a difference. Instead of starting with sets of final demands related to technical coefficients to construct interindustry transaction for all the sectors, the centrally planned economy starts with sets of total outputs which include final demands. To deal with the political obstacles, the technique manages to identify the sectors in the input output table with a Marxian classification. Total output is made up of value of capital used in a given period, value of labour used in the production process and the surplus value of profit.

Development

Economic development requires, among other things, a restructuring of the economy. For this the analytical tool of input output table is quite useful as it can show the effect of expanding a sector or adding to the existing number of sectors. The growth of certain sectors can set a chain reaction to generate expansion in other sectors of the economy. Suitable comparisons with the input output tables of other economies can be made to study economic development in a proper perspective. The input output technique does not bring about economic development but can help in achieving faster progress at a lower cost of labour, capital and human sacrifice. As Leontief remarks, "much sweat and tears go into the actual realisation even of the best conceived development plan."

Developing countries usually do not satisfy all the requirements of input output technique which expects considerable stability of the economic structure, relatively subordinate external trade factor for the national economy to remain important, high level of industrialisation, adequate data and a well functioning government machinery working for forecasts etc. Despite some constraints many developing countries have made good use of input output techniques. These countries must allow for certain shortcomings like nonconsistancy of returns to scale, nonhomogeneity of sectoral outputs, frequent price changes, input substitution, altering relative costs etc. All these may affect the constancy of technological coefficients and hence of the technology even in the short run.

Nonconstancy of returns to scale may be overcome by using marginal instead of average coefficients. This requires a considerable amount of data. In fact, inadequate data constitute another problem of developing countries where the coefficients are more like allocation coefficients rather than optimal adjustment indicators of output mix which is what they should be.

CHAPTER 15

National Income and Economic Accounting

15.1 NATIONAL INCOME

An understanding of an economic system and its effect on people becomes easy if we have a general description of the economic activity of the region in the form of a numerical statement. Estimates of national income, expenditure and output give a convenient and concise view of an economic system.

National income, expressed in money terms, is the flow of goods and services produced in the economy in a given period of time. A national income estimate gives a money measure of the net aggregate of goods and services produced as a result of the economic activities of the people of a country in a given period of time. Usually the period taken is one year.

Three sets of flows in an economy

(1) The *flow* of productive services from owners to firms, industries, agriculture and productive enterprises, *which creates*

(2) the *flow* of money incomes, rents, wages etc. from firms etc. to owners of productive services, *which results in*

(3) the *flow* caused by investment, government activity, foreign trade, consumption etc.

Three measures of national income

(*i*) *National Product* (*N.P.*). The sum of the values of net outputs arising in all the sectors of the economy in a given period. This is the production phase and includes incomes from abroad.

(*ii*) *National Dividend or National Income* (*N.I.*). The sum of all incomes, dividends rents, wages, etc. received in cash or kind, accruing to the production factors in the given period. This is the distribution phase.

(*iii*) *National Expenditure* (*N.E*). The sum of all expenditures on goods and services by consumers and government and net addition to capital. This is the consumption phase.

We may write the identity of national income flows as:

$$NP = NI = NE$$

Circular flow. National income accounts relate to flow variables and not to stock variables. A stock variable has a given numerical value at a particular time, *e.g.,* the wealth of an individual or of a country. A flow variable is measured by units which are cumulative over time, *e.g.,* the level of incomes or expenditures over a year.

Income Viewpoint

Gross National Product (GNP). It is the market value of all final goods and services produced during a given year. Final goods are goods for final use and not for resale or reprocessing. This avoids double counting.

Net National Product (NNP). The whole of GNP is not added to the wealth of the economy. Replacement expenses and depreciation must be subtracted from GNP to get net national product.

$$\text{NNP} = \text{GNP} - \text{Depreciation}$$

National Income (NI). Indirect taxes are deducted from government subsidies added to NNP to arrive at national income or national product at factor cost.

NI at factor cost = NNP – Indirect taxes + Subsidies.

Personal Income (PI). NI does not give the total income of all the persons. Suitable adjustments have to be made to arrive at PI.

PI = NI – Social security contributions–Undistributed profits

– Corporate taxes + Transfer payments (pensions etc.)

Disposable Income (DI). Deduction of income taxes from PI yields DI.

Consumption and Savings (C and S). The income used for purchases of final goods and services is consumption. The remainder is saving.

$$\text{PI} - \text{Personal taxes} = \text{DI} = C + S$$

Expenditure Viewpoint

Consumption (C). The purchases of final goods and services constitute consumption expenditure.

Investment (I). Total capital expenditures to make up for depreciation and capital growth constitutes the level of private investment.

Unsold goods form a part of final output or GNP.

Government expenditure (G). Taxes and other revenues from the income stream are used in the expenditure stream by the government.

Foreign trade (W). Deduction of expenditure on imports from the incomes obtained from sales of final goods and services to foreigners constitutes net foreign investment *W.*

Ex. 1. Calculation of GNP, NNP etc. for country *X*.

		Rs. (million)
GNP at market prices		5000
less depreciation allowances		– 200
NNP at market prices		4800
less indirect taxes		– 400
plus subsidies		+ 100
NNP at factor cost		4500
less social security contributions	1200	
undistributed profits	500	– 1700
plus dividends from firms	800	
benefits from government	100	
interest and rent	300	+ 1200
Personal income		4000

Ex. 2. Compute GNP from the following:

	Rs. (million)
Depreciation allowances	50
Indirect taxes	100
Corporate income	300
Surplus of state enterprises	20
Interest paid by firms	30
Wages paid by firms	50
Interest and wages paid by others	100
Direct corporate taxes	

Ans. GNP is the total of all these items: 800

Difficulties in the Estimation of National Income

There are many conceptual and statistical difficulties to be faced in the measurement of national income. The obstacles in the way of measurement are all the greater in developing countries.

Theoretical or conceptual difficulties are with regard to

(*i*) definitions of the terms 'nation' and 'income',

(*ii*) method and concept of national income to be used in calculations,

(*iii*) the purpose of estimation, *e.g.*, whether it is to study the economic power of the economy or to study distribution of income and so on, (*iv*) the nature of commodities and services to be included in national income,

(*v*) the method of treatment of the government sector, foreign firms and international transactions,

(*vi*) non-monetised sectors and barter transactions,

(*vii*) suitability of money as a measure of national income.

Statistical and other difficulties are

(*i*) inadequacy and inaccuracy of data

(*ii*) regional diversities

(*iii*) inadequacy of efficient and trained personnel to handle statistical work

(*iv*) defective and improper occupational classification

(*v*) indifference of people

(*vi*) inadequate attention by government, particularly in developing countries.

15.2 ECONOMIC ACCOUNTING

Business accounts consolidate large unorganised data and give information about the actual position of a firm. They can be utilised to improve efficiency and for comparison with past performances. Superfluous parts of the business and wasteful expenditure may be checked with their help.

Economic accounting, for which the broader term social accounting is often used, is based on the double entry system of business accounting though on a much larger scale. It is a method of presenting a comprehensive and systematic record in the form of balancing statements of the activities and operations of an economy over successive periods. The statistical cum accounting framework of the aggregates of a number of important variables and relationships gives an excellent idea about the way in which a modern complex economy functions. Economic accounting is a useful structure for analytical purposes; it is a closed synthetic system showing all conceptual and statistical transactions and their interrelationships. Here it differs from business accounting and provides an empirical content to test the theories and concepts of economics and even contributes to the refinement and development of existing ideas which in turn, contribute to the improvement of the structuring of economic accounts themselves.

15.3 UN SYSTEM OF NATIONAL ACCOUNTS (SNA)

The transactions in SNA are classified into various sectors.

(*a*) The *Production Account* helps in analysing the demand and cost structure of industries.

(*b*) The *Consumption Account* pertains to goods and services, domestic or otherwise. This account provides the link (*i*) between sectors in which value added emerges and (*ii*) between sectors of origin and receipt of incomes.

(*c*) *The Capital Account* is concerned with capital formation and capital finance.

(*d*) *Imputed Transactions* are bifurcated from monetary transactions and can help in testing the worth of monetary policy.

The transactions are carried by the transactors which may be classified as (*a*) consumers (*b*) producers as industries (*c*) producers of government services and (*d*) profit services, (*e*) inter-household services.

Uses of Economic Accounts

The uses of national income accounts run parallel to those of business accounts:

(*i*) Economic accounts give a clear picture of the state of affairs in the economy. They consolidate large masses of unorganised data, give continuous information about the economy and can be utilised to improve results. Geographical and temporal comparisons are made possible by these accounts.

(*ii*) Greater efficiency and stability in the economy can be brought about with their help. While giving the annual incomes of various sectors and the extent and nature of consumption, production and capital formation they also reflect the trends in various sectors, the mode of distribution of economic resources, the purchasing power of the people, the presence of inflationary and deflationary forces, and so on.

(*iii*) These accounts can suggest measures for economic welfare, reducing inequalities, cutting down wastes and minimising costs in the economy.

(*iv*) National income accounts reveal the extent of the contributions of, and the interrelationships between different sectors of the economy. They show how each sector is functioning on its own and as compared with other sectors.

(*v*) They provide a comprehensive approach to economic analysis and method to deal with the changing nature of economic policy. National income accounts are not sufficient for formulating economic policy nor do they give a complete answer to the problem. But they give, in quantitative terms, a basis for the formulation of economic policy.

Preparation of Economic Accounts

To obtain useful information about the working of the economy, the classification of transactions has to be done carefully. The system used is similar to that in business *i.e.*, double entry book keeping. Each transaction is in money terms and recorded twice, once on the receipt side of one account and again on the expenditure side of another account.

Transactions may be *real* involving payments for goods and services or they may be *transfer payments* involving gifts, borrowings, pensions and other financial transactions without any transfer of goods and services.

Depreciation and consumption of one's own products are real economic transactions even though they do not seem to involve exchange in terms of money. Their value, however, can be recorded in terms of money.

Sectors

Based on their behavioral similarities, the entities which undertake economic transactions are usually divided into five sectors.

The producing sector. It organises production of goods and services with the object of earning profits. If makes use of factors provided by the household sector and pays for their use.

The household sector. These owners of factors of production are also consumers who pay for the goods and services produced by the producing sector. Often producers and households are the same persons but acting in different capacities at different times.

The government sector. Its revenues come from state enterprise, taxation etc. Its expenditure takes the forms of road construction, defence, law and order, etc. The government sector in modern times plays an important role in the redistribution of incomes.

The rest of the wold sector. It considers the role of exports and imports *i.e.*, economic transactions of the country with the rest of the world.

The capital sector. It symbolises the process of financial investment and savings.

Given above is the scheme of *institutional* sectoring. Alternatively, we may have sectoring by industries.

Two Sectors

Consider a two-sector model for an economy having

(*i*) the producing sector (business), and

(*ii*) the households sector (consumers).

The system tells us only this : aggregate income from production equals aggregate expenditure on consumption and *vice versa.*

Given below are the economic accounts for such a system:

B. Business Account

Receipts from		*Payments to*	
consumers (for sales)	x	consumers (for services)	y
Total	x		y

$x = y$

C. Households Account

Receipts from		*Payments to*	
business (for services)	y	business (for purchases)	x
Total	y		x

$y = x$

A matrix showing these transactions can present the relationship between the flows in the accounts given above.

Receipts by \ *Payments by*	B	C	Total
B	—	x	x
C	y	—	y
Total	y	x	$x + y$

Ex. 3. Consider the accounts of a 3-sector economy.

B. Business Account

Receipts from		*Payments to*	
consumers (sales)	500	households (purchase of services)	700
gross investment	200		
Total	700		700

C. Household Account

Receipts from		*Payments to*	
business (for services)	700	business (for goods)	500
		savings	200
Total	700		700

I. Capital Account

Receipts from		*Payments to*	
households (savings)	200	gross investment	200
Total	700		700

The matrix for the above transactions is shown below.

Payment by

Receipts by	*B* *Business*	*C* *Households*	*I* *Capital*	*Total*
B. Business	—	500	200	700 (*i*)
C. Households	700	—	—	700 (*iii*)
I. Capital	—	200	—	200 (*v*)
Total	700 (*ii*)	700 (*iv*)	200 (*vi*)	1600

(*i*) *B* row total: Expenditure on consumption + Exp. on investment

$$= 500 + 200 = 700$$

This is *gross national expenditure* which equals *gross national product.*

(*ii*) B column total: Incomes of households = 700

This is *gross national income* which equals *gross national product.*

(*iii*) C row total : 700. This is *personal income.*

(*iv*) C column total : Expenditure by consumers + savings

$$= 500 + 200 = 700$$

(*v*) I row total : 200. These are savings of households.

(*iv*) I column total : 200. This is business investment.

A More General Case : Five Sector Model

Given below is the systematic presentation of the economic accounting scheme for five sectors.

We have used the following notation for the purpose.

x_Y : receipt from sector X by sector Y

which equals X_y : payment by sector X to sector Y

x_Y occurs on the receipt side of Y account

X_y occurs on the payment side of X account

e.g., g_B : receipt from government (G) by firms (B)

which equals G_b : payment by government (G) to firms (B)

SECTORWISE NATIONAL INCOME ACCOUNTS

B. Producing Sector (Business)

Receipts from	*Payments to*
Households or consumers c_B	Households (for services) B_c
Government g_B	Government (taxes etc.) B_g
Net exports w_B	Savings (for investment) B_i
Gross domestic investment i_B	
Total $c_B + g_B + w_B + i_B = B_c + B_g + B_i$	

C. Households (Consumers)

Receipts from	*Payments to*
Business (for services) b_c	Business (for purchases) C_b
Government (transfer payments) g_c	Government (investment) C_g
	Savings (for investment) C_i
Total $b_C + g_C = C_b + C_g + C_i$	

G. Government

Receipts from	*Payments to*
Households (taxes etc.) c_G	Business (purchases) G_b
Business (taxes etc.) b_G	Households (tr. payments) G_c
	Savings (for investment) G_i
Total $c_G + b_G = G_b + G_c + G_i$	

I. Capital (Investment and savings)

Receipts		*Payments*
Savings : Firms b_I		Gross domestic investment I_b
: Households c_I		Net investment abroad I_w
: Government g_I		
Total	$b_I + c_I + g_I = I_b + I_w$	

W. Rest of the World

Receipts from		*Payments to*
Net investment abroad i_W		Business for net exports W_b
Total	$i_W = W_b$	

MATRIX OF ECONOMIC ACCOUNTS

Payments by / *Receipts by*	B *Business firms*	C *Households*	G *Govern-ment*	I *Capital*	W *Rest of the world*
B. Business firms	—	$c_B = C_b$	$g_B = G_b$	$i_B = I_b$	$w_B = W_b$
C. House-holds	$b_C = B_c$	—	$g_C = G_c$	—	—
G. Govern-ment	$b_G = B_g$	$c_G = C_g$	—	—	—
I. Capital	$b_I = B_i$	$c_I = C_i$	$g_I = G_i$	—	—
W. Rest of the World	—	—	—	$i_W = I_w$	—

Each sector has one row and one column. Each cell shows the equality of the receipts (x_Y) by one sector from another and the payments (X_y) by another sector to it.

Ex. 4. Construct the matrix for the following accounts:

B. Business Accounts

Receipts		*Payments*	
Sales to consumers	160	Purchases from household	170
Investment	30	Imports	20
Exports	10	Savings	10
Total	200		200

C. Households

Receipts		*Payments*	
Sales of services to business	170	Purchases from business	190
		Savings	10
Total	170		170

I. Capital Account

Receipts		*Payments*	
Savings : firms	10	Gross domestic investment	30
Savings : households	10		
Net disinvestment abroad	10		
Total	30		30

W. Rest of the World

Receipts		*Payments*	
Sales of imports to business	20	Purchases of exports	10
		Net disinvestment abroad	10
Total	20		20

Ans.

Payments by / *Receipts by*	*B*	*C*	*I*	*W*	*Total*
B	—	160	30	–10	180
C	170	—	—	—	170
I	10	10	—	—	20
W	—	—	–10	—	–10
Total	180	170	20	–10	360

Interpret the row and column totals.

Ex. 5. Prepare economic accounts from the following matrix of transactions.

Payments by / *Receipts by*	*B*	*C*	*G*	*I*	*W*	*Total*
B	—	160	30	5	15	210
C	175	—	10	—	—	185
G	20	20	—	—	—	40
I	20	5	—	—	—	25
W	—	—	—	20	—	15
Total	210	185	40	25	15	475

Interpret the row and column totals.

Ans. *e.g.*, Row 1 total 210 is gross national expenditure at market prices etc.

15.4 INPUT OUTPUT TABLE AND ECONOMIC ACCOUNTING

An input output table records transactions among various industries or sectors. Being a comprehensive record of the economy it can be used to derive national income aggregates and their relationships with other transactions. Purchases by any industry from other industries constitute *intermediate* output. Purchases of goods by sectors (other than industries) for final consumption constitute final demand.

Exclusion of intermediate transactions

To derive GNP free from errors of multiple counting using an input output table we must exclude intermediate goods and services that enter into the production of other goods and services. Such transactions as constitute purchases by productive units from other pródductive units should be omitted and only final demand needs to be considered.

The four quadrants of an input output table

The input output table may be divided into 4 parts

(*i*) Top left: inter-industry transactions

(*ii*) Top right: transactions to meet final demand

(*iii*) Bottom left: payments for imports etc

(*iv*) Bottom right: final demand met from imports etc., re-exports and indirect taxes at the final demand stage.

Consider an economy with only three industries A_1, A_2, A_3. From the table we can see a number of things.

Transactions matrix for an economy (Rs. millions)

Input from Industries	*Output to Industries*			*Total*	*Final Demand*				*Total Gross Output*
					Export	*Capital*	*Govt*	*personal consump-tion*	
	A_1	A_2	A_3		W	I	G	C	
A_1	—	3	14	17	5	2	3	43	70
A_2	2	—	5	7	3	8	10	52	80
A_3	10	6	—	16	12	20	8	34	90
Total	12	9	19	40	20	30	21	129	240
Govt.	7	3	10	20	—	6	7	18	51
Imports	6	12	17	15	10	5	9	75	114
Primary Input	45	66	54	165	—	—	—	—	165
Total Gross Input	70	80	90	240	30	41	37	222	570

Rows

(*i*) Where does the output of A_1 go?

Row 1 : 17 mil. goes to intermediate products (3 to A_2 and 14 to A_3) and the balance (53 mil.) to satisfy final demand. Gross total output of industry A_1 is $17 + 53 = 70$.

(*ii*) How does industry A_1, satisfy final demand?

Exports: 5; Capital formation : 2; Government consumptions: 3; Personal consumption : 43; Total 53.

(*iii*) What does row 1 show?

It shows the distribution of output. We see that personal consumption accounts for a large share of the total output. Similarly for rows 2 and 3.

Columns

(*iv*) What does column 1 show?

It shows inputs coming from various industries (12) and from imported materials (6). Other items are indirect taxes (less subsidies): 7; value added *i.e.,* payments to primary factors or the sum of wages, profits etc. and depreciation : 45.

Total gross input of A_1 is $12 + 58 = 70$

For any industry, row total = column total because total gross output = total gross input.

(*v*) How to find gross national product from the table?

The payments to primary factors constitute gross national product at factor prices: 165.

Total resources available include GNP and imports: $114 + 51 + 165 = 330$ which is also the difference $570 - 240 = 330$.

Disposal of resources gives the sum of national expenditure and exports. It is the sum of payments to primary factors and imports. It is also the sum of capital formation, consumption and government expenditure

$$W + I + G + C = 30 + 41 + 37 + 222 = 330$$

which is the same as the total resources showing that national income equals national expenditure.

Note that interindustry transactions (40) do not enter into national income accounting.

15.5 SAM

A social accounting matrix or SAM organises information about the social and economic structure of a region in a given period, usually any particular year. This static picture of the economy can be analysed to examine the nature of the economy, to forecast the effects of policy charges and for economic modelling.

The information contained in a SAM is similar to but more detailed than that in national accounts. However, instead of double entries there are only single entries in the form of a matrix whose size depends on the details required. The information in the matrix shows the influence on one another of various constituents of the matrix. These constituents may be socioeconomic groups enabling us for instance, to study issues of poverty and basic needs or they may be productive factors and institutions enabling us to study the effects of say, fiscal incentives, and so on.

The SAM principle is that of double entry book keeping achieved by only single entries in a matrix.

A SAM for Country X (million rupees)

		1*a*	1*b*	2*a*	2*b*	2*c*	2*d*	3	4*a*	4*b*	4*c*	4*d*	4*e*	4*f*	5	Total
1*a*	Labour Income								2015	561	245	909	564	1275		5559
1*b*	Capital Income								2009	909	734	1423	829			5904
2*a*	Household Current a/c	5569	4216		644	248									18	10695
2*b*	Corporation Current a/c		1575			294									15	1854
2*c*	Government Current a/c			447	376		1368	43								2234
2*d*	Indirect Taxes			119				270	80	504	66	130	76	29	94	1368
3	Combined Capital a/c			1337	834	43									425	2639
4*a*	Agriculture			2861				104	1191	354	3		62	40	1288	5903
4*b*	Industry			1824				109	254	815	417	172	66	74	335	4066
4*c*	Construction							1595		1		7	50	92		1745
4*d*	Trade and Transport			1606				154	135	344	206	96	42	59	203	2845
4*e*	Private Services			1410					37	4	9	38	55	37	287	1877
4*f*	Government Services					1649										1649
5	Rest of the World		113	1091				364	182	574	65	70	133	43		2635
	Total	5569	5904	10695	1854	2234	1368	2639	5903	4066	1745	2845	1877	1649	2635	

A SAM can be used to upgrade or improve statistical information. It can form the basis of the economic mode of an economy. Some parts of a SAM may be more aggregated and others less aggregated. Use may be made of one or more parts or the whole SAM. The input output matrix may form a part of a SAM and may be used in conjunction with other parts.

The multiplier analysis in a SAM enables us to examine the effect of changes in one or more elements on the other parts of the SAM.

In a SAM, the rows represent incomings and the columns represent outgoings. In a sufficiently simplified model of an economy, it is not difficult to notice the working of the circular process in the SAM or at least in its core.

In the table above the numbering of columns and rows is identical.

1a, *1b* are for factors of production (labour, capital); *2a*, *2b*, *2c*, *2d* are for institutions (households, corporations, government and indirect taxes)

3 is for combined capital account; *4a*, *4b*, *4c*, *4d*, *4e*, *4f* are for industries (agriculture, industry, construction, trade and transport, private services and government services)

5 is for rest of the world account.

The subaccounts for factors of production show incomes of labour and capital. The box corresponding to columns *2a*, *2b*, *2c*, *3d* shows payments between institutional sectors. These are transfer payments arising from asset ownership, direct taxes etc. and do not constitute income from production.

Account 3 combines the capital accounts of households, corporations, government and the rest of the world.

The production account made of six subsectors *4a* to *4f* is contained in the box formed by these rows and columns and forms an input output matrix which captures the relation between the production and other accounts.

In addition to these accounts those for financial accounts and flow of capital funds into investment may also be included in a SAM.

NIPA

National Income and Product Accounts (NIPA) provide information on the identification and extent of surplus or deficit of various sectors of the economy.

NIPA help in realising the need to fill the saving investment gap and promote efforts for growth on the domestic fronts. Even the need for foreign aid becomes more clear by examining NIPA.

NIPA deal with the final activities of the economic system.

These are:

(*a*) Production, covering goods and services.

(*b*) Consumption of goods and services and transfers and distribution of factor incomes.

(*c*) Capital formation, gross and fixed.

(*d*) Rest of the world transactions including exports and imports of visible and invisible goods and services.

There are conceptual problems regarding such issues as expenditures on education, durable goods and military, environment pollution etc. These can be incorporated into the national income concept by widening its scope and definition.

Ex. 6. UN system of National Accounts : Policy Areas

	Opening assets	Production	Consumption	Accumulation	Rest of the world	Revaluation	Closing assets
Opening Assets				Analysis of national wealth, productivity			
Production		I–O, production analysis	Consumer demand, Govt. Exp.	Models of stock, fixed K-formation, invest. policy	Export demand analysis		
Consumption		Prod. Ins Factor shares, productivity	Fiscal Policy Distribution & Redis. of income	Depreciation analysis: invests allowances	Return on Foreign invest. double taxation		
Accumulation	Net worth		Saving behaviour	Monetary policy Liquidity preference	International finance & liquidity; Foreign aid (Long term)	Capital gains/ losses & tax poivy	Net worth
Rest of the world		Import demand analysis	Short term foreign aid policy	International finance & liquidity Foreign aid policy (Long terms)	BOT & BOP		
Revaluation				Capital revaluation			
Closing Assets				National wealth & productivity			

15.6 FLOW OF FUNDS ACCOUNTING

Flow of funds accounting is a system of accounting more comprehensive than that of national income accounting and covers the entire complex of transactions including property ownership, debts, insurance benefits, sales of assets etc.

This system, being concerned with the sources and use of funds has sectors like consumers, farms, banking, insurance, state and local government investors, rest of the world and so on. National income accounting, although sufficiently comprehensive, assumes homogeneity of products in industries and cannot reveal the sources and uses of funds in the system.

The columns are used for institutional rather than functional sector classification in flow of funds accounting and the rows show the kinds of transactions which are current non-financial, capital non-financial and financial. The information thus made available can be used to analyse the effect of different monetary policies on the functioning of the economy. The effects of other changes on the structure of transaction flows can be measured similarly.

Flow of funds accounting is not concerned with the technological aspects of the economic system. Its sectors and transactions are interlocked. The presentation enforces consistency of analysis in view of all economic transactions entered therein. Some details of the flow of funds system are given below.

1. National income accounting is concerned with the working of the economy. The money flow accounts enable us to understand the working of the economy. Flow of funds accounting reviews the economy as a monetary, credit and financial entity which produces as well as purchases goods.

The flow of funds accounts have the following characteristics.

The generalised balance sheet gives the net worth after subtracting all the liabilities of an entity from all the financial and nonfinancial assets at a particular point of time. Nonfinancial transactions pertain to flow of goods and services and unilateral transfers like taxes and donations. Financial flows partain to borrowing and debt repayments and accumulations.

2. The Flow of Funds System of Accounts as a component of national accounts presents an economy's intersectoral financial accounts and links savings and investment components with other sectors. It differs from national income accounts which measure a country's incomes, expenditures, investments and savings in money terms. It presents current productive and distributive activities among factors of production. Input output accounts are concerned with interindustry production and technological relations between physical inputs and outputs. Sectoral production rather than the economy's financial output is examined in input output analysis. The intersectoral financial behaviour of flow of funds accounts can be put in the form of an input output table.

Each sector plays a dual role for lending and borrowing investment funds. The transactions matrix, for example, from a given sectoral saving pattern can be used to study the unique pattern of investment.

3. The Flow of Funds System shows the transactions among economic decision makers grouped into institutionally homogeneous sectors in the form of a matrix of interlocking balanced sectoral accounts. There is no indication of cause and effect relations for performance.

The matrix of accounts shows sectoral surpluses, deficits, range and magnitude of financial sectors.

We can examine the general trend, structure and other aspects of different financial instruments and sectors. It is possible to link financial development with economic development through the causality pattern of these variables.

4. These accounts can be useful in studying monetary and fiscal policies which cannot be analysed from national income accounts except indirectly.

Flow of Funds Accounts are also useful in planning and projection.

5. The Flow of Funds matrix analyses each sector's surplus or deficit, range and relative magnitude of financial claims and provides charges on savings, physical and financial and physical assets while identifying the magnitude and pattern of financial flows.

6. These accounts can be used to explain the demand and supply of bank deposits and their variations so that they are useful in examining the role of short term and long term interest rates.

They can be used to study state outlay function and to analyse the cyclical pattern of the nonbanking sector.

Flow of Funds Accounts : Some Applications

(*a*) Policy formulation to locate the sources conducive to expansion and contraction of the economy.

(*b*) Estimation of the effects of changes in policy.

(*c*) Assessment of the efficacy of fiscal and monetary policies.

(*d*) Forecasting and model building.

(*e*) Provision of data on financial transactions to be used, for instance, in capital market analysis along with data on nonfinancial transactions for the same sectors.

Balance of Payments Accounts

The balance of payments accounts show the transactions between a country and the rest of the world, under various heads. There being important economic relationships between economic accounts and balance of payments, the two sets of accounts are often studied together.

15.7 NATIONAL BALANCE SHEETS

Like the business balance sheets, a balance sheet of any sector of the economy shows the net values of assets and inventories and also assets in the form of financial claims less liabilities. A national balance sheet gives the net worth of a sector at the end of a period which equals the net worth at the beginning plus any additions or deductions during the period. The structural analysis of wealth estimates, industrial concentrations, distribution of incomes and property etc. is the logical outcome of a study of national balance sheets whose preparation is usually coordinated with that of economic accounts.

Other methods of presentation of economic accounts are:

(*i*) flow diagrams, illustrating the basic flows in the economy.

(*ii*) single (instead of double) entry balance statements.

(*iii*) algebraic equations showing equalities of debits and credits.

(*iv*) matrices with rows and columns showing incomings and outgoings as in the case of interindustry relationships.

Different systems of estimating national income have their own peculiarities. Each can be used to meet specific requirements.

Ex. 6.

		Export of goods W	*Inv.* I	*Govt.* G	*Households* C	*GNE = GNP*
Y_{fc} at factor cost		export income W_1	Depr. $-D$			NNP
	Corporate savings S_c	Current deficit F		Govt. savings S_g	Hh savings S_h	Finance for capital formation
	Wages, profits. etc. N					Household incomes
Indirect taxes T_i	Corporate taxes T_c				Hh taxes T_h	Govt. revenue
Imports W_2						Disposal of receipts
GNP at market price	NI	Current receipts	Capital formation	Disposal Revenue		

Show that this matrix of aggregates is composed of the Keynesian identities:

$$Y = C + G + I + W$$

$$Y - D = C + G + (I - D) + W$$

$$NNP = Y_{fc} - D + W_1 = N$$

$$I = S_c + S_g + S_h - F$$

$$T_i + T_c + T_h = G + S$$

$$W + W_1 = W_2 + F \text{ etc.}$$

Prepare fictitious accounts for production, national income, investments, households, government and the rest of the world conforming to the above matrix.

Ex. 7. Interpret the following matrix of social accounts:

Receipts from / *Payments to*	*Transaction*	*I*	*II*	*III*	*IV*	*V*	*Total*
I	Sales of goods and services	—	800	190	250	260	1500
	Less imports	—	—	—	—	–245	1255 (*i*)
	Less depreciation	—	—	—	–125	—	1130 (*ii*)
	Less net transfers	—	—	–130	—	—	1000 (*iii*)
II	Factor in–comes	990	—	—	—	—	—
	transfers	—	—	–90	—	—	900 (*iv*)
III	Factor incomes	10	—	—	—	—	—
	Less foreign transfers	—	—	—	—	–5	5
IV	Saving	—	100	35		–10	125 (*v*)
Total		1000	900	5	125	0	—

Ans. (*i*) GNP
(*ii*) NNP at market prices
(*iii*) NNP at factor prices
(*iv*) Disposable income
(*v*) Net saving.

Ex. 8. Analyse the following input output table.

Outputs / *Inputs*	*Industry* 1	2	3	*Inter-mediate outputs*	*Personal income account*	*Govt. income account*	*National capital account*	*Exports etc.*	*Total final output*	*Total output*
Industry 1	62	134	—	196	20	—	5	50	75	271
2	46	601	247	896	450	100	95	130	775	1669
3	27	250	115	398	230	300	100	20	650	1042
Factor Payments										
Wages	36	334	385	—	—	—	—	—	—	755
Profits	38	125	—	—	—	—	—	—	245	
Net expenditure taxes	5	84	41	—	—	—	—	—	—	130
Capital consumption allowances	7	77	40	—	—	—	—	—	—	124
Imports etc.	5	150	90	—	—	—	—	—	—	245
Total inputs	271	1669	1042	1490	700	200	200	200	1500	

Ex. 9. Interpret the following input output table:

	Manufacturer A	B	C	*Total of intermediate output*	*Final buyers* House holds	Govt.	Cap-ital	exp-orts	*Total final output*	*Total output*
A	—	4	0	4	23	0	0	5	28	32
B	3	–	2	5	0	3	10	7	20	25
C	4	5	–	9	3	11	0	0	14	23
Imports	3	9	0	12	5	3	0	0	8	20
Total	10	18	2	30	31	17	10	12	70	100
Wages	9	2	7	18	0	0	0	0	0	18
Profits	7	3	6	16	0	0	0	0	0	16
Taxes	6	2	8	16	0	0	0	0	0	16
Total input	32	25	23	80	31	17	10	12	70	150

Ex. 10. Consider the gross output of a multisectoral economy.

The real value of gross output is equal to total expenditure or demand.

(*i*) Gross output=Intersectoral outflows+[Investment+Consumption+Govt. expenditure+Exports]

i.e., Gross output=Intersectoral outflows+Final demand

Again, the value of gross output is equal to its factor and other shares, as viewed from the distribution side.

(*ii*) Gross output=Intersectoral inflows
+[Wages+Profits+Taxes+Imports]

i.e., Gross output=Intersectoral inflows+GNP by distributive shares

Comment on the above result.

Ans. (*i*) emphasises the revenue aspect of interindustrial transactions.

(*ii*) emphasises their cost aspect.

The two identities clarify the interdependent aspects of general economic activity from the spending and distributive points of view.

From an analytical point of view the above identities may be considered as equations to be satisfied for conditions of equilibrium to be established.

Ex. 11. "Economic accounting may be used for testing feasibility of overall economic projections." Discuss.

Ex. 12. "Economic accounting provides no automatic solution to problems but helps chart the method of attack." Comment.

CHAPTER 16

Linear Programming

16.1 PROGRAMMING IN ECONOMICS

Programming means planning. In economics, programming means planning of economic activities to get optimal solutions to problems.

A firm can produce goods by using certain quantities of labour and capital. The firm will use more labour or more capital depending upon the costs involved. There are restrictions on the quantities of labour and capital available. Under these circumstances the firm has to maximise its profits.

What is the solution? Is there one solution? Or are there many solutions to the problem? If so, what is the optimal feasible solution? This is a programming problem. It requires, for instance, the integration and coordination of economics and engineering in an economic problem faced with a choice between various technologies.

A consumer buying a number of goods is restricted by the prices, his preferences and his income. His nutritional requirements introduce further restrictions. Subject to all these constraints the consumer has to minimise his costs so that he gets at least the minimum of the nutrients. The standard diet, which is a feasible solution, may be obtained in many ways by buying foods in different quantities. Programming methods help us to obtain an optimum feasible solution.

16.2 OPTIMISATION (MAXIMISATION OR MINIMISATION)

The problem of optimisation of some quantity subject to certain restrictions or constraints is a common feature of economics, industry, defence, etc. The usual method of maximising or minimising a function involves constraints in the form of equations. Thus utility may be maximised subject to the budget constraint of fixed income, given in the form of an equation. The minimisation of cost is a familiar problem to be solved subject to some minimum standards.

If the constraints are in the form of equations, methods of calculus can be useful. But if the constraints are inequalities instead of equations and we have an objective function to be optimised subject to these inequalities, we use the method of mathematical programming.

For simplicity, we employ linear objective functions and linear inequalities and the technique we use is called *linear programming,* a special case of mathematical programming.

Linear Programming

Linear programming is a practical tool of analysis which yields the optimum solution for a linear objective function subject to constraints in the form of linear inequalities. The technique is, in many ways, superior to the conventional mathematical techniques.

A firm wants to maximise its profits, and has limited resources (resources *less than* certain given amounts : hence *inequalities*). Linear programming gives a non-negative optimum solution to such a problem subject to the availability of resources.

An economy may have some goals. A good plan suggesting the optimum allocation of resources could be formulated with the help of programming methods.

A linear programming problem has the following essential parts.

(*i*) *A set of non-negative restraints.* This is to avoid getting negative solutions in which we are not interested, *e.g.*, negative price or negative output do not mean anything.

(*ii*) *A set of linear constraints* representing the limitations and extensions of the factors available.

(*iii*) *An objective function* to be optimised subject to the above constraints.

16.3 APPLICATIONS OF LINEAR PROGRAMMING METHODS

There is a wide variety of problems to which linear programming methods have been successfully applied.

(*i*) *Diet problems.* To determine the minimum requirement of nutrients subjects to availability of foods and their prices.

(*ii*) *Transportation problems.* To decide the routes, number of units, the choice of factories so that the cost of operation is the minimum.

(*iii*) *Manufacturing problems.* To find the number of items of each type that should be made so as to maximise the profits.

(*iv*) *Production problems,* subject to sales fluctuations. To decide the production schedule to satisfy demand and minimise cost in the face of fluctuating rates and storage expenses.

(*v*) *Assembling problems.* To have the best combination of basic components to produce goods according to certain specifications.

(*vi*) *Purchasing problems.* to have the least cost objective in, say, the processing of goods purchased from outside and varying in quantity, quality, and prices.

(*vii*) *Job assigning problems.* To assign jobs to workers for maximum effectiveness and optimum results subject to restrictions of wages and other costs.

16.4 ASSUMPTIONS OF LINEAR PROGRAMMING

The LP problems are solved on the basis of some assumptions which follow from the nature of the problem.

1. *Linearity*

   The objective function to be optimised and the constraints involve only linear relations. They should be linear in their variables. If they are not, alternative techniques to solve the problem have to be found. Linearity implies proportionality between activity levels and resources.Constraints are rules governing the process.

2. *Non-negativity*

   The decision variables should necessarily be non-negative.
3. *Additivity and Divisibility*

   Resources and activities must be additive and divisible.
4. *Alternatives*

   There should be alternative courses of action with a well defined objective function to be maximised or minimised.
5. *Finiteness*

   Activities, resources, constraints should be finite and known.
6. *Certainty*

   Prices and various coefficients should be known with certainty.

Some Limitations

The computations required in complex problems may be enormous. The assumption of divisibility of resources may often be not true. Linearity of the objective function and constraints may not be a valid assumption. In practical work there can be several objectives, not just a single objective as assumed in LP.

16.5 FORMULATION OF LP PROBLEMS

The formulation has to be done in an appropriate form. We should have

1. An objective function to be maximised or minimised. It will have n decision variables $x_1, x_2, \ldots, x_n$ and is written in the form

   max z or min $z = c_1x_1 + c_2x_2 + \ldots\ldots + c_nx_n$

   where each c_j is a constant which stands for per unit contribution of profit (in the maximisation case) or cost (in the minimisation case) to each x_j
2. The constraints in the form of linear inequalities

$$\left.\begin{array}{l} a_{11}x_1 + a_{12}x_2 + \ldots + a_{1n}x_n \lesseqgtr b_1 \\ a_{21}x_1 + a_{22}x_2 + \ldots + a_{2n}x_n \lesseqgtr b_2 \\ \ldots\ldots\ldots\ldots\ldots\ldots\ldots\ldots\ldots\ldots\ldots \\ a_{m1} + a_{m2}x_2 + \ldots + a_{mn}x_n \lesseqgtr b_m \end{array}\right\} \begin{array}{l} \text{Briefly written} \\ \Sigma a_{ij}x_j \lesseqgtr b_i \\ i = 1, 2, \ldots, n \end{array}$$

where b_i stands for the ith requirement or constraint.

The non-negativity constraints are

$$x_1, x_2, \ldots\ldots, x_n \geq 0$$

In *matrix notation* we can write: max or min $z = cx$

Subject to constraints $Ax \leq b$ or $Ax \geq b$

and the non-negativity conditions $x \geq 0$ where $c = [c_1c_2\ldots c_n]$. Here

$$A = \begin{bmatrix} a_{11} & a_{12} & \ldots & a_{1n} \\ a_{21} & a_{22} & \ldots & a_{2n} \\ \ldots & \ldots & \ldots & \ldots \\ a_{m1} & a_{m2} & \ldots & a_{mn} \end{bmatrix}, \quad b = \begin{bmatrix} b_1 \\ b_2 \\ \vdots \\ b_m \end{bmatrix}, \quad x = \begin{bmatrix} x_1 \\ x_2 \\ \vdots \\ x_n \end{bmatrix}$$

Ex. 1. A firm has two factories F_1, F_2 and two selling shops S_1, S_2. Factory F_1 can produce 700 units of a commodity X and F_2 900 units. S_1 must have 500 units of X and S_2 1000 units of X for sale. The transportation costs per unit of X are Rs 6 from F_1 to S_1, Rs 5 from F_1 to S_2, Rs 4 from F_2 to S_1 and Rs 8 from F_2 to S_2. Show a linear programme to minimise the total cost of transportation.

Let x_{11} be the number of units transported from F_1 to S_1, x_{12} from F_1 to S_2, x_{21} from F_2 to S_1 and x_{22} from F_2 to S_2. The problem can be written

$$\min z = 6x_{11} + 5x_{12} + 4x_{21} + 8x_{22}$$

Subject to

$$x_{11} + x_{12} \le 700$$

$$x_{21} + x_{22} \le 900$$

$$x_{11} + x_{21} \ge 500$$

$$x_{12} + x_{22} \ge 1000$$

$$x_{11}, x_{12}, x_{21}, x_{22} \ge 0$$

Ex. 2. A diet has to be decided to fulfil the daily requirements of proteins, fats and carbohydrates at the minimum cost. The choice is to be made from four different foods whose yields per unit are given below from which the LP model has to be formulated.

	Yield Per Unit			
Food	*Proteins*	*Fats*	*Carbohydrates*	*Cost/unit*
1	3	2	6	45
2	4	4	4	40
3	8	7	7	85
4	6	5	4	65
Minimum Requirement	800	200	700	

(1) To determine units of food (4 types): $x_1, x_2, x_3, x_4 \ge 0$

(2) To minimise $C = 45x_1 + 40x_2 + 85x_3 + 65x_3$

(3) *Constraints* Proteins: $3x_1 + 4x_2 + 8x_3 + 6x_4 \ge 800$

Fats: $2x_1 + 4x_2 + 7x_3 + 5x_4 \ge 200$

Carbohydrates: $6x_1 + 4x_2 + 7x_3 + 4x_4 \ge 700$

Ex. 3. A financial expert gives advice to invest a sum of one million rupees for maximum returns. There are six alternative investment types:

Inv. Type	A_1	A_2	B_1	B_2	C_1	C_2
Yield per cent	3	2.5	3.5	4	5	4.5

At least 40 per cent must be invested in Type A and not more than 35 per cent in any single one of the other two types. Suggest a L.P. Investment model.

$$\text{Max. } R = .03X_{11} + .025x_{12} + .035x_{21} + .04x_{22} + .05x_{31} + .045x_{32}$$

Subject to

$$x_{11} + x_{12} \ge 40\% \text{ of 1 mil.}$$

$$x_{21} + x_{22} \le 35\% \text{ of 1 mil.}$$

$$x_{31} + x_{32} \le 35\% \text{ of 1 mil.}$$

$$x_{11} + x_{12} + x_{21} + x_{22} + x_{31} + x_{33} = 1 \text{ million}$$

$$x_{11}, x_{12}, x_{21}, x_{22}, x_{31}, x_{33} \geq 0$$

Ex. 4. A firm has to produce a tea mixture of 200 kg using x_1 tea at a cost of Rs 30 per kg and x_2 tea at Rs 50 per kg. Not more than 80 kg of x_1 tea and at least 60 kg of x_2 must be used. Minimise the mixture cost.

$$\text{Min } C = 30x_1 + 50x_2 \text{ subject to}$$

$$x_1 + x_2 = 200;\ x_1 \leq 80;\ x_2 \geq 60;\ x_1, x_2 \geq 0$$

16.6 SOLVING OF LINEAR PROGRAMMING PROBLEMS

There are several methods of solving linear programming problems. Many algorithms or step by step procedures have been suggested. Among these the Simplex method is the most convenient and efficient. We may use the graphical method in simple cases.

16.7 GRAPHICAL SOLUTION

In this method, which requires the graphing of inequalities we examine only those points which lie on the intersection set of all admissible sets defined by the constraints. Any point lying in the intersection set of all the constraints is a feasible solution. From the polygon of the feasible solutions we determine the feasible solution or solutions which optimise the given objective function.

An optimising solution is always a vertex or corner of the polygon of feasible solutions. In case of more than one solution, at least two of the solutions must be at the vertices. Thus the optimum feasible solutions, in the graphical method, are at corner points or somewhere in the boundary of the feasible region.

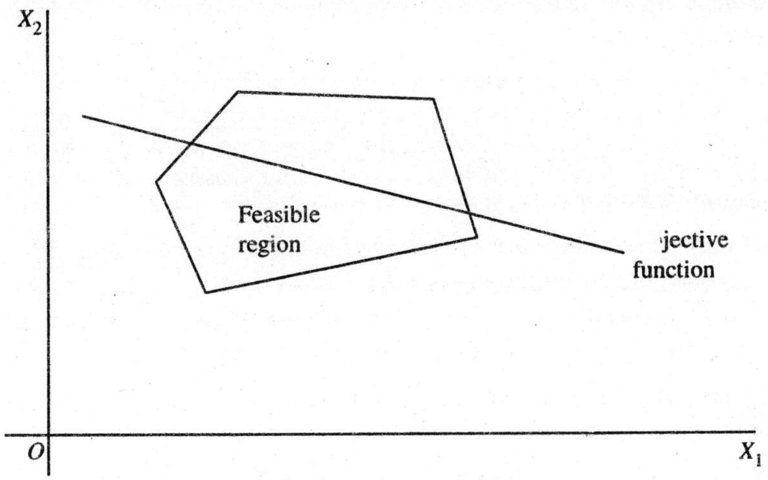

The objective is to determine the optimal solution which is a feasible solution with the best value for the objective function.

A feasible solution in the form of a set of values of decision variables is that which satisfies all non-negativity conditions and all structural constraints.

There may be an infinite number of feasible solutions which can be within the polygon of the feasible region or on its edges. There is a finite number of solutions at the corner points or vertices. The optimal solution is usually one of the corner points of the feasible region which should be a convex set.

Note. A convex set is such that any line joining any two points in the region lies entirely within the region. A linear function defined over a convex set takes on its maximum or minimum value at an extreme point.

16.8 LP STEPS IN THE GRAPHICAL METHOD

The following steps may be followed in the graphical solution of a linear programming problem:

Part A

A 1. Plot each constraint as an equality. On the basis of these constraints, identify the region which satisfies all the constraints. This is the feasible region.

A 2. In the objective function $z = c_1x_1 + c_2x_2$ (assuming there are only two variables), take an arbitrary value of z and plot it on the same graph.

If there is no feasible solution, it is not necessary to proceed. If the problem has a feasible solution, identify the region of feasible solution and move a line parallel to the z line so that it touches at least one extreme point

(*a*) as far way from the origin as possible for a maximisation problem.

(*b*) as close to the origin as possible for a minimisation problem.

A 3. The extreme point gives the optimal solution.

Part B

B 1. Find if the optimal s 'ution is (*a*) definite and unique (*b*) multiple or infinite (*c*) unbounded or (*d*) non-existent.

If the objective function touches only one extreme point for maximum or minimum (See Diagrams Ex. 7 below), then there is a unique solution. In the case of multiple solutions the objective function touches the extreme points as well as all the points between these. If the objective function suggests an unbounded solution it implies one or more constraints have not been taken into account.

B 2. If the optimal solution is unique or of multiple type, determine it.

Find the value of z by substituting the optimum solution values of x_i. Interpret the solution.

Part C

C. 1. Sensitivity Analysis may be carried out if necessary. This is to find how the optimal solution changes as a result of variations in

(*i*) the coefficients of variables in the objective function or constraints;

(*ii*) the number of constraints or the number of decision variables;

(*iii*) the right hand values of constraints.

This provides the decision maker with a tool to extract additional information without necessarily being required to re-solve the linear programming problem.

Ex. 5. A firm can produce a good either by (*i*) a labour intensive technique, using 8 units of labour and 1 unit of capital or (*ii*) a capital intensive technique using 1 unit of labour and 2 of capital.

The firm can arrange up to 200 units of labour and 100 units of capital. It can sell the good at a constant *net* price *p i.e., p* is obtained after subtracting costs. Obviously we have simplified the problem because in this way *p* becomes profit per unit. Let $p = 1$.

Let x_1 and x_2 be the quantities of the goods produced by the processes 1 and 2 respectively. To maximise the profit $px_1 + px_2$ we write the objective function $\pi = x_1 + x_2$ (Since $p = 1$). The problem becomes

$$\text{Max } \pi = x_1 + x_2$$

Subject to the labour constraint $8x_1 + x_2 \le 200$

and the capital constraint $x_1 + 2x_2 \le 100$

and the non-negativity conditions $x_1 \ge 0, x_2 \ge 0$

This is a problem in linear programming. There is a *linear* objective function of non-negative variables, to be maximised subject to certain linear inequalities.

Since there are two variables involved we can easily use the graphical method to arrive at the solution.

The procedure is first to draw the graphs of the linear constraints treating them as equations.

$$8x_1 + x_2 = 200$$
$$x_1 + 2x_2 = 100$$

Each of these lines, giving the boundaries or borders for labour and for capital divides the quadrant into two non-overlapping regions.

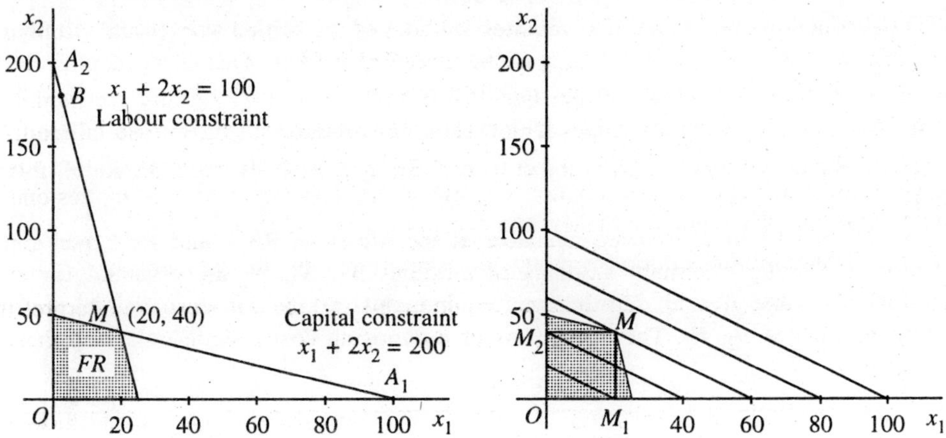

We have to satisfy simultaneously two restraints of the ≤ type. The acceptable points are, on the southwest side where, therefore, the feasible region (FR) is located. (shown shaded)

Each pair of values (x_1, x_2) in this region is a feasible solution. The feasible region includes of *all* the points *inside the region* and *even those on the boundary lines.*

The kinked boundary of the feasible region is made of part segments of the constraint lines and the axes.

There are three corners of extreme points (0, 50), (25, 0) and (20, 40). The solution is usually at one of these corner points

For the point (0, 50), $\pi = x_1 + x_2 = 0 + 50 = 50$

(25, 0), $\pi = 25 + 0 = 25$

(20, 40), $\pi = 20 + 40 = 60$

The optimal feasible solution is $x_1 = 20$, $x_2 = 40$ so that $\pi_{max} = 60$

More appropriately, a series of isoprofit lines are constructed giving different values to k in the profit function:

$$k = x_1 + x_2 \quad \text{or} \quad x_2 = k - x_1 \qquad \text{(replace } \pi \text{ by a parameter } k\text{)}$$

We superimpose a series of parallel isoprofit lines (obtained from $x_2 = k - x_1$) on the feasible region.

To maximise profit we select the highest possible isoprofit line while still remaining in the feasible region.

$M(20, 40)$ meets both these tests and is, therefore, the *optimal feasible solution* of the linear program.

The Optimal Feasible solution: $x_1 = 20$, $x_2 = 40$ gives $\pi = x_1 + x_2 = 20 + 40 = 60$

A combination of 20 units of the good produced by the first technique and 40 by the second technique gives the maximum profit 60.

The solution meets all the requirements and both the production techniques are used.

There is an infinite number of solutions if the profit line happens to be parallel to A_1M (or A_2M). Then every point on A_1M (or A_2M) provides a solution.

Ex. 6. A handloom firm produces two types of cloth, x_1, x_2 and makes a profit of Rs 20 per unit on x_1 and Rs 25 per unit on x_2. Three types of workers (skilled, semiskilled, unskilled) are available each for 8 hours a day.

The production of one unit of x_1 requires 1/4 hour of the skilled worker and 1/3 hour of the semiskilled worker and 1/2 hour of the unskilled worker. That of x_2 requires 1/3 hour of the semi-skilled and 1/2 of the unskilled worker.

What combination of the two types of cloth should be produced to maximise total profit?

(*Hint.* Maximise $20x_1 + 25x_2$ subject to $x_1 \leq 32$, $x_2 \leq 8$, $4x_1 + 3x_2 \leq 48$ and $x_1 \geq 0$, $x_2 \geq 0$)

Ex. 7. Two foods F_1, F_2 are available at the prices of Re 1 and Rs 2 per unit respectively. Certain minimum amounts of nutrients N_1, N_2, N_3 are essential for an individual. The table gives these minimum requirements and the nutrients available from one unit of each of F_1, F_2. The question is of minimising cost t while satisfying these requirement.

	Minimum Requirements	*one unit of* F_1	*one unit of* F_2
N_1	17	9	2
N_2	19	3	4
N_3	15	2	5
Price per unit		1	2

Total cost $C = p_1x_1 + p_2x_2$ (x_1, x_2 quantities of F_1, F_2)

where $p_1 = 1, p_2 = 2$

We therefore have to minimise

$$C = x_1 + 2x_2$$

subject to the minimum nutrient requirement constraints,

$$9x_1 + 2x_2 \geq 17$$
$$3x_1 + 4x_2 \geq 19$$
$$2x_1 + 5x_2 \geq 15$$

and $x_1 \geq 0, x_2 \geq 0$

First draw the graphs of the equations

$$9x_1 + 2x_2 = 17$$
$$3x_1 + 4x_2 = 19$$
$$2x_1 + 5x_2 = 15$$

These represent the borders for the nutrients as the right hand sides in these equations represent the minimum requirements.

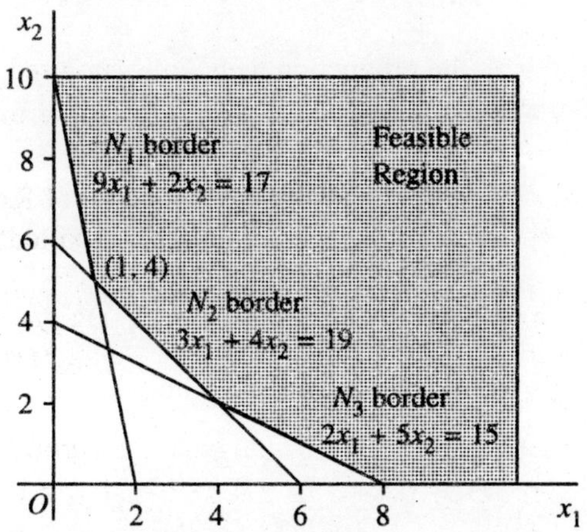

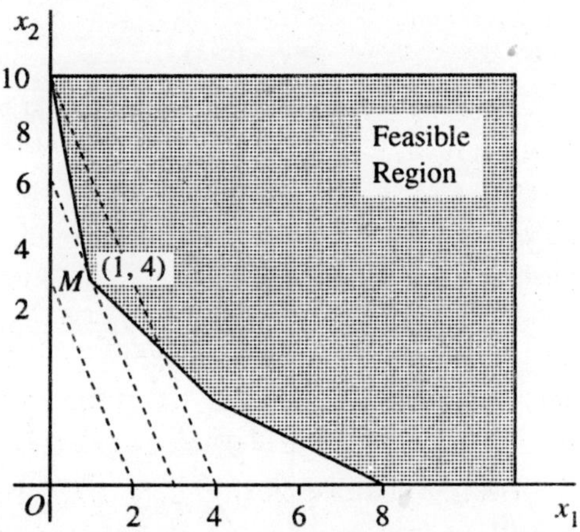

The constraints of the $\geq$ type suggest that the shaded portion in the diagram satisfies these constraints simultaneously and hence is the feasible region. This gives us a set of feasible solutions. Corner points can be used to arrive at the solution. Alternatively cost lines may be drawn.

To minimise cost, draw a set of parallel cost lines:

$$k_1 = x_1 + 2x_2 \text{ or } x_2 = k - \tfrac{1}{2}x_1$$

with slope $-\frac{1}{2}$, on the feasible region.

The minimum cost corresponds to the lowest isocost which is seen to pass through (1, 4).

The optimal feasible solution, therefore, is $x_1 = 1$, $x_2 = 4$ involving the minimum cost:

$$C = x_1 + 2x_2 = 1 + 2 \times 4 = 9.$$

The optimum combination is 1 unit of F_1 and 4 units of F_2.

Ex. 8. In Ex. 7 what will happen if the isocost line were

(*a*) $9x_1 + 2x_2 = k$?

(*b*) $3x_1 + 4x_2 = k$

Ans. In either case there will be an infinity of solutions lying on a whole border line. Draw the relevant diagrams and check the results.

Ex. 9. Draw the diagrams for Ex. 5 if the isoprofit line is

(*a*) $8x_1 + x_2 = k$ (*b*) $x_1 + 2x_2 = k$

interpret the results.

Ex. 10. In Ex. 5, what will happen if the profit from the product of the first process is 1 and that from the second is 2?

Ans. There is an infinity of optimal solutions.

Ex. 11. Max $z = 5x + 2y$

subject to

$$2x + 3y \geq 6$$
$$x - 2y \leq 2$$
$$3x + 2y \leq 12$$
$$-3x + 2y \leq 3$$
$$x, y \geq 0$$

The dotted dash lines show the objective function $z = 5x + 2y$

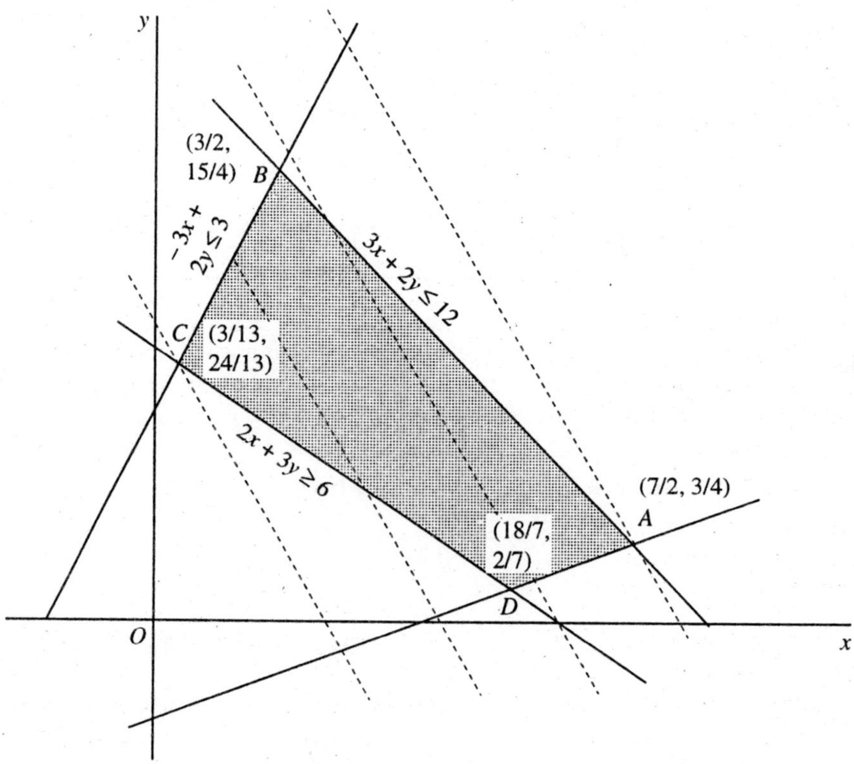

The feasible region is $ABCD$. The highest point cut by the objective function on the feasible region is A where $x = \frac{7}{2}$, $y = \frac{3}{4}$. The value of the objective function can be found by substituting these.

Assuming that the answer lies at one of the corners A, B, C, D we may find the value of z at each of these:

$$z_A = 5x + 2y = 5 \times \tfrac{7}{2} + 2 \times \tfrac{3}{4} = 19$$

$$z_B = 5 \times \tfrac{3}{2} + 2 \times \tfrac{15}{4} = 15$$

$$z_C = 5 \times \tfrac{3}{13} + 2 \times \tfrac{24}{13} = \tfrac{63}{13}$$

$$z_D = 5 \times \tfrac{15}{7} + 2 \times \tfrac{2}{7} = \tfrac{94}{7}$$

We can see that $z_{max} = 19$. Also $z_{min} = 63/13$

Ex. 12. Min $z = 3x + 5y$

subject to
$$x + y \geq 3$$
$$x - 2y \leq 2$$
$$-2x + y \leq 4$$
$$x, y \geq 0$$

FR, the shaded portion is the feasible region.

$$z_A = 3x + 5y$$
$$= 3 \times \tfrac{8}{3} + 5 + \tfrac{1}{3} = \tfrac{29}{3}$$
$$z_B = 3 \times 0 + 5 \times 5 = 20$$
$$z_C = 3 \times 0 + 5 \times 3 = 15$$

The minimum value of z is at A where $z_A = \tfrac{29}{3}$

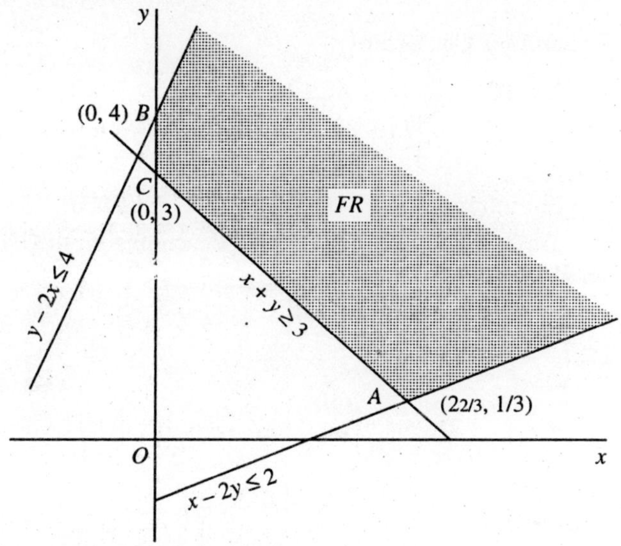

Ex. 13. Min $u = -x_1 + 2x_2$

subject to
$$-x_1 + 3x_2 \leq 10$$
$$x_1 + x_2 \leq 6$$
$$x_1 - x_2 \leq 2$$
$$x_1, x_2 \geq 0$$

At the four vertices A_1, A_2, A_3, A_4 of the polygon

$$u_1 = 6.67,\ u_2 = 6$$
$$u_3 = 0,\ u_4 = -2$$
$$\therefore u_{min} = -2$$

corresponding to

$$x_1 = 2,\ x_2 = 0$$

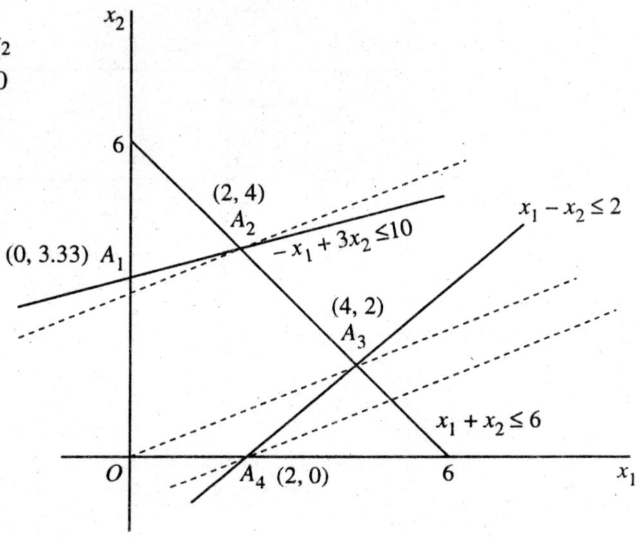

Ex. 14. Max $u = 3x_1 + 2x_2$

subject to $-2x_1 + 3x_2 \leq 9$

$3x_1 - 2x_2 \leq -20$

$x_1, x_2 \geq 0$

No overlapping in first quadrant and hence no common point. There is no solution.

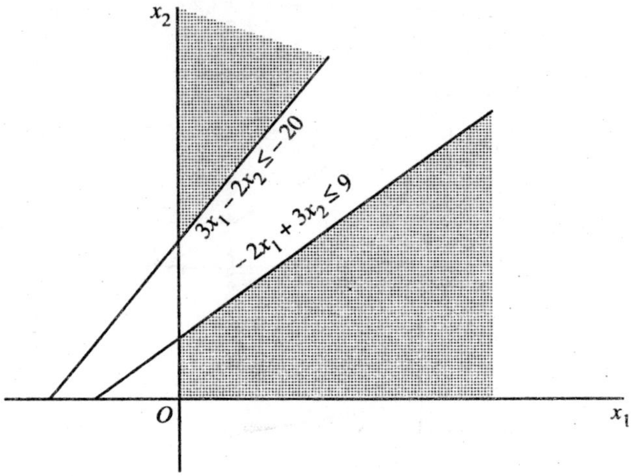

Ex. 15. Show that there is no solution to: Min $u = 2x_1 + x_2$ subject to $1.5x_1 + x_2 \leq 6$; $x_1 \leq 2$; $x_1 + x_2 \geq 7$; $-x_1 + x_2 \geq 4$; $x_1, x_2 > 0$

Redundant Constraint

Ex. 16. $5x + 6y \leq 1$ (1)

$2x + 3y \geq 6$ (2)

$y \geq 2, x \geq 0$

For minimisation, constraint (1) is redundant.

The solution may be at *A* or *B* depending on the objective function. *FR* is the shaded feasible region.

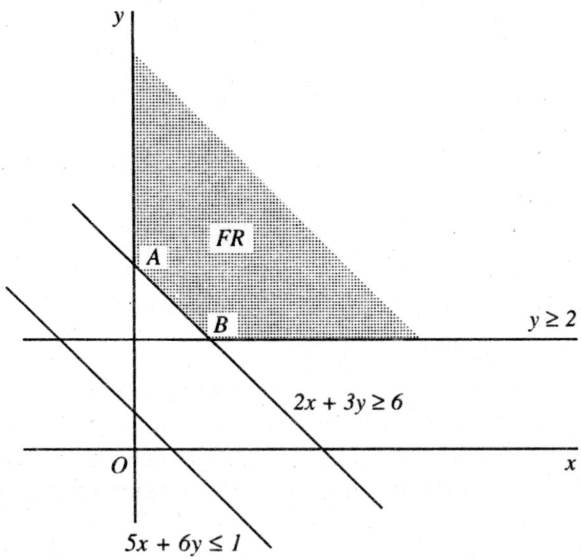

Ex. 17. Note the nature of constraints:

$3x + y \geq 3$ (1)

$x + 4y \geq 4$ (2)

$x + y \leq 6$ (3)

$x \leq 2$ (4)

$y \leq 4$ (5)

$x \geq 0, y \geq 0$

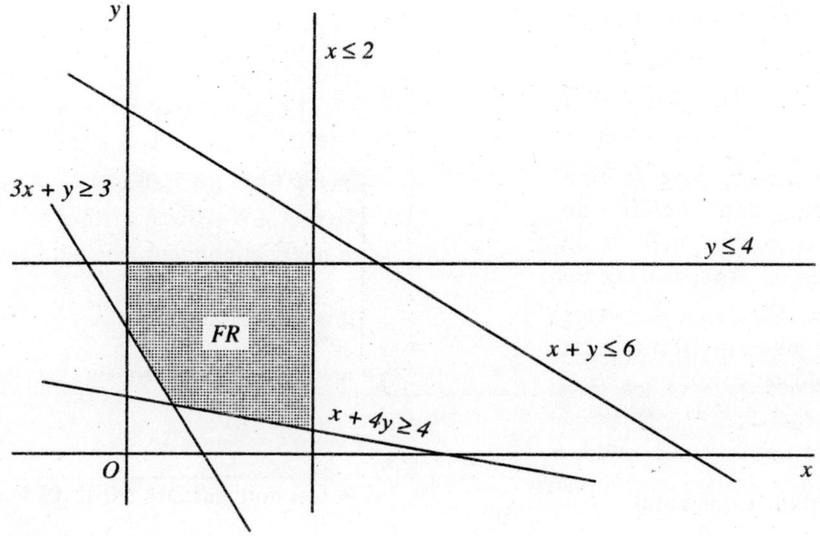

Ex. 18. Max $z = 2x_1 + 3x_2$

subject to

$$2x_1 + x_2 \le 6$$
$$x_1 + 2x_2 \le 8$$
$$x_1 - x_2 \le 1$$
$$x_1 \le 2$$
$$x_1, x_2 \ge 0$$

Ans. $x_1 = \frac{4}{3}, x_2 = \frac{10}{3}, z = \frac{38}{3}$

Ex. 19. A firm produces bread and biscuits and sells them at prices of Rs 5 and Rs. 3 per kg. The flour required for a unit of bread is twice that for biscuits. The supply of flour is sufficient for only 1000 kg of biscuits per day. Bread requires a special additive unit of which only 400 are available per day. For biscuits, additive units are also required and 700 of them are available per day. Find the production combination for maximum profit.

Hint. max $z = 5x_1 + 3x_2$ $\quad x_1 \ge 0$ bread

subject to $2x_1 + x_2 \le 1000$ $\quad x_2 \ge 0$ biscuits

$$x_1 \le 400$$
$$x_2 \le 700$$

Ans. $z_{max} = 2850$ for $x_1 = 150, x_2 = 700$

Ex. 20. Two plants of a firm produce pens, pencils and exercise books. Plant X_1 produces daily 1500, 3000 and 2000 units (in '000) of pens, pencil and books, plant X_2 produces 1500, 1000 and 5000 units of these. The operating costs at the two plants are 600 and 400 money units daily. The monthly demand for pens, pencils and books is 20000, 40000 and 44000 units. How many days should the plants work to minimise product cost and still meet most of the demand.

Hint. Min $z = 600x_1 + 400x_2$ $\quad$ No. of days x_1, x_2

$x_1 \ge 0$ plant X_1

subject to $1500x_1 + 1500x_2 \ge 20000$ $\quad x_2 \ge 0$ plant X_2

$$3000x_1 + 1000x_2 \ge 40000$$
$$2000x_1 + 5000x_2 \ge 44000$$

Ans. $z_{min} = 8800$ at $x_1 = 12, x_2 = 4$

Ex. 21. Which constraints are redundant in the following with $x_1, x_2 \geq 0$?

$$4x_1 + 3x_2 \leq 12 \quad (1)$$
$$-x_1 + x_2 \geq 1 \quad (2)$$
$$x_1 + x_2 \leq 4 \quad (3)$$
$$x_1 + x_2 \leq 6 \quad (4)$$

Ans. The last two

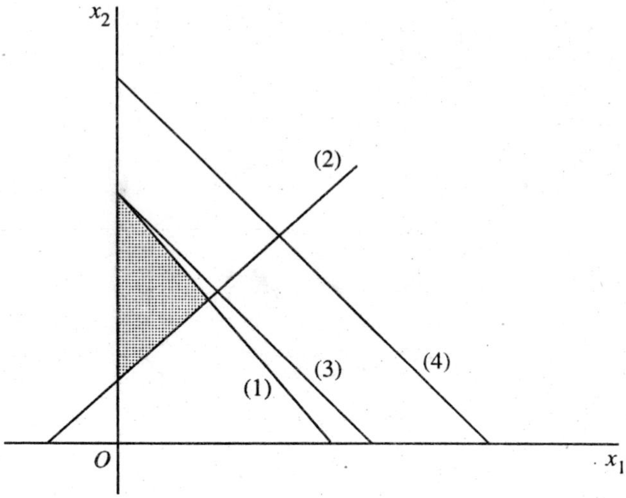

Ex. 22. A firm produces two blends *X*, *Y* of tea using two processes *A*, *B* whose requirements and whose production are (in appropriate units)

Process	*Requirements*		*Blend Production*	
	I	*II*	*X*	*Y*
A	5	3	5	8
B	4	5	4	4

The maximum amounts of I & II raw materials available are 200 and 150 units. The minimum demands for blends *X*, *Y* are 100 and 80 units. The profits are 300 and 400 money units per production run from process 1 and process 2. What can be the maximum profit?

Hint. *Max* $z = 300x_1 + 400x_2$

Subject to
$$5x_1 + 4x_2 \leq 200$$
$$3x_1 + 5x_2 \leq 150$$
$$5x_1 + 4x_2 \geq 100 \qquad x_1 \geq 0 \text{ Process } A$$
$$8x_1 + 4x_2 \geq 80 \qquad x_2 \geq 0 \text{ Process } B$$

Ans. $z_{max} = \dfrac{180000}{13}$ at $x_1 = \dfrac{400}{13}$, $x_2 = \dfrac{150}{13}$

Note that the constraint $8x_1 + 4x_2 \geq 80$ is redundant.

Ex. 23. *Min* $z = 200x_1 + 400x_2$

Subject to
$$x_1 + x_2 \geq 200 \quad x_1, x_2 \geq 0$$
$$x_1 + 3x_2 \geq 400$$
$$x_1 + 2x_2 \leq 350$$

Ans. $z_{min} = 60000$ at $x_1, x_2 = 100$

(More problems in later sections)

16.9 THE SIMPLEX ALGORITHM

The graphic solution can be useful only in very simple cases. In actual practice there are many complications and we may be required to improve the solution until the optimal solution is obtained. If there are three or more than three variables, then the technique of *simplex algorithm* helps in locating the optimal solution, if any, in a finite number of steps, by using a convergent iterative computational process based on matrix algebra, leading to successively better solutions. The iteration goes on improving the solution till the attainment of such a value of the objective function as cannot be improved any further.

An initial solution is obtained, in terms of *slack* variables which are non-negative quantities taking away the slacks in the constraints, *i.e.*, removing the *slacks* in the inequalities ≤ and converting them into equalities. The data are arranged in a simplex tableau. The feasible solution obtained is checked for optimality and the process repeated until the optimal solution is obtained.

There are m basic variables and $n - m$ non-basic variables.

Each iteration examines whether the objective function can be improved by exchanging a basic variable for a non-basic variable.

LP Search for the Optimum

The optimum has to be searched at extreme points. The LP method moves from one extreme point to another in search of a better solution. The simplex method finds the optimal solution (whatever extreme point is chosen initially) in a finite number of steps.

In the context of use of the simplex algorithm some definitions given below should be noted.

Some Definitions and Notes

A solution of a linear programming (LP) problem is a set of values that satisfy the equality and inequality constraints. This set may not satisfy non-negativity constraints.

A feasible solution satisfies all constraints.

If there are m consistent and independent constraint equations or inequalities and n non-negative variables we have $n - m$ non-basic variables chosen to be zero and m basic variables for which there is a unique solution subject to the constraints.

An optimal solution, say for a maximum, is feasible when the objective function has a finite value greater than or equal to the values for all other feasible solutions.

By making a search among basic feasible solutions we finally get an optimal solution, if any. The value of the objective function is improved as new solutions are attempted. We have the optimal solution when the objective function cannot be improved any further.

The possibility of search for improved solutions makes the techniques useful in many situations. This results in numerous different applications of the technique. Thus, plans can be constructed for capital investment proposals. If the existing capacity or market demand acts as a constraint against production or profitability, the impact of provision of additional facilities can be judged. The technique to handle the market for higher profitability can be considered and so on.

16.10 PROCEDURE IN SIMPLE CASES

The simplex method is amenable to use on computers and can be applied to a variety problems incorporating many variables. Its applications to maximisation and minimisation problems are analogous since that can be done by changing the signs of coefficients. Thus z_{max} implies $-z_{min}$. The inequality constraints should be appropriately changed. Thus $\Sigma a_{ij} x_j \leq b_i$ implies $-\Sigma a_{ij} x_j \geq -b_i$.

Max $z = 7x_1 + 3x_2$ is equivalent to *Min* $-z = -7x_1 - 3x_2$ *i.e.*, for the same set of constraints the optimum values of x_1, x_2, are identical in both cases. The numerical values of z are opposite in sign.

Ex. 24. Consider Ex. 5. The problem is to maximise $z = x_1 + x_2$ subject to $8x_1 + x_2 \leq 200$, $x_1 + 2x_2 \leq 100$ and $x_1 \geq 0$, $x_2 \geq 0$.

Using slack variables S_1, S_2 the inequalities become equalities which should be written in the form

$$8x_1 + x_2 + S_1 + 0S_2 = 200$$
$$x_1 + 2x_2 + 0S_1 + S_2 = 100$$

To maximise $z = x_1 + x_2 + 0S_1 + 0S_2$ form the initial tableau with zero profit as the solution corresponding to zero production. This provides us with the initial feasible solution.

Row		c_j	*1*	*1*	*0*	*0*		
	c_B	*Basic Variable*	x_1	x_2	S_1	S_2	b_i	*Ratio*
R_1	0	S_1	8	1	1	0	200	200/1 = 200
R_2	0	S_2	1	2	0	1	100	100/2 = 50
		z_i	0	0	0	0		
	$z_j - c_j$		–1	–1	0	0		

Step 1. First determine the optimal column. The row $z_j - c_j$ shows the net profit when one unit of the variable is added. Any negative value of $z_j - c_j$ implies the solution can be improved.

Between x_1, x_2, the *coming in* variable is that which contributes most to the profit. Here, since both x_1, x_2 contribute equally we may take either, say, x_2, as the coming in variable. The x_2 column is the optimal column.

Step 2. Consider the ratios obtained by dividing the b_i entries in R_1, R_2 by the corresponding entries in the optimal column

$$\frac{200}{1} = 200, \frac{100}{2} = 50$$

The *going out* variable is the one corresponding to the smaller ratio. Here S_2 is the going out variable to be replaced by x_2 in the new tableau. The largest quantity of x_2 that can be taken is 50.

	c_j	*1*	*1*	*0*	*0*		
c_B	*Basic Variable*	x_1	x_2	S_1	S_2	b_i	*Ratio*
0	S_1	$7\frac{1}{2}$	0	1	$-\frac{1}{2}$	50	$150/7\frac{1}{2} = 200$
1	x_2	$\frac{1}{2}$	1	0	$\frac{1}{2}$	50	$50/\frac{1}{2} = 100$
	z_j	$\frac{1}{2}$	1	0	$\frac{1}{2}$	50	
	$z_j - c_j$	$-\frac{1}{2}$	0	0	$\frac{1}{2}$		

To find the elements of the rows the following formula is used:

Old row element—old row element × corresponding element in optimal column in the coming in row.

Thus $200 - 1 \times 50 = 150$, $8 - 1 \times \frac{1}{2} = 7\frac{1}{2}$, $1 - 1 \times 1 = 0$ etc.

Step 3. The negative profit per unit in the $z_j - c_j$ rows suggests the need for further improvement with the help of x_1

We therefore repeat the steps. Between $\dfrac{150}{7\frac{1}{2}} = 20$ and $\dfrac{50}{\frac{1}{2}} = 100$, the smaller ratio corresponds to S_1 which is now the going out variable to be replaced by x_1.

	c_j	*1*	*1*	*0*	*0*	
c_B	*Basic*	x_1	x_2	S_1	S_2	b_i
1	x_1	1	0	$\frac{2}{15}$	$-\frac{1}{15}$	20
1	x_2	0	1	$-\frac{1}{15}$	$\frac{8}{15}$	40
z_j		1	1	$\frac{1}{15}$	$\frac{7}{15}$	60
$z_j - c_j$		0	0	$\frac{1}{15}$	$\frac{7}{15}$	$\frac{7}{15}$

There is no negative $z_j - c_j$ now so that the optimal solution is $x_1 = 20, x_2 = 40$ which agrees with the graphical solution obtained earlier. The method used here can be presented in a somewhat lengthier but more convenient form as will be done in some of the solved exercises later where these same steps have been described in greater detail.

The exercises given earlier for graphical solution can be attempted with the help of the simplex method.

Ex. 25. Show that if we have to maximise $z = 6x_1 + 9x_2$

subject to $7x_1 + 12x_2 \leq 120,\ 10x_1 + 8x_2 \leq 120,\ x_1, x_2 \geq 0$

the simplex tableau will ultimately take the form:

	c_j	6	9	0	0	
c_B	Solution	x_1	x_2	S_1	S_2	b_i
9	x_2	0	1	$\frac{5}{32}$	$-\frac{7}{64}$	45/8
6	x_1	1	0	$-\frac{1}{8}$	$-\frac{3}{16}$	15/2
	$z_j - c_j$	0	0	$\frac{21}{32}$	$\frac{9}{64}$	

suggesting that for z_{max}, we have $x_1 = \frac{15}{2}$, $x_2 = \frac{45}{8}$, $z_{max} = 765/8$

Ex. 26. Show that the maximisation of $z = 5x_1 + 2x_2$ subject to $x_1 + x_2 \leq 5$, $2x_1 - x_2 \leq 7$, $x_1, x_2 \geq 0$ results in the final simplex tableau:

	c_j	5	2	0	0	
c_B	Solution	x_1	x_2	S_1	S_2	b_i
2	x_2	0	1	$\frac{2}{3}$	$-\frac{1}{3}$	1
6	x_1	1	0	$\frac{1}{3}$	$\frac{1}{3}$	4
	$z_j - c_j$	0	0	–3	–1	

suggesting that for for z_{max}, we have $x_1 = 4$, $x_2 = 10$, $z_{max} = 5 \times 4 + 2 \times 10 = 40$

Ex. 27. A firm manufactures two types of cloth, *A*, *B* using three different colours of wool. The table shows how much wool of each colour is available at a given time:

Require-ment (in dgm)		*For 1 meter length of type (in dgm)*	*Wool available for manufacture (in dgm)*
A	*B*		
4	5	of red wool	1000
5	2	of green wool	1000
3	8	of yellow wool	1200

Assuming that he can make a profit of Rs. 5 on one meter of type *A*, and Rs. 3 on one meter of type *B*, the manufacturer wants to know how to maximise his total profit. Find the optimum solution.

Ex. 28. A watch dealer wishes to buy new watches and has two models M_1, M_2 to choose from. Model M_1 costs Rs. 100 and M_2 costs Rs. 200. His showcase has space for 30 watches and he has Rs. 4000 to spend.

The watch dealer can make a profit per watch of Rs. 20 in M_1 and Rs. 50 in M_2. How many of each model should he buy to obtain the maximum profit?

Ex. 29. Maximise $z = \frac{1}{4}(7x_1 + 6x_2)$

Subject to
$$8x_1 + 5x_2 \leq 320$$
$$4x_1 + 5x_2 \leq 200$$
$$x_1 \geq 0, x_2 \geq 0$$

Ans. $x_1 = 30$, $x_2 = 16$, $z_{max} = 76.5$

Draw suitable diagrams.

What will be the result if the non-negatively conditions are replaced by $x_1 \geq 15$, $x_2 \geq 10$?

Ans. Same solution $x_1 = 30$, $x_2 = 16$

Ex. 30. (*i*) Minimise $t = 2x_1 + x_2$

Subject to
$$4x_1 + 7x_2 \geq 300$$
$$3x_1 + 4x_2 \geq 200$$
$$x_1 \geq 0, x_2 \geq 0$$

Ans. $x_1 = 40$, $x_2 = 20$

(*ii*) Maximise $z = 6x_1 + 21x_2$

Subject to
$$x_1 + 2x_2 \leq 3$$
$$x_1 + 4x_2 \leq 4$$
$$3x_1 + x_2 \leq 3$$
$$x_1 \geq 0, x_2 \geq 0$$

Ans. $x_1 = 24/33$, $x_2 = 27/33$, $z_{max} = 711/33$

Ex. 31. Maximise $z = x_1 + x_2$

Subject
$$x_1 + 3x_2 \leq 90$$
$$10x_1 + 7x_2 \geq 440$$
$$x_1 \geq 0, x_2 \geq 0$$

Ans. $z_{max} = 50$

Ex. 32. Maximise $z = 2x + 3x_2$

Subject to
$$x_1 + 2x_2 \leq 41$$
$$8x_1 + 5x_2 \leq 185$$
$$x_1 \geq 5, x_2 \geq 5$$

Ans. $x_1 = 15$, $x_2 = 13$, $z_{max} = 69$

Ex. 33. A machine producing *A* and *B* can produce *A* by using 2 units of chemical C_1 and 1 unit of chemical C_2 and can produce *B* by using 1 unit of chemical C_1 and 2 units of C_2. Only 800 units of C_1 and 100 units of C_2 are available. What will be the optimum allocation between *A* and *B* to maximise the total profit if the profits available per unit of *A* and *B* are 30 and 20 respectively?

Ans. *A* : 200, *B* : 400, profit = 14000

16.11 LINEAR PROGRAMMING: A GENERAL FORMULATION

To find non-negative numbers, or choice variables,

$$x_1, x_2, \ldots, x_n$$

which maximise (or minimise) the linear objective function of these variables

$$z = c_1x_1 + c_2x_2 + \ldots + c_nx_n$$

subject to m linear constraints

$$a_{11}x_1 + a_{12}x_2 + \ldots + a_{1n}x_n \leq b_1$$
$$a_{21}x_1 + c_{22}x_2 + \ldots + a_{2n}x_n \leq b_2$$
$$\ldots\ldots$$
$$a_{m1}x_1 + a_{m2}x_2 + \ldots + a_{mn}x_n \leq b_m$$

and the non-negativity restrictions

$$x_1 \geq 0, x_2 \geq 0, \ldots\ldots, x_n \geq 0.$$

(*i*) There are m constraints and n variables.

(*ii*) The subscript in a_{ij} is useful in locating the position of a coefficient.

(*iii*) $b_1, b_2, \ldots, b_m$ are restrictions imposed on the program.

(*iv*) Any set of non-negative numbers $x_1, x_2, \ldots, x_n$ which satisfies the above inequalities gives a feasible solution.

(*v*) A feasible solution which gives the maximum or minimum value of the objective function is an optimal solution giving $z_{\max}$ (or $z_{\min}$) which is the value of the programme.

We can formulate the problem with the sigma notation

To maximise $\quad z = \sum_{j=1}^{n} c_j x_j$

Subject to $\quad \sum_{j=1}^{n} a_{ij} x_j \leq b_i \qquad i = 1, 2, \ldots, m$

and $\quad x_j \geq 0 \qquad j = 1, 2, \ldots, n$

We can write it in the matrix form: Define the column vectors

$$c = [c_1 c_2 \ldots c_n], x = \begin{bmatrix} x_1 \\ x_2 \\ \vdots \\ x_n \end{bmatrix}, \; b = \begin{bmatrix} b_1 \\ b_2 \\ \vdots \\ b_m \end{bmatrix}.$$

and the $m \times n$ matrix

$$A = \begin{bmatrix} a_{11}\; a_{12} \cdots a_{1n} \\ a_{21}\; a_{22} \cdots a_{2n} \\ \cdots\;\cdots\;\cdots \\ a_{m1}\; a_{m2} \cdots a_{mn} \end{bmatrix}$$

The problem is to maximise the objective function $z = cx$ subject to $Ax \leq b$ and $x \geq 0$.

The minimisation problem can be framed similarly.

In the standard form, all constraints are written as equations *i.e.*, $Ax = b$. This can be done by introducing slack or surplus variables.

16.12 MAXIMISATION PROBLEM: DETAILED SIMPLEX PROCEDURE

The problem is written

$$\begin{aligned}
\text{Max} \quad & z = c_1x_1 + c_2x_2 + \ldots + c_nx_n \\
\text{Subject} \quad & a_{11}x_1 + a_{12}x_2 + \ldots + a_{1n}x_n \le b_1 \\
& a_{21}x_1 + a_{22}x_2 + \ldots + a_{2n}x_n \le b_2 \\
& \ldots\ldots\ldots\ldots\ldots\ldots\ldots\ldots \\
& a_{m1}x_1 + a_{m2}x_2 + \ldots + a_{mn}x_n \le b_m \\
& x_1, x_2, \ldots, x_n \ge 0
\end{aligned}$$

In matrix notation

$$\begin{aligned}
& \text{Max } z = cx \\
\text{Subject to} \quad & Ax \le b \\
& x \ge 0
\end{aligned}$$

Since all m inequalities are of the $\le$ type we introduce m slack variables $S_1, S_2, \ldots, S_m$.

(If there are inequalities of $\ge$ type then surplus variables would have to be introduced with a minus sign).

The set of $m + 1$ equations has n decision variables $x_1, x_2, \ldots, x_n$ and m slack variables $S_1, S_2, \ldots, S_m$. The first equation is the objective function and the rest the constraint equations

$$\begin{aligned}
z - c_1x_1 - c_2x_2 - .. - c_nx_n + 0S_1 + 0S_2 + .. + 0S_m &= 0 \\
a_{11}x_1 + a_{12}x_2 + .. + a_{1n}x_n + 1S_1 + 0S_2 + .. + 0S_m &= b_1 \\
a_{21}x_1 + a_{22}x_2 + .. + a_{2n}x_n + 0S_1 + 1S_2 + .. + 0S_m &= b_2 \\
\ldots\ldots\ldots\ldots\ldots\ldots\ldots\ldots\ldots\ldots & \\
a_{m1}x_1 + a_{m2}x_2 + .. + a_{mn}x_n + 0S_1 + 0S_2 + .. + 1S_m &= b_m
\end{aligned}$$

These can be put in the form of a table with each line in its proper place as shown below:

Standard Form of a Simplex Tableau

	c_j		c_1	c_2	...	c_n	0	0	...	0	*Resources*	*Ratio*
			x_1	x_2	...	x_n	S_1	S_2	...	S_m	b_i	b_i/a_{ij}
Row		*Basic*	*(non-basic variables)*				*(slack variables)*					
	c_B	*variables*	*Coefficient matrix*				*Identity Matrix*					
R_1	0	S_1	a_{11}	a_{12}	...	a_{1n}	1	0	...	0	b_1	b_1/a_{1j}
R_2	0	S_2	a_{21}	a_{22}	...	a_{2n}	0	1	...	0	b_2	b_2/a_{2j}
		...			...				...			
R_m	0	S_m	a_{m1}	a_{m2}	...	a_{mn}	0	0	...	1	b_m	b_m/a_{mj}
	$z_j = \Sigma\, c_{Bj}\, a_{ij}$		0	0	...	0	0	0	...	0	$z = \Sigma\, c_{Bi}\, b_i$	
$z_j - c_j$			$-c_1$	$-c_2$	...	$-c_n$	0	0	...	0	(Index Row)	

The c_B column shows coefficients of basic variables in the objective function. In the initial tableau these basic variables are slacks S_1, S_2, ..., S_m shown in the adjoining column.

The matrix under non-basic variables is the coefficient matrix and that under slack variables is the identity matrix.

Each z_j is the sum of products of c_B values by the corresponding elements of each column. Thus for the j th column, $z_j = \Sigma\, c_{Bj}\, a_{ij}$. Also $z_j - c_j$ is the index row and the value of the objective function is given by $z = \Sigma\, c_{Bi}\, b_i$

16.13 SIMPLEX ALGORITHM: STEPS FOR SOLUTION IN DETAIL

We shall consider here the maximisation problem.

1. Convert the inequality constraints into equalities by adding slack variables S_i for each $\leq$ inequality and subtracting S_i for each $\geq$ inequality. Thus

$$a_{i1}x_1 + a_{i2}x_2 + \ldots + a_{in}x_n \leq b_i \text{ becomes}$$
$$a_{i1}x_1 + a_{i2}x_2 + \ldots + a_{in}x_n + S_i = b_i$$
$$a_{i1}x_1 + a_{i2}x_2 + \ldots + a_{in}x_n \geq b_i \text{ becomes}$$
$$a_{i1}x_1 + a_{i2}x_2 + \ldots + a_{in}x_n - S_i = b_i$$
$$S_i \geq 0 \quad i = 1, 2, \ldots m$$

2. Introduce all $S_i (i = 1, 2, \ldots, m)$ into the objective function which should be written $\max z = x_1 + c_2x_2 + \ldots + c_n x_n + 0S_1 + 0S_2 + \ldots + 0S_m$

The initial basic feasible solution is written by putting $x_1 = x_2 = \ldots = x_n = 0$

Note: (*i*) The slack variable appearing in any inequality will not appear in any other inequality.

(*ii*) The coefficients of slack variables in the objective function will be all zero.

3. The c_j row in the simplex tableau shows the profit or cost per unit. The z_j row shows the opportunity costs of bringing in one unit of a variable into the solution. The $z_j - c_j$ row called the index row shows the net benefit (or loss) arising from bringing in one unit of a variable into the solution.

   Name the rows R_1, R_2, ..., R_m.

4. To find the optimal solution check the index row. If all its element are positive the current solution is optimum. If there is some negative element, the solution can be improved by the procedure in the steps given below.
5. Find the largest negative element in the $z_j - c_j$ row. The column containing that element is the pivotal column. The pivotal column is the incoming vector.
6. Divide each element in the b_i column by the corresponding element in the pivotal column. The row corresponding to the minimum of these ratios is the pivotal row. Omit non-positive cases. The pivotal row is the outgoing vector.
7. The number lying at the intersection of the pivotal row and pivotal column is called the *pivot* element which has to be greater than zero. Put a circle around the pivotal element.
8. The pivotal column under the incoming variable is the incoming column. The pivotal row corresponding to the going out variable is the pivot equation.

The Gauss Jordan method given below leads to a change of basis

New pivot row = Old pivot row ÷ pivot element

$$\left.\begin{array}{l}\text{Every Other row}\\ \text{including } z \text{ row}\end{array}\right\}\text{New row} = \begin{array}{c}\text{old}\\ \text{row}\end{array} - \begin{pmatrix}\text{incoming}\\ \text{column}\\ \text{coefficient}\end{pmatrix}\times\begin{pmatrix}\text{new}\\ \text{pivot}\\ \text{row}\end{pmatrix}$$

Note. Adjustments in a row imply adjustments in its elements

A tableau is optimal if none of the non-basic variables has a negative coefficient in the z equation. The solution has to be improved if one or more coefficients are negative.

9. To improve the solution, the procedure is as follows. Begin with Tableau 1.

Divide each element in the pivotal row R_p by the pivotal element which now becomes unity. Rewrite the tableau with this new pivotal row replacing the old row. Let the new pivotal row be R'_p.

In each row perform the following transformations and obtain new values of elements in the row. Denote the new row R'_i if the old row is R_i.

$$R'_i = R_i - kR_p$$

where k is the element in the ith row of the pivotal column and R_p the old pivotal row.

Do this for the whole tableaux so that the pivotal column will now consist of the pivotal element 1 and all other elements zero. The values of R'_p provide us with Tableau 2 to be worked upon in an identical manner till we find that none of the elements in the index row or z equation is negative.

16.14 MINIMISATION PROBLEM

The problem is written

$$\begin{aligned}\text{Min} \quad & z = c_1x_1 + c_2x_2 + \ldots + c_nx_n\\ \text{Subject to} \quad & a_{11}x_1 + a_{12}x_2 + \ldots + a_{1n}x_n \geq b_1\\ & a_{21}x_1 + a_{22}x_2 + \ldots + a_{2n}x_n \geq b_2\\ & \ldots\ldots\ldots\ldots\ldots\ldots\ldots\ldots\\ & a_{m1}x_1 + a_{m2}x_2 + \ldots + a_{mn}x_n \geq b_m\\ & x_1, x_2, \ldots, x_n \geq 0\end{aligned}$$

The inequalities usually being of the ≥ type, negative slack variables $S_1, S_2, \ldots, S_m$ are introduced. Since this would lead to a non-feasible solution, we introduce simultaneously artificial variables $A_1, A_2, \ldots, A_m$ in the inequalities so that we get a system of m equations in $n + 2m$ variables (made of n decision variables, m surplus and m artificial variables). Each artificial variables is given a large value M and each slack variable a value zero in the objective function so that the artificial variables should be driven out of the solution when they become nonbasic.

This simplex method is called the penalty technique in which the penalty is M for a minimisation problem and $-M$ for a maximisation problem whenever there is a constraint of = or ≥ type.

The problem takes the following form in which the first equation is the objective function followed by equations corresponding to constraints.

$$\text{Min} \quad z = c_1x_1 + c_2x_2 + \ldots + c_nx_n + 0S_1 + 0S_2 + \ldots + 0S_m + MA_1 + MA_2 + \ldots + MA_m$$

Subject to

$$a_{11}x_1 + a_{12}x_2 + \ldots + a_{1n}x_n - S_1 + A_1 = b_1$$

$$a_{21}x_1 + a_{22}x_2 + \ldots + a_{2n}x_n - S_2 + A_2 = b_2$$

$$\ldots\ldots\ldots\ldots\ldots\ldots\ldots\ldots\ldots$$

$$a_{m1}x_1 + a_{m2}x_2 + \ldots + a_{mn}x_n - S_m + A_m = b_m$$

$$x_1, x_2, \ldots, x_n \geq 0$$

$$S_1, S_2, \ldots, S_m \geq 0; A_1, A_2, \ldots A_m \geq 0$$

The procedure to deal with the minimisation case is almost the same as that for maximisation.

16.15 NOTES AND PRECAUTIONS IN SIMPLEX ALGORITHMS

1. All variables must be non-negative.
2. Every equality constraint can be replaced by two inequalities.
3. The problem max u subject to a set of constraints is exactly the same as the problem, min $(-u)$ subject to the same set of constraints and $-$ max u = min $(-u)$.
4. The optimatily conditions for the maximisation and minimisation problems are different. The feasibility conditions for the maximisation and minimisation problems are not different.
5. It is not true to say that the entering variable selected from among the current nonbasic variables as the one with the most negative objective coefficient guarantees the largest increase in the objective value in the next iteration.
6. If the going out variable does not correspond to the minimum ratio, at least one basic variable becomes negative in the following iteration.
7. The greater the number of constraints, the greater the computation work. The dual can help.
8. The pivot element should not be zero or negative.
9. An artificial variable can be dropped altogether once it becomes non-basic. Degeneracy can be avoided if redundant constraints are removed.
10. The solution is not feasible if an artificial variable appears positive in the optimal solution.
11. If a unit worth of resource is positive, the resource is scarce.
12. A model is said to have no feasible solution if the constraints cannot be satisfied simultaneously. With all $\leq$ type constraints this will not happen but with $\geq$ or $=$ constraints, the use of artificial variables may not allow a feasible solution. The penalty method may help provided the model has a feasible space. An infeasible space means incorrect model formulation or conflicting constraints or non-simultaneously satisfied constraints in which case other methods of solution are needed.

13. The value of the objective function is calculated after each iteration. It will indicate any positive or negative improvement that may have taken place in it.
14. The coefficients in the matrix of a_{ij} values indicate the marginal rate of exchange between row variables and column variables.
15. The quantities in the index row are shadow prices. If a shadow price is positive, it means an algebraic reduction in the objective function resulting from one unit of the variable in that column being introduced in the solution. If a shadow price is negative the converse is true.
16. In the final solution, if some slack variables appear, this would imply that there is some unutilised capacity or resource. If no slack variables appear, there is no unutilised capacity or resource.
17. If there is no feasible solution to a linear programming problem, this is a case of pseudo-optimality. Here the optimality criterion is satisfied but all the constraints are not satisfied. A solution of this type may contain artificial variables in the basis.
18. The complete simplex tableau can be computed from a knowledge of the associated inverse matrix.Only the feasibility of the optimal tableau is affected by changes on the right sides of the constraints. But changes in any coefficients of a linear programme other than those on the right sides may not affect the optimality of the solution.
19. The optimality and feasibility of a solution no more remain if both the right sides of the constraints and the coefficients of the objective function are changed.
20. If a new activity is added the objective value improves. A new constraint does not improve the objective value.
21. In a maximisation problem, the optimal profit equals the worth of used resources.

In the optimal solution, changes in the objective coefficient of a basic variable can change all the coefficients of non-basic variables in the optimal tableau. Again, in the optimal tableau, changes in the original objective coefficient of a non-basic variable can affect only its objective coefficients.

16.16 KEY COLUMN : TIES

While doing the simplex iteration, we move from one basic solution to another. We determine the incoming variable depending on the largest entering variable. If the $z_j - c_j$ have the same largest value in two or more cases, any one may be taken arbitrarily but to reduce the number of possible interactions the leftmost column may be selected if there is a *tie*.

In the case of a tie between a decision variable and a slack variable, the decision variable should be selected as the incoming variable.

In the case of a tie between two or more slack variables, the choice is arbitrary though for uniformity, the leftmost column may be selected first.

Key Row

The going out variable is decided as follows. We find the ratios of the right side values to the key column coefficients. That variable is the going out variable for which the ratio is the minimum.

Sometimes there may be two or more ratios with the same minimum value *i.e.*, there is a tie for determining the key row. This presents a situation of *degeneracy*. In the case of degeneracy one or more basic variables can take a zero value. The number of basic variable in the solution is not equal to the number of constraints or, the number of zero variables is not equal to the number of decision variables.

To break the tie, divide each slack variable coefficient by the corresponding coefficient in the key column. If the ratios do not break the tie, find the same ratios or decision variable coefficients and select the row with the smallest ratio. This is the key row. Now proceed for the optimal solution.

Alternative Optimals

More than one optimal solutions are possible when the objective function line is parallel to one of the constraints. In the graphical method the tie can be noticed by observing the corner parts. In the simplex algorithm the alternative optimal is noticed when the top row coefficient of a non-basic variable becomes zero. The non-basic variable can become a basic variable without the optimal changing.

16.17 MULTIPLE UNBOUNDED NULL SOLUTION

Sometimes the slopes of the objective function and one of the constraints may coincide giving more than one optimal solutions. In the final table $z_j - c_j$ becomes zero for a non-basic variable which may be a decision or slack variable.

On the other extreme, there can be a situation of a non-feasible solution because of incompatible constraints.

We have an unbounded solution where the value of the objective function goes on increasing indefinitely (without violation of feasibility conditions) because of one or more decision variables.

Ex. 35. max $z = 3x_1 + 2x_2$

Subject to

$$x_1 + 2x_2 \le 6$$
$$2x_1 + x_2 \le 8$$
$$-x_1 + x_2 \le 1$$
$$x_2 \le 2,\ x_1 \ge 0,\ x_2 \ge 0$$

Write Max $z = 3x_1 + 2x_2 + 0S_1 + 0S_2 + 0S_3 + 0S_4$

$$x_1 + 2x_2 + S_1 + 0S_2 + 0S_3 + 0S_4 = 6$$
$$2x_1 + x_2 + 0S_1 + S_2 + 0S_3 + 0S_4 = 8$$
$$-x_1 + x_2 + 0S_1 + 0S_2 + S_3 + 0S_4 = 1$$
$$0x_1 + x_2 + 0S_1 + 0S_2 + 0S_3 + S_4 = 2$$
$$x_1, x_2, S_1, S_2, S_3, S_4 \ge 0$$

The tableaus to solve this problem are given below

Tableau I

Row	c_B	c_j / Basic Variables	x_1 (+3)	x_2 (+2)	S_1 (0)	S_2 (0)	S_3 (0)	S_4 (0)	Resources b_i	
R_1	0	S_1	1	2	1	0	0	0	6	6/1 = 6
R_2	0	S_2	(2)	1	0	1	0	0	8	8/2 = 4
R_3	0	S_3	– 1	1	0	0	1	0	1	
R_4	0	S_4	0	1	0	0	0	1	2	
		z_j	0	0	0	0	0	0		
		$z_j - c_j$	– 3	– 2	0	0	0	0		

Initial solution with $x_1 = 0, x_2 = 0$

$$S_1 = 6, S_2, = 8, S_3 = 1, S_4 = 2$$

To proceed for the optimal solution, note the most negative value of $z_j - c_j$ is – 3.

$\therefore$ x_1 column is the key column and x_1 is the incoming variable.

The minimum ratio in the last column is 4

$\therefore$ R_2 is the key row and S_2 is the outgoing variable. The pivot element is 2. Put a circle around 2. Rewrite with R_2 row divided throughout by the pivot element 2 so that the pivot is 1.

Perform the following operations to get Tableau 2. The new rows denoted by R_i' replace old rows R_i

$R_1' = R_1 - 1R_2$ This will make all elements

$R_3' = R_3 - (-1)R_2$ in the pivot column zero

$R_4' = R_4 - 0R_2$ (except the pivot element 1)

Thus $R_1' : 1 - 1 \times 1 = 0, 2 - 1 \times \frac{1}{2} = \frac{3}{2}, 1 - 1 \times 0 = 1$ etc.

$R_3' : -1 + 1 \times 1 = 0, 1 + 1 \times \frac{1}{2} = \frac{3}{2}$ etc. (All values now in Tableau 2)

Replace the variable S_2 by x_1 in the basic variable column

Tableau 2

Row	c_B	c_j B.V.	3 x_1	2 x_2	0 S_1	0 S_2	0 S_3	0 S_4	b_i	
R_1'	0	S_1	0	(3/2)	1	– 1/2	0	0	2	$2/\frac{3}{2} = 4/3$
R_2'	3	x_2	1	1/2	0	1/2	0	0	4	$4/\frac{1}{2} = 8$
R_3'	0	S_3	0	3/2	0	1/2	1	0	5	$5/\frac{3}{2} = 10/3$
R_4'	0	S_4	0	1	0	0	0	1	2	$2/1 = 2$
		z_j	3	3/2	0	3/2	0	0		
		$z_j - c_j$	0	– 1/2	0	3/2	0	0		

Here each z_j is obtained by multiplying each c_B element by the corresponding element in each column. Thus in x_1 column,

$$z_1 = 0 \times 0 + 3 \times 1 + 0 \times 0 + 0 \times 0 = 3 \text{ and so on.}$$

The solution is $z = 0 \times 2 + 3 \times 4 + 0 \times 5 + 0 \times 2 = 12$

This is not optimal since there is a negative element –1/2 in $z_j - c_j$ row so that x_2 is the incoming variable. Since the least value in the last column is 4/3, S_1 is the outgoing variable. The pivot is 3/2. Divide R_1 by 3/2.

Proceeding as before we now have the next tableau 3 where

$$R_2'' = R_2' - \tfrac{1}{2} \times R_1' \text{ etc.}$$

$$R_3'' = R_3' - \tfrac{3}{2} \times R_1' \text{ etc.}$$

$$R_4'' = R_4' - 1 \times R_1' \text{ etc.}$$

Tableau 3

Row	c_B	c_j B.V.	3 x_1	2 x_2	0 S_1	0 S_2	0 S_3	0 S_4	b_i
R_1''	2	x_2	0	1	2/3	– 1/3	0	0	4/3
R_2''	3	x_1	1	0	– 1/3	2/3	0	0	10/3
R_3''	0	S_3	0	0	–1	1	1	0	3/1
R_4''	0	S_4	0	0	– 2/3	1/3	0	1	2/3
		z_j	3	2	1/3	4/3	0	0	
		$z_j - c_j$	0	0	1/3	4/3	0	0	

The solution is $z = 2 \times \frac{4}{3} + 3 \times \frac{10}{3} + 0 \times 3 + 0 \times \frac{2}{3}$

$= 38/3$

This is the optimal solution since there is no negative value in the $z_j - c_j$ row. This tableau is optimal.

Ex. 36. In the following problem the objective function remains the same as in the last exercise but the constraints are slightly different. Solve and compare.

Max $z = 3x_1 + 2x_2$

Subject to
$$-x_1 + 2x_2 \leq 4$$
$$3x_1 + 2x_2 \leq 14$$
$$x_1 - x_2 \leq 3$$
$$x_1, x_2 \geq 0$$

The problem is max $z = 3x_1 + 2x_2 + 0S_1 + 0S_2 + 0S_3$

Subject to
$$-x_1 + 2x_2 + 1S_1 + 0S_2 + 0S_3 = 4$$
$$3x_1 + 2x_2 + 0S_1 + 1S_2 + 0S_3 = 14$$
$$x_1 - x_2 + 0S_1 + 0S_2 + S_3 = 3$$
$$x_1, x_2, S_1, S_2, S_3 \geq 0$$

Show that the final tableau is as follows with the optimum solution at $x_1 = 4$, $x_2 = 1$, $S_1 = 6$ and $z_{max} = 14$

	c_j	3	2	0	0	0	b_i
c_B	*Basis*	x_1	x_2	S_1	S_2	S_3	
0	S_1	0	0	1	– 1/5	8/5	6
2	x_2	0	1	0	1/5	– 3/5	1
3	x_1	1	0	0	1/5	2/5	4
	z_j	3	2	0	1	0	
	$z_j - c_j$	0	0	0	1	0	

Ex. 37. (*i*) Min $z = 8x_1 + 3x_2 + 5x_3$

Subject to
$$29x_1 + 2x_2 + 13x_3 \geq 194$$
$$14x_1 + 8x_2 + 10x_3 \geq 152$$
$$11x_1 + 2x_2 + 7x_3 \geq 98$$
$$4x_1 + x_2 + 2x_3 \geq 34$$
$$x_1, x_2, x_3 \geq 0$$

Ans. $x_1 = 5$, $x_2 = 4$, $x_3 = 5$, $z_{min} = 77$

(*ii*) Max $z = 2x_1 + 3x_2$

Subject to $x_1 + 2x_2 \leq 8$ $\quad x_1 \geq 0$

$2x_1 + x_2 \leq 10$ $\quad x_2 \geq 0$

We can write this problem in standard form by adding slack variables

$$z = 2x_1 + 3x_2 + 0S_1 + 0S_2$$
$$x_1 + 2x_2 + S_1 + 0S_2 = 12$$
$$2x_1 + x_2 + 0S_1 + S_2 = 10$$

The procedure for Tableau 1 is given below. Proceed further and show that the optimum solution is $x_1 = 4$, $x_2 = 2$ use the graphical method for verification.

Tableau 1

	c_j	2	3	0	0		
c_B	*Basic Variable*	x_1	x_2	S_1	S_2	b	
0	S_1	1	(2)	1	0	8	$8 \div 2 = 4$
0	S_2	2	1	0	1	10	$10 \div 1 = 10$
	$z_j - c_j$	– 2	– 3	0	0	0	

1. The most negative value in the last row (the index row) is – 3. Hence x_2 column is the key column.
2. Divide each number in the b column by the corresponding number in the key column. The smallest ratio gives the key row. Here S_1 row is the key row.
3. The number at the crossing of x_2 column and S_1 row is 2. It is encircled. This is the pivot element.

Ex. 38. Interpret the concept of opportunity cost using the following problem:

Max $\quad z = 15x_1 + 45x_2$

Subject to

$$x_2 \le 50$$
$$x_1 + \frac{8}{5}x_2 \le 240$$
$$\frac{x_1}{2} + x_2 \le 62$$
$$x_1, x_2 \ge 0$$

Using slack variables, the problem becomes

$$0x_1 + x_2 + S_1 + 0S_2 + 0S_3 = 50$$
$$x_1 + \frac{8x_2}{5} + 0S_1 + S_2 + 0S_3 = 240$$
$$\frac{x_1}{2} + x_1 + 0S_1 + 0S_2 + S_3 = 10$$

Tableau 1

c_B	c_j Basic Variables	15 x_1	45 x_2	0 S_1	0 S_2	0 S_3	b	b_i/a_{ij}
0	S_1	0	1	1	0	0	50	$\frac{50}{1} = 50$
0	S_2	1	8/5	0	1	0	240	$\frac{240}{8/5} = 150$
0	S_3	1/2	2	0	0	1	162	$\frac{162}{2} = 81$
	z_j 0	0	0	0	0	0		
	$z_j - c_j$	– 15	– 45	0	0	0		

Optimal column x_2

Least ratio 50

Going out variables S_1. Show that the final simplex tableau is as follows:

Final Tableau

c_B	c_j B.V.	15 x_1	45 x_2	0 S_1	0 S_2	0 S_3	b
45	x_2	0	1	0	– 5/12	– 5/6	35
0	S_2	0	0	1	+ 5/12	– 5/6	15
15	x_1	1	0	0	+ 10/6	– 8/6	184
	z_j	15	45	0	6.25	17.5	4335
	$z_j - c_j$	0	0	0	6.25	17.5	

Ans. The optimal solution is

$$z_{max} = 45 \times 35 + 15 \times 184$$
$$= 1575 + 2760$$
$$= 4335$$

where $x_1 = 15, x_2 = 45$

is the optimal production level for maximum profit.

Note. The last row can be interpreted. It gives the *shadow prices* for the constraints on related variables: (See Sec. 16.19 and 16.22)

$$S_1 = 0, S_2 = 6.25, S_3 = 17.50$$

The z_j terms indicate the *opportunity* cost *i.e.*, if one unit of x_1 is not utilised the loss of profit is 15 and the non-utilisation of one unit of x_2 means a loss of profit of 45. The coefficients of non-basic variables S_1, S_2, S_3 represent shadow costs amounting to a decrease in the optimal value of the objective function resulting from a unit increase in a non-basic variable.

Ex. 39. Max $z = 2x_1 - x_2 + x_3$

Subject to
$$3x_1 + x_2 + x_3 \leq 60$$
$$x_1 - x_2 + 2x_3 \leq 10$$
$$x_1 + x_2 - x_3 \leq 20$$
$$x_1, x_2, x_3 \geq 0$$

Rewriting with slack variables S_1, S_2, S_3
$$z - 2x_1 + x_2 - x_3 - 0S_1 - 0S_2 - 0S_3 = 0$$
$$3x_1 + x_2 + x_3 + 1S_1 + 0S_2 + 0S_3 = 0$$
$$x_1 - x_2 + 2x_3 + 0S_1 + 1S_2 + 0S_3 = 0$$
$$x_1 + x_2 - x_3 + 0S_1 + 0S_2 + 1S_3 = 0$$

Check and complete the skeleton of steps given below:

Tableau 1

		+2	−1	+1	0	0	0		
		x_1	x_2	x_3	S_1	S_2	S_3	b	
0	S_1	3	1	1	1	0	0	60	$\frac{60}{3}$
0	S_2	(1)	−1	2	0	1	0	10	$\frac{10}{1}$
0	S_3	1	1	−1	0	0	1	20	$\frac{20}{1}$
	z_j	0	0	0	0	0	0		
	$z_j - c_j$	−2	1	−1	0	0	0		

Tableau 2

0	S_1	0	4	−5	1	−3	0	30	$\frac{30}{4}$
2	x_1	1	−1	2	0	1	0	10	
0	S_3	0	(2)	−3	0	−1	1	10	$\frac{10}{2}$
	$z_j - c_j$	0	−1	−3	0	2	0		

Tableau 3

0	S_1	0	0	1	1	– 1	– 2	10
2	x_1	1	0	1/2	0	1/2	1/2	15
– 1	x_3	0	1	– 3/2	0	– 1/2	1/2	5
	$z_j - c_j$	0	0	3/2	0	3/2	1/2	25

$x_1 = 15$ $\quad z_{max} = 2 \times 15 - 1 \times 5 + 1 \times 0$

$x_2 = 5$ $\quad = 25$

$x_3 = 0$

Ex. 40. Max $z = 3x_1 + 5x_2 + 2x_3$

Subject to $\quad -2x_2 + x_3 \geq 2$

$$x_1 + 4x_2 + 2x_3 = 5$$

$$x_1, x_2, x_3 \geq 0$$

Using slack variables S_1, S_2 and artificial variables A_1, A_2 the problem becomes

$$z - 3x_1 - 5x_2 - 2x_3 - 0S_1 - 0A_1 - 0A_2 = 0$$

$$0x_1 - 2x_2 + x_3 - 1S_1 + 1A_1 + 0A_2 = 0$$

$$x_1 + 4x_2 + 2x_3 + 0S_1 + 0A_1 + 1A_2 = 0$$

Check the steps below and complete the skeleton tableaus:

Tableau 1

	x_1	x_2	x_3	S_1	A_1	A_2		
A_1	0	– 2	(1)	– 1	1	0	2	2/1
A_2	1	4	2	0	0	1	5	5/2
$z_j - c_j$	$-M-3$	$-2M-5$	$-3M-2$	M	0	0	$-7M$	

Tableau 2

x_3	0	– 2	1	– 1	1	0	2	
A_2	1	(8)	0	2	– 2	1	1	1/8
$z_j - c_j$	$-M-3$	$-8M-9$	0	$-2M-2$	$3M+2$	0	$-M+4$	

Tableau 3

x_3	1/4	0	1	– 1/2	1/2	1/4	9/4
x_2	1/8	1	0	1/4	– 1/4	1/8	1/8
$z_j - c_j$		0	0	1/4	$M - 1/4$	$M + 9/8$	41/8

Tableau 4

x_3	0	– 2	1	– 1	1	0	2
x_1	1	8	0	2	– 2	1	1
$z_j - c_j$	0	15	0	4	$M - 4$	$M + 3$	7

$$x_1 = 1, x_2 = 0, x_3 = 2, z_{opt} = 7$$

Ex. 41. Minimise $z = x_1 + 2x_2$

s.t. $$-2x_1 + x_2 \geq 1$$
$$-x_1 + x_2 \geq 2 \quad x_1, x_2 \geq 0$$

The problem can be converted into:

Maximise $-z = -x_1 - 2x_2$

by using two slack variables S_1, S_2 and two artificial variables A_1, A_2. We rewrite the problem as

max $u = -z - MA_1 - MA_2$ where M is a large positive number. Alternatively we write:

$$x_1 + 2x_2 + MA_1 + MA_2 + u = 0$$

s.t. $$-2x_1 + x_2 - S_1 + A_1 = 1$$
$$-x_1 + x_2 - S_2 + A_2 = 2$$

check and complete the tableaus given below

Tableau 1

	1	2	*0*	*0*	$-M$	$-M$	
	x_1	x_2	S_1	S_2	A_1	A_2	b
A_1	– 2	(+ 1)	– 1	0	1	0	0
A_2	– 1	1	0	– 1	0	1	2

Tableau 2

	x_1	x_2	S_1	S_2	A_1	A_2	b
x_1	– 2	1	– 1	0	1	0	1
A_2	1	0	1	– 1	– 1	1	1

Tableau 3

	x_1	x_2	S_1	S_2	A_1	A_2	
x_2	– 1	1	0	– 1	0	1	2
S_2	1	0	1	– 1	– 1	1	1

The optimal solution is $x_1 = 0, x_2 = 2, u_{max} = 4 = z_{min}$

Ex. 42. Max $z = 4x_1 + 2x_2$ s.t. $x_1 + x_2 \leq 15, 2x_1 + x_2 \leq 20; x_1\ x_2 \geq 0$

Write
$$z = 4x_1 + 2x_2 + 0S_1 + 0S_2$$
$$x_1 + x_2 + 1S_1 + 0S_2 = 15$$
$$2x_1 + x_2 + 0S_1 + 1S_2 = 20$$

Tableau 1

		4	2	0	0		b/a_{ij}
		x_1	x_2	S_1	S_2	b	
0	S_1	1	1	1	0	15	15/1 = 15
0	S_2	(2)	1	0	1	20	20/2 = 10
	$z_j - c_j$	– 4	– 2	0	0	0	

The largest negative value in the index row – 4 corresponds to the lowest value 10 in the last column. The pivot 2 is encircled.

Tableau 2

		4	2	0	0		
		x_1	x_2	S_1	S_2	b	b/a_{ij}
0	S_1	0	1	1/2	0	5	5/1/2 = 10
4	x_1	0	1/2	1	–1/2	10	10/12 = 20
	$z_j - c_j$	0	0	0	2	40	

The optimal solution is $z_{max} = 5 \times 0 + 10 \times 4 = 40$ corresponding to $x_1 = 10$, $S_1 = 5$

But since there is zero coefficient for x_2 an alternative solution may exist.

The pivot row is 1/2

Tableau 3

		4	2	0	0	
		x_1	x_2	S_1	S_2	b
2	x_2	0	1	2	– 1	10
4	x_1	1	0	– 1	1	5
$z_j - c_j$		0	0	0	2	40

Again $z_{max} = 40$ at $x_1 = 5$, $x_2 = 10$

Ex. 43. A firm has 3 machines *A, B, C* with available working hours as follows:

Machine *A* 250 hours, *B* 150 hours, *C* 50 hours. The firm can manufacture three products *X, Y, Z*. Their machine requirement are

		Hours Required	
Machine/Product	*X*	*Y*	*Z*
A	8	2	3
B	4	3	–
C	2	–	1

If the profits per unit on *X, Y, Z* are Rs. 10, 3 and 4 respectively, what will be the maximum profit?

Hint. Let *x, y, z* be the quantities of the three products. The problem becomes

Max $u = 10x + 3y + 4z$

s.t. $8x + 2y + 3z \leq 250$

$4x + 3y \leq 150$

$2x + z \leq 50$

$x, y, z \geq 0$

Ex. 44. A firm produces products *X, Y, Z* on which profits per unit are Rs. 3, 5, 4 respectively. These products require the use of raw materials *A, B, C* whose available capacities are 8, 10, 15 respectively.

A unit of *X* requires 2 units of *A*, 3 of *B* to produce, *Y* requires 2 of *B* and 5 of *C* and *Z* requires 3 of *A*, 2 of *B* and 4 of *C*. Show how the firm can maximise its profit.

Hint: max $u = 3x + 5y + 4z$

Subject to
$$2x + 3z \le 8$$
$$3x + 2y + 2z \le 10$$
$$5y + 4z \le 15$$
$$x, y, z \ge 0$$

Ex. 45. Minimise $z = 3x_1 + 6x_2$

Subject to
$$4x_1 + x_2 \ge 20$$
$$x_1 + x_2 \le 20$$
$$x_1 + x_2 \ge 10$$
$$x_1, x_2 \ge 0$$

Using the slack variables S and artificial variables A, the constraints can be written
$$4x_1 + x_2 - S_1 + A_1 \doteq 20$$
$$x_1 + x_2 + S_3 = 0$$
$$x_1 + x_2 - S_2 + A_2 = 0$$
$$x_1, x_2, S_1, S_2, S_3, A_1 A_2 \ge 0$$

The equations for the simplex tableau are

Min $z = 3x_1 + 6x_2 + 0S_1 - MA_1 + 0S + 0S_2 + MA_2$

Subject to
$$4x_1 + x_2 - 1S_1 + 1A_1 + 0S_3 + 0S_2 + 0A_2 = 20$$
$$x_1 + x_2 + 0S_1 + 0A_1 + 1S_3 + 0S_2 + 0A_2 = 20$$
$$x_1 + x_2 + 0S_1 + 0A_1 + 0S_3 - 1S_2 - 1A_2 = 10$$

Tableau 1

	3	6	0	–M	0	0	–M	0	
	x_1	x_2	S_1	A_1	S_3	S_2	A_2	b_i	
A_1	4	+ 1	– 1	1	0	0	0	20	(1)
S_3	1	1	0.	0	1	0	0	20	(2)
A_2	1	1	0	0	0	– 1	– 1	10	(3)

Beginning with this tableau, show that the optimal solution is
$$z_{min} = 30 \text{ when } x_1 = 10, S_3 = 10, S_1 = 20$$

Ex. 46. max $z = x_1 + 2x_2 + 3x_3 - 4x_4$

Subject to
$$x_1 + 2x_2 + 3x_3 = 15$$
$$2x_1 + x_2 + 5x_3 = 20$$
$$x_1 + 2x_2 + x_3 + x_4 = 10$$
$$x_1, x_2, x_3, x_4 \ge 0$$

Hint: Since there are equality constraints we can write
$$z = x_1 + 2x_2 + 3x_3 - 4x_4 + MA_1 + MA_2 + MA_3$$

Subject to $x_1 + 2x_2 + 3x_3 + 0x_4 + MA_1 + 0A_2 + 0A_3 = 15$

$$2x_1 + x_2 + 5x_3 + 0x_4 + 0A_1 + MA_2 + 0A_3 = 20$$

$$x_1 + 2x_2 + x_3 + x_4 + 0A_1 + 0A_2 + MA_3 = 10$$

Hint: Start with the basic feasible solution

$$A_1 = 15, A_2 = 20, A_3 = 10$$

Ans. $z_{max} = 15$ with $x_1 = \frac{5}{2}, x_2 = \frac{5}{2}, x_3 = \frac{5}{2}, x_4 = 0$

Ex. 47. A firm has to buy 200 kg of an alloy containing gold and silver. The condition is that the alloy must have at least 100 kg of gold and no more than 35 kg of silver. The firm buys the alloy from two sources I and II at prices of 200 and 400 money units per kg. Source I gives the alloy with gold and silver proportions at 25 and 10 per cent. Source II gives these proportions as 75 and 20 per cent respectively. To minimise the cost what quantities should be bought from the two sources? Call the quantities x_1, x_2.

Hint: Min $z = 200x_1 + 400x_2$

Subject to $x_1 + x_2 \geq 200$

$$0.25x_1 + 0.75x_2 \geq 100$$

$$0.10x_1 + 0.20x_2 \leq 35, x_1, x_2 \geq 0$$

Write $z = 200x_1 + 400x_2 + 0S_1 + 0S_2 + 0S_3 + MA_1 + MA_2$

Subject to $x_1 + x_2 - 1S_1 + 0S_2 + 0S_3 + 1A_1 + 0A_2 = 200$

$$x_1 + 3x_2 + 0S_1 - 1S_2 + 0S_3 + 0A_1 + 1A_2 = 400$$

$$x_1 + 2x_2 + 0S_1 + 0S_2 + 1S_3 + 0A_1 + 0A_2 = 350$$

Ans. $z_{min} = 60,000$ at $x_1 = 100 = x_2$

Ex. 48. Min $z = x_1 + x_2 - 3x_3$

Subject to $-2x_1 + x_2 + x_3 \leq 11$

$$x_1 + 2x_2 - 4x_3 \geq 3$$

$$-x_2 + 2x_3 = -1$$

$$x_1, x_2, x_3 \geq 0$$

The problem may be written

Min $z = x_1 + x_2 - 3x_3 + 0S_1 + OS_2 + MA_1 + MA_2$

Subject to $-2x_1 + x_2 + x_3 + 1S_1 + 0S_2 + 0A_1 + 0A_2 = 11$

$$x_1 + 2x_2 - 4x_3 + 0S_1 - 1S_2 + 1A_1 + 0A_2 = 3$$

$$0x_1 + x_2 - 2x_3 + 0S_1 + 0S_2 + 0A_1 + 1A_2 = 1$$

$$x_1, x_2, x_3 S_1, S_2, A_1, A_2 \geq 0$$

Show that the (*a*) initial and (*b*) final tableaus are as follows with the solution given below:

(*a*)

	c_j	*1*	*1*	*– 3*	*0*	*0*	*M*	*M*	b_i
c_B	*Basic Variables*	x_1	x_2	x_3	S_1	S_2	A_1	A_2	
0	S_1	– 2	1	1	1	0	0	0	11
M	A_1	1	2	– 4	0	– 1	1	0	3
M	A_2	0	1	– 2	0	0	0	1	1
	c_j	M	3M	– 6*M*	0	– *M*	M	0	
	$z_j - c_j$	*M* – 1	3*M* – 1	– 6*M* + 3	0	– *M*	0	0	

(*b*)

	c_j	*1*	*1*	*– 3*	*0*	*0*	*M*	*M*	b_i
c_B	*Basic Variables*	x_1	x_2	x_3	S_1	S_2	A_1	A_2	
– 3	x_3	0	0	1	1/3	– 2/3	2/3	– 5/3	4
1	x_1	1	0	0	0	– 1	1	– 2	1
1	x_2	0	1	0	2/3	– 4/3	4/3	– 7/3	9
	z_j	1	1	– 1	– 1/3	– 1/3	1/3	2/3	
	$z_j - c_j$	0	0	0	– 1/3	– 1/3	1/3 – *M*	2/3 – *M*	

Ans. $x_1 = 1, x_2 = 9, x_3 = 4, \min z = -2$

16.18 SENSITIVITY ANALYSIS

Sensitivity analysis examines the degree of sensitivity of the linear programming solution as a result of minor variations in such parameters of the problem as the coefficients of the objective function, the b_i constraints and so on. The minor changes may result in (*i*) a major change in the optimal solution, (*ii*) a minor change or (*iii*) no change at all depending upon the nature of change in the parameter.

It is possible to incorporate the minor changes in the problem and solve it from the beginning to arrive at the optimal solution. Sensitivity analysis helps cut down these computations and also provides guidelines for improvement of results.

We shall examine the changes in the optimal solution resulting from the following:

(*a*) Changes in the coefficients c_j of the objective function

(*i*) If the coefficient c_k of the *non-basic* variable x_k changes, $z = c_B\, x_B$ being independent of c_k, both the objective function and the optimal solution remain unaffected.

(*ii*) If the coefficient c_k of a *basic* variable x_k changes by Δc_{BK}, the new objective function becomes $z_1 = z + \Delta c_{BK} x_{BK}$ but the optimal solution remains unchanged.

(*b*) Changes in the resource vector b_i

This does not change the optimality condition. It affects the basic variable values x_{Bi} and the objective function $z = c_B x_B$ as b_k becomes $b_k \div \Delta b_k$

(*c*) Changes in the coefficients a_{ij} of matrix A

(*i*) If the a_{ij} of a basic column is changed, the new column can no longer form the basis of the simplex tableau. Both the optimality condition and the feasibility conditions will be affected. One can find the range of changes in a_{ij} for these conditions to remain unaffected.

(*ii*) If the a_{ij} of a non-basic column is changed, the feasibility of the solution remains unaffected but the optimality of the solution is affected. The optimality can, however, be preserved by determining the appropriate change in a_{ij}.

(*d*) Addition of a New Variable

The additional variable x_{n+1} with unit cost c_{n+1} will add an extra column a_{n+1} in A. The current solution

(*i*) is optimal if $c_{n+1} - z_{n+1} \leq 0$

(*ii*) can be improved if $c_{n+1} - 1 - z_{n+1} > 0$

(*e*) Addition of a New Constraint

The new constraint $a_{m+1} x \leq b_{m+1}$ may be

(*i*) satisfied by the optimal solution x_B so that x_B remains feasible as well as optimal

(*ii*) not satisfied by x_B, so that the optimal solution of the modified problem may be obtained. The optimal solution may be obtained by the dual simplex.

16.19 DUALITY

Every linear programming problem has associated with it another linear programming problem which is the dual problem of the original or *primal* problem.

This is the principle of duality in linear programming.

Between a primal and its dual any one may be called the primal and the other its dual.

Duality exists because of the latent symmetries in a linear programming problem. Corresponding to every linear programming maximisation problem there is an associated minimisation problem and *vice versa.*

Ex. 49.

Primal	*Dual*
x_1, x_2 non-negative	y_1, y_2 non-negative
$4x_1 + 7x_2 \geq 300$	$4y_1 + 3y_2 \leq 2$
$3x_1 + 4x_1 \geq 200$	$7y_1 + 4y_2 \leq 1$
Minimise $u = 2x_1 + 1x_2$	Maximise $z = 300y_1 + 200y_2$

These two problems, obtained by transposing the primal matrix, as shown, are duals of each other. It may be noted that the rows and columns have been interchanged and the directions of inequalities reversed. The direction of optimisation has been reversed too.

Linear programming problems are always in pairs.

If the primal problem has a finite optimal solution then the dual problem has a finite optimal solution and the values of the objective functions are equal. Thus $u_{min} = z_{max}$.

This gives us the *fundamental theorem of linear programming* which states that if a linear programming problem and its dual are both feasible, then they both have optimal solutions and each problem has the same value.

It is always possible to give an economic interpretation to the dual problem. Thus in the dual of the profit maximising problem we may consider the costs of inputs $y_1, y_2, \ldots, y_m$ which do not enter directly into the profit calculations but which do result in these profits. The firm may impute values to the efforts of these inputs. These *imputed values* or *shadow prices* do not necessarily have any relation with the market prices.

Note. If the primal and dual are maximisation and minimisation problems, in whatever order, their optimisation concepts are opposite. The dual is formulated from information in the primal whose solution leads to the solution of the dual.

The number of primal variables equals the number of constraints in the dual and *vice versa.* The primal technological coefficients become the transpose of those in the dual.

Symmetric Dual Program

Primal	*Dual*
$\max z = c_1x_1 + \ldots + c_nx_n$	$\min u = b_1y_1 + \ldots + b_my_m$
s.t. $a_{11}x_1 + a_{12}x_2 + \ldots + a_{in}x_n \le b_1$	s.t. $a_{11}y_1 + a_{21}y_2 + \ldots + a_{m1}y_m \ge c_1$
$a_{21}x_1 + a_{22}x_2 + \ldots + a_{2n}x_n \le b_2$	$a_{12}y_1 + a_{22}y_2 + \ldots + a_{m2}y_m \ge c_2$
... ...	... ...
$a_{m1}x_1 + a_{m2}x_2 + \ldots + a_{mn}x_n \le b_m$	$a_{1n}y_1 + a_{2n}y_2 + \ldots + a_{mn}y_m \ge c_n$
$x_1, x_2, \ldots, x_n \ge 0$	$y_1, y_2, \ldots, y_m \ge 0$
i.e., $\max z = cx$	$\min u = yb$
s.t. $Ax \le b,\ x \ge 0$	s.t. $yA \ge c,\ y \ge 0$

Weak Duality. The value of the maximum problem for any feasible solution is always greater than or equal to the value of the minimum problem.

If the primal has a finite optimal solution then the dual also has a finite optimal solution. The two optimal objective values are equal.

i.e.,

$$z_{max} = u_{min}$$

16.20 DUAL SIMPLEX

In the dual simplex tableau,

$$z_j = \sum_{i=1}^{m} a_{ij}\, y_i$$

stands for the *imputed* cost of used resources per unit of activity j

Notes on the Dual

1. The dual can be formulated with the help of information contained in the primal. When the dual is solved it yields all essential information about the primal also. The structural characteristics of the primal and dual will tell us which of the two would be easier to solve.

2. The optimisation directions of the primal and the dual are opposite.

3. The number of variables in the primal equals the number of constraints in the dual.

4. The number of constraints in the primal equals the number of variables in the dual.

5. The objective function coefficients of the primal become the right hand sides of the constraints in the dual and the right hand constraints of the primal become the coefficients of the objective function of the dual.

6. The row coefficients in the primal become the column coefficients in the dual.

7. An equality constraint in the primal corresponds to an unrestricted variable (*i.e.*, positive, negative or zero) in the dual.

8. Suitable techniques are used to incorporate into the problem different types of constraints.

9. The dual objective function can be interpreted as the total value of the limited inputs of the primal *i.e.*, the maximum value of the primal objective function. The dual amounts to finding a set of shadow prices that will minimise the total value of the restricted inputs. We are then in a position to pinpoint the profit from each activity arising from inputs used therein.

10. If the primal has few activities and a large number of constraints implying a large number of variables it would be preferable to solve the dual, now with fewer constraints and yet get all the required information.

11. If a primal constraint does not require any slack or surplus variable for conversion into an equation, the dual is necessarily unrestricted. (*i*) If a slack variable requires the primal constraint, the dual variable will be restricted to non-negative values for a maximisation primal and non-positive values for a minimisation primal. (*ii*) If a surplus variable is so required the dual variable will not be unrestricted.

12. An unrestricted primal variable has a dual constraint with equality. The dual constraint corresponding to an artificial variable in the primal leads to redundant constraints in the dual.

13. When the primal is non-optimal, the dual is non-feasible. If the primal is infeasible, the dual may not necessarily have an unbounded optimum. But if the primal has an unbounded optimal, the dual is always infeasible.

14. In the dual simplex procedure (*i*) the starting solution must be optimal and infeasible and (*ii*) if none of the constraint coefficients of the going out variable equation is negative the program has no feasible solution.

Dual Advantages

Duality is a general concept and is noticed in economics and other social and physical sciences. Production and cost, input and output systems are examples of duals and are useful in long-term planning.

It is often useful to employ the dual if there are more constraints than the number of decision variables. This is because the number of constraints usually determines the number of iterations or tableaus required to solve the problem. If there are too many constraints it is preferable to solve the dual.

The dual concept is useful in sensitivity analysis or postoptimality analysis which helps in strategic decision making.

Dual Formation Procedure

Primal	*Dual*
Number of constraints	Number of variables
Number of variables	Number of constraints
ith variable coefficient in objective from	ith constraint right hand constant
jth constraint right hand constant	jth variable coefficient in objective form
$\leq$ constraint	non-negative variable
$\geq$ constraint	non-positive variable
= constraint	unrestricted variable
Non-negative variable	$\geq$ constraint
Non-positive variable	$\leq$ constraint
Unrestricted variable	= constraint

Ex. 50. Show that the dual of

Max $z = 20x_1 + 15x_2 + 18x_3 + 10x_4$

Subject to

$$5x_1 - 3x_2 + 10x_3 + 4x_4 \leq 60$$
$$x_1 + x_2 + x_3 = 25$$
$$-x_2 + 4x_3 + 7x_4 \geq 35$$
$$x_1, x_2, x_3 \geq 0,\ x_4 \text{ unrestricted}$$

is given by

Min $u = 60y_1 + 25y_2 + 35y_3$

$$5y_1 + y_2 \geq 20$$
$$-3y_1 + y_2 - y_3 \geq 15$$
$$10y_1 + y_2 + 4y_3 \geq 18$$
$$4y_1 + 7y_3 = 10$$
$$y_1 \geq 0,\ y_3 \leq 0$$

y_2 unrestricted.

Primal-Dual Table

Primal variables	x_1	x_2	...	x_j	...	x_n		
Right side of dual constraints	c_1	c_2	...	c_j	...	c_n	Dual objective	Dual variables
Left side of	a_{11}	a_{12}	...	a_{1j}	...	a_{1n}	b_1	y_1
Dual constraints	a_{21}	a_{22}	...	a_{2j}	...	a_{2n}	b_2	y_2
	...	...	...	...	...	...	...	...
	a_{m1}	a_{m2}	...	a_{mj}	...	a_{mn}	b_m	y_m

Dual Constraints

A Note on the Objective Value for any feasible solution in the Maximisation and Minimisation problems.

1. obj. value in the max. problem ≤ obj. value in the min. problem for any *feasible* solutions of the primal and the dual.

2. obj. value in the max. problem = obj. value in the min. problem for the *optimum* solution of both the problems.

3. Optimal z equation coefficient of a starting primal variable = Difference between the left and right sides of the dual constraint of the starting variable.

4. When the max. and min. procedures reach an equilibrium point, further improvement in the objective value is not possible.

Ex. 51. Formulate the dual

(*a*) Min $z = 8x_1 + 4x_2 + 5x_3 + 6x_4$

Subject to
$$3x_1 - 2x_2 + 4x_3 + x_4 \le 125$$
$$x_1 + x_3 + x_4 = 75$$
$$2x_1 + x_2 - 3x_3 \le 150$$
$$x_1, x_2, x_3 \ge 0$$
$$x_4 \text{ unrestricted.}$$

The dual is Max $u = 125y_1 + 75y_2 + 150y_3$

Subject to
$$3y_1 + y_2 + 2y_3 \ge 0$$
$$-2y_1 + y_3 \ge 0$$
$$4y_1 + y_2 - 3y_3 \ge 0$$
$$y_1 + y_2 = 6$$
$$y_1, y_3, \le 0, y_2 \text{ unrestricted.}$$

(*b*) Max $z = 20x_1 + 15x_2 + 18x_3 + 10x_4$

s.t.
$$5x_1 - 3x_2 + 10x_3 + 4x_4 \le 60$$
$$x_1 + x_2 + x_3 = 25$$
$$-x_2 + 4x_3 + 7x_4 \ge 35$$
$$x_1, x_2, x_3 \ge 0$$
$$x_4 \text{ unrestricted}$$

The dual is min $u = 60y_1 + 25y_2 + 35y_3$

s.t.
$$5y_1 + y_2 \ge 20$$
$$-3y_1 + y_2 - y_3 \ge 15$$
$$10y_1 + y_2 - y_3 \ge 18$$
$$4y_1 + 7y_3 = 10$$
$$y_1 \ge 0, y_3 \le 0, y_2 \text{ unrestricted.}$$

Ex. 52. Find the dual of

Max $z = 10x_1 + 12x_2$

Subject to
$$x_1 + 2x_2 = 5$$
$$-x_1 + 5x_2 \ge 3$$
$$4x_1 + 7x_2 \le 8$$
$$x_1 \text{ unrestricted}; \quad x_2 \ge 0$$

Since there is an equality constraint, we should rewrite it by putting $x_1 = x_{11} - x_{12}$; x_1, x_{11}, $x_{12} \geq 0$. The problem becomes

Max $z = 10x_{11} - 10x_{12} + 12x_2$

Subject to
$$x_{11} - x_{12} + 2x_2 = 5$$
$$-x_{11} + x_{12} + 5x_2 - 1S_1 + 0S_2 = 3$$
$$4x_{11} - 4x_{12} + 7x_2 + 0S_1 + 1S_2 = 8$$
$$x_{11}, x_{12}, x_2, S_1, S_2 \geq 0.$$

The dual is

Min $t = 5y_1 + 3y_2 + 8y_3$

Subject to
$$y_1 - y_2 + 4y_3 \geq 10$$
$$-y_1 + y_2 - 4y_3 \geq -10$$
$$2y_1 + 5y_2 + 7y_3 \geq 12$$
$$-y_2 \geq 0$$
$$y_3 \geq 0$$
y_1 unrestricted.

Ex. 53. Find the dual of

Max $z = 7x_1 + 9y_2$

Subject to
$$x_1 + 2x_2 \leq 15$$
$$-x_1 + 3x_2 \leq 12$$
$$x_1, x_2 \geq 0$$

Ans. Min $t = 15y_1 + 2y_2$

Subject to $y_1 - y_2 \geq 7$
$$2y_1 + 3y_2 \geq 9$$
$$y_1, y_2 \geq 0$$

Ex. 54. Write the dual of max $u = 4x_1 - 2x_2$

Subject to $x_1 + x_2 = 3$, $-3x_1 + x_2 \leq -4$, $x_1 \geq 0$, x_2 unrestricted.

Write $x_2 = x_{21} - x_{22}$; $x_{21}, x_{22} \geq 0$
$$x_1 + x_2 \leq 3$$
$$x_1 + x_2 \geq 3 \text{ or } -x_1 - x_2 \leq -3$$

The primal is max $u = 4x_1 - 2(x_{21} - x_{22})$

Subject to
$$x_1 + (x_{21} - x_{22}) \leq 3$$
$$-x_1 - (x_{21} - x_{22}) \leq -3$$
$$-3x_1 + (x_{21} - x_{22}) \leq -4$$
$$x_1, x_{21}, x_{22} \geq 0$$

Dual. min $z = 3y_1 - 3y_2 - 4y_3$

Subject to
$$y_1 - y_2 - 3y_3 \geq 4$$
$$y_1 - y_2 + y_3 \geq -2$$
$$-y_1 + y_2 - y_3 \geq 2$$
$$y_1, y_2, y_3 \geq 0$$

or if we write $y = y_1 - y_2$, min $z = 3y - 4y_3$

Subject to $y - 3y_3 \geq 4$, $y + y_3 \geq -2$, $-y - y_3 \geq 2$ with y unrestricted, $y_3 \geq 0$.

Ex. 55. The dual of max $c'x$ subject to $Ax = b, x \geq 0$ is min $b'y$ subject to $A'y \geq c$ and y unrestricted.

Ex. 56. Show that the optimal solution for the following duals is given by $x = 10/3 = y$; $p = 0$, $q = 5/3$, $r = 1/3$.

Min $u = 2x + 3y$

Subject to $\quad 2x + 3y \leq 30 \qquad x, y \geq 0$

$x + 2y \geq 10$

$x - y \geq 0$

Max $v = 30p + 10q$

Subject to $\quad 2p + q + r \leq 2 \qquad q, r \geq 0$

$3p + 2q - r \leq 3 \qquad p \leq 0$

Note. 1. The objective functions

$$u_{\min} = 2 \times \frac{10}{3} + 3 \times \frac{10}{3} = 50/3$$

and

$$v_{\max} = 30 \times 0 + 10 \times \frac{5}{3} = 50/3$$

have the same value.

2. Check. In the index row, it will be found that in the final tableau, the absolute values of $z_j - c_j$ under S_1, S_2, S_3 in the primal table will be the same as entries in the resources column b in the dual tableau.

Primal	*Dual*
Values of basic variables	$-(z_j - c_j)$ values in the non-basic variable column
$z_j - c_j$ values in the non-basic variable column	values of basic variables

Ex. 57. Product X_1 requires 3 units of a raw material and 2 units of labour and contributes a profit of 7 money units. X_2 requires one unit of the raw material and one of labour to contribute 5 money units. A total of 48 units of material and 40 units of labour are available.

Solve the dual and get the shadow prices of material and labour.

Primal	*Dual*
max $z = 7x_1 + 7x_2$	min $u = 48y_1 + 40y_2$
st $\quad 3x_1 + x_2 \leq 48$	st $\quad 3y_1 + 2y_2 \geq 7$
$2x_1 + x_2 \leq 40$	$y_1 + y_2 \geq 5$
$x_1, x_2 \geq 0$	$y_1, y_2 \geq 0$

The solution to the primal can be found by introducing slacks S_1, S_2:

max $z = 7x_1 + 7x_2 + 0S_1 + 0S_2$; $x_1, x_2, S_1, S_2 \geq 0$

s.t. $\quad 3x_1 + x_2 + 1S_1 + 0S_2 = 48$

$2x_1 + x_2 + 0S_1 + 1S_2 = 40$

We will find the solution to the dual which will include surplus variables S_3, S_4 and artificial variable A_1, A_2

$$\min z = 48y_1 + 40y_2 + 0S_1 + 0S_2 + MA_1 + MA_2$$

$$\text{st. } 3y_1 + 2y_2 - 1S_1 + 0S_2 + MA_1 + 0A_2 = 7$$

$$y_1 + y_2 + 0S_1 - 1S_2 + 0A_1 + MA_2 = 5$$

$$y_1, y_2, S_1, S_2, A_1, A_2 \geq 0$$

Tableau 1

	c_j	48	40	0	0	M	M	b_i	Ratio
c_B	B.V.	y_1	y_2	S_1	S_2	A_1	A_2		
M	A_1	(3)	2	-1	0	1	0	7	7/3
M	A_2	1	1	0	-1	0	1	5	5/1
	z_j	4M	3M	$-M$	$-M$	M	M		
	$z_j - c_j$	$\begin{pmatrix} 4M \\ -48 \end{pmatrix}$	$\begin{pmatrix} 3M \\ -40 \end{pmatrix}$	$-M$	$-M$	0	0		

The minimum ratio is 7/3 and the pivot is 3 in the column corresponding to $z_j - c_j = 4M - 48$. Thus y_1 is the coming in and A_1 the going out variable.

Tableau 2

	c_j	48	40	0	0	M	M	b_i	Ratio
c_B	B.V.	y_1	y_2	S_1	S_2	A_1	A_2		
48	y_1	1	2/3	$-1/3$	0			7/3	7/2
M	A_2	0	1/3	1/3	-1			8	24
	z_j	48	$32 + M/3$	$-16 + M/3$	$-M$				
	$z_j - c_j$	0	$-8 + M/3$	$-16 + M/3$	$-M$				

Tableau 3

	c_j	48	40	0	0	M	M	b_i	Ratio
c_B	B.V.	y_1	y_2	S_1	S_2	A_1	A_2		
40	y_2	3/2	1	$-1/2$	0			7/2	—
M	A_2	$-1/2$	0	1/2	-1			3/2	3
	z_j	$60 - M/2$	40	$-20 + M/2$	$-M$				
	$z_j - c_j$	$12 - M/2$	0	$-20 + M/2$	$-M$				

Tableau 4

	c_j	48	40	0	0	M	M	b_i
c_B	B.V.	y_1	y_2	S_1	S_2	A_1	A_2	
40	y_2	1	1	0	– 1			5
0	S_1	– 1	0	1	– 2			0
	z_j	40	40	0	– 40			
	$z_j - c_j$	– 8	0	0	– 40			

The u_{min} corresponds to $y_2 = 5$ and $y_1 = 0$

Also $S_1 = 0$, $x_2 = 40$ which are the values of the primal decision variables.

∴ For optimal solution, produce 0 units of product x_1 and 5 unity of x_2

$$z_{max} = 7 \times 0 + 5 \times 40 = 200$$

The shadow price of raw material is $y_1 = 0$ and that of labour is $y_2 = 5$

Ex. 58. Show that the following two problems are duals of each other.

(*a*) Max $z = 45y_1 + 24y_2 + 20y_3$

Subject to
$$y_1 + 3y_2 \le 5$$
$$y_1 + 5y_2 \le 5$$
$$y_1 + 7y_3 = 2$$
$$y_1 - 2y_2 - 5y_3 \le 3$$
$$y_1 + 3y_3 = 0$$
$$y_1 \text{ unrestricted}, y_2 \le 0, y_3 \ge 0$$

(*b*) Min $u = 4x_1 + 5x_2 + 2x_3 + 3x_4 + x_5$

Subject to
$$x_1 + x_2 + x_3 + x_4 + x_5 = 45$$
$$3x_1 + 5x_2 - 2x_4 \le 24$$
$$7x_3 - 5x_4 + 3x_5 \ge 20$$

$x_1, x_2, x_4, \ge 0$, x_3, x_5 unrestricted

Ex. 59. Show that for the problem min $u = 5x + 6y + 3z$

Subject to
$$x + y - z \ge 20, \quad x, y, z \ge 0$$
$$5x + 5y + 5z \ge 35$$
$$5x + 5y + 3z \ge 50$$
$$7x + 6y - 9z \ge 30$$
$$2x + 4y - 15z \ge 10$$
$$12x + 10y \ge 90$$
$$y - 10z \ge 20$$

The solution is $x_1 = 0$, $x_2 = 20$, $x_3 = 0$, $u_{min} = 120$. Solve the dual.

Ex. 60. Write the dual of max $z = 2x_2 - 5x_3$

Subject to
$$x_1 + x_3 \ge 2$$
$$2x_1 + x_2 + 6x_3 \le 6$$
$$x_1 - x_2 + 3x_3 = 0$$
$$x_1, x_2, x_3 \ge 0.$$

Solve this to find the dual solution.

The Dual is Min $u = 2y_1 + 6y_2$

Subject to
$$y_1 + 2y_2 + y_3 \geq 0$$
$$y_2 - y_3 \geq 2$$
$$y_1 + 6y_2 + 3y_3 \geq -5$$
$$y_1 \leq 0$$
$$y_2 \geq 0$$
y_3 unrestricted. **Ans.** $y_1 = 0,\ y_2 = 2/3,\ y_3 = -4/3$

Ex. 61. Each of the given two plants has capacities to produce three products x_1, x_2, x_3. Plant A has capacities for x_1 3000, x_2 1000, x_3 2000 with daily operating costs 600. Plant B has capacities for x_1 1000, x_2 1000, x_3 6000 with daily operating costs 400. The maximum demands are for x_1 24000, x_2 16000, x_3 48000 units per month. How many days should the plants work to minimise the product cost?

If d_1, d_2 are the number of working days then min $z = 600\, d_1 + 400\, d_2$

Subject to
$$3000\, d_1 + 1000\, d_2 \geq 24000$$
$$1000\, d_1 + 1000\, d_2 \geq 16000$$
$$2000\, d_1 + 6000\, d_2 \geq 48000$$
$$d_1, d_2 \geq 0$$

Prove that $z_{min} = 7200$ corresponding to $d_1 = 4$, $d_2 = 12$ and solve the dual: max $z = 24000\, y_1 + 16000\, y_2 + 40000\, y_3$

Subject to
$$3000\, y_1 + 1000\, y_2 + 2000\, y_3 \leq 600$$
$$1000\, y_1 + 1000\, y_2 + 6000\, y_3 \leq 400$$
$$y_1, y_2, y_3 \geq 0$$

Ex. 62. For the primal, min $z = x_1 + 4x_2 + 3x_4$

Subject to
$$x_1 + 2x_2 - x_3 + x_4 \geq 3$$
$$-2x_1 - x_2 + 4x_3 + x_4 \geq 2$$
$$x_1, x_2, x_3, x_4 \geq 0$$

Show that $z_{min} = 7$ corresponding to $x_1 = 7$, $x_2 = 0$, $x_3 = 4$, $x_4 = 0$. Find the solution for the dual.

Ex. 63. A diet must contain a minimum number of units of three ingredients: 4000 units of G_1, 50 of G_2, 1400 of G_3. Food F_1 containing 200 units of G_1, 1 of G_2 and 40 of G_3 costs Rs. 4 per unit. Food F_2 containing 100 units of G_1, 2 of G_2 and 40 of G_3 costs Rs 3 per unit. Minimise the cost by combining the two foods. Can the dual help?

One may solve the primal, min $z = 4x_1 + 3x_2$

Subject to
$$200x_1 + 100x_2 \geq 4000$$
$$x_1 + 2x_2 \geq 50$$
$$40x_1 + 40x_2 \geq 1400 \qquad x_1, x_2 \geq 0.$$

and get $x_1 = 5$, $x_2 = 30$; $z_{min} = 4 \times 5 + 3 \times 30 = 110$ or it may be found convenient to use the dual

$$\max z = 4000y_1 + 50y_2 + 1400y_2$$

Subject to
$$200y_1 + y_2 + 40y_3 \le 4$$
$$100y_1 + 2y_2 + 40y_3 \le 3 \qquad y_1, y_2, y_3 \ge 0.$$

16.21 DUALITY AND INPUT-OUTPUT PROBLEMS

In the minimisation problem, the objective value is an upper bound on the objective value in the maximisation problem. The two problems thus have an equilibrium point which cannot be improved and which is reached when the two objective values are equal.

If we consider the limited resources available as *inputs* and the contribution of an activity to the objective values as *output* the problem becomes an input output system.

16.22 SHADOW PRICE

When the optimal solution of the primal problem is obtained, the optimal in the dual is simultaneously attained. The two solutions are identical.

The primal problem deals with the primary relationship of decision variables in the objective function and the constraints. It is concerned with the allocation of resources to fulfil the objective of the problem. The dual problem is economically meaningful and is concerned with the pricing of these allocated resources. The dual may not always be interpretable in words.

The values assigned to different resources can be looked upon as their operational costs called *shadow prices* or *imputed values* which might influence the product price. The addition of individual resources multiplied by their shadow prices gives the total shadow price.

If in the optimal solution, the imputed cost of an activity is more than its marginal profit, the level of the activity is zero.

The unit worth of a resource or its shadow price or imputed price is also called the dual price. The unit worth implies the rate of improvement in the optimal solution following an increase in the available amount of that resource.

The shadow price is used to decide the extent of expansion of any particular resource facility and the determination of the range of variation in the resource availability that would result in the shadow price obtained in the optimal tableau. The extent of changes in marginal profits or marginal costs can also be similarly determined.

Notes

1. A shadow price indicates the value of having an additional unit of a resource *i.e.*, it is the amount by which the optimal value of the objective function would change if any constraint is changed marginally.

2. A shadow price thus indicates the marginal return or change in the value of the objective, given a unit change in a constraint.

3. A shadow price is positive for any constraint which has been satisfied as an equality (*i.e.*, with no slack) in the optimal solution.

4. A shadow price is zero for any constraint which is not satisfied as an equality (*i.e.*, there is a slack).

5. Shadow prices or imputed values can be considered in terms of the costs of restrictions on the amounts of inputs available, say, to a firm. For example, the potential change in profit resulting from the increase by one unit of an input is the imputed value of that unit and can be looked upon as the highest price that the firm would be willing to pay for that one unit.

6. The imputed value or shadow price is not the same as the cost of that input nor is it the market price of that input. The total value of the input used can be found by multiplying the input's shadow unit price by its total amount used. It may be noted that (1) the sum total of all the imputed values of inputs equals the gross profit minus fixed cost and (2) when the optimal solution requires less than the maximum amount available of some input then the imputed value of that input is zero. The imputed value arises in a situation of restrictions.

7. The shadow idea is useful when an input is not purchased in the market or in which the connection between that input and the output is a distant one. The shadow price then is that which the decision maker would have to pay for this input in a competitive market which may not even exist. In a socialist economy the planner may use shadow prices since markets may not exist and valuation may not be easy.

8. In business, shadow price estimation may prove to be a useful exercise. Also in sensitivity analysis, if the shadow price of current production capacity of a plant is found to be higher than the cost of increasing the capacity it may be worth while to reexamine the scope for additional profits.

Ex. 64. How does a linear programming framework contribute to a development programme?

(It suggests a proper allocation of resources and projection of production levels, criteria for choice of investments, balance of payments considerations, etc. subject to constraints of employment, available resources, etc.

It also suggests a technique of production etc. in each sector of an underdeveloped economy depending on its requirements.)

Ex. 65. Give some points of difference between linear programming and input output analysis.

(*LP* assumes most efficient adjustments can be made according to changing conditions in the economy. In analysing the actual working of the economic system *IO* may be more useful.)

CHAPTER 17

Probability and Statistics

17.1 THE CONCEPT OF PROBABILITY

Chance and luck play an important role in our lives. Such events as marriages, deaths etc. depend substantially on chance. The birth of an individual depends on a biological chance resulting in a male or a female, a white or a black, a genius or an idiot.

One may try to measure the extent of chance or probability of any such event as is often done in gambling, weather forecasts, business etc. A right decision may mean, for instance, the choice of the best possible bet, irrespective of the ultimate actual result. But experience can help a person make fairly accurate decisions and predictions.

Any attempt to investigate an economic phenomenon is likely to have to reckon with the element of probability.

Measurement in economics makes use of statistical theory, which in turn is based on the laws of probability. Statistics and probability are fundamentally related. The theory of probability describes statistical phenomena and analyses them. They are used to study correlation and regression, sampling methods, business decisions, etc.

In everyday conversation, the term probability is used loosely. Statements of the following type are common:

It is likely to rain today.

It will probably rain today.

There is a fair chance of it raining today.

In each case the sense conveyed is almost the same and a certain degree of confidence is shown in the outcome of the event.

A mathematical measure for the degree of confidence is provided by the theory of probability.

If we toss a coin, the event, 'head' or the event 'tail' may occur. Both these possibilities are equally likely. We say the probability or chance of a head is 50 per cent or $\frac{1}{2}$ and the probability of a tail is also $\frac{1}{2}$.

If we roll a die, any one of the six faces (1, 2, 3, 4, 5, 6) is equally likely to show up. The probability of any one of these events is $\frac{1}{6}$.

We shall express the probability p of an event as a ratio. It can never be less than zero and can never exceed one *i.e.*, $0 \leq p \leq 1$.

17.2 SOME DEFINITIONS

Events. An event is said to have occurred in an experimental if the result of the experiment is a specified outcome E. If this specified outcome is observed when the experiment is performed then E is said to have occurred.

Trial. A procedure or an experiment to collect any statistical information is called a trial.

Exhaustive events. A set of events is exhaustive if all possible events associated with a trial are included in the set. The set of events {head, tail} exhausts all possibilities in the toss of a thin coin since nothing else can occur. In the roll of a die, (1, 2, 3, 4, 5, 6) is an exhaustive set of events.

Favourable events. Such cases as result in the happening of an event are said to be cases favourable to that event.

Elementary or simple events. Such events as contain only a single outcome are called elementary events.

Impossible events. An event which can never occur is an impossible event.

Certain events. An event which is certain to occur is a certain event.

The occurrence of number 10 when a die is rolled is an impossible event. The death of a person is a certain event.

Complementary events. Events E_1, E_2 are complementary if whenever E_1 occurs, E_2 does not; whenever E_2 occurs, E_1 does not.

Occurrences of head and tail in the toss of a coin are complementary events.

The complement of an event E is written E'.

17.3 A PRIORI PROBABILITY

If an event E can happen in m cases out of a total of n possible equally likely cases then the a priori probability or briefly, the probability of the event E is defined as

$$p(E) = \frac{m}{n}$$

If we can find

n : the total number of possible cases

m : the number of favourable cases

then the probability of the event is the ratio m/n.

Here $n - m$ cases are favourable to E' *i.e.*, to the nonoccurrence of E and E and E' are *complementary* events.

$$p(E') = \frac{n-m}{n}$$

We can see that

$$p(E) + p(E') = 1 \quad \text{or} \quad p + q = 1$$

where $p = p(E)$, $q = p(E')$ so that $q = 1 - p$.

Ex. 1. What is the probability of getting a six when a die is rolled?

Number of possible outcomes $n = 6$

Number of favourable outcomes $m = 1$

$$\therefore \quad p = \tfrac{1}{6}.$$

Ex. 2. What is the probability of getting a white ball from an urn containing 2 white and 3 red balls? (The draw is made without looking into the urn and all the balls are absolutely similar in all respects.)

Number of possible outcomes $n = 5$ (five balls)

Number of favourable outcomes $m = 2$ (two white balls)

$$\therefore \quad p = \tfrac{2}{5}$$

Ex. 3. A box contains 4 red, 2 white, 3 black balls. When a ball is drawn at random what is the probability that it will be (*i*) red (*ii*) white (*iii*) black?

Ans. (*i*) $\tfrac{4}{9}$ (*ii*) $\tfrac{2}{9}$ (*iii*) $\tfrac{3}{9}$.

Ex. 4. What is the probability of getting a 5 or 6 in the rolling of a die?

Total number of cases $n = 6$

Total number of favourable cases $m = 2$ (*i.e.*, 5 or 6)

$$\therefore \quad p = \tfrac{2}{6} = \tfrac{1}{3}$$

Total number of unfavourable cases = 4 (*i.e.*, 1, 2, 3, 4)

The probability of not getting a 5 or 6 is given by

$$q = \tfrac{4}{6} = \tfrac{2}{3}$$

Check:

$$p + q = 1$$

$$\tfrac{1}{3} + \tfrac{2}{3} = 1$$

Alternatively if we know $p = \tfrac{1}{3}$ then $q = 1 - \tfrac{1}{3} = \tfrac{2}{3}$.

Ex. 5. If two coins are tossed at random what is the probability of getting (*a*) 2 heads, (*b*) two tails, (*c*) one head, one tail?

Ans. (*a*) Total number of cases = 4 (*HT, TH, HH, TT*)

Total number of favourable cases for 2 heads = 1 (*HH*)

$\therefore \; p = \tfrac{1}{4}$ (*b*) $\tfrac{1}{4}$ (*c*) $\tfrac{1}{2}$.

Ex. 6. 100 bulbs were found to be defective in a lot of 800 bulbs. Find the probability that a bulb selected from the lot is not defective.

The probability that a bulb selected from the lot is defective is given by

$$p = \tfrac{100}{800} = \tfrac{1}{8}$$

The probability that a bulb selected is not defective is

$$q = 1 - p = 1 - \tfrac{1}{8} = \tfrac{7}{8}$$

Ex. 7. An urn contains r red, w white balls. A sample of two balls is drawn. Find the probability that they are (*a*) both red (*b*) both white (*c*) one red, one white.

Probability of the sample containing 2 red $= \dfrac{r}{(r+w)} \times \dfrac{(r-1)}{(r+w-1)}$,

Probability of the sample containing 2 white $= \dfrac{w}{(r+w)} \times \dfrac{(w-1)}{(r+w-1)}$

Consider the probability of the sample containing one red and one white ball. This can happen in two mutually exclusive ways, 1st ball red, 2nd ball white or 1st ball white, 2nd ball red.

Thus, the probability of the sample containing 1 white and 1 red ball

$$= \frac{r}{(r+w)} \times \frac{w}{(r+w-1)} + \frac{w}{(r+w)} \times \frac{r}{(r+w-1)}$$

Note. The sum of probabilities in the three cases equals 1.

Ex. 8. In a group of ten people where six are males and four are females, what is the chance that a committee of four, formed from the group with random selection, has

(*a*) four females?

(*b*) three females and one male?

Ans. (*a*) Probability of a committee with four females

$$= \frac{4}{10} \times \frac{3}{9} \times \frac{2}{8} \times \frac{1}{7} = 0.0048$$

(*b*) A committee of three females and one male can be formed in four mutually exclusive ways: (*i*) MFFF, (*ii*) FMFF, (*iii*) FFMF, (*iv*) FFFM

(*i*) The probability of each arrangement is the same:

$$p(i) = \frac{6}{10} \times \frac{4}{9} \times \frac{3}{8} \times \frac{2}{7} = 0.0286$$

$$p(ii) = \frac{4}{10} \times \frac{6}{9} \times \frac{3}{8} \times \frac{2}{7} = 0.0286 \text{ and so on.}$$

Total $p(i) + p(ii) + p(iii) + p(iv) = 4 \times 0.0286 = 0.114$

Ex. 9. From a consignment containing 100 items of which 10% are defective, a random sample of 10 is drawn.

What is the probability that

(*i*) the sample contains 0 defective items?

(*ii*) the sample contains exactly 1 defective?

(*i*) Probability of 0 defective arises in only one way. (NNNNNNNNNN)

$$p(0) = \frac{90}{100} \times \frac{89}{99} \times \frac{88}{98} \times \ldots \times \frac{81}{91} = 0.33$$

N = Non-defective, D = Defective

(*ii*) Exactly one defective in the sample can arise in 10 mutually exclusive ways: (each gives the same result)

1	D	N	N	...	N
2	N	D	N	...	N
3	N	N	D	...	N
:					
:	....	....	....	....	....
10	N	N	N	...	D

The probability of 1 defective in 10 items sampled is 10 times any of these

$$p(1) = 10 \times \frac{10}{100} \times \frac{90}{99} \times \frac{89}{98} \times \ldots \times \frac{82}{91}$$

17.4 PROBABILITY: DIFFERENT APPROACHES

There are several approaches to the definition of the probability concept. They are complementary and not competitive.

Classical Approach : The probability $p(E)$ of an event E is defined as

$$p(E) = \frac{\text{Number of outcomes favourable to the event}}{\text{Total number of possible outcomes}}$$

This is an priori approach and requires no experimentation. It is based on the assumption that all outcomes are equally likely.

Personal Approach : The decision maker, in a given situation may assign a certain value to the probability of an event depending on the degree of his belief in the occurrence of that event. Different persons may assign different probability values to the same event. This happens in marketing, managerial, and economics related situations.

Ex. 10. A, B, C toss a coin in that order. The first to get a head wins. The game continues till this happens. What are the chances of each wining?

$$p(\text{head}) = \frac{1}{2} \text{ and } p(\text{not getting a head}) = \frac{1}{2}$$

Begin with A. He gets a head in toss 1 or, if he does not get a head in 1, he can win if B does not get it in toss 2, C does not get it in 3 and then A gets it in toss 4 and so on. Thus A can win if he gets a head in

Toss 1, $p_1 = \frac{1}{2}$

Toss 4, $p_2 = p(\overline{A})\, p(\overline{B})\, p(\overline{C})\, p(A) = \frac{1}{2} \cdot \left[\frac{1}{2}\cdot\frac{1}{2}\cdot\frac{1}{2}\right] = \left(\frac{1}{2}\right)^4$

Toss 7, $p_3 = \left(\frac{1}{2}\right)^7$ and so on. This gives an infinite geometric series.

$$\therefore p(A \text{ wining}) = \frac{1}{2} + \left(\frac{1}{2}\right)^4 + \left(\frac{1}{2}\right)^7 + \ldots = \frac{\frac{1}{2}}{1-(1/2)^3} = \frac{4}{7} \quad \text{(use GP)}$$

Similarly $p(B) = \frac{1}{2}\cdot\frac{1}{2}+\frac{1}{2}\cdot\left(\frac{1}{2}\right)^4+\frac{1}{2}\left(\frac{1}{2}\right)^7+\ldots$

$$= \frac{1}{2}\cdot\left[\frac{1}{2}+\left(\frac{1}{2}\right)^4+\left(\frac{1}{2}\right)^7+\ldots\right] = \frac{1}{2}\cdot\frac{4}{7} = \frac{2}{7}$$

$$p(C) = \frac{1}{2}\left(\frac{1}{2}\right)^2+\frac{1}{2}\left(\frac{1}{2}\right)^5+\frac{1}{2}\left(\frac{1}{2}\right)^8+\ldots$$

$$= \frac{1}{2}\cdot\frac{1}{2}\left[\frac{1}{2}+\left(\frac{1}{2}\right)^2+\left(\frac{1}{2}\right)^7+\ldots\right] = \frac{1}{2}\cdot\frac{1}{2}\cdot\frac{4}{7} = \frac{1}{7}$$

A posteriori probability

In the case of mathematical or a *priori* probability all the possibilities of the occurrence are considered in advance. The events are equally likely and mutually exclusive.

There is another aspect of probability. The statistical probability, also called empirical or *a posteriori* probability, is determined experimentally rather than theoretically. It is concerned with problems in which past experiences are influential. To arrive at the conclusion a large number of cases must be observed.

If m is the number of successes in n trials and if the sequence of relative frequencies m/n obtained for larger and larger values of n approaches a limit p *i.e.*, when a large number of trials is made, we define the a posteriori probability of the event as

$$p = \lim_{n\to\infty}\frac{m}{n}$$

17.5 SETS AND PROBABILITY

Events can be treated as subsets of an outcome space. All the relations in set theory apply without change to events. For this we merge the two definitions of probability into one. Observations of uncontrolled events in nature or those under controlled conditions will be assumed to be resulting from a process and an *experiment* so that our approach will become direct and systematic.

We shall consider an event to be a collection of points which is a subject of the outcome space S of the experiment.

Probability Axioms.

1. An *impossible* event is that which equals the empty set ϕ: $p(\phi) = 0$.

2. A *certain* event equals the outcome space : $p(S) = 1$.

3. For any event E, $p(E) \geq 0$, $p(E) \leq 1$ *i.e.*, $0 \leq p(E) \leq 1$.

4. The complement E', of an event E, is the collection of all sample points in S but not in E.

$$p(E) + p(E') = 1,\ p(E) = 1 - p(E')$$

17.6 COMPOUND EVENTS

Most events occurring in practice are compound *i.e.*, combinations of two or more events.

Let E_1, E_2 be two events. Then we define

1. *Union event* : $E_1 \cup E_2$ or $E_1 + E_2$ *i.e.*, either E_1 or E_2 or both E_1, E_2 occur.

2. *Intersection event* : $E_1 \cap E_2$ or E_1E_2, *i.e.*, both E_1 and E_2 occur.

3. *Inclusion event* : $E_1 \subset E_2$ *i.e.*, if E_1 occurs then so does E_2.

4. *Difference event* : $E_1 - E_2$ or $E_1E'_2$ *i.e.*, if E_1 occurs then E_2 does not occur.

5. *Mutually exclusive events* : $E_1 \cap E_2 = \phi$ *i.e.*, E_1, E_2 are mutually exclusive if the event E_1E_2 contains no sample points.

If there are two events E_1, E_2, then $E_1 = (E_1E_2) + (E_1E'_2)$. Here E_1E_2 and $E_1 E'_2$ are mutually exclusive.

$$\therefore \quad p(E_1) = p(E_1E_2) + p(E_1 E'_2)$$

Similarly $p(E_2) = p(E_2E_1) + p(E_2E'_1)$

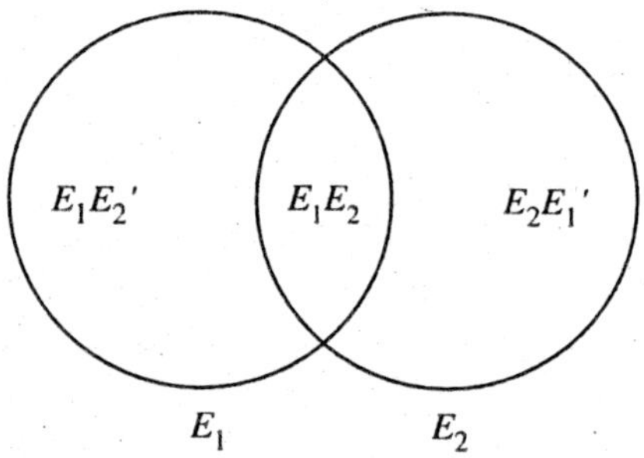

6. *Independent Events*

Two events E_1, E_2 are said to be *independent* if

$$p(E_1/E_2) = p(E_1)$$

or

$$p(E_2/E_1) = p(E_2)$$

It may be easily seen that if $p(E_1/E_2) = p(E_1)$ then

$$p(E_2/E_1) = p(E_2)$$

Note that E_1, E_2 are said to be mutually exclusive if they cannot occur together.

7. *Dependent Events*

When the occurrence of an even depends on the nature of occurrence of one or more events that go before it, it is said to be *dependent* on these events.

17.7 CONDITIONAL PROBABILITY

The conditional probability of event E_1, given that E_2 has already occurred, is defined by

$$p(E_1/E_2) = \frac{p(E_1E_2)}{p(E_2)}$$

The probability of E_2, given E_1 is

$$p(E_2/E_1) = \frac{p(E_1E_2)}{p(E_1)}$$

Ex. 11. Consider the following data:

Activity Participation in a college

	Yes	*No*	*Total*
Boys	120	80	200
Girls	60	40	100
Total	180	120	300

Let E_1 = Yes, E_2 = boys, E'_1 = No, E'_2 = Girls

Then $p(E_1) = \frac{180}{300}$, $p(E_2) = \frac{200}{300}$, $p(E_1E_2) = \frac{120}{300}$

$$\therefore \quad p(E_1/E_2) = \frac{p(E_1E_2)}{p(E_2)} = \frac{120/300}{200/300} = \frac{3}{5}$$

$$p(E_2/E_1) = \frac{p(E_1E_2)}{p(E_1)} = \frac{120/300}{180/300} = \frac{2}{3}.$$

Ex. 12. Two dice are rolled. Let E_1 be the event that the first die shows an even face and E_2 that the second die shows 1 or 2.

$$p(E_1) = \tfrac{3}{6} = \tfrac{1}{2},\ p(E_2) = \tfrac{2}{6} = \tfrac{1}{3}$$

$$p(E_1E_2) = \tfrac{1}{2} \times \tfrac{1}{3} = \tfrac{1}{6}$$

$$\therefore \ p(E_1/E_2) = \frac{p(E_1E_2)}{p(E_2)} = \frac{\frac{1}{6}}{\frac{1}{3}} = \frac{1}{2}$$

$$p(E_2/E_1) = \frac{p(E_1E_2)}{p(E_1)} = \frac{\frac{1}{6}}{\frac{1}{2}} = \frac{1}{3}.$$

Ex. 13. A factory has some automatic and some mechanical machines. Some of these machines are new and some old. A machine is selected at random and found to be new. Using the data below, show the probability that it is automatic is $\frac{4}{7}$.

	Auto.	*Mech.*	*Total*
New	40	30	70
Old	20	10	30
Total	60	40	100

Diagrammatic representation of two or more events.

Two mutually exclusive events

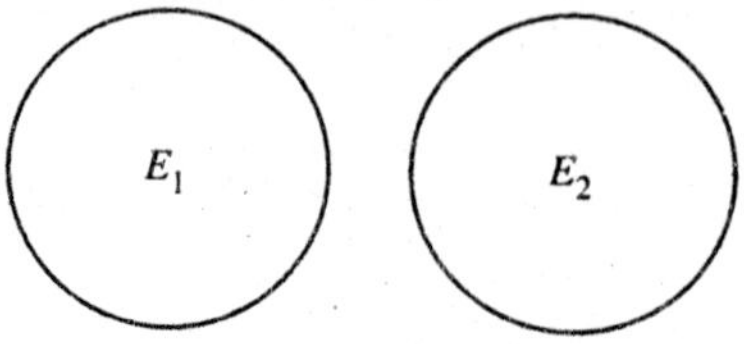

Two compatible events

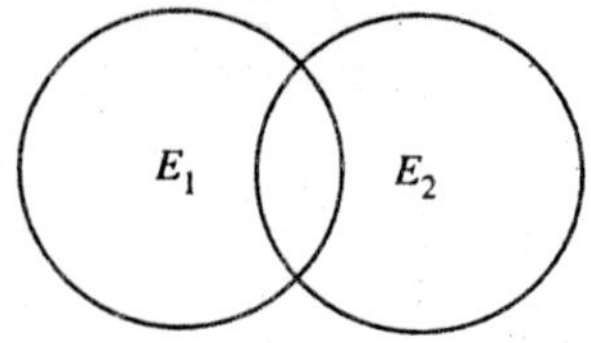

E_1 or E_2

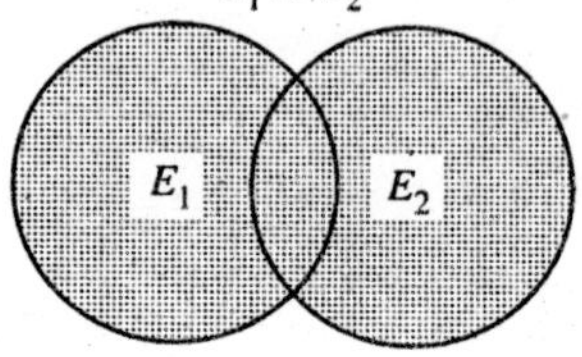

E_1 or E_2 but not both

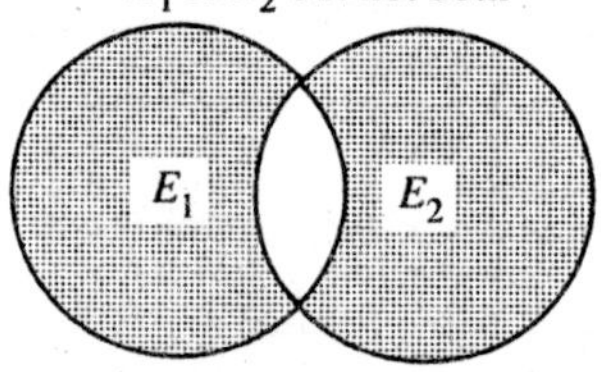

E_1 and E_2

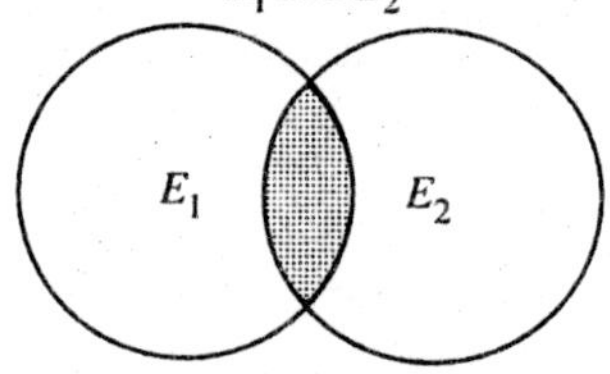

E_1 but not E_2

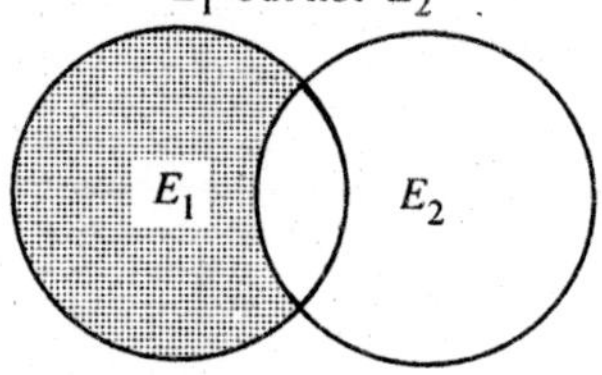

E_2 but not E_1

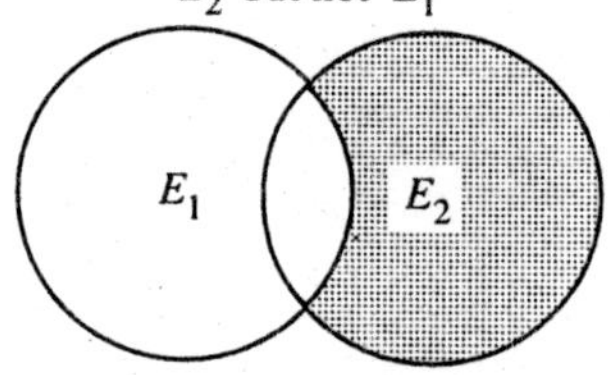

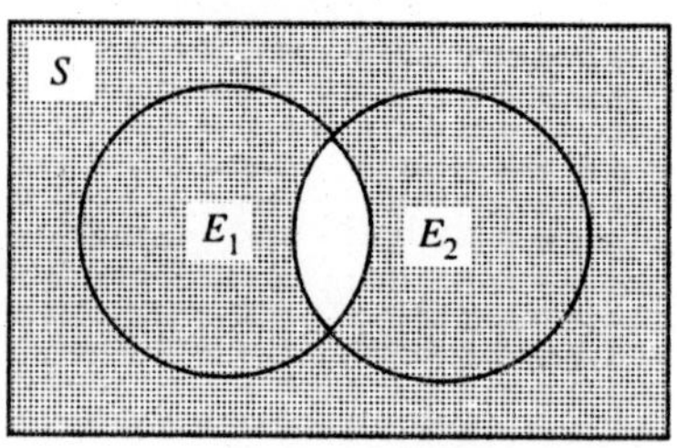

E_1 or E_2 or
neither but not both

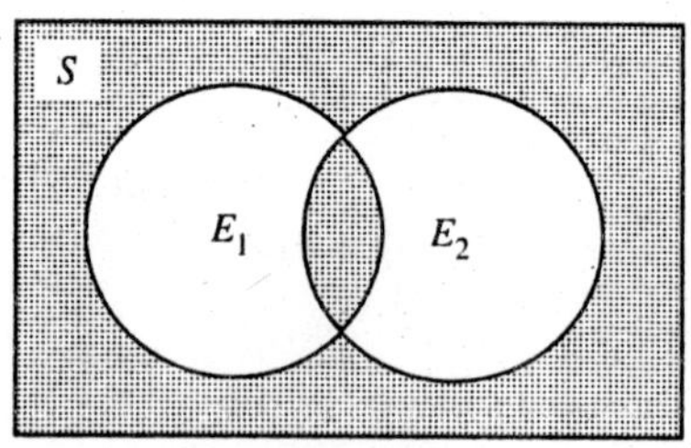

Either E_1 and E_2
both or niether

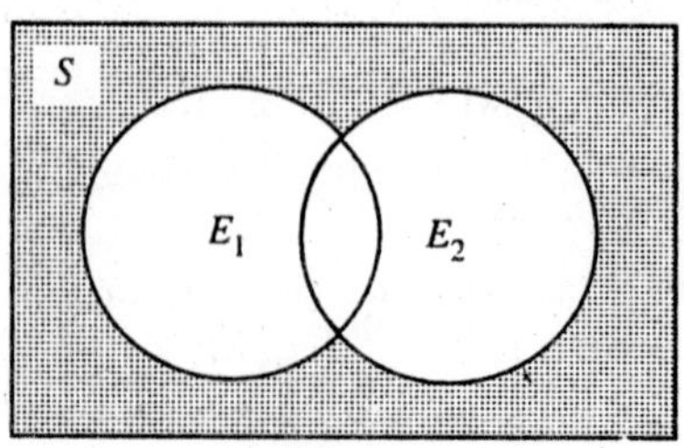

Nither E_1 nor E_2

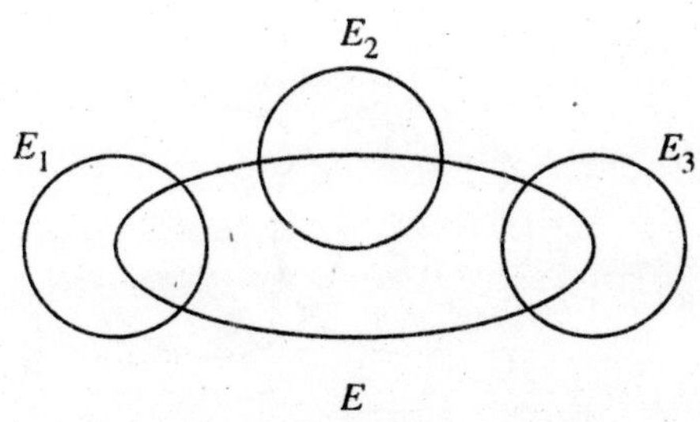

E_1, E_2, E_3 mutually exclusive but each compatible with E

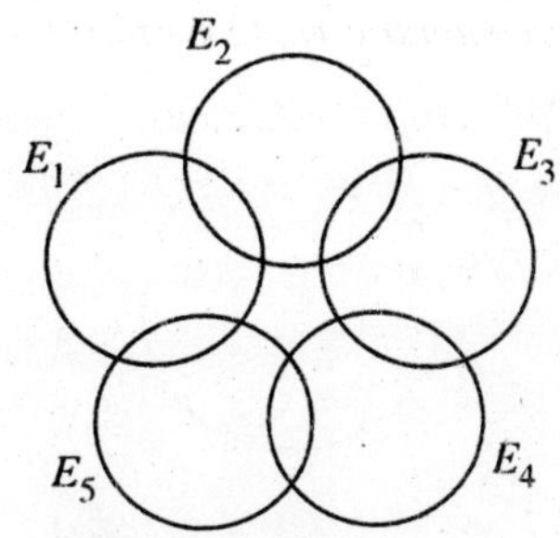

Each E_i compatible with 2 others but incompatible with the rest

17.8 FUNDAMENTAL LAWS OF PROBABILITY

The Law of Addition of Probabilities

(*i*) The probabilities of the union event $E_1 + E_2$ is given by

$$p(E_1 + E_2) = p(E_1) + p(E_2) - p(E_1E_2)$$

(*ii*) If E_1, E_2 are mutually exclusive events *i.e.*, $p(E_1E_2) = 0$ then

$$p(E_1 + E_2) = p(E_1) + p(E_2)$$

(*iii*) If E_1, E_2, E_3 are any three events

$$p(E_1 + E_2 + E_3) = p(E_1) + p(E_2) + p(E_3) - p(E_1E_2) - p(E_2E_3) - p(E_3E_1) + p(E_1E_2E_3)$$

If E_1, E_2, E_3 are mutually exclusive then

$$p(E_1 + E_2 + E_3) = p(E_1) + p(E_2) + p(E_3)$$

The results follow from set theory.

The Law of Multiplication of Probabilities

The probabilities of the intersection event E_1E_2 is given by

$$p(E_1E_2) = p(E_1)\, p(E_2/E_1) = p(E_2)\, p(E_1/E_2)$$

If E_1, E_2 are independent

$$p(E_1E_2) = p(E_1)\, p(E_2)$$

For three events E_1, E_2, E_3

$p(E_1E_2E_3) = p(E_1)\, p(E_2/E_1)\, p(E_3/E_1, E_2)$ and so on.

17.9 THE PRINCIPLE OF INVERSE PROBABILITY: BAYE'S THEOREM

Often we come across problems in which after noticing the outcome of an experiment we want to find the probability that the outcome was due to a particular one of the various possible causes of the outcome.

Suppose an event E can occur with one and only one of events $F_1, F_2, \ldots, F_n$ *i.e.*, if E occurs some F_i must occur. What is the probability that if E has occurred then it is, say, F_1, that has occurred in conjunction with E. This gives rise to Baye's theorem.

To take a simple two event case suppose F can occur as F_1 or F_2

Since $$E = (EF_1) + (EF_2)$$

$$\therefore \quad p(E) = p(EF_1) + p(EF_2) = p(F_1)\, p(E/F_1) + p(F_2)\, p(E/F_2)$$

From the definition of conditional probability, we have

$$p(F_1/E) = \frac{p(F_1)p(E/F_1)}{p(E)} = \frac{p(F_1)p(E/F_1)}{p(EF_1)+p(EF_2)}$$

$$\therefore \quad p(F_1/E) = \frac{p(F_1)\,p(E/F_1)}{p(F_1)\,p(E/F_1)+p(F_2)\,p(E/F_2)}$$

Similarly $$p(F_2/E) = \frac{p(F_2)\,p(E/F_2)}{p(F_1)\,p(E/F_1)+p(F_2)\,p(E/F_2)}.$$

Three Event Case. If F can occur as F_1 or F_2 or F_3, the denominator in the formula consists of three terms.

Each $$p(F_i/E) = \frac{p(F_i)\,p(E/F_i)}{p(F_1)\,p(E/F_1)+p(F_2)\,p(E/F_2)+p(F_3)\,p(E/F_3)} \quad (i = 1, 2, 3)$$

***n* Event Case.** In general if F can occur as $F_1, F_2, \ldots, F_n$ then

$$p(F_i/E) = \frac{p(F_i)\,p(E/F_i)}{\sum_{i=1}^{n} p(F_i)\,p(E/F_i)} \quad (i = 1, 2, \ldots, n)$$

Ex. 14. 90 per cent students of a class are well prepared (F_1) and 10 per cent unprepared (F_2) for an examination in mathematics. If the probability of passing with preparedness (E/F_1) is 0.85 and that of not passing with unpreparedness is (E'/F_2) is 0.60, find the probability that a student selected at random

(*i*) who is found to have passed must have been well prepared: $p(F_1/E)$

(*ii*) who is found to have failed must have been unprepared: $p(F_2/E)$.

Note that $p(E/F_2) + p(E'/F_2) = 1$

$\therefore\ p(E/F_2) = 1 - p(E'/F_2) = 1 - 0.60 = 0.40.$

Using Baye's theorem, we have

$$p(F_1/E) = \frac{p(F_1)\,p(E/F_2)}{p(F_1)\,p(E/F_1)+p(F_2)\,p(E/F_2)}$$

$$= \frac{0.90\times 0.85}{0.90\times 0.85+0.10\times 0.40} = 0.95$$

$$p(F_2/E') = \frac{p(F_2)\,p(E'/F_2)}{p(F_1)\,p(E'/F_1)+p(F_2)\,p(E'/F_2)}$$

$$= \frac{0.10\times 0.60}{0.90\times 0.15+0.10\times 0.60} = 0.31$$

We can similarly show that

$$p(F_2/E) = 0.05,\ p(F_1/E') = 0.69.$$

Ex. 15. Each of three classes has a variety of students belonging to three grades: Class A: 4(I), 3(II), 1(III), Class B : 1(I), 3(II), 2(III), Class C : 2(I), 2(II), 4(III). A class is chosen and a student selected is found to have grade I. What is the probability that he is from Class A?

Ex. 16. Machines A, B, C contribute to 20, 35, 45 per cent production of bolts in a factory. These produce 8, 6, 5 per cent defectives respectively. One bolt is picked up at random and found defective. What is the probability it was produced by machine C?

Ans. 45/119

Ex. 17. In a class 40 per cent students are working and 60 per cent non-working. Of those working it is found that 15 per cent secure a first division and of the non-working 20 per cent secure a first division. What is the probability that a student selected at random and having a first division is a working student? **Ans.** $\frac{1}{3}$

Ex. 18. A biased die which always shows a six when rolled, is mixed with 5 fair dice. A die chosen at random is rolled thrice and a six appears each time. Show that the probability that the die chosen is the biased one is given by $\dfrac{(\frac{1}{6})1}{(\frac{1}{6})1+\frac{5}{6}(\frac{1}{6})^3}$.

Laws of Total and Compound Probabilities: Alternative Proofs.

Ex. 19. If in a space of n elements, n_1 elements are favourable to E_1 and n_2 to E_2 and n_{12} elements are common to both E_1 and E_2 then the probability that a selected element belongs to E_1 or E_2 is given by

$$p(E_1+E_2)=\frac{n_1+n_2-n_{12}}{n}=\frac{n_1}{n}+\frac{n_2}{n}-\frac{n_{12}}{n}=p(E_1)+p(E_2)-p(E_1E_2)$$

This is an alternative proof of the law of addition or total of probabilities.

If E_1, E_2 are mutually exclusive events (*i.e.*, they cannot occur simultaneously) so that $p(E_1E_2) = 0$ then we have

$$p(E_1 + E_2) = p(E_1) + p(E_2)$$

Thus, the probability of one or the other of two mutually exclusive events is equal to the sum of the probability of each occurring separately.

In general, the probability of one of the mutually exclusive events E_1, E_2, ..., E_m is the sum of the probabilities of each occurring separately

$$p(E_1 + E_2 + \ldots + E_m) = p(E_1) + p(E_2) + \ldots + p(E_m)$$

If n is the total number of all possible and equally likely cases of which

n_1 are favourable to E_1

n_2 are favourable to E_2 ...

n_m are favourable to E_m, then

$$p(E_1+E_2+\ldots+E_m)=\frac{n_1+n_2+\ldots+n_m}{n}=\frac{n_1}{n}+\frac{n_2}{n}+\ldots+\frac{n_m}{n}$$

$$= p(E_1) + p(E_2) + \ldots + p(E_m).$$

For instance, the probability of drawing a king or a queen or an ace from a deck is $\frac{1}{13}+\frac{1}{13}+\frac{1}{13}=\frac{3}{13}$.

Ex. 20. A proof of the theorem of compound probability can also be given as follows:

Let n be the total number of all possible and equally likely cases of which n_1 are favourable to E_1.

Let m be the number of cases favourable to E_1 and E_2. The m cases favourable to E_1 and E_2 are to be found among the n_1 cases favourable to E_1.

$$p(E_1E_2) = \frac{m}{n} \text{ which can be written as } \frac{n_1}{n} \cdot \frac{m}{n_1}$$

Now $$p(E_1) = \frac{n_1}{n} \text{ and } p(E_2/E_1) = \frac{m}{n_1}$$

because, assuming E_1 has already occurred, there are only n_1 cases left out of which m are favourable to E_2.

$$\therefore \quad p(E_1E_2) = p(E_1) \cdot p(E_2/E_1)$$

Similarly $$p(E_1E_2) = p(E_2) \cdot p(E_1/E_2)$$

Ex. 21. If $E_1 \subset E_2$ *i.e.*, the event E_1 is contained in E_2

Then $$p(E_1/E_2) = \frac{p(E_1)}{p(E_2)}$$

The probability axioms are satisfied, *i.e.*,

$$p(E_1/E_1) = 1,\ p(E_1/E_2) \geq 0$$

Ex. 22. $p(E_3/E_2) = \dfrac{p(E_3E_2)}{p(E_2)}$

We usually write the multiplicative law of probability as

$$p(E_1E_2) = p(E_2)\ p(E_1/E_2)$$

Extensions:

$$\begin{aligned} p(E_1E_2E_3) &= p(E_1/E_2E_3)\ p(E_2E_3) \\ &= p(E_1/E_2E_3)\ p(E_3)\ p(E_2/E_3) \\ &= p(E_3)\ p(E_2/E_3)\ p(E_1/E_2E_3) \end{aligned}$$

$$\begin{aligned} p(E_1E_2E_3E_4) &= p(E_1/E_2E_3E_4)\ p(E_2/E_3E_4)\ p(E_3/E_4)\ p(E_4) \\ &= p(E_4)\ p(E_3/E_4)\ p(E_2/E_3E_4)\ p(E_1/E_2E_3E_4) \end{aligned}$$

Ex. 23. From an urn containing 8 red and 12 blue balls, 5 balls are drawn. A ball is drawn, its colour noted and kept aside (not put back). Then another ball is drawn and the process repeated a total of 5 times. Find the probability that the order of colours is red, blue, blue, red, blue *i.e.*, *RBBRB*.

$$p(RBBRB) = p(R)\ p(B/R)\ p(B/RB)\ p(R/RBB)\ p(B/RBBR)$$

$$= \frac{8}{20} \cdot \frac{12}{19} \cdot \frac{11}{18} \cdot \frac{7}{17} \cdot \frac{10}{16}$$

MISCELLANEOUS

Ex. 24. What is the probability of obtaining a double six in two rolls of a die (or one roll of two dice)? (**Note.** E_1 is independent of E_2)

$$p(E_1E_2) = p(E_1)\ p(E_2) = \tfrac{1}{6} \times \tfrac{1}{6} = \tfrac{1}{36}.$$

Ex. 25. What is the probability of getting four heads in a single toss of 4 coins?

(or four tosses of one coin) **Ans.** $(\frac{1}{2})^4 = \frac{1}{16}$

Ex. 26. What is the probability of getting 2 aces in two draws from a deck of cards?

If the first card is replaced before the second draw,

$$p(E_1 E_2) = p(E_1)\, p(E_2) = \frac{4}{52} \times \frac{4}{52} = \frac{1}{13} \cdot \frac{1}{13} = \frac{1}{169}$$

If the first card is not replaced

$$p = \frac{4}{52} \cdot \frac{3}{51} = \frac{1}{121}$$

Ex. 27. From an urn containing 3 white (W) and 2 black (B) balls, two balls are drawn without replacement. What is the probability that the first ball drawn is white and the second black?

$$p(WB) = p(W)\, p(B/W) = \frac{3}{5}\frac{2}{4} = \frac{8}{20}.$$

Ex. 28. From a pack of 52 cards, a card is drawn and found to be a face card. What is the probability that it is a queen?

$$p(\text{Queen/Face card}) = \frac{p(\text{Queen})}{p(\text{Face card})} = \frac{4/52}{12/52} = \frac{1}{3}.$$

Ex. 29. If a die is rolled what is the probability of an odd number coming up?

There are three odd numbers 1, 3, 5. There are 3 favourable cases in a total of six possible outcomes.

$$p = \frac{3}{6} = \frac{1}{2}$$

Ex. 30. If two dice are rolled what is the probability that a sum of 7 will be obtained?

Total number of cases = 6 × 6 = 36

Total number of favourable cases showing seven ≐ 6

First die	1	2	3	4	5	6
Second die	6	5	4	3	2	1

These 6 are the only possible combinations for a total of 7,

$$p = \frac{6}{36} = \frac{1}{6}$$

Ex. 31. What is the probability of getting (*i*) 9 (*ii*) 10 (*iii*) 11 points with 3 dice?

Ans. (*i*) $\frac{25}{216}$ (*ii*) $\frac{27}{216}$ (*iii*) $\frac{27}{216}$.

Ex. 32. 40 cards remain after removing the 12 face cards from a deck. What is the probability that 4 cards drawn out of these belong to different suits? **Ans.** $\frac{1000}{9139}$

Ex. 33. If 4 coins are tossed what is the probability of obtaining 2 heads and 2 tails?

Ans. $\frac{3}{8}$

Ex. 34. Two cards are drawn from a deck. What is the probability that both the cards drawn are queens? **Ans.** $\frac{12}{221}$

Ex. 35. Ten cards are drawn from a deck. What is the probability of finding among them at least (*i*) one king, (*ii*) two kings? **Ans.** $\frac{349}{595}$, (*ii*) $\frac{1257}{7735}$

Ex. 36. If the probability of A winning a game is $\frac{1}{5}$ and that of B winning is $\frac{1}{6}$, find the probability of

(*i*) both A, B winning **Ans.** (*i*) $\frac{1}{5}\cdot\frac{1}{6}=\frac{1}{20}$

(*ii*) both A, B losing (*ii*) $\frac{4}{5}\cdot\frac{5}{6}=\frac{2}{3}$

(*iii*) either A or B winning (*iii*) $\frac{1}{5}\cdot\frac{5}{6}+\frac{1}{6}\cdot\frac{4}{5}+\frac{1}{5}\cdot\frac{1}{6}=\frac{1}{3}$ or $1-\frac{2}{3}=\frac{1}{3}$

(*iv*) A winning and B losing (*iv*) $\frac{1}{5}\cdot\frac{5}{6}=\frac{1}{6}$.

Ex. 37. In a biased coin the probability of tossing a head is $\frac{1}{3}$. In another biased coin the probability of tossing a tail is $\frac{1}{4}$. If both coins are tossed what is the probability of getting, (*i*) two heads, (*ii*) two tails, (*iii*) no tails, (*iv*) no heads, (*v*) exactly one head, (*vi*) exactly one tail, (*vii*) at least one head.

Ans. (*i*) and (*iii*) $\frac{1}{4}$, (*ii*) and (*iv*) $\frac{1}{6}$, (*v*) and (*vi*) $\frac{7}{12}$, (*vii*) $\frac{5}{6}$.

Ex. 38. An urn contains 4 black and 2 white balls. Two balls are drawn at random. What is the probability that (*i*) the balls are of the same colour? (*ii*) the balls are of different colours? (*iii*) one of the balls is black? (*iv*) one of the balls is white when it is known that the other is white? **Ans.** (*i*) $\frac{7}{15}$, (*ii*) $\frac{1}{15}$, (*iii*) $\frac{14}{15}$, (*iv*) $\frac{1}{5}$.

Ex. 39. If a single card is drawn from a deck of 52 cards, what is the probability that it will be a (*i*) queen of hearts? (*ii*) a heart or a 5?

Ans. (*i*) $\frac{1}{52}$ (*ii*) $\frac{1}{4}+\frac{1}{13}-\frac{1}{52}=\frac{4}{13}$

Ex. 40. If a card is drawn from a deck what is the probability of obtaining a diamond or a face card (*i.e.*, king, queen and jack)? **Ans.** $\frac{11}{26}$

Ex. 41. If three dice are rolled what is the probability of

(*i*) 3 sixes **Ans.** (*i*) $(\frac{1}{6})^3=\frac{1}{216}$

(*ii*) exactly 2 sixes (*ii*) $3(\frac{1}{6}\frac{1}{6}\frac{5}{6})=\frac{15}{216}$

Ex. 42. What is the probability of getting

(*i*) two 3's in two successive rolls of a die?

(*ii*) three 3's in three successive rolls of a die? **Ans.** (*i*) $\frac{1}{36}$ (*ii*) $\frac{1}{216}$

Ex. 43. Find the probability of drawing a queen from a deck of cards 4 times in succession. **Ans.** $(\frac{4}{52})^4$ If the cards are replaced every time.

$\frac{4}{52}\cdot\frac{3}{51}\cdot\frac{2}{50}\cdot\frac{1}{49}$ if the cards are not replaced.

Ex. 44. Four cards are drawn from a deck without replacement. Find the probability that all the cards are (*i*) spades (*ii*) of the same suit (*iii*) of different suits.

Ans. (*i*) $\frac{11}{4165}$ (*ii*) $\frac{44}{4165}$ (*iii*) $\frac{2197}{20825}$

Ex. 45. An urn contains 3 white and 5 black balls: (*i*) One ball is drawn. What is the probability that it is black? **Ans.** $\frac{5}{8}$

(*ii*) One ball is drawn and its colour unnoted laid aside. Then another ball is drawn. What is the probability that it is (*a*) black, (*b*) white? **Ans.** (*a*) $\frac{5}{8}$, (*b*) $\frac{3}{8}$

Ex. 46. Three identical boxes have two drawers each. The first box contains a gold coin in each drawer, the second contains a silver coin in each drawer, the third contains a gold coin in one and a silver coin in the other. (*a*) A box is chosen at random. What is the probability that it contains coins of different metals? **Ans.** $\frac{1}{3}$

(*b*) A box is chosen, one of its drawers opened and a gold coin found. What is the probability that the other drawer contains a silver coin?

The answer to (*b*) is not $\frac{1}{2}$ as might be thought to be the case. It is not true that there are now three equally likely cases: box 1, box 2, and box 3. Rather there are six equally likely cases (if we number the drawers 1 to 6) while opening the first drawer.

As soon as the gold coin is discovered three cases become impossible. In only one of the remaining 3 cases there is the possibility of a silver coin. The answer therefore is $\frac{1}{3}$.

17.10 REPEATED TRIALS

Let p be the probability of success of an event and q the probability of its failure.

Let the event be repeated n times.

What is the probability of r successes? It is p^r

What is the probability of $n - r$ failures? It is q^{n-r}

What is the probability of the first r successes, remaining $n - r$ failures? It is $p^r \cdot q^{n-r}$

Now, out of n trials, r successes are possible (*i.e.*, $n - r$ failures are possible) in ${}^nC_r$ ways.

$\therefore$ The probability of r successes in n trials is ${}^nC_r p^r q^{n-r}$.

Ex. 47. The probability of getting two heads in 5 tosses of a coin is

(since $p = \frac{1}{2}$, $q = \frac{1}{2}$)

$$\text{prob}(2) = p(2) = {}^5C_2 \left(\tfrac{1}{2}\right)^2 \left(\tfrac{1}{2}\right)^3 = \frac{5!}{2!\ 3!} \cdot \frac{1}{32} = \frac{5}{16}$$

Ex. 48. The probability of getting two heads twice in 5 tosses of two coins is

$$\text{prob}(2) = p(2) = {}^5C_2 \left(\tfrac{1}{4}\right)^2 \left(\tfrac{3}{4}\right)^3 \qquad \text{(since } p = \tfrac{1}{4}, q = \tfrac{3}{4}\text{)}$$

$$= 10 \cdot \frac{27}{4^5} = \frac{270}{1024}.$$

Ex. 49. What is the probability of getting at least two heads in 5 tosses of a coin?

It means either 2 heads or 3 heads or 4 heads or 5 heads

The answer $= p(2) + p(3) + p(4) + p(5)$

$$= {}^5C_2 \left(\tfrac{1}{2}\right)^2 \left(\tfrac{1}{2}\right)^3 + {}^5C_3 \left(\tfrac{1}{2}\right)^3 \left(\tfrac{1}{2}\right)^2 + {}^5C_4 \left(\tfrac{1}{2}\right)^4 \left(\tfrac{1}{2}\right) + {}^5C_5 \left(\tfrac{1}{2}\right)^5$$

$$= ({}^5C_2 + {}^5C_3 + {}^5C_4 + {}^5C_5) \tfrac{1}{32} = \tfrac{13}{16}.$$

Ex. 50. 10 per cent of items produced on a machine are usually found to be defective. What is the probability that in a random sample of 12 items, (*i*) none, (*ii*) one, (*iii*) two, (*iv*) at the most two (*v*) 9 (*vi*) at least 9 items are found defective?

$$p = \frac{1}{10},\ q = \frac{9}{10}$$

(*i*) $p(0) = {}^{12}C_0 \left(\frac{1}{10}\right)^0 \left(\frac{9}{10}\right)^{12} = \left(\frac{9}{10}\right)^{12}$

(*ii*) $p(1) = {}^{12}C_1 \left(\frac{1}{10}\right)^1 \left(\frac{9}{10}\right)^{11} = 12 \times \frac{9^{11}}{10^{12}}$

(*iii*) $p(2) = {}^{12}C_2 \left(\frac{1}{10}\right)^2 \left(\frac{9}{10}\right)^{10}$

(*iv*) p (at the most 2 items being defective) $= p(0) + p(1) + p(2)$

(*v*) $p(9) = {}^{12}C_9 \left(\frac{1}{10}\right)^9 \left(\frac{9}{10}\right)^3$

(*vi*) p (at least 9 items are defective) $= p(9) + p(10) + p(11) + p(12)$.

Ex. 51. What is the probability that neither a doublet nor a 9 will appear in a single roll of two dice? **Ans.** $\frac{13}{18}$

Ex. 52. A student can solve on an average one half the number of a problem set. To pass the examination 7 out of 10 problems must be solved. What is the probability that he passes? **Ans.** $\frac{11}{64}$

Ex. 53. A coin is tossed three times giving a total of two heads. What is the conditional probability of a head on the first trial? **Ans.** $\frac{2}{3}$

Ex. 54. A person wishes to make a telephone call but cannot remember the first digit in the number. What is the probability of getting the correct number if a digit is dialled at random? **Ans.** $\frac{1}{9}$

Ex. 55. 500 men attend a show and have tickets numbered 1 to 500. One of these selected at random is to get a prize. Find the probability of the winning ticket having (*i*) a number ending with 0, (*ii*) a number less than 100, (*iii*) the number 399.

Ans. $\frac{1}{10}, \frac{99}{500}, \frac{1}{500}$

Ex. 56. The probability that a person selected is fat is 0.5. The probability is 0.7 that he is short, and 0.3 that he is both. Out of 500 persons satisfying the above situation, how many approximately are likely to be either fat or short? **Ans.** 450.

Ex. 57. You have good mood on some days, better on some other days, and the best on still other days. If the probability of good mood is $\frac{3}{20}$ and of better mood is $\frac{3}{5}$, what is the probability that on a certain day you do not have the best mood? **Ans.** 0.75

Ex. 58. The probability that a majority of students will attend a meeting is 0.5 and the probability that if a majority attend and the resolution will be passed is 0.9; what is the probability that a majority will attend and the resolution will be passed? **Ans.** 0.45.

Ex. 59. Three applicants are to be selected at random out of 4 boys and 6 girls, what is the probability of selecting (*i*) all girls, (*ii*) all boys (*iii*) at least one boy?

Ans. $\frac{5}{30}, \frac{1}{30}, \frac{49}{180}$

Ex. 60. The probability that a student attends a class is 0.5 and the probability, if he attends the class, of his understanding the subject is 0.9. What is the probability that he will attend the class and understand the subject? It may be noted that the probability of his not attending the class and understanding the subject is 0.5. If a student understands the subject what is the probability that he attended the class? **Ans.** 0.45, 0.64.

Ex. 61. Three machines A, B, C produce 30 per cent, 45 per cent and 25 per cent of total output and their percentages of defectives are : 1 per cent, 2 per cent, 1.5 per cent respectively. An item is selected at random from the combined output of the three and is found to be defective. What is the probability that it was produced by A? **Ans.** $\frac{19}{100}$.

Ex. 62. Three boys B_1, B_2, B_3 join a marathon. The odds in fouvour of their reaching their destinations are 2 : 5, 3 : 7, 6 : 11. Find the chance they will reach their destinations

$$p(B_1) = \text{Prob. of } B_1 \text{ reaching} = \frac{2}{2+5} = \frac{2}{7}$$

$$p(B_2) = \frac{3}{3+7} = \frac{3}{10}$$

$$p(B_3) = \frac{6}{6+11} = \frac{6}{17}$$

$$p(B_1 B_2 B_3) = \frac{2}{7}\,\frac{3}{10}\,\frac{6}{17} = \frac{18}{595}.$$

Ex. 63. Each of the boys B_1, B_2, B_3 has the same probability of getting the first rank. Girls G_1, G_2 each have the same probability of getting the first rank. The girls, however, are twice as likely to get the first rank as the boys. Find the probability that a boy gets the first rank. If B_1, G_1 are brother and sister, find the probability that one of them gets the first rank.

Since each girl has double the probability of a boy, take each girl as two cases making a total of 4; 3 boys correspond to 3 cases. Toal no. of cases = 7.

Favourable cases *i.e.*, $p(\text{Boy getting rank}) = \frac{3}{7}$

In the second part 1 girl = 2 favourable cases and 1 boy = 1 favourable case.

$$p(B_1 \text{ or } G_1) = \frac{2}{7} + \frac{1}{7} = \frac{3}{7}$$

Ex. 64. A's chance of hitting a target is 0.6. Find the probability that A hits one or the other of two independent targets. **Ans.** $0.6 + 0.6 - 0.36 = 0.84$

Ex. 65. In a lottery k^2 tickets are sold. A person buys k tickets. What is the probability of his winning a prize if there are k prizes in all? We may add $p(1)$, $p(2)$, ..., $p(k)$, the probabilities of his winning 1 or 2 or ... or k prizes.

Alternatively, the probability of his not winning a prize is

$$p(0) = \frac{k^2 - k}{k^2} = \frac{k(k-1)}{k^2} = \frac{k-1}{k}$$

The compound probability of his not winning any prize $=\left(\frac{k-1}{k}\right)^k$

The probability of his winning 1 or more prizes $=1-\left(\frac{k-1}{k}\right)^k$

Ex. 66. A person rolls a die. How many times he must roll the die so that his chance of getting a six at least once is $\frac{1}{2}$?

If he tries x times, the probability of his not getting a six is $(5/6)^x$. The probability of his getting a six at least once is $1-(5/6)^x$

$$\therefore \quad 1-\left(\frac{5}{6}\right)^x=\frac{1}{2} \text{ i.e., } \left(\frac{5}{6}\right)^x=\frac{1}{2}$$

$$\therefore \quad x(\log 5-\log 6)=\log 2$$

$$\therefore \quad x=\frac{\log 2}{\log 6-\log 5}=\frac{0.3010}{0.0792}\simeq 4$$

Ex. 67. There are 4 envelopes into which 4 letters are to be inserted each meant for a particular envelope. Find the probability that all the letters are not inserted into their right envelopes.

4 letters into 4 envelopes can be put in $4\times 3\times 2\times 1=4!$ ways.

All letters can be put in the right envelope in only 1 way our of 4! ways.

p(all going into right envelopes) = 1/4!

p(none going into right envelopes) $=1-\frac{1}{4!}$

Ex. 68. A, B cast two dice alternately.

A wins if he gets 6 before B gets 7

B wins if he gets 7 before A gets 6

If A begins, find the probability of his winning.

p(getting a 6 from 2 dice) $=\frac{5}{36}$

p(getting a 7 from 2 dice) $=\frac{6}{36}$

If A starts, his probability of winning is

$$\frac{5}{36}+\frac{31}{36}\times\frac{5}{6}\times\frac{5}{36}+\frac{31}{36}\times\frac{5}{6}\times\frac{31}{36}\times\frac{5}{6}\times\frac{5}{36}+\ldots=30/61$$

Ex. 69. In a college there is an equal number of boys and girls. 10% of the boys and 5% of the girls have offered history and the rest economics. If a history student is selected at random what is the probability that the selected person is (*i*) a boy (*ii*) a girl .

Suppose a person is selected and found to have taken history. What will be the conditional probability that the selected person is (*iii*) a boy (*iv*) a girl?

	Proportional with Economics	History	Total	Notation
				B Boy
Boys	0.450	0.050	0.5	*G* Girl
Girls	0.475	0.025	0.5	*H* History
Total	0.925	0.075	1	*E* Economics

(*i*) $P(BH) = P(B/H)\, P(H)$

$$\therefore \quad P(B/H) = \frac{p(BH)}{p(H)} = \frac{0.050}{0.075} = \frac{2}{3}$$

$$\text{Selection from history students of a boy} = \frac{p(\text{all history boys})}{p(\text{history students})}$$

(*ii*) Similarly $p(G/H) = \dfrac{p(GH)}{p(H)} = \dfrac{0.025}{0.075} = \dfrac{1}{3}$

(*iii*) $p(B/F) = \dfrac{p(BE)}{p(E)} = \dfrac{0.050}{0.925} = \dfrac{18.}{37}$

(*iv*) $p(G/E) = \dfrac{p(GE)}{p(E)} = \dfrac{0.475}{0.925} = \dfrac{19}{37}$

Ex. 70. 35% of those visiting a fair did shopping over Rs. 5000, 40% of those who did this shopping were teenagers, 20% of visitors who did not do such shopping were teenagers. If a visitor happens to be a teenager what is the probability that he will do shopping over Rs 5000? What is the probability that a nonteenager will do such shopping?

Ans. 14/27, 21/73

Ex. 71. The following is a weekly salary record of 400 workers in a factory

Salary	Undergraduates	Graduates	Total
Below 300	180	20	200
300 – 400	40	60	100
over 400	80	20	100
Total	300	100	400

A worker is selected at random. Find the probability that

(*i*) he is an undergraduate

(*ii*) he is graduate given that his income is over 400.

(*iii*) his income is below 300 given that he is a graduate.

Ans. $\dfrac{300}{400}, \dfrac{20}{100}, \dfrac{180}{300}$

17.11 PROBABILITY, LOGIC, SETS

Consider a set of logical outcomes of an experiment in the sample space *E*. Any subset of *E* is an event. Two events E_1, E_2 are mutually exclusive if $E_1 \cap E_2 = \phi$.

Let *r* be a statement relative to a logically possible outcome of this experiment. Let *R* be the truth set of *r i.e.*, *R* is that subset of logical possibilities for which *r* is true.

If a sample space is used to determine the probability of a statement it must contain the truth set of that statement.

The probability of r, given additional information s, is the conditional probability written

$$p(r/s) = \frac{n(R \cap S)}{n(S)} = \frac{p(r \wedge s)}{p(s)}$$

where R, S are truth sets of r, s respectively and $n(S)$ is the probability measurement of S.

Since the probability of a statement is the probability measure of its truth set it follows that

(*i*) $0 \le p(r) \le 1$ for any r

(*ii*) $p(\sim r) = 1 - p(r)$

(*iii*) $p(r \vee s) = p(r) + p(s) - p(r \wedge s)$ for any r and s

$= p(r) + p(s)$ if $R \cap S = \phi$

(*iv*) If $p(r/s) = p(r)$ then r, s are independent statements.

17.12 RANDOM EXPERIMENTS

An experiment is said to be *random* if we cannot predict the outcome before the experiment is carried out. A random experiment is one which can be repeated, practically or theoretically, any number of times. We can toss a coin or roll a die any number of times and study the outcomes. The result of any outcome may or may not influence that of succeeding outcomes. Thus any throw of a coin or die is independent of all earlier throws. But if a card is drawn from a deck of cards and not replaced the experiment consisting of a second draw will be influenced by the result of the first.

The consideration of the random character of a phenomenon is inevitable because of the complicated characteristics of the laws of nature, ignorance about the relevant laws and the theoretical or practical difficulty allowing for the effect of a large number of factors under consideration or disturbing factors or other shortcomings of our tools of work.

A random variable is a *probability variable.* A specified degree of probability is *attached* to every value of a random variable. What is meant by this statement is that when we conduct an experiment, say by tossing a coin, we cannot definitely say what will come up. Random forces will decide the outcome, although each outcome may have an equal chance or probability of showing up. A random variable is a quantity which in different observations can assumes different values.

17.13 EXPERIENCE

In practice, if an event has a negligibly small chance of occurring it may not occur in a single or in a few observations. It dos not mean that the event cannot occur thus. It is only that we feel that the event being highly unlikely to occur we may safely not expect its occurrence for practical purposes. Thus we are almost sure that blind, random typing for one hour is not likely to produce a poem by Eliot.

If an event with a negligible probability does not occur in a single observation it agrees with theory. If it does it is not against the theory. But a single observation can never prove or disprove a statistical theory. There should be more observations for a convincing proof.

A reasonable person assigns a probability to an event depending on his experience with events in the real world. Two reasonable persons with nearly similar experiences are likely to assign nearly the same probability to a certain kind of event.

17.14 THE LAW OF STATISTICAL REGULARITY

A sample of observed values of a random variable can be assumed to be distributed according to a population or theoretical distribution function. In other words, if from a large group known as the population, a small group or a sample of observations is chosen at random it tends to possess the characteristics of the population.

The sample should be of a reasonable size, neither too large nor too small. Randomness in selection assures each observation an equal chance of being selected.

The law of intertia of large numbers. This is a corollary of the law of statistical regularity.The average of a large number of independent observations should almost be equal to the theoretical mean value (of the population). In other words the difference between the theoretical mean value and the observed mean of a number of random variables becomes as small as desired if the number of observations is sufficiently large.

This means that aggregates of observations tend to be more stable as the number of observations increases, so that movements will tend to compensate one another.

Experience and the law of statistical regularity help a person, directly or indirectly, to assign a probability p to a certain event under a given set of conditions. The assessment of this probability, however, will be substantially influenced by the quality of observations and the relative importance attached to these observations.

17.15 PROBABILITY, STATISTICS, EXPERIMENTS

Statistical theory rests upon the laws of probability. By studying the ideas of probability, chance and likelihood of events, we can pave the way to the study of statistics.

Statistical methods are used in almost every branch of learning: in physical, social and biological sciences, in market and consumer research, industrial quality control and so on. The applications in each case depend upon the actual or conceptual possibility of repeated observations or experiments. From these we may seek to find predictable relationships between some events and the causes of those events. We may be able to control a number of causes and still there might remain several uncontrolled causes, unassignable disturbances or random errors which bring in the concept of probability.

Phenomena in social sciences and management are the results of a number of variables in action. We may not be able to *conduct* an experiment in the usual sense of the term but we can try to assign causes to an event under consideration. The large number of unassignable causes or random errors require us to make use of the concept of probability.

Statistics deals with uncertainties and the tendency of outcomes to vary under identical experimental conditions. Statistical laws are not exact in the sense in which physical or natural laws are. They are true on the average or in the long run.

Development of Statistics

From the earliest times, statistics or statistical data are found to have been compiled, mostly for military and fiscal reasons, to provide government heads with data to manage the affairs of the state. Gradually statistics of wealth, taxes, deaths, marriages, etc. also began to be compiled regularly. Mathematical studies of problems involving probability, vital statistics, insurance and economic statistics were carried out in the seventeenth century. It was in the nineteenth century that a study of the significance of large numbers was made. The ideas of correlation and regression, their application to studies in social sciences and sampling techniques were developed.

In modern times, no economist, engineer, manager or scientist of any description can afford to work without statistical data. Sales and purchases, prices, production, profits etc. on the one hand and tables, diagrams, reports on the other, all these are directly or indirectly connected with statistical work.

17.16 DEFINITION OF STATISTICS

The word *statistics* may be used either in the plural or in the singular.

Statistics *are* numerical facts systematically collected.

Statistics *is* the science dealing with the collection, organisation, analysis and interpretation of numerical data. It is a systematic body of knowledge.

As statisticians we may collect statistics (plural) *i.e.,* statistical or quantitative data. We may study the science of statistics (singular).

An interesting Marxist definition of statistics is that it is an independent social science which studies the quantitative aspect of mass social phenomena, regularities of social development, social production and the influence of natural and technical factors on quantitative changes in social life.

Secrist defines statistics as "aggregates of facts affected to a marked extent by a multiplicity of causes, numerically expressed, enumerated or estimated according to reasonable standards of accuracy, collected in a systematic manner for a predetermined purpose and placed in relation to each other".

17.17 CHARACTERISTICS AND FUNCTIONS OF STATISTICS

These follow from the definition:

Numerical Expression. Statistics deals with quantitative data, with numbers as expressions of meaningful relationships. Statistical data are not abstract numbers. They constitute concrete material which represents objects and their measurements.

Accuracy. The degree of accuracy depends on circumstances. Based on the nature and object of an inquiry, a reasonable standard of accuracy has to be maintained.

Systematic collection of data. There should be a definite method and purpose in the collection of data. A systematic procedure should be followed.

Aggregates of facts. Isolated figures are not statistics. Masses of facts in social sciences contain the laws of general behaviour. Statistical methods extract meaningful information from a mass of data.

Multiplicity of causes. There are various unassignable causes and forces acting and reacting with one another when we consider any phenomenon.

Collection of data for a predetermined purpose. The statistician should proceed with a specific object in view. Vagueness and ambiguity of purpose can lead to waste of time, energy and money.

Comparison of data. Relationships between and comparisons of variables are an important part of the study of statistics.

Hypothesis testing, prediction and formulation of policies. These are parts of any scientific study. Statistical methods are useful in formulating and testing hypotheses, forecasting future events and framing suitable policies.

Statistics is a science as it describes facts objectively without delivering value judgements. It is an art in that it applies certain techniques to available facts, sometimes in a subjective manner. Being a systematic body of knowledge statistics is considered a science though not one as exact as physical sciences.

17.18 ROLE OF STATISTICS IN VARIOUS FIELDS

In business and management, planning, organising, supervision and control and decision making are necessary at every stage. At all these stages an analysis of data collected becomes easier with the help of statistical methods.

In physical and natural sciences, statistics has been used at various stages. Research work becomes easier with the help of statistical tools if quantitative data are employed.

Banks, insurance companies, trade associations, governments, etc. have their departments of statistics. In each case the statistical approach is adjusted to suit the specific requirements of the field.

Students of economics often come across statistics of savings, interest rates, demand, costs, production, stocks, taxes and so on.

Social scientists, politicians, biologists, research workers, all are making useful applications of statistical methods in their work.

Economic Statistics. Economic statistics is a systematic body of knowledge concerned with the application of statistical methods to theoretical economic arguments. It is a science that checks the ideas in economic theory with the help of quantitative and historical information and suggests methods of improvements where necessary. In economics, national consumption is related to national income. We cannot conduct an experiment to check this but we can study, with the help of statistics, the relation between the cause (income) and effect (consumption) and may come to the conclusion that there may be other causes influencing consumption. These causes may be known or unassignable. Some developments in economies have been made possible because of the use of statistics.

Economists use statistical information to build sound reasoning. Extensive studies have been made with the help of statistics in the fields of business cycles, national income, economic growth, price trends and inflation, money markets, banking, credit, public finance and so on.

Food and land economics, population and price theories, revenues, costs and profits of firms, use statistical analysis at some stage or other.

17.19 STATISTICAL INFERENCE

As a result of historical development statistics may be classified into two parts.

Descriptive statistics is concerned with data processing *i.e.,* organisation and presentation of classified facts efficiently and in a convenient form.

Inferential statistics is concerned with deductive reasoning. It leads us from the particular to the general, from one or several causes to universal situations. We pick up a *sample* from a *population*, take into account the way in which the sample was selected and express our generalisation in specific probability terms, *i.e.,* we make a judgement about the population from which the sample comes. The population may or may not be observed either because it is infinite or it is destructive to go through it. As we shall see the purpose of *sampling* and the business of statistical inference is to make judgements about population *parameters* on the basis of sample *statistics.*

Decision theory under conditions of uncertainty, used in management and other social sciences uses techniques of statistical inference.

Misuse and Distrust of Statistics

There are many ways in which statistics are likely to be used improperly and wrongly.

Often, the truth is not obvious. It has to be searched and even researched. It may come in the form of a paradox. Sometime great efforts have to be made to arrive at the truth. Misuse of statistics leads to distortions. This demands a keener search for the truth.

1. *Errors of Context.* Facts, otherwise true, may be quoted or presented out of context in such a way as to misrepresent the real state of affairs. Politicians, journalists and others may sometimes do so.

2. *Errors of Generalisation.* A generalisation or conclusion based on incomplete or inadequate or unrepresentative data can lead to wrong inferences. On the basis of very poor marks obtained by two or three students we cannot conclude that all the students from a college are equally bad.

3. *Errors of Deduction.* A general result may be wrongly applied to a specific case. It is a natural human trait to take, a tendency generally noted, for a law. If the students of a particular class have been showing good results every year in the past it does not mean they will necessarily do so this year too.

4. *Bias.* Conscious or unconscious entry of bias in statistical work is common. It may enter consciously when it is associated with a purpose. The ruling party in the parliament may show an upward bias while maintaining that the per capita income has increased. The opposition group may show a downward bias in the matter. A careful study of the data and the accompanying circumstances may reveal the truth in such cases. It is more difficult to discover the unconscious bias which has a chance of entering at one or more of all the stages of statistical work. Some methods are available which can help detect the bias and the probable extent of bias which can be taken into account while analysing or interpreting the results.

Bias may sometimes be introduced by exaggeration of quantities or through the use of colourful high sounding words.

5. *Errors of Noncomparability.* Wrong, improper, meaningless or untimely comparisons can lead to faulty conclusions. A comparison of index numbers of different regions based on different years or different commodities can be meaningless.

6. *Errors of Oversimplication.* Sweeping statements, omission of essential details and quick half truths are misleading. The mention of a rise in national income without reference to a rise in population, cannot give a correct idea of changes in per capita income.

7. *Errors of Spurious Accuracy.* As against oversimplification, there may be found an irrelevant attempt at showing high accuracy. It is not necessary to give the result beyond a certain degree of accuracy especially when chances of errors in the preceding steps cannot be ruled out.

To give the size of a population in millions correct to the unit's or ten's place or, to speak of a person's income running into thousands correct to the last paisa does not make much sense.

8. *Errors of Calculation.* Mistakes in the calculation of ratios, percentages and in the application of mathematical operations are common and should be avoided.

9. *Errors of Assuming Causation.* The assumption of a wrong cause and effect relationship may lead to nonsensical results. If the movements of two time series, one showing the number of sick persons in a city in India and the other showing the production of wheat in Russia are found to be correlated by means of a formula it does not follow that there is a cause and effect relationship between the two. Interesting examples of the search for cause and effect relationship are to be found in the study of business cycles, cancer, common cold, and so on.

10. *Abuse.* Abuse of statistics is common. It is often the result of ignorance rather than design. Differences in definitions, methods, sampling procedures, sources of data, etc., give rise to differences in results leading to misunderstanding and misinterpretation.

An insight into basic statistical concepts and a proper application of these can be of immense use in removing misunderstandings about statistical results.

Distruct of statistics results from the abuse of statistics and statistical concepts. "There are three kinds of lies, *viz.* lies, damned lies and statistics" —statements like these should be understood in the appropriate context since they refer to cases where misuse of statistics may have been made. And statistics is not always misused. Simply because some people, to serve their immediate purposes, happened to have twisted statistical facts it does not mean that the whole science of statistics is a bundle of lies. As a matter of fact, such sweeping statements can be made in connection with any science.

There is nothing wrong with statistical tools. The fault lies with the user of the science and not with the science.

17.20 LIMITATIONS OF STATISTICS

Statistical data should be collected carefully and systematically. Errors are likely to creep into facts collected and also into the use of statistical tools which may be wrongly or improperly applied. And there is every possibility of drawing wrong conclusions.

Statistical methods, cannot, in general, be used when data are in a qualitative form. Statistical methods are applicable mainly to quantitative data. There are special methods for qualitative data, too.

Statistics deals with aggregates. Individual facts do not constitute statistics. We have to have a sufficiently large amount of data to study and draw conclusions.

Statistical results are true only on an average. They are not universally true.

To get good results an experiment may have to be repeated a large number of times.

Statistics can be misused. It is possible to suppress some facts and emphasize others or to misrepresent some parts and represent others partially. This may be done to establish misleading conclusions.

Some problems may not be amenable to statistical analysis either because it may be difficult to do so or it may be because of ignorance on the part of the analyst, inadequacy of tools or inappropriate data.

Ex. 72. Comment on the following statements:

(*i*) Statistics is the science of averages.

(*ii*) Statistics can prove anything.

(*iii*) Statistics are like clay of which you can make a god or a devil, as you please.

(*iv*) Statistics affects everybody and touches life at many points.

(*v*) Statistics is a science and an art.

(*vi*) Statistics is a body of method for making wise decisions in the face of uncertainty.

(*vii*) Statistics is the science of measurement of the social organism regarded as a whole in all its manifestations.

(*viii*) Statistics only furnishes tools, necessary though imperfect, which are dangerous in the hands of those who do not know their uses and deficiencies.

(*ix*) The science of statistics is a most useful servant but only of great value to those who understand its proper use.

(*x*) Statistics helps business by its employment as a tool of industrial research and as an important factor in scientific research.

(*xi*) Probability is based on the degree of rational belief.

(*xii*) In spite of the diversity of philosophical concepts, at the mathematical level there is almost no disagreement about the superstructure of probability.

(*xiii*) Two events are statistically independent if the occurrence of one in some way affects the occurrence of the other.

(*xiv*) A quantity whose value is determined by the outcome of an experiment of change is known as a random variable.

Ex. 73. What is economic statistics?

Ex. 74. What are the uses and abuses of statistics?

Ex. 75. Discuss the scope, utility and limitations of statistics.

Ex. 76. What are the different definitions of probability? Explain with examples.

Ex. 77. State and prove the laws of total and compound probability.

Ex. 78. Comment on the following advice for statistical analysis.

(*i*) Search for those facts that show events and relationships most truly. Numbers are but a shorthand convention for describing past events and forecasting trends.

(*ii*) Do not try to *prove* a thing by manipulating data. Then there will be no value left in figures. Do not interpret the results to confirm any preconceived hypothesis.

(*iii*) Use the best judgement. Make valid probability tests. Give warning of possible fallacies or errors.

(*iv*) Do not think statistical methods can replace the thought process as a way of avoiding pitfalls. They may help us to bridge them. .

(*v*) Statistics are an aid to, but not a substitute for, commonsense.

CHAPTER 18

Frequency Distributions

18.1 ARRANGING DATA

Large masses of data can be handled and interpreted more conveniently if they are arranged and grouped in a suitable manner. They may be summarised in such a way as to reveal the important characteristics of the data at a glance or to enable the derivation of its essential features with little effort. Frequency distributions and graphs of various kinds are useful for the purpose.

Frequency Distributions

The *frequency* of an observation in a distribution is the number of times it appears in the distribution.

Ex. 1. In the *distribution* of 14 observations showing marks obtained by students in a test:

12, 12, 34, 12, 45, 12, 45, 12, 34, 45, 12, 12, 34, 34

the frequency of 12 is 7 (*i.e.*, 12 occurs 7 times)

the frequency of 34 is 4 (*i.e.*, 34 occurs 4 times)

the frequency of 45 is 3 (*i.e.*, 45 occurs 3 times)

Total frequency is 14 ($\because$ 7 + 4 + 3 = 14)

Let the symbol for observations be x and the symbol for frequencies be f. Then we have the *frequency distribution* of the given data:

x	f	
12	7	
34	4	
45	3	
Total	14	observations

Here x is a discrete variable which takes different values: 12, 34, 45, in this case with frequencies 7, 4, 3, respectively.

18.2 FREQUENCY DISTRIBUTIONS

A frequency distribution is a table in which the values of a variable are classified according to size. Extensive data in ungrouped form cannot be as useful as grouped data in the form of a frequency table. It is necessary, however, that data be homogeneous and comparable.

To form a distribution we tally each observation with a mark, every fifth mark across the first four marks giving us a frequency of five at a time. For the given example:

x	*Tally Marks*	*Frequency*
12	~~IIII~~ II	7
34	IIII	4
45	III	3
Total		14

The number of observations normally is large. It is better, therefore, to take *class intervals* which may represent all the values between certain limits called class limits.

Ex. 2. The class interval 10—20 contains all the values between 10 and 20. Its midpoint is 15. The lower class limit is 10, the upper class limit is 20.

Ex. 3. The class interval 0.05—0.06 contains all the values between 0.05 and 0.06. The midpoint is 0.055. The class limits are 0.05 (lower), 0.06 (upper).

The following steps give the procedure for the *construction* of a frequency distribution.

First, the range or the difference between the highest *i.e.* x_{max} and the lowest *i.e.* x_{min} values of all the observations should be found.

Then the number and size of class intervals should be decided. The number of class intervals should be neither too large nor too small. Usually, anywhere between 10 and 15 intervals are taken. The size or width of the class intervals should be reasonable and convenient.

The tally marks are recorded one at a time, every fifth mark shown by crossing the last block of four.

It may be noted that odd numbers are frequently used for the size of the intervals giving us a whole number for the midpoint of the interval.

Ex. 4. A class was given an arithmetic test. Set up a frequency distribution for the following scores resulting from the test:

55 44 45 48 69 30 47 50 66 38 58 44 42 49 70 32 48 60
46 47 53 54 35 50 62 49 82 40 46 29 42 39 46 61 35 51
75 25 46 28 43 50 47 62 30 56 48 30 47 53

$$\text{Range} = x_{max} - x_{min} = 82 - 25 = 57.$$

If we take class intervals of size 10 there will be only 6 classes for 57 observations. A class interval of size 3 will give us 20 classes. We want neither too many nor too few classes. We may take a class interval of size 5 which gives us 12 classes.

Class Interval	*Tally Marks*	*f*
25—29	I I I	3
30—34	I I I I	4
35—39	I I I I	4
40—44	𝍸 I	6
45—49	𝍸 𝍸 I I I I	14
50—54	𝍸 I I	7
55—59	I I I	3
60—64	I I I I	4
65—69	I I	2
70—74	I	1
75—79	I	1
80—84	I	1
	Total	50

We can see that the unintelligible mass of data has become more meaningful. Suppose in an examination the results are to be decided as follows:

35 marks and above : third division
45 marks and above : second division
60 marks and above : first division

there are 7 students failing
9 getting third division
25 getting second division
9 getting first division
Total 50

Ex. 5. If we take class intervals of size 10 for the data above we have the frequency distributions follows.

Class Interval	*Tally Marks*	*f*
25—34	𝍸 I I	7
35—44	𝍸 I I I I	9
45—45	𝍸 𝍸 𝍸 𝍸 I I	22
55—74	𝍸 I I	7
65—74	I I I	3
75—84	I I	2
	Total	50

Ex. 6. For the same data construct frequency distributions with class intervals of size (*i*) 6 (*ii*) 7 (*iii*) 8 (*iv*) 9.

Ex. 7. What per cent of students obtained marks between (*i*) 40 and 49, (*ii*) 30 and 49, (*iii*) more than 70, (*iv*) less than 34? **Ans.** (*i*) 40% (*ii*) 56% (*iii*) 6% (*iv*) 15%.

Ex. 8. Is a score of 45 high, average or low? **Ans.** average.

Ex. 9. Construct a frequency distribution for the following data showing the weights in kgs. of 40 students.

40	50	45	38	49	60	55	40	42	41	43	38	48	59
54	40	43	42	44	39	47	58	53	40	41	47	47	40
46	57	52	41	39	46	50	42	45	56	51	42		

Continuous and Discrete Classes.

Continuous data move from one class to another without a break. The variable values can be put on a continuous scale.

Discrete data move from one class to another with a break.

18.3 CLASS INTERVALS AND CLASS LIMITS

The class interval is the range of values of each class. It is the difference between the lower boundaries of any two successive classes.

Grouping of data amounts to (*i*) dividing the data into a certain number of class intervals and (*ii*) finding how many observations fall into each interval. The frequency of an interval is the number of observations falling into that interval.

An interval is defined by its limits. Adjacent groups should not overlap. In the illustrative example we have taken the intervals like 25 – 29, 30 – 34, and so on. The first number in any interval is the lower limit and the second number is the upper limit.

Alternatively we could write them thus:

25 – 29.9, 30 – 34.9, 35 – 39.9, and so on.

Class limits should be located in such a way as to minimise the errors of grouping which are similar to errors of rounding off numbers in general.

The above mentioned class intervals may be put in any one of the following alternative forms:

25 – 29	25 and under 29	25 to 29
30 – 34	30 and under 34	30 to 34
35 – 39	35 and under 39	35 to 39
etc.	etc.	etc.

The size of the class interval is a matter of convenience in representing facts. The class interval should be small enough so that all the units in one class interval may be treated as identical for practical purposes. It should be large enough to avoid making the frequency table too long and laborious to handle.

Often the class interval is given in the form of *group* boundaries to maintain *continuity*. This is particularly useful in drawing certain diagrams.

Thus the continuous group boundaries in the above case will be

25—30	or alternatively	24.5—29.5
30—35		29.5—34.5
35—40		34.5—39.5 etc.

This is done to eliminate the gaps between any two groups. To avoid ambiguity it is understood while constructing the frequency distribution, that the value 25 will belong to the interval 25—30. Similarly 30 will be in the interval 30—35 and so on. The distance between a pair of group boundaries is called group width.

A		B	C
25	Group width	30	35

A		B	C
24.5	Group width	29.5	34.5

It is desirable that all intervals used in a frequency distribution have the same width so that the results are easier to find and understand. Differences in widths or unequal class intervals are likely to cause confusion. But sometimes, unequal class intervals are used, as in grouping incomes:

Income Rs.
under 100
100 — 200
200 — 500
500 — 2500
2500 — 10000
10000 and above.

Unequal intervals should be dealt with according to requirements.

In this case, the first and the last classes are open end classes at one extreme each to show very small and very large incomes since either it is difficult to include more classes or it is not useful to include them.

Midpoints or Class Marks

The midpoint of a class interval is halfway between the class limits. It is denoted by x. It is found by taking the average of the class limits.

It is possible to represent all the cases within a class interval by a single point. The midpoint is convenient and appropriate for the purpose.

Ex. 10.

Class interval	*Midpoint*	*Explanation*
25—29	27	$\frac{25+29}{2} = 27$
30—34	32	$\frac{30+34}{2} = 32$
40—44	42	etc.

The class interval here is not 4 but 5, often the midpoint is found by considering the data implicitly and comprehensively.

Ex. 11.

Class interval	*Midpoint*	*Explanation*
25—30	27.5	$\frac{25+30}{2} = 27.5$
30—35	32.5	$\frac{30+35}{2} = 32.5$
35—40	37.5	etc.
40—45	42.5	

Ex. 12.

Class interval	*Midpoint*	*Explanation*
1.03 – 1.04	1.035	$\frac{1.03+1.04}{2} = 1.035$
1.04 – 1.05	1.045	etc.
10.5 – 1.06	1.055	

Ex. 13. If the class marks for some data are 0.25, 0.27, 0.29, 0.31, 0.33, 0.35, find the class interval, size, and the class limits. **Ans.** 0.24—0.26, 0.26—0.28 etc.

Note. Statistical information can be classified, if necessary, according to qualitative characteristics, *i.e.*, unlike quantitative classes, we have mutually exclusive characteristics like regions, physical features, etc.

Class of Person in an office	*Frequency*
Managing Director	1
Director	5
Managers	10
Officers	18
Clerks	25
Attendants	11
Others	10
Total	70

Ex. 14. A perfume manufacturer promises to give 7 cc. of perfume in his bottles. 50 bottles are checked. The following data gives in cc the quantity of perfume found in them. Make class intervals of (*i*) size 5 (*ii*) size 10. (*iii*) How many bottles have more than or less than 7 cc of perfume.

5.5	7.4	6.2	6.8	6.0	7.3	6.6	7.9
7.4	9.2	5.6	4.2	7.7	6.6	5.6	5.1
6.0	8.2	7.5	5.4	6.5	4.7	6.7	7.0
6.5	6.8	8.4	9.8	5.1	8.2	8.0	9.9
6.9	6.4	6.1	8.3	7.6	7.8	6.8	6.6
7.1	8.8	5.5	8.7	5.1	9.1	3.8	4.4
6.5	8.3						

Ans. (*i*)

Class Interval	frequency
3.5—3.9	1
4.0—4.4	2
4.5—4.9	1
5.0—5.4	4
5.5—5.9	4
6.0—6.4	5
6.5—6.9	11
7.0—7.4	5
7.5—7.9	5
8.0—8.4	6
8.5—8.9	2
9.0—9.5	2
9.5—10.0	2
	50

(*ii*)

Class	Frequency
3.0—3.9	1
4.0—4.9	3
5.0—5.9	8
6.0—6.9	16
7.0—7.9	10
8.0—8.9	8
9.0—9.9	4
	50

(*iii*)
22 bottles more and 28 bottles less than 7 cc

18.4 THE HISTOGRAM

A histogram is a set of adjoining vertical bars whose areas are proportional to the frequencies represented by the bars., A histogram is drawn by taking the class intervals on the x-axis and the frequencies on the y-axis.

When the class intervals are of equal width, the height of a bar corresponds to the frequency in that class.

The class intervals with the highest frequency will have the highest bar and the highest concentration of observations. This class will be called the modal class.

Usually there is only one modal class and the rest of the frequencies on either side taper off. Sometimes, however, there are two or more modal classes.

The scales of axes are normally taken in such a way that the length of the y-axis is about $1\frac{1}{2}$ times that of the x-axis. This will result in a decent looking graph.

Ex. 15. For the following data showing the scores obtained in an arithmetic test by 50 students, a histogram can be drawn.

Class interval	*Midpoint* x	f
25—30	27.5	3
30—35	32.5	4
35—40	37.5	4
40—45	42.5	5
45—50	47.5	15
50—55	52.5	7
55—60	57.5	3
60—65	62.5	4
65—70	67.5	2
70—75	72.5	1
75—80	77.5	1
80—85	82.5	1
	Total	50

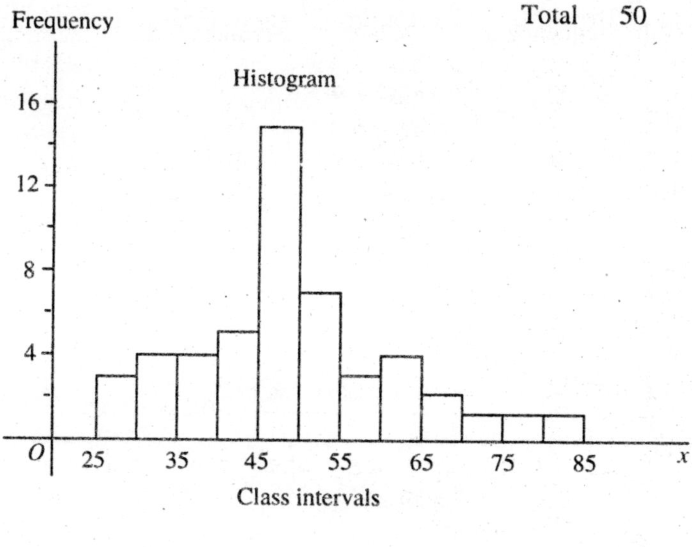

18.5 THE FREQUENCY POLYGON

The frequency polygon is the most frequently encountered graphic device in statistics. It is easy to construct and simple to interpret.

The frequency polygon is a line chart plotted in the same way as the histogram. On the x-axis class midpoints are taken and frequencies along the y-axis are represented by points which are joined by straight lines to give us the frequency polygon. The first end point is joined to the x-axis to a midpoint showing zero frequency just before the first class interval, and the last end joined to the one after the last class interval as shown in the diagram.

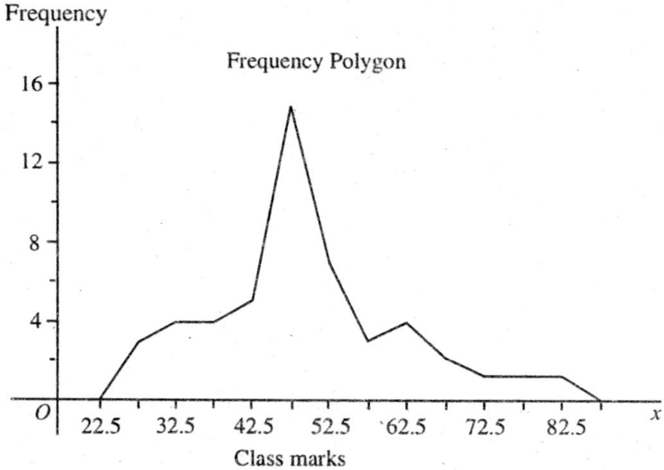

The first midpoint 22.5 corresponds to zero frequency and the last, *i.e.* 87.5 also has zero frequency.

18.6 FREQUENCY CURVES

A limiting form of the histogram or frequency polygon is the frequency curve. A smooth curve instead of broken lines joining the points corresponding to the frequencies gives us the frequency curve of the data. It is a continuous curve and provides a frequency value for every value of x. The curve need not necessarily pass through each and every point.

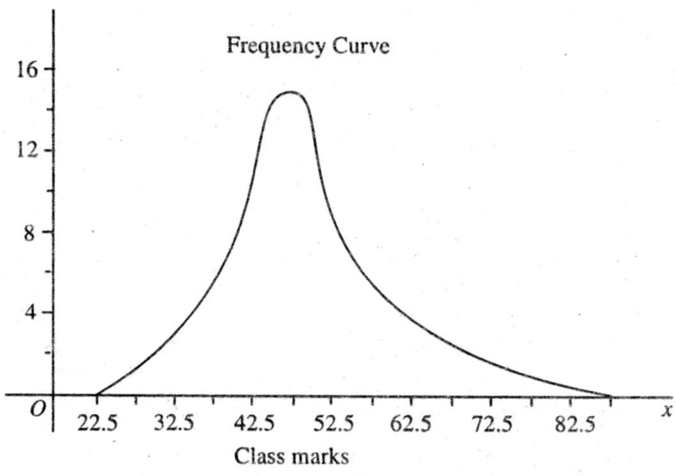

A frequency curve smooths out sampling errors particularly evident in small samples.

If a histogram or frequency polygon is drawn from a large amount of data with a large number of classes the resulting diagram approximates a frequency curve.

Frequency polygon and histogram

Each has its advantages and disadvantages. Both represent, however, the same story.

The histogram has bars, each representing the exact number of observations in a class. The polygon does not show this.

The class intervals in a histogram are easily noticed and understood and the classes are more clear than in a frequency polygon which shows only the midpoints.

The bars could be separated in a histogram if required to show gaps in a discontinuous distribution.

Unequal widths of class intervals are more accurately recorded in a histogram.

A histogram is usually better than a frequency polygon where the number of classes is small. For a large number of classes a frequency polygon is preferable.

A frequency polygon is easier to draw and is less time consuming.

A frequency polygon gives a better idea about the flow of data. It is more convenient for comparisons and theoretical studies.

Two or more frequency polygons can be easily drawn together and studiesd simultaneously. Histograms are not easy to draw this way.

18.7 RELATIVE FREQUENCY DISTRIBUTIONS

In a frequency distribution, if the frequency in each class interval is converted into a proportion by dividing by the total frequency we get a series of proportions called *relative frequencies.* A distribution presented with relative frequencies rather than actual frequencies is called a *relative frequency* distribution. The sum of all relative frequencies in a distribution is 1.

Ex. 16.

Class interval	*Frequency*	*Relative frequency*	*Explanation*
25—35	7	0.14	$\frac{7}{50} = 0.14$
35—45	9	0.18	$\frac{9}{50} = 0.18$
45—55	22	0.44	etc.
55—65	7	0.14	
65—75	3	0.06	
75—85	2	0.04	
Total	50	1.00	

The concept of relative frequencies is useful in sampling theory. It can also be used to compare two frequency distributions with unequal total frequency though with the same series of class intervals as in the following example.

Ex. 17.

Class interval	f_1	f_2	*Rel. freq.* f_1	*Rel. freq.* f_2
10—20	5	12	0.20	0.12
20—30	10	24	0.40	0.24
30—40	6	30	0.24	0.30
40—50	3	19	0.12	0.19
50—60	1	15	0.04	0.15
Total	25	100	1.00	1.00

A direct visual comparison of two frequency distributions can be made by drawing their frequency polygons.

Ex. 18. Draw frequency polygons for the above relative frequency distributions.

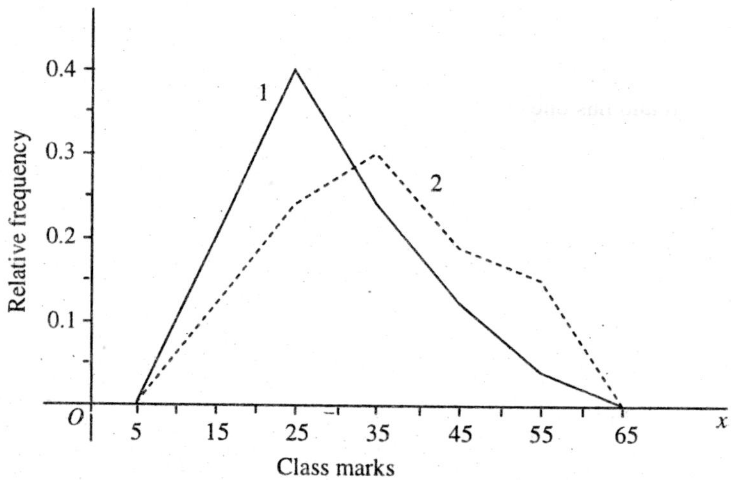

Relative Frequencies and Area

If the area under a curve is taken to be 1 then we can easily establish the correspondence between this area and the total of relative frequencies which is also 1.

The proportion of area between any two points is equal to the proportion or relative frequency of cases falling between those points.

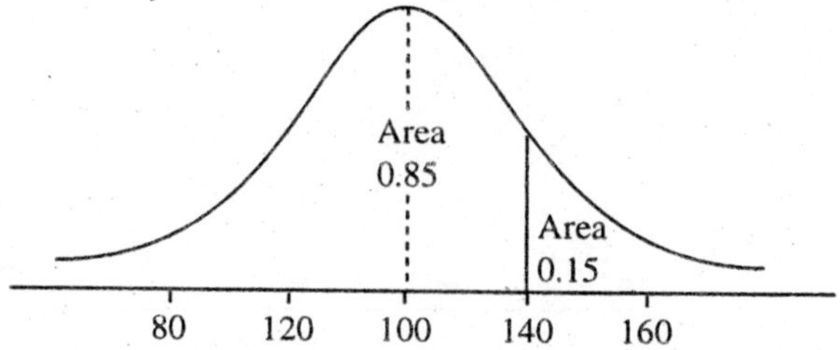

Ex. 19. In the frequency curve given in the diagram the area to the left of the point 140 is 0.85 and beyond 140 the area is 0.15.

Ex. 20. The area to the left or to the right of the point 100 (which is at the centre) is 0.5.

18.8 TYPES OF FREQUENCY CURVES

A frequency polygon or a frequency curve gives us an idea of the way in which the observations are spread out or distributed along the x-axis. It helps us to observe the pattern of the distribution.

Frequency curves can have various shapes which can be classified on the basis of

(*i*) degree of asymmetry or skewness

(*ii*) central tendency and amount of spread (or dispersion or variability)

(*iii*) extent of peakedness

(*iv*) number of peaks

(*v*) other characteristics.

The standard curve is the normal curve corresponding to the normal distribution. It is bell shaped, symmetrical and has one peak. It has various special characteristics which we shall consider later.

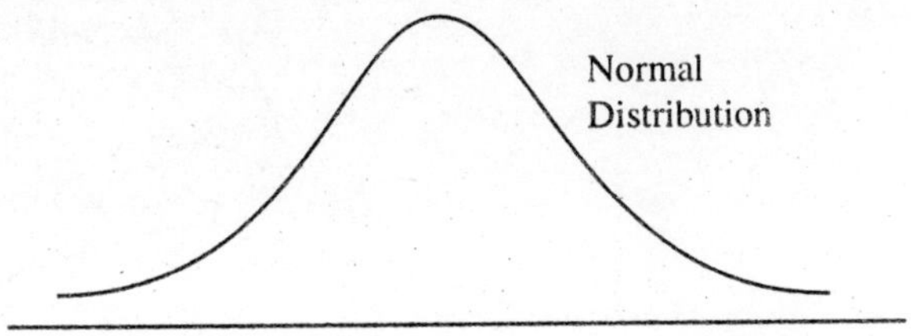

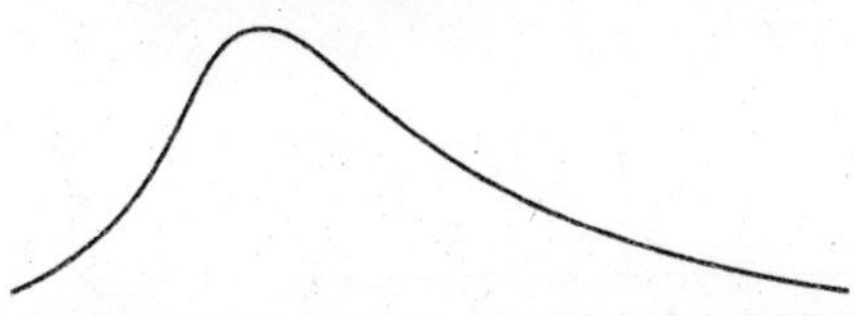

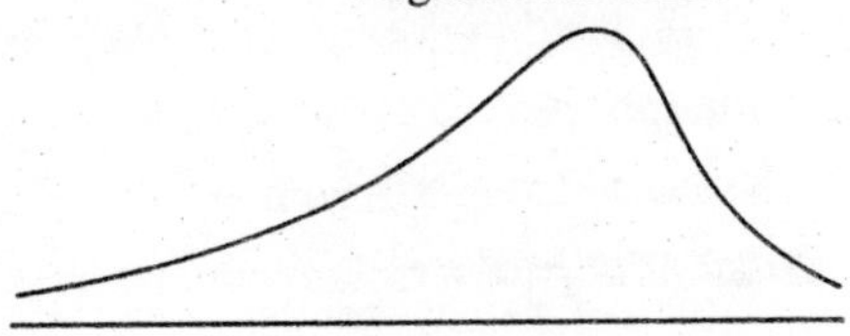

(*i*) *Skewness*

If a curve is not symmetrical along a vertical line passing through the centre of the distribution it is either positively skew or negatively skew depending on whether the tail is on the right or left respectively. Skewness shows the degree of asymmetry of a distribution.

(*ii*) *Central Tendency and Variability*

Distributions may differ with regard to central tendency and may not differ with regard to variability or spread.

Distributions may differ with regard to variability and may not differ with regard to central tendency.

We say that the dispersion or variability is low if the values in the distribution are clustered together around some central value. Widely differing values give a more spread out curve, *i.e.*, one with a higher dispersion.

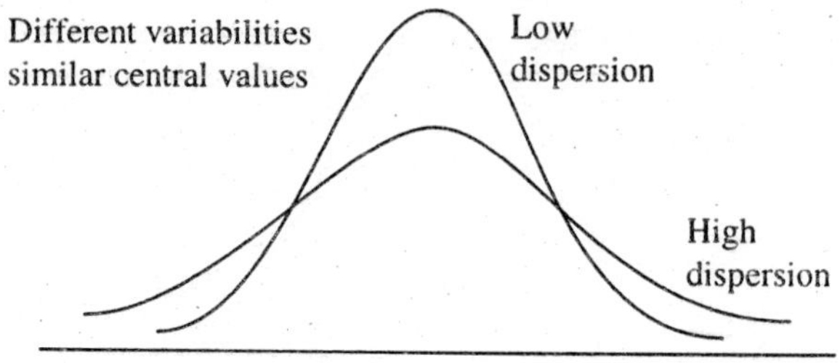

(*iii*) *Kurtosis*

Kurtosis is the measure of the degree of peakedness of a distribution.

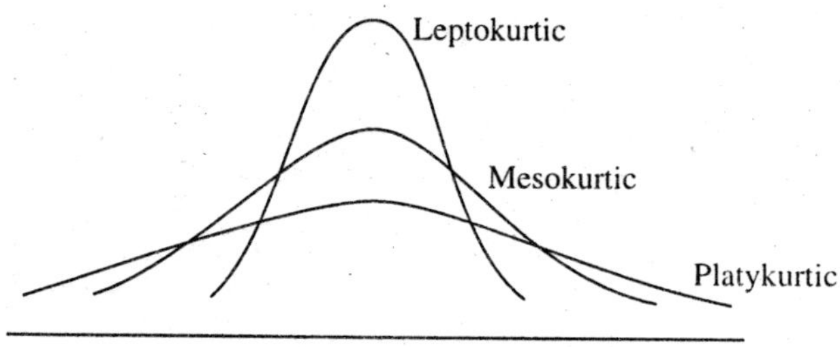

A flat topped curve is called *platykurtic.*

A sharply peaked curve is called *leptokurtic.*

The normal curve is *mesokurtic.*

(*iv*) *Number of peaks*

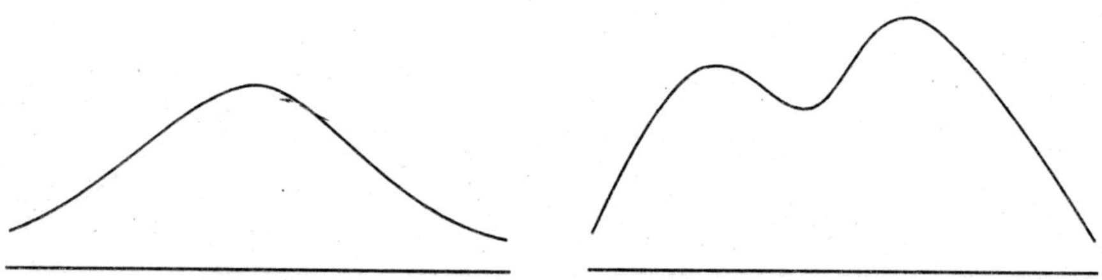

Unimodal

Bimodal

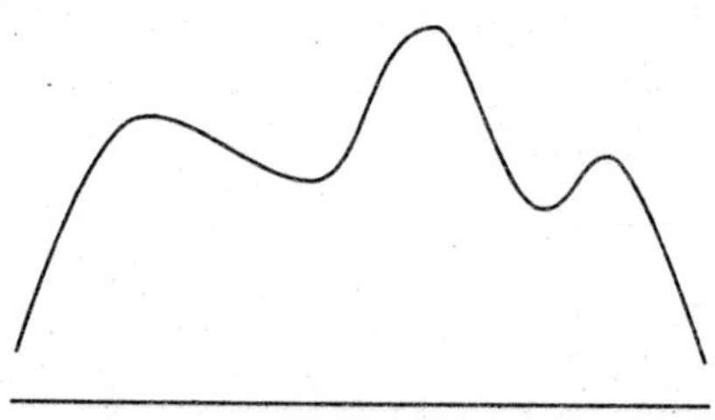

Multimodal

A curve with a single peak or modal value is called unimodal. There can be two or more peaks giving us bimodal or multimodal curves.

(*v*) *Other characteristics*

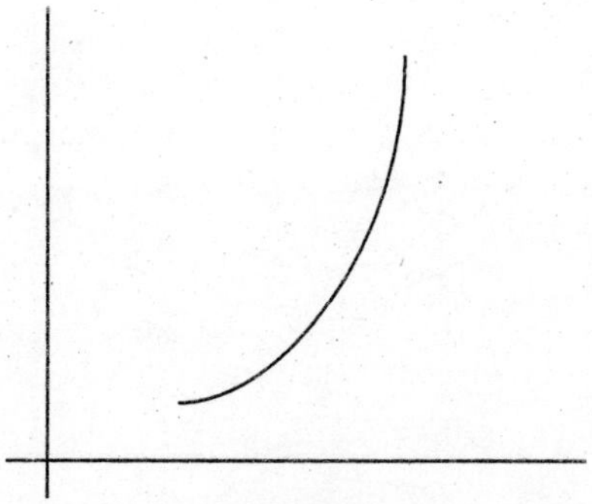

J-shape

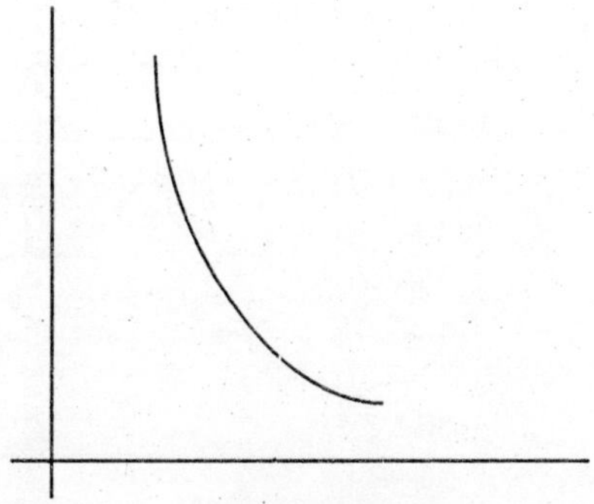

Reverse J-shape

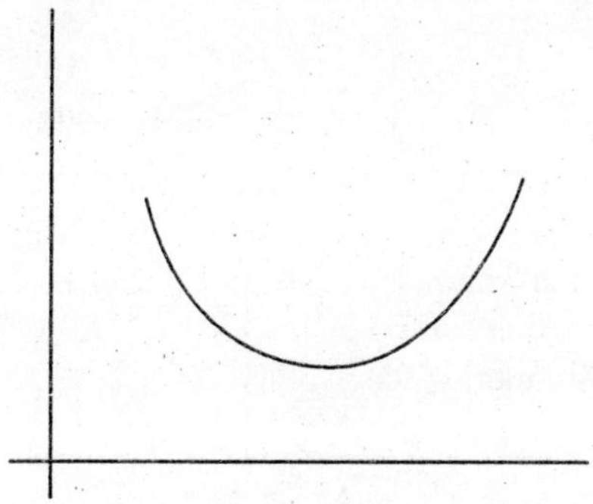

U-shape

In practice we come across a large variety of shapes of curves. Some of these have been shown here.

18.9 CUMULATIVE FREQUENCY CURVES : OGIVES

The cumulative frequency of any class in a frequency distribution is the sum of the frequencies of all the classes, beginning with the first, to that class.

If f_1, f_2, f_3, f_4 are the frequencies of the first four classes, the cumulative frequencies of these classes are $f_1, f_1 + f_2, f_1 + f_2 + f_3$ and $f_1 + f_2 + f_3 + f_4$ respectively.

The cumulative frequency of the kth class is $f_1 + f_2 + \ldots + f_k$ and may be written

$$\text{cum } f_k = \sum_{i=1}^{k} f_i$$

The cumulative frequency curve is obtained by drawing the cumulative frequencies against the boundary values or end points of the class intervals.

If we go on adding the frequencies as above we get a "less than" table and a "less than" cumulative frequency curve. If the process is done from the last frequency backwards we have a "more than" frequency curve.

A "less than" frequency curve is an ascending curve and a "more than" frequency curve is a descending curve. These curves are called *ogives*.

Ex. 21. Consider the ogives for the following data.

Class Interval	*f*	*Less than Cum. f*	*More than Cum. f*
25—30	3	3	50
30—35	4	7	47
35—40	4	11	43
40—45	5	16	39
45—50	15	31	34
50—55	7	38	19
55—60	3	41	12
60—65	4	45	12
65—70	2	47	5
70—75	1	48	3
75—80	1	49	2
80—85	1	50	1

Meaning of "Less than" cum f : The first cum $f = 3$ shows there are 3 students getting less than 30 marks; the next cum $f = 7$ (obtained by addition : $3 + 4 = 7$) shows there are 7 students getting less than 35; the next cum $f = 11$ (obtained by addition : $7 + 4 = 11$) shows the number of students getting less than 40 and so on.

"More than" cum. f : The first cum. $f = 50$ shows there are 50 students getting more than 25 marks; the next cum. $f = 47$ students getting more than 30; the next cum. $f = 43$ (obtained by subtracting $47 - 4 = 43$) shows the number of students getting more than 35 and so on.

The ogives show graphically the number of students getting less than or more than the marks listed in the data.

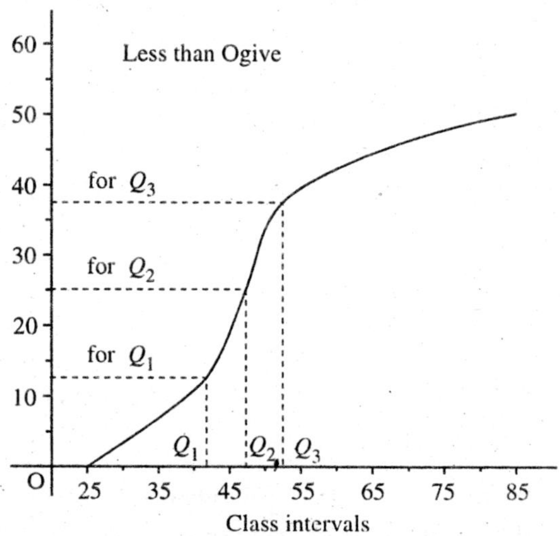

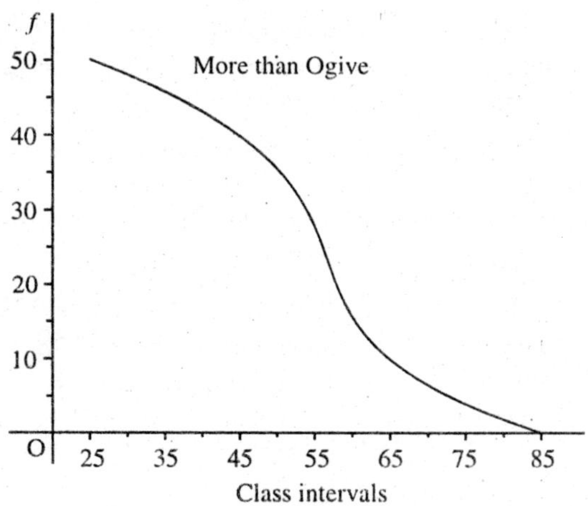

OGIVES

An ogive is the graph of a cumulative frequency distribution.

In drawing a "less than" ogive the upper limit of each of the intervals is used and the points are placed above these. In the case of a "more than" ogive we place the points above the lower limits as has been done in the diagrams.

An ogive permits an easy interpolation of values between the plotted points.

Ex. 22. From the ogives given above, find how many students get less than 45 marks. (*Hint* : Either draw a perpendicular above the point 45 on the x-axis in the "less than" ogive and read the value along the y-axis or find the cumulative frequency straight from the table showing cumulative frequencies. **Ans.** 16

Ex. 23. How many students get more than 45 marks? (From either ogive or from the table we get the answer.) **Ans.** 34.

Ex. 24. What percentage of students gets marks between 40 and 60? There are 30 students getting marks between 40 and 60. The percentage, therefore, is 60.

Ex. 25. What are the marks obtained by the 25th student?

This cannot be found from the table. In the ogive, from the frequency 25 draw a horizontal line. From the point where it meets the curve draw a perpendicular on the x-axis. The point where it cuts the x-axis gives the marks. **Ans.** 48

Ogives and frequency curves. An ogive has some advantages over a frequency curve:

Its shape does not depend on the size of the class interval.

The ogive can be drawn even if the class intervals are of unequal sizes. It is for this reason that we can easily compare the ogives of different data. An ogive is suitable for the determination of the number of cases falling within any given interval, not necessarily coinciding with the intervals given in the data.

18.10 DECILES, PERCENTILES, QUARTILES

An ogive is useful in finding the proportion of a set of measurements which exceed a given value or in finding the range of measurements containing a given proportion of the total. The ogive provides a quick and simple graphical method for approximately determining the *percentiles.*

The *percentiles* are the values of the measurements which divide the distribution into 100 equal parts. *Deciles* divide the distribution into 10 equal parts. The first decile is the same as the tenth percentile and so on.

There are three *quartiles,* written Q_1, Q_2, Q_3 which divide the population into four equal parts. The first or the lowest quartile Q_1, divides the distribution into two parts in the ratio 25 : 75. The third or the upper quartile, Q_3, divides it in the ratio 75 : 25.

The second quartile Q_2, commonly known as the *median*, divides the distribution into two equal parts, ratio 50 : 50. The median is the same as the fifth decile or the fiftieth percentile. The lower and the upper quartiles are the 25th and 75th percentiles respectively.

An ogive is useful in finding deciles, percentiles, quartiles.

From the less than ogive the median can be obtained by reading the value of the measurement corresponding to the middle of the total frequency. Q_1 is obtained by reading the value corresponding to 25 per cent frequency and Q_3 by reading the value corresponding to 75 per cent frequency. The more than ogive can be used similarly.

Ex. 26. Find Q_1, Q_2, Q_3 from the less than ogive given above. By drawing the lines corresponding to 25%, 50% and 75% frequencies, as shown we get, approximately,

$$Q_1 = 42,$$

$$Q_3 = 54$$

and the median $$Q_2 = 48.$$

Ex. 27. The *interquartile range* is the difference between the upper and lower quartiles:

$$Q_3 - Q_1 = 54 - 42 = 12.$$

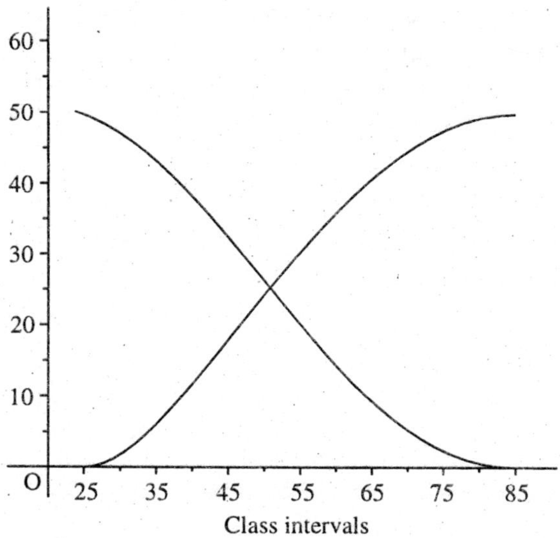

Ex. 28. Draw both the ogives by superimposing them on the same graph. Show that the point of intersection gives the median.

Ex. 29. Find the quartiles for the following data:

Wages	*No. of Workers*
80—100	5
100—120	7
120—140	10
140—160	15
160—180	14
180—200	9
200—220	8
220—240	7

Relative or percentage cumulative frequency distributions.

If cumulative frequencies are divided by the total frequency we get relative cumulative frequencies.

Ex. 30.

Production of steel in tonnes	*Number of months*	*Cum. f.*	*Relative c. f.*
570—600	3	3	0.05
600—630	7	10	0.17
630—660	24	34	0.57
660—690	20	54	0.90
690—720	3	56	0.95
720—750	2	59	0.98
750—780	1	60	1.00

Explanation: Relative c.f. 0.05 shows that in 5 per cent of the total number of months production was less than 600; relative c.f. 0.90 shows that in 90 per cent of months production was less than 690 and so on.

The data with relative cumulative frequencies shown gives us an example of a relative cumulative frequency distribution.

The shape of the curve of this distribution is similar to that of the ogive.

18.11 LORENZ CURVE OF CONCENTRATION

The Lorenz curve is concerned with the distribution of any characteristic among the items possessing that characteristics. The extent of concentration of the characteristic in a few items can be examined by the use of the Lorenz curve. Also see Ch. 12.

In many countries a very large proportion of wealth is generally concentrated in the hands of a very small proportion of the total population. A large proportion of workers is employed in a very small proportion of factories and so on.

Lorenz curve helps us to demonstrate this common economic phenomenon of disparities on a graph.

Consider the incomes received by some individuals, their numbers arranged according to class intervals. The classes are arranged in ascending order of income size and the cumulative totals for incomes and the number of individuals are noted.

A Lorenz curve is plotted by taking the cumulative percentages of total incomes on the y-axis and the cumulative percentages of income earners on the x-axis.

Ex. 31.

Income range Rs.	*No. of persons (thousands) f*	*Total income (Rs. million) y*	*Cum. fr. total cum. f*	*Cum. total of y cum. y*	*% of cum. f*	*% of cum. y*
Over 5000	50	790	50	790	5	52.7
2000—5000	120	300	170	1090	17	72.7
1000—2000	200	220	370	1310	37	87.3
500—1000	180	100	550	1410	55	94.1
200—500	250	60	800	1470	80	98.0
100—200	200	30	1000	1500	100	100.0

The Line of Equal Distribution: If all groups had been receiving comparable proportions income, *e.g.*, 25 per cent people getting 25 per cent incomes, 60 per cent people getting 60 per cent incomes and so on, then the graph would be a straight line joining the points (0, 0) (100, 100). This is the line of equal distribution.

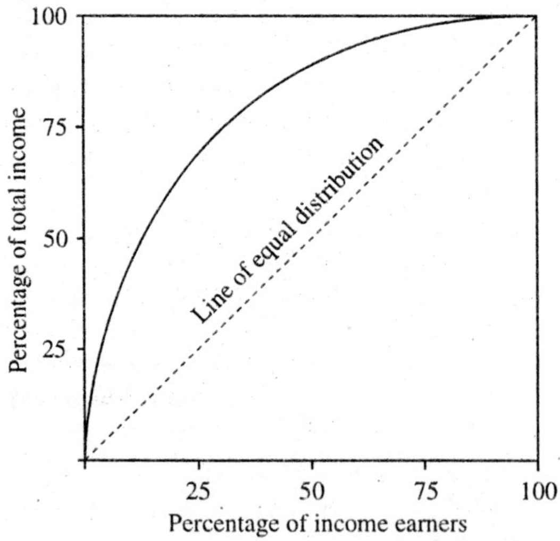

A Lorenz Curve

Uses and Interpretation of the Curve: The degree to which a Lorenz curve deviates from the line of equal distribution is a measure of the *inequality* of distributions of incomes. See Sec. 12.21.

The farther the curve moves away from this line the greater is the inequality.

The degree of this inequality at any stage is indicated by the distance from the equal distribution line of the curve.

Comparisons of different classes of earners can be made by superimposing their Lorenz curves on the same graph. Similarly we can compare the distributions of incomes in different countries.

Lorenz curves can be used to measure or compare inequalities in a variety of cases like industrial outputs, examination marks, payments of taxes and so on.

Ex. 32. Draw a Lorenz curve to compare the distribution of sales of eggs in two towns.

	Town A		*Town B*	
Sales (Rs.)	*Number of shops*	*Total sales*	*Number of shops*	*Total sales*
under 100	12	1000	18	1500
100—200	20	4000	30	6000
200—400	35	12500	58	18700
400—650	40	23500	50	28300
650—1000	23	22800	32	26500
Over 1000	10	19200	12	22000

The two curves should be drawn on the same graph. The curve farther from the line of equal distribution will have a higher degree of variability.

Ex. 33. Construct a frequency curve, a frequency polygon, and a histogram for the following data showing the salaries of 100 employees in a firm. Find the median and quartiles.

Salary (Rs.)	*Number of employees*
200—250	10
250—300	11
300—350	15
350—400	23
400—450	20
450—500	12
500—550	7
550—600	2

Ans. Med = 380.5, $Q_1 = 313.3$, $Q_3 = 440$

Ex. 34. Construct "less than" and "more than" ogives for the above data. How many employees earn more than

(*i*) Rs. 200 p.m. (*ii*) Rs. 500 p.m. (*iii*) Rs. 550 p.m.?

How many employees earn less than

(*iv*) Rs. 300 p.m. (*v*) Rs. 500 p.m. (*vi*) Rs. 600 p.m.?

Ans. (*i*) 100 (*ii*) 9 (*iii*) 2

(*iv*) 21 (*v*) 91 (*vi*) 100.

Ex. 35. Construct the frequency curves and histograms for the following data:

	Students in	
Marks obtained	*Group A*	*Group B*
25—40	10	38
40—55	35	72
55—70	27	40
70—85	23	30
85—100	5	20

(*i*) Draw the ogives for groups *A* and *B* on the same graph and find from them the values of Q_1, Q_2, Q_3. **Ans.** A: 46.4, 57.8, 72; B: 42.5, 53, 70

(*ii*) Draw the relative frequency polygons.

Ex. 36. The following data show the number of hours of study per week of two classes of students.

Hours of study	f_A	f_B
2—5	1	5
5—8	2	17
8—11	4	28
11—14	10	32
14—17	12	40
17—20	15	35
20—23	20	28
23—26	24	10
26—29	10	3
29—32	2	2
Total	100	200

(*i*) Draw their frequency polygons and find which group is more variable.
(*Note.* The polygon with a greater spread has, in general, a higher variability)

(*ii*) Draw the histograms and the frequency curves.

(*iii*) Draw their relative frequency curves and compare them.

(*iv*) Find the quartiles from the ogives and calculate the interquartile range.

18.12 PROBABILITY DISTRIBUTIONS

We know that the relative frequency of any event is a number between 0 and 1. The sum of the relative frequencies of all possible events is 1.

This means that we can visualise a relationship between relative frequencies and probabilities. We can use the probability theory to do this.

A random variable is one that can assume a number of values with given probabilities.

The probability distribution of a random variable x is the complete statement giving different values of the variable and the respective probabilities of these values. It may be considered as the limiting form of a relative frequency distribution when n, the number of observations, is large.

18.13 DISCRETE AND CONTINUOUS DISTRIBUTIONS

Distributions are either discrete or continuous.

Discrete distribution : If a function $p(x)$ takes the values $p_1, p_2, \ldots, p_n$ when x takes the values $x_1, x_2, \ldots\ldots, x_n$ then $p(x)$ is the discrete probability distribution of x. Its properties are

$$0 \le p(x) \le 1, \qquad \sum_{i=1}^{n} p(x_i) = 1, \qquad p(E) = \sum_{E} p(x)$$

The graphs of discrete probability distributions are histograms or bar charts.

Continuous distributions : We denote by $p(x)$ the continuous probability density function (*pdf*) in the continuous sample space of the random variable x. Its properties are

$$p(x) \ge 0 \qquad -\infty < x < \infty, \qquad \int_{-\infty}^{\infty} p(x)dx = 1$$

$$p(E) = \int_{E} p(x)\,dx$$

The graph of a *pdf* is a smooth curve.

$\sum_{i=1}^{n} p(x) = 1$ or $\int_{-\infty}^{\infty} p(x) = 1$ means that the area under the histogram or the curve (as the case may be) is unity.

Often we are given the interval in which a distribution exists.The constant of a distribution can be found. Thus if $p(x) = C(2x + 1) \quad 0 < x < 2$

where C is a constant which can be found by equating to unity the area under the curve.

$$\int_0^2 C(2x+1)\,dx = [Cx^2 + Cx]_0^2 = 4C + 2C = 1$$

i.e., $\qquad 6C = 1$ or $C = \frac{1}{6}$.

$\therefore \qquad p(x) = \frac{1}{6}(2x + 1)$

Ex. 37. Find C if (*i*) $p(x) = C(3 - x) \qquad 0 < x < 3$

(*ii*) $p(x) = Ce^{-x} \qquad 0 < x < 1 \qquad$ **Ans.** $\frac{2}{9}, 1 - \frac{1}{e}$.

Ex. 38. Are the following probability distributions?

(*i*) $p(x) = \frac{x}{10}$ for $x = 1, 2, 3, 4$ **Ans.** Yes

(*ii*) $p(x) = \frac{1}{20}$ for $x = 1, 2, 3, 4, \ldots, 20$ Yes

(*iii*) $p(x = 1) = \frac{1}{4}$, $p(x = 2) = \frac{1}{2}$, $p(x = 3) = \frac{1}{4}$ Yes

(*iv*) $p(3) = \frac{1}{3}$, $p(4) = \frac{1}{5}$. No

Note. (*i*) $\Sigma p(x) = \frac{1}{10} + \frac{2}{10} + \frac{3}{10} + \frac{4}{10} = 1$

(*ii*) $\Sigma p(x) = \frac{1}{20} + \frac{1}{20} + \ldots + \frac{1}{20} = 1$

since $p(x) = \frac{1}{20}$ for all x.

(*iii*) $\Sigma p(x) = \frac{1}{4} + \frac{1}{2} + \frac{1}{4} = 1$

(*iv*) $\Sigma p(x) = \frac{1}{3} + \frac{1}{5} \neq 1$

18.14 JOINT PROBABILITY DISTRIBUTIONS

We have considered the case of probability distributions of one random variable. Several variables, however, may be involved in many experiments. They give rise to joint distributions of two or more variables.

18.15 BIVARIATE PROBABILITY DISTRIBUTIONS

The joint probability function of two random variables X and Y is written $p(x, y)$ where X has assumed a value x and Y has assumed a value y.

If X takes the values $x_1, x_2, \ldots\ldots, x_n$
and Y takes the values $y_1, y_2, \ldots\ldots, y_n$
then the bivariate probability distribution of X, Y is

$$p(X = x_i \text{ and } Y = y_i) \text{ or briefly } p(x, y)$$

18.16 INDEPENDENT DISTRIBUTIONS

If the individual probability distributions of X and Y are $p_1(x)$ and $p_2(y)$ then $p_1(x)$ and $p_2(y)$ are independent if and only if

$$p(x, y) = p_1(x)\, p_2(y) \text{ for all } x, y.$$

18.17 MARGINAL DISTRIBUTIONS

A probability distribution for a single discrete variable obtained by summing the appropriate values in a joint distribution is called a marginal probability distribution. The marginal probability distributions of X and Y are

$$p(x) = \sum_y p(x, y), \quad p(y) = \sum_x p(x, y)$$

18.18 CONDITIONAL DISTRIBUTIONS

The conditional probability distribution of Y when X is fixed is given by

$$p(y/x) = \frac{p(x, y)}{p(x)} \qquad p(x) \neq 0$$

Similarly $$p(x/y) = \frac{p(x, y)}{p(y)} \qquad p(y) \neq 0$$

The ideas and formulae for joint, marginal and conditional distributions for *continuous* random variables can be derived on the same lines by using integrals in place of summations where the distributions are discrete.

18.19 DESCRIPTION OF A FREQUENCY DISTRIBUTION : DESCRIPTIVE STATISTICS

A frequency distribution can be described with the help of the following measures.

1. *Averages* : These are measures of central tendency of the data. They are also called measures of location because they determine the location of values.

2. *Measures of dispersion* : These measures (variously called measures of variation, scatter or spread) determine the degree of dispersion about the central tendency of the data.

3. *Measures of skewness* : These measures determine the degree of asymmetry of the distribution on the two sides of the point of the greatest concentration in the data.

4. *Measures of kurtosis* : They measure the degree of peakedness showing the extent of concentration at the central value.

CHAPTER 19

Measures of Central Tendency

19.1 AVERAGE

Quantitative data arranged in the form of frequency distributions generally exhibit a common characteristic: they have a tendency to concentrate at certain values, usually somewhere near the centre of the distribution. This tendency of observations to cluster near the central part of the distribution is known as *Central Tendency* and can be measured statistically.

A measure of central tendency or an average is regarded as the most *representative* value of *the given data.* This is because it is determined at the point where the concentration of items or values is the greatest, *i.e.,* the frequency is the highest on a distribution scale. An average is thus a precise yet simple expression representing a series of divergent individual values, or in other words, it is the consolidated essence of a complex distribution.

Averages serve the following purposes :

1. They provide a condensed or consolidated formulation of a large quantity of numerical data which ordinarily cannot be easily interpreted. For example, it is not easy to memorize the individual performances of college students, but quite simple to remember their average performance.

2. They afford a basis for comparison with other similar grouped data. For instance, it is impossible to compare the height of each individual in one state with that of each inhabitant of another, but it is possible to compare the mean heights in the two states.

The types of averages in general use are:

I. Mathematical Averages

1. Arithmetic Mean
2. Geometric Mean
3. Harmonic Mean
4. Quadratic Mean (or root mean square)

II. Positional Averages (based on their position in the data)

5. Median
6. Mode

The averages in common use are the arithmetic mean, median and mode.

19.2 THE ARITHMETIC MEAN

It is the best known and most commonly used form of average.

The arithmetic mean is obtained by dividing the sum of values of observations by the number of observations. It is usually denoted by $\bar{x}$.

Simple Arithmetic Mean

Ex. 1. For the observations 5, 4, 9, 12, 10, the arithmetic mean is given by

$$\bar{x} = \frac{5+4+9+12+10}{5} = \frac{40}{5} = 8$$

It x is a variable having n values

$$x_1, x_2, \ldots\ldots, x_n$$

then its simple arithmetic mean may be written

$$\bar{x} = \frac{x_1 + x_2 + \ldots\ldots\ldots x_n}{n} = \frac{\sum_{i=1}^{n} x_i}{n} \quad \text{or simply} \quad \frac{\sum x}{n}$$

Arithmetic Mean of Grouped Data

When observations repeat themselves, we may reduce our work by considering their frequencies.

Ex. 2. For the observations 5, 5, 5, 4, 4, 3, 3, 3, 3, the arithmetic mean may be written

$$\bar{x} = \frac{3\times5+2\times4+4\times3}{3+2+4} = \frac{35}{9} = 3.9$$

If x is a variable having values

$$x_1, x_2, \ldots\ldots, x_k$$

occurring with frequencies

$$f_1, f_2, \ldots\ldots, f_k$$

then its arithmetic mean may be written

$$\bar{x} = \frac{f_1x_1 + f_2x_2 + \ldots + f_kx_k}{f_1 + f_2 + \ldots + f_k} = \frac{\sum_{i=i}^{k} f_ix_i}{\sum_{i=1}^{k} f_i} = \frac{\sum_{i=i}^{k} f_ix_i}{n} \quad \text{or} \quad \frac{\sum fx}{n}$$

where $\sum_{i=1}^{k} f_i = n$

Often we have data in the form of frequency distributions in which there are regular class intervals. If we assume that each class interval can be represented by its midpoint then the method used above can be easily applied and the procedure can be simplified. Unequal class intervals can be dealt with appropriately.

The first step is to find the midpoints (denoted by x_i) of all the class intervals and put them under x. The frequencies f_i of these are given under f. (as in Ex. 4 below)

The arithmetic mean is given by

$$\bar{x} = \frac{\Sigma fx}{n}$$

Note. If in any exercise there are no frequencies, this implies that each frequency $f_i = 1$ so that above formula can be written $\bar{x} = \frac{\Sigma x}{n}$.

Ex. 3.

Wages x	*Number of Workers* f	fx
5	2	10
7	4	28
9	5	45
11	6	66
13	2	26
15	1	15
Total	20	190

It means 20 workers receive a total 190 in the form of wages and their arithmetic mean is $\frac{190}{20} = 9.50$.

We write this with the help of the formula

$$\bar{x} = \frac{\Sigma fx}{n} = \frac{190}{20} = 9.50$$

Arithmetic mean of wages = 9.50

In a frequency distribution we use the midpoints as representatives of the class intervals and calculate the arithmetic mean accordingly.

Ex. 4.

Class interval of marks	*Midpoints* x	*Number of students* f	fx
0—10	5	12	60
10—20	15	15	225
20—30	25	28	700
30—40	35	25	875
40—50	45	20	900
Total		100	2760

Explanation
f_1x_1 : $12 \times 5 = 60$
f_2x_2 : $15 \times 15 = 225$
etc.

$$\bar{x} = \frac{\Sigma fx}{n} = \frac{2760}{100} = 27.6$$

Short cut method

Instead of x use a new variable

$$u = \frac{x - x_0}{c}$$

where c is the class interval of the data and x_0 is any one of the x values picked up from the data. Usually a value of x_0 near the middle of the data is taken.

Here $c = 10$; take $x_0 = 15$. The new variable is

$$u = \frac{x - 15}{10}$$

when $\quad x = 5, \quad u = \frac{5 - 15}{10} = -1$

$\quad x = 15, \quad u = \frac{15 - 15}{10} = 0$

$\quad x = 25, \quad u = \frac{25 - 15}{10} = 1$ and so on

This can be noted in the table given below.

First find the mean of u values.

Marks	x	f	$u = \frac{x - 15}{10}$	fu
0—10	5	12	−1	−12
10—20	15	15	0	0
20—30	25	28	1	28
30—40	35	25	2	50
40—50	45	20	3	60
		100		126

$$\bar{u} = \frac{\Sigma fu}{\Sigma f} = \frac{126}{100} = 1.26$$

To find the arithmetic mean of the data use the following formula

$$\bar{x} = x_0 + c\bar{u}$$

where x_0 is the number subtracted (here 15) and c is the dividing number (here 10)

$$\therefore \quad \bar{x} = 15 + 10 \times 1.26$$
$$= 15 + 12.6$$
$$= 27.6, \text{ the same answer as in Ex. 4 above.}$$

There is no restriction on what number is taken as the origin. But it is preferable to take one of the x numbers, preferably the number near the centre of the data. The dividing number should be the size of the class interval c.

Ex. 5. We now take a different origin for the same data

Marks	f	x	$u = \dfrac{x-25}{10}$	fu
0—10	12	5	– 2	– 24
10—20	15	15	– 1	– 15
20—30	28	25	0	0
30—40	25	35	1	25
40—50	20	45	2	40
Total	100			$\Sigma fu = 26$

Explanation

col. 4

$$\frac{5-25}{10} = \frac{-20}{10} = -2$$

$$\frac{15-25}{10} = \frac{-10}{10} = -1$$

$$\frac{25-25}{10} = \frac{0}{10} = 0 \text{ etc.}$$

$$x_0 = 25 \quad c = 10, n = 100$$

$$\bar{u} = \frac{\Sigma fu}{n} = \frac{26}{100} = 0.26$$

$$\bar{x} = x_0 + c\bar{u} = 25 + 10 \times 0.26$$

$$= 25 + 2.6 = 27.6,$$

the same answer as obtained earlier.

Note. For any data of this type note the following procedure:

Write the midpoints of class intervals.

Subtract from x-values a number which is nearest the centre of the data, roughly, the midpoint corresponding to the highest frequency. Then divide the result by the class interval c to obtain the u values. Calculate $\bar{u}$.

Multiply $\bar{u}$ by c and add it to x_0.

We get the mean $\bar{x} = x_0 + c\bar{u}$.

Ex. 6. For the data given below find the arithmetic mean of daily wages.

Wages	*Number of Workers*
3.75—3.85	2
3.85—3.95	23
3.95—4.05	49
4.05—4.15	63
4.15—4.25	45
4.25—4.35	25
4.35—4.45	3
4.45—4.55	4

Ans. 4.11.

Ex. 7. The arithmetic mean of the following data has been calculated by the direct as well as the indirect method. Check.

Direct				*Short cut* $x_0 = 30$, $c = 5$	
Monthly wages	*Midpoint x*	*Freq. f*	*fx*	$u = \frac{x-30}{5}$	*fu*
12.5 – 17.5	15	2	30	– 3	– 6
17.5 – 22.5	20	22	440	– 2	– 44
22.5 – 27.5	25	19	475	– 1	– 19
27.5 – 32.5	30	14	420	0	0
32.5 – 37.5	35	3	105	1	3
37.5 – 42.5	40	4	160	2	8
42.5 – 47.5	45	6	270	3	18
47.5 – 52.5	50	1	50	4	4
		$x = 71$	$\Sigma fx = 1950$		$\Sigma fu = -36$

By the direct method $\bar{x} = \frac{\Sigma fx}{n} = \frac{1950}{71} = 27.46$

By the short method $\bar{x} = x_0 + c\bar{u} = 30 + 5 \times \frac{-36}{71} = 27.46.$

Ex. 8. Find the mean height of 50 men from the data given below:

Height in inches	73	72	71	70	.69	68	67	66	65
Number of men	2	4	6	10	11	7	5	4	1

Height x	*f*	$u = x - 69$	*fu*
73	2	+ 4	+ 8
72	4	+ 3	+ 12
71	6	+ 2	+ 12
70	10	+ 1	+ 10
69	11	0	0
68	7	– 1	– 7
67	5	– 2	– 10
66	4	– 3	– 12
65	1	– 4	– 4
	$n = 50$		$\Sigma fu = 9$

$$x_0 = 69,\ n = 50,\ c = 1,\ \bar{u} = \frac{\Sigma fu}{n} = \frac{9}{50} = 0.18$$

$$\bar{x} = x_0 + c\bar{u} = 69 + 1 \times 0.18 = 69.18$$

If we had done it directly we would get

$$\Sigma fx = 3459 \text{ (check)}$$

$$\bar{x} = \frac{\Sigma fx}{n} = \frac{3459}{50} = 69.18$$

Properties of the Arithmetic Mean

(*i*) The arithmetic mean is affected by all the items in the series which is good in so far as it makes $\bar{x}$ a very representative average. It is not a good representative if there are large deviations from the central value.

(*ii*) It is easy to calculate and is capable of algebraic manipulations.

(*iii*) It is determinate.

(*iv*) It is a typical representative value of all items in the data.

(*v*) Its value may be substituted for the value of each item without changing the total: $\Sigma fx = \Sigma f\bar{x} = n\bar{x}$

The algebraic sum of the deviations of a set of numbers from their arithmetic mean is zero: $\Sigma f(x - \bar{x}) = \Sigma fx - \bar{x}\Sigma f = n\bar{x} - n\bar{x} = 0$.

If there are no frequencies we can write

$$\Sigma x = \Sigma\bar{x} = n\bar{x}$$

$$\Sigma(x - \bar{x}) = n\bar{x} - n\bar{x} = 0$$

Ex. 9. The mean of 3, 10, 15, 8 is 6

$\therefore$ $(3 - 6) + (10 - 6) + (15 - 6) + (8 - 6) = 0$

i.e., $\Sigma(x - \bar{x}) = 0$.

(*vi*) The sum of squared deviations of a set of numbers from their arithmetic mean is the minimum *i.e.*,

$$\Sigma(x - \bar{x})^2 \text{ is a minimum}$$

For any $A \neq \bar{x}$, $\Sigma(x - A)^2 > \Sigma(x - \bar{x})^2$

(*vii*) A combined mean can be found from component means.

If $\bar{x}_1$, $\bar{x}_2$ are means of n_1, n_2 observations respectively the combined mean is

$$\bar{x} = \frac{n_1\bar{x}_1 + n_2\bar{x}_2}{n_1 + n_2}.$$ See Sec. 20.5 for details.

Ex. 10. If the mean of marks of one class of 50 students is 30 and that of another class of 100 students is 40, what is the mean of all the 150 students?

$$n_1 = 50,\ n_2 = 100,\ \bar{x}_1 = 30,\ \bar{x}_2 = 40$$

$$\bar{x} = \frac{50 \times 30 + 100 \times 40}{50 + 100} = \frac{5500}{150} = 36.7$$

In general $\bar{x} = \dfrac{n_1\bar{x}_1 + n_2\bar{x}_2 + \ldots\ldots + n_k\bar{x}_k}{n_1 + n_2 + \ldots\ldots + n_k}$

for k components.

(*viii*) If x_0 is an assumed mean or origin and $u = x - x_0$ the arithmetic mean can be written

$$\bar{x} = x_0 + c\,\bar{u}$$

Merits of Arithmetic Mean

The arithmetic mean is the most widely used measure of central tendency.

(*i*) Its definition is clear and precise. It corresponds to the centre of gravity of the observations.

(*ii*) It is simple to understand and easy to compute. Easier still by the short cut method.

(*iii*) It uses each and every item in the data.

(*iv*) It has a determinate value and is rigidly defined.

(*v*) It can be subjected to further algebraic treatment and advanced statistical theory is based on it.

(*vi*) It can be found even if only the total of values is known and the individual values are not known.

(*vii*) It provides a good standard of comparison since extreme values can cancel each other out when the number of observations is large.

Demerits of Arithmetic Mean

(*i*) It may be greatly influenced by unrepresentative values. The mean of 49, 50, 51 is 50. But the mean of 1, 50, 99 is also 50. In such cases the representative character of the mean is lost.

(*ii*) It gives greater importance to larger and less importance to smaller values. It has an upward bias.

(*iii*) It cannot be calculated if one or more items in the data are missing.

(*iv*) It cannot be located by inspection (like the mode and median).

(*v*) It may conceal facts and may lead to distorted conclusions.

Thus, if the performance of two students in three terms is given by

(*a*) 30%	40%	50%	marks
(*b*) 50%	40%	30%	marks

although the means are equal we cannot know from the means alone that one student is showing improved results, and the other deteriorating. Similarly inequalities of incomes are concealed by the per capita income and the average result of a college may tell nothing about the general performance of the students.

19.3 WHAT IS WEIGHTING?

In the computation of arithmetic mean we assume that all items are of equal importance. It may not be so.

Importance of different items can be shown by attaching suitable weights to them relative to their importance.

If $w_1, w_2, \ldots\ldots, w_k$ are the weights assigned to the values $x_1, x_2, \ldots\ldots, x_k$ respectively according to their relative importance the weighted mean is

$$\bar{x} = \frac{\Sigma wx}{\Sigma x}$$

Ex. 11. A teacher assigns weights 4, 2, 3 respectively to seminar work, class and monthly tests of students. The average academic performance may be found by the weighted mean.

Work	*Marks out of 100* x	*Weights* w	wx
Seminar	45	4	180
Class work	62	2	124
Monthly test	52	3	156
		$\Sigma w = 9$	$\Sigma wx = 460$

$$\bar{x} = \frac{\Sigma wx}{\Sigma w} = \frac{460}{9} = 51.1.$$

The weights show that seminar work is twice as important as class work from the teacher's point of view and so on.

A weight indicates the number of times a given item is to be counted. It corresponds to the concept of frequency. The sum of weights is like the sum of frequencies. The usual formula for the mean can therefore be utilised.

Weights may be actual or arbitrary.

Actual weights are estimated on the strength of the available data.

Arbitrary weights are assigned by experts who use their experience and intuition, of course, with chances of going wrong sometimes.

19.4 WHEN SHOULD WEIGHTS BE USED?

Weights should be used

(*i*) when items fall into different grades or, the classes of the same group show considerable variation,

(*ii*) when the averages of different grades constituting a group of observations and the number of observations falling in each grade are known, and

(*iii*) when the relative proportion of the number of items changes with a change in the size of items.

The weighted mean possesses all the properties and advantages of the simple mean. It is usually applied in the calculation of

(*i*) birth, marriage, and death rates

(*ii*) spatial and temporal comparisons

(*iii*) index numbers.

Difference between simple and weighted averages : The simple average is a particular case of the weighted average with weights 1 for all items.

In the case of a weighted average, the weights are chosen according to a systematic plan.

If larger items are associated with smaller weights and smaller items with larger weights the weighted average, in general, will be influenced by the latter and will be less than the simple average. In the opposite case the weighted average will be larger.

If there is no such connection between items and weights the difference between simple and weighted averages may be insignificant.

Ex. 12. Two colleges show the following results. Which one is better on an average?

Percentage results

	College A	*College B*
First year	70% (200 students)	80% (150 students)
Second year	60% (150 students)	60% (100 students)
Final year	80% (100 students)	80% (50 students)

We should calculate the weighted means for both the colleges.

For college A : $\bar{x}_A = \dfrac{200\times 70+150\times 60+100\times 80}{200+150+100}$

$$= \frac{31000}{450} = 68.9$$

For college B : $\bar{x}_B = \dfrac{150\times 80+100\times 60+50\times 80}{150+100+50}$

$$= \frac{22000}{300} = 73.3$$

College B has a better average than A.

19.5 THE GEOMETRIC MEAN

The geometric mean (G) is the nth root of the product of n values.

$$G = \sqrt[n]{x_1 \times x_2 \times \ldots \times x_n}$$

The G.M. of 2, 4, 8 is the cube root of their product.

$$G = \sqrt[3]{2\cdot 4\cdot 8} = \sqrt[3]{64} = 4$$

If the frequencies of $x_1, x_2, \ldots\ldots, x_k$ are respectively $f_1, f_2, \ldots\ldots, f_k$ $(\Sigma f = n)$

$$G = \sqrt[n]{x_1^{f_1}\cdot x_2^{f_2} \ldots\ldots x_k^{f_k}}$$

Logarithms may be used in the calculation of G.M.

$$\text{Log } G = \frac{1}{n}[f_1 \log x_1 + f_2 \log x_2 + \ldots + f_k \log x_k] = \frac{\Sigma f \log x}{n}$$

$$G = \text{Antilog } \frac{1}{n}\Sigma f \log x$$

If there are no frequencies $G = (x_1 x_2 \ldots x_n)^{\frac{1}{n}}$ and $\log G = \frac{1}{n} \Sigma \log x$

Ex. 13. To compute G.M. of some x values we proceed as follows:

x	$\log x$
4.5	0.6532
250.5	2.3979
12.0	1.0792
119.5	2.0789
30.0	1.4771
42.0	1.6232
75.0	1.8751
35.4	1.5490
	$12.7336 = \Sigma \log x$

$$G = \text{Antilog } \frac{1}{n}\Sigma \log x = \text{Antilog } \tfrac{1}{8} \times 12.7336$$

$$= \text{Antilog } 1.5917 = 39.81.$$

Ex. 14.

Class interval	f	x	$\log x$	$f \log x$
2 – 4	20	3	0.4772	9.542
4 – 6	40	5	0.6990	27.660
6 – 8	30	7	0,8451	25.353
8 – 10	10	9	0.9542	9.542
	$\Sigma f = 100$			$\Sigma f \log x = 72.097$

$$G = \text{Antilog } \frac{1}{n}\Sigma f \log x$$

$$= \text{Antilog } \tfrac{1}{100}(72.097) = \text{Antilog } 0.72097$$

$$G = 5.279.$$

Ex. 15. Find the G.M. for Ex. 3.

Merits and Uses of Geometric Mean

Most of the properties and merits of G.M. resemble those of A.M.

(*i*) The GM takes into account all the items in the data and condenses them into one representative value.

(*ii*) It has a downward bias. It gives more weight to smaller values than to larger values.

(*iii*) It is determinate. For the same data there cannot be two geometric means.

(*iv*) It balances the ratios of the values on either side of the data. It is ideally suited to average rates of change such as index numbers and ratios between measures and percentages.

(*v*) It is amenable to algebraic manipulations like the A.M.

Demerits of Geometric Mean

(*i*) It is difficult to use and to compute.

(*ii*) It is determined for positive values and cannot be used for negative values or zero. A zero will convert the whole product into zero.

19.6 THE HARMONIC MEAN

The harmonic mean is the reciprocal of the arithmetic mean of the reciprocals of a series of observations. In other words, the total number of items of a variable in a series, divided by the total of the reciprocals of the items gives the H.M.

The H.M. is computed by using the formula

$$H = \frac{n}{\frac{1}{x_1} + \frac{1}{x_2} + \ldots + \frac{1}{x_n}} = \frac{n}{\Sigma \frac{1}{x}}$$

If we have a frequency distribution, $H = \frac{n}{\Sigma \frac{f}{x}}$.

Ex. 16. The harmonic mean of 0.5 and 0.25 is

$$H = \frac{2}{\frac{1}{0.5} + \frac{1}{0.25}} = \frac{2}{2+4} = \frac{1}{3} \text{ or } 0.33.$$

Ex. 17. To calculate the harmonic mean for the data below we proceed as shown.

Class interval	*Midpoint* x	$\frac{1}{x}$	f	$f \times \frac{1}{x}$
2 – 4	3	0.333	20	6.67
4 – 6	5	0.200	40	8.00
6 – 8	7	0.143	30	4.29
8 – 10	9	0.111	10	1.11
			$\Sigma f = 100$	$\Sigma f \times \frac{1}{x} = 20.07$

$$H = \frac{n}{\Sigma \frac{f}{x}} = \frac{100}{20.07} = 4.98.$$

Ex. 18. Calculate the H.M. for the following data:

Class interval	*Frequency*
5 – 10	2
10 – 15	7
15 – 20	11
20 – 25	9
25 – 30	8
30 – 35	3

Merits and Uses of the Harmonic Mean

(*i*) As a computed average like the A.M. and G.M., the H.M. also takes into account all the observations in the data.

(*ii*) It gives more weight to smaller items *i.e.*, it has a downward bias. It is of great utility when weights are attached to smaller items.

(*iii*) It measures rates of change and can be adapted to problems involving time and certain types of ratios and rates.

(*iv*) It is amenable to algebraic manipulations.

Demerits of Harmonic Mean

(*i*) It is difficult to compute when the number of items is large.

(*ii*) Some people may find it difficult to understand. It is not in common use as an average.

(*iii*) It assigns too much weight to the smaller items and thus has limited scope. It may be noted that its value is always less than that of the G.M. which is usually smaller than the A.M. The relative positions of A.M., G.M. and H.M. may be symbolically presented as follows:

$$\bar{x} \geq G \geq H$$

The three are identical if all the items in a given data have the same value.

19.7 THE QUADRATIC MEAN OR THE ROOT MEAN SQUARE

It is the square root of the mean of the squared values of x in a series of k observations:

$$M_q = \sqrt{\frac{x_1^2 + x_2^2 + \ldots + x_k^2}{k}} = \sqrt{\frac{\Sigma x^2}{k}}$$

The weighted quadratic mean with weights or frequencies $f_1, f_2 \ldots f_k$ for the values $x_2, x_2, \ldots, x_k$ respectively, where $\Sigma f = n$, is given by

$$M_q = \sqrt{\frac{\Sigma f x^2}{n}}$$

It can be shown that

$$M_q > \bar{x} > G > H$$

Ex. 19. We may write the generalised formula for averages as

$$\sqrt[k]{\frac{\Sigma x^k}{n}}$$

An infinite number of averages can be obtained by giving different values to k,

$$k = -1 \text{ gives the H.M.} = \frac{n}{\Sigma \frac{1}{x}}$$

$$k = 1 \text{ gives the A.M.} = \Sigma \frac{x}{n}$$

$$k = 2 \text{ gives the Q.M.} = \sqrt{\frac{\Sigma x^2}{n}}$$

Ex. 20. Find M_q for 3, 2, 4, 5, 1, 7

$$M_q = \sqrt{\frac{1}{6}(3^2 + 2^2 + 4^2 + 5^2 + 1^2 + 7^2)} = \sqrt{\frac{104}{6}} = 4.16.$$

19.8 THE MEDIAN

When observations are arranged in increasing or decreasing order of magnitude, the middle observation is the median.

Ex. 21. For the seven observations 1, 2, 5, 7, 9, 10, 12. Median = 8, this is the value of the 4th observation.

Ex. 22. If the number of observations is even, as in 1, 2, 5, 7, 9, 10, 12, 14, then

$$\text{Median} = \frac{7+9}{2} = 8$$

the average of the middle two items. The position of the median is such that 50 per cent of items are below and 50 per cent above it.

If there are n observations, the median is the value of the $\left(\frac{n+1}{2}\right)^{th}$ item.

In Ex. 21, $(n + 1)/2 = (7 + 1)/2 = 4$

The value of the median cannot be found from a disorderly series. The series should be first arranged in order.

Ex. 23. Find the median for the following data:

0, 5, – 100, – 20, 80

Arranging the data in ascending order

– 100, – 20, 0, 5, 80

we have the median = 0.

Ex. 24. Find the median height of 9 students with heights: (in inches)

59, 72, 63, 65, 62, 61, 70, 69, 10 **Ans.** 63.

Note. The median is not affected by changes in the extreme observations. Each of the sets

2, 5, 10, 12, 15

2, 5, 10, 12, 100

– 100, 5, 10, 12, 2000

has the same median 10

Ex. 25. The median of

2, 5, 10, 12, 15, 20

is
$$\frac{10+12}{2} = 11$$

Median of Grouped Data

In a frequency distribution, the median has to be interpolated in the class interval containing the median.

The class which contains the median is called the median class. It corresponds to the middle of the total frequency.

If the cumulative frequencies are calculated for the data the median class can be easily located.

Let us use the following notation:

n total number of observations

l_o lower limit of the median class

f_o frequency of the median class

F cumulative frequency of the pre-median class

c size of class interval

We have the formula for the median

$$Me = l_o + \frac{\frac{n+1}{2} - F}{f_o} \times c$$

The other method of finding the median is by drawing the graph of cumulative frequencies or the orgive.

Ex. 26. The method to calculate the median for a frequency distribution is as follows:

Class interval marks	*Number of students* f	*Cumulative frequency c.f.*
30 – 40	2	2
40 – 50	18	20
50 – 60	24	44
60 – 70	20	64
70 – 80	8	72
80 – 90	3	75

Explanation

$$n = 75, \quad \frac{n+1}{2} = \frac{76}{2} = 38$$

The middlemost observation *i.e.*, the 38th observation lies in the class interval 50 – 60 which, therefore, is the median class

Lower Limit $l_o = 50$

$l_o = 50,\ c = 10, f_o = 24$

F = c.f. of pre-median class = 20

$$Me = l_o = \frac{\frac{n+1}{2}}{f_o} \times c = 50 + \frac{38-20}{24} \times 10$$

$$= 50 + \frac{180}{24} = 57.5$$

Ex. 27. Find the median of the following data:

Class interval	*f*	*Class interval*	*f*
7.5 – 10.5	12	22.5 – 25.5	14
10.5 – 13.5	10	25.5 – 28.5	3
13.5 – 16.5	15	28.5 – 31.5	0
16.5 – 19.5	19	31.5 – 34.5	4
19.5 – 22.5	12	34.5 – 37.5	1

Ans. 17.82

Ex. 28. Compare the medians of groups *A* and *B* in the following data:

	Workers in	
Wages	*factory A*	*factory B*
45 – 50	1	1
50 – 55	0	0
55 – 60	1	1
60 – 65	5	5
65 – 70	6	15
70 – 75	12	10
75 – 80	14	9
80 – 85	7	6
85 – 90	2	2
90 – 95	1	1
95 – 100	0	0
100 – 105	1	0
Total	50	50

Ans. 75.2, 71.75

Ungrouped Data

If the data are ungrouped the median is easier to find. Observations are arranged in ascending or descending order with their respective frequencies. The middle item gives the median.

Ex. 29. Find the median of the following data:

x	6.7	6.2	8.4	6.9	7.1
f	2	5	6	8	6

Arranging in increasing order

x	f	cf
6.2	5	5
6.7	2	7
6.9	8	15
7.1	7	22
8.4	7	29
$n = 29$		

$\frac{n+1}{2} = \frac{29+1}{2} = 15$ and the median is 6.9, the value of the 15th observation.

Alt. arrange in full:

62. 6.2 62. 6.2 6.2 6.7 6.7 6.9 6.9 6.9 6.9 6.9 6.9 6.9 6.9
7.1 7.1 7.1 7.1 7.1 7.1 7.1 7.1 8.4 8.4 8.4 8.4 8.4 8.4 8.4

The median is the value of the 15th item which is 6.9.

Ex. 30. Find the median sale size of a commodity sold in different sizes as given below:

Size	3	2	1	6	5	4	7
Sales	25	10	30	37	5	19	29

Ans. 3.

Properties of the Median

(*i*) Geometrically, the median divides a histogram or a frequency curve into two parts with equal areas.

(*ii*) It remains unaffected by the magnitude of the extreme deviations from the average.

(*iii*) It may be located even when the items in a series cannot be measured so long as they can be ranked or arranged in any order.

(*iv*) The sum of absolute deviations of observations from their median is the least *i.e.*, $\Sigma|x - \text{Med}|$ is minimum.

Merits and Uses of the Median

(*i*) It can be easily understood and its computation is simple.

(*ii*) It can be computed even for incomplete data. It is concerned only with a few central observations.

(*iii*) It balances the number of items in a distribution. It is useful in describing scores, ratios and grades.

(*iv*) It is useful in the case of skewed distributions like those of incomes and prices.

(*v*) It can be used for qualitative data.

(*vi*) In the case of open end classes the median can be calculated but the mean cannot.

(*vii*) In can be easily determined graphically (see ogives Ch. 18)

Demerits of the Median

(*i*) Rearrangement of data may be necessary to compute the median.

(*ii*) The median is not easily capable of algebraic manipulations. As such it is not much used in advanced studies.

(*iii*) The empirical formula for the median based on interpolation may not always give correct results.

(*iv*) It may ignore significant extreme values.

(*v*) Weighting cannot be used in the case of the median. The scope of operations is thus narrowed.

(*vi*) It cannot be computed as exactly as the mean.

19.9 THE MEDIAN AND QUARTILES

The median divides the data into two equal parts so that 50% of the items lie on either side of it. For this reason it is also called the second quartile.

Quartiles are the values of the items that divide the series into 4 equal parts. The median is the second quartile or Q_2.

The first quartile (Q_1) is the value of the item which divides the lower half of the distribution into two equal parts. As such it is the value of the $\frac{n+1}{4}$th item in the series.

The third quartile (Q_3) is the value of the item which divides the upper half of the distribution into two equal parts *i.e.*, it is the value of the $3\left(\frac{n+1}{4}\right)$th item in the series.

Q_1, Q_2, Q_3 can be calculated from the ogive (see Ch. 18) of the data. We shall now consider their computation procedure with the help of formulae.

We shall use the same notation and formulae for Q_1, Q_3 as for the median or Q_2. It should, however, be remembered that the Q_1-class corresponds to the $(n + 1)/4$th item and the Q_3-class corresponds to the $\frac{3}{4}(n+1)$th item. Sometimes for simplicity in calculations, n is used in place of $n + 1$.

Median

$$Q_2 = l_0 + \frac{\frac{n+1}{2} - F}{f_0} \times c \quad \left[\begin{array}{l} l_0 = \text{lower class limit of med. class} \\ f_0 = \text{frequency of med. class} \\ F = c.f. \text{ of pre-median class} \end{array}\right.$$

First Quartile

$$Q_1 = l_0 + \frac{\frac{n+1}{4} - F}{f_0} \times c \quad \left[\begin{array}{l} l_0 = \text{lower limit of } Q_1 \text{ class} \\ f_0 = \text{frequency of } Q_1 \text{ class} \\ F = c.f. \text{ of pre-}Q_1 \text{ class} \end{array} \right.$$

Third Quartile

$$Q_3 = l_0 + \frac{\frac{3(n+1)}{4} - F}{f_0} \times c \quad \left[\begin{array}{l} l_0 = \text{lower limit of } Q_3 \text{ class} \\ f_0 = \text{frequency of } Q_3 \text{ class} \\ F = c.f. \text{ of pre-}Q_3 \text{ class.} \end{array} \right.$$

Ex. 31. The computation procedure for Q_1, Q_2, Q_3 has been shown in the following data:

Class interval	*f*	*c.f.*	*Explanation*
2 – 6	3	3	$\frac{n+1}{2} = \frac{59+1}{2} = 30$
6 – 10	5	8	∴ 14 – 18 is the median class
$Q_1 \rightarrow$ 10 – 14	12	20	$\frac{n+1}{4} = 15$
$Me \rightarrow$ 14 – 18	18	38	∴ 10 – 14 is the Q_1 – class
$Q_3 \rightarrow$ 18 – 22	10	48	$3\left(\frac{n+1}{4}\right) = 45$
22 – 20	9	57	∴ 18 – 22 is the Q_3 class
26 – 30	2	59	

For *median* $l_0 = 14, f_0 = 18, F = 20, c = 4$

$$Q_2 = l_0 + \frac{\frac{n+1}{2} - F}{f_0} \times c = 14 + \frac{30 - 20}{18} \times 4 = 16.2$$

This is the middlemost value which divides the data into two equal parts

For *first quartile* $l_0 = 10, f_0 = 12, F = 8, c = 4$

$$Q_1 = l_0 + \frac{\frac{n+1}{4} - F}{f_0} \times c = 10 + \frac{15 - 8}{12} \times 4 = 12.3$$

One fourth or 25% of data lies to the left of Q_1 or 12.3, three fourth or 75% to the right

For *third quartile* $l_0 = 18, f_0 = 10, F = 38, c = 4$

$$Q_3 = l_0 + \frac{\frac{3(n+1)}{4} - F}{f_0} \times c = 18 + \frac{45 - 38}{10} \times 4 = 20.8.$$

75% of the data lies to the left of $Q_3 = 20.8$ and 25% to the right.

Ex. 32. Draw the ogive for the above data and compute Q_1, Q_2, Q_3. Compare the results. (Identical answers will be found by the use of less than or more than ogive.)

Ex. 33. Find Q_1, Q_2 and Q_3 for the following data:

Wages	*No. of Workers*
0 – 500	50
500 – 1000	72
1000 – 1500	104
1500 – 2000	120
2000 – 2500	96
2500 – 3000	53
3000 – 3500	42
3500 – 4000	46

Ans. $Q_1 = 1096.1$
$Q_2 = 1741.7$
$Q_3 = 2416.3$

Ex. 34. Find Q_1, Q_2 and Q_3 for the following data:

Item	*Frequency*
0 – 4	7
5 – 9	9
10 – 14	13
15 – 19	18
20 – 24	14
25 – 29	10

Ans. $Q_1 = 11.1$
$Q_2 = 17.6$
$Q_3 = 21.9$

Ex. 35. Find the mean and median for the following data:

Number of absent days	*Number of students*
Less than 5	29
" 10	224
" 15	465
" 20	582
" 25	634
" 30	644
" 35	650
" 40	653
" 45	655

These data should be rearranged first. The given frequencies are *less than* cumulative frequencies. To calculate the frequencies in different class intervals subtract each *cumulative frequency* from the one immediately following. The frequency table is given below.

Number of absent days	*Number of Students*
0 – 5	29
5 – 10	195
10 – 15	241
15 – 20	117
20 – 25	52
25 – 30	10
30 – 35	6
35 – 40	3
40 – 45	2
Total	655

Explanation

In the class 0 – 5, $cf_1 = 29$

In the class 5 – 10, $cf_2 = 224$

$\therefore$ For the class 5 – 10

$f_2 = cf_2 - cf_1 = 224 - 29 = 195$

Similarly for the class 10 – 15

$f_3 = cf_3 - cf_2 = 465 - 224 = 241$ etc.

The usual procedure gives $\bar{x} = 12.5$, Med = 12.1.

Ex. 36. The above data might as well be given in the following form:

Number of absent days	*Number of students*
More than 0	655
5	626
10	431
15	190
20	73
25	21
30	11
35	5
40	2

The procedure is just the opposite of that used above. The frequencies given are *more than* cumulative frequencies. To get the frequencies subtract every *c.f.* from the preceeding *c.f.*

Ans. same.

Deciles, Percentiles, etc.

We can calculate other position measures like deciles and percentiles in the same way as quartiles with appropriate changes in the formulae given above. Thus the 60th percentile (P_{60}) is the value of the $\frac{60}{100}$th item. The seventh decile (D_7) is the value of the $\frac{7}{10}$ $(n + 1)$th item. These values can be easily found from the ogive.

19.10 THE MODE

The mode is the value which occurs most often in the data. It is the value around which there is the greatest degree of concentration.

The mode means 'norm' or 'fashion'. The mode is thus a typical measure of central tendency inasmuch as it is the most probable value in a series. This concept is often in the minds of some people when they speak of averages. The modal size of shoes is the size bought by most people, the modal consumer is the so called 'average' consumer.

Ex. 37. The mode in the series 3, 7, 3, 3, 5, 3, 1, 2 is 3.

When there are two values having the same maximum frequencies the mode is difficult to find. It may imply two modes.

The mode is the value of the x variable in a frequency distribution that occurs the greatest number of times *i.e.,* corresponds to the maximum ordinate of the curve.

The class in which the mode falls is called the modal class.

It corresponds to the highest frequency.

The class midpoint of a modal class is called a crude mode. The true mode is easy to define but difficult to determine.

A distribution with one mode is called unimodal; with two modes it is bimodal. With many modes we have a multimodal distribution.

The mode may be found by

(*i*) grouping

(*ii*) using interpolation formula

(*iii*) using mean and median

(*iv*) graphical methods.

Mode by grouping

If the concentration of frequencies is at more than one place the point of maximum concentration can be found by the grouping method.

There are generally six columns of frequencies to be calculated from the given data. The steps are explained in the example below.

Ex. 38. To determine the modal wages for the following data:

Wages	5	6	7	8	10	12	14	15	16	18
Number of Persons	3	2	7	10	9	6	5	4	3	1

We arrange the data as shown below.

Wages	*Frequencies*	(1)	(2)	(3)	(4)	(5)	(6)
					Columns		
5	f_1	3					
			5				
6	f_2	2					
				8	11		
7	f_3	6					
			17			19	
8	f_4	11					26
				20			
10	f_5	9					
			15		26		
12	f_6	6					
				11		20	
14	f_7	5					15
			9		12		
15	f_8	4					
				7		8	
16	f_9	3					
			4				
18	f_{10}	1					

Explanation of Columns:

(1) Original frequencies

(2) $f_1 + f_2 = 3 + 2 = 5, f_3 + f_4 = 6 + 11 = 17$ etc.

(3) $f_2 + f_3 = 2 + 6 = 8, f_4 + f_5 = 11 + 9 = 20$ etc.

(4) $f_1 + f_2 + f_3 = 3 + 2 + 6 = 11, f_4 + f_5 + f_6 = 11 + 9 + 6 = 26$ etc.

(5) $f_2 + f_3 + f_4 = 2 + 6 + 11 = 19, f_5 + f_6 + f_7 = 9 + 6 + 5 = 20$ etc.

(6) $f_3 + f_4 + f_5 = 6 + 11 + 9 = 26, f_6 + f_7 + f_8 = 6 + 5 + 4 = 15$ etc.

Most of the *highest* frequencies in different columns correspond to the wage 8 which therefore is the mode.

This can be seen by drawing the following table:

No.	5	6	7	8	10	12	14	15	16	18
					Wages					
1				1						
2			1	1						
3				1	1					
4				1	1	1				
5				1	1	1	1			
6			1	1	1					
	—	—	2	6	4	2	1	—	—	—

The value 8 has been repeated the maximum (6) number of times and hence is the mode.

Mode by interpolation

This involves two steps. The first is to find the modal class by grouping. Then the mode is found by using the formula

$$\text{Mode} = l_0 + \frac{f_0 - f_1}{2f_0 - f_1 - f_2} \times C$$

where l_0 is the lower limit of the modal class
c is the size of the class interval
f_0 is the frequency of the modal class
f_1 is the frequency of the class immediately above the modal class
f_2 is the frequency of the class immediately below the modal class

The modal class is generally the class with the highest frequency.

Ex. 39.

Class interval	*Frequency*
4 – 6	17
7 – 9	39
10 – 12	27
13 – 15	16

The modal class here is the 7 – 9 class which has the highest frequency 39 so that $l_0 = 7, f_0 = 39, f_1 = 17, f_2 = 27$.

$$Mo = l_0 + \frac{f_0 - f_1}{2f_0 - f_1 - f_2} \times c = 7 + \frac{39 - 17}{78 - 17 - 27} \times 2 = 7 + 1.29 = 8.29.$$

Ex. 40. Find the mode for the following data :

Height (in inches)	56	57	58	59	60	61	62	63	64	65	66
No. of boys	14	17	23	25	44	52	65	67	61	57	50

Ans. 63

Ex. 41. Find the mode for the following data :

Size of shoes	12	11	10	9	8	7	6	5	4	3
Number of persons	9	21	28	36	23	9	7	3	2	2

Ans. Size 9

Ex. 42. Find the mode for the following data :

Weights of students x	*frequency* f
90 – 100	10
100 – 110	37
110 – 120	65
120 – 130	80
130 – 140	51
140 – 150	35
150 – 160	18
160 – 170	4

premodal class : $f_1 = 65$
modal class : $f_0 = 80, l_0 = 120$
postmodal class : $f_2 = 51$
class interval size : $c = 10$

$$M_0 = l_0 + \frac{f_0 - f_1}{2f_0 - f_1 - f_2} \times c$$

$$= 120 + \frac{80 - 65}{160 - 65 - 51} \times 10$$

$$= 123.4$$

Mode from mean and median

An empirical, approximate relationship has been shown to exist between the mean, median and mode.

$$\text{Mode} = 3 \text{ Median} - 2 \text{ Mean}$$

We can calculate the mode if we know the values of the mean and the median. The mode value calculated by this method need not necessarily be the same as that found by using any other method.

Ex. 43. In the last problem check

$$\bar{x} = 125.73 \qquad \text{Med.} = 124.75$$

$$\text{Mo} = 3 \text{ Med.} - 2\bar{x} = 3 \times 124.75 - 2 \times 125.73$$

$$= 124.79$$

We had found $Mo = 123.4$ by direct calculation.

Graphical Location of Mode

It is possible to locate the mode from a histogram of the given data. This is done by drawing two diagonally intersecting lines joining each upper corner of the modal class bar to each opposite upper corner of the adjacent bars. The perpendicular drawn from the point of intersection on the x-axis gives the modal value on the x-axis.

Merits and Uses of the Mode

(*i*) It is simply defined and computed. When graphed it can be located almost by inspection.

(*ii*) It is a popular average in the sense that it is the one that most people use without being aware of it *e.g.* when they speak of the average number of car accidents, bus breakdowns etc.

(*iii*) Extreme values have no effect on the mode and it can be calculated when complete data are not available.

(*iv*) It is the most typical in the sense that it denotes the most probable value in the series.

Demerits of the Mode

(*i*) In the case of bimodal or multimodal distributions, it is not possible to pinpoint any one value as the mode.

(*ii*) It is not rigidly defined and thus cannot be called an ideal average.

(*iii*) It is often indeterminate when the distribution is highly irregular.

(*iv*) It is not based on all the observations in the data and hence is not an ideal measure of central tendency. It lays too much emphasis on the modal group and thus may not be fully representative of the whole series.

(*v*) It is not easily amenable to algebraic manipulation.

Despite these disadvantages the mode is widely used and is particularly useful to proprietors of shops and others who wish to know which items of food or other things must be most heavily stocked so as to be able to meet the demand for them. The mode is also employed in anthropometric and biological statistics where there is a definite characteristic of a race or community from which the measurements of individuals or groups show some deviations.

19.11 IS THERE SUCH A THING AS AN IDEAL AVERAGE?

The answer is 'probably not' because there is no ideal average for all time and for all occasions. But what do we mean by ideal? It is generally agreed that the title should be given to that average which possesses the following characteristics :

1. It is rigidly defined and easily computed.
2. It is simple enough to be understood by the layman.
3. It takes into account all the items in the series and gives them equal weights. It is not unduly influenced by a few extreme values.
4. Its values are not greatly affected by sampling fluctuations.
5. It is amenable to algebraic manipulations.

From the points listed above it is obvious that each measure of central tendency discussed in the preceding pages possesses some of these characteristics and the arithmetic mean satisfies them to a greater degree than any other measure. The arithmetic mean is rigidly defined, is invariably determinate and quite easily computed. It is in common use, takes into account all items and its value is not much influenced by sampling fluctuations. Most important, it is amenable to algebraic manipulations. Its only major drawback is that it is influenced by a few very low or high values in the extremities. For all that, it is the average most often opted for when a choice has to be made.

Choice of an Average

There are three factors that have to be kept in mind when deciding what average to use.

1. The purpose for which the average is being used.
2. The nature, characteristics and properties of the average.
3. The nature and characteristics of the data: in particular, the degree of homogeneity in the data.

The Arithmetic Mean is generally applicable to all kinds of data, and hence is really an all-purpose average which is part of the reason for its immense popularity.

The Geometric Mean is generally used in the study of socio-economic phenomena, especially when it is desired to give more weight to small items or when relative rather than absolute differences have to be taken into account rather than absolute ones.

The Harmonic Mean has applications to data relating to time series since its special advantage is that it averages ratios more accurately than other measures. It has a downward bias and is based on all items in a series.

The Median is generally used when attributes of data are not directly measurable. In case the data are incomplete it is definitely preferable to the mean.

The Mode has this in common with the median that it can be calculated even when data are not fully presented. It should be used by preference when the data have definite characteristics from which measurements of items show deviations as in anthropometry and biometric studies.

19.12 SOME LIMITATIONS OF MEASURES OF CENTRAL TENDENCY

(*a*) Any measure of central tendency averages and summarises several details of a series in a concise form and thus becomes a representative of the series. Unless handled carefully it may lead to wrong and absurd conclusions. Some examples may be noted.

(*i*) On an average the number of accidents occurring in the middle of the road is 5 per thousand. The number of deaths at other places is 30 per thousand. Therefore it is safer to walk in the middle of the road.

An absurd conclusion. How many people walk in the middle of the road?

(*ii*) The depth of a river at four different points is 2, 7, 5, 6 feet respectively. The average depth is 5 feet. Therefore, all the people with height above 5 feet can cross it.

A wrong conclusion.

(*iii*) The average of marks obtained by one class of students is 30 per cent. Therefore every student is hopeless.

Can there not be several brilliant students in the class?

(*iv*) The average of marks of one class of students is greater than that of another. Therefore the second class has very bad students.

That there is no student in the second class better than any one in the first class may be a wrong conclusion.

Distrust of statistics results from bad instances like these.

(*b*) An average does not reveal all the information about the data.

(*i*) It does not indicate if the series is rising or falling. It shows the central tendency, not the tendency of the data.

(*ii*) Widely differing series may have the same average.

The mean of 0 and 1000 is 500

The mean of 450 and 550 is 500.

The median of 2, 4, 8 is 4.

The median of – 1000, 4, 15000 is 4.

(*iii*) An average is not necessarily a member of the series. Sometimes this may give funny results like 5.4 students.

A measure of central tendency does not give information about the form of a series and the distribution of frequencies. Measures of dispersion show the degree to which numerical data tend to spread about an average. In the next chapter we shall study measures of dispersion.

Notes:

(*i*) If a distribution is absolutely symmetrical, the mean, median and mode coincide. In a moderately symmetrical distribution the median lies between the mean and the mode, approximately one third of the distance along the x-axis from the mean towards the median.

Mean–Mode $\simeq$ 3 (Mean–Median)

(*ii*) The mean, mode and median may be found in the same or different class intervals. There is no hard and fast rule about it.

(*iii*) The median and quartiles need not necessarily be in different class intervals. One or more of them may be found together in the same class interval.

Ex. 44. (*i*) Find the mean, median and mode for the following data:

Marks	*Number of students*
30 – 35	12
35 – 40	18
40 – 45	22
45 – 50	27
50 – 55	17
55 – 60	23
60 – 65	19
65 – 70	8

(*ii*) Also find the mode by using the formula

Mode = 3 Med. – $2\bar{x}$. $\bar{x}$ *Me* *Mo*

Ans. (*i*) 49.5, 49.0, 46.1, (*ii*) 48.0.

It is not necessary that the value of mode calculated directly from the data and that from the empirical formula should be the same.

Ex. 45. At a self service canteen students' waiting time in the queue is shown below:

Waiting minutes	*Number of students*
No waiting	10
Less than 1	12
1 – 2	27
2 – 4	35
4 – 7	10
7 – 10	4
10 and above	2
	100

How long do the middle 50 per cent students wait?

What will be the mean waiting time if the maximum waiting time is 15 minutes?

(For the first case Q_1, Q_3 may be calculated either directly or from an ogive. The mean can be calculated even if there are unequal class intervals whose midpoints may be appropriately taken).

CI	*x*	*f*	*fx*	*cf*
0	0	10	0	10
0 – 1	0.5	12	6	22
1 – 2	1.5	27	40.5	49
2 – 4	3.0	35	105.0	84
4 – 7	5.5	10	55.0	94
7 – 10	8.5	4	34.0	98
10 – 15	12.5	2	25.0	100
			265.5	

$$Q_1 = 1 + \frac{25-22}{27} \times 1 = 1.11$$

$$Q_3 = 2 + \frac{75-49}{35} \times 2 = 3.49$$

For Q_1: $c = 1$; for Q_3: $c = 2$

The middle 50% wait for

1.11 to 3.49 minutes

$$\bar{x} = \frac{265.5}{100} = 2.655 \text{ mins.}$$

Ex. 46. Four teachers of economics reported grades of 80, 90, 50 and 60 respectively for 30, 40, 50 and 60 students. What is the average grade? **Ans.** 60.2.

Ex. 47. Find the mean and quartiles for the following data:

Weekly earnings	*Number of weeks*
110 – 114	4
115 – 119	20
120 – 124	38
125 – 129	24
130 – 134	10
135 – 139	4

Ans. $\bar{x} = 123.9$, $Q_1 = 120.2$, $Q_2 = 123.5$, $Q_3 = 127.9$.

Ex. 48. Find the mean, median and mode for the following data:

Sales	*Number of days*
over 1500	0
1400	8
1300	10
1200	16
1100	28
1000	43
900	55
800	65
700	72
600	77
500	80

Ans. $\bar{x} = 1017.5$, Med $= 1023.3$, $M_o = 1050.0$.

Ex. 49. What is the purpose served by a statistical average? What are its desirable properties?

Ex. 50. What are the principal criteria for a satisfactory average? State giving reasons the circumstances in which it would be preferable to use (1) the mean, (2) the median, (3) the mode, (4) the geometric mean, (5) the harmonic mean.

Ex. 51. Distinguish between simple and weighted means and state the circumstances under which each is used.

Ex. 52. Comment on the following statements :

(*i*) An average is a substitute for a complex group of variables or for a group of large numbers.

(*ii*) An average does not reveal the whole story about the items which it represents.

(*iii*) No particular average can be regarded as the best at all times.

(*iv*) The average is that value of a distribution which represent the entire data.

(*v*) An average provides a comprehensive formulation of a large quantity of numerical data.

CHAPTER 20

Dispersion, Descriptive Measures, Expectation, Moments

20.1 MEASURES OF DISPERSION

An average condenses into one figure, certain characteristics of data. This gives adequate information about the data but in the process some information is concealed.

In particular, the nature and extent of differences between the items themselves and their differences or deviations from their average are not at all revealed by a measure of central tendency.

Variability is an important consideration in statistical description and a basic concept in statistical inference.

A measure of variation or dispersion in any data shows the extent to which the numerical values tend to spread about an average.

Different series may possess different dispersions of items around the average. We may not know this by merely looking at the averages, especially if they happen to be identical.

Ex. 1.

Subject	*Marks of Student A*	*Marks of Student B*
Mathematics	52	97
History	55	60
Geography	56	39
English	57	37
French	55	42
Total	275	275
A.M.	55	55

Here the means are identical but variations in the performance of the two students are large. A is definitely superior to B on the whole though B may have other commendable characteristics. The difference can be perceived conveniently by using suitable measures of dispersion.

Measures of central tendency are averages of the *first order.*

Measures of dispersion are averages of the *second order.*

There are averages of higher order also.

A measure of dispersion gives an idea about the extent of lack of uniformly in the sizes and qualities of the items in a series. It helps us to know the degree of uniformity and consistency in the series.

If the difference between items is large the dispersion or variation is large and *vice versa.*

If the difference between items is small, the average represents and describes the data adequately. For large differences it is proper to supplement information by calculating a measure of dispersion in addition to an average.

It is useful to know the dispersion in any data because its knowledge helps us :

(*i*) to compare the current results with the past results,

(*ii*) to compare two or more sets of observations,

(*iii*) to suggest methods to control variations in the data.

The average number of passes this year in a college may be the same as last year but there may be significant variation in performance.

A similar comparison with the results of other colleges may be made.

If variations are large, the authorities may move to do something to improve the results.

A study of variations helps us in knowing the extent of uniformity or consistency in any data.

Uniformity in production is an essential requirement of industry.. Quality control methods are based on the laws of dispersion.

Absolute and Relative Dispersion

A precise measure of dispersion is one which gives the magnitude of the variations in a series *i.e.,* it measures, in numerical terms, the extent of the scatter of the values around the average.

When dispersion is measured in terms of the original units of a series, it is absolute dispersion or variability. It is difficult to compare absolute values of dispersion in different series, especially when the series are in different units or have different sets of values.

Relative dispersion or variation is in the form of a ratio or percentage and is independent of original units. It is useful in comparing of data of different types.

A good measure of dispersion should have properties similar to those described for a good measure of central tendency.

Measures of Dispersion

Absolute Variability

The Range
The Quartile Deviation
The Mean Deviation
The Mean Difference
The Standard Deviation
Graph Based

Relative Variability

Relative Range
Relative Quartile Deviation
Relative Mean Deviation
Coefficient of Variation

20.2 THE RANGE

The range (R) of a set of numbers is the difference between the maximum and minimum values. It indicates the limits within which the values fall.

If in the series $x_1, x_2, \ldots\ldots, x_n$

$$x_{max} = \text{maximum value among all observations}$$

$$x_{min} = \text{minimum value}$$

then the range $R = x_{max} - x_{min}$

Ex. 2. The range of the series 5, 15, 17, 20, 24 is

$$R = 24 - 5 = 19$$

Relative range or the coefficient of range is defined by the ratio

$$RR = \frac{x_{max} - x_{min}}{x_{max} - x_{min}} = \frac{24 - 5}{24 + 5} = 0.66$$

For grouped data, the range is the difference between the upper limit of the last class interval and the lower limit of the first class interval.

Merits and Uses of the Range

It is simple to compute and easy to understand.

It is time saving and widely used in industrial quality control where the range of each batch of production is calculated to determine its variability.

Variations in rates of exchange are studied with the help of the range. Weather forecasts give a measure of the range of temperatures etc.

Demerits of the Range

It is not a precise measure.

It depends only on the extreme values and is not based on all the items in a series. Nor does it give any consideration to the arrangement of values in a series.

It is greatly influenced by the presence of one or two exceptionally large or small values and is subject to considerable fluctuations from one sample to another.

The range of two widely different series may be the same. The range of the series 1, 2, 5, is 3 and the range of the series, 5001, 5000, 5003 is also 3.

The range cannot be found for open end distributions.

It is not amenable to algebraic manipulations.

On the whole it is an unsatisfactory measure of dispersion but it has several uses.

20.3 THE QUARTILE DEVIATION

The *quartile deviation* (Q.D.) is a type of range based on the quartiles. The quartile deviation or *semi-inter-quartile-range* is half the distance along the x-axis between the lower (Q_1) and upper (Q_3) quartiles.

The greater the concentration of the distribution, the smaller is this interval which, therefore, gives an absolute measure of dispersion.

$$\text{Q.D.} = \frac{Q_3 - Q_1}{2}$$

The relative measure is given by the

$$\text{Coefficient of Q.D.} = \frac{Q_3 - Q_1}{Q_3 + Q_1}$$

Ex. 3. For a given data $Q_1 = 5$, $Q_3 = 10$.

$$\text{Q.D.} = \frac{10-5}{2} = 2.5$$

$$\text{Coef. of Q.D.} = \frac{10-5}{10+5} = \frac{5}{15} = \frac{1}{3}.$$

Ex. 4. Find the Q.D. and coefficient of Q.D. for the following data:

Marks	*No. of students*
35 – 40	4
40 – 45	8
45 – 50	12
50 – 55	7
55 – 60	2

Ans. 38.3, 0.35.

Ex. 5.

Earnings	f	*c.f.*
10 – 20	42	42
20 – 30	25	67
30 – 40	58	125
40 – 50	42	167

$$Q_1 = 10 + \frac{42-0}{42} \times 10 = 20.00$$

$$Q_3 = 40 + \frac{126-125}{42} \times 10 = 40.23$$

$$\text{Q.D.} = \frac{40.23 - 20.00}{2} = 10.115$$

Interquartile Range = $Q_3 - Q_1 = 40.23 - 20.00 = 20.23$
Note that Q.D. is one half of the interquartile range

$$\text{Coef. of Q.D.} = \frac{Q_3 - Q_1}{Q_3 + Q_1} = \frac{20.23}{60.23} = 0.29$$

Merits and Uses of Q.D.

It is easy to compute and simple enough for quick comprehension.

It is not affected by extreme values.

It is useful in measuring variations in the case of open end distributions.
It considers the most important part, the central 50%, of the data.

Demerits of Q.D.
It is not based on all the observations in the data.
It is not amenable to algebraic manipulations.
It is greatly influenced by sampling fluctuations.
It gives only an approximate and vague idea of the dispersion in the data. This is a general drawback of empirical measures of dispersion. This is overcome by the method of averaging deviations, viz. the mean deviation and the standard deviation.

20.4 THE MEAN DEVIATION

The mean deviation of a series of values is the arithmetic mean of their absolute deviations.

The deviations may be taken from any average. It is better to take deviations from the median as the sum of the absolute deviations of values in a series from their median is the minimum. But in practice it is convenient to take deviations from the arithmetic mean.

We define mean deviation from the arithmetic mean as follows:

$$\text{M.D.} = \frac{\Sigma|x - \bar{x}|}{n}$$

For grouped data $\text{M.D.} = \dfrac{\Sigma f|x - \bar{x}|}{n}$

Coefficient of $\text{M.D.} = \dfrac{\text{M.D.}}{\bar{x}}$

Mean Deviation from the median $= \dfrac{\Sigma\,|\,x - Me\,|}{n}$

or for grouped data $= \dfrac{\Sigma f\,|\,x - Me\,|}{n}$.

If the M.D. has been calculated with reference to the median, the coefficient of M.D. is M.D./Median.

Ex. 6. Find the M.D. from the mean for the series 5, 7, 10, 12, 6.

We have $\bar{x} = 8$.

x	$\lvert x - \bar{x}\rvert$
5	3
7	1
10	2
12	4
6	2

$$\Sigma\,|x - \bar{x}| = 12$$

$$\text{M.D.} = \frac{\Sigma|x - \bar{x}|}{n} = \frac{12}{5} = 2.4$$

$$\text{Coefficient of M.D.} = \frac{\text{M.D.}}{\bar{x}} = \frac{2.4}{8} = 0.3.$$

Ex. 7. The M.D. from the median and the coefficient of M.D. have been calculated for the data given below:

Age (Yrs.)	*No. of persons* f	x	$\lvert x - Me\rvert$	$f\lvert x - Me\rvert$
1 – 5	7	3	17	119
6 – 10	10	8	12	120
11 – 15	16	13	7	112
16 – 20	32	18	2	64
21 – 25	24	23	3	72
26 – 30	18	28	8	144
31 – 35	10	33	13	130
36 – 40	5	38	18	90
41 – 45	1	43	23	23
	123			$\Sigma f\lvert x - Me\rvert = 874$

By the usual method we find (c.f. not shown in the table)

$$Me = l_o + \frac{\frac{n+1}{2} - F}{f_o} \times c = 20 \text{ (check)}$$

Absolute values of deviations from the median are given by $\lvert x - Me\rvert = \lvert x - 20\rvert$

For example, in the fourth column,

$$\lvert 3 - 20\rvert = 17, \lvert 8 - 20\rvert = 12 \text{ etc.}$$

These values are multiplied by the corresponding frequencies.

We have the M.D. from the median

$$= \frac{\Sigma f\lvert x - Me\rvert}{n} = \frac{874}{123} = 7.11$$

$$\text{Coefficient of M.D.} = \frac{\text{M.D.}}{\text{Median}} = \frac{7.11}{20} = 0.356$$

Similarly M.D. from the mean may be found.

Ex. 8. (*i*) Calculate the mean deviation from the mean for the following data. What light does it throw on the social conditions of the community?

Data showing differences in ages of husbands and wives.

Difference in years	*Frequency*
0 – 5	499
5 – 10	705
10 – 15	507

15 – 20	281
20 – 25	109
25 – 30	52
30 – 35	164

(*ii*) The age distribution of 100 life insurance policy holders is as follows:

Age	*No. of policy holders*
17 – 19	9
20 – 25	16
26 – 35	12
36 – 40	26
41 – 50	14
51 – 55	12
56 – 60	5
61 – 70	5

Calculate the mean deviation from the median age.

Ex. 9. Calculate the mean deviation from the mean and median and their coefficients for the following data.

Size of shoes	:	3	6	11	2	4	10	5	7	8	9
No. of pairs sold	:	10	15	25	6	4	3	2	8	9	4

Ex. 10. Compute the mean deviation from the mean and median and their coefficients from the following data.

Age	:	15 – 25	25 – 35	35 – 45	45 – 55
No. of persons	:	150	325	200	125

Merits and demerits of the mean deviation

It is rigidly defined, easy to understand and calculate.

It is based on all the items in a series and as such it measures the dispersion in the entire range of the data.

It is a useful measure of dispersion but is difficult to handle.

It is not greatly affected by the values of extreme items.

Its great shortcoming is that in the interests of simplicity it ignores the signs of deviations from the average. This is not mathematically defensible. It is because of this that further algebraic manipulation is not easy. Despite this it is used in certain economic studies such as the distribution of personal wealth in a community which has both rich and poor members.

The mean difference

The mean difference of a distribution is the arithmetic mean of the absolute values of all possible pairs of its values including pairs consisting of identical values.

20.5 THE STANDARD DEVIATION

It is the most important and the most used measure of dispersion. It is an improvement over the mean deviation and is free from some defects of other measures of dispersion.

The standard deviation is the square root of the average of the squared deviations from their mean of a set of observations.

If μ is the mean of the observations of a population of size, its standard deviation is denoted by σ and defined

$$\sigma = \sqrt{\sum_{i=1}^{n}(x_i - \mu)^2 / n}$$

The sample standard deviation (s.d.) is usually denoted by s where

$$s = \sqrt{\frac{\Sigma(x-\bar{x})^2}{n}} \quad \text{or} \quad s = \sqrt{\frac{\Sigma x^2 - n\bar{x}^2}{n}} \quad \text{if } n \text{ is large, } n > 30.$$

For grouped data $s = \sqrt{\dfrac{\Sigma f(x-\bar{x})^2}{n}}$ or $\sqrt{\dfrac{\Sigma fx^2 - n\bar{x}^2}{n}}$

In the formula for s.d. of a sample, the denominator is $n - 1$ if the sample size is small, $n \le 30$.

For the following distribution of a sample, the s.d. has been calculated. Note the procedure.

Ex. 11.

x	x^2	*Alternatively*	x	$x - \bar{x}$	$(x - \bar{x})^2$
2	4		2	– 3	9
5	25		5	0	0
7	49		7	2	4
6	36		6	1	1
$\bar{x} = 5$	114	$n = 4$	$\bar{x} = 5$		14

$$s = \sqrt{\frac{\Sigma x^2 - n\bar{x}^2}{n-1}} = \sqrt{\frac{114 - 4\times 5\times 5}{3}} = 2.16 \qquad s = \sqrt{\frac{\Sigma(x-\bar{x})^2}{n-1}} = \sqrt{\frac{14}{3}} = 2.16$$

For grouped data with frequencies the procedure is as follows:

Ex. 12.

x	f	fx	fx^2	*Alternatively*	$x - 7$	$(x - 7)^2$	$f(x - 7)^2$
3	2	6	18		– 4	16	32
5	1	5	25		– 2	4	4
7	4	28	196		0	0	0
9	1	9	81		2	4	4
11	2	22	242		4	16	32
	10	70	562				72

$$\bar{x} = \frac{\Sigma fx}{n} = \frac{70}{10} = 57$$

$$s = \sqrt{\frac{\Sigma f(x-\bar{x})^2}{n-1}}$$

$$= \sqrt{\frac{72}{9}}$$

$$= \sqrt{8}$$

$$s = \sqrt{\frac{\Sigma fx^2 - n\bar{x}^2}{n-1}}$$

$$= \sqrt{\frac{562 - 10\times 7\times 7}{9}}$$

$$= \sqrt{\frac{72}{9}} = \sqrt{8}$$

Short Cut (Analogous to the Arithmetic Mean Short Cut)

For a frequency distribution of a sample, with $n > 30$

$$s = \sqrt{\frac{\Sigma f(x-\bar{x})^2}{n}} = \sqrt{\frac{\Sigma fx^2 - n\bar{x}^2}{n}}$$

If we substitute $u = (x - x_o)/c$ where x_0 is the assumed mean

c = size of class interval, n = size of sample

We have the formula for practical work

$$s = c\sqrt{\frac{\Sigma f(u-\bar{u})^2}{n}} = c\sqrt{\frac{\Sigma fu^2 - n\bar{u}^2}{n}}$$ (See Ex. 14 below for procedure)

The s.d. remains unchanged by change of origin. The denominator is $n - 1$ if $n \le 30$.

Note. If μ is the population mean, the *population s.d.* is denoted by the Greek letter σ and its formula is

$$\sigma = \sqrt{\frac{\Sigma(x-\mu)^2}{n}} = \sqrt{\frac{\Sigma x^2 - n\bar{\mu}^2}{n}}$$

where n is the total number of observations in *the population.* Note carefully that we have used μ here in place of $\bar{x}$ which is used in sample s.d.

The standard deviation is an absolute measure of dispersion. The relative measure is obtained by dividing it by the mean.

$$\text{CV} = \text{Coefficient of Variation} = \frac{s}{\bar{x}}\times 100$$

If we do not take percentages, we have the

$$\text{Coefficient of Standard Deviation} = \frac{s}{\bar{x}}.$$

Ex. 13. Find the S.D. and C.V. of the sample observations: 2, 5, 7, 6.

We have $\bar{x} = 20/4 = 5$, $n = 4$ which is less than 30.

x	$x - \bar{x}$ *i.e.*, $x - 5$	$(x - 5)^2$
2	− 3	9
5	0	0
7	2	4
6	1	1
		$\Sigma(x - 5)^2 = 14$

$$s = \sqrt{\frac{\Sigma(x - \bar{x})^2}{n - 1}} = \sqrt{\frac{14}{3}} = 2.16$$

$$\text{Coefficient of Variation} = \frac{s}{\bar{x}} \times 100 = \frac{2.16}{5} \times 100 = 43.2\%$$

If we take an origin or assumed mean x_0 say 4, the steps follow the same lines.

x	$u = x - 4$	u^2
2	− 2	4
5	1	1
7	3	9
6	2	4
	$\Sigma u = 4$, $\bar{u} = 1$	$\Sigma u^2 = 18$ $n = 4$

$$s = \sqrt{\frac{\Sigma u^2 - n\bar{u}^2}{n - 1}} = \sqrt{\frac{18 - 4 \times 1}{3}} = 2.16$$

Ex. 14. We now calculate S.D. for grouped data by the short cut method.

Class	x	*Frequency*	$u = \dfrac{x - x_0}{c}$	fu	fu^2
5 − 15	10	3	− 3	− 9	27
15 − 25	20	7	− 2	− 14	28
25 − 35	30	9	− 1	− 9	9
35 − 45	40	23	0	0	0
45 − 55	50	15	1	15	15
55 − 65	60	8	2	16	32
65 − 75	70	6	3	18	54
75 − 85	80	4	4	16	64
		$n = 75$		$\Sigma fu = 33$	$\Sigma fu^2 = 229$

Any value may be taken as x_0, preferably, one of the values which is somewhere in the middle of the x column. Here we take $x_0 = 40$.

Here, the number of observations n is large ($n > 30$). Use the s.d. formula with n in the denominator.

We have : $x_0 = 40, c = 10, n = 75, \bar{u} = \Sigma fu/n = 33/75$

$$s = c\sqrt{\frac{\Sigma fu^2 - n\bar{u}^2}{n}} = 10\sqrt{2.86} = 10 \times 1.691 = 16.91$$

$$\bar{x} = x_0 + c\bar{u} = 40 + 10 \times \frac{33}{75} = 44.4$$

$$\text{CV} = \frac{s}{\bar{x}} \times 100 = \frac{16.91}{44.4} \times 100 = 38.1\%$$

Ex. 15. Find s and C.V. for the following frequency distribution:

Variable :	2	3	14	5	9	10	12	13	15
Frequency :	25	37	44	59	68	43	31	23	12

Ex. 16. Find s and C.V. for the following data:

Wages	*No. of workers*
0 – 5	2
5 – 10	5
10 – 15	7
15 – 20	13
20 – 25	21
25 – 30	16
30 – 35	8
35 – 40	3

20.5 (*a*) COMBINED MEAN

Combined Standard Deviation

The combined arithmetic mean for two sets of data with arithmetic means $\bar{x}_1$, $\bar{x}_2$ and number of observations n_1, n_2 is given by

$$\bar{x} = \frac{n_1\bar{x}_1 + n_2\bar{x}_2}{n_1 + n_2}, \text{ No. of observatios here} = n_1 + n_2$$

If there are k data sets with means $\bar{x}_1, \bar{x}_2, \ldots, \bar{x}_k$ and number of observations $n_1, n_2, \ldots, n_k$, the combined mean is given by

$$\bar{x} = \frac{n_1\bar{x}_1 + n_2\bar{x}_2 + \ldots + n_k\bar{x}_k}{n_1 + n_2 + \ldots + n_k}$$

We can similarly find the combined standard deviation of two or more data sets whose standard deviations s_1, s_2 etc. are known.

We consider two data sets.

Let the means of the two be $\bar{x}_1$, $\bar{x}_2$ and the number of observations n_1, n_2. The individual standard deviations are s_1, s_2.

If the combined mean is $\bar{x}$, let $d_1 = \bar{x}_1 - \bar{x}$ and $d_2 = \bar{x}_2 - \bar{x}$

The combined standard deviation of the two series is given by (with n_1, n_2 large)

$$s = \sqrt{\frac{n_1 s_1^2 + n_2 s_2^2 + n_1(\bar{x}_1 - \bar{x})^2 + n_2(\bar{x}_2 - \bar{x})^2}{n_1 + n_2}}$$

or
$$s = \sqrt{\frac{n_1 s_1^2 + n_2 s_2^2 + n_1 d_1^2 + n_2 d_2^2}{n_1 + n_2}}$$

Note. For large n use $s^2 = \dfrac{\Sigma(x-\bar{x})^2}{n}$

The combined variance is (See Sec. 20.6 below)

$$s^2 = \frac{n_1 s_1^2 + n_2 s_2^2 + n_1 d_1^2 + n_2 d_2^2}{n_1 + n_2}$$

$$= \frac{n_1(s_1^2 + d_1^2) + n_2(s_2^2 + d_2^2)}{n_1 + n_2}.$$

Ex. 17. An analysis of the monthly wages paid to workers of 2 firms A and B belonging to the same industry gives the following results:

	Firm A	Firm B
Number of wage earners (n)	586 (n_1)	648 (n_2)
Average monthly wage ($\bar{x}$)	Rs. 52.5 ($\bar{x}_1$)	Rs. 47.5 ($\bar{x}_2$)
Standard deviation (s)	Rs. 10 (s_1)	Rs. 11 (s_2)

To calculate s for the whole series we first compute $\bar{x}$.

$$\bar{x} = \frac{n_1\bar{x}_1 + n_2\bar{x}_2}{n_1 + n_2} = \frac{586 \times 52.5 + 648 \times 47.5}{586 + 648} = \text{Rs. } 49.9$$

$$s = \sqrt{\frac{n_1 s_1^2 + n_1(\bar{x}_1 - \bar{x})^2 + n_2 s_2^2 + n_2(\bar{x}_2 - \bar{x})^2}{n_1 + n_2}}$$

$$= \sqrt{\frac{586 \times 100 + 586(52.5 - 49.9)^2 + 648 \times 121 + 648(47.5 - 49.9)^2}{586 + 648}}$$

$$= \sqrt{\frac{58600 + 3984.6 + 78408 + 3693.6}{1234}}$$

$$= \sqrt{\frac{144686.2}{1234}} = \sqrt{117.2} = 10.8.$$

Ex. 18. Given below are the frequencies, means and standard deviations of two series. Find the s.d. of the combined series.

	Series I	Series II
Frequency	35	45
Mean	30	20
Standard deviation	10	5

Some Properties of Standard Deviation

(*i*) The standard deviation is extremely useful in estimating the *representativeness* of the mean. A small standard deviation means a high degree of uniformity in the observations and of homogeneity in a series. Conversely a large s.d. means just the opposite. Thus if we have 2 or more comparable series with identical means, one with the smaller s.d. has the more representative mean.

(*ii*) The standard deviation, being concerned with the mean of squared deviations, increases with an increase in these deviations and decreases with a decrease in them.

(*iii*) It is based on all the items of a series and gives them proper weights.

(*iv*) For a symmetrical distribution the mean added to multiples of s.d. covers certain fixed percentages of areas. We take the area under a frequency curve as equal to 1. The following results may be noted:

The area between $\bar{x} - s$ and $\bar{x} + s = 0.6827$

The area between $\bar{x} - 2s$ and $\bar{x} + 2s = 0.9545$

The area between $\bar{x} - 3s$ and $\bar{x} + 3s = 0.9973$

The area between $\bar{x} - 1.96s$ and $\bar{x} + 1.96s = 0.95$

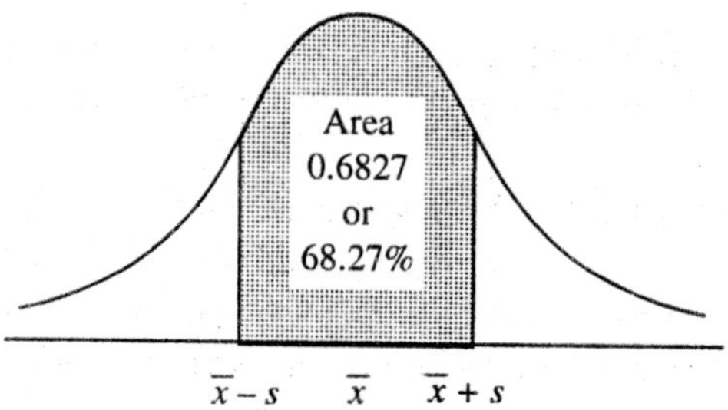

In the shaded area between $\bar{x} \pm s$ there are 68.27% items

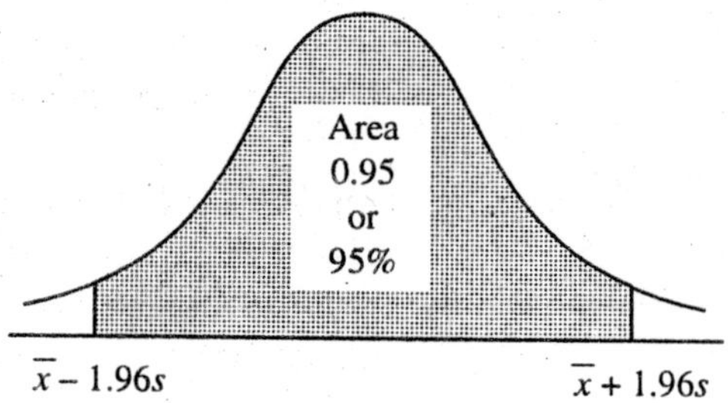

In the shaded area between $\bar{x} \pm 1.96s$ there are 95% items

(*v*) In the case of moderately symmetrical distributions, we have the following empirical relationships:

Mean Deviation MD $= 0.7979s$

Quartile Deviation QD $= 0.6745s$

(*vi*) For a given data the standard deviation is greater than the mean deviation measured from the mean or the median.

Merits and Importance of Standard Deviation

1. It is a reliable and convenient measure of dispersion. It is based on every item and is rigidly defined.

2. It is not as much affected by sampling fluctuations as other measures of dispersion.

3. The *range* is unstable and unreliable, the *quartile deviation* considers only the middle part of the data, the *mean deviation* neglects signs. The *standard deviation* is free from all these defects.

4. It measures the consistency and uniformity of data.

5. It is useful in finding the proportion of items falling within given distances (multiples of s) from the mean of a distribution.

6. It is useful in obtaining further descriptive measures of a distribution, like skewness, kurtosis, etc.

7. It is of great significance in testing the reliability of measures calculated from samples, the difference between such measures and in comparing the extent of fluctuations in two or more series.

8. Its analytical significance rests on its relation to an important theoretical distribution, known as the normal distribution which is defined by its mean and s.d. The normal distribution is symmetrical and property (*iv*) of the normal distribution given above applies to it. This property enables one to find the probability of a specified value in the sample, if the distribution of the population approximates to the normal and its mean and s.d. are known.

An important use of s.d. is as a standard unit of measure. It measures the deviation of a given observation from the mean. Thus we consider a given x values as being a given number of standard deviations away from the mean. This value is denoted by z which is often called the standard deviate or standard score:

$$z = \frac{x - \mu}{\sigma}$$

where μ is the mean of the population and σ its s.d.

Some Limitations of the Standard Deviation

The computational work is heavy in the case of s.d.

The s.d. gives more weight to extreme items and less to those near the mean.

These limitations do not come in the way of its regular employment and importance in theory and practice.

Comparison of variation in two or more sets of data

We normally use the coefficient of s.d. $\left(i.e., \frac{s}{\bar{x}}\right)$ or coefficient of variation $\left(i.e., \frac{s}{\bar{x}} \times 100\right)$ for such a comparison. We may, however, use other measures like M.D., Q.D. or Range.

That series is more uniform which has a lower value of C.V.

Ex. 29. Series 1 Series 2.

x_1	x_2	$x_1 - \bar{x}_1$	$(x_1 - \bar{x}_1)^2$	$x_2 - \bar{x}_2$	$(x_2 - \bar{x}_2)^2$
3	7	−2	4	0	0
5	9	0	0	2	4
4	10	− 1	1	3	9
8	2	3	9	− 5	25

$\bar{x}_1 = \frac{20}{4} = 5 \quad \bar{x}_2 = \frac{28}{4} = 7 \quad \Sigma(x_1 - \bar{x}_1)^2 = 14 \quad \Sigma(x_2 - \bar{x}_2)^2 = 38$

$$s_1 = \sqrt{\frac{\Sigma(\bar{x}_1 - \bar{x}_1)^2}{n_1 - 1}} = \sqrt{\frac{14}{3}} = 2.16$$

$$s_2 = \sqrt{\frac{\Sigma(x_2 - \bar{x}_2)^2}{n_2 - 1}} = \sqrt{\frac{38}{3}} = 3.58$$

$$\text{CV}_1 = \frac{s_1}{\bar{x}_1} \times 100 = \frac{2.16}{5} \times 100 = 43.2\%$$

$$\text{CV}_2 = \frac{s_2}{\bar{x}_2} \times 100 = \frac{3.56}{7} \times 100 = 50.9\%$$

The first series is more uniform.

Ex. 20. The fluctuations in the market rates of the shares of two mills during the year 1997 are recorded as follows:

A	:	280,	282,	284,	285,	281,	280,
		284,	285,	284,	288,	286,	285.
B	:	49,	44,	41,	45,	46,	48,
		42,	41,	43,	48,	49,	50.

Which of the two shares shows greater variability?

Ex. 21. The C.V. for each of the two classes of students has been calculated in the following distributions:

Marks	x	Class A f_1	Class B f_2	$u = \frac{x-55}{10}$	f_1u	f_1u^2	f_2u	f_2u^2
30 – 40	35	3	10	– 2	– 6	12	– 20	40
40 – 50	45	8	11	– 1	– 8	8	– 11	11
50 – 60	55	12	23	0	0	0	0	0
60 – 70	65	7	6	1	7	7	6	6
		30	50		– 7	27	– 25	57

$$s_1 = c\sqrt{\frac{\Sigma f_1 u^2}{n} - \left(\frac{\Sigma f_1 u}{n}\right)^2} = 10\sqrt{\frac{27}{30} - \left(\frac{-7}{30}\right)^2} = 9.22 \qquad n_1 \geq 30$$

$$s_2 = c\sqrt{\frac{\Sigma f_2 u^2}{n} - \left(\frac{\Sigma f_2 u}{n}\right)} = 10\sqrt{\frac{57}{50} - \left(\frac{-25}{50}\right)^2} = 11.09 \qquad n_2 \geq 30$$

$$\bar{x}_1 = x_0 + c\,\bar{u}_1 = 55 + 10 \times \left(\frac{-7}{30}\right) = 52.7$$

$$\bar{x}_2 = x_0 + c\bar{u}_2 = 55 + 10 \times \left(\frac{-25}{30}\right) = 46.7$$

$$\text{CV}_1 = \frac{s_1}{\bar{x}_1} \times 100 = \frac{9.22}{52.7} \times 100 = 17.5\%$$

$$\text{CV}_2 = \frac{s_2}{\bar{x}_2} \times 100 = \frac{11.09}{46.7} \times 100 = 23.8\%$$

The class *B* students show more variable results.

20.6 THE VARIANCE

The variance of a set of numbers is the square of the standard deviation

$$s^2 = \frac{\Sigma f(x - \bar{x})^2}{n}$$

We distinguish between sample and population variance by using s^2 for the sample and σ^2 for the population. The sample variance has $n - 1$ in its denominator. ($n \leq 30$)

We use $\bar{x}$ for sample mean and μ for population mean.

We generally use Roman alphabets for sample characteristics and Greek letters for population characteristics.

The Graphical Method to Find Dispersion

The Lorenz curve can be used to find the dispersion in the data graphically.

The extent of departure of an actual curve from the line of equal distribution gives a measure of dispersion. A greater departure implies a greater dispersion and *vice versa*. See Sec. 12.21.

Ex. 22 (1) What is dispersion? Explain the importance of the concept. What are the different measures of dispersion?

(2) Explain and illustrate any three measures of dispersion. Why cannot the arithmetic mean be used as a measure of dispersion? How does the mean deviation or the standard deviation help in this regard?

(3) Distinguish between absolute and relative measures of dispersion.

(4) Distinguish between the measures of central tendency and dispersion with reference to their properties and uses.

(5) Show that the standard deviation is independent of any change of origin but is dependence on the change of scale.

(6) True or false, comment:

(*i*) The coefficient of variation serves the same person as the standard deviation.

(*ii*) Th standard deviation is invariant under change of scale and origin.

(*iii*) If for a group of observations $x_1, x_2, \ldots, x_n$ the mean is $\bar{x}$ and s.d. is s then for the observations $-x_1, -x_2, \ldots, -x_n$, the mean is given by $-\bar{x}$ and s.d. by $-s$.

Ex. 23. Find the s.d. for the following data:

Wages	*Number of men*
5 – 7	15
7 – 9	18
9 – 11	27
11 – 13	10
13 – 15	6

Ans. $s = 2.33$

Ex. 24. A distribution consists of three parts as follows:

Part	*Number of items*	*Mean*	*s.d.*
1	200	5	3
2	250	10	4
3	300	15	5

Show that the mean of the whole distribution is 16 and its s.d. 7.2

Ex. 25. Find the C.V. if for a given data

(*i*) s.d. $= 3.5$, $n = 10$, $\Sigma x = 145$

(*ii*) $s^2 = 148.6$, $\bar{x} = 40$. **Ans.** (*i*) 24.14 (*ii*) 30.50.

Ex. 26. The scores of two batsmen *A*, *B* in 10 innings are given below. Find who is more consistent in scoring.

A	32	28	47	63	71	39	10	60	96	14
B	19	31	48	53	67	90	10	62	40	80.

Ans. $CV_A = 55.4$, $CV_B = 48.8$; *B* is more consistent.

Ex. 27. If the first batsman in the above problem consistently scores 20 runs more in each innings, what will be the conclusion?

Ans. The same conclusion. His inconsistency does not change by an addition of the same number throughout.

Ex. 28. Compare the variability of expenditures of families in two towns as given below:

Expenditure	*No. of families in town A*	*No. of families in town B*
20 – 30	3	2
30 – 40	61	14
40 – 50	132	20
50 – 60	153	27
60 – 70	140	28
70 – 80	51	7
80 – 90	2	2

Ex. 29. For a set of n values of x find n if it is known that

$$\Sigma x^2 = 1360, \Sigma x = 60, ns^2 = 1280.$$

Ans. Use the relation $s^2 = \dfrac{\Sigma x^2}{n} - \left(\dfrac{\Sigma x}{n}\right)^2$; $n = 45$.

Ex. 30. If the average of class A is 10, of class B is 20 and the combined average 16, what is the ratio of the total frequency in class A to that in B? **Ans.** 2 : 3

Some Bases for Comparison of Measures of Dispersion

Simplicity : The range is the simplest to calculate and understand. The s.d., in spite of computational difficulties, is more useful. Other measures lie in between these two in the matter of simplicity.

Sampling Stability : The range varies considerably from sample to sample and is the most unstable in the sampling sense. For unimodal distributions that we usually employ, the s.d. is the most stable.

Description of a Distribution :·A distribution is described, among other things, by measures of central tendency and dispersion. These should go in proper form, *e.g.*, mean with s.d. and median with Q.D.

Type of a Distribution : For an open end distribution, the Q.D. is the best. Other measures are difficult to use. In the usual cases, the s.d. is the best.

Algebraic Manipulations : The s.d. is amenable to algebraic manipulations. As such it is used in advanced statistical work.

The mean, s.d. and variance are central concepts in statistical theory.

20.7 MOMENTS OF A DISTRIBUTION

With the help of moments we can measure the central tendency, variability and other features of a distribution. Moments are denoted by μ_r where r shows the order of the moment.

The Mean is the first moment, written $\mu = \dfrac{1}{n} = \Sigma fx$ (not μ_1)

The variance (s^2) is the second moment. It is written

$$s^2 = \mu_2 = \frac{1}{n}\Sigma f(x - \bar{x})^2$$

The third moment is written $\mu_3 = \dfrac{1}{n}\Sigma f(x - \bar{x})^3$

The fourth moments is written $\mu_4 = \dfrac{1}{n}\Sigma f(x - \bar{x})^4$

In general we have the rth moment $\mu_r = \dfrac{1}{n} \Sigma f(x - \bar{x})^r$

These are moments about the mean. If instead of the mean we take some other number A, we get moments about A.

For probability distributions we may write the corresponding results in the discrete and continuous cases as follows:

$$\text{Mean } \mu = \sum_{i=1}^{n} x_i p(x_i) \quad \text{or} \quad \int_{-\infty}^{\infty} x p(x) dx$$

$$\text{Variance } \mu_2 = \sum_{i=1}^{n} (x_i - \mu)^2 p(x_i) \quad \text{or} \quad \int_{-\infty}^{\infty} (x - \mu)^2 p(x) dx \text{ etc.}$$

See Sec. 20.12.

20.8 SKEWNESS

Skewness refers to lack of symmetry in a distribution. In a symmetrical distribution, the mean, median and mode coincide.

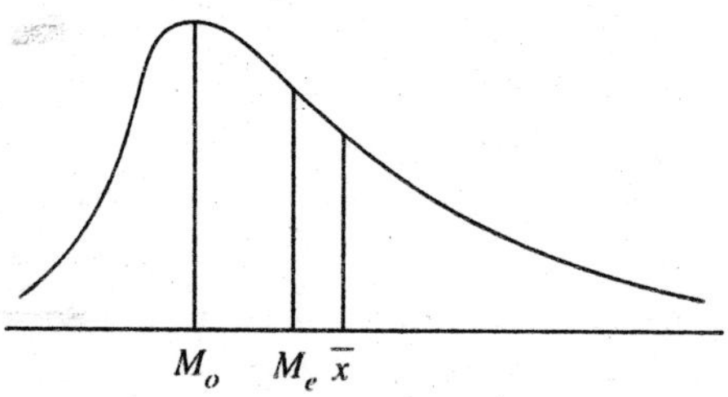

Positive Skewness

Symmetry

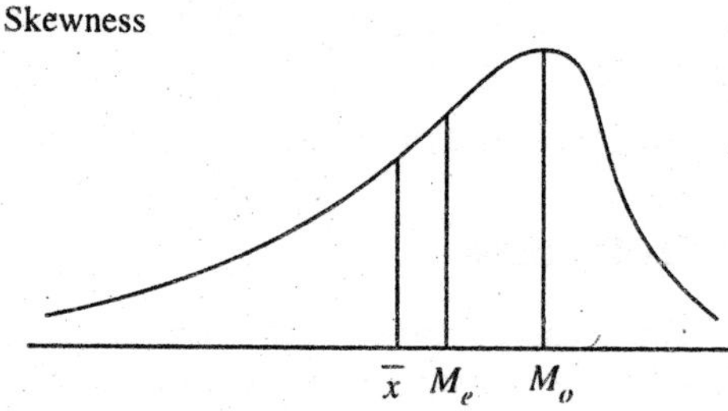

Negative Skewness

In a positively skew distribution the longer tail is on the right side and the mean is on the right of the median.

In a negatively skew distribution the longer tail is on the left and the mean is on the left of the median.

In a skew distribution the distance between the mean and median is nearly one-third of that between the mean and mode.

How to check the presence of skewness in a distribution ?

Skewness is present in the data

(*i*) if the graph is not symmetrical

(*ii*) if the mean, median and mode do not coincide

(*iii*) if the quartiles are not equidistant from the mean

(*iv*) if the sum of positive and negative deviations from the median is not zero

(*v*) if frequencies are not similarly distributed on either side of the mode.

Measures of Skewness

A measure of skewness gives a numerical expression for and the direction of asymmetry in a distribution. It gives information about the shape of the distribution and the degree of variation on either side of the central value.

We consider some relative measures of skewness.

(*i*) Pearson's Coefficient of Skewness

$$PSk = \frac{\bar{x} - Mo}{s} = \frac{3(\bar{x} - Me)}{s}$$

It may have any value. But usually it lies between –1 and +1.

Ex. 31. If for a given data it is found that

$\bar{x}$ = 10, Mode = 8, s = 4, we have

$$PSk = \frac{\bar{x} - Mo}{s} = \frac{10 - 8}{4} = 0.5$$

(*ii*) Bowley's Coefficient of Skewness

$$BSk = \frac{Q_3 + Q_1 - 2Me}{Q_3 - Q_1}$$

Its value lies between –1 and +1.

Ex. 32. If for a given data $Q_1 = 2$, $Q_3 = 8$, $Me = 5$

$$BSk = \frac{Q_3 + Q_1 - 2Me}{Q_3 - Q_1} = \frac{8 + 2 - 5}{8 - 2} = 0.83.$$

(*iii*) Kelley's Coefficient of Skewness

$$KSk = P_{50} - \tfrac{1}{2}(P_{10} + P_{90})$$

where P_{10}, P_{50} and P_{90} are the 10th, 50th, and 90th percentiles of the data.

(*iv*) *Method of Moments* : If μ_2, μ_3 are moment about the mean we have the

$$\text{Coefficient of Skewness } \beta_1 = \frac{\mu_3^2}{\mu_2^3} = \mu_3^2/\sigma^6$$

Sometimes we define the coefficient of skewness

$$\gamma_1 = \sqrt{\beta_1} = \sqrt{\frac{\mu_3^2}{\mu_2^3}} = \frac{\mu_3}{\sigma^3}$$

20.9 KURTOSIS

Kurtosis is a measure of peakedness of a distribution. It shows the degree of convexity of a frequency curve.

If the normal curve is taken as the standard, symmetrical, bell shaped curve, kurtosis gives a measure of departure from the normal convexity of a distribution.

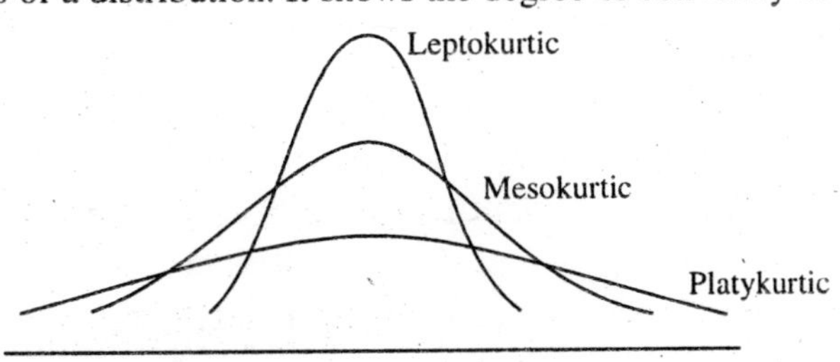

The normal curve is *mesokurtic.* It is of intermediate peakedness.

The flat-topped curve, broader than the normal, is termed *platykurtic.*

The slender, highly peaked curve is termed *leptokurtic.*

Measures of Kurtosis

(*i*) Moment Coefficient of Kurtosis : $\beta_2 = \dfrac{\mu_4}{\mu_2^2}$.

Instead of β_2 we often use $\gamma_2 = \beta_2 - 3$ which is positive for a leptokurtic distribution, negative for a platykurtic distribution and zero for the normal distribution.

Percentile Coefficient of Kurtosis $k = \dfrac{Q}{P_{90} - P_{10}}$

where $Q = \frac{1}{2}(Q_3 - Q_1)$ is the semi interquartile range.

Ex. 33. Find the skewness and kurtosis for the following distribution by the method of moments:

Number of hours worked	*Number of days*
1 – 3	3
3 – 5	5
5 – 7	1
7 – 9	1

For the data we have $\bar{x} = \dfrac{\Sigma fx}{n} = \dfrac{40}{10} = 4$

Write $d = x - \bar{x}$

x	f	d	d^2	fd^2	fd^3	fd^4
2	3	– 2	4	12	– 24	48
4	5	0	0	0	0	0
6	1	2	4	4	8	16
8	1	4	16	16	64	256
Total	10			32	48	320

$$s^2 = \mu_2 = \frac{1}{n}\Sigma fd^2 = \frac{1}{10} \times 32 = 3.2 \quad \therefore \; s = 1.79$$

$$\mu_3 = \frac{1}{n}\Sigma fd^3 = \frac{1}{10} \times 48 = 4.8$$

$$\mu_4 = \frac{1}{n}\Sigma fd^4 = \frac{1}{10} \times 320 = 32.0$$

Skewness : $\beta_1 = \dfrac{\mu_3^2}{\mu_2^3} = \dfrac{4.8^2}{3.2^3} = \dfrac{9}{128} = 0.0703$

Kurtosis : $\beta_2 = \dfrac{\mu_4}{\mu_2^2} = \dfrac{32.0}{3.2^2} = 3.125$

Alternatively $\gamma_1 = \sqrt{\beta_1} = \sqrt{0.0703} = 0.265$

$\gamma_2 = \beta_2 - 3 = 0.125$

The skewness is positive and the distribution is leptokurtic as $\gamma_2 > 0$.

Ex. 34. Find the skewness and kurtosis for the above problem by other methods. Compare the results (which need not be the same).

Ex. 35. Show that the following distribution is platykurtic. Find the kurtosis by any method known to you.

x:	10	11	12	13	14	15	16
f:	5	10	20	30	20	10	5

Ans. $\beta_2 = \mu_4/\mu_2^2 = 265$

Ex. 36. Calculate the mean, median and s.d. for the following data and find its skewness:

Class interval	*Frequency*
50 – 60	4
60 – 70	16
70 – 80	36
80 – 90	52
90 – 100	64
100 – 110	40
110 – 120	32
120 – 130	11

Ans. First find $\bar{x} = 53$, $Me = 43$, $s = 16.1$, $Sk = \dfrac{3(\bar{x} - Me)}{s} = \dfrac{30}{16.1} = 18.5$.

Ex. 37. Find the quartiles in the above problem and hence determine the coefficient of skewness.

Ex. 38. Comment on the following statements:

(*i*) Measures of variation help us in studying some important characteristics of a distribution.

(*ii*) The range is not based on the whole data.

(*iii*) Quartile deviation ignores 50% of the data.

(*iv*) Standard deviation is superior to mean deviation.

(*v*) Averages, measures of dispersion, skewness and kurtosis are complementary to one another in understanding a frequency distribution.

Ex. 39. What are the various measures of dispersion of a distribution? Discuss their relative advantages. Why is the s.d. regarded as superior to other measures of dispersion?

20.10 EXPECTATION OR EXPECTED VALUE

If x is a discrete variable assuming discrete values

$$x_1, x_2, \ldots, x_n$$

with probabilities $p_1, p_2, \ldots, p_n$

we define the Expectation of x

$$E(x) = x_1p_1 + x_2p_2 + \ldots + x_np_n = \sum_{i=1}^{x} x_i p_i$$

$$= \Sigma xp$$

Write $p_i = \frac{f_i}{n}$, the relative frequency. Then

$$E(x) = \Sigma x_i \frac{f_i}{n} = \frac{\Sigma f_i x_i}{n} = \bar{x}, \text{ arithmetic mean}$$

$E(x)$ is the mean of the probability distribution of X

In general for any function $g(x)$

$$Eg(x) = \Sigma g(x)\, p(x)$$

If k is a constant, $E(k) = k$

$$E(kx) = kE(x)$$

$$E(x - \bar{x}) = E(x) - E(\bar{x}) = \bar{x} - \bar{x} = 0$$

Expectation of a sum

If there are two or more variables,

The expectation of their sum is the sum of their expectations.

$$E(x + y) = E(x) + E(y)$$

$$E(x + y + z) = E(x) + E(y) + E(z)$$

$$E(ax + by) = aE(x) + bE(y)$$

$$E(x - y) = E(x) - E(y)$$

$$E(-x) = -E(x)$$

$$|E(x)| \leq E(|x|) \text{ since } -x \leq |x|,\ x \leq |x|$$

Ex. 40. If $x = x_1 + x_2 + \ldots + x_u$, $E(x) = E(x_1) + E(x_2) + \ldots + E(x_u)$

Ex. 41. The expectation of the sum of points on a die is

$$E(x = 1) + E(x = 2) + \ldots + E(x = 6) = \frac{1}{6} \cdot 1 + \frac{1}{6} \cdot 2 + \frac{1}{6} \cdot 3 + \frac{1}{6} \cdot 4 + \frac{1}{6} \cdot 5 + \frac{1}{6} \cdot 6 = \frac{7}{2}$$

Expected Value of a Product

If x, y are independent random variables assuming values $x_1, x_2, \ldots, x_n$ with probabilities $p_1, p_1, \ldots, p_n$ and values $y_1, y_2, \ldots, y_n$ with probabilities $p_1', p_2', \ldots, p_n'$ the expectation of their product is the product of their expectation $E(xy) = E(x)\, E(y)$

Similarly $E(xyz) = E(x)\, E(y)\, E(z)$

Ex. 42. Find the joint expectation of (1) the sum of numbers x on a normal die and (2) the following

y	– 1	0	1	2
p_i	1/3	1/6	1/6	1/3

1. For a die $E(x) = \frac{1}{6} \times 1 + \frac{1}{6} \times 2 + \ldots + \frac{1}{6} \times 6 = \frac{7}{2}$

2. In this case, $E(y) = \frac{1}{3}(-1) + \frac{1}{6}(0) + \frac{1}{6}(1) + \frac{1}{3}(2) = \frac{1}{2}$

$$E(xy) = E(x)\,E(y) = \frac{7}{2} \cdot \frac{1}{2} = \frac{7}{14}$$

Variance

$$\begin{aligned} V(x) &= E(x - \bar{x})^2 = E(x^2) - 2E(x\bar{x}) + E\bar{x}^2 \\ &= E(x^2) - 2\bar{x}\,E(x) + \bar{x}^2 \\ &= Ex^2 - 2\bar{x}^2 + \bar{x}^2 = E(x^2) - \bar{x}^2 \\ &= E(x^2) - \{E(x)\}^2 \end{aligned}$$

$V(ax) = a^2V(x)$, $E(bx) = bE(x)$, $E(a + bx) = a + bE(x)$

$V(a + bx) = b^2V(x) = V(a - bx)$ since $V(a) = 0$ (A constant has no variance)

$V(x + y) = V(x) + V(y) = V(x - y)$ if x, y are independence random variables

$V(ax + by) = a^2V(x) + b^2V(y)$ if x, y are independent.

$= a^2V(x) + b^2V(y) + 2ab\,\text{Cov}(x, y)$ if x, y are not independent.

Ex. 43. A box contains 8 tickets. 3 tickets each carry a prize of Rs. 5, the remaining 5 each carry a prize of Rs. 2 (*i*) If one ticket is drawn what is the expectations of the prize? (*ii*) If two are drawn, what is the expectation? **Ans.** 3.125, 6.25

Covariance

The covariance between random variables x, y is defined as

$$\text{Cov.}(x, y) = E(xy) - E(x)\,E(y)$$

$$\text{Cov.}(aX, bY) = ab\,\text{Cov}(X, Y)$$

Correlation between X, Y

$$\text{Corr}(x, y) = \frac{\text{Cov}(x, y)}{\sqrt{V(x)}\sqrt{V(y)}}$$

Standard Deviation of X

$$\text{S.D.}(x) = \sqrt{\text{Var}(x)} = \sqrt{V(x)}$$

20.11 EXPECTATION : CONTINUOUS VARIABLES

If x is a continuous random variable we define

$$E(x) = \int_{-\infty}^{\infty} x\,p(x)\,dx = \mu$$

where $p(x)$ is the density function of x

The expectation of a function $h(x)$ is

$$Eh(x) = \int_{-\infty}^{\infty} h(x)\, p(x)\, dx$$

The rth moment about the mean is

$$E(x-\mu)^r = \int_{-\infty}^{\infty} (x-\mu)^r p(x)\, dx$$

Ex. 44. A newspaper earns Rs. 100 a day if there is suspense in the news. He loses Rs. 10 a day if it is an eventless newspaper. What is the expectation of his earnings if the probability of suspense news is 0.4?

$$\begin{aligned} E(x) &= p_1x_1 + p_2x_2 \\ &= 0.4 \times 100 - 0.6 \times 10 \\ &= 40 - 6 = 34 \end{aligned}$$

Ex. 45. A player tossing three coins, earns Rs. 10 for 3 heads, Rs. 6 for 2 heads and Re. 1 for 1 head. He loses Rs. 25, if 3 tails appear. Find his expectation.

Probability of 3 heads is given by

$$p(HHH) = \frac{1}{2}\cdot\frac{1}{2}\cdot\frac{1}{2} = \frac{1}{8} = p_1 \text{ say}$$

$$p(HHT) = {}^3C_2 \cdot \frac{1}{2}\cdot\frac{1}{2}\cdot\frac{1}{2} = \frac{3}{8} = p_2 \text{ say} \qquad \text{2 heads, 1 tail}$$

$$p(HTT) = {}^3C_1 \cdot \frac{1}{2}\cdot\frac{1}{2}\cdot\frac{1}{2} = \frac{3}{8} = p_3 \text{ say} \qquad \text{1 head, 2 tails}$$

$$p(TTT) = \frac{1}{2}\cdot\frac{1}{2}\cdot\frac{1}{2} = \frac{1}{8} = p_4 \text{ say} \qquad \text{3 tails}$$

$$\begin{aligned} E(x) &= p_1x_1 + p_2x_2 + p_3x_3 - p_4x_4 \\ &= \frac{1}{8}\times 10 + \frac{3}{8}\times 6 + \frac{3}{8}\times 2 - \frac{1}{8}\times 25 \\ &= \frac{9}{8} = \text{Rs. } 1.125 \end{aligned}$$

Ex. 46. A and B roll a die. Whoever gets a 6 first wins Rs. 550. Find their individual expectations if A makes a start. What will be the answer if B makes a start?

A may get 6 in 1st trial, $p_1 = \dfrac{1}{6}$

A may not get in 1st, B may not get in 2nd and A may get in 3rd,

$$p_3 = \frac{5}{6}\times\frac{5}{6}\times\frac{1}{6} = \left(\frac{5}{6}\right)^2 \frac{1}{6} \quad \text{and so on}$$

$$A\text{'s winning chance} = \frac{1}{6} + \left(\frac{5}{6}\right)^2 \frac{1}{6} + \left(\frac{5}{6}\right)^4 \frac{1}{6} + \ldots\ldots$$

$$= \frac{1}{6} \cdot \frac{1}{1-\left(\frac{5}{6}\right)^2} = \frac{6}{11}$$

$$\left(\text{Geometric progression with common ratio } \left(\frac{5}{6}\right)^2\right)$$

$\therefore$ B's winning chance $= 1 - \frac{6}{11} = \frac{5}{11}$

A wins Rs. 550 with probability $p = \frac{6}{11}$

A gets nothing if he loses with prob. $q = \frac{5}{11}$

Expectation of $A = p \cdot x + q \cdot 0$

$$= \frac{6}{11} \times 550 + \frac{5}{11} \times 0 = 300$$

Similarly B wins with $p = \frac{5}{11}$

B gets nothing if he loses with $q = \frac{6}{11}$

Expectation of $B = \frac{5}{11} \times 550 + \frac{6}{11} \times 0 = 250$

If B starts first his expectation would be 300 and A's would be 250.

Ex. 47. A box contains 15 bulbs of which 5 are defectives. A person selects 3 bulbs at random. Find the expected number of defective bulbs. **Ans.** 1

Ex. 48. If it rains a person can earn Rs. 100 per day by hiring out his vehicle. If it is fair weather he loses Rs. 10. The probability of raining is 0.4. Find his expectation.

Ans. 34

Ex. 49. x takes the values 0, 1, 3, 9 with probability 0.4, 0.2, 0.3, 0.1.

$$\mu = E(x) = \Sigma(x_i p_i) = 0. \times 0.4 + 1 \times 0.2 + 3 \times 0.3 + 9 \times 0.1$$

$$= 2.0$$

$$E(x^2) = \Sigma x_i^2 p_i = 0^2 \times 0.4 + 1^2 \times 0.2 + 3^2 \times 0.3 + 9^2 \times 0.1$$

$$= 11.0$$

$$V(x) = E(x^2) - \mu^2 = 11 - 2 = 9$$

$$\text{s.d.}(x) = \sqrt{9} = 3.$$

Ex. 50. There are 3 tickets numbered 0, 2, 3. One ticket is selected and replaced; another ticket is selected and replaced. A third ticket is selected and replaced once again. If X stands for the sum of the 3 ticket numbers, construct the probability distributive of X and find μ and σ^2.

Ex. 51. The purchase of some shares can give a profit of Rs. 400 with probability 1/100 and Rs. 300 with probability 1/20. Comment on a fair price of the share.

Expected value $E(x) = \Sigma x_i p_i = 400 \times \frac{1}{100} + 300 \times \frac{1}{20} = 19.$

Ex. 52. A car insurance policy offers Rs. 1000 after an accident the probability of whose occurrence is 0.04. If the expected gain is to be zero what should be the premium?

Ans. Rs. 41

Ex. 53. Find the variance and standard deviation of the probability distribution

x_i	1	2	3	4
p_i	0.1	0.3	0.2	0.4

$$E(x) = \Sigma p_i x_i = 0.1 \times 1 + 0.3 \times 2 + 0.2 \times 3 + 0.4 \times 4$$
$$= 2.9$$

$$\text{Variance } (x) = V(x) = E(x^2) - \bar{x}^2$$
$$= \Sigma p_i x_i^2 - \bar{x}^2 = 0.1 \times 1^2 + 0.3 \times 2^2 + 0.2 \times 3^2 + 0.4 \times 4^2 - 2.9^2$$
$$= 0.1 + 1.2 + 1.8 + 6.4 - 8.41 = 1.09$$

$$\text{S.D. } (x) = \sqrt{V(x)} = \sqrt{1.09} = 1.044$$

Ex. 54. The variance of a constant is zero.

If k is a constant, it has no variability.

$$V(k) = 0$$

If the constant k is attached to a variable x then

$$V(kx) = k^2 V(x)$$
$$V(2x) = 4V(x)$$
$$V(2 + 3x) = V(3x) = 9\ V(x)$$

Ex. 55. For the probability distribution

x	− 1	− 2	3	7.5	8
p	0.2	0.15	0.5	0.1	0.05

$E(x) = 2.15$ $V(x) = 9.5$

Ex. 56. In a queue the probability of the number of people joining per minute is given below.

Number of persons joining the queue	0	1	2	3
Probability	0.4	0.3	0.2	0.1

What is the expected number of persons joining the queue?

Ans. 1

Ex. 57. A box contain 2^n tickets of which ${}^nC_i$ tickets bear the number i ($i = 1, 2, 3,\ldots, n$). If a set of m tickets is drawn find the expected value of the sum of their numbers.

To find $E(A)$ where $A = x_1 + x_2 + \ldots + x_m$. Here x_i can take values 0, 1, 2,..., n with probabilities

$$\frac{{}^nC_0}{2^n}, \frac{{}^nC_1}{2^n}, \ldots, \frac{{}^nC_n}{2^n}$$

$$E(x_i) = \Sigma p_i x_i = \frac{{}^nC_0}{2^n}\cdot 0 + \frac{{}^nC_1}{2^n}\cdot 1 + \frac{{}^nC_2}{2^n}\cdot 2 + \ldots + \frac{{}^nC_n}{2^n}\cdot n$$

$$= \frac{1}{2^n}[{}^nC_1 + 2\,{}^nC_2 + \ldots + n\,{}^nC_n]$$

$$= \frac{n}{2^n}[1 + {}^{n-1}C_1 + {}^{n-1}C_2 + \ldots + {}^{n-1}C_{n-1}] \text{ (take } n \text{ common)}$$

$$= \frac{n}{2^n}(1+1)^{n-1} = \frac{n}{2} \text{ [Expand } (1+1)^{n-1} \text{ and check]}$$

$$\therefore \quad E(A) = \sum_{i=1}^{m} E(x_i) = \frac{mn}{2}$$

20.12 MORE ON MOMENTS

The expected value of a vandom variable X about a point m is called the moment of that variable about m. This is the first moment about A and written

$$\mu_1' = E(x - A)$$

The kth moment about A is written

$$\mu_k' = E(x - A)^k$$

Moments are often found about the mean μ. the kth moment about the mean is

$$\mu_k = E(x - \mu)^k \text{ (Not } \mu_k'\text{)}$$

The second moment about the mean is

$$\mu_2 = E(x - \mu)^2$$

This can be seen to be the variance.

The third moment about the mean can be used as a measure of skewness:

$$\mu_3 = E(x - \mu)^3$$

The fourth moment about the mean can be used as a measure of kurtosis.

$$\mu_4 = E(x - \mu)^4 \text{ and so on.}$$

These are all *central moments* since they are moments about the mean.

The kth moments about any other value is written μ_k' (as distinguished from μ_k)

The first moment about the mean is zero.

$$\mu_1 = E(x - \mu) = E(x) - \mu = 0$$

i.e., $\quad E(x) = \mu$

Moments can be used to describe a distribution.

The rth moment about any point A is written

$$\mu_r' = \frac{1}{n}\sum_{i=1}^{k} f_i(x_i - A)^r$$

The rth moment about the mean μ is written

$$\mu_r = \frac{1}{n}\Sigma f(x-\mu)^r$$

Note : $\mu_0' = \frac{1}{n}\Sigma f(x-A)^\circ = \frac{1}{n}\Sigma f = \frac{n}{n} = 1$

Since $\mu_1 = \frac{1}{n}\Sigma f(x-\mu) = \frac{1}{n}\Sigma fx - \frac{1}{n}\Sigma f\,\mu = 0$

where $\frac{1}{n}\Sigma fx = \mu$

$$\therefore \quad \mu_1' = \frac{1}{n}\Sigma f(x-A) = \frac{1}{n}\Sigma fx - \frac{A}{n}\Sigma f$$

$$= \mu - A = d \text{ say.}$$

If $A = 0$, *i.e.*, the moments are taken about zero,

$$\mu_1' = \frac{1}{n}\Sigma fx = \mu, \text{ the arithmetic mean}$$

The second moment about the mean is

$$\mu_2 = \mu_2' - 2\mu^2 + \mu^2 = \mu_2' - \mu^2 = \mu_2' - \mu_1'^2, \text{ the variance}$$

It can be shown that the third and fourth moments are

$$\mu_3 = \mu_3' - 3\mu_1'\mu_2' + 2\mu_1'^3$$

$$\mu_4 = \mu_4 - 4\mu_3'\mu_1' + 6\mu_2'\mu_1'^2 - 3\mu_1'^4$$

Ex. 50. Define the measures of central tendency for a continuous distribution $p(x)$

(*a*) A.M. : $\bar{x} = \mu_1' = E(x) = \int_{-\infty}^{\infty} x p(x)\,dx$

(*b*) G.M. : G defined by $\log G = E(\ln x) = \int_{-\infty}^{\infty} \ln x\, p(x)dx$

(*c*) H.M. : H defined by $\frac{1}{H} = E\left(\frac{1}{x}\right) = \int_{-\infty}^{\infty} \frac{1}{x} p(x)\,dx$

(*d*) Mode is that value of x for which p is the maximum

(*e*) Median M divides total frequency into two equal parts *i.e.*, $p(x \geq M) = p(x \leq M)$

or $$\int_{-\infty}^{M} p(x)\,dx = \int_{M}^{\infty} p(x)\,dx = \frac{1}{2}\int_{-\infty}^{\infty} p(x)\,dx = \frac{1}{2}$$

(*f*) First Quartile Q_1 : $\int_{-\infty}^{Q_1} p(x)\,dx = \frac{1}{4}$

(*g*) Third Quartile Q_3 : $\int_{Q_3}^{\infty} p(x)\,dx = \frac{1}{4}$

Ex. 59. Define the measures of dispersion for a continuous probability distribution.

The mean deviation from the mean $\bar{x}$ is defined by

1. $MD = \int_{-\infty}^{\infty} |x - \bar{x}| f(x) dx$

2. The variance is defined by

$$\mu_2 = \int_{-\infty}^{\infty} (x - \bar{x})^2 f(x) dx$$

3. The *r*th moment about the mean is

$$\mu_r = \int_{-\infty}^{\infty} (x - \bar{x})^r f(x) dx$$

Ex. 60. For $f(x) = e^{-x}$, $0 \le x \le \infty$

$$\mu_r' = \int_0^{\infty} x e^{-x} dx = [-xe^{-x} + e^{-x}]_0^{\infty} = 1$$

The *r*th moment about the mean (*i.e.*, 1 in this case) is $\int_0^{\infty} (x-1)^r e^{-x} dx$ which can be obtained by expanding $(x - 1)^r$.

For instance $\mu_2 = \int_0^{\infty} (x-1)^2 e^{-x} dx$

$$\therefore \quad \mu_2 = \int_0^{\infty} x^2 e^{-x} dx - 2\int_0^{\infty} x e^{-x} dx + \int_0^{\infty} e^{-x} dx$$

$$= [-x^2 e^{-x} + 2(-xe^{-x} + e^{-x})]_0^{\infty} - 2.1$$

$$= 2 - 2 + 1 = 1.$$

CHAPTER 21

Distributions I: Some Theoretical Distributions

21.1 THEORETICAL DISTRIBUTIONS

The frequency distributions in the form of data considered earlier refer to samples drawn from a population. Measures of central tendency and dispersion etc. are the main features of a frequency distribution but their knowledge is not sufficient if complete information about the distribution is needed. If we take a lrrge number of observations of an actual distribution, the law of that distribution can be established. The variate values follow some mathematical law which can be found.

It is possible to fit such a theoretical distribution to the given observations if its characteristics and laws are known. We may be able to discover additional features of the observed values with its help.

In statistics we come across many distributions either experimentally or theoretically.

The most important theoretical distributions are:

(*i*) The Binomial Distribution

(*ii*) The Poisson Distribution

(*iii*) The Normal Distribution

21.2 THE BINOMIAL DISTRIBUTION

A binomial distribution is one is which the frequencies are proportional to the successive terms of the binomial expansion. This theoretical distribution is very useful in practice. It can be applied to a number of commonly occurring situations.

We know that the probability of r successes in n repeated trials is given by

$$^nC_r p^r q^{n-r}$$

Here p is the probability of success

q is the probability of failure; $q = 1 - p$

Prob. of r successes, p^r must necessarily accompany the

Prob. of $n - r$ failures *i.e.*, q^{n-r}. This is possible in ^nC_r ways

See Repeated Trials in Chapter on *Probability*.

Note. The binomial distribution is discrete in which the random variable r assumes the values 0, 1, 2,..., n where

(*i*) n is finite

(*ii*) trials are repeated n times

(*iii*) Either a success or a failure results in each trial.

(*iv*) Prob. of success in each trial is constant.
It does not change from trial to trial.

(*v*) The trials are independent
If the independent trials are repeated k times, the expected frequency of r successes is

$$k\,{}^nC_r\, p^r q^{n-r}$$

In n trials we can write the probability of

0	success	$P(0) = {}^nC_0 p^0 q^n$
1	...	$P(1) = {}^nC_1 p^1 q^{n-1}$
2	successes	$P(2) = {}^nC_2 p^2 q^{n-2}$
3	...	$P(3) = {}^nC_3 p^3 q^{n-3}$
...	...	... ...
r	...	$P(r) = {}^nC_r p^{n-r}$
...	...	... ...
n	...	$P(n) = {}^nC_r p^n q^0$

These terms are the successive terms of the binomial expansion

$$(q+p)^n = C_0 p^0 q^n + {}^nC_1 p^1 q^{n-1} + {}^nC_2 p^2 q^{n-2} + \ldots + {}^nC_n p^n q^0$$

$$= \sum_{r=0}^{n} {}^nC_r p^r q^{n-r}$$

The successive terms on the right show the probabilities of 0, 1, 2,..., n successes.

Definition. If an event E has probability p of occurring in each of n independent trials then the probability that it will occur *exactly* r times in n trials is given by

$$f(r) = {}^nC_r\, p^r q^{n-r}$$

This probability distribution is called the *binomial probability distribution.* The binomial distribution is a discrete distribution with parameters n and p. If p, q are equal it is symmetrical, otherwise it is nonsymmetrical. Some characteristics features of the binomial distribution are given below:

Mean $\mu = np$ (it coincides with the greatest term of the Binomial distribution)

Standard deviation $\sigma = \sqrt{npq}$

Variance $\sigma^2 = npq$

Coefficient of skewness $\sqrt{\beta_1} = \dfrac{q-p}{\sqrt{npq}}$, $\beta_1 = \dfrac{(q-p)^2}{npq}$

Coefficient of kurtosis $\beta_2 = 3 + \frac{1-6pq}{npq}$, $\gamma_2 = \beta_2 - 3 = \frac{1-6pq}{npq}$

The binomial distribution is useful in sampling theory and tests of significance.

Ex. 1. A machine is known to be producing 20 per cent defective items. Construct the probability distribution for 5 items selected at random from the production of this machine.

This is a case of the binomial distribution with $p = 0.2$, $q = 0.8$, $n = 5$. We know that the probability of x successes out of 5 is given by ${}^5C_x\, p^x q^{n-x}$.

Binomial Probability Distribution:

$$p(x = 0) = {}^5C_0\,(0.2)^0\,(0.8)^5 = 0.3277$$
$$p(x = 1) = {}^5C_1\,(0.2)^1\,(0.8)^4 = 0.4096$$
$$p(x = 2) = {}^5C_2\,(0.2)^2\,(0.8)^3 = 0.2048$$
$$p(x = 3) = {}^5C_3\,(0.2)^3\,(0.8)^2 = 0.0512$$
$$p(x = 4) = {}^5C_4\,(0.2)^4\,(0.8)^1 = 0.0064$$
$$p(x = 5) = {}^5C_5\,(0.2)^5\,(0.8)^0 = 0.0003$$

Total 1

If the experiment is repeated 10000 times we can calculate the expected frequencies obtained by multiplying the individual probabilities by 10000. Thus $0.3277 \times 10000 = 3277$ etc. We have the following frequency distribution, showing expected frequencies:

x	$p(x)$	*expected frequency f*
0	.3277	3277
1	.4096	4096
2	.2048	2048
3	.0521	512
4	.0064	64
5	.0003	3
Total	1.0000	10000

The mean and variance of this binomial distribution can be easily found.

Mean $\mu = np = 5 \times 0.2 = 1$

i.e., 1 defective is expected in every 5.

Variance $\sigma^2 = npq = 5 \times 0.2 \times 0.8 = 0.8$

Standard deviation $\sigma = \sqrt{npq} = \sqrt{0.8} = 0.894$.

Ex. 2. In the above problem find the probability of getting (*i*) exactly one defective (*ii*) at least 3 (*iii*) all (*iv*) at most 3 (*v*) no defectives.

Ans. (*i*) $p(1) = 0.4096$. (*ii*) At least 3 means 3, 4 or 5. The required probability is $p(3) + p(4) + p(5) = .0512 + .0064 + .0003 = 0.0579$ (*iii*) $p(5) = 0.0003$. (*iv*) $p(0) + p(1) + p(2) + p(3) = 0.9933$ (*v*) $p(0) = 0.3277$.

Ex. 3. A die is rolled thrice and a 5 or 6 is considered a success. Find the probability of (*i*) no success (*ii*) at least two successes (*iii*) at least one but not more than 2 successes. **Ans.** (*i*) $\frac{8}{27}$ (*ii*) $\frac{7}{27}$ (*iii*) $\frac{18}{27}$

Ex. 4. The probability is $\frac{2}{5}$ that a batsman scores 100 runs in a cricket match. What is the probability that in 5 matches the batsman will score 100 runs in at least 3 matches? **Ans.** 0.317

Ex. 5. A variate takes values 0, 1, 2,..., n with frequencies proportional to the binomial coefficients $^nC_0$, $^nC_1$, $^nC_2$,..., ^nC_n find the mean and variance of the distribution.

Total frequency = $^nC_0 + {^nC_1} + \ldots + {^nC_n} = (1 + 1)^n = 2^n$

Probabilities of 0, 1, 2,..., n values are $\frac{^nC_0}{2^n}, \frac{^nC_1}{2^n}, \frac{^nC_2}{2^n}, \ldots, \frac{^nC_n}{2^n}$

These can be written $^nC_0\left(\frac{1}{2}\right)^0\left(\frac{1}{2}\right)^n, {^nC_1}\left(\frac{1}{2}\right)^1\left(\frac{1}{2}\right)^{n-1}, \ldots, {^nC_n}\left(\frac{1}{2}\right)^n\left(\frac{1}{2}\right)^0$

This is a binomial expansion of $(p + q)^n$ where $p = q = \frac{1}{2}$

$\therefore$ Mean $= np = n \times \frac{1}{2} = \frac{n}{2}$ variance $= upq = \frac{n}{4}$

Ex. 6. One half of the population of a town consists of smokers. 100 investigators each check 10 persons for smokers. How many investigators may be expected to find that 3 or less people are smokers?

Probability of anyone being a smoker $= \frac{1}{2}$

Probability of 3 or less being found smokers is

$p(x \le 3) = p(x = 0) + p(x = 1) + p(x = 2) + p(x = 3)$

$$= {^{10}C_0}\left(\frac{1}{2}\right)^{10} + {^{10}C_1}\left(\frac{1}{2}\right)^1\left(\frac{1}{2}\right)^9 + {^{10}C_2}\left(\frac{1}{2}\right)^2\left(\frac{1}{2}\right)^8 + {^{10}C_3}\left(\frac{1}{2}\right)^3\left(\frac{1}{2}\right)^7$$

$$= \frac{1}{2^{10}} \times 176 = \frac{176}{1024}$$

The number reporting 3 or less $= 100 \times \frac{176}{1024} \simeq 17$

Ex. 7. If $p = 0.01$ how many trials are necessary in order that the probability of at least one success is greater than $\frac{1}{2}$? **Ans.** 69

Ex. 8. The incidence of a disease in an industry is such as to create a 20% chance of suffering from it. What is the probability that out of 6 workers 4 get the disease?

Ans. $\frac{53}{3125}$

Ex. 9. Bring out the fallacy: The mean of a Binomial is 5 and its s.d. is 3.

$np = 5$ and $npq = 3^2 = 9$ $\therefore$ $q = \frac{9}{5} > 1$ which is not possible

21.3 THE POISSON DISTRIBUTION

When the probability of success of an event is very small *i.e.*, the event is rare in a large number of trials over a continuum of time, the resulting distribution is known as Poisson distribution. The number of deaths from horsekicks in battles, the number of printing errors per page in a book, the number of blind children born in a city etc., provide examples of the Poisson distribution which has many uses in industry, demography and elsewhere.

Poisson distribution is a discrete distribution given by

$$p(x) = \frac{m^x e^{-m}}{x!}$$

where $m = np$ is the mean number of successes. This distribution can be obtained as a limit of the binomial (B) when n is very large, p is very small and $np = m$ is small.

$$\lim_{\substack{n\to\infty \\ p\to 0}} B \xrightarrow[\text{const.}]{np} \text{Poisson}$$

As the distribution describes the behaviour of rare events it is called *the law of improbable events.* There is only one parameter m in the distribution which does not involve n at all. The distribution is skew.

Some characteristic features of the Poisson distribution are:

Mean $\mu = m$

Variance $\sigma^2 = m$ (*Note* : mean $\simeq$ variance)

$\mu_3 = m$ and $\mu_4 = 3m^2 + m$

Standard deviation $\sigma = \sqrt{m}$

Coefficient of skewness $\gamma_1 = 1/\sqrt{m}$

Coefficient of kurtosis $\beta_2 = 3 + \frac{1}{m}$, $\gamma_2 = \beta_2 - 3 = \frac{1}{m}$

Mode = Integral number between $m - 1$ and m if m is not an integer

= $m - 1$ and m if m integer (two modes)

Note: 1. $\gamma_1\gamma_2\, \sigma m = 1$

2. The Poisson is the limiting from of the Binomial when $n \to \infty$, $p \to 0$, $np = m$. However, this reference to its limiting property is not necessary since it has numerous independent applications as in queuing theory.

3. Mean $\simeq$ variance.

Ex. 10. If 3 per cent items manufactured by a machine are defective, find the probability that 3 items out of 100 taken at random will be defective.

Here it is preferable to use Poisson rather than the binomial distribution since $p(= 0.03)$ is very small, n is large and $m = np = 100 \times .03 = 3$

$$p(x = 3) = \frac{m^x e^{-m}}{x!} = \frac{3^3 e^{-3}}{3!} = 0.2241.$$

Find the probability of (*i*) at the most 2 defectives (*ii*) at least 3 defectives (*iii*) no defective (*iv*) not less than 2 defectives.

[Calculate (*i*) $p(0) + p(1) + p(0)$ (*ii*) $1 - p(0) - p(1) - p(2)$

(*iii*) $p(0)$ (*iv*) $1 - [p(0) - p(1)]$

Ex. 11. There is a chance of 1/500 of a blade being defective in each packet of 10. Find the number of packets containing 0, 1, 2 defectives in a consignment of 10000 blades.

$$p = \frac{1}{500}, \quad n = 10, \quad m = np = \frac{10}{500} = \frac{1}{50}, \quad e^{-m} = 0.9802$$

$$p(0 \text{ defective}) = \frac{e^{-0.02}(0.02)^0}{0!} = 0.9802$$

$$p(1) = \frac{e^{-.02}(.02)^1}{1!} = 0.0.0196$$

$$p(2) = \frac{e^{-.02}(.02)^2}{2!} = .000196$$

Out of 10000, the number of packets with 0 defective is 9802; with 1 defective is 196; with 2 defectives is $1.96 \simeq 2$

Ex. 12. A Poisson distribution has a mode at $x = 4$ and another at $x = 5$. Find the probability that x will have either of these values.

Prob. $(x = 4)$ = Prob. $(x = 5)$ *i.e.*, $\dfrac{e^{-m} m^4}{4!} = \dfrac{e^{-m} m^5}{5!}$ whence $m = 5$

Prob. $(x = 4 \text{ or } 5)$ = Prob. $(x = 4)$ + Prob. $(x = 5)$ = 2 Prob. $(x = 4)$

$$= 2 \times \frac{e^{-5} \times m^4}{4!} = 2 \times \frac{0.006738 \times 5^4}{4!} = 0.035$$

Ex. 13. In a precision bombing attack there is a 5% chance of hitting the target. Two direct hits are required to destroy the target. How many bombs must be dropped to give a 99% chance or more to destroy the target? **Ans.** 11

Ex. 14. If the probability of success = 0.01 find how many trials are necessary so that the probability of at least one success exceeds $\frac{1}{2}$. **Ans.** 69

21.4 THE NORMAL DISTRIBUTION

The normal distribution provides a standard pattern with which other distributions may be compared and which provides a common feature of interest among economists, sociologists, psychologists etc.

It has been noticed that empirical distributions of various types of observations in natural and social sciences are often very close to the normal distribution. In statistical analysis the distributions of observations is frequently assumed to be approximately normal. In statistical estimation and testing of hypotheses the normal distribution plays an important role.

The normal distribution is defined by the equation:

$$y = \frac{1}{\sigma\sqrt{2\pi}} e^{-\frac{1}{2}\left(\frac{x-\mu}{\sigma}\right)^2}$$

where μ is the mean value of x in the population and σ is the standard deviation of x in the population. μ and σ are the *parameters* of the equation and completely define the distribution.

y is the height of the curve at any particular value of x :

$e(= 2.7183)$ and π (= 3.1416) are constants.

For applications of the normal distributions it is very convenient to standardise it *i.e.*, to construct it in such a way that it has a zero mean and standard deviation 1. For this we define a random variable (called the standardised normal variate)

$$z = \frac{x-\mu}{\sigma} \text{ so that } y = \frac{1}{\sqrt{2\pi}} e^{-\frac{1}{2}z^2}$$

Now the mean of z is zero and its standard deviation is 1.

Some Properties of the Normal Distribution

1. The normal distribution is a bell shaped, symmetrical distribution, asymptotic in both directions to the x-axis.

2. It is a continuous distribution, can be expressed by an equation given above. The curve corresponding to this equation is smooth.

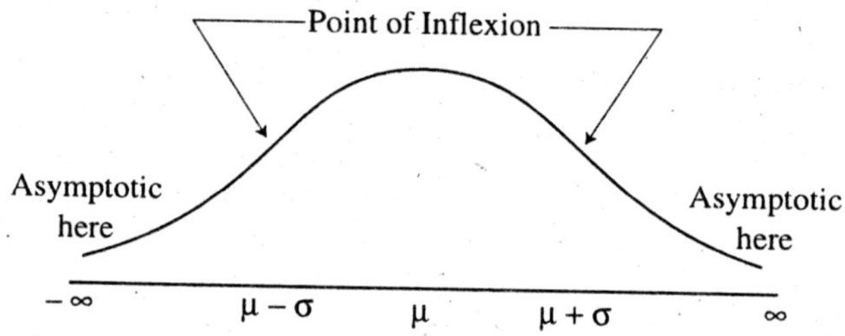

The normal curve

3. Depending on the particular values of the parameters μ and σ there is a whole family of normal distributions.

μ, the central value of the distribution fixes its general level and σ, the dispersion decides the extent to which it is spread out. A smaller value of σ means a lower spread and a lower value of σ means a larger spread as is evident from the diagram.

Two normal curves
$N_1 : \mu = 20, \sigma = 5$
$N_2 : \mu = 25, \sigma = 10$

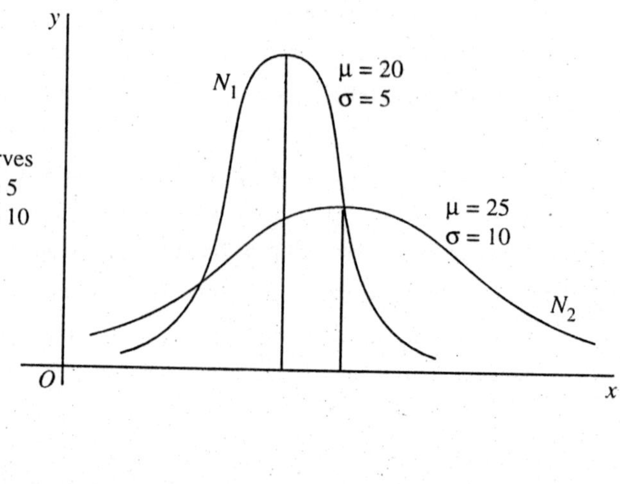

The shapes of curves look different if different scales are taken but the properties remain the same.

4. If any distribution in practice can be approximated by a normal curve then that distribution will have approximately the same properties as the best fitting normal curve.

5. A normal curve is symmetric about its own mean so that one half of the area falls on each side of mean. In the normal distribution, the median and mode coincide with the mean.

6. A normal curve is asymptotic to the horizontal axis *i.e.,* if the value of x is increased (on either side) indefinitely the curve comes closer and closer to the horizontal axis but never quite reaches it for any value of x. The range of the normal distribution is from $-\infty$ to $+\infty$.

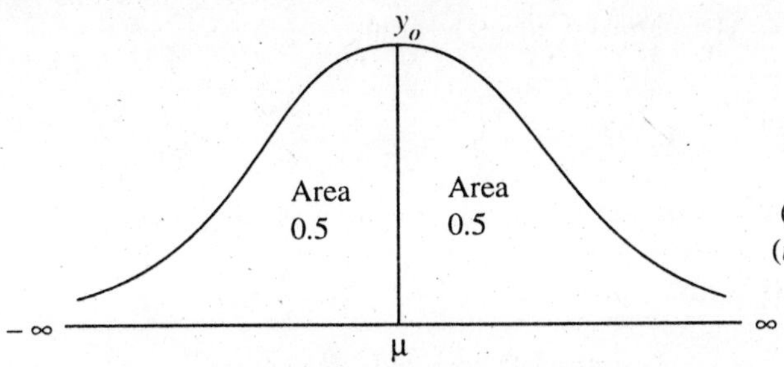

(*i*) For the normal curve the mean, median, mode coincide at μ
(*ii*) Range $(-\infty, \infty)$
(*iii*) Highest frequency y_o

For $z = -\infty$, $p(z) = 0$ i.e., area to the left of $-\infty$ is 0
For $z = +\infty$, $p(z) = 1$ i.e, area to the left of $+\infty$ is 1

7. The standardised normal distribution is one for which the random variable $z = \dfrac{x-\mu}{\sigma}$ so that the mean of $z = 0$, s.d. of $z = 1$. All normal distributions, when written in this form, will have the same shape and the same equation with mean 0, s.d. 1.

To convert any value of a normally distributed variable x to the corresponding standard curve with the standardised variable z we have simply to write

$$z = \frac{x - mean}{\text{s.d.}}$$

The table showing normal curve areas is based on z values. Use the table to check the areas mentioned in the following property.

8. The total area under a normal curve is equal to 1. It can be found that

(*i*) The area between μ and $\mu + \sigma$ is 0.3413 (See area between 0, 1)
The area between μ and $\mu - \sigma$ is 0.3413 (See area between 0, −1)
The area between $\mu - \sigma$ and $\mu + \sigma$ is 0.6827 (area between ± 1)

(*ii*) The area between μ and $\mu + 2\sigma$ is 0.4772 and between μ and $\mu - \sigma$ is 0.4772
The area between $\mu - 2\sigma$ and $\mu + 2\sigma$ is 0.9545. (area between ± 2)

(*iii*) The area between $\mu - 3\sigma$ and $\mu + 3\sigma$ is 0.9973. (area between ± 3)
If we use the standardised variate z we may write the above more conveniently.

(*i*) The area between -1 and $+1$ is 0.6827

(*ii*) The area between – 2 and + 2 is 0.9545

(*iii*) The area between – 3 and + 3 is 0.9973.

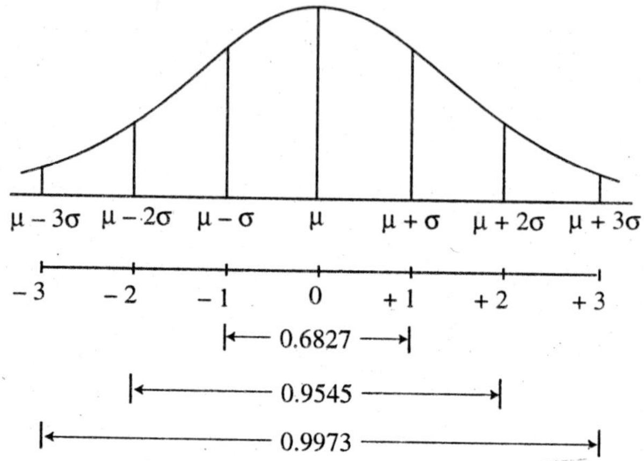

The numerical value of the area at any value x_0 (or z_0) is equal to the numerical value of the area under the curve to the left of x_0 (or z_0).

The above results can be written

(*i*) $p(\mu - \sigma \leq x \leq \mu + \sigma) = 0.6827$

or $p(-1 \leq z \leq +1) = 0.6827$

i.e., the probability that z lies between – 1 and + 1 is 0.6827

Similarly (*ii*) $p(-2 \leq z \leq 2) = 0.9545$

(*iii*) $p(-3 \leq z \leq 3) = 0.9973$

In general, the probability that z lies between any two values z_1, z_2 can be written

$$p\ (z_1 \leq z \leq z_2).$$

9. The curve has points of inflextion at $\mu - \sigma$ and $\mu + \sigma$ where the curvature of the curve changes its direction.

10. The probable error of a normal population is the amount which is exceeded by the errors of one half of all the observations. It is equal to $0.6745\ \sigma$. *i.e.*, P.E. $= \dfrac{Q_3 - Q_1}{2} = 0.6745\sigma$. It means that the probability is $\frac{1}{2}$ that a variable selected at random will have a value between

$$\bar{x} \pm 0.6745\ \sigma.$$

11. The cumulative distribution corresponding to the normal frequency function $f(x)$ is called the normal distribution function $F(x)$. It may be noted that $f(x)$ is the derivative of $F(x)$. This curve is the ogive of the normal curve. The ordinates of this curve correspond to the areas under the normal curve:

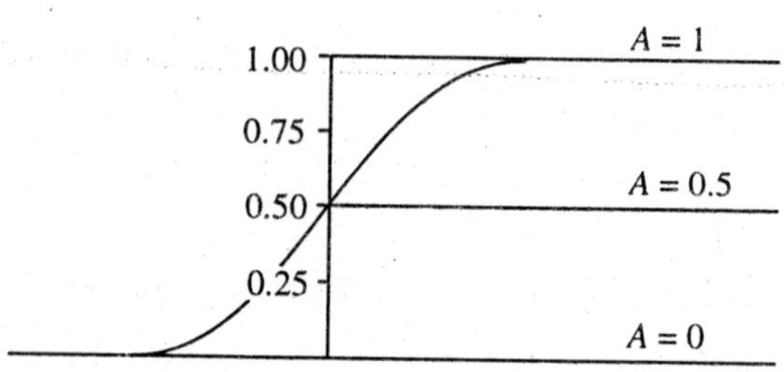

The normal distribution function

As x varies from $-\infty$ to $+\infty$, the area varies from 0 to 1. The ordinate of the ogive is 0 at $x = -\infty$, 0.5 at $x = 0$ and the maximum 1 at $x = \infty$.

12. The following are the theoretical descriptive measures of the normal distribution:

Mean μ (In the standard form : $\mu = 0$)

Standard deviation σ (In the standard form : $\sigma = 1$)

Variance σ^2

Coefficient of skewness $\beta_1 = 0$

Coefficient of kurtosis $\beta_2 = 3$ or $\gamma_2 = 0$

Mean Deviation about the mean $\sigma\sqrt{\dfrac{2}{\pi}} = 0.7979\sigma \simeq \dfrac{4}{5}\sigma$

Quartiles $Q_1 = \mu - 0.6745\ \sigma$, $Q_3 = \mu + 0.6745\ \sigma$

Quartile deviation $\dfrac{Q_3 - Q_1}{2} = 0.6745\ \sigma$

The height of the maximum ordinate in the standard form is 0.3989.

The mean, median, mode coincide

The ratio $QD : MD : SD$

$\simeq 10 : 12 : 15$

Ex. 13. Find the standard scores of students receiving the grades:

(*i*) 40, (*ii*) 50, (*iii*) 60, (*iv*) 80, (*v*) 100

if the mean is 60 and s.d. 10.

(*i*) $x = 40,\ \bar{x} = 60,\ s = 10$

$$z = \frac{x - \bar{x}}{s} = \frac{40 - 60}{10} = -2$$

(*ii*) $z = \dfrac{50 - 60}{10} = -1$ (*iii*) $z = \dfrac{60 - 60}{10} = 0$

(*iv*) $z = \dfrac{80 - 60}{10} = 2$ (*v*) $z = \dfrac{100 - 60}{10} = 4.$

Ex. 14. For the above problem find the grade corresponding to the standardised values (*i*) $z = 1.5$, (*ii*) $z = -2$

We have $z = \dfrac{x - \bar{x}}{s}$ or $x = \bar{x} + zs$

(*i*) $1.5 = \dfrac{x - 60}{10}$ $\therefore\ x - 60 = 15$ or $x = 75$

(*ii*) $-2 = \dfrac{x - 60}{10}$ $\therefore\ x - 60 = -20$ or $x = 40$

21.5 THE CENTRAL LIMIT THEOREM

The normal distribution arises in many contexts. Many distributions connected with problems in sociology, psychology, biology, industry etc. approximate to the normal.

Distributions of heights, weights, marks obtained by students, etc. all follow the normal law. The reason is to be found in a basic law of nature, *the central limit theorem* :

If a large number of equal sized random samples (each containing a large enough number of observations) are taken from an infinite population and the sum of each of these samples is found, it will be observed that the distribution of these sums is normal.

21.6 THE NORMAL CURVE TABLE

It is possible to find the area under the normal curve between any limits with the help of the table prepared for the purpose. This table gives the proportion of the area under the curve corresponding to a given range of values of z *i.e.*, it gives the probability that a given value z will be observed.

For the problems given below the normal curve table should be used. In each case it is preferable to draw the diagram.

Ex. 15. (*i*) To find the area between $z = 0$ and $z = 1.5$ see the value against 1.5 from the table. It is 0.4332.

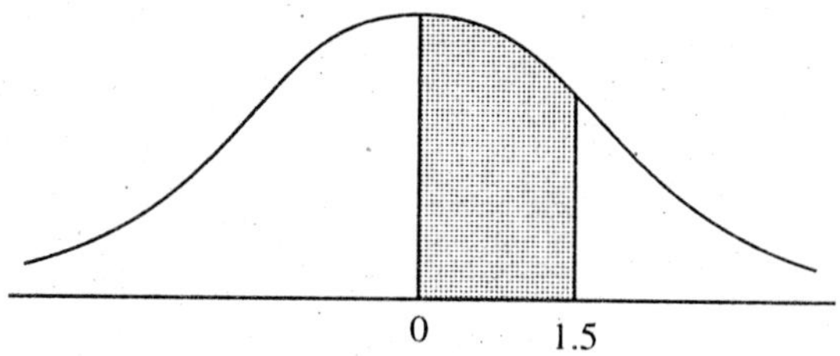

This area represents the probability that z lies between 0 and 1.5. We write $p(0 \leq z \leq 1.5) = 0.4332$.

(*ii*) The area between $z = -1.5$ and $z = 0$ is the same as that between 0 and 1.5 because of symmetry. The table gives values on one side only. It is not necessary to have a table for values on the other side.

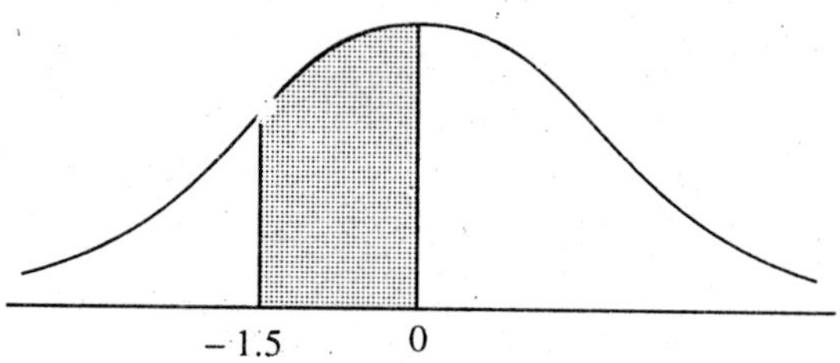

$$p(-1.5 \leq z \leq 0) = 0.4332$$

(*iii*) The area between

$z = -1.5$ and $z = +1.5$ is the sum of the two areas in (*i*) and (*ii*)

$$p(-1.5 \leq z \leq +1.5) = 0.8664$$

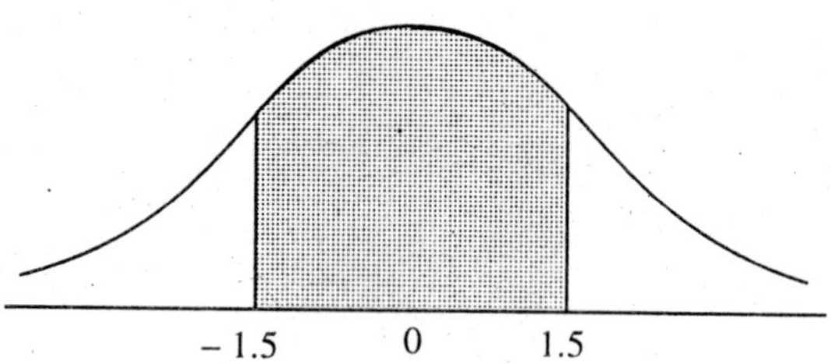

Ex. 16. The area between $z = -0.75$ and $z = 1.2$ is given by the sum of (the area between $z = -.75$ and $z = 0$) and (the area between $z = 0$ and $z = 1.2$).

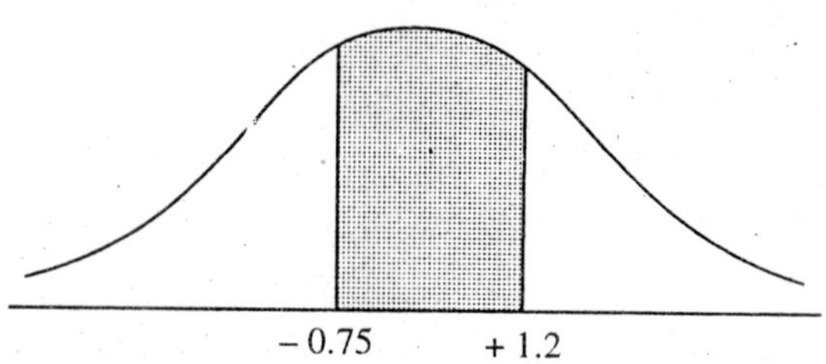

i.e., $p(-0.75 \leq z \leq 0)$

$$+ p(0 \leq z \leq 1.2)$$

$$= (0 \leq z \leq 0.75) + p(0 \leq z \leq 1.2)$$

$$= 0.2734 + 0.3849$$

$$= 0.6583$$

Ex. 17. The area between $z = 0.46$ and $z = 1.28$

$= (\text{Area between } z = 0 \text{ and } z = 1.28)$

$- (\text{Area between } z = 0 \text{ and } z = 0.46)$

$= 0.3997 - 0.1772$

$= 0.2225$

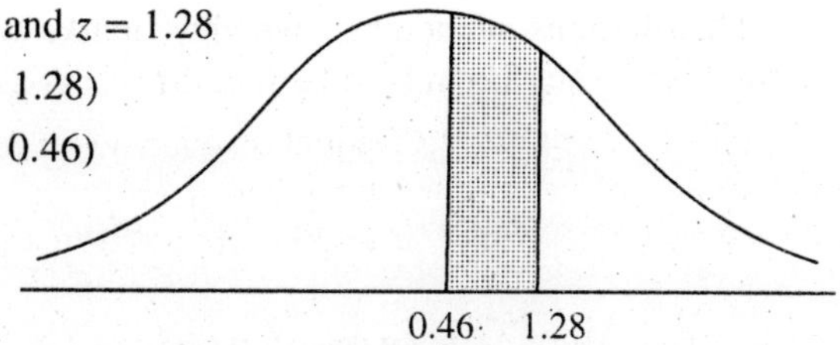

Ex. 18. The area to the right of $z = 0$ is 0.5.

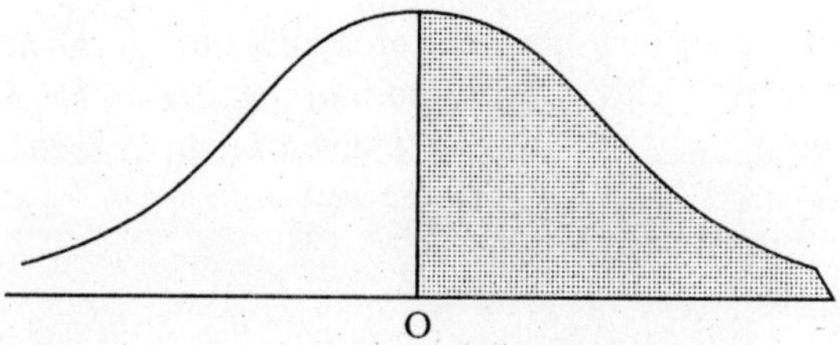

Ex. 19. The area to the right of $z = 1.2$ is

$= (\text{Area to the right of } z = 0) -$

$(\text{Area between } z = 0 \text{ and } z = 1.2)$

$= 0.5 - 0.3849$

$= 0.1151.$

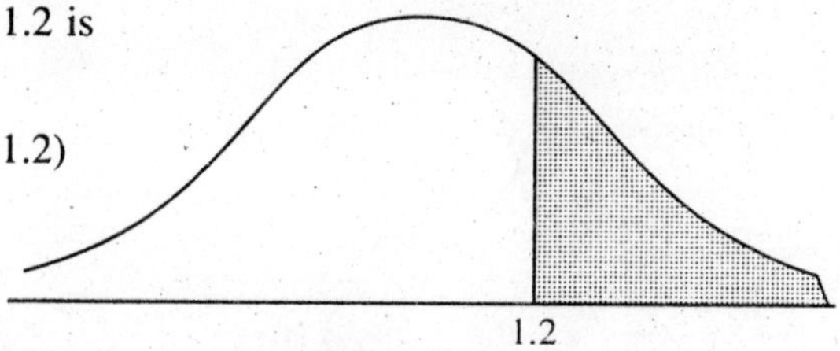

Ex. 20. The area to the left of $z = 0$ is 0.5.

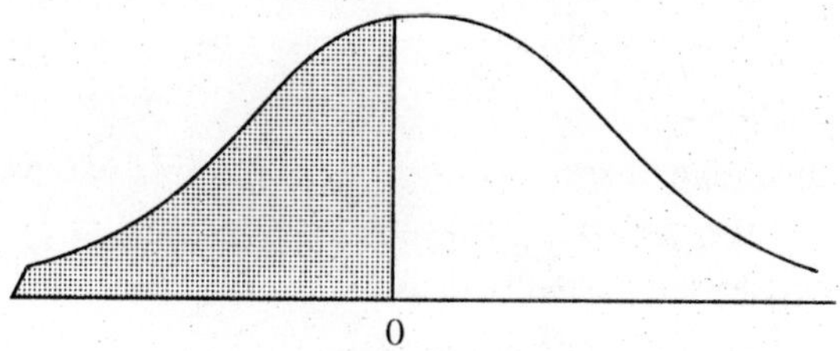

Ex. 21. The area between

$z = -1.96$ and $z = -1.16$

is the same as the area between $z = 1.16$ and $z = 1.96$

i.e., (Area bet. $z = 0$ and $z = 1.96$) –

(Area bet. $z = 0$ and $z = 1.16$)

$= 0.4750 - 0.3770 = .0980.$

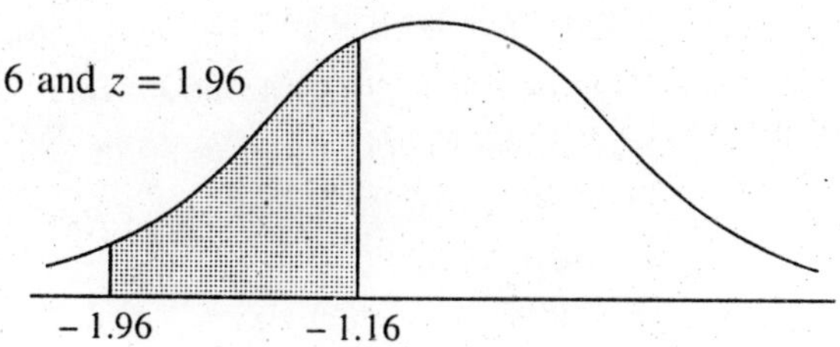

Ex. 22. The area to the left of $z = -1.96$ is $0.5 - 0.4750 = 0.0250$.

Ex. 23. The sum of the areas to the left of $z = -1.96$ and to the right of $z = +1.96$ is

$0.0250 + 0.0250 = 0.05.$

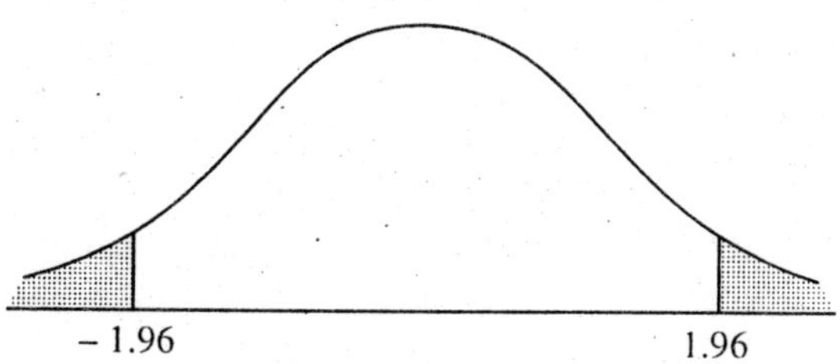

Ex. 24. Find the area between $z = \pm 1.16$. **Ans.** 0.7540

Ex. 25. Show that the area between $z = -1.44$ and $z = 2.05$ is 0.9049.

Ex. 26. Find z if the area between 0 and z is

(*i*) 0.4887 (*ii*) 0.4549, (*iii*) 0.8612 to the left of z.

(*i*) (*ii*) (*iii*)

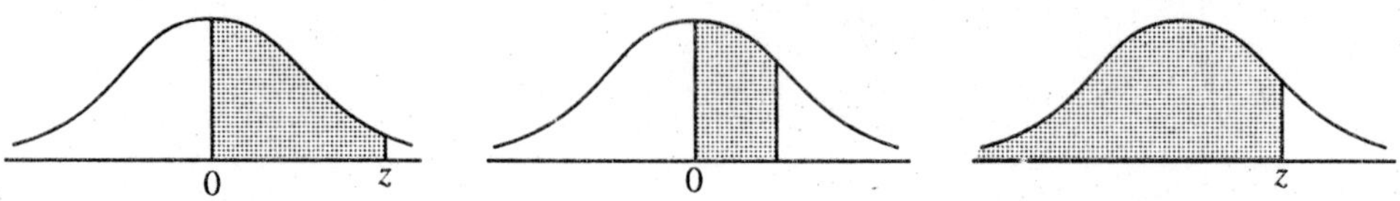

Ans. (*i*) $z = 3$, (*ii*) $z = 1.5$, (*iii*) $z = 1.09$

Ex. 27. A normal curve has $\bar{x} = 10$, $s = 2$. Find the area between $x_1 = 8$ and $x_2 = 14$

$$z_1 = \frac{x_1 - \bar{x}}{s} = \frac{8-10}{2} = -1$$

$$z_2 = \frac{x_2 - \bar{x}}{s} = \frac{14-10}{2} = 2$$

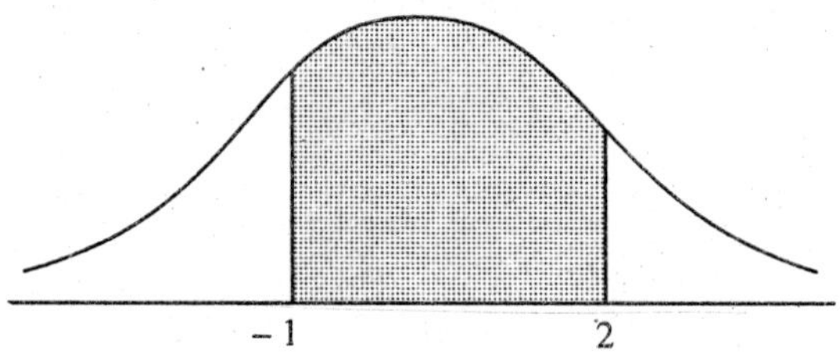

(Area between $z_1 = -1$ and $z_2 = 2$)

= (Area between $z_1 = -1$ and 0)

+ (Area between 0 and $z_2 = 2$)

= 0.3413 + 0.4772 = 0.8185

Show that if $x_1 = 9$, $x_2 = 14$ the area is 0.6687.

Ex. 28. The mean height of 1000 men is 68 inches and their s.d. 5 inches. If the heights are normally distributed find how many men have heights between

(*i*) 67 and 69, (*ii*) between 62 and 64,

(*iii*) above 80 inches (*iv*) less than 60 inches,

(*v*) less than or equal to 60 inches,

(*vi*) more than or equal to 80 inches.

(*i*) Assuming that the heights have been recorded to the nearest inch, the heights between 67 and 69 inches can actually have any values between $x_1 = 66.5$ and $x_1 = 69.5$.

We do this for the sake of continuity. It should be remembered the normal distribution is a continuous distribution and the data being discrete have to be adjusted for continuity.

$$z_1 = \frac{x_1 - \bar{x}}{s} = \frac{66.5 - 68}{5} = -0.3$$

$$z_2 = \frac{x_2 - \bar{x}}{s} = \frac{69.5 - 68}{5} = +0.3$$

Area between z_1, z_2 = (Area between – 0.3 and 0) + (Area between 0 and 0.3)

= 0.1179 + 0.1179 = 0.2358

This is the required proportion of men. The number of men having heights between 67 and 69 inches is

$$1000 \times 0.2358 = 23.58 \text{ say } 23$$

(*ii*) To find the number of men with heights between 62 and 64 inches we write

$$x_1 = 61.5 \qquad x_2 = 64.5 \qquad \text{(for continuity)}$$

$$z_1 = \frac{61.5 - 68}{5} = -1.3$$

$$z_2 = \frac{64.5 - 68}{5} = -0.7$$

(Area between $z_1 = -1.3$ and $z_2 = -0.7$)

= (Area between $z_1 = -1.3$, $z = 0$) – (Area between $z = 0$, $z_2 = -0.7$)

= (Area between $z = 0$, $z_1 = +1.3$) – (Area between $z = 0$, $z_2 = 0.7$)

= 0.4032 – 0.2580 =0.1450

The number of men with heights between 62 and 64 inches = 1000 × 0.1450 = 145

(*iii*) For men to have heights above 80 they must have at least 80.5 inches heights.

$$z = \frac{80.5 - 68}{5} = 2.50$$

(Area to the right of $z = 2.5$)

= (Area to the right of $z = 0$) – (Area between $z = 0$ and $z = 2.5$)

= 0.5 – 0.4938 = .0062

The number of men with heights above 80 inches is

$$1000 \times 0.0062 = 6.2, \text{ way } 6$$

(*iv*) Heights less than 60 means less than 59.5

$$z = \frac{59.5 - 68}{5} = -1.7$$

(Area to the left of $z = -1.7$)

= (Area to the right of $z = 1.7$)

= 0.5 – 0.4554 =. 0446

The number with heights less than 60 is

$$1000 \times 0.0446 = 44.6, \text{ say } 44$$

(*v*) The number of men having heights less than or equal to 60 inches means they must have heights less than 60.5.

Compare with (*iv*)

$$z = \frac{60.5 - 68}{5} = -1.5$$

(Area to the left of $z = -1.5$)

= 0.5 – 0.4332 = 0.0668

The required number = 1000 × .0668 = 66.8 say 66

(*vi*) More than or equal to 80 implies more than 79.5

$$z = \frac{79.5 - 68}{5} = 2.3$$

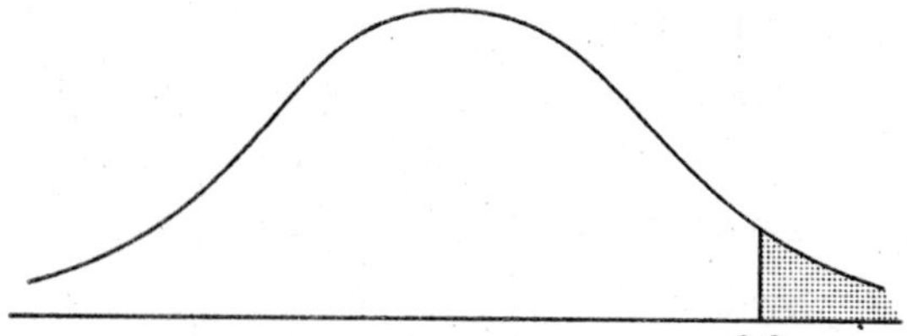

(Area to the right of $z = 2.3$)

$= 0.5 - 0.4893 = 0.0107$

The required number

$= 1000 \times 0.0107$

$= 10.7$ say 10.

In the probability form, all the above results can be written

(*i*) $p(67 \le x \le 69) = p(-0.3 \le z \le +0.3) = 0.2358$

(*ii*) $p(62 \le x \le 64) = p(-1.3 \le z \le -0.7) = 0.1450$

(*iii*) $p(x > 80) = p(z \ge 2.5) = 0.0062$

(*iv*) $p(x < 60) = p(z \le -1.7) = 0.0446$

(*v*) $p(x \le 60) = p(z \le -1.5) = 0.0668$

(*vi*) $p(x \ge 80) = p(z \ge 2.3) = 0.0107$

Whenever we approximate the continuous normal to a given discrete distribution we should make the adjustment for continuity.

Ex. 29. In the preceding problem how many students have the height 72 inches?

Men with height 72 are to be taken as men having heights between $x_1 = 71.5$ and $x_2 = 72.5$

$$z_1 = \frac{71.5 - 68}{5} = 0.7$$

$$z_2 = \frac{72.5 - 68}{5} = 0.9$$

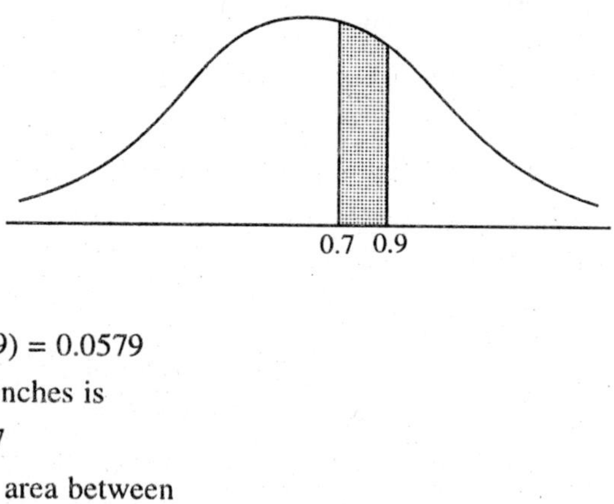

(Area between $z_1 = 0.7$ and $z_2 = 0.9$)

$= 0.3159 - 0.2580 = 0.0579$

Alternatively we write $p(0.7 \le z \le 0.9) = 0.0579$

The number of men with heights 72 inches is

$1000 \times 0.0579 = 57.9$ say 57

Ex. 30. In the same problem find the area between

(*i*) $\bar{x} \pm s$ (*ii*) $\bar{x} \pm 2s$ (*iii*) $\bar{x} \pm 3s$

(*i*) To find the area between $\bar{x} - s$ and $\bar{x} + s$ *i.e.*, between $68 - 5 = 63$ and $68 + 5 = 73$,

write $z_1 = \frac{63 - 68}{5} = -1$, $z_2 = \frac{73 - 68}{5} = +1$

We have to find the area between $z_1 = -1$, $z_1 = +1$.

As a matter of fact this step is not necessary when the area is in terms of x + some multiple of s, *e.g.*, the area between $\bar{x} \pm 2s$ is equal to the area between $z_1 = -2$ and $z_2 = +2$ and so on.

The area between $\bar{x} - s$ and $\bar{x} + s$ is equal to the (area between $z_1 = -1$ and $z_2 = +1$) *i.e.*,

(Area between $z_1 = -1$ and $z = 0$)

+ (Area between $z = 0$ and $z_2 = +1$)

$= 0.3413 + 0.3413 = 0.6826$

Similarly the area between

(*ii*) $\bar{x} \pm 2s$ is 0.9545 (*iii*) $\bar{x} \pm 3s$ is 0.9973

We can write these results as follows:

$P(\bar{x} - s \le z \le \bar{x} + s) = 0.6826$, $P(\bar{x} - 2s < z < \bar{x} + 2s) = 0.9545$ etc.

Ex. 31. The mean weight of 500 students at a certain college is 151 1b. and the s.d. is 15 1b. Assuming that the weights are normally distributed, how many students weigh:

(*i*) between 120 and 155 1b. **Ans.** (*i*) 300

(*ii*) more than 185 1b. (*ii*) 5

(*iii*) less than 128 1b. (*iii*) 29

(*iv*) 128 1b. (*iv*) 4

(*v*) less than or equal to 128 1b. (*v*) 33

Ex. 32. In an examination the mean and s.d. of marks are 80 and 10. If 500 students appeared for the examination, how many of them got marks (*i*) more than 85 (*ii*) between 70 and 90 (*iv*) more than 80? Assume normality of the distribution and find

(*i*) $p(\bar{x} - s < x < \bar{x} + s)$ (See Ex. 30)

(*ii*) $p(\bar{x} - 2s < x < \bar{x} + 2s)$

(*iii*) $p(\bar{x} - 3s < x < \bar{x} + 3s)$

Ex. 33. In an examination; the marks obtained were found to follow the normal distribution with $\bar{x} = 58$, $s = 10$. What is the chance of a student getting marks (*i*) less than 63, (*ii*) between 41 and 63, (*iii*) exactly 68?

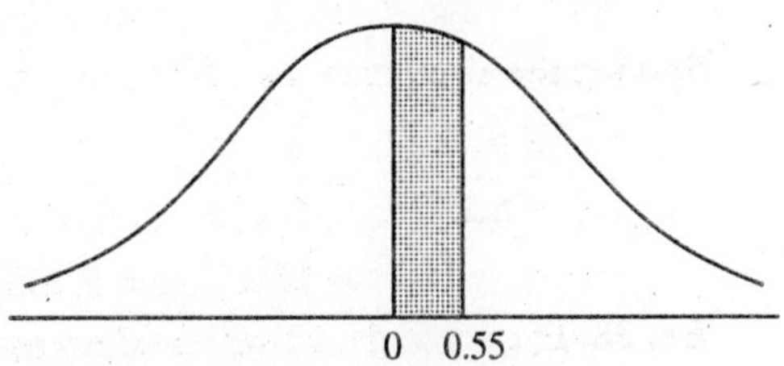

Ans. (*i*) $p(x_1 \le 63) = p(z_1 \le 0.55) = 0.7088$

(*ii*) $p(-1.75 \le z \le 0.55) = 0.6687$

(*iii*) $p(0.95 \le z \le 1.05) = 0.0242$.

Ex. 34. Find the area under the normal curve

(*i*) between $z = -2.25$ and $z = -0.40$ **Ans.** (*i*) 0.3324

(*ii*) between $z = -1.5$ and $z = 2.6$ (*ii*) 0.9285

(*iii*) to the left of $z = 0.48$ (*iii*) 0.6844

(*iv*) to the right of $z = 0.48$ (*iv*) 0.3156

Ex. 35. If a set of measurements is normally distributed what is the proportion which differs from the mean by

(*i*) more than $\frac{1}{4}s$ (*ii*) less than $\frac{1}{2}s$?

Ans. (*i*) $1 - p\left(\bar{x} - \frac{s}{4} \le x \le \bar{x} + \frac{s}{4}\right) = 1 - p(-\frac{1}{4} \le z \le \frac{1}{4})$

$= 1 - 0.1974 = 0.8026.$

(*ii*) 0.3830.

Ex. 36. If 49 per cent of a normal population lies between $\bar{x} = 83$ and a value of $x = 99$, find s.

$$z = \frac{99 - 83}{s} = \frac{16}{s}$$

$$P\left(0 \le z \le \frac{16}{s}\right) = 0.49.$$

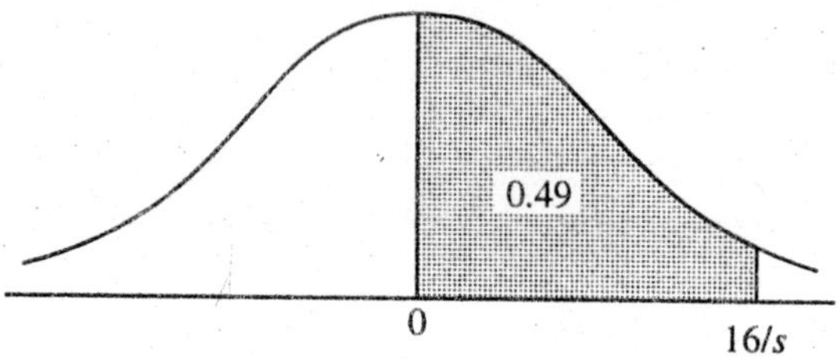

From the table see the value of z which corresponds to the area 0.49. This value is 2.33.

$$\therefore \quad \frac{16}{s} = 2.33 \quad \therefore \quad s = \frac{16}{2.33} = 6.9.$$

Ex. 37. In a normal population, 15 per cent values are less than 12 and 40 per cent above 16.2. Find the mean and s.d.

From the table, the area

(*i*) 0.15 corresponds to $z_1 = 1.037$

(*ii*) 0.40 corresponds to $z_2 = 0.253$

$$\frac{\bar{x} - 12}{s} = 1.037, \quad \frac{16.2 - \bar{x}}{s} = 0.253$$

Solving the simultaneous equations we get

$$\frac{\bar{x} - 12}{16.2 - \bar{x}} = \frac{1.037}{0.253}$$

$$\bar{x} = 15.375, \ s = 3.256.$$

Ex. 38. For some data distributed normally, the mean is 50 and 5% of the values are greater than 60. Find the standard deviation.

$$z = \frac{\bar{x} - \mu}{s} = \frac{60 - 50}{s} = \frac{10}{s}$$

$$P(x \ge 60) = 0.05$$

i.e., $$P\left(z \ge \frac{10}{s}\right) = 0.05$$

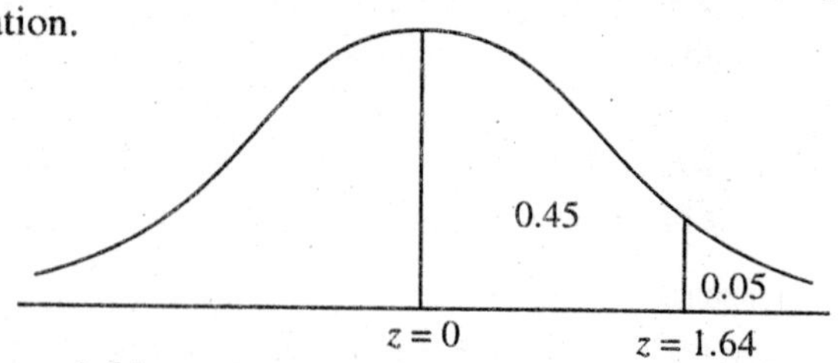

From the table, this corresponds to $z = 1.64$

Since the area between $z = 0$ and $z = 1.64$ is

$0.5 - 0.05 = 0.45$

$$\therefore \quad \frac{10}{s} = 1.64 \quad \text{or} \quad s = \frac{10}{1.64} = 6.1$$

Ex. 39. In an intelligence test administered to 1000 children the average score is 42 and s.d. 24. Find the number of children (*i*) exceeding the score 60, (*ii*) with scores between 20 and 40. Assume normal distribution.

Ex. 40. In a normally distributed population of size 2000, the mean is 45 and s.d. 5. Below what score do (*i*) 65 per cent of the cases fall? (*ii*) 600 of the cases fall?

Ans. (*i*) $z = 0.39$ or $x = 45 + 1.95 = 46.95$. (*ii*) $x = 42.40$.

Ex. 41. The mean grade on an examination is 72 and s.d. 9. The top 10 per cent of the students are to receive grade one. What is the minimum grade a student must get in order to receive grade one? **Ans.** 84.

Ex. 42. If scores are normally distributed and a student's score in standard units is 0.8 what percentage of students is expected to score higher than him? **Ans.** 21 per cent.

Ex. 43. If the heights of college students are normally distributed with mean 69 and s.d. 3 find the probability that (*i*) $x < 65$ and (*ii*) $65 \le x \le 70$.

Ex. 44. A minimum eligibility height for service is to be prescribed to enable 60% of candidates are selected. The heights are normally distributed with mean 60.6 inches and S.D. 2.55. Determine the minimum eligibility.

Let x = minimum height

$$z = \frac{x - 60.6}{2.55}$$

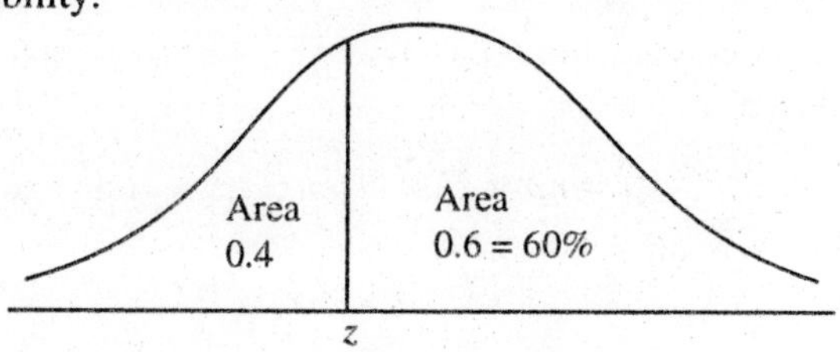

To find x corresponding to the area on the left of x being 0.4

This is possible when $z = -0.2533$

(From the tables)

$$\therefore \qquad \frac{x - 60.6}{2.55} = -0.2533$$

i.e., $\qquad x = 59.95$ inches.

Ex. 45. The net weight of a tinned item is guaranteed to be at least 16 ounces.The manufacturer sets up an average of 3 per cent over the guaranteed amount. If the s.d. is 2 ounces what fraction of his output actually fails to support his claim? Assume normality of fills about the overrun value. **Ans.** 0.0032.

21.7 RELATIONSHIP BETWEEN THE NORMAL, BINOMIAL AND POISSON DISTRIBUTIONS

Under certain conditions many distributions approximate the normal.

(*i*) The Poisson distribution can be derived from the binomial if $n \to \infty$, $p \to 0$ and $np = m$, a small number

$$\lim_{\substack{n \to \infty \\ p \to 0 \\ np \to m}} {}^nC_x p^x (1-p)^{n-x} = \frac{e^{-m} m^x}{x\,!}$$

The Poisson distribution is easy to use and hence replaces the binomial when n is large and p is small. An obvious characteristic of the Poisson distribution is that its mean and variance are nearly equal.

(*ii*) If n is large and p not very small, the binomial distribution is approximately normal with the standardised variable $z = \frac{x - np}{\sqrt{npq}}$, since the mean and s.d. of the binomial are np and $\sqrt{npq}$.

(*iii*) If $m \to \infty$, the Poisson distribution also tends to be normal with $z = \frac{x - m}{\sqrt{m}}$ since the mean and s.d. of the Poisson are m and $\sqrt{m}$.

Ex. 46. Show that if the probability of success in an experiment is 0.2 then, by using the binomial distribution the probability of getting 5 successes in 25 trials is 0.579.

Check $p(x \geq 5) = 1 - p(0) - p(1) - p(2) - p(3) - p(4)$

$= 0.579$

Compare the result with the normal approximation. For the normal approximation we have

$$\mu = np = 25 \times 0.2 = 5, \ \sigma = \sqrt{npq} = \sqrt{25 \times 0.2 \times 0.8} = 2$$

Allowing for continuity

$$z = \frac{4.5 - 5}{2} = -0.25$$

Using the normal tables we get

$$p(x \geq 5) = p(z \geq -0.25) = 0.5987$$

which is a rough approximation. The approximation would be much better in the case of values of p nearer 0.5.

Ex. 47. If $p = 0.03$, $n = 100$ and we want to find the probability of say 4 successes. The binomial is very difficult to use here and the normal approximation is not good since p is small. The Poisson approximation which is very convenient gives

$$p(x = 4) = \frac{e^{-m} m^x}{x!} = \frac{e^{-3} 3^4}{4!} = 0.168.$$

Ex. 48. If $p = 0.06$ what is the probability of getting at least 10, but not more than 20 successes in 250 trials? **Ans.** 0.8584 (Use Normal Approx.)

Ex. 49. Use the binomial distribution and also the normal and Poisson approximations to find the probability of finding (*i*) 2 (*ii*) less than 4 defectives in a batch of 20 produced by a machine known to be producing 10 per cent defectives.

Ans. (*i*) $B : 0.2852, N : 0.2906, P : 0.2707$

(*ii*) $B : 0.8671, N : 0.8369, P : 0.85771.$

Ex. 50. In a telephone exchange the probability of a wrong call is 0.001. Find the probability of 4 wrong calls if 2000 calls are made.

What is the minimum number of calls required to ensure a probability 0.9 that at least one of the calls is wrongly connected? **Ans.** 0.09, 2303.

Ex. 51. The number of persons entering a petrol pump centre in any 5 minute period follows a Poisson with mean 0.3. Find the probability that during a 5 minute period, 4 customers enter the centre.

$m = 0.3$ and $x = 4$

$$p(x) = \frac{e^{-m} m^x}{x!}$$

$$p(4) = \frac{e^{-0.3} (0.3)^4}{4!} = 0.00025$$

Ex. 52. A Life Insurance Company which has insured 350 people knows from past record that between the ages 30 – 40, 1% people die. Find the probability that less than 3 people in this age group die.

$$p(x < 3) = p(x = 0) + p(x = 1) + p(x = 2)$$
$$= 0.32$$

Ex. 53. The items produced by a factory are known to have 2% defectives. Find the probability that in a random sample of 200 items there are

(*i*) 3 or more are defective

(*ii*) 4 or less are defective

This is a case of Poisson distribution

$$p = 0.02,\ n = 200,\ m = np = 200 \times 0.02 = 4$$

If x is the number of defective items

$$p(x \geq 3) = 1 - \{p(x = 0) + p(x = 1) + p(x = 2)\}$$

$$= 1 - \left\{\frac{e^{-m} m^0}{0!} + \frac{e^{-m} m^1}{1!} + e^{-m} \frac{m^2}{2\,!}\right\}$$

$$= 1 - e^{-m}\left\{1 + \frac{4}{1} + \frac{4^2}{2}\right\}$$

$$= 1 - 0.0183\ (13) = 1 - .2379 = 0.7621$$

$$P(x \leq 4) = \sum_{x=0}^{4} \frac{e^{-m} m^x}{x\,!} = 0.6283$$

Ex. 54. Between 10 and 11 a.m. the number of calls received by an operator follows the Poisson with mean 5. Find the probability that 2 calls are received on a day between 10 and 11 a.m.

$$p(x = 2) = \frac{e^{m} m^2}{2\,!} = \frac{e^{-5} \times 25}{2} = \frac{0.0067 \times 25}{2}$$

$$= \frac{.1675}{2} = 0.08375$$

Ex. 55. If for a random variable following Poisson distribution $p(x = 1) = p(x = 2)$ find $p(x = 4)$

$$\frac{e^{-m}m^1}{1!} = \frac{e^{-m}m^2}{2!} \quad \therefore\ 1 = \frac{m}{2} \text{ or } m = 2$$

$$p(x = 4) = \frac{e^{-2}2^4}{4!} = 0.1353 \times \frac{16}{24} = 0.0902$$

21.8 FITTING OF DISTRIBUTIONS TO GIVEN DATA

On the basis of probability reasoning or experience, it is often possible to fit a theoretical distribution to a given sample frequency distribution. In doing so we assume that the given frequency distribution has the characteristics of the fitted theoretical distribution. We make use of characteristics like the mean and s.d. of the sample to estimate those for the population which is assumed to be represented by the fitted theoretical distribution.

Fitting a Binomial Distribution

Ex. 56. A machine is known to be producing 20% defective items. 5 items are selected at random and the number of defectives noted. The procedure is repeated 10000 times and the following results are obtained:

Number of defectives	*Observed frequency*	*Expected frequency*	
0	3120	3277	(Ex. 1)
1	4107	4096	
2	2202	2048	
3	510	512	
4	59	64	
5	2	3	

Fit a binomial distribution to the data.

The procedure amounts to calculating the probabilities for 0, 1, 2, ...defectives and calculating the expected frequencies as in Ex. 1 where this problem is worked out. By comparing the observed and expected frequencies, we may say that the *fit* seems to be good. (This can be tested by using χ^2 distribution; sec next chapter)

Fitting a Poisson Distribution

The procedure here is similar to that in the case of the binomial, *viz.* the calculation of probabilities followed by finding the expected frequencies.

Ex. 57. Fit a Poisson distribution to the following data showing the number of deaths per year (x) resulting from the kick of a horse, over a period of 20 years.

x	0	1	2	3	4
f	109	65	22	3	1

$$m = \frac{\Sigma fx}{n} = \frac{122}{200} = 0.61$$

$$p(x) = \frac{e^{-m} m^x}{x!}$$ for the Poisson distribution. We have

$$p(0) = \frac{e^{-0.61}(0.61)^0}{0!} = e^{-0.61} = 0.5434$$

$$p(1) = \frac{e^{-0.61}(0.61)^1}{1!} = p(0) \times \frac{0.61}{1} = 0.3315$$

$$p(2) = \frac{e^{-0.61}(0.61)^2}{2!} = p(1) \times \frac{0.61}{2} = 0.1010$$

$$p(3) = \frac{e^{-0.61}(0.61)^3}{3!} = p(2) \times \frac{0.61}{3} = 0.0205$$

$$p(4) = \frac{e^{-0.61}(0.61)^4}{4!} = p(3) \times \frac{0.61}{4} = 0.0035$$

The expected frequencies are obtained by multiplying each of the probabilities by 200. They are 108.7, 66.3, 20.2, 4.1 and 0.7.

Ex. 58. Fit (*i*) a binomial, (*ii*) a Poisson distribution to the following data and show that the expected frequencies are as shown.

Observed		*Expected f*	
x	f	B	N
0	30	32	41.7
1	62	60	53.4
2	46	43	34.2
3	10	13	14.6
4	2	2	4.7

Fitting a Normal Distribution

(A) *The method of ordinates*

This method uses the tables of ordinates of the standard normal curve. The steps are as follows:

(*i*) Find the midpoints x of classes.

(*ii*) For each x find $z = \frac{x - \overline{x}}{s}$

(*iii*) Find ordinates at each of these distances z from the table.

(*iv*) Multiply each of these by 0.3989, the mean ordinate height. These are the heights of ordinates or frequencies corresponding to the class midpoints.

(B) *The method of areas*

We find z for each class interval and the area for each class is found by using the tables. Expected frequencies are found by multiplying the areas by the total frequency.

Ex. 59. The following problem shows the method of fitting a normal distribution to the given data by the method of areas.

Heights in inches	*Frequency*
60—62	5
63—65	18
65—67	42
67—69	27
69—71	8
Total	100

We find $\bar{x} = 67.45$ inches, $s = 2.92$ inches, $n = 100$

We should revise the class limits to adjust for continuity:

Heights	*Class Boun-daries* x	$z = \frac{x - \bar{x}}{s}$	*normal area from 0 to z*	*Area for each class or Rel. freq.*	*Expected freq. = n × Rel. fr.*	*Obs. freq.*
59.5—62.5	59.5	—2.72	0.4967	.4967—.4554 = .0413	4.1	5
62.5—65.5	62.5	—1.70	0.4554	.4554—.2486 = .2068	20.7	18
65.5—68.5	65.3	—0.67	0.2486	.2486—.1406 = .3892	38.9	42
68.5—71.5	68.5	0.36	0.1406	4177—1406 = .2771	27.7	27
71.5—74.5	71.5	1.31	0.4177	.4920—.4177 = .0743	7.4	8
	74.5	2.41	0.4920			

In can be seen that the expected frequencies agree well with the observed frequencies.

(C) *The method of fitting a cumulative normal distribution*

If the cumulative frequency polygon of a distribution has a shape which is fairly closely approximated by the graph of the cumulative normal distribution we have another method of fitting a normal distribution.

In fitting the cumulative normal distribution only the following values of z are taken: 0, ± 1, ± 2, ± 3, and the corresponding values of $F(x)$ for the given data are calculated and a smooth curve drawn through them giving us the fitted cumulative normal distribution.

We can calculate the fitted frequencies from the cumulative fitted frequencies.

Ex. 60. Fit a normal distribution to the following data :

Class interval	*Frequencies*
0 — 10	1 (0)
10 — 20	1 (2)
20 — 30	4 (5)

30 — 40	8 (9)
50 — 60	13 (12)
60 — 70	5 (7)
70 — 80	2 (3)
80 — 90	0 (1)
90 — 100	1 (0)

Show that the expected frequencies are as shown within the brackets in each case.

Ex. 61. In a certain examination the percentages of passes and distinctions were 46 and 9 respectively. Estimate the average marks obtained by the candidates, given that the minimum pass and distinction marks are 40 and 75.

What will be the minimum qualifying marks for admission to a re-examination of the failed candidates if only the best 25% of them are to be given another opportunity to be examined ?

Ans. $\bar{x}$ = 37.2 ; x = 36.

Ex. 62. In a normal distribution 31% items are under 45 and 8% ever 64. find $\bar{x}$, s.

Ans. 50, 10

Ex. 63. An examination was taken by 1000 students who are to be classified into subgroups *A, B, C, D, E* according to ability, the range of ability being equal in subgroups. On the assumption that the trait measured is normally distributed, how many students should be placed in each group ?

Ans. Divide the range of ability $\bar{x} - 3s$ to $\bar{x} + 3s$ *i.e.*, $6s$ into 5 equal parts, each of interval $1.2s$.

Group	*Areas*		*Number of students*
A	$P(\bar{x} - 3s \le x \le \bar{x} - 1.8s)$	$\doteq 0.037$	37
B	$P(\bar{x} - 1.8s \le x \le \bar{x} - 0.6s)$	$= 0.238$	238
C	$P(\bar{x} - 0.6s \le x \le \bar{x} + 0.6s)$	$= 0.450$	450
D	$P(\bar{x} - 0.6s \le x \le \bar{x} - 1.8s)$	$= 0.038$	238
E	$P(\bar{x} - 1.8s \le x \le \bar{x} - 3s)$	$= 0.037$	37
Total 1,000		1000	

Verify the results.

Ex. 64. 2000 electric bulbs with an average life of 1000 hours and s.d. 200 hours are installed in a town. What number might be expected to fail in the first 700 burning hours? After what period of burning hours might we expect that 10% of the bulbs would have failed? Assume the lives of bulbs to be normally distributed.

Ans. 134 bulbs, 744 hours.

Ex. 65. The following distribution gives the frequencies of occurrence of a variate x between certain limits:

Variate x	*Frequency*
Less than 40	30
40 or more but less that 50	33
50 or more	37

The distribution is exactly normal. Find the frequency between $x = 50$ and $x = 60$.

Ans. 26

Ex. 66. We use s as a yardstick in the scaling of a distribution on either side of the mean μ.

Can we use any other measure for the purpose?

Yes. For instance, instead of s. we may use Q.D. = $0.6745\ s$

(Area from $x = 0$ to $z = \bar{x} + 0.6745s$)

= (Area from $z = 0$ to $z = 0.6745$) = 0.25

Similarly (Area from $z = -0.6745$ to $z = 0$) = 0.25

The area between $\bar{x} \pm 0.6745\sigma$ is therefore the middle 50%.

Similarly find the area

(*i*) between $\bar{x} \pm 2$QD (*ii*) between $\bar{x} \pm 3$QD.

Ex. 67. If $\mu = 75$, $\sigma = 15$, $n = 100$ for a normal distribution find Q_1, Q_3. Also find how many values lie between $x = 20$ and $x = 30$.

Ex. 68. Prove that for the normal distribution, the Q.D., mean deviation and s.d. are approximately in the ratio 10 : 12 : 15.

Ex. 69. Comment on the following:

(*i*) Everybody believes in the normal law of errors: the experimenter because he thinks it is a mathematical theorem, the mathematician because he thinks it is an experimental fact.

(The normal distribution has important and useful mathematical as well as experimental properties. It fascinates both the mathematician and the statistician.)

(*ii*) The law of large numbers states: Whenever the outcome of a random process arises as the sum of a large number of independent effects no one of which predominates, outcomes tend to be normally distributed.

(*iii*) The law of large numbers provides theoretical justification for the great homage rendered to the normal distribution throughout the history of statistics.

(*iv*) A sampling distribution is a theoretical probability distribution that shows the functional relation between the possible values of some summary characteristics of n cases drawn at random and the probability associated with each value over all possible samples of size n from a particular population.

Ex. 70. Describe the main characteristics of the normal distribution. Discuss its importance in statistical theory.

CHAPTER 22

Distributions II : Discrete and Continuous Distributions Exact Sampling Distributions

22.1 DISCRETE PROBABILITY DISTRIBUTIONS

When a random variable x takes discrete values $x_1, x_2, \ldots, x_n$ with probabilities $p_1, p_2, \ldots, p_n$, we have a Discrete Probability Distribution of X.

The function $p(x)$ which for $X = x_1, x_2, \ldots, x_n$ takes values $p_1, p_2, \ldots, p_n$ is the Probability Function of X.

The variable is discrete because it does not assume all values. Its properties are

$$\begin{aligned} p(x_i) &= \text{Probability that } X \text{ assumes the value } x_i \\ &= \text{Prob } (x = x_i) = p_i \\ p(x) &\geq 0, \ \Sigma p(x) = 1 \end{aligned}$$

Ex. 1. Four coins are tossed and the number of heads X noted. X can take value 0, 1, 2, 3, 4 heads.

$$p(X=0) = \left(\frac{1}{2}\right)^4 = \frac{1}{16}$$

$$p(X=1) = {}^4C_1\left(\frac{1}{2}\right)\left(\frac{1}{2}\right)^3 = \frac{4}{16}$$

$$p(X=2) = {}^4C_2\left(\frac{1}{2}\right)^2\left(\frac{1}{2}\right)^2 = \frac{6}{16}$$

$$p(X=3) = {}^4C_3\left(\frac{1}{2}\right)^3\left(\frac{1}{2}\right) = \frac{4}{16}$$

$$p(X=4) = {}^4C_4\left(\frac{1}{2}\right)^4\left(\frac{1}{2}\right)^0 = \frac{1}{16}$$

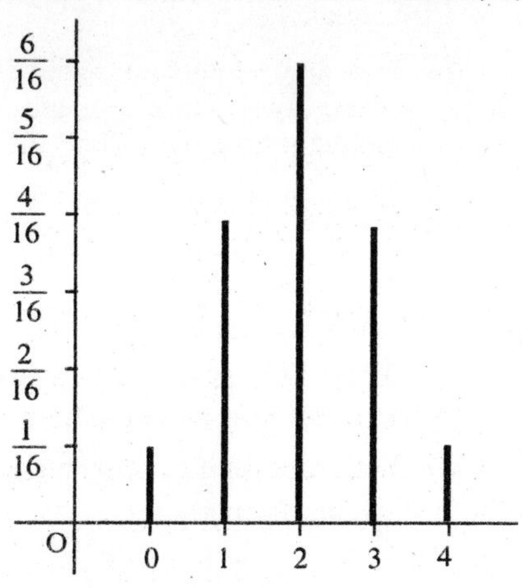

$$\sum_{x=0}^{4} p(x) = \frac{1}{16} + \frac{4}{16} + \frac{6}{16} + \frac{4}{16} + \frac{1}{16} = 1$$

This is a discrete probability distribution.

If a discrete variable X has the following probability function, find (*i*) a (*ii*) $p(X \leq 3)$ (*iii*) $p(X \geq 3)$

x_i	$p(x_i)$
0	0
1	a
2	$2a$
3	$2a^2$
4	$4a^2$
5	$2a$

Since $\Sigma p(x) = 1$, $0 + a + 2a + 2a^2 + 4a^2 + 2a = 1$

$\therefore \quad 6a^2 + 5a - 1 = 0$ so that $(6a - 1)(a + 1) = 0$

$$a = \frac{1}{6} \quad \text{or} \quad a = -1 \text{ (not admissible)}$$

For $a = \frac{1}{6}$, $p(X \leq 3) = 0 + a + 2a + 2a^2 = 2a^2 + 3a = \frac{5}{9}$

$$p(X \geq 3) = 4a^2 + 2a = \frac{4}{9}.$$

Discrete Distributions

There are several discrete distributions. The Binomial and Poisson which have been discussed are used very widely. Some other discrete distributions are described below.

22.2 UNIFORM OR RECTANGULAR DISTRIBUTION

Each possible value of the random variable x has the same probability in the uniform distribution. If x takes values $x_1, x_2, \ldots, x_k$, then

$$p(x_i,\ k) = \frac{1}{k}$$

The numbers on a die follow the uniform distribution

$$p(x_i,\ 6) = \frac{1}{6} \quad \text{(Here } x = 1, 2, 3, 4, 5, 6\text{)}$$

Bernoulli Trials

In a Bernoulli experiment an event E either happens or does not happen (E'). Examples are: getting a head on tossing a coin, getting a six on rolling a die and so on.

The Bernoulli random variable is written

$$\begin{aligned} X &= 1 \text{ if } E \text{ occurs} \\ &= 0 \text{ if } E' \text{ occurs} \end{aligned}$$

Since there are two possible values it is a case of a discrete variable where

Probability of success = $p = p(E)$

Profitability of failure = $1 - p = q = p\ (E')$

We can write

For $k = 1, f(k) = p$

For $k = 0, f(k) = q$

For $k = 0$ or $1, f(k) = p^k q^{1-k}$

22.3 NEGATIVE BINOMIAL

In this distribution the variance is larger than the mean.

Suppose the probability of success p in a series of independent Bernoulli trials remains constant.

Suppose the rth success occurs after x failures in $x + r$ trials.

(1) The probability of the success of the last trial is p

(2) The number of remaining trials is $x + r - 1$ in which there should be $r - 1$ successes. The probability of $r - 1$ successes is given by

$${}^{x+r-1}C_{r-1}\, p^{r-1}\, q^x$$

The combined probability of (1) and (2) happening together is

$$p(x) = px\ {}^{x+r-1}C_{r-1}\, p^{r-1}\, q^x \qquad x = 0, 1, 2, \ldots\ldots$$

This is the Negative Binomial distribution. We can write it in an alternative form

$$p(x) = {}^{-r}C_x\, p^r\, (-q)^x \qquad x = 0, 1, 2, \ldots\ldots$$

We can sum up as follows :

In an infinite series of Bernoulli trials the probability that $x + r$ trials will be required to get r successes is the Negative Binomial

$$p(x) = {}^{x+r-1}C_{r-1}\, p^r q^x \qquad r \geq 0$$

If $r = 1$, it becomes the Geometric distribution

If $p \to 0$, $r \to \infty$, $rp = m$ a constant, then the negative binomial tends to the Poisson distribution.

22.4 GEOMETRIC DISTRIBUTION

Suppose the probability of success p in a series of independent trials remains constant.

Suppose the first success occurs after x failures, *i.e.*,there are x failures preceding the first success. The probability of this event will be given by $p(x) = q^x p$ $(x = 0, 1, 2, \ldots\ldots)$

This is the Geometric distribution and can be derived from the Negative Binomial. If we put $r = 1$ in the Negative Binomial distribution:

$$p(x) = {}^{x+r-1}C_{r-1}\, p^r q^x$$

we get the Geometric distribution

$$p(x) = {}^{x}C_o\, p^1 q^x = pq^x$$

$$\Sigma p(x) = \sum_{n=0}^{p} q^x p = \frac{p}{1-q} = 1$$

$$E(x) = \text{Mean} = \frac{q}{p}$$

$$\text{Variance} = q/p^2$$

$$\text{Mode} = \left(\frac{1}{2}\right)^x$$

Ex 2. Find the expectation of the number of failures preceding the first success in an infinite series of independent trials with constant probability p of success.

The probability of success in

1st trial $= p$ (success at once)

2nd trial $= qp$ (one failure then success)

3rd trial $= q^2p$ (two failures then success and so on)

The expected number of failures preceding the success

$$E(x) = 0 \cdot p + 1 \cdot qp + 2q^2p + \ldots\ldots\ldots\ldots$$

$$= qp(1 + 2q + 3q^2 + \ldots\ldots)$$

$$= qp \frac{1}{(1-q)^2} = qp \frac{1}{p^2} = \frac{q}{p} \quad (\text{since } p = 1 - q)$$

22.5 HYPERGEOMETIC DISTRIBUTION

From a finite population of size N, a sample of size n is drawn without replacement.

Let there be N_1 successes out of N

The number of failures is $N_2 = N - N_1$

The distribution of the random variable X which is the number of successes obtained in the above case, is called the Hypergeometic distribution

$$p(x) = \frac{{}^{N_1}C_x \, {}^{N}C_{n-x}}{{}^{N}C_n} \quad (X = 0, 1, 2, \ldots, n)$$

Here x is the number of successes in the sample and $n - x$ is the number of failures in the sample.

It can be shown that

$$\text{Mean} : E(X) = n\frac{N_1}{N}$$

$$\text{Variance} : \text{Var}(X) = \frac{N-n}{N-1}\left(\frac{nN_1}{N} - \frac{nN_1^2}{N}\right)$$

Ex. 3. There are 20 lottery tickets with three prizes. Find the probability that out of 5 tickets purchased exactly two prizes are won.

We have $N_1 = 3$, $N_2 = N - N_1 = 17$, $x = 2$, $n = 5$

$$p(2) = \frac{{}^3C_2 \, {}^{17}C_3}{{}^{20}C_5}$$

The probability of no prize $p(0) = \frac{{}^3C_0 \, {}^{17}C_5}{{}^{20}C_5}$

The probability of exactly 1 prize $p(1) = \frac{{}^3C_1 \, {}^{17}C_4}{{}^{20}C_5}$

Ex. 4. From an urn containing 20 black and 30 red balls, 6 balls are drawn at random. Find the probability that no red ball is selected. **Ans.** $p(X = 0) = {}^{20}C_6/{}^{50}C_6$

Ex. 5. Examine the nature of the distribution if r balls are drawn, one at a time without replacement, from a bag containing m white and n black balls.

It is the hypergeometric distribution. It corresponds to the probability that x balls will be white out of r balls so drawn and is given by

$$p(x) = \frac{{}^mC_x \, {}^nC_{r-x}}{{}^{m+n}C_r}$$

22.6 THE MULTINOMIAL

There are k possible outcomes of trials viz $x_1, x_2, \ldots, x_k$ with probabilities $p_1, p_2, \ldots, p_k$. n independent trials are performed. The multinomial distribution gives the probability that out of these n trials, x_1 occurs n_1 times, x_2 occurs n_2 times and so on. This is given by

$$\frac{n!}{n_1!\, n_2! \ldots n_k!} \; p_1^{n_1} p_2^{n_2} \ldots p_k^{n_k}$$

where $$\sum_{i=1}^{k} n_i = n$$

Ex. 6. What are the characteristic features of the Binomial distribution?

1. It is a discrete distribution.
2. It gives the probability of x successes and $n - x$ failures in a specific order.
3. The experiment consists of n repeated trials.
4. Each trial results in a success or a failure.
5. The probability of success remains constant from trial to trial.
6. The trials are independent.
7. The success probability p of any outcome remains constant over time. This condition is usually not fully satisfied in situations involving management and economics, *e.g.*, the probability of response from successive informants is not the same. However, it may be assumed that the condition is reasonably well satisfied in many cases and that the outcome of one trial does not depend on the outcome of another. This condition too may not be fully satisfied in many cases. An investigator may not approach a second informant with the same set-up of mind as used for the first informant.

8. The binomial distribution depends on two parameters n and p. Each set of different values of n, p has a different binomial distribution.

9. If $p = 0.5$ the distribution is *symmetrical*. For a symmetrical distribution, in n trials

$$\text{Prob. } (X = 0) = \text{Prob } (X = n)$$

i.e., the probabilities of 0 or n successes in n trials will be the same. Similarly

$$\text{Prob. } (X = 1) = \text{Prob}(X = n - 1) \text{ and so on.}$$

If $p > 0.5$, the distribution is not symmetrical. The probabilities on the right are larger than those on the left. The reverse case is when $p < 0.5$.

When n becomes large the distribution becomes bell shaped. Even when n is not very large but $p \simeq 0.5$, it is fairly bell shaped.

10. The binomial distribution can be approximated by the normal. As n becomes large and p is close to 0.5 the approximation becomes better.

Ex. 7. If the ratio n/N, *i.e.*, sample size to population size is small, the results given by the Binomial may not be reliable. Comment.

When the distribution is binomial, each successive trial, being independent of other trials, has a constant probability of success. If the sampling of n items is without replacement from a population of size N, the probability of success of any event depends upon what happened in the previous events. In that case the Binomial cannot be used unless the ratio n/N is small. Even then there is no guarantee of getting accurate results.

The Binomial should be used only if the ratio $\frac{n}{N}$ is very small, say less that 0.05.

Ex. 8. Explain the concept of a Discrete Probability Distribution.

If a random variable x assumes n discrete values $x_1, x_2, \ldots\ldots\ldots, x_n$ with respective probabilities $p_1, p_2, \ldots\ldots, p_n$ $(p_1 + p_2 + \ldots + p_n = 1)$ then the distribution of values x_i with probabilities p_i (= 1, 2,...n) is called the Discrete Probability Distribution of x.

The frequency function or frequency distribution of x is defined by $p(x)$ which for different values $x_1, x_2, \ldots, x_n$ of x gives the corresponding probabilities:

$p(x_i) = p_i$ where $p(x) \geq 0$, $\Sigma p(x) = 1$

Ex. 9. For the following probability distribution find $p(x > 4)$ and $p(x \leq 4)$

x	0	1	2	3	4	5
$p(x)$	0	a	$a/2$	$a/2$	$a/4$	$a/4$

Since $\Sigma p(x) = 1$, $0 + a + \frac{a}{2} + \frac{a}{2} + \frac{a}{4} + \frac{a}{4} = 1$

$$\therefore \quad \frac{5}{2}a = 1 \quad \text{or} \quad a = \frac{2}{5}$$

$$p(x > 4) = p(x = 5) = \frac{9}{4} = \frac{1}{10}$$

$$p(x \leq 4) = 0 + a + \frac{a}{2} + \frac{a}{2} + \frac{a}{4} = \frac{9a}{4} = \frac{9}{10}$$

Ex. 10. A fair coin is tossed 400 times. Find the mean number of heads and the corresponding standard deviation.

This is a case of Binomial distribution with $p = q = \frac{1}{2}$, $n = 400$

The mean number of heads is given by $\mu = np = 400 \times \frac{1}{2} = 200$ and S.D.

$\sigma = \sqrt{npq} = \sqrt{400 \times \frac{1}{2} \times \frac{1}{2}} = 10.$

Ex. 11. A manager has thought of 4 planning strategies each of which has an equal chance of being successful. What is the probability that at least one of his strategies will work if he tries them in 4 situations? Here $p = \frac{1}{4}$, $q = \frac{3}{4}$

The probability that none of the strategies will work is given by

$$p(0) = {}^4C_0 \left(\frac{1}{4}\right)^0 \left(\frac{3}{4}\right)^4 = \left(\frac{3}{4}\right)^4$$

The probability that at least one will work is given by $1 - \left(\frac{3}{4}\right)^4 = \frac{175}{256}$

Ex. 12. Suppose the proportion of people preferring a car C is 0.5. Let X denote the number of people out of a set of 3 who prefer C. The probabilities of 0, 1, 2, 3 of them preferring C are

$$p(X = 0) = {}^3C_0(0.5)^0 (0.5)^3 = \frac{1}{8}$$

$$p(X = 1) = {}^3C_1(0.5)^1 (0.5)^2 = \frac{3}{8}$$

$$p(X = 2) = {}^3C_1(0.5)^2 (0.5)^1 = \frac{3}{8}$$

$$p(X = 3) = {}^3C_3(0.5)^3 (0.5)^0 = \frac{1}{8}$$

$$\mu = E(X) = \Sigma x_i p_i = 0 \times \frac{1}{8} + 1 \times \frac{3}{8} + 2 \times \frac{3}{8} + 3 \times \frac{1}{8} = 1.5$$

$$\sigma^2 = E(X - \mu)^2 = E(X^2) - \mu^2 = \Sigma x_i^2 p_i - \mu^2$$

$$= 0^2 \times \frac{1}{8} + 1^2 \times \frac{3}{8} + 2^2 \frac{3}{8} + 3^2 \times \frac{1}{8} - 1.5^2$$

$$= 0.75$$

Ex. 13. For the Poisson distribution, write the probabilities of 0, 1, 2, ... successes.

x	$p(x) = e^{-m} \dfrac{m^x}{x!}$
0	$p(0) = e^{-m}\, m^0/0! = e^{-m}$
1	$p(1) = e^{-m} \dfrac{m}{1!} = p(0) \cdot m$

2	$e^{-m} \dfrac{m^2}{2!} = p(2) = p(1) \cdot \dfrac{m}{2}$
3	$e^{-m} \dfrac{m^3}{3!} = p(3) = p(2) \cdot \dfrac{m}{3}$
⋮	

and so on.

Total of all probabilities $\Sigma p(x) = 1$

Ex. 14. What are the raw moments of Poisson distribution?

First raw moment $\mu_1' = m$

Second raw moment $\mu_2' = m^2 + m$

Third raw moment $\mu_3' = m^3 + 3m^2 + m$

Ex. 15. 20% of bolts in a factory are defective. Deduce the probability distribution of the number of defectives in a sample of 5.

Ans.

x	$p(x)$
0	1024/3125
1	256/625
2	128/625
3	32/625
4	4/625
5	1/3125

Total of all probabilities = 1

Ex. 16. For a Poisson distribution $p(0) = p(1)$. Find $p\ (x > 0)$.

We have $e^{-m}m^0/0! = e^{-m}m/1!$ so that $m = 1$

$\therefore$ and $p(0) = e^{-1} = 1/2.718 = 0.37$

$$p(x > 0) = 1 - p(0) = 0.63$$

Ex. 17. In 122 ancient battles the number of men killed by a horse kick is shown below. Find the expected frequencies using the Poisson distribution.

No. killed	Frequencies of battles	
x	f	fx
0	109	0
1	65	65
2	22	44
3	3	9
4	1	4
		122

$$m = \frac{122}{200} = 0.61 \text{ (Check : For the Poisson, mean } m = \text{variance)}$$

Find variance and check that it is approximately equal to m, the mean. This is the characteristic of a Poisson distribution. The probabilities of 0, 1, 2, 3, 4 killed are

$$e^{-0.61}, \quad \frac{e^{-0.61} \times .61}{1!}, \quad \frac{e^{-0.61} \times (.61)^2}{2!}, \quad \frac{e^{-0.61} \times (.61)^3}{3!}, \quad \frac{e^{-0.61} \times (.61)^4}{4!}$$

or 0.5435 0.3315 0.1110 0.0205 0.0085

Multiplying each by 200 we get the expected frequencies

	f_e	108.7,	66.3,	22.2	4.1,	1.7
Compare	f_o	109,	65	22,	3	1

The observed frequencies are reasonably close to the expected frequencies.

Ex. 18. Find the probability that at the most 5 defectives will be found in a box of 200 if 2% items are usually known to be defective? (Assume Poisson distribution)

Hint. $p = \frac{1}{50}$, $n = 200$, $e^{-4} = 0.0183$ **Ans.** 0.785

Ex. 19. (*a*) It is claimed that the TV branded *M* is demanded by 70% of customers. If *X* is the number of TVs demanded find the probability distribution of *X* when there are four customers.

If all the four demand *M* then $p(4) = 0.7^4 = 0.2401$. The prob. dist. is

X	4	3	2	1	0
p	0.2401	.4116	.2646	.0756	.0081

These values may be plotted.

(*b*) Comment on the accuracy of the following :

(*i*) For a Binomial distribution (*a*) mean = 4, variance = 3 (*b*) mean = 3, S.D. = 2

(*ii*) For a Poisson, mean = 10, S. D. = 5.

(*iii*) For a normal distribution, mean = 50, median = 52.

Hint. (*i*) (*a*) For a Binomial mean = $np = 4$ variance = $npq = 3$

$$\frac{\text{Variance}}{\text{Mean}} = \frac{npq}{np} = q = \frac{3}{4}.$$ This is possible since q should be less than 1.

(*i*) (*b*) $SD^2/\text{mean} = npq/np = q = 4/3$. Results inaccurate

(*ii*) For a Poisson, mean $\simeq$ variance. Here mean = 10, variance = 25. Results inaccurate.

(*iii*) The Normal is symmetrical so that mean = median = mode

Results inaccurate.

Note. Poisson Approximation to the Binomial. When the number of trials *n* is large and the probability *p* of a success is small, the binomial distribution can be approximated by the Poisson. This approximation is useful is practice.

If $p = 0.002$, $n = 1000$, $m = np = 2$

The probability of 2 successes in 1000 trials is

$$p(X = 2) = {}^nC_xp^xq^{n-x} = 1000\ C_2(0.002)^2\ (0.998)^{998}$$

Similarly, $$p(X = 3) = {}^{1000}C_3(0.002)^3\ (0.998)^{997} \text{ etc.}$$

These terms are difficult to calculate. If we employ the Poisson, we have a much easier task before us.

$$m = np = 1000 \times .002 = 2$$

$$p(X = 2) = \frac{e^{-m} m^x}{x!} = \frac{e^{-2} 2^2}{2!} = 0.1353 \times 2 = 0.2706$$

$$p(X = 3) = \frac{e^{-2} 2^3}{3!} = 0.1804 \text{ etc.}$$

Ex. 20. One in every 100 items made on a machine is defective. Out of 25 items picked, find the probability of 1 item being defective.

$$p = .01, q = .99, n = 25, np = 0.25$$

$$\text{Binomial} : p(1) = {}^{25}C_1 \, (0.1)^1 \, (0.99)^{24} = 0.1964$$

$$\text{Poisson} : p(1) = \frac{e^{-.25} (0.25)^1}{1!} = 0.1947$$

22.7. CONTINUOUS PROBABILITY DISTRIBUTIONS

When a random variate can take any value in the given interval $a \leq x \leq b$, it is a continuous variate and its distribution is a *Continuous Probability Distribution.*

Theoretical distributions are often continuous. They are useful in practice because they are convenient to handle mathematically. They can serve as good approximations to discrete distributions.

The range of the variate may be finite or infinite.

A continuous random variable can take all values in a given interval.

A continuous probability distribution is represented by a smooth curve.

The total area under the curve for a probability distribution is necessarily unity. The curve is always above the x axis because the area under the curve for any interval represents probability and probabilities cannot be negative.

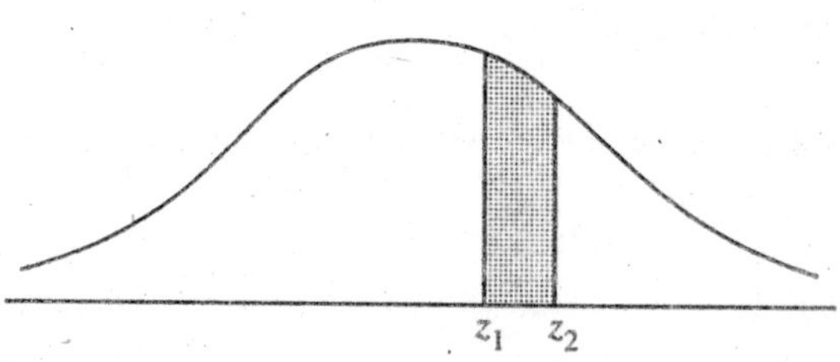

If X is a continuous variable, the probability of X falling in an interval with the end points z_1, z_2 may be written $p(z_1 \leq X \leq z_2)$

This probability corresponds to the shaded area under the curve.

A function is a probability density function if $\int_{-\infty}^{\infty} p(x)dx = 1, \; p(x) \geq 0, -\infty < x < \infty,$

i.e., the area under the curve $p(x)$ is 1 and the probability of x lying between two values a, b, *i.e.*, $p(a < x < b)$ is positive. The most prominent example of a continuous probability function is the normal distribution.

Cumulative Probability Function (cpf)

The cpf shows the probability that x takes a value less than or equal to, say, z and corresponds to the area under the curve up to z.

$$p(x \leq z) = \int_{-\infty}^{z} p(x)dx$$

This is denoted by $F(x)$.

22.8. DEGREES OF FREEDOM

To compute a sampling statistic we have to make use of sample observations. We have also to make use of population parameters which being usually unknown, have to be estimated from the same. The computation of k such parameters from the sample implies the loss of k free observations.

We use the term degrees of freedom to denote the number of independent observations in a sample of size n from which k population parameter have been estimated. The number of degrees of freedom is written

$$\nu = n - k$$

In the case of t statistic if the population mean μ is estimated from the sample, then $\nu = n - 1$

22.9. EXACT SAMPLING DISTRIBUTIONS

Sampling distributions are fundamental in the study of statistics. If $x_1, x_2, \ldots, x_n$ is a random simple from a normal population, we can study the *exact* sampling distributions of

(1) $\chi^2 = \Sigma x^2$

(2) $t = \dfrac{(\overline{x} - \mu)}{s / \sqrt{n}}$ and

(3) $F = s_1^2 / s_2^2$

22.10 THE χ^2 DISTRIBUTION (Read chi-square)

χ^2 distribution is a continuous probability distribution

If $z = \dfrac{x - \mu}{\sigma} \sim N(0, 1)$ *i.e.*, z is normal with mean = 0, SD = 1

Then $$z = \left(\frac{x - \mu}{\sigma}\right)^2 \sim \chi^2 \text{ with 1 d.f.}$$

Note. ~ stands for "is distributed as'

N for normal, $N(0, 1)$ for normal with 0 mean and SD = 1

Properties

1. If $x_1, x_2, \ldots x_n$ are independent normal variates with means $\mu_1, \mu_2, \ldots, \mu_n$ and s.d. $\sigma_1, \sigma_2, \ldots, \sigma_n$ then

$$\chi^2 = \sum_{i=1}^{n} \left(\frac{x - \mu_i}{\sigma_i}\right)^2 \sim \chi^2 \text{ with d.f.n.}$$

2. It can be shown that $p(\chi^2) = k\ (\chi^2)^{\frac{n}{2}-1}\ e^{-\chi^2/2}$ (Probability function of χ^2). Here k is a constant, $v = d.f.$, n = no. of observations. $0 < \chi^2 < \infty$

3. Mean = v, Mode = $v - 2$, SD = $\sqrt{2v}$

4. If $\chi_1^2 \sim \chi^2$ with n_1 $d.f.$, $\chi_2^2 \sim \chi^2$ with n_2 $d.f.$ then $\chi_1^2 + \chi_2^2 \sim \chi^2$ with $n_1 + n_2$ $d.f.$

5. χ^2 can be used for both large and small samples.

The χ^2 distribution, (read chi – square distribution) is a continuous probability distribution. This distribution is useful in testing the independence between observed and expected frequencies, goodness of fit and other purposes.

The χ^2 distribution is highly skewed on the right for small numbers of degrees of freedom. As this number starts going beyond 10, the distribution curve tends to become symmetrical. For large n, the χ^2 distribution is approximately normal.

22.11 *t* DISTRIBUTION

If $x_1, x_2, \ldots, x_n$ is a random sample from a normal population with mean μ, variance σ^2, then

$$t = \frac{\bar{x} - \mu}{s/\sqrt{n}} \text{ follows the } t \text{ distribution with } n - 1 \ d.f.$$

1. We can write $p(t) = k(1 + t^2/v)^{-\frac{v}{2}-\frac{1}{2}}$

This is the probability function of t where v stands for $d.f.$ and $-\infty < t < \infty$

Here, k is a constant

$$s^2 = \frac{\Sigma(x - \bar{x})^2}{n - 1} \text{ is an unbiassed estimate of } \sigma^2$$

The t distribution is widely used in small sample tests.

2. The curve has a maximum ordinate at $t = 0$

3. The curve is asymptotic to the t axis at each end.

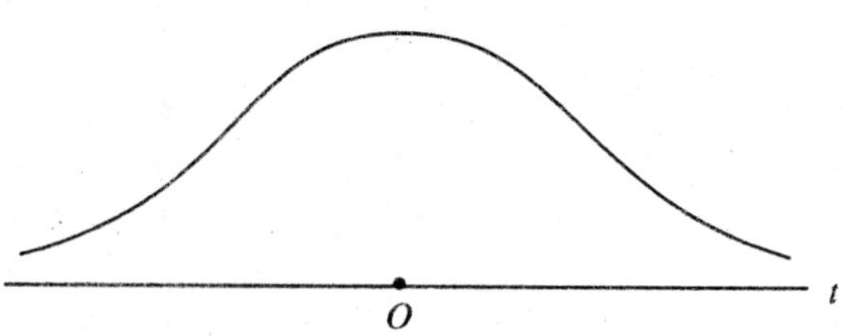

22.12 F DISTRIBUTION

If χ_1^2, χ_2^2 are independent χ^2 variates with v_1, v_2 $d.f.$ then

$$F = \frac{\chi_1^2/v_1}{\chi_2^2/v_2}$$

follows F distribution with (v_1, v_2) $d.f.$

1. The probability density function of F is $k\ F^{\frac{\nu_1}{2}-1}\left(1+\frac{\nu_1 F}{\nu_2}\right)^{-\frac{\nu_1}{2}-\frac{\nu_2}{2}}$ $\quad 0 \leq F < \infty$

Here k is a constant; ν_1, ν_2 are *d.f.* for numerator and denominator.

2. $AM = \frac{\nu_2}{\nu_2 - 2}$, $SD = \frac{\nu_2}{\nu_2 - 2}\sqrt{\frac{2(\nu_1 + \nu_2 - 2)}{\nu_1(\nu_2 - 4)}}$

3. The F curve is positive skew.

4. If $\nu_1 = \nu_2 = h - 1$ then A.M. = H.M. = $\frac{h-1}{h-3}$

The median is at $F = 1$.

Ex. 21. If $\log_{10} x$ is normally distributed with mean 4 and variance 4 find the probability of $1.202 < x < 83,180,000$

$$z = \frac{\log_{10} x - \mu}{\sigma}$$

$$z_1 = \frac{(\log_{10} 1.202) - 4}{2} = \frac{0.08 - 4}{2} = -1.96$$

$$z_2 = \frac{\log_{10} 8318000 - 4}{2} = \frac{7.92 - 4}{2} = +1.96$$

Area between $-$ 1.96 and 1.96 = Prob $(-1.96 \leq z \leq 1.96) = 0.95$

(Using the Normal Tables.)

CHAPTER 23

Correlation and Regression

23.1 RELATIONSHIPS

We are familiar with functional relationships like those between pressure and volume of a gas, radius and circumference of a circle and so on. These are fixed or perfect relationships, frequent in natural sciences.

Often, there are sets of observations found in pairs or groups which tend to change together. There is a certain degree of association but not of a fixed or functional type. For example, heights and weights of individuals are highly associated or correlated but there is no definite mathematical relationship between the two. Demand for a commodity and its price are correlated but no mathematically exact relationship can be established between them. Almost any association which is concerned directly or indirectly with an underlying human element is likely to be imperfect and cannot be accurately determined like a functional relationship. Inadequacy of information and inaccuracies of various types may not let us determine the degree of association properly. But it is possible to measure the degree of such relationships by means of correlation and regression techniques.

Functional relationships are usefully employed in physical sciences to study the degree of association between variables and the laws connecting them. In social sciences, functional relationships cannot be utilised since exact relationships between statistical variables do not exist in general.

Correlation and regression are statistical devices to measure the amount of similarity and variations *i.e.*, the degree of association between series of pairs or sets of observations of two or more variables. These tools are useful in finding how much a variable changes corresponding to the average amount of change in other variables in the data.

Bivariate population

If for every value of a variable x there is a corresponding value of another variable y, and there are n observations, the set of pairs of values may be written as follows :

x	y
x_1	y_1
x_2	y_2
x_3	y_3
$\vdots$	$\vdots$
x_n	y_n

The set of pairs of observations $(x_1, y_1), (x_2, y_2), \ldots, (x_n, y_n)$ constitutes a bivariate population.

Is there a relationship between the two variables?

Can this relationship be measured?

Can it be expressed by means of a functional relation?

What is the degree of relationship, if any?

Related Measurements : Correlation and Regression

Often we want to find the possible relationship between two series of observations such that the values in one series are affected by the values in the other.

A relationship between two such sets of observations is called *correlation.*

The procedure to use one variable to make predictions about another is called *regression.* These relations are not functional and the degree of relationship can be measured approximately.

23.2 PEARSON'S COEFFICIENT OF CORRELATION

Alternatively called *total correlation* or *product moment correlation*, the coefficient of correlation is a measure of the degree of relationship between two variables x and y. It gives, in numerical terms, the extent to which sample observations on x are correlated with sample observations on y.

The coefficient of correlation, denoted by r, is defined in such a way as to be a pure number so that the units of x and y do not affect the result. The value of r lies between -1 and $+1$.

$r = +1$ shows a perfect positive correlation

$r = -1$ shows a perfect negative correlation

$r = 0$ shows absence of correlation

r nearly equal to $+1$ shows a strong or significant positive correlation

r nearly equal to zero shows an insignificant correlation.

A formula to estimate the coefficient of correlation given by

$$r = \frac{\Sigma(x-\bar{x})(y-\bar{y})}{\sqrt{\Sigma(x-\bar{x})^2}\sqrt{\Sigma(y-\bar{y})^2}}$$

which can be written in the following alternative form for practical purposes :

$$r = \frac{\Sigma xy - n\bar{x}\bar{y}}{\sqrt{\Sigma x^2 - n\bar{x}^2}\sqrt{\Sigma y^2 - ny^2}}$$

Ex. 1. Find the coefficient of correlation for the following 10 sample observations:

x :	5	7	8	4	9	3	2	5	4	3
y :	2	4	5	5	6	5	4	4	3	2

The procedure shown below should be followed to calculate Σx^2, Σy^2, Σxy. Note that $n = 10$.

x	y	x^2	y^2	xy
5	2	25	4	10
7	4	49	16	28
8	5	64	25	40
4	5	16	25	20
9	6	81	36	54
3	5	9	25	15
2	4	4	16	8
5	4	25	16	20
4	3	16	9	12
3	2	9	4	6
$\Sigma x = 50$	$\Sigma y = 40$	$\Sigma x^2 = 298$	$\Sigma y^2 = 176$	$\Sigma xy = 213$

$$r = \frac{\Sigma xy - n\bar{x}\,\bar{y}}{\sqrt{\Sigma x^2 - n\bar{x}^2}\ \sqrt{\Sigma y^2 - n\bar{y}^2}}$$

$$= \frac{213 - 10 \times 5 \times 4}{\sqrt{298 - 10 \times 25}\ \sqrt{176 - 10 \times 16}}$$

$$= \frac{13}{\sqrt{48}\ \sqrt{16}} = 0.47$$

showing that x and y are positively correlated. The correlation between x, y does not seem to the strong.

23.3 SCATTER DIAGRAM

When the observations in both the series increase (or decrease) together the two series are positively correlated.

If the observations in one series increase while those in the other decrease, the correlation is negative.

A simple representation of the relationship between two variables can be done by means of a scattergram which is merely a graphic presentation of the set of n given pairs of observations on x and y in a coordinate system.

Ex. 2. Consider the scattergrams for the following data :

Demand x	*Price* y	*Supply* x	*Price* y
16	3	4	3
10	4	5	4
8	6	8	5
5	5	9	6
4	7	10	7
2	8	12	8
1	11	15	11

In the demand price relation we note that there is a tendency for small values of x to be associated with large values of y and *vice versa.*

We also see that the general trend of the scatter is nearly a straight line showing a linear relationship between x and y.

This is a case of a nearly linear *inverse* relationship between x and$y$ (demand and price). See Diagram below.

The supply price relation presents a case of nearly linear, *direct* relationship between x and y (supply and price).

If the points all lie on a straight line we have a perfect positive or negative linear relationship ($r = \pm 1$).

The relationships between the two variables x and y has been represented by means of a scatter diagram which is a visual pattern of pairs of data shown as points between the coordinate axes.

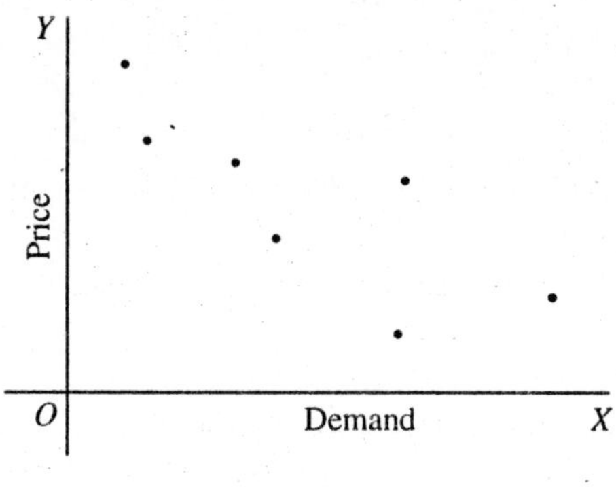

The scatter diagram shows the presence of a negative relationships between demand and price.

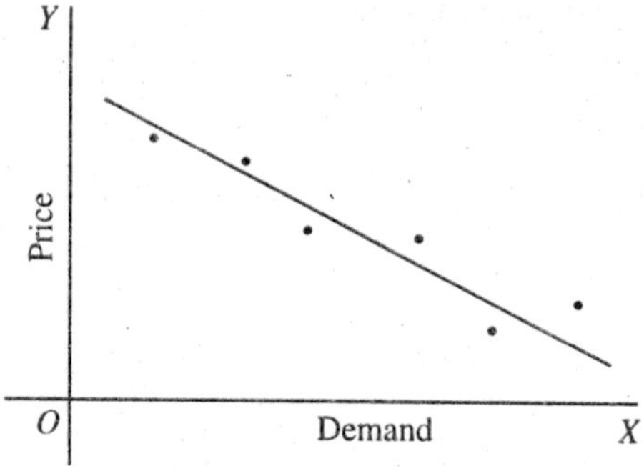

Inverse regression relationship between demand and price assumed to be linear.

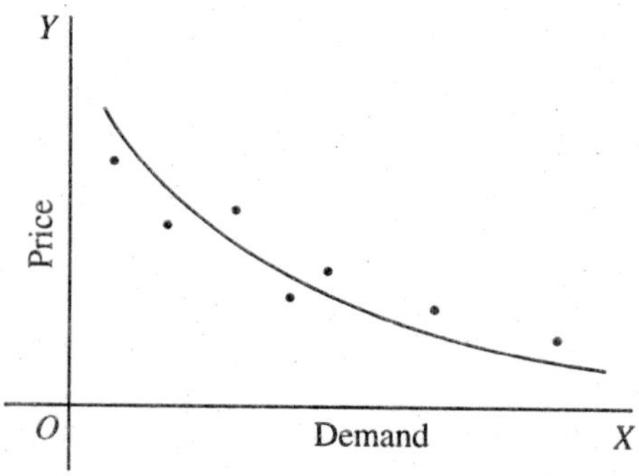

Inverse regression relationship between demand and price assumed to be curvilinear

The relationship between the variables x and y can be linear or curvilinear *i.e.*, it may be assumed that a straight light passes through the points or a curve passes through the points. All points need not necessarily lie on the line or the curve. They should be well balanced on both sides of the line or the curve.

Ex. 3. Consider the scatter diagram for the supply-price data:

Supply	x	4	5	8	9	10	12	15
Price	y	3	4	6	5	7	8	11

The relationship is direct and can be seen to be approximately linear

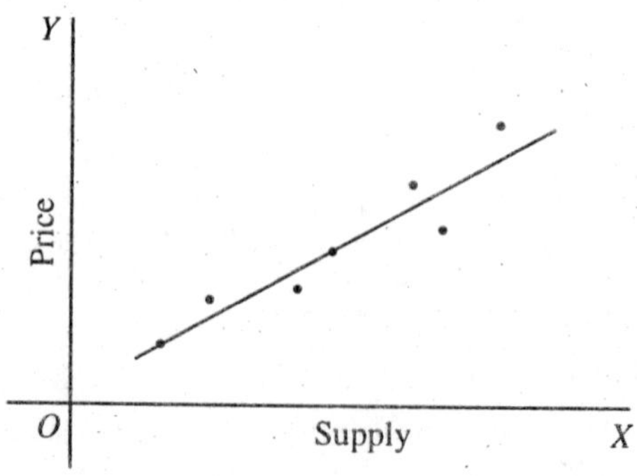

The following diagrams show some possible relations in practice.

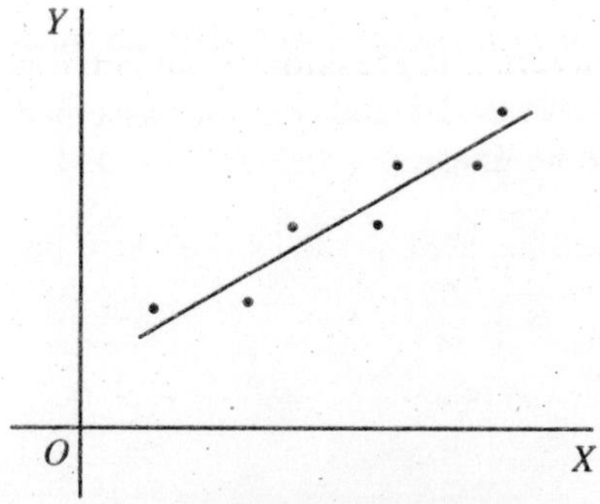

Direct Linear
High positive correlation
r, high positive

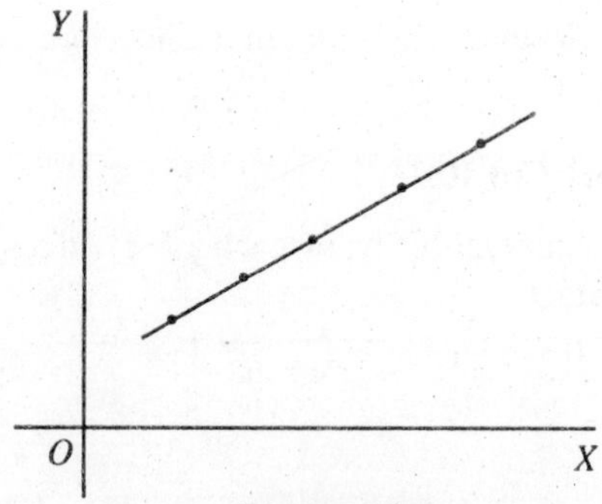

Direct Linear
Perfect positive correlation
$r = +1$

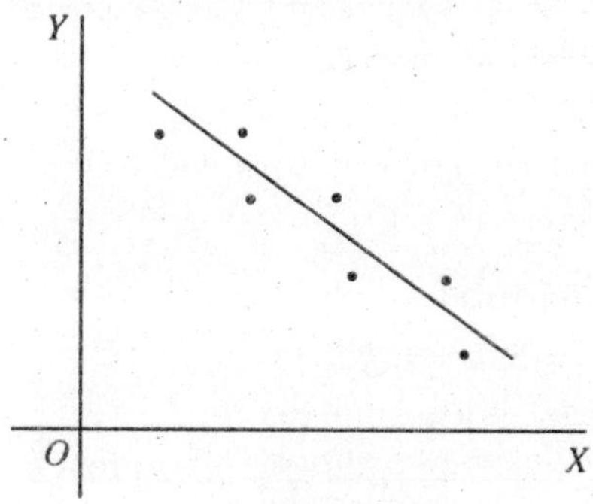

Inverse Linear
High negative correlation
r, high negative

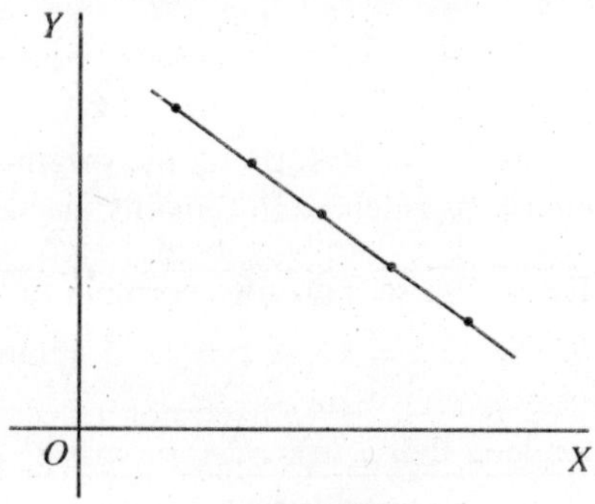

Inverse Linear
Perfect negative correlation
$r = -1$

Absence of any relationship between variables is reflected by scattergrams in which all points are haphazardly scattered without showing any definite relationship. In practice, relationships, if any, are never perfect. They are approximately linear or curvilinear.

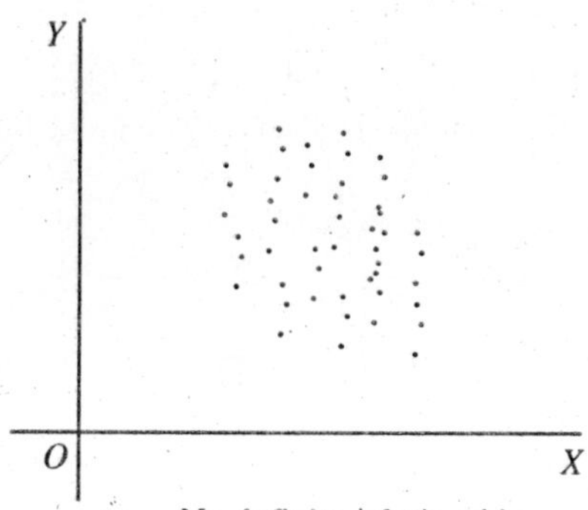

No definite relationship

Change of origin and scale

Computational simplification.

The value of r is independent of origin of reference and the units of measurement. It means r remains unaffected by the addition (or subtraction) of a constant to, and/or by multiplication (or division) by a constant, of the values of either or both variables. In practice it means that if we subtract any number from x values and any number from y

values and divide the resulting quantities by any number and then find r for the resulting new set of observations, the value of r remains unchanged.

Whether the units of x and y are comparable or not, r will have a definite meaning provided x, y are logically related.

Short Cut Rule

Subtract x_0 from each of x values and y_0 from each of y values. Let h, k be any constants.

If we write

$$u = \frac{x - x_0}{h}, \quad u = \frac{y - y_0}{k}$$

it can be proved that the formula for r given above becomes

$$r = \frac{\Sigma(u - \bar{u})(v - \bar{v})}{\sqrt{\Sigma u^2 - n\bar{u}^2}\sqrt{\Sigma v^2 - n\bar{v}^2}}$$

$$= \frac{\Sigma u\, v - n\bar{u}\,\bar{v}}{\sqrt{\Sigma u^2 - n\bar{u}^2}\ \sqrt{\Sigma v^2 - n\bar{v}^2}}$$

which has the same form as the original except that x is replaced by u and y by v. The procedure for calculation remains the same.

Ex. 4. We will use this formula to calculate r for Ex. 1.

We write $u = x - 4$, $v = y - 5$. (Here $x_0 = 4$, $y_0 = 5$, $h = 1$, $k = 1$)

(We need not necessarily *divide* by any quantities.)

x	y	$u = x - 4$	$v = y - 5$	u^2	v^2	uv
5	2	1	– 3	1	9	– 3
7	4	3	– 1	9	1	– 3
8	5	4	0	16	0	0
4	5	0	0	0	0	0
9	6	5	1	25	1	5
3	5	– 1	0	1	0	0
2	4	– 2	– 1	4	1	2
5	4	1	– 1	1	1	– 1
4	3	0	– 2	0	4	0
3	2	– 1	– 3	1	9	3
		$\Sigma u = 10$	$\Sigma v = -10$	$\Sigma u^2 = 58$	$\Sigma v^2 = 26$	$\Sigma uv = 3$

$$\bar{u} = \frac{\Sigma u}{n} = \frac{10}{10} = 1,\ \bar{v} = \frac{\Sigma v}{n} = \frac{-10}{10} = -1$$

$$r = \frac{\Sigma uv - n\bar{u}\bar{v}}{\sqrt{\Sigma u^2 - n\bar{u}^2}\sqrt{\Sigma v^2 - n\bar{v}^2}} = \frac{3 - 10 \times 1 \times (-1)}{\sqrt{58 - 10 \times 1}\sqrt{26 - 10 \times 1}}$$

$$= \frac{3+10}{\sqrt{48}\sqrt{16}} = \frac{13}{\sqrt{48 \times 16}}\ 0.47.$$ Same value as obtained earlier.

Take a different origin *i.e.*, subtract different numbers from x and y, *e, g.* Take (*i*) $u = x - 3$, $v = y - 6$ (*ii*) $u = x - 5$, $v = y - 4$ and check if r remains the same in all the cases.

Ex. 5. Calculate r for demand-price and supply-price data given above.

Ex. 6. For the following data showing the salaries and allowances of 12 persons, calculate r :

Salary	*Allowance*	*Salary*	*Allowance*
439	148	245	106
467	148	465	143
535	190	456	149
580	130	453	135
388	101	428	165
102	203	477	182

Write x for salary, y for allowances

Put $u = x - x_0$, $v = y - y_0$ where

x_0, y_0 are suitable numbers.

In this case, as in any other case, it is preferable to take x_0, y_0 nearest the mean values of x, y respectively.

Take $x_0 = 450$, $y_0 = 150$ and show that $r = 0.57$.

Ex. 7. Find r for the following data :

Age of husband :	23	27	28	29	30	31	33	35	36	39
Age of wife :	18	22	23	24	25	26	28	22	30	32

Ans. 0.90

Ex. 8. The marks obtained by 10 students in two tests (out of 100 marks) are shown below. Show that $r = 0.67$.

Test I :	10	9	8	7	6	5	4	3	2	1
Test II :	7	5	9	3	7	5	7	4	1	2

23.4 SPURIOUS CORRELATION

$r = 0$ means the variables x, y are not linearly related.

r nearly equal to 1 means there is a strong linear relationship between x, y.

Other values of r show different degrees of correlation.

A high value of r does not necessarily imply a cause and effect relationship between x, y.

A low value of r does not necessarily imply the measurements are totally unrelated.

The reason is that r on its own gives a purely mathematical relationship between x, y. There need not be any cause and effect relationship between them. We are interested in r which relates variables with cause and effect connection.

It may so happen that we may find a high value of r for two series even if there is no true relationship between them. This may be because the two series may be rising and falling in the same manner for any third set of reasons. We may say that they are both moving similarly with a third variable time. This dependence on time does not give us any cause and effect relationship. The resulting correlation is called *spurious* or nonsensical correlation.

If it is found that the correlation between the number of students passing the university examinations and the production of steel is 0.95, it is spurious correlation.

If there is a strong correlation between tax revenue and the number of dealings in black money, it may not be spurious correlation.

A high correlation between increasing number of hours available for study and decreasing number of hours devoted to study can be a meaningful correlation.

To get meaningful information about relationships between variables handle r with care.

23.5 *r* FROM CORRELATION TABLE

Coefficient of Correlation for Coded Scattergram

A correlation table is a two way frequency table showing the frequencies f in each class interval.

The formula to be used in this case is

$$r = \frac{\Sigma f xy - \Sigma f x \, \Sigma f y / n}{\sqrt{\Sigma f x^2 - \frac{(\Sigma f x)^2}{2}} \sqrt{\Sigma f y^2 - \frac{(\Sigma f y)^2}{n}}}$$

If we use, instead of x, y,

$$u = \frac{x - x_0}{h}, \quad v = \frac{y - y_0}{k}$$

where h, k are class intervals of x series and y series respectively, we have

$$r = \frac{\Sigma f uv - \frac{\Sigma f u \Sigma f v}{n}}{\sqrt{\Sigma f u^2 - \frac{(\Sigma f u)^2}{n}} \sqrt{\Sigma f v^2 - \frac{(\Sigma f v)^2}{n}}}$$

Ex. 9. The following table gives the marks in mathematics and statistics obtained by students in an examination. Find r.

Marks in Stats. \ Marks in Maths.	15–35	35–55	55–75	75–95
20—30	1	3		
30—40		2	2	2
40—50		4	1	1
50—60			3	
60—70			1	

$\Sigma fv = -6$ $\Sigma fv^2 = 28$

$\Sigma fu = -4$ $\Sigma fu^2 = 16$

$\Sigma fuv = 13$ $n = 20$ (See table below)

Explanation

(*i*) $u = \frac{x-60}{20}$, *e.g.* $\frac{20-60}{20} = -2$ etc.

$v = \frac{y-45}{10}$, *e.g.* $\frac{55-45}{10} = 1$ etc.

(*ii*) f_v : sum of frequencies for v *e.g.* $1 + 3 = 4$

f_u : sum of frequencies for u *e.g.* $2 + 1 + 3 + 1 = 7$

(*iii*) vf, is obtained by multiplying v and f_v values *e.g.* $-2 \times 4 = -8$ etc. Similarly $ufu = u \times f_u$

(*iv*) $v^2 f_v$ is obtained by multiplying v by vf_v *e.g.* $(-2) \times (-8) = 16$. Similarly $u^2 f_u = u \times u\, f_u$

(*v*) Bracketed numbers in cells are obtained by multiplying the frequency in the cell by the corresponding u and v values. *e.g.* in the cell corresponding to

(*a*) row 1, col. 1, $u = -2, v = -2, f = 1, \therefore fuv = 4$

(*b*) row 1, col. 2, $fuv = 3 \times (-2)(-1) = 6$

(*c*) row 3, col, 2, $fuv = 4 \times (0)(-1) = 0$

$$r = \frac{\Sigma fuv - \frac{\Sigma fu \Sigma v}{n}}{\sqrt{\Sigma fu^2 - \frac{(\Sigma fu)^2}{n}}\sqrt{\Sigma fv^2 - \frac{(\Sigma fv)^2}{n}}}$$

$$= \frac{13 - (-4)(-6)/20}{\sqrt{16 - \frac{(-4)^2}{20}}\sqrt{28 - \frac{(-6)^2}{20}}}$$

$$= \frac{13 - 1.2}{\sqrt{16 - 0.8}\sqrt{28 - 1.8^2}} = \frac{11.8}{\sqrt{15.2}\sqrt{26.2}}$$

$$= \frac{11.8}{3.90 \times 5.12} = 0.58.$$

		Marks in Maths. x	15—35	35—56	55—75	75—95				
			20	40	60	80				
	y	$v \backslash u$	–2	–1	0	1	f_v	vf_v	v^2f_v	f_{uv}
20—30	25	–2	1 (4)	3 (6)			4	–8	16	10
30—40	35	–1		2 (2)	2 (0)		4	–4	4	2
40—50	45	0		4 (0)	1 (0)	2 (0)	7	0	0	0
50—60	55	1			3(0)	1 (1)	4	4	4	1
60—70	65	2			1 (0)		1	2	4	0
		f_u	1	9	7	3	20	–6	28	13
		uf_u	–2	–9	0	3	–4			
		u^2f_u	4	9	0	3	16			
		f_{uv}	4	8	0		13			

Origin for x is $x_0 = 60$, C.I. $= h = 10$, $u = (x - 60)/10$

Origin for y is $y_0 = 45$, $k = 10$, $v = (x - 45)/10$

f_u frequecies of u vlaues, f_v the frequencies of v values.

Ex. 10. Calculate the coefficient of correlation for the following data ($n = 260$) and show that $r = 0.4$.

Age of husband

	y \ x	15–20	20–25	25–30	30–35	35–40	40–45	45–50	50–55
	12—16	2	3	3	4	1	1		
	16—20	2	5	5	8	8	6	1	
	20—24	2	1	5	6	16	8	5	
Age	24—28	1	3	12	11	7	12	7	1
of	28—32		2	1	10	12	9	8	2
Wife	32—36			5	7	8	11	8	7
	36—40			1	3	1	4	4	4
	40—44					2	3	2	

23.6 SPEARMAN'S RANK CORRELATION

If observations on two variables are given in the form of ranks and not numerical values, it is possible to compute what is known as *rank correlation between* the two series.

The rank correlation, written ρ, is a descriptive index of agreement between ranks over individuals. It is the same as the ordinary coefficient of correlation computed on ranks but its formula is simpler.

$$\rho = 1 - \frac{6\Sigma D_i^2}{n(n^2 - 1)}$$

where n is the number of observations and Di the positive difference between ranks associated with the individual i.

Like r, the rank correlation lies between – 1 and + 1.

Ex. 11. The ranks given by two judges to 10 individuals are given below.

Individual	*Rank given by* Judge I x	Judge II y	D $= x - y$	D^2
1	1	7	6	36
2	2	5	3	9
3	7	8	1	1
4	9	10	1	1
5	8	9	1	1
6	6	4	2	4
7	4	1	3	9
8	3	6	3	9
9	10	3	7	49
10	5	2	3	9
				$\Sigma D^2 = 128$

The Rank Correlation is given by

$$\rho = 1 - \frac{6\Sigma D^2}{n^3 - n} = 1 - \frac{6 \times 128}{10^3 - 10} = 1 - 0.776 = 0.224.$$

The value of ρ = 0.224 shows the agreement between the judges is not high.

Ex. 12. In the above case, compute r and compare.

The simple coefficient of correlation r for the above data is calculated below.

x	y	x^2	y^2	xy
1	7	1	49	7
2	5	4	25	10
7	8	49	64	56
9	10	81	100	90
8	9	64	81	72
6	4	36	16	24
4	1	16	1	4
3	6	9	36	18
10	3	100	9	30
5	2	25	4	10
$\Sigma x = 55$	$\Sigma y = 55$	$\Sigma x^2 = 385$	$\Sigma y^2 = 385$	$\Sigma xy = 321$

$$r = \frac{321 - 10 \times \frac{55}{10} \times \frac{55}{10}}{\sqrt{385 - 10 \times \left(\frac{55}{10}\right)^2} \sqrt{385 - 10 \times \left(\frac{55}{10}\right)^2}} = \frac{18.5}{\sqrt{82.5 \times 82.5}} = \frac{18.5}{82.5} = 0.224$$

This shows that the Spearman ρ for any two sets of ranks is the same as the Pearson r for the sets of ranks. But it is much easier to compute ρ.

Often the ranks are not given. Instead the numerical values of observations are given. In such a case we must attach the ranks to these values to calculate ρ.

Ex. 13.

Marks in Maths.	*Marks in Stats.*	*Rank in Maths.* x	*Rank in Stats.* y	D	D^2
45	60	4	2	2	4
47	61	3	1	2	4
60	58	1	3	2	4
38	48	5	4	1	1
50	46	2	5	3	9
					$\Sigma D^2 = 22$

$$\rho = 1 - \frac{6\Sigma D^2}{n^3 - n} = 1 - \frac{6 \times 22}{125 - 5} = -0.1$$

This shows a negative, though small, correlation between the ranks.

If two or more observations have the same value their ranks are equal and obtained by calculating the means of the various ranks.

If in the above data, marks in maths. are 45 for each of the first two students, the rank of each would be $\frac{3+4}{2}$ = 3.5. Similarly if the marks of each of the last two students in statistics are 48, their ranks would be $\frac{4+5}{2}$ = 4.5.

The problem takes the following shape :

Marks in Maths.	*Mark in Stats.*	*Rank* x	y	D	D^2
45	60	3.5	2	1.5	2.25
45	61	3.5	1	2.5	6.25
60	58	1	3	2	4.00
38	48	5	4.5	1.5	2.25
50	48	2	4.5	2.5	6.25

$$\rho = 1 - \frac{6\Sigma D^2}{n^3 - n} = 1 - \frac{6 \times 21}{120} = -0.05$$

An elaborate formula which can be used in cases of equal ranks is

$$\rho = 1 - \frac{6}{n^3 - n}\left[\Sigma D^2 + \tfrac{1}{12}\,\Sigma(m^3 - m)\right]$$

where $\frac{1}{12}(m^3 - m)$ is to be added to ΣD^2 for *each* group of equal ranks, m being the number of equal ranks *each* time.

For the given data, we have

for x series, number of equal ranks $m = 2$

For y series, also, $m = 2$ so that

$$\rho = 1 - \frac{6}{5^3 - 5}\left[21 + \tfrac{1}{12}(2^3 - 2) + \tfrac{1}{12}(2^3 - 2)\right]$$

$$= 1 - \tfrac{6}{120}\left[21 + \tfrac{6}{12} + \tfrac{6}{12}\right] = 1 - \frac{6 \times 22}{120} = -0.1$$

Ex. 14. Find the rank correlation for the following rankings of 6 students in two tests:

Test I	4	3	5	6	2	1
Test II	5	4	3	1	2	6

Ex. 15. For the following results showing marks obtained by 15 students show that the rank correlation is 0.89.

Marks in Maths.	50	50	40	39	38	37	36	35	34	33	32	31	30	29	28
Marks in Stats.	50	49	51	52	43	47	42	40	44	40	30	41	32	33	31

Ex. 16. Show that 0.86 is the rank correlation as well as the simple correlation for the following data showing ranks given to 12 qualities of roses by two judges.

Quality	1	2	3	4	5	6	7	8	9	10	11	12
Judge I	4	10	6	5	9	3	12	11	7	8	2	1
Judge II	5	9	7	8	11	2	12	10	6	4	1	3

Ex. 17. Show by means of diagrams various cases of scatter epressing correlation between x, y.

(*a*)

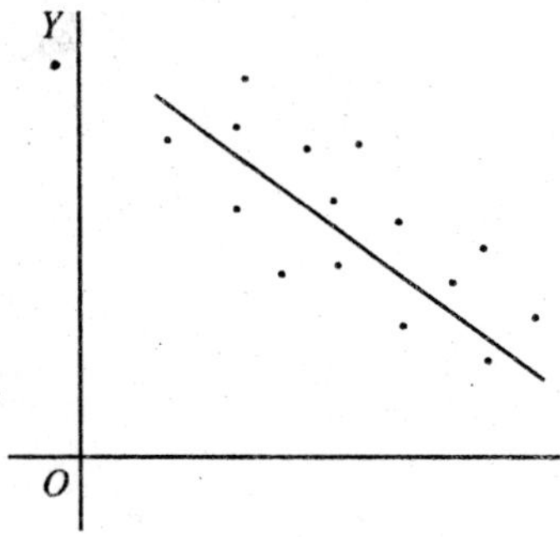

Negative slope
Inverse linear relationship
High scatter
r low, negative

(*b*)

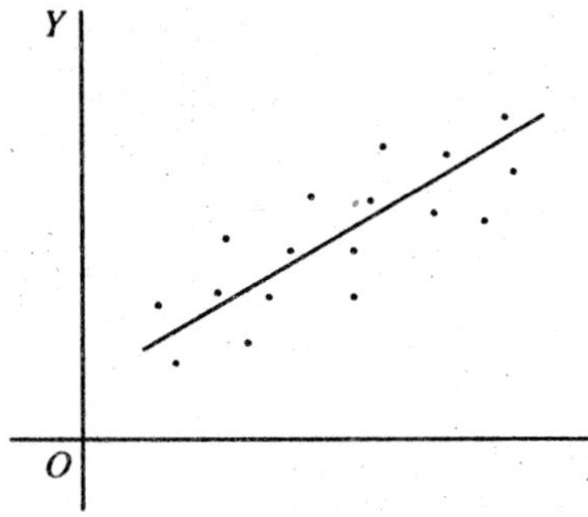

Positive slope
Direct linear relationship
High scatter
r low, positive

(*c*)

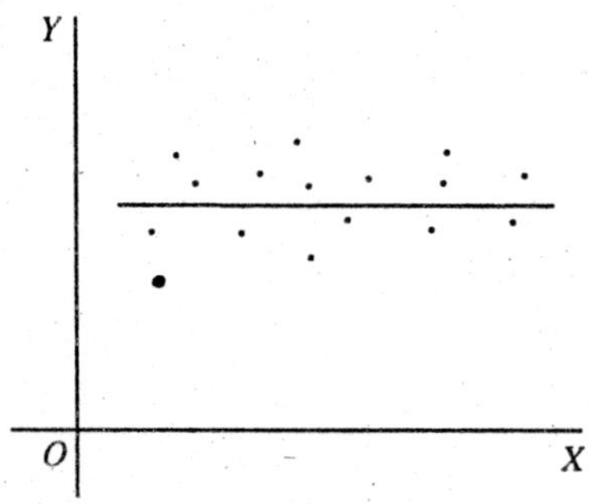

Slope = 0
$r = 0$

(*d*)

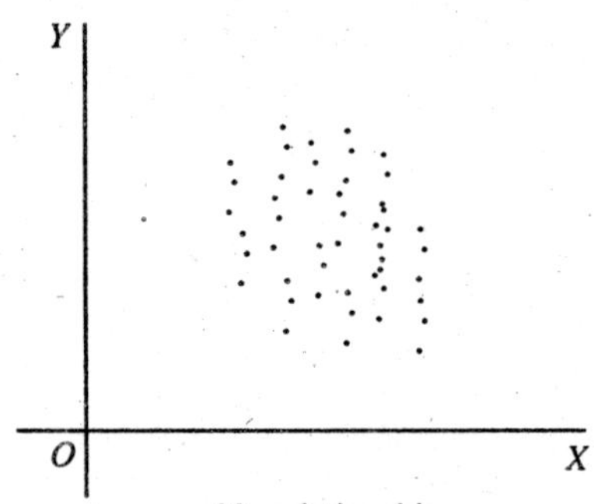

No relationship
$r \simeq 0$

(*e*)

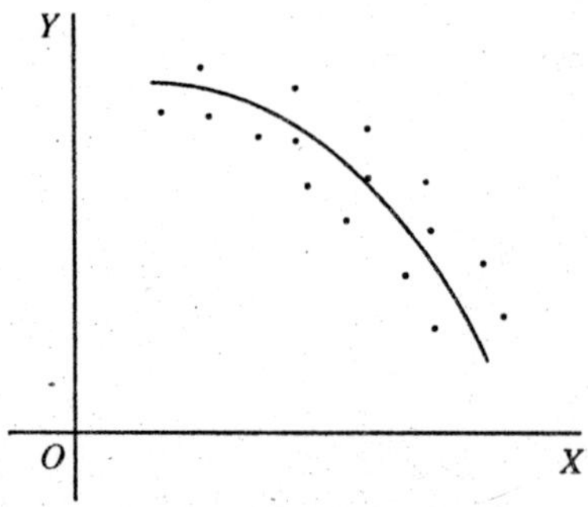

Inverse
Curvilinear relationship

(*f*)

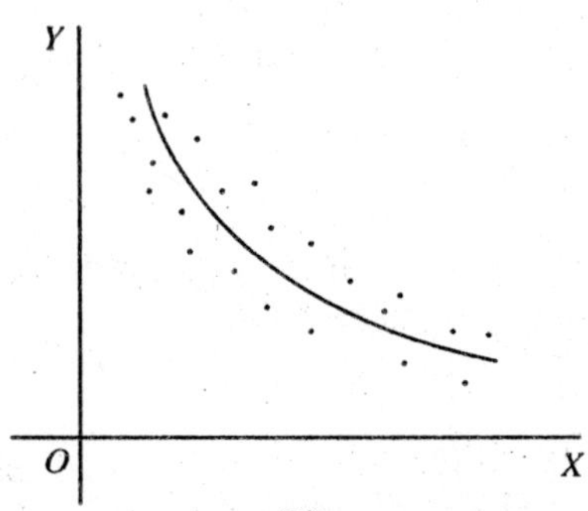

Direct
Curvilinear relationship

(*e*)

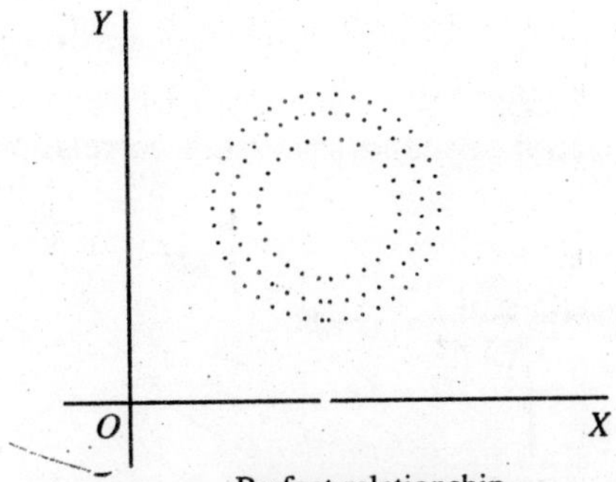

Perfect relationship
But r = 0 because of
nonlinear relation

Ex. 18. Differentiate between Correlation Analysis and Regression Analysis.

Correlation analysis helps us in determining the degree to which two or more variables are related to each other.

When there are only two variables we can determine the degree to which one variable is linearly related to the other. Regression analysis helps in determining the pattern of relationship between one or more independent variables and a dependent variable. This is done by an equation estimated with the help of data.

23.7 COEFFICIENT OF DETERMINATION

The square of the correlation coefficient is called the *coefficient of determination.* The coefficient of determination = r^2 is the proportion of total variation in y that is explained by the linear relation between x and y.

Ex. 19. If $r = 0.6$, the coefficient of determination is 0.36.

23.8 COEFFICIENT OF ALIENATION

The absence of relationship between two variables can be expressed by an index of lack of relationship which may be written

$$\text{Coefficient of Alienation} = \sqrt{1-r^2}$$

which is the opposite of the coefficient of correlation.

Ex. 20. If $r = 0.6$, the coefficient of alienation is

$$\sqrt{1-r^2} = \sqrt{1-0.36} = \sqrt{0.64} = 0.8.$$

Standard Error of *r*

Probable Error of *r*

If r is the coefficient of correlation between a pair of values of x, y then the standard error (SE) of r is given by

$$SE(r) = \frac{1-r^2}{\sqrt{n}}$$

The probable error of r is given by

$$PE(r) = 0.6745\ SE(r) = 0.6745\ \frac{1-r^2}{\sqrt{n}}$$

Ex. 21. If $r = 0.75$ $\qquad SE = \frac{1-0.75^2}{\sqrt{100}} = 0.04375$

$$PE(r) = 0.6745\ SE(r) = 0.0295$$

$PE(r)$ is the limit such that the probability of r lying between $r \pm PE(r)$ is exactly $\frac{1}{2}$ provided r is normally distributed. $PE(r)$ is never negative.

23.9 REGRESSION

Correlation and regression analysis is an important technique used to estimate the relationships among economic (and other) variables.

Regression is the *estimation* of unknown values or the *prediction* of one variable from known values of other variables.

Simple linear regression involves a relationship between two variables only.

Multiple regression (and correlation) analyses consider the relationship between three or more variables.

Economic concepts and policy issues and forecasting problems are often concerned with the relationships between two or more variables. The analysis is simplified by taking into account only two variables as in the case of demand and supply functions, income and expenditure functions, cost and production functions and so on. These relationships are not exact functional relationships. Regression analysis attempts to measure the degree of relationships among the variables.

It is possible that, while we consider the relationship between two variables only, there may be other variables which strongly influence the dependent variable under consideration. In such cases we should make use of multiple regression analysis.

Linear Regression

If we assume linear regression between two variables, it means that if the set of observations on these variables are plotted on a graph paper, a straight line can be approximately drawn through the scattergram. This straight line reasonably reflects the information available in the data.

The linear relationship between Y and X can be written in the form of a straight line equation

$$Y = a + bX$$

Where a, b are constants determining the position of line, b being its slope and a the intercept on the Y-axis.

It may be noted that all the points do not lie on the line. The relationship

$$Y = a + bX$$

shows only a general tendency of association or average relationship between Y and X values.

This relationship is called the linear regression of Y on X. The constant b is called the regression coefficient of Y on X.

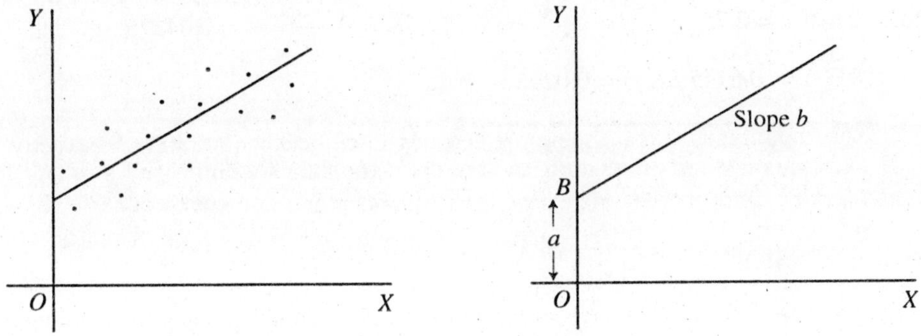

The line $Y = a + bX$ has inercept $a = OB$ on the y-axis and slope b

The line $Y = a + bX$ has intercept $a = OB$ on the Y axis and slope b

Note. A particular line is *completely determined* if in the equation corresponding to the line

$$Y = a + bX$$

the constants a, b are known.

Ex. 22. What is meant by the Line of Best Fit ?

We should determine a, b in such a way that the points lie as close to the line as possible. This will give us the line of best fit.

This line will give, corresponding to each X value, a Y-value which may not necessarily be the same as the actual Y-value. This estimated Y-value which will in general be different from the actual Y will be denoted by Y'.

To obtain the line of best fit, we use the method of *least squares*.

23.10. METHOD OF LEAST SQUARES

For each value of X, we have the actual Y and the estimated Y'. The error of prediction $Y - Y'$ may be positive or negative.

To find a, b we minimise this difference and a good way to do that is to minimise the sum of all the squared differences $\sum_i (Y_i - Y_i')^2$.

The least squares method of fitting a line to a set of n points consists in minimising

$$\sum_{i=1}^{n} (Y_i - Y_i')^2, \text{ i.e. } \sum_{i=1}^{n} [Y_i - (a + bX)]^2$$

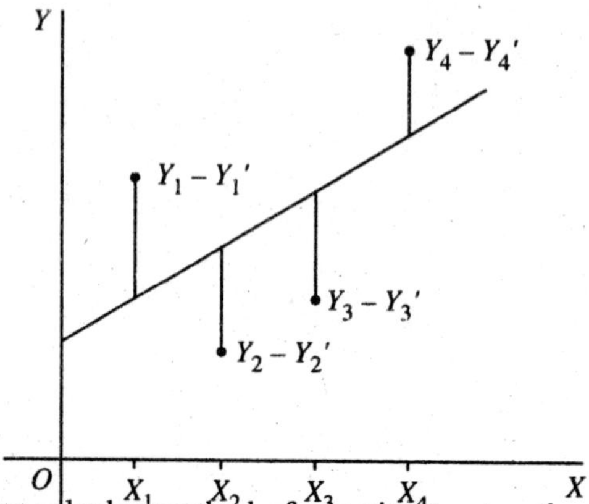

The short vertical distances show $Y - Y'$ (Positive or negative) corresponding to each value of X

The calculus method of equating to zero the differentials with respect to a and b in turn and further simplification yields the least squares regression coefficients

$$a = \overline{Y} - b\,\overline{X}$$

$$b = \frac{\Sigma XY - n\overline{X}\,\overline{Y}}{\Sigma X^2 - n\overline{X}^2}$$

Normal Equations : Least Squares

Let $(X_1, Y_1), (X_2, Y_2), \ldots, (X_n, Y_n)$
be n pairs of observationso on X, Y.

If we use the Principle of Least Squares to fit the regression line

$$Y = a + bX$$

to these data we get the following *normal equations* :

$$\Sigma Y = na + b\Sigma X \qquad \ldots(1)$$

$$\Sigma XY = a\Sigma X + b\Sigma X^2 \qquad \ldots(2)$$

(1) Can be written $n\overline{Y} = n\,a + bn\,\overline{X}$

$$\therefore \quad a = \overline{Y} - b\overline{X}$$

$$= \frac{\Sigma Y}{n} - b\frac{\Sigma X}{n}$$

Substitute in (2),

$$\Sigma XY = \left(\Sigma\frac{Y}{n} - b\frac{\Sigma X}{n}\right)\Sigma X + b\,\Sigma X^2$$

$$\therefore \quad \Sigma XY - \frac{\Sigma X \Sigma Y}{n} = b\left(\Sigma X^2 - \frac{\Sigma X \Sigma X}{n}\right)$$

$$\therefore \quad b = \frac{\Sigma XY - \Sigma X \Sigma Y / n}{\Sigma X^2 - (\Sigma X)^2 / n}$$

$$= \frac{\Sigma XY - n\overline{X}\,\overline{Y}}{\Sigma X^2 - n\overline{X}^2}$$

$$a = \overline{Y} - b\overline{X} \qquad b = \frac{\Sigma XY - n\overline{X}\,\overline{Y}}{\Sigma X^2 - n\overline{X}^{2}}$$

Ex. 23. The regression of Y (price) on X (demand) has been determined for the data given below :

Demand X	*Price* Y	X	Y^2	XY
10	4	100	16	40
8	6	64	36	48
5	5	25	25	25
4	6	16	49	28
2	8	4	64	16
1	9	1	81	9
30	39	210	271	166

$n = 6 \qquad \overline{X} = \frac{30}{6} = 5 \qquad \overline{Y} = \frac{39}{6} = 6.5$

$\Sigma X^2 = 210 \qquad \Sigma Y^2 = 271 \qquad \Sigma XY = 166$

$$b = \frac{\Sigma XY - n\overline{X}\,\overline{Y}}{\Sigma X^2 - nX^2} = \frac{166 - 6 \times 5 \times 6.5}{210 - 6 \times 5 \times 5} = -\frac{29}{60} = -0.48$$

$$a = \overline{Y} - b\overline{X} = 6.5 - \left(-\tfrac{29}{60}\right)5 = 6.5 + \tfrac{29}{12} = 6.5 + 2.42 = 8.92$$

The regression of Y on X is written $Y = a + bX$

Here $a = 8.92$, $b = -0.48$

The regression of Y on X for the given data is

$$Y = 8.92 - 0.48X.$$

Predict *Y* given *X*

If $X = 3$ can we predict the corresponding Y?

By substituting $X = 3$ in the regression equation, we have

$$Y' = 8.92 - 0.48 \times 3 = 7.48$$

i.e., if $X = 3$, Y is likely to be 7.48. This is not a definite result. This is a likely result.

If $X = 12$. $Y = 8.92 - 0.48 \times 12 = 8.92 - 5.76 = 3.16$, a likely value of Y when $X = 12$.

We can predict the Y value corresponding to any X value within a *reasonable range.* We cannot go far beyond the values in the data because the nature of data can change.

Now suppose $X = \overline{X} = 5$(the mean of X) then

$$Y = 8.92 - 0.48 \times 5 = 6.5 = \overline{Y}$$

The regression of Y on X can not be used to predict X, given Y. See Sec. 23.13.

When the mean of X is substituted in the regression equation, the predicted value of Y is the mean of Y.

A regression equation is sometimes called a prediction equation as it helps in the making of prediction.

What are *exact* and *inexact* relationships?

If there is a certain degree of association or relationship between the demand for a commodity and its price or if the sales of a particular brand are influenced by the expenses on advertisement of the brand, these relationships are not exact. The extent of relationship between two or more variables (like demand, price, advertisement expenses etc.) and the degree of this relationship can be determined by the techniques of correlation analysis and regression analysis.

Relationships between variables in physical sciences are exact and can be expressed by means of mathematical functions.

In social sciences exact functional relationships cannot be employed because the variables occurring in these sciences do not have mathematically exact relations. Inadequacy of complete information and inaccuracies of various types do not permit us to determine the exact degree of association between variables in social sciences.

Mathematical methods can be usefully employed, though with some precaution, in correlation and regression analysis.

What is Correlation and Regression Analysis?

What are Direct and Indirect relationships?

What are logical relationships?

Correlation and regression analyses are two related statistical devices to measure the degree of similarity of variations between two or more related variables, the observations on which are available in the form of series. The easiest case is one involving two variables and hence two series of observations. Correlation and regression analyses can also be employed for prediction.

Direct and Inverse Relationships

The relationship between two variables X, Y can be positive (or direct) implying that the values of observations in the two series rise or fall together. It can be negative (or inverse) implying that the values in one series rise while the corresponding values in the other series fall.

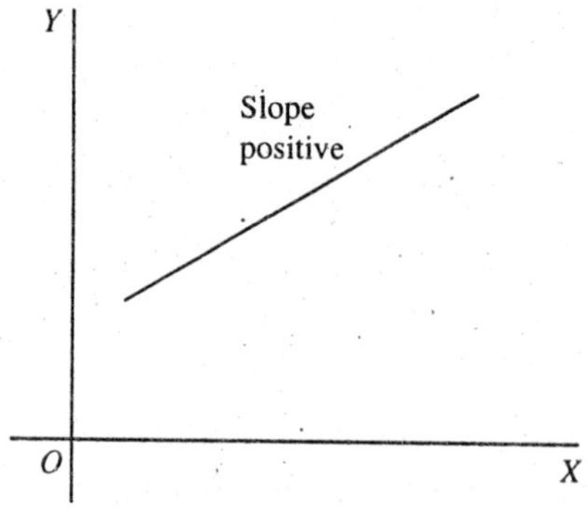

Direct relationship

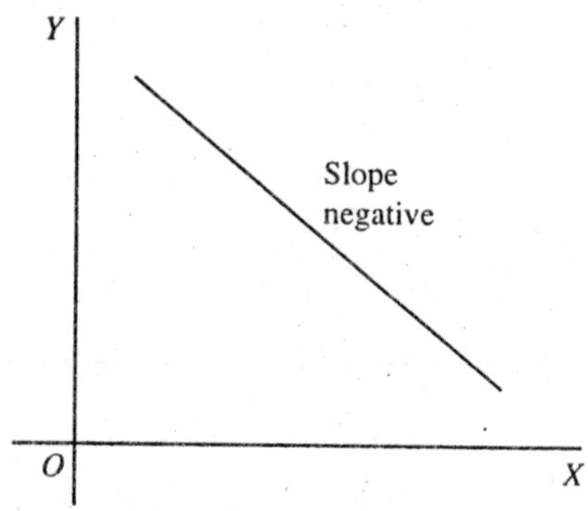

Inverse relationship

Causal or Logical Relationships

A causal relationship implies the presence of cause and effect between variables. Correlation and regression analyses may help us to infer causality provided there is a basis for the existence of this causality. For instance, the relationship between the increase in the sales of a commodity and expenditure on its advertisement is causal but not that between the increase in potato production in Russia and the increase in human population in India.

23.11 PREDICTION

A *prediction* is a guess about the value of an observation that is to be drawn from a population.

If we know nothing about the population it is blind prediction.

Some knowledge about the population is necessary to make meaningful predictions.

A *predicted variable* is one about which predictions are made.

A *predictor variable* is one that provides relevant information for prediction.

The technique of using one variable to make prediction about another is called *regression* analysis. In regression analysis, we make use of known values of one variable to predict unknown values of the other. The regression curve is useful for prediction.

23.12 REGRESSION EQUATION : EQUIVALENT FORMS : RELATION WITH CORRELATION

In the regression equation

$$Y = a + bX \quad \text{or} \quad Y = a + b_{yx}X \qquad \ldots(1)$$

where b_{yx} stands for b to suggest regression coefficient of y on x.

b is the slope of the *best fitting* line in the scattergram of the data. b is called the regression coefficient of Y on X. It shows the amount by which Y changes for a unit change in X. It can be shown that

$$b = b_{yx} = r\frac{s_y}{s_x}$$

where s_x, s_y are standard deviations of x and y respectively.

Thus the regression equation can also be written

$$Y = a + r\frac{s_y}{s_x}X \qquad \ldots(2)$$

or

$$Y - \overline{Y} = r\frac{s_y}{s_z}(X - \overline{X}) \qquad \ldots(3)$$

since

$$a = \overline{Y} - b\overline{X}$$

Ex. 24. Given $\overline{X} = 5$, $\overline{Y} = 7$, $r = 0.4$, $s_x = 2$, $s_y = 3$ the regression of Y on X can be written

$$Y - 7 = 0.4 \times \tfrac{3}{2}(X - 5)$$

i.e.,

$$Y - 7 = 0.6(X - 5) \text{ or } Y = 0.6X - 3 + 7$$

The required equation is

$$Y = 4 + 0.6X.$$

Alternatively we have $b = r\,\dfrac{s_y}{s_x} = 0.4 \times \frac{3}{2} = 0.6$

$$a = \overline{Y} - b\overline{X} = 7 - 0.6 \times 5 = 4 \text{ so that}$$
$$Y = 4 + 0.6X$$

Regression methods are more flexible than correlation methods. Regression is the basic technique in the study of the relationship between two variables, correlation is a part of it.

23.13 LINEAR REGRESSION OF *X* ON *Y* (COMPARE WITH REGRESSION OF *Y* ON *X*) : PREDICT *X* GIVEN *Y*.

If X is treated as the independent variable and Y the dependent variable we have the regression of Y on X.

$$Y = a + bX \qquad \text{or} \qquad Y = a + b_{yx}X$$

If we take Y as the independent variable and X as the dependent variable, we have the regression of X on Y.

$$X = a' + b'Y \qquad X = a' + b_{xy}Y \qquad \ldots(1')$$

This can be done provided the dependence of X on Y is feasible; it may not be possible in all cases.

Note. b' is the regression coefficient of X on Y and a' the intercept on the x-axis.

a' and b' are determined by the method of least squares and the formulae are similar to those for a and b with X and Y interchanged.

$$a = \overline{X} - b'\overline{Y} \text{ and } b' = \frac{\Sigma XY - n\overline{X}\overline{Y}}{\Sigma Y^2 - n\overline{Y}^2} = b_{xy}$$

Compare these with a, b.

We may write b_{xy} or $b' = r\,\dfrac{s_x}{s_y}$ and we have the alternative forms of regression of X on Y:

$$X = a' + r\frac{s_x}{s_y}(Y - \overline{Y}) \qquad \ldots(2')$$

or

$$X - \overline{X} = r\,\frac{s_x}{s_y}(Y - \overline{Y}) \qquad \ldots(3')$$

Ex. 26. Find the regression of X on Y for the price demand data (Ex. 23)

$$n = 6,\ \overline{X} = 5,\ \overline{Y} = 6.5,\ \Sigma Y^2 = 271,\ \Sigma XY = 166$$

$$b' = \frac{\Sigma XY - n\overline{X}\overline{Y}}{\Sigma Y^2 - n\overline{Y}^2} = \frac{166 - 195}{271 - 253.5} = \frac{-29}{17.5} = -1.66$$

$$a' = \overline{X} - b'\,\overline{Y} = 5 - (-1.66) \times 6.5 = 15.77$$

The regression of X on Y is $X = a' + b'Y$

We have $a' = 15.77$, $b' = -1.66$ so that for our data, the regression of X on Y is $X = 15.77 - 1.66Y$.

To predict X, given a value of Y we use the relation $X = a' + b' Y$. It would be wrong to use $Y = a + bX$ for the purpose.

23.14 TWO REGRESSION LINES

The relationship between Y and X is not perfect. The *average relationship* of Y and X in which X is the independent and Y the dependent variable is *not* the same as the average relationship of X and Y in which Y is the independent and X the dependent variable.

The two regression lines, which best describe these two average relationships, are given by the regression equations:

$$Y = a + bX \qquad \text{...(1)}$$

$$X = a' + b'Y \qquad \text{...(1')}$$

a, b in (1) are obtained by minimising $\Sigma(Y - Y')^2$

a', b' in (1') are obtained by minimising $\Sigma(X - X')^2$

$$b = \frac{\Sigma XY - n\overline{X}\overline{Y}}{\Sigma X^2 - n\overline{X}^2}$$

$$b' = \frac{\Sigma XY - n\overline{X}\overline{Y}}{\Sigma Y^2 - n\overline{Y}^2}$$

$$a = Y - b\overline{X}$$

$$a' = \overline{X} - b'\overline{Y}$$

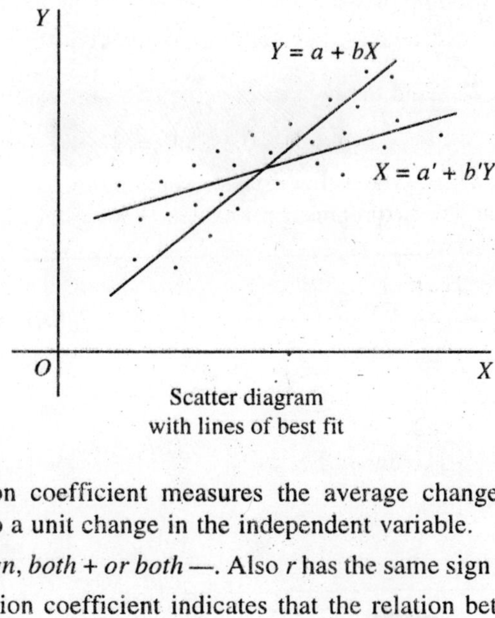

Scatter diagram with lines of best fit

The signs of b and b' indicate whether the slopes of the lines of best fit are positive or negative. It may be recalled that the regression coefficient measures the average change in the dependent variable corresponding to a unit change in the independent variable.

b and b' must have the same sign, both + or both —. Also r has the same sign as b, b'.

A positive value of the regression coefficient indicates that the relation between X and Y is direct.

A negative value shows an inverse relationship between X and Y, *i.e.*, high and low values are paired together.

Since $\quad b = r\dfrac{s_y}{s_x} = b_{yx} \qquad b' = r\dfrac{s_x}{s_y} = b_{xy}$

We have $bb' = r^2$ or $b_{yx}\,b_{xy} = r^2$

That r, b, b' have the same sign gives us an alternative definition of r. It is the square root of the product of b and b' and has the *same* sign as b or b'.

$$r = \sqrt{bb'} \text{ or } \sqrt{b_{yx}b_{xy}}$$

The product bb' can never exceed 1 because r cannot numerically exceed 1. This result can be used to distinguish between two regression lines for the same data.

Ex. 27. We have $b = -0.48$, $b' = -1.66$

$$\therefore \qquad r = \sqrt{(-0.48)(-1.66)} = -0.89.$$

If the regression line of Y on X, *i.e.*, $Y = a + bX$ exists it does not imply that the regression of X on Y necessarily exists.

If X is time, Y sales, the regression of sales on time is expressed by $Y = a + bX$.

But there is no question of regression (or dependence) of time on sales.

23.15. CAN THE TWO REGRESSION LINES COINCIDE?

The two regression lines are identical if and only if all the points in the scattergram lie on one straight line, *i.e.*, if the correlation is perfect, $r = 1$.

What is the point of intersection of the two regression lines?

$(\overline{X}, \overline{Y})$ is the only point common to both and hence the point of intersection. If we solve the two regression equations simultaneously we get $\overline{X}$, $\overline{Y}$.

Ex. 28. Find the means from the two regression equations for a certain data :

$Y = 5 + 0.2\,X$, $X = 3 + Y$ **Ans.** By solving $\overline{X} = 10$, $\overline{Y} = 7$

Ex. 29. For the following data showing index numbers of prices and production for 5 years find the two regression lines and show that $bb' = r^2$.

Year	*Index Numbers of*	
	Production	*Prices*
1961	100	107
1962	101	123
1963	106	133
1964	99	109
1965	97	128

Estimate the index number of prices when it is known that the index number of production is 110. Predict the index number of production when that of prices is known to be 120. Use X for production, Y for prices.

Subtract 100 from each of X values and 120 from each of Y values.

Note. r does not change by change of scale and origin *i.e.* by subtraction and division etc. b changes by division (or multiplication) of observations by any number. But b does not change by subtraction as done in this exercise. In simple regression problems such subtraction may be avoided.

Thus $u = X - 100$, $v = Y - 120$.

u	v	u^2	v^2	uv
0	– 13	0	169	0
1	3	1	9	3
6	13	36	169	78
– 1	– 11	1	121	11
– 3	9	9	81	– 27
$\Sigma u = 3$	$\Sigma v = 1$	$\Sigma u^2 = 47$	$\Sigma v^2 = 549$	$\Sigma uv = 65$

$$\bar{u} = \frac{\Sigma u}{n} = \frac{3}{5} = 0.6, \bar{v} = \frac{\Sigma v}{n} = \frac{1}{5} = 0.2$$

(*i*)
$$r = \frac{\Sigma uv - n\bar{u}\bar{v}}{\sqrt{\Sigma u^2 - n\bar{u}^2}\sqrt{\Sigma v^2 - n\bar{v}^2}}$$

$$= \frac{65 - 5 \times 0.6 \times 0.2}{\sqrt{47 - 5 \times .6 \times .6}\sqrt{549 - 5 \times .2 \times .2}} = \frac{64.4}{\sqrt{45.2}\sqrt{548.8}}$$

$$= \frac{64.4}{6.3 \times 23.4} = 0.41.$$

(*ii*) For the regression of Y on X, no additional work is necessary. b does n change by change of origin only. It changes by change of scale.

$$b = \frac{\Sigma uv - n\bar{u}\bar{v}}{\Sigma u^2 - n\bar{u}^2} = \frac{64.4}{45.2} = 1.42$$

Now
$$\bar{X} = \frac{\Sigma X}{n} = \frac{503}{5} = 100.6 \text{ and } \bar{Y} = \frac{\Sigma Y}{n} = \frac{601}{5} = 120.2$$

$\therefore$
$$a = \bar{Y} - b\bar{X} = 120.2 - 1.42 \times 100.6 = -14.25$$

The regression of Y on X is given by : $Y = -14.25 + 1.42X$

(*iii*) To find the regression of X on Y

$$b' = \frac{\Sigma uv - n\bar{u}\bar{v}}{\Sigma v^2 - n\bar{v}^2} = \frac{64.4}{548.8} = 0.12$$

$$a' = \bar{X} - b'\bar{Y} = 100.6 - 0.12 \times 120.2 = 68.58$$

The regression of X on Y is given by

$$X = 68.58 + 0.12\, Y$$

(*iv*) $\sqrt{bb'} = \sqrt{1.42 \times 0.12} = 0.41 = r \qquad \because bb' = r^2$

(*v*) To predict Y from X substitute $X = 110$ in the regression of Y on X

Predicted $Y = -14.25 + 1.42 \times 110 = 141.95$

To predict X from Y substitute $Y = 120$ in the regression of X on Y.

Predicted $X = 68.58 + 0.12 \times 120 = 82.98$

Ex. 30.

Price of a commodity Y	*Its supply* X
5	90
28	409
10	120
9	155
20	372
15	153
12	130
26	452
18	250
12	199

Find the elasticity of supply at $(\overline{X}, \overline{Y})$ for the above deta

Ans. The regression of X on Y is $X = 234.27 + 0.176\ Y$

$b = 0.176$, $b' = 0.513$, $r = 0.95$, $\overline{Y} = 16.5$, $\overline{X} = 237$

Elasticity of Supply at $(\overline{X}, \overline{Y})$ is $\eta_s = \dfrac{dX}{dY} \cdot \dfrac{\overline{Y}}{\overline{X}} = 0.176 \times \dfrac{16.5}{237} = 0.012.$

Ex. 31. A firm doubles the number of its employees and profit increases significantly. Does it imply profit depends on the number of employees?

It is likely that the increase in the employee number has come along with increase in capital, efficiency or other development. The increase in employee number need not be the basic cause of profit increase.

If X and Y have nothing to do with each other logically but the observations on X and Y happen to move according to a pattern, the resulting regression equation, even if it is well fitted and has significant coefficients, is spurious and meaningless.

Ex. 32. Show that if five students get respectively 1, 2, 3, 4, 5 marks (X) out of 10 in subject A and 3, 3, 5, 7, 6 marks (Y) out of 10 in subject B, the regression of subject B on subject A can be written $Y = 1.8 + X$.

Ex. 33. Show that for 10 students $\rho = 0.79$ for marks given by

Judge I	23	20	19	17	16	28	24	25	27	22
Judge II	30	28	27	41	36	45	46	44	43	39

Ex. 34. For the following results showing average increase in hours of sleep gained by using two drugs on 10 persons, find r and ρ.

Drug A	1.1	1.0	0.9	0.8	0.7	1.9	1.6	1.6	1.2	1.1
Drug B	1.2	0.9	0.9	0.6	0.7	1.6	1.7	1.4	1.3	1.2

Ans. $r = 0.93$, $\rho = 0.96$

Ex. 35. In a linear regression analysis of 60 observations, the two lines of regression are

$$1000\,Y = 768X - 3608 \text{ and } 5X = 6Y + 24$$

What is the coefficient of correlation in the data?

Show that the ratio of the coefficient of variation of X to that of Y is $\frac{5}{24}$.

From the data we have $b = \dfrac{768}{1000}$, $b' = \frac{6}{5}$

Coef. of corln. $r = \sqrt{bb'} = \sqrt{0.922} = 0.96$

If we solve the two equations we get $\overline{X} = 6$, $\overline{Y} = 1$

Now since $b = r\dfrac{s_y}{s_x}$ $\therefore$ $\dfrac{s_x}{s_y} = \dfrac{r}{b} = \dfrac{0.96}{0.768} = 1.25$

Coefficient of variation of X is $\dfrac{s_x}{\overline{X}} \times 100$

Coefficient of variation of Y is $\dfrac{s_y}{\overline{Y}} \times 100$

Their ratio is $\dfrac{s_x/\overline{X}}{s_y/\overline{Y}} = \dfrac{s_x}{s_y}\,\dfrac{\overline{Y}}{\overline{X}} = 1.25 \times \frac{1}{6} = \frac{5}{24}$

Ex. 36. What if $bb' > 1$?

Since $r^2 \not> 1$, $\therefore$ If $bb' > 1$, interchange dependent and independent variables in the two regression lines.

Ex. 37. In a regression analysis problem the following data are given: $s_x = 3$ and the regression lines $8X - 10Y + 66 = 0$, $40X - 18Y = 214$.

Find $\overline{X}$, $\overline{Y}$, s_y, r,   **Ans.** $\overline{X} = 13$, $\overline{Y} = 17$, $r = \sqrt{\dfrac{8}{10} \times \dfrac{14}{40}} = 0.6$, $s_y = 4$.

Ex. 38. In each of the cases given below find (*i*) r, (*ii*) the two regression equations and predict the values as asked. Draw the scattergrams and plot the lines.

(*a*)

Year	*Index of industrial Production X*	*National income Y (Rs. million)*
1941	100	500
1942	105	520
1943	107	580
1944	98	600
1945	109	610
1946	110	600
1947	114	650

Predict Y if $X = 120$. Predict X if $Y = 700$

(*b*)

X	Y
10	0
10	– 1
11	– 2
11	4
12	– 2
15	0
16	– 2
11	2

Predict Y if $X = \frac{5}{2}$ and predict X if $Y = 18$.

Ex. 39. Find the equation of the demand curve for a commodity whose prices and the corresponding quantities demanded are shown below. Assume linearity. Find η_d at mean price and quantity.

Year	*Quantity demanded* x	*Price* p
1951	51	6.9
1952	54	3.0
1953	35	7.9
1954	63	7.3
1955	75	6.6
1956	69	2.5
1957	88	1.5
1958	97	2.8
1959	76	3.5
1960	58	7.5
1961	20	8.5
1962	38	5.0
1963	44	4.3
1964	42	7.6
1965	28	6.5
1966	70	4.5

$b' = -6.5,\qquad a' = \overline{X} - b'\overline{p} = 56.8 - (-6.5) \times 5.4 = 91.9$

The regression of X on p is the demand curve

$$X = 91.9 - 6.5\,p$$

The elasticity of demand can be calculated at the mean price and mean quantity by writing

$$\eta_d = \frac{dX}{dp} \cdot \frac{\overline{p}}{\overline{X}} = 6.5 \times \frac{5.4}{56.8} = -0.62 \text{ (minus sign inserted)}$$

23.16 TRANSFORMATION TO LINEARITY

Relationships between variables are often not linear, *i.e.*, the relationships may be curvilinear or in any other nonlinear forms. In such cases if we can make a suitable transformation of either or both the variables so as to convert the relationship into a linear form, the work of prediction and analysis can become easy.

Many kinds of transformations can be affected but here we confine ourselves to converting nonlinear into linear relationships.

The standard form of a linear equation is

$$Y = a + bX$$

If Y is plotted against X, b is the slope of the line and a its intercept on the Y-axis.

1. Suppose $\quad y = \frac{b}{x} + a$ is the given nonlinear relation.

If we write $\quad X = \frac{1}{x}$ the equation becomes $y = bX + a$

which is a linear equation and can be handled by the usual method. In the data we convert X values into their reciprocals and use the new variable x as the independent variable. An example of this is the Philips model in which y = wage rate and x = unemployment rate

2. $\quad Y = \frac{A}{B+X}$

Transformation : $\quad y = \frac{1}{Y}, a = \frac{B}{A}, b = \frac{1}{A}$

The equation becomes

$$\frac{1}{Y} = \frac{B+X}{A} = \frac{B}{A} + \frac{1}{A} X \text{ or } y = a + bx$$

3. $\quad Y = \frac{AX}{B+X}$

Transformation $\quad y = \frac{1}{Y}, x = \frac{1}{X}, b = \frac{B}{A}, \; a = \frac{1}{A}$

The equation can be written

$$\frac{1}{Y} = \frac{B+X}{XA} = \frac{B}{AX} + \frac{1}{A}$$

or

$$y = a + bx$$

4. $\quad Y = (A + BX)^c$

Transformation $\quad y = \log Y, x = \log (A + BX)$

The equation becomes

$$\log Y = c \log (A + BX)$$

or $y = cx$. Additional work will be needed here.

5. The power curve $Y = AX^B$ (Log linear model)

Taking logarithms, we get

$$\log Y = \log A + B \log X$$

This is a linear relationship in log Y and log X although the original relationship was not linear.

log A is constant $= a$, say, Let $y = \log Y$, $x = \log X$.

Then we can write the equation

$$y = a + bx$$

which means we may plot log Y against log X and the curve becomes a straight line.

Since $b = \frac{d \ln y}{d \ln x} = \frac{dy}{dx}\frac{x}{y}$, it is the elasticity of y with respect to x

6. Semi log model $\quad Y = AB^X$

$$\ln Y = \ln A + X \ln B$$

Put $\ln Y = y$, $\ln A = a$, $\ln B = b$ then

$$y = a + bX$$

$$b = \log B = \frac{d \ln y}{dx} = \frac{1}{y}\frac{dy}{dx}$$

= relative change in y for an absolute change in x. The model gives a constant rate of growth.

Alternatively we have the model

$$e^{(Y-a)/b} = X \text{ or } Y - a = b \ln X; \text{ This is the same as } Y = a + b \ln X$$

Here b measures the ratio of absolute change in Y with respect to a relative change in Y.

$$b = \frac{dY}{dX/X} = X\frac{dY}{dX}$$

Ex. 40. Fit a growth function $y = AB^X$ to the following data on national income.

Year (*X*)	*N.I.* (*y*)	*X*	*Y* = log *y*	X^2	*XY*
1981	250	– 3	2.40	9	– 7.20
1982	251	– 2	2.40	4	– 4.80
1983	279	– 1	2.45	1	– 2.45
1984	319	0	2.50	0	0
1985	361	1	2.56	1	2.56
1986	439	2	2.64	4	5.28
1987	547	3	2.74	9	8.22
			17.69	28	1.61

The function $y = AB^X$, upon log transformation becomes

$$\log y = \log A + X \log B$$

Putting $Y = \log y,\ \beta_1 = \log A,\ \beta_2 = \log \beta$

We have to estimate the relation

$$Y = \beta_1 + \beta_2 X \text{ where } X = x - 1984$$

$$\beta_2 = \frac{\Sigma xy - n\overline{xy}}{\Sigma x^2 - n\bar{x}^2} = \frac{1.61 - 7 \times 0 \times 2.53}{28 - 7 \times 0} = 0.0575$$

$$i.e.,\ \beta = \text{antilog } \beta_2 = 1.14$$

$$\beta_1 = \bar{Y} - \beta_1 \bar{X} = 2.53$$

$$A = \text{antilog } \beta_1 = 336$$

The required relation is

$$Y = 2.53 + 0.0575X$$

or $y = (336)\ 1.14^{x - 1984}$

6. The exponential $Y = A.\ 10^{bx}$

Since $\log Y = \log A + bx.$

let $\log Y = y$, $\log A = a$ and the equation takes the usual form

$$y = a + bx$$

7. The logarithmic curve $y = a + b \log X$ becomes $y = a + bx$ if we simply write $\log X = x$.

8. $Y = AX^{bX}$

Transformation $y = \log Y,\ x = X \log X,\ a = \log A$

The equation can be written

$$\log Y = \log A + bX \log X$$

or $y = a + bx$

10. $Y = Ae^{B/X}$

Transformation $y = \log Y,\ x = \dfrac{1}{X},\ a = \log A,\ \ b = B \log e$

$$\log Y = \log A + \frac{B}{X} \log e$$

or $y = a + bx.$

23.17 RELIABILITY OF THE REGRESSION EQUATION

The closer to the estimated regression line are the points in the scatter diagram the better will the line be in accurately representing the data.

The reliability of the estimated equation can be measured by S_y which is called the standard error of estimate

$$S_y = \sqrt{\frac{\Sigma(Y - \hat{Y})^2}{n - 2}}$$

where Y and $\hat{Y}$ are, respectively, the observed and estimated values of the dependent variable. We can simplify the formula by writing

$$S_y = \sqrt{\frac{\Sigma Y^2 - a\Sigma Y - b\Sigma XY}{n-2}}$$

If S_y is large it means the scatter or variability of data points around the regression line is high.

If $S_y = 0$, the regression line is a perfect estimator of the variable Y.

Note. We assume that the observed Y values are normally distributed about each estimated Y. The distribution variance is assumed to be constant.

We can predict a value of Y if we can find the specific value X_o of X. The prediction interval limits are

$\hat{Y} \pm t\, S_y$ where the t value corresponds to the confidence level

Note. The standard error of prediction is given by

$$s_p = S_y\sqrt{1 + \frac{1}{n} + \frac{(\overline{X} - X_0)^2}{\Sigma X^2 - n\overline{X}^2}}$$

Coefficient of Determination

The degree or strength of *linear* (only linear) association between the variables X, Y can be measured by r^2, the coefficient of determination. It can be proved that

$$r^2 = 1 - \frac{\Sigma(Y - \hat{Y})^2}{\Sigma(Y - \overline{Y})^2}$$

$\hat{Y}$ is the estimated value of Y using the regression relation

The value of r^2 lies between 0 and 1,

$$0 \le r^2 \le 1$$

The value of r^2 close to zero implies the absence of any significant correlation between X, Y and r^2 close to 1 implies a strong correlation between X, Y.

The formula for r^2 can also be written

$$r^2 = \frac{a\Sigma Y - n\overline{Y}^2 + b\Sigma XY}{\Sigma Y^2 - n\overline{Y}^2}$$

Note. This r^2 is also the square of the coefficient of correlation r,

where
$$r = \frac{\Sigma XY - n\overline{X}\overline{Y}}{\sqrt{\Sigma X^2 - n\overline{X}^2}\sqrt{\Sigma Y^2 - n\overline{Y}^2}}$$

Explained Variation, Unexplained Variation

The total variation $\Sigma(Y - \overline{Y})^2$ can be broken into two parts :

Explained variation $\Sigma(\hat{Y} - \overline{Y})^2$

Unexplained variation $\Sigma(Y - \hat{Y})^2$

Thus $\Sigma(Y - \overline{Y})^2 = \Sigma(\hat{Y} - \overline{Y})^2 + \Sigma(Y - \hat{Y})^2$

$$\therefore \quad \frac{\Sigma(Y-\overline{Y})^2}{\Sigma(Y-\overline{Y})^2} = \frac{\Sigma(\hat{Y}-\overline{Y})^2}{\Sigma(Y-\overline{Y})^2} + \frac{\Sigma(Y-\hat{Y})^2}{\Sigma(Y-\overline{Y})^2}$$

$$\therefore \quad 1 = r^2 + \frac{\Sigma(Y-\hat{Y})^2}{\Sigma(Y-\overline{Y})^2}$$

$$\therefore \quad r^2 = 1 - \frac{\Sigma(Y-\hat{Y})^2}{\Sigma(Y-\overline{Y})^2}$$

$$= 1 - \frac{\text{Unexplained fraction of total variation}}{\text{Total variation}}$$

For the regression line $Y = a + b\,X$, show that

$$r^2 = \frac{a\Sigma Y + b\Sigma XY - n\overline{Y}^2}{\Sigma Y^2 - n\overline{Y}^2}$$

Ex. 41. The calculations for r^2 are shown here.

X	Y	X^2	Y^2	XY	$\hat{Y}$	$Y - \hat{Y}$
3	11	9	121	33	8.46	2.54
7	18	49	324	126	15.30	2.70
4	9	16	81	36	10.17	– 1.17
2	4	4	16	8	6.75	– 2.75
0	7	0	49	0	3.33	3.67
4	6	16	36	24	10.17	– 4.17
1	3	1	9	3	5.04	– 2.04
2	8	4	64	16	6.75	1.25
23	66	99	700	246		

$$b = \frac{\Sigma XY - n\overline{X}\overline{Y}}{\Sigma X^2 - n\overline{X}^2} = \frac{246 - 8 \times \frac{23}{8} \times \frac{66}{8}}{99 - 8 \times \frac{23}{8} \times \frac{23}{8}}$$

Use $\hat{Y}$ values calculated below

$$= \frac{56.25}{32.87} = 1.71$$

$$a = \overline{Y} - b\,\overline{X} = \frac{66}{8} - 1.71 \times \frac{23}{8} = 3.33$$

The regression line of Y on X is given

$Y = 3.33 + 1.71\,X$

For $X = 3,\ \hat{Y}_1 = 3.33 + 1.71 \times 3 = 8.46$

$X = 7,\ \hat{Y}_2 = 3.33 + 1.71 \times 7 = 15.30$

$$X = 4,\ \hat{Y}_3 = 3.33 + 1.71 \times 4 = 10.17$$

$$X = 2,\ \hat{Y}_4 = 3.33 + 1.71 \times 2 = 6.75$$

$$X = 0,\ \hat{Y}_5 = 3.33 + 1.71 \times 0 = 3.33$$

$$X = 4,\ \hat{Y}_6 = 3.33 + 1.71 \times 4 = 10.17$$

$$X = 1,\ \hat{Y}_7 = 3.33 + 1.71 \times 1 = 5.04$$

$$X = 2,\ \hat{Y}_8 = 3.33 + 1.71 \times 2 = 6.75$$

The coefficient of determination is given by

$$r^2 = 1 - \frac{\Sigma(Y - \hat{Y})^2}{\Sigma(Y - \overline{Y})^2} = 1 - \frac{49.8}{155.5}$$

$$= 0.68$$

$$r = \sqrt{r^2} = 0.79$$

Alternatively, $r^2 = \dfrac{a\Sigma Y + b\Sigma XY - n\overline{Y}^2}{\Sigma Y^2 - n\overline{Y}^2}$

$$= \frac{3.33 \times 66 + 1.71 \times 246 - 8 \times (66/8)^2}{700 - 8(66/8)^2}$$

$= 0.68$. This procedure may be preferred.

Note. r^2 is a measure of goodness of fit. It shows how one variable describes another variable. r^2 should *not* be used to imply a percentage change in Y caused by a change in X.

23.18 CAUSE AND EFFECT

Correlation and Regression analyses can be used to predict the values of the dependent variable Y for various possible or future values of the independent variable X.

The relation $Y = a + bX$

can be used to predict Y if X is given. It cannot be used to predict X if Y is given. This is because the relation assumes X to be the cause and Y the effect. It is a one way relation from X to Y.

To predict X from Y, one must make sure that the relation from Y to X is possible. For example if Y is sales, X is advertisement expenses, we can have $Y = a + bX$ (to predict Y from X) and $X = a' + b'Y$ (to predict X from Y)

Note. 1. For predictions, one should not go too far beyond the range of the data. This is because there is no guarantee of the trends of observations continuing on the same pattern.

2. The relation $Y = a + bX$ does not *necessarily* imply X is directly the cause and Y its effect. It is possible that both are caused by other common factors. Also the relation on its own may amount to nothing very important because there may be other relations or equations that may need to be considered along with the given relation.

23.19 MULTIPLE REGRESSION

Multiple regression allows the use of two or more independent variables. Thus it uses more information to estimate the dependent variable Y. The technique can be used to fit nonlinear as well as linear relations. It can be manipulated to use qualitative variables.

If X_1, X_2 are independent variables which explain Y we have the regression of Y on X_1, X_2, given by

$y = a + b_1 X_1 + b_2 X$ where a, b_1, b_2 can be estimated by the principle of least squares. For this, solve the normal equations.

$$\Sigma Y = na + b_1 \Sigma X_1 + b_2 \Sigma X_2$$

$$\Sigma X_1 Y = a\Sigma X_1 + b_1 \Sigma X_1^2 + b_2 \Sigma X_1 X_2$$

$$\Sigma X_2 Y = a\Sigma X_2 + b_1 \Sigma X_1 X_2 + b_2 \Sigma X_2^2$$

23.20 MULTIPLE CORRELATION, PARTIAL CORRELATION

The correlation between a variable X_1 with the group of variables X_2, X_3 is measured by the coefficient of *multiple correlation*, written $R_{1.23}$

Variable 1 corresponds to the dependent variable; 2, 3 correspond to the independent variables.

If r_{12} stands for the correlation between x_1, x_2 (same as r_{21})

r_{23} between x_2, x_3 (same as r_{32})

r_{31} between x_3, x_1 (same as r_{13})

We can show that the multiple correlation coefficient

$$R_{1.23} = \sqrt{\frac{r_{12}^2 + r_{13}^2 - 2_{12} r_{23} r_{31}}{1 - r_{23}^2}} \quad (0 \leq R_{1.23} \leq 1)$$

We can similarly define $R_{2.31}$, $R_{3.12}$. In the case of n variables, the multiple correlation between X_1 and X_2, X_3, ..., X_{n-1} can be written $R_{1.23...n-1}$.

The correlation between x_1, x_2 keeping x_3 constant is expressed by the *partial correlation* coefficient

$$r_{12.3} = \frac{r_{12} - r_{13} r_{23}}{\sqrt{1 - r_{13}^2}\sqrt{1 - r_{23}^2}} \quad (-1 \leq r_{12.3} \leq 1)$$

$$r_{23.1} = \frac{r_{23} - r_{12} r_{13}}{\sqrt{1 - r_{12}^2}\sqrt{r - r_{13}^2}} \text{ and so on. } (r_{21} = r_{12},\ r_{31} = r_{13})$$

The partial correlation between x_1, x_2 in the presence of variables x_3, x_4, ..., x_n is written $r_{12.34...n}$

Find $R_{1.23}$ and $r_{12.3}$ given $r_{12} = .2$, $r_{13.} = .3$, $r_{23} = .4$

$$R_{1.23} = \sqrt{\frac{.04 + .09 - 2 \times .2 \times .3 \times .4}{1 - .16}} = 0.31$$

$$r_{12.3} = \frac{0.2 - 0.3 \times 0.4}{\sqrt{1 - .09}\sqrt{1 - .16}} = 0.092$$

Ex. 42. $r_{12} = 0.7$, $r_{13} = 0.6$, $r_{23} = 0.4$, $R_{1.23} = 0.78$, $r_{23.1} = 0.035$

CHAPTER 24

Sampling

24.1 SAMPLING

The theory of sampling establishes the basis for statistical estimation. A scientist takes a number of observations for the measurement of a certain phenomenon and repeats his experiment a number of times. Every time he depends on a *sample* of observations to be able to make some general statements about his results.

Inductive inference is the central theme of statistics. Theoretical propositions are generalisations referring to all conceivable observations on the subject matter under consideration. In practice, we cannot study all the conceivable cases. Sometimes it is senseless to test all the possibilities. To test the length of life of a bulb it is subjected to a strain. The bulb is destroyed in the process but its probable life can be calculated. If all the bulbs in a factory were checked thus no bulb would be left intact. A sample of bulbs subjected to the test will give a fairly good idea about the quality of the whole population of bulbs being produced.

Samples and Populations

A *sample* is a set of measurements taken from a process or series of experiments. It can be regarded as having been drawn from a large population of measurements covering a large number of observations. The word population is a technical term in statistics, not necessarily referring to people. A *population* or a universe is a set of large number of objects or the total number of conceivable observations of any kind or possessing some common specified characteristics. The existence of a population is often required as an assumption.

The set of observations that form a population need not necessarily be measurements, they may as well correspond to some attributes or characteristics whose presence or absence may be under consideration.

Ex. 1. 5 out of 100 items produced on a machine are examined. The sample size is 5 and the population size is 100.

Ex. 2. A coin is tossed 50 times. The number of tosses at a time constitutes a sample and the observations are the heads or tails that result from it. On the basis of the sample, an inference can be made about the population. Repeated tossed of a coin comprise a population in which the observation is either a head or a tail.

24.2 VARIABLES : QUANTITATIVE AND QUALITATIVE

When a characteristic takes different values depending on circumstances it is called a variable. Variables that can be measured are called quantitative variables. Variables that cannot be measured but only categorised as belonging to some group or other are called qualitative variables and give information regarding some attribute.

Statistical data pertaining to any social science are based on individual observations which are not necessarily uniform in nature. Each observation corresponds to a characteristic which is subject to variation. This variable quantity is called a variable and each observation a variate.

The prices of a share on different days, the heights or marks of a set of students are examples of variables.

In statistics we usually deal with quantative variables, *i.e.*, those which can be measured.

We also deal with qualitative variables which are distinguished by quality or attribute. For example eye colour, flower quality etc. are qualitative variables. Thus, we may consider the number of people with a certain skin colour.

24.3 WHAT IS A RANDOM VARIABLE

A random variable is a quantity whose value is determined by a chance experiment. It is also called a chance variable or stochastic variable. The term random does not imply equally likely outcomes; rather, it implies that the quantity under consideration cannot be predicted from the experiment. However, ordinarily we say that each independent random event has an equally likely chance of being selected. In technical terms, a random variable is defined as a measurable function determined on a probability space.

X is a *discrete random variable* if it can assume only a finite number of values.

X is a *continuous random variable* if it can assume all the values in an interval. Unlike the discrete variable, a continuous variable takes an infinite number of values.

Probability based sampling techniques maintain the conditions of randomness.

Note. The term *independent events* implies that the probability of occurrence of one event remains constant whether the other event has occurred or not.

It is not true that after a long series of successes a failure is inevitable. Similarly it is not true that after a long series of failures, a success must occur. This is known as Monte Carlo fallacy.

24.4 FINITE AND INFINITE POPULATIONS

If there is a limit to the number of possible observations, we have a *finite* population. An unlimited number of possible observations constitutes an *infinite* population.

In practice we deal with finite populations. Often the differences in the results from finite and infinite populations are not very significant. For relatively small samples, the results obtained with the help of infinite population assumptions are applicable to finite populations.

An *existent* population is an aggregate of concrete objects, *e.g.*, students of a college.

A *hypothetical* population consists of all the conceivable ways in which an event may occur, *e.g.* the rolling of a die an infinite number of times and noting all the results.

24.5 POPULATION PARAMETERS AND SAMPLE STATISTICS

A measure that differentiates a population quantitatively from all other populations of the same form is called a *parameter*. It is a *characteristic* of the population under consideration.

A sample is a partial collection or a limited number of observations out of all possible observations constituting the population. From a knowledge of the characteristics of the sample, information about the unknown population has to be obtained.

A random sample is such that each observation in the population has an equal chance of being included in the random sample. Every observations is as likely to be considered as any other. Further it means that if a sufficiently large number of samples were selected in the same manner from the population each observation in the population would be accounted for in the population of the samples.

In a biased sample, certain observations have a greater chance of being included than others.

A population is described by its *parameters*. Correspondingly a sample is described by its characteristics each of which is called a *statistic* (plural : statistics) not to be confused with the word *statistics* meaning data or statistics meaning the science of statistics.

The mean or standard deviation of a sample is a statistic. The same characteristic of a population is a parameter. A sample statistic is a measurable characteristic derived from a sample.

24.6. WHY DRAW A SAMPLE?

Advantages of Sampling

A sample provides partial information about the population and yet, as we have seen, we have to depend, for various reasons, on the information available from the sample.

Is there an alternative? Not always. Sometimes it may be feasible to obtain more complete or accurate information by taking a complete enumeration of data or what is known as the census method. But there are some definite advantages of sampling as compared with the census method.

(*i*) *Practicability.* Often a complete census is impracticable. It may, therefore, not be done at all. Why not, therefore, resort to sampling techniques?

(*ii*) *Flexibility.* Sampling has greater scope and flexibility in the matter of types of information to be obtained.

(*iii*) *Accuracy.* More efficient workers, intensive training to a small number of workers and comparatively less work result in greater accuracy of results. Beyond a certain point the improvement in information from additional data is not worth the cost. There is no guarantee that complete enumeration will give decidedly better results, or that a complete census could even be carried out. Often sampling methods give more accurate results than a complete survey.

(*iv*) *Speed.* The collection and analysis of data can be done more quickly if the data are not excessive. Time and energy are saved.

(*v*) *Superfluity Avoided.* Often there are many items of the same type. A complete enumeration will consider each one of them. That amounts to waste of time and energy.

(vi) *Reduced Cost.* Expenditure on a sample is likely to be very small as compared with that on a complete enumeration.

According to the Law of Statistical Regularity a reasonable sized sample randomly selected from a population exhibits such composition and characteristics as would be found in the population. As per the Law of Inertia of Large numbers, the larger the sample size, the greater the accuracy of results.

Precision in Sampling

The precision of a sample is the difference between the sample result and that of a complete census, theoretically taken under the same conditions.

Sampling theory plays an important role in many fields of scientific activity. Its purpose is to make sampling more efficient. Attaining a specified precision at minimum cost is the main intention of sampling theory.

It is possible to predict the precision of a sampling method. To find the extent of the error, the frequency distribution of the estimate is useful. We usually assume the frequency distribution to be normal.

After we have decided the possible nature of the frequency distribution we can proceed to make the needed inference about the population from the sample.

24.7 SAMPLING DISTRIBUTIONS

The sampling distribution of a statistic is the frequency distribution of the statistic calculated from all possible samples of a given size, drawn with or without replacement, from a given population.

Let μ and σ denote the mean and standard deviation of a population of size N and all possible samples of size n are drawn. It can be shown that if the population is infinite or if sampling is with replacement the sampling distribution of the mean is approximately normal with mean μ and standard deviation $\dfrac{\sigma}{\sqrt{n}}$.

If the population is finite and sampling is without replacement then the sampling distribution of the mean is approximately normal with mean μ and s.d. $\sqrt{\dfrac{N-n}{N-1}}\dfrac{\sigma}{\sqrt{n}}$.

Standard Error. Since the standard deviation of the sampling distribution of a statistic cannot be found exactly we use instead, the *standard error* which is an estimate of the standard deviation.

Thus if σ is not known we may use instead, the standard error given by $s/\sqrt{n}$

where $$s = \sqrt{\frac{\Sigma(x-\bar{x})^2}{n}} \text{ if } n \text{ is large}$$

and $$s = \sqrt{\frac{\Sigma(x-\bar{x})^2}{n-1}} \text{ if } n \text{ is small.}$$

Confidence interval for the mean. For a normal distribution 95 per cent values fall in the interval $\bar{x} \pm 1.96$ S.E. or $\bar{x} \pm 1.96\ \sigma/\sqrt{n}$.

i.e., we can say with 95 per cent confidence that the population mean might lie between these two limits.

Similarly we can say that the population mean lies between $\bar{x} \pm 3$ S.E. with 99.73 per cent confidence, and so on.

24.8 STATISTICAL INFERENCE AND DECISION MAKING

The process of inferring information about a population from a sample is known as statistical inference which deals with two main problems: Estimation and Testing of Hypotheses.

Estimation provides us with an estimate of the population parameter based on the sample statistic.

Point estimation assigns a single true value to the parameter being estimated.

Interval estimation specifies the upper and lower limits within which the parameter lies with a given probability or degree of belief. The two limits thus obtained are called *confidence limits*.

A *statistical hypothesis* specifies properties of a distribution of hypothetical parameters with which we compare statistics derived from actual samples, *i.e.*, we try to determine, on the basis of a sample statistic, whether the sample could have been drawn from a population with a certain hypothetical parameter. We use *tests of significance* of the difference between the actual statistic and the corresponding parameter for the purpose.

If we find the data or the sample size insufficient to have enough confidence in the conclusions we may collect additional data and carry out more tests.

It may be noted, however, that on the basis of a sample, one can never be sure about one's decisions regarding the population. But an effort should be made to arrive at the right decisions, to avoid wrong decisions and to get an idea about the extent of possible errors in them. Wrong decisions can be costly and the minimisation of these costs is another aspect of decision making with the help of samples.

24.9. THE ESTIMATE OF SAMPLE SIZE

The size of a sample to be taken in a survey is a problem that requires our attention because it is related to the time, cost and efficiency factors. If we take a very large ample it involves a waste of resources and time. A very small sample may not give enough information.

Sampling theory suggests methods of estimating sample sizes in particular situations depending upon:

(*i*) the desired precision and the limits of error to be allowed

(*ii*) the properties of the population

(*iii*) the kind of sampling to be used

(*iv*) the purpose

(*v*) the resources, time and labour, available and the costs to be incurred.

24.10 SAMPLING TECHNIQUES

The accuracy of statistical inference based on samples depends on the adequacy of samples and sampling methods. The problem of estimating the characteristics of a population would be very simple if the data were uniform and having the same pattern as the population. We know that in practice it it not so.

A sample may be drawn from the whole population or, by dividing the population into *strata,* from different strata of the population. A sample may be *systematically* drawn or it may be a cluster from some part of the population.

If each member of the population is allowed to be selected any number of times, we have sampling *with replacement.* Sampling *without replacement* implies that each member cannot be selected more than once.

There are two principal methods of drawing a sample from a population.

Probability and Non-Probability Samples

Probability samples. Here each observation in the population has an equal chance of being selected to become a part of the sample. Simple random sampling, stratified sampling, cluster sampling etc. are based on probability sampling. Personal judgement has no role to play in probability sampling. The precision of a probability sample increases with the size of the sample. In large scale surveys probability sampling is commonly used.

Nonprobability samples. Here there is no provision for any probability that each individual will be included in the sample. Quota sampling, incidental sampling and judgement sampling are examples of nonprobability samples. They are sometimes more convenient and economical.

PROBABILITY SAMPLING

24.11 SIMPLE RANDOM SAMPLING

A random sample is a sample selected from the population in such a way that every item in the population has an equal chance of being included in the sample. Random sampling or the process of drawing a random sample should be free from bias and the selection of any item should depend completely on chance.

Random sampling does not imply haphazard selection. If there are N units in the population, simple random sampling consists of selecting n units out of N in such a way that every one of ${}^{N}C_{n}$ samples that can be drawn from the population, has an equal chance of being selected.

Methods of Drawing Random Samples

For valid statistical inference from random samples we have to ensure that every element in the population has an equal chance of being taken for the sample. Haphazard procedures can hardly ever produce random samples.

Every method of random sampling will necessarily require a listing of all the elements in the population. For a small population, slips corresponding to the element numbers may be placed in a container and we may form a sample by drawing the required number of slips from it. However, it is usual to obtain random samples with the help of tables of random numbers generated by a random process, mechanically or electronically. The digits in all directions, rows or columns, are in random order in such tables which are available in a variety of publications, for instance, those of Indian Statistical Institute, Tippett or Kendall and Smith and others.

Random Sampling with Replacement

If each selected unit is returned to the population after each drawing, every item continues to have the same chance of being drawn. If the population is of a finite size N, this procedure may be employed if necessary.

Random Sampling without Replacement

If each selected unit is not returned back into the population and after each draw each of the remaining items has an equal chance of being drawn we have sampling without replacement. This method is utilised when the population size is infinitely large.

A Table of Random Numbers

Row	*Column*									
	1	*2*	*3*	*4*	*5*	*6*	*7*	*8*	*9*	*10*
1	6	9	1	8	3	3	7	5	9	6
2	7	0	0	3	8	1	3	4	7	9
3	3	7	2	0	8	1	5	6	8	0
4	2	7	0	0	0	6	5	0	6	5
5	3	0	7	0	7	8	4	9	4	2
6	6	2	9	3	3	1	7	7	5	2
7	5	4	9	2	1	4	8	5	7	0
8	0	3	7	0	1	7	3	8	0	3
9	9	3	6	6	2	2	0	9	7	2
10	2	9	5	6	9	9	5	5	6	9

The procedure to use a random number table. The elements in the population are numbered 1 to N and the n elements needed for the sample are drawn from the random number tables by starting at any point on it and proceeding to note the required numbers till the required sample size is obtained.

Ex. 3. To select a sample of size 8 from a population of size 800, let us start, say, from the sixth row, second column, taking 3 digit numbers and omitting such numbers as do not have a serial number in the population.

293 317 752 549 214 857 (omit) 003 701 738

If sampling is without replacement, any reappearing number is to be ignored. A sample randomly selected may happen to be unrepresentative though the probability of such a thing happening is small.

Ex. 4. Draw a 10 per cent random sample from a population of size 100 (Let the numbers be 00, 01, ..., 99. Take 10 two digit numbers from the tables)

Ex. 5. Select by using random numbers, 25 random samples (with replacement) of 4 workers each from the frequency distribution of the wages of 100 workers given below. Find their means and standard deviations and compare them by calculating the actual values.

Wages	*Frequency*	*Sampling Number*
100—120	4	00—03
120—140	20	05—24
140—160	32	25—56
160—180	39	57—95
180—200	5	96—99
Total	100	

(**Hint.** Attach the sampling numbers corresponding to the frequency as shown in the last column).

Estimates of mean and s.d. in simple random sampling

Let x_1, x_2 , ..., x_n be a simple random sample of size n drawn from a finite population of size N. Then the estimate of the population mean is given by

$$\bar{x} = \frac{\sum_{i=1}^{n} x_i}{n}$$

If S^2 is the population variance, the variance of the estimate is given by

$$V(\bar{x}) = \frac{S^2}{n} \cdot \frac{N-n}{N-1}$$

The standard error of the estimate is given by

$$\text{S.E.} = \frac{S}{\sqrt{n}}\sqrt{\frac{N-n}{N-1}}$$

If either the population is infinite or the sampling fraction $\frac{n}{N}$ small, we have $S.E. = \frac{S}{\sqrt{n}}$.

Merits of Simple Random Sampling

Simple Random sampling is based on chance. Personal bias, therefore, cannot influence the results. The technique becomes more reliable as the sample size increases.

Some Limitations. Use of random number tables or numbered slips and the need to catalogue the data are tedious jobs.

The area covered by random sampling is sometimes very large geographically and a lot of time and cost may be wasted in sampling. It is for this reason that stratified random sampling (discussed below) is often suggested in such cases.

The size of sample used in random sampling can be made large. Yet, at times, some areas may not be considered at all by chance and others may be over-represented by chance.

24.12 STRATIFIED RANDOM SAMPLING

Simple random sampling is often not convenient or accurate particularly when the population is heterogeneous. There is also a probability of some parts of the population being kept out by chance. Simple random sampling is more commonly used in the case of homogeneous populations.

If a population can be divided into sub-populations or strata the elements in each of which are of a fairly uniform nature, a simple random sample can be drawn from each stratum and the results thus obtained from all the strata are pooled together to obtain a precise estimate for the population. This is stratified sampling which can also be used when specific information from different parts of the population is needed. These parts can be treated as populations in their own right. It is sometimes possible that administratively it may be more convenient to deal with strata rather than a whole population or that a different approach to different strata may be desirable because of non-uniformity.

The size of a stratified sample may be either proportional or non-proportional to the stratum size. We may draw, for instance, a 5 per cent or 10 per cent sample from each stratum. A non-proportional sample, however, may be more efficient in many cases.

If the observations in some strata are cheaper to select and to survey or if the observations are highly variable in some strata it may be advisable to take larger samples from such strata.

Estimation of population parameters from non-proportional sample statistics can be done by weighting each subsample statistic in proportion to the size of the corresponding subpopulation.

The chief merits of stratified sampling are low cost, greater accuracy, allowance for geographical concentration and a better coverage.

The method requires a careful division of the population into strata and a truly random sampling in each of these strata.

Stratification

Stratification in employed when a heterogeneous population can be subdivided into such strata as are nearly homogeneous themselves. A prior knowledge of the population definitely helps.

Stratified sampling amounts to taking a number of simple random samples from subpopulations which are the strata of the given population. Stratified sampling gives the most efficient results when the variability within each stratum is the least, *i.e.*, each stratum is more or less homogeneous within itself.

Let $\bar{x}_1, \bar{x}_2, ..., \bar{x}_L$, be the means of samples of sizes $n_1, n_2, ..., n_L$ respectively drawn from the L strata of sizes $N_1, N_2, ..., N_L$ where $\sum_{i=1}^{L} N_i = N$, the population size.

The estimate of the population mean is given by $\frac{\Sigma N_i \bar{x}_i}{N}$

There are various ways in which the sampling fractions from different strata may be taken. Let n_h be the sample size drawn from the stratum of size N_L whose s.d. is S_h.

Proportional allocation consists in taking a uniform sampling fraction $\frac{n_n}{N_h}$ from each stratum resulting in self weighting samples.

Allocation proportional to s.d. consists in taking samples such that $\frac{n_h}{S_h}$ remains the same for each stratum.

Optimum allocation consists in n_h being proportional to both the size and s.d. of each stratum, *i.e.*, $\frac{n_h}{N_h S_h}$ remains the same.

The last two are cases of variable sampling fractions. The size of the sampling fraction depends on the purpose of the sampling inquiry. Proportional allocation is commonly used though optimum allocation can give the maximum overall gain.

24.13 MULTISTAGE SAMPLING

When a survey is to be carried out over a wide area involving heavy travel expenses, use can be made of multistage sampling where

(*i*) the population is first divided into large groups or first stage units; one of the first stage units is first selected to be used in the second stage.

(*ii*) this is divided into smaller, second stage units, and a sample selected out of these; further we may go to third stage units and so on till we arrive at a reasonable sized group to be utilised for the selection of the ultimate sample.

24.14 CLUSTER SAMPLING OR SUBSAMPLING

The use of a sampling unit which consists of a group or cluster of the elements in the population is known as cluster sampling. The selection of the sample is done in two stages. First certain groups or clusters are selected from the population. These are *primary sampling units.* From each of these clusters, elementary sampling units are drawn. From this follow alternative names for this method of sampling viz., *subsampling* or *two stage sampling* or block or area sampling. Of course, the process could be continued for *multistage sampling.*

Cluster sampling permits groups of observations for easier coverage. Travel expenses may be significantly reduced and thus the cost per elementary sampling unit becomes much less.

The results of a cluster sample may not be as precise as those of a random sample but they can be made more precise by taking a larger sample size.

24.15 SYSTEMATIC SAMPLING

Systematic sampling is a particular case of cluster sampling in which the sample is a single cluster. Here a sample is selected at a predecided sampling interval. Thus, every 10th unit in the population gives us a 10 per cent sample. In general every n^{th} unit may be selected. This method has an advantage over random sampling if neighbouring observations resemble one another. The method serves almost the same purpose as stratification in distributing the sampling units more evenly over the entire population.

Systematic sampling being quasi random in nature, may sometimes give misleading results when for instance, every n^{th} unit selected happens to be of the same type. Precaution can help, however. Reliability of information obtained from systematic sampling, in general, is adequate.

The merit of systematic sampling lies in that it is time and money saving if it is handled carefully. A biased sample may mar the utility of the result.

24.16 SEQUENTIAL SAMPLING

Sequential sampling consists in testing a small number of items from a lot of units and accepting and rejecting the whole lot depending on the decision taken with the help of the sample. If no decision can be taken from the first sample, more samples are taken till a decision is possible. Sequential sampling is used mostly in production units to sample manufactured products.

NONPROBABILITY SAMPLING

24.17 JUDGEMENT SAMPLING

Personal judgement can sometimes be more useful than a probability sample in pilot surveys or small scale surveys. Judgement sampling can be resorted to if the person in charge is experienced and is known to have a good power of judgement. In the construction of index numbers, judgements of experts in the field are more useful than random methods.

Judgement sampling can be biased and there is no objective test to determine the validity of its results. Its utility may not necessarily increase with the increase in sample size. Judgement sampling is sometimes recommended in the case of small samples and some special cases.

24.18 QUOTA SAMPLING

In quota sampling, probability methods are not used. Each investigator is instructed to collect information from an assigned number or quota of individuals having specified characteristics. The investigator is free to choose any individuals in the group.

There is great scope for bias in quota sampling but in many cases it has proved to be a very convenient and less costly method. Quota sampling is popular in market surveys and public opinion polls.

24.19 CONVENIENCE SAMPLING

When a sample is selected from available municipal lists or directories etc. according to some convenience factor, it is a case of convenience sampling. It may be employed in pilot studies, public opinion surveys etc.

24.20 PRECAUTIONS IN AND LIMITATIONS OF PRACTICAL PLANNING OF SURVEYS

The nature and type of sampling to be used in a particular situation depends on various considerations like time, cost, efficiency desired and other factors. Any of the above methods or a combination of these may be used depending on requirements.

A sample survey should be carefully planned and executed to avoid inaccuracies. In the absence of experts there is the possibility of inexperienced persons ruining a survey. Greater care has to be taken as the size of the survey and its coverage become larger.

Often a *pilot survey* is conducted to get a rough idea of the problems to be faced and to plan their solution. This may give experience, training and preparation for better results.

24.21 SIMULATION

Problems in real life are often very complex and cannot be easily expressed in the form of comprehensible mathematical equations. The simulation technique can be employed to arrive at the solution of such problems.

The technique of simulation consists of a series of organised trial-error experiments with the model of a system to enable one to derive certain results when the mathematical format of the model is not easy. Simulation helps in prediction and decision making under conditions of uncertainty to provide solutions to problems arising in a variety of business and managerial situations where values of variables are not available or not easily available or when analytical and algebraic methods do not help because of complications of various types.

Simulation Methodology

The operation of the system is described by a mathematical model which more or less is known to fit the reality. Sampling experiments are performed on the system. The method commonly used for the purpose is called the Monte Carlo method which does not give the Optimal solution but tends to give a nearly optimal solution as the number of simulation trials is increased.

Another method for the purpose is System Simulation method which works with a replica of the operating environment. This method consumes more time, cost and energy than the Monte Carlo method which is more commonly used.

Note. 1. Monte Carlo simulation can be applied only to models under uncertainty. It requires a continuous observation of the system over a long period of time.

2. Monte Carlo method is based on the use of random numbers for which tables are availables. Each random number has an equal chance of being selected.

3. Each selected random number is converted into an observation as if it were drawn from the probability distribution specified by the model describing the system.

24.22 ERRORS

In statistical measurements, hypothesis testing and decision making, etc. there is usually a possibility of errors.

Some errors are systematic and are independent of the size of the sample. They may be related to defects in planning stages, procedures, methods of collecting data, faulty questionnaire design, nonresponse and other deficiencies. The term conventionally used for systematic error is *bias*.

The other type of error is random or sampling error which arises from differences between the results of sample observations and universe values. The means of different samples drawn from the same population differ although the procedure of drawing the sample is the same for all samples. The size of this error indicates the reliability or precision of the experiment. As the sample size increases, this error decreases and hence larger samples are considered more reliable than smaller samples.

The true value of any measurement is thus associated with Bias (Systematic Error) + Random Error each of which may be positive or negative.

Ex. 6. With a larger sample size which of the two measurement errors decreases ? Which error can be eliminated?

(Random errors decrease on the average with increasing sample size. Random errors may usually compensate one another. Systematic errors can be discovered and minimised).

Ex. 7. Is the error in the following random or systematic?

A selected number of 1000 people were asked in quetionnaires whether they thought the politicians to be corrupt, honest or otherwise. Of the 50 people that replied, a majority did not consider them honest. But the majority of the people are known to consider them corrupt.

Ans. Both random and systematic errors

Systematic error because of the attitudinal differences between a mere 50 respondents and a much larger original sample ; A "selection", apart from a small sample size, may introduce bias.

Ex. 8. Consider a population of Size $N = 5$. The five values are 3, 6, 9, 12, 15. Find the means of all possible samples of size 3 and compare the mean of these samples with the population mean

$$\text{Population mean} = \frac{3+6+9+12+15}{5} = 9$$

${}^5C_3 = 10$ Samples of size 3 are possible.

Sample No.	Sample values	Mean
1	3, 6, 9	6
2	3, 6, 12	7
3	3, 6, 15	8
4	6, 9, 12	9
5	6, 9, 15	10
6	9, 12, 15	12
7	3, 9, 12	8
8	3, 9, 15	9
9	3, 12, 15	10
10	6, 12, 15	11

Mean of sample means = $\frac{90}{10} = 9$

This equals the population mean because all possible samples have been considered.

Ex. 9. A sample of size 5 is drawn without replacement from a population of size 41. If the population S.D. is 6.25 find the SE of the sample mean. Note. $N = 41$, $n = 5$.

$$\text{SE}(\bar{x}) = \frac{\sigma}{\sqrt{n}}\sqrt{\frac{N-n}{N-1}} = \frac{6.25}{\sqrt{5}}\sqrt{\frac{41-5}{41-1}} = 2.65$$

Ex. 10. Comment on the following:

(*i*) Sampling is necessary under certain conditions.

(*ii*) A population is any collection of objects.

(*iii*) Probability sampling is different from purposive sampling.

Ex. 11. Differentiate between

(*i*) Simple and stratified random sampling

(*ii*) Systematic and judgement sampling

(*iii*) Sequential and cluster sampling

Ex. 12. Explain the following terms.

(*i*) Sample (*ii*) Population (*iii*) Statistic (*iv*) Parameter

Ex. 13. What is the importance of Sample size?

Sample Size

The size of a sample plays an important role in any statistical exercise. In sampling theory it is necessary to have an idea of sampling error whose estimates are based on probability theory which is the basis of simple random sampling. This estimation procedure is not applicable to nonprobability samples.

For small samples, as the sample size increases, the variance reduces rapidly. For large samples the variance does not fall rapidly with increasing sample size.

Note. The sampling error of any subgroup in a survey will depend on the size of that subgroup and not on the total sample size.

Ex. 14. Answer the following questions:

(*i*) What is a random sample?

(*ii*) What is the best sample size?

(*iii*) What are the precautions that should be taken in sampling?

(*iv*) What are the advantages of sampling?

Ex. 15. Is there any source of bias in the following:

(*i*) A sample of unemployed persons obtained by taking every fiftieth name from a register of the unemployed.

(*ii*) A survey of the incomes of the residents of a locality by taking every tenth house.

(*iii*) Estimation of per capita income by interviewing car owners.

(*iv*) A sampling of mangoes by taking a few from the top of the basket.

Ans. (*i*) and (*ii*) are cases of systematic sampling. Fairly unbiased, if there is no periodicity.

(*iii*) Biased. Car owners may generally be richer than non-car owners.

(*iv*) Biased. The best fruit is often placed on the top.

Ex. 16. How can a pilot survey help?

It is a preliminary and exploratory exercise. It can reveal defects at various stages and suggest the need for improvements. It can give an idea of the possible responses, the nature of response and provide useful insights and guidance.

Ex. 17. What are Sampling and Non-sampling Errors?

A sample, though characterising the population cannot extract full information about the population. The resulting difference between each statistic and the corresponding parameter stands for sampling error. Errors arising at the stage of data collection, aggregation etc. are non-sampling errors. Efforts should be made to minimise these errors.

24.23 SAMPLING DISTRIBUTION OF A STATISTIC

Ex. 18. Explain the concept of Sampling Distribution of a statistic.

Populations are usually large, sometimes inaccessible and difficult to handle in their entirety. To get information about a population and to draw inferences about it we make use of samples. In other words, we make inferences about the population parameters with the help of statistics calculated from a sample of random observations drawn from the population. A statistic is a descriptive measure computed from sample observations.

The researcher usually has the option of a single sample from which he calculates a statistic and uses it to estimate the population parameter with the help of an appropriate sampling distribution.

Sampling Distributions

A sampling distribution is the probability distribution of all values of a sample statistic computed from all possible random samples from a population. This is a theoretical statement. In actual practice, the researcher employs a single sample with the help of the sampling distribution of the statistic. It is possible to make inferences from a sample about the population so that it can be found if the sample estimate is reasonable.

Every statistic (*e.g.* mean, standard deviation, coefficient of correlation etc.) has a sampling distribution. If, $x_1, x_2, \ldots, x_n$ are values of observations in a sample, the statistic t in the functional relation $t = t(x_1, x_2, \ldots, x_n)$

amounts to the reduction of the data to a single number. Some population information is lost but that may not be relevant to the problem under consideration.

If $t_1, t_2, \ldots, t_k$ are values of the statistic t in k samples, each value occurring with a certain probability then the resulting probability distribution is the sampling distribution of the statistic t.

24.24 CENTRAL LIMIT THEOREM

I. For the Proportion

If p is the proportion of a certain characteristic in a sample drawn from a population having this characteristic in proportion P then the sampling distribution of this proportion p is approximately normal with mean P and S.D. = SE (p) provided the sample size n is sufficiently large.

II. For the Mean

If a reasonably symmetrical population has mean μ and standard deviation σ and large samples of size $n \geq 30$ are taken from the population with means $\bar{x}_1$, $\bar{x}_2$, ... etc. then it can be shown that

(*i*) the sampling distribution of the means is approximately normal and the mean of this normal distribution is equal to the population mean, *i.e.*, Mean $(\bar{x}_i) = \mu$.

(*ii*) the standard error of these means equals $\frac{1}{\sqrt{n}}$ times the population standard deviation:

$$SE\,(\bar{x}_i) = \frac{\sigma}{\sqrt{n}}$$

where n is the sample size.

If the original population has a normal distribution then the distribution of $\bar{x}$ is exactly normal irrespective of the sample size. Even when the original population is not normal, $\bar{x}$ is approximately normal if $n \geq 30$.

The Central limit Theorem is useful in the theory of estimation and in testing of hypotheses about the population mean μ. This is because regardless of the population distribution form the distribution of $\bar{x}$ can be assumed to approach the normal as the sample size increases.

Ex. 19. What are the characteristics of the sampling distribution of (*a*) p, (*b*) $\bar{x}$.

Population Proportion p

The proportion of units with a particular characteristic in a sample is the best estimator of the corresponding population proportion P. This estimator is unbiased, consistent, efficient and sufficient.

Ex. 20. If the proportion of defectives in a sample from a consignment is found to be 0.2 we can say that the defective proportion in the population is also 0.2.

If the population proportion P is known, the variance of the proportion in a large sample is given by

$$\sigma_p^2 = \frac{PQ}{n} \quad \text{where } Q = 1 - P$$

The standard error of the proportion p is given by

$$\sigma_p = \sqrt{\frac{PQ}{n}}$$

If the population proportion P is not known we can use p instead and write

$$\sigma_p = \sqrt{\frac{pq}{n}} \text{ (where } q = 1 - p)$$

The mean of the sampling distribution of the proportion p of successes is given by $\mu = np$.

The sample mean = np

$$S.E.\ (np) = \sqrt{nPQ} \text{ if } P \text{ is known and } Q = 1 - P$$

$$S.E.\ (np) = \sqrt{npq} \text{ if } P \text{ is not known}$$

For the sample mean $\bar{x}$, on the same lines, $S.E.\ (\bar{x}) = \frac{\sigma}{\sqrt{n}}$

24.25 FINITE POPULATION: SAMPLING WITHOUT REPLACEMENT

In theory we assume that sampling is from an infinite population. In that case, $SE\ (\bar{x}) = \frac{\sigma}{\sqrt{n}}$.

If the population is of finite size N and the sample size $n > 5\%$ of N, we must apply a correction factor $\sqrt{\frac{N-n}{N-1}}$ to the stand error. Thus

the corrected $SE(\bar{x}) = \sqrt{\frac{N-n}{N-1}}\ \frac{\sigma}{\sqrt{n}}$.

Sometimes N is large so that the correction factor may be written $\sqrt{\frac{N-n}{N}}$. This is because in that case it does not matter much whether we divide by $N - 1$ or N.

Similarly in a finite population, $S.E.$ of sample proportion, $SE\ (p) = \sqrt{\frac{PQ}{n}}\sqrt{\frac{N-n}{N-1}}$.

Ex. 21. Estimate the mean and S.D. of the population as well as the estimate of the SE of the sample mean for the random sample of 5 observations given by 14, 19, 17, 20, 25.

$$\bar{x} = \frac{\Sigma x}{n} = \frac{95}{5} = 19$$

$$S^2 = \frac{\Sigma(x-\bar{x})^2}{n-1} = \frac{\Sigma(x-19)^2}{5-1}$$

$$= \frac{66}{5-1} = 16.5$$

x	$(x - 19)^2$
14	25
19	0
17	4
20	1
25	36
95	66

$S = \sqrt{16.5} = 4.1$ (Hereafter we have used S instead of s when the denominator is $n - 1$)

Note. $S = \sqrt{\frac{\Sigma(x-\bar{x})^2}{n-1}}$, $s = \sqrt{\frac{\Sigma(x-\bar{x})^2}{n}}$

The estimate of μ is 19

The estimate of σ is 4.1

$$SE(\bar{x}) = \frac{\sigma}{\sqrt{n}} \simeq \frac{S}{\sqrt{n}} \quad \text{(using } S \text{ as an estimate of } \sigma)$$

$$= \frac{\sqrt{16.5}}{\sqrt{5}} = 1.8$$

Ex. 22. From a population of size 128 units and S.D. 4 a sample of 64 units was drawn. If the sample mean is 28 find the 95% confidence limits for the population mean.

The population is finite : $N = 128$

The population SD is known, $\sigma = 4$

$$SE = \frac{\sigma}{\sqrt{n}}\sqrt{\frac{N-n}{N-1}} = \frac{4}{\sqrt{64}}\sqrt{\frac{128-64}{128-1}} = 0.35$$

The 95% confidence limits are $\bar{x} \pm 1.96$ SE = 28 ± 1.96 × .35 = 28.69 and 27.31

Ex. 23. From a population of 1000 a random sample of size 100 is taken and is found to have a mean of 4.8 and SD. 1.1. Find the 95% confidence limits for the population means. (For Ex. 22, 23, see Sec. 25.11, 25.12, 25.13)

Ans. 4.6, 5

Sampling Distribution of the Mean : Summing Up

The arithmetic mean (or for that matter any other characteristic) of any sample drawn from a population is called a statistic. Different samples from the same population will have different values of these statistics though the difference, in general, may be small. These values will be distributed around the population mean which is called a parameter (being one of the many other characteristics or parameters of the population).

The sampling distribution of the mean is a theoretical distribution which is helpful in carrying out statistical inference.

If samples of a given size are drawn from a normal population, the sampling distribution of the mean $\bar{x}$ is also normal. If the population mean is μ and standard deviation σ then the sampling distribution of $\bar{x}$ is normal and has its

(*i*) mean $\mu_{\bar{x}}$ equal to the population mean μ

(*ii*) standard deviation $\sigma_{\bar{x}} = \dfrac{\sigma}{\sqrt{n}}$ where n is the sample size.

Note. $\sigma_{\bar{x}}$ is a measure of the precision of the sample mean $\bar{x}$ in estimating the population mean μ. It is referred to as the standard error.

Since $\sigma_{\bar{x}}$ varies inversely with the square root of sample size n, quadrupling the sample size cuts the standard error to half and so on.

What happens if the population is not exactly normal as it may be the case in situations pertaining to economics and management? The central limit theorem states that the sampling distribution of x for almost all types of populations is nearly normal or can be taken to be so.

For a non-normal population, the sampling distribution of (sample means) $\bar{x}$ may be considered to be approximately normal for large samples.

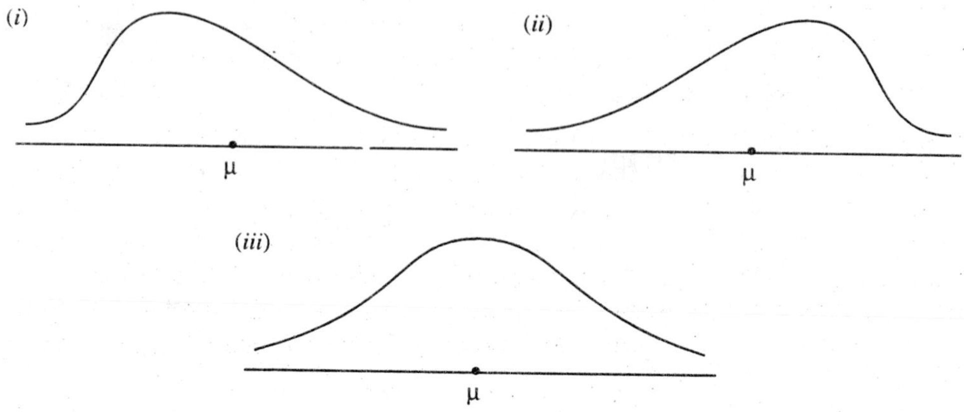

The sampling distribution in (*iii*) is approximately normal for sample means drawn from non-normal distributions (*i*) or (*ii*)

Note. The following standard errors of different statistics may be noted

1. S.E. $(p) = \sqrt{\frac{PQ}{n}}$ (p sample proportion, $q = 1 - p$) (P population proportion, $Q = 1 - P$)

2. S.E. $(\bar{x}) = \frac{\sigma}{\sqrt{n}}$ $\bar{x}$ sample mean, σ population standard division

3. S.E. $(p_1 - p_2) = \sqrt{\frac{P_1Q_1}{n_1} + \frac{P_2Q_2}{n}}$ p_1, p_2 sample proportions from samples of sizes, n_1, n_2; P_1, P_2 corresponding population proportions

4. S.E. $(\bar{x}_1 - \bar{x}_2) = \sqrt{\frac{\sigma_1^2}{n_1} + \frac{\sigma_2^2}{n_2}}$ $\bar{x}_1$, $\bar{x}_2$ means from two populations with S.D. σ_1, σ_2

Also S.E. (Median) $= \frac{1.2533\sigma}{\sqrt{n}}$

S.E. $(SD) = \frac{\sigma}{\sqrt{2n}}$

S.E. $(r) = \frac{1-\rho^2}{\sqrt{n}}$ (ρ = population correlation coefficient)

If the population is finite of size N

$$\text{S.E. }(p) = \sqrt{\frac{PQ}{n}}\sqrt{\frac{N-n}{N-1}}$$

$$\text{S.E. }(\bar{x}) = \frac{\sigma}{\sqrt{n}}\sqrt{\frac{N-n}{N-1}}$$

CHAPTER 25

Statistical Inference : Estimation

25.1 INFERENCE AND INFORMATION

Statistical inference involves

1. Estimation of population parameters by using sample data with a desired degree of precision.

2. Hypothesis testing and decision making to choose from various alternatives while controlling and minimising risks arising from wrong decisions.

The estimation of population parameters is necessary because complete enumeration of a population to get the required information could be very expensive and difficult as in the case of consumers of a commodity, percentage of defective items in production and so on.

Notes. 1. There are two types of estimates: *point* estimates and *interval* estimates.

2. Estimation is done with the help of information obtained from a sample which is drawn from a population.

For instance, we calculate a statistic, say the arithmetic mean, from a sample. This statistic may be inferred to be an approximation of the corresponding population parameter, in this case, the population arithmetic mean. Parameters can be *estimated* with the help of certain sample statistics. Some of the commonly estimated parameters are population mean, proportion, variance etc.

3. We try to get information about the population or estimate a population parameter with the help of sample data.

25.2 ESTIMATION IN DECISION MAKING

Decision making is usually marked by uncertainty because the available information is often incomplete or inadequate. Often one has to resort to some kind of estimation which may be (*i*) guesswork, (*ii*) something based on experience or (*iii*) something evolved by the use of scientific methods. Decisions made by a manager on the basis of estimates can influence the working of a large organisation and that made by an economist can influence the working of an economy.

In quantitative work, statistical techniques are used to infer or estimate the characteristics or *parameters* of a population with the help of the characteristics of a sample selected from the population.

A sample characteristic is a *statistic* which is used to estimate a population parameter. Such a statistic is called the *estimator* which gives an *estimate* of a given population parameter. It is the specific numerical value of the estimator, calculated from the sample which is the basis of the observed numerical value of the statistic to be used in the estimation exercise. If the decision maker does not have prior information about the state of nature he can make use of or can employ estimation from sample information.

Ex. 1. In a city it is decided to estimate people's preference for a certain product. A scientifically selected sample of the people of the city can be used to find the nature of their preference which can be used as an estimate of the preference of all the people in the city.

Ex. 2. 30% people consume a given product. This is a point estimate. The percentage of people consuming a given product could be anywhere between 20 and 40. This is an interval estimate.

25.3. POINT ESTIMATION AND CONFIDENCE INTERVAL

Population parameters can be estimated either as point estimates or as interval estimates *i.e.*, as single values or as intervals.

The sample mean $\bar{x}$ is a point estimate of the population mean μ. This unique value may not be equal to or may not be the representative of the population mean. The interval estimate provides the limits within which the true population value may lie with a certain degree of confidence or probability.

25.4 ESTIMATORS

A point estimate indicates a single value which can be used to estimate the population parameter.

An interval estimate is a *range* of values which may include the population parameter, with a certain degree of confidence. The probability of the true parameter lying in a wider range is greater than that lying in a narrow range. This probability indicates the degree of confidence.

The procedure or formula that enables us to arrive at the estimate is called an estimator. There can be more than one estimator for a parameter.

Some criteria are available to decide which of these several estimators may be used in a given situation.

If $x_1, x_2, \ldots, x_n$ are the sample observations from a population, *i.e.* if a sample of size n is taken from a population then the estimator of the population parameter is a function that shows how to use $x_1, x_2, \ldots, x_n$ to estimate the parameter. The value thus obtained is the estimate.

If we have many samples, each of size n, the means of these samples will give us what is called the "sampling distribution" of the mean.

The possible range within which the population mean may lie is called "confidence interval"

25.5 CONFIDENCE

The confidence (in terms of a percentage) attached to an interval estimate implies the extent of our confidence that the population parameter is included in the range of the interval estimate. This is expressed in terms of a probability statement to indicate the degree of confidence in the estimated interval.

25.6 PROPERTIES OF GOOD ESTIMATES

A given population parameter can be estimated by means of several different estimators. In a given situation we can make a proper selection of an estimator on the basis of certain criteria.

25.7 UNBIASED ESTIMATORS

If t is an unbiased estimator of the parameter θ. Then $E(t) = \theta$. A sample statistic is an *unbiased estimator* of the corresponding population parameter θ if the expected value (*i.e.*, the mean) of the sampling distribution of the statistic equals the corresponding population parameter. Thus if the statistic t, which is a function of sample value, *i.e.*, $t = t(x_1, x_2, \ldots, x_n)$, is an unbiased estimator of the population parameter θ then

$$E(t) = \theta$$

If $E(t) \neq \theta$, the estimator t is biased. We can measure the bias. We write Bias $= E(t) - \theta$.

Ex. 3. (*i*) It can be proved that the sample mean $\bar{x}$ is an unbiased estimator of the population mean μ, *i.e.*, $E(\bar{x}) = \mu$.

(*ii*) The sample variance s^2 is not an unbiased estimator of population variance σ^2.

But $\frac{n}{n-1}s^2$ is an unbiased estimator of σ^2. It can be shown that

$$E\left(\frac{n}{n-1}s^2\right) = \sigma^2 \quad \text{where } s^2 = \frac{\Sigma(x-\bar{x})^2}{n}$$

(*iii*) The arithmetic mean is an unbiased estimator of the population median. This is because the average of the values taken by the mean would equal the population median.

(*iv*) An unbiased estimator neither overestimates nor underestimates the population parameter, *i.e.*, its average (or expected) value equals the population parameter.

(*v*) The sample proportion P is an unbiased estimator of the proportion p.

25.8 CONSISTENT ESTIMATORS

If a statistic t comes closer and closer to the population parameter θ as the sample size becomes large, *i.e.*, $t \to \theta$ as $n \to \infty$ then t is said to be a consistent estimator of θ.

For the same population parameter θ, it is possible to have one or more consistent estimators.

Consistency is a limiting property based on the sample size n. An estimator may become consistent as n increases. It implies that an increase in the sample size n goes on adding information about the population parameter.

Ex. 4. The sample mean $\bar{x}$ is a consistent estimator of the population mean μ.
i.e., as $n \to \infty$, $\bar{x} \to \mu$.

25.9 EFFICIENT ESTIMATORS

There can be several consistent estimators say t_1, t_2, t_3 for the same parameter θ.

The most efficient or the best estimator of these is the one which has the least variance (or least standard error)

If $V(t_1) < V(t_2) < V(t_3)$ or if $SE(t_1) < SE(t_2) < SE(t_3)$

then t_1 is more efficient than t_2 and t_2 is more efficient than t_3.

Ex. 5. If $V(\bar{x}) = \dfrac{\sigma^2}{n}$, $V(\text{median}) = \dfrac{\pi}{2n}\sigma^2$, which is the more efficient between $\bar{x}$ and the median as an estimator of the population mean? $\bar{x}$ is more efficient because $V(\bar{x}) < V(\text{median})$

Note. $\dfrac{\sigma^2}{n} < \dfrac{\pi}{2n}\sigma^2$ since $1 < \dfrac{\pi}{2}$ or $\dfrac{11}{7}$

25.10 MEAN SQUARE ERROR (MSE)

The performance of an estimator is measured by

$$\begin{aligned} \text{MSE} &= E(t-\theta)^2 \\ &= \text{Var}(t) + (E(t) - \theta)^2 \\ &= \text{Var}(t) + B^2 \end{aligned}$$

where the bias $B = E(t) - \theta$ (Sec. 25.7)

In the case of the sample mean used as an estimator of the population mean,

$$B = E(\bar{x}) - \mu = \mu - \mu = 0$$

$$\text{MSE} = E(\bar{x} - \mu)^2 = \text{Var}(\bar{x}) = \frac{\sigma^2}{n}$$

Again, for the sample variance s^2, we have with $s^2 = \dfrac{\Sigma(x-\bar{x})^2}{n}$

$$E(s^2) = \frac{n-1}{n}\sigma^2 \text{ (there is a negative bias)}$$

It can be shown that

$$\text{MSE}(s^2) = \text{Var}(s^2) + (E(s^2) - \sigma^2)^2$$

Ex. 6. In a certain situation there are three estimators α, β, γ. Other things being equal which one of these should be preferred if

$$\text{Var}(\alpha) < \text{Var}(\beta) < \text{Var}(\gamma)$$

Ans. α because it has the least variance.

Note. In the case of a normally distributed population, $\bar{X}$ is an unbiased estimator of μ and has the least variance. Hence $\bar{X}$ is a Minimum Variance Unbiased Estimator (MVUE) of μ.

Sufficiency

If an estimator extracts from the sample such an amount of information as no other estimator does, it is said to be a sufficient estimator.

An estimator may or may not possess any number of the above population.

Ex. 7. Explain the term confidence. Find the confidence interval for the mean.

Ans. It is not possible to specify any specific value or *a point estimate* for the population mean based on a single sample. Instead, we can specify a range of values or *an interval estimate* within which the hypothesis may work. This interval is called a *confidence interval* and its extreme values are called *confidence limits*.

The *confidence* is the level of probability with which we place our confidence in the estimate.

For a large sample with $z = \left|\frac{\bar{x}-\mu}{s/\sqrt{n}}\right|$, the upper and lower limits of the confidence interval for μ are found from

$$\bar{x} \pm z\frac{s}{\sqrt{n}}$$

$$\left(\text{Since } z = \left|\frac{\bar{x}-\mu}{s/\sqrt{n}}\right| \therefore z\,\frac{s}{\sqrt{n}} = |\bar{x}-\mu| \therefore \bar{x}-\mu = \pm z\,\frac{s}{\sqrt{n}}\right)$$

For small samples with $t = \left|\frac{\bar{x}-\mu}{s/\sqrt{n-1}}\right|$, similarly, the confidence interval for μ is

$$\bar{x} \pm t\,\frac{s}{\sqrt{n-1}} \qquad \left(\textbf{Note. } s = \sqrt{\frac{\Sigma(x-\bar{x})^2}{n}}\right)$$

The values of z and t are taken according to the degree of confidence desired.

Ex. 8. If a company makes a claim of an estimated mean life of 200 hours for the bulbs manufactured by it, it is giving a point estimate.

But this does not describe the uncertainty associated with the estimate. The actual mean bulb life may be anywhere between two given values, say 160 and 240 hours. This statement has to be associated with a certain degree of confidence which can be determined with the help of some assumptions.

25.11 CONFIDENCE INTERVAL ESTIMATE AND PROBABILITY STATEMENTS

The area under the normal curve between – 1.96 S.E. and + 1.96 S.E. of the population mean is known to be 0.95. We can describe this by saying that if the sample data are normally distributed, the probability of the mean lying between ± 1.96 S.E. is 0.95. In other words, the chance of the population mean lying between ± 1.96 S.E. is 95% or 0.95. Our confidence is 95% that the interval contains the population mean.

The probability (in the above case 0.95 or 95%) associated with an interval estimate is the level of confidence.

The usually employed confidence intervals are

(*i*) 95% corresponding to ± 1.96 S.E.

(*ii*) 95.45% corresponding to ± 2 S.E.

(*iii*) 99% corresponding to ± 2.58 S.E.

(*iv*) 99.73% corresponding to ± 3 S.E.

Other confidence levels can be employed as required by using the normal tables. For instance the 90% confidence interval corresponds to ± 1.64 S.E. The value corresponding to + 1.64 S.E. gives the upper confidence limit or UCL and that corresponding to – 1.64 S.E. gives the lower confidence limit or LCL.

The wider the confidence limits the greater is the certainty of the statement. But very wide intervals may not be very useful in practice.

Ex. 9. From a large consignment a sample of 100 items is selected and found to contain 10% defective items. Find 95.45% and 99.73% confidence limits for defectives.

$$p = 0.1,\ q = 0.9,\ \text{S.E. }(p) = \sqrt{\frac{PQ}{n}} \simeq \sqrt{\frac{pq}{n}} = \sqrt{\frac{0.1 \times 0.9}{400}} = 0.015$$

where P, Q being not known p, q are used.

The 95.45% limits are $0.1 \pm 2 \times 0.015 = 0.13, 0.07$.

The 99.73% limits are $0.1 \pm 3 \times 0.015 = 0.145, 0.055$

Ex. 10. The A.M. of marks of 100 students is 60 with S.D. 10. Find 95% and 99% confidence limits

$$\bar{x} = 60,\ \text{S.E. }(\bar{x}) = \frac{\sigma}{\sqrt{n}} = \frac{10}{\sqrt{100}} = 1$$

The 95% confidence limits for the population mean are $\bar{x} \pm 1.96$ S.E., *i.e.*, $60 \pm 1.96 \times 1 = 61.96, 58.04$. The 99% confidence limits are $\bar{x} \pm 2.58$ SE $(\bar{x})$, *i.e.*, $60 \pm 2.58 \times 1 = 62.58, 57.42$.

In this exercise if the population size is known to be $N = 200$, then $\frac{n}{N} = 0.5$ cannot be neglected.

$$\text{SE}(\bar{x}) = \frac{\sigma}{\sqrt{n}}\sqrt{\frac{N-n}{N-1}} = \frac{10}{\sqrt{100}}\sqrt{\frac{200-100}{200-1}} = 0.7$$

The 95% confidence limits are $60 \pm 1.96 \times .7 = 61.37, 58.63$.

The 99% confidence limits are $60 \pm 2.58 \times .7 = 61.81, 58.19$.

Ex. 11. If $p = 0.1$, $n = 100$, $N = 400$, a finite population

$$\text{SE }(p) = \sqrt{\frac{pq}{n}}\sqrt{\frac{N-n}{N-1}} = \sqrt{\frac{0.1 \times 0.9}{100}}\sqrt{\frac{400-100}{400-1}} = 0.026$$

The 95.45 confidence limits for p are

$$p \pm \sqrt{\frac{pq}{n}}\sqrt{\frac{N-n}{N-1}} = 0.1 \pm 2 \times 0.026 = 0.152, 0.048$$

Ex. 12. A sample of 100 items taken from a batch of 2696 is found to contain 5 defectives. Find the 95% confidence interval for the proportion of defectives $p = 0.05$.

Ans. 0.0923, 0.0077

Ex. 13. In a random sample of 400 items taken from a large consignment, 52 are found to be defective. Show that the S.E. of the proportion of defectives is 0.01%.

Ex. 14. The values x of a characteristic of a population of size six are given by 2, 6, 5, 1, 7, 3.

Take all possible samples of size two and verify that the mean of the sample means is equal to the population mean. (See Ex. 8 Sec. 24.22)

Ex. 15. A random sample of 100 units taken from a batch of 2000 has the average size = 0.354 cm and SD = 0.048. Find the 95% confidence interval for the average.

Ans. Use $\bar{x} \pm 1.96$ SE to get 0.3632, 0.3448

Ex.16. What is estimation? What are the tests for good estimators?

The data consist of observations $x_1, x_2, \ldots, x_n$ with a probability distribution $f(x, \theta)$ which is a function of known form depending on the unknown parameter θ. There may be one or more such parameters.

The estimation problem consists of finding methods of using sample observations to construct estimates of the parameters.

An estimator is a function of observations $x_1, x_2, \ldots x_n$ with values in the parameter space. An estimator is a statistic and we want to know how close it can be to the parameter to be estimated. If a statistic t is an estimator of the parameter θ, the *distance* or *error* is measured by $t - \theta$ and the bias $B = E(t) - \theta$.

If $B = 0$, the estimator is unbiased.

Ex. 17. Comment on the qualities of the arithmetic mean as an estimator of the population mean.

The sample mean is an unbiased estimator of the population mean.

It is efficient because it has a low standard error.

It is consistent because as the sample size n becomes large, the sample mean becomes closer and closer to the population mean.

The population mean is denoted by μ and the sample mean $\bar{x}$ is the best estimator of μ.

Note. The median or the mode does not have all those desirable qualities of the best estimator of μ.

25.12 CONFIDENCE INTERVAL FOR POPULATION PROPORTION

In a lot of an item we may want to know the interval within which the percentage or proportion p of defectives in the lot may lie.

If a sample of size n is drawn with replacement from a population having a proportion p of a certain attribute and a proportion $q = 1 - p$ of not having that attribute then the expected value of the proportion is p and

$$\text{Variance} = \frac{pq}{n}, \textit{ i.e.}, \text{ S.D.} = \sqrt{\frac{pq}{n}}$$

Since the population P is not known we use the sample p (notation for sample estimate) and

$$\text{S.E. } (p) = \sqrt{pq/n}$$

The 95% confidence interval for P can be derived from

$$-1.96 \le \frac{p - P}{\sqrt{pq/n}} \le 1.96$$

or $\quad p \pm 1.96 \text{ SE } (p) = p \pm 1.96 \sqrt{\dfrac{p(1-p)}{n}}$

The 99% confidence interval for p is $p \pm 2.58$ S.E.
The 99.73 confidence interval for p is $p \pm 3$ S.E.

25.13 CONFIDENCE INTERVAL FOR THE MEAN

To find the confidence interval for a statistic, say the mean, we compute $z = \dfrac{\bar{x} - \mu}{\sigma / \sqrt{n}}$

If we want 99.73% confidence limits, the confidence area lies between $z = \pm 3$ so that

$$-3 \le \frac{\bar{x} - \mu}{\sigma \sqrt{n}} \le +3$$

i.e., this inequality holds over 99.73% of the data and does not hold the remaining 0.27% of the data. This can be written $\bar{x} - \dfrac{3\sigma}{\sqrt{n}} \le \mu \le \bar{x} + \dfrac{3\sigma}{\sqrt{n}}$

These are the confidence limits for the population mean μ.

The 95.45% confidence limits for μ would be $\bar{x} - 2\dfrac{\sigma}{\sqrt{n}} \le \mu \le \bar{x} + 2\dfrac{\sigma}{\sqrt{n}}$

Note that in each case we have written the appropriate value of z in

$\bar{x} - z \text{ SE } (\bar{x}) \le \mu \le \bar{x} + z \text{ SE } (\bar{x})$

In general, if θ is the parameter and t the statistics, the limits are

$t - z \text{ SE } (t) \le \theta \le t + z \text{ SE } (t)$

Ex. 18. Write 95% confidence limits for the proportion p where $p = \dfrac{110}{200} = 0.55$. (getting 110 heads in 200 tosses of a coin)

The limits for np are $np \pm 1.96 \sqrt{npq}$

$$\begin{aligned} &= 110 \pm 1.96 \sqrt{200 \times \tfrac{1}{2} \times \tfrac{1}{2}} \\ &= 110 \pm 1.96 \times 7.1 \\ &= 110 \pm 13.86 = 123.86, 96.14 \text{ or } 124 \text{ and } 96 \text{ app.} \end{aligned}$$

Thus, the 95% limits for the heads are 124 and 96

The 95% confidence limits for p are

$$\begin{aligned} & p \pm 1.96 \sqrt{\frac{pq}{n}} \\ &= 0.55 \pm 1.96 \sqrt{\frac{\frac{1}{2} \times \frac{1}{2}}{200}} \\ &= 0.55 \pm 1.96 \times .0355 \\ &= 0.55 \pm .07 = 0.62 \text{ and } 0.48 \end{aligned}$$

Ex. 19. In a factory producing 10000 shirts every month. 2% defectives are found in a sample of 500 shirts. Find the 95% confidence limits for 10000 shirts. (2% means $p = 0.02$)

$$\text{S.E. } (p) = \sqrt{\frac{pq}{n}} = \sqrt{\frac{0.02 \times .98}{500}} = 0.0063$$

The 95% limits are $p \pm 1.96 \sqrt{pq/n}$

i.e. $0.02 \pm 1.96 \times 0.0063 = 0.0077$ and $.0323$

The number of defectives lie between 10000×0.0077 and 10000×0.0323

i.e., between 77 and 323

Confidence Limits for Differences

We can write the confidence limits for difference of proportions $p_1 - p_2$ and difference of means $\bar{x}_1 - \bar{x}_2$ on similar lines. Note the following:

$z =$ 1.96 for 95% confidence limits
$=$ 2 for 95.45% limits
$=$ 2.58 for 99% limits
$=$ 3 for 99.73% limits
and other values from the normal table

Summary of Confidence Limits

For population mean μ

$\bar{x} \pm z \text{ SE } (\bar{x})$

For population proportion p

$\bar{x} \pm z \text{ SE}(p)$

For the difference $\mu_1 - \mu_2$ of two populations

$\bar{x}_1 - \bar{x}_2 \pm z \text{ SE } (\bar{x}_1 - \bar{x}_2)$

For the difference $P_1 - P_2$ of two populations

$p_1 - p_2 \pm z \text{ SE } (p_1 - p_2)$ Also see Sec. 26.11, 26.12

Ex. 20. 50 items taken from a batch of 2000 are found to contain 5 defectives. Find the 95.75% confidence interval for defectives in the whole batch.

$$p = \frac{5}{50} = 0.1,$$

$$\text{SE } (p) = \sqrt{\frac{pq}{n}}\sqrt{\frac{N-n}{N-n}} = \sqrt{\frac{0.1 \times 0.9}{50}}\sqrt{\frac{1950}{1999}} = 0.042$$

The 95% confidence limits are

$p \pm 2\text{SE}(p)$ or $0.1 \pm 2 \times 0.042$

i.e., 0.184 and 0.016

CHAPTER 26

Testing of Hypothesis

26.1 WHAT IS A HYPOTHESIS?

A hypothesis is a statement or assumption concerning a population. For the purpose of decision making, a hypothesis has to be verified and then accepted or rejected. This is done with the help of observations. We test a sample and make a decision on the basis of the result obtained.

Decision making situations arise in marketing, industry, management etc.

Statistical Decision Making

Testing a statistical hypothesis on the basis of a sample enables us to decide whether the hypothesis should be accepted or rejected. The sample data enable us to accept or reject the hypothesis. Since the sample data give incomplete information about the population the result of the test need not be considered to be final or unchallengeable.

The procedure which, on the basis of sample results, enables us to decide whether a hypothesis is to be accepted or rejected is called Hypothesis Testing or Test of Significance.

Note 1. A test provides evidence, if any, against a hypothesis, usually called a *null hypothesis*. The test cannot prove the hypothesis to be correct. It can give some evidence against it.

The hypothesis makes some assumption about the density function of the random variate. The sampling distribution is fundamental to this subject.

The test of a hypothesis means a procedure to decide whether to accept or reject a hypothesis.

If a sample is found to have an untenable probability (of occurrence) level (called the significance level), we reject the hypothesis. Usually the probability levels of 0.05 and 0.01 are taken. They are called 5% and 1% significance levels.

Note 2. The acceptance of a hypothesis implies there is no evidence from the sample that we should believe otherwise.

The rejection of a hypothesis leads us to conclude that it is false. This way of putting the problem is convenient because of the uncertainty inherent in the problem. In view of this we must always briefly state a hypothesis that we *hope to reject.*

A hypothesis stated in the hope of being rejected is called a *null hypothesis* and is denoted by H_0.

If H_0 is rejected, it may lead to the acceptance of an alternative hypothesis denoted by H_1.

Ex. 1. A new fragrance soap is introduced in the market. The null hypothesis H_0 which may be rejected, is that the new soap is not better than the existing soap.

Ex. 2. A die is suspected to be loaded. Roll the die a number of times to test.

The Null Hypothesis $H_0 : p = 1/6$ for showing six.

The Alternative $H_1 : p \neq 1/6$.

Ex. 3. Skulls found at an ancient site may all belong to race X or race Y on the basis of their diameters. We may test the hypothesis that the mean is μ of the population from which the present skulls came. We have the hypotheses

$$H_0 : \mu = \mu_x, \ H_1 : \mu = \mu_y$$

Here we should not insist on calling either hypothesis null and the other alternative since the reverse could also be true.

Committing Errors : Type I and Type II

In how many ways can we commit errors?

We reject a hypothesis when it may be true. This is Type I Error.

We accept a hypothesis when it may be false. This is Type II Error.

The other true situations are desirable:

We accept a hypothesis when it is true. We reject a hypothesis when it is false.

	Accept H_0	Reject H_0
H_0 True	Accept True H_0 Desirable	Reject true H_0 Type I Error
H_0 False	Accept False H_0 Type II Error	Reject False H_0 Desirable

The level of significance implies the probability of Type I error. A five per cent level implies that the probability of committing a type I error is 0.05. A one per cent level implies 0.01 probability of committing type I error.

Lowering the significance level and hence the probability of type I error is good but unfortunately it would lead to the undesirable situation of committing type II error.

To sum up :

(*a*) **Type I Error.** Rejecting H_0 when H_0 is true

(*b*) **Type II Error.** Accepting H_0 when H_0 is false.

Note. The probability of making a Type I error is the level of significance of a statistical test. It is denoted by α

where

$$\alpha = \text{Prob. (Rejecting } H_0/H_0 \text{ true)}$$

$$1 - \alpha = \text{prob. (Accepting } H_0/H_0 \text{ true)}$$

The probability of making a Type II error is denoted by β where

$$\beta = \text{Prob. (Accepting } H_0/H_0 \text{ false)}$$

$$1-\beta = \text{Prob. (Rejecting } H_0/H_0 \text{ false)} = \text{Prob. (The test correctly rejects } H_0 \text{ when } H_0 \text{ is false)}$$

$1 - \beta$ is called the power of the test. It depends on the level of significance α, sample size n and the parameter value.

26.2 NULL AND ALTERNATIVE HYPOTHESES

The hypothesis to be tested is called the Null Hypothesis and is denoted by H_0.

This is to be tested against other possible states of nature called alternative hypotheses. The alternative is usually denoted by H_1.

The null hypothesis implies that there is no difference between the statistic and the population parameter. To test whether there is no difference between the sample mean $\bar{x}$ and the population μ, we write the null hypothesis.

$$H_0 : \bar{x} = \mu$$

The alternative hypothesis would be

$$H_1 : \bar{x} \neq \mu$$

This means $\bar{x} > \mu$ or $\bar{x} < \mu$. This is called a two tailed hypothesis.

The alternative $H_1 : \bar{x} > \mu$ is right tailed

The alternative $H_1 : \bar{x} < \mu$ is left tailed

These are one sided or one tailed alternatives.

Note 1. The alternative hypothesis H_1 implies all such values of the parameter which are not specified by the null hypothesis H_0.

Note 2. Testing a statistical hypothesis is a rule which leads to a decision to accept or reject a hypothesis.

A one tailed test requires rejection of the null hypothesis when the sample statistic is greater than the polulation value or less than the population value at a certain level of significance.

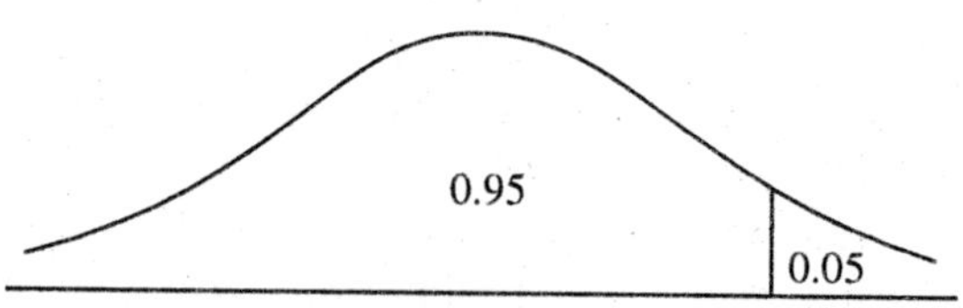

$H_0 : \bar{x} > \mu$

(*i*) We may want to test if the sample mean $\bar{x}$ exceeds the population mean μ. Then the null hypothesis is

$$H_0 : \bar{x} > \mu$$

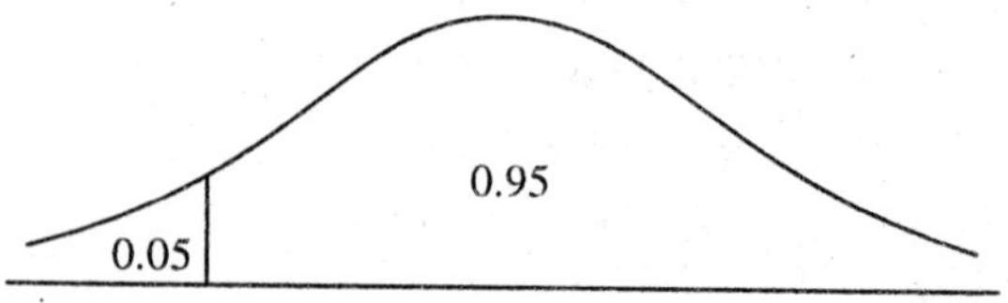

$H_0 : \bar{x} < \mu$

(*ii*) In the other case the null hypothesis could be

$$H_0 : \bar{x} < \mu$$

Each of these two situations leads to a one tailed test and has to be dealt with in the same manner as the two tailed test.

Here the critical rejection is on one side only, right for $\bar{x} > \mu$ and left for $\bar{x} > \mu$ Each diagram here shows a five per cent level of test of significance.

Ex. 4. A minister in a certain government has an average life of 11 months without being involved in a scam. A new party claims to provide ministers with an average life of more than 11 months without scam. We would like to test if, on the average the new ministers last longer than 11 months. We may write the null hypothesis $H_0 : \mu = 11$ and the alternative $H_1 : \mu > 11$.

Ex. 5. A drug is known to benefit one half of its users. A new drug is claimed to do better. We have

$$H_0 : p = 0.5$$

$$H_1 : p > 0.5 \text{ (One tailed hypothesis)}$$

A company claims to give at least 12 gm of chocolate in its chocobar. Another company claims it is less than 12 gm. We have

$$H_0 : \mu \geq 12$$

$$H_i : \mu < 12$$

Ex. 6. The normal sales of a commodity are likely to be 10,000 units. As a result of a campaign the sales go up. If the standard error is known we may want to know if sales have increased decisively. This involves a one tailed test.

In an examination, a certain batch of candidates scores (*i*) unusually high marks (*ii*) unusually low marks. Is each of these a proper case for a one tailed test ?

Ans. Yes, one tailed.

26.3 CRITICAL REGION (REGION OF HYPOTHESIS REJECTION)

We shall consider test problems arising out of Type I Error.

The *level of significance* of a test is the maximum probability with which we are willing to take a risk of Type I Error.

If we take a 5% significance level ($p = 0.05$) we are 95% confident ($p = 0.95$) that a right decision has been made.

A 1% significance level ($p = 0.01$) makes us 99% confident ($p = 0.99$) about the correctness of the decision.

The critical region is the area of the sampling distribution in which the test statistic must fall for the null hypothesis to be rejected.

We can say that the critical region corresponds to the range of values of the statistic which according to the test requires the hypothesis to be rejected.

Two Tailed Test

If the level of significance is α then in a two tailed test, a part of the critical region (say half, *i.e.*, $\alpha/2$) is placed on the right and another part (*i.e.*, $\alpha/2$) on the left.

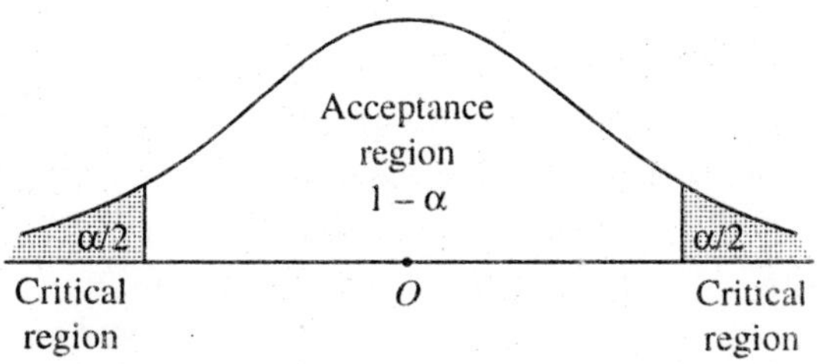

This enables us to reject H_0 *if* the z value is very large either on the right or on the left.

Ex. 7. What will happen if the acceptance region is made larger?

α will decrease. It will be more easily possible to accept H_0 when H_0 is false (Type II error), *i.e.*, it will lower the probability of making a Type I error but raise that of β, Type II error.

Note. α, β are probabilities of making an error; $1-\alpha$, $1-\beta$ are probabilities of making correct decisions.

Ex. 8. Can we say $\alpha + \beta = 1$?

No. Each is concerned with a different type of error. But both are not independent of each other.

26.4 PENALTY

Ex. 9. Which error type is worse?

Usually type II error is considered the worse of the two though it is mainly the circumstances of a case that decide the answer to this question.

If type I error means accepting the hypothesis that a guilty person is innocent and if type II error means accepting the hypothesis that an innocent person is guilty then type II error would be dangerous. The penalties and costs associated with an error determine the balance or trade off between Type I and Type II errors.

Usually type I error is shown as the shaded area, say 5% of a normal curve which is supposed to represent the data.

If the sample statistic, say the sample mean, falls in the shaded area the hypothesis is rejected at 5 per cent level of significance.

Ex. 10. Is a particular significance level important?

A significance level is arbitrary and there is nothing like a scientific or fixed level of significance in all situations. Each particular case has to be considered in the light of circumstances.

26.5 STANDARD ERROR

The concept of standard error of statistic is used to test the precision of a sample and provides the confidence limits for the corresponding population parameter.

The statistic may be the sample arithmetic mean $\bar{x}$, the sample proportion p etc.

The standard error (SE) of any such statistic is the standard deviation of the sampling distribution of the statistic.

Given below are SEs in common use. No proof is given. The number of observations is n

Notation : $\bar{x}$ Sample mean
μ population mean
s sample standard deviation
σ population standard deviation
p sample proportion, $q = 1 - p$
P population proportion , $Q = 1 - p$

$$\text{SE}(\bar{x}) = \frac{\sigma}{\sqrt{n}} \qquad \text{SE}(p) = \sqrt{\frac{PQ}{n}}$$

SE of difference between two means $\bar{x}_1$, $\bar{x}_2$ or two proportions p_1, p_2 with sample sizes n_1, n_2

$$\text{SE}(\bar{x}_1 - \bar{x}_2) = \sqrt{\frac{\sigma_1^2}{n_1} + \frac{\sigma_2^2}{n}}$$

$$\text{SE}(p_1 - p_2) = \sqrt{\frac{P_1Q_1}{n_1} + \frac{P_2Q_2}{n_2}}$$

26.6 DECISION RULE

Suppose u is distributed normally with mean 0 and S.D.1. Briefly we write $u \sim N(0, 1)$.

If the expected value of u is written $E(u)$ the standardised normal variate is

$$z = \frac{u - E(u)}{SE(u)}$$

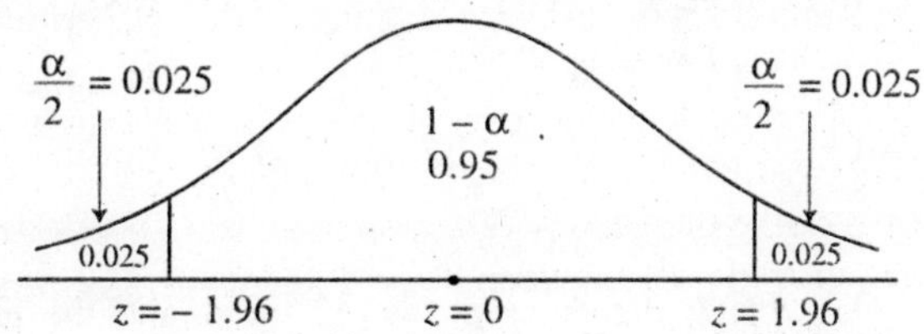

The total area under the normal curve is 1 (corresponding to 100%).

The area between $z = -1.96$ and $z = +1.96$ is 0.95. This is the region of acceptance with 95% confidence . We write

$$p(-1.96 \le z \le 1.96) = 0.95$$

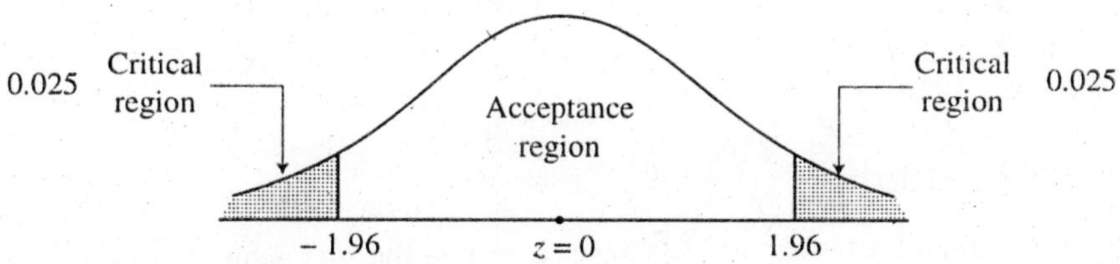

The size of the critical region is 0.05 (shaded area 0.025 on the left and 0.025 on the right). This is a two tailed test.

If $|z|$ remains between the range $\pm$ 1.96, we are in the hypothesis acceptance region. The two values – 1.96 and 1.96 are the 5% critical values.

If $|z| > 1.96$ we are in the critical region, *i.e.*, the region of rejection of the hypotheses.

26.7 LARGE SAMPLE TESTS

Two Tailed Decision Rule

At 5% level of significance.

Accept the null hypothesis if $|z| \le 1.96$ Using Normal Tables

Reject the null hypothesis if $|z| > 1.96$ ...(1)

At 1% level of significance

Accept if $|z| \leq 2.58$ ...(2)

Reject if $|z| > 2.58$

At 4.55% level of significance

Accept if $|z| \leq 2$ ...(3)

Reject if $|z| > 2$

At 0.27% level of significance

Accept if $|z| \leq 3$ ...(4)

Reject if $|z| > 3$

For other levels of significance, normal tables should be used.

The values $|z| \lesseqgtr 2$ and $|z| \lesseqgtr 3$ in (3) and (4) above are popularly used.

One Tailed Decision Rule

Right Tailed Test $H_0 : \mu > \mu_0$, Accept if $z < 2.33$ at 1%

Accept if $z < 1.645$ at 5%

Left Tailed Test $H_0 : \mu < \mu_0$, Accept if $z < -2.33$ at 1%

Accept if $z < -1.645$ at 5%

Ex. 11. A die is rolled 49152 times and the number of times it shows 4, 5 or 6 is 25149. Test the hypothesis that the die is unbiased.

$H_0 : P = 1/2$ (The probability of getting 4, 5 or 6)

$H_1 : P \neq 1/2$

This is a two tailed test with SE = $\sqrt{\frac{PQ}{n}}$

$$z = \left|\frac{p-P}{SE(p)}\right| = \left|\frac{\frac{25149}{49152} - \frac{1}{2}}{\sqrt{\frac{\frac{1}{2} \times \frac{1}{2}}{49152}}}\right| = 5.4 > 3$$

The die is biased. We reject the null hypothesis at 0.27% level of confidence.

Ex. 12. A die is rolled 9000 times. Of these 3220 times it shows 3 or 4. Test the hypothesis that the die is unbiased.

Hint. $P = \frac{1}{3}$, $1 - P = \frac{2}{3}$ $z > 3.89$, Difference Significant

Hypothesis rejected at 0.27% level.

Ex. 13. A coin is tossed 200 times. It shows heads 116 times. Can you say the coin is biased?

$$z = \frac{np - nP}{\sqrt{nPQ}} = \frac{116 - 100}{\sqrt{200 \times \frac{1}{2} \times \frac{1}{2}}} = 2.26 > 1.96$$

The difference is significant at 5% level. But since $z < 2.58$, it is not significant at 1% level.

We can also proceed thus

$$z = \frac{p - P}{\sqrt{PQ/n}}$$

Test for a Sample Proportion *p*

To test the null hypothesis that the population proposition has a specified value p_0, *i.e.*, $H_0 : p = p_0$

For large n, if H_0 is true then

$$z = \frac{p - p_0}{SE(p)} \text{ is approximately normal.}$$

The critical region for z, depending on the desired level of significance, can be found out

We can also write $z = \dfrac{np - np_0}{SE(np)} = \dfrac{np - np_0}{\sqrt{nPQ}} = \dfrac{p - p_0}{\sqrt{PQ/n}}$

Ex. 14. A die was rolled 400 times. It showed a 6 coming up 80 times. Can you say the die is unbiased? $p = 80/400 = 1/5 \quad P = 1/6$

$$z = \frac{1/5 - 1/6}{\sqrt{\frac{1}{6} \times \frac{5}{6} \times \frac{1}{400}}} = \frac{1/30}{\sqrt{5}/120} = 1.77 < 1.96$$

The result suggests an unbiased die.

Ex. 15. A company supplied 500 units of an item. The number of defectives was found to be 42 as against the company's conviction that 6% items could be defective.

Examine the tenability of the company claim. **Ans.** Untenable at 5% level.

Note. If the requirements are given it is possible to determine the sample size using the formula $n = z_{\alpha/2}^2 \, pq/e^2$ when e is the error permitted.

Ex. 16. It is desired to find the percentage of citizens in a town wanting iodised salt. How large a sample will be needed with 95% confidence that the estimate is within 1% of the true percentage given? Given

(*a*) a prior estimate of 25% want the scheme

(*b*) no estimate of those wanting it

(*a*) use the formula $n = z_{\alpha/2}^2 \, pq/e^2$ when e = 1% error = 0.01

$$\therefore \quad n = \frac{1.96^2 \times 0.25 \times 0.75}{(0.01)^2} \simeq 7200$$

(*b*) If no estimate of p is available, take $n = \dfrac{z_{\alpha/2}^2}{4e^2}$

$$= \frac{1.96^2}{4\times(0.01)^2} \simeq 9600$$

26.8 TEST FOR A SAMPLE MEAN $\bar{x}$

We have to test the null hypothesis that the population mean has a specified value μ, *i.e.*, $H_0 : \bar{x} = \mu$. For large n, if H_0 is true then

$z = \left|\dfrac{\bar{x}-\mu}{SE(\bar{x})}\right|$ is approximately nominal. The theoretical region for z depending on the desired level of significance can be found out.

Ex. 17. A factory produces items, each weighing 5 kg with variance 4. Can a random sample of size 900 with mean weight 4.45 kg. be justified as having been taken from this factory?

$n = 900 \quad \bar{x} = 4.45 \quad \mu = 5 \quad \sigma = \sqrt{4} = 2$

$$z = \left|\frac{\bar{x}-\mu}{SE}\right| = \left|\frac{\bar{x}-\mu}{\sigma/\sqrt{n}}\right| = \left|\frac{4.45-5}{2/30}\right| = 8.25$$

We have $z > 3$. The null hypothesis is rejected. The sample may not be regarded as originally from the factory at 0.27% level of significance (corresponding to 99.73% acceptance region).

Ex. 18. A sample of 400 persons has a mean height of 171.38 cm. Can the sample be regarded as having been drawn from a population with mean height of 171.17 cm and S.D. 3.3?

Ans. Yes, since $z = 1.27$

26.9 TESTS FOR EQUALITY OF TWO PROPORTIONS

If p_1, p_2 are proportions of some characteristic of two samples of sizes n_1, n_2 drawn from populations with proportions P_1, P_2, then we have $H_0 : P_1 = P_2$ vs $H_1 : P_1 \neq P_2$

Case (*a*) If H_0 is true then let $P_1 = P_2 = p$

where p can be found from the data ;

$$p = \frac{n_1 p_1 + n_2 p_2}{n_1 + n_2} \text{ and } q = 1 - p$$

p is the mean of the two proportions

$$SE(p_1 - p_2) = \sqrt{pq\left(\frac{1}{n_1}+\frac{1}{n_2}\right)}$$

$$z = \frac{p_1 - p_2}{SE(p_1 - p_2)} \text{ is approx. normal } (0, 1)$$

We write $z \sim N(0, 1)$

The usual rules for rejection or acceptance are applicable here.

Case (*b*) If it is assumed that the proportion under question is not the same in the two populations from which the samples are drawn and that p_1, p_2 are the true proportions, we write

$$SE\,(p_1 - p_2) = \sqrt{\frac{p_1 q_1}{n_1} + \frac{p_2 q_2}{n_2}}$$

We can also write the confidence interval for $p_1 - p_2$.

For 2 independent samples of sizes n_1, n_2 selected from two binomial populations, the 100 (1 – α) % confidence limits for $p_1 - p_2$ are

$$(p_1 - p_2) \pm z_{\alpha/2} \sqrt{\frac{p_1 q_1}{n_1} + \frac{p_2 q_2}{n_2}}$$

The 90%, confidence limits would be [with $\alpha = 0.1$, $100\,(1 - \alpha) = 0.90$]

$$p_1 - p_2 \pm 1.645 \sqrt{\frac{p_1 q_1}{n_1} + \frac{p_2 q_2}{n_2}}$$

Ex. 19. Out of 5000 interviewees, 2400 are in favour of a proposal, out of another set of 2000 interviewees 1200 are in favour. Is the difference significant?

$$p_1 = \frac{2400}{5000} = 0.48,\ p_2 = \frac{1200}{2000} = 0.6\ n_1 = 5000,\ n_2 = 2000$$

$$SE = \sqrt{\frac{.48 \times 52}{5000} + \frac{.6 \times .4}{2000}} = 0.013 \quad \text{Case } (b)$$

$$z = \left|\frac{p_1 - p_2}{SE}\right| = \frac{0.12}{0.013} = 9.2 > 3$$

The difference is highly significant at 0.27% level.

Ex. 20. 360 persons out of 600 are found to suffer from pollution induced bronchitis in one city. In another, 400 out of 500 are found to suffer from bronchitis. Is there any significant difference in the incidence of bronchitis?

$$p_1 = \frac{360}{600} = 0.6 \quad p_2 = \frac{400}{800} = 0.5 \quad \text{Case } (a)$$

$$p = \frac{n_1 p_1 + n_2 p_2}{n_1 + n_2} = \frac{360 + 400}{1400} = \frac{760}{1400} = 0.543$$

$$q = 1 - 0.543 = 0.457$$

$$SE = pq\sqrt{\frac{1}{n_1} + \frac{1}{n_2}} = \sqrt{0.543 \times 0.457}\sqrt{\frac{1}{600} + \frac{1}{800}}$$

$$z = \frac{0.6 - 0.5}{.027} = 3.7 > 2.58$$

The difference is significant at 1% level of significances.

Ex. 21. The number of defectives produced by a machine is 20 in a batch of 400. Another machine produced 10 defectives in a batch of 300.

(*i*) Is there any difference?

(*ii*) If the second machine is the same as the first machine after overhauling, would you say the machine has improved?

(*i*) $H_0 : P_1 = P_2, H_1 : P_1 \neq P_2$ two tailed test

$$P = \frac{20}{400} = \frac{1}{20},\quad p_2 = \frac{10}{300} = \frac{1}{30},\qquad \text{Case } (a)$$

$$p = \frac{n_1 p_1 + n_2 p_2}{n_1 + n_2} = \frac{20+10}{400+300} = \frac{3}{70},\quad q = \frac{67}{70}$$

$$SE\,(p_1 - p_2) = \sqrt{\frac{3}{70} \times \frac{67}{70}\left(\frac{1}{400} + \frac{1}{300}\right)} = 0.0155$$

$$z = \frac{p_1 - p_2}{SE(p_1 - p_2} = \frac{\frac{1}{20} - \frac{1}{30}}{0.0155} = 1.1 < 1.96$$

The difference is not significant at 5% level

(*ii*) $H_0 : p_1 > p_2$. This is one tailed test

$$z = 1.1 < 1.645$$

There seems to be no significant improvement (at 5% level).

Ex. 22. In two large populations there are 30% and 25% smokers. Is this difference likely to be hidden in samples of 1200 and 900 from the populations?

Hint. Take $SE = \sqrt{\frac{p_1 q_1}{n_1} + \frac{p_2 q_2}{n_2}}$ and get $z = 2.55$

Ans. The difference is not likely to be hidden

Ex. 23. Of 2000 men aged 25 working in a factory, 400 die before they are 50. Out of 1000 men of age 25 in another factory 175 die before they are 50, Test if there is a difference in the proportion of deaths. **Ans.** No, $z = 1.64$

Ex. 24. 450 men are found to be smokers in a sample of 600 men. In another sample of 900 men, 450 men are smokers. Are the two groups different in the matter of smoking?

$$p_1 = \frac{450}{600} = 0.75,\quad p_2 = \frac{450}{900} = 0.5$$

If H_0 is true then $P_1 = P_2 = P$ so that

$$SE\,(p_1 - p_2) = \sqrt{pq\left(\frac{1}{n_1} + \frac{1}{n_2}\right)} \quad \text{where } p = \frac{n_1 p_1 + n_2 p_2}{n_1 + n_2}$$

Here $$p = \frac{600 \times 0.75 + 900 \times 0.5}{600 + 900} = \frac{2}{5} = 0.4$$

$$z = \frac{p_1 - p_2}{SE(p_1 - p_2)} = \frac{0.75 - 0.50}{\sqrt{0.4 \times 0.6\left(\frac{1}{600} + \frac{1}{900}\right)}} = 9.5 > 1.96$$

The difference is significant at 5% level of confidence.

26.10 LARGE SAMPLE TEST FOR EQUALITY OF TWO MEANS $\bar{x}_1$, $\bar{x}_2$

Suppose two samples of sizes n_1, n_2 are drawn from populations having means μ_1, μ_2 and standard deviations σ_1, σ_2

To test the equality of means $\bar{x}_1$, $\bar{x}_2$, we write

$$H_0 : \mu_1 = \mu_2$$

$$H_1 : \mu_1 \neq \mu_2$$

If we assume H_0 is true then

$$z = \frac{\bar{x}_1 - \bar{x}_2}{SE(\bar{x}_1 - \bar{x}_2)}$$

$$= \frac{\bar{x}_1 - \bar{x}_2}{\sqrt{\frac{\sigma_1^2}{n_1} + \frac{\sigma_2^2}{n_2}}}$$ Approximately normally distributed with mean 0, S.D. 1.

We write $z \sim N(0, 1)$

As usual, if $|z| > 2$ we reject H_0 at 4.55% level of significance and so on.

Ex. 25. Two groups of sizes 121 and 81 are subjected to tests. Their means are found to be 84 and 81 and standard directions 10 and 12. Test for the significance of difference between the groups.

$$\bar{x}_1 = 84,\ \bar{x}_2 = 81,\ n_1 = 121,\ n_2 = 81,\ \sigma_1 = 10,\ \sigma_2 = 12$$

$$z = \frac{\bar{x}_1 - \bar{x}_2}{\sqrt{\frac{\sigma_1^2}{n_1} + \frac{\sigma_2^2}{n_2}}} = \frac{84 - 81}{\sqrt{\frac{100}{121} + \frac{144}{81}}} = 1.86 < 1.96$$

The difference is not significant at the 5% level of confidence.

Ex. 26. The average score of two groups *A*, *B* were found to be 25 and 22 with S.D. 4 and 5.5 respectively. Test for the equality of the two group scores. Given $n_1 = n_2 = 400$.

$$H_0 : \mu_1 = \mu_2 \text{ and } H_1 : \mu_1 \neq \mu_2$$

$$SE(\bar{x}_1 - \bar{x}_2) = \sqrt{\frac{\sigma_1^2}{n_1} + \frac{\sigma_2^2}{n_2}} = \sqrt{\frac{s_1^2}{n_1} + \frac{s_2^2}{n_2}}$$

$$= \sqrt{\frac{4^2}{400} + \frac{5.5^2}{400}} = 0.34$$

$$z = \frac{25 - 22}{0.34} = 8.8 > 3$$

The difference is highly significant at 0.27% level of significance.

Ex. 27. 800 ore pieces from a mine were found to contain an average of 74.5 gm of gold. From a nearby mine 1600 pieces had 75 gm gold. Test the equality of the averages from the two mines, each having a S.D. of 2.4.

Ans. $z = 4.8$; significant difference

Ex. 28. To test the goodness of a coin, it is tossed 5 times. It is considered a bad coin if more than 4 heads show up. What is the probability of Type I error? (*b*) If the probability of a head is 0.2, what is the probability of Type II error?

Ans. (*a*) The rejection region is prob. (5 heads) = 1/32

(*b*) $1 - (0.2)^5$

Ex. 29. In a sample of 500 people, 280 are tea takers and the rest coffee takers. Are tea and coffee equally popular?

Ans. No. $z = 2.7$, significant at 1% level.

Ex. 30. A factory claims to produce a lower average number of defectives compared to the prevalent average of 30.5. A random sample of 100 defective articles from this factory showed the following distribution.

Class limits	Frequency
16—20	12
21—25	22
26—30	20
31—35	30
36—40	16

Is the claim sustained at 5% level of significance?

Hint. $H_0 : \mu = 30.5$, $H_1 : \mu < 30.5$ Calculate $\bar{x} = 28.8$, $s = 6.4$

Use the left side one tailed test with

$$z = \frac{28.8 - 30.5}{6.4/\sqrt{100}} = -2.7 < -1.645$$

H_0 rejected; claim sustained.

Ex. 31. One population has the same mean as the second but has S.D. twice that of the second. (*a*) Show that in samples of size 500 from each, the difference between means will in all probability not exceed 0.3σ where σ is the smaller S.D. (*b*) Assuming the distribution of difference of means to be normal, find the probability that it exceeds half that amount

$$SE = \sqrt{\frac{\sigma^2}{500} + \frac{(2\sigma)^2}{500}} = \frac{\sigma}{10}$$

At 0.27% level we must have $\dfrac{\bar{x}_1 - \bar{x}_2}{SE} < 3$

i.e., $\bar{x}_1 - \bar{x}_2 < 3\ SE$

i.e., $< 3 \times \dfrac{\sigma}{10} = 0.3\sigma$

(*b*) Prob $|\bar{x}_1 - \bar{x}_2| > 0.15\sigma$ = area to the right of $z = 0.15$

$= 0.4404.$

26.11 SMALL SAMPLE TESTS OF SIGNIFICANCE

The sampling distribution of many statistics for large samples is approximately normal.

For small samples with $n < 30$, the normal distribution, as shown above, can be used only if the sample is from a normal population with known σ.

If σ is not known we can use student's t distribution instead of the normal. We then replace σ by sample standard deviation s with some modification, given below.

Let $x_1, x_2, \ldots, x_n$ be a random sample of size n drawn from a normal population with mean μ and S.D. σ. We define

$t = \dfrac{\overline{x} - \mu}{s / \sqrt{n-1}}$. Here t follows the students' t distribution with $n - 1$ degrees of freedom.

Note. For small samples of $n < 30$ the term $\sqrt{n-1}$ in $SE = s/\sqrt{n-1}$ corrects the bias resulting from the use of sample standard deviation as an estimator of σ

Also $\dfrac{s^2}{S^2} = \dfrac{n-1}{n}$ or $s = S\sqrt{\dfrac{n-1}{n}}$

Procedure : Small Samples

To test the null hypothesis $H_0 : \mu = \mu_0$
against the alternative $H_1 : \mu \neq \mu_0$

Calculate $|t| = \dfrac{\overline{x} - \mu}{SE(\overline{x})}$ and compare it with the table value with $n - 1$ d.f. at level of significance $\alpha\%$.

If this value > table value, reject H_0
If this value < table value, accept H_0
(Significance level idea same as for large samples)
We can also find the 95% (or any other) confidence limits for μ.

For the two tailed test (use the same rules as for large samples ; substitute t for z) the 95% confidence limits are

$$\overline{x} \pm t_\alpha\, s/\sqrt{n-1} \quad \alpha = .025$$

Rejection Region. At $\alpha\%$ level for *two tailed* test if $|t| > t_{\alpha/2}$ reject
For *one tailed* test (right) if $t > t_\alpha$ reject
(left) if $t < -t_\alpha$ reject

At 5% level the three cases are		
	if $|t| > t_{.025}$ reject	two tailed
	if $t > t_{.05}$ reject	one tailed right
	if $t < -t_{.05}$ reject	one tailed left

For proportions, the same procedure is to be followed.

Ex. 32. A firm produces tubes of diameter 2 cm. A sample of 10 tubes is found to have a diameter of 2.01 cm and variance 0.004. Is the difference significant? Given $t_{0.05,\,9} = 2.26$

$$t = \frac{\overline{x} - \mu}{s/\sqrt{n-1}} = \frac{2.01 - 2}{\sqrt{.004/10-1}} = \frac{.01}{.021} = 0.48$$

Since $|t| < 2.26$ the difference is not significant at 5% level.

Ex. 33. 10 persons randomly selected are found to have heights 63, 63, 66, 67, 68, 69, 70, 71, 71, 71 inches. Discuss the suggestion that the mean height in the population is 66 inches.

Hint. Calculate $\quad \bar{x} = 67.8, \ \dfrac{s}{\sqrt{n-1}} = \sqrt{\dfrac{\Sigma(x - \bar{x})^2}{n}} / \sqrt{n-1} = 0.95$

$$t = \frac{67.8 - 66}{s/\sqrt{n-1}} = 1.9 < t_{9,0.05} = 2.26$$

26.12 SMALL SAMPLE TEST

Test for Equality of Two Means $\bar{x}_1, \bar{x}_2$

σ_1, σ_2 unknown ; $n_1 < 30, n_2 < 30$

S_1^2, S_2^2 are sample variances

$H_0 : \mu_1 = \mu_2$, Alt $H_1 : \mu_1 \neq \mu_2$

The variances are assumed equal

An unbiased estimate of the common variance of the two small samples is given by

$$S_0^2 = \frac{n_1 s_1^2 + n_2 s_2^2}{n_1 + n_2 - 2}$$

To test the null hypothesis H_0, use

$$t = \frac{\bar{x}_1 - \bar{x}_2}{S_0\sqrt{\dfrac{1}{n_1} + \dfrac{1}{n_2}}} \text{ with } n_1 + n_2 - 2 \text{ d.f.}$$

$$\left[\begin{array}{l} \text{Note } s^2 = \dfrac{\Sigma(x - \bar{x})^2}{n} \\ \quad\ \ S^2 = \dfrac{\Sigma(x - \bar{x})^2}{n-1} \end{array}\right.$$

The 95% confidence limits for the difference of population means are (for two tailed test)

$$(\bar{x}_1 - \bar{x}_2) \pm t_\alpha S_0\sqrt{\frac{1}{n_1} + \frac{1}{n_2}} \quad \text{(Here } \alpha = 0.025)$$

Ex. 34. 5 students attend a class and get respectively 42, 39, 48, 60, 41 marks. 7 other students attending another class get marks 38, 42, 56, 64, 68, 69, 62.

Is there any significance difference between the classes?

$\bar{x}_1 = 46, \ \bar{x}_2 = 57, \qquad s_1^2 = \dfrac{\Sigma(x_1 - \bar{x}_1)^2}{n_1} = 58, \ s_2^2 = \dfrac{(x_2 - \bar{x}_2)^2}{n_2} = 132.3$

$n_1 = 5, n_1 = 7 \qquad S^2 = \dfrac{n_1 s_1^2 + n_2 s_2^2}{n_1 + n_2 - 2} = 121.5$

$$|t| - \left|\frac{\bar{x}_1 - \bar{x}_2}{S\sqrt{\dfrac{1}{n_1} + \dfrac{1}{n_2}}}\right| = \left|\frac{46 - 507}{\sqrt{121.5\sqrt{\dfrac{1}{5} + \dfrac{1}{7}}}}\right| = 1.7$$

From the table $t_{10,0.05} = 2.23$

The observed value is less than the table value.

The difference is not significant.

Ex. 35. A sample of 12 pigs is given food. *A* and another of 12 given food *B*. The gain in weights are given below. Is there a significant difference in gains in weights?

A 31	34	29	26	32	35	38	34	30	29	32	31
B 26	24	28	29	30	29	32	26	31	29	32	28

If we assume that the variances of the samples are unequal *i.e.*, we assume what is called heteroscedasticity, then

$$t = \frac{\bar{x}_1 - \bar{x}_2}{\sqrt{\frac{S_1^2}{n_1} + \frac{S_2^2}{n_2}}}$$

26.13 PAIRED OBSERVATIONS

t test for the difference of means

Let $(x_1, y_2), (x_2, y_2), \ldots, (x_n, y_n)$ be the pairs of values for the same subjects *e.g.*, Sales data before (x) and after an advertisement campaign (y)

Performance of candidates before (x) and after training (y)

We have to test the significance of the difference between x, y values.

For each pair (x_i, y_i), find $d_i = x_i - y_i$

$H_0 : \mu_1 = \mu_2$, *i.e.*, no difference before and after and $H_i : \mu_1 \neq \mu_2$

We find the mean $\bar{d}$ of d values and use the statistic

$$t = \frac{\bar{d}}{S/\sqrt{n}} \quad \text{where } S = \sqrt{\frac{\Sigma(d-\bar{d})^2}{n-1}}$$

Ex. 36. 11 students were given a test and their marks noted. After training, their marks in a second test were noted.

Do the marks indicate any benefit from training?

Student	1	2	3	4	5	6	7	8	9	10	11
x	23	20	19	21	18	20	18	17	23	16	19
y	24	19	22	18	20	22	20	20	23	20	17
d_i	–1	1	–3	3	–2	–2	–2	–3	0	–4	Z

$$\bar{d} = \frac{-11}{11} = -1, \quad \sqrt{\frac{\Sigma(d-\bar{d})^2}{n-1}} = 2.24, \text{ d.f.} = 11 - 1 = 10$$

$$t = \frac{|\bar{d}|}{2.24/\sqrt{11}} = \frac{\sqrt{11}}{2.24} = 1.5$$

Table value $t_{10,\,.05} = 2.23$

The difference is not significant.

Ex. 37. A diet produced the following increases in the body weight of 12 cattle : 6, 3, 8, –2, 3, 0, –1, 1, 1, 6, 0, 5, 4. Can you say the diet is effective in increasing weight?

Hint. These value are d values already given

Table value $t_{11, .05} = 2.2$ **Ans.** Yes.

26.14 TO TEST THE SIGNIFICANCE OF AN OBSERVED CORRELATION COEFFICIENT *r*

Let $(x_1, y_1), (x_2, y_2), \ldots, (x_n, y_n)$ be a random sample from a normal population.

To test the hypothesis that the population coefficient of correlation is zero, use

$$t = \frac{r\sqrt{n-2}}{\sqrt{1-r^2}} \text{ with } n-2 \text{ degrees of freedom}$$

If $|t| >$ table value of $t_{n-2, .05}$ then r is significant

If $|t| < t_{n-2, .05}$ then r is not significant

Ex. 37. For 18 observations it was found that $r = -0.6$. Is this significant for the existence of correlation in the population?

$$|t| = \frac{0.6\sqrt{18-2}}{\sqrt{1-.36}} = 2.67$$

Table value $t_{16, 0.05} = 2.12$

The observed value > table value.

The coefficient of correlation is significant.

Note. In the linear regression model.

$$Y = a + bX$$

It can be easily proved that the estimators $\hat{a}, \hat{b}$ are

1 Linear
2. Unbiased
3. Efficient (because they have the minimum variance)
4. Consistent: (they approach the population value as n becomes large.)

Ex. 38. Can you test the significance of the regression coefficients a, b in the linear regression model $Y = a + bX$?

Yes. Since the estimators $\hat{a}, \hat{b}$ are normally distributed we can use the statistic

$$t = \frac{\hat{a}-a}{SE(\hat{a})}, \; t = \frac{\hat{b}-b}{SE(\hat{b})}$$

where a, b are the population values of the coefficients

It can be shown that if σ is the standard deviation of regression and $x = X - \overline{X}$

$$SE(\hat{a}) = \sigma^2\sqrt{\frac{\Sigma X^2}{n\Sigma x^2}}, \; SE(\hat{b}) = \sigma^2\sqrt{\frac{1}{\Sigma x^2}}$$

26.15 CHI-SQUARE (χ^2) TEST

χ^2 distribution is used for the following tests :

1. To test the goodness of fit, *i.e.*, to test the difference between theoretical and observed frequencies.
2. To test the independence of attributes : Contingency Tables.
3. To test if the population variance has a specified value.

Goodness of Fit

This is to test how well an observed frequency distribution fits a theoretical distribution.

For this test, we shall make use of the statistic Chi-squared written

$$\chi^2 = \Sigma \frac{(O-E)^2}{E} = \Sigma \frac{O^2}{E} - n$$

O is observed frequency
E is expected frequency

where n is the total frequency and d.f. $= n - 1$

If the calculated value of $\chi^2 >$ table value of χ^2 at a given level of significance, we reject the null hypothesis H_0 that there is no significant difference between observed and expected frequencies.

Ex. 39. There are 100 slips in an urn supposed to have all digits in equal numbers. A check gives the following result. Is the result consistent with the hypothesis that the urn contains all digits in equal numbers *i.e.* 10?

Problem		*Worked Problem*	
Digit	*O (Observed frequencies)*	*E (Expected frequencies)*	$(O-E)^2/E$
0	9	10	1/10
1	11	10	1/10
2	12	10	4/10
3	12	10	4/10
4	10	10	0
5	8	10	4/10
6	7	10	9/10
7	10	10	0
8	9	10	1/10
9	12	10	4/10
			28/10 = 2.8

$$\chi^2 = \Sigma \frac{(O-E)^2}{E} = 2.8 \text{ and d.f. } = n - 1 = 10 - 1 = 9$$

From the tables $\chi^2_{9, 0.05} = 16.92$

Since the observed χ^2 is less than the table value, we accept the null hypothesis of no difference between observed and expected values.

Note. The χ^2 test is valid if total frequency is large and is greater than 50. Also if the individual theoretical frequencies are small we pool the smaller frequencies which are less than 5.

Ex. 40. Five dice were rolled 96 times and the frequencies of numbers 4, 5 or 6 shown were noted :

No. of dice showing 4, 5 or 6	0	1	2	3	4	5
Frequency	3	8	24	35	19	7

These are observed frequencies. The expected frequencies are obtained from the binomial distribution

$$N\ {}^{n}C_x p^x q^{n-x} \text{ i.e. } 96 \times {}^5C_x \left(\frac{1}{2}\right)^x \left(\frac{1}{2}\right)^{n-x} = 3, 15, 30, 30, 15, 3 \quad \text{for } x = 0, 1, 2, 3, 4, 5$$

$\left(\text{probability of getting 4, 5 or 6 is } \frac{1}{2}\right)$

Since the first and the last observed frequencies are small, we combine them with adjoining frequencies (3 + 15 = 18, 15 + 3 = 18)

O	E
11	18
24	30
35	30
26	18

$\chi^2 = (O - E)^2/E = 8.31$

d.f. = 4 – 1 = 3

Table value

$\chi^2_{3,0.5} = 7.81$

The difference between observed and expected frequencies is significant since observed $\chi^2 >$ table value.

Ex. 41. 200 digits are taken from random number tables. Their frequencies are as follows :

Digit	0	1	2	3	4	5	6	7	8	9
O	18	19	23	21	16	25	22	20	21	15

Test if the digits are distributed equally.

Ans. $\chi^2 = 4.3$, $d.f. = 9$. H_0 appears to be correct

Ex. 42. Of 160 offsprings of a cross of pigs, 102 were red, 24 black and 34 white. According to a genetic model the probabilities of red, black and white are $\frac{9}{16}, \frac{3}{16}, \frac{4}{16}$.

Test the consistency of the model. Given $\chi^2_{2,0.05} = 5.99$

Ans. $\chi^2 = 3.7$; model supported

Ex. 43. Fit a Poisson distribution to the following and test the goodness of fit (x stands for the number of printing errors in a book) and O stands for the number of pages, *i.e.*, observed frequency

x	$O = f$	E (calculations shown below)
0	211	209.3
1	90	92.1
2	19	20.3
3	5	3.0
4	0	0.3

Note: $\Sigma O = \Sigma f = 325$

Note. In this exercise $d.f. = n - 2$

where n is the number of observations after adjusting for small frequencies.

$m = \frac{\Sigma fx}{n} = 0.44$ (Check)

	Expected frequency $= 325 \times p(x)$
For $x = 0,\ p(0) = \frac{e^{-m}m^0}{0!} = 0.644$	209.3
$x = 1,\ p(1) = \frac{e^{-m}m^1}{11} = 0.644 \times .44$	92.1
$x = 2,\ p(2) = \frac{e^{-m}m^2}{2!} = 0.644 \times \frac{0.44^2}{21}$	20.3
$x = 3,\ p(3) = \frac{e^{-m}m^3}{3!} = 0.644 \times \frac{0.44^3}{3!}$	3.0
$x = 4,\ p(4) = \frac{e^{-m}m^4}{4!} = 0.644 \times \frac{(0.44)^4}{4!}$	0.3

The last two frequencies 3.0 and 0.3 are small. Combine them with 20.3. We then have

x	O	E	$(O-E)^2/E$
0	211	209.3	0.014
1	90	92.1	0.048
2, 3, 4	24	23.6	0.007
			0.069

$$\chi^2 = \frac{\Sigma(O-E)^2}{E} = 0.069$$

d.f. $= n - 2 = 3 - 2 = 1$ since only 3 observations remain now.

Table $\chi^2_{1.\ 0.05} = 3.84$.

Since observed $\chi^2 <$ table value, the fit is good (difference not significant).

26.16 χ^2 TEST OF INDEPENDENCE

Contingency Tables

Suppose the frequencies in the data are classified according to attribute A into r classes (rows) and according to attribute B into c classes (columns) as follows :

Class	B_1	B_2	....	B_c	Total
A_1	O_{11}	O_{12}	....	O_{1c}	(A_1)
A_2	O_{21}	O_{22}	....	O_{2c}	(A_2)
⋮					⋮
A_r	O_{r1}	O_{r2}	....	O_{rc}	(A_r)
Table	(B_1)	(B_2)	....	(B_c)	N

The totals of row and column frequencies are (A_i), (B_j).

To test if there is any relation between A, B we set up the null hypothesis of *independence* between A, B.

The expected frequency in any cell is calculated by using the formula :

$$E_{ij} = \frac{(A_i)(B_j)}{N}$$

Use
$$\chi^2 = \Sigma \frac{(O_{ij} - E_{ij})^2}{E_{ij}} \quad \text{with d.f.} = (r-1)(c-1)$$

Ex. 44. From the following test if there is any association between sex and education. Given $\chi^2_{2,0.05} = 5.99$, $\chi^2_{2,0.01} = 9.2$

Observed Frequencies

	School	*College*	*University*	*Total*
Boys	10	15	25	50
Girls	25	10	15	50
Total	35	25	40	100

Expected Frequencies

Boys	17.5	12.5	20	5
Girls	17.5	12.5	20	50
Total	35	25	40	100

e.g. $\frac{35 \times 50}{100} = 17.5$

$\frac{25 \times 50}{100} = 12.5$

$\frac{40 \times 50}{100} = 20$ etc.

d.f. $= (2-1)(3-1) = 2$

$\chi^2 = \Sigma (O-E)^2/E = 9.9$

This is greater than the table value. It is not true that education does not depend on sex, *i.e.* the two are not independent.

Ex. 45. Usually we come across 2×2 contingency tables. The cells show observed frequencies a, b, c, d.

A, B are attributes ; $\bar{A}$, $\bar{B}$ show the absence of the attributes.

Class	B	$\bar{B}$	Total
A	a	b	$a+b$
$\bar{A}$	c	d	$c+d$
Total	$a+c$	$b+d$	$a+b+c+d$

Expected frequency for a is $(a+c)(a+b)/N = a'$

for b is $(b+d)(a+b)/N = b'$

for c is $(a+c)(c+d)/N = c'$

for d is $(b+d)(c+d)/N = d'$

$$\chi^2 = \frac{(a-a')^2}{a'} + \frac{(b-b')^2}{b'} + \frac{(c-c')^2}{c'} + \frac{(d-d')^2}{d'}$$

It can be proved that this becomes

$$\chi^2 = \frac{N(ad-bc)^2}{(a+b)(c+d)(a+c)(b+d)} \quad \text{with } df = (2-1)(2-1) = 1$$

Note. Yate's correction. In this formula:

If $ad > bc$, replace a by $a - \frac{1}{2}$ and d by $d + \frac{1}{2}$

b by $b + \frac{1}{2}$ and c by $c + \frac{1}{2}$

If $ad < bc$, replace a by $a + \frac{1}{2}$ and d by $d + \frac{1}{2}$

b by $b - \frac{1}{2}$ and c by $c - \frac{1}{2}$

Alt. use $\chi^2 = \frac{(ad-bc-N/2)^2 N}{(a+b)(c+d)(a+c)(b+d)}$ or $\frac{(ad-bc+N/2)^2 N}{(a+b)(c+d)(a+c)(b+d)}$

According as $ad - bc > 0$ or $ad - bc < 0$

Ex. 46. In a town with 8000 graduates there are 800 female graduates. Out of 1500 graduate employees, 120 are females. Is there any distinction in employment on sex basis?

Given $\chi^2_{1,0.05} = 3.84$ **Ans.** Yes. Since $\chi^2 = 13.9$

Ex. 47.

	Got Typhoid	Did not get	Total
Inoculated	2	10	12
Non inoculated	8	4	12
Total	10	14	24

Can the inoculation be regarded effective against typhoid? Given $\chi^2_{1,0.05} = 3.84$.

Applying Yate's correction ($ad < bc$)

	Got Typh.	Did not get	Total
Inoculated	2.5	9.5	12
Not inoculated	7.5	4.5	12
	10	14	24

$$\chi^2 = \frac{(2.5\times 4.5 - 9.5\times 7.5)^2 24}{12\times 12\times 10\times 14} = 4.3$$

This is greater than the table value.

∴ The inocultion seems to be effective against typhoid.

Ex. 48.

Father's eye colour	Son's eye colour Dark	Light	Total
Dark	230	148	378
Light	151	471	622
Total	381	619	1000

Are the eye colour of fathers and sons associated?

Ans. They are associated *d.f.* = 1

26.17 TEST FOR A GIVEN POPULATION VARIANCE

To test the hypothesis that a sample $x_1, x_2, \ldots, x_n$ of size n has a specified variance $\sigma^2 = \sigma_0^2$.

Null hypothesis $H_0 : \sigma^2 = \sigma_0^2$ or $\sigma = \sigma_0$

$$H_1 : \sigma^2 > \sigma_0^2$$

Test statistic $\chi^2 = \dfrac{ns^2}{\sigma_0^2} = \dfrac{\Sigma(x - \bar{x})^2}{\sigma_0^2}$

If χ^2 is greater than the table value we reject the null hypothesis.

26.18 F TEST

If there are two independent random samples from normal populations we have to test the hypothesis that the population variances σ_1^2, σ_2^2 are the same

$$H_0 : \sigma_1^2 = \sigma_2^2 \text{ and } H_1 : \sigma_1^2 \neq \sigma_2^2$$

We find $S_1^2 = \dfrac{\Sigma(x_1 - \bar{x}_1)^2}{n_1 - 1}$, estimate of σ_1^2

$$S_2^2 = \frac{\Sigma(x_2 - \bar{x}_2)^2}{n_2 - 1}, \text{ estimate of } \sigma_2^2$$

To carry out the test of signficance, find

$$F = \frac{S_1^2}{S_2^2} \text{ if } S_1^2 > S_2^2$$

or

$$\frac{S_2^2}{S_1^2} \text{ if } S_2^2 > S_1^2$$

See the F tables for $n_1 - 1$, $n_2 - 1$ degree of freedom ($n_1 - 1$ corresponds to the numerator of F with the greater variance, $n_2 - 1$ for the denominator) at 5% level of significance.

If the observed F is less than the table value we assume the two populations have a common variance.

At 1% level of significance, the same procedure is to be followed.

Ex. 49. One sample of 10 bulbs gives a S.D. of 9 hours of life and another sample of 11 bulbs gives a S.D. of 10 hours of life. Can you say the variances are different at 1% level of significance? (Note : here $s_1 = 9$, $s_2 = 10$)

$$S_1^2 = \frac{n_1}{n_1 - 1} s_1^2 = \frac{10}{10 - 1} \times (9)^2 = 90 \qquad F = \frac{s_2^2}{s_1^2} = \frac{110}{90} = 1.22 < \text{table value}$$

$$S_2^2 = \frac{n_2}{n_2 - 1} s_2^2 = \frac{11}{11 - 1} \times (10^2) = 110$$

(Table value $F_{9,10,\,00.1} = 4.94$)

We accept the null hypothesis. The population variances may not be different.

Ex. 50. Two samples of size 9 and 8 give the sum of squared deviations from means as follows

$\Sigma\,(x_1 - \bar{x}_1)^2 = 160$, $\Sigma\,(x_2 - \bar{x}_2)^2 = 91$

Can they be considered as drawn from the same normal populations? $F_{8,7,0.05} = 3.4$

Ans. $F = 20/13 = 1.54$

The two samples can be considered as having been drawn from the same normal population with the same variance.

Can their means be regarded as equal?

Yes; this can be checked by t test.

Ex. 51. Check whether the following samples from normal populations have the same variance.

Sample 1 : 0 5 11 14 16 22 25 25 27

Sample 2 : 1, 6, 7, 25, 18, 3, 25, 26, 28, 31

Ans. $F = 1.47$ Table value $F_{9,7,.05} = 3.68$ The two variances may be regarded as equal.

26.19 NONPARAMETRIC TESTS

In hypothesis testing we make some inference about population parameters like the mean, the proportion etc. We usually make the assumption that our sample data come from a normal population.

What if the population is not normal or the normality assumption is not proper? We then cannot use the parametric testing method. Instead we use nonparametric testing method which is *distribution free*, *i.e.*, it does not depend on any assumption about the population distribution. These methods are not very efficient and may lose some information but they are easy to understand and are capable of wide applications.

Distribution free methods make use of ordered samples, *i.e.*, the sample observations are in ascending order of magnitude. Here the median, quartiles etc. are preferred to mean, standard deviation etc.

26.20 A NOTE ON STATISTICAL DECISION MAKING

Statistical Decisions have to be made in the presence of uncertainty.

In testing of hypothesis, the choice is between H_0 and H_1. In estimation, there are several choices available. The design of experiments requires one to choose between the nature and extent of observations. All this has to be done in the presence of uncertainty.

A decision function $D(x)$ assigns to every possible outcome a unique action. This may result in loss, positive or negative, depending on an unknown parameter w

The loss function $L\ (w, D)$

which depends on the outcome x, is a random variable. Its expected value is called the Risk Function.

CHAPTER 27

Analysis of Variance (ANOVA)

27.1 INTRODUCTION

To test the significance of difference between two sample means we use the t distribution.

When there are three or more sample means assumed to have been drawn from the same population, they have to be tested for differences. For this we use the technique of Analysis of Variance (ANOVA) which separates "the variance ascribable to one group of causes from the variance ascribable to other groups." By this method we can test whether the differences between the means of three or more populations are significant or not.

The ANOVA technique is also used to test the linearity of the regression line fitted to the data.

If there is only one factor assumed to be affecting the observations we examine the significance of differences between the class means resulting from the *variability between different classes* based on that one factor. This is called One Way Classification.

The analysis carried out in situations where there are two factors affecting the variables is called Two Way Classification.

Other classifications with more factors and interactions can also be analysed.

One Way Classification. The hypothesis is to test the equality (or homogeneity) of k population means $\mu_1, \mu_2, \mu_3, \ldots, \mu_k = \mu$.

Null Hypothesis $H_0 : \mu_1 = \mu_2 = \mu_3 = \ldots = \mu_k = \mu$

Alternative H_1 : All μ's are not equal.

Assumptions (*a*) The samples are independently drawn from their populations

(*b*) The variances of all populations are equal

(*c*) The populations are normally distributed

Computation

Procedure for One Way Classification. This is based on the partition of total variation into two parts :

Total variation = Variation between classes + Variation within classes

(1) (2) (3)

The observations are classified according to one factor or criterion and we calculate (1), (2), (3) below.

(1) **Total.** This is calculated from Total Sum of Squares (TSS)

$$TSS = \Sigma\Sigma\,(x_{ij} - \bar{x})^2$$

x_{ij} are observations, $\bar{x}$ the mean of all observations.

If T is the total of all n observations it can be shown that

$$TSS = \Sigma\Sigma\, x_{ij}^2 - \frac{T^2}{n} \quad \text{(d.f.} = n-1)$$

(2) **Between Sample Variation.** This is calculated from the Sum of Squares Between Samples (BSS)

$$BSS = \sum_{i=1}^{k} n_i(\bar{x}_i - \bar{x})^2$$

There are k classes (or criteria), $\bar{x}_i$ is the mean of the ith class with n_i observations. ($d.f. = k - 1$)

If T_i is the total of each class it can be shown that

$$BSS = \Sigma\, x_{ij}^2 - \Sigma\frac{T_i^2}{n_i}$$

(3) **Within Sample Variation (WSS).** This is calculated from the difference.

$$WSS = TSS - BSS$$

or

$$WSS = \Sigma\Sigma\, x_{ij}^2 - \Sigma\frac{T_i^2}{n_i} \quad (d.f. = n - k)$$

Mean Square

BSS divided by degrees of freedom $k - 1$ gives Between Variance or Mean Square Between (MSB). Similarly WSS divided by degrees of freedom $n - k$ gives Within Variance or Mean Square Within (MSW)

We have the following results :

$$MSB = \frac{BSS}{k-1}$$

$$MSW = \frac{WSS}{n-k}$$

To carry out the test for significance of difference between the two variances we calculate

$$F = \frac{MSB}{MSW} \text{ with } k-1,\ n-k\ d.f.$$

See the F table corresponding to $k - 1$, $n - k$ $d.f.$ under 5% (or 1% if required) level of significance.

If the observed value of $F >$ table value, the null hypothesis is rejected at the given level of significance.

Note. It is essential that F calculated as above should be greater than 1, *i.e.*,

Numerator should be > Denominator. If it so happens that MSW > MSB then write

$$F = \frac{MSW}{MSB}$$

In other words, while calculating F, the larger variance should be in the numerator. The results are put in the form of an ANOVA table.

Note. The results follow the same pattern in the *two way classification*.

27.2 ANOVA : ONE WAY CLASSIFICATION

Source of Variation	*Sum of Squares*	*Degrees of Freedom*	*Mean square MS*	*Variance Ratio F*
Between classes	$\Sigma \frac{T_i^2}{n_i} - \frac{T^2}{n}$	$k-1$	BSS/k – 1 = MSB	$\frac{\text{MSB}}{\text{MSW}}$ or $\frac{\text{MSW}}{\text{MSB}}$
Within classes	$\Sigma\Sigma x_{ij}^2 - \Sigma \frac{T_i^2}{n_i}$	$n-k$	WSS/n – k = MSW	whichever is greater than 1.
Total	$\Sigma\Sigma x_{ij}^2 - \frac{T^2}{n}$	$n-1$		

Ex. 1. The number of washing machines sold by 4 salesmen over three weeks is shown in the following chart.

Use ANOVA to test the difference between salesmen

Week \ Salesmen	A	B	C	D
I	6	13	10	10
II	14	15	13	9
III	7	14	9	11

Note. The result will remain unchanged if we simplify the data by subtracting some number (preferly closer to the overall mean) from all observations. We shall subtract 11

A	*B*	*C*	*D*	Total
–5	2	–1	1	
3	4	2	–2	
–4	3	–2	0	
$T_1 = -6$	$T_2 = 9$	$T_3 = -1$	$T_4 = -3$	$\Sigma T_i = 1$
$\frac{T_1^2}{3} = 12$	$\frac{T_2^2}{3} = 27$	$\frac{T_3^2}{3} = \frac{1}{3}$	$\frac{T_4^2}{3} = 3$	$\Sigma \frac{T_i^2}{n_i} = 42\frac{1}{3}$

$$T = -1, \Sigma\Sigma x_{ij}^2 = 93, T/n = 1/12$$

$$\text{TSS} = \Sigma x_{ij}^2 - \frac{T^2}{n} = 93 - \frac{1}{12} = 92.91$$

$$\text{BSS} = \Sigma \frac{T_i^2}{n_i} - \frac{T^2}{n} = 42\frac{1}{3} - \frac{1}{12} = 42.25$$

$$\text{WSS} = 92.91 - 42.25 = 50.66$$

ANOVA

Source	SS	DF	M	F
Between	42.25	3	14.08	2.22
Within	50.66	8	6.33	
Total	92.91	11		

Table $F_{3, 8, 0.05} = 4.07$

Observed $F <$ Table value

Conclusion. No significant difference between salesmen.

Ex. 2. Carry out the analysis of variance of the following data showing production quantities by three machines :

A	*B*	*C*
4	4	8
6	3	6
8	5	6
	12	9
		10

It is convenient to subtract 7 from the data. (*Note* : The data may be used as they are. But this subtraction simplifies calculation work.)

A	B	C
–3	–3	1
–1	–4	–1
1	–2	–1
	5	2
		3
$T_1 = -3$	$T_2 = -4$	$T_3 = 4$
$\frac{T_1^2}{3} = 3$	$\frac{T_2^2}{4} = 4$	$\frac{T_3^2}{5} = 3.2$

$$\Sigma\Sigma x_{ij}^2 = (3)^2 + (-1)^2 + (1)^2 + (-3)^2 + \dots + 2^2 + 3^2 = 81$$

$$T = -3, T^2/n = 9/12 = 0.75$$

$$\Sigma \frac{T_i^2}{n_i} = 3 + 4 + 3.2 = 10.2$$

$$TSS = \Sigma\Sigma x_{oj}^2 - T^2/n = 81 - 0.75 = 80.25$$

$$\text{BSS} = \Sigma T_i^2/n_i - T^2/n = 10.2 - 0.75 = 9.45$$

ANOVA

Source	SS	DF	MS	F
Between	9.45	2	4.725	$\frac{7.87}{4.725}$
Within	70.80	9	7.87	= 1.67
Total	80.25	11		

Table $F_{9,\,2,\,0.05} = 4.26$

Observed F is less than the table value. The difference between the three machines is not significant.

Ex. 3. The sales figures of an item in four towns *A, B, C, D* are given below. Test if the differences are significant

Town	*A*	*B*	*C*	*D*
	72	74	78	67
	69	72	74	67
	63	70	70	66
	59	69	58	64
	59	69	58	62
	53	58	56	58
	51	52	56	54

Ans. $F = \frac{60.25}{31.66} = 1.47$ Table value $F_{24,3} = 3.01$. Differences not significant

Ex. 4. Explain One Way Classification in a tabular form

Suppose there are n observation $\{x_{ij}\}$

$$i = 1, 2, \ldots, k \ (k \text{ classes each of size } n_i)$$

$$j = 1, 2, \ldots, n_i \left(n = \sum_{i=1}^{k} n_i \right)$$

The total and mean of each class is denoted by T_i and $\bar{x}_i$ in the following data on $\{x_{ij}\}$.

Class					Class Total	Class Mean
1	x_{11}	x_{12}	...	x_{1n1}	T_1	$\bar{x}_1$
2	x_{11}	x_{12}	...	x_{2n2}	T_2	$\bar{x}_2$
	...	...	...	...		
i	x_{i1}	x_{i2}	...	x_{in_i}	T_i	$\bar{x}_i$
	...	...	...	...		
k	x_{k1}	x_{k2}	...	x_{kn_k}	T_k	$\bar{x}_k$
				Total	T	

Notation

T = Total of all observations

(1) $\bar{x}_i$ = Mean of the ith class $= \frac{T_i}{n_i} = \frac{\sum_{j=1}^{n_i} x_{ij}}{n_i}$

(2) $\bar{x}$ = Mean of all observations = $\dfrac{T}{n}$

$$= \frac{1}{n}\sum_{i=1}^{k} n_i \bar{x}_i = \frac{1}{n}\sum_{j=1}^{n_i} x_{ij}$$

(3) SSE = Error Sum of Squares or Within SS (WSS)

$$= \sum_i \sum_j \left(x_{ij} - \bar{x}_i\right)^2 = \sum_i \sum_j x_{ij}^2 - \sum_i \left(\frac{T_i^2}{n_i}\right)$$

(4) SSB = Sum of squares due to factor variations or Between SS

$$= \sum_i n_i (\bar{x}_i - \bar{x})^2 = \sum_i \left(\frac{T_i^2}{n_i}\right) - \frac{T^2}{n}$$

(5) TSS = Total sum of squares

$$= \Sigma\Sigma(x_{ij} - \bar{x})^2 = \sum_i \sum_j x_{ij}^2 - \frac{T^2}{n}$$

(6) Degrees of Freedom (*df*)

TSS based on n observations has $n - 1$ *df*. One degree of freedom is lost due to one constraint of the total. BSS has $n - k$ *df* since there are k constraints

ESS has $k - 1$ *df*

Note. $n - 1 = (n - k) + (k - 1)$

Ex. 5. A firm keeps record of the number of effective customers per week as shown below. Can you say there is a significance in the number over a period of 3 weeks given below?

Weeks					
I	300	400	300	500	0
II	600	300	300	400	—
III	700	300	400	600	500

Given $F_{2, 11, 0.05} = 3.98$

Ans. $F = 1.83$, The differences are not significant

Ex. 6. The yields of 6 varieties of a crop are given below. Is there a significant difference between the yields?

Variety 1	2	3	4	5	6
72	75	82	70	77	80
77	74	80	76	76	76
76	81	74	77	78	74
80	82	76	71	80	75
71	78	78	72	75	75
74		83	72	76	76
		79		77	77
		80			74
					77

given $F_{5,30,0.05} = 2.53$

Ans. $F = \dfrac{30.4}{7.2} = 4.2$ This is greater than the table value.

The difference is significant at 5% level of significance.

CHAPTER 28

Index Numbers

28.1 MEANING AND TYPES OF INDEX NUMBERS

An index number is a device which measures the relative change in the magnitude of a group of related, distinct variables in two or more situations. An average represents the general level of magnitude of a particular variable. An index number represents the general level of magnitude of changes between two or more situations or times.

Index numbers may be used to measure changes in prices, wages, production, employment, national income and so on, over a period of time. They may also be used to study differences between places or categories.

An index number in a given situation is calculated by taking into account the variables influencing that situation. If we want to compare the standards of living at two different periods of time we must consider the consumption pattern of the people concerned at these two periods of time. This is done by studying the variations in the prices of commodities used by those people.

The method of construction of an index number and the items to be included in it are matters decided by the purpose for which it is to be used.

Index numbers used in economics and business may be classified as follows :

(*i*) *Index numbers of prices* : *e.g.* cost of living index, wholesale price index, stock price index etc.

(*ii*) *Index numbers of quantities or production* : *e.g.* index of industrial production, business activity index, production or employment index etc.

(*iii*) *Index numbers of value* : *e.g.* index of gross national product, index of retail sales, index of personal incomes etc.

(*iv*) *Index numbers of qualitative changes* and other index numbers.

We shall consider here price index numbers in detail. The procedure for other types follows the same lines.

28.2 PRICE INDEX NUMBERS

A study of changes in prices is of special significance in economics. While changes in the price of a single commodity can be explained by the forces of demand and supply, the changes in the prices of several commodities may be due to various reasons. One reason may be the value of money itself. This may be associated with other causes.

Prices of different commodities form a heterogeneous group subject to some general and special forces from within or without the economy. Index numbers take into account the effects of both the types of forces. The effects of special forces normally tend to cancel one another with little influence on the average; the general forces tend to either increase or decrease the quantities concerned. Index numbers consider the segregation of the effect of the general forces in the form of rise or fall in quantities from the special forces on these quantities.

A price index number may be regarded as a summary statistic for the frequency distribution of price relatives. It is the weighted arithmetic mean of the distribution of successive years. A quantity index may be similarly treated as a statistic for a frequency distribution of quantity relatives. We rarely regard the prices and quantities of commodities included in two different index numbers as functionally related.

A price index number measures changes in the general level of prices of a group of commodities. It thus measures changes in the purchasing power of money.

Price index numbers are generally used to make comparisons over time though sometimes they can be used for comparisons over regions.

Price index numbers may be used to compare wholesale prices, retail prices, agricultural prices, industrial prices and so on.

Construction

The *construction* of price index numbers (or any index number) involves the consideration of the following problems :

(1) The choice of items to be included

(2) The choice of the base period to calculate price relatives

(3) The choice of average to be adopted

(4) The choice of weighting systems to be used.

(1) Choice of Items

(*a*) *Nature of Commodities Selected* :

In the construction of price index numbers it is not possible to include all commodities and their quotations. We therefore select a few items which are

(*i*) representative of the tastes, habits or requirements of the people concerned.

(*ii*) easily identifiable,

(*iii*) not varying in quality or deign or contents over the given period of time.

(*b*) *Number of Commodities* :

A reasonable number of commodities should be taken. A very large number may be difficult to manage, a small number may not be adequate to give a good idea about the nature of movements. The number of commodities is decided by the nature of the economy and the extent of accuracy desired.

(*c*) *Quotations* :

Price quotations should be obtained from some selected markets, from leading firms or from newspapers. For the wholesale price index number, quotations should be obtained from wholesale dealers. For cost of living index numbers, retailers should be contacted and so on.

(*d*) *Quality of Commodities* :

Out of several varieties of a commodity the most popular or common varieties should be used in the base period as well as in the current period. Differences in varieties at different periods may give wrong signals through the index numbers.

(*e*) *Grouping of Commodities* :

Homogeneous groups of similar commodities may be formed. Thus all food items may be put in one sub-group and so on.

(2) Choice of Base

The comparison of prices in any year has to be made with the prices in the base year which in general is an earlier, ideal period of time during which prices may have remained more or less stable.

With the help of the current year prices and the base year prices we calculate the price relatives which are the percentages of prices in the current year as compared with the prices in the base year.

There are two methods of calculating the price relatives.

(*i*) Fixed base method

(*ii*) Chain base method

What is a Price Relative?

A price relative is a pure number, free of any unit of measure. It is usually calculated as a percentage of the price of a commodity in any year as compared with the base year price.

We define the price relative for the current year for the commodity as

$$P.R. = \frac{\text{Commodity price in current year}}{\text{Commodity price in base year}} \times 100 = \frac{p_1}{p_0} \times 100$$

where p_1 is the price in current year

p_0 is the price in base year

Fixed Base Method :

In the fixed base method, a definite base year or the average of a certain number of years is chosen and the price relatives are calculated with the fixed base.

The base year should be a reasonably stable and normal period. It should not be too distant a period from the data under study. An abnormal base year may lead to abnormal conclusions. Care should be taken to see that the selected base year is more or less free from economic fluctuations, wars, famines etc.

Chain Base Method :

In the chain base method every price relative is chained to the preceding year as base. Thus the price relative for each year has a different base, the preceding year.

A chain base provides a direct comparison between each year and the next rather than between remote years as in the fixed base method. Moreover, it facilitates the entry of new items and removal of old items.

(3) Choice of Average

To calculate the index number, the average of the price relatives for different commodities should be calculated. For this purpose we may use the arithmetic mean, the geometric mean or the median. While the geometric mean and median have some specific advantages, the arithmetic mean has been found to be more useful in practice as well as in theory. The geometric mean is reversible, the arithmetic mean is not. The geometric mean is sometimes preferred by experts but the arithmetic mean is more intelligible and easier to calculate.

(4) Choice of Weights

The items selected to construct an index number being of unequal importance, a simple average will not be truly representative of the state of affairs. Suitable weights, therefore, should be attached to the items, depending on their relative importance.

A high degree of accuracy in weighting is neither practical nor necessary since errors in weighting have little effect compared with variations in prices.

Arbitrary or chance weighting is subjective. Rational weighting based on knowledge and experience is more useful.

Opinions differ regarding the use of weights. According to some, weighting does not usually make much difference and almost the same result is obtained by different systems of weighting. Others emphasise its use because of the heterogeneous character of various series of prices and their unequal importance.

In the construction of certain types of index numbers, weighting is definitely useful. Approximate weights selected on the basis of the economic importance of a commodity or group can give a better index. Often to avoid the search for weights physical quantities of items consumed in the base or current year are taken as weights. Values include the effect of prices and are generally not recommended as weights.

Weights are usually kept constant over a reasonably long period of time to facilitate comparisons.

Summing up

The construction of an index number requires the choice of a suitable number of representative items, an ideal base year and appropriate weights to be processed by a good average. It may be noted that changes in fashions, tastes etc. bring about the replacement or disappearance of many items so that a base year close to the current year should normally be taken. A more recent base year can accommodate various changes in weights and other aspects of index numbers.

The process of calculation of index numbers is simplified by the use of suitable formulae depending on the accuracy desired or the information available. The form of index number to be used depends on individual situations.

Index numbers are not determinate quantities. An index number is trustworthy in its tendencies without being reliable to the last digit.

Methods of constructing index numbers

(A) *Simple index numbers* : Aggregates of prices and price relative. Here we have, for a series of time intervals, the prices per unit of a number of items.

(*a*) *A simple index number* is the per cent of each of a set of price relatives to some base year price. In the following case, the price of only one item is considered.

Ex. 1. (*a*) *Year*	*Price*	*Index number* (1961 = 100)
1961	20	100
1962	25	125
1963	30	150
1964	40	200

Explanation
Base 1961
(25/20) × 100 = 125
(30/20) × 100 = 150
(40/20) × 100 = 200

(*b*) *A simple aggregate price index* is the sum of the prices per unit of each of the items in the current year relative to the sum of the prices per unit of the items in the base year.

If the prices per unit of n commodities in the current year 1 are

$$p_{11}, p_{12}, p_{13}, \ldots\ldots\ldots, p_{1n}$$

and in the base year 0 they are

$$p_{o1}, p_{o2}, p_{o3}, \ldots\ldots\ldots, p_{on}$$

then the simple aggregate price index is written

$$I = \frac{p_{11} + p_{12} + \ldots\ldots + p_{1n}}{p_{o1} + p_{o2} + \ldots\ldots\ldots + p_{on}} = \frac{\sum_{1}^{n} p_{1i}}{\sum_{1}^{n} p_{oi}}$$

Briefly, we write, $I = \frac{\Sigma p_1}{\Sigma p_o}$. This index is not of much help. It suffers from various defects.

Ex. 1. (*b*)

Commodity	*Unit*	*Price in* Base year p_o	*Price in* Current year p_1
x	kg	5	10
y	m	7	12
z	doz	10	11
	22	33	

$$I = \frac{\Sigma p_1}{\Sigma p_o} \times 100 = \frac{33}{22} \times 100 = 150$$

The aggregate price level has gone up by 50 per cent.

(c) *A simple average of price relatives* is an index obtained by taking the average of the price relatives of all the items in the series. The average normally used is the arithmetic mean. If the number of items is n we have the index,

$$1 = \frac{1}{n}\Sigma\frac{p_1}{p_0} \times 100$$

Ex. 1. (c)

Commodity	Unit	p_0	p_1	Price relative $\frac{p_1}{p_0}\times 100$	Explanation
x	kg	5	10	200.0	$\frac{10}{5} \times 100 = 200.0$
y	m	7	12	171.4	$\frac{12}{7} \times 100 = 171.4$
z	doz	10	11	110.0	$\frac{11}{10} \times 100 = 110.0$
	Total			481.4	

$$I = \frac{\sum \frac{p_1}{p_o}\times 100}{n} = \frac{481.4}{3} = 160.5$$

This method shows that the aggregate price level has gone up by 60.5%.

Composite Index Numbers

(B) In practice, composite index numbers are used. Simple or unweighted index numbers are defective either because they ignore units of measurements or do not consider the relative importance of different items.

Weighted Index Numbers : *Weighted aggregates of prices and price relatives*

An index number which takes into account the weight attached to each of the items is called a weighted index number. A weighted index number is a measure of the change in the prices of items and not in the weights of items.

If $w_1, w_2, \ldots, w_k$ are the weights of k items, then their cost in the base period $= \sum_1^k p_o w$ and their cost in the current period $= \sum_1^k p_1 w$ (written briefly $\Sigma p_0 w$, $\Sigma p_1 w$)

(a) *The weighted aggregate price index number* is given by :

$$I = \frac{\Sigma p_1 w}{\Sigma p_o w}\times 100$$

If the base year quantity q_o is used as weight, we have

$$I = \frac{\Sigma p_1 q_0}{\Sigma p_o q_0}\times 100$$

Ex. 2. (*a*)

Commodity	*Unit*	p_o	p_1	*Weight* w	p_0w	p_1w
x	kg	5	10	4	20	40
y	m	7	12	7	49	84
z	doz	10	11	3	30	33
				Total	99	157

$$I = \frac{\Sigma p_1 w}{\Sigma p_0 w} \times 100 = \frac{157}{99} \times 100 = 158.6$$

The aggregate price level is seen to have gone up by 58.6 per cent.

(*b*) *The weighted aggregate of price relatives* is obtained by taking the average of weighted price relatives.

If $R = \frac{p_1}{p_0} \times 100$ stands for a price relative and w the corresponding weights we write the weighted aggregate of price relatives as :

$$I = \frac{\Sigma Rw}{\Sigma w}$$

Ex. 2. (*b*)

Commodity	*Unit*	p_0	p_1	w	$R = \frac{p_1}{p_o} \times 100$	Rw
x	kg	5	10	4	200.0	800
y	m	7	12	7	171.4	1200
z	doz	10	11	3	110.0	330
			Total	14		2330

$$I = \frac{\Sigma Rw}{\Sigma w} = 166.4$$

The aggregate price level is up by 66.4 per cent.

Any other set of weights can also be used in this index number. Sometimes, instead of using w as weights, value weights are attached.

If V is the value of goods purchased (either in the base year or in the current year) the index number can be written

$$I = \frac{\Sigma RV}{\Sigma V}$$

Normally the base year value $V = p_0 q_0$ is used.

$$I = \frac{\Sigma RV}{V} = \frac{\sum \frac{p_1}{p_0} p_0 q_0 \times 100}{\Sigma p_0 q_0} = \frac{\Sigma p_1 q_o}{\Sigma p_o q_o} \times 100$$

This is known as Laspeyre's index number. It has base year quantities q_0 as weights.

With current year quantities q_1 as weights we get Paasche's index number : $\frac{\Sigma p_1 q_1}{\Sigma p_0 q_1} \times 100.$

Chain Base Technique

This technique provides the comparison of any given year's price with the previous year's price.

Each year's price is taken as the base in the next year. Thus 1941 is the base for 1942 (Here the first year 1941 = 100). Then 1942 is the base for 1943 etc.

(*a*) The Chain Base or Link Index Number (CBIN) gives a relation with the immediate past rather than a distant past.

(*b*) This avoids seasonal variations since the time gap is narrowed down.

(*c*) Outdated or unwanted items can be dropped and new items added if required.

The procedure for conversion from Fixed Base (FB) to chain base (CB) can be seen from the exercises below.

Note. Instead of taking a single year base price, one may take the average of a few years as base to avoid the effect of abnormalities in a single selected year.

A base cannot be a permanent feature because of changes in tastes, habits, technology, incomes and availability of resources. It is usual to shift the base to a period closer to the period of study.

Ex. 3. Calculate (*a*) fixed base and (*b*) chain base index numbers for the following:

Year	*Price*
1941	10
1942	12
1943	15
1944	21
1945	21

(*a*)

Year	*Fixed Base Index Number* (1941 = 100)
1941	100
1942	120
1943	150
1944	210
1945	210

If 1941 is the base year, each year's index number is obtained by dividing the year's price by base year price and multiplying by 100.

Explanation

$\left(\frac{12}{10} \times 100 = 120\right)$

$\left(\frac{15}{10} \times 100 = 150\right)$

$\left(\frac{21}{10} \times 100 = 210\right)$

$\left(\frac{21}{10} \times 100 = 210\right)$

(b) Year	Chain Base Index Number	Explanation
1941	100	
1942	120	$\left(\frac{12}{10}\times 100=120\right)$
1943	125	$\left(\frac{15}{12}\times 100=125\right)$
1944	140	$\left(\frac{21}{15}\times 100=140\right)$
1945	100	$\left(\frac{21}{21}\times 100=100\right)$

In chain base, each year is the base of the next. Divide each year's price by the preceding year's price and multiply by 100.

Ex. 4. Convert (*i*) the fixed base index numbers (FBIN) into chain base index numbers (CBIN) and (*ii*) chain base index numbers into fixed base index numbers in the previous exercise.

(*i*) Year	F.B.I.N.	C.B.I.N.
1941	100	100
1942	120	$\frac{120}{100}\times 100 = 120$
1943	150	$\frac{150}{120}\times 100 = 125$
1944	210	$\frac{210}{150}\times 100 = 140$
1945	210	$\frac{210}{210}\times 100 = 100$

Explanation

To convert FBIN into CBIN treat the given index numbers as prices and proceed as usual

(*ii*) Year	CBIN	FBIN
1941	100	100
1942	120	$\frac{120\times 100}{100} = 120$
1943	125	$\frac{125\times 120}{100} = 150$
1944	140	$\frac{140\times 150}{100} = 210$
1945	100	$\frac{100\times 210}{100} = 210$

Explanation

To calculate FBIN from CBIN multiply each CBIN by the previous FBIN and divide by 100

Link Relatives

A Link Relative of an item in any period is the ratio of its price in that period divided by its price in the just preceding period, multiplied by 100.

For each year, the link relative is averaged over the year to arrive at the Link Index for each year. These link indices for all the years are converted into chain indices.

$$\text{Chain Index for any year} = \frac{\begin{matrix}\text{Link Index} \\ \text{for the year}\end{matrix} \times \begin{matrix}\text{Chain Index} \\ \text{of previous year}\end{matrix}}{100}.$$

Commodity \ Year	0	1	2	...	t
1	p_{01}	p_{11}	p_{21}		p_{t1}
2	p_{02}	p_{12}	p_{22}		p_{12}
⋮	⋮	⋮	⋮		⋮
n	p_{0n}	p_{1n}	p_{2n}		p_{tn}

Link Relatives

Commodity \ Year	0	1	2	...	t
1	100	$100p_{11}/p_{01}$	$100p_{21}/p_{11}$	...	$100p_{t1}/p_{t-1,1}$
2	100	$100p_{12}/p_{02}$	$100p_{22}/p_{12}$	...	$100p_{t2}/p_{t-1,2}$
⋮	⋮				
n	100	$100p_{1n}/p_{0n}$	$100p_{2n}/p_{1n}$	...	$100p_{tn}/p_{t-1,n}$
Link Index for Year 100 (= Average for column)		$\frac{100}{n}\sum\frac{p_1}{p_0} = L_{01}$	$\frac{100}{n}\sum\frac{p_2}{p_1} = L_{12}$	...	$100\sum\frac{p_t}{p_{t-1}}$, $= L_{t-1,t}$

∴ Chain Index for year 0 is 100

Chain Index for year 1 is $I_{01} = L_{01}$

Chain Index for year 2 is $I_{02} = I_{01}\,L_{12}/100$

Chain Index for year 3 is $I_{03} = I_{02}\,L_{23}/100$

Ex. 5. The calculation scheme for the fixed base relative index numbers and chain relative index numbers for the wholesale prices of certain commodities is given below :

Prices in Rs. per quintal

Commodity	1941	1942	1943	1944	1945
A	2	4	5	6	8
B	3	4.5	9	9	12
C	10	12	15	21	21
D	8	6	8	10	16

Fixed base relative index numbers for the given data (1941 = 100)

Commodity	1941	1942	1943	1944	1945
A	100	200	250	300	400
B	100	150	300	300	400
C	100	120	150	210	210
D	100	75	100	125	200
Total of Relatives	400	545	800	935	1210
Average of Relatives	100	136	200	234	303

Chain Relative Index Numbers for the given data :

Commodity	1941	1942	1943	1944	1945
A	100	200	125	120	133
B	100	150	200	100	133
C	100	120	125	140	100
D	100	75	133	125	160
Total of Link Relatives	400	545	583	485	526
Average Link Relatives	100	136	146	121	132

28.3 COST OF LIVING INDEX NUMBERS OR CONSUMER PRICE INDEX NUMBERS

A consumer price index number is a statistical measure of changes in the prices of the goods and services bought by families of wage-earners. The prices collected are retail prices of items like food, clothing, rent, fuel, medicine, transport, electricity, entertainment etc.

This index studies the effect of changes in prices of certain chosen combinations of goods on the standard of living of people considered as consumers. General price index numbers cannot do this.

Cost of living index numbers work as economic indicators in the same way as wholesale prices index numbers in measuring the purchasing power of money.

These index numbers can be compiled separately for :

(*i*) different classes of people

(*ii*) different regions or homogeneous zones

(*iii*) different time periods

(*iv*) different occupations.

The reason for this is that each class or region has its own choice of commodities whose qualities and quantities consumed differ from time to time and place to place. Moreover retail prices which are the basis of living index numbers differ from time to time and place to place.

Construction

The steps to be followed in the *construction* of cost of living index numbers are nearly the same as those for general index numbers :

(*i*) The class of people for whom this index number is to be constructed is decided first.

(*ii*) The sample family budget inquiry of the class is conducted.

(*iii*) The items covered are those which are in common use by the class concerned.

(*iv*) Price quotations for these items are to be obtained from the locality in which these people reside or from where they make their purchases.

(*v*) Suitable weights are attached to different commodities depending on their importance as revealed by family budget inquiries or otherwise. The cost of living index number is then calculated by the weighted aggregate of actual prices or the weighted average of price relatives.

Limitations of Cost of Living Index Numbers

(*i*) Possibility of selection of wrong or non-representative items.

(*ii*) No measure of changes in the quality of living

(*iii*) Inadequate reflection of inflationary effect, black market prices, change in quality, goods, etc.

(*iv*) Arbitrary assignment of weights.

(*v*) Nonrepresentation of experience of families.

(*vi*) Frequent need for recalculation.

Ex. 6. Construct the cost of living index number for the following data for 1958 with 1950 as base. Use as weights, (*i*) the quantities consumed in the base year, (*ii*) the values in the base year.

Article	*Quantity*	*Prices*	
	1950	1950	1958
Food	3 mds	12.00	18.00
Cloth	12 yds	1.00	0.90
Electricity	40 units	0.20	0.25
Rent	3 rooms	25.00	23.00
Miscellaneous	34 units	0.40	0.50

(*i*) *Aggregate Expenditure Method (quantities q_0 as weights)*

Article	q_0	p_0	p_1	p_0q_0	p_1q_0
F	3	12.00	18.00	36.00	54.00
C	12	1.00	0.90	12.00	10.80
E	40	0.20	0.25	8.00	10.00
R	3	25.00	23.00	75.00	69.00
M	30	0.40	0.50	12.00	15.00
				143.00	158.80

$$I = \frac{\Sigma p_1q_0}{\Sigma p_0q_0} \times 100 = \frac{158.80}{143.00} \times 100 = 111.5$$

(*ii*) *Family Budget Method*

Article	q_0	p_0	p_1	$R = \frac{p_1}{p_0} \times 100$	$V = p_0q_0$	*RV*
F	3	12.00	18.00	150.00	36	5400
C	12	1.00	0.90	90.00	12	1080
E	40	0.20	0.25	125.00	8	1000
R	3	25.00	23.00	92.00	75	6900
M	30	0.40	0.50	125.00	12	1500
					143	15880

$$I = \frac{\Sigma RV}{\Sigma V} = \frac{15880}{143} = 111.5$$

The results of both the methods are identical. Why ?

Ex. 7. An inquiry into the budgets of middle class families in a city gave the following information. Comment on the results. **Ans.** $I = \Sigma \, wp_1/p_0 = 97.9$

Expenses on :	*Food*	*Rent*	*Clothing*	*Fuel*	*Miscellaneous*
	35%	15%	20%	10%	20%
Prices 1928	150	30	75	25	40
Prices 1929	145	30	65	23	45

Ex. 8. Find the cost of living index number for January 1953 and state the percentage by which it is higher or lower than the 1939 level which is taken here as the base year.

Group	*Group Index Number for Jan. 1953*	*Weights*
Food	152	48
Fuel	110	6
Clothing	130	8
Rent	100	12
Miscellaneous	90	26

Ans. 125.4

Ex. 9. The procedure for deflating per capita income by the cost of living index is shown below. (This will give real P.C.I.)

Year	*Per capita Income (Rs.)*	*Cost of living Index base 1951 = 100*	*Deflated (real) P.C.I.*	*Explanation*
1951	120	100	120	$\frac{120}{100} \times 100 = 120$
1952	140	110	127	$\frac{140}{110} \times 100 = 127$
1953	150	120	125	$\frac{150}{120} \times 100 = 125$
1954	180	140	129	$\frac{180}{140} \times 100 = 129$
1955	200	150	133	$\frac{200}{150} \times 100 = 133$

To deflate a value divide it by the index number and multiply by 100. The deflated values represent the purchasing power.

It may be noted that changes in real P.C. I. are small as compared with those in money incomes showing that although money incomes seem to have increased, purchasing power has only marginally increased.

28.4 SOME INDEX NUMBER FORMULAE WITH SYSTEMATIC WEIGHTS

Representative index numbers must employ weights truly reflecting the relative importance of the commodities included.

Laspeyre's Index Number. This is a weighted aggregate price index number and uses base year quantities q_0 as weights. This type of index number is likely to have an upward bias and may be said to give an upper limit to the price change. Laspeyre's index number may be written :

$$L = \frac{\Sigma p_1 q_0}{\Sigma p_0 q_0} \times 100$$ Laspeyre's formula uses base year weights.

2. *Paasche's Index Number.* This index number, also a weighted aggregate, uses current year quantities q_1 as weights. It is likely to have a downward bias and may provide a lower limit to the price change. Paasche's index number may be written

$$P = \frac{\Sigma p_1 q_1}{\Sigma p_0 q_1} \times 100$$ Paasche's formula uses current year weights.

3. *Fisher's Ideal Index Number.* This index number is the geometric mean of Laspeyre's and Paasche's Index numbers.

$$F = \sqrt{L \times P}$$

$$= \sqrt{\frac{\Sigma p_1 q_0}{\Sigma p_0 q_0} \times \frac{\Sigma p_1 q_1}{\Sigma p_0 q_1}} \times 100$$ Fisher's formula is G.M. of L and P.

4. *Marshall-Edgeworth Index Number.* This is a composite index number and has no bias in either direction.

$$M = \frac{\Sigma p_1 (q_0 + q_1)}{\Sigma p_0 (q_0 + q_1)} \times 100$$

5. Bowley's Index Number $= \frac{1}{2}(L + P) = \frac{1}{2}\left(\frac{\Sigma p_1 q_0}{\Sigma p_0 q_0} + \frac{\Sigma p_1 q_1}{\Sigma p_0 q_1}\right) \times 100$

28.5 A NOTE ON LASPEYRE'S (L) AND PAACHE'S (P) NUMBERS

There are many formulae suggested for the calculation of index numbers. But Laspeyre's and Paasche's index numbers are the most used in statistics and economics. The two give different results in general but if the change in prices (or quantities) is in the same ratio they will be equal.

Prices and quantities in practice do not move in the same ratio. It has been found that P shows a smaller rise than L when prices are rising and a greater fall when prices are falling. If the periods being compared are not very distant L and P are usually fairly close.

L assigns to each price relative a weight which is appropriate for the base year. The weight for each commodity is then held constant for a number of years. P applies to each price relative a weight which is appropriate for that specific year. P thus measures pure price change only between the base year and that given year to which a particular set of weights applies.

Both L, P being aggregates of prices weighted by quantities with comparable biases and similarities, it is a difficult task to make a choice between them. A number of compromise formulae which are cross weight index numbers are sometimes used.

Ex. 10. Calculate L, P, F, M for the following data :

Commodity	*Base year Price*	*Base year quantity*	*Current year Price*	*Current year quantity*
A	6	50	10	56
B	2	100	2	120
C	4	60	6	60
D	10	30	12	24
E	8	40	12	36

Production Index Numbers

The index numbers of production or quantity define changes in the physical volume of output either by comparing quantities directly or by deflating values by suitable price index numbers.

The p's and q's in the formulae for price and production index numbers are perfectly symmetrical and can be interchanged. Thus Laspeyre's and Paasche's *quantity* index numbers may be written

$$L_q = \frac{\Sigma q_1 p_0}{\Sigma q_0 p_0} \times 100, \quad p_q = \frac{\Sigma q_1 p_1}{\Sigma q_0 p_1} \times 100.$$

Productivity Index Numbers

Productivity is measured by comparing change in the volume of production x with change in the volume of input of factor of production y.

The index number of productivity $= \dfrac{x}{y}$

The reciprocal $\dfrac{y}{x}$ is the index of factor requirements per unit of output.

Index Number of Unit Labour requirements

$$IULR = \frac{\text{Total of manhours spent in the current year}}{\text{Total of manhours spent in the base year}} \times 100$$

$$= \frac{\Sigma x_1 r_1}{\Sigma x_0 r_0} \times 100$$

where x, r represent the output and the unit labour requirements of individual commodities, enterprises or industries.

28.6 INDEX NUMBERS FOR COMBINED GROUPS

Sometimes index numbers are calculated separately for each of various groups of commodities. Thus we may have the index number I_1 for food items, I_2 for clothing items, and so on. If $w_1, w_2, \ldots$ are the weights attached to the group index numbers, we have the index number for the combined groups of commodities.

$$I_g = \frac{\Sigma I_i w_i}{\Sigma w_i}$$

Value Index Numbers

A value index in its simplest form compares the total value of all commodities in the current period with the total value in the base period.

$$I_v = \frac{\Sigma p_1 q_1}{\Sigma p_0 q_0}$$

Weighted value index numbers may be calculated by attaching suitable weights according to the relative importance of the items.

28.7 CHANGE OF THE BASE PERIOD

Sometimes it becomes necessary to change the base of a given index number from the year, say a to the year b. For this purpose we may construct the whole series of index numbers with the new base. This procedure being tedious, we may use the alternative method of obtaining the series of new base index numbers by dividing each index number of the old series by the index number of the new time period selected for base.

Ex. 11. For the following data the base period is shifted from 1951 to 1953.

Year	*Index Numbers* 1951 = 100	*Index Numbers* 1953 = 100	*Explanation*
1951	100	50	1951 : $\frac{100}{200} \times 100 = 50$
1952	160	80	1952 : $\frac{160}{200} \times 100 = 80$
1953	200	100	
1954	180	90	1954 : $\frac{180}{200} \times 100 = 90$
1955	220	110	

28.8 SPLICING TWO SERIES OF INDEX NUMBERS

An old index number series may be discontinued because of obsolete items included in it or other reasons. If a new series is constructed with a discontinuation year of the first as base, the two series so obtained, having different bases are not comparable. But the two series can be spliced together into one continuous series by multiplying each index number of the new series by the index number of the discontinuation year of the old series.

Ex. 12. If 1953 is the discontinuation year for the old series in the following data the new series can be constructed as shown:

Year	*Old series* (1)	*New series* (2)	*Spliced series old base* (3)	*Spliced series new base* (4)
1951	100	—	100	50
1952	160	—	160	80
1953	200	100	200	100
1954		90	180	90
1955		105	210	105
1956		165	210	105

Explanation

	Col. 3 Old Series			Col. 4 New Series
1951		100	$\frac{100}{200} \times 100 \rightarrow$	50
1952		160	$\frac{100}{200} \times 160 \rightarrow$	80
1953		200	$\frac{100}{200} \times 200 \rightarrow$	100
1954	$\frac{200}{100} \times 90 \rightarrow$	180	90	
1955	$\frac{200}{100} \times 110 \rightarrow$	220	110	
1956	$\frac{200}{100} \times 105 \rightarrow$	210	105	

28.9 COMPARISON OF COSTS OF PRODUCTION OF TWO FIRMS

Let $c_{11}, c_{12}, \ldots, c_{1n}$ be the costs of production of goods produced in quantities $q_{11}, q_{12}, \ldots, q_{1n}$ by one firm.

Let $c_{21}, c_{22}, \ldots, c_{2n}$ be the costs of production of goods produced in quantities q_{21}, $q_{22}, \ldots, q_{2n}$ by the second firm.

A comparison of their total cost $T_1 = \Sigma c_1 q_1$ and $T_2 = \Sigma c_2 q_2$ does not enable us to realise the differences in prices of inputs. But if we take as standard weights, the quantities produced by either firm say the first, we have an index number for comparison of costs of production.

$$I_1 = \frac{\Sigma c_2 q_1}{\Sigma c_1 q_1}$$

If the second firm quantities are taken as the standard, we have

$$I_2 = \frac{\Sigma c_1 q_2}{\Sigma c_2 q_2}.$$

Ex. 13. The procedure for the calculation of various price and quantity index numbers is illustrated below by taking only two commodities.

Item	p_0	q_0	p_1	q_1	$p_0 q_0$	$p_1 q_0$	$p_1 q_1$	$p_0 q_1$	$p_0(q_0+q_1)$	$p_1(q_0+q_1)$
A	1	10	2	5	10	20	10	5	15	30
B	1	5	4	2	5	20	8	2	7	28
Total					15	40	18	7	22	58

Index Numbers of Prices

$$L = \frac{\Sigma p_1 q_0}{\Sigma p_0 q_0} \times 100 = \frac{40}{15} \times 100 = 266.7$$

$$P = \frac{\Sigma p_1 q_1}{\Sigma p_0 q_1} \times 100 = \frac{18}{7} \times 100 = 257.1$$

$$M = \frac{\Sigma p_0 (q_0 + q_1)}{\Sigma p_0 (q_0 + q_1)} \times 100 = \frac{58}{22} \times 100 = 263.6$$

$$F = \sqrt{L \times P} = 266.7 \times 257.1 = 261.8.$$

Index Numbers of Production : No additional computational work is necessary for quantity index numbers.

$$L_q = \frac{\Sigma q_1 p_0}{\Sigma q_0 p_0} \times 100 = \frac{7}{15} \times 100 = 46.7$$

$$P_q = \frac{\Sigma q_1 p_1}{\Sigma q_0 p_1} \times 100 = \frac{18}{40} \times 100 = 45.0 \text{ etc.}$$

Ex. 14. Construct an appropriate index number for the following data :

Year	*Wheat*		*Rice*		*Sugar*	
	Price	*Quantity*	*Price*	*Quantity*	*Price*	*Quantity*
1961	5	30	6	20	4	10
1962	7	25	8	27	7	12

Ex. 15. Show that $L_q = 87.5$, $P_q = 88.2$ and $F_q = 87.8$ for the following data :

	1961		*1971*	
Commodity	*Quantity*	*Value*	*Quantity*	*Value*
X	4202	1954	3809	1973
Y	256	134	147	88
Z	45	2	23	1

Ex. 16. Show that for the following data *L*, *P*, *F*, *M* each equals 95.3.

	1949		*1958*	
Commodity	*Price*	*Quantity*	*Price*	*Quantity*
A	1.39	237	1.24	470
B	1.24	3238	1.15	3800
C	0.72	1220	0.65	1422
D	0.086	4077	0.097	4702
E	1.42	18.1	1.27	32.5
F	2.24	1098	2.23	1462

Ex. 17. The price relative for the year 1956 with 1958 as base is 62.5 while the price relative for the year 1957 with 1956 as base is 133.33. Show that the price relative for 1958 (*i*) with 1957 as base is 120 and (*ii*) with 1956-57 as base is 137.

Ex. 18. The price of a commodity in 1960 increased by 50% over that in 1950 and the production decreased by 30% during the same period. What is the percentage change in the value of money?

Ans. 5% increase.

Ex. 19. For the following data show that although the nominal per capita income is as shown the real per capita income from 1951 to 1958 is given by 360, 385, 417, 325, 227, 203, 203 respectively:

Year	*Per capita income*	*Cost of living index (1949 = 100)*
1951	360	100
1952	400	104
1953	480	115
1954	520	160
1955	550	210
1956	590	260
1957	610	300
1958	650	320

Ex. 20. Construct a full series with (*i*) base 1948 (*ii*) base 1952 from the following data :

Year	1951	1952	1953	1954	1955	1956	1957	1958
Index Numbers 1951 = 100	100	104	115	160	210	—	—	—
Index Numbers 1955 = 100	—	—	—	—	100	124	143	152

Ex. 21. The relative importance of eight commodities is food 348, clothing 97, rent 85, fuel 65, durables 71, entertainment 35, education 79, other 217. The increase in the prices of these commodities over a period of 10 years is 25, 22, 1, 18, 14, 11, 4 per cent respectively. What is the percentage increase in the general price level of these commodities? **Ans.** 115.37

28.10 TESTS OF INDEX NUMBER FORMULAE

There is no perfect formula for measuring changes over time and no index number can adequately represent all the changes taking place from time to time. It is difficult to lay down any definite rule to be followed in the construction of a formula for an ideal index number.

Fisher, however, has, suggested two tests, that should be met by a good index number formula. There is no mathematical basis of these tests of consistency and any conclusions drawn with their help about any formula need not be considered as adequate or final.

I. Time Reversal Test

The formula for the index number should be such that the product of the index number by another index number based on the same data with time interchanged should be equal to one.

If P_{01} is the price index number of the year 1 with base year 0 and P_{10} is the price index number of the year 0 with base year 1 then the test requires that

$P_{01} \times P_{10} = 1$ (ignoring multiplication by 100)

This test is not satisfied by L and P. Thus writing P_{01} for L in Laspeyre's formula, we have

$$P_{01} = \frac{\Sigma p_1 q_0}{\Sigma p_0 q_0}, \; P_{10} = \frac{\Sigma p_0 q_1}{\Sigma p_1 q_1}$$

$$P_{01} \times P_{10} = \frac{\Sigma p_1 q_0}{\Sigma p_0 q_0} \times \frac{\Sigma p_0 q_1}{\Sigma p_1 q_1} \neq 1$$

Laspeyre's formula does not satisfy the time reversal test. Paasche's formula also does not satisfy this test. Fisher's index number satisfies time reversal test. Writing now P_{01} for F,

$$P_{01} = \sqrt{\frac{\Sigma p_1 q_0}{\Sigma p_0 q_0} \times \frac{\Sigma p_1 q_1}{\Sigma p_0 q_1}}, \; P_{10} = \sqrt{\frac{\Sigma p_0 q_1}{\Sigma p_1 q_1} \times \frac{\Sigma p_0 q_0}{\Sigma p_1 q_0}}$$

$$P_{01} \times P_{10} = \sqrt{\frac{\Sigma p_1 q_0}{\Sigma p_0 q_0} \times \frac{\Sigma p_1 q_1}{\Sigma p_0 q_1} \times \frac{\Sigma p_0 q_1}{\Sigma p_1 q_1} \times \frac{\Sigma p_0 q_0}{\Sigma p_1 q_0}} = 1$$

Note. If for any index number $P_{01} \, P_{10} \neq 1$ it may be interpreted to mean that there is a *time bias* or error in index formula given by $B_T = P_{01} P_{10} - 1$.

2. Factor Reversal Test

The formula for the index number should be such that the product of the index number by another index number based on the same data with price and quantity interchanged should be equal to the ratio of aggregate value in the current to the aggregate value in the base year.

If P_{01} is the index number for year 1 with year 0 as base and Q_{01} is the index number for year 1 with year 0 as base (obtained by interchanging p's and q's) then the test requires that

$$P_{01} \times Q_{01} = \frac{\Sigma p_1 q_1}{\Sigma p_0 q_0} = \frac{\text{Value in current year}}{\text{Value in base year}} = V_{01}$$

The bias is given by $B_F = \dfrac{P_{01} Q_{01}}{V_{01}} - 1$

L and P fail in this test. Thus if we write P_{01} for L

$$P_{01} = \frac{\Sigma p_1 q_0}{\Sigma p_0 q_0}, \; Q_{01} = \frac{\Sigma q_1 p_0}{\Sigma q_0 p_0}$$

$$P_{01} \times Q_{01} = \frac{\Sigma p_1 q_0}{\Sigma p_0 q_0} \times \frac{\Sigma q_1 p_0}{\Sigma q_0 p_0} \neq \frac{\Sigma p_1 q_1}{\Sigma p_0 q_0}$$

The factor reversal test is satisfied by Fisher's formula :

$$P_{01} = \sqrt{\frac{\Sigma p_1 q_0}{\Sigma p_0 q_0} \cdot \frac{\Sigma p_1 q_1}{\Sigma p_0 q_1}}, \; Q_{01} = \sqrt{\frac{\Sigma q_1 p_0}{\Sigma q_0 p_0} \cdot \frac{\Sigma q_1 p_1}{\Sigma q_0 p_1}}$$

$$P_{01} Q_{01} = \sqrt{\frac{\Sigma p_1 q_0}{\Sigma p_0 q_0} \cdot \frac{\Sigma p_1 q_1}{\Sigma p_0 q_1} \cdot \frac{\Sigma q_1 p_0}{\Sigma q_0 p_0} \cdot \frac{\Sigma q_1 p_1}{\Sigma q_0 p_1}} = \sqrt{\frac{\Sigma p_1 q_1}{\Sigma p_0 q_0} \cdot \frac{\Sigma p_1 q_1}{\Sigma p_0 q_0}} = \frac{\Sigma p_1 q_1}{\Sigma p_0 q_0}$$

Fisher's index number is called an ideal index number because it satisfies both the time reversal and factor reversal tests.

3. A third test known as *circular test* is an extension of time reversal test. If

P_{01} is the price index of year 1 with base year 0,

P_{12} is the price index of year 2 with base year 1,

P_{20} is the price index of year 0 with base year 2, then

$$P_{01} \times P_{12} \times P_{20} = 1 \quad \text{(with 3 periods)}$$

In general if there are n periods

$$P_{01} \times P_{12} \times P_{23} \times \ldots \times P_{n0} = 1$$

This test is not satisfied by most index numbers.

Ex. 22. Which of the following index numbers satisfy the above tests?

(*i*) $\dfrac{\Sigma p_1}{\Sigma p_0}$ (*ii*) $\dfrac{\Sigma p_1 q_1}{\Sigma p_0 q_0}$ (*iii*) $\dfrac{\Sigma q_1}{\Sigma q_0}$

(*iv*) $\dfrac{\Sigma p_1 (q_0 + q_1)}{\Sigma p_0 (q_0 + q_1)}$ (*v*) $\dfrac{\Sigma p_1 \sqrt{q_0 q_1}}{\Sigma p_0 \sqrt{q_0 q_1}}$.

Other Tests of Index Number Formulae

Index numbers may be regarded as indicators of a change in a group of statistical series allowing us to split the total change into its components called sub-indices $I_1, I_2, \ldots, I_n$. The total index I is the product of the sub-indices:

$$I = I_1 \times I_2 \ldots \times I_n$$

4. *Test of Proportionality.* If all sub-indices are equal to A the total index equals A.

5. *Test of Definiteness.* If one of the sub-indices is zero or infinity, the total index must not become zero or infinity or indeterminate.

6. *Test of Commensurability.* If the unit of measurement is changed the value of the index must not change.

28.11 USES OF INDEX NUMBERS

(*i*) They reflect movements of prices, quantities and other variables over time.

(*ii*) They reflect simultaneous movements in the value of money and hence suggest the need to control inflation or depression as the case may be.

(*iii*) They indicate the rise and fall in real wages, whatever is suggested by money wages.

(*iv*) Index numbers of industrial activity show the progress of general industrialisation.

(*v*) Index numbers of business activity and investment show the conditions in the business and stock markets. They describe seasonal patterns in business.

(*vi*) Index numbers of employment, national income etc. show the movements in these variables.

Limitations of Index Numbers

(*i*) Index numbers are useful only when data can be expressed in quantitative terms.

(*ii*) An index number being based on samples and on some parts of data may not necessarily be truly representative of the actual state of affairs.

(*iii*) An index only roughly indicates relative changes. The accuracy cannot be ascertained.

(*iv*) If there is any inaccuracy in the primary data from which an index number is calculated it is not easy to locate it.

(*vi*) There being widely differing methods of index number construction, one may remain uncertain about the choice of method to be used.

(*vii*) Differences in units, in qualities and quantities of goods, in tastes, fashion, technology, habits etc. are not reflected properly by index numbers.

In spite of some limitations, an index number is a useful tool in the hands of the statistician and the economist.

Ex. 23. Explain the uses of index numbers. What is the procedure followed in the preparation of an index number?

Ex. 24. How is the cost of living index number prepared? What are its limitations?

Ex. 25. (*i*) The percentage increases in certain food prices and the weights attached to different constituents are shown below. Find the index number for food. (Calculations shown in the same table)

Food Item	*Percentage increase in price*	*Weights* w	*Index No.* I	Iw
Wheat	180	12	250	3000
Rice	60	5	160	800
Milk	100	3	200	600
Vegetables	20	6	120	720
Others	50	4	150	600
		30		5720

Note. A percentage increase of 150 means a total of 250 ($\because$ 100 + 150 = 250) which is the index number for wheat. Similarly for rice, 100 + 60 = 160 and so on.

$$\text{Food Index} = \frac{\Sigma Iw}{\Sigma w} = \frac{5720}{30} = 190.7$$

(*ii*) If clothing items, Rent and Miscellaneous items have index numbers given by 200, 180, 190 with weights 10, 5, 15, find the cost of living index number including food above.

Item	*Weight*	*Composite Index No.*	*Iw*
Food	70	190.7	13350
Clothing	10	200	2000
Rent	5	180	900
Miscellaneous	15	190	2850
	100		19100

$$\text{Cost of Living Index Number} = \frac{19100}{100} = 191$$

Ex. 26. A person spends Rs. 350 in a month on certain consumer items. The cost of living index in that month is 136, find the amounts he spent on items *A* and *B* using the following data

Item	*Expenditure*	*Index*
Food	140	180
A	x	150
B	y	100
Rent	56	110
Miscellaneous	63	80

Ans. $x = 42$, $y = 49$

Ex. 27. Write a note on Laspeyre's and Paasche's index numbers.

Ex. 28. What are the tests for an ideal index number?

Ex. 29. "Index numbers seek to set aside the irregularity of individual instances and replace it by the regularity of big numbers." Discuss.

Ex. 30. What is a weighted index number? Should weighting be left to chance?

Ex. 31. Find the cost of living index number for the following :

Items	*Weight*	*Group Index*
Food	58.5	91.4
Clothing	5.4	106.5
Find	6.2	102.2
Housing	9.6	100.0
Miscel.	20.3	100.3

An. 95.5

Ex. 32. (*i*) Calculate the real wage index numbers (RWI) for the following.

(*ii*) How much increase in money wages is needed to keep the real wages intact?

Year	Wages (Rs.)	Price Index	RWI	Increase needed
1957	200	100	100	0
1958	240	150	80	60
1959	350	200	87.5	50

1960	360	220	81.8	80
1961	360	230	78.3	100
1962	380	250	76.0	120
1963	400	250	80.0	100

Explanation (*i*) Real wage in any period is obtained by dividing the money wage by the index number of that period.

$$\text{With } 1957 = 100,\ I_{1958} = \frac{240}{150} \times \frac{100}{200} = 80$$

$$I_{1979} = \frac{350}{200} \times \frac{100}{200} = 87.5,\ I_{1960} = \frac{360}{120} \times \frac{100}{200} = 81.8 \text{ etc.}$$

(*ii*) Price index 1958 increases from 100 to 150

∴ Wages should increase from 200 to 300

∴ Deficit = 300 – 240 = 60.

Price index 1959 increases from 100 to 200

∴ Wages should increase from 200 to 400

Deficit = 400 – 350 = 50 and so on.

Ex. 33. The cost of living index number rose from 160 to 200. What should be the new wages with stability of a person getting Rs. 800? **Ans.** Rs. 1000

Ex. 34. (*a*) The average prices of rice, wheat, jowar are given below. Find the price relatives with 1964 as base and get the composite index numbers.

Year	*Rice*	*Wheat*	*Jawar*	(prices in Rs./Quintal)
1962	150	82	80	
1963	175	72	72	
				PR base 1964
1964	200	90	80	**Ans.** 88.3, 85.8, 100.0, 113.3, 99.7
1965	220	99	96	
1966	198	81	88	

(*b*) In 1966, the average price of a commodity was 20% more than in 1965 and 20% less than in 1964. It was also 50% more than in 1967. Reduce the data to price relatives with 1965 as base.

If $R_{1965} = 100$ then $PR_{1966} = 120$ (20% more) = 80% of PR of 1964

$$\therefore\ PR \text{ of } 1964 = 120 \times \frac{100}{800} = 150$$

$$PR_{1976} = 150\% \text{ of } PR_{1967}$$

$$\therefore \; PR_{1967} = \frac{120 \times 100}{150} = 80$$

Price Relatives are **Ans.** $\begin{cases} 1964 & 150 \\ 1965 & 100 \\ 1966 & 120 \\ 1967 & 80 \end{cases}$

Ex. 35. From the following data

	Rice		Wheat		Jawar	
	qty.	*Price*	*qty.*	*price*	*qty.*	*price*
1960	10	9	11	6	5	5
1970	9	3	10	4	3	3

Check if Marshall Edgeworth and Fisher Index Numbers are nearly the same.

CHAPTER 29

Time Series

29.1 REGULARITY AND ORDER IN DATA

A time series is a sequence of observations on a variable made at regular points or intervals of time and arranged in chronological order. The importance of regularity and order in which the observations are made is explicitly recognised.

Economic and business activities provide us with various examples of time series. Prices, production, employment and national income data, noted at regular intervals of time, are familiar cases of time series.

When we consider situations which exist in a theoretical state of equilibrium *i.e.* when time is of no importance we have a static analysis. Systems trying to reach their equilibrium over time permit us to carry out a dynamic analysis of variations over time. Time series analysis enables us to determine useful patterns in the data with the help of statistical tools.

Ex. 1. Price(y) of wheat per quintal 1921—1945

Year	*Price*	*Year*	*Price*	*Year*	*Price*	*Year*	*Price*	*Year*	*Price*
1921	3	1926	5	1931	5	1936	10	1941	9
1922	4	1927	8	1932	8	1937	12	1942	10
1923	9	1928	10	1933	12	1938	14	1943	14
1924	8	1929	4	1934	10	1919	12	1944	16
1925	7	1930	2	1935	7	1940	11	1945	12

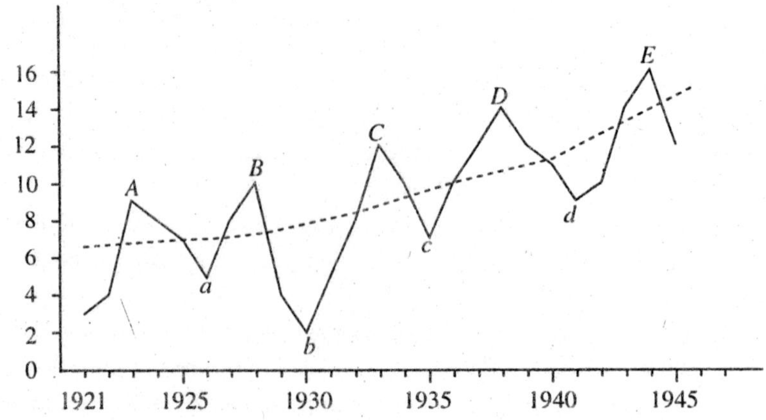

The observations $y_1, y_2, y_3, \ldots,$ forming a time series have been made at equidistant intervals of time. The variable y_t (standing for price of wheat in period t) may be continuous or discontinuous over time.

It is convenient to represent a time series graphically either by joining the points by straight lines or by a smooth curve. In either case the major characteristics of the time series become evident.

A time series moves in a fairly regular manner with time. The observations are highly time dependent and reflect the economic history of the variable under consideration. The dotted curve in the diagram shows the approximate trend of the data.

29.2 TIME SERIES AND FREQUENCY DISTRIBUTIONS

Frequency distributions are subject only to random factors while time series are subject to some non-random factors also.

In frequency distributions the observations are statistically independent. In time series they may be dependent and the dependency may depend on the position of the observation in the sequence at any time. A probabilistic mechanism producing a time series cannot be ruled out.

Functional Relation

We may represent a time series by a functional relation of the type

$$y_t = f(t)$$

which shows the value of the variable y_t at time t.

If, at times $t_1, t_2, \ldots, t_k$, the variable y has values

$$y_1, y_2, \ldots, y_k$$

then these values constitute a time series. The time interval may be a day, a year or any other period.

The fluctuations in the time series are usually due to a number of forces. We call them *components* of the time series.

29.3 OBJECTIVES AND USES OF TIME SERIES ANALYSIS

A time series is the result of the simultaneous interaction of various forces, some recurring and others nonrecurring at intervals. Many theories have been put forth to explain various types of fluctuations, particularly trade cycles, arising in economics and other sciences. An analysis of time series can test the validity of these theories.

In economics, an explanation of the forces of inflation and depression is sought to be given with the help of time series analysis.

The analysis can be used to forecast the probable economic events in the near future based on past knowledge. Budgetting and planning of economic activities depend considerably on accurate economic forecasts.

The analysis can be used to control the process producing the series. For instance, it may suggest methods to check inflationary forces before they raise their heads.

Time series analysis can help in understanding the mechanism generating the series. It can be useful in the description of the features of an economy.

29.4 DECOMPOSITION OF TIME SERIES : FOUR COMPONENTS

Time series may be affected by factors which are

(*i*) recurring or nonrecurring

(*ii*) periodic or random

There are four major types of variation or *components* of a time series :

(*i*) Secular trend T

(*ii*) Cyclical variations C

(*iii*) Other periodic changes, particularly seasonal changes S

(*iv*) Random or irregular movements R

All these four components, jointly or severally, are found in a time series.

We may *assume* a multiplicative or additive relationship among these components. In the first case, the observed value (y_t) at any time is assumed to be the resultant *product* of the play of all the four forces:

$$y_t = T \times C \times S \times R$$

In the second case, it is the resultant *sum* of the impact of the four forces

$$y_t = T + C + S + R$$

The multiplicative hypothesis amounts to a log-additive relation. Since

$$y_t = T.C.S.R$$

$$\therefore \quad \ln y_t = \ln T + \ln C + \ln S + \ln R$$

Note. Different components are not necessarily independent of one another.

Mixed additive or multiplicative combinations of different components give rise to alternative models.

The objective of our analysis is a breakdown of the series into its major components. The magnitude of each of these components will show the make up of the time series.

T : Secular Trend

The trend is the general, smooth, long term movement of the series. It is a gradual movement over the whole period showing a persistent growth, decline, or any other long run tendency.

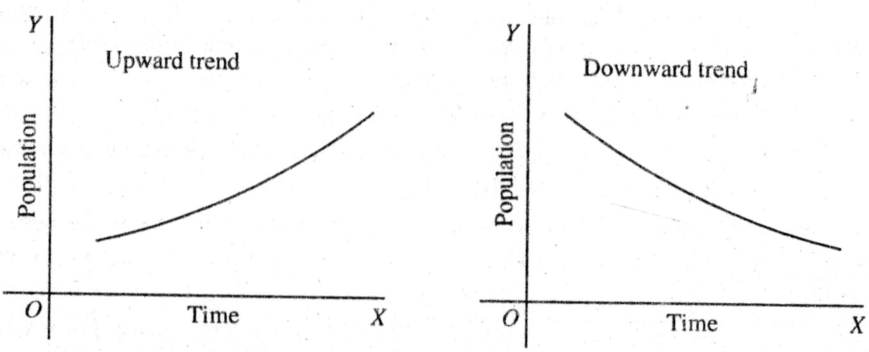

The trend ignores sudden or short term fluctuations. It moves steadily in a certain direction for a fairly long period of time. The term long period is relative. One year or even a month may be a long period in some cases. In other cases, as in agriculture, even 50 years may be inadequate sometimes. The movement may be slow or fast.

Population trends often show an upward movement. Death rates may show a downward trend.

It is possible that upward and downward movements as well as stabilty may be noticed in different parts of a trend.

Time trend is relative to the period of reference. Changes in population, technology, capital formation and other factors give a shape to the trend. Social, economic and political institutions are the basic elements underlying the trend.

C: Cyclical Variations

These are long term movements which represent constantly recurring rises and falls in the variable under consideration. Fluctuations over a period of one or more years at nearly regular intervals in economic activity give rise to business cycles which show alternating periods of depression and prosperity in the economy. Cycles of economic variables like wages, employment, prices, production etc. are the result of interaction between and the cumulative act of political, economic, social and other forces.

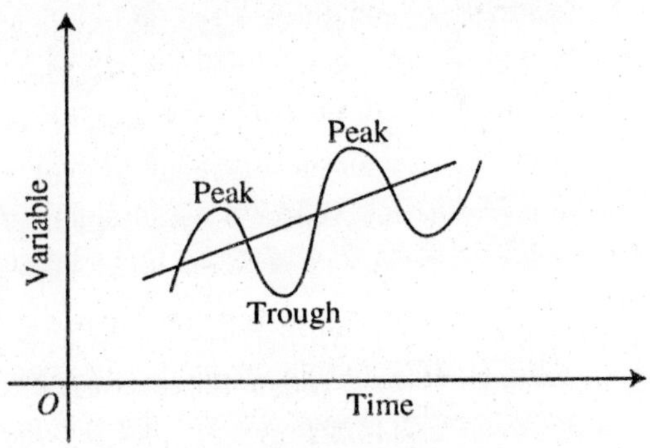

Peaks and throughs are turning points

The amplitude and the period of the oscillating cycle may not be very regular but a perspective view of the graph of a time series in general shows a more or less regular pattern of cycles recurring roughly every three, four or more years and sometimes even 18 years.

An observation y_t of a time series is a peak if its value is greater than the immediately preceding (y_{t-1}) and the immediately following (y_{t+1}) observations.

$$\text{For a peak } y_{t-1} < y_t > y_{t+1}$$

$$\text{for a trough } y_{t-1} > y_t < y_{t+1}$$

Phase : The interval between two successive turning points is a phase of the time series.

Period : The interval between two successive peaks (or troughs) is the period of oscillation. One complete period gives rise to a cycle which is a four phase cycle made of, for instance, a boom, a recession, a depression and recovery in a business cycle.

S : Seasonal Changes

These are variations that occur regularly at specific intervals of time. Seasonal movements are cycles of relatively short duration, mostly one year, and related to seasonal factors and customs.

Every year seasons follow practically the same pattern and seasonal variations follow the seasons. So do the customs and other rhythmic forces.

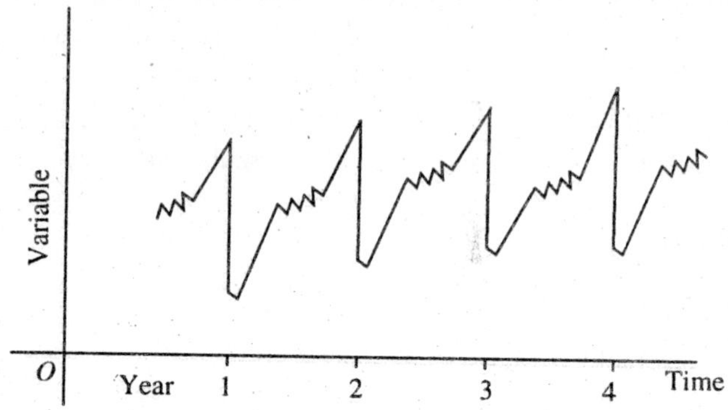

Rainfall, temperatures, eggs laid by hens, harvests, certain types of investments, etc. are examples showing seasonal variations.

Seasonal variations occur due to *natural* forces.

They may also occur as a result of man made customs, rites, festivals and conventions.

In a time series, seasonal variations may sometimes be misinterpreted to mean other types of variations. For a better understanding of the components of a time series, the data should be corrected for seasonal variation *i.e.* the series should be deseasonalised by writing y_t/S_t or $y_t - S_t$ (depending on the assumption of multiplication or additivity of components). Here S_t stands for seasonal effect.

Seasonal variations are considered either for their own sake or for being removed, thus allowing us to concentrate on the remaining variation. It is difficult to make a study of business cycles without removing seasonal variations. Planning in business and elsewhere depends much on seasonal factors. Production or purchase of certain commodities takes place according to the nature of the season as in the case of woollens in winter or umbrellas in the rainy season.

R : Random movements

These occur in a completely unpredictable manner. They do not follow any definite pattern like the trend, cyclical or seasonal movements. Irregular variations are strictly random movements which keep on operating all the time. Episodic or catastrophic variations occur from time to time in the form of strikes, wars, floods, political events etc.

The minute to minute changes in the prices of stocks give us a good example of random movements. Random changes have been subjects of intensive studies.

29.5 TIME SERIES ANALYSIS

Preliminaries : The object of time series analysis is to discover the magnitude and direction of (*i*) the trend that may exist in the series. (*ii*) the nature and amplitude of cycles, (*iii*) the effect of seasonal change and (*iv*) the size of random movements.

Before proceeding with an analysis, often, corrections in the time series for changes in various factors like population, prices etc. are made.

The four components of a time series are not necessarily mutually independent. Any one of them may be under the influence of one or more of the remaining forces thus complicating the work of analysis.

Sometimes one or more of these components may not be recognised owing to a mix up. Lack of regularity in the components may be highly confusing.

29.6 THE TREND

A study of the trend may be useful in itself and also in the separation of deviations of periodic movements. It may be useful for comparison with the past and present trends and particularly helpful in having a peep into the probable future *i.e.*, forecasting.

The trend may be determined by the following methods :

(*i*) Free hand drawing.

(*ii*) The method of moving averages.

(*iii*) High and low midpoints method.

(*iv*) The method of semi-averages.

(*v*) The method of mathematical functions.

(*i*) Freehand drawing

To get the general nature of the trend in a time series, a smooth curve may be drawn through the graph of the series.

The method is simple, quick and useful in determining any kind of trend, linear or otherwise. But even an expert with a sound judgement may not be able to give an exact graphical trend for any data. The result is rough, approximate and subjective. See Ex. 1.

(*ii*) The method of moving averages

The method of moving averages consists in finding the average of a certain number of terms of the time series and taking this average as the trend value for the middle of the period covered in the calculation of the average which is the period of the moving average. Successive moving averages are calculated for the data by omitting every time the beginning observation and adding the one immediately after the last.

An odd number for the period of the moving average is preferred as it is easier to centre it. The moving average of an even number period cannot be centred and in such a case it is recentred between pairs. The procedure will be clear from Ex. 2.

A moving average is useful in smoothing out periodic fluctuations of the cyclical type, particularly when the period of the moving average is carefully selected.

The longer the period of the moving average the smoother is the trend likely to be provided the period coincides with the period of the cycle.

The method of moving averages can also be used in the context of seasonal variations where the period is 12 months or 4 seasons or some other division of the year.

It may be noted that the method of moving averages has a descriptive rather than an analytical value.

Ex. 2. Same data as for Ex. 1.

Year	*Price*	*3 year moving average*
1921	3	
1922	6	6.0
1923	9	7.7
1924	8	8.0
1925	7	6.7
1926	5	6.7
1927	8	7.7
1928	10	7.3
1929	4	5.3
1930	2	3.7
1931	5	5.0
1932	8	8.3
1933	12	10.0
1934	10	9.7
1935	7	9.0
1936	10	9.7
1937	12	12.0
1938	14	12.7
1939	12	12.3
1940	11	10.7
1941	9	10.0
1942	10	11.3
1943	15	13.7
1944	16	14.3
1945	12	

Explanation

$$1921-23 : \frac{3+6+9}{3}=\frac{18}{3}=6.0$$

placed opposite 1922

$$1922-25 : \frac{6+9+8}{3}=\frac{23}{3}=7.7$$

placed opposite 1923

$$1923-25 : \frac{9+8+7}{3}=\frac{24}{3}=8.0$$

placed opposite 1924

$$1924-26 : \frac{8+7+5}{3}=\frac{20}{3}=6.7$$

placed opposite 1925 and so on.

Each moving average thus calculated is centered by placing it opposite the middle of the period, *e.g.* 6.0 against 1922 which is the middle of 1921—1923; 7.7 against 1923 and so on.

Note that the first and the last years of the data have no moving average.

The moving average series is smooth and more or less regular. It has been shown graphically as a dotted curve in the diagram for Ex. 1.

Period of the Moving Average

If the period of the moving average is equal to the period of the cycle or is a multiple of it, the smoothing is perfect and we have a straight line trend. Otherwise, the smoothing will not be good.

If the period of the cycle is not fixed because of fluctuations, the trend will be a curve also marked by fluctuations though of a smaller magnitude.

If the original series is without any fluctuations the trend will reproduce the original series.

The moving average may remove the curvature if any, in the original series.

The moving average cannot remove such fluctuations in a series as are irregular throughout. A moving average with a longer period can remove these fluctuations to a greater extent.

A shorter period is more useful in averaging out the cycles though, in general, the period must be selected on the basis of the actual data and taking into account all relevant considerations.

The method of moving averages cannot be applied with success to every series but when it is successfully applied the trend may be stretched free hand on either side for prediction purposes.

The procedure to calculate the moving average for a *period* with *even* number of years is given below.

Ex. 3. Consider a part of the above data for a 4-year moving average.

Year	*Price*	*4 year moving average*	*Centred 4 year moving average*
1921	3		
1922	6	6.5	
1923	9		7.0
		7.5	
1924	8		7.4
		7.25	
1925	7		7.1
		7.0	
1926	5	7.5	7.3
1927	8		7.1
		6.75	
1928	10		6.4
		6.0	
1929	4	5.25	5.6
1930	2	4.75	5.0
1931	5		
1932	8		

Explanation

Column 3

1921—24 : $\frac{3+6+9+8}{4}=\frac{26}{4}=6.5$

inserted between 1922 and 1923,

1922, – 25 : $\frac{6+9+8+7}{4}=\frac{30}{4}=7.5$

inserted between 1923 and 1924 etc.

Column 4

These are averages of pairs :

$$\frac{6.5+7.5}{2}=7.5,$$

put against 1923.

$$\frac{7.5+7.25}{2}=7.4$$

put against 1924 etc.

The first two and the last two moving averages do not exist.

Ex. 4. Determine a (*i*) 5 year (*ii*) 6 year moving average trend for the data of Ex. 1.

Short Term Fluctuations

The deviations of original value in the time series from trend values are a combination of the remaining components *viz.* cyclical, seasonal, and irregular fluctuations. These deviations are called short term fluctuations.

Ex. 5. Short term fluctuations with a 4 year M.A.

Year	*Production* y_t	*4 yr M.A.*	*Centred 4 yr M.A.*	*Short term fluctuations*
1971	77			
1972	88			
		86.0		
1973	94		87.75	+ 6.25
		89.5		
1974	85		90.75	– 5.75
		92.0		
1975	91		91.50	– 0.50
		91.0		
1976	98			
1977	90			

Note. 94—87.75 = 6.25

85—90.75 = – 5.75

91—91.50 = – 50

Ex. 6. Short term fluctuations with a 3 year of M.A.

Year	*Price*	*3 year moving average*	*Short term fluctuations*
1921	3		
1922	6	6.0	0
1923	9	7.7	1.3
1924	8	8.0	0
1925	7	6.7	0.3
1926	5	6.7	–1.7
1927	8		

Ex. 7. For the following data, fit a 4 year M.A. and find short term fluctuations

1971	1972	1973	1974	1975	1976	1977	1978
238	252	251	279	318	361	439	548

Ex. 8. Obtain the trend of sale in the following data by a 5 year moving average and calculate the short term fluctuations :

Year	*Sales*	*Year*	*Sales*	*Year*	*Sales*
1941	120	1948	118	1955	130
1942	122	1949	125	1956	131
1943	124	1950	126	1957	135
1944	125	1951	129	1958	130
1945	120	1952	124	1959	129
1946	120	1953	123	1960	140
1947	119	1954	122	1961	140

(*iii*) **The high low midpoints method**

In this method we connect by straight lines the high points of each cycle and determine by interpolation the value on these lines for each year of the times series. The same is done with the low points. The average of each pair of high and the corresponding low points is obtained by taking the midpoints. The trend is represented by the lines joining these midpoints. In Ex. 1, join *AB*, *BC*, *CD* etc. Similarly join *ab*, *bc*, *cd*, etc.

Any other points besides the high and low points may also be taken.

(*iv*) **The method of semi-averages**

The time series is divided into two (or more) convenient, preferably equal parts for each of which an average is found. The line joining the points corresponding to these averages gives the trend. The method should be applied only when the trend is linear or approximately linear.

If there is an even number of years, say 10 years from 1961 to 1970, the two parts will be made of 1961 – 65 and 1966 – 70.

If there is an odd number of years as in 1961 – 71 omit the middle year 1966. The two parts are 1961 – 65, 1967 – 71.

Compute the arithmetic mean of each point and plot against the central year of each part. The line joining the two points is the required trend. The linearity of trend is an assumption which may not necessarily hold in practice.

Ex. 9. Year	1961	1962	1963	1964	1965	1966	1967	1968
Production	380	400	650	720	690	600	870	930

Average of first 4 yrs. (1961–64) $\bar{y}_1 = \dfrac{2150}{4} = 537.5$

Average of the last 4 years (1965–68) $\bar{y}_2 = \dfrac{3090}{4} = 772.5$

Plot 537.5 against July 1962. Plot 772.5 against July 1966. The join of $\bar{y}_1$, $\bar{y}_2$ is shown by a dotted line which is the trend determined by the method of semi averages.

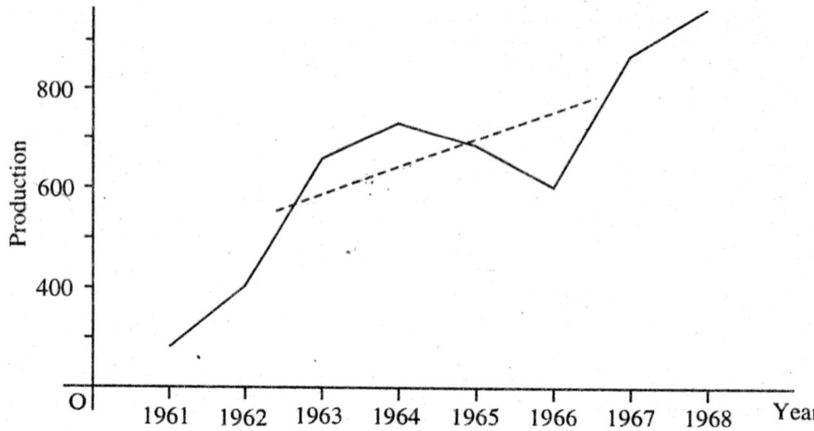

(v) The method of mathematical functions

A mathematical function describes a trend as well as shows the law of development of a time series. A mathematical function is very useful for prediction purposes though the selection of a proper function can be difficult.

Several simple functions are available. One of the simplest is a straight line function : $y = a + bx$ where time is represented by x which is the independent variable and y is the dependent variable. The values of the parameters a, b are determined by the principle of *least squares* as explained in the chapter on regression. The procedure for a fictitious data set of 5 years is shown below.

Ex. 10.

Year t	*Production* y	*Time Deviation* $x = t - 1963$	x^2	xy	*Trend values*	*Short term fluctuation*
1961	6	– 2	4	– 10	5.4	0.6
1962	7	– 1	1	– 7	6.7	0.3
1963	10	0	0	0	8.0	2.0
1964	6	1	1	6	9.3	– 3.3
1965	12	2	4	24	10.6	1.4

$n = 5$, $\Sigma y = 40$ $\Sigma x = 0$ $\Sigma x^2 = 10$, $\Sigma xy = 13$

(Calculations shown below)

$$\bar{x} = \frac{\Sigma x}{n} = \frac{0}{5} = 0 \qquad \bar{y} = \frac{\Sigma y}{n} = \frac{40}{5} = 8$$

$$b = \frac{\Sigma xy - n\bar{x}\bar{y}}{\Sigma x^2 - n\bar{x}^2} = \frac{13 - 5 \times 0 \times 8}{10 - 5 \times 0} = \frac{13}{10} = 1.3$$

$$a = \bar{y} - b\bar{x} = 8 - 1.3 \times 0 = 8$$

The trend is given by the regression of y on x, estimated by the principle of least squares :

$$y = 8 + 1.3x$$

To compute the *trend values* substitute the value of x corresponding to years

1961	$x = -2$	$y = 8 + 1.3\,(-2) = 5.4$
1962	$x = -1$	$y = 8 + 1.3\,(-1) = 6.7$
1963	$x = 0$	$y = 8 + 1.3\,(0) = 8.0$
1964	$x = 1$	$y = 8 + 1.3\,(1) = 9.3$
1965	$x = 2$	$y = 8 + 1.3\,(2) = 10.6$

The computed trend values and short term fluctuations obtained by subtracting the trend values from the observed values are shown in separate columns.

Can we forecast the future trend?

We *can* forecast the trend though for a not very remote period.

Thus for 1966, $x = 3$ and $y = 8 + 1.3\,(3) = 11.9$

for 1970, $x = 7$ and $y = 8 + 1.3\,(7) = 17.1$ etc.

In fitting a straight line trend to time series data in the above problem we have taken time deviations $x = t - 1963$. This is done for convenience in calculations.

The rule that should be followed is that when the number of years is odd, take the middle year as the origin, like 1963 here so that the x – values are $-2, -1, 0, 1, 2$. This results in $\sum x = 0$.

If the number of years is even, say 6, the origin should be taken at the middle of the middle two years *i.e.*, the x values are $-5, -3, -1, 1, 3, 5$. This also results in $\sum x = 0$.

Nonlinear trends. A linear trend may not always be a good fit to the data. Often the rise and fall in the data may be better expressed by a parabolic trend of the type :

$$y = a + bx + cx^2$$

which is a second degree curve. We can find the values of a, b, c by solving the normal equations obtained by the principle of least squares:

$$\sum y = na + b\sum x + c\sum x^2$$

$$\sum xy = a\sum x + b\sum x^2 + c\sum x^3$$

$$\sum x^2 y = a\sum x^2 + b\sum x^2 + c\sum x_4$$

Usually a straight line trend is fitted to the data. If after plotting the data we notice the presence of a second degree, log-linear or any other trend we can still use the method of least squares which (*a*) is free from subjective bias (*b*) provides annual rate of growth in a linear trend and (*c*) can help predict the values of the variable for any period in future or over the time series.

Note. This method ignores the presence of cyclical, seasonal and random fluctuations in the data and is based on long term variations in the trend.

It is sensitive to any change in the data.

If we have taken the time deviations according to the rule given above *i.e.* if $x = t -$ (middle of all years) then $\sum x = 0$, $\sum x^3 = 0$.

These normal equations take the simple form.

$$\Sigma y = na + c\Sigma x^2$$

$$\Sigma xy = b\Sigma x^2 \quad \text{or} \quad b = \frac{\Sigma xy}{\Sigma x^2}$$

$$\Sigma x^2 y = a\Sigma x^2 + c\Sigma x^4$$

The first and third equations give the values of a, c.

Many other types of trends can be fitted to time series data depending on the pattern of their movement. Nonlinear trends can usually be reduced to linear forms by suitable transformations for convenience in calculations.

Ex. 11. Fit a second degree curve to the data in Ex. 8 and Ex. 10.

Ex. 12. The following are the annual profits in thousands of rupees in a certain business :

Year	1951	1952	1953	1954	1955	1956	1957
Profit	63	72	75	65	80	85	95

(*i*) Use the method of least squares to fit a straight line trend and make an estimate of the profits in 1958. **Ans.** $y = 76 + 4.86x$

(*ii*) Calculate the short term oscillations for the above data.

(*iii*) Fit a parabolic trend.

29.7 CYCLICAL EFFECTS

A study of cyclical effects in a time series is useful for forecasting, for contra-cyclical measures and for a study of forces affecting the time series.

There are various method of determining the cyclical effects.

1. *The Graphical Method*

It consists of directly plotting the raw data of the time series on a graph paper and drawing and comparing the cycles for several years. The seasonal fluctuations are more or less eliminated in the process.

2. *The Residual Method*

After eliminating the trend (T) and seasonal fluctuations (S) the residual is the sum of cyclical (C) and irregular fluctuations (R).

The residual is likely to contain a part of T and S for which suitable precaution may be taken. The random fluctuations may be approximately eliminated by the method of moving averages. The resulting cyclical effect C is a very approximate quantity from which we may calculate the amplitude of the cyclical movement.

3. *The Method of Harmonic or Periodogram Analysis*

The method takes a time series (from the point of view of a cycle) as made up of the sum of a number of periodic functions, *i.e.* a number of pure harmonic movements.

An alternative is to consider the time series as an oscillatory system incorporating a continuous stream of external shocks influencing both the period and the amplitude of the periodic functions.

Comparison of cyclical movements

A comparison of cyclical movements of time series is useful for economic forecasting.

How to compare cycles with different amplitudes?

This may be done by taking different scales on the graph paper for different series. Thus a series with an amplitude double that of another may be drawn with a scale twice that of the other. Alternatively deviations from the trend in units of the standard deviations may be taken for each series.

29.8 BUSINESS CYCLES

These are important cyclical movements to be found in market economies which are based on free enterprise.

The aggregative economic activity in such economies is often found to follow almost a regular pattern of oscillations, A period of prosperity or expansion is followed by a period of depression or contraction and then again a revival into the expansion phase of the next cycle. The period, anywhere between two to about fourteen years, is not regular for this recurrent sequence of changes.

Business cycles are not true periodicities. As a matter of fact, each cycle is a historical phenomenon in itself differing from others in various respects.

Each cycle is a product of the interaction of various endogeneous forces like investments, interest rates, bank credit etc. and exogenous forces like wars, innovations, population changes etc. A study of these cycles involves the study of all these forces and contracyclical economic measures may be based on their study.

A cycle with a period between two or three years is called a *kitchen* cycle in economics.

A cycle of ten to twelve years is called a *Jugular* cycle.

The length of a *Kondratieff* cycle may be about fifty years.

Cycles in general business activity are called *reference* cycles.

A full cycle is measured between successive troughs (or peaks).

AB : Expansion Phase : It is the phase between the initial trough and the following peak.

BC : Contraction Phase : It is the phase between a peak and the following trough.

The extent of conformity of any given cycle to a reference cycle can be estimated.

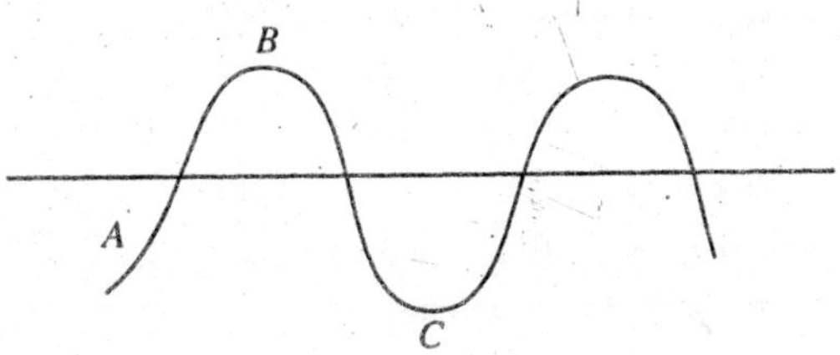

It may be noted that the turning points in a business cycle do not affect all industries at the same time. A recovery or recession begins first in some sectors of the economy, then spreads with a cumulative effect.

29.9 SEASONAL EFFECTS

Seasonal movements recur at regular intervals in a more or less constant pattern. Seasonal patterns are true periodicities noticed in most economic time series.

The presence of seasonal effects should be first verified by plotting the data on a graph paper and after a clear evidence is obtained the following methods may be used to measure seasonal variations.

1. Ratios to monthly averages.
2. Ratios to 12-month moving averages.
3. Ratios to trend values.
4. Link relatives.

It may be noted that any seasonal pattern is valid only for the period under consideration and may not be reliable beyond it. Changes in the seasonal pattern being frequent, a regular check and revision are helpful.

The removal of seasonal effect from the data, *i.e.*, deseasonalisation consists in dividing each observation in the original data by the appropriate seasonal index.

1. Ratios to monthly averages

The monthly average for each given month is calculated and then written as a percentage of the average of all the monthly averages, *e.g.*

$$\text{Seasonal variation index for January} = \frac{\text{Monthly average for January}}{\text{Average of monthly averages}} \times 100$$

Ex. 9. The method has been illustrated for the following data on monthly exports.

Months	*Exports 1941*	*(Rs. 1942*	*Crores) 1943*	*3 Year average for each month*	*Seasonal Variation Index*
Jan.	6	6	5	5.7	71.25
Feb.	2	3	4	3.0	37.5
Mar.	2	2	4	2.7	33.8
Apr.	6	7	8	7.0	87.5
May	5	8	10	7.7	96.3
Jun.	10	12	14	12.0	150.0
Jul.	8	9	10	9.0	111.1
Aug.	9	9	12	10.0	125.0
Sep.	6	7	10	7.7	96.3
Oct.	9	8	12	9.7	121.3
Nov.	6	9	14	9.7	121.3
Dec.	10	11	14	11.7	146.3
			Total	95.9	
			Average	8.0	

Explanation

The last column shows percentages for months

Jan. $\frac{5.7}{8.0} \times 100 = 71.25$

Feb. $\frac{3.0}{8.0} \times 100 = 37.5$ etc.

The procedure would be similar if weekly or quarterly instead of monthly data were available

2. Ratios to 12-month moving averages

Seasonal Variations. The data may be given for seasons or months. If the data are monthwise, a 12-month moving average may be found and each term subtracted from the corresponding value of the original data. The averages of these differences taken monthwise give seasonal variations.

Indices of Seasonal Variations. In this case, the percentages of the original values with reference to the moving averages are to be found and the average percentage of each month expressed as a percentage of the total of all the months gives the index of seasonal variations.

This method removes the influence of the trend and cyclical movements provided the periodicity and amplitude of the original data are more or less constant.

Ex. 13. Find the seasonal variation index for each quarter for the following data on value added (in Rs. millions) of a firm. Data given on quarterly basis.

Year \ *Quarter*	*I*	*II*	*III*	*IV*
1967	8	11	9	16
1968	10	14	12	16
1969	12	15	10	16
1970	10	12	11	19
4 yr. AV.	10.0	13.0	10.5	16.5

Note. $\frac{10}{12.5} \times 100 = 80$

$\frac{13}{12.5} \times 100 = 104$

$\frac{10.5}{12.5} \times 100 = 84$

$\frac{16.5}{12.5} \times 100 = 132$

Mean of 4 years = 12.5

Seasonal variation Index	80	104	84	132

Seasonal index for Quarter I 80
Quarter II 104
Quarter III 84
Quarter IV 132

Ex. 14. The method of seasonal variation indices has been illustrated below by taking seasonal (quarterly) data so that instead of a 12-month moving average we will find 4 quarterly moving averages.

Quarter	(*a*) *Sales* (*000*)	(*b*) *4 quarter moving average*	(*c*) *Moving average centred*	(*d*) *Short term flucua-tions* (*a*)—(*c*)	(*e*) *Average seasonal variation* (*see table below*)	(*f*) *Residual variations* (*d*)—(*e*)
1961 I	2					
II	3					
		3.5				
III	4		3.5	0.5	0.9	– 0.4
		3.5				
IV	5		3.6	1.4	1.9	– 0.5
		3.8				
1962 I	2		3.9	– 1.9	– 2.2	0.3
		4.0				
II	4		4.1	– 0.1	– 0.4	0.3
		4.3				
III	5		4.3	0.7	0.9	– 0.2
IV	6		4.3	2.3	1.9	0.4
		4.3				
1963 I	2		4.4	– 2.4	– 2.2	– 0.2
II	4		4.5	– 0.5	– 0.4	– 0.1
		4.5				
III	6		4.6	1.4	0.9	0.5
		4.8				
IV	7		5.0	2.0	1.9	0.1
		5.3				
1964 I	3		5.4	– 2.4	– 2.2	– 0.2
		5.5				
II	5		5.6	– 0.6	– 0.4	– 0.2
		5.8				
III	7					
IV	8					

Computation of Seasonal Variation

Yr. / Quarter	Seasonal Variation Deviation from trend (d) 1961	1962	1963	1964	Average seasonal variations (e)	Explanation : Last column (e) transferred to table above
I	—	–1.9	–2.4	–2.4	–2.2	I : $\frac{-1.9-2.4-2.4}{3} = -2.2$
II	—	–0.1	–0.5	–0.6	– 0.4	II: $\frac{0.5+0.7+1.4}{3} = 0.2$
III	0.5	0.7	1.4	—	0.9	
IV	1.4	2.3	2.0	—	1.9	etc.

If instead of absolute value in Col. (*d*) we take percentages of actual value to trend value (*e.g.* 1961 III : $\frac{4}{3.5} \times 100 = 114.3$ etc.) and prepare a table of these percentages we get seasonal variation indices. (Ratio to 4-quarter moving averages).

The ratios to 12 month moving averages can be found on the same lines if monthly data are given.

	Quarter	Sales (*a*)	M.A. (*b*)	Ratio = $\frac{(a)}{(c)}$
1961	I	2		
	II	3		
	III	4	3.5	114.3
	IV	5	3.6	138.9
1962	I	2	3.9	51.3
	II	4	4.1	97.5
	III	5	4.3	116.3
	IV	6	4.3	141.4
1963	I	2	4.4	45.5
	II	4	4.5	88.9
	III	6	4.6	130.4
	IV	7	5.0	140.0
1964	I	3	54	55.6
	II	5	54	55.6
	III	7	5.6	89.2
	IV	8		

Computation of Seasonal Indices

Qr. \ Yr.	1961	1962	1963	1964
I		51.3	45.5	55.6
II		97.5	88.9	89.2
III	114.3	116.3	130.4	
IV	138.9	141.4	140.0	

3. Ratios to trend values

The methods given above compare the monthly data with 12-month averages or moving averages. Instead we may compare them with trend values and average out over several years the resulting ratios, thus getting seasonal indices.

The trend values may also be obtained by fitting a curve by the principle of least squares and expressing each original observation as a percentage of the corresponding trend value.

The resulting series contains cyclical and random components which can be more or less eliminated by averaging the series for different months or different quarters.

The final series is obtained by multiplying the results from the previous step by 1200/total of indices (for monthly data) or by 400/total of indices (for quarterly data). The ratio to trend method gives results for each month of the data without loss of observations.

The ratio to trend method is inferior to the method of moving averages if cyclical swings are strong. In that case deviations from moving averages provide us with better seasonal variations than those from trend values.

4. Link relatives

The value for the observation in any month expressed as a percentage of the preceding month gives us the link relative for the month. For each month, the link relatives, thus obtained, are averaged out. Each such monthly average is then linked to the average of the preceding month by successive multiplication. The results are adjusted for trend by deduction and the preliminary seasonal indices thus obtained are adjusted further by making their sum total 1,200. This give us the seasonal indices.

The method is complicated. The seasonal indices calculated by this method may include cyclical and trend effects. But they are not likely to be much influenced by irregular fluctuations of more than one month duration. An advantage in this method is that the monthly data are utilised more completely.

29.10 IRREGULAR FLUCTUATIONS

If we succeed in separating the trend, cyclical and seasonal effects, the residue left may consist of irregular or random movements. It is very difficult to isolate them and often they are left associated with cyclical fluctuations after the trend and seasonal effects are removed.

It may be be noted that the random effect can be written in the multiplicative relationship

$$R = \frac{y_t}{T \times C \times S},$$

and in the additive relationship

$$R = y_t - T - C - S.$$

Ex. 15. For the following data showing average wages of some workers during 1949—1960 find the trend by (*i*) graphical method, (*ii*) 3-year moving averages, (*iii*) least squares method.

Find the trend values for the years listed.

Year	1949	1950	1951	1952	1953	1954	1955	1956	1957	1958	1959	1960
Wages	100	120	130	140	160	170	180	200	210	220	230	250

Ex. 16. Find the seasonal variations and their indices for the following data showing sales of a commodity, the amounts being in thousands of rupees.

Month	1951	1952	1953	1954
Jan.	27	22	25	20
Feb.	12	14	12	15
Mar.	10	12	12	15
Apr.	8	1C	12	18
May.	15	16	14	17
Jun.	1	4	6	8
July	2	6	5	8
Aug.	4	4	8	8
Sept.	5	6	9	10
Oct.	7	4	5	9
Nov.	15	17	20	21
Dec.	20	28	30	31

Obtain 5-month and 7-month moving averages for these data. Construct the graphs and interpret the results.

Ex. 17. Use a 5-year moving average to find the trend in the following data and plot the trend on the graph of the data : 40, 39, 37, 38, 41, 42, 47, 42, 41, 36, 45, 47, 50, 52, 48, 45, 40, 54, 56, 55, 50, 45, 40, 48, 66, 60, 60, 56, 52, 50, 60.

29.11 BUSINESS FORECASTING

Anticipation is a prominent feature of life in general and business life in particular. And no prediction is the final word.

Prediction may fail in which case we must reconsider our method and our observations.

Prediction may succeed in which case we need not be too sure of our method or the circumstances which brought about the success.

Whether we fail or succeed we try to eliminate the uncertainty of the future by forecasting. If there were mystic oracles in the past, we have some scientific methods now. Neither of the two techniques can claim perfection but a scientific study has certain advantage we should like to explore.

Uncertainty about the future is faced by a firm, an industry, an economy and all human institutions. Mere guesswork cannot help in this age of automation and dynamism. Quick decisions about production, sales, costs, profits, etc. have to be made. A systematic method of getting definite estimates of future conditions is what is called forecasting.

Forecasting is only a method of reducing the future uncertainty. It does not replace uncertainty by certainty.

Past versus Present

If most circumstances remain the same the future is likely to follow the pattern of the past. Successful prediction depends on the possible continuity from the past through the present to the future. Experience and familiarity with the problem under consideration are additional tools.

We may study the past for historical evidence, experiment in the present and make a forecast about the future using the totality of experience gained from all the sources. The use of historical evidence is made either on the general rhythmic pattern or on the basis of some specific historical analogy.

Another basis used for forecasting is that of a knowledge and study of all the factors having a bearing on the given situation. Surveys are conducted to get a fuller information and any changes in factors, technology, political conditions, demand and supply etc. are noted and incorporated in the analysis. Naturally, history gets a secondary position here.

In general, all forecasting methods make use of the past evidence as well as the present circumstances.

Method of business forecasting—I

The model building approach

In the model building method, a set of relationships or equations describing the economic activity is established and the interrelations between various aspects of the economy are studied.

A model thus built may well fit the past data so that a forecast can be made. Discredited hypotheses may be dropped and replaced by better ones following a more comprehensive understanding of the economy and learning from experience.

The method can be useful to a firm desirous of linking its forecasting with tactical and strategic plans for the future. It may not be feasible for every firm to construct its own model for forecasting since the effect and cost involved are substantial.

The model building approach, considered as an econometric approach, instead of bothering to extract the components of a time series considers it as a result of movements on certain causal variables which are discoverable. Economic knowledge and experience play an important role in this approach.

It can be seen that statistical techniques may be used as a means of constructing and testing hypotheses and also as a means of forecasting.

Forecasting with index numbers

Index numbers are business barometers. They can be used in many ways for forecasting purposes.

A combined study of index numbers of prices, employment, production, etc. may help us in getting a general business activity index. This and other specific indices may be used for business forecasting purposes.

Index numbers may not necessarily give reliable results. It is preferable to use them in conjunction with other methods of forecasting.

Time lag forecasting

Different series of measurements influenced by the same factors, often react to them in different ways and with different speeds. There is a time lag between different movements along with different stages. For example, deficit financing may lead to inflationary pressures, the rise in prices may lead to a fall in demand and so on. The speed of these phenomena is greater in the share market.

The forecast by this method can be improved by choosing proper variables and an accurate time lag. Correlation methods are commonly used for the purpose.

Extrapolation and forecasting from the trend

Extrapolation means estimation of probable values for a function outside the limits between which values are known. It is based on the expectation of continuity of the past pattern into the future.

Long term trend projection. If a trend has been fitted to a given time series then we may use the extrapolation method to study its future movements.

The trend may be fitted by any of the methods discussed earlier. Some trends can also be fitted by using special mathematical functions suited for specific purposes.

For example we have :

The straight line trend	:	$y = a + bx$
Higher degree polynomials	:	$y = a + bx + cx^2$
The exponential curve	:	$y = ab^x + c$
The Gompertz curve	:	$\log y = (\log a)\, b^x + \log c$
The logistic curve	:	$\dfrac{1}{y} = ab^x + c$

and so on.

We have seen that by a suitable transformation, each of these can be converted into the more familiar straight line trend. These curves may also be fitted to the data as they are. The extrapolation method of forecasting can be used in either case.

The trend is a purely statistical means of forecasting without considering economic or other forces which are operating to make the time series.

Short term trend projection. Different method have to be used when the forecasting time period is short, say less than a year. Most of these methods make use of the moving average method with some alterations. For example, instead of taking a simple moving average over certain months, we may take a weighted moving average, weighted variously, often exponentially, in such a way that the maximum importance is given to the most recent measurements.

The method is sometimes known as exponential smoothing method or *exponential forecasting*. To apply the method we have to know only the last-value in the series and the previous estimate, thus saving a lot of time.

Ex. 18. The following data show GNP and per capita income of a country. Plot the series on the same graph by taking the scale for GNP in Rs. million. Comment on the similarity between the two trends.

Also find the trend (*i*) approximately by a scatter diagram (*ii*) by least squares method and predict the values for five years in the future.

Year	*GNP* (*in million*)	*PCI*	*Year*	*GNP*	*PCI*
1958	400	200	1965	537	290
1959	430	210	1966	570	300
1960	460	230	1967	650	320
1961	500	240	1968	565	330
1962	500	250	1969	575	350
1963	510	270	1970	580	380
1964	520	280	1971	600	400

Is there any improvement in the situation over years?

Ex. 19. Forecast the production in January and February 1960 by fitting a trend to the following data :

1959	Jan	Feb	Mar	Apr	May	Jun	Jul	Aug	Sep	Oct	Nov	Dec
Prod.	120.3	120.9	128	127	129	133	134	135	137	138	140	141.

Ex. 20. What is meant by seasonal variation ? How will you separate it from time series data?

Ex. 21. What is meant by analysis of time series?

Ex. 22. What is a moving average? How does it differ from an arithmetic mean?

Ex. 23. What are the uses of time series? What is the difference between a time series and a frequency distribution?

Ex. 24. What is forecasting? What are the different methods of forecasting?

CHAPTER 30

Compilation and Presentation of Statistics

30.1 STATISTICAL INQUIRIES

Statistical methods apply to any situation where numerical information is available and the data are of a variable nature. Statistical analysis can be used in any social science for measurements showing a mixture of controlled and uncontrolled variability.

A study of the essence of the pattern of the development of socio-economic phenomena is more complete if we make a study of their quantitative aspects. This may help in the control and verification of plan targets, unused potentialities in the economy, a distorted growth pattern and so on. In general it may help establish various economic balances.

A *statistical inquiry* is an investigation or search, with the help of quantitative data, conducted in accordance with certain statistical techniques.

Statistical methods are not, in general, applicable to cases in which quantitative expressions are not available. Questions like the average level of sincerity of members of parliament or the mean degree of beauty in Indian women cannot be answered easily with statistical techniques.

A statistical inquiry emphasises *aggregates* characterised by certain *uniformities* as opposed to individual or subjective data.

The assumption of uniformities based on *the law of large numbers*, is indispensable. It enables us to generalise with the help of a limited number of observations.

Statistical Units

The statistical unit is the person, object, event etc. whose characteristics are to be described in the course of a statistical inquiry.

The reporting unit or the unit of *collection* is that from which information is obtained.

The tabulation or *compilation* unit is that for which the required statistics are to be tabulated.

The unit of *analysis and interpretation* is that which helps in a better understanding of the data and making comparisons with reference to time or space through rates, ratios, percentages, coefficients, etc.

Units may be simple or composite

A *simple unit* is one which represents a single condition without qualifications or a combination of characteristics that occur together.

A *composite unit* is subject to one or more qualifications attached to it.

30.2 FRAME AND DESIGN

A list or description of all the units in a statistical population is called the *frame* of the inquiry.

The *design* of a statistical inquiry determines

(*i*) the time when it is made,

(*ii*) the reporting period,

(*iii*) the period of survey operation,

(*iv*) the method of collection of data.

(*v*) the coverage (i.e. complete or incomplete enumeration), and

(*vi*) the sources utilised.

Instruction sheets and guidance at all stages falling within the scope of a statistical inquiry should be available regularly. All results should be recorded on prescribed forms.

30.3 NATURE AND TYPES OF INQUIRIES

A decision about the nature and type of an inquiry depends on its object and scope.

A *special* inquiry may collect information on a single topic. A *multipurpose* inquiry may be concerned with various aspects of the subject and may collect relatively summary information.

An inquiry collecting original information is *primary*, an inquiry based on official and other records is *secondary*.

An inquiry may be official, semi-official or non-official depending on the agency conducting it.

A *census* inquiry considers the units of the entire population. A *sample* inquiry considers only some representative units selected by the methods of sampling. This is the type of inquiry mostly used in practice.

An *initial* inquiry is one which is conducted for the first time. There may be a *repetitive* inquiry which is a continuation of the previous inquiries.

A *regular* inquiry is on a set periodical basis over a stretch of time. An *ad hoc* inquiry is undertaken whenever the need for the same arises.

An inquiry may be confidential or open, direct or indirect, specific or general, brief or detailed and so on. The title in each case clarifies the meaning.

30.4 STAGES OF A STATISTICAL INQUIRY

A statistical inquiry passes through several stages before it is completed. It begins with its conception, followed by some preliminaries and execution of the inquiry and ends with the actual report.

For convenience we group the stages as follows:

A. Planning the inquiry
B. Executing the inquiry
C. Collection and Editing of data
D. Organisation: Classification and Tabulation of data
E. Diagrammatic Representation of data
F. Analysis and Interpretation through computation of relevant statistical measures and derivatives like averages, coefficient of correlation, etc.
G. Writing up a report embodying the conclusions.

30.5 STAGE A : PLANNING THE INQUIRY

Without proper planning, time, money and energy are often expended in vain.

Planning a statistical inquiry requires a consideration of various aspects of the inquiry.

(*i*) Object and scope
(*ii*) Sources of data and technique of collection
(*iii*) Units of data collection
(*iv*) Ascertaining the degree of accuracy desired

A (i) Object and scope

Every statistical inquiry is conducted with a predetermined purpose in view and its scope is usually defined in advance.

The scope and object should be explicit and unambiguous. This diminishes the possibility of irrelevant data being collected and thus saves time and expenses.

The object may be general or specific. The cost of living index number for the whole economy corresponds to a general object; that for a particular region corresponds to a specific object.

The variables used in the inquiry should be carefully defined and the data collected should relate to them.

The scope should fix the limits of the inquiry and should be decided with reference to

(*a*) the *area* limit that best serves the purpose,
(*b*) the *time* which should be reasonable,
(*c*) the *quantum* of data to be collected wich should be neither too unwieldy nor too small to bias the results, and
(*d*) the *nature* and type of inquiry.

Ex. 1. Explain the concepts of data and observations.

The raw material of statistical science is in the form of data which is made of observations resulting from measurement or counting.

Data can be collected by conducting surveys. Sometimes data are obtained from experiments.

Data need not be collected from outside if they are already available in routinely kept and other records of institutions, data banks, published reports etc.

Observations

Values resulting from measurements of characteristics are called observations pertaining to those characteristics. Data on heights of individuals are observations or measurements pertaining to the variable height.

A (ii) Sources of data and technique of their collection

Original data collected in the process of investigation are called *primary* data. *Secondary* data are obtained from published or unpublished data collected by others.

A source of data may be primary and secondary at the same time.

The decision about the source of data depends on the nature and scope of the inquiry.

The technique of collection depends on the

(*a*) nature and scope of the problem

(*b*) cost, time and resources

(*c*) degree of accuracy desired.

A (iii) Units of data collection

The units of data collection should be clearly and unambiguously defined and should be stable and uniform throughout the inquiry. The unit of national income is money and the amounts should be mentioned in a certain currency throughout.

A statistical unit may be used in a conventional sense or in a special sense. In each case it should be defined.

A statistical unit should be

(*a*) precise and definite

(*b*) homogeneous and uniform

(*c*) appropriate for the purpose of the inquiry

A (iv) Degree of accuracy

Absolute accuracy in a statistical inquiry is unattainable because of the very nature of the subject and the tools used. Moreover, the attainment of accuracy, if at all possible, may not be worth the time and costs involved.

The necessary degree of accuracy should, therefore, be specified and this depends on

(*a*) the nature and scope of the inquiry,

(*b*) cost, time and resources,

(*c*) other factors.

30.6 STAGE B: EXECUTING THE INQUIRY

For supervision, control, follow up and proper handling of a statistical inquiry, the execution work must go on smoothly.

Either the existing administrative centres may be used or suitably situated centres may be established. There should be uniformity in the matter of designing of forms, questionnaires, training and supervision of field investigators for the control of accuracy of work. In cases of non-response there should be a proper follow-up.

The last stages of executing the inquiry consist of the statistical analysis of the collected information and preparation of reports for which plans should be made at initial stages.

30.7 STAGE C: COLLECTION AND EDITING OF DATA

The nature of data required may be determined by the statistician himself.

Data may be obtained from primary or secondary sources.

Primary data are collected by or under the direct supervision of the statistician. This requires a familiarity with the conditions under which data are collected so that errors could be avoided.

Secondary data are all other data with lower degrees of control by the statistician. The degree of accuracy of these data depends on the degree of control. See *A* (*ii*).

The distinction between the two is a matter of degree. Primary data for someone may become secondary for another person. Population census data or national income statistics are primary data to those collecting it and secondary to those using it.

Reliability of data

Relevance and meaning of data depend on their reliability which in turn depends on:

Who collected them and how and when?

What was measured and what were the units used?

In practice, very often, secondary data have to be used. They should be treated with caution and care.

Collection of secondary data

Secondary data are obtained from official, semi-official or unofficial sources. It may be in a published or unpublished form.

Published form. Various publications of government, local bodies, chambers of commerce, journals, newspapers etc. appearing with different periodicities.

Unpublished form. Manuscripts, records of government and other bodies, business houses, etc.

Collection of primary data

One or more of the following methods may be adopted:

(*i*) Direct personal investigation

(*ii*) Indirect investigation

(*iii*) Information from correspondents and local reports

(*iv*) Questionnaires.

C (i) Direct investigation

The informants are directly contacted by the statistician and a first hand information is obtained. The method is likely to be more accurate and there is a possibility of getting supplementary information which may prove to be useful. The statistician can adapt his language and technique to the nature and status of the informant.

The experienced statistician can get the best and the maximum possible information by this method.

The method may, however, prove to be costly and time consuming unless the area or the number of items to be covered is small. Personal bias is likely to enter the results. The method, if properly used, is convenient for some intensive surveys.

C (ii) Indirect investigation

Third parties or witnesses having information may be interviewed in this case either because direct sources do not exist or do not respond for some reason or other.

C (iii) Information from correspondents and local reports

Newspapers and some government departments use this method which is adopted when information is to be collected at regular intervals from wide areas, *e.g.*, crop estimates.

C (iv) Questionnaires

A questionnaire is a list of queries designed to elicit the required information in the form of short answers filled into the blank spaces against the questions either by the interviewer or by the informant.

Postal inquiry. The questionnaires may be sent by post. The informant is requested to fill up questionnaire and send it back by post within a certain period of time. There is no need of any prior appointment with the informant.

This method is used when the area to be covered is vast. It is cheap and quick.

There are limitations, however. Usually, the response is poor. Very few questionnaires may come back. Moreover, the accuracy of the information supplied is difficult to verify.

Often the results are biased in this method as only a certain type of people may supply the information.

Postal inquiries are not recommended unless there is a legal obligation or the nonresponders are subsequently interviewed personally.

Sending questionnaires through enumerators. The enumerator may help the informant to fill the questionnaire or may fill it himself after asking questions.

In this case the nonresponse is less and the information received is more reliable. The informant may not refuse to answer questions. Even difficult questions can be answered by him with the enumerator's help.

The method can be used for illiterate informants.

The enumerator should be well trained, experienced and tactful for the success of this method. The interview in this method is of greater importance than the questionnaire.

30.8 DESIGNING A QUESTIONNAIRE

This is a highly skilled job which should be undertaken only by persons specially trained for it. Since the success of an inquiry may depend significantly upon the questionnaire (especially in the case of the postal inquiry method), great care must be taken when designing it and framing the questions. In general the following points should be observed:

(*i*) The purpose of the questionnaire should always be stated clearly and unambiguously at the head of the form to indicate to the informant the nature of the material required.

(*ii*) Details such as name, address, age, etc. should be specified where necessary.

(*iii*) The questions should be simple and brief.

(*iv*) They should be unambiguously phrased in easily understandable terms.

(*v*) The best kinds of questions are those which allow a pre-printed answer to be ticked as in the case of multiple choice questions. This is especially the case when most of the possible answers or choices are known and their number is small.

(*vi*) The questions should neither be irrelevant nor too personal.

(*vii*) Leading questions should not be asked. A *leading question* is one that suggests the desired answer, *e.g.*, the question "don't you agree that all people with taste buy Dumdum sarees?" may suggest the answer "yes".

(*viii*) The questions should be designed so that they fall into a logical sequence. This will make it easier for the respondent to pay attention to their significance and the quality of answers may be improved.

Ex 2. We now frame a model questionnaire whose object is to find out the present occupation and career (or profession) of all adult persons of Indian extraction (*i.e.*, having one grandparent or parent born in India) who were born in Africa but have now returned to India.

QUESTIONNAIRE

Object : Collection of information regarding present occupation, career or profession of all adult persons of Indian extraction who were born in Africa but have now returned to settle in India.

1. Name : Shri/Smt/Kumari ....
   (In block capitals, starting with surname)
2. Present Address: ....
   Permanent address if other than Present Address
   ....
3. Age : .... (years completed)
4. Date of birth : .... Place of birth : ....
5. Names of Indian parents(s) or grandparent (s)
   ....
6. Date(s) of birth of Indian parent(s) or grandparents(s)

....

7. Date of departure of parent(s) or grandparent(s) for Africa

   ....

8. Place where they settled in Africa and length of period of stay

   .... and ....

9. Occupation of parents(s) in Africa ....

10. Average annual income of parent(s) in Africa ....

11. Date of your departure from Africa ....

12. Date of your arrival in India ....

13. Are you employed—or do you have your own business?
    (Tick one, unless No. 16 applies).

14. (*a*) Where are you employed? ....
    (*b*) Since when? ....
    (*c*) What position do you hold ? ....
    (*d*) What remuneration do you get? . . ..

15. (*a*) What is the nature of your business? ....
    (*b*) When did you begin it? ....
    (*c*) Have you any previous experience or training for this line?
    ....
    (*d*) If so, what? ....
    (*e*) Have you any partners in business? ....
    How many? ....
    (*f*) Of what nationality? ....

16. (*a*) Are you a student? ....
    (*b*) Which degree are you studying for? ....
    (*c*) Which university or college? ....
    (*d*) What level of education did you have in Africa? ....
    .... (name, degree obtained, if any)

17. What career do you intend to take up after finishing your studies?
    ....

18. Do you intend to remain in India permanently? ....

19. How many members of your family who have left Africa, have settled in countries other than India? ....

30.9 CHECKING AND EDITING OF DATA

To check various types of errors and irregularities in the collected data a proper scrutiny of the same should be carried out. Data with large errors, inconsistencies and omissions should be rejected.

Editing primary data. Editing has to be done for various reasons :

(*a*) *For completeness.* All schedules and questionnaires should be complete in all respects. Efforts should be made, in particular, to get the answers of vital questions even at this stage.

(*b*) *For accuracy.* Reliability and validity of data can be checked by an experienced editor. Faulty conclusions result from inaccurate data.

(*c*) *For uniformity.* Questions may be misunderstood or misinterpreted by the informants. The editor should remove nonuniformities, *e.g.*, if the monthly income has been wrongly taken for annual income, the editor can verify the figures.

Editing secondary data. Secondary data should not be accepted as they are. They may contain errors of various types or may be inadequate. Secondary data should be edited for

(*a*) reliability, from the point of view of degree of accuracy, source, method of collection, etc.

(*b*) adequacy and suitability for the purpose of investigation. *Whether primary or secondary*, the collected data should be carefully checked for

(*i*) completeness and adequacy

(*ii*) logical consistency for plausibility and freedom from internal logical contradictions

(*iii*) arithmetical consistency for correct figures, totals and sub-totals etc.

30.10 APPROXIMATIONS

For a better understanding of data and for comparison purposes, approximations may be resorted to. The usual method is that of rounding off figures according to the degree of accuracy desired.

Statistical errors

In planning and executing a statistical inquiry both sampling and non-sampling errors can occur.

Sampling errors

Sampling errors may arise from chance. These are *unbiased sampling errors.*

They may arise from the deliberately employed sampling methods in which case they are *errors of bias.*

What is bias?

Bias is a cumulative or non-compensating error which does not become less as the sample size increases.

Random sampling error decreases with increasing sample size.

Bias may arise from prejudice, vanity or other reasons resulting in a cumulatively wrong effect in some direction. It may enter at various stages.

(*a*) *Sampling stage*: Bias may enter because of

(*a*) wilful selection of a sample with a purpose

(*b*) substitution of one item for another.

(*b*) *Collection stage*: Bias may enter due to

(*a*) faulty collection of data

(*b*) wrong definitions or incomplete understanding

(*c*) nonresponse

(*d*) response associated with vanity etc.

Bias is possible at any stage of a statistical inquiry. Bias can be and should be avoided and checked at all the stages of an inquiry.

Unbiased errors

Unbiased errors are subject to the law of statistical regularity and tend to eliminate one another if the sample size is increased.

Non-sampling errors

These errors result from lack of training on the part of enumerators, wrong statistical units, approximations and so on. These errors are likely in any survey, census or sample. Sampling errors can arise when sampling techniques are used.

Measurement of errors

Absolute error is the difference between the actual value (A) and estimated value (E).

Absolute error, $A — E$ may be positive or negative.

Relative error is the ratio

$$RE = \frac{A - E}{E}$$

It can be used for comparison. It may be positive or negative and is often expressed as a percentage.

Ex. 3. If actual value = 11 and estimated value = 10

$$RE = \frac{11 - 10}{10} = \frac{1}{10} = 0.1 \text{ or } 10\%$$

30.11 STAGE D: ORGANISATION OF DATA: CLASSIFICATION AND TABULATION

D (*i*) CLASSIFICATION

The collected data should be arranged systematically to give it shape, form and meaning.

The division of the data into homogeneous groups according to their characteristics, recorded in a statistical inquiry, is called *classification*.

Classification should be done in such a way as (*i*) to reflect the salient features of the data (*ii*) to condense the unorganised mass of data in homogeneous classes, (*iii*) to enable the making of comparisons and (*iv*) to eliminate unnecessary details.

Classification along with tabulation is a preliminary to the presentation of statistical results. It should be :

(*i*) unambiguous, to avoid confusion,

(*ii*) stable, to facilitate comparisons,

(*iii*) flexible enough, to allow incorporation of new and varying situations,

(*iv*) able to cover all possibilities; there should be a class for every item of the data.

We may consider various forms of classification.

I. Simple and Cross Classification

A simple classification is made according to one characteristic. A cross classification is made according to several characteristics. *i.e.* with classes distinguished according to additional characteristics after sub-division. It may also be called manifold classification.

Ex. 4. A case of manifold classification showing radio listeners of Karnataka (*K*), North Indian (*N*) and Western (*W*) music.

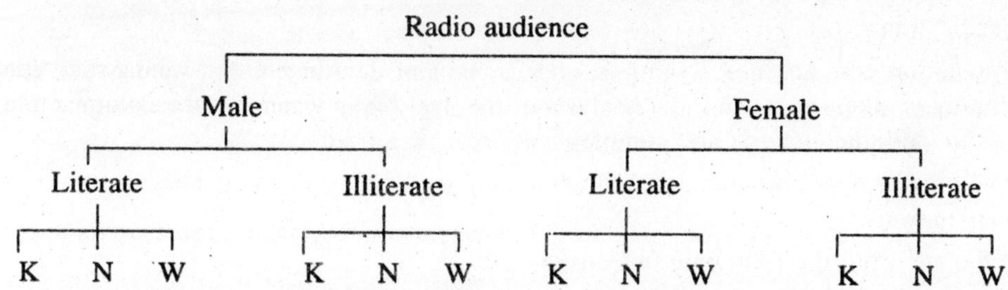

II. Classifications according to Attributes and Variables

These may be called qualitative and quantitative classifications.

An attribute is a characteristic which can be described only qualitatively and not numerically. Attributes may be indirectly measured, counted and classified according to some criteria.

The example given above shows classification according to attributes.

Quantitative classification resulting in frequency distributions is on the basis of measurable characteristics like marks in an examination, heights of men and so on. This method of classification has been discussed in the chapter on frequency distributions.

III. Geographical and Chronological Classifications

When data are classified on the basis of geography or location we have geographical classification.

Ex. 5. A case of geographical classification.

Name of Country	*Population in million*
A	500
B	750
C	220

Chronological classification shows series of quantitative data over time. Time series have been discussed in a separate chapter.

IV. Reclassification or Secondary Classification

This consists in the formation of new groups on the basis of an existing classification.

V. Series

The classification of data may be done on the bases mentioned above, giving rise to, mainly, three type of series:

(*i*) Time series based on time

(*ii*) Spatial series based on space

(*iii*) Condition series based on some condition.

Frequency distributions are examples of condition series. Any phenomenon other than time and location can give rise to a condition series.

D (*ii*) TABULATION

Classification of data is usually followed by tabulation which is the mechanical part of classification.

Tabulation consists in a systematic arrangement of data in columns and rows, The opportunity is taken of starting the analysis of the data by so arranging the columns and rows as to facilitate analysis and comparisons.

The processes of classification and tabulation need not be distinct. They are steps to the final analysis.

Tabulation has the following objectives:

(*i*) Simplicity. The removal of unnecessary details gives a clear and concise picture of the data

(*ii*) Economy of space and time

(*iii*) Ease in comprehension and remembering

(*iv*) Facility of comparisons. Comparisons within a table and with other tables may be made

(*v*) Ease in handling of totals, analysis, interpretation etc.

Construction of Tables

A table is constructed depending on the type of information to be presented and the requirements of statistical analysis. The following are the essential features of a table:

(*i*) *Title.* It should have a clear *title* which describes the contents of the table. The title should be brief and self-explanatory.

(*ii*) *Stubs and captions.* It should have clear headings and sub-headings. Column headings are called *captions* and row headings are called *stubs.* The stubs are usually wider than captions.

(*iii*) *Units.* It should indicate all the *units* used.

(*iv*) *Body.* The *body* of the table should contain all information arranged according to description.

(*v*) *Headnote.* The *headnote* or prefatory note placed just below the title, in a less prominent type, gives some additional explanation about the table. Sometimes the headnote consists of the unit of measurement.

(*vi*) *Footnotes.* A *footnote* at the bottom of the table may clarify some omissions of special features.

A source note gives information about the source used, if any.

(*vii*) *Arrangement of data.* Data may be arranged according to requirements in chronological, alphabetical, geographical, or any other other.

(*viii*) *Emphasis.* The items to be emphasized may be put in different print or marked suitably.

(*ix*) *Other details.* Percentages, ratios, etc. should be shown in separate columns. Thick and thin lines should be drawn at proper places.

A table should be easy to read and should contain only the relevant details. If the aim of clarification is not achieved the table should be redesigned.

Types of Tables

Depending on the nature of the data and other requirements, tables may be divided into various types.

General tables or reference tables. They contain detailed information for general use and reference, *e.g.*, tables published by government agencies.

Specific purpose or derivative tables. They are usually summarised from general tables and are useful for comparison and analytical purposes. Averages, percentages etc. are incorporated along with information in these tables.

Simple and complex tables. A table showing only one characteristic is a simple table. The more common tables are complex and show two or more characteristics or groups of items.

Ex. 6. A simple table.

Cinema attendance among Adult Male Factory Workers in Bombay
March 1972

Frequency	*Number of workers*
Less than once a month	3780
1 to 4 times a month	1652
More than 4 times a month	926

Ex. 7. A Complex table. The table given below is the result of a survey into the cinema going habits of adult factory workers.

Cinema Attendance among Adult Male Factory Workers in Bombay
March 1972

Cinema	*Single*		*Married*	
Attendance	*Under 30*	*Over 30*	*Under 30*	*Over 30*
Less than once a month	122	374	1404	1880
1—4 times a month	1046	202	289	115
More than 4 times a month	881	23	112	10
Total	2049	599	1805	2005

It is obvious that the tabular form of classification of data is a great improvement on the narrative form.

Frequently, table construction involves deciding which attribute should be taken as primary and which as secondary. For the above table, we might ask whether it would be improved if "under 30" and "30 and over" had been the main column headings and "single" and "married" the sub-headings. The answer depends on the purpose of the table. If the activities of *age groups* are to be compared, it is best left as it stands. But if a comparison between men of different *marital status* is required the change would be an improvement.

Advantages of tabulation of data

(*i*) Tabulated data can be more easily understood and grasped than untabulated data.

(*ii*) A table facilitates comparisons between sub-divisions and with other tables.

(*iii*) It enables the required figures to be located easily.

(*iv*) It reveals patterns within the figures, which otherwise might not have been obvious, *e.g.*, in the table above we can at once conclude that regular and frequent cinema attendance is mainly confined to younger men.

(*v*) It makes the summation of items and the detection of errors and omissions easier.

(*vi*) It obviates repetition of explanatory phrases and headings and hence takes less space.

Ex. 8. In 1962, 1090 million passenger journeys were made on the Western Railway, yielding Rs. 137.6 million in receipts. Of the journeys, 351 million were at full fares; 426 mil. at reduced fares and an estimated 313 mil. by season ticket holders. By 1964, full fare journeys had fallen by 78 million, reduced fare ones were up by 9 million and season ticket holders up by 4 million. During the same period, receipts from full fares went up from Rs. 79.7 mil. to Rs. 100 mil., those from reduced fare increased from Rs. 40 mil. to Rs. 54.1 mil., and season ticket receipts increased by Rs. 5.9 mil.

Draw up tables which will summarize the above data and bring out the principal changes between the two years. Also compute the derived statistics.

Ex. 9. The number of electors in the 575 parliamentary constituencies of a country are in the range of 0.5 million to 1.5 million. Suggest a suitable classification.

Ex. 10. Present the following information in a suitable tabular form.

In 1945 out of a total of 1750 workers in a factory, 1200 were members of a trade union.

The number of women was 200 of which 175 did not belong to a trade union.

In 1950, the number of union workers increased to 1580 of which 1290 were men and the number of non-union workers fell to 208 of which 180 were men.

In 1955 there were 1800 workers of whom 1600 belonged to a trade union. Of the 300 women there were 8 who did not belong to a trade union.

Categories of factory workers according to their membership of a trade union (M) und non-membership (N)

	Men			Women			Total of Men and Women		
Year	M	N	Total	M	N	Total	M	N	All
1945	1175	375	1500	25	175	200	1200	500	1750
1950	1290	180	1470	290	28	318	1580	208	1788
1955	1308	192	1500	292	8	300	1600	200	1800

Ex. 11. In a survey of 10048 persons, of whom 2703 were children under 16 years of age, 605 retired men and 2212 housewives, the following information was obtained.

Of those above 16, there were 2720 males above 16 and under 60, and 2931 females of this age group.

There were 227 males between 60 and 65 and 482 males above 65 and 270 and 515 females in these age ranges respectively.

Construct tables showing the distribution of the sample by work status and of the adult population by age and sex. Use percentages to help in making comparisons. Give your comment on the data.

30.12 STAGE E: DIAGRAMMATIC PRESENTATION OF DATA

Statistical results can be presented in an appealing and vivid form with the help of charts.

An appropriate chart can give a clear, truthful and easily understandable picture of the facts contained in the data. A carelessly drawn inappropriate chart can mislead and waste the labour undertaken.

The drawing of an *effective* chart requires talent, imagination and a grasp of the subject under consideration.

Objectives of constructing charts

(*i*) To get a quick view of the data. For understanding and remembering the nature and trend of the data.

(*ii*) To compare with other data and charts.

(*iii*) To analyse the data, for further study and planning.

(*iv*) To check the accuracy of certain computational results. Incompatibilities can be noticed from the trend in the chart.

(*v*) To make a step by step visual study of development.

(*vi*) To present the data in an alterate form.

Chart Norms

A chart should be planned so as to possess the following fundamental features.

(*i*) *Simplicity and neatness*. The chart should not include unnecessary information and should deal with the relevant data for the relevant purpose only. The form should be attractive and easily understandable.

(*ii*) *Size and proportion*. It should be neither too large nor too small. The size depends on the data and other details. The proportion of its height to width should be reasonable. Often it is taken as 1 : 1.5.

(*iii*) *Accuracy*. Conscious and unconscious misrepresentation should be avoided.

(*iv*) *Emphasis*. Colours, heavy, light and dotted lines may be suitably used. The materials should be used carefully and skilfully.

Construction of a chart

The above norms should be kept in mind when constructing a chart. With proper scales on the x and y axes it is possible to show useful information in a convenient form.

A title, subtitles, scale headings and proper labelling should accompany every chart.

Charts versus tabular presentation of data

Data may be presented in the form of tables or charts.

A chart has an eye appeal. A table is likely to be skipped by a reader.

A table gives more information but requires detailed and minute study.

Exact values can be read from a table. But inexact values may be concealed by a table and revealed by a chart.

A chart involves less work in preparation. But to construct a chart tables are needed first.

Charts arouse interest. They do not replace the data. They provide a qualitative appreciation of quantitative data.

Both may be presented together to give the best of both.

Charts in Economics

Graphs and charts are extensively used in economics. The attempt to avoid the use of mathematics in economics has resulted in a great use of geometrical and other types of representations in economics.

A graphical presentation of economic ideas has been found to be extremely useful. Economic laws are statements of tendencies and any curves and lines drawn with reference to an economic argument need not necessarily have any definite scales because the pattern can be clear without them. Charts in economics are, therefore, rough sketches of behavioural relationships.

Limitations of charts

A chart can easily portray two or three characteristics of a set of data. An attempt to represent more characteristics may make it complicated. Sometimes a chart may add new meaning to the facts but in general it only illustrates the data and shows a different way of expressing facts.

A chart can present only a limited amount of information which may be approximate.

A chart is liable to be misinterpreted or misunderstood.

A good chart requires skill and labour to produce and care should be taken in using and interpreting it.

Types of charts

There are many types of charts. We consider below some types in common use:

E_1 Graphs

(*a*) Scattergrams and functional graphs

(*b*) Time graphs

(*c*) Band charts and Cartograms

(*d*) Semilogarithmic graphs

(*e*) Analytical graphs

E_2 Bar charts

E_3 Pictograms

E_4 Maps

E_5 Area charts

E_6 Planning charts

E_1 Graphs.

In a graph the independent variable should always be placed on the horizontal axis and the dependent variable on the *y*-axis.

Too many curves should not be shown simultaneously.

(*a*) *Scattergrams.* A scattergram shows the statistical relationship between two variables for both of which data are recorded. The existence of some kind of a functional relationship between two variables may be reflected in it.

(*b*) *Time graphs.* Different values of one or more variables changing over time are shown in a *historigram* or time series.

(*c*) *Band charts.* Several curves each showing one stage in the build up of a cumulative total, takes the form of bands running across the graphs.

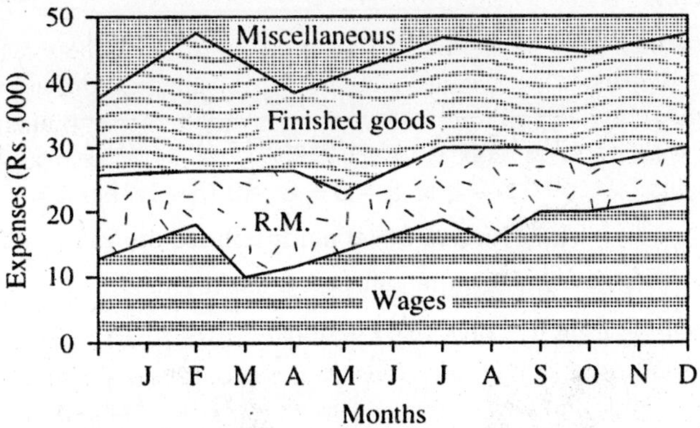

Band chart

Expenses over months

A percentage band chart shows the variation of the relative proportions of the components over time.

A *Z-chart* is a special type of time graph which represents in a single graph in one place the following :

Monthly figures,

Cumulative monthly figures,

12-month moving totals,

and at times, a target.

On the vertical axis, the scale for the monthly figure should be larger than that for cumulative figures and for moving totals to contain the graph within reasonable bounds.

(*d*) *Semilogarithmic or ratio graphs.* A semilog or ratio chart has a logarithmic scale on the vertical axis and a normal i.e. arithmetical scale on the x-axis. A ratio chart shows ratios in their proper form.

This graph is useful in making an analysis of relative changes and to compare series of widely differing magnitudes.

The slope of the curve indicates the rate at which the figures are increasing (or decreasing). For straight line graphs, the rate of change is constant.

A semilog graph is the presentation of a logarithmic function transformed into a linear (or any other) function.

A semilog graph may be drawn by

(*i*) using a slide rule

(*ii*) taking logarithms of the values of the variable

(*iii*) using a semi-log graph paper.

(*e*) *Analytical graphs.* Most diagrams serve an illustrative function. Some may serve an analytical function. Scattergrams, logarithmic graphs, Lorenz curves are examples of diagrams serving analytical functions.

Histograms, frequency polygons and curves, cumulative frequency curves are charts of frequency distributions whose analysis and description are important parts of statistics.

E_2 Bar Charts

Bar charts are drawn almost in the same way as graphs. Data are represented by a series of bars, the heights of each bar showing the size of the observation represented.

When a few observations of one or two series are to be shown, bar charts are preferable to line charts which should be used when there are many observations or several series.

Bars reflect the individuality of each observation. Lines emphasise the general trend. The two may be sometimes used together.

Bar charts (*i*) are easy to construct, (*ii*) can show the data accurately, (*iii*) emphasise the individual items, and (*iv*) may show their component parts more clearly.

While drawing the bar charts:

(*i*) the widths of the bars should be kept uniform, and

(*ii*) the gaps between successive bars should remain the same.

Bars may be drawn horizontally or vertically. Vertical bars are preferred as they allow an easy comparison with other bars.

Bar charts may be of various types:

(*i*) Simple bar charts
(*ii*) Component bar charts
(*iii*) Percentage component bar charts
(*iv*) Multiple bar charts
(*v*) Deviation bars.

(*i*) *Simple bar charts*

They represent only one variable or one classification. There is a series of bars representing the data.

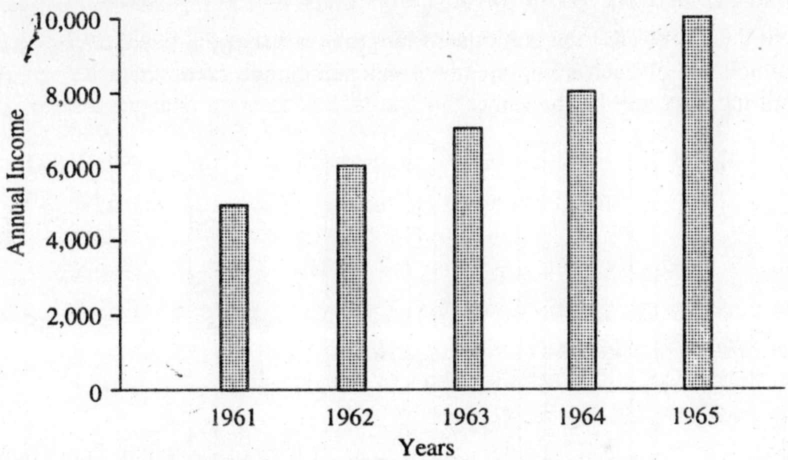

Simple bar chart showing annual income of Mr. X

(*ii*) *Component bar charts*

When each figure is made up of two of more component figures the bars may be subdivided into components. Too many components should not be shown.

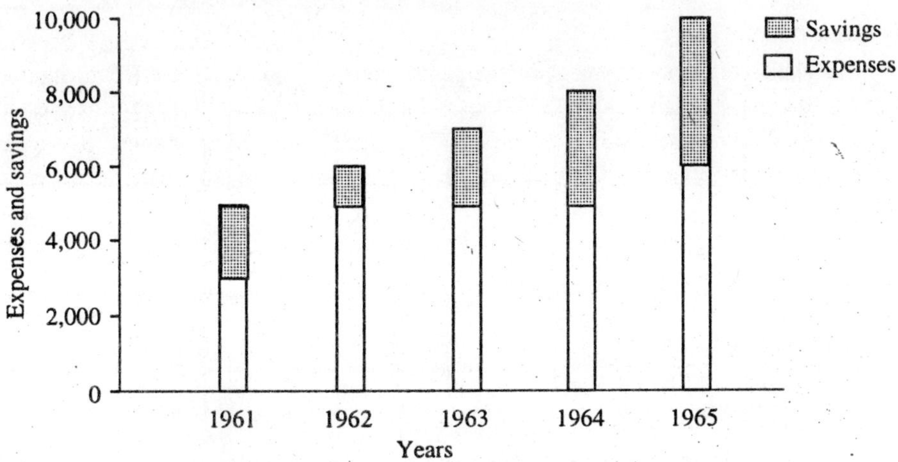

Component bar chart showing expenses and savings of Mr. X

Table showing annual income, expenses and savings of Mr. X.

Year	Amounts in Rs. of			Percentages of		
	Income	*Expenses*	*Savings*	*Income*	*Expenses*	*Savings*
1961	5000	3000	2000	100.0	60.0	40.0
1962	6000	5000	1000	100.0	83.3	16.7
1963	7000	5000	2000	100.0	71.4	28.6
1964	8000	5000	3000	100.0	62.5	37.5
1965	10000	6000	4000	100.0	60.0	40.0

(iii) Percentage component bars or divided bar charts

When in the above case the component lengths represent the percentages (instead of the actual amounts) of each component we get percentage component bar charts. The heights of all the bars will be the same.

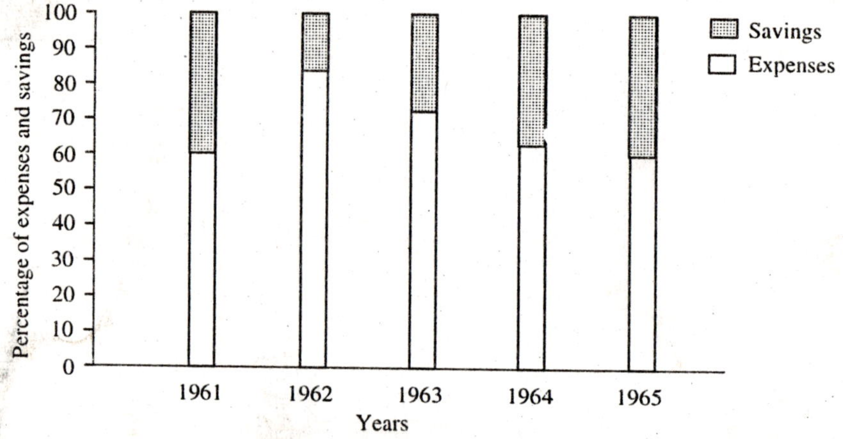

Percentage component bar chart
showing expenses and savings of Mr. X

(iv) Multiple bar charts

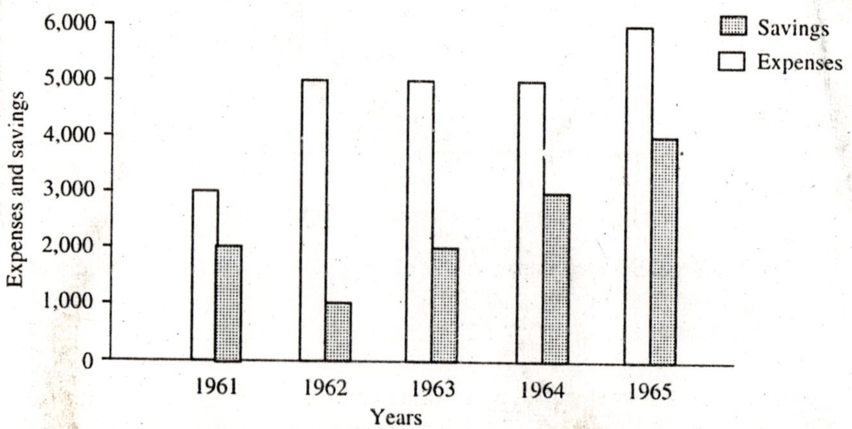

Multiple bar chart showing expenses and savings of Mr. X

Here the interrelated component parts are shown in adjoining bars, coloured or marked differently, thus allowing comparison between different parts.

These charts can be used if the overall total is not required. Some charts given earlier show totals also.

(*v*) *Deviation charts*

Positive and negative values may be shown in the form of bars in deviation charts, either horizontally or vertically.

Trade between countries A and B, 1961–64

(Amounts in million rupees)

Year	*Exports*	*Imports*	*Balance of trade*
1961	40	50	– 10
1962	60	100	– 40
1963	80	50	+ 30
1964	90	70	+ 20

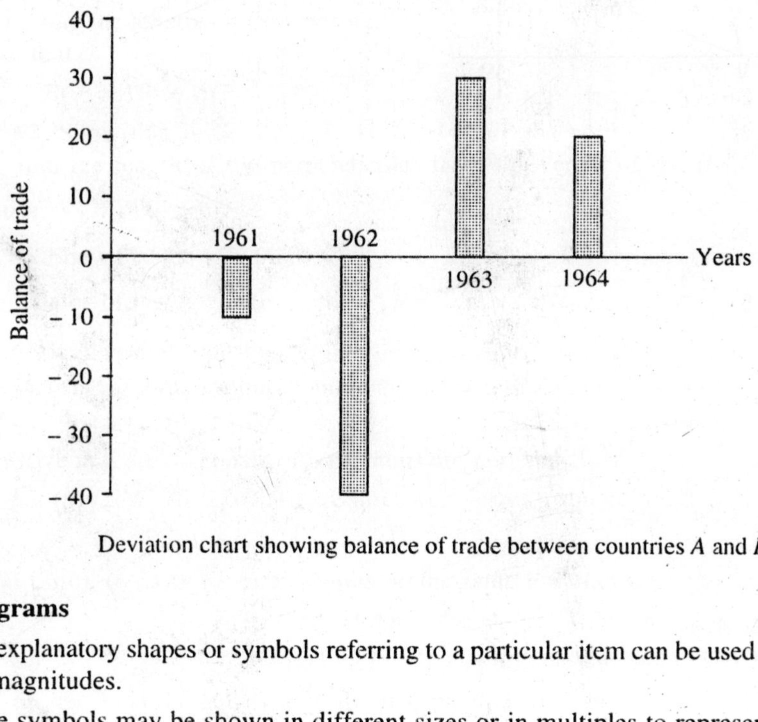

Deviation chart showing balance of trade between countries *A* and *B*.

E_3. Pictograms

Self-explanatory shapes or symbols referring to a particular item can be used sometimes to show magnitudes.

These symbols may be shown in different sizes or in multiples to represent data.

Cartograms are statistical maps which classify data on a geographical basis. The quantities are shown on the maps in differing shades or colours.

E_4. Flow maps

The size of flows between different points can be shown by flow maps by varying the width of the principal features of the data more emphatically.

Alternatively the flow lines may be shown in different shades, the darker ones representing higher density and the lighter ones lower density.

E_5. Area charts

Two-dimensional charts may be shown in the form of rectangles, squares or circles.

Rectangles are used when two sets of figures are to be represented. Both the length and breath of the rectangle can be varied.

Squares are used to compare quantities widely differing in magnitude. They are used for comparisons. Circles may be used for the same purpose.

Ex. 12. Show in the form of squares or circles the per capita income in three countries as given below:

Country	*P.C.I. (Rs.)*
A	200
B	400
C	800

(*i*) For squares take the square roots of the values of various items and then take a suitable scale.

(*ii*) In the case of circles the radii are proportional to the square roots of the areas.

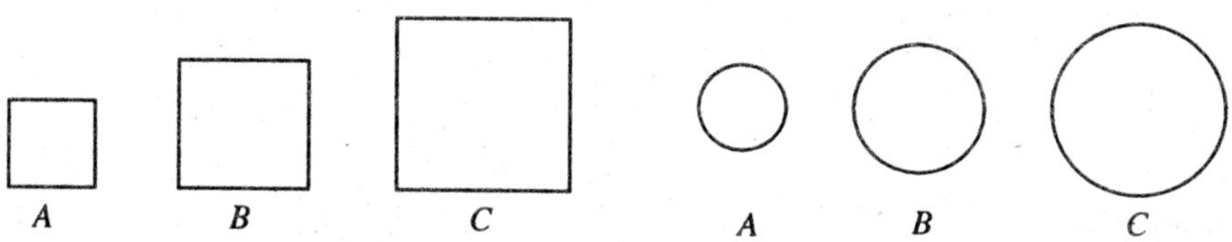

P.C.I. of countries *A, B, C.*

Three-dimensional or volume diagrams may be similarly used.

Pie charts or angular charts. A pie chart is a circle divided by radial lines into sections or slices so that the area of each section is proportional to the size of the amount represented. It is a convenient way of showing the component amounts in proportion to the total.

The sum of angles at a point being 360°, the component parts of the data are expressed as proportions of 360° and the sectors of a circle represent these parts. The pie chart follows the same pattern as percentage divided bar chart. In the case of a pie chart the degrees corresponding to components are to be drawn by means of a protractor.

It is not easy to compare different pie charts.

Ex. 13. Expenses of family with income Rs. 200.

Item	*Rs.*	*Percentages*	*Degrees*
Food	80	40	144
Clothing	40	20	72
Rent	50	25	90
Others	30	15	54
Total	200	100	360

(*Explanation*—Last column: Degrees are obtained by dividing the amount for each item by the total and multiplying by 360).

Divided
Bar diagram

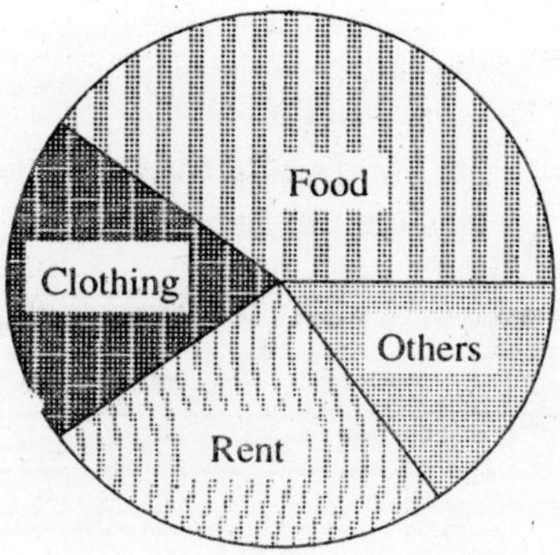

Pie chart

Expenses of family with income of Rs. 200 shown in the form of a divided bar diagram and a pie chart.

E_6. Planning charts

One or more of the above mentioned types of charts may be used in planning of projects and elsewhere. Planning charts may make use of statistical ideas and methods depending upon individual cases.

Gantt charts. This chart consists of a series of columns each corresponding to a certain period of time or other variable with a proportional width.

In the Gantt progress chart each column corresponds to a working day with a width proportional to the amount of work expected.

A horizontal line within the columns shows the relationship between this amount of work and the amount actually done. The resulting chart is the linear graph of plan fulfilment.

The Gantt chart is a form of bar chart and can be used for the planning and review of projects.

Charts like *population pyramids* can be used to represent cross classifications. A population pyramid is often used to show the distribution of population according to sex and age.

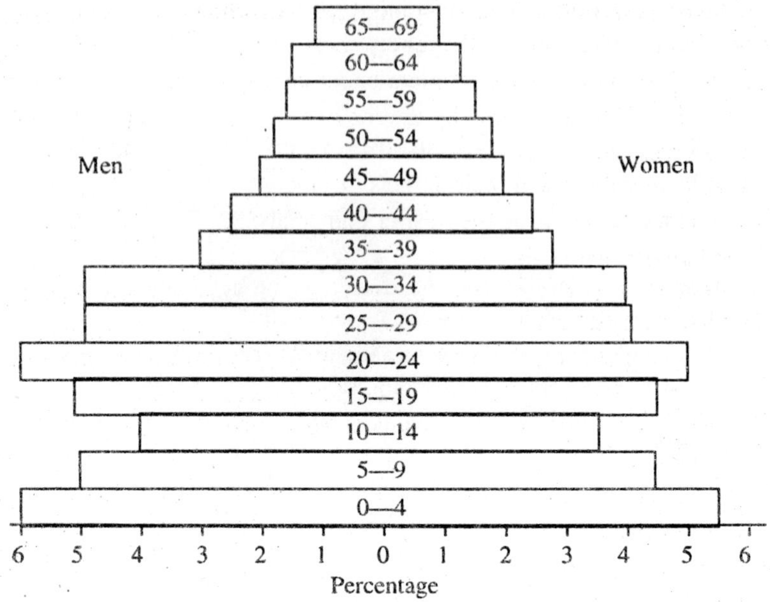

Population pyramid showing cross classification of age and sex.
Data fictitious.

Network planning. A plan or a project can be broken down into a set of subprojects or activities in a number of ways. These activities may be related in certain ways and they may have to be completed in a certain order.

The first sub-division is usually a broad one, say at the director level in management studies. Further sub-divisions indicate interrelationships at lower levels of management and so on.

The network method of graphing helps in the scheduling of projects, improves on the planned timings and enables us to derive further information. A carefully chosen network model can be immensely useful in the study of a system and its various functions. It can also help increase efficiency and reduce cost in the carrying out of complex government and industrial planning projects.

30.13 STAGE F: ANALYSIS AND INTERPRETATION OF DATA

The analysis and interpretation of data consists in classifying and summarising it and studying its various aspects.

There are several methods of analysing statistical material. After characteristics like the mean, standard deviation etc. have been determined we should try to draw inferences from the facts before us.

Wrong interpretations and false generalisations should be avoided and an effort should be made to determine the extent of error in the results.

30.14 STAGE G: THE FINAL REPORT

The conclusions of a statistical inquiry are embodied in a consolidated report which is the last step of an inquiry giving various details from the beginning to end.

The final report is a result of the integration and summary of all the relevant features of the inquiry. Among other details, the report gives

(*i*) a brief description of the problem under consideration,

(*ii*) the period of time covered by the inquiry.

(*iii*) the sources, nature and amounts of data collected, the difficulties encountered in collecting them and their defects, if any,

(*iv*) the techniques used in organising and analysing the data together with the assumptions made,

(*v*) a summary of findings and the extent of the accuracy and reliability of the results,

(*vi*) a critical appraisal and overall evaluation of the inquiry along with its relation to the findings of other similar inquiries,

(*vii*) suggestions for possible improvements that could be made in similar inquiries in future,

(*viii*) the uses to which the inquiry could be profitably put,

(*ix*) credits to the participants and contributors, and

(*x*) an appendix showing copies of various forms, schedules and instructions used in the inquiry.

Ex. 14. The following table shows the monthly expenditure of three families. Represent the data by a suitable diagram and write a report.

		Family	
Items	*A*	*B*	*C*
Food	43	83	120
Clothing	8	17	25
Entertainment	3	10	12
Education	5	9	15
Rent	10	21	17
Miscellaneous	6	15	17

Ex. 15. Show by means of a bar diagram the production of a factory using the following information:

Quarter of the year	*Finished goods*	*Semifinished goods*	*Total*
I	200	100	300
II	300	200	500
III	150	100	250
IV	450	300	750

Ex. 16. A rupee spent on khadi is distributed as follows:

	Paise
Farmer	19
Spinner	35
Weaver	28
Washerman	8
Administration	10
	100

Present the data in a pie chart and a bar diagram.

Ex. 17. Show the following data in a suitable chart:

Continent	*Area (million sq. miles)*
Africa	11.7
Asia	10.4
Europe	1.9
N. America	9.4
Oceania	3.3
S. America	6.9
USSR	7.9
Total	51.5

Ex. 18 Construct a suitable chart for the following data showing the outlay of public sector in the second and third five year plans:

	Expenditure (Rs. crores)	
	Second plan	*Third plan*
Agriculture	530	1068
Irrigation and power	865	1662
Village and small industries	175	265
Organised industry	900	1520
Transport and communication	1300	1486
Social services and misc.	830	1500
Total	4600	7500

Ex. 19. Distinguish between primary and secondary data. What are the various methods of collecting primary data?

Ex. 20. State the preliminary steps you would take in planning a statistical inquiry.

Ex. 21. Define classification and tabulation. What are the various methods of classification used in statistics.

Ex. 22. What are the tests of a good diagram? Briefly explain the uses and limitations of diagrams in presenting statistical data.

Ex. 23. You are required to plan a sample survey to study the problems of indebtedness among the rural agricultural population of India. Suggest a suitable plan. Draft a suitable questionnaire that may be used in this connection.

Ex. 24. Comment on the following:

(*i*) It is never safe to take published statistics at their face value without knowing their meaning and limitations.

(*ii*) In collection and tabulation, commonsense is the chief requisite and experience the chief teacher.

(*iii*) Diagrams do not add anything to the meaning of statistics but when drawn and studied intelligibly they bring to view the salient characteristics of graphs and series.

(*iv*) The job of the business statistician is to look into the future. He is employed to furnish those in positions of top control of his firm with a basis for their thinking and acting.

(*v*) The fundamental gospel of statistics is to push back the domain of ignorance, prejudice, rule of thumb, arbitrary or premature decisions, tradition and dogmatism and to increase the domain in which decisions are made and principles are formulated on the basis of analysed quantitative facts.

Select References

ALLEN R.G.D., *Mathematical Analysis for Economists*, Macmillan

————, *Mathematical Economics*, Macmillan

AITKENS, A.C., *Determinants and Matrices*, Oliver and Boyd.

BAUMOL W., *Economic Theory and Operation Analysis*, Prentice Hall.

BIRKHOFF, G. AND MACLANE, S., *A Survey of Modern Algebra*, Macmillan

CHENERY, H. AND CLARK; P., *Interindustry Economics*, John Wiley.

COCHRAN, W.G., *Sampling Techniques*, Wiley Eastern.

CAMERON, E.A., *College Algebra*, Henry H. & Co.

CROXTON, E.F. AND COWDEN, D. I. *Applied General Statistics*, Prentice Hall.

DICKINSON, G.C., *Statistical Mapping and Presentation of Statistics*, Eduard Arnold

DORFMAN, R., SAMUELSON, P.A. AND SOLOW, R. M., *Linear Programming and Economic Analysis*, McGraw Hill.

EZEKIEL, *Methods of Correlation Analysis*, John Wiley

FRAZER, R.A., DUNCAN, W.J. AND COLLAR, A.R., *Elementary Matrices*, Macmillan

HICKS, J. R., *A Revision of Demand Theory*, Oxford University Press.

HOEL, P.G. *Introduction to Mathematical Statistics*, John Wiley

HENDERSON, J.M. AND QUANDT, R.E., *Microeconomic Theory, A Mathematical Approach*, McGraw Hill.

JOHNSTON, J., *Econometric Method*, McGraw Hill.

KENDALI, M.G. AND STUART, A., *The Advanced Theory of Statistics*, Charles Griffin.

KLEIN, L.A., *A Texbook of Econometrics*, Prentice Hall.

LEONTIEF, W.W., *The Structure of American Economy 1919—1939*, Oxford University Press.

MOOD, A.M., *Introduction to the Theory of Statistics*, McGraw Hill.

SAMUELSON, P.A., *Foundations of Economic Analysis*, Harvard University Press.

SPIEGEL M.R., *Theory and Practice of Statistics*, Schaum Series.

STOLL, R., *Set Theory and Logic*, W.H. Freeman & Co.

USPENSKY, J.V., *Introduction to Mathematical Probability*, McGraw Hill.

WEATHERBURN, C.E., *A First Course in Mathematical Statistics*, Cambridge.

WIDDER, D.V., *Advanced Calculus*, Prentice Hall.

WILKS, S.S., *Mathematical Statistics*, Princeton University.

YAMANE, T., *Mathematics for Economics*, Prentice Hall.

Statistical Tables

TABLE I

Value or e^{-m}

m	e^{-m}	m	e^{-m}	m	e^{-m}	m	e^{-m}
0.1	0.90484	2.6	0.07427	5.1	0.00610	7.6	0.00050
0.2	0.81873	2.7	0.06721	5.2	0.00552	7.7	0.00045
0.3	0.74042	2.8	0.06081	5.3	0.00499	7.8	0.00041
0.4	0.67032	2.9	0.05502	5.4	0.00452	7.9	0.00037
0.5	0.60653	3.0	0.04979	5.5	0.00409	8.0	0.00034
0.6	0.54881	3.1	0.04505	5.6	0.00370	8.1	0.00030
0.7	0.49659	3.2	0.04076	5.7	0.00335	8.2	0.00027
0.8	0.44933	3.3	0.03688	5.8	0.00303	8.3	0.00025
0.9	0.40657	3.4	0.03337	5.9	0.00274	8.4	0.00022
1.0	0.36788	3.5	0.03020	6.0	0.00248	8.5	0.00020
1.1	0.33287	3.6	0.02732	6.1	0.00224	8.6	0.03018
1.2	0.30119	3.7	0.02472	6.2	0.00203	8.7	0.00017
1.3	0.27253	3.8	0.02237	6.3	0.00184	8.8	0.00015
1.4	0.24660	3.9	0.2024	6.4	0.00166	8.9	0.00014
1.5	0.22313	4.0	0.01832	6.5	0.00150	9.0	0.00012
1.6	0.20190	4.1	0.01657	6.6	0.00136	9.1	0.00011
1.7	0.18268	4.2	0.01500	6.7	0.00123	9.2	0.00010
1.8	0.16530	4.3	0.01357	6.8	0.00111	9.3	0.00009
1.9	0.14957	4.4	0.01228	6.9	0.00101	9.4	0.00008
2.0	0.13534	4.5	0.01111	7.0	0.00091	9.5	0.00007
2.1	0.12246	4.6	0.01005	7.1	0.00083	9.6	0.00007
2.2	0.11080	4.7	0.00910	7.2	0.00075	9.7	0.00006
2.3	0.10026	4.8	0.00823	7.3	0.00068	9.8	0.00006
2.4	0.09072	4.9	0.00745	7.4	0.00061	9.9	0.00005
2.5	0.08208	5.0	0.00674	7.5	0.00055	10.0	0.00005

TABLE 2

Values of e^x

x	e^x	x	e^x
0.0	1.000		
0.1	1.105	5.1	164.0
0.2	1.221	5.2	181.3
0.3	1.350	5.3	200.3
0.4	1.492	5.4	221.4
0.5	1.649	5.5	244.7
0.6	1.822	5.6	270.4
0.7	2.014	5.7	298.9
0.8	2.226	5.8	330.3
0.9	2.460	5.9	365.0
1.0	2.718	6.0	403.4
1.1	3.004	6.1	445.9
1.2	3.320	6.2	492.8
1.3	3.669	6.3	541.6
1.4	4.055	6.4	601.8
1.5	4.482	6.5	665.1
1.6	4.953	6.6	735.1
1.7	5.474	6.7	812.4
1.8	6.050	6.8	897.8
1.9	6.686	6.9	992.3
2.0	7.389	7.0	1096.6
2.1	8.166	7.1	1212.0
2.2	9.025	7.2	1339.4
2.3	9.974	7.3	1480.3
2.4	11.023	7.4	1636.0
2.5	12.18	7.5	1808.0
2.6	13.46	7.6	1998.2
2.7	14.88	7.7	2208.3
2.8	16.44	7.8	2440.6
2.9	18.17	7.9	2697.3
3.0	20.09	8.0	2981.0

x	e^x	x	e^x
3.1	22.20	8.1	3294.5
3.2	24.53	8.2	3641.0
3.3	27.11	8.3	4023.9
3.4	29.96	8.4	4447.1
3.5	33.12	8.5	4914.8
3.6	36.60	8.6	5431.7
3.7	40.45	8.7	6002.9
3.8	44.70	8.8	6634.2
3.9	49.40	8.9	7332.0
4.0	54.60	9.0	8103.1
4.1	60.34	9.1	8955.3
4.2	66.69	9.2	9897.1
4.3	73.70	9.3	10938
4.4	81.45	9.4	12088
4.5	90.02	9.5	13360
4.6	99.48	9.6	14765
4.7	109.95	9.7	16318
4.8	121.51	9.8	18034
4.9	134.29	9.9	19930
5.0	148.4		

TABLE 3

AREA UNDER THE NORMAL CURVE BETWEEN *o*, *z*

$P\{0 \le z_1 \le z\}$

z	*.00*	*.01*	*.02*	*.03*	*.04*	*.05*	*.06*	*.07*	*.08*	*.09*
0.0	.000	.0040	.0080	.0120	.0160	.0199	.0239	.0279	.0319	.0359
0.1	.0398	.0438	.0478	.0517	.0557	.0596	.0636	.0675	.0714	.0753
0.2	.0793	.0832	.0871	.0910	.0948	.0987	.1026	.1064	.1103	.1141
0.3	.1179	.1217	.1255	.1293	.1331	.1368	.1406	.1443	.1480	.1517
0.4	.1554	.1591	.1628	.1664	.1700	.1736	.1772	.1808	.1879	.1879
0.5	.1915	.1950	.1985	.2019	.2054	.2088	.2123	.2157	.2190	.2224
0.6	.2257	.2291	.2324	.2357	.2389	.2422	.2454	.2486	.2517	.2549
0.7	.2580	.2611	.2642	.2673	.2704	.2734	.2764	.2794	.2823	.2852
0.8	.2881	.2910	.2939	.2967	.2995	.3023	.3051	.3078	.3106	.3133
0.9	.3159	.3186	.3212	.3238	.3264	.3289	.3315	.3340	.3365	.3389
1.0	.3413	.3438	.3461	.3485	.3508	.3531	.3554	.3577	.3599	.3621
1.1	.3643	.3665	.3686	.3708	.3729	.3749	.3770	.3790	.3810	.3830
1.2	.3849	.3869	.3888	.3907	.3925	.3944	.3962	.3980	.3997	.4015
1.3	.4032	.4049	.4066	.4082	.4099	.4115	.4121	.4147	.4162	.4177
1.4	.4192	.4207	.4222	.4236	.4251	.4265	.4279	.4292	.4306	.4319
1.5	.4332	.4345	.4357	.4370	.4382	.4394	.4406	.4418	.4429	.4441
1.6	.4452	.4463	.4474	.4484	.4495	.4505	.4515	.4525	.4535	.4545
1.7	.4554	.4564	.4573	.4582	.4591	.4599	.4608	.4616	.4625	.4633
1.8	.4641	.4649	.4656	.4664	.4671	.4678	.4686	.4693	.4699	.4706
1.9	.4713	.4719	.4726	.4732	.4738	.4744	.4750	.4756	.4761	.4767
2.0	.4772	.4778	.4783	.4788	.4793	.4798	.4803	.4808	.4812	.4817
2.1	.4821	.4826	.4830	.4834	.4838	.4842	.4846	.4850	.4854	.4857
2.2	.4861	.4864	.4868	.4871	.4875	.4878	.4881	.4884	.4887	.4890
2.3	.4893	.4896	.4898	.4901	.4904	.4906	.4909	.4911	.4913	.4916
2.4	.4918	.4920	.4922	.4925	.4927	.4929	.4931	.4932	.4934	.4936
2.5	.4938	.4940	.4941	.4943	.4945	.4946	.4948	.4949	.4951	.4952
2.6	.4953	.4955	.4956	.4957	.4959	.4960	.4961	.4962	.4963	.4964
2.7	.4965	.4966	.4967	.4968	.4969	.4970	.4971	.4972	.4973	.4974
2.8	.4974	.4975	.4976	.4977	.4977	.4978	.4979	.4979	.4980	.4981
2.9	.4981	.4982	.4982	.4983	.4984	.4984	.4985	.4985	.4986	.4986
3.0	.4987	.4987	.4987	.4988	.4988	.4989	.4989	.4989	.4990	.4990

TABLE 4

CRITICAL VALUES OF *t* (for two tailed tests)

Degrees of freedom	*Level of Significance*		
	1%	5%	10%
1	63.657	12.706	6.314
2	9.925	4.303	2.920
3	5.841	3.182	2.353
4	4.694	2.776	2.132
5	4.032	2.571	2.015
6	3.707	2.447	1.943
7	3.499	2.365	1.895
8	3.355	2.306	1.860
9	3.250	2.262	1.833
10	3.169	2.228	1.812
11	3.106	2.201	1.796
12	3.055	2.179	1.782
13	3.012	2.160	1.771
14	2.977	2.145	1.761
15	2.947	2.131	1.753
16	2.921	2.120	1.746
17	2.898	2.110	1.740
18	2.878	2.101	1.734
19	2.861	2.093	1.729
20	2.845	2.086	1.725
21	2.831	2.080	1.721
22	2.819	2.074	1.717
23	2.807	2.069	1.714
24	2.797	2.064	1.711
25	2.787	2.060	1.708
26	2.779	2.056	1.706
27	2.771	2.052	1.703
28	2.763	2.048	1.701
29	2.756	2.045	1.699
α	2.576	1.960	1.645

For one tailed test at 5% take the *t* value at 10%.

TABLE 5
CRITICAL χ^2 VALUES

Degrees of Freedom	*Level of Significance*		
	10%	*5%*	*1%*
1	2.706	3.841	6.635
2	4.605	5.991	9.210
3	6.251	7.815	11.345
4	7.779	9.488	13.277
5	9.236	11.071	15.086
6	10.645	12.592	16.812
7	12.017	14.067	18.475
8	13.362	15.507	20.090
9	14.684	16.919	21.666
10	15.987	18.307	23.209
11	17.275	19.675	24.724
12	18.549	21.026	26.217
13	19.812	22.362	27.688
14	21.064	23.685	29.141
15	22.307	24.996	30.578
16	23.542	26.296	32.000
17	24.769	27.587	33.409
18	25.989	28.869	34.805
19	27.204	30.144	36.191
20	28.412	31.410	37.566
21	29.615	32.671	38.932
22	30.813	33.924	40.289
23	33.007	35.172	41.638
24	33.196	36.415	42.980
25	34.382	37.652	44.314
26	35.363	38.885	45.642
27	36.741	40.113	46.963
28	37.916	41.337	48.278
29	38.087	42.567	49.588
30	40.256	42.773	50.892

TABLE 6

Critical Values of F

Significance Level 1 per cent

v_2 \ v_1	*1*	*2*	*3*	*4*	*5*	*6*	*7*	*8*	*12*	*24*	∞
1	4052	5000	5403	5625	5764	5859	5982	6106	6235	6366	
2	98.50	99.00	99.17	99.25	99.30	99.33	99.37	99.42	99.46	99.50	
3	34.12	30.82	29.46	28.71	28.24	27.91	27.49	27.05	26.60	26.13	
4	21.20	18.20	16.69	15.88	15.52	15.21	14.80	14.37	13.93	13.45	
5	16.26	13.27	12.06	11.39	10.97	10.67	10.29	9.89	9.47	9.02	
6	13.75	10.92	9.78	9.15	8.75	8.47	8.10	7.72	7.31	6.88	
7	12.25	9.55	8.45	7.85	7.46	7.19	6.84	6.47	6.07	5.65	
8	11.26	8.65	7.59	7.01	6.63	6.37	6.03	5.67	5.28	4.86	
9	10.56	8.02	6.99	6.42	6.06	5.80	5.47	5.12	4.73	4.31	
10	10.04	7.56	6.55	5.99	5.64	5.39	5.06	4.71	4.33	3.91	
11	9.65	7.21	6.22	5.87	5.32	5.07	4.74	4.40	4.02	3.60	
12	9.33	6.93	5.95	5.41	5.06	4.82	4.50	4.16	3.78	3.36	
13	9.07	6.70	5.74	5.21	4.86	4.62	4.30	3.96	3.59	3.17	
14	8.86	6.51	5.56	5.04	4.69	4.46	4.14	3.80	3.43	3.00	
15	8.68	6.36	4.42	4.89	4.56	4.32	4.00	3.67	3.29	2.87	
16	8.53	6.23	5.29	4.77	4.44	4.20	3.89	3.55	3.18	2.75	
17	8.40	6.11	5.18	4.67	4.34	4.10	3.79	3.46	3.08	2.65	
18	8.29	6.01	5.09	4.58	4.25	4.01	3.71	3.37	3.00	3.59	
19	8.18	5.93	5.01	4.50	4.17	3.94	3.63	3.30	3.92	2.49	
20	8.10	5.85	4.94	4.43	4.10	3.87	3.56	3.23	2.86	2.42	
21	8.02	5.78	4.87	4.37	4.04	3.81	3.51	3.17	2.80	2.36	
22	7.95	5.72	4.82	4.31	3.99	3.76	3.45	3.12	2.75	2.31	
23	7.88	5.66	4.76	4.26	3.94	3.71	3.41	3.07	2.70	2.26	
24	7.82	5.61	4.72	4.22	3.90	3.67	3.36	3.03	2.66	2.21	
25	7.77	5.57	4.68	4.18	3.85	3.63	3.32	2.99	2.62	2.17	
26	7.72	5.53	4.64	4.14	3.82	3.59	3.20	2.96	2.58	2.10	
27	7.68	5.49	4.60	4.11	3.78	3.56	3.26	2.93	2.45	2.13	
28	7.64	5.45	4.57	4.07	3.75	3.53	3.23	2.90	2.52	2.06	
29	7.60	5.42	4.54	4.04	3.73	3.50	3.20	2.87	2.49	2.03	
30	7.56	5.39	4.51	4.02	3.70	3.47	3.17	2.84	2.47	2.01	
40	7.31	5.18	4.31	3.83	3.51	3.29	2.99	2.66	2.29	1.80	
60	7.08	4.98	4.13	3.65	3.34	3.12	2.82	2.50	2.12	1.60	
120	6.85	4.79	3.95	3.48	3.17	2.96	2.66	2.34	1.95	1.38	
∞	6.64	4.60	3.78	3.32	3.02	2.80	2.51	2.18	1.79	1.00	

v_1 = Degrees of freedom for the larger variance.

v_2 = Degrees of freedom for the smaller variance.

TABLE 7

CRITICAL VALUES OF F

SIGNIGICANCE LEVEL 5 PER CENT

v_2 \ v_1	1	2	3	4	5	6	8	12	24	∞
1	161.4	199.5	215.7	224.6	230.2	234.0	238.9	243.9	249.0	254.3
2	18.51	19.00	19.16	19.25	19.30	19.33	19.37	19.41	19.45	19.50
3	10.13	9.55	9.28	9.12	9.01	8.94	8.84	8.74	8.64	8.53
4	7.71	6.94	6.59	6.39	6.26	6.16	6.04	5.91	5.77	5.63
5	6.61	5.79	5.41	5.19	5.05	4.95	4.82	4.68	4.53	4.36
6	5.99	5.14	4.76	4.53	4.39	4.28	4.15	4.00	3.84	3.67
7	5.59	4.74	4.35	4.12	3.97	3.87	3.73	3.57	3.41	3.23
8	5.32	4.46	4.07	3.84	3.69	3.58	3.44	3.28	3.12	3.93
9	5.12	4.26	3.86	3.63	3.48	3.37	3.23	3.07	2.90	2.71
10	4.96	4.10	3.71	3.48	3.33	3.22	3.07	2.91	2.74	2.54
11	4.84	3.98	3.59	3.36	3.20	3.09	2.95	2.79	2.61	2.40
12	4.72	3.88	3.49	3.26	3.11	3.00	2.85	2.69	2.50	2.30
13	4.67	3.80	3.41	3.18	3.02	3.92	2.77	2.60	2.42	2.21
14	4.60	3.74	3.34	3.11	2.96	2.85	2.70	2.53	2.35	2.13
15	4.54	3.68	3.29	3.06	2.90	2.79	2.64	2.48	2.29	2.07
16	4.49	3.63	3.24	3.01	2.85	2.74	2.59	2.42	2.24	2.01
17	4.45	3.59	3.20	2.96	2.81	3.70	2.55	2.38	2.19	1.96
18	4.41	3.55	3.16	2.93	2.77	2.66	2.51	2.34	2.15	1.92
19	4.38	3.52	3.13	2.90	2,74	2.63	2.48	2.31	2.11	1.88
20	4.35	3.49	3.10	2.87	2.71	2.60	2.45	2.28	2.08	1.84
21	4.32	3.47	3.07	2.84	2.68	2.57	2.42	2.25	2.05	1.81
22	4.30	3.44	3.05	2.82	2.66	2.55	2.40	2.23	2.03	1.78
23	4.28	3.42	3.03	2.80	2.64	2.53	2.38	2.20	2.00	1.76
24	4.26	3.40	3.01	2.78	2.62	2.51	2.36	2.18	1.98	1.73
25	4.24	3.38	2.99	2.76	2.60	2.49	2.34	2.16	1.96	1.71
26	4.22	3.37	2.98	2.74	2.59	2.47	2.32	2.15	1.95	1.69
27	4.21	3.35	2.96	2.73	2.57	2.46	2.30	2.13	1.93	1.67
28	4.20	3.34	2.95	2.71	2.56	2.44	2.29	2.12	1.91	1.65
29	4.18	3.33	2.93	2.70	2.54	2.43	2.28	2.10	1.90	1.64
30	4.17	3.32	2.92	2.69	2.53	2.42	2.27	2.09	1.89	1.62
41	4.08	3.23	2.84	2.61	2.45	2.34	2.18	2.00	1.79	1.51
60	4.00	3.15	2.76	2.52	2.37	2.25	2.10	1.92	1.70	1.89
120	2.92	3.07	2.68	2.45	2.29	2.17	2.02	1.83	1.61	1.25
∞	3.84	2.99	2.60	2.37	2.21	2.09	1.94	1.75	1.52	1.00

v_1 = Degree of freedom for the larger variance.
v_2 = Degrees of freedom for the smaller variance.